ELECTRIC MACHINES

Fifth Edition

About the Authors

D P Kothari obtained his BE (Electrical) in 1967, ME (Power Systems) in 1969 and PhD in 1975 from BITS, Pilani, Rajasthan. He was involved in teaching and development of several courses at BITS Pilani, from 1969 to 1977. Currently, he is the Dean (R&D) at J D College of Engineering & Management, Nagpur. Earlier, he served as Vice Chancellor at VIT, Vellore. He was also the Director in-charge and Deputy Director (Administration) as well as Head at the Centre of Energy Studies at Indian Institute of Technology, Delhi. He also served as Principal at VRCE, Nagpur. During 1982–83 and 1989, he was a visiting professor at the Royal Melbourne Institute of Technology, Melbourne, Australia. He taught at Melbourne University for one semester in 1989. He was also NSF Fellow at Perdue University, USA, in 1992.

Dr Kothari, who is a recipient of the most Active Researcher Award, has published and presented 812 research papers in various national and international journals, conferences, guided 50 PhD scholars and 68 M.Tech. students, and authored 50 books in various allied areas. He has delivered several keynote addresses and invited lectures in both national and international conferences. He has also delivered 42 video lectures on YouTube with maximum of 40,000 hits!

Dr Kothari is a Fellow of the National Academy of Engineering (FNAE), Fellow of Indian National Academy of Science (FNASc), Fellow of Institution of Engineers (FIE), Fellow IEEE, Hon. Fellow ISTE and Fellow IETE.

His many awards include the National Khosla Award for Lifetime Achievements in Engineering from IIT, Roorkee, in 2005. The University Grants Commission (UGC), Government of India, has bestowed the UGC National Swami Pranavandana Saraswati Award for his outstanding scholarly contributions in the field of education, in 2005.

He is also a recipient of the Lifetime Achievement Award conferred by the World Management Congress, New Delhi, for his contribution in the areas of educational planning and administration, in 2009. Recently, he received Excellent Academic Award at IIT Guwahati by NPSC, 2014.

In 2016, he received six Life Time Achievement awards by various agencies and Living Legend Award at Chennai Conference.

I J Nagrath was an Adjunct Professor at BITS Pilani, from where he retired in July 1999 as Professor of Electrical Engineering and Deputy Director. He was actively engaged in writing books related to his long teaching and research experience.

Prof. Nagrath obtained his BE with Honours in Electrical Engineering from Birla Engineering College in 1951, and MS from the University of Wisconsin, USA, in 1956.

He had co-authored several successful books, which include *Electric Machines*, 4/e, *Modern Power System Analysis,* 4/e, *Power System Engineering,* 2/e, *Electrical Machines*, 2/e (Sigma Series) and *Basic Electrical Engineering*, 3/e (all published by MHEI); and also co-authored *Control System Engineering* and authored *Electronics: Analog and Digital*. He had also published several research papers in prestigious national and international journals.

ELECTRIC MACHINES

Fifth Edition

D P Kothari

Dean (R&D), J D College of Engineering & Management, Nagpur
Fellow-INAE, Fellow-INASc, Fellow-IEEE, Hon. Fellow-ISTE, FIE (India), Fellow-IETE, MCSI
Former Director Research, MVSR Engineering College, Hyderabad
Former Director General, J B Group of Educational Institutions, Hyderabad
Director General, Raisoni Group of Institutions, Nagpur
Former Director General, Vindhya Institute of Technology and Science, Indore
Former Vice Chancellor, VIT University, Vellore
Former Director-Incharge, IIT Delhi
Former Principal, VRCE, Nagpur
Visiting Fellow, RMIT, Melbourne, Australia

I J Nagrath

Adjunct Professor and Former Deputy Director
Birla Institute of Technology and Science
Pilani

McGraw Hill Education (India) Private Limited

CHENNAI

McGraw Hill Education Offices

Chennai New York St Louis San Francisco Auckland Bogotá Caracas
Kuala Lumpur Lisbon London Madrid Mexico City Milan Montreal
San Juan Santiago Singapore Sydney Tokyo Toronto

McGraw Hill Education (India) Private Limited

Published by McGraw Hill Education (India) Private Limited
444/1, Sri Ekambara Naicker Industrial Estate, Alapakkam, Porur, Chennai 600 116

Electric Machines, 5/e

Print Edition
ISBN (13): 978-93-5260-640-5
ISBN (10): 93-5260-640-X

e-Book Edition
ISBN (13): 978-93-5260-641-2
ISBN (10): 93-5260-641-8

Managing Director: *Kaushik Bellani*

Director—Science & Engineering Portfolio: *Vibha Mahajan*
Senior Manager Portfolio—Science & Engineering: *Hemant Jha*

Senior Manager—Content Development: *Shalini Jha*
Content Developer: *Sachin Kumar*

Production Head: *Satinder S Baveja*
Copy Editor: *Taranpreet Kaur*
Assistant Manager—Production: *Suhaib Ali*

General Manager—Production: *Rajender P Ghansela*
Manager—Production: *Reji Kumar*

Typeset at Text-o-Graphics, B-1/56, Aravali Apartment, Sector-34, Noida 201 301, and printed at

Magic International Pvt. Ltd., Plot No. 26E, Sector-31,(Industrial), Site-IV, Greater Noida 201306

DZADCRBMDZZZB

Visit us at: www.mheducation.co.in

Dedicated
to

Shobha

—D P Kothari

and

Pushpa

—I J Nagrath

Contents

Preface to the Fifth Edition

This revised edition is intended for third and fourth year undergraduate students and first year postgraduate students of electrical engineering programs. It emphasizes on the fundamental principles and their applications in day to day life. The mathematical complexity of the book is kept at a level that is well within the grasp of engineering students. The book can be used by IAS, IES, GATE aspirants and practicing engineers.

It caters to a one-semester course, which is common for all engineering disciplines and a one-semester topping off course for those majoring in electrical engineering.

Highlights

- Content presentation supports outcomes based learning approach
- Coverage on technological trends like losses in dc machines, use of electric machines in electric drives, etc.
- In-depth coverage of important topics—Rotating Electric Machines, Armature Reaction of ac and dc Machines, Speed Control of dc and ac Motors, Performance Characteristics of dc and ac Machines and Special Motors
- MATLAB examples within the chapters facilitates problem solving skills

What is New in the Fifth Edition?

The fifth edition is student-centric, and follows the '**Outcomes based Learning**' approach. This is an educational process that emphasizes on developing engineering skills in student and testing the outcomes of the study of a course, as opposed to rote learning. It contains 12 chapters each containing a set of learning objectives.

Learning Tools

Learning Objectives

Each chapter is organized into multiple Learning Objectives (LOs) that provide a quick reference to the chapter's key aspects. These help students better anticipate what they will be studying and help instructors measure students' understanding. Solved examples, chapter-end questions, and summary points are tagged with respective LO.

CHAPTER 2

Magnetic Circuits and Induction

LEARNING OBJECTIVES

After studying this chapter, students will be able to

- **LO1** Explain magnetic circuits and magnetization characteristics
- **LO2** Identify the different magnetic materials and illustrate their properties
- **LO3** Develop expressions for magnetically induced emf and force and understand the ac operation of magnetic circuits
- **LO4** Define hysteresis and classify the eddy current losses
- **LO5** Classify permanent magnets and list their applications

2.1 | INTRODUCTION

The electromagnetic system is an essential element of all rotating electric machinery and electromechanical devices as well as static devices like the transformer. The role of the electromagnetic system is to establish and control electromagnetic fields for carrying out conversion of energy, its processing and transfer. Practically all electric motors and generators, ranging in size from fractional horsepower units found in domestic appliances to the gigantic several thousand kW motors employed in heavy industry and several hundred megawatt generators installed in modern generating stations, depend upon the magnetic field as the coupling medium allowing

Pedagogy as per Blooms Taxonomy

The pedagogy is arranged as per levels of difficulty which is derived from Blooms' Taxonomy.

Note: Difficulty Level ⟶ ○○● — Level 1 and Level 2 Category; ○●● — Level 3 and Level 4 Category; ●●● — Level 5 and Level 6 Category

○○● Knowledge and Comprehension based easy-to-solve problems

○●● Application and Analysis based medium-difficulty problems

●●● Synthesis and Evaluation based high-difficulty problems

Recall It is, therefore, seen that an ideal transformer changes (transforms) voltages in *direct ratio* of the number of turns in the two windings.

In terms of rms values Eq. (3.16) implies

$$\frac{V_1}{V_2} = \frac{E_1}{E_2} = \frac{N_1}{N_2} = a \quad \text{(same as Eq. (3.9))} \qquad (3.17)$$

!Attention In Fig. 3.13 the current $\bar{I}_1$ flowing in the primary of the semi-ideal transformer can be visualized to comprise two components as below:

(i) Exciting current $\bar{I}_0$ whose magnetizing component $\bar{I}_m$ creates mutual flux $\bar{\phi}$ and whose core-loss component $\bar{I}_i$ provides the loss associated with alternation of flux.

(ii) A load component $\bar{I}_2'$ which counterbalances the secondary mmf $\bar{I}_2 N_2$ so that the mutual flux remains constant independent of load, determined only by $\bar{E}_1$. Thus

$$\bar{I}_1 = \bar{I}_0 + \bar{I}_2' \qquad (3.28)$$

Important! The efficiency of a transformer is in the range of 96–99%.

It is of no use trying to determine it by measuring the output and input under load conditions because the wattmeter readings are liable to have an error of 1–2%. The best and accurate method of determining efficiency would be to find the losses from the OC and SC tests. With this data efficiency can then be calculated as

$$\eta = \frac{\text{output}}{\text{output + losses}} = 1 - \frac{\text{losses}}{\text{output + losses}} \qquad (3.54)$$

Emphasis on Crucial Concepts

Keywords and concepts have been highlighted throughout the book to aid ease of understanding.

Integrated Assessment Tools

Within the Chapter Exercises

More than 350 Concept Check questions, within the narrative, prompt the students to apply the material as they read. This offers great retention through looping mechanism. More than 200 solved examples are given to develop a clear understanding of the concepts. They illustrate step-by-step procedure to solve a problem.

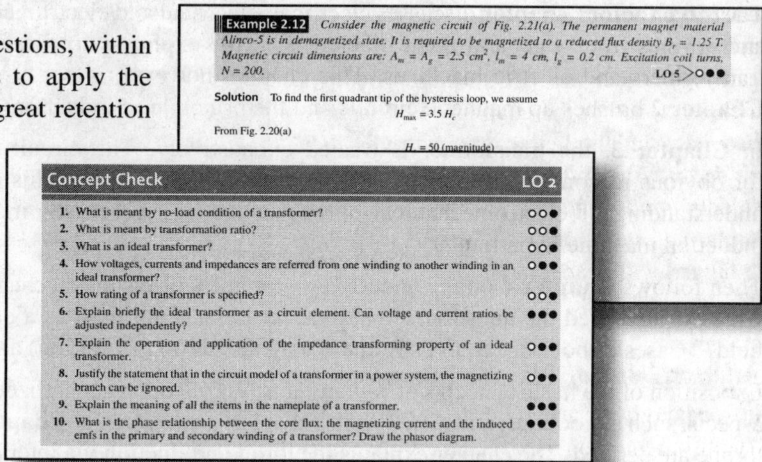

Example 2.12 *Consider the magnetic circuit of Fig. 2.21(a). The permanent magnet material Alinco-5 is in demagnetized state. It is required to be magnetized to a reduced flux density B_r = 1.25 T. Magnetic circuit dimensions are: $A_m = A_g$ = 2.5 cm², l_m = 4 cm, l_g = 0.2 cm. Excitation coil turns, N = 200.* LO 5

Solution To find the first quadrant tip of the hysteresis loop, we assume

$$H_{max} = 3.5 H_c$$

From Fig. 2.20(a)

H_c = 50 (magnitude)

Concept Check LO 2

1. What is meant by no-load condition of a transformer?
2. What is meant by transformation ratio?
3. What is an ideal transformer?
4. How voltages, currents and impedances are referred from one winding to another winding in an ideal transformer?
5. How rating of a transformer is specified?
6. Explain briefly the ideal transformer as a circuit element. Can voltage and current ratios be adjusted independently?
7. Explain the operation and application of the impedance transforming property of an ideal transformer.
8. Justify the statement that in the circuit model of a transformer in a power system, the magnetizing branch can be ignored.
9. Explain the meaning of all the items in the nameplate of a transformer.
10. What is the phase relationship between the core flux; the magnetizing current and the induced emfs in the primary and secondary winding of a transformer? Draw the phasor diagram.

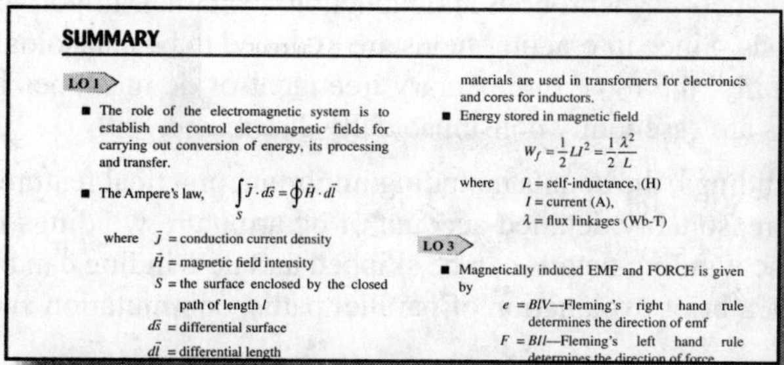

SUMMARY

LO 1

- The role of the electromagnetic system is to establish and control electromagnetic fields for carrying out conversion of energy, its processing and transfer.
- The Ampere's law, $\int_S \vec{J}\cdot d\vec{s} = \oint \vec{H}\cdot d\vec{l}$

where $\vec{J}$ = conduction current density
$\vec{H}$ = magnetic field intensity
S = the surface enclosed by the closed path of length *l*
$d\vec{s}$ = differential surface
$d\vec{l}$ = differential length

materials are used in transformers for electronics and cores for inductors.

- Energy stored in magnetic field

$$W_f = \frac{1}{2}LI^2 = \frac{1}{2}\frac{\lambda^2}{L}$$

where L = self-inductance, (H)
I = current (A),
λ = flux linkages (Wb-T)

LO 3

- Magnetically induced EMF and FORCE is given by

e = BlV—Fleming's right hand rule determines the direction of emf
F = BIl—Fleming's left hand rule determines the direction of force

Summary

A detailed chapter-end summary related to LOs covered in the chapter is provided for a quick review of the important concepts.

Chapter-end Exercises

More than 500 chapter-end exercises have been carefully constructed to enhance the knowledge of the students. These are categorized into Multiple-Choice Questions and Practice Problems alongwith 21 solved examples using MATLAB.

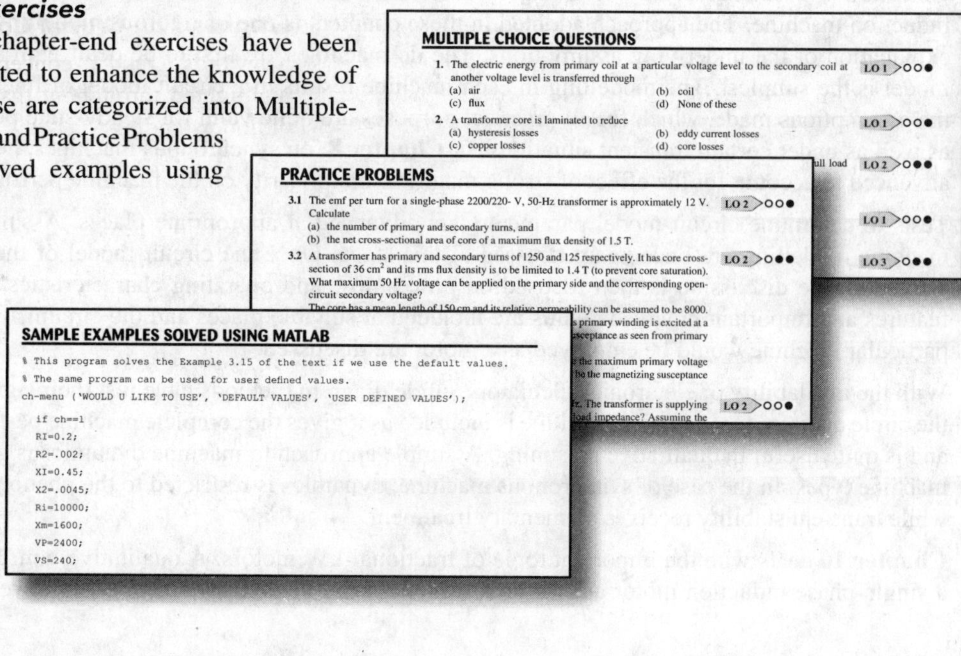

MULTIPLE CHOICE QUESTIONS

1. The electric energy from primary coil at a particular voltage level to the secondary coil at another voltage level is transferred through LO 1
(a) air (b) coil
(c) flux (d) None of these
2. A transformer core is laminated to reduce LO 1
(a) hysteresis losses (b) eddy current losses
(c) copper losses (d) core losses

PRACTICE PROBLEMS

3.1 The emf per turn for a single-phase 2200/220- V, 50-Hz transformer is approximately 12 V. Calculate LO 2
(a) the number of primary and secondary turns, and
(b) the net cross-sectional area of core of a maximum flux density of 1.5 T.

3.2 A transformer has primary and secondary turns of 1250 and 125 respectively. It has core cross-section of 36 cm² and its rms flux density is to be limited to 1.4 T (to prevent core saturation). What maximum 50 Hz voltage can be applied on the primary side and the corresponding open-circuit secondary voltage? LO 2

SAMPLE EXAMPLES SOLVED USING MATLAB

```
% This program solves the Example 3.15 of the text if we use the default values.
% The same program can be used for user defined values.
ch=menu ('WOULD U LIKE TO USE', 'DEFAULT VALUES', 'USER DEFINED VALUES');
if ch==1
R1=0.2;
R2=.002;
X1=0.45;
X2=.0045;
Ri=10000;
Xm=1600;
VP=2400;
VS=240;
```

Organization of the Book

The **introductory chapter** discusses electrical–electrical and electromechanical energy conversion processes and devices from a general point of view with the explicit purpose of motivating the reader for studying transformers and electric machines. This chapter, however, is not a prerequisite for the rest of the book. **Chapter 2** brushes up magnetic circuits and the principle of induction.

In **Chapter 3**, the transformer is treated exhaustively. The circuit-model approach is emphasized and for obvious reasons the role of phasor diagram is underplayed. This chapter lays the groundwork for the understanding of electromechanical energy conversion processes in machines and the circuit model of induction machine in particular.

Then follows **Chapter 4** on the underlying principles of electromechanical energy conversion, at the end of which is answered the question, "Why is electric field not used as a coupling medium instead of magnetic field?" Cases of both linear and nonlinear magnetization (saturation) are treated.

Exposition of the basic concepts of rotating machines from a generalized point of view as well as engineering aspects, such as cooling, rating and load mechanics is advanced in **Chapter 5**. General expressions for emf and torque are derived. The chapter explains the torque production via interaction of two magnetic fields assumed to be sinusoidally distributed. An alternative current-sheet approach is also given for interested readers. Then follows the elementary treatment of specific machine types—synchronous and induction—and their important characteristics are visualized on a field-interaction basis. Since interacting fields are assumed to be sinusoidal, which is justified in these two classes of machines only, the most rudimentary treatment of dc machines is given here because the fields in this class of machines are essentially nonsinusoidally distributed.

While Chapter 5 gives an essential treatment of ac windings, the details including important practical features are dealt in **Chapter 6** on ac armature windings. A reasonably detailed account of dc armature windings is given in **Chapter 7**. When time is a limiting factor, ac winding details can be skipped and dc winding can be directly introduced via two developed diagrams with a brief explanation of parallel paths, commutation and brush location.

Chapters 7–9 extensively cover the three basic machine types—dc machine, synchronous machine and induction machine. The approach adopted in these chapters is one of rigorous modelling with due stress on explanation of the underlying assumptions. The dc machines are first to be dealt with as their steady-state model is the simplest. The modelling in each machine results in a circuit model of linear kind by virtue of the assumptions made, which for all practical purposes are quite valid for steady-state performance analysis as well as under certain transient situations. In **Chapter 8**, on synchronous machines, heuristic methods are advanced to account for the effect of strong magnetic nonlinearity on the machine performance.

Tests to determine circuit-model parameters are advanced at appropriate places. Assumptions involved in machine modelling are once again stressed at this stage. Once the circuit model of the machine has been arrived at, the discussion is then focused on power flow and operating characteristics. The constructional features and important practical details are included at suitable places and the circumstances under which a particular machine would be employed as a motor are discussed.

With the availability of electronic calculators, circle diagram methods have lost their significance. However, the circle diagram for induction machine is included as it gives the complete machine performance at a glance and is quite useful in qualitative reasoning. A simple approach to machine dynamics is given in all the three machine types. In the case of synchronous machine, dynamics is restricted to the phenomenon of "hunting", while transient stability receives elementary treatment.

Chapter 10 deals with the important topic of fractional-kW motors. A qualitative-cum-heuristic analysis of a single-phase induction motor and its circuit model are followed by a rigorously developed circuit model for a two-winding motor. This rigorous coverage may be skipped when time does not permit it. A variety of single-phase induction, synchronous and series commutator types of motors are treated. Comprehensiveness is imparted to this chapter by the inclusion of stepper motors, ac servomotor and ac tachometer; the latter two follow simply as a corollary from the two-winding motor analysis.

Chapter 11 deals with the generalized theory of electrical machines.

Probably the most significant development in recent years in the allied area of motor control is the use of power semiconductors—diodes, power transistors and thyristors. The growth in this area has already qualified for a separate undergraduate level course. However, for the sake of completeness, a comprehensive chapter is included in this book. This, in our view, is a better approach than to burden the previous chapters by spreading out the relevant details. **Chapter 12** on this topic has a wide coverage and includes all the three varieties of SCR (silicon controlled rectifier) circuitry, namely—converters, choppers and inverters. The contents and effects of non-smooth dc and nonsinusoidal ac outputs of these control equipment on the circuit behaviour and on machine performance are beyond the scope of this book.

With phenomenal developments in SCR circuitry for power control, cross-field machines and ac commutator machines have become almost obsolete. However, to fulfil the need of such universities which still include these topics in their curriculum, fairly detailed appendices (IV and V) on these topics are added.

A number of cross-sectional views of built-up machines and their parts are included and the student is exhorted to carefully study these to help him visualize the physical picture of the machine being modelled. Laboratory exercises associated with a machines course will further aid this process.

A large number and variety of illustrative examples are spread throughout the book. These would greatly help in imprinting a clear physical picture of the devices and associated physical reasoning on the student's mind. Concept check questions are given at the end of each section within the chapters. This will help the readers to assess their learning after each section. An equally large number of unsolved problems are given at the end of each chapter. Answers to selected unsolved problems are also given at the end of each chapter. Some of these problems are devised to illustrate some points beyond what is directly covered in the text.

International Standard (SI) units are used throughout the book. The list of symbols is necessarily large. Apart from being illustrated at the point of occurrence, the symbols used are listed in the beginning of the book.

Web Supplements

The text is supported by a website accessible at *http://www.mhhe.com/electmach5e* with the following supplements:

- For Instructors
  - Solutions Manual
  - PowerPoint lecture slides
- For students
  - Additional Reading Material
- Links to animation videos

Acknowledgements

This edition of the book owes its existence to many wonderful people. While revising the text, I have had the benefit of valuable advice and suggestions from many teachers, students and readers who used the earlier editions of this book during last 32 years. All these individuals have influenced this edition and hence my heartfelt thanks and appreciation to them. I hope this support would continue in future as well.

I would like to acknowledge the special contributions made by following colleagues:

S Kalyani	*Kamaraj College of Engineering & Technology, Virudhunagar, Tamil Nadu*
Shubhajit Pal	*University of Engineering & Management, Jaipur, Rajasthan*
R Shanthi	*Tamil Nadu Generation and Distribution Corporation, Chennai*
A Jagadeeshwaran	*Sona College of Technology, Salem, Tamil Nadu*

My sincere thanks to Mr Praveen Kumar from Deshya Technologies Pvt. Ltd. (incubated at IIT Guwahati), who helped us with the e-learning resources and the authorities of GPGI, Nagpur, SBJITMR, Nagpur, and GEC, Nagpur, for their cooperation and the facilities provided while writing this edition.

At SBJITMR, I am also thankful to Mr Nilesh and Mr Nandkishor for helping in typing some portions of the revised book. I extend my deepest thanks to my family and Late Professor Nagrath's family who supported us all through and for providing all possible help since the time this book is into existence.

The help and support provided by the entire editorial and production team at McGraw Hill Education (India) are highly appreciated.

My sincere thanks are due to the following reviewers for their valuable comments and suggestions.

Mukhtar Ahmad	*Zakir Husain College of Engineering & Technology, Aligarh, Uttar Pradesh*
Anurag Agrawal	*Shri Ramswaroop Memorial Group of Professional Colleges, Lucknow, Uttar Pradesh*
Harvinder Kumar	*Greater Noida Institute of Technology, Greater Noida, Uttar Pradesh*
Pooja Rani	*ABES Engineering College, Ghaziabad, Uttar Pradesh*
P M Tiwari	*Amity School of Engineering and Technology, Manesar, Haryana*
R K Jarial	*National Institute of Technology, Hamirpur, Himachal Pradesh*
Aritras Chakraborty	*Swami Vivekanand Institute of Engineering & Technology, Chandigarh, Punjab*
Sudhir Y Kumar	*College of Engineering and Technology, Lakshmangarh, Rajasthan*
Shubhajit Pal	*University of Engineering & Management, Jaipur, Rajasthan*
Vinay Gupta	*Manipal University Jaipur, Jaipur, Rajasthan*
Akash Talwariya	*JECRC University, Jaipur, Rajasthan*
Manas Ranjan Nayak	*Institute of Technical Education and Research, Bhubaneswar, Odisha*
S Kalyani	*Kamaraj College of Engineering & Technology, Virudhunagar, Tamil Nadu*
S Periyanayagi	*Ramco Institute of Technology, Rajapalayam, Tamil Nadu*
R Rajasekaran	*SNS College of Technology, Coimbatore, Tamil Nadu*
Ajitha	*Sri Sairam College of Engineeing, Bengaluru, Chennai, Tamil Nadu*
M G Umamaheswari	*Velammal Engineering College, Chennai, Tamil Nadu*
A Jagadeeshwaran	*Sona College of Technology, Salem, Tamil Nadu*
N Chandrasekaran	*PSNA College of Engineering and Technology, Dindigul, Chennai*
Shivasharanappa	*Don Bosco Institute of Technology, Bangalore, Bengaluru, Karnataka*

D P KOTHARI
I J NAGRATH

Publisher Note

McGraw Hill Education (India) invites suggestions and comments from you, all of which can be sent to *info.india@mheducation.com* (kindly mention the title and author name in the subject line).

Piracy-related issues may also be reported.

Preface to the Third Edition

This book is designed to supplement textbooks on electric machines. The material relates to courses like Basic Electrical Engineering, Electrical Science I, Electric Machines I and II. The basic structure of each chapter, apart from the first introductory chapter comprises six items.

To begin with, the concept review helps the student to quickly brush up the fundamentals, and refer to important expressions and relations of electric machines.

Numerous worked out examples are given in the text to illustrate how the theory can be applied to obtain quantitative results and to emphasize the order of magnitude of various quantities. These are followed by supplementary problems for the student to solve; answers are provided for a check.

In almost all numerical problems, realistic parameters, values and specifications have been chosen. Considerable care has been taken in the development of these problems, which are an integral and important part of electric machines theory.

In addition to concepts, solved and unsolved problems, there are large number of objective questions—multiple choice, true-false type and fill-in the blanks. These are provided for the student's quick recall of concepts, modelling analysis and design of all type of electric machines.

Student's understanding of the fundamentals of electric machines cannot be evaluated through numerical questions alone. Therefore, most tests and examinations in this area focus attention on certain qualitative write ups as well. The review questions in each chapter are designed to acquaint the student with these type of questions along with the qualitative write-ups as answers to these review questions.

The repertoire of objective questions and concept reviews will help the reader in understanding the subject and serve as a guide to students studying on their own for competitive examinations such as the AMIE Sections, GATE, graduate IETE, UPSC (IAS/IES) and the screening examinations of various recruiting agencies. Besides, the book can also be used by teachers for infusing variety in tests and quizzes.

While writing the text, we have been greatly encouraged by many colleagues and students. We appreciate the suggestions made by them and thank them for the same. We would like to put on record the valuable suggestions and cooperation of the entire editorial and production teams of Tata McGraw-Hill in bringing out this book. We hope to receive suggestions for improving the text in future also.

D P KOTHARI
I J NAGRATH

CHAPTER 1

Introduction

LEARNING OBJECTIVES

After studying this chapter, students will be able to

- **LO1** Discuss the importance of prime mover, electric generators, electric motors and transformers in practical power system networks
- **LO2** Explain the construction, working principle, types and applications of basic rotating dc and polyphase ac machines
- **LO3** Summarize the recent trends and developments in the field of electric machines

1.1 | INTRODUCTION

Electric machines are employed in almost every industrial and manufacturing process. Pages can be filled in listing the applications of electric machines right from giant-size generators (500 MW and above), industrial motors ranging up to a few megawatts to fractional-kW domestic appliances and to sophisticated aerospace applications requiring stringent reliability in operation.

This book deals with the important topic of electric machines, the indepth understanding of which is necessary to tackle the problems of energy, pollution and poverty that presently confront the whole of mankind.

Since Thomas Alva Edison developed an electric generator, more than hundred years ago, engineers have continually strived and successfully reduced the size and revised upwards the efficiencies of electric machines by the use of improved materials and optimal design strategies. We appear to have reached close to the upper limit imposed by nature.

Industry Relevance

The electrical machinery sector in India primarily caters to the power sector. India produces the full range of electric power generation and transmission machinery. The electrical machinery industry consists of four key product categories based on their use. They are as follows:

- Generation machinery which includes generators, boilers and turbines.
- Transmission machinery includes different types of transformers and transmission towers.
- Distribution machinery which includes circuit breakers, switch gears and control gears.
- Others include electric motors, wires and cables.

The growth of the electrical machinery industry is directly related to the development of power generation and distribution. India's power generation capacity of 2,300 MW in 1950 has expanded to over 128,182 MW at present.

The industry has been gearing up to meet the increase in demand owing to this increase in capacity. India has the capability to manufacture transmission and distribution machinery up to 400 Kilovolts (kV) ac and high voltage dc (HVDC). Industry players are now engaged in upgrading transmission to the next higher voltage system of 765 kV. Thus, the electrical machinery sector in India is on a high growth path.

1.2 ELECTRIC MACHINES

LO 1 Discuss the importance of prime mover, electric generators, electric motors and transformers in practical power system networks

Electricity does not occur naturally in usable form and it also cannot be stored* in usefully large quantities. Therefore, it must be generated continuously to meet the demand (of power) at all times. An efficient and convenient way to generate electric power is by conversion of mechanical power into electrical form in a *rotating* device† called a *generator*. In the process a small part of power is lost in the generator (efficiencies in large generators are above 90%). The mechanical power is itself obtained from heat power by thermodynamical means in a steam turbine (efficiency in the range of 40–50% as the present upper limit) or by conversion of potential energy of water in a hydraulic turbine with very little loss. The basic source of mechanical power—steam/hydraulic turbine is called the *prime mover*. Electricity can also be generated directly from hot gases in *plasma* form, obviating the need of converging heat power to intermediate mechanical power. This process‡ is still in an experimental stage. The electromechanical process of electric power generation is shown schematically in Fig. 1.1. Under steady conversion conditions, T_{PM} (prime mover) = T_G (generator) and the turbine and generator run at steady speed.

Other than lighting and heating§, the major use of electric energy is made by converting it back to the mechanical form to run the wheels of industry as well as tiny household appliances. The intermediary, the electric power, permits the use of large efficient central generating stations, while it is easily transported to

* Attempts are on to store a sizeable amount of electric energy in large superconducting coils. While these attempts are not likely to succeed in the near future, this stored energy would only be sufficient to meet sharp load peaks.

† The device always has an outer stationary member (refer to Sec. 1.2).

‡ The process is known as *magnetohydrodynamics* (MHD) which uses the Hall effect to generate electric power. The process is inefficient because the outlet gases are at high temperature. By utilizing the hot gases in a convenient gas turbine, the composite process could be made more efficient than the conventional steam turbine.

§ It is expensive to use electricity for heating except in special processes (e.g. electric arc furnaces) and where highly accurate controlled heating is required (e.g. induction heating).

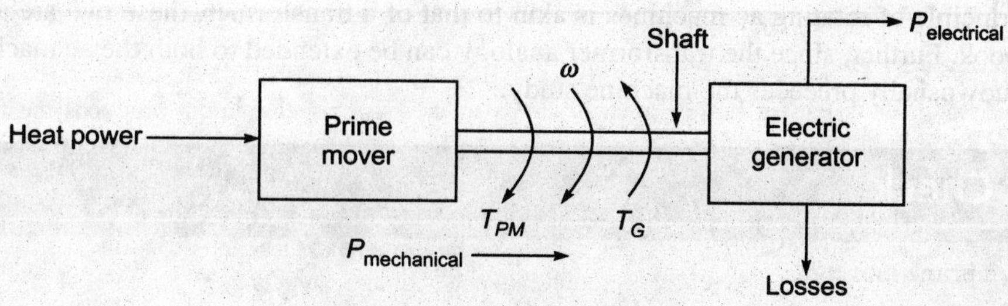

Fig. 1.1 *Electric generator*

the myriads of use points. The electromechanical energy conversion process is a *reversible* one and simple adjustment of mechanical shaft and electrical conditions reverses the flow of power as illustrated in Fig. 1.2. In this mode of operation, the electromechanical device, in general called the *electric machine*, is known as the *motor* and the machine is said to be in the motoring mode. Under steady speed operation, again T_M (motor) = T_L(load). Both in generating and motoring modes, losses occur in the electric machine but the overall conversion efficiencies are very high (close to or above 90%).

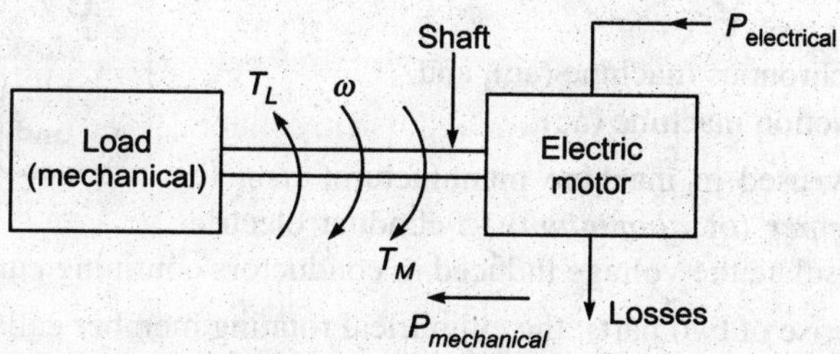

Fig. 1.2 *Motoring mode of operation of an electric machine*

A *transformer* is a static device that transforms electric energy from one ac voltage level to another. It is this device that has made the electric system almost universally ac. The electric power is generated at relatively low voltages (up to a maximum of 33 kV) which then is raised to very high voltages (e.g. 756 kV) by means of a transformer and then transmitted. High voltages are associated with low currents and reduced transmission losses. Geographically close to the use points, the electric power is transformed back to safe low utility voltages (400/231 V). A transformer consists basically of two coils (three sets of coil pairs for a 3-phase system) tightly coupled by means of magnetic (steel) core. Figure 1.3(a) gives the symbolic representation of a transformer and Fig. 1.3(b) shows a simple electric power generation transmission and reception system. A practical electric power system is an integrated one, far more complex than the simple diagrammatic representation of Fig. 1.3(b), and is in the form of an interconnected network for reasons of economy, operational efficiency and reliability.

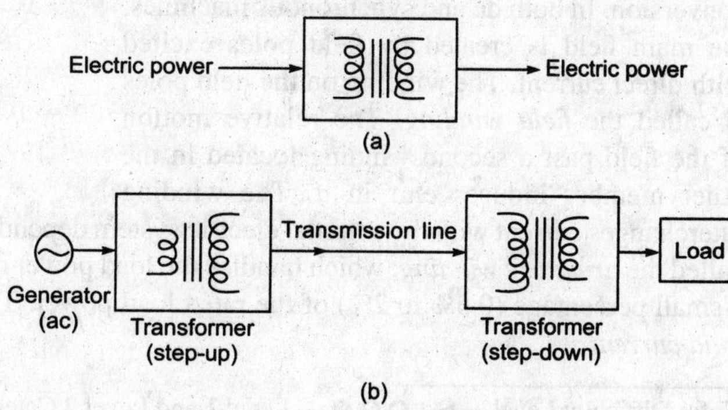

Fig. 1.3 *(a) Transformer, (b) Simple electric power system*

Because the principle of rotating ac machines is akin to that of a transformer, these two are always studied together in a book. Further, since the transformer analogy can be extended to both the ac machine types, the transformer study usually precedes the machine study.

Concept Check — LO 1

1. What is a prime mover? ○○●
2. What is an electric generator? ○○●
3. What is an electric motor? ○○●
4. What is a transformer? ○○●

1.3 BASIC PRINCIPLE, TYPES AND CONSTRUCTIONAL FEATURES OF ELECTRIC MACHINES

LO 2 Explain the construction, working principle, types and applications of basic rotating dc and polyphase ac machines

There are three basic rotating electric machine types, namely

1. the dc machine,
2. the polyphase synchronous machine (ac), and
3. the polyphase induction machine (ac).

Three materials are mainly used in machine manufacture; *steel* to conduct magnetic flux, *copper* (or *aluminium*) to conduct electric current and *insulation* to insulate the voltage induced in conductors confining currents to them.

All electric machines comprise of two parts: the cylindrical rotating member called the *rotor* and the annular stationary member called the *stator* with the intervening air-gap as illustrated in Fig. 1.4. The rotor has an axial *shaft* which is carried on bearings at each end located in *end covers* bolted to the stator. The shaft extends out of the end cover usually at one end and is coupled to either the prime mover or the load.

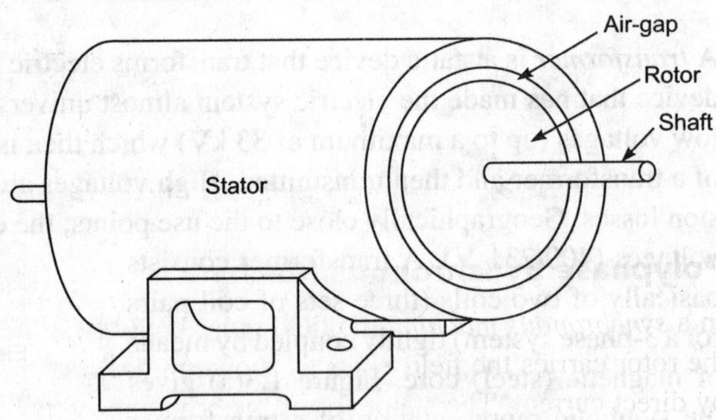

Fig. 1.4 *An electric machine*

The stator and rotor are both made of magnetic material (steel) which conducts the magnetic flux upon which depends the process of energy conversion. In both dc and synchronous machines, the main field is created by field poles excited with direct current. The winding on the field poles is called the *field winding*. The relative motion of the field past a second winding located in the other member induces emf in it. The winding interchanges current with the external electric system depending upon the circuit conditions. It is this winding, called the *armature winding*, which handles the load power of the machine, while the field winding consumes a small percentage (0.5% to 2%) of the rated load power. The load dependent armature current is known as *load current*.

Note: Difficulty Level ⟶ ○○● — Level 1 and Level 2 Category; ○●● — Level 3 and Level 4 Category; ●●● — Level 5 and Level 6 Category

dc Machine

In a *dc machine* the field poles are on the stator while the rotor is the armature as shown in the cross-sectional view of Fig. 1.5. The field poles are symmetrical and are even in number, alternately north and south. As the armature rotates, alternating emf and current induced in the armature winding are rectified to dc form by a rotating mechanical switch called the *commutator*, which is tapped by means of stationary carbon *brushes*. The commutator is cylindrical in shape and comprises several wedge-shaped copper segments bound together while they are insulated from each other. The armature is made of laminated steel with slots cut out on the periphery to accommodate the insulated armature winding. The ends of each armature coil are connected to the commutator segments to form a closed winding. The armature when carrying current produces stationary poles (same as number of field poles) which interact with the field poles to produce the electromagnetic torque.

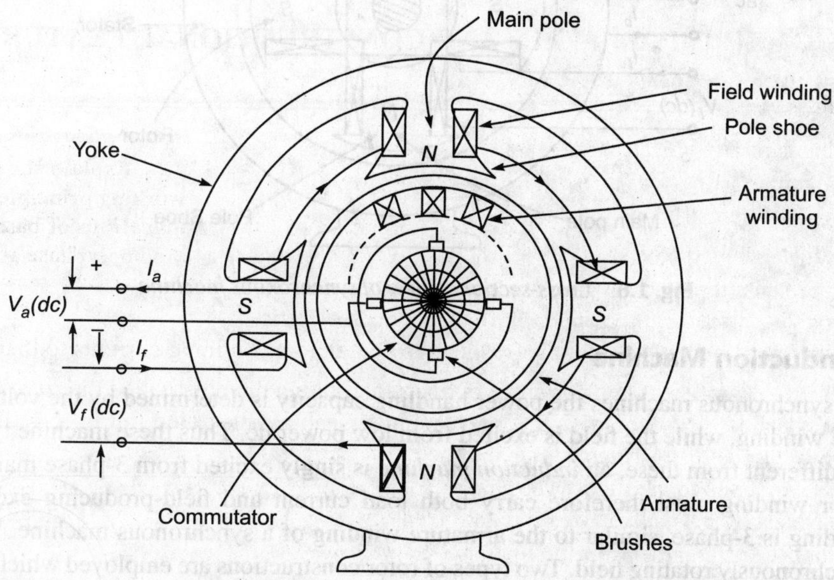

Fig. 1.5 *Cross-sectional view of dc machine*

Polyphase Synchronous Machine

In a *synchronous machine* the field poles could be either on the stator or rotor, but in all practical machines the rotor carries the field poles as shown in the cross-sectional view of Fig. 1.6. The field poles are excited by direct current. The stator forms the armature carrying a 3-phase winding wound for the same number of poles as the rotor. All the three phases have identical windings with the same angular displacement between any pair of phases. When the rotor rotates, it produces alternating emf in each phase forming a balanced set with frequency given by

$$f = \frac{nP}{120} \tag{1.1}$$

where f = frequency in Hz

n = rotor speed in rpm

P = number of field poles

For a given number of poles, there is a fixed correspondence between the rotor speed and the stator frequency; the rotor speed is therefore called the *synchronous speed.* When balanced 3-phase currents are allowed to flow in the armature winding, these produce a synchronously rotating field, stationary with respect to the rotor field as a result of which the machine produces torque of electromagnetic origin. The synchronous motor is, however, *nonselfstarting*.

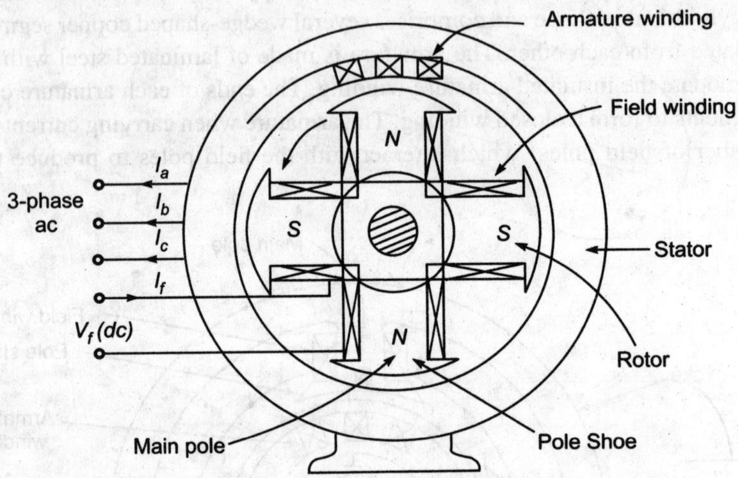

Fig. 1.6 *Cross-sectional view of synchronous machine*

Polyphase Induction Machine

In both dc and synchronous machines the power handling capacity is determined by the voltage and current of the armature winding, while the field is excited from low power dc. Thus these machine types are doubly excited. Quite different from these, an *induction machine* is singly excited from 3-phase mains on the stator side. The stator winding must therefore carry both load current and field-producing excitation current. The stator winding is 3-phase, similar to the armature winding of a synchronous machine. When excited it produces a synchronously rotating field. Two types of rotor constructions are employed which distinguish the type of induction motor.

1. Squirrel-cage rotor Here the rotor has copper (or aluminium) bars embedded in slots which are short-circuited at each end as shown in Fig. 1.7(a). It is a rugged economical construction but develops low starting torque.

2. Slip-ring (or wound-rotor) rotor The rotor has a proper 3-phase winding with three leads brought out through slip-rings and brushes as shown in Fig. 1.7(b). These leads are normally short-circuited when the motor is running. Resistances are introduced in the rotor circuit via the slip-rings at the time of starting to improve the starting torque.

The rotating field created by the stator winding moves past the shorted rotor conductors inducing currents in the latter. These induced currents produce their own field which rotates at the same speed (synchronous) with respect to the stator as the stator-produced field. Torque is developed by the interaction of these two relatively stationary fields. The rotor runs at a *speed close* to synchronous but always slightly lower than it. At the synchronous speed no torque can be developed as zero relative speed between the stator field and the rotor implies no induced rotor currents and therefore no torque.

Single-phase ac motors are employed for low-voltage, low-power applications—fractional-kW motors. They operate on the same basic principles as the 3-phase motor, but the pulsating single-phase field produces additional losses, reducing motor torque and the pulsating torque component increases the noise level of the motor.

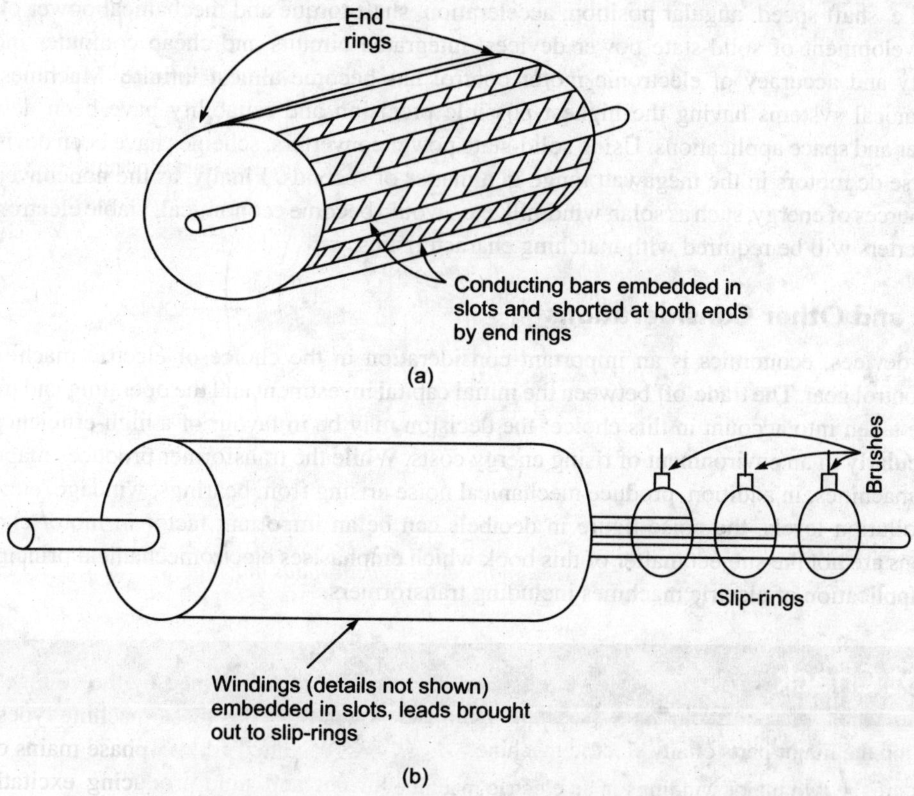

Fig. 1.7 *Schematic diagram: (a) A squirrel-cage rotor, (b) A wound rotor*

An induction machine connected to the mains when driven at *supersynchronous* speed behaves as a generator feeding power into the electric system. It is used in small hydroelectric stations and wind and aerospace applications.

The insulation of a machine (or transformer) is its most vulnerable part because it cannot be stressed beyond a certain temperature. For a given frame size, the steady temperature rise is determined by the machine loading, the associated power loss (this appears in the form of heat) and the cooling provided. Thus the maximum machine loading called its *rating* for a given frame size is limited by the permissible temperature rise which is dependent upon the class of insulation used. In the case of high-speed dc machines poor *commutation* (reversal of current in armature coils) may become a limiting factor even before the temperature limit is reached. The speed itself may be a limiting factor in very high-speed machines on account of the centrifugal forces developed. This limit is more stringent in dc machines with complicated armature construction than in the rugged rotor induction motor. Because of their high thermal capacity, machines are quite capable of withstanding a fair amount of overloads for short durations.

Motor Control

There is great diversity and variety in the components and systems used to control rotating machines. The purpose of a motor control may be as simple as start/stop or the control of one or more of the motor output parameters, i.e. shaft speed, angular position, acceleration, shaft torque and mechanical power output. With the rapid development of solid-state power devices, integrated circuits and cheap computer modules, the range, quality and accuracy of electronic motor control has become almost infinite. Machines and other electromechanical systems having the highest possible precision and reliability have been developed for nuclear power and space applications. Using solid-state power converters, schemes have been devised to start, stop or reverse dc motors in the megawatt range in a matter of seconds. Finally, as the nonconventional and renewable sources of energy, such as solar, windmill, etc., would become economical, viable electromechanical energy converters will be required with matching characteristics.

Economic and Other Considerations

As in other devices, economics is an important consideration in the choice of electric machines and the associated control gear. The trade-off between the initial capital investment and the operating and maintenance cost must be taken into account in this choice; the decision may be in favour of a high-efficiency high-cost motor, particularly in an environment of rising energy costs. While the transformer produces magnetic noise, the rotating machines, in addition, produce mechanical noise arising from bearings, windage, etc. In present-day noise-pollution levels, the noise figure in decibels can be an important factor in motor choice. These considerations are not the subject matter of this book which emphasises electromechanical principles and the theory and application of electric machines including transformers.

Concept Check — LO 2

1. Mention the major parts of any electric machine. ○○●
2. What are the two major windings in an electric machine? ○○●
3. What is a commutator? ○●●
4. What is meant by synchronous speed? ○○●
5. List the two types of rotor constructors employed in an induction motor. ○○●
6. What are the various output parameters to be monitored and controlled in a motor? ○●●

1.4 RECENT TRENDS IN RESEARCH AND DEVELOPMENTS IN ELECTRIC MACHINES

LO 3 Summarize the recent trends and developments in the field of electric machines

Design and operation of electrical machines become easier and cheaper with suitable electric drives. This electric drive converts and feeds the input energy to the machine according to the desired operation. A power electronic converter constitutes the heart of the drive system which uses the power semiconductor devices. These converters help to convert the power from one form to another. Various advancements in converter topologies, and control methods have been proposed to convert and control the energy efficiently. Other intelligent techniques such as Neural Network, Artificial Intelligence, Expert system, Fuzzy logic and Evolutionary Computing are used to make the control most accurate and fast.

Research in converter topologies has also improved the power quality at the supply end. Various multipulse and multilevel power electronic converters have been developed for power-quality improvement along with cost-effective ac-dc converters for power factor improvement in electric drive system.

Electrical energy offers the most flexible, economic and efficient mode of power generation, transmission, distribution and utilization compared to other forms of energy systems. Most of the power required for human activities round the globe continues to come from electrical machines from the very large generators installed in power stations to the very small motors in automotive control systems. The rapid depletion and the increased cost of conventional fuels have given a thrust in the research on isolated asynchronous generators as alternative power sources, converting from wind energy, biogas, hydro units and biomass. Asynchronous generators operated in isolated systems for supplying electricity to the remote areas, where grid supply is not accessible, are best options because of having certain advantages such as low cost, less maintenance and brushless constructions.

In a wind-energy conversion system, the voltage and frequency variation at the generator terminals is due to varying consumer loads as well as change in wind speeds. Therefore, the controller should have the capability to control the voltage and frequency of isolated generators under dynamic conditions. Various types of voltage and frequency controllers are proposed for constant-speed, constant-power applications.

Friction, vibration and noise can be eliminated in industrial drive by using permanent magnet direct-drive technology. Both ac and dc *PM* motors are more suitable for high performance and wide speed variation applications. Motors with higher torque and low speed are highly appreciated. PM brushless motors are of this category.

The electric machines market is rapidly growing because of various developments and emerging areas such as wind energy, marine, traction and offshore.

The materials used for making the electrical machines play a major role in their performance. Particularly, maximum temperature rise of the material (permanent magnet and insulation materials) used will affect the rated torque of an electrical machine. The operating temperature strongly affects the performances of electrical machines. Finite Element Analysis (FEA) can be used to compute the temperature of the material and machines. A thermal network of the electrical machines is considered and input to this network is calculated from magnetic and electrical loading of the machines. An Object Oriented Program (OOP) is used to develop thermal network and it allows a convenient organization of thermal analysis process.

Modern commercial electrical steels can be grouped under non-oriented, grain-oriented and rapidly quenched alloy types, of which the first two dominate the applications. Because of the limitations on the shape of the magnetic path, grain-oriented types are used predominantly in large power and distribution transformers, while non-oriented ones are used in rotating machines and small apparatus. The silicon content is critical to the performance of electrical steels because it increases resistivity but it, also reduces the anisotropy and so reduces losses and magnetostriction and permeability and makes the material brittle. In non-oriented type, this content varies between 2.9% to 3.2%.

The continuing development of high permeability silicon-steel, metallic glass, ferrites, aluminum ceramics and high temperature insulating materials, permanent magnetic materials like Neomax (Nd-Fe-B) and rare-earth cobalts has been influencing the construction and design of many large and small machines and apparatus from large generators to small step motors. Not only the cost of material per kVA has come down in many cases, more reliable and highly efficient machines have been successfully designed.

Areas which have revolutionized the growth of research in electrical machines could be grouped as

(a) Design of power electronic converters for motor drive and with better efficiency and control
(b) Increase in rating of power semiconductor solid-state devices
(c) Development of *cheap* digital signal processing controllers for operation and control
(d) The development of evolutionary computing techniques, artificial intelligent techniques for machine design, operation and control
(e) Power quality and power factor improvement using improved power quality converters for various motor-drive applications
(f) Design of motor drives of electric and hybrid electric vehicles
(g) Condition monitoring of electrical machines using Artificial Intelligence techniques
(h) Design of electric machines using CAD/CAM techniques and validation through *FE* analysis

Thyristor is still unbeaten in high voltage and current rating among power semiconductor devices with voltage and current rating of 12 kV and 6 kV, respectively. New devices such as IGCT, GCT, power MOSFET, power IGBT have opened up new vistas for the electronic switching and control of energy-converting devices; ac/dc, dc/ac, ac/ac and dc/dc converters are at their peak in various applications and totally changed the way of operating the machines from the conventional mode of operation. Development of self-commutated devices overcome the drawbacks of line commutated converters and it is used in both Voltage Source Converters (VSC) and Current Source Converters (CSC). High voltage dc transmission (HVDC), Flexible ac Transmission System (FACTS), variable frequency operation of machines, voltage control of dc machines, Switched Mode Power Supply (SMPS), offer a wide variety of applications for converters and for power engineers to develop new and realiable system configurations. The most common application is the variable speed drives using dc and ac machines. Power electronics appear to have shifted the emphasis of electrical engineers from the design of special types of variable speed machines to the use of special electronic circuits to make an existing machine to give the desired variable speed characteristics and performances.

Electric vehicles and hybrid electric vehicles are one of the most recent and potential applications of electrical machines. These vehicles use the motor drives in association with power electronic converters. Application of electrical drives in *system automobile* began with 6 V in the first quarter of the 20th century. Initially, it was used only for basic and necessary functions such as ignition, cranking and lighting loads. Since then, there has been a constant increase in power demand and now the whole vehicle is *driven* by the electrical system. Electrical system has replaced all mechanical, pneumatic and hydraulic systems, thereby increasing the efficiency and performance of the automobile system.

Present voltage level being used is 12 V, and in the future it is expected to switch over to higher levels such as 42 V and 300 V. There is a immense challenge in design of power electronic converters and suitable high-efficiency machines for hybrid electric vehicles.

To make the squirrel-cage induction motor run like a separately excited dc motor has remained a dream of generations of electrical engineers. Direct Torque Control (DTC) has fulfilled this dream and simplified the control circuit of induction motor to a great extent. The availability of *cheap* microprocessor-based system has further improved the control of electrical *devices* of all forms for obtaining the desired steady state, transient as well as dynamic characteristics from the existing drive machine with the help of suitable designed solid-state control circuits.

Electrical machines and drive systems are subjected to *many different* faults. They may include stator faults, rotor electrical faults and rotor mechanical faults, failure of power electronics system and damage of mechanical parts. Most of industrial processes demand continuous operation. This is mainly influenced by the condition monitoring leading to fault diagnosis and prediction of performance of electrical machines and drives. Fast and accurate diagnosis of machine faults results in prevention of failure and avoiding processes interrupt and reduce the idle time of machines. It also helps in reducing financial loss, avoids harmful effects and devastation of the system.

For many years, the manufactures and users had relied on protective relays such as over-current and over-voltage relays to trip fault machines. This scheme may lead to machine damage and other harmful effects. Various intelligence techniques such as Artificial Neural Networks (ANN), Expert System (ES), Fuzzy System (FS) are being used for machines analysis for monitoring and control to make the process continuous, fast and accurate.

One of the latest developments in ac motor research has been in the direction of field-oriented control or making ac motors perform like dc motors with highly accurate torque and power control. Another recent development in ac motor drives is a system called Direct Torque Control and Direct Self-Control. There is no modulator and no need for an encoder to feedback information about motor shaft speed and position. The DTC sensorless-type control incorporates fast digital signal processing hardware, resulting in a torque response which is ten times faster than any ac or dc drive.

One of the reasons why ac technology will continue to make inroads into dc dominance is reduced power consumption. Brushless dc will survive eventually but it is not dc at all but a completely different technology. It is called dc because of the concept of external commutation.

Now there are increased levels of customer support. With the emphasis on just-in-time production, downtime is unacceptable. With today's systems, if the machine goes down one can, via modem, have a technician at a remote site use system diagnosis software to troubleshoot the entire system from anywhere in the world.

Large Rating Machines

The innovative design features as direct water-cooled armature windings, gap-pickup rotor winding cooling, Micapal II stator insulation, Class F rotor and stator insulation, advanced Tetraloc stator endwinding support systems, and the side ripple-spring armature bar slot support structure were developed for heavy rating machines. The Finite Element Model (FEM) is exercised to interrogate the generator assembly for all loads encountered during assembly and operation. Electromagnetic force is the most likely force to cause structural issues. This force is cyclic and acts in the radial direction at the inside diameter of the stator core with a magnitude of one million pounds. The resulting stator core vibration and the transmission to the generator structure and foundation is a significant design consideration. Forced harmonic response *analyses* are performed to ensure that the electromagnetic forces cannot excite the machine's natural frequencies. The structural design of the stationary components also must be considered when calculating the dynamic behavior of the rotor, since the rotor is supported on bearings located in the end shields of the machine. In this load configuration, the structural vibratory loads caused by the rotor, and the loading caused by stator vibration that drives rotor behavior, are interrogated. Once again, a forced harmonic analysis is performed to understand and optimize the interactions. Lastly, the structural design has a major impact on the overall producibility and serviceability of the generator. The complexity of the fabrication determines the unit's machining cycles as well as its accessibility for thoroughly cleaning the inner cavities of the machine before shipping.

Generator 2 pole machine running at 3000/3600 rpm and 4 pole machine running at 1500/1800 rpm have become common for large outputs or 1000 MW of more. The main dimension of 1000 MVA machine is the output coefficient C which will be 2 MVA s/m^3 and D^2L will be 10 m^3. The higher value of output coefficient is made possible by enhanced cooling techniques. The excitation current is 7 kA at 650 V for 1500 MVA machine and 5.7 kA, 640 V for 1000 MVA machine respectively.

The stator core, made from grain-oriented silicon steel for low loss and high permeability, is mounted rigidly on the inner frame. Isolation of the core vibration from the remainder of the structure is accomplished through the use of flexible pads between the feet on the inner frame and the base structure. The end windings are secured to a strong structure of insulating materials to be used. A solid cone of filament-wound resin bonded fibre glass is used to generate strength and long-term rigidity. The coils are bedded to the structure with comfortable packing material. The complete structure is bolted to the end of the core. Axial movement may be allowed to accommodate expansion of coils relative to the core. Low-loss stator core, of grain-oriented silicon steel, minimizes electrical losses within the core to increase machine efficiency.

The insulation between turns is usually provided by interleaves of resin-bonded glass fabric material. The stator winding arrangements for 1000 MVA should be four parallel paths in 2-pole machines and in 4-pole machines, it can be in parallel or series parallel combination. The efficiency of 1000–1500 MVA is normally very high. Due to this megawatt loss, good cooling medium is required. Direct hydrogen cooling of rotors is developed for 1000 MVA. The cooling of the rotor can be done by 1). Each coil is wound with continuous length of copper strap bent on edge at the four corners, 2). Larger section conductors of silver bearing copper-containing grooves and holes to provide the passage of gas. Cool, deionized water, supplied by a closed-loop auxiliary system, flows through copper strands in the stator winding, and the warm water is discharged at the turbine end of the generator. The hydrogen-cooled machines can be designed with a higher electrical loading than the air-cooled machines due to the better cooling, and tend to have a larger subtransient reactance than an air-cooled machine.

The generator's performance is heightened by optimization of its bar strand configuration, including hollow-to-solid-strand ratio. Spring bar stator core support system isolates vibration of the stator core, to minimize vibration transmitted to the foundation. Stator winding support features top wedges and ripple springs to secure stator bars in the slot and eliminate bar vibration. This maximizes insulation life and reduces maintenance requirements. Core-end cooling is enhanced through proven design concepts to control core temperatures and minimize eddy current losses. Approaches include split tooth, stepped core, flux shields and non-magnetic materials. Retaining rings of 18-Mangaanese/18-Chromium, non-magnetic stainless steel resists stress-corrosion cracking. Internal view of a 1000 MW generator is shown in Fig. 1.8.

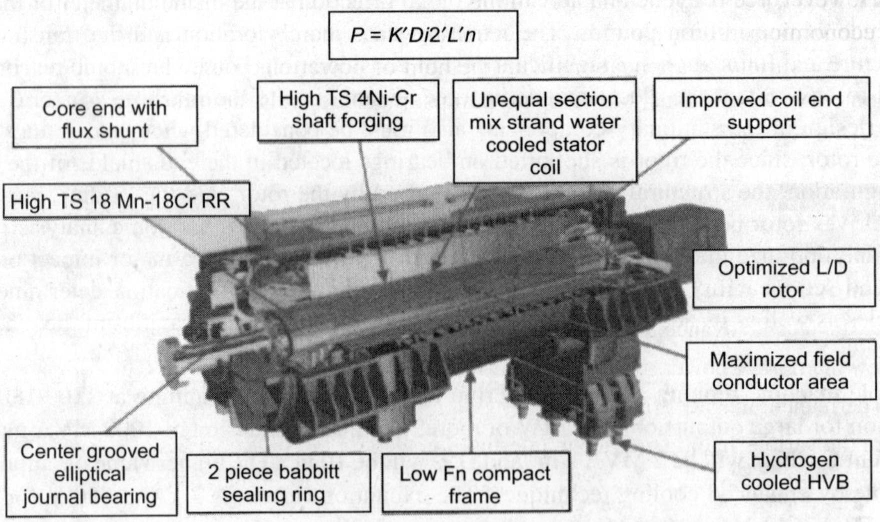

Fig. 1.8 *Internal view of 1000 MW generator (Courtesy, Toshiba Corporation)*

Table 1.1 Generator Parameters (600 MW)

Parameters	Unit	
Rated capacity	MW	600
Rated power factor		0.85
Synchronous reactance	Xd%	240.313
Transient reactance	Xd′%	28.281
Sub-transient reactance	Xd″%	21.582
Nagative-seq reactance	X2%	21.295
Zero-seq reactance	X0%	10.131
Direct axis time constant		
Open circuit time constant	Tdo′ Sec	8.724
Short circuit time constant	Td′ Sec	1.027
Open circuit sub-transinet	Tdo″ Sec	0.046
Short circuit sub-transient	Td″ Sec	0.035
Qudrature axis time constant		
Open circuit time constant	Tdo′ Sec	0.969
Short circuit time constant	Td′ Sec	0.169
Open circuit sub-transinet	Tdo″ Sec	0.068
Short circuit sub-transient	Td″ Sec	0.035

Applications of Electric Machines

The term 'electric machines' is generally used for electric generators and motors in the context of electrical engineering. Generators normally act as a source and motors as loads. These electric machines may be powered by direct current or by alternating current which leads to the two main classifications: dc machines and ac machines. dc generators are used for general lighting, charging battery, giving excitation to alternators, as boosters in feeders, for supplying field excitation current in dc locomotives for regenerative braking, arc welding, etc. However, at present, the applications of dc generators have become very limited because of technical and economic reasons. Nowadays, electric power is mainly generated in the form of alternating current and used for variety of applications with the help of power electronic devices. Electric motors find wide applications in industrial fans, blowers and pumps, machine tools, household appliances, power tools, and disk drives.

Concept Check — LO 3

1. What is an electric drive?
2. What are the recent advancements in converter topologies?
3. Name few alternative power sources for conventional fuels.
4. What is permanent magnet drive technology?
5. What are the various intelligence techniques used in electric machine analysis?

SUMMARY

LO 1

- Prime mover is basically an engine that converts energy from a source into mechanical energy thereby imparting power or motion to another device.
- Electric machines may be categorized into three major devices, namely, electric generator, electric motor and transformer. Generators normally acts as a source and motors as loads. These electric machines may be powered by direct current or by alternating current which leads to the two main classifications: dc machines and ac machines.

LO 2

- All electric machines comprise of two parts: the cylindrical rotating member called the rotor and the annular stationary member called the stator with the intervening air-gap.
- Induction motors employ two types of rotor constructions, viz., Squirrel Cage rotor and Slip-ring or Wound rotor.

LO 3

- An electric drive is basically a power electronic converter system which converts and feeds energy to the electric machine according to the desired operation.
- Both ac and dc Permanent Magnet motors are highly suitable for high performance and wide speed variations.

MULTIPLE CHOICE QUESTIONS

1. The field coils of a dc generator are usually made of **LO 2**
   (a) mica (b) copper
   (c) cast iron (d) steel
2. Which of the following applications require high starting torque? **LO 2**
   (a) Lathe machine (b) Centrifugal pump
   (c) Locomotive (d) Air blower
3. Which of the following does not change in a transformer? **LO 2**
   (a) Voltage (b) Current
   (c) Frequency (d) All of the above
4. The commercial sources of energy are **LO 2**
   (a) solar, wind and biomass
   (b) fossil fuels, hydropower and nuclear energy
   (c) wood, animal wastes and agricultural wastes
   (d) None of the above
5. Electric power is generated at **LO 2**
   (a) 33 kV (b) 110 kV
   (c) 400 kV (d) 12 kV

Answers

1. (b) **2.** (c) **3.** (c) **4.** (b) **5.** (a)

CHAPTER 2

Magnetic Circuits and Induction

LEARNING OBJECTIVES

After studying this chapter, students will be able to

LO1 Explain magnetic circuits and magnetization characteristics

LO2 Identify the different magnetic materials and illustrate their properties

LO3 Develop expressions for magnetically induced emf and force and understand the ac operation of magnetic circuits

LO4 Define hysteresis and classify the eddy current losses

LO5 Classify permanent magnets and list their applications

2.1 | INTRODUCTION

The electromagnetic system is an essential element of all rotating electric machinery and electromechanical devices as well as static devices like the transformer. The role of the electromagnetic system is to establish and control electromagnetic fields for carrying out conversion of energy, its processing and transfer. Practically all electric motors and generators, ranging in size from fractional horsepower units found in domestic appliances to the gigantic several thousand kW motors employed in heavy industry and several hundred megawatt generators installed in modern generating stations, depend upon the magnetic field as the coupling medium allowing interchange of energy in either direction between electrical and mechanical systems. A transformer though not an electromechanical conversion device, provides a means of transferring electrical energy between two electrical ports via the medium of a magnetic field. Further, transformer analysis runs parallel to rotating machine analysis and greatly aids in understanding the latter. It is, therefore, seen that all electric machines including transformers use the medium of magnetic field for energy conversion and transfer. The study of these devices essentially involves electric and magnetic circuit analysis and their interaction. Also, several other essential devices like relays, circuit breakers, etc. need the presence of a confined magnetic field for their operation.

The purpose of this chapter is to review the physical laws governing magnetic fields, induction of emf and production of mechanical force, and to develop methods of magnetic-circuit analysis. Simple magnetic circuits and magnetic materials will be briefly discussed. In the chapters to follow, how the concepts of this chapter are applied in the analysis of transformers and machines will be shown.

Industry Relevance

Electromagnetic induction is an incredibly useful phenomenon with a wide variety of applications. Induction is used in power generation and power transmission. Practically all electric motors and generators, ranging in size from fractional horsepower units found in domestic appliances to the gigantic several thousand kW motors employed in heavy industries and several hundred megawatt generators installed in modern generating stations, depend upon the magnetic field as the coupling medium allowing interchange of energy in either direction between the electrical and mechanical systems.

Some of the most important applications of magnetic materials and induction are as follows:

- Power conversion from electrical to mechanical in motors, generators, and electromagnets.
- Power adaption in transformers.
- Signal transfer in transformers in both LF and HF.
- Magnetic field screening.
- Permanent magnets are used in loudspeakers, small generators, small motors, and sensors.
- Data storage analog in audio and video tapes.
- Data storage analog in ferrite core memory drum, hard disc, floppy disc, and bubble memory.
- In quantum devices like GMR reading head and MRAM.

2.2 MAGNETIC CIRCUITS

Maxwell's Equations

LO 1 Explain magnetic circuits and magnetization characteristics

The exact description of the magnetic field is given by the Maxwell's equations* and the constitutive relationship of the medium in which the field is established.

Such description apart from being highly complex is otherwise not necessary for use in electric machines wherein the fields (magnetic and electric) are slowly varying (fundamental frequency being 50 Hz) so that the *displacement current* can be neglected. The magnetic field can then be described by Ampere's law and is solely governed by the *conduction current*. This law is in integral form

Probe

Derive Maxwell's equations governing the electric and magnetic fields.

*Maxwell's equations governing the electric and magnetic fields are

$$\vec{\nabla}\cdot\vec{B} = 0 \quad \text{and} \quad \vec{\nabla}\cdot\vec{D} = \rho$$

$$\vec{\nabla}\times\vec{H} = \vec{J} + \frac{\partial \vec{D}}{\partial t} \quad \text{and} \quad \vec{D} = \varepsilon_0 \vec{E}\,;\ \varepsilon_0 = 8.85 \times 10^{-12}$$

wherein $\vec{J}$ = conduction current density and $\vec{D}$ = displacement current density, negligible for slowly-varying fields $\vec{D} = \varepsilon_0 \vec{E}\,;\ \varepsilon_0 = 8.85 \times 10^{-12}$ F/m.

and is easily derivable from the third Maxwell's equation (by ignoring displacement current) by means of well-known results in vector algebra. The Ampere's law is reproduced as follows:

$$\int_s \vec{J} \cdot d\vec{s} = \oint \vec{H} \cdot d\vec{l} \tag{2.1}$$

wherein $\vec{J}$ = conduction current density

$\vec{H}$ = magnetic field intensity

s = the surface enclosed by the closed path of length l

$d\vec{s}$ = differential surface

$d\vec{l}$ = differential length

Simple Electromagnetic System

Consider the example of a simple electromagnetic system comprising an exciting coil and *ferromagnetic* core as shown in Fig. 2.1. The coil has N turns and carries a constant (dc) current of iA. The magnetic field is established in the space wherein most of the total magnetic flux set up is confined to the ferromagnetic core for reasons which will soon become obvious. Consider the flux path through the core (shown dotted) which in fact is the *mean path* of the core flux. The total current piercing the surface enclosed by this path is as follows:

$$\int_s \vec{J} \cdot d\vec{s} = Ni$$

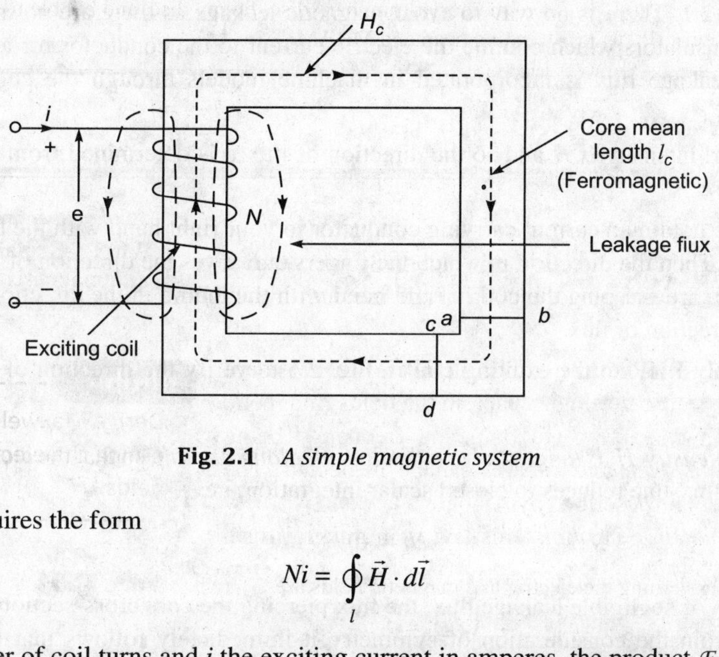

Fig. 2.1 *A simple magnetic system*

Hence Eq. (2.1) acquires the form

$$Ni = \oint_l \vec{H} \cdot d\vec{l} \tag{2.2}$$

Since N is the number of coil turns and i the exciting current in amperes, the product $\mathcal{F} = Ni$ has the units of *ampere-turns* (AT) and is the cause of establishment of the magnetic field. It is known as the *magnetomotive force* (mmf) in analogy to the electromotive force (emf) which establishes current in an electric circuit.

Probe

Define magnetomotive force.

Magnetic Field Intensity

The magnetic field intensity H causes a flux density B to be set up at every point along the flux path which is given by

$$B = \mu H = \mu_0 \mu_r H \qquad \text{(for flux path in core)} \qquad (2.3a)$$

and

$$B = \mu_0 H \qquad \text{(for flux path in air)} \qquad (2.3b)$$

The units of flux density are weber (Wb)/m^2 called *tesla* (T). The term μ_0 is the *absolute permeability* of free space and has a value of

$$\mu_0 = 4\pi \times 10^{-7} \qquad \text{henry (H)/m}$$

The permeability $\mu = \mu_0\mu_r$ of a material medium is different from μ_0 because of a certain phenomenon occurring in the material. The term μ_r is referred to as *relative permeability* of a material and is in the range of 2000-6000 for ferromagnetic materials (see Sec. 2.3). It is, therefore, seen that for a given H, the flux density B and, therefore, the flux over a given area

$$\phi = \oint_s \vec{B} \cdot d\vec{s}$$

will be far larger in the magnetic core in Fig. 2.1 than in the air paths. Hence, it is safe to assume that the magnetic flux set up by mmf Ni is mainly confined to the ferromagnetic core and the flux set up in air paths is of negligible value. The flux set up in air paths is known as the *leakage flux* as if it leaks through the core; some of the leakage flux paths are shown chain-dotted in Fig. 2.1. There is no way to avoid magnetic leakage as there are no magnetic insulators in contrast to electric insulators which confine the electric current to the conductor for all practical purposes. The effect of the leakage flux is incorporated in machine models through the concept of the *leakage inductance*.

> **Probe**
>
> Define leakage inductance.

The direction of field intensity *is* H and so the direction of flux ϕ is determined from the Right Hand Rule (RHR). It is stated as:

Imagine that you are holding a current carrying conductor in your right hand with the thumb pointing in the direction of current. Then the direction in which the fingers curl gives the direction of flux. In case of a coil you imagine that you are grasping the coil in right hand with the thumb in the direction of current; then the fingers curl in the direction of flux.

The reader may apply RHR to the exciting coil in Fig. 2.1 to verify the direction of flux as shown in the figure.

The magnetic field intensity $\vec{H}$ *is tangential to a flux line all along its path*, so that the closed vector integration in Eq. (2.2) along a flux-line reduces to closed scalar integration, i.e.

$$Ni = \oint_l H \cdot dl \qquad (2.4)$$

With the assumption of negligible leakage flux, the flux piercing the core cross-section at any point remains constant. Further, from the consideration of symmetry it immediately follows that the flux density over straight parts of the core is uniform at each cross-section and remains constant along the length; such that H is constant along the straight parts of the core. Around the corners, flux lines have different path lengths between magnetic equipotential planes (typical ones being *ab* and *cd* shown in Fig. 2.1) so that H varies from a high value along inner paths to a low value along outer paths. It is reasonable to assume that H shown dotted along the mean path will have the same value as in straight parts of the core (this *mean path technique* renders simple the analysis of magnetic circuits of machines and transformers).

It has been seen previously that the magnetic field intensity along the mean flux path in the core can be regarded constant at H_c. It then follows from Eq. (2.4) that

$$\mathcal{F} = Ni = H_c l_c \tag{2.5}$$

where $\mathcal{F}$ = **mmf** in AT and l_c = mean core length (m)

From Eq. (2.5)

$$H_c = \frac{Ni}{l_c} \quad \text{AT/m} \tag{2.6}$$

If one now imagines that the exciting current i varies with time, Eq. (2.6) would indicate that H_c will vary in unison with it. Such fields are known as *quasi-static* fields in which the field pattern in space is fixed but the field intensity at every point varies as a replica of the time variation of current. This simplified field picture is a consequence of negligible displacement current in slowly-varying fields as mentioned earlier. In a quasi-static field, the field pattern and field strength at a particular value of time-varying exciting current will be the same as with a direct current of that value. In other words, a field problem can be solved with dc excitation and then any time variation can be imparted to it.

Probe

What are quasi-static fields?

Now, the core *flux density* is given by,

$$B_c = \mu_c H_c \quad \text{tesla (T)}$$

and core flux (assumed to be total flux) is given by,

$$\phi = \oint_s \vec{B} \cdot d\vec{s} = B_c A_c \quad \text{Wb}$$

where A_c = cross-sectional area of core and flux in the limbs is oriented normal to cross-sectional area. Then from Eq. (2.6)

$$\phi = \mu_c H_c A_c = \frac{Ni}{\left(\dfrac{l_c}{\mu_c A_c}\right)} \quad \text{or} \quad \phi = \frac{\mathcal{F}}{\mathcal{R}} = \mathcal{F}\mathcal{P} \tag{2.7}$$

where $\mathcal{R} = \dfrac{\mathcal{F}}{\phi} = \dfrac{l_c}{\mu_c A_c}$ = ***reluctance***[†] of the magnetic circuit (AT/Wb) (2.8)

and $\mathcal{P} = 1/\mathcal{R}$ = *permeance* of the magnetic circuit. It is, therefore, seen that by certain simplifying assumptions and field symmetries, it has been possible to lump the distributed magnetic system into a lumped magnetic circuit described by Eq. (2.7) which is analogous to Ohm's law in dc circuits. The electrical circuit analog of the magnetic system (now reduced to a magnetic circuit) is shown in Fig. 2.2 wherein $\mathcal{F}$ (mmf) is analogous to E (emf), $\mathcal{R}$ (reluctance) is analogous to R (resistance) and ϕ (flux) is analogous to i (current).

Probe

Give the electrical analogy of a simple magnetic circuit.

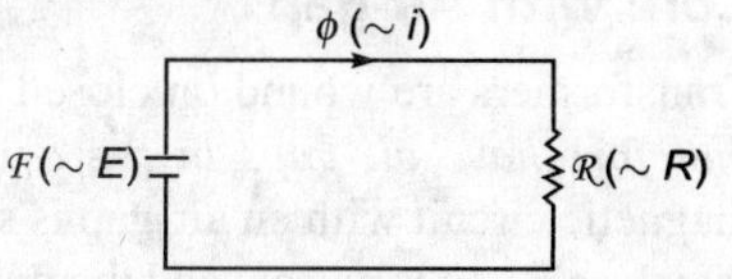

Fig. 2.2 *Electrical analog of the simple magnetic circuit of Fig. 2.1*

[†] Unit of reluctance is AT/Wb and will not be specified every time in examples.

The analogy though useful is, however, not complete; there being two points of difference: (i) magnetic reluctance is nondissipative of energy unlike electric resistance, (ii) when $\mathcal{F}$ is time-varying, the magnetic circuit still remains resistive as in Fig. 2.2, while inductive effects are bound to appear in an electric circuit. This is because there is no time-lag between the exciting current and the establishment of magnetic flux (quasi-static field).

The lumped magnetic circuit and its electrical analog are useful concepts provided the permeability (μ) of the core material and, therefore, the core reluctance is constant as is tacitly assumed above. This, however, is not the case with ferromagnetic materials, but when air-gaps are involved, the assumption of constant reluctance is generally valid and leads to considerable simplicity in magnetic circuit analysis.

In more complicated magnetic circuits—with multiple excitations and series-parallel core arrangement—the general theorems of electric circuits apply, i.e. Kirchhoff's voltage (mmf) law and Kirchhoff's current (flux) law. This is illustrated in Example 2.3.

B-H Relationship (Magnetization Characteristic)

In free space (also nonmagnetic materials), the permeability μ_0 is constant so that *B-H* relationship is linear. This, however, is not the case with ferromagnetic materials used in electric machines, wherein the *B-H* relationship is strictly nonlinear in two respects—*saturation* and *hysteresis*. Hysteresis non-linearity is the double valued *B-H* relationship exhibited in cyclic variation of *H* (i.e. exciting current). This nonlinearity is usually ignored in magnetic circuit calculations and is important only when current wave shape and power loss are to be accounted for. This is discussed in Sections 2.3 and 2.6. A typical normal *B-H* relationship (magnetization characteristic) for ferromagnetic materials is shown in Fig. 2.3. It has an initial nonlinear zone, a middle almost linear zone and a final saturation zone in which *B* progressively increases less rapidly with *H* compared to the linear zone. In the deep saturation zone, the material behaves like free space.

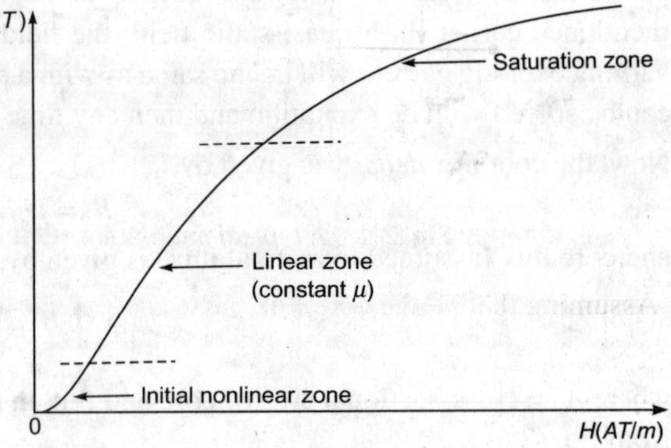

Fig. 2.3 *Typical normal magnetization curve of ferromagnetic material*

Due to considerations dictated by economy, electric machines and transformers are designed such that the magnetic material is slightly saturated (i.e. somewhat above the linear zone). In exact magnetic circuit calculations the nonlinear magnetization curve has to be used necessitating graphical/numerical solutions.

Core with Air-gap

Transformers are wound on closed cores as in Fig. 2.1. *Rotating machines have a moving element and must therefore have air-gaps in the cores out of necessity*. A typical magnetic circuit with an air-gap is shown in Fig. 2.4. It is assumed that the air-gap is narrow and the flux coming out of the core passes straight down the air-gap such that the flux density in the air-gap is the same as in the core. Actually as will soon be seen, that the flux in the gap fringes out so that the gap flux density is somewhat

> **Probe**
> What is the need for air-gaps in the cores of rotating machines?

less than that of the core. Further, let the core permeability μ_c be regarded as constant (linear magnetization characteristic).

The mmf Ni is now consumed in the core plus the air-gap. From the circuit model of Fig. 2.4(b) or directly from Fig. 2.4(a)

$$Ni = H_c l_c + H_g l_g \tag{2.9a}$$

or

$$Ni = \frac{B_c}{\mu_c} l_c + \frac{B_g}{\mu_0} l_g \tag{2.9b}$$

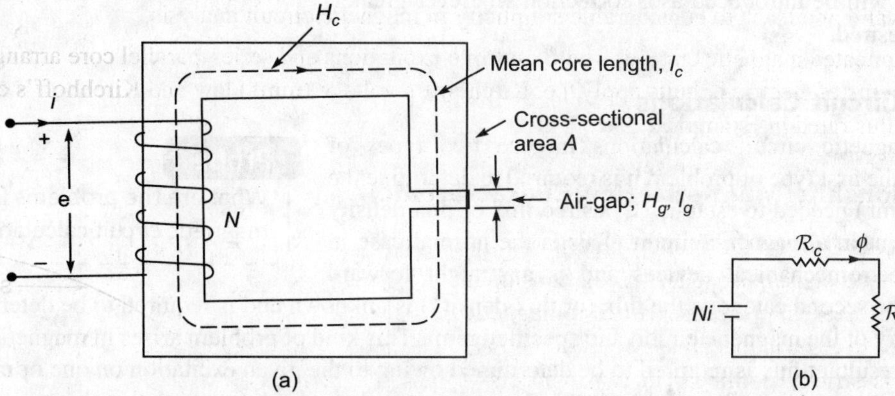

Fig. 2.4 *A typical magnetic circuit with air-gap and its equivalent electric circuit*

Assuming that all the core flux passes straight down the air-gap (it means no fringing (see Fig. 2.5))

$$B_g = B_c$$

∴

$$\phi = B_c A = B_g A \tag{2.10}$$

Substituting Eq. (2.10) in Eq. (2.9b)

$$Ni = \phi\left(\frac{l_c}{\mu_c A}\right) + \phi\left(\frac{l_g}{\mu_0 A}\right) \tag{2.11}$$

Recognizing various quantities in Eq. (2.11)

$$\mathcal{F} = \phi(\mathcal{R}_c + \mathcal{R}_g) = \phi\mathcal{R}_{eq} \tag{2.12}$$

where

$$\mathcal{R}_c = \frac{l_c}{\mu_c A} = \text{core reluctance}$$

$$\mathcal{R}_g = \frac{l_g}{\mu_0 A} = \text{air-gap reluctance}$$

From Eq. (2.12)

$$\phi = \frac{\mathcal{F}}{\mathcal{R}_c + \mathcal{R}_g} = \frac{\mathcal{F}/\mathcal{R}_g}{1 + \mathcal{R}_c/\mathcal{R}_g} \tag{2.13}$$

But

$$\frac{\mathcal{R}_c}{\mathcal{R}_g} = \frac{\mu_0 l_c}{\mu_c l_g} \ll 1$$

because μ_c is 2000 to 6000 times μ_0 in ferromagnetic materials. The permeability effect predominates the usual core and air-gap dimensions even though $l_c \gg l_g$. It then follows from Eq. (2.13), that

$$\phi \approx \mathcal{F}/\mathcal{R}_g \tag{2.14}$$

which means that in a magnetic circuit with air-gap(s), core reluctance may be neglected with no significant loss of accuracy. This assumption will be generally made in modelling rotating machines. The effect of core saturation (reduction of core permeability) will be introduced as a correction wherever greater accuracy is desired.

Probe
What is the assumption made in modelling rotating machines?

Magnetic Circuit Calculations

Normally magnetic circuit calculations involve two types of problems. In the first type of problem it is required to determine the excitation (mmf) needed to establish a desired flux or flux density at a given point in a magnetic circuit. This is the normal case in designing electromechanical devices and is a straight forward problem. In the second category, the flux (or flux density) is unknown and is required to be determined for a given geometry of the magnetic circuit and specified mmf. This kind of problem arises in magnetic amplifiers wherein this resultant flux is required to be determined owing to the given excitation on one or more control windings. A little thought will reveal that there is no direct analytical solution to this problem because of the non-linear *B-H* characteristic of the magnetic material. Graphical/numerical techniques have to be used in obtaining the solution of this problem.

Probe
What are the problems faced in magnetic circuit calculations?

Leakage Flux

In all practical magnetic circuits, most of the flux is confined to the intended path by use of magnetic cores but a small amount of flux always leaks through the surrounding air. This stray flux as already stated is called the *leakage flux*, Leakage is characteristic of all magnetic circuits and can never be fully eliminated. Calculations concerning the main magnetic circuit are usually carried out with the effect of leakage flux either ignored or empirically accounted for. Special studies of leakage must be made for ac machines and transformers since their performance is affected by it.

Fringing

At an air-gap in a magnetic core, the flux fringes out into neighbouring air paths as shown in Fig. 2.5; these being of reluctance comparable to that of the gap. The result is *nonuniform* flux density in the air-gap (decreasing outward), enlargement of the effective air-gap area and a decrease in the average gap flux density. The fringing effect also disturbs the core flux pattern to some depth near the gap. The effect of fringing increases with the air-gap length. Corrections for fringing in short gaps (as used in machines) are empirically made by adding one gap length to each of the two dimensions making up its area. For the example of the core with the air-gap previously presented, the gap reluctance would now be given by

Probe
With what value does the effect of fringing increases?

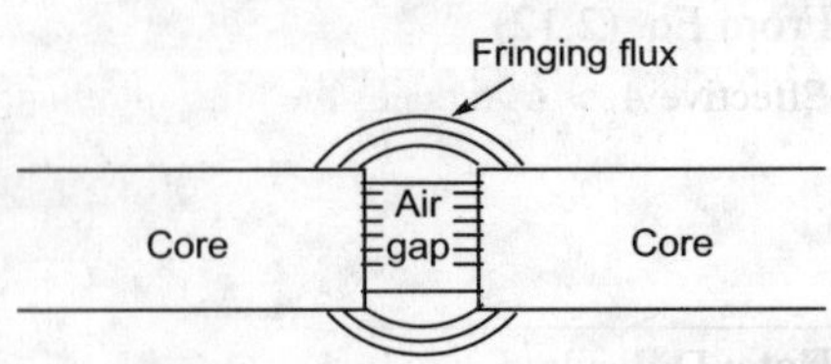

Fig. 2.5 *Flux fringing at air-gap*

$$\mathcal{R}_g = \frac{l_g}{\mu_0 A_g}$$

which will be less than the previous value as $A_g > A$.

It can be shown theoretically that the magnetic flux leaves and enters the surface of an infinitely permeable material normally. This will be nearly so in ferromagnetic materials which have high permeability. In electric machines a small amount of the tangential flux component present at iron surfaces will be neglected.

Stacking Factor

Magnetic cores are made up of thin, lightly insulated (coating of varnish) *laminations* to reduce power loss in cores due to the eddy-current phenomenon (explained in Sec. 2.6). As a result, the net cross-sectional area of the core occupied by the magnetic material is less than its gross cross-section; their ratio (less than unity) being known as the stacking factor. Depending upon the thickness of laminations, stacking factor may vary from 0.5–0.95, approaching unity as the lamination thickness increases.

Example 2.1 *The magnetic circuit of Fig. 2.4(a) has dimensions: $A_c = 4 \times 4$ cm^2, $l_g = 0.06$ cm, $l_c = 40$ cm; N = 600 turns. Assume the value of $\mu_r = 6000$ for iron. Find the exciting current for $B_c = 1.2$ T and the corresponding flux and flux linkages.*

LO 1 ○●●

Solution From Eq. (2.9), the ampere-turns for the circuit are given by

$$Ni = \frac{B_c}{\mu_0 \mu_r} l_c + \frac{B_g}{\mu_0} l_g \qquad \text{(i)}$$

Neglecting fringing

$$A_c = A_g \qquad \text{therefore} \qquad B_c = B_g$$

Then

$$i = \frac{B_c}{\mu_0 N}\left(\frac{l_c}{\mu_r} + l_g\right)$$

$$= \frac{1.2}{4\pi \times 10^{-7} \times 600}\left(\frac{40}{6000} + 0.06\right) \times 10^{-2} \qquad \text{(ii)}$$

$$= 1.06 \text{ A}$$

The reader should note that the reluctance of the iron path of 40 cm is only $\left(\frac{2/3}{6}\right) = 0.11$ of the reluctance of the 0.06 cm air-gap.

$$\phi = B_c A_c = 1.2 \times 16 \times 10^{-4} = 19.2 \times 10^{-4} \text{ Wb}$$

Flux linkages,

$$\lambda = N\phi = 600 \times 19.2 \times 10^{-4} = 1.152 \text{ Wb-turns}$$

If fringing is to be taken into account, one gap length is added to each dimension of the air-gap constituting the area. Then

$$A_g = (4 + 0.06)\,(4 + 0.06) = 16.484 \text{ cm}^2$$

Effective $A_g > A_c$ reduces the air-gap reluctance. Now

$$B_g = \frac{19.2 \times 10^{-4}}{16.484 \times 10^{-4}} = 1.165 \text{ T}$$

Note: Difficulty Level ⟶ ○○● — Level 1 and Level 2 Category; ○●● — Level 3 and Level 4 Category; ●●● — Level 5 and Level 6 Category

From Eq. (i)

$$i = \frac{1}{\mu_0 N}\left(\frac{B_c l_c}{\mu_r} + B_g l_g\right) \tag{iii}$$

$$= \frac{1}{4\pi \times 10^{-7} \times 600}\left(\frac{1.2 \times 40 \times 10^{-2}}{6000} + 1.165 \times 0.06 \times 10^{-2}\right)$$

$$= 1.0332\,\text{A}$$

Example 2.2 *A wrought iron bar 30 cm long and 2 cm in diameter is bent into a circular shape as shown in Fig. 2.6. It is then wound with 600 turns of wire. Calculate the current required to produce a flux of 0.5 mWb in the magnetic circuit in the following cases:*

(i) no air-gap;

(ii) with an air-gap of 1 mm; μ_r (iron) = 4000 (assumed constant); and

(iii) with an air-gap of 1 mm; assume the following data for the magnetization of iron:

H in AT/m	*2500*	*3000*	*3500*	*4000*
B in T	*1.55*	*1.59*	*1.6*	*1.615*

LO 1

Solution

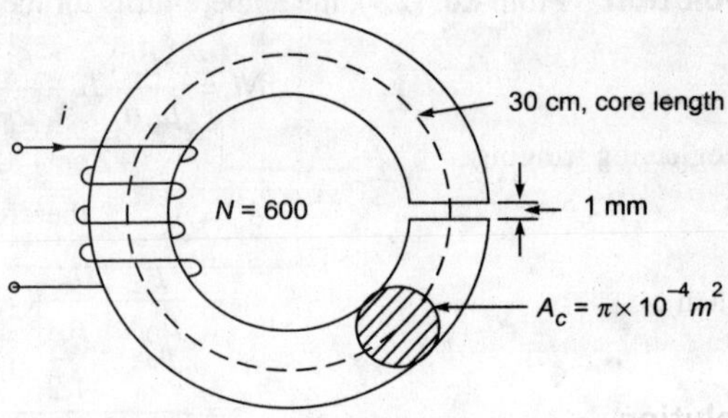

Fig. 2.6

(i) No air-gap

$$\mathcal{R}_c = \frac{30 \times 10^{-2}}{4000 \times 4\pi \times 10^{-7} \times \pi \times 10^{-4}} = 1.9 \times 10^5$$

$$Ni = \phi \mathcal{R}_c$$

or $$i = \phi \mathcal{R}_c / N = \frac{0.5 \times 10^{-3} \times 1.9 \times 10^5}{600} = 0.158\,\text{A}$$

(ii) Air-gap = 1 mm, μ_r (iron) = 4000

$$\mathcal{R}_c = 1.9 \times 10^5 \text{ (as in part (i)}$$

$$\mathcal{R}_g = \frac{1 \times 10^{-3}}{4\pi \times 10^{-7} \times \pi \times 10^{-4}} = 25.33 \times 10^5$$

$$\mathcal{R}(\text{total}) = \mathcal{R}_c + \mathcal{R}_g = 27.1 \times 10^5$$

$$\therefore \quad i = \frac{0.5 \times 10^{-3} \times 27.1 \times 10^5}{600} = 2.258\,\text{A}$$

(iii) Air-gap = 1 mm; *B-H* data as given

$$B_c = B_g = \frac{0.5 \times 10^{-3}}{\pi \times 10^{-4}} = 1.59 \text{ T (fringing neglected)}$$

$$H_g = \frac{B_g}{\mu_0} = \frac{1.59}{4\pi \times 10^{-7}}$$

$$\text{AT}_g = H_g l_g = \frac{1.59 \times 1 \times 10^{-3}}{4\pi \times 10^{-7}} = 1265$$

From the given magnetization data (at $B_c = 1.59$ T),

$$H_c = 3000 \text{ AT/m}$$

$$\text{AT}_c = H_c l_c = 3000 \times 30 \times 10^{-2} = 900$$

$$\text{AT (total)} = \text{AT}_c + \text{AT}_g$$

$$= 900 + 1265 = 2165$$

$$i = \frac{2165}{600} = 3.61 \text{ A}$$

Example 2.3 *The magnetic circuit of Fig. 2.7 has cast steel core with dimensions as shown:*

Mean length from A to B through either outer limb = 0.5 m

Mean length from A to B through the central limb = 0.2 m

In the magnetic circuit shown it is required to establish a flux of 0.75 mWb in the air-gap of the central limb. Determine the mmf of the exciting coil if for the core material (a) $\mu_r = \infty$ (b) $\mu_r = 5000$. Neglect fringing.

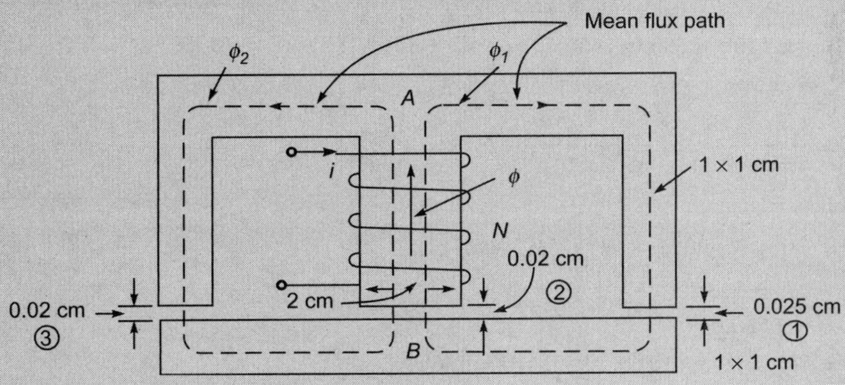

Fig. 2.7

LO 1

Solution

(a) $\mu_r = \infty$, i.e. there are no mmf drops in the magnetic core. It is easy to see from Fig. 2.7 that the two outer limbs present a parallel magnetic circuit. The electrical analog of the magnetic circuit is drawn in Fig. 2.8(a). Various gap reluctances are:

$$\mathcal{R}_{g1} = \frac{0.025 \times 10^{-2}}{4\pi \times 10^{-7} \times 1 \times 10^{-4}} = 1.99 \times 10^6$$

$$\mathcal{R}_{g2} = \frac{0.02 \times 10^{-2}}{4\pi \times 10^{-7} \times 1 \times 10^{-4}} = 1.592 \times 10^6$$

$$\mathcal{R}_{g3} = \frac{0.02 \times 10^{-2}}{4\pi \times 10^{-7} \times 2 \times 10^{-4}} = 0.796 \times 10^6$$

From Fig. 2.8(b),

$$Ni = 0.75 \times 10^{-3} (\mathcal{R}_{g3} + \mathcal{R}_{g1} \parallel \mathcal{R}_{g2})$$

$$= 0.75 \times 10^{-3} (0.796 + 0.844) \times 10^6$$

$$= 1230 \text{ AT}$$

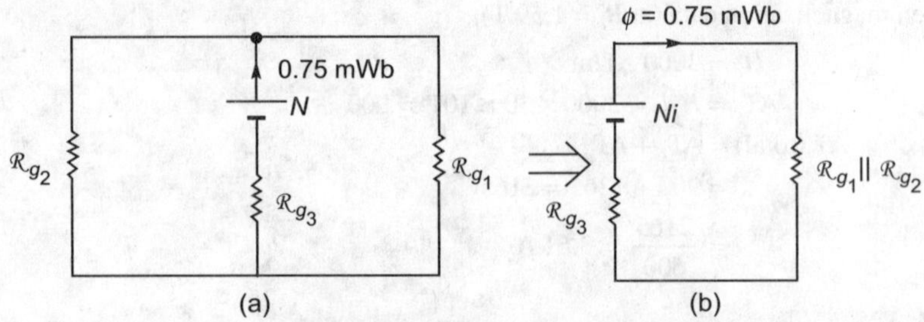

Fig. 2.8 *Electrical analog of Fig. 2.7*

(b) μ_r = 5000. This means that the reluctance of magnetic core must be taken into consideration. The analogous electric circuit now becomes that of Fig. 2.9.

Since gap lengths are negligible compared to core lengths, various core reluctances can be calculated as follows:

$$\mathcal{R}_{c_1} = \frac{0.5}{4\pi \times 10^{-7} \times 5000 \times 1 \times 10^{-4}} = 0.796 \times 10^6$$

$$\mathcal{R}_{c_2} = \mathcal{R}_{c_1} = 0.796 \times 10^6$$

$$\mathcal{R}_{c_3} = \frac{0.2}{4\pi \times 10^{-7} \times 5000 \times 2 \times 10^{-4}} = 0.159 \times 10^6$$

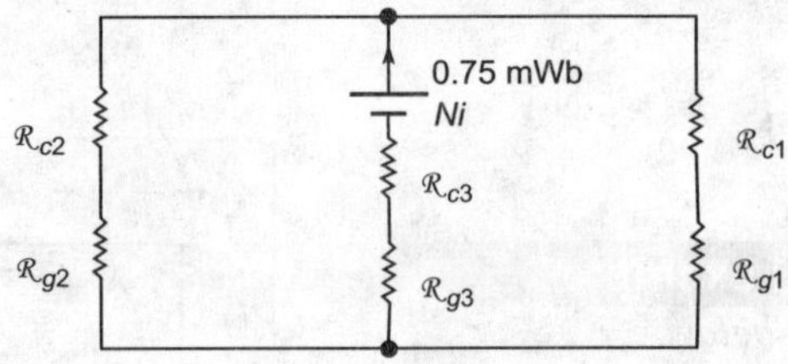

Fig. 2.9

The equivalent reluctance is

$$\mathcal{R}_{\text{eq}} = (\mathcal{R}_{c_1} + \mathcal{R}_{g_1}) \| (\mathcal{R}_{c_2} + \mathcal{R}_{g_2}) + \mathcal{R}_{c_3} + \mathcal{R}_{g_3})$$

$$= \frac{27.86 \times 23.86}{51.72} \times 10^6 + 0.955 \times 10^6 = 1.955 \times 10^6$$

Now $Ni = \phi \mathcal{R}_{\text{eq}} = 0.75 \times 10^{-3} \times 1.955 \times 10^6 = 1466$ AT

Example 2.4 *The magnetic circuit of Fig. 2.10 has cast steel core. The cross-sectional area of the central limb is 800 mm² and that of each outer limb is 600 mm². Calculate the exciting current needed to set up a flux of 0.8 mWb in the air gap. Neglect magnetic leakage and fringing. The magnetization characteristic of cast steel is given in Fig. 2.16.*

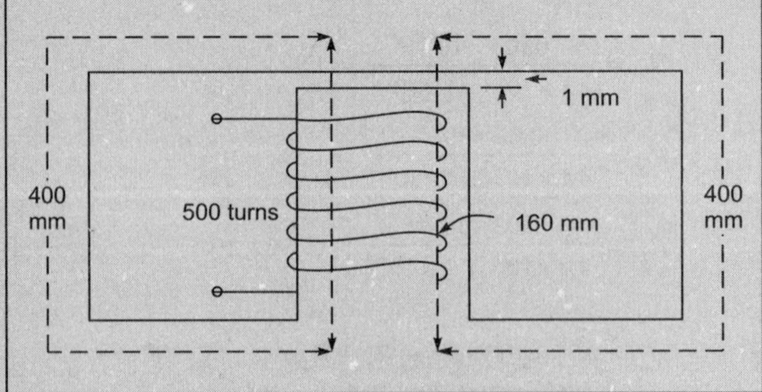

Fig. 2.10

LO 1

Solution

Air gap $\quad B_g = \dfrac{0.8}{800} \times \dfrac{10^{-3}}{10^{-6}} = 1\text{ T} \quad \text{and} \quad H_g = \dfrac{1}{4\pi \times 10^{-7}}\text{ AT/m}$

$$\mathcal{F}_g = \frac{1}{4\pi \times 10^{-7}} \times 1 \times 10^{-3} = 796\text{ AT}$$

Central limb $\quad B_c = B_g = 1\text{ T}$

From Fig. 2.16 $\quad H_c = 1000\text{ AT/m}$

$$\mathcal{F}_c = 1000 \times 160 \times 10^{-3} = 160\text{ AT}$$

Because of symmetry, flux divides equally between the two outer limbs. So

$$\phi\text{ (outer limb)} = 0.8/2 = 0.4\text{ mWb}$$

$$B\text{ (outer limb)} = \frac{0.4 \times 10^{-3}}{600 \times 10^{-6}} = 0.667\text{ AT}$$

$$\mathcal{F}\text{ (outer limb)} = 375 \times 400 \times 10^{-3} = 150\text{ AT}$$

$$\mathcal{F}\text{ (total)} = 796 + 160 + 150 = 1106\text{ AT}$$

$$\text{Exciting current} = 1106/500 = 2.21\text{ A}$$

Example 2.5 *The magnetic circuit of Fig. 2.11 has a cast steel core whose dimensions are given below:*

Length (ab + cd) = 50 cm	*Cross-sectional area = 25 cm²*
Length ad = 20 cm	*Cross-sectional area = 12.5 cm²*
Length dea = 50 cm	*Cross-sectional area = 25 cm²*

Determine the exciting coil mmf required to establish an air-gap flux of 0.75 m Wb. Use the B-H curve of Fig. 2.13.

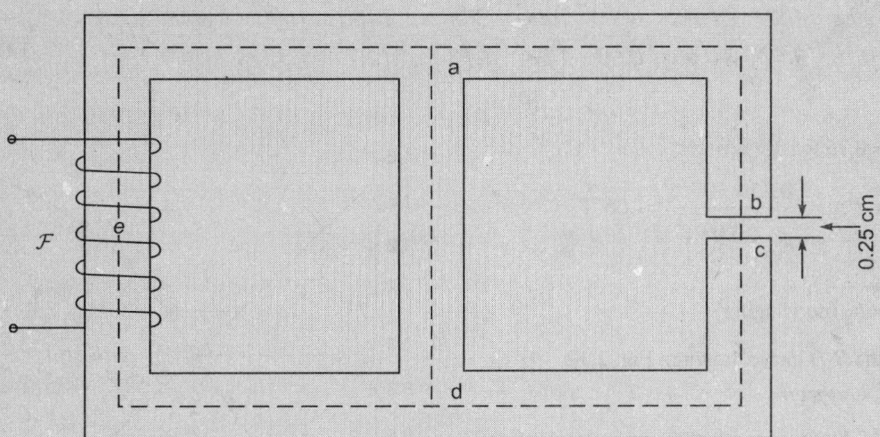

Fig. 2.11

LO 1

Solution Assuming no fringing the flux density in the path *abcd* will be same, i.e.

$$B = \frac{0.75 \times 10^{-3}}{25 \times 10^{-4}} = 0.3\text{ T}$$

$$\mathcal{F}_{bc} = \frac{B}{\mu_0} l_{bc} = \frac{0.3 \times 0.25 \times 10^{-3}}{4\pi \times 10^{-7}} = 60 \text{ AT}$$

$H_{ab} = H_{cd}$ (from Fig. 2.16 for cast steel for $B = 0.3$ T) = 200 AT/m

$$\mathcal{F}_{ab+cd} = 200 \times 50 \times 10^{-2} = 100 \text{ AT}$$

$$\therefore \quad \mathcal{F}_{ad} = 60 + 100 = 160 \text{ AT}$$

$$H_{ad} = \frac{160}{20 \times 10^{-2}} = 800 \text{ AT/m}$$

B_{ad} (from Fig. 2.16) = 1.04 T

$$\phi_{ad} = 1.04 \times 12.5 \times 10^{-4} = 1.3 \text{ mWb}$$

$$\phi_{dea} = 0.75 + 1.3 = 2.05 \text{ mWb}$$

$$B_{dea} = \frac{2.05 \times 10^{-3}}{25 \times 10^{-4}} = 0.82 \text{ T}$$

H_{dea} (from Fig. 2.16) = 500 A T/m

$$\mathcal{F}_{dea} = 500 \times 50 \times 10^{-2} = 250 \text{ AT}$$

$$\mathcal{F} = \mathcal{F}_{dea} + \mathcal{F}_{ad} = 250 + 160 = 410 \text{ AT}$$

Example 2.6 *A cast steel ring has a circular cross-section of 3 cm in diameter and a mean circumference of 80 cm. A 1 mm air-gap is cut out in the ring which is wound with a coil of 600 turns.*

(a) Estimate the current required to establish a flux of 0.75 mWb in the air-gap. Neglect fringing and leakage.

(b) What is the flux produced in the air-gap if the exciting current is 2A? Neglect fringing and leakage.

Magnetization data:

H (AT/m)	*200*	*400*	*600*	*800*	*1000*	*1200*	*1400*	*1600*	*1800*	*2020*
B (T)	*0.10*	*0.32*	*0.60*	*0.90*	*1.08*	*1.18*	*1.27*	*1.32*	*1.36*	*1.40*

LO 1

Solution

$$\phi = 0.75 \times 10^{-3} \text{ Wb}$$

$$B_g = \phi/A = \frac{0.75 \times 10^{-3}}{\pi \times \left(\frac{0.03}{2}\right)^2} = 1.06 \text{ T}$$

$B_c = B_g$ (no fringing)

Reading from the *B-H* curve drawn in Fig. 2.12,

$H_c = 900$ AT/m

$l_c = 0.8$ m (air-gap length can be neglected)

$$\text{AT}_c = H_c l_c = 900 \times 0.8 = 720$$

$$\text{AT}_g = \frac{1.06}{4\pi \times 10^{-7}} \times 10^{-3} = 843$$

$$Ni = \text{AT}_c + \text{AT}_g = 720 + 843 = 1563$$

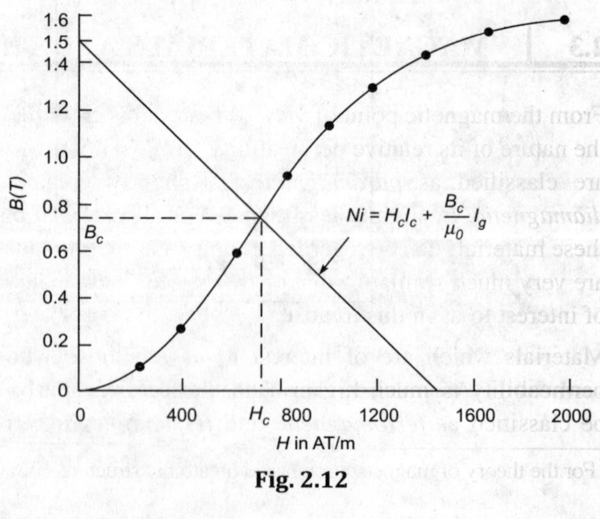

Fig. 2.12

Therefore $i = \dfrac{1563}{600} = 2.6\,\text{A}$

(b) The excitation is now given and the flux is to be determined from the *B-H* curve given. The problem must, therefore, be solved numerically/graphically. It is solved here graphically. Now

$$Ni = \frac{B_2}{\mu_0} l_g + H_c l_c;\ (B_g = B_c) \tag{i}$$

This is a linear equation in B_c and H_c; the second equation is the nonlinear *B-H* curve. The intersection of the two for a given *Ni* will yield the solution. For this problem

$$Ni = 600 \times 2 = 1200 \text{ AT}$$

Substituting various values in Eq. (i)

$$1200 = \frac{B_c}{4\pi \times 10^{-7}} \times 10^{-3} + 0.8\, H_c \tag{ii}$$

This equation is plotted in Fig. 2.12, by locating the points

$$H_c = 0, \quad B_c = 1.5$$
$$B_c = 0, \quad H_c = 1500$$

The intersection gives the result

$$B_c = 0.78 \text{ T}$$
$$\phi = B_c A = 0.78 \times \frac{\pi}{4}(0.03)^2 = 0.55 \text{ mWb}$$

Concept Check — LO 1

1. State Ohm's law for magnetic circuits.
2. Define magnetic reluctance.
3. Explain why a ferromagnetic material exhibits its typical B-H behaviour.
4. Explain the practical use made of magnetic saturation.

2.3 MAGNETIC MATERIALS AND THEIR PROPERTIES

LO 2 Identify the different magnetic materials and illustrate their properties

From the magnetic point of view[‡] a material is classified according to the nature of its relative permeability (μ_r). All nonmagnetic materials are classified as *paramagnetic*, μ_r slightly greater than 1, and *diamagnetic*, μ_r slightly less than 1. For all practical purposes, μ_r of these materials can be regarded as unity, i.e. their magnetic properties are very much similar to that of free space. Such materials are not of interest to us in this treatise.

Probe

On what basis is a magnetic material classified?

Materials which are of interest to us are those whose relative permeability is much higher than that of free space. These can be classified as *ferromagnetic* and *ferrimagnetic*. Ferromagnetic

[‡]For the theory of magnetization based on atomic structure of materials a suitable book on material science may be consulted.

materials can be further subdivided as *hard* and *soft*. Hard ferromagnetic materials include permanent magnet materials, such as alnicos, chromium steels, certain copper-nickel alloys and several other metal alloys. Soft ferromagnetic materials are iron and its alloys with nickel, cobalt, tungsten and aluminium. Silicon steels and cast steels are the most important ferromagnetic materials for use in transformers and electric machines. Ferrimagnetic materials are the *ferrites* and are composed of iron oxides—MeO. Fe_2O_3, where Me represents a metallic ion. Ferrites are also subgrouped as hard (permanent magnetic) and soft (nickel-zinc and manganese-zinc) ferrites. Soft ferrites are quite useful in high frequency transformers, microwave devices, and other similar high-frequency operations. There is a third category of magnetic materials, known as *superparamagnetic*, made from powdered iron or other magnetic particles. These materials are used in transformers for electronics and cores for inductors. *Permalloy* (molybdenum-nickel-iron powder) is the best known example of this important category of magnetic materials.

Properties of Magnetic Materials

Magnetic materials are characterized by high permeability and the nonlinear *B-H* relationship which exhibits both saturation and hysteresis. The physics of these properties is explained by the domain theory of magnetization usually taught in junior level courses.

The *B-H* relationship for cyclic *H* is the *hysteresis loop* shown in Fig. 2.13 for two values of maximum flux density. It is easily observed from this figure that *B* is a symmetrical two-valued function of *H*; at any given *H*, *B* is higher if *H* is reducing compared to when *H* is increasing. This is the basic hysteresis property in which *B* lags behind *H*. It can also be recognized as a *memory-type non-linearity* in which the material remembers its previous history. Further, it is observed that the hysteresis loop becomes wider for increasing maximum flux densities. The dotted curve drawn through the positive and negative tips of the hysteresis loops with increasing maximum flux densities is the *normal magnetization curve* and is obtainable in virgin (unmagnetized) material by increasing the dc magnetization in either direction. The normal magnetization curve exhibits the saturation phenomenon as discussed in Sec. 2.2.

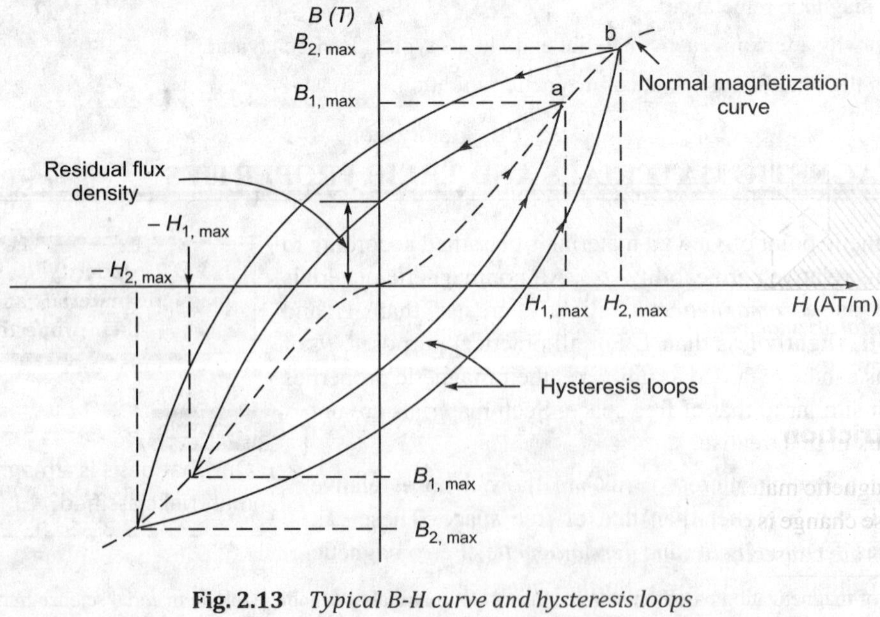

Fig. 2.13 *Typical B-H curve and hysteresis loops*

!Attention It is easily seen that because of hysteresis and saturation, the magnetic characteristic of a given material cannot be described by a few overall parameters but must be expressed in the form of a set of curves. It will be shown in Sec. 2.6 that the area of the hysteresis loop is the energy loss (it appears in the form of heat energy) per unit volume in one cycle of magnetization. This loss depends upon the quality of material and the maximum flux density at which the material is operated. Hysteresis loop as such is of little use in engineering applications except in illustrating the waveform of the exciting current. It is the normal (dc) magnetization curve which is of direct application in magnetic circuit calculations and design. In short form it will be referred as the *magnetization curve*. The hysteresis loss represented by the loop area is usually lumped with the eddy-current loss (Sec. 2.6) and the two together are known as the *core* (*or iron*) *loss* which is parameterized by material thickness and frequency and is expressed as loss per unit volume (specific loss).

Sheet Steels

Nearly all transformers and certain parts of electric machines use silicon steel in the form of sheets, thinly insulated on each side. Silicon is added to steel to increase its resistivity thereby reducing the eddy-current loss. Building the core out of thin sheets (laminations) greatly aids in reduction of the eddy-current loss (see Sec. 2.6).

The crystalline structure of silicon steel is body centred cubic—an atom at each corner of the cube and an atom at the cube centre. The cubic structure presents considerable ease of magnetization (high μ_r) along the cube edge as compared to the diagonal on the cube side and the diagonal through the cube body which is most difficult to magnetize. Therefore, if silicon steel crystals are aligned so that the cube edges lie parallel to the direction of magnetization, the material has a much higher relative permeability and can be operated at much higher flux densities with moderate exciting currents and also has superior core-loss qualities. This crystal arrangement is illustrated in Fig. 2.14. The crystal arrangement is practically achieved by *cold-rolling* steel sheets. The material is then termed as *cold-rolled grain-oriented steel* (*crgos*) and is invariably employed in electromechanical devices. Annealing of silicon steel sheets further helps in proper alignment of crystals. The saturation flux density and electric resistivity are of course independent of the grain orientation.

Probe

Describe the structure of silicon steel.

Probe

What is cold-rolled grain-oriented steel?

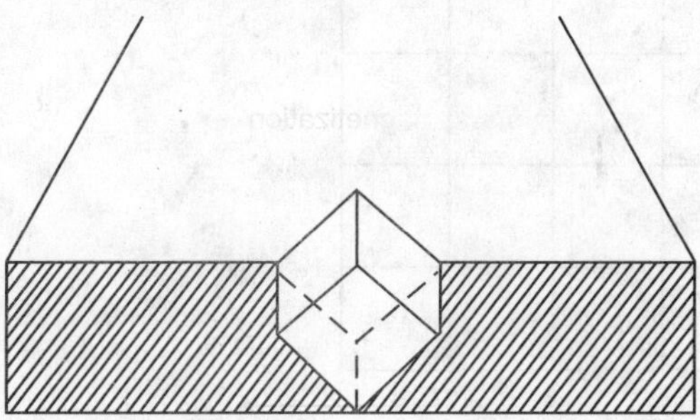

Fig. 2.14 *Crystal arrangement in grain-oriented steel*

Magnetostriction

When ferromagnetic materials are subjected to magnetizing mmf, these undergo small changes in dimensions. The lengthwise change is of the order of 10^{-5} m and is accompanied by transverse changes of the opposite sign. These changes are caused by magnetostriction. Their nature is hysteric with consequent dissipation of energy when magnetization is alternating. Further, there are associated mechanical stresses which produce noise in audible bandwidth which can be a nuisance for high flux densities employed in transformers in modern practice. Magnetostriction noise may, therefore, be the subject of limiting specifications in transformers.

Magnetization Curves

Typical magnetization curves of transformer steel are given in Fig. 2.15 and a comparative set of magnetization curves are given in Fig. 2.16.

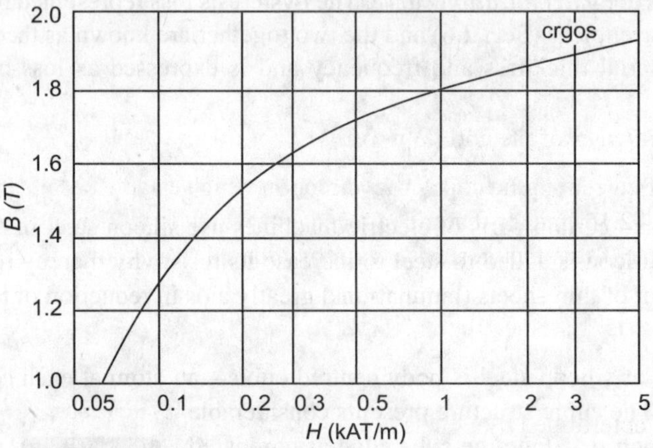

Fig. 2.15 *Magnetization curves: transformer steel*

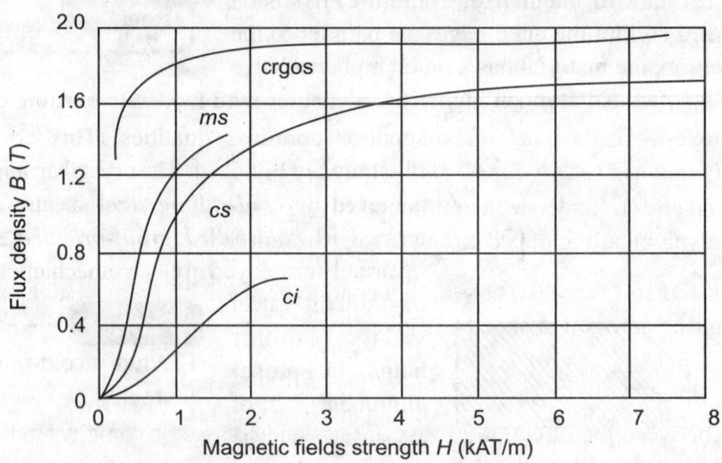

Fig. 2.16 *Magnetization curves: comparative [Cold-rolled grain-oriented steel (crgos), mild steel, cast steel, cast iron]*

Concept Check LO 2

1. Explain the origin of magnetostriction noise in ferromagnetic materials.

2.4 MAGNETICALLY INDUCED EMF AND FORCE

LO 3 Develop expressions for magnetically induced emf and force and understand the ac operation of magnetic circuits

Faraday's law of induction, which is the integral form of the fourth Maxwell's equation, is given as

$$\oint \vec{E} \cdot d\vec{l} = -\int \frac{\partial \vec{B}}{\partial t} \cdot d\vec{s} \tag{2.15}$$

The integrated form of Eq. (2.15) for a coil of N turns is

$$e = -N\frac{d\phi}{dt} = -\frac{d\lambda}{dt} \tag{2.16}$$

where $\lambda = N\phi =$ *flux linkages* of the coil (Wb-Turns)

The positive direction of current in the coil is that direction which establishes positive flux and flux linkages. The negative sign in Eq. (2.16) means that the induced emf owing to an increase in λ is in opposite direction to that of positive current. If this fact is separately remembered (it is known as *Lenz's law*), Eq. (2.16) may be written

$$e = N\frac{d\phi}{dt} = \frac{d\lambda}{dt} \tag{2.17}$$

with the sign of the emf determined by Lenz's law.

Change in flux linkages of a coil may occur in three ways:

Probe

What are the ways in which change in flux linkages of a coil may occur?

(i) The coil remains stationary with respect to flux, but the flux through it changes with time. The emf induced is known as *statically induced emf*.

(ii) Flux density distribution remains constant and stationary but the coil moves relative to it. The emf induced is known as *dynamically induced* (*or motional*) *emf*.

(iii) Both changes (i) and (ii) may occur simultaneously, i.e. the coil moves through time-varying flux. Both statically and dynamically induced emfs are then present in the coil.

The dynamically induced emf (case (ii) above) in a conductor of length l placed at 90° to a magnetic field of flux density B and cutting across it at speed v is given by

$$e = \left|\vec{v} \times \vec{B}\right| l = Blv \sin\theta \tag{2.18a}$$

where θ is the angle between the direction of flux density and conductor velocity, and l the conductor along which the flux density is assumed uniform. In electric machines $\theta = 90°$, so that

$$e = Blv \tag{2.18b}$$

This is known as the *flux-cutting rule* and the direction of emf is given by $\vec{v} \times \vec{B}$ or by the well-known *Fleming's right-hand rule*[§].

[§]Extend the thumb, first and second fingers of the right hand so that they are mutually perpendicular to each other. If the thumb represents the direction of v (conductor with respect to B) and the first finger the direction of B, then the second finger represents the direction of emf along l.

Inductance

Equation (2.17) may be written as

$$e = N\frac{d\phi}{dt} = N\frac{d\phi}{di}\frac{di}{dt} = L\frac{di}{dt} \tag{2.19a}$$

where
$$L = N\frac{d\phi}{di} = \frac{d\lambda}{di} \quad \text{H} \tag{2.19b}$$

is the *self-inductance* of the circuit. For a magnetic circuit having a linear *B-H* relationship (constant permeability of material) or with a dominating air-gap, the inductance L is a constant, independent of current and depends only on the geometry of circuit elements and permeability of the medium. In this case Eq. (2.19) can also be expressed as

$$L = \frac{\lambda}{i} \tag{2.20}$$

The inductance can be written in terms of field quantities as

$$L = \frac{N^2 BA}{Hl} = N^2\mu\frac{A}{l} = \frac{N^2}{\mathcal{R}} = N^2\mathcal{P} \quad \text{H} \tag{2.21}$$

Thus self-inductance is proportional to N^2.

The inductance concept is easily extendable to the *mutual inductance* of two coils sharing a common magnetic circuit. Thus,

> **Probe**
> Write the equation for inductance in terms of field quantities.

$$\left.\begin{aligned} M_{12} &= \frac{\lambda_{12}}{i_2} \quad \text{H} \\ M_{21} &= \frac{\lambda_{21}}{i_1} \quad \text{H} \end{aligned}\right\} \tag{2.22}$$

where
λ_{12} = flux linkages of coil 1 due to current in coil 2
λ_{21} = flux linkages of coil 2 due to current in coil 1

For a *bilateral** magnetic circuit,

$$M = M_{12} = M_{21}$$

It can also be shown that for *tight coupling*, i.e. all the flux linking both the coils (no leakage)

$$M = \sqrt{L_1 L_2} \tag{2.23a}$$

In general,

$$M = k\sqrt{L_1 L_2} \tag{2.23b}$$

where
k = *coupling coefficient* (which can be at most unity)

From Eq. (2.20)
$$\lambda = Li$$

In static magnetic configuration, L is fixed independent of time so that the induced emf is given by Eq. (2.19a). In rotating devices both L and i vary with time giving the induced emf

$$e = \underbrace{L\frac{di}{dt}}_{\substack{\text{Statically}\\ \text{induced emf}}} + \underbrace{i\frac{dL}{dt}}_{\substack{\text{Dynamically}\\ \text{induced emf}}} \tag{2.24}$$

Force

Force of electromagnetic origin is given by the *Lorentz force equation*

$$d\vec{F} = I d\vec{l} \times \vec{B} \tag{2.25}$$

where I is the current flowing in the differential conductor of length dl. Integrating over the conductor length along which B is assumed uniform, the total force is obtained as

$$\vec{F} = I\vec{l} \times \vec{B} = BIl \sin\theta \, \vec{a}_F \quad \text{N} \tag{2.26a}$$

where θ is the angle between the direction of conductor and the magnetic field and $\vec{a}_F$ is unit vector in the direction defined by the cross product. For $\theta = 90°$ which is used in most machine configurations, Eq. (2.26a) reduces to

$$F = BIl \quad \text{N} \tag{2.26b}$$

The direction of force being given by $I\vec{l} \times \vec{B}$ or by the well-known Fleming's lefthand rule[¶]. It immediately follows from Eq. (2.26b) that B can be imagined to have unit of N/Am.

Example 2.7 *For the magnetic circuit of Fig. 2.17 find the self and mutual inductances between the two coils. Core permeability = 1600.*

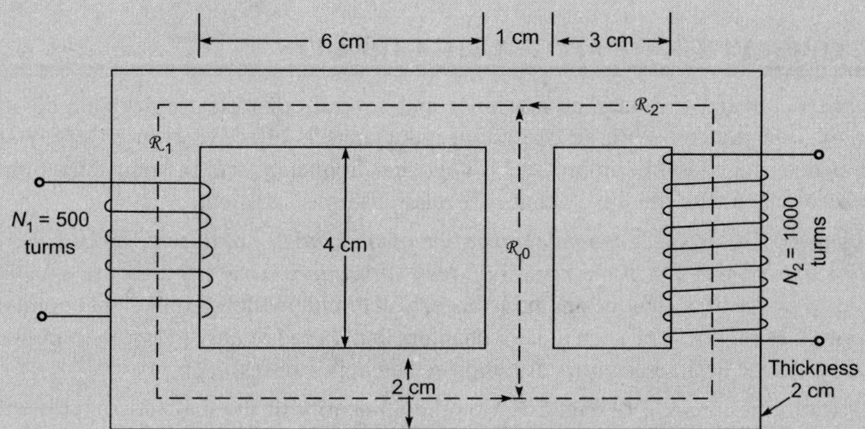

Fig. 2.17

LO 3

Solution

$$l_1 = (6 + 0.5 + 1) \times 2 + (4 + 2) = 21 \text{ cm}$$
$$l_2 = (3 + 0.5 + 1) \times 2 + (4 + 2) = 15 \text{ cm}$$
$$l_0 = 4 + 2 = 6 \text{ cm}$$

$$\mathcal{R}_1 = \frac{21 \times 10^{-2}}{4\pi \times 10^{-7} \times 1600 \times 2 \times 2 \times 10^{-4}} = 0.261 \times 10^6$$

[¶]Extend the thumb, first and second fingers of the left hand so that they are mutually perpendicular to each other. If the first finger represents the direction of B and the second finger the direction of I, then the thumb points in the direction of force on the conductor.

$$\mathcal{R}_2 = \frac{15 \times 10^{-2}}{4\pi \times 10^{-7} \times 1600 \times 2 \times 2 \times 10^{-4}} = 0.187 \times 10^6$$

$$\mathcal{R}_0 = \frac{6 \times 10^{-2}}{4\pi \times 10^{-7} \times 1600 \times 1 \times 2 \times 10^{-4}} = 0.149 \times 10^6$$

(i) Coil 1 excited with 1A

$$\mathcal{R} = \mathcal{R}_1 + \mathcal{R}_0 \parallel \mathcal{R}_2 = 0.261 + 0.1871 \parallel 0.149 = 0.344 \times 10^6$$

$$\phi_1 = (500 \times 1)/(0.344 \times 10^6) = 1.453 \text{ mWb}$$

$$\phi_{21} = \phi_2 = 1.453 \times 0.149/(0.149 + 0.187) = 0.64 \text{ mWb}$$

$$L_{11} = N_1\phi_1 = 500 \times 1.453 \times 10^{-3} = 0.7265 \text{ H}$$

$$M_{21} = N_2\phi_{21} = 1000 \times 0.649 \times 10^{-3} = 0.64 \text{ H}$$

(ii) Coil 2 excited with 1A

$$\mathcal{R} = \mathcal{R}_2 + (\mathcal{R}_0\mathcal{R}_1)/\mathcal{R}_0 + \mathcal{R}_1) = [0.187 + (0.149 \times 0.281)/(0.149 + 0.281)] \times 10^6 = 0.284 \times 10^6$$

$$\phi_2 = (1000 \times 1)/(0.284 \times 10^6) = 3.52 \text{ mWb}$$

$$L_{22} = N_2\phi_2 = 1000 \times 3.52 \times 10^{-3} = 3.52 \text{ H}$$

$$M_{12} = M_{21} \text{ (bilateral)} = 0.65 \text{ H}$$

2.5 AC OPERATION OF MAGNETIC CIRCUITS

LO 3

The magnetic circuits of transformers, ac machines and several other electromagnetic devices are excited from ac rather than dc sources. With ac operation, inductance is effective even in steady-state operation. Often, the flux is determined by the impressed voltage and frequency, and the magnetization current has to adjust itself in accordance with the flux so that *B-H* relationship is satisfied.

Except when linearity is desirable, economic utilization of material demands that working flux density should lie in the nonlinear zone (but not in the region of deep saturation). Exact and accurate analysis, therefore, cannot be predicted on the basis of constant inductance. Still circuit models (equivalent circuits) with constant parameters are often used. It will be seen in later chapters that these not only provide simplified approach but at the same time yield the desired accuracy for engineering applications.

Consider the *N*-turn iron-core coil of Fig. 2.1. Complete linearity of the magnetic circuit will be assumed. Magnetic flux ϕ is produced by the exciting current *i*. Let the current and so the flux vary sinusoidally with time. Then

$$\phi = \phi_{\max} \sin \omega t \tag{2.27}$$

where $\phi_{\max}$ = maximum value of flux in core (Wb)

$\omega = 2\pi f$, where *f* is frequency in Hz

The induced emf in the coil as per Faraday's law (Eq. (2.17)) is

$$e = N\frac{d\phi}{dt} = \omega N\phi_{\max} \cos \omega t \text{ V} \tag{2.28}$$

Probe

Derive an expression for induced emf in the coil as per Faraday's law and it's rms value.

and its rms value is

$$E = \frac{2\pi}{\sqrt{2}} fN\phi_{\max} = 4.44\, fN\phi_{\max} \tag{2.29}$$

$$E = 444\, fNA_c B_{max} \tag{2.30}$$

where B_{max} is the maximum value of the flux density and A_c is the core's area of cross-section.

The polarity of the emf must, in accordance with Lenz's law, oppose the flux change and, therefore, is as shown in Fig. 2.1 when the flux is increasing. Since the current produces the flux instantaneously and in proportion to it (quasi-static field), they are in phase. From Eqs. (2.27) and (2.28), it is found that the induced emf leads the flux (hence the current) by 90°. The induced emf and coil resistance drop oppose the impressed voltage. However, resistance drop in many ac electromagnetic devices is quite small and may be neglected to a close approximation.

The electric power input into the magnetic circuit of Fig. 2.1 through the coil terminals is

$$p = ie = i\frac{d\lambda}{dt} \tag{2.31}$$

The electric energy input which gets stored in the magnetic field** in the time interval t_1 to t_2 is

$$W_f = \int_{t_1}^{t_2} p\,dt = \int_{\lambda_1}^{\lambda_2} i\,d\lambda \tag{2.32}$$

where W_f = increase in field energy as the coil flux linkages change from λ_1 to λ_2. In field quantities

$$W_f = \int_{B_1}^{B_2} \left(\frac{H_c l_c}{N}\right)(A_c N)dB_c = A_c l_c \int_{B_1}^{B_2} H_c\ dB_c \tag{2.33}$$

wherein $A_c l_c$ is the volume of the core. Thus the energy density in the field is given by

$$w_f\,(\text{density}) = \int_{B_1}^{B_2} H_c\ dB_c \quad \text{J/m}^3 \tag{2.34}$$

Example 2.8 *For the magnetic circuit of Example 2.1 and Fig. 2.4(a), find the following:*

(a) Induced emf e for B_c = 1.2 sin 314t T,
(b) reluctance $\mathcal{R}_c$ and $\mathcal{R}_g$
(c) coil inductance, L and
(d) magnetic field energy at B_c = 1.2 T

LO 3

Solution

(a) In Example 2.1 the value of λ was found as 1.152 Wb-T for B_c = 1.2 T. Therefore, for sinusoidal variation of B_c,

$$\lambda = 1.152 \sin 314t \text{ Wb-T}$$

The emf is

$$e = \frac{d\lambda}{dt} = 361.7 \cos 314t \quad \text{V}$$

(b)

$$\mathcal{R}_c = \frac{l_c}{\mu_0 \mu_r A_c} = \frac{0.4}{4\pi \times 10^{-7} \times 6000 \times 16 \times 10^{-4}}$$

$$= 3.317 \times 10^4$$

** It will be seen in Chapter 4 that a part of this energy is converted to mechanical form if mechanical motion is permitted between parts of the magnetic system.

$$\mathcal{R}_g = \frac{l_g}{\mu_0 A_g} = \frac{6 \times 10^{-4}}{4\pi \times 10^{-7} \times 16 \times 10^{-4}} = 29.856 \times 10^4$$

(c) From Example 2.1

$$i = 1.06 \text{ A}$$

$$\therefore \quad L = \frac{\lambda}{i} = \frac{1.152}{1.06} = 1.09 \text{ H}$$

It can also be found by using Eq. (2.21). Thus

$$L = N^2 P = \frac{N^2}{\mathcal{R}} = \frac{N^2}{\mathcal{R}_c + \mathcal{R}_g} = \frac{(600)^2}{(3.316 + 29.84) \times 10^4}$$

$$= 1.08 \text{ H}$$

(d) The energy stored in the magnetic field is from Eq. (2.32)

$$W_f = \int_0^{\lambda} i d\lambda = \int_0^{\lambda} \frac{\lambda}{L} d\lambda = \frac{1}{2}\frac{\lambda^2}{L}$$

$$= \frac{1}{2} \times \frac{(1.152)^2}{1.08} = 0.6144 \text{ J}$$

Concept Check — LO 3

1. What is the phase angle between flux and induced emf in an ac excited coil wound on an iron core? ○●●
2. Write the expression for inducted emf (rms) in an ac excited coil wound in an iron core. Use standard symbols. ○●●
3. Write the expression for the self inductance of a coil wound on an iron core. ○○●

2.6 HYSTERESIS AND EDDY-CURRENT LOSSES

LO 4 Define hysteresis and classify the eddy current losses

When a magnetic material undergoes cyclic magnetization, two kinds of power losses occur in it—hysteresis and eddy-current losses—which together are known as *core-loss*. The core-loss is important in determining heating, temperature rise, rating and efficiency of transformers, machines and other ac run magnetic devices.

Hysteresis Loss

Figure 2.18 shows a typical hysteresis loop of a ferromagnetic material. As the mmf is increased from zero to its maximum value, the energy stored in the field per unit volume of material is

$$\int_{-B_r}^{B_b = B_m} H dB = \text{area } ofabgo$$

As H is now reduced to zero, dB being negative, the energy is given out by the magnetic field (from the exciting coil back to the voltage source) and has a value

$$\int_{B_b = B_m}^{B_c} H dB = \text{area } cbg$$

The net energy unrecovered in the process is area *ofabco* which is lost irretrievably in the form of heat and is called the hysteresis loss. The total hysteresis loss in one cycle is easily seen to be the area of the complete loop (*abcdefa*) and let it be indicated as w_h (hysteresis loss/unit volume). Then hysteresis loss in volume V of material when operated at f Hz is

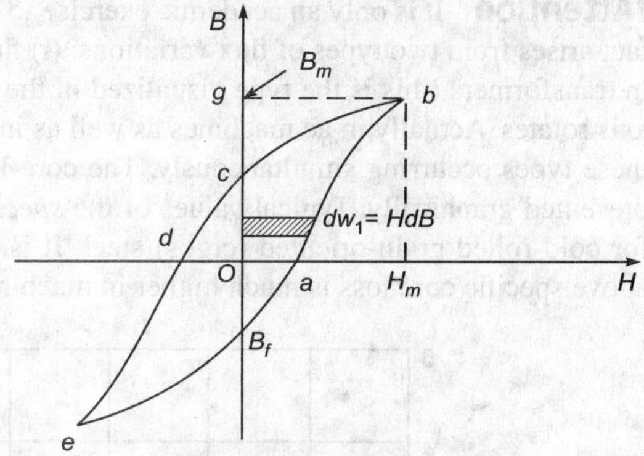

Fig. 2.18 *Hysteresis loss*

$$P_h = w_h V f \quad \text{W} \tag{2.35}$$

Probe

Derive an expression for hysteresis loss in volume V of material when operated at f Hz.

In order to avoid the need for computation of the loop area, Steinmetz gave an empirical formula for computation of the hystersis loss based on experimental studies according to which

$$P_h = k_h f B_m^n \quad \text{W/m}^3 \tag{2.36}$$

where k_h is a characteristic constant of the core material, B_m is the maximum flux density and n, called the Steinmetz exponent, may vary from 1.5 to 2.5 depending upon the material and is often taken as 1.6.

Eddy-current Loss

Probe

How is high resistivity achieved?

When a magnetic core carries a time-varying flux, voltages are induced in all possible paths enclosing the flux. The result is the production of circulating currents in the core (all magnetic materials are conductors). These currents are known as eddy-currents and have power loss (i^2R) associated with them called eddy-current loss. This loss, of course, depends upon the resistivity of the material and lengths of the paths of circulating currents for a given cross-section. Higher resistivity and longer paths increase the effective resistance offered by the material to induced voltages resulting in reduction of eddy-current loss. High resistivity is achieved by adding silicon to steel and hence silicon steel is used for cores conducting alternating flux. Dividing up the material into thin *laminations* along the flow of flux, with each lamination lightly insulated (varnish is generally used) from the adjoining ones, increases the path length of the circulating currents with consequent reduction in eddy-current loss. The loss in fact can be shown to depend upon the square of lamination thickness. The lamination thickness usually varies from 0.3 to 5 mm for electromagnetic devices used in power systems and from about 0.01 to 0.5 mm for devices used in electronic applications where low core-loss is desired.

The eddy-current loss can be expressed by the empirical formula

$$p_e = k_e f^2 B^2 \quad W/m^3 \tag{2.37}$$

wherein

$$k_e = K'_e d^2 / \rho \tag{2.38}$$

d being the thickness of lamination and ρ the resistivity of material.

!Attention It is only an academic exercise to split the core-loss into its two components. The core loss in fact arises from two types of flux variations: (i) flux that has a fixed axis and varies sinusoidally with time as in transformers (this is the type visualized in the above discussion), (ii) flux density is constant but the flux axis rotates. Actually in ac machines as well as in armature of dc machines the flux variation comprises both these types occurring simultaneously. The core-loss is measured experimentally on material specimen and presented graphically. Typical values of the *specific core-loss* (W/kg of material) are displayed in Fig. 2.19 for cold-rolled grain-oriented (crgos) steel. It is easy to see from these figures that for reasons mentioned above specific core loss is much higher in machines than in transformers.

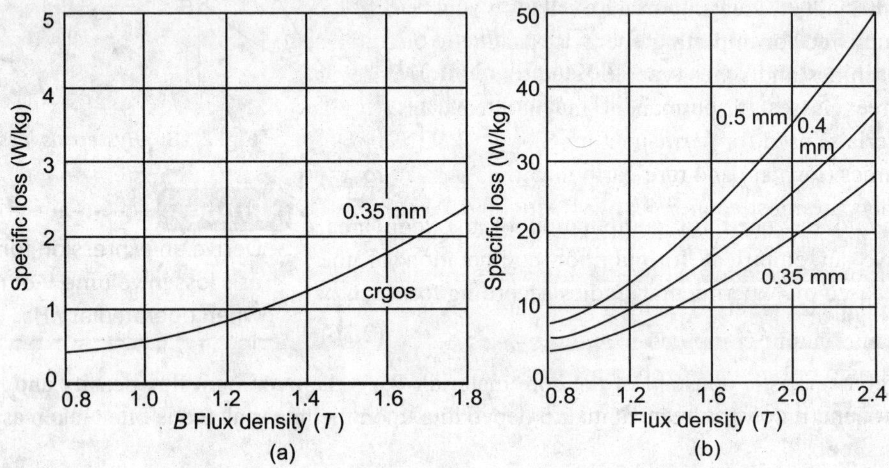

Fig. 2.19 *Core-loss at 50 Hz: (a) transformers, (b) machines*

Example 2.9 *The total core loss of a specimen of silicon steel is found to be 1500 W at 50 Hz. Keeping the flux density constant the loss becomes 3000 W when the frequency is raised to 75 Hz. Calculate separately the hysteresis and eddy current loss at each of those frequencies.* LO 4

Solution From Eqs. (2.36) and (2.37) for constant flux density, total core loss can be expressed as

$$P = Af + Bf^2 \quad \text{or} \quad P/f = A + Bf$$

$$1500/50 = A + 50\,B \quad \text{or} \quad 30 = A + 50\,B \qquad \text{(i)}$$

$$3000/75 = A + 75\,B \quad \text{or} \quad 40 = A + 75\,B \qquad \text{(ii)}$$

Solving Eqs. (i) and (ii), we get $A = 10$, $B = 2/5$

Therefore $$P = 10f + 2/5\,f^2 = P_h + P_e \qquad \text{(iii)}$$

At 50 Hz $$P_h = 10 \times 50 = 500 \text{ W}$$

$$P_e = 2/5 \times 2500 = 1000 \text{ W}$$

At 75 Hz $$P_h = 10 \times 75 = 750 \text{ W}$$

$$P_e = 2/5 \times (75)^2 = 2250 \text{ W}$$

Concept Check LO 4

1. Advance a qualitative explanation for reduction of eddy current loss by using a core composed of silicon steel laminations. ●●●

2.7 | PERMANENT MAGNETS

LO 5 Classify permanent magnets and list their applications

The permanent magnet is an important excitation source (life long) commonly employed for imparting energy to magnetic circuits used in rotating machines and other types of electromechanical devices. There are three classes of permanent magnet materials (or hard magnetic materials) used for permanent magnet dc (PMDC) motors: Alnicos, ceramics (ferrites) and rare-earth materials. Alnico magnets are used in motors up to 200 kW, while ceramic magnets are most economical in fractional kW motors. The rare-earth magnetic materials are very costly, but are the most economic choice in very small motors. Latest addition is neodymium-iron boron (Nd FeB). At room temperature, it has the highest energy product (to be explained later in this section) of all commonly available magnets. The high permeance and coercivity allow marked reductions in motor frame size for the same output compared to motors using ferrite (ceramic) magnets. For very high temperature applications Alnico or rare-earth cobalt magnets must be used.

Two important qualities of a permanent magnet (PM) are defined below with reference to the second quadrant of its hysteresis loop.

Permanent Magnetization or Residual Flux Density (B_r)

It is the flux density trapped in closed magnetic structure if the applied mmf (and therefore the magnetic field intensity, H) were reduced to zero.

Coercivity

It is the measure of mmf (or H) which, when applied to the magnetic circuit, would reduce its flux density to zero, i.e. it would demagnetize the material. Its value is negative and in units of kA/m.

The second quadrant of the hysteresis loops for Alnico 5 and M-5 steel are shown respectively in Figs. 2.20(a) and (b). Their residual flux densities and coercivities are given below:

Alnico 5 : $B_r \approx 1.25\ T,\ H_c \approx -50$ kA/m

M-5 steel : $B_r \approx 1.4\ T,\ H_c \approx -6$ kA/m

It is therefore observed that while B_r of M-5 steel is higher than that of Alnico 5 but the latter (Alnico 5) has a far greater coercivity. As we shall see below that materials with high coercivity qualify as PM materials.

An important measure of the capability of permanent magnet is known as its *maximum energy product*. This corresponds to the largest *BH* product, $(BH)_{max}$, which is a point on the second quadrant of the hysteresis loop; see Fig. 2.20(a). It has the dimensions of energy density (J/m^3) and it can be shown that operation of a given PM material at this point will result in the minimum volume of material required to produce a given flux density in the air gap.

Probe

Explain maximum energy product.

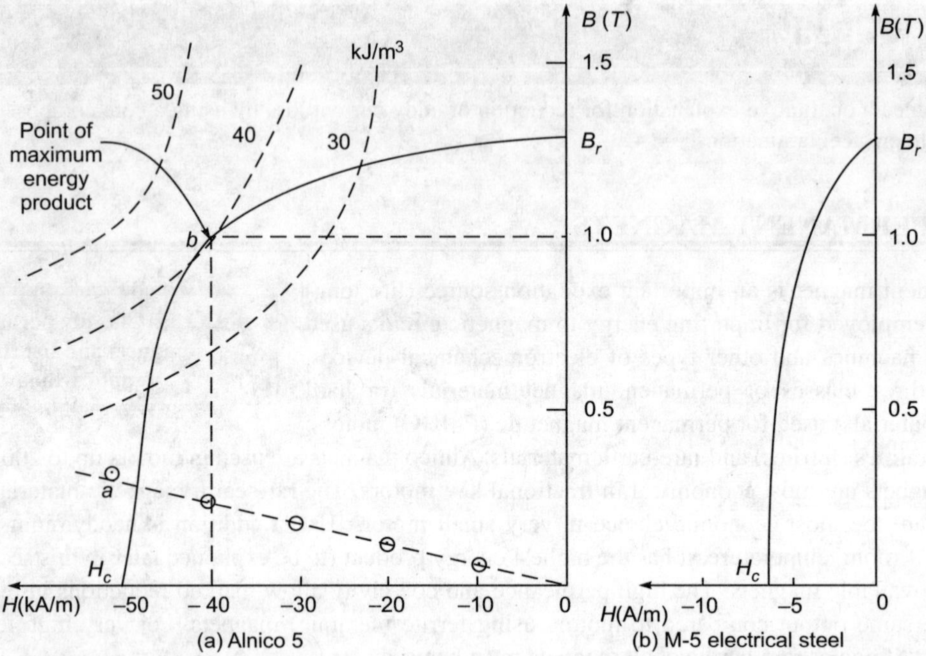

Fig. 2.20 *Second quadrant of hysteresis loop for (a) Alnico 5 and (b) M-5 electrical steel*

Example 2.10 *A magnetic circuit (Fig. 2.21a) consists of a core of very high permeability, an air-gap length of l_g = 0.4 cm and a section of permanent magnet (made of Alnico 5) of length l_m = 2.4 cm. Assume μ of core = ∞.*

Calculate the flux density B_g in the air-gap. Given: A_m = 4 mm^2.

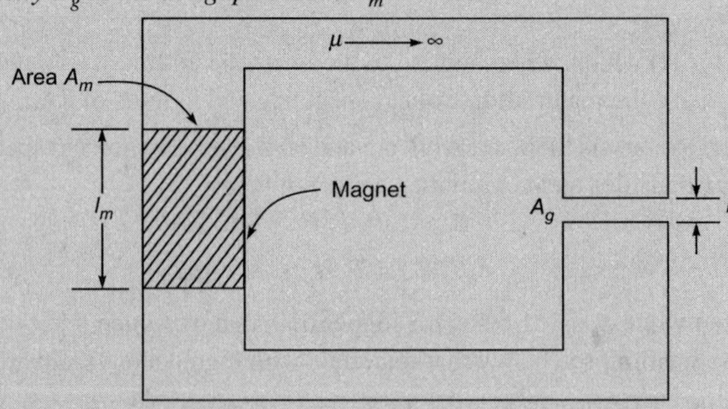

Fig. 2.21(a) *A Magnetic circuit with a PM*

LO 5

Solution

$$\mu_{core} = \infty \Rightarrow H_{core} = 0$$

From Ampere's circuital law

$$H_m l_m + H_g l_g = 0 = \mathcal{F} \tag{2.39}$$

or

$$H_g = -\left(\frac{l_m}{l_g}\right) H_m \tag{2.40}$$

where H_g and H_m are the magnetic field intensities in the air-gap and the PM respectively. Thus the existence of an air-gap is equivalent to the application of a negative field to the PM material.

As the flux must be continuous around the path

$$\phi = B_m A_m = B_g A_g \tag{2.41}$$

Also

$$B_g = \mu_0 H_g$$

We obtain from Eqs. (2.40) and (2.41)

$$B_m = -\mu_0 \left(\frac{A_g}{A_m}\right)\left(\frac{l_m}{l_g}\right) H_m \tag{2.42}$$

Substituting values we get

$$B_m = -6\mu_0 H_m = -7.54 \times 10^{-6} H_m \tag{2.43}$$

This is a straight line (also called *load line*) shown in Fig. 2.20(a), where its intersection with the demagnetization curve at point *'a'* gives the solution for B_m.

Thus $$B_g = B_m = 0.33 \text{ T}$$

Note: If we repeat the above problem for M-5 electrical steel, it is easy to find the answer since the load line is the same as given by Eq. (2.43). It can be shown that $B_m = 4 \times 10^{-5}$ T. This is much less than the value of B_m for Alnico 5.

From Eq. (2.40) we can get the expression for B_g as

$$B_g = \mu_0 H_g = -\mu_0 \left(\frac{l_m}{l_g}\right) H_m \tag{2.44}$$

From Eqs. (2.41) and (2.44) we get

$$B_g^2 = \mu_0 \left(\frac{l_m A_m}{l_g A_g}\right)(-H_m B_m)$$

$$= \mu_0 \left(\frac{Vol_m}{Vol_g}\right)(-H_m B_m) \tag{2.45}$$

$$Vol_m = \frac{B_g Vol_g}{\mu_0(-H_m B_m)} \text{; a positive value as } H_m \text{ is negative (Fig. 2.20)} \tag{2.46}$$

Thus, *to produce a flux density B_g in an air-gap of volume Vol_g, minimum volume of magnet material would be required if the material is operated in the state represented by the maximum value of the product $B_m H_m$.*

From Eq. (2.46) it may appear that one can get an arbitrarily large air-gap flux density just by reducing the air-gap volume. However, in practice this cannot be achieved because the on increasing flux density in the magnetic circuit beyond a given point, the magnetic core gets saturated and the assumption of infinite core permeability becomes invalid.

It may be noted that in Fig. 2.20(a) a set of constant *BH* product curves (hyperbolas) is also plotted.

Example 2.11 *For the magnetic circuit of Fig. 2.21(a) if A_g = 3.0 cm^2 find the minimum magnet volume required to produce an air-gap flux density of 0.7 T.*

LO 5

Solution The smallest magnet volume will be obtained with the magnet operating at point as shown in Fig. 2.20(a) which corresponds to B_m = 1.0 T and H_m = –40 kA/m.

From Eq. (2.41)

$$A_m = B_g A_g / B_m = (0.7 \times 3)/1 = 2.1 \text{ cm}^2$$

From Eq. (2.39)

$$l_m = -\frac{H_g l_g}{H_m} = -\frac{B_g l_g}{\mu_0 H_m}$$

$$= -\frac{0.7 \times 0.4}{4\pi \times 10^{-7} \times (-40 \times 10^3)} = 5.57 \text{ cm}$$

$$\text{Minimum magnet volume} = 2.1 \times 5.57 = 11.7 \text{ cm}^3$$

Examples 2.10 and 2.11 depict the operation of hard (PM) magnetic materials. However, the situation is more complex as discussed in the next section.

Now consider the case when an exciting coil is placed on the core of the permanent magnet. Circuit of Fig. 2.21(a) with *Ni* ampere-turns. The circuit is shown in Fig. 2.21(b).

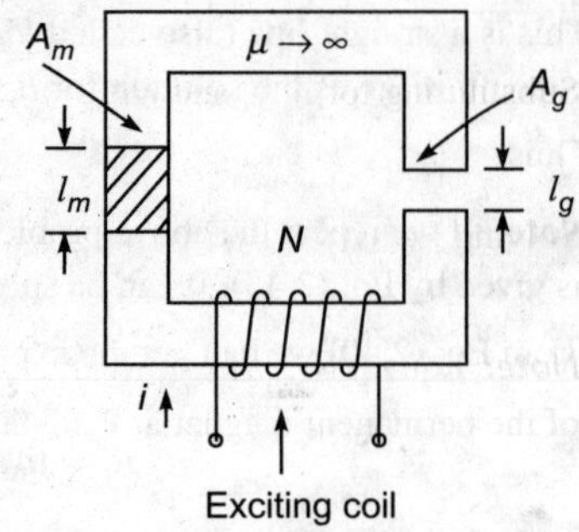

Fig. 2.21(b)

From the magnetic circuit

$$\mathcal{F} = Ni = H_m l_m + H_g l_g \quad (2.47)$$

As flux lines are continuous and no leakage is assumed.

So, $$B_m A_m = B_g A_g = \mu_0 H_g A_g \quad (2.48)$$

Equation (2.47) can be written as follows:

$$Ni = H_m l_m + \frac{B_m A_m}{\mu_0 A_g} \quad (2.49)$$

Important! This can be reorganized as follows:

$$B_m = -\mu_0 \left(\frac{A_g}{A_m}\right)\left(\frac{l_m}{l_g}\right) H_m + \mu_0 \left(\frac{A_g}{A_m}\right)\frac{1}{l_g}(Ni) \quad (2.50)$$

This equation is the general form of the load line. The intersection of this line gives the operational point *B, H*. By adjusting the current in the exciting coil, the permanent magnet can be brought to desired magnetization state. This is accomplished as per steps below.

- Select the *B-H* (hysteresis) curve with desired value of B_r (residual magnetization).
- From the corresponding *B-H* curve, find B_{max} and H_{max} at the hysteresis loop tip.
- Substitute these values in Eq. (2.50) to obtain i_{max}. Usually the desired *B-H* curve may not be available. So from B_r we have to estimate B_{max} and H_{max}. This can be done by extrapolating the *B-H* curve from B_r to about 4 times H_c (into positive *H*-side).

While magnetizing a permanent magnet the exciting current is raised to i_{max} and then reduced to zero gradually. The exciting coil may then be removed.

Example 2.12 *Consider the magnetic circuit of Fig. 2.21(a). The permanent magnet material Alinco-5 is in demagnetized state. It is required to be magnetized to a reduced flux density B_r = 1.25 T. Magnetic circuit dimensions are: $A_m = A_g = 2.5\ cm^2$, l_m = 4 cm, l_g = 0.2 cm. Excitation coil turns, N = 200.*

LO 5

Solution To find the first quadrant tip of the hysteresis loop, we assume

$$H_{max} = 3.5\ H_c$$

From Fig. 2.20(a)

$$H_c = 50 \text{ (magnitude)}$$

So, $$H_{max} = 170 \text{ kA/m}$$

Extrapolating *B-H* curve into first quadrant, we get

$$B_{max} \approx 2T$$

Substituting values in Eq. (2.50)

$$B_{max} = \mu_0 \left[-\left(\frac{2}{2}\right)\left(\frac{4}{0.2}\right) H_{max} + \left(\frac{2}{2}\right) \frac{1 \times 200\, i}{0.2 \times 10^{-2}} \right]$$

$$\mu_0 = 4\pi \times 10^{-7}$$

$\therefore$ $$B_{max} = -2.51 \times 10^{-5}\, H_{max} + 12.57 \times 10^{-2}\, i$$

Substituting for B_{max} and H_{max}

$$2 = -2.51 \times 10^{-5} \times 170 \times 10^3 + 12.57 \times 10^2\, i \qquad (2.51)$$

Solving Eq. (2.51), we get

$$i = 49.86 \text{ A}$$

Note: Equation (2.51) represents a straight line in *B-H* plane for a given *i*. Its intersection with *B-H* curve gives the state of the permanent magnet at that value of exciting current.

2.8 | APPLICATION OF PERMANENT MAGNET MATERIALS

LO 5

Typical values of the properties of four classes of PM materials are given in Table 2.1. These properties are: residual flux density B_r, coercive magnetizing force H_c, maximum energy stored $(BH)_{max}$ and resistivity.

Probe
Explain the four classes of PM materials.

Table 2.1

Type		Material	B_r (T)	H_c kA/m	BH_{max} kJ/m^3	$\frac{\rho}{\mu_m}$
A	Ceramic	Barium or Strontium ferrite carbonate powders	0.39	200	30	High
B	Metallic	Alcomax	1.25	60	45	500
C	Metallic	Hycomax	0.8	100	35	500
D	Rare-earth	Somarium-cobalt	0.75	600	130	60

Type A material is cheap but heavy, and suitable for low-rated production-run motors. Types B and C materials are hard and can be given simple shapes only. Type D material can be easily moulded and machined and is used for most electric motors but is costly. Materials with higher coercivities are much less prone to demagnetization. Figure 2.22 shows dc magnetization curves for a few commonly used materials.

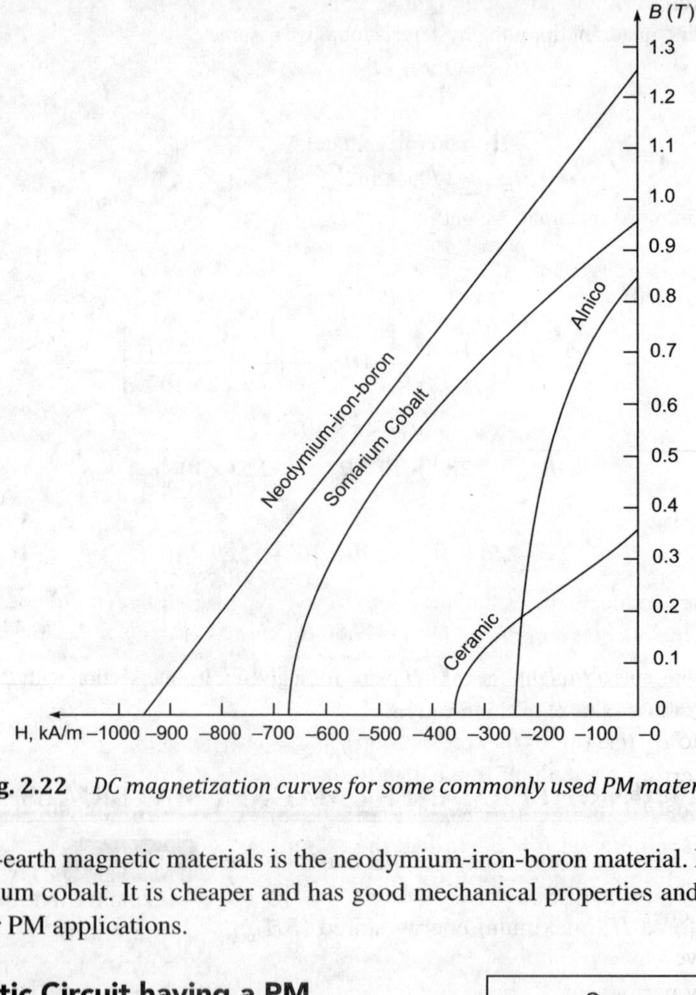

Fig. 2.22 *DC magnetization curves for some commonly used PM materials*

The latest of the rare-earth magnetic materials is the neodymium-iron-boron material. It has larger B_r, H_c and $(BH)_{max}$ than Somarium cobalt. It is cheaper and has good mechanical properties and this is expected to be used in a big way for PM applications.

Practical Magnetic Circuit having a PM

Consider the magnetic circuit of Fig. 2.23 consisting of a section of PM material in a core of highly permeable soft magnetic material with an N-turn exciting winding.

Figure 2.24 shows that the PM is initially unmagnetised and current is applied to the exciting winding. As core is of infinite permeability, the X-axis represents both i and H.

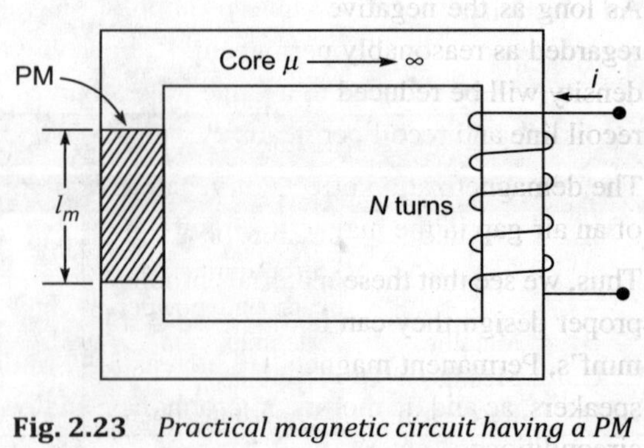

Fig. 2.23 *Practical magnetic circuit having a PM*

As current i is raised to its maximum value (saturation), the BH trajectory rises from the origin to its maximum value at point '*a*'. Now if the current is reduced to zero, the *BH* characteristic starts forming a hysteresis loop meeting Y-axis at point *b* at which $B = B_r$ and $H = 0$. As the current is further decreased to a negative value the *BH* curve continues to follow a hysteresis loop as shown in Fig. 2.24. For $i = -i_1$, the operating point is *c*. It may be noted that the same operating point *c* would be arrived (see Ex. 2.10) if the material were to start at point *b* with zero excitation and air-gap length of $l_m\ (A_g/A_m)\ (-\mu_0\ H_1/B_1)$ (Eq. (12.42)) were then inserted in the core.

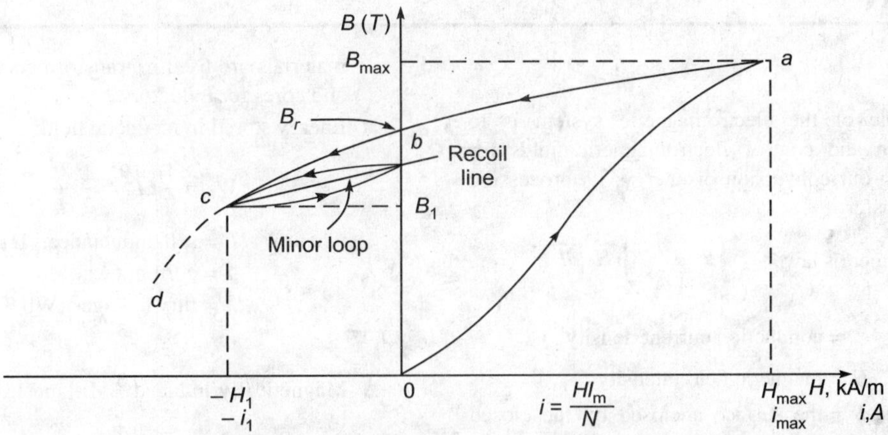

Fig. 2.24 *Demagnetization curve with recoil line*

If the current is further reduced, the trajectory would further trace the *BH* curve toward *d* as shown in Fig. 2.24. However, if the current is reduced to zero, the trajectory does not normally retrace the loop toward point *b*. Instead it starts to trace out a minor hysteresis loop, as shown in Fig. 2.24. As the current varies from 0 to i_1. The minor hysteresis loop may usually be replaced with little error by a straight line called the *recoil line*.

> **Probe**
> Explain recoil permeability with the help of a diagram.

This line has a slope called the *recoil permeability* μ_{rec}, which is approximately the same as that of the original *BH* curve at $H = 0$, i.e., at $B = B_r$. In fact the recoil line is essentially tangent to the *BH* curve for a large portion of the useful operating region for many materials such as Somarium Cobalt, Ceramic 7 etc. having large values of coercivity.

As long as the negative value of applied magnetic field intensity does not exceed H_1, the magnet may be regarded as reasonably permanent. If, however, a negative field intensity greater than H_1 is applied, the flux density will be reduced to a value lower than B_1 and a new and lower minor loop will be created with a new recoil line and recoil permeability.

The demagnetization effects of negative excitation which have been discussed above are equivalent to those of an air-gap in the magnetic circuit.

Thus, we see that these materials produce enough magnetic flux even in magnetic circuits with air-gaps. With proper design they can be operated stable even when subjected to a wide range of destabilizing forces and mmf's, Permanent magnets are increasingly finding greater applications in many small devices such as loud speakers, ac and dc motors, microphones, analog electric meters, driving, windshield wipers, radio antennas, airconditioners, etc.

Concept Check — LO 5

1. What are the important qualities of a PM? ○○●

SUMMARY

LO 1

- The role of the electromagnetic system is to establish and control electromagnetic fields for carrying out conversion of energy, its processing and transfer.
- The Ampere's law, $\int_S \vec{J} \cdot d\vec{s} = \oint \vec{H} \cdot d\vec{l}$

  where $\vec{J}$ = conduction current density
  $\vec{H}$ = magnetic field intensity
  S = the surface enclosed by the closed path of length l
  $d\vec{s}$ = differential surface
  $d\vec{l}$ = differential length
- In all practical circuits most of the flux is confined to the intended path by use of magnetic cores but a small amount of flux always leaks through the surrounding air. The stray flux is called the leakage flux.
- The effect of the fringing field is to increase the effective cross-sectional area A_g of the air-gap.
- Magnetic circuit law

$$\phi = \frac{\mathcal{F}}{\mathcal{R}} = \mathcal{F}\mathcal{P} = \text{flux (Wb)}$$

$$\mathcal{F} = mmf \text{ (AT)}$$

$$\mathcal{R} = reluctance = \frac{l_c}{\mu_c A_c} \text{ (AT/Wb)}$$

$$\mathcal{P} = permeance = 1/\mathcal{R}$$

- Electrical analog of magnetic circuit

$$\mathcal{F} \sim E, \quad \mathcal{R} \sim R, \quad \phi \sim i$$

LO 2

- Super paramagnetic materials are made from powdered iron or other magnetic particles. These materials are used in transformers for electronics and cores for inductors.
- Energy stored in magnetic field

$$W_f = \frac{1}{2}LI^2 = \frac{1}{2}\frac{\lambda^2}{L}$$

  where L = self-inductance, (H)
  I = current (A),
  λ = flux linkages (Wb-T)

LO 3

- Magnetically induced EMF and FORCE is given by

  $e = BlV$—Fleming's right hand rule determines the direction of emf
  $F = BIl$—Fleming's left hand rule determines the direction of force
  B = flux density (Wb/m^2 or T)
  V = conductor speed m/s relative to flux
  l = conductor length (m)
  I = conductor current (A)

LO 4

- Hysteresis loss,

$$p_h = k_h f B_m^n \quad \text{W/m}^3$$

  n = Stenmetz exponent, 1.5 to 2.5 upto taken as 1.6
- Eddy current loss, $p_e = k_e f^2 B_m^2$ W/m^3
- Magnetic cores are made up of thin, lightly insulated (coating of varnish) laminations to reduce power loss in cores due to the eddy current phenomenon. As a result, the net cross-sectional area of the core occupied by the magnetic material is less than its gross cross-section, their ratio (less than unity) being known as *stacking factor*.

LO 5

- Permanent Magnet – B_r = residual flux; Coercerlity – H (negative) needed to reduce; B to zero

PRACTICE PROBLEMS

Note: Unless otherwise specified, neglect leakage and fringing.

2.1 A square loop of side $2d$ is placed with two of its sides parallel to an infinitely long conductor carrying current I. The centre line of the square is at distance b from the conductor. Determine the expression for the total flux passing through the loop. What would be the loop flux if the loop is placed such that the conductor is normal to the plane of the loop. Does the loop flux in this case depend upon the relative location of the loop with respect to the conductor? LO 1

2.2 For the magnetic circuit of Fig. P.2.2, find the flux density and flux in each of the outer limbs and the central limbs. Assume the relative permeability of iron of the core to be (a) ∞, (b) 4500. LO 1

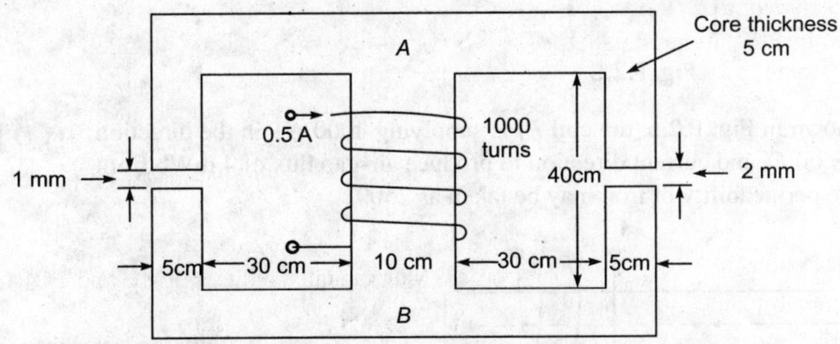

Fig. P.2.2

2.3 For the magnetic circuit shown in Fig. P.2.3, calculate the exciting current required to establish a flux of 2 mWb in the air-gap. Take fringing into account empirically. Use the *B-H* curve of Fig. 2.15. LO 1

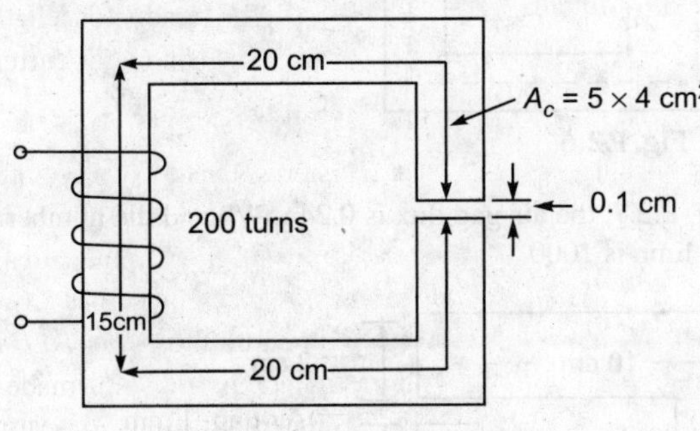

Fig. P.2.3

2.4 A steel ring has a mean diameter of 20 cm, a cross-section of 25 cm^2 and a radial air-gap of 0.8 mm cut across it. When excited by a current of 1 A through a coil of 1000 turns wound on the ring core, it produces an air-gap flux of 1 m Wb. Neglecting leakage and fringing, calculate (a) relative permeability of steel, and (b) total reluctance of the magnetic circuit. LO 1

2.5 The core made of cold rolled silicon steel (*B-H* curve of Fig. 2.15) is shown in Fig. P.2.5. It has a uniform cross-section (net iron) of 5.9 cm^2 and a mean length of 30 cm. Coils *A*, *B* and *C* carry 0.4, 0.8 and 1 A respectively in the directions shown. Coils *A* and *B* have 250 and LO 1

500 turns respectively. How many turns must coil C have to establish a flux of 1 mWb in the core?

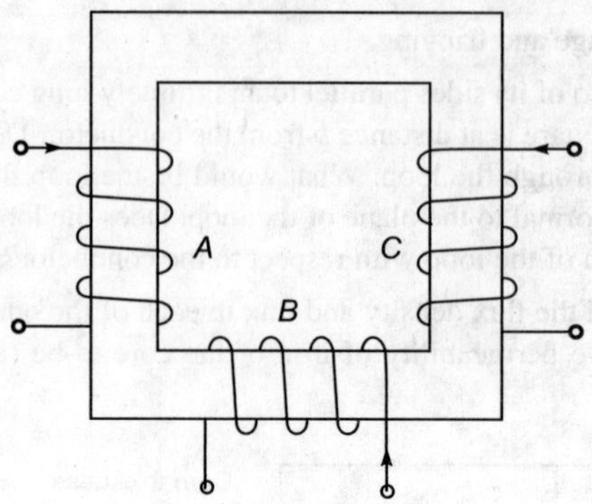

Fig. P.2.5

2.6 In the magnetic circuit shown in Fig. P.2.6, the coil F_1 is supplying 4000 AT in the direction indicated. Find the AT of coil F_2 and current direction to produce air-gap flux of 4 mWb from *top to bottom*. The relative permeability of iron may be taken as 2500. LO 1

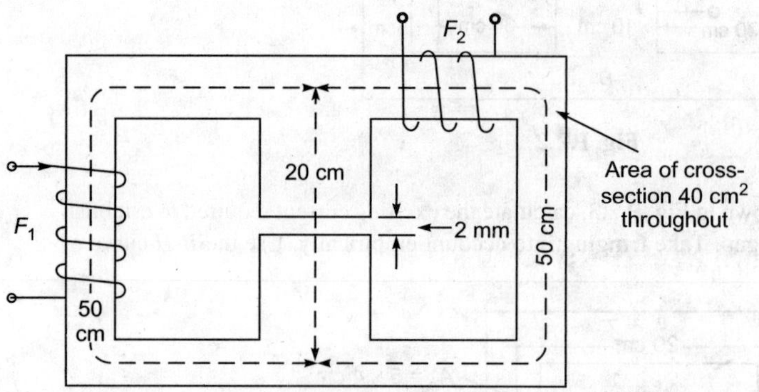

Fig. P.2.6

2.7 For the magnetic circuit shown in Fig. P.2.7, the air-gap flux is 0.24 mWb and the number of turns of the coil wound on the central limb is 1000. LO 1

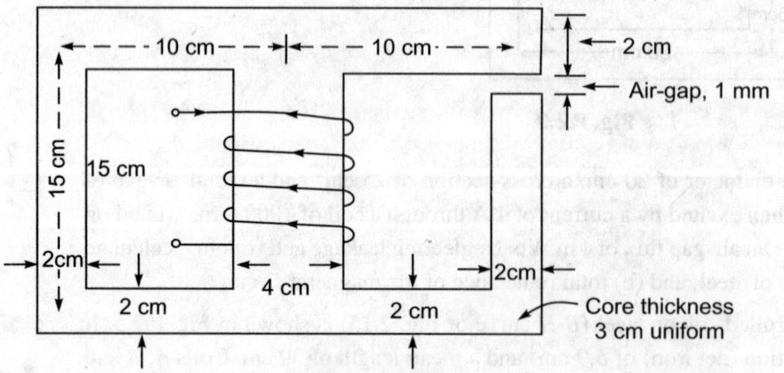

Fig. P.2.7

Calculate (a) the flux in the central limb, (b) the current required. The magnetization curve of the core is as follows:

H(AT/m)	200	400	500	600	800	1060	1400
B(T)	0.4	0.8	1.0	1.1	1.2	1.3	1.4

2.8 The magnetic circuit shown in Fig. P.2.8 has a coil of 500 turns wound on the central limb which has an air-gap of 1 mm. The magnetic path from A to B via each outer limb is 100 cm and via the central limb 25 cm (air-gap length excluded). The cross-sectional area of the central limb is 5 cm × 3 cm and that each outer limb is 2.5 cm × 3 cm. A current of 0.5 A in the coil produces an air-gap flux of 0.35 mWb. Find the relative permeability of the medium. LO 1

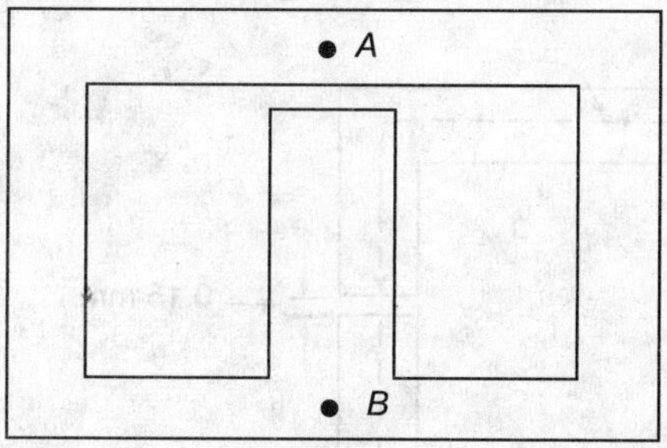

Fig. P.2.8

2.9 A cast steel ring has an external diameter of 32 cm and a square cross-section of 4 cm side. Inside and across the ring a cast steel bar 24 × 4 × 2 cm is fitted, the butt-joints being equivalent to a total air-gap of 1 mm. Calculate the ampere-turns required on half of the ring to produce a flux density of 1 T in the other half. Given: LO 1

H(AT/m)	0	200	400	600	800	1000	1200	1400	1600
B(T)	0	0.11	0.32	0.6	0.8	1.0	1.18	1.27	1.32

2.10 In Prob. 2.2 the *B-H* curve of the core material is characterized by the data given below. Find now the flux and flux densities in the three limbs of the core. LO 1

H(AT/m)	50	100	150	200	250	300	350
B(T)	0. 14	0.36	0.66	1.00	1.22	1.32	1.39

Hint: This problem can be solved by the graphical-cum-iterative technique.

2.11 A ring of magnetic material has a rectangular cross-section. The inner diameter of the ring is 20 cm and the outer diameter is 25 cm, its thickness being 2 cm. An air-gap of 1 mm length is cut across the ring. The ring is wound with 500 turns and when carrying a current of 3 A produces a flux density of 1.2 T in the air-gap. Find (a) magnetic field intensity in the magnetic material and in the air-gap. (b) relative permeability of the magnetic material, and (c) total reluctance of the magnetic circuit and component values. LO 2

2.12 For the magnetic ring of Prob. 2.11, the exciting current is again 3 A. Find the following: LO 3

(a) Inductance of the coil,

(b) energy stored in the magnetic material and in the air-gap, and

(c) rms emf induced in the coil when it carries alternating current of 3 sin 314t.

2.13 Assume that the core of the magnetic circuit of Fig. P.2.3 has $\mu_r = 2500$. **LO 3**

(a) Calculate the energy stored in the core and in the air-gap for an excitation current of 5 A. What will be these values if $\mu_r = \infty$?

(b) What will be the excitation current to produce a sinusoidally varying flux of 0.5 sin 314*t* mWb in the air-gap?

(c) Calculate the inductance of the coil. What will be the inductance if $\mu_r = \infty$?

2.14 The magnetic circuit of Fig. P.2.14 has a magnetic core of relative permeability 1600 and is wound with a coil of 1500 turns excited with sinusoidal ac voltage, as shown. Calculate the maximum flux density of the core and the peak value of the exciting current. What is the peak value of the energy stored in the magnetic system and what percentage of it resides in the air-gap? **LO 3**

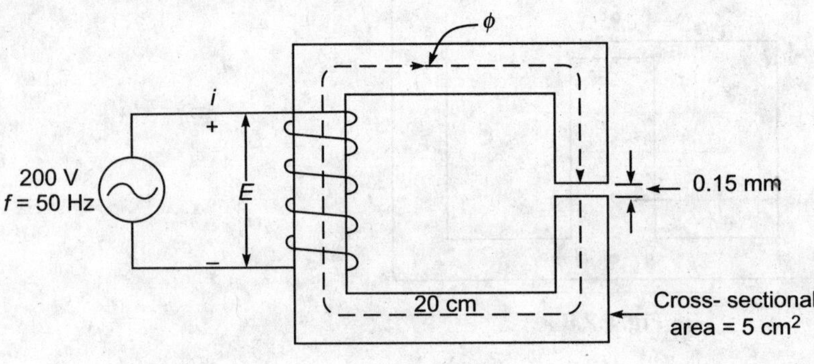

Fig. P 2.14

2.15 The material of the core of Fig. P.2.15, wound with two coils as shown, is sheet steel (*B-H* curve of Fig. 2.15). Coil 2 carries a current 2 A in the direction shown. What current (with direction) should coil 1 carry to establish a flux density of 1.4 T in the core in the indicated direction? **LO 3**

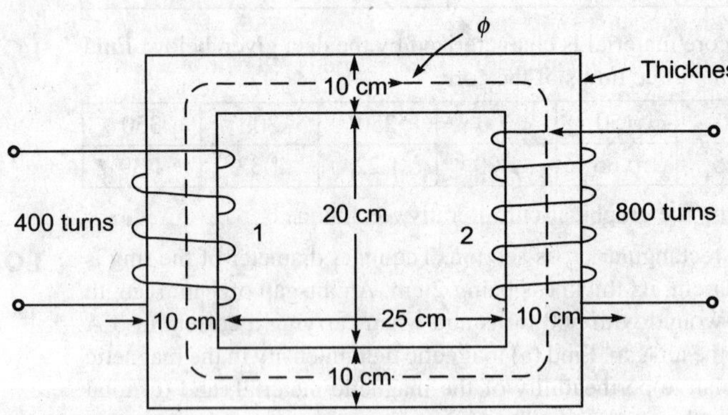

Fig. P 2.15

2.16 The flux in a magnetic core is alternating sinusoidally at a frequency of 600 Hz. The maximum flux density is 2 T and the eddy-current loss is 15 W. Find the eddy-current loss in the core if the frequency is raised to 800 Hz and the maximum flux density is reduced to 1.5 T. **LO 4**

2.17 The core-loss (hysteresis + eddy-current loss) for a given specimen of magnetic material is found to be 2000 W at 50 Hz. Keeping the flux density constant, the frequency of the supply is raised to 75 Hz resulting in core-loss of 3200 W. Compute separately hysteresis and eddy-current losses at both the frequencies. *Hint*: $P_L = P_c + P_k = k_e f^2 B_m^2 V + k_h f B_m^n V$; V = fixed core volume **LO 4**

Since B_m remains constant

$$P_L = k'_e f^2 + k'_h f$$

or

$$P_L / f = k'_e f + k'_h$$

which gives a straight line from which k'_e and k'_h can be determined.

2.18 A permanent magnet (PM) made of neodymium-iron boron alloy is placed in the magnetic circuit of Fig. P.2.18. Given **LO 5**

$$A_g = 4 \text{ cm}^2,\ l_g = 0.4 \text{ cm}$$

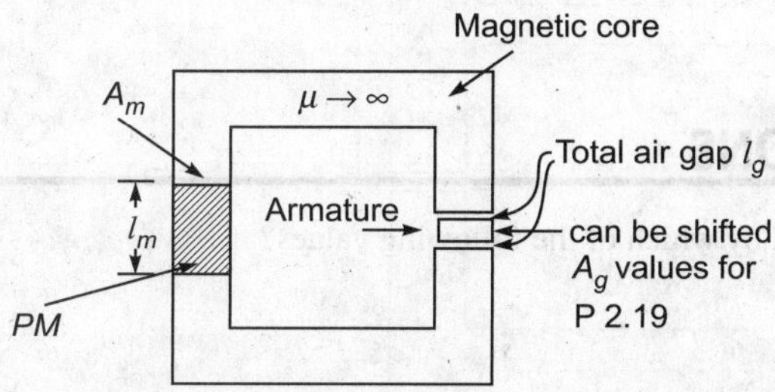

Fig. P 2.18

It is desired to have air gap flux density $B_g = 0.5$ T. For optimum design (minimum volume of PM) determine l_m,

Note $A_m = A_g$

2.19 In the PM circuit of Fig. P.2.18, A_g is reduced to 2 cm^2. Determine B_g and B_m. **LO 5**

2.20 The armature in the PM circuit Fig. 2.18 is now taken out and its height reduced so that when it is placed back in the circuit the air gap length l_g is now 0.5 cm. Determine B_g and B_m. **LO 5**

2.21 On the core of Fig. P.2.18 an exciting coil is wound with 200 turns and is fed with an exciting current of 1A. Determine air-gap flux density B_g. Note that direction of exciting current is such that it aids magnetization. **LO 5**

Answers

2.1 $\dfrac{\mu_0 + Id}{\pi} \ln\left(\dfrac{b+d}{b-d}\right)$ Wb

2.2 (a) 0.314 T, 0.785 mWb; 0.628 T, 1.571 mWb; 0.471 T, 2.356 mWb

(b) 0.261 T, 0.653 mWb; 0.468 T, 1.17 mWb; 0.365 T, 1.823 mWb

2.3 3.973 A

2.4 268

2.5 350 turns

2.6 530 AT

2.7 (a) 1.026 mWb (b) 0.453 A

2.8 3612

2.9 980.8

2.10 0.317 T

2.11 (a) 771.2, 9.55×10^5 AT/m (b) 1238.3 (c) $(1.592 + 0.909) \times 10^4 = 2.5 \times 10^4$ AT/Wb

2.12 (a) 0.1 H (b) 0.1635, 0.2865 J (c) 94.2 cos 314 *t* V

2.13 (a) 0.2, 0.869 J; 0, 1.31 J (b) 1.17 sin 314 *t* A (c) 0.0855, 0.1051 H

2.14 0.158 J, 54.4%

2.15 5.463 A

2.16 15 W

2.17 At 50 Hz; 1,733, 266.7 W; at 75 Hz; 2,600, 600 W

MULTIPLE CHOICE QUESTIONS

1. The magnetic field is solely governed by which of the following values? LO 1
(a) Displacement Current
(b) Conduction current
(c) Eddy current
(d) Magnetic current

2. Magnetomotive force is analogous to which of the following electric fields mentioned below? LO 1
(a) Electric field (b) Electromotive force
(c) Electrostatic field (d) Electric force

3. Choose the true statement among the four options given below. LO 1
(a) Magnetic energy is dissipative of energy.
(b) Magnetic reluctance is non-dissipative of energy unlike electric resistance.
(c) When F is time-varying, the magnetic circuit does not remain resistive.
(d) Magnetic reactance is dissipative.

4. What are magnetic cores made of? LO 1
(a) Thin, heavily insulated (coating of varnish) laminations
(b) Thick, lightly insulated (coating of varnish) laminations
(c) Thick, heavily insulated (coating of varnish) laminations
(d) Thin, lightly insulated (coating of varnish) laminations

5. All nonmagnetic materials are classified as *paramagnetic* having μ_r value LO 2
(a) slightly less than 1 (b) slightly greater than 1
(c) equal to 1 (d) equal to 0

6. Which of the following materials is the best known example of *superparamagnetic* materials? LO 2
(a) Perialloy (b) Permalloy
(c) Magnetic alloy (d) Superparamagnetic alloy

7. Magnetic materials are classified by LO 2
   (a) low permeability and linear B-H relationship
   (b) high permeability and non-linear B-H relationship
   (c) low permeability and non-linear B-H relationship
   (d) high permeability and linear B-H relationship
8. When coil remains stationary with respect to flux, but the flux through it changes with time, the emf induced is known as LO 3
   (a) dynamically induced emf
   (b) statically induced emf
   (c) induced emf
   (d) emf
9. In ac operation of magnetic circuits, except when linearity is desirable, economic utilization of material demands that working flux density should lie in a LO 3
   (a) linear zone (but not in the region of deep saturation)
   (b) non-linear zone (but not in the region of deep saturation)
   (c) linear zone (but in the region of deep saturation)
   (d) non-linear zone (but in the region of deep saturation)
10. What is trapped in a closed magnetic structure if the applied mmf (and therefore the magnetic field intensity, H) were reduced to zero? LO 5
    (a) Flux (b) Flux-density
    (c) Density (d) Magnetic flux

Answers

1. (b) **2.** (b) **3.** (b) **4.** (d) **5.** (a) **6.** (b) **7.** (b) **8.** (b) **9.** (b) **10.** (b)

CHAPTER

3 Transformers

LEARNING OBJECTIVES

After studying this chapter, students will be able to

- **LO1** Explain the constructional details, underlying working principle and types of single phase transformers
- **LO2** Discuss the performance of an ideal and practical transformer under no-load and load condition and hence deduce the equivalent circuit of a transformer
- **LO3** Outline various losses occurring in a transformer and testing methods used to determine the performance characteristics of a transformer and hence compute the losses
- **LO4** Summarize the concept of per-unit system and its advantages
- **LO5** Analyze the performance of a transformer by determining its efficiency, voltage regulation and the effect of loading
- **LO6** Assess the excitation phenomenon in transformers
- **LO7** Discuss the working principle of different varieties of transformers
- **LO8** Interpret the necessary conditions to be satisfied for parallel operation of transformers
- **LO9** Explain in brief about the three-winding transformers and tap-changing transformers and hence deduce their equivalent circuits
- **LO10** Discuss special types of transformers used for different applications

3.1 | INTRODUCTION

A transformer is a static device comprising coils coupled through a magnetic medium connecting two ports at different voltage levels (in general) in an electric system allowing the interchange of electrical energy between the ports in either direction via the magnetic field. The transformer is one of the most important component of a variety of electrical circuits ranging from low-power, low-current electronic and control circuits to ultra high-voltage power systems. Transformers are built in an astonishing range of sizes from the tiny units used in communication systems to monsters used in high-voltage transmission systems, weighing hundreds of tons. A circuit model and performance analysis of transformers is necessary for understanding of many electronic and control systems and almost all power systems. The transformer being an electromagnetic device, its analysis greatly aids in understanding the operation of electromechanical energy conversion devices which also use magnetic field but the interchange of energy is between electrical and mechanical ports.

The most important tasks performed by transformers are: (i) changing voltage and current levels in electric power systems, (ii) matching source and load impedances for maximum power transfer in electronic and control circuitry, and (iii) electrical isolation (isolating one circuit from another or isolating dc while maintaining ac continuity between two circuits). Transformers are used extensively in ac power systems because they make possible power generation at the most desirable and economical level (10–20 kV), power transmission at an economical transmission voltage (as high as 400–1000 kV) and power utilization at most convenient distribution voltages (230/400 V) for industrial, commercial and domestic purposes but in industrial applications voltages may have to be as high as 3.3, 6.6 or 11 kV for large motors. In communication and electronic systems where frequency ranges from audio to radio and video, transformers are used for a wide variety of purposes. For instance input/output transformers (used to connect the microphone to the first amplifying stage/to connect the last amplifying stage to the loudspeaker) and interstage transformers are to be found in radio and television circuits. Indeed the transformer is a device which plays an important and essential role in many facets of electrical engineering.

A transformer, in its simplest form, consists essentially of two insulated windings interlinked by a common or mutual magnetic field established in a core of magnetic material. When one of the windings, termed the *primary*, is connected to an alternating-voltage source, an alternating flux is produced in the core with an amplitude depending on the primary voltage, frequency and number of turns. This mutual flux links the other winding, called the *secondary*. A voltage is induced in this secondary of the same frequency as the primary voltage but its magnitude depends on the number of secondary turns. When the number of primary and secondary turns are properly proportioned, almost any desired voltage ratio, or *ratio of transformation* can be achieved. The subscript "1" will be associated with the primary and "2" with the secondary. The reader should note that these are arbitrary terms and in no way affect the inherent properties of a transformer.

Basic Types of Transformers

If the secondary voltage is greater than the primary value, the transformer is called a *step-up* transformer; if it is less, it is known as a *step-down* transformer; if primary and secondary voltages arc equal, the transformer is said to have a one-to-one ratio. One-to-one transformers are used to electrically isolate two parts of a circuit. Any transformer may be used as a step-up or step-down depending on the way it is connected.

Transformers are also categorized as electronic transformers and power transformers. Electronic transformer's operating voltages are very low and are rated at low power levels. These are used in consumer electronic equipment like televisions, personal computers, CD/DVD players, and other devices. The term power transformer is referred to the transformers with high power and voltage ratings. These are extensively used in power generation, transmission, distribution and utility systems to increase or decrease the voltage levels. Various applications use wide variety of transformers including power, instrumentation and pulse transformers.

In order to ensure the largest and most effective magnetic linkage of the two windings, the core, which supports them mechanically and conducts their mutual flux, is normally made of highly permeable iron or steel alloy (cold-rolled, grain oriented sheet steel). Such a transformer is generally called an *iron-core transformer*. Transformers operated from 25–400 Hz are invariably of iron-core construction. However, in special cases, the magnetic circuit linking the windings may be made of nonmagnetic material, in which case the transformer is referred to as an *air-core transformer*. The air-core transformer is of interest mainly in radio devices and in certain types of measuring and testing instruments. An intermediate type, exemplified by a type of induction coils and by small transformers used in speech circuits of telephone systems, utilizes a straight core made of a bundle of iron wires on which the primary and secondary coils are wound in layers.

Industry Relevance

A number of industry applications require **Specific Industrial Transformers** due to the usage of power (current) as a major resource for production. Electric Arc Furnaces (EAF), Ladle Furnaces (LF) and high-current **Rectifier and Converter Transformers** need a specific design to supply the necessary power at a low voltage level. These transformer types (**EAF Transformers and Ladle Furnace Transformers**), as well as converter transformers for large drive applications are called special-purpose or **Industry Transformers**, whose design is tailor-made for high-current solutions for industry applications.

Electronic transformers, operating at low voltages and low power levels, are used in consumer electronic equipments like televisions, personal computers, CD/DVD players, and other devices. **Power transformers** with high power and voltage ratings are extensively used in power generation, transmission, distribution and utility systems to increase or decrease the voltage levels. Various applications use wide variety of transformers including power, instrumentation and pulse transformers.

3.2 TRANSFORMER CONSTRUCTION AND PRACTICAL CONSIDERATIONS

LO 1 Explain the constructional details, underlying working principle and types of single phase transformers

The type of construction adopted for transformers is intimately related to the purpose for which these are to be used; winding voltage, current rating and operating frequencies. The construction has to ensure efficient removal of heat from the two seats of heat generation—core and windings, so that the temperature rise is limited to that allowed for the class of insulation employed. Further, to prevent insulation deterioration, moisture ingress to it must not be allowed. These two objectives are simultaneously achieved in power transformers, other than those in very small sizes, by immersing the built-up transformer in a closed tank filled with noninflammable insulating oil called *transformer oil.*

Transformer Cooling Process

To facilitate natural oil circulation and to increase the cooling surface exposed to the ambient, tubes or fins are provided on the outside of tank walls. In large-size transformers tubes may be forced-cooled by air. For still larger installations the best cooling system appears to be that in which the oil is circulated by pump from the top of the transformer tank to a cooling plant, returning when cold to the bottom of the tank. In small sizes the transformers are directly placed in a protective housing or are encased in hard rubber moulding and are air-cooled. Figures 3.1. (a), (b) and (c) show the constructional details of practical transformers.

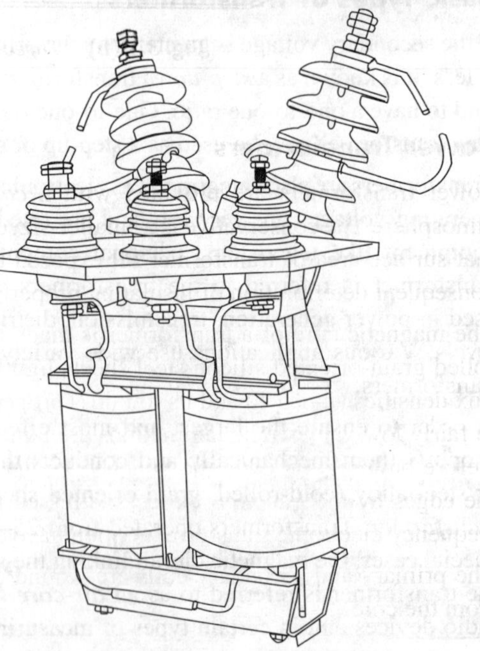

Fig. 3.1 (a) *Single-phase transformer core and windings*

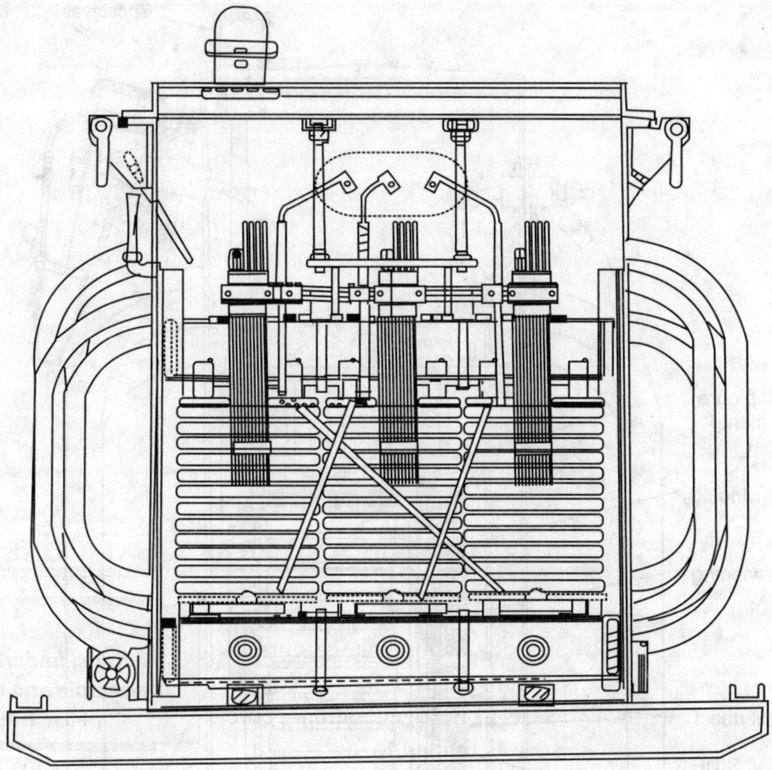

Fig. 3.1 (b) *Three-phase transformer core and windings*

Power Transformers

Power transformers are provided with a *conservative* through which the transformer breathes into the atmosphere The conservative is a smaller-sized tank placed on top of the main tank. This arrangement ensures that surface area of transformer oil exposed to atmosphere is limited so as to prevent fast oxidization and consequent deterioration of insulating properties of the oil.

The magnetic core of a transformer is made up of stacks of thin laminations (0.35 mm thickness) of cold-rolled grain-oriented silicon steel sheets lightly insulated with varnish. This material allows the use of high flux densities (1–1.5 T) and its low-loss properties together with laminated construction reduce the core-loss to fairly low values. The laminations are punched out of sheets and the core is then built of these punching. Before building the core, the punched laminations are annealed to relieve the mechanical stresses set in at the edges by the punching process; stressed material has a higher core-loss. Pulse transformers and high-frequency electronic transformers often have cores made of *soft ferrites*.

The primary and secondary coils are wound on the core and are electrically insulated from each other and from the core.

Important! Two types of cores are commonly employed in practice—*core-type* and *shell-type.*

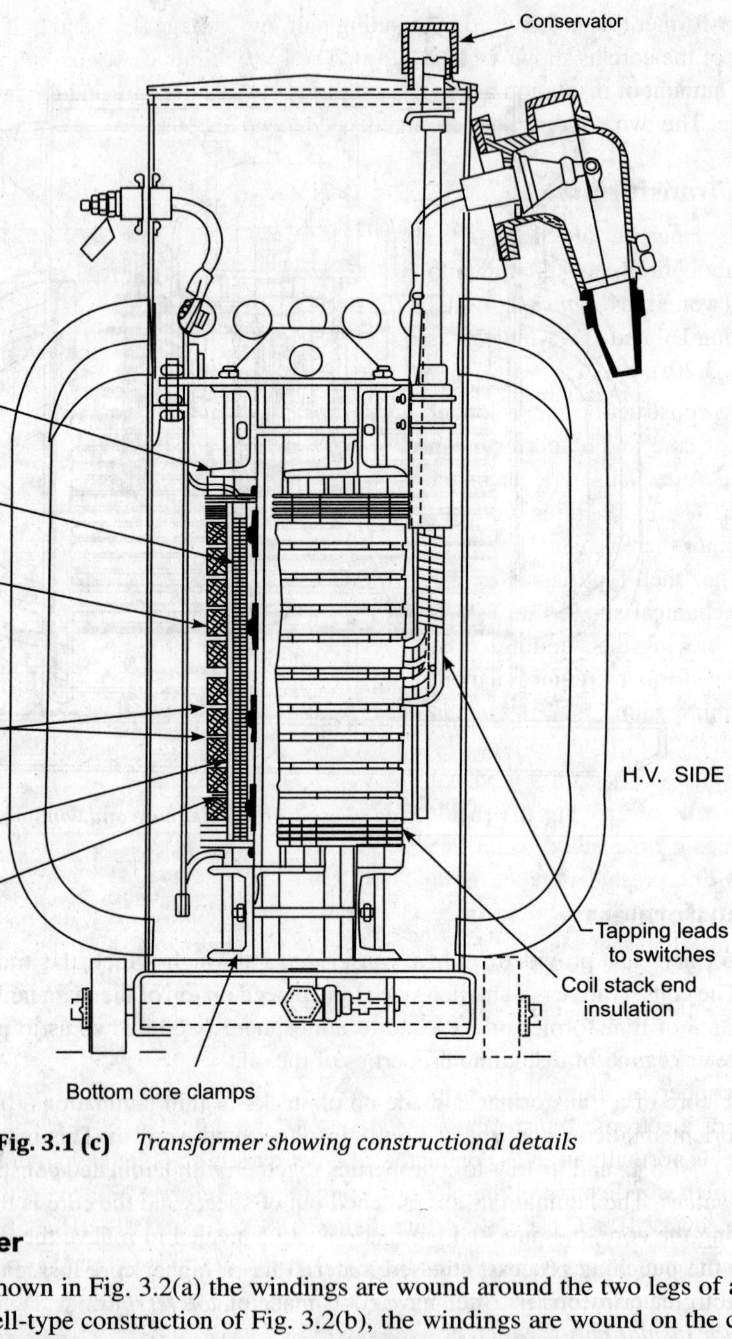

Fig. 3.1 (c) *Transformer showing constructional details*

Core-Type Transformer

In core-type construction shown in Fig. 3.2(a) the windings are wound around the two legs of a rectangular magnetic core, while in shell-type construction of Fig. 3.2(b), the windings are wound on the central leg of a three-legged core. Though most of the flux is confined to a high permeability core, some flux always leaks through the core and embraces paths which partially lie in the air surrounding the core legs on which the coils are wound. This flux which links one of the windings without linking the other, though small in magnitude, has a significant effect on the transformer behaviour. Leakage is reduced by bringing the two coils closer. In a core-type transformer this is achieved by winding half low-voltage (LV) and half high-voltage (HV) winding on each limb of the core as shown in Fig. 3.2(a). The LV winding is wound on the inside and HV on outside to reduce the amount of insulation needed. Insulation between the core and the inner winding is then stressed to low voltage. The two windings are arranged as *concentric* coils.

Shell-type Transformer

In shell-type construction leakage is reduced by subdividing each winding into subsections (wound as *pancake* coils) and interleaving LV and HV windings as shown in Fig. 3.2(b).

The core-type construction has a longer mean length of core and a shorter mean length of coil turn. This type is better suited for EHV (extra high voltage) requirement since there is better scope for insulation. The shell-type construction has better mechanical support and good provision for bracing the windings. The shell-type transformer requires more specialized fabrication facilities than core-type, while the latter offers the additional advantage of permitting visual inspection of coils in the case of a fault and ease of repair at substation site. For these reasons, the present practice is to use the core-type transformers in large high-voltage installations.

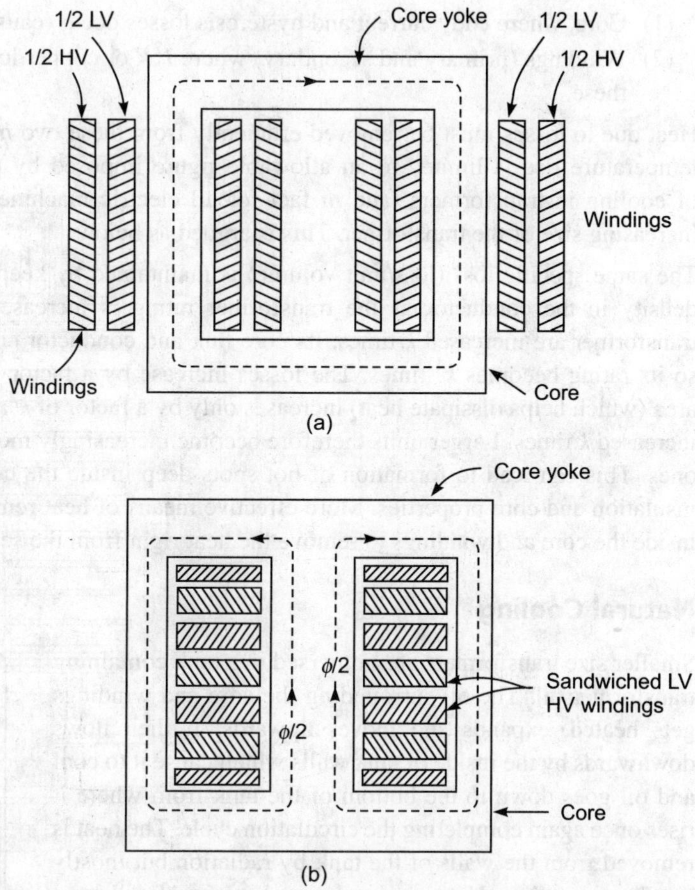

Fig. 3.2 *(a) Core-type transformer, (b) Shell-type transformer*

Transformer Windings

Transformer windings are made of solid or stranded copper or aluminium strip conductors. For electronic transformers, "magnet wire" is normally used as conductor. Magnet wire is classified by an insulation class symbol, A, B, C, F and H, which is indication of the safe operating temperature at which the conductor can be used. Typical figures are the lowest 105 °C for class-A and highest 180 °C for class-H.

The windings of huge power transformers use conductors with heavier insulation (cloth, paper, etc.) and are assembled with greater mechanical support and the winding layers are insulated from each other—this is known as *minor insulation* for which pressed board or varnished cloth is used. *Major insulation*, insulating cylinders made of specially selected pressed board or synthetic resin bounded cylinders, is used between LV and core and LV and HV. Insulating barriers are inserted between adjacent limbs when necessary and between coils and core yokes.

Transformer Cooling (Large Units)

!Attention Some idea of transformer losses, heating and cooling has been presented above. Details of transformer losses will be presented in Section 3.6. Basically, there are two seats of losses in a transformer namely:

(1) Core, where eddy current and hysteresis losses occur (caused by alternating flux density).

(2) Windings (primary and secondary) where I^2R or copper loss occurs because of the current flowing in these.

Heat due to losses must be removed efficiently from these two main parts of the transformer so that steady temperature rise is limited to an allowable figure imposed by the class of insulation used. The problem of cooling in transformers (and in fact for all electric machinery) is rendered increasingly difficult with increasing size of the transformer. This is argued as below:

The same specific loss (loss/unit volume) is maintained by keeping constant core flux density and current density in the conductor as the transformer rating is increased. Imagine that the linear dimensions of transformer are increased k times. Its core flux and conductor current would then increase by k^2 times and so its rating becomes k^4 times. The losses increase by a factor of k^3 (same as volume), while the surface area (which helps dissipate heat) increases only by a factor of k^2. So the loss per unit area to be dissipated is increased k times. Larger units therefore become increasingly more difficult to cool compared to the smaller ones. This can lead to formation of hot spots deep inside the conductors and core which can damage the insulation and core properties. More effective means of heat removal must therefore be adopted with ducts inside the core and windings to remove the heat right from the seats of its generation.

Natural Cooling

Smaller size transformers are immersed in a tank containing transformer oil. The oil surrounding the core and windings gets heated, expands and moves upwards. It then flows downwards by the inside of tank walls which cause it to cool and oil goes down to the bottom of the tank from where it rises once again completing the circulation cycle. The heat is removed from the walls of the tank by radiation but mostly by air convection. Natural circulation is quite effective as the transformer oil has large coefficient of expansion. Still for large sizes, because of the arguments presented earlier, the cooling area of the tank must be increased by providing cooling fins or tubes (circular or elliptical) as shown in Fig. 3.3. This arrangement is used for all medium size transformers.

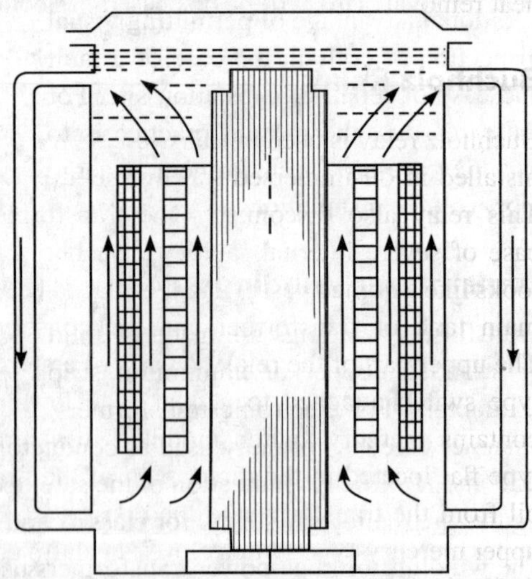

Fig. 3.3 *Natural cooling in transformers*

Forced Cooling

For transformer sizes beyond 5 MVA additional cooling would be needed which is achieved by supplementing the tank surface by a separate radiator in which oil is circulated by means of a pump. For better cooling oil-to-air heat exchanger unit is provided as shown in Fig. 3.4(a). For very large size transformers cooling is further strengthened by means of oil-to-water heat exchanger as shown in Fig. 3.4(b).

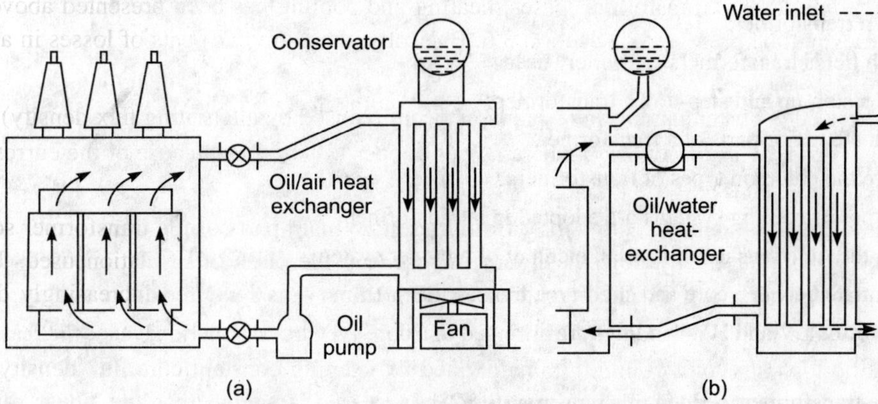

Fig. 3.4 *Forced cooling in transformers*

As already pointed out, ducts are provided in core and windings for effective heat removal by oil. Vertical flow is more effective compared to horizontal flow but for pancake coils some of the ducts will have to be horizontal.

The problem of cooling in transformers is more acute than in electric machines because the rotating member in a machine causes forced air draft which can be suitably directed to flow over the machine part for efficient heat removal. This will be discussed in Section 5.10.

Buchholz Relay

Buchholz relay is used in transformers for protection against all kinds of faults. It is a gas-actuated relay and installed in oil-immersed transformers. It will give an alarm in case of incipient faults in the transformer. This relay also disconnects the transformer in case of severe internal faults. A Buchholz relay looks like a domed vessel and it is placed between main tank of transformer and the conservator. The upper part of the relay consists of a mercury-type switch attached to a float. The lower part contains mercury switch mounted in a hinged-type flat located in the direct path of the flow of oil from the transformer to the conservator. The upper mercury-type switch closes an alarm circuit during incipient fault, whereas the lower mercury switch is used to trip the circuit breaker in the case of sever faults. The Buchholz relay is shown in Fig. 3.4 (c).

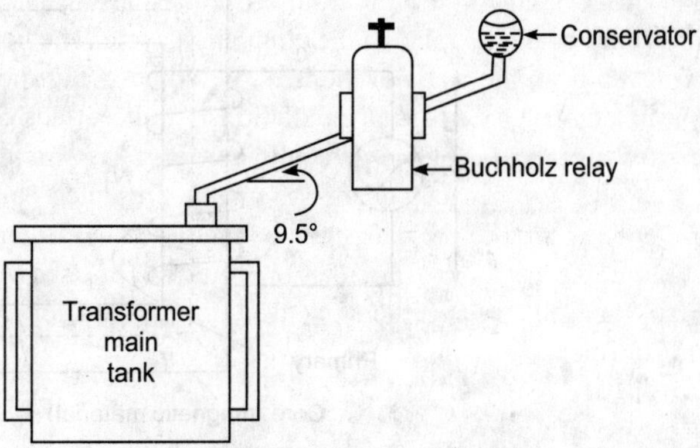

Fig. 3.4 (c) *Buchholz relay set up*

Concept Check LO 1

1. What is a transformer? ○○●
2. In which fields transformers are widely used? ○●●
3. What is a step-up and step-down transformer? ○○●
4. Mention the major parts of a transformer. ○○●
5. What are the common types of transformers? ○○●
6. What types of cooling systems are adopted in a transformer? ○●●
7. Explain the functions it fulfils as an element of a power system. ○●●
8. Differentiate between core and shell-type transformers. ○●●
9. State how the LV and HV windings are arranged in a core-type transformer. Advance the reason why? ●●●
10. Why are transformers needed in a power system? ●●●
11. Why are transformers placed in oil-filled tanks? ●●●

3.3 | TRANSFORMER ON NO-LOAD

LO 2 Discuss the performance of an ideal and practical transformer under no-load and load condition and hence deduce the equivalent circuit of a transformer

Figure 3.5 shows the schematic diagram of a two-winding transformer on no-load, i.e. the secondary terminals are open while the primary is connected to a source of constant sinusoidal voltage of frequency f Hz. The simplifying assumption that the resistances of the windings are negligible, will be made.

The primary winding draws a small amount of alternating current of instantaneous value i_0, called the *exciting current*, from the voltage source with positive direction as indicated on the diagram. The exciting current establishes flux ϕ in the core

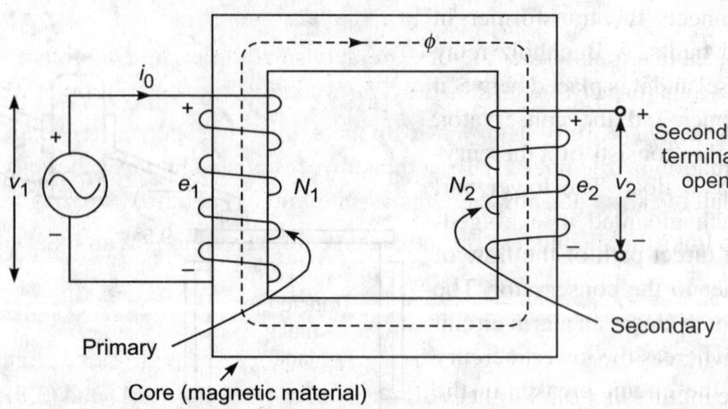

Fig. 3.5 *Transformer on no-load*

Note: Difficulty Level ⟶ ○○● — Level 1 and Level 2 Category; ○●● — Level 3 and Level 4 Category; ●●● — Level 5 and Level 6 Category

(positive direction marked on diagram) all of which is assumed confined to the core i.e., there is no leakage of flux. Consequently the primary winding has flux linkages,

$$\lambda_1 = N_1\phi$$

which induces emf in it is given by

$$e_1 = \frac{d\lambda_1}{dt} = N_1\frac{d\phi}{dt} \tag{3.1}$$

As per Lenz's law, the positive direction of this emf opposes the positive current direction and is shown by + and – polarity marks on the diagram. According to Kirchhoff's law,

$$v_1 = e_1 \text{ (winding has zero resistance)} \tag{3.2}$$

and thus e_1 and therefore ϕ must be sinusoidal of frequency f Hz, the same as that of the voltage source. Let

$$\phi = \phi_{max} \sin \omega t \tag{3.3}$$

where ϕ_{max} = maximum value of core flux

$\omega = 2\pi f$ rad/s (f = frequency of voltage source)

The emf induced in the primary winding is

$$e_1 = N_1\frac{d\phi}{dt} = \omega N_1 \phi_{max} \cos \omega t \tag{3.4}$$

From Eqs (3.3) and (3.4) it is found that the induced emf leads the flux by 90°*. This is indicated by the phasor diagram of Fig. 3.6. The rms value of the induced emf is

$$E_1 = \sqrt{2}\,\pi f N_1 \phi_{max} = 4.44\, f N_1 \phi_{max} \tag{3.5}$$

Since $E_1 = V_1$ as per Eq. (3.2),

$$\phi_{max} = \frac{E_1(=V_1)}{4.44\, f N_1} \tag{3.6}$$

Even if the resistance of the primary winding is taken into account,

$$E_1 \approx V_1 \tag{3.7}$$

as the winding resistances in a transformer are of extremely small order. It is, therefore, seen from Eq. (3.6) that maximum flux in a transformer is determined by V_1/f (voltage/frequency) ratio at which it is excited.

According to Eq. (3.6) the flux is fully determined by the applied voltage, its frequency and the number of winding turns. This equation is true not only for a transformer but also for any other electromagnetic device operated with sinusoidally varying ac and where the assumption of negligible winding resistance holds.

All the core flux ϕ also links the secondary coil (no leakage flux) causing in it an induced emf of

$$e_2 = N_2\frac{d\phi}{dt} \tag{3.8}$$

The polarity of e_2 is marked + and – on Fig. 3.5 according to Lenz's law (e_2 tends to cause a current flow whose flux opposes the mutual flux ϕ). Further, it is easily seen from Eqs. (3.1) and (3.8) that e_1 and e_2 are in phase. This is so indicated by phasor diagram of Fig. 3.6 where $\bar{E}_1$, and $\bar{E}_2$ are the corresponding phasors in terms of the rms values. As the secondary is open-circuited, its terminal voltage is given as

---

* cos ωt leads sin ωt by 90°

$$v_2 = e_2$$

From Eqs (3.1) and (3.8) we have the induced emf ratio of the transformer windings as

$$\frac{e_1}{e_2} = \frac{N_1}{N_2} = a$$

$$\frac{E_1}{E_2} = \frac{N_1}{N_2} = a \text{ ratio of transformation} \tag{3.9}$$

This indeed is the transformation action of the transformer. Its current transformation which is in inverse ratio of turns will be discussed in Section 3.4.

Exciting Current

The value of exciting current i_0 has to be such that the required mmf is established so as to create the flux demanded by the applied voltage (Eq. (3.6)). If a linear *B-H* relationship is assumed (devoid of hysteresis and saturation), the exciting current is only magnetizing in nature and is proportional to the sinusoidal flux and in phase with it. This is represented by the phasor $\bar{I}_m$, in Fig. 3.6, lagging the induced emf by 90°. However, the presence of hysteresis and the phenomenon of eddy-currents, though of a different physical nature, both demand the flow of active power into the system and as a consequence the exciting current $\bar{I}_0$ has another component $\bar{I}_i$ in phase with $\bar{E}_1$. Thus, the exciting current lags the induced emf by an angle θ_0 slightly less than 90° as shown in the phasor diagram of Fig. 3.6. Indeed it is the hysteresis which causes the current component $\bar{I}_i$ leading $\bar{I}_m$ by 90° and eddy-currents add more of this component. The effect of saturation nonlinearity is to create a family of odd-harmonic components in the exciting current, the predominant being the third harmonic; this may constitute as large as 35–40% of the exciting current. While these effects will be elaborated in Sec. 3.10, it will be assumed here that the current $\bar{I}_o$ and its magnetizing component $\bar{I}_m$ and its core-loss component $\bar{I}_i$ are sinusoidal on equivalent rms basis. In other words, $\bar{I}_m$ is the *magnetizing current* and is responsible for the production of flux, while $\bar{I}_i$ is the *core-loss current* responsible for the active power[†] being drawn from the source to provide the hysteresis and eddy-current loss.

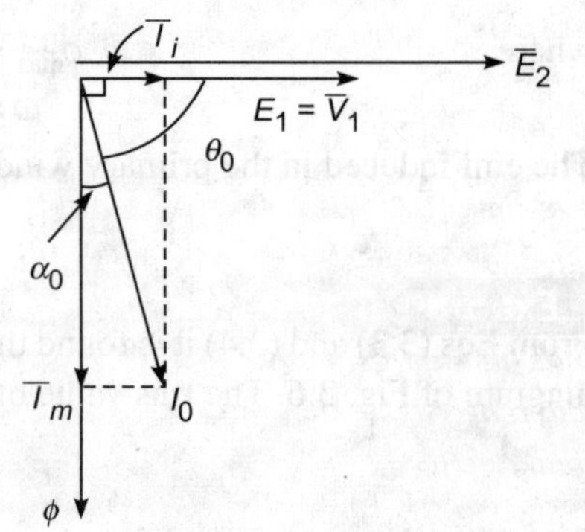

Fig. 3.6 *Phasor relationship of induced emf, flux and exciting current*

To account for the harmonics, the exciting current $|\bar{I}_o|$ is taken as the rms sine wave equivalent of the actual non-sinusoidal current drawn by the transformer on no-load. Since the excitation current in a typical transformer is only about 5% of the full-load current, the net current drawn by the transformer under loaded condition is almost sinusoidal.

From the phasor diagram of Fig. 3.6, the core-loss is given by

$$P_i = E_1 I_0 \cos \theta_0 \tag{3.10}$$

[†] Suffix *i* is used as this current provides the core-loss which occurs in the iron core and is also referred as *iron-loss*.

In a practical transformer, the magnetizing current (I_m) is kept low and the core-loss is restrained to an acceptable value by use of high permeability silicon-steel in laminated form.

From the no-load phasor diagram of Fig. 3.6, the parallel circuit model[‡] of exciting current as shown in Fig. 3.7 can be easily imagined wherein conductance G_i accounts for core-loss current $\bar{I}_i$ and inductive susceptance B_m for magnetizing current I_m. Both these currents are drawn at induced emf $\bar{E}_1 = \bar{V}_1$ for resistance-less, no-leakage primary coil; even otherwise $\bar{E}_1 \approx \bar{V}_1$.

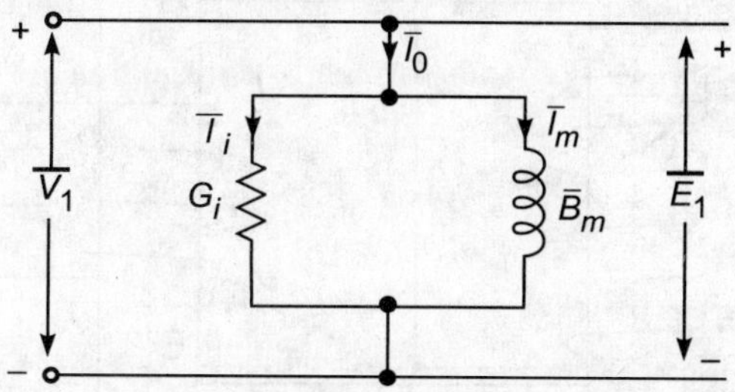

Fig. 3.7 *Circuit model of transformer on no-load (exciting current)*

Example 3.1 *A transformer on no-load has a core-loss of 50 W, draws a current of 2 A (rms) and has an induced emf of 230 V (rms). Determine the no-load power factor, core-loss current and magnetizing current. Also calculate the no-load circuit parameters of the transformer. Neglect winding resistance and leakage flux.*

LO 2

Solution

Power factor, $$\cos\theta_0 = \frac{50}{2 \times 230} = 0.108 \text{ lagging};$$

$$\theta_0 = 83.76°$$

Magnetizing current, $$I_m = I_0 \sin\theta_0 = 2 \sin(\cos^{-1} 0.108) = 1.988 \text{ A}$$

Since $\theta_0 \approx 90°$, there is hardly any difference between the magnitudes of the exciting current and its magnetizing component.

$$\text{Core-loss current, } I_i = I_0 \cos\theta_0 = 2 \times 0.108 = 0.216 \text{ A}$$

In the no-load circuit model of Fig. 3.7 core loss is given by

$$G_i V_1^2 = P_i$$

or $$G_i = \frac{P_i}{V_1^2} = \frac{50}{(230)^2} = 0.945 \times 10^{-3} \text{ ℧}$$

Also $$I_m = B_m V_1$$

or $$B_m = \frac{I_m}{V_1} = \frac{1.988}{230} = 8.64 \times 10^{-3} \text{ ℧}$$

‡Series circuit is equally possible but not convenient for physical understanding.

Example 3.2 *The BH curve data for the core of the transformer shown in Fig. 3.8 is given in Problem 2.10. Calculate the no-load current with the primary excited at 200 V, 50 Hz. Assume the iron loss in the core to be 3 W/kg. What is the pf of the no-load current and the magnitude of the no-load power drawn from the mains? Density of core material = 7.9 g/cc.*

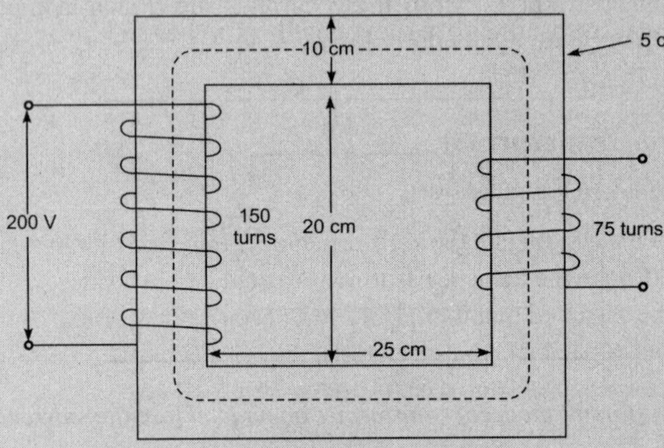

LO 2

Fig. 3.8

Solution Substituting values in Eq. (3.5)

$$200 = 4.44 \times 50 \times 150 \times \phi_{max}$$

or

$$\phi_{max} = 6.06 \text{ mWb}$$

$$B_{max} = \frac{6.06 \times 10^{-3}}{10 \times 5 \times 10^{-4}} = 1.212 \text{ T}$$

From the data of BH curve of Problem 2.10, we get

$$H_{max} = 250 \text{ AT/m}$$

$$AT_{max} = 250\, l_c = 250 \times 2\,(30 + 35) \times 10^{-2} = 325$$

$$I_{m(max)} = \frac{325}{150} = 2.17 \text{ A}$$

$$I_{m(rms)} = \frac{2.17}{\sqrt{2}} = 1.53 \text{ A}$$

$$\text{Core volume} = 2(20 \times 10 \times 5) + 2(45 \times 10 \times 5) = 6500 \text{ cm}^3$$

$$\text{Weight of core} = 6500 \times 7.9 \times 10^{-3} = 51.35 \text{ kg}$$

$$\text{Core loss} = 51.35 \times 3 = 154.7 \text{ W}$$

$$I_i = \frac{154.7}{200} = 0.77 \text{ A}$$

Referring to the phasor diagram of Fig. 3.7

$$\bar{I}_0 = 0.77 - j\,2.17 = 2.3 \angle -70.5^\circ$$

$$I_0 = 2.3 \text{ A (no-load current)}$$

$$\text{No-load } \quad pf = \cos 71.5^\circ = 0.334 \text{ lagging}$$

3.4 | IDEAL TRANSFORMER

LO 2

In order to visualize the effect of flow of secondary current in a transformer, certain idealizing assumptions will be made which are close approximations for a practical transformer. A transformer possessing these ideal properties is hypothetical (has no real existence) and is referred to as the *ideal transformer*. It possesses certain essential features of a real transformer but some details of minor significance are ignored which will be reintroduced at a refined stage of analysis.

Assumptions for an Ideal Transformer

The idealizing assumptions made are listed below:

(i) The primary and secondary windings have zero resistance. It means that there is no ohmic power loss and no resistive voltage drop in the ideal transformer. An actual transformer has finite but small winding resistances. It will also be assumed that there is no stray capacitance, though the actual transformer has inter-turn capacitance and capacitance between turns and ground but their effect is negligible at 50 Hz.

(ii) There is no leakage flux so that all the flux is confined to the core and links both the windings. An actual transformer does have a small amount of leakage flux which can be accounted for in detailed analysis by appropriate circuit modelling.

(iii) The core has infinite permeability so that zero magnetizing current is needed to establish the requisite amount of flux (Eq. (3.6)) in the core.

(iv) The core-loss (hysteresis as well as eddy-current loss) is considered zero.

Figure 3.9 shows an ideal transformer having a primary of N_1 turns and a secondary of N_2 turns on a common magnetic core. The voltage of the source to which the primary is connected is

$$v_1 = \sqrt{2}\, V_1 \cos \omega t \tag{3.11}$$

while the secondary is initially assumed to be an open circuited. As a consequence, flux ϕ is established in the core (Eq. (3.6)) such that

$$e_1 = v_1 = N_1 \frac{d\phi}{dt} \tag{3.12}$$

but the exciting current drawn from the source is zero by virtue of assumption (iii) above. The flux ϕ which is wholly mutual (assumption (ii) above) causes an emf

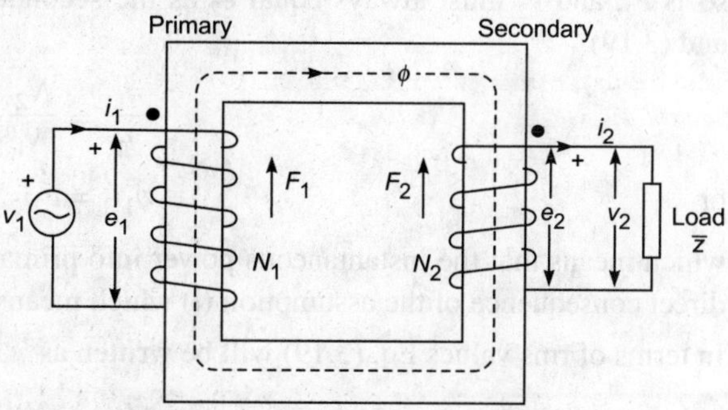

Fig. 3.9 *Ideal transformer on load*

$$e_2 = N_2 \frac{d\phi}{dt} \tag{3.13}$$

to be induced in the secondary of polarity[§] marked on the diagram for the winding direction indicated. The dots marked at one end of each winding indicate the winding ends which simultaneously have the same polarity due to emfs induced. From Eqs (3.12) and (3.13)

$$\frac{e_1}{e_2} = \frac{N_1}{N_2} = a \tag{3.14}$$

[§] The reader may check these polarities by applying Lenz's law while assuming that flux ϕ is increasing.

Since a, the *transformation ratio*, is a constant, e_1 and e_2 are in phase. The secondary terminal voltage is

$$v_2 = e_2 \tag{3.15}$$

Hence
$$\frac{v_1}{v_2} = \frac{e_1}{e_2} = \frac{N_1}{N_2} = a \tag{3.16}$$

Recall It is, therefore, seen that an ideal transformer changes (transforms) voltages in *direct ratio* of the number of turns in the two windings.

In terms of rms values Eq. (3.16) implies

$$\frac{V_1}{V_2} = \frac{E_1}{E_2} = \frac{N_1}{N_2} = a \quad \text{(same as Eq. (3.9))} \tag{3.17}$$

Now let the secondary be connected to a load of impedance Z_2 so that the secondary feeds a sinusoidal current of instantaneous value i_2 to the load. Due to this flow of current, the secondary creates mmf $F_2 = i_2N_2$ opposes the flux ϕ. However, the mutual flux ϕ cannot change as otherwise the (v_1, e_1) balance will be disturbed (this balance must always hold as winding has zero leakage and resistance). The result is that the primary draws a current i_1 from the source so as to create mmf $F_1 = i_1 N_1$ which at all time cancels out the load caused mmf $i_2 N_2$ so that ϕ is maintained constant independent of the load current flow, Thus

$$i_1 N_1 = i_2 N_2 \tag{3.18}$$

or
$$\frac{i_1}{i_2} = \frac{N_2}{N_1} = \frac{1}{a} \tag{3.19}$$

Obviously i_1 and i_2 are in phase for positive current directions marked on the diagram (primary current *in* at the dotted terminal and secondary current *out* of the dotted terminal). Since flux ϕ is independent of load, so is e_2, and v_2 must always equal e_2 as the secondary is also resistanceless. Therefore, from Eqs (3.17) and (3.19)

$$\frac{i_1}{i_2} = \frac{N_2}{N_1} = \frac{v_2}{v_1}$$

or
$$v_1 i_1 = v_2 i_2 \tag{3.20}$$

which means that the instantaneous power into primary equals the instantaneous power out of secondary, a direct consequence of the assumption (i) which means a *loss-less* transformer.

In terms of rms values Eq. (3.19) will be written as

$$\frac{I_1}{I_2} = \frac{N_2}{N_1} = \frac{1}{a} \tag{3.21}$$

which implies that currents in an ideal transformer transform in *inverse ratio* of winding turns. Equation (3.20) in terms of rms values will read

$$V_1 I_1 = V_2 I_2 \tag{3.22}$$

i.e. the VA output is balanced by the VA input.

Figure 3.10(a) shows the schematic of the ideal transformer of Fig. 3.9 with dot marks identifying similar polarity ends. It was already seen above that $\bar{V}_1$ and $\bar{V}_2$ are in phase and so are $\bar{I}_1$ and $\bar{I}_2$. Now

$$\frac{\bar{V}_1}{\bar{V}_2} = \frac{N_1}{N_2} \tag{3.23a}$$

and
$$\frac{\overline{I}_1}{\overline{I}_2} = \frac{N_2}{N_1} \tag{3.23b}$$

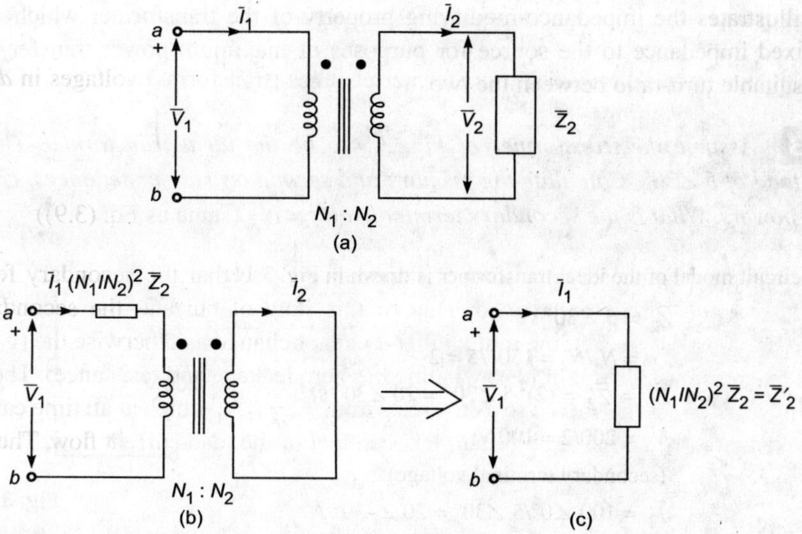

Fig. 3.10 *Ideal transformer; referring impedance from secondary to primary*

Dividing Eq. (3.23a) by Eq. (3.23b)

$$\frac{\overline{V}_1/\overline{V}_2}{\overline{I}_1/\overline{I}_2} = \frac{N_1/N_2}{N_2/N_1}$$

or
$$\frac{\overline{V}_1}{\overline{I}_1} = \left(\frac{N_1}{N_2}\right)^2 \frac{\overline{V}_2}{\overline{I}_2} = \left(\frac{N_1}{N_2}\right)^2 \overline{Z}_2 \tag{3.24}$$

or
$$\overline{Z}_1 = \left(\frac{N_1}{N_2}\right)^2 \overline{Z}_2 = a^2\overline{Z}_2 = \overline{Z}'_2 \tag{3.25}$$

Impedance on One Side Referred to Another Side

It is concluded from Eq. (3.25) that the impedance on the secondary side when seen (*referred to*) on the primary side is transformed in the direct ratio of square of turns. Equivalence of Eqs (3.24) and (3.25) to the original circuit of Fig. 3.10(a) is illustrated through Figs 3.10(b) and (c). Similarly an impedance $\overline{Z}_1$ from the primary side can be referred to the secondary as

$$\overline{Z}'_1 = \left(\frac{N_2}{N_1}\right)^2 \overline{Z}_1 = \frac{1}{a^2}\overline{Z}_1 \tag{3.26}$$

Transferring an impedance from one side of a transformer to the other is known as *referring the impedance* to the other side. Voltages and currents on one side have their counterpart on the other side as per Eqs (3.23(a) and (b)).

In conclusion it may be said that in an ideal transformer voltages are transformed in the direct ratio of turns, currents in the inverse ratio and impedances in the direct ratio squared; while power and VA remain unaltered.

Equation (3.25) illustrates the impedance-modifying property of the transformer which can be exploited for matching a fixed impedance to the source for purposes of maximum power transfer by interposing a transformer of a suitable turn-ratio between the two.

Example 3.3 *Assume the transformer of Fig. 3.8 to be the ideal transformer. The secondary is connected to a load of 5 ∠30°. Calculate the primary and secondary side impedances, current and their pf, and the real powers. What is the secondary terminal voltage?*

LO 2

Solution The circuit model of the ideal transformer is drawn in Fig. 3.11.

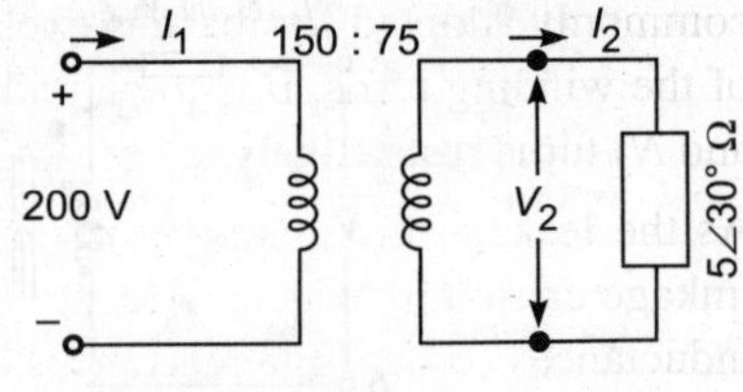

Fig. 3.11

$$\bar{Z}_2 = 5 \angle 30° \ \Omega$$

$$a = N_1/N_2 = 150/75 = 2$$

$$\bar{Z}_1 = \bar{Z}_2' = (2)^2 5 \angle 30° = 20 \angle 30° \ \Omega$$

$$V_2 = 200/2 = 100 \text{ V};$$

(secondary terminal voltage)

$$\bar{I}_2 = 100 \angle 0°/5 \angle 30° = 20 \angle -30° \text{ A}$$

or $$I_2 = 20 \text{ A; pf} = \cos 30° = 0.866 \text{ lagging}$$

$$\bar{I}_1 = \bar{I}_2' = 20 \angle 30°/2 = 10 \angle -30° \text{ A}$$

or $$I_1 = 10 \text{ A; pf} = \cos 30° = 0.866 \text{ lagging}$$

$$P_2 \text{ (secondary power output)} = (20)^2 \times \text{Re } 5 \angle 30° = 400 \times 4.33 = 1.732 \text{ kW}$$

$$P_1 \text{ (primary power input)} = P_2 \text{ (as the transformer is lossless)} = 1.732 \text{ kW}$$

or otherwise $$P_1 = V_1 I_1 \cos \theta_1 = 200 \times 10 \times 0.866 = 1.732 \text{ kW}$$

3.5 | REAL TRANSFORMER AND EQUIVALENT CIRCUIT

LO 2

Figure 3.12 shows a real transformer on load. Both the primary and secondary have finite resistances R_1 and R_2 which are uniformly spread throughout the winding; these give rise to associated copper (I^2R) losses. While a major part of the total flux is confined to the core as mutual flux ϕ linking both the primary and secondary, a small amount of flux does leak through paths which lie mostly in air and link separately the individual windings.

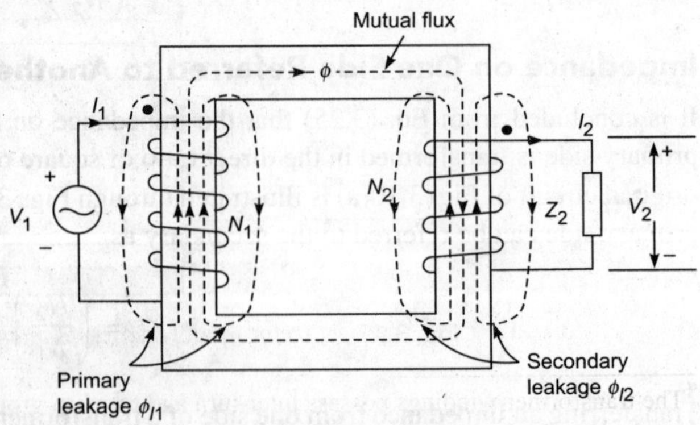

Fig. 3.12 *Real transformer*

MMF of Primary and Secondary Windings

In Fig. 3.12 with primary and secondary for simplicity assumed to be wound separately on the two legs of the core, leakage flux ϕ_{l1} caused by primary mmf I_1N_1 links primary winding itself and ϕ_{l2} caused by I_2N_2 links the secondary winding, thereby causing self-linkages of the two windings. It was seen in Sec. 3.2 with reference to Fig. 3.2(a) that half the primary and half the secondary is wound on each core leg. This reduces the leakage flux linking the individual windings. In fact it can be found by tracing flux paths that leakage flux is now confined to the annular space between the halves of the two windings on each leg.

Leakage Flux Reduction

In shell-type construction, the leakage will be still further reduced as LV and HV pancakes are interleaved. Theoretically, the leakage will be eliminated if the two windings could be placed in the same physical space but this is not possible; the practical solution is to bring the two as close as possible with due consideration to insulation and constructional requirements. Shell-type construction though having low leakage is still not commonly adopted for reasons explained in Sec. 3.2. Actually some of the leakage flux will link only a part of the winding turns. It is to be understood here that ϕ_{l1} and ϕ_{l2} are equivalent leakage fluxes linking all N_1 and N_2 turns respectively.

As the leakage flux paths lie in air for considerable part of their path lengths, winding mmf and its self-linkage caused by leakage flux are linearly related in each winding; therefore contributing constant leakage inductances (or leakage reactances corresponding to the frequency at which the transformer is operated) of both primary and secondary windings. These leakage reactances¶ are distributed throughout the winding though not quite uniformly.

Both resistances and leakage reactances of the transformer windings are series effects and for low operating frequencies at which the transformers are commonly employed (power frequency operation is at 50 Hz only), these can be regarded as lumped parameters. The real transformer of Fig. 3.12 can now be represented as a semi-ideal transformer having lumped resistances R_1 and R_2 and leakage reactances symbolized as X_{l1} and X_{l2} in series with the corresponding windings as shown in Fig. 3.13. The semi-ideal transformer draws magnetizing current to set up the mutual flux ϕ and to provide for power loss in the core; it, however, has no winding resistances and is devoid of any leakage. The induced emfs of the semi-ideal transformer are E_1 and E_2 which differ respectively from the primary and secondary terminal voltages V_1 and V_2 by small voltage drops in winding resistances and leakage reactances (R_1, X_{l1} for primary and R_2, X_{l2} for secondary). The ratio of transformation is

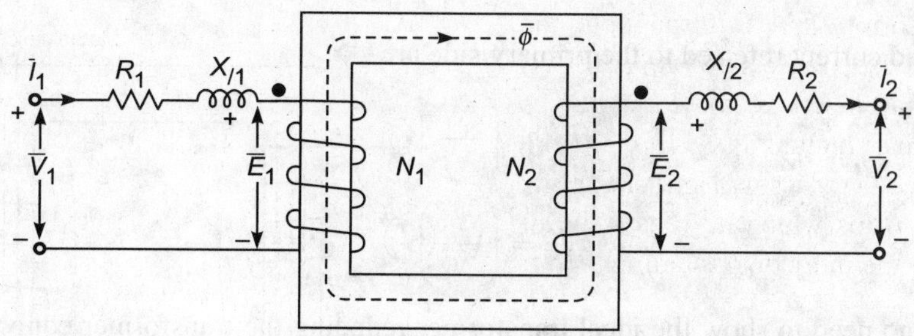

Fig. 3.13 *Circuit model of transformer employing semi-ideal transformer*

¶The transformer windings possess inter-turn and turns-to-ground capacitances. Their effect is insignificant in the usual low-frequency operation. This effect must, however, be considered for high-frequency end of the spectrum in electronic transformers. In low-frequency transformers also, capacitance plays an important role in surge phenomenon caused by switching and lightning.

$$a = \frac{N_1}{N_2} = \frac{E_1}{E_2} \approx \frac{V_1}{V_2} \tag{3.27}$$

because the resistances and leakage reactance of the primary and secondary are so small in a transformer that $E_1 \approx V_1$ and $E_2 \approx V_2$.

Equivalent Circuit

!Attention In Fig. 3.13 the current $\bar{I}_1$ flowing in the primary of the semi-ideal transformer can be visualized to comprise two components as below:

(i) Exciting current $\bar{I}_0$ whose magnetizing component $\bar{I}_m$ creates mutual flux $\bar{\phi}$ and whose core-loss component $\bar{I}_i$ provides the loss associated with alternation of flux.

(ii) A load component $\bar{I}'_2$ which counterbalances the secondary mmf $\bar{I}_2 N_2$ so that the mutual flux remains constant independent of load, determined only by $\bar{E}_1$. Thus

$$\bar{I}_1 = \bar{I}_0 + \bar{I}'_2 \tag{3.28}$$

where
$$\frac{\bar{I}'_2}{\bar{I}_2} = \frac{N_2}{N_1} \tag{3.29}$$

The exciting current $\bar{I}_0$ can be represented by the circuit model of Fig. 3.7 so that the semi-ideal transformer of Fig. 3.13 is now reduced to the true ideal transformer. The corresponding circuit (*equivalent circuit*) modelling the behaviour of a real transformer is drawn in Fig. 3.14(a) wherein for ease of drawing the core is not shown for the ideal transformer.

The impedance $(R_2 + jX_{l2})$ on the secondary side of the ideal transformer can now be referred to its primary side resulting in the equivalent circuit of Fig. 3.14(b) wherein

$$X'_{l2} = \left(\frac{N_1}{N_2}\right)^2 X_{l2} \tag{3.30a}$$

$$R'_2 = \left(\frac{N_1}{N_2}\right)^2 R_2 \tag{3.30b}$$

The load voltage and current referred to the primary side are

$$\bar{V}'_2 = \left(\frac{N_1}{N_2}\right)\bar{V}_2 \tag{3.31a}$$

$$\bar{I}'_2 = \left(\frac{N_2}{N_1}\right)\bar{I}_2 \tag{3.31b}$$

Therefore there is no need to show the ideal transformer reducing the transformer equivalent circuit to the T-circuit of Fig. 3.14(c) as referred to side 1. The transformer equivalent circuit can similarly be referred to side 2 by transforming all impedances (resistances and reactances), voltages and currents to side 2. It may be noted here that *admittances* (conductances and susceptances) are transformed in the inverse ratio squared in contrast to impedances (resistances and reactances) which as already shown in Sec. 3.4 transform in direct ratio squared. The equivalent circuit of Fig. 3.14(c) referred to side 2 is given in Fig. 3.14(d) wherein

$$\bar{V}_1' = \frac{N_2}{N_1}\bar{V}_1$$

$$\bar{I}_1' = \frac{N_1}{N_2}\bar{I}_1$$

$$G_i' = \left(\frac{N_1}{N_2}\right)^2 G_i;\ B_m' = \left(\frac{N_1}{N_2}\right)^2 B_m$$

$$R_2' = \left(\frac{N_2}{N_1}\right)^2 R_2;\ X_{l1}' = \left(\frac{N_2}{N_1}\right)^2 X_{l1}$$

With the understanding that all quantities have been referred to a particular side, a superscript dash can be dropped with a corresponding equivalent circuit as drawn in Fig. 3.14(d).

Effect of Hysteresis and Eddy-Current Losses

In the equivalent circuit of Fig. 3.14(c) if G_i is taken as constant, the core-loss is assumed to vary as E_1^2 or $\phi_{\max}^2 f^2$ (Eq. (3.6)). It is a fairly accurate representation as core-loss comprises hysteresis and eddy-current loss expressed as $(K_h \phi^{1.6}{}_{\max} f + K_e \phi_{\max}^2 f^2)$. The magnetizing current for linear *B-H* curve varies proportional to $\phi_{\max} \propto \frac{E_1}{f}$. If inductance (L_m) corresponding to susceptance B_m is assumed constant.

$$I_m = \frac{E_1}{2\pi f L_m} = B_m E_1$$

It therefore is a good model except for the fact that the saturation effect has been neglected in which case B_m would be a nonlinear function of E_1/f. It is an acceptable practice to find the shunt parameters G_i, B_m at the rated voltage and frequency and assume these as constant for small variations in voltage and frequency.

The passive lumped T-circuit representation of a transformer discussed above is adequate for most power and radio frequency transformers. In transformers operating at higher frequencies, the interwinding capacitances are often significant and must be included in the equivalent circuit. The circuit modification to include this parameter is discussed in Sec. 3.12.

The equivalent circuit given here is valid for a sinusoidal steady-state analysis. In carrying out transient analysis all reactances must be converted to equivalent inductances.

The equivalent circuit developed above can also be arrived at by following the classical theory of magnetically coupled circuits, section 3.22. The above treatment is, however, more instructive and gives a clearer insight into the physical processes involved.

Phasor Diagram of Exact Equivalent Circuit of Transformer [Fig. 3.14(a)]

The KVL equations for the primary and secondary circuits are

$$\bar{V}_1 = \bar{E}_1 + \bar{I}_1 R_1 + j\bar{I}_1 X_1$$

$$\bar{V}_2 = \bar{E}_2 - \bar{I}_2 R_2 - j I_2 X_1$$

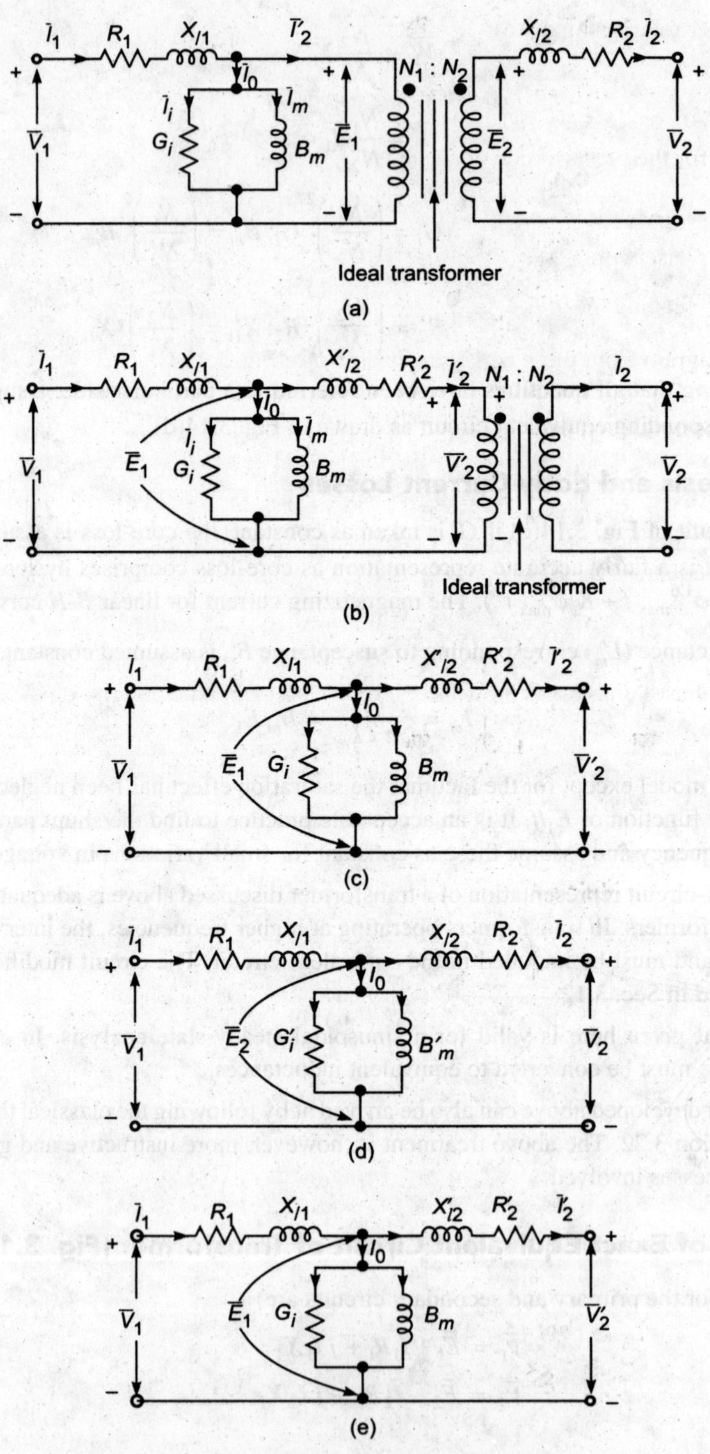

Fig. 3.14 *Evolution of transformer equivalent circuit*

The ideal transformer relationships are

$$\frac{\bar{E}_1}{\bar{E}_2} = a\,;\ \frac{\bar{I}'_2}{\bar{I}_2} = \frac{1}{a}$$

The nodal equation for the primary side currents is

$$\bar{I}_1 = \bar{I}'_2 + \bar{I}'_0 = \bar{I}'_2 + \left(\bar{I}_i + \bar{I}_m\right)$$

where $\bar{I}_i$ is in phase with $\bar{E}_1$

I_m is 90° lagging $\bar{E}_1$

The exact phasor diagram from these equations is drawn in Fig. 3.15.

Alternative Phasor Diagram

Alternatively if we use $e = -\dfrac{d\lambda}{dt}$; direction of $\bar{E}_1$ and $\bar{E}_2$ will reverse in Fig. 3.13 and these will lag the flux phasor by 90°. The direction of secondary current is now into the dotted terminal. So for mmf balance $\bar{I}'_2 = -\bar{I}_2$. The KVL equation on the primary side is

$$\bar{V}_1 = (-\bar{E}_1) + \bar{I}'_2 R_2 + j\,\bar{I}'_2 X_2$$

The corresponding phasor diagram is drawn in Fig. 3.16. Note that the polarity of V_2 reverses in Fig. 3.13 but it is of no consequence.

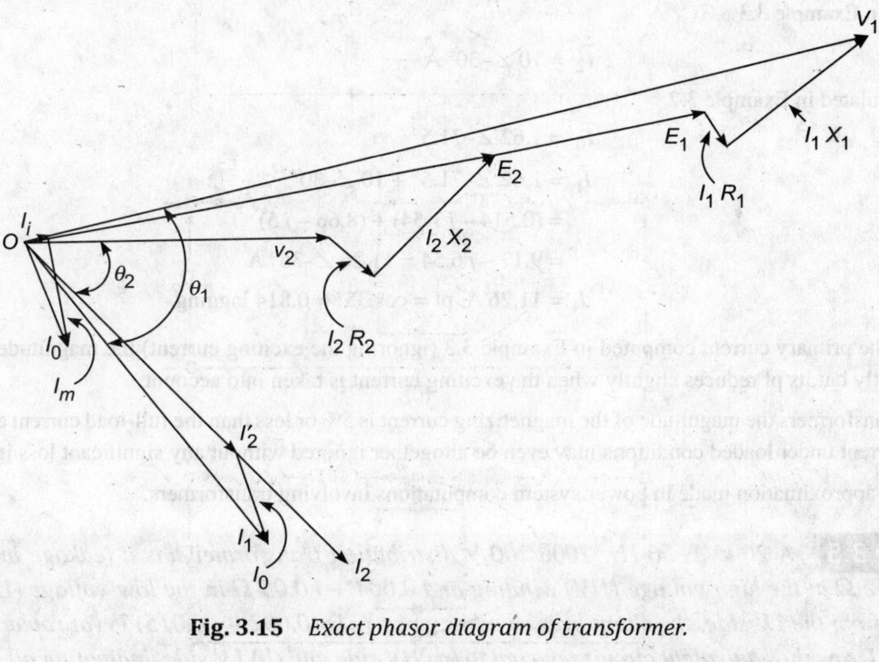

Fig. 3.15 *Exact phasor diagram of transformer.*

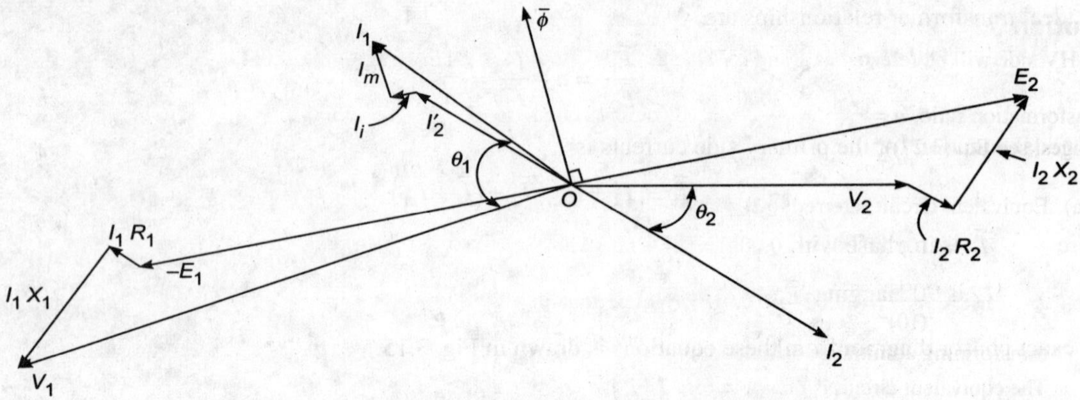

Fig. 3.16 *Alternative exact phasor diagram of transformer*

Example 3.4 *Consider the transformer of Example 3.2 (Fig. 3.8) with load impedance as specified in Example 3.3. Neglecting voltage drops (resistive and leakage reactive drops), calculate the primary current and its pf. Compare with the current as calculated in Example 3.3.* LO 2

Solution As per Eqs (3.28) and (3.29)

$$\bar{I}_1 = \bar{I}_0 + \bar{I}_2' \quad \text{and} \quad \frac{\bar{I}_2'}{\bar{I}_2} = \frac{N_2}{N_1}$$

As calculated in Example 3.3

$$\bar{I}_2' = 10\,\angle{-30^\circ}\text{ A}$$

Further as calculated in Example 3.2

$$\bar{I}_0 = 1.62\,\angle{-71.5^\circ}$$

Hence

$$\bar{I}_1 = 1.62\,\angle{-71.5^\circ} + 10\,\angle{-30^\circ}$$

$$= (0.514 - j\,1.54) + (8.66 - j\,5)$$

$$= 9.17 - j\,6.54 = 11.26\,\angle{-35^\circ}\text{ A}$$

$$I_1 = 11.26\text{ A, pf} = \cos 35^\circ = 0.814 \text{ lagging}$$

Compared to the primary current computed in Example 3.3 (ignoring the exciting current) the magnitude of the current increases slightly but its pf reduces slightly when the exciting current is taken into account.

In large size transformers the magnitude of the magnetizing current is 5% or less than the full-load current and so its effect on primary current under loaded conditions may even be altogether ignored without any significant loss in accuracy.

This is a usual approximation made in power system computations involving transformers.

Example 3.5 *A 20-kVA, 50-Hz, 2000/200-V distribution transformer has a leakage impedance of $0.42 + j\,0.52\ \Omega$ in the high-voltage (HV) winding and $0.004 + j\,0.05\ \Omega$ in the low-voltage (LV) winding. When seen from the LV side, the shunt branch admittance Y_0 is $(0.002 - j\,0.015)$ ℧ (at rated voltage and frequency). Draw the equivalent circuit referred to (a) HV side and (b) LV side, indicating all impedances on the circuit.* LO 2

Solution

The HV side will be referred as 1 and LV side as 2.

Transformation ratio, $a = \frac{N_1}{N_2} = \frac{2000}{200} = 10$ (ratio of rated voltages; see Eq. (3.27))

(a) Equivalent circuit referred to HV side (side 1)

$$\bar{Z}_2' = (10)^2\,(0.004 + j\,0.005) = 0.4 + j\,0.5\ \Omega$$

$$\bar{Y}_0' = \frac{1}{(10)^2}\,(0.002 - j\,0.015)$$ (Notice that in transforming admittance is divided by a^2)

The equivalent circuit is drawn in Fig. 3.17(a).

(b) Equivalent circuit referred to LV side (side 2).

$$\bar{Z}_1' = \frac{1}{(10)^2}\,(0.42 + j\,0.52) = 0.0042 + j\,0.0052$$

The equivalent circuit is drawn in Fig. 3.17(b).

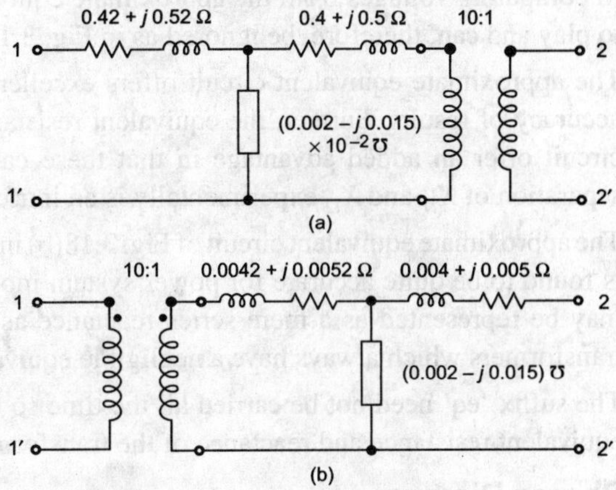

Fig. 3.17

Approximate Equivalent Circuit

In constant frequency (50 Hz) power transformers, approximate forms of the exact T-circuit equivalent of the transformer are commonly used. With reference to Fig. 3.14(c), it is immediately observed that since winding resistances and leakage reactances are very small, $\bar{V}_1 \approx \bar{E}_1$ even under conditions of load. Therefore, the exciting current drawn by the magnetizing branch ($G_i \parallel B_m$) would not be affected significantly by shifting it to the input terminals, i.e. it is now excited by $\bar{V}_1$ instead of $\bar{E}_1$ as shown in Fig. 3.18(a). It may also be observed that with this approximation, the current through R_1, X_{l1} is now I_2' rather than $\bar{I}_1 = \bar{I}_0 + \bar{I}_2'$. Since $\bar{I}_0$ is very small (less than 5% of full-load current), this approximation changes the voltage drop insignificantly. Thus it is basically a good approximation. The winding resistances and reactances being in series can now be combined into equivalent resistance and reactance of the transformer as seen from the appropriate side (in this case side 1). Remembering that all quantities in the equivalent circuit are referred either to the primary or secondary dash in the referred quantities and suffixes l, 1 and 2 in equivalent resistance, reactance and impedance can be dropped as in Fig. 3.18(b).

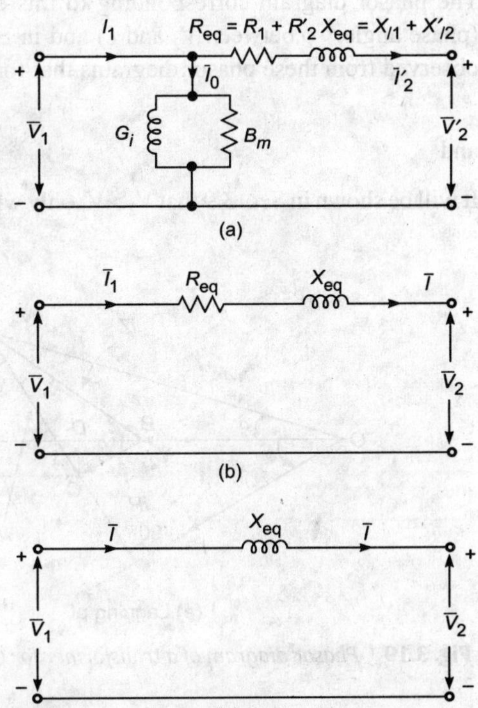

Fig. 3.18 *Approximate equivalent circuits of a transformer*

Here R_{eq} (equivalent resistance) $= R_1 + R_2$

X_{eq} (equivalent reactance) $= X_{l1} + X_{l2}$

Z_{eq} (equivalent impedance) $= R_{eq} + j\,X_{eq}$

In computing voltages from the approximate equivalent circuit, the parallel magnetizing branch has no role to play and can, therefore, be ignored as in Fig. 3.18(b).

The approximate equivalent circuit offers excellent computational ease without any significant loss in the accuracy of results. Further, the equivalent resistance and reactance as used in the approximate equivalent circuit offer an added advantage in that these can be readily measured experimentally (Sec. 3.7), while separation of X_{l1} and X_{l2} experimentally is an intricate task and is rarely attempted.

The approximate equivalent circuit of Fig. 3.18(b) in which the transformer is represented as a series impedance is found to be quite accurate for power system modelling [7]. In fact in some system studies, a transformer may be represented as a mere series reactance as in Fig. 3.18(c). This is a good approximation for large transformers which always have a negligible equivalent resistance compared to the equivalent reactance.

The suffix 'eq' need not be carried all the time so that R and X from now onwards will be understood to be equivalent resistance and reactance of the transformer referred to one side of the transformer.

Phasor Diagram

For the approximate equivalent circuit of Fig. 3.18(b), (suffix 'eq' is being dropped now),

$$\bar{V}_2 = \bar{V}_1 - \bar{I}\,\bar{Z} \tag{3.32a}$$

or

$$\bar{V}_1 = \bar{V}_2 + \bar{I}\,(R + jX) \tag{3.32b}$$

The phasor diagram corresponding to this equation is drawn in Fig. 3.19(a) for the lagging power factor (phase angle** ϕ between V_2 and I) and in Fig. 3.19(b) for the leading power factor (pf). It is immediately observed from these phasor diagrams that for the phase angle indicated

$V_2 < V_1$ for lagging pf

and $V_2 > V_1$ for leading pf

It will be shown in Sec. 3.9 that $V_2 > V_1$ only when leading phase angle ϕ is more than $\tan^{-1}(R/X)$ (see Eq. (3.67).

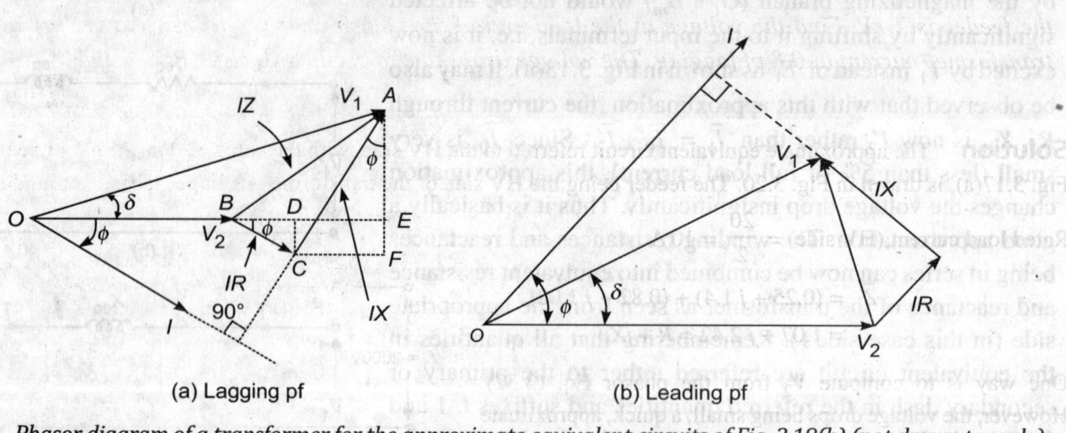

Fig. 3.19 *Phasor diagram of a transformer for the approximate equivalent circuits of Fig. 3.18(b) (not drawn to scale)*

In the plasor diagrams of Figs. 3.19(a) and (b) (these are not drawn to scale), the angle δ is such that V_1 leads V_2. This is an indicator of the fact that real power flows from side 1 to side 2 of the transformer (this is proved

** Phase angle ϕ should not be confused with flux though the same symbol has been used.

in Section 8.9). This angle is quite small and is related to the value of the equivalent reactance: resistance of the transformer being negligible.

Name Plate Rating

The voltage ratio is specified as V_1 (rated)/V_2 (rated). It means that when voltage V_1 (rated) is applied to the primary, the secondary voltage on fullload at specified pf is V_2 (rated). The ratio V_1 (rated)/V_2 (rated) is not exactly equal to N_1/N_2, because of voltage drops in the primary and secondary. These drops being small are neglected and it is assumed that for all practical purposes

$$\frac{V_1(\text{rated})}{V_2(\text{rated})} = \frac{N_1}{N_2} \tag{3.33}$$

The rating of the transformer is specified in units of VA/kVA/MVA depending upon its size.

$$\text{kVA(rated)} = \frac{V(\text{rated}) \times I(\text{full-load})}{1000} \tag{3.34}$$

where V and I are referred to one particular side. The effect of the excitation current is of course ignored.

!Attention The transformer name plate also specifics the equivalent impedance, but not in actual ohm. It is expressed as the *percentage voltage drop* (see Sec. 3.8) expressed as

$$\frac{I(\text{full-load})Z}{V(\text{rated})} \times 100\%$$

where all quantities must be referred to anyone side.

Example 3.6 *The distribution transformer described in Example 3.5 is employed to step down the voltage at the load-end of a feeder having an impedance of 0.25 + j 1.4 Ω. The sending-end voltage of the feeder is 2 kV. Find the voltage at the load-end of the transformer when the load is drawing rated transformer current at 0.8 pf lagging. The voltage drops due to exciting current may be ignored.*

LO 2

Solution The approximate equivalent circuit referred to the HV side, with the value of transformer impedance from Fig. 3.17(a), is drawn in Fig. 3.20. The feeder being the HV side of the transformer, its impedance is not modified.

Rated load current (HV side) = $\frac{20}{2} = 10$ A

$$\bar{Z} = (0.25 + j\,1.4) + (0.82 + j\,1.02)$$
$$= 1.07 + j\,2.42 = R + jX$$

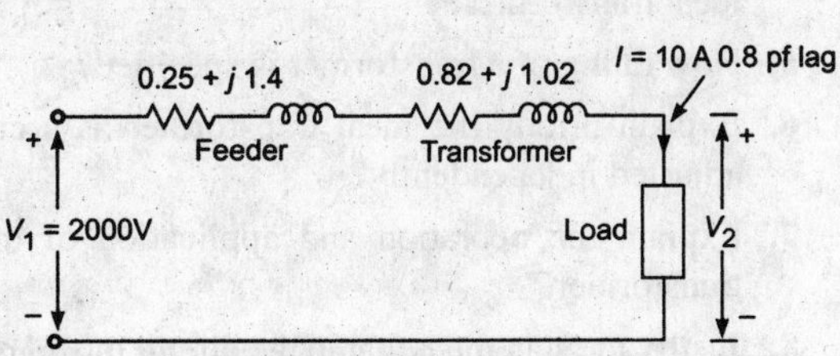

Fig. 3.20

One way is to compute V_2 from the phasor Eq. (3.32). However, the voltage drops being small, a quick, approximate but quite accurate solution can be obtained from the phasor diagram of Fig. 3.19(a) without the necessity of carrying out complex number calculations.

From Fig. 3.19(a)

$$OE = \sqrt{(OA)^2 - (AE)^2}$$

From the geometry of the phasor diagram

$$AE = AF - FE$$
$$= IX \cos \phi - IR \sin \phi$$
$$= 10(2.42 \times 0.8 - 1.07 \times 0.6) = 12.94 \text{ V}$$

Now $$OE = \sqrt{(2000)^2 - (12.94)^2} = 1999.96 \text{ V}$$

It is therefore seen that

$$OE \approx OA = V_1 \text{ (to a high degree of accuracy; error is 2 in } 10^5\text{)}$$

V_2 can then be calculated as

$$V_2 = OE - BE$$
$$\approx V_1 - BE$$

Now $$BE = BD + DE$$
$$= I(R \cos \phi + X \sin \phi)$$
$$= 10(1.07 \times 0.8 + 2.42 \times 0.6) = 23.08 \text{ V}$$

Thus $$V_2 = 2000 - 23.08 = 1976.92 \text{ V}$$

$$\text{Load voltage referred to LV side} = \frac{1976.92}{10} = 197.692 \text{ V}$$

Remark It is noticed that to a high degree of accuracy, the voltage drop in transformer impedance can be approximated as

$$V_1 - V_2 = I(R \cos \phi + X \sin \phi); \text{ lagging pf} \tag{3.35a}$$

It will soon be shown that

$$V_1 - V_2 = I(R \cos \phi - X \sin \phi); \text{ leading pf} \tag{3.35b}$$

Concept Check LO 2

1. What is meant by no-load condition of a transformer? ○○●
2. What is meant by transformation ratio? ○○●
3. What is an ideal transformer? ○○●
4. How voltages, currents and impedances are referred from one winding to another winding in an ideal transformer? ○●●
5. How rating of a transformer is specified? ○○●
6. Explain briefly the ideal transformer as a circuit element. Can voltage and current ratios be adjusted independently? ●●●
7. Explain the operation and application of the impedance transforming property of an ideal transformer. ○●●
8. Justify the statement that in the circuit model of a transformer in a power system, the magnetizing branch can be ignored. ○●●
9. Explain the meaning of all the items in the nameplate of a transformer. ●●●
10. What is the phase relationship between the core flux; the magnetizing current and the induced emfs in the primary and secondary winding of a transformer? Draw the phasor diagram. ●●●

11. How can we refer the transformer winding resistance and leakage reactance from one side to the other?

12. From the percentage impedance given as the nameplate, find the voltage to be applied for full load current to flow in SC test.

3.6 | TRANSFORMER LOSSES

LO 3 Outline various losses occurring in a transformer and testing methods used to determine the performance characteristics of a transformer and hence compute the losses

Important! The transformer has no moving parts so that its efficiency is much higher than that of rotating machines. The various losses in a transformer are enumerated as follows:

Core-loss

These are hysteresis and eddy-current losses resulting from alternations of magnetic flux in the core. Their nature and the remedies to reduce these have already been discussed at length in Sec. 2.6. It may be emphasized here that the core-loss is constant for a transformer operated at constant voltage and frequency as are all power frequency equipment.

Copper-loss (I^2R-loss)

This loss occurs in winding resistances when the transformer carries the load current; varies as the square of the loading expressed as a ratio of the full-load.

Load (Stray)-loss

It largely results from leakage fields inducing eddy-currents in the tank wall, and conductors.

Dielectric-loss

The seat of this loss is in the insulating materials, particularly in oil and solid insulations.

Important! The major losses are by far the first two: P_i, the constant core (iron)-loss and P_c, the variable copper-loss. Therefore, only these two losses will be considered in further discussions.

It will be seen in Sec. 3.7 that transformer losses and the parameters of its equivalent circuit can be easily determined by two simple tests without actually loading it.

3.7 | TRANSFORMER TESTING

LO 3

Two chief difficulties which do not warrant the testing of large transformers by direct load test are: (i) large amount of energy has to be wasted in such a test, (ii) it is a stupendous (impossible for large transformers) task to arrange a load large enough for direct loading.

!Attention Performance characteristics of a transformer must be computed from a knowledge of its equivalent circuit parameters which, in turns, are determined by conducting simple tests involving very little power consumption, called *nonloading tests*. In these tests the power consumption is simply that which is needed to supply the losses incurred. The two nonloading tests are the Open-circuit (OC) test and Short-circuit (SC) test.

In both these tests voltage, current and power are measured from which the resistance and reactance of the input impedance can be found, as seen in each test. Thus only four parameters can be determined which correspond to the approximate equivalent circuit of Fig. 3.16(a).

Before proceeding to describe OC and SC tests, a simple test will be advanced for determining similar polarity ends on the two windings of a transformer.

Polarity Test

Similar polarity ends of the two windings of a transformer are those ends that acquire simultaneously positive or negative polarity of emfs induced in them. These are indicated by the dot convention as illustrated in Sec. 3.4. Usually the ends of the LV winding are labelled with a small letter of the alphabet and are suffixed 1 and 2, while the HV winding ends are labelled by the corresponding capital letter and are suffixed 1 and 2 as shown in Fig. 3.21. The ends suffixed 2 (a_2, A_2) have the same polarity and so have the ends labelled 1 (a_1, A_1).

In determining the relative polarity of the two-windings of a transformer the two windings are connected in series across a voltmeter, while one of the windings is excited from a suitable voltage source as shown in Fig. 3.21. If the polarities of the windings are as marked on the diagram, the voltmeter should read $V = V_1 \sim V_2$. If it reads $(V_1 + V_2)$, the polarity markings of one of the windings must be interchanged.

Applications of Polarity Test

The above method of polarity testing may not be convenient in field testing of a transformer. Alternatively the polarity testing can be easily carried out by a dc battery, switch and dc voltmeter (permanent magnet type which can determine the polarity of a voltage) as shown in the simple setup of Fig. 3.21(b). As the switch on the primary side is closed, the primary current increases and so do the flux linkages of both the windings, inducing emfs in them. The positive polarity of this induced emf in the primary is at the end to which the battery positive is connected (as per Lenz's law). The end of secondary which (simultaneously) acquires positive polarity (as determined by the dc voltmeter) is the similar polarity end. The reverse happens when the switch is opened, i.e. the similar polarity end of the secondary is that end which acquires negative potential.

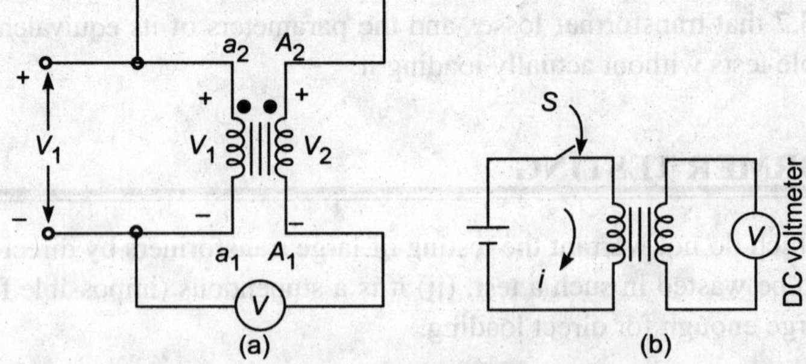

Fig. 3.21 *(a) Polarity test on two-winding transformer; (b) A simple polarity test by a battery*

Open-Circuit (OC) or No-Load Test

!Attention The purpose of this test is to determine the shunt branch parameters of the equivalent circuit of the transformer (Fig. 3.14(c)). One of the windings is connected to supply at rated voltage, while the other winding is kept open-circuited. From the point of view of convenience and availability of supply the test is usually performed from the LV side, while the HV side is kept open circuited as shown in Fig. 3.22. If the transformer is to be used at voltage other than rated, the test should be carried out at that voltage. Metering is arranged to read.

$$\text{voltage} = V_1;\ \text{current} = I_0 \text{ and power input} = P_0 \tag{3.36}$$

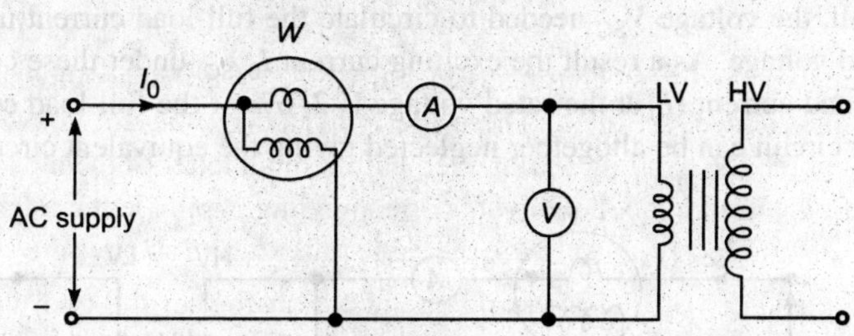

Fig. 3.22 *Connection diagram for open-circuit test*

Figure 3.23(a) shows the equivalent circuit as seen on open-circuit and its approximate version in Fig. 3.23(b). Indeed the no-load current I_0 is so small (it is usually 2–6% of the rated current) and R_1 and X_1 are also small, that V_1 can be regarded as $= E_1$ by neglecting the series impedance. This means that for all practical purposes the power input on no-load equals the core (iron) loss i.e.,

$$P_0 = P_i \text{ (iron-loss)} \tag{3.37}$$

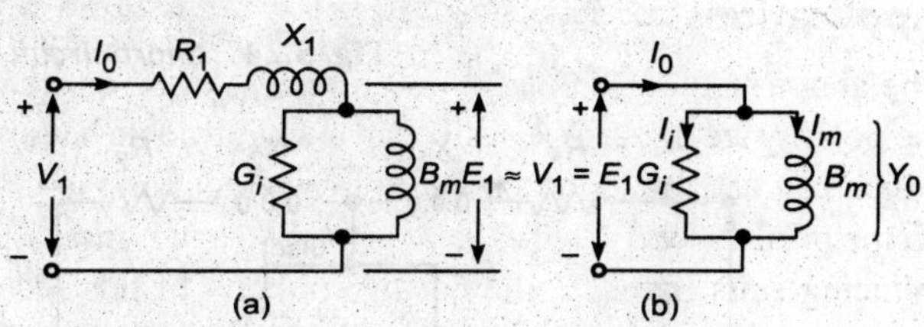

Fig. 3.23 *Equivalent circuit as seen on open-circuit*

The shunt branch parameters can easily be determined from the three readings (Eq. (3.36)) by the following circuit computations and with reference to the no-load phasor diagram of Fig. 3.6.

$$\bar{Y}_0 = G_i - jB_m \tag{3.38}$$

Now

$$Y_0 = \frac{I_0}{V_1} \tag{3.39}$$

$$V_1^2 G_i = P_0$$

or

$$G_i = \frac{P_0}{V_1^2} \tag{3.40}$$

It then follows that

$$B_m = \sqrt{Y_0^2 - G_i^2} \tag{3.41}$$

These values are referred to the side (usually LV) from which the test is conducted and could easily be referred to the other side if so desired by the inverse square of transformation ratio. The transformation ratio if not known can be determined by connecting a voltmeter on the HV side as well in the no-load test.

It is, therefore, seen that the OC test yields the values of core-loss and parameters of the shunt branch of the equivalent circuit.

Short-circuit (SC) Test

This test serves the purpose of determining the series parameters of a transformer. For convenience of supply arrangement[††] and voltage and current to be handled, the test is usually conducted from the HV side of the transformer, while the LV is short-circuited as shown in Fig. 3.24. The equivalent circuit as seen from the HV under short-circuit conditions is drawn in Fig. 3.25(a). Since the transformer resistances and leakage reactances are very small, the voltage V_{SC} needed to circulate the full-load current under short-circuit is as low as 5–8% of the rated voltage. As a result the exciting current $I_{0\,(SC)}$ under these conditions is only about 0.1 to 0.5% of the full-load current (I_0 at the rated voltage is 2–6% of the full-load current). Thus the shunt branch of the equivalent circuit can be altogether neglected giving the equivalent circuit of Fig. 3.25(b).

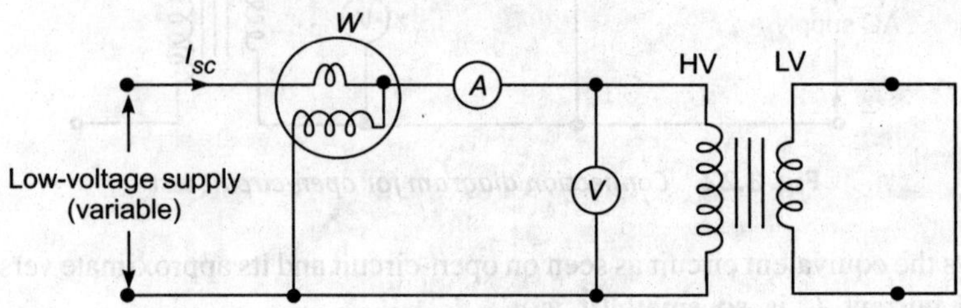

Fig. 3.24 *Short-circuit test on transformer*

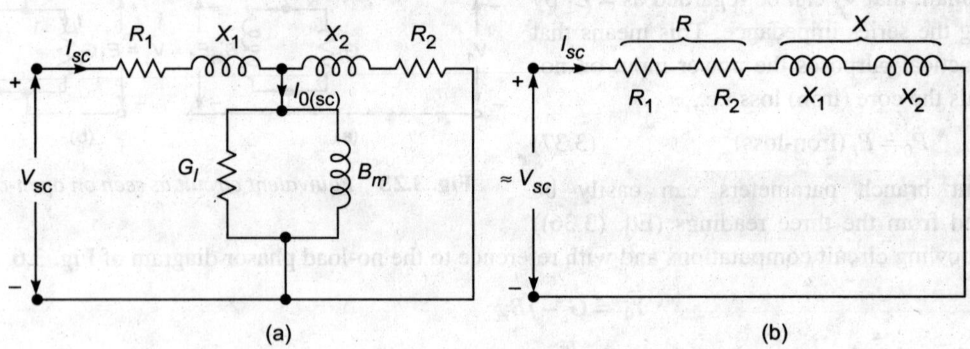

Fig. 3.25 *Equivalent circuit under short-circuit conditions*

[††]Voltage needed for the SC test is typically 5% of the rated value. For a 200 kVA, 440/6600-V transformer, test on the HV side would require

$$\frac{6600 \times 5}{100} = 330 \text{ V} \qquad \text{and} \qquad \frac{200 \times 1000}{6600} = 30 \text{ A supply}$$

while if conducted from the LV side it would need

$$\frac{440 \times 5}{100} = 22 \text{ V} \qquad \text{and} \qquad \frac{200 \times 1000}{440} = 445 \text{ A supply}$$

Low-voltage, high-current supply needed for conducting the SC test from the LV side is much more difficult to arrange than the supply required for the same test from the HV side.

While conducting the SC test, the supply voltage is gradually raised from zero till the transformer draws full-load current. The meter readings under these conditions are:

voltage = V_{SC}; current = I_{SC}; power input = P_{SC}

Since the transformer is excited at very low voltage, the iron-loss is negligible (that is why shunt branch is left out), the power input corresponds only to the copper-loss, i.e.

$$P_{SC} = P_c \text{ (copper-loss)} \tag{3.42}$$

From the equivalent circuit for Fig. 3.25(b), the circuit parameters are computed as below:

$$Z = \frac{V_{SC}}{I_{SC}} = \sqrt{R^2 + X^2} \tag{3.43}$$

Equivalent resistance,
$$R = \frac{P_{SC}}{(I_{SC})^2} \tag{3.44}$$

Equivalent reactance,
$$X = \sqrt{Z^2 - R^2} \tag{3.45}$$

These values are referred to the side (HV) from which the test is conducted. If desired, the values could be easily referred to the other side.

It is to be observed that the SC test has given us the equivalent resistance and reactance of the transformer; it has not yielded any information for separating[‡‡] these into respective primary and secondary values.

It was observed that OC and SC tests together give the parameters of the approximate equivalent circuit of Fig. 3.16(a) which as already pointed out is quite accurate for all important computations.

Example 3.7 *The following data were obtained on a 20 kVA, 50 Hz, 2000/200 V distribution transformer: Draw the approximate equivalent circuit of the transformer referred to the HV and LV sides respectively.*

LO 3

Table 3.1

	Voltage (V)	Current (A)	Power (W)
OC test with HV open-circuited	200	4	120
SC test with LV short-circuited	60	10	300

Solution

OC test (LV side)

$$Y_0 = \frac{4}{200} = 2 \times 10^{-2}\ ℧;\ G_i = \frac{120}{(200)^2} = 0.3 \times 10^{-2}\ ℧$$

$$B_m = \sqrt{Y_0^2 - G_i^2} = 1.98 \times 10^{-2}\ ℧$$

[‡‡]Resistances could be separated out by making dc measurements on the primary and secondary and duly correcting these for ac values. The reactances cannot be separated as such. Where required, these could be equally apportioned to the primary and secondary, i.e.

$$X_1 = X_2 \text{ (referred to anyone side)}$$

This is sufficiently accurate for a well-designed transformer.

SC test (HV side)

$$Z = \frac{60}{10} = 6\ \Omega;\ R = \frac{300}{(10)^2} = 3\ \Omega$$

$$X = \sqrt{Z^2 - R^2} = 5.2\ \Omega$$

Transformation ratio, $$\frac{N_H}{N_L} = \frac{2000}{200} = 10$$

Equivalent circuit referred to the HV side:

$$G_i\,(\text{HV}) = 0.3 \times 10^{-2} \times \frac{1}{(10)^2} = 0.3 \times 10^{-4}\ \mho$$

$$B_m\,(\text{HV}) = 1.98 \times 10^{-2} \times \frac{1}{(10)^2} = 1.98 \times 10^{-4}\ \mho$$

The equivalent circuit is drawn in Fig. 3.26(a).

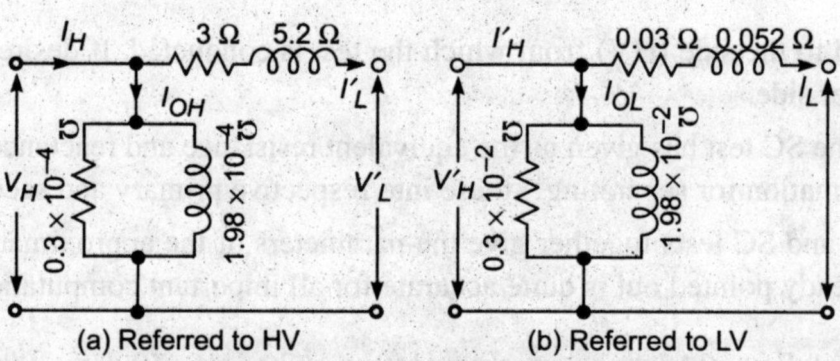

Fig. 3.26 *Equivalent circuit*

Equivalent circuit referred to the LV side:

$$R(\text{LV}) = 3 \times \frac{1}{(10)^2} = 0.03\ \Omega$$

$$X(\text{LV}) = 5.2 \times \frac{1}{(10)^2} = 0.052\ \Omega$$

The equivalent circuit is drawn in Fig. 3.26(b).

Example 3.8 *The parameters of the equivalent circuit of a 150-kVA, 2400/240-V transformer are:*

$R_1 = 0.2\ \Omega$ $\quad R_2 = 2 \times 10^{-3}\ \Omega$

$X_1 = 0.45\ \Omega$ $\quad X_2 = 4.5 \times 10^{-3}\ \Omega$

$R_i = 10\ k\Omega$ $\quad X_m = 1.6\ k\Omega$ *(as seen from 2400-V side)*

Calculate:

(a) Open-circuit current, power and pf when LV is excited at rated voltage

(b) The voltage at which the HV should be excited to conduct a short-circuit test (LV shorted) with full-load current flowing. What is the input power and its pf?

LO 3

Solution

Note

$$R_i = \frac{1}{G_i}, \quad X_m = \frac{1}{B_m}$$

$$\text{Ratio of transformation, } a = \frac{2400}{240} = 10$$

(a) Referring the shunt parameters to LV side

$$R_i\,(\text{LV}) = \frac{10 \times 1000}{(10)^2} = 100\ \Omega$$

$$X_m\,(\text{LV}) = \frac{1.6 \times 1000}{(10)^2} = 16\ \Omega$$

$$\bar{I}_0\,(\text{LV}) = \frac{240\angle 0^\circ}{100} - j\,\frac{240\angle 0^\circ}{16}$$

$$= 2.4 - j\,15 = 15.2\angle{-80.9^\circ}\ \text{A}$$

or

$$I_0 = 15.2\ \text{A, pf} = \cos 80.9^\circ = 0.158 \text{ lagging}$$

(b) LV shorted, HV excited, full-load current flowing: Shunt parameters can be ignored under this condition. Equivalent series parameters referred to HV side:

$$R = 0.2 + 2 \times 10^{-3} \times (10)^2 = 0.4\ \Omega$$

$$X = 0.45 + 4.5 \times 10^{-3} \times (10)^2 = 0.9\ \Omega$$

$$\bar{Z} = 0.4 + j\,0.9 = 0.985\ \angle 66^\circ\ \Omega$$

$$I_{fl}\,(\text{HV}) = \frac{150 \times 1000}{2400} = 62.5\ \text{A}$$

$$V_{SC}\,(\text{HV}) = 62.5 \times 0.958 = 59.9\ \text{V or } 60\ \text{V (say)}$$

$$P_{SC} = (62.5)^2 \times 0.4 = 1.56\ \text{kW}$$

$$pf_{SC} = \cos 66^\circ = 0.406 \text{ lagging}$$

Sumpner's (Back-to-Back) Test

While OC and SC tests on a transformer yield its equivalent circuit parameters, these cannot be used for the 'heat run' test wherein the purpose is to determine the steady temperature rise if the transformer was fully loaded continuously; this is so because under each of these tests the power loss to which the transformer is subjected is either the core-loss or copper-loss but not both. The way out of this impasse without conducting an actual loading test is the Sumpner's test which can only be conducted simultaneously on two identical transformers[§§].

Circuit Connections for Sumpner's Test

In conducting the Sumpner's test the primaries of the two transformers are connected in parallel across the rated voltage supply (V_1), while the two secondaries are connected in phase opposition as shown in Fig. 3.27. For the secondaries to be in phase opposition, the voltage across T_2T_4 must be zero otherwise it will be double the rated secondary voltage in which case the polarity of one of the secondaries must be reversed. Current at low voltage (V_2) is injected into the secondary circuit at T_2T_4. The supply (1) and supply (2) are from the same mains.

[§§] In very large sizes two identical transformers may not be available as these are custom-built.

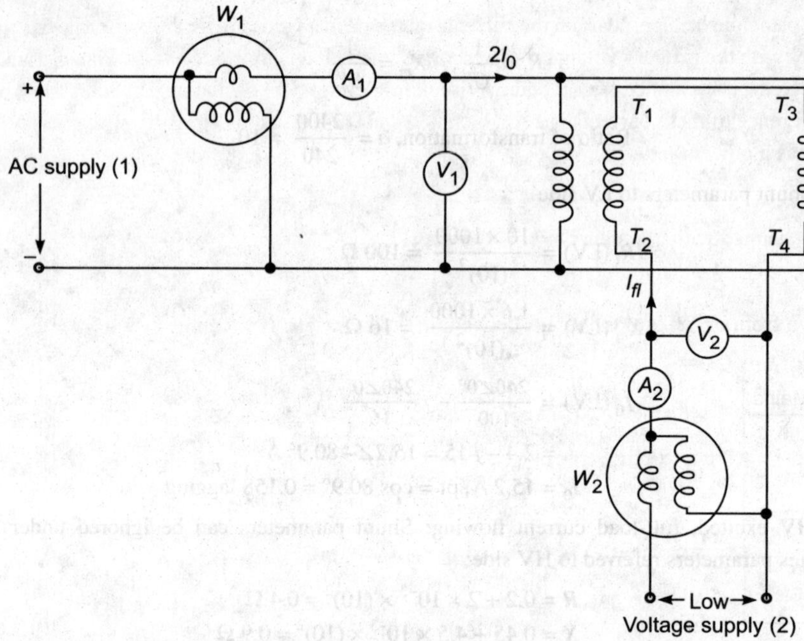

Fig. 3.27 *Sumpner's test on two identical single-phase transformers*

As per the superposition theorem, if V_2 source is assumed shorted, the two transformers appear in open-circuit to source V_1 as their secondaries are in phase opposition and therefore no current can flow in them. The current drawn from source V_1 is thus $2I_0$ (twice the no-load current of each transformer) and power is $2P_0$ ($= 2P_i$, twice the core-loss of each transformer). When the ac supply (1) terminals are shorted, the transformers are series-connected across V_2 supply (2) and are short-circuited on the side of primaries. Therefore, the impedance seen at V_2 is $2Z$ and when V_2 is adjusted to circulate full-load current (I_{fl}), the power fed in is $2P_c$ (twice the full-load copper-loss of each transformer). Thus in the Sumpner's test while the transformers are not supplying any load, full iron-loss occurs in their cores and full copper-loss occurs in their windings; net power input to the transformers being $(2P_0 + 2P_c)$. The heat run test could, therefore, be conducted on the two transformers, while only losses are supplied.

In Fig. 3.27 the auxiliary voltage source is included in the circuit of secondaries; the test could also be conducted by including the auxiliary source in the circuit of primaries.

The procedure to connect a 3-phase transformer for the back-to-back test will be explained in Sec. 3.12.

Example 3.9 *Two transformers of 20 kVA each with turn-ratios respectively of 250 : 1000 and 250 : 1025 are connected in back-to-back test; the two primaries being fed from a 250 V supply and secondaries being connected in phase opposition. A booster transformer connected on primary side to the same 250 V supply is used to inject voltage in the circuit of secondaries such as to circulate a current of 20 A. The core losses of each transformer are 350 W and each transformer has a reactance 2.5 times its resistance. Calculate the possible readings of the wattmeter connected to measure the input to the primaries.*

LO 3

Solution Using the principle of superposition the currents on the primary side are first found, caused by the circulating current in the secondaries with the primary voltage source shorted; the voltage injected on the secondary side being intact. The primary currents necessary to balance the secondary circulating current are shown in Fig. 3.28; these being in phase with each other. The difference of these currents is 2 A which flows in the lines connecting the primaries to the main (refer to figure). This current has a power factor of

$$\cos \tan^{-1} 2.5 = 0.371$$

Therefore, the power exchanged with the mains by this current is

$$250 \times 2 \times 0.371 = 185.5 \text{ W}$$

This power will be drawn from or fed into the mains depending on the polarity of the injected voltage.

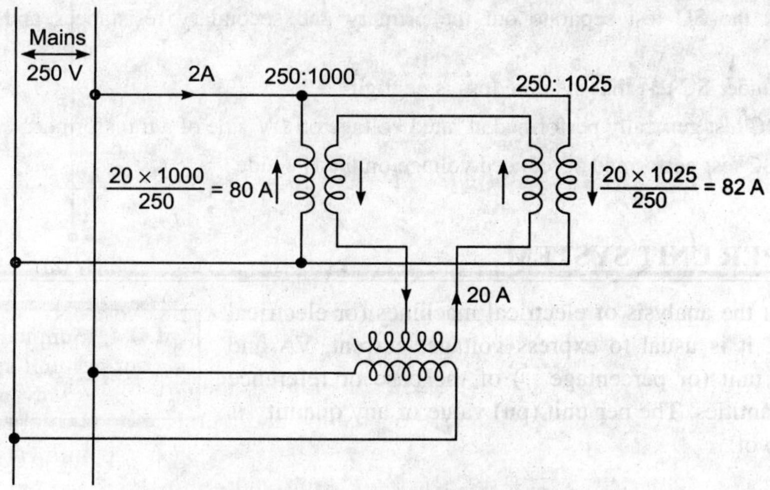

Fig. 3.28 *Circuit for Example 3.9*

Consider now the currents owing to the voltage source connected to the primaries with the secondary injected voltage source shorted. The primaries now draw the magnetizing currents from the mains with associated core-loss of both the transformers equal to 2 × 350 = 700 W. The currents in the secondaries because of the small voltage unbalance (transformers have a slightly different turn ratio) would be small with very little associated loss.¶

Hence power drawn from the mains is

$$700 \pm 185.5 = 885.5 \text{ W or } 514.5 \text{ W}$$

¶ The unbalanced voltage in secondary circuit because of unequal turn ratio is 25 V, while the rated secondary voltage is 1000 V. Assuming a high voltage side transformer impedance of 40 Ω, the circulating current caused by the unbalanced voltage would be

$$\frac{25}{2 \times 40} = 0.3125 \text{ A}$$

The ratio of losses because of this current to the losses caused by the 20A current is

$$\left(\frac{0.3125}{20}\right)^2 = 2.44 \times 10^{-4}$$

This being of negligible order, there is no error of consequence in neglecting the effect of unequal turn ratio.

Concept Check LO 3

1. Name the various losses in a transformer. ○○●
2. What are the two non-loading tests? ○○●
3. What is a polarity test? ○○●
4. Why open-circuit test is called no-load test? ○●●
5. What is a short-circuit test? ○○●
6. What is the significance of open-circuit and short-circuit tests in a transformer? ○●●
7. What is the significance of Sumpner's (Back-to-Back) test? ○●●
8. Why cannot the SC test separate out the primary and secondary resistances and leakage inductances? ○●●
9. Justify that under SC test that the core loss is negligible. ●●●
10. Why is an OC test generally performed at rated voltage on LV side of a transformer? ○○●
11. Why is the SC test performed at reduced voltage on the HV side? ●●●

3.8 | THE PER UNIT SYSTEM

LO 4 Summarize the concept of per-unit system and its advantages

While carrying out the analysis of electrical machines (or electrical machine systems), it is usual to express voltage, current, VA and impedance in per unit (or percentage***) of the base or reference values of these quantities. The per unit (pu) value of any quantity is defined as the ratio of:

$$\frac{\text{The actual value in any units}}{\text{The base or reference value in the same units}}$$

While the base values can be selected arbitrarily, it is normal to choose the rated value of the device as its base values.

There are two important advantages that accrue from the use of the pu system. First, the parameters of transformers as well as rotating machines lie roughly in the same range of numerical values irrespective of their ratings if expressed in per unit of their ratings; correctness of analysis becomes immediately obvious in this system.

Second, the pu system is most convenient in power systems as it relieves the analyst of the need to refer circuit quantities to one or other side of the transformers. It is a universal practice to use the pu system for modelling real-life large integrated power systems and in computer simulation of machine systems for their transient and dynamic analysis.

Per-Unit for Single Phase System

Base values of various quantities are related to each other by the usual electrical laws. For a single-phase system,

*** Per cent values are not preferred as a factor of 100 has to be carried.

$$P_{base}, Q_{base}, (VA)_{base} = V_{base}I_{base} \tag{3.46}$$
$$R_{base}, X_{base}, Z_{base} = V_{base}/I_{base} \tag{3.47}$$
$$G_{base}, B_{base}, Y_{base} = I_{base}/V_{base} \tag{3.48}$$

Always, $(VA)_{base}$ and V_{base} are first selected and their choice automatically fixes the other base values as per Eqs (3.46)–(3.48). It immediately follows from these equations that

$$Z_B = \frac{V_B^2}{(VA)_B}$$

Then
$$Z(\text{pu}) = \frac{Z(\Omega) \times (VA)_B}{V_B^2} \tag{3.49}$$

In large devices and systems it is more practical to express the bases in kVA/ MVA and kV. Then Eq. (3.49) is written as

$$Z(\text{pu}) = \frac{Z(\Omega) \times (\text{kVA})_B}{1000(\text{kV})_B^2} \tag{3.50a}$$

or
$$Z(\text{pu}) = \frac{Z(\Omega) \times (\text{MVA})_B}{(\text{kV})_B^2} \tag{3.50b}$$

When $(\text{MVA})_B$ and $(\text{kV})_B$ are modified, the new pu impedance is given by

$$Z(\text{pu})_{\text{new}} = Z(\text{pu})_{\text{old}} \frac{(\text{MVA})_{B,\text{new}}}{(\text{MVA})_{B,\text{old}}} \times \frac{(\text{kV})_{B,\text{old}}^2}{(\text{kV})_{B,\text{new}}^2} \tag{3.51}$$

Important! It can be easily shown that in a transformer equivalent circuit using pu notations, the need for an ideal transformer is eliminated because the pu impedance of a transformer is the same whether computed from the primary or secondary side so long as the voltage bases on the two sides are selected in the ratio of transformation (see Example 3.10).

A procedure universally adopted is to translate all quantities to pu values for carrying out analysis and to convert the results obtained back to actual units.

As has been mentioned earlier the pu parameters of transformers (and electric machines) lie within a narrow range. For example, the magnetizing current normally lies between 0.02 and 0.05 pu, the equivalent resistance between 0.005 pu (large transformers) and 0.02 pu (small transformers) and the equivalent reactance usually varies from 0.05 (large) to 0.1 pu (small transformers). This information helps a great deal in comparing units of a given size made by various manufacturers.

Per-Unit for 3-phase System

In the 3-phase system, the bases are chosen as

$$(\text{MVA})_B = \text{3-phase MVA}$$
$$(\text{kV})_B = \text{line-to-line kV}$$

Assuming star connection (equivalent star can always be found),

$$Z_B = \frac{((\text{kV})_B/\sqrt{3})^2}{\frac{1}{3}(\text{MVA})_B} = \frac{(\text{kV})_B^2}{(\text{MVA})_B}$$

Then $$Z(\text{pu}) = \frac{Z(\Omega) \times (\text{MVA})_B}{(\text{kV})_B^2} \tag{3.52}$$

which relationship is indeed the same as Eq. (3.50) for the single-phase system.

Consider three impedances, Z each, connected in delta. Then with 3-phase $(\text{MVA})_B$ and line-to-line $(\text{kV})_B$,

$$Z_B(\Delta) = \frac{(\text{kV})_B^2}{(\text{MVA})_B/3} = \frac{3(\text{kV})_B^2}{(\text{MVA})_B}$$

Therefore $$Z(\text{pu}) = \frac{(Z/3)(\text{MVA})_B}{(\text{kV})_B^2}$$

Since $Z/3$ is the equivalent star impedance, the pu impedance for delta or its equivalent star is the same for a given 3-phase MVA base and line-to-line kV base.

Example 3.10 *The exciting current was found to be 3 A when measured on the LV side of a 20-kVA, 2000/200- V transformer. Its equivalent Impedance (referred to the HV side) is 8.2 + j 10.2 Ω. Choose the transformer rating as the base.*

(a) Find the exciting current in pu on the LV as well as HV side.

(b) Express the equivalent impedance in pu on the LV as well as HV side.

LO 4

Solution

$$V_B(\text{HV}) = 2000 \text{ V} \qquad V_B(\text{LV}) = 200 \text{ V}$$
$$I_B(\text{HV}) = 10 \text{ A} \qquad I_B(\text{LV}) = 100 \text{ A}$$

$$Z_B(\text{HV}) = \frac{2000}{10} = 200\ \Omega$$

$$Z_B(\text{LV}) = \frac{200}{100} = 2\ \Omega$$

(a) $$I_0(\text{LV}) = \frac{3}{100} = 0.03 \text{ pu}$$

The exciting current referred to the HV side is 0.3 A

$$I_0(\text{HV}) = \frac{0.3}{10} = 0.03 \text{ pu}$$

(b) $$Z(\text{HV})_{(\text{pu})} = \frac{8.2 + j10.2}{200} = 0.041 + j\,0.051$$

$$Z(\text{LV}) = \frac{8.2 + j10.2}{(10)^2} = 0.082 + j\,0.102$$

$$Z(\text{LV})_{(\text{pu})} = \frac{0.082 + j0.102}{2} = 0.041 + j\,0.051$$

Remark The earlier remark is, therefore, confirmed that pu values referred to either side of the transformer are the same so long as voltage bases on the two sides are in the ratio of transformation of the transformer.

Concept Check LO 4

1. What is the per-unit value of any quantity? ○○●
2. What are the advantages of per-unit system? ○○●
3. What are the applications of per-unit system? ○●●
4. Mention the four quantities normally considered in per-unit system. ○○●
5. State and prove the condition from maximum efficiency of a transformer. ○●●
6. Prove that in the system, if the voltage bases are selected in the ratio of transformation, the pu impedance of the transformer is same on either side. ○●●

3.9 | EFFICIENCY AND VOLTAGE REGULATION

LO 5 Analyze the performance of a transformer by determining its efficiency and voltage regulation and the effect of loading

Power and distribution transformers are designed to operate under conditions of constant rms voltage and frequency and so the efficiency and voltage regulation are of prime importance.

Efficiency

The rated capacity of a transformer is defined as the product of rated voltage and full load current on the output side. The power output depends upon the power factor of the load.

The efficiency η of a transformer is defined as the ratio of the useful power output to the input power. Thus

$$\eta = \frac{\text{output}}{\text{input}} \tag{3.53}$$

Important! The efficiency of a transformer is in the range of 96–99%.

It is of no use trying to determine it by measuring the output and input under load conditions because the wattmeter readings are liable to have an error of 1–2%. The best and accurate method of determining efficiency would be to find the losses from the OC and SC tests. With this data efficiency can then be calculated as

$$\eta = \frac{\text{output}}{\text{output + losses}} = 1 - \frac{\text{losses}}{\text{output + losses}} \tag{3.54}$$

In Eq. (3.54) the effect of meter readings is confined to losses only so that the overall efficiency as obtained from it is far more accurate than that obtained by direct loading.

Various losses in a transformer have already been enumerated in Sec. 3.6 and the two important losses (iron-loss P_i and copper-loss P_c) are shown in Fig. 3.29 (a) of a loaded transformer.

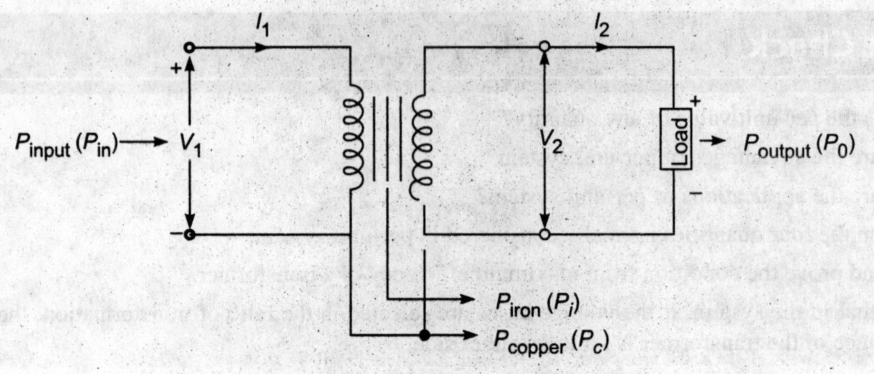

Fig. 3.29(a) *Transformer on load*

Effect of Load and Load Power Factor

The iron (core) losses depend upon the flux density and so on the induced emf. As $E_1 \approx V_1$ at all loads, these losses can be regarded as constant (independent of load) for constant primary voltage.

Copper losses in the two windings are

$$P_c = I_1^2 R_1 + I_2^2 R_2$$
$$= I_2^2 R_{eq}(2)$$

where $R_{eq}(2)$ = equivalent resistance referred to the secondary side. Thus it is found that copper losses vary as the square of the load current.

$$\text{Transformer output, } P_0 = V_2 I_2 \cos\theta_2$$
$$\cos\theta_2 = \text{load pf}$$

From Eq. (3.54)

$$\eta = \frac{V_2 I_2 \cos\theta_2}{V_2 I_2 \cos\theta_2 + P_i + I_2^2 R_{eq}(2)} \tag{3.55}$$

where V_2 is the rated secondary voltage. It varies slightly with the load but the variation is so small (about 3-5%) that it can be neglected for computing efficiency.

Equation (3.55) shows that for a given power factor, efficiency varies with load current. It can be written as

$$\eta = \frac{V_2 \cos\theta_2}{V_2 \cos\theta_2 + \left(\frac{P_i}{I_2} + I_2 R_{eq}(2)\right)} \tag{3.56}$$

For maximum value of η for given $\cos\theta_2$ (pf), the denominator of Eq. (3.56) must have the least value. The condition for maximum η, obtained by differentiating the denominator and equating it to zero, is

$$I_2^2 R_{eq}(2) = P_i \tag{3.57}$$

or Copper-loss (variable) = core-loss (constant)

It means that the efficiency is maximum at a load when the copper-loss (variable loss) equals the core-loss (constant loss). Thus for maximum efficiency,

$$I_2^2 = \frac{P_i}{R_{eq}(2)}$$

Dividing by I_{2fl}^2 on both sides

$$\left(\frac{I_2}{I_{2fl}}\right)^2 = \frac{P_i}{I_{2fl}^2 R_{eq}(2)} = k^2$$

or
$$k = \sqrt{\frac{P_i}{P_c(fl)}} \tag{3.58}$$

where
$$k = I_2/I_{2fl} \tag{3.59a}$$

and
$$P_c(fl) = \text{full-load copper-loss}$$

Thus the efficiency is maximum at a fractional load current given by Eq. (3.59a). Multiplying the numerator and denominator of Eq. (3.59a) by rated V_2

$$k = \frac{V_2 I_2}{V_2 I_{2fl}} \tag{3.59b}$$

Thus the maximum efficiency is given at a load $k(V_2 I_{2fl})$ or kS_2, where S_2, is VA (or kVA) rating of the transformer. The expression for maximum efficiency is given by

$$\eta_{max} = \frac{kS_2 \cos\theta_2}{kS_2 \cos\theta_2 + 2P_i} \tag{3.60}$$

Inference It can be easily observed from Eq. (3.60) that η_{max} increases with increasing pf (cos θ_2) and is the highest at unity pf. Also $\eta_{max} = 0$ when cos $\theta = 0$, i.e. at zero pf (purely reactive load). Therefore, knowledge of transformer losses is as important as its efficiency.

Power transformers used for bulk power transmission are operated near about full load at all times and are therefore designed to have maximum efficiency at full-load. On the other hand, the distribution transformers supply load which varies over the day through a wide range. Such transformers are, therefore, designed to have maximum efficiency at about three-fourths the full load.

Recall Normally transformer efficiency is maximum when the load power factor is unity.

From Fig. 3.29(b), it is seen that the maximum efficiency occurs at same load current independent of power factor, because the total core loss P_c, and equivalent resistance $R_{eq}(2)$ are not affected by load power factor. Any way reduction of load power factor reduces the transformer output and the transformer efficiency also reduced.

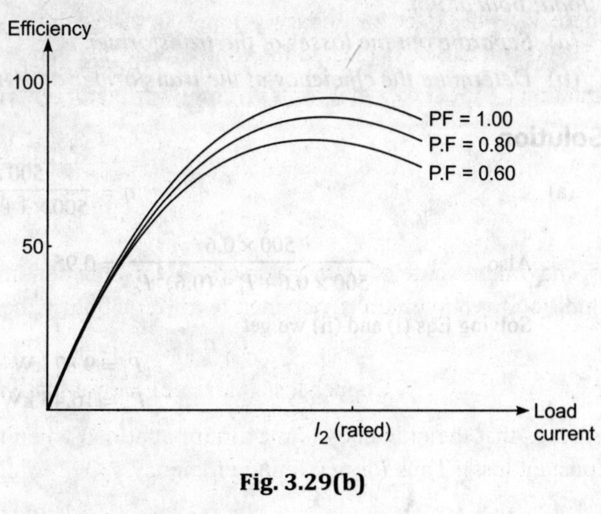

Fig. 3.29(b)

All-day (Energy) Efficiency

The all-day efficiency of a transformer is the ratio of the total energy output (kWh) in a 24-h day to the total energy input in the same time. Since the core losses are constant independent of the load, the all-day efficiency of a transformer is dependent upon the load cycle; but no prediction can be made on the basis of the load factor (average load/peak load). It is an important figure of merit for distribution transformers which feed daily load cycle varying over a wide load range. Higher energy efficiencies are achieved by designing distribution transformers to yield maximum (power) efficiency at less than full load (usually about 70% of the full load). This is achieved by restricting the core flux density to lower values by using a relatively larger core cross-section. (It means a larger iron/copper weight ratio.)

Example 3.11 *For the transformer of Example 3.7 calculate the efficiency if the LV side is loaded fully at 0.8 power factor. What is the maximum efficiency of the transformer at this power factor and at what pu load would it be achieved?*

LO 5

Solution

$$\text{Power output} = V_2 I_2 \cos\theta_2$$

$$= 200 \times 100 \times 0.8 = 16000 \text{ W (independent of lag/lead)}$$

$$\text{Total loss } P_L = P_i + k^2 P_c$$

$$= 120 + 1 \times 300 = 420 \text{ W}$$

$$\therefore \qquad \eta = 1 - \frac{P_L}{P_0 + P_L} = 1 - \frac{420}{16000 + 420} = 97.44\%$$

For maximum efficiency

$$k = \sqrt{\frac{P_i}{P_c}} = \sqrt{\frac{120}{300}} = 0.632$$

i.e. at 0.632 pu load (this is independent of power factor). Now

$$\eta_{max}\,(\cos\theta_2 = 0.8) = 1 - \frac{2P_i}{P_0 + 2P_i} = 1 - \frac{2 \times 120}{16000 \times 0.632 + 2 \times 120} = 97.68\%$$

Example 3.12 *A 500 kVA transformer has an efficiency of 95% at full load and also at 60% of full load; both at upf.*

(a) Separate out the losses of the transformer.

(b) Determine the efficiency of the transformer at 3/4th full load.

LO 5

Solution

(a)
$$\eta = \frac{500 \times 1}{500 \times 1 + P_i + P_c} = 0.95 \qquad \text{(i)}$$

Also
$$\frac{500 \times 0.6}{500 \times 0.6 + P_i + (0.6)^2 P_c} = 0.95 \qquad \text{(ii)}$$

Solving Eqs (i) and (ii) we get

$$P_i = 9.87 \text{ kW}$$
$$P_c = 16.45 \text{ kW}$$

(b) At 3/4th full load upf

$$\eta = \frac{500 \times 0.75}{500 \times 0.75 + 9.87 + (0.75)^2 \times 16.45} = 95.14\%$$

Example 3.13 *A transformer has its maximum efficiency of 0.98 at 15 kVA at upf. Compare its all-day efficiencies for the following load cycles:*

(a) Full load of 20 kVA 12 hours/day and no-load rest of the day.

(b) Full load 4 hours/day and 0.4 full-load rest of the day.

Assume the load to operate on upf all day.

LO 5

Solution

$$\eta_{max} = \frac{P_0}{P_0 + 2P_i} \quad \text{or} \quad 0.98 = \frac{15}{15 + 2P_i}$$

or $\qquad P_i = 0.153$ kW

Now $\qquad k^2 = \frac{P_i}{P_c} \quad$ or $\quad \left(\frac{15}{20}\right)^2 = \frac{0.153}{P_c}$

or $\qquad P_c = 0.272$ kW

(a)

P_0	Time, h	W_0	$P_{in} = P_0 + P_i + k^2 P_c$	W_{in}
20	12	240	$20 + 0.153 + 0.272 = 20.425$	245.1
0	12	0	$0 + 0.153 = 0.153$	1.8
		240 kWh		246.9 kWh

$$\eta_{allday} = \frac{\Sigma W_0}{\Sigma W_{in}} = \frac{240}{246.9} = 97.2\%$$

(b)

P_0	Time, h	W_0	$P_{in} = P_0 + P_i + k^2 P_c$	W_{in}
20	4	80	$20 + 0.153 + 0.272 = 20.425$	81.7
8	20	160	$8 + 0.153 + \left(\frac{8}{20}\right)^2 \times 0.272 = 8.196$	163.9
		240 kWh		245.6 kWh

$$\eta_{allday} = \frac{240}{245.6} = 97.7\%$$

Remark Even though the load factor is the same in each case [(240/24)/20 = 0.5] the all day efficiencies still differ because of the difference in the nature of the two load cycles.

Voltage Regulation

Constant voltage is the requirement of most domestic, commercial and industrial loads. It is, therefore, necessary that the output voltage of a transformer must stay within narrow limits as the load and its power factor vary. This requirement is more stringent in distribution transformers as these directly feed the load centres. The voltage drop in a transformer on load is chiefly determined by its leakage reactance which must be kept as low as design and manufacturing techniques would permit.

The figure of merit which determines the voltage drop characteristic of a transformer is the voltage regulation. It is defined as the change in magnitude of the secondary (terminal) voltage, when full-load (rated load) of specified power factor supplied at rated voltage is thrown off, i.e. reduced to no-load with primary voltage (and frequency) held constant, as percentage of the rated load terminal voltage. In terms of symbols

$$\% \text{ Voltage regulation} = \frac{V_{20} - V_{2.fl}}{V_{2.fl}} \times 100 \tag{3.61}$$

where $V_{2,fl}$ = rated secondary voltage while supplying full load at *specified power factor*

V_{20} = secondary voltage when load is thrown off.

Figure 3.30(a) shows the transformer equivalent circuit[†††] referred to the secondary side and Fig. 3.30(b) gives its phasor diagram. The voltage drops *IR* and *IX* are very small in a well-designed transformer (refer Example 3.6). As a result the angle δ between V_1 and V_2 is of negligible order, so that[‡‡‡]

$$V_1 \approx OE$$

$$V_1 - V_2 \approx BE = I(R\cos\phi + X\sin\phi);\ \phi \text{ lagging} \tag{3.62a}$$

$$= I(R\cos\phi - X\sin\phi);\ \phi \text{ leading} \tag{3.62b}$$

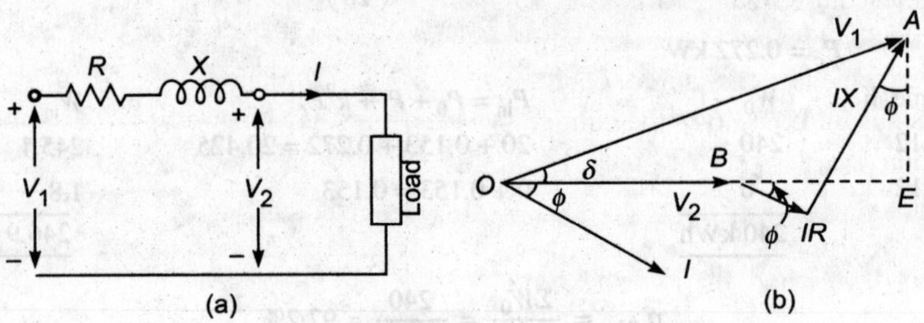

Fig. 3.30 *(a) Equivalent circuit referred to secondary, (b) Phasor diagram (not proportional)*

When the load is thrown off

$$V_{20} = V_1$$

$$\therefore \quad V_{20} - V_2 = I(R\cos\phi \pm \sin\phi) \tag{3.63}$$

where I is the full-load secondary current and V_2, the full-load secondary voltage (equal to the value of V_2 (rated)). Thus

$$\% \text{ Voltage regulation, Reg} = \frac{V_{20} - V_2}{V_2} \times 100$$

$$= \frac{I(R\cos\phi \pm X\sin\phi)}{V_2} \times 100 \tag{3.64}$$

[†††] In approximate equivalent circuit, the magnetizing shunt branch plays no role in determining voltages and hence is left out.

[‡‡‡] $\bar{V}_1$ could be calculated from the phasor equation

$$\bar{V}_1 = \bar{V}_2 + I\angle\phi\,(R + jX)$$

but the approximate method is very much quicker and quite accurate (see Example 3.6).

Recognizing that $\dfrac{IR}{V_2} = R(\text{pu})$ and $\dfrac{IX}{V_2} = X(\text{pu})$

We have Per unit voltage regulation = $R(\text{pu}) \cos \phi \pm X(\text{pu}) \sin \phi$ (3.65)

Effect of Load Power Factor on Voltage Regulation

It is seen from Eq. (3.64) that the voltage regulation varies with power factor and has a maximum value when

$$\frac{d\,\text{Reg}}{d\phi} = 0 = -R \sin \phi + X \cos \phi$$

or $$\tan \phi = \frac{X}{R}$$

or $$\cos \phi = \frac{R}{\sqrt{R^2 + X^2}}\text{; lagging} \quad (3.66)$$

Equation (3.66) implies that voltage regulation is the maximum when the load power factor (lagging) angle has the same value as the angle of the equivalent impedance. From Eq. (3.64), the voltage regulation is zero when

$$R \cos \phi - X \sin \phi = 0$$

or $$\tan \phi = \frac{R}{X}\text{; leading} \quad (3.67)$$

For ϕ (leading) larger than that given by Eq. (3.67), the voltage regulation is negative (i.e. the secondary full-load voltage is more than the no-load voltage).

The complete variation of % regulation with power factor is shown in Fig. 3.31.

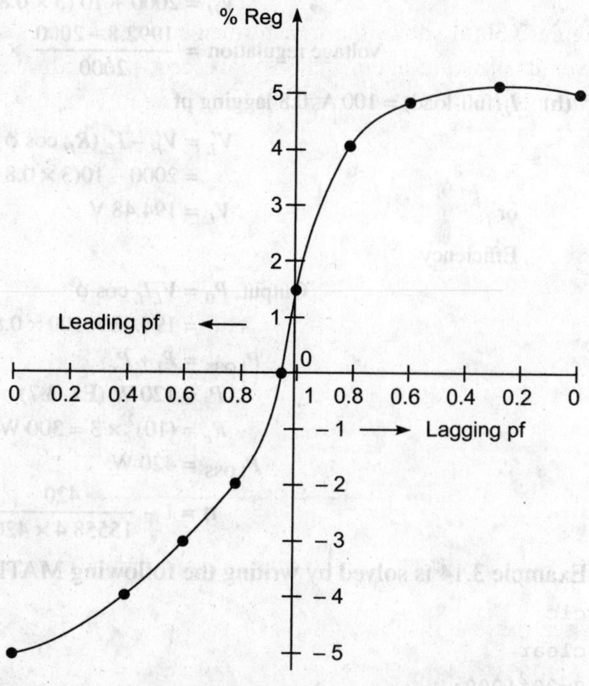

Fig. 3.31 *Percentage regulation versus power factor; R = 0.015 pu, X = 0.05 pu*

Example 3.14 *Consider the transformer with data given in Example 3.7.*

(a) With full-load on the LV side at rated voltage, calculate the excitation voltage on the HV side. The load power factor is (i) 0.8 lagging, (ii) 0.8 leading. What is the voltage regulation of the transformer in each case?

(b) The transformer supplies full-load current at 0.8 lagging power factor with 2000 V on the HV side. Find the voltage at the load terminals and the operating efficiency.

LO 5

Solution

(a) The HV side equivalent circuit of Fig. 3.26(a) will be used.

$$V_L = 200\text{ V}, \qquad I_L = \frac{200 \times 1000}{200} = 100\text{ A}$$

$$V'_L = 2000 \text{ V}, \qquad I'_L = 10 \text{ A}$$

Now $$V_H = V'_L + I'_L (R_H \cos\phi \pm X_H \sin\phi); \qquad R_H = 30\ \Omega$$

$$X_H = 5.2\ \Omega$$

(i) $$\cos\phi = 0.8 \text{ lagging}, \sin\phi = 0.6$$

$$V_H = 2000 + 10(3 \times 0.8 + 5.2 \times 0.6) = 2055.2 \text{ V}$$

$$\text{Voltage regulation} = \frac{2055.2 - 2000}{2000} \times 100 = 2.76\%$$

(ii) $$\cos\phi = 0.8 \text{ leading}, \sin\phi = 0.6$$

$$V_H = 2000 + 10\,(3 \times 0.8 - 5.2 \times 0.6) = 1992.8 \text{ V}$$

$$\text{Voltage regulation} = \frac{1992.8 - 2000}{2000} \times 100 = -0.36\%$$

(b) I_L(full-load) = 100 A, 0.8 lagging pf

$$V'_L = V'_H - I'_L (R_H \cos\phi + X_H \sin\phi)$$

$$= 2000 - 10(3 \times 0.8 + 5.2 \times 0.6) = 1944.8 \text{ V}$$

or $$V_L = 194.48 \text{ V}$$

Efficiency

$$\text{Output, } P_0 = V_L I_L \cos\phi$$

$$= 194.48 \times 100 \times 0.8 = 15558.4 \text{ W}$$

$$P_{\text{LOSS}} = P_i + P_c$$

$$P_i = 120 \text{ W (Ex. 3.7)}$$

$$P_c = (10)^2 \times 3 = 300 \text{ W}$$

$\therefore$ $$P_{\text{LOSS}} = 420 \text{ W}$$

$$\eta = 1 - \frac{420}{15558.4 \times 420} = 97.38\%$$

Example 3.14 is solved by writing the following MATLAB code.

```
clc
clear
S=20*1000;
V1=200;
V2=2000;
I1=S/V1;
I2=S/V2;
RH=3;
XH=5.2;
Cosine-phi=0.8;
Sin-phi=0.6;
VH=V2+I2*(RH*cosine-phi+XH*sin-phi)
Vreg=(VH-V2)*100/V2
%% case2
VH=V2+I2*(RH*cosine-phi-XH*sin-phi)
Vreg=(VH-V2)*100/V2
```

```
Il=100;
Vll=V2-I2*(RH*cosine-phi+XH*sin-phi);
Vl=Vll/10
Ploss=120+10*10*3;
Pop=Vl*Il*cosine-phi;
eff=(1-(Ploss/(Ploss+Pop)))*100
```

Answer:

VH = 2.0552e+003

Vreg = 2.7600

VH = 1.9928e+003

Vreg = –0.3600

Vl = 194.4800

eff = 97.3715

Example 3.15 *For the 150 kVA, 2400/240 V transformer whose circuit parameters are given in Example 3.8, draw the circuit model as seen from the HV side. Determine therefrom the voltage regulation and efficiency when the transformer is supplying full load at 0.8 lagging pf on the secondary side at rated voltage. Under these conditions calculate also the HV side current and its pf.* LO 5

Solution

$$R(\text{HV}) = 0.2 + 2 \times 10^{-3} \times (10)^2 = 0.4\ \Omega$$

$$X(\text{HV}) = 0.45 + 4.5 \times 10^{-3} \times (10)^2 = 0.9\ \Omega$$

The circuit model is drawn in Fig. 3.32.

$$I_{2(fl)} = \frac{150 \times 1000}{240} = 625 \text{ A, 0.8 pf lagging}$$

$$V_2 = 240 \text{ V}$$

$$I_2 = \frac{625}{10} = 62.5 \text{ A, 0.8 pf lagging}$$

$$V_2' = 2400 \text{ V}$$

$$\text{Voltage drop} = 62.5(0.4 \times 0.8 + 0.9 \times 0.6) = 53.75 \text{ V}$$

$$\text{Voltage regulation} = \frac{53.75}{2400} \times 100 = 2.24\%$$

$$V_1 = 2400 + 53.75 = 2453.75 = 2454 \text{ V}$$

$$P_{(\text{out})} = 150 \times 0.8 = 120 \text{ kW}$$

$$P_{c(\text{copper loss})} = (62.5)^2 \times 0.4 = 1.56 \text{ kW}$$

$$P_{i(\text{core loss})} = \frac{(2454)^2}{10 \times 1000} = 0.60 \text{ kW}$$

$$P_L = P_i + P_c = 0.60 + 1.56 = 2.16 \text{ kW}$$

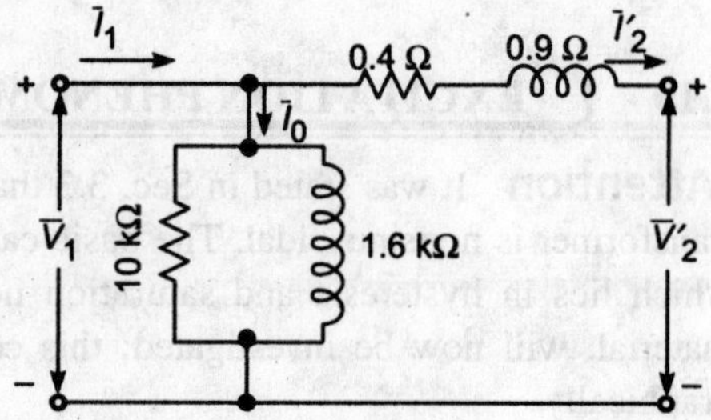

Fig. 3.32

$$\eta = \frac{120}{120 \times 2.16} = 98.2\%$$

$$\bar{I}_0 = \frac{2454\angle 0^\circ}{10 \times 1000} - j\frac{2454\angle 0^\circ}{1.6 \times 1000} = 0.245 - j\,1.53 \text{ A}$$

$$\bar{I}_2' = 62.5\,(0.8 - j\,0.6) = 50 - j\,37.5 \text{ A}$$

$$\bar{I}_1 = \bar{I}_0 + \bar{I}_2 = 50.25 - j\,39.03 = 63.63\,\angle{-37.8^\circ} \text{ A}$$

or $$I_1 = 63.63 \text{ A, pf} = 0.79 \text{ lagging}$$

Concept Check LO 5

1. What is rated capacity of a transformer? ○○●
2. Define efficiency of a transformer? ○○●
3. What is the typical range of efficiency of a transformer? ○○●
4. What is the accurate method of determining the efficiency of a transformer? ○●●
5. What is the effect of load power factor on the efficiency of a transformer? ○●●
6. Define all-day efficiency. ○○●
7. What is voltage regulation of a transformer? ○○●
8. When does voltage regulation become zero? ○●●
9. Draw the phasor diagram of a transformer as seen from any one side for zero voltage regulation. ○●●
10. From the phasor diagram of Question 9, derive the approximate condition for zero voltage regulation. ○●●
11. State and prove the condition from maximum efficiency of a transformer. ○●●
12. Describe how the primary current adjusts itself as the load on a transformer is increased. ●●●
13. Explain why the core flux in a transformer is almost independent of load current. ●●●

3.10 | EXCITATION PHENOMENON IN TRANSFORMERS

LO 6 Assess the excitation phenomenon in transformers

!Attention It was stated in Sec. 3.3 that the no-load current in a transformer is nonsinusoidal. The basic cause for this phenomenon, which lies in hysteresis and saturation non-linearities of the core material, will now be investigated; this can only be accomplished graphically.

Assume that the voltage v_1 applied to the transformer of Fig. 3.5 is sinusoidal. Since the ohmic drop ($r_1 i_0$) is assumed negligible compared to the magnitude of the applied voltage, the induced emf which balances the applied voltage must also be sinusoidal and so must be the flux established in the core (see Eqs (3.3) and (3.4)). Further, the flux must lag the induced emf by 90° as shown in the emf and flux waveforms drawn in Fig. 3.33. The current necessary to set up sinusoidal flux can be obtained graphically by looking up the hysteresis curve (ϕ-i_0 curve) also drawn in Fig. 3.33.

Assume that the steady-state operation has been reached so that hysteresis loop of Fig. 3.33 is being repeated in successive cycles of the applied voltage. Consider the instant when the flux has a value $-\phi_1$, the corresponding exciting current being zero. When the flux becomes zero (at time instant t_1), the current is a small positive value i_{01}. When the flux has a positive value ϕ_2 as shown in the figure, there are two possible values of current, i_{02} when the flux is on the increasing part of the hysteresis loop and i_{03} when the flux is on the decreasing part of the loop; $i_{02} > i_{03}$. The flux maximum $+\phi_{max}$ coincides with the current maximum $+i_{0\,max}$. The current becomes zero once again for flux $+\phi_1$. So far the positive half of exciting current has been traced out; the negative half will be symmetrical (odd symmetry) to it because of the inherent symmetry of the magnetic hysteresis loop. The complete cycle of the exciting current is sketched in Fig. 3.33.

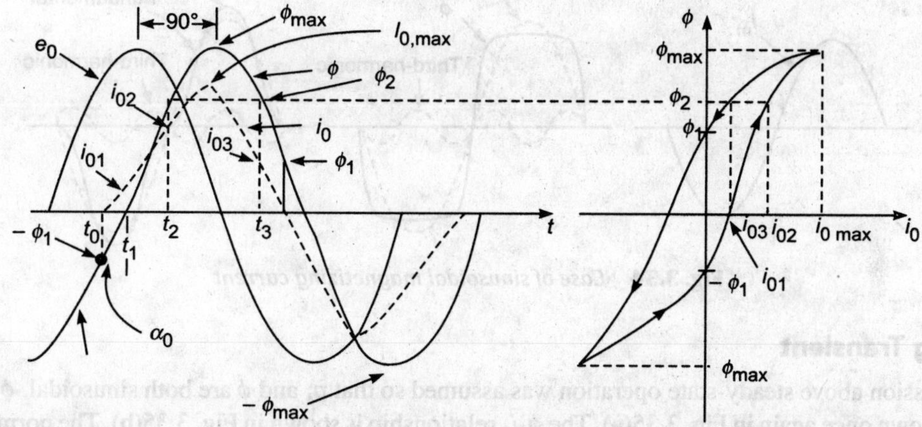

Fig. 3.33 *Effect of hysteresis and saturation on waveform of exciting current (flux sinusoidal)*

Exciting Current Wave Shape

From the exciting current wave shape of Fig. 3.33, it is observed that it is nonsinusoidal and peaky[§§§]. While odd symmetry is preserved and the current and flux maximas occur simultaneously, the current zeros are advanced[¶¶¶] in time with respect to the flux wave shape. As a consequence the current has fundamental and odd harmonics, the strongest being the third harmonic which can be as large as 40% of the fundamental. Further, the fundamental of the exciting current leads the flux by a small angle α_0 (also refer Fig. 3.6); so that the current fundamental has a component in phase with flux (I_m of Fig. 3.6) and a much smaller component in quadrature to the flux (leading) or in phase with voltage (I_i of Fig. 3.6). While I_m is responsible for creation of core flux, I_i accounts for the power lost in the core due to hysteresis.

Current I_i must of course be modified to account for the eddy-current loss in the core. The corresponding current component apart from being in phase with V_1 is sinusoidal in nature as it balances the effect of sinusoidal eddy-currents caused by the sinusoidal core flux. It is, therefore, seen that eddy-currents do not introduce any harmonics in the exciting current.

Under Linear-Loads

When the transformer feeds current to a linear load, the load current is sinusoidal and being much larger than the excitation current would 'swamp out' the nonsinusoidalness in the resultant primary current; as a consequence the primary current on load is sinusoidal for all practical purposes.

[§§§] Peakiness in exciting current is due to saturation phenomenon and would be present even if hysteresis were absent.

[¶¶¶] This shift in phase is contributed by the hysteric nature of ϕ-i_0 curve.

In certain 3-phase transformer connections, third-harmonic current cannot flow (Sec. 3.12), as a result the magnetizing current i_m is almost sinusoidal. To satisfy the *B-H* curve, the core flux must then be nonsinusoidal; it is a flat-topped wave. This can be verified by assuming a sinusoidal i_m and then finding out the ϕ wave shape from the ϕ-i_m relationship, the normal magnetizing curve****. Since the flux is flattopped, the emf which is its derivative will now be peaky with a strong third-harmonic content. The various waveforms are illustrated in Fig. 3.34.

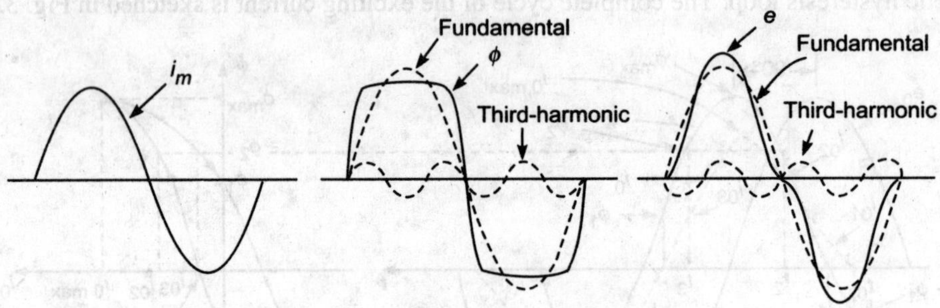

Fig. 3.34 *Case of sinusoidal magnetizing current*

Switching Transient

In the discussion above steady-state operation was assumed so that v_1 and ϕ are both sinusoidal, ϕ lagging v_1 by 90° as shown once again in Fig. 3.35(a). The ϕ-i_0 relationship is shown in Fig. 3.35(b). The normal exciting current under these conditions is about 0.05 pu if the transformer is designed with B_{max} about 1.4 T.

Doubling Effect

When the voltage v_1 is switched on to the transformer, the core flux and the corresponding exciting current undergo a transient before reaching steady-state values. The severity of the switching transient is related to the instant when the voltage wave is switched on; the worst conditions being when the applied voltage has zero value at the instant of switching as shown in Fig. 3.35(c). It is assumed here that the initial flux in the transformer at the instant of switching has zero value. It is seen from this figure that the steady-state value of flux demanded at this instant is $-\phi_m$, while the flux can only start with zero value (in the inductive circuit). As a consequence, a transient flux component (off-set flux) $\phi_t = \phi_m$ originates so that the resultant flux is $(\phi_t + \phi_{ss})$ which has zero value at the instant of switching. The transient component ϕ_t will decay according to the circuit time constant (*L*/*R*) which is generally low in a transformer. If the circuit dissipation (core-loss) is assumed negligible, the flux transient will go through a maximum value of $2\phi_m$, a phenomenon called *doubling effect*.

Inrush Current

The corresponding exciting current will be very large as the core goes into deep saturation region of magnetization ($B_m = 2 \times 1.4 = 2.8$ T); it may indeed be as large as 100 times the normal exciting current, i.e, 5 pu (normal exciting current being 0,05 pu) producing electromagnetic forces 25 times the normal. This is why the windings of large transformers must be strongly braced. In subsequent half-periods ϕ_t gradually decays till it vanishes and the core flux acquires the steady-state value, Because of the low time constant of

**** Hysteresis contributes i_i in phase with v_1 which is not being considered here.

the transformer circuit, distortion effects of the transient may last several seconds. The transformer switching transient is referred to as the *inrush current*.

The initial core flux will not be zero as assumed above but will have some residual value ϕ_r because of retentivity. As shown in Figs. 3.35(e) and (f), the transient will now be even more severe; $\phi_t = \phi_m + \phi_r$ and the core flux will now go through a maximum value of $(2\phi_m + \phi_r)$.

It is observed from Figs. 3.35(c) and (e) that the offset flux is unidirectional so that the transient flux and exciting current are unidirectional in the initial stage of the transient. A typical oscillogram of the *inrush current* is shown in Fig. 3.36.

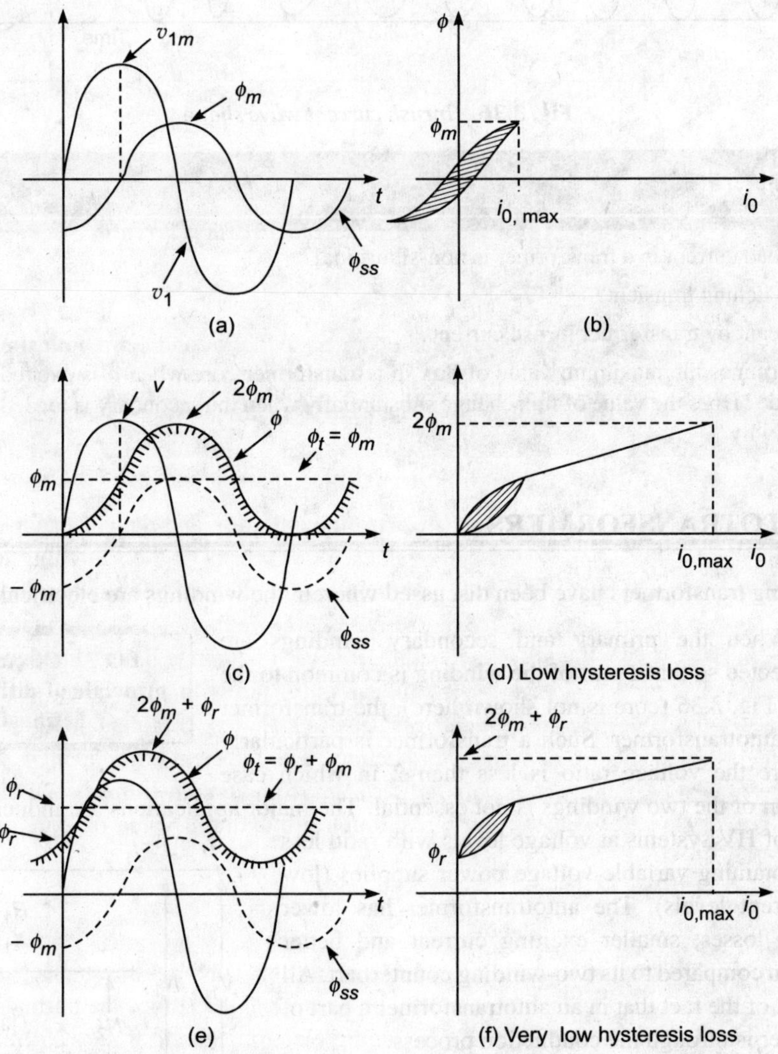

Fig. 3.35 *Transformer inrush current*

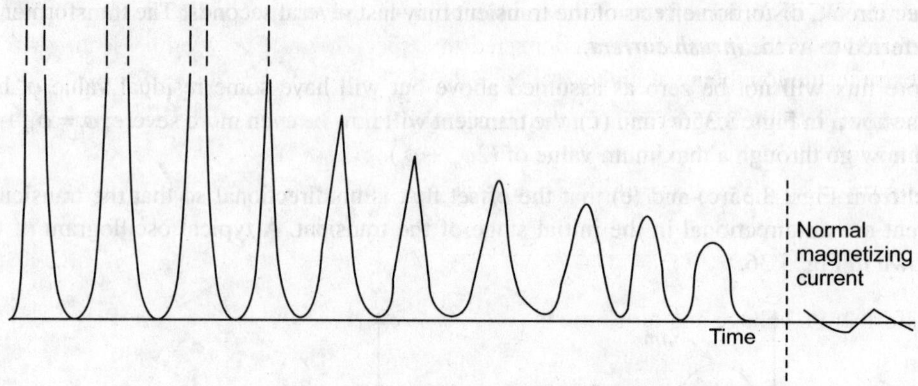

Fig. 3.36 *Inrush current wave shape*

Concept Check LO 6

1. Why no-load current in a transformer is non-sinusoidal?
2. What is switching transient?
3. What is meant by transformer inrush current?
4. What determines the maximum value of flux in a transformer core when it is excited from the primary side? Does the value of flux change substantially when the secondary is loaded? Explain the reason why.

3.11 | AUTOTRANSFORMERS

So far two-winding transformers have been discussed wherein the windings are electrically isolated.

LO 7 Discuss the working principle of different varieties of transformers

Important! When the primary and secondary windings are electrically connected so that a part of the winding is common to the two as shown in Fig. 3.36 (core is not shown here), the transformer is known as an autotransformer. Such a transformer is particularly economical where the voltage ratio is less than 2 in which case electrical isolation of the two windings is not essential. The major applications are induction motor starters, interconnection of HV systems at voltage levels with ratio less than 2, and in obtaining variable voltage power supplies (low voltage and current levels). The autotransformer has lower reactance, lower losses, smaller exciting current and better voltage regulation compared to its two-winding counterpart, All this is on account of the fact that in an autotransformer a part of the energy transfer is through the conduction process.

Figure 3.37 shows a single-phase autotransformer having N_1 turns primary with N_2 turns tapped for a lower voltage secondary.

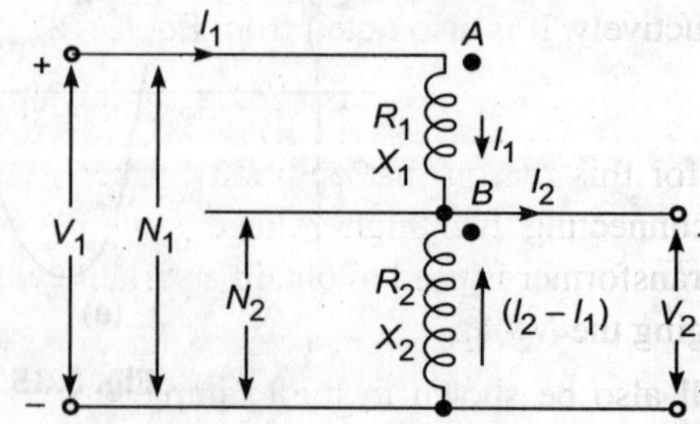

Fig. 3.37 *Autotransformer*

The winding section BC of N_2 turns is common to both primary and secondary circuits. In fact it is nothing but a conventional two-winding transformer connected in a special way. The winding section AB must be provided with extra insulation, being at higher voltage.

It will be assumed here that the magnetizing current is negligible; but it can easily be determined by a no-load test and accounted for.

With reference to Fig. 3.37 the two-winding voltage and turn-ratio is

$$a = \frac{V_1 - V_2}{V_2} = \frac{N_1 - N_2}{N_2}\,;\, N_1 > N_2 \tag{3.68}$$

As an autotransformer its voltage and turn-ratio is

$$a' = \frac{V_1}{V_2} = \frac{N_1}{N_2} > 1 \tag{3.69}$$

It is easy to see that Eqs (3.68) and (3.69) are related as

$$a' = 1 + a \tag{3.70}$$

Visualizing that in Fig. 3.37 a two-winding transformer is connected as an autotransformer, let us compare the VA ratings of the two. As a two-winding transformer

$$(VA)_{TW} = (V_1 - V_2)I_1 = (I_2 - I_1)V_2 \tag{3.71}$$

When used as an autotransformer

$$(VA)_{\text{auto}} = V_1 I_1 = V_2 I_2 \tag{3.72}$$

Equation (3.71) can be written as

$$(VA)_{TW} = \left(1 - \frac{V_2}{V_1}\right)(V_1 I_1) = \left(1 - \frac{N_2}{N_1}\right)(VA)_{\text{auto}}$$

or

$$(VA)_{\text{auto}} = \left[\frac{1}{1-(1/a')}\right](VA)_{TW};\; a' = N_1/N_2 > 1 \tag{3.73}$$

It immediately follows that

$$(VA)_{\text{auto}} > (VA)_{TW} \tag{3.74}$$

It is therefore seen that a two-winding transformer of a given *VA* rating when connected as an autotransformer can handle higher *VA*. This is because in the autotransformer connection (Fig. 3.37) part of the *VA* is transferred conductively. It is also noted from Eq. (3.74) as $a' = N_1/N_2$ (the autotransformation ratio) approaches unity,

$$(VA)_{\text{auto}} >> (VA)_{TW} \tag{3.75}$$

It is for this reason that autotransformer is commonly used when turn-ratio needed is 2 or less, like in interconnecting two high-voltage systems at different voltage levels. For low voltage, low *VA* rating autotransformer is used to obtain a variable voltage supply for testing purposes. Here $a' = N_1/N_2$ is varied by changing the N_2-tap.

It will also be shown in the example that follows that an autotransformer compared to its two-winding counterpart has a higher operating efficiency.

Reduction in Volume of Conductor Material

Let us see the problem from the design point of view by comparing winding copper needed for a given voltage ratio and VA rating for a two-winding transformer and an autotransformer. Assuming constant conductor current density, we can write

$$\frac{G_{\text{auto}}}{G_{TW}} = \frac{I_1(N_1 - N_2) + (I_2 - I_1)N_2}{I_1N_1 + I_2N_2}$$

$$= 1 - \frac{2I_1N_2}{2I_1N_1} \qquad (\because I_1N_1 = I_2N_2)$$

$$= 1 - \frac{N_2}{N_1} = 1 - \frac{V_2}{V_1} \tag{3.76}$$

where G stands for weight of winding material. It then follows from Eq. (3.76) that

$$G_{TW} - G_{\text{auto}} = \frac{1}{a'}G_{TW}$$

$$= \text{saving of conductor material in using autotransformer}$$

If $a' = 10$, saving is only 10% but for $a' = 1.1$, saving is as high as 90%. Hence the use of autotransformer is more economical when the turn-ratio is close to unity.

!Attention The interconnection of EHV systems (e.g. 220 kV and 132 kV) by the autotransformers results in considerable saving of bulk and cost as compared to the conventional two-winding transformers. Of course, a 3-phase autotransformer will be required.

It can be easily shown with reference to Fig. 3.37 that

$$R_2' = \left(\frac{N_1}{N_2} - 1\right)^2 R_2; \; X_2' = \left(\frac{N_1}{N_2} - 1\right)^2 X_2 \tag{3.77}$$

as seen on the primary side.

Example 3.16 *The 2000/200-V, 20-kVA transformer of Ex. 3.7 is connected as a step-up autotransformer as in Fig. 3.38 in which AB is 200 V winding and BC is 2000-V winding. The 200-V winding has enough insulation to withstand 2200-V to ground. Calculate (a) the LV and HV side voltage ratings of the autotransformer; (b) its kVA rating; (c) kVA transferred inductively and conductively; (d) its efficiency at full-load 0.8 pf.*

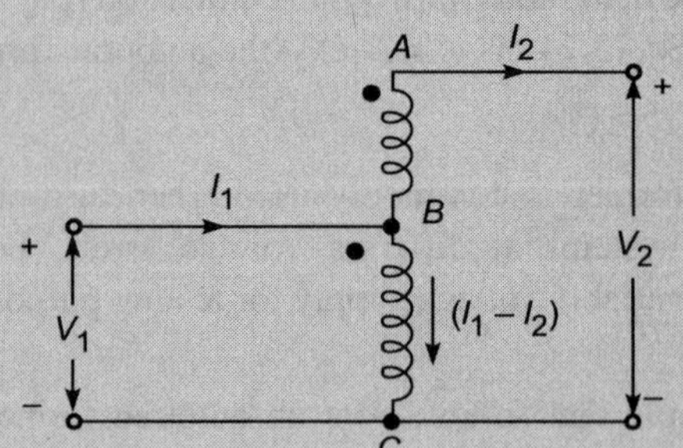

Fig. 3.38

LO 7

Solution

(a) $$V_1 = 2000\text{ V};$$
$$V_2 = 2000 + 200 = 2200\text{ V}$$

(b) $$I_2 = \frac{20 \times 1000}{200} = 100\text{ A}$$
$$I_1 - I_2 = 10\text{ A};\ I_1 = 110\text{ A}$$
$$\text{kVA rating} = \frac{2200 \times 100}{1000} = 220$$

It is, therefore seen that a 20-kVA two-winding transformer has a rating of 220 kVA as autotransformer, an 11 times increase.

(c) $$\text{kVA transferred inductively} = \frac{V_1(I_1 - I_2)}{1000} = \frac{2000 \times 100}{1000}$$

kVA transferred coductively = 220 – 20 = 200

(d) With data given in Ex. 3.7;

Core-loss (excitation voltage 2000 V) = 120 W

Full-load copper loss 300 W

(I_2 = 100 A, $I_1 - I_2$ = 10 A)

$\overline{420}$ W (Total loss)

$$\text{Full-load output} = 2200 \times 100 \times 0.8 = 176\text{ kW}$$
$$\eta = 1 - \frac{420}{176000} = 99.76\%$$

It was shown in Ex. 3.11 that this transformer as a two-winding transformer has a full-load efficiency of 97.44%. The reason for such high efficiency (99.76%) for the autotransformer is its higher output for the same excitation voltage and winding currents i.e., for the same losses.

Example 3.17 *A 240V/120V, 12 kVA transformer has full-load unity pf efficiency of 96.2%. It is connected as an auto-transformer to feed a load at 360 V. What is its rating and full-load efficiency at 0.85 pf lagging?*

LO 7

Solution 240 V/120 V, 12 kVA has rated currents of 50 A/100 A. It's connection as an autotransformer as shown in Fig. 3.39.

$$\text{Auto-transformer rating} = 360 \times 100 \times 10^{-3} = 36\text{ kVA}$$

It is 3-times 2-winding connection.

As 2-winding connection,

$$\text{Output, } P_0 = 12 \times 1 = 12\text{ kW}$$
$$\eta = \frac{P_0}{P_0 + P_L} = \frac{1}{1 + \dfrac{P_L}{P_0}} = 0.962$$

from which find full-load loss

$$1 = 0.962 + 0.962\left(\frac{P_L}{P_0}\right)$$

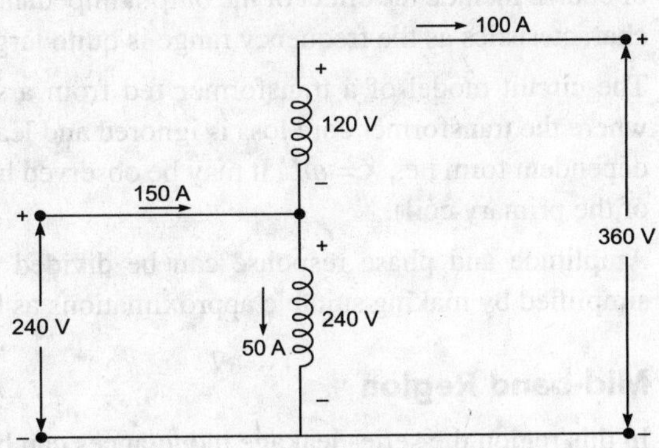

Fig. 3.39

or $$\frac{P_L}{P_0} = \frac{0.038}{0.962}; P_L = 12 \times \frac{0.038}{0.962} \; 0.474 \text{ kW}$$

In auto connection full-load loss remains the same. At 0.85 pf

$$P_0 = 36 \times 0.85 = 30.6 \text{ kW}$$

$$\eta = \frac{1}{1 + \dfrac{0.474}{30.6}} = 0.985 \quad \text{or} \quad 98.5\%$$

3.12 | VARIABLE FREQUENCY TRANSFORMER

LO 7

So far we have considered transformers which operate at fixed frequency (50 Hz).

!Attention Their purpose is to transform electric power from one voltage level to another; their performance measures being high efficiency and low voltage regulation. Small transformers (usually iron-cored) are used for coupling purposes in electronic circuits for communication, measurement and control. These transformers process signals which contain a wide band of frequencies (the width of band depends upon the signal measurement and control, audio, video, etc).

Applications of VF Transformers

The two basic applications of these transformers are:

- Coupling of load to the source so as to maximize power delivered to the load. This application exploits the impedance transforming property of the transformer. Under condition of impedance matching the over-all efficiency of the system is as low as 50%. But in electronic circuit applications the performance criterion is the maximum power unlike the maximum efficiency in power system applications. Such transformers are known as output transformers while in audio applications these are known as *audio-transformers*.
- Providing a path for dc bias current through the primary while keeping it out of the secondary circuit.

An important requirement of these transformers is that the amplitude voltage gain (ratio of output/input voltage amplitude) should remain almost constant over the range of frequencies (bandwidth) of the signal. Further, it is desirable that the phase shift of output signal from the input signal over the signal bandwidth be small. We shall now investigate the gain and phase frequency characteristics of the transformer. This would of course include the effect of the output impedance (resistance) of the electronic circuit output stage. In these characteristics as the frequency range is quite large the frequency scale used is logarithmic.

The circuit model of a transformer fed from a source of finite output resistance is drawn in Fig. 3.40(a) where the transformer core loss is ignored and leakage and magnetizing effects are shown in their frequency dependent form i.e., $X = \omega L$. It may be observed here that L_m (megnetizing inductance) = L_{11} (self inductance of the primary coil).

Amplitude and phase response can be divided into three regions wherein the response calculations are simplified by making suitable approximations as below.

Mid-band Region

In this region the series leakage inductances can be ignored (as these cause negligible voltage drops) and the shunt inductance (magnetizing inductance) can be considered as open circuit. With these approximations the equivalent circuit as seen on the primary side is drawn in Fig. 3.40(b). It immediately follows from the circuit analysis that V_L and V_S are in phase, the circuit being resistive only. As for the amplitude gain, it is given as

$$V'_L = V_S\left[\frac{R'_L}{R+R'_L}\right];\ R = R_S + R_1 + R'_2$$

$$\left[\frac{N_1}{N_2}\right]V_L = V_S\left[\frac{R'_L}{R+R'_L}\right]$$

$$A_0 = \frac{V_L}{V_S} = \left[\frac{N_2}{N_1}\right]\left[\frac{R'_L}{R+R'_L}\right];\ \angle A_0 = 0 \tag{3.78}$$

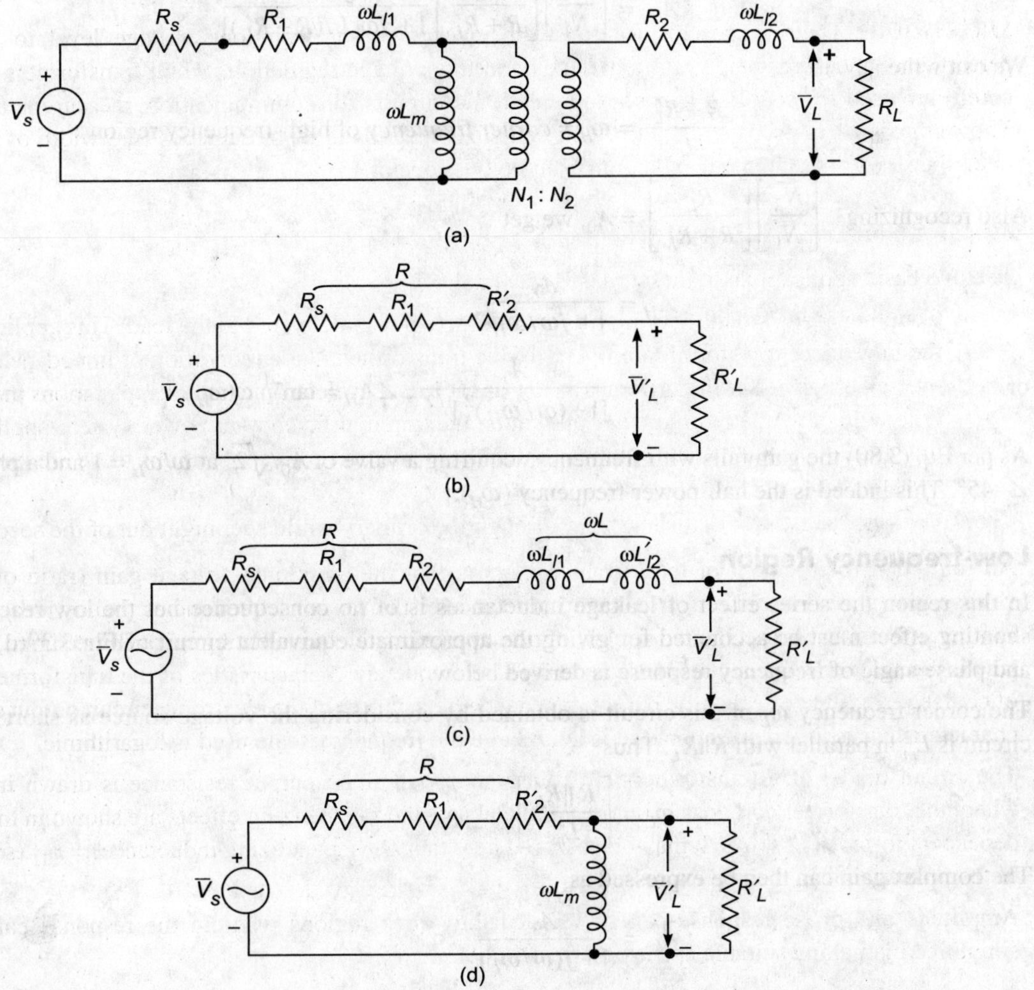

Fig. 3.40 *(a) Circuit model of transformer, (b) Approximate circuit model in mid-frequency region, (c) Approximate circuit model in high-frequency region, (d) Approximate circuit model in low-frequency region*

High-frequency Region

In this region the series inductances must be taken into account but the shunt inductance is an effective open circuit yielding the approximate equivalent circuit of Fig. 3.40(c). Amplitude and phase angle as function of frequency are derived below.

$$\bar{A}_H = \frac{\bar{V}_L}{\bar{V}_S} = \left[\frac{N_2}{N_1}\right]\frac{R'_L}{(R+R'_L)+j\omega L}$$

where $L = L_{l1} + L_{l2}$ = total leakage inductance as seen on primary side.

Further rearrangement leads to

$$\bar{A}_H = \left[\frac{N_2}{N_1}\right]\left[\frac{R'_L}{R+R'_L}\right]\frac{1}{1+j\omega[L/(R+R'_L)]}$$

We can write

$$\frac{R+R'_L}{L} = \omega_H = \textit{corner frequency} \text{ of high-frequency region}$$

Also recognizing $\left[\frac{N_2}{N_1}\right]\left[\frac{R'_L}{R+R_L}\right] = A_0$, we get

$$A_H = \frac{A_0}{1+j\omega/\omega_H} \tag{3.79}$$

or

$$A_H = \frac{A_0}{[1+(\omega/\omega_H)^2]^{1/2}}\,;\ \angle A_H = \tan^{-1}\,\omega/\omega_H \tag{3.80}$$

As per Eq. (3.80) the gain falls with frequency acquiring a valve of $A_0/\sqrt{2}$ at $\omega/\omega_H = 1$ and a phase angle of $\angle -45°$. This indeed is the half power frequency (ω_H).

Low-frequency Region

In this region the series effect of leakage inductances is of no consequence but the low reactance (ωL_m) shunting effect must be accounted for giving the approximate equivalent circuit of Fig 3.39(d). Amplitude and phase angle of frequency response is derived below.

The corner frequency ω_L of this circuit is obtained by considering the voltage source as short circuit. This circuit is L_m in parallel with $R \| R'_L$. Thus

$$\omega_L = \left[\frac{R\,\|\,R'_L}{L_m}\right]$$

The complex gain can then be expressed as

$$\bar{A}_L = \frac{A_0}{1+j(\omega/\omega_L)} \tag{3.81}$$

or

$$A_L = \frac{A_0}{[1+(\omega/\omega_L)^2]^{1/2}}\,;\ \angle\bar{A}_L = \tan^{-1}\,\omega/\omega_L \tag{3.82}$$

Again the lower corner frequency is the half power frequency.

The complete amplitude and phase response of the transformer (with source) on log frequency scale are plotted in Fig. 3.41. At high frequencies the interturn and other stray capacitances of the transformer windings begin to play a role. In fact the capacitance-inductance combination causes parallel resonance effect on account of which an amplitude peak shows up in the high-frequency region of the frequency response. No reasonably accurate modelling of these effects is possible and best results are obtained experimentally. The frequency response of Fig. 3.41 gives a general guidance as to its nature.

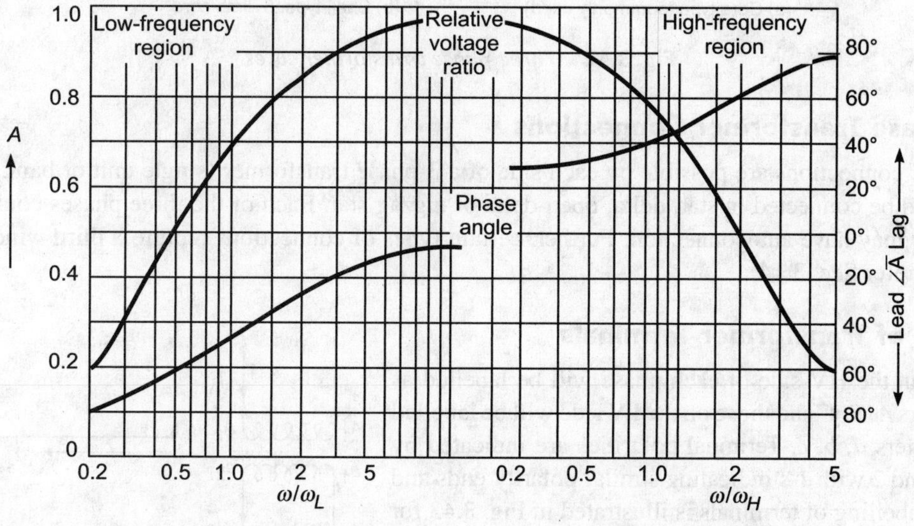

Fig. 3.41 *Normalized frequency characteristic of output transformers*

3.13 | THREE-PHASE TRANSFORMERS

LO 7

Important! In generation, transformation, transmission and utilization of electric energy it can be shown that it is economical to use the three-phase system rather than the single-phase. For three-phase transformation, three single-phase transformers are needed. Two arrangements are possible: a bank of three single-phase transformers or a single three-phase transformer with the primary and secondary of each phase wound on three legs of a common core. The three-phase transformer unit costs about 15% less than that of a bank and furthermore, the single unit occupies less space. There is little difference in reliability, but it is cheaper to carry spare stock of a single-phase rather than a three-phase transformer. In underground use (mines) a bank of single-phase units may be preferred as it is easier to transport these units. The bank also offers the advantage of a derated open-delta operation when one single-phase unit becomes inoperative. Reduced cost being an overweighing consideration, it is common practice to use a three-phase transformer unit.

Core Types of Three-Phase Transformers

In a three-phase bank the phases are electrically connected but the three magnetic circuits are independent. In the more common three-phase, 3-limb core-type transformer (Fig. 3.42(a)), the three magnetic circuits are also linked. Where delinking of the magnetic circuits is desired in a three-phase unit, a 5-limb shell type transformer could be used (Fig. 3.42(b)).

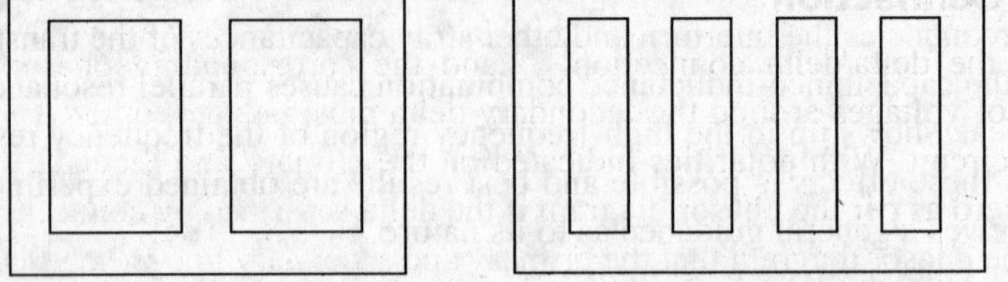

(a) Core type (commonly used) (b) Shell type, 5 -limb core

Fig. 3.42 *Three-phase transformer cores*

Three-phase Transformer Connections

A variety of connections are possible on each side of a 3-phase transformer (single unit or bank). The three phases could be connected in star, delta, open-delta or zigzag star. Each of the three phases could have two windings or may have autoconnection. Further, certain types of connections require a third winding known as *tertiary* (refer Sec. 3.14).

Labelling of Transformer Terminals

Terminals on the HV side of each phase will be labelled as capital letters A, B, C and those on the LV side will be labelled as small letters a, b, c. Terminal polarities are indicated by suffixes 1 and 2 with 1's indicating similar polarity ends and so do 2's Labelling of terminals is illustrated in Fig. 3.43 for phase a. Assuming the transformer to be ideal, $V_{A_2A_1}$ (voltage of terminal A_2 with respect to A_1) is in phase with $V_{a_2a_1}$ and I_A is in phase with I_a.

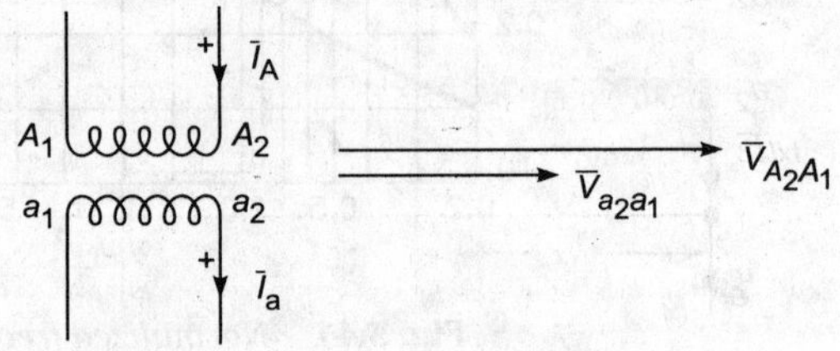

Fig. 3.43 *Labelling of transformer terminals*

Star/Star (Y/Y) Connection

Star connection is formed on each side by connecting together phase winding terminals suffixes 1 as in Fig. 3.44(a). The phasor diagram is drawn in Fig. 3.44(b) from which it is easily seen that the voltages of the corresponding phases (and therefore of the corresponding lines) are in phase. This is known as the *0°-connection*. The letters within brackets on the phasor diagram indicate the lines, to which the terminals are connected. If the winding terminals on secondary side are reversed, the *180°-connection* is obtained.

It is also observed from Fig. 3.44 that if the phase transformation ratio is $x : 1$, the line transformation (line-to-line voltages, line currents) the ratio is also $x : 1$.

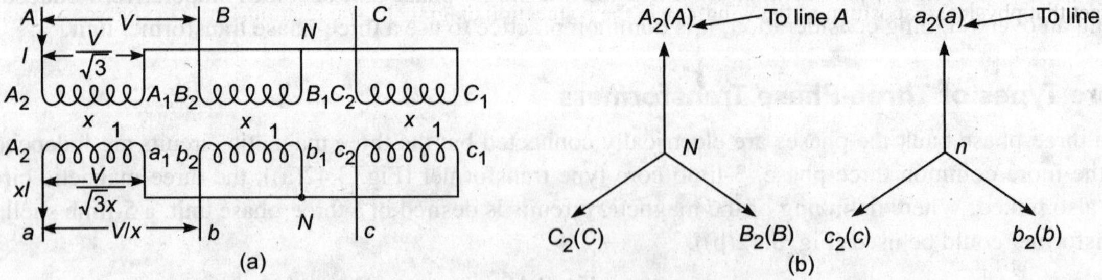

Fig. 3.44 *Star/star 0°-connection*

Delta/Delta (Δ/Δ) Connection

Figure 3.45(a) shows the delta/delta connection†††† and the corresponding phasor diagram is given in Fig. 3.45(b). The sum of voltages around the secondary delta must be zero; otherwise delta, being a closed circuit, means a short circuit. With polarities indicated on the primary and secondary sides, voltages V_{a2a1}, V_{b2b1} and V_{c2c1} add to zero as per the phasor diagram if the delta is formed by connecting a_1b_2, b_1c_2 and c_1a_2. It is easily seen from the phasor diagram that the primary and secondary line voltages are in phase so it is the *0°-connection*. However, if the secondary leads *a*, *b*, *c* are taken out from the delta nodes a_1b_2, b_1c_2, c_1c_2, the secondary voltages are in phase opposition to the primary voltages can be visualized from the phasor diagram of Fig. 3.45(b). This is the *180°-connection*.

It is also seen from Fig. 3.45(a) that if the phase transformation ratio is $x : 1$, the transformation ratio for line quantities is also $x : 1$.

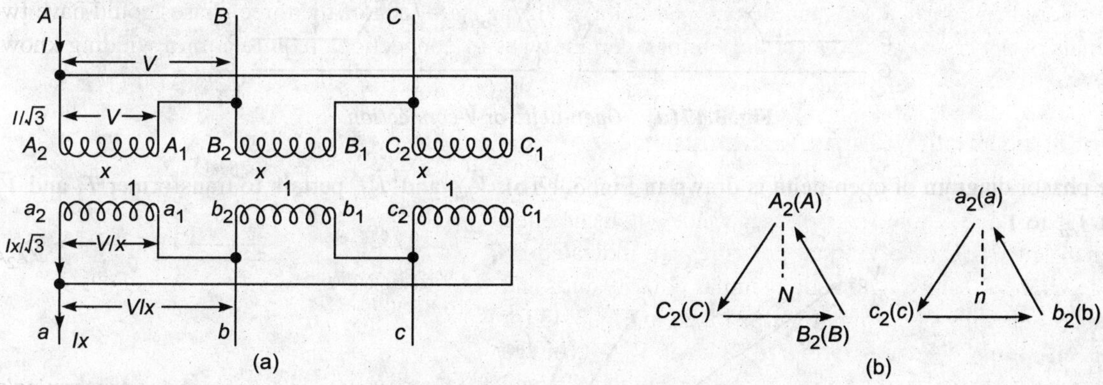

Fig. 3.45 *Delta/delta connection*

Open-Delta Connection

In the delta/delta connection if one of the transformers is disconnected, the resulting connection is known as *open-delta*. Supposing the *b*-phase transformer in Fig. 3.47(a) is removed, and the open-delta is excited from balanced 3-phase supply, then it easily follows from the phasor diagram of Fig. 3.47(b) that the voltage V_{b2b1} $= V_{bc}$ does not change as it equals $-(V_{ca} + V_{ab})$; thus the voltages on the secondary side still remain balanced

†††† The star and delta connections in later parts of the book will generally be indicated as in Fig. 3.46. The style temporarily adopted here is for the sake of clarity of identifying the primary and secondary of each phase. Furthermore, it also stresses the fact that physical disposition of the windings in the connection diagram has no relationship to the phasor diagram.

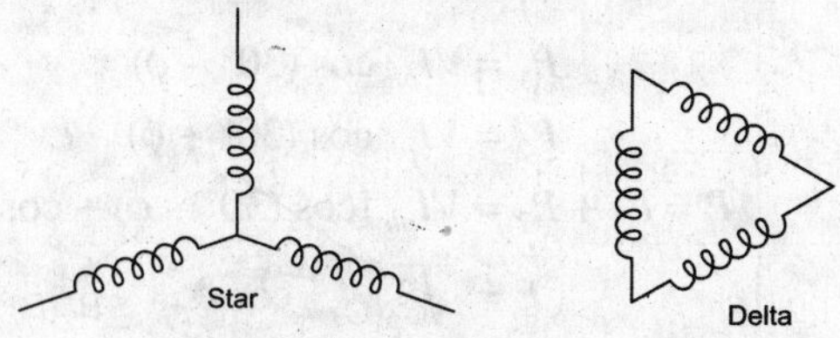

Fig. 3.46

3-phase. The open-delta connection supplying a balanced load is shown in Fig. 3.47(a). If the maximum allowable secondary phase current is I_{ph}, the transformer can handle VA of

$$S_{\text{open-delta}} = \sqrt{3}\, VI_{ph}\, ; I_{ph} = I_{\text{line}}$$

which for normal delta/delta connection is

$$S_{\text{delta}} = 3\, VI_{ph}$$

Thus the open-delta connection has a VA rating of $1/\sqrt{3} = 0.58$ of the rating of the normal delta/delta connection.

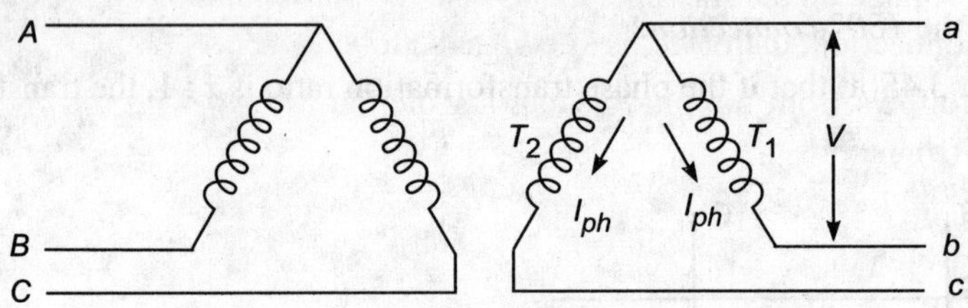

Fig. 3.47(a) *Open-delta or V-connection*

The phasor diagram of open-delta is drawn in Fig. 3.47(b). $\bar{V}_{ba}$ and $\bar{I}_{ab}$ pertain to transformer T_1 and $\bar{V}_{ca}$ and $\bar{I}_{ca}$ to T_2.

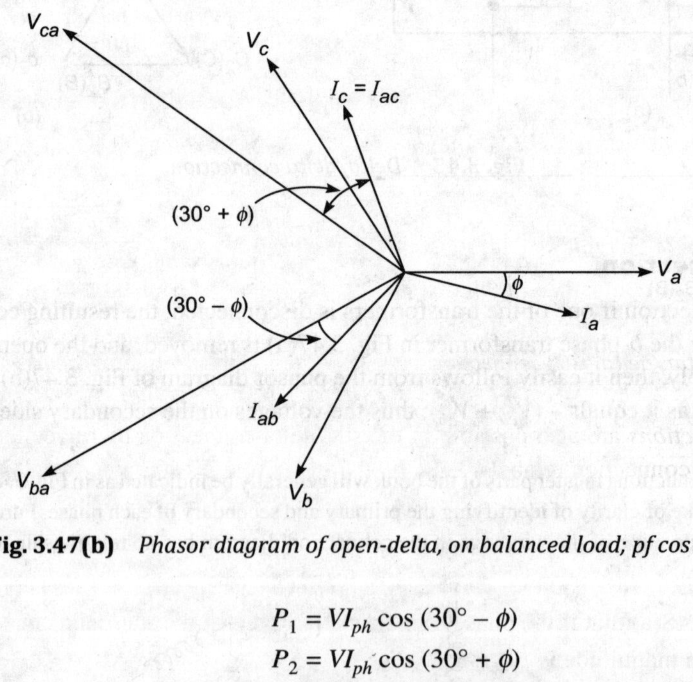

Fig. 3.47(b) *Phasor diagram of open-delta, on balanced load; pf cos ϕ*

Power output of T_1 $\quad P_1 = VI_{ph} \cos (30° - \phi)$

and that of T_2 $\quad P_2 = VI_{ph} \cos (30° + \phi)$

Total power delivered $\quad P = P_1 + P_2 = VI_{ph} [\cos (30° - \phi) + \cos (30° + \phi)]$

Upon simplification, we find $\quad P = \sqrt{3}\, VI_{ph} \cos \phi$

The two transformers supply equal power at upf, i.e., $\phi = 0$.

Star/Delta (Y/Δ) Connection

Star connection is formed on primary side by connecting together 1 suffixed terminals; 2 suffixed terminals being connected to appropriate lines; the delta is formed by connecting c_1a_2, a_1b_2 and b_1c_2 with the lines connected to these junctions being labelled as a, b and c respectively as shown in Fig. 3.48(a). The phasor diagram is drawn in Fig. 3.48(b). It is seen from the phasor diagram on the delta side that the sum of voltages around delta is zero. This is a must as otherwise closed delta would mean a short circuit. It is also observed from the phasor diagram that phase a to neutral voltage (equivalent star basis) on the delta side lags by –30° to the phase-to-neutral voltage on the star side; this is also the phase relationship between the respective line-to-line voltages. This connection, therefore, is known as *–30°-connection*.

The *+30°-connection* follows from the phasor diagram of Fig. 3.49(a) with the corresponding connection diagram as in Fig. 3.49(b).

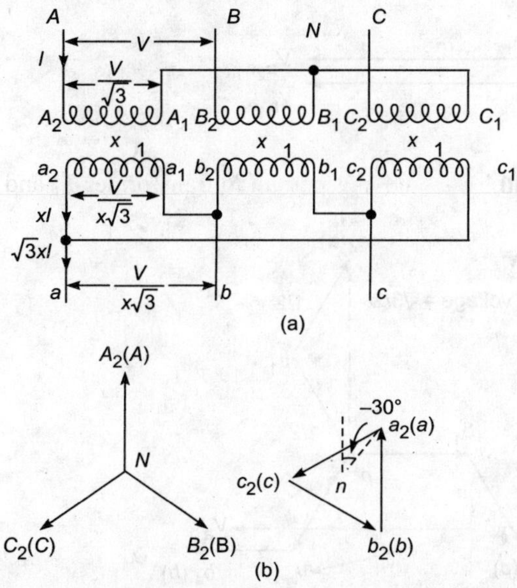

Fig. 3.48 *Star/delta –30°-connection*

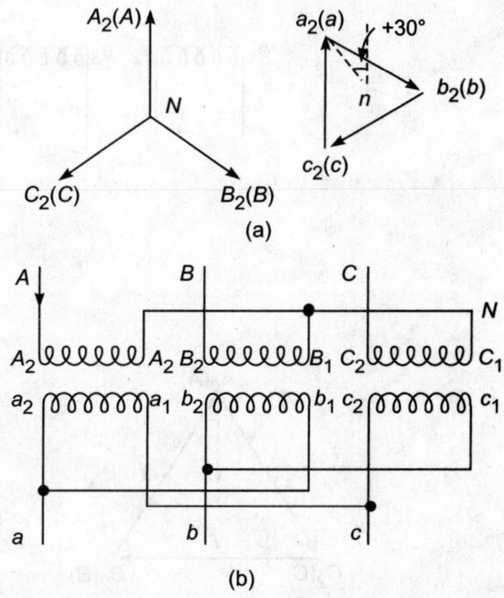

Fig. 3.49 *Star/delta +30°-connection*

Similarly ±*90°-connections* are also possible in the star/delta connection by relabelling the delta side lines. For example for +90° connection relabel $c \rightarrow a$, $b \rightarrow c$ and $a \rightarrow b$. Reader may work out relabelling for –90° connection.

In Indian and British practices ± *30°-connections* are employed. The American practice is to use ±90°-connections.

It follows from Fig. 3.48(a) that if the phase transformation ratio of the star/delta connection is x : 1, the line transformation ratio in magnitude is $\sqrt{3}x$: 1.

Delta/Star (Δ/Y) Connection

This connection is simply the interchange of primary and secondary roles in the star/delta connection. One just interchanges capital and small letter suffixings in Figs 3.48 and 3.49. Of course what was the

–30°-connection will now be the +30°-connection and vice versa. If the phase transformation ratio is $x : 1$ (delta/star), the transformation ratio for line quantities will be $(x/\sqrt{3}) : 1$.

Delta/Zig-zag Star Connection

The winding of each phase on the star side is divided into two equal halves with labelling as in Fig. 3.50. Each leg of the star connection is formed by using halves from two different phases. The phasor diagram for this connection is given in Fig. 3.51 from which the connection diagram easily follows. Obviously it is the 0°-connection. Reversal of connections on the star side gives us the 180°-connection.

$$\text{Phase transformation} = x : 1$$

$$\text{Line transformation} = x : \frac{\sqrt{3}}{2} \text{ or } \frac{2}{\sqrt{3}}x : 1$$

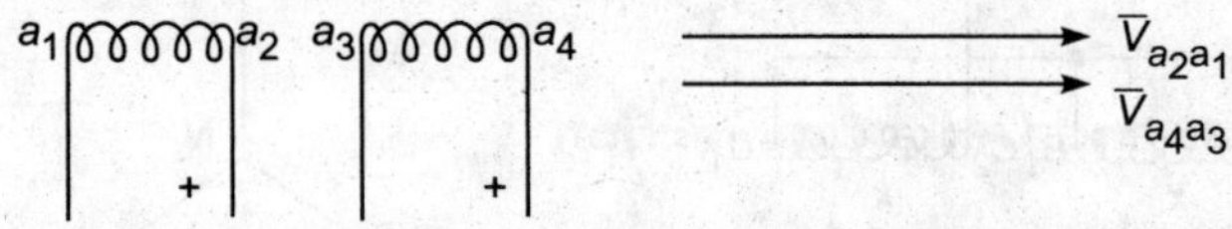

Fig. 3.50

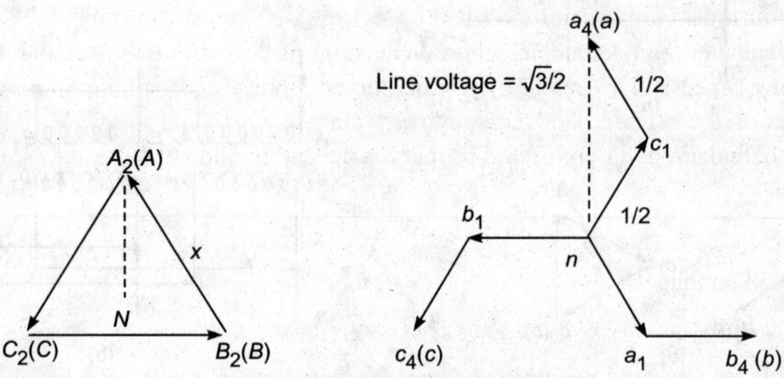

Fig. 3.51 *Delta/zig-zag star 0°-connection*

Star/Zig-zag Star

The connection is indicated by the phasor diagram of Fig. 3.52.

$$\text{Phase transformation} = x : 1$$

$$\text{Line transformation} = \sqrt{3}\,x : \frac{3}{2} \text{ or } \frac{2}{\sqrt{3}}x : 1$$

Phase Groups

Various transformer connections with the same phase shift are grouped together. Thus there are Group I (0°), Group II (180°), Group III (30°) and Group IV (–30°).

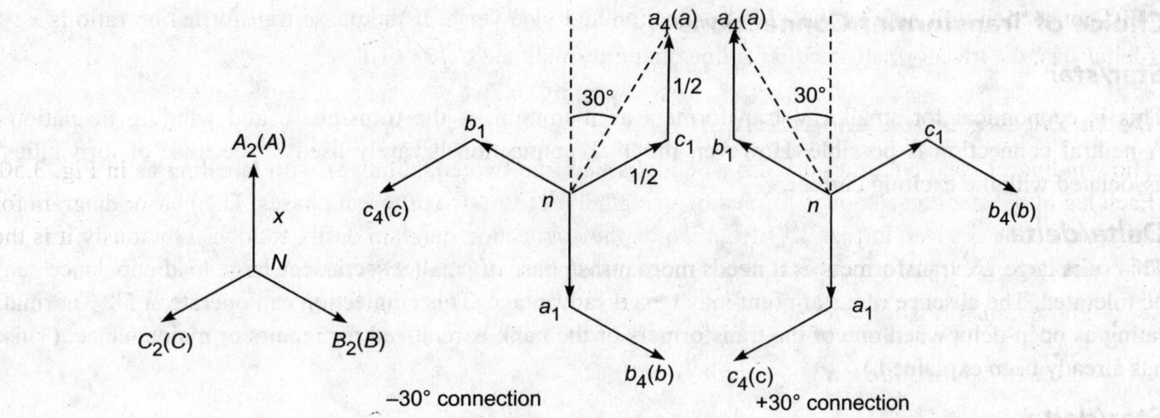

Fig. 3.52 *Star/zig-zag star*

Choice between Star and Delta Connection

Recall In star connection with earthed neutral, the maximum voltage of the phase winding to ground is $1/\sqrt{3}$ or 58% of the line voltage, while in delta connection this is equal to the line voltage in case of earthing of one of the lines during a fault. Therefore, for very high voltage transformers the star connection on the HV side is about 10% cheaper than delta connection on account of insulation cost. A delta-connected primary is necessary for a star-connected LV secondary feeding mixed 3-phase and 1-phase (line-to-neutral) loads. This is because the lines on the primary side can only carry current which add to zero. In the case of unbalanced 1-phase loads on secondary, delta-connected primary is needed to allow the flow of *zero sequence* current

$$\bar{I}_0 = \frac{\bar{I}_n}{3} = \frac{1}{3}(\bar{I}_a + \bar{I}_b + \bar{I}_c)$$

as shown in Fig. 3.53 so that

$$\bar{I}_A + \bar{I}_B + \bar{I}_C = 0$$

This means that only *positive* and *negative sequence* currents flow in the lines on the delta side.

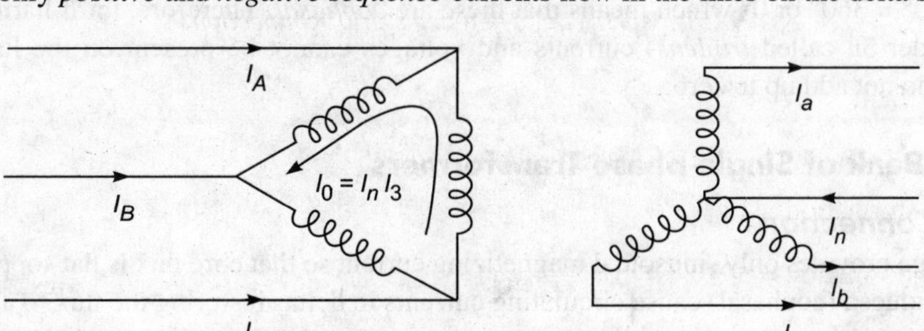

Fig. 3.53

This could also be achieved by star-connected primary provided the primary and secondary star points are grounded. But this is not recommended on account of flow of ground current for unbalanced secondary loads.

Choice of Transformer Connections

Star/star

This is economical for small HV transformers as it minimizes the turns/phase and winding insulation. A neutral connection is possible. However, the *Y* /*Y* connection is rarely used[‡‡‡‡] because of difficulties associated with the exciting current.

Delta/delta

This suits large LV transformers as it needs more turns/phase of smaller section. A large load unbalance can be tolerated. The absence of a star point may be a disadvantage. This connection can operate at 58% normal rating as open-delta when one of the transformers of the bank is removed for repairs or maintenance. (This has already been explained.)

Star/delta

This is the most commonly used connection for power systems. At transmission levels star connection is on the HV side, i.e. Δ/Y for step-up and Y/Δ for step-down. The neutral thus available is used for grounding on the HV side. At the distribution level the Δ/Y transformer is used with star on the LV side which allows mixed 3-phase and 1-phase loads, while delta allows the flow of circulating current to compensate for neutral current on the star side (Fig. 3.53).

The Y/Δ connection has an associated phase shift of ± 30° which must be accounted for in power system interconnections.

Harmonics

It was seen in Sec. 3.10 that when the third-harmonic current is permitted to flow, by circuit conditions, along with the sinusoidal magnetizing current in a transformer, the core flux is sinusoidal and so is the induced emf. On the other hand, when the circuit does not permit the flow of the third-harmonic current, i.e. the magnetizing current is sinusoidal, the flux is flat-topped containing "depressing" third-harmonic and as a consequence third-harmonic voltages are present in the induced emfs. This problem in 3-phase transformers will now be examined.

It is to be observed here that the phase difference in third-harmonic currents and voltages on a 3-phase system is $3 \times 120° = 360°$ or 0° which means that these are *cophasal.* Therefore, third-harmonic (in general harmonics of order $3n$ called *triplens*) currents and voltages cannot be present on the lines of a 3-phase system as these do not add up to zero.

Three-phase Bank of Single-phase Transformers

Delta/delta Connection

The supply voltage provides only sinusoidal magnetizing current so that core flux is flat-topped; but the third-harmonic emfs induced (cophasal) cause circulating currents in deltas restoring the flux to almost sinusoidal. The third-harmonic voltages are very small as the transformer offers low impedance to third-harmonic currents.

Star/delta and delta/star Connection

Because of one delta connection the same conditions are obtained as in Δ/Δ connection except that the impedance offered to the flow of third-harmonic currents in delta is now larger and so are third-harmonic voltages.

[‡‡‡‡] Recently a favourable trend is developing for reasons of economy.

Star/star Connection

In the case of isolated neutrals, third-harmonic voltages are present in each phase as explained earlier. Further, since these voltages are cophasal, no third-harmonic voltages are present between lines. The voltage of phase *a* to neutral can now be expressed as

$$e_{aN} = e_a \sin \omega t + e_{a3} \sin 3\omega t$$

While fundamental frequency voltages in the three phases have a relative phase difference of 120°, the third-harmonic voltages in them are cophasal (with respect to each other), but their phase with respect to the fundamental frequency (voltage changes at the rate of 2 ω, twice the fundamental frequency). This situation is illustrated in the phasor diagram of Fig. 3.54 from which it is immediately observed that the voltage of the neutral point oscillates at frequency 2ω. The phenomenon is known as *oscillating neutral* and is highly undesirable because of which the star/star connection with isolated neutrals is not used in practice.

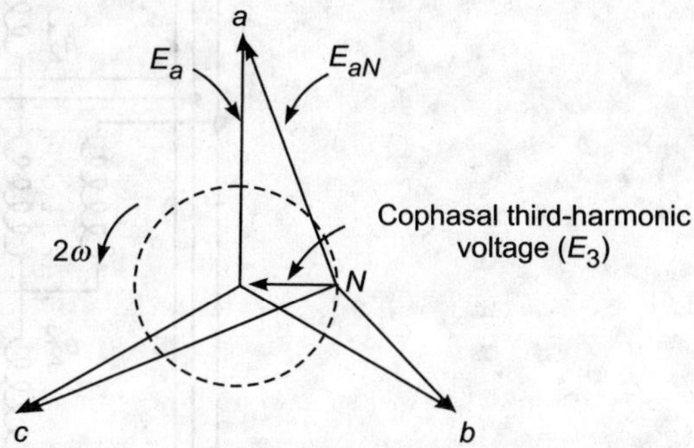

Fig. 3.54 *Oscillating neutral*

If the neutrals are connected, it effectively separates the three transformers. Third-harmonic currents can now flow via the neutrals.

Three-phase Transformer

In core type transformer (Fig. 3.42(a)), the third-harmonic fluxes in all the three limbs are simultaneously directed upwards or downwards so that this flux must return through air (high-reluctance path). The high-reluctance path tends to suppress the third-harmonic flux. The phenomenon gets more complex now and at core densities exceeding 1.5 T, the total harmonic content (particularly fifth) is very marked in the magnetizing current (fifth harmonic currents can flow on lines as their relative phase difference is 5 × 120° = 600° or 120°).

To reduce the strong fifth harmonic in the magnetizing current for the star/star connection with isolated neutral, a path must be provided through iron for the third-harmonic flux. Hence, the use of a 5-limb core as in Fig. 3.42(b).

Back-to-Back Test on Three-phase Transformers

Fig. 3.55(a) shows the connection arrangement for the back-to-back test on two identical 3-phase transformers. The two secondaries must be connected in phase opposition and in proper phase sequence. The auxiliary transformer for circulating full-load current is included in the circuit of the two secondaries; it could also be included in the circuit of the primaries. Thus with only losses (core-loss and copper-loss) supplied from the mains, a "heat run" test could be conducted on the transformers.

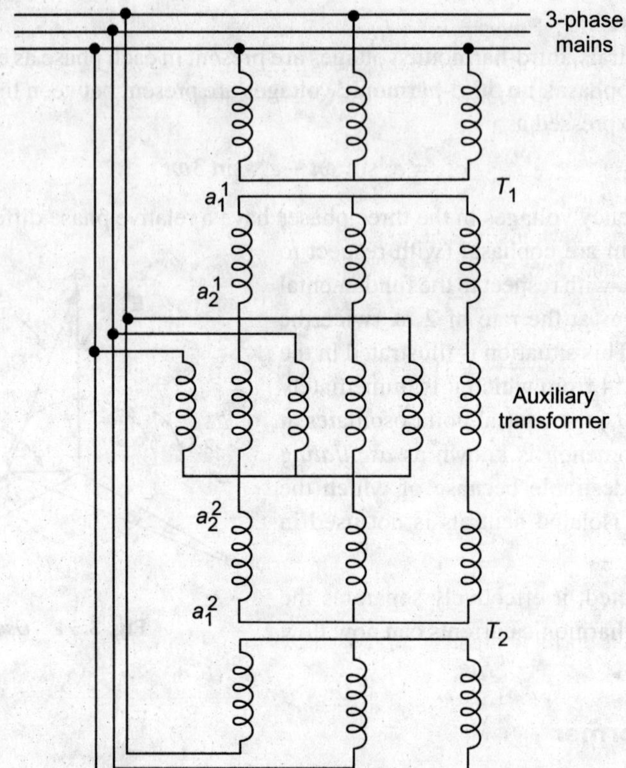

Fig. 3.55(a) *Back-to-back test on 3-phase transformers*

Delta/Delta Connected Transformers

The primaries are normally excited from the mains. Each secondary delta is opened at one junction and a single-phase transformer can be employed to circulate full-load current in both the deltas as shown in Fig. 3.55(b).

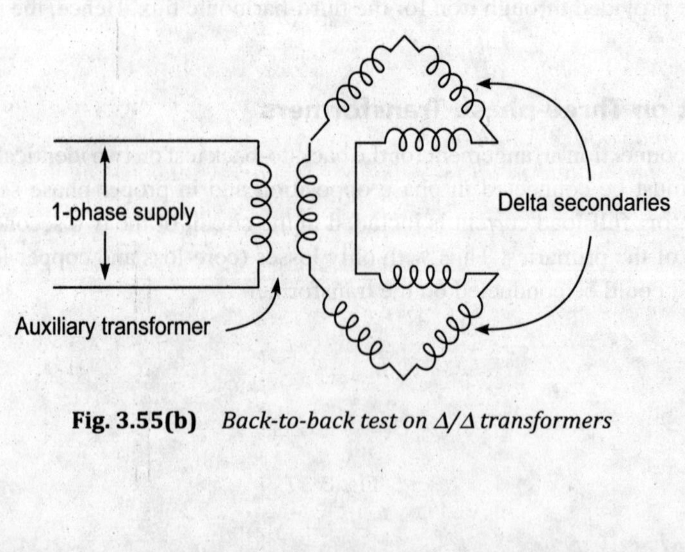

Fig. 3.55(b) *Back-to-back test on Δ/Δ transformers*

Example 3.18 *A 3-phase transformer bank consisting of three 1-phase transformers is used to step-down the voltage of a 3-phase, 6600 V transmission line. If the primary line current is 10 A, calculate the secondary line voltage, line current and output kVA for the following connections: (a) Y/Δ and (b) Δ/Y. The turns ratio is 12. Neglect losses.* LO 7

Solution

(a) The Y/Δ connection is drawn in Fig. 3.56(a).

$$V_{PY} = \frac{6600}{\sqrt{3}}$$

$$V_{P\Delta} = V_{L\Delta} = \frac{6600}{\sqrt{3} \times 12} = 317.55 \text{ V}$$

$$I_{P\Delta} = 10 \times 12 = 120 \text{ A}$$

$$I_{L\Delta} = 120\sqrt{3} = 207.84 \text{ A}$$

$$\text{Output kVA} = \sqrt{3} \times \frac{6600}{\sqrt{3} \times 12} \times 120\sqrt{3} \times \frac{1}{1000}$$

$$= 66\sqrt{3} = 114.3$$

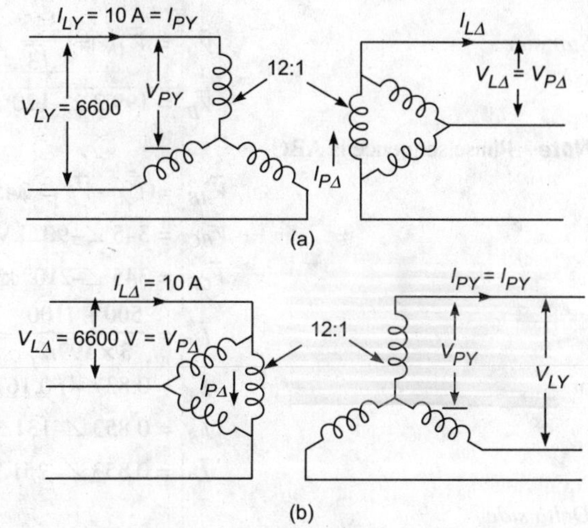

Fig. 3.56

(b) The Δ/Y connection is drawn in Fig. 3.56(b).

$$I_{P\Delta} = \frac{10}{\sqrt{3}} \text{ A}$$

$$I_{LY} = \frac{12 \times 10}{\sqrt{3}} = 69.28 \text{ A}$$

$$V_{PY} = \frac{6600}{12} \text{ V}$$

$$V_{LY} = \frac{6600\sqrt{3}}{12} \quad 925.6 \text{ V}$$

$$\text{Output kVA} = \sqrt{3} \times \frac{6600\sqrt{3}}{12} \times \frac{120}{\sqrt{3}} \times \frac{1}{1000} = 114.3 \text{ (same as in part (a))}$$

Example 3.19 *A Δ/Y connected 3-phase transformer as shown in Fig. 3.57 has a voltage ratio of 22 kV (Δ)/345 kV(Y) (line-to-line). The transform is feeding 500 MW and 100 MVAR to the grid (345 kV). Determine the MVA and voltage rating of each unit (single-phase). Compute all currents and voltages of both magnitude and phase angle in all the windings (primaries and secondaries). Assume each single-phase transformer to be ideal.* LO 7

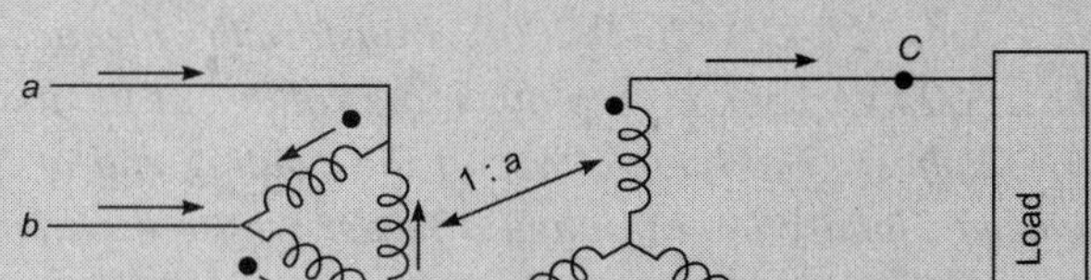

Fig. 3.57

Solution

$$\text{Load MVA, } \bar{S} = 500 + j\,100$$

$$S = 510$$

$$\text{MVA rating of each (single phase) transformer} = 510/3 = 170$$

$$\text{Voltage rating of each transformer} = \frac{345/\sqrt{3}}{22} = 9.054$$

Let us choose voltage of star phase A as reference then

Star side

$$\bar{V}_A = \bar{V}_{AN} = \frac{345}{\sqrt{3}} \angle 0° = 199.2 \angle 0° \text{ kV}$$

$$\bar{V}_B = 199.2 \angle -120° \text{ kV}, \; \bar{V}_C = 199.2 \angle -240° \text{ kV}$$

Note Phase sequence is ABC

$$\bar{V}_{AB} = \bar{V}_A - \bar{V}_B = 345 \angle 30° \text{ kV}$$

$$\bar{V}_{BC} = 345 \angle -90° \text{ kV}$$

$$\bar{V}_{CA} = 345 \angle -210° \text{ kV}$$

$$\bar{I}_A^* = \frac{500 + j100}{3 \times 199.2} = 0.837 + j\,0.167 \text{ kA; as } \bar{S} = VI^*$$

or

$$\bar{I}_A = 0.837 - j\,0.167 = 0.853 \angle -11.3° \text{ kA}$$

$$\bar{I}_B = 0.853 \angle -131.3° \text{ kA}$$

$$\bar{I}_C = 0.853 \angle -251.3° \text{ kA}$$

Delta side

$$\bar{V}_{ab} = \frac{\bar{V}_A}{a} = \frac{199.2}{9.054} \angle 0° = 22 \angle 0° \text{ kV}$$

$$\bar{V}_{bc} = 22 \angle -120° \text{ kV}$$

$$\bar{V}_{ca} = 22 \angle -240° \text{ kV}$$

$$\bar{I}_{ab} = 9.054 \times 0.853 \angle -11.3° = 7.723 \angle -11.3° \text{ kA}$$

$$\bar{I}_{bc} = 7.723 \angle -131.3° \text{ kA}$$

$$\bar{I}_{ca} = 7.723 \angle -251.3° \text{ kA}$$

$$\bar{I}_a = \bar{I}_{ab} - \bar{I}_{bc} = \sqrt{3} \times 7.723 \angle(-11.3° - 30°) = 13.376 \angle -41.3°$$

$$\bar{I}_b = 13.376 \angle(-120° - 11.3°) = 13.376 \angle -131.3° \text{ kA}$$

$$\bar{I}_c = 13.376 \angle(-240° - 11.3°) = 13.376 \angle -251.3° \text{ kA}$$

Note It is easily observed from above that line voltages and currents on the star side lead those on the delta side by 30°.

Example 3.20 *Three 1-phase 20-kVA, 2000/200-V transformers identical with that of Ex. 3.3 are connected in Y/Δ in a 3-phase, 60 kVA bank to step-down the voltage at the load end of a feeder having impedance of 0.13 + j 0.95 Ω/phase. The line voltage at the sending-end of the feeder is 3464 V. The transformers supply a balanced 3-phase load through a feeder whose impedance is 0.0004 + j 0.0015 Ω/phase. Find the load voltage (line-to-line) when it draws rated current from transformers at 0.8 lagging power factor.*

LO 7

Solution Figure 3.58 gives the circuit diagram of the system. The computations will be carried out on per phase-Y basis by referring all quantities to the HV (Y-connected) side of the transformer bank.

LV feeder impedance referred to the HV side is

$$\left(\frac{2000\sqrt{3}}{200}\right)^2 (0.0004 + j\,0.0015) = 0.12 + j\,0.45\ \Omega/\text{phase}$$

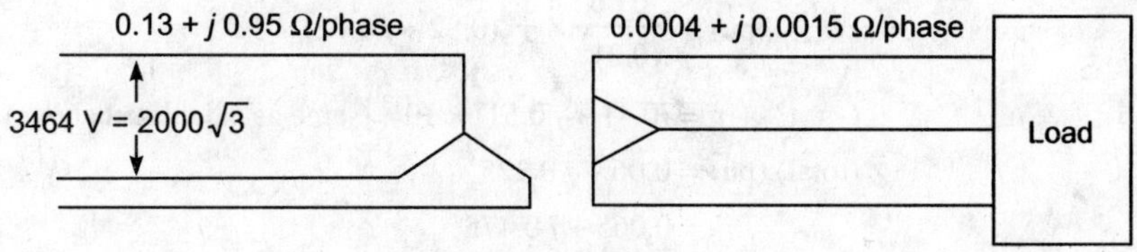

Fig. 3.58

The total series impedance of the HV and LV feeders referred to the HV side is

$$Z_F = (0.13 + j\,0.95) + (0.12 + j\,0.45)$$

$$= 0.25 + j\,1.4\ \Omega/\text{phase}$$

From Ex. 3.5, the equivalent impedance of the transformer bank is referred to the HV side

$$Z_T = 0.82 + j\,1.02\ \Omega/\text{phase Y}$$

$$\text{Sending-end feeder voltage} = \frac{3464}{\sqrt{3}} = 2000\ \text{V/phase Y}$$

Load current on the HV side = rated current of transformer

= 10 A/phase Y

It is now seen that the equivalent circuit for one phase referred to the Y-connected HV side is exactly the same as in Ex. 3.6, Fig. 3.20. Thus the load voltage referred to the HV side is 197.692 V to neutral. The actual load voltage is 197.688 V, line-to-line (since the secondaries are Δ-connected).

PU method In such problems it is convenient to use the pu method. We shall choose the following base values:

$$(MVA)_B = \frac{3 \times 20}{1000} = 0.06$$

Voltage base on HV side = $2\sqrt{3}$ kV (line-to-line)

Voltage base on LV side = 0.2 kV (line-to-line)

Note The voltage base values are in the ratio of line-to-line voltages (same as phase voltages on equivalent star basis).

$$\bar{Z}_1\ (\text{LV line})\ (\text{pu}) = (0.0004 + j\,0.0015) \times \frac{0.06}{(0.2)^2} = (0.06 + j\,0.225) \times 10^{-2}$$

$$\bar{Z}_2\ (\text{HV line})\ (\text{pu}) = (0.13 + j\,0.95) \times \frac{0.06}{(2\sqrt{3})^2} = (0.065 + j\,0.475) \times 10^{-2}$$

$$\bar{Z}_T\ (\text{star side})\ (\text{pu}) = (0.82 + j\,1.02) \times \frac{0.06}{(2\sqrt{3})^2} = (0.41 + j\,0.51) \times 10^{-2}$$

Note Suppose the transformer impedance was given on the delta side

$$\bar{Z}_T \text{ (delta side)} = \left(\frac{200}{2000}\right)^2 \times (0.82 + j\,1.02)$$

$$= (0.82 + j\,1.02) \times 10^{-2}\ \Omega \text{ (delta phase)}$$

$$\text{Equivalent star impedance} = \frac{1}{3}(0.82 + j\,1.02) \times 10^{-2}\ \Omega$$

$$\bar{Z}_T \text{ (pu)} = \frac{0.06}{(0.2)^2} \times \frac{1}{3}(0.82 + j\,1.02) \times 10^{-2}$$

$$= (0.41 + j\,0.51) \times 10^{-3} \text{ (same as calculated above)}$$

$$\bar{Z} \text{ (total) (pu)} = 0.06 + j\,0.225$$

$$0.065 + j\,0.475$$

$$0.41 + j\,0.51$$

$$\overline{(0.535 + j1.21) \times 10^{-2}}$$

$$V_1 \text{ (sending–end voltage)} = 2\sqrt{3}\ \text{kV (line) or 1 pu}$$

$$I_1 \text{ (= rated current)} = 1 \text{ pu; pf= 0.8 lag}$$

$$V_2 \text{ (load voltage)} = 1 - 1 \times (0.535 \times 0.8 + 1.2 \times 0.6) \times 10^{-2}$$

$$= 0.98846 \text{ pu}$$

$$= 0.98846 \times 200 = 197.692 \text{ V (line)}$$

Example 3.21 *Three transformers, each rated 20 kVA, 2 kV/200 V, are connected Δ/Δ and are fed through a 2000 V (line-to-line) 3-phase feeder having a reactance of 0.7 Ω/phase. The reactance of each transformer is 0.0051 pu. All resistances are to be ignored. At its sending-end the feeder receives power through the secondary terminals of a 3-phase Y/Δ connected transformer whose 3-phase rating is 200 kVA, 20/2 kV (line-to-line). The reactance of the sending-end transformer is 0.06 pu on its own rating. The voltage applied to the primary terminals is 20 kV (line-to-line).*

A 3-phase fault takes place at the 200 V terminals of the receiving-end transformers. Calculate the fault current in the 2 kV feeder lines, in the primary and secondary windings of the receiving-end transformers and at the load terminals (200 V terminals).

LO 7

Solution Choose a common 3-phase base of 60 kVA. Line-to-line voltage bases are in ratio of transformation 20 kV : 2 kV : 200 V. It is observed that

$$X_T \text{ (sending-end}^{\S\S\S\S}) = 0.06 \times \frac{60}{200} = 0.018 \text{ pu}$$

For the 2 kV feeder

$$V_B = \frac{2000}{\sqrt{3}} = 1154.7 \text{ V (line-to-neutral)}$$

$$I_B = \frac{60 \times 1000}{\sqrt{3} \times 2000} = 17.32 \text{ A, phase Y}$$

§§§§ Transformer impedance in pu is independent of the connection.

$$Z_B = \frac{1154.7}{17.32} = 66.6\ \Omega\text{/phase Y}$$

$$X_{\text{feeder}} = \frac{0.7}{66.6} = 0.0105 \text{ pu}$$

$$X_T\ (\text{receiving-end}^{\P\P\P\P}) = 0.0051 \text{ pu}$$

Total reactance from the sending-end to the fault point (on the secondary side of the receiving-end transformer) = 0.018 + 0.0105 + 0.0051 = 0.0336 pu

$$\text{Sending-end voltage} = \frac{20}{20} = 1.0 \text{ pu}$$

$$\text{Fault current} = \frac{1.0}{0.0336} = 29.76 \text{ pu}$$

The current in any part of the system can be easily computed as below:

$$\text{Current in 2 kV feeder} = 29.76 \times 17.32 = 515.4 \text{ A}$$

$$\text{Current in 2 kV winding of } \Delta/\Delta \text{ transformer} = 515.3\sqrt{3} = 297.56 \text{ A}$$

$$\text{Current in 200 V winding of } \Delta/\Delta \text{ transformer} = 297.56 \times 10 = 2975.6 \text{ A}$$

$$\text{Current at load terminals} = 2975.6\sqrt{3} = 5154 \text{ A}$$

Example 3.22 *A 3-phase bank of three single-phase transformer are fed from 3-phase 33 kV (line-to-line). It supplies a load of 6000 kVA at 11 kV (line-to-line). Both supply and load are 3-wire. Calculate the voltage and kVA rating of the single-phase transformer for all possible 3-phase transformer connection.*

LO 7

Solution

1. Star-Star connection

$$\text{Primary-side phase voltage, } V_{P1} = \frac{33}{\sqrt{3}} = 19.05 \text{ kV}$$

$$\text{Secondary-side phase voltage, } V_{P2} = \frac{11}{\sqrt{3}} = 6.35 \text{ kV}$$

$$\text{Transformer voltage rating} = 19.05/6.35 \text{ kV}$$

$$\text{kVA rating} = \frac{6000}{3} = 2000$$

2. Star-Delta connection

$$V_{P1} = 19.05 \text{ kV},\ V_{P2} = 11 \text{ kV}$$

$$\text{Transformer rating} = 19.05/11 \text{ kV, 2000 kVA}$$

3. Delta-Star connection

$$\text{Transformer rating} = 33/6.35 \text{ kV, 2000 kVA}$$

4. Delta-Delta connection

$$\text{Transformer rating} = 33/11 \text{ kV, 2000 kVA}$$

¶¶¶¶Transformer impedance in pu is independent of the connection.

Example 3.23 *A 6.6 kV/400 V, 75 kVA single-phase transformer has a series reactance of 12% (0.12 pu).*

(a) Calculate the reactance in ohms referred to LV and HV sides.

(b) Three such transformers are connected in Star-Star, calculate (i) the line voltage and kVA rating, (ii) pu reactance of the bank, (iii) series reactance in ohms referred to HV and LV sides

(c) Repeat part (b) if the bank is connected star on HV side and delta on LV side. LO 7

Solution

(a)
$$X(\text{pu}) = \frac{X(\Omega)\,\text{MVA}}{(\text{kV})^2}$$

HV side

$$0.12 = \frac{X(\Omega)\times 75\times 10^{-3}}{(6.6)^2}$$

$$X(\Omega) = \frac{0.12\times(6.6)^2}{75\times 10^{-3}} = 69.696\ \Omega$$

LV side

$$0.12 = \frac{X(\Omega)\times 75\times 10^{-3}}{(0.4)^2}$$

$$X(\Omega) = 0.256\ \Omega$$

(b) *Star-Star connection*

(i) Line voltage

$$\text{HV}\quad 6.6\sqrt{3} = 11.43\ \text{kV}$$

$$\text{LV}\quad 400\sqrt{3} = 692.8\ \text{V}$$

$$\text{Rating} = 3\times 75 = 225\ \text{kVA}$$

(ii)
$$X(\text{pu}) = \frac{X(\Omega)\,\text{MVA}\,(3-\text{phase})}{(\text{kV}\,(\text{line}))^2} = \frac{69.696\times 225\times 10^{-3}}{(6.6\sqrt{3})^2} = 0.12$$

(iii) HV side

$$X(\Omega) = 69.696\ \Omega/\text{phase}$$

LV side

$$X(\Omega) = 0.256\ \Omega/\text{phase}$$

(c) *Star-Delta*

(i) Line voltages

$$\text{Star side } 6.6\sqrt{3} = 11.43\ \text{kV}$$

$$\text{Delta side} = 400\ \text{V}$$

$$\text{Rating} = 3\times 75 = 225\ \text{kVA}$$

(ii)
$$X(\text{pu}) = 0.12$$

(iii) Star side
$$X = 69.69\ \Omega/\text{phase}$$

Delta side
$$X = 0.256\ \Omega/\text{phase}$$

X(pu), calculated from delta side

$$X(\text{pu}) = \frac{(0.256/3)\times 225\times 10^{-3}}{(0.4)^2} = 0.12$$

Concept Check LO 7

1. How a two-winding transformer can be converted to an autotransformer?
2. List the practical applications of an autotransformer.
3. What are variable frequency transformers?
4. What are the applications of variable frequency transformers?
5. How the frequency regions are divided for amplitude and phase response?
6. What are three-phase transformers?
7. Mention the connections of a three-phase transformer.
8. Which type of winding connection is preferred for three-phase transformers used in power systems?
9. Where is an autotransformer employed in a power system? Why?
10. In a transmission system the star side of a star/delta transformer is HV side, while in a distribution system the star side is the LV side. Explain.

3.14 | PARALLEL OPERATION OF TRANSFORMERS

LO 8 Interpret the necessary conditions to be satisfied for parallel operation of transformers

When the load outgrows the capacity of an existing transformer, it may be economical to install another one in parallel with it rather than replacing it with a single larger unit. Also, sometimes in a new installation, two units in parallel, though more expensive, may be preferred over a single unit for reasons of reliability—half the load can be supplied with one unit out. Further, the cost of maintaining a spare is less with two units in parallel. However, when spare units are maintained at a central location to serve transformer installations in a certain region, single-unit installations would be preferred. It is, therefore, seen that parallel operation of the transformer is quite important and desirable under certain circumstances.

Conditions for Parallel Operation

The satisfactory and successful operation of transformers connected in parallel on both sides requires that they fulfil the following conditions:

(i) The transformers must be connected properly as far as their polarities are concerned so that the net voltage around the local loop is zero. A wrong polarity connection results in a dead short circuit.

(ii) Three-phase transformers must have zero relative phase displacement on the secondary sides and must be connected in a proper phase sequence. Only the transformers of the same phase group can be paralleled. For example, Y/Y and Y/Δ transformers cannot be paralleled as their secondary voltages will have a phase difference of 30°. Transformers with +30° and –30° phase shift can, however, be paralleled by reversing the phase-sequence of one of them.

(iii) The transformers must have the same voltage-ratio to avoid no-load circulating current when transformers are in parallel on both primary and secondary sides. Since the leakage impedance is low, even a small voltage difference can give rise to considerable no-load circulating current and extra I^2R loss.

(iv) There should exist only a limited disparity in the per-unit impedances (on their own bases) of the transformers. The currents carried by two transformers (also their kVA loadings) are proportional to their ratings if their ohmic impedances (or their pu impedances on a common base) are inversely proportional to their ratings or their per unit impedances on their own ratings are equal. The ratio of equivalent leakage reactance to equivalent resistance should be the same for all the transformers. A difference in this ratio results in a divergence of the phase angle of the two currents, so that one transformer will be operating with a higher, and the other with a lower power factor than that of the total output; as a result, the given active load is not proportionally shared by them.

Parallel Transformers on No-load

The parallel operation of transformers can be easily conceived on a per phase basis. Figure 3.59 shows two transformers paralleled on both sides with proper polarities but on no-load. The primary voltages V_1 and V_2 are obviously equal. If the voltage-ratio of the two transformers are not identical, the secondary induced emf's, E_1 and E_2 though in phase will not be equal in magnitude and the difference $(E_1 - E_2)$ will appear across the switch S. When secondaries are paralleled by closing the switch, a circulating current appears even though the secondaries are not supplying any load. The circulating current will depend upon the total leakage impedance of the two transformers and the difference in their voltage ratios. Only a small difference in the voltage-ratios can be tolerated.

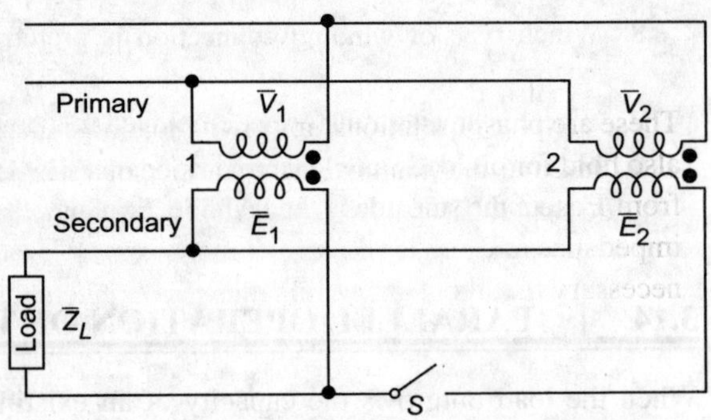

Fig. 3.59 *Transformer connection for parallel operation. Subscripts 1 and 2 respectively denote the transformer numbers*

Division of Load between Transformers in Parallel

Equal Voltage-ratios

When the transformers have equal voltage ratio, $E_1 = E_2$ in Fig. 3.59, the equivalent circuit of the two transformers would then be as shown in Fig. 3.60 on the assumption that the exciting current can be neglected in comparison to the load current. It immediately follows from the sinusoidal steady-state circuit analysis that

$$\bar{I}_1 = \frac{\bar{Z}_2}{\bar{Z}_1 + \bar{Z}_2}\bar{I}_L \tag{3.83}$$

and

$$\bar{I}_2 = \frac{\bar{Z}_1}{\bar{Z}_1 + \bar{Z}_2}\bar{I}_L \tag{3.84}$$

Of course

$$\bar{I}_1 + \bar{I}_2 = \bar{I}_L \tag{3.85}$$

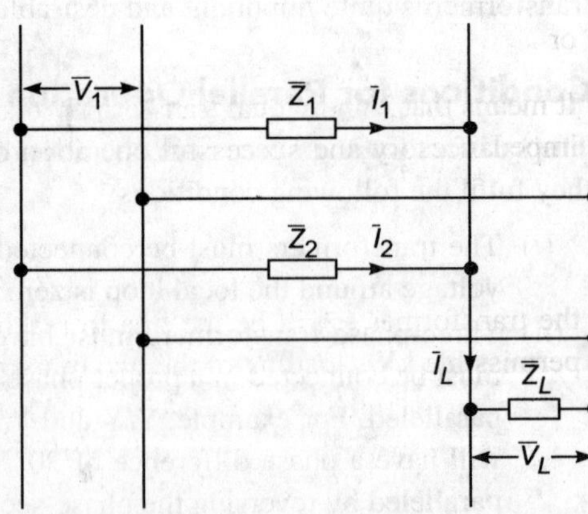

Fig. 3.60 *Approximate equivalent circuit for two transformers in parallel (equal turn-ratio)*

Taking $\bar{V}_L$ as the reference phasor and defining complex power as $\bar{V}^* \bar{I}$, the multiplication of $\bar{V}_L^*$ on both sides of Eqs (3.83) and (3.84) gives

$$\bar{S}_1 = \frac{\bar{Z}_2}{\bar{Z}_1 + \bar{Z}_2} \bar{S}_L \tag{3.86}$$

$$\bar{S}_2 = \frac{\bar{Z}_1}{\bar{Z}_1 + \bar{Z}_2} \bar{S}_L \tag{3.87}$$

where
$$\bar{S}_1 = \bar{V}_L^* I_1$$
$$\bar{S}_2 = \bar{V}_L^* I_2$$
$$\bar{S}_L = \bar{V}_L^* I_L$$

These are phasor relationships giving loadings in the magnitude and phase angle. Equations (3.86) and (3.87) also hold for pu loads and leakage impedances if all are expressed with reference to a common base. It is seen from Eqs (3.83) and (3.84) that the individual currents are inversely proportional to the respective leakage impedances. Thus, if the transformers are to divide the total load in proportion to their kVA ratings, it is necessary that the leakage impedances be inversely proportional to the respective kVA ratings, i.e.

$$\frac{Z_1}{Z_2} = \frac{S_2(\text{rated})}{S_1(\text{rated})} \tag{3.88}$$

This condition is independent of the power factor of the total load. The condition of Eq. (3.88) can be written as

$$\frac{Z_1}{Z_2} = \frac{V_L I_2(\text{rated})}{V_L I_1(\text{rated})}$$

or
$$\frac{Z_1 I_1(\text{rated})}{V_L} = \frac{Z_2 I_2(\text{rated})}{V_L}$$

or
$$Z_1(\text{pu}) = Z_2(\text{pu});\ \text{on own rating} \tag{3.89}$$

It means that if individual transformer loadings are to be in the ratio of their respective kVA ratings, their pu impedances (on their own ratings) should be equal. If

$$Z_1 < Z_2 \frac{S_2(\text{rated})}{S_1(\text{rated})} \tag{3.90a}$$

the transformer 1 will be the first to reach its rated loading as the total kVA load is raised. The maximum permissible kVA loading of the two in parallel without overloading anyone is then given by

$$S_1(\text{rated}) = \frac{Z_2}{|\bar{Z}_1 + \bar{Z}_2|} S_L(\text{max})$$

or
$$S_L(\text{max}) = S_1(\text{rated}) \frac{|\bar{Z}_1 + \bar{Z}_2|}{Z_2} \tag{3.90b}$$

Similarly if

$$Z_2 < Z_1 \frac{S_1(\text{rated})}{S_2(\text{rated})} \tag{3.91a}$$

then

$$S_L(\max) = S_2(\text{rated}) \frac{|\bar{Z}_1 + \bar{Z}_2|}{Z_1} \tag{3.91b}$$

In either case (Eq. (3.90a) or (3.91a))

$$S_L(\max) < S_1(\text{rated}) + S_2(\text{rated}) \tag{3.92}$$

Unequal Voltage Ratios

It has already been mentioned that a small difference in voltage ratios can be tolerated in the parallel operation of transformers. Let E_1 and E_2 be the no-load secondary emfs of two transformers in parallel. With reference to Fig. 3.58, if a load current $\bar{I}_L$ is drawn at voltage $\bar{V}_L$, two mesh voltage balance equations can be written as

$$\bar{E}_1 = \bar{I}_1\bar{Z}_1 + \bar{I}_L\bar{Z}_L = \bar{I}_1\bar{Z}_1 + (\bar{I}_1 + \bar{I}_2)\bar{Z}_L \tag{3.93}$$

and

$$\bar{E}_2 = \bar{I}_2\bar{Z}_2 + \bar{I}_L\bar{Z}_L = \bar{I}_2\bar{Z}_2 + (\bar{I}_1 + \bar{I}_2)\bar{Z}_L \tag{3.94}$$

$\therefore$

$$\bar{E}_1 - \bar{E}_2 = \bar{I}_1\bar{Z}_1 - \bar{I}_2\bar{Z}_2 \tag{3.95}$$

On no-load $\bar{I}_L = 0$, so that the circulating current between the two transformers is given by

$$\bar{I}_1 = -\bar{I}_2 = \frac{\bar{E}_1 - \bar{E}_2}{\bar{Z}_1 + \bar{Z}_2} \tag{3.96}$$

On short-circuit

$$\bar{I}_1 = \frac{\bar{E}_1}{\bar{Z}_1}, \bar{I}_2 = \frac{\bar{E}_2}{\bar{Z}_2} \tag{3.97}$$

On loading

$$\bar{I}_1 = \frac{(\bar{E}_1 - \bar{E}_2) + \bar{I}_2\bar{Z}_2}{\bar{Z}_1}$$

Substituting for I_1: in Eq. (3.94) we get

$$\bar{E}_2 = \bar{I}_2\bar{Z}_2 + \left[\frac{(\bar{E}_1 - \bar{E}_2) + \bar{I}_2\bar{Z}_2}{\bar{Z}_1} + \bar{I}_2\right]\bar{Z}_L$$

$\therefore$

$$\bar{I}_2 = \frac{\bar{E}_2\bar{Z}_1 - (\bar{E}_1 - \bar{E}_2)\bar{Z}_L}{\bar{Z}_1\bar{Z}_2 + \bar{Z}_L(\bar{Z}_1 + \bar{Z}_2)} \tag{3.98}$$

Similarly

$$\bar{I}_1 = \frac{\bar{E}_1\bar{Z}_2 - (\bar{E}_1 - \bar{E}_2)\bar{Z}_L}{\bar{Z}_1\bar{Z}_2 + \bar{Z}_L(\bar{Z}_1 + \bar{Z}_2)} \tag{3.99}$$

Normally $\bar{E}_1$ and $\bar{E}_2$ are in phase or their phase difference is insignificant. Severe results of paralleling transformers not belonging to the same phase-groups (say Y/Y and Y/Δ transformers) are immediately obvious from Eq. (3.96) for no-load circulating current. When many transformers are in parallel, their load sharing can be found out using the Millman theorem [2].

Example 3.24 *A 600-kVA, single-phase transformer with 0.012 pu resistance and 0.06 pu reactance is connected in parallel with a 300-kVA transformer with 0.014 pu resistance and 0.045 pu reactance to share a load of 800 kVA at 0.8 pf lagging. Find how they share the load (a) when both the secondary voltages are 440 V and (b) when the open-circuit secondary voltages are respectively 445 V and 455 V.*

LO 8

Solution

(a) The pu impedances expressed on a common base of 600 kVA are

$$\bar{Z}_1 = 0.012 + j\,0.06 = 0.061\ \angle 79°$$

$$\bar{Z}_2 = 2(0.014 + j\,0.045) = 0.094\ \angle 73°$$

$$\bar{Z}_1 + \bar{Z}_2 = 0.04 + j\,0.15 = 0.155\ \angle 75°$$

The load is

$$\bar{S}_L = 800(0.8 - j\,0.6) = 800\ \angle{-37°}\ \text{kVA}$$

From Eqs (3.86) and (3.87)

$$\bar{S}_1 = 800\ \angle{-37°} \times \frac{0.094\angle 73°}{0.155\angle 75°} = 485\ \angle{-39°} = 377 - j\,305.2$$

$$\bar{S}_2 = 800\ \angle{-37°} \times \frac{0.061\angle 79°}{0.155\angle 75°} = 315\ \angle{-33°} = 264 - j171.6$$

It may be noted that the transformers are not loaded in proportion to their ratings. At a total load of 800 kVA, the 300 kVA transformer operates with 5% overload because of its pu impedance (on common kVA base) being less than twice that of the 600 kVA transformer.

The maximum kVA load the two transformers can feed in parallel without any one of them getting overloaded can now be determined. From above it is observed that the 300 kVA transformer will be the first to reach its full-load as the total load is increased. In terms of magnitudes

$$\frac{0.061}{0.155} S_L(\text{max}) = S_2(\text{rated}) = 300\ \text{kVA}$$

$$\therefore \qquad S_L(\text{max}) = \frac{300 \times 0.155}{0.061} = 762.3\ \text{kV A}$$

while the sum of the ratings of the two transformers is 900 kVA. This is consequence of the fact that the transformer impedances (on common base) are not in the inverse ratio of their ratings.

(b) In this case it is more convenient to work with actual ohmic impedances. Calculating the impedances referred to secondary

$$\bar{Z}_1\ (\text{actual}) = (0.012 + j\,0.06) \times \frac{440}{\dfrac{600 \times 1000}{440}} = 0.0039 + j\,0.0194 = 0.0198\ \angle 79°$$

$$\bar{Z}_2\ (\text{actual}) = (0.028 + j\,0.09) \times \frac{440}{\dfrac{600 \times 1000}{440}} = 0.009 + j\,0.029 = 0.0304\ \angle 73°$$

$$\bar{Z}_1 + \bar{Z}_2 = 0.0129 + j\,0.0484 = 0.05\ \angle 75°$$

The load impedance $\bar{Z}_L$ must also be estimated. Assuming an output voltage on load of 440 V,

$$\bar{V}_L^* \bar{I}_L \times 10^{-3} = (\bar{V}_L^2/\bar{Z}_L) \times 10^{-3} = 800 \angle -37°$$

$$\therefore \qquad \bar{Z}_L = \frac{(440)^2}{800 \times 10^3 \angle -37°} = 0.242 \angle 37° = 0.1936 + j\,0.1452$$

From Eqs (3.98) and (3.99)

$$\bar{I}_1 = \frac{445 \times 0.0304\angle 73° - 10 \times 0.242\angle 37°}{0.0198\angle 79° \times 0.0304\angle 73° + 0.242\angle 37° \times 0.05\angle 75°} = 940 \angle -34° \text{ A}$$

$$\bar{I}_2 = \frac{445 - 0.0198\angle 79° - 10 \times 0.242\angle 37°}{0.0198\angle 79° \times 0.0304\angle 73° + 0.242\angle 37° \times 0.05\angle 75°} = 883 \angle -44° \text{ A}$$

The corresponding kV As are

$$\bar{S}_1 = 440 \times 940 \times 10^{-3} \angle -34° = 413.6 \angle -34°$$
$$\bar{S}_2 = 440 \times 883 \times 10^{-3} \angle -4444° = 388 \angle -44°$$

The total output power will be

$$413.6 \cos 34° + 388 \cos 44° \qquad = 621.5 \text{ kW}$$

This is about 3% less than $800 \times 0.8 = 640$ kW required by the load because of the assumption of the value of the output voltage in order to calculate the load impedance.

The secondary circulating current on no-load is

$$\frac{(\bar{E}_1 - \bar{E}_2)}{|\bar{Z}_1 + \bar{Z}_2|} = \frac{-10}{0.05} = -200 \text{ A}$$

which corresponds to about 88 kVA and a considerable waste as copper-loss.

Concept Check — LO 8

1. What are the advantages of parallel operation of transformers? ○○●
2. What are the necessary conditions to be satisfied for parallel operation of transformers? ○●●
3. When would the load division among the transformers be proportional to the respective KVA ratings? ●●●

3.15 THREE-WINDING TRANSFORMERS

LO 9 Explain in brief about the three-winding transformers and tap-changing transformers and hence deduce their equivalent circuits

Transformers may be built with a third winding, called the *tertiary*, in addition to the primary and secondary. Various purposes which dictate the use of a tertiary winding are enumerated below:

(i) **!Attention** To supply the substation auxiliaries at a voltage different from those of the primary and secondary windings.

(ii) Static capacitors or synchronous condensers may be connected to the tertiary winding for reactive power injection into the system for voltage control.

(iii) A delta-connected tertiary reduces the impedance offered to the zero sequence currents thereby allowing a larger earth-fault current to flow for proper operation of protective equipment. Further, it limits voltage imbalance when the load is unbalanced. It also permits the third harmonic current to flow thereby reducing third-harmonic voltages.

(iv) Three windings may be used for interconnecting three transmission lines at different voltages.

(v) Tertiary can serve the purpose of measuring voltage of an HV testing transformer.

When used for purpose (iii) above the tertiary winding is called a *stabilizing* winding.

Stabilization by Tertiary Winding

The star/star transformer comprising single-phase units or a single unit with a 5-limb core offers high reactance to the flow of unbalanced load between the line and neutral. Any unbalanced load can be divided into three 3-phase sets (positive, negative and zero sequence components). The zero-sequence component (cophasal currents on three lines, $I_0 = I_n/3$) caused by a line-to-neutral load on the secondary side cannot be balanced by primary currents as the zero-sequence currents cannot flow in the isolated neutral star-connected primary. The zero-sequence currents on the secondary side therefore set up magnetic flux in the core. Iron path is available for the zero sequence flux***** in a bank of single-phase units and in the 5-limb core and as a consequence the impedance offered to the zero-sequence currents is very high (0.5 to 5 pu) inhibiting the flow of these currents. The provision of a delta-connected tertiary permits the circulation of zero-sequence currents in it, thereby considerably reducing the zero-sequence impedance. This is illustrated in Fig. 3.61.

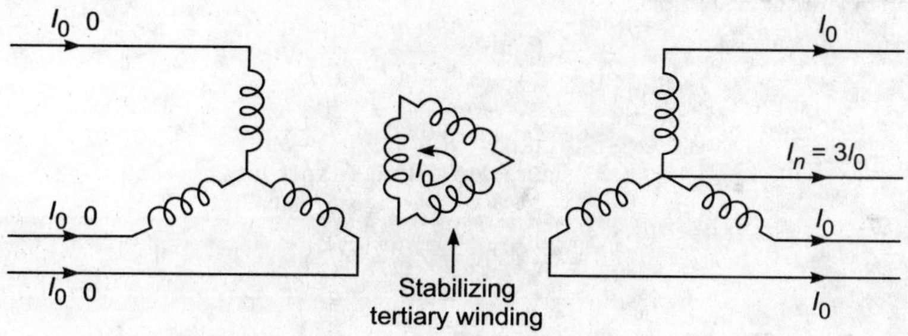

Fig. 3.61

Equivalent Circuit

The equivalent circuit of a 3-winding transformer can be represented by the single-phase equivalent circuit of Fig. 3.62 wherein each winding is represented by its equivalent resistance and reactance. All the values are reduced to a common rating base and respective voltage bases. The subscripts 1, 2 and 3 indicate the primary, secondary and tertiary respectively. For simplicity, the effect of the exciting current is ignored in the equivalent circuit. It may be noted that the load division between the secondary and tertiary is completely arbitrary. Three external circuits are connected between terminals 1, 2 and 3 respectively and the common terminal labelled 0. Since the exciting current is neglected, $\bar{I}_1 + \bar{I}_2 + \bar{I}_3 = 0$.

The impedance of Fig. 3.62 can be readily obtained from three simple short-circuit tests. If $\bar{Z}_{12}$ indicates the SC impedance of windings 1 and 2 with winding 3 open, then from the equivalent circuit,

$$\bar{Z}_{12} = \bar{Z}_1 + \bar{Z}_2 \tag{3.100}$$

***** In a 3-limb core the zero-sequence flux (directed upwards or downwards in all the limbs) must return through the air path, so that only a small amount of this flux can be established; hence a low zero-sequence reactance.

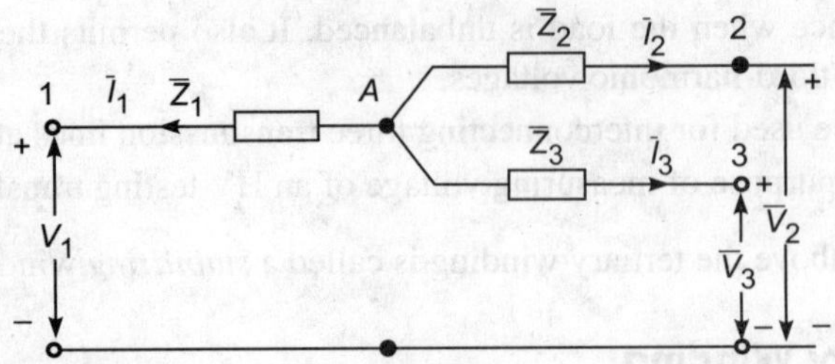

Fig. 3.62 *Single-phase equivalent circuit of 3-winding transformer*

Similarly
$$\bar{Z}_{23} = \bar{Z}_2 + \bar{Z}_3 \tag{3.101}$$
$$\bar{Z}_{13} = \bar{Z}_1 + \bar{Z}_3 \tag{3.102}$$

where $\bar{Z}_{23}$ = SC impedance of windings 2 and 3 with winding 1 open.

$\bar{Z}_{13}$ = SC impedance of windings 1 and 3 with winding 2 open.

All the impedances are referred to a common base.

Solving Eq. (3.100) to Eq. (3.102) yields

$$\bar{Z}_1 = \frac{1}{2}(\bar{Z}_{12} + \bar{Z}_{13} - \bar{Z}_{23}) \tag{3.103}$$

$$\bar{Z}_2 = \frac{1}{2}(\bar{Z}_{23} + \bar{Z}_{12} - \bar{Z}_{13}) \tag{3.104}$$

$$\bar{Z}_3 = \frac{1}{2}(\bar{Z}_{13} + \bar{Z}_{23} - \bar{Z}_{12}) \tag{3.105}$$

The open-circuit test can be performed on anyone of the three windings to determine the core-loss, magnetizing impedance and turn-ratio.

Example 3.25 *The primary, secondary and tertiary windings of a 50-Hz, single-phase, 3-winding transformer are rated as 6.35 kV, 5 MVA; 1.91 kV, 2.5 MVA; 400 V, 2.5 MVA respectively. Three SC tests on this transformer yielded the following results:*

(i) Secondary shorted, primary excited: 500 V, 393.7 A
(ii) Tertiary shorted, primary excited: 900 V, 393.7 A
(iii) Tertiary shorted, secondary excited: 231 V, 21 312.1 A

Resistances are to be ignored.

(a) Find the pu values of the equivalent circuit impedances of the transformer on a 5 MVA, rated voltage base.

(b) Three of these transformers are used in a 15 MVA, Y-Y-Δ, 3-phase bank to supply 3.3 kV and 400 V auxiliary power circuits in a generating plant. Calculate the pu values of steady-state short-circuit currents and of the voltage at the terminals of the secondary windings for a 3-phase balanced short-circuit at the tertiary terminals. Use 15 MVA, 3-phase rated voltage base.

LO 9

Solution

(a) Let us first convert the SC data to pu on 5 MVA base/phase.

For primary, $V_B = 6.35$ kV

$$I_B = \frac{5000}{6.35} = 787.4 \text{ A}$$

For secondary, $V_B = 1.91$ kV

$$I_B = \frac{5000}{1.91} = 2617.8\text{A}$$

Converting the given test data to pu yields:

Test No.	Windings involved	V	I
1	P and S	0.0787	0.5
2	P and T	0.1417	0.5
3	S and T	0.1212	0.5

From tests 1,2 and 3, respectively. 0.0787

$$X_{12} = \frac{0.0787}{0.5} = 0.1574 \text{ pu}$$

$$X_{13} = \frac{0.1417}{0.5} = 0.2834$$

$$X_{23} = \frac{0.1212}{0.5} = 0.2424 \text{ pu}$$

From (3.103) – (3.105)

$$X_1 = 0.5(0.1574 + 0.2834 - 0.2424) = 0.0992 \text{ pu}$$
$$X_2 = 0.5(0.2424 + 0.1574 - 0.2834) = 0.05825 \text{ pu}$$
$$X_3 = 0.5(0.2834 + 0.2424 - 0.1574) = 0.1842 \text{ pu}$$

(b) The base line-to-line voltage for the Y-connected primaries is $\sqrt{3} \times 6.35 = 11$ kV, i.e. the bus voltage is 1 pu. From Fig. 3.62, for a short-circuit at the terminals of the tertiary, $V_3 = 0$. Then

$$I_{SC} = \frac{V_1}{X_1 + X_3} = \frac{V_1}{X_{13}}$$

$$= \frac{1.00}{0.2834} = 3.53 \text{ pu}$$

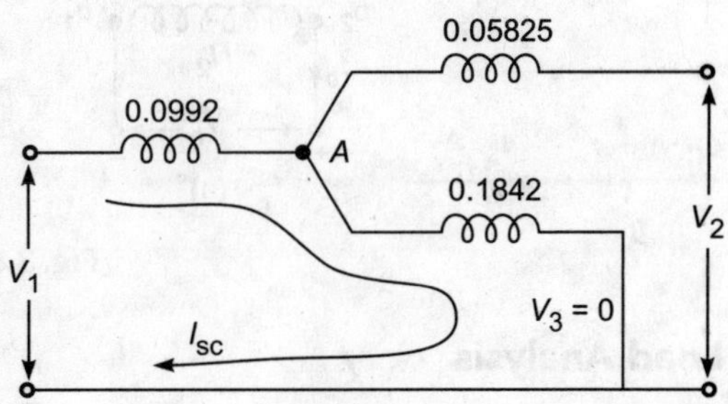

Fig. 3.63

SC current primary side = $3.53 \times 787.4 = 2779.5$ A

SC current tertiary side = $3.53 \times \dfrac{5000 \times 1000}{400/\sqrt{3}}$

= 76424 A (line current)

Neglecting the voltage drops due to the secondary load current, the secondary terminal voltage is the voltage at the junction point A (Fig. 3.63), i.e

$$V_A = I_{SC} X_3 = 3.53 \times 0.1842 = 0.6502 \text{ pu}$$
$$V_A(\text{actual}) = 0.6502 \times 1.91\sqrt{3} = 2.15 \text{ kV (line-to-line)}$$

3.16 | PHASE CONVERSION

LO 9

Three/Two-phase Conversion (Scott Connection)

Important! Phase conversion from three to two phase is needed in special cases, such as in supplying 2-phase electric arc furnaces.

The concept of 3/2-phase conversion follows from the voltage phasor diagram of balanced 3-phase supply shown in Fig. 3.64(b). If the point M midway on V_{BC} could be located, then V_{AM} leads V_{BC} by 90°. A 2-phase supply could thus be obtained by means of transformers; one connected across AM, called the *teaser transformer* and the other connected across the lines B and C. Since $V_{AM} = (\sqrt{3}/2)\, V_{BC}$, the transformer primaries must have $\sqrt{3}\, N_1/2$ (teaser) and N_1 turns; this would mean equal voltage/turn in each transformer. A balanced 2-phase supply could then be easily obtained by having both secondaries with equal number of turns, N_2. The point M is located midway on the primary of the transformer connected across the lines B and C. The connection of two such transformers, known as the Scott connection, is shown in Fig. 3.64(a), while the phasor diagram of the 2-phase supply on the secondary side is shown in Fig. 3.64(c).

The neutral point on the 3-phase side, if required, could be located at the point N which divides the primary winding of the teaser in the ratio 1 : 2 (refer Fig. 3.64(b)).

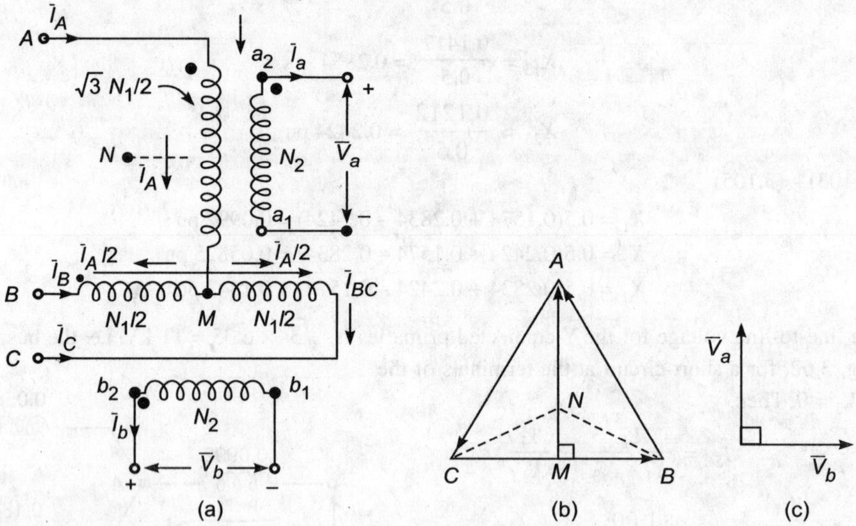

Fig. 3.64 *Scott connection*

Load Analysis

If the secondary load currents are $\bar{I}_a$ and $\bar{I}_b$, the currents can be easily found on the 3-phase side from Fig. 3.64(a).

$$\bar{I}_A = \frac{2N_2}{\sqrt{3}N_1}\bar{I}_a = \frac{2}{\sqrt{3}}\bar{I}_a \text{ (for } N_1/N_2 = 1)$$

$$\bar{I}_{BC} = \frac{N_2}{N_1}\bar{I}_b = \bar{I}_b \text{ (for } N_1/N_2 = 1)$$

$$\overline{I}_B = \overline{I}_{BC} - \overline{I}_A/2$$
$$\overline{I}_C = -\overline{I}_{BC} - \overline{I}_A/2$$

Phasor Diagram

The corresponding phasor diagram for balanced secondary side load of unity power factor is drawn in Fig. 3.65 from which it is obvious that the currents drawn from the 3-phase system are balanced and cophasal with the star voltages. The phasor diagram for the case of an unbalanced 2-phase load is drawn in Fig. 3.66.

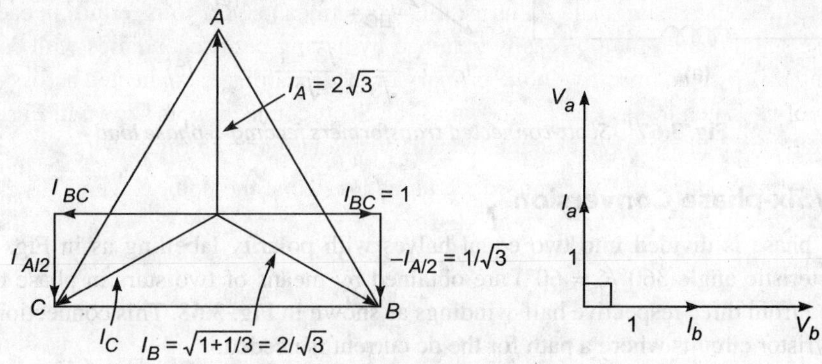

Fig. 3.65 *Scott connection, balanced load (unity power factor)*

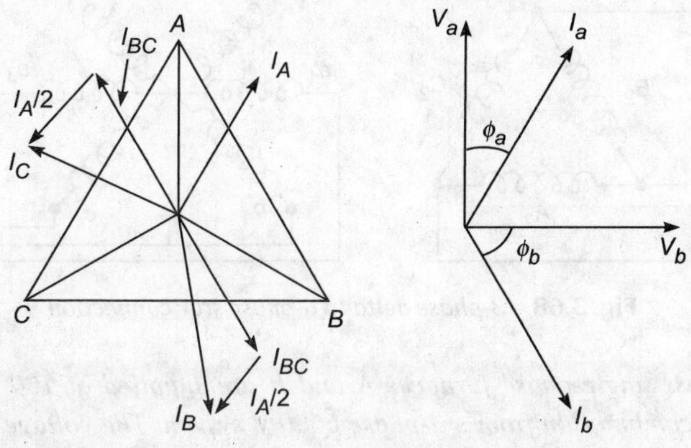

Fig. 3.66 *Scott connection, unbalanced load*

Three/One-phase Conversion

A single-phase power pulsates at twice the frequency, while the total power drawn by a balanced 3-phase load is constant. Thus a 1-phase load can never be transferred to a 3-phase system as a balanced load without employing some energy-storing device (capacitor, inductor or rotating machine). Suitable transformer connections can be used in distributing a 1-phase load on all the three phases though not in a balanced fashion. For large 1-phase loads, this is better than allowing it to load one of the phases of a 3-phase system.

A variety of transformer connections are possible. Figure 3.67(a) shows how Scott-connected transformers could be used for this purpose and Fig. 3.67(b) shows the corresponding phasor diagram.

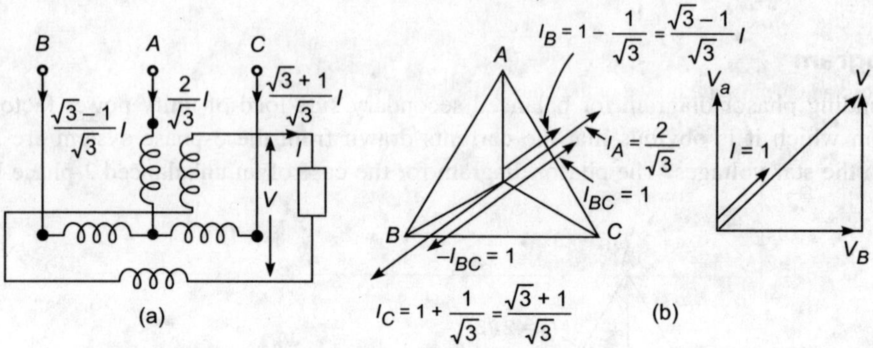

Fig. 3.67 *Scott-connected transformers feeding 1-phase load*

Three Phase/Six-phase Conversion

Each secondary phase is divided into two equal halves with polarity labelling as in Fig. 3.50. Six-phase voltages (characteristic angle 360°/6 = 60°) are obtained by means of two stars in phase opposition, each star being formed from three respective half-windings as shown in Fig. 3.68. This connection is employed in rectifiers and thyristor circuits where a path for the dc current is needed.

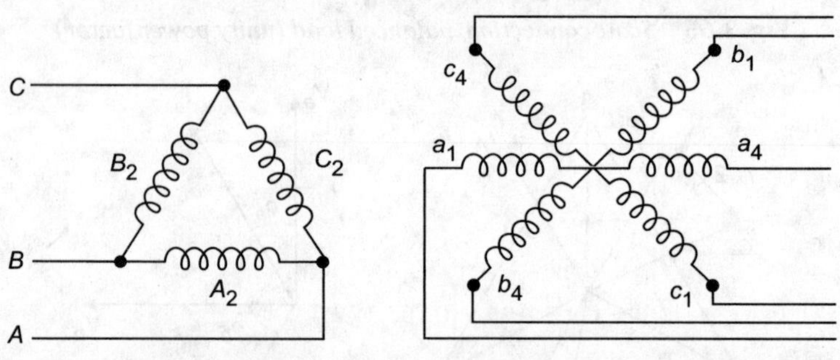

Fig. 3.68 *3-phase delta to 6-phase star connection*

Example 3.26 *Two single-phase furnaces A and B are supplied at 100 V by means of a Scott-connected transformer combination from a 3-phase 6600 V system. The voltage of furnace A is leading. Calculate the line currents on the 3-phase side, when the furnace A takes 400 kW at 0.707 pf lagging and B takes 800 kW at unity pf.*

LO 9

Solution With reference to Fig. 3.64(a)

$$\frac{N_1}{N_2} = \frac{6600}{100} = 66$$

$$\therefore \qquad \frac{\frac{\sqrt{3}}{2} N_1}{N_2} = 57.16$$

Furnace currents are

$$I_a = \frac{400 \times 1000}{100 \times 0.707} = 5658\text{ A}; \qquad \phi_a = 45° \text{ lagging}$$

$$I_b = \frac{800 \times 1000}{100 \times 1} = 8000\text{ A}; \qquad \phi_b = 0°$$

Furnace voltages and currents are drawn in the phasor diagram of Fig. 3.69(a).

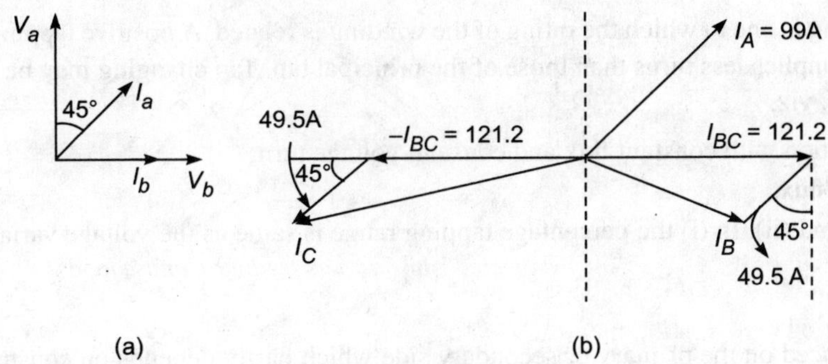

Fig. 3.69

On the 3-phase side

$$I_A = \frac{5658}{57.16} = 99\text{ A}$$

$$I_{BC} = \frac{8000}{66} = 121.2\text{ A}$$

From the phasor diagram of Fig. 3.69(b)

$$\bar{I}_B = 121.2 - 49.5(0.707 + j\,0.707)$$

$$= 86.2 - j\,35$$

or $$\bar{I}_B = 93\text{ A}$$

$$\bar{I}_C = 121.2 + 49.5\,(0.707 - j\,0.707)$$

$$= 156.2 - j\,35$$

or $$I_C = 160\text{ A}$$

3.17 TAP CHANGING TRANSFORMERS

LO 9

Voltage variation in power systems is a normal phenomenon owing to the rapid growth of industries and distribution network. System voltage control is therefore essential for:

(i) Adjustment of consumers' terminal voltage within prescribed limits.
(ii) Control of real and reactive power flow in the network.
(iii) Periodical adjustment (1–10%) to check off-set load variations.

Adjustment is normally carried out by off-circuit tap changing, the common range being 5% in 2.5% steps. Daily and short-time control or adjustment is carried out by means of on-load tap changing gear.

Besides the above, tapping are also provided for one of the following purposes:

(i) For varying the secondary voltage.
(ii) For maintaining the secondary voltage constant with a varying primary voltage.
(iii) For providing an auxiliary secondary voltage for a special purpose, such as lighting.
(iv) For providing a low voltage for starting rotating machines.
(v) For providing a neutral point, e.g. for earthing.

The principal tapping is one to which the rating of the winding is related. A positive tapping means more, and a negative tapping implies less turns than those of the principal tap. Tap changing may be achieved in one of the three conditions, viz.

(i) voltage variation with constant flux and constant voltage turn,
(ii) with varying flux,
(iii) a mix of (i) and (ii). In (i) the percentage tapping range is same as the voltage variation.

Location

The taps may be placed on the primary or secondary side which partly depends on construction. If tappings are near the line ends, fewer bushings insulators are required. If the tappings are placed near the neutral ends, the phase-to-phase insulation conditions are eased.

For achieving large voltage variation, tappings should be placed near the centres of the phase windings to reduce magnetic asymmetry. However, this arrangement cannot be put on LV windings placed next to the core (as in core type transformer) because of accessibility and insulation considerations. The HV winding placed outside the LV winding is easily accessible and can, thus, be tapped easily.

It is not possible to tap other than an integral number of turns and this may not be feasible with LV side tappings. For example 250 V phase winding with 15 V/turn cannot be tapped closer than 5%. It is therefore essential to tap the HV windings which is advantageous in a step-down transformer.

Some of the methods of locating tappings are depicted in Fig. 3.70(a) and (b).

(a) Taps at one end for small transformers (b) Large transformer taps centrally placed for both delta and star transformer

Fig. 3.70 *Location of transformer tappings*

Axial mmf unbalance is minimized by thinning out the LV winding or by arranging parts of the winding more symmetrically. For very large tapping ranges a special tapping coil may be employed.

Tap changing causes changes in leakage reactance, core loss, I^2R loss and perhaps some problems in parallel operation of dissimilar transformers.

No-load (off-load or off-circuit) Tap Changing

The cheapest method of changing the turn ratio of a transformer is the use of off-circuit tap changer. As the name indicates, it is required to deenergize the transformer before changing the tap. A simple no-load tap changer is shown in Fig. 3.71. It has eight studs marked one to eight. The winding is tapped at eight points. The face plate carrying the suitable studs can be mounted at a convenient place on the transformer such as upper yoke or located near the tapped positions on the windings. The movable contact arm A may be rotated by handwheel mounted externally on the tank.

If the winding is tapped at 2% intervals, then as the rotatable arm A is moved over to studs 1, 2; 2, 3; … …6, 7; 7, 8 the winding in circuit reduces progressively by it from 100% with arm at studs (1, 2) to 88% at studs (7,8).

The stop *F* which fixes the final position of the arm A prevents further anticlockwise rotation so that stud 1 and 8 cannot be bridged by the arm. Adjustment of tap setting is carried out with transformer deenergized. For example, for 94% tap the arm is brought in position to bridge studs 4 and 5. The transformer can then be switched on.

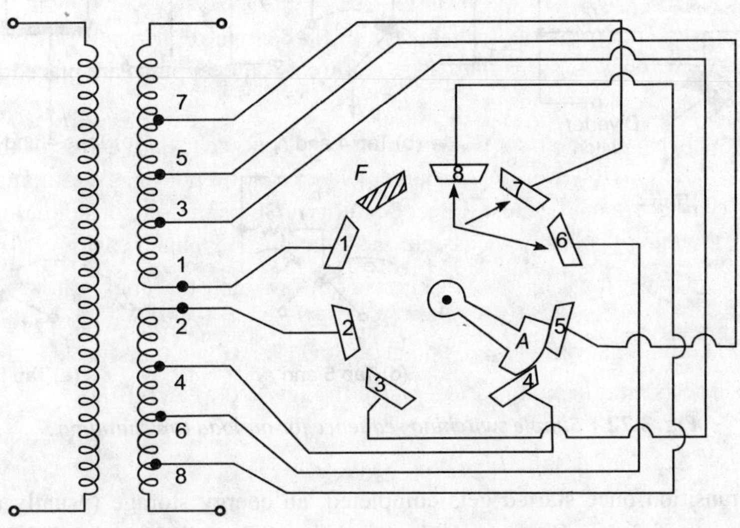

Fig. 3.71 *No-load tap changer*

To prevent unauthorized operation of an off-circuit tap changer, a mechanical lock is provided. Further, to prevent inadvertent operation, an electromagnetic latching device or microswitch is provided to open the circuit breaker so as to deenergize the transformer as soon as the tap changer handle is moved; well before the contact of the arm with the stud (with which it was in contact) opens.

On-load Tap Changing

On-load tap changers are used to change the turn ratio of transformer to regulate system voltage while the transformer is delivering load. With the introduction of on-load tap changer, the operating efficiency of electrical system gets considerably improved. Nowadays almost all the large power transformers are fitted with on-load tap changer. During the operation of an on-load tap changer the main circuit should not be opened to prevent (dangerous) sparking and no part of the tapped winding should get short-circuited. All forms of on-load tap changing circuits are provided with an impedance, which is introduced to limit short-circuit current during the tap changing operation. The impedance can either be a resistor or centre-tapped reactor. The on-load tap changers can in general be classified as resistor or reactor type. In modern designs the current limiting is almost invariably carried out by a pair of resistors.

On-load tap changing gear with resistor transition, in which one winding tap is changed over for each operating position, is depicted in Fig. 3.72. The figure also shows the sequence of operations during the transition from one tap to the next (adjoining) (in this case from tap 4 to tap 5). Back-up main contractors are provided which short-circuit the resistor for normal operation.

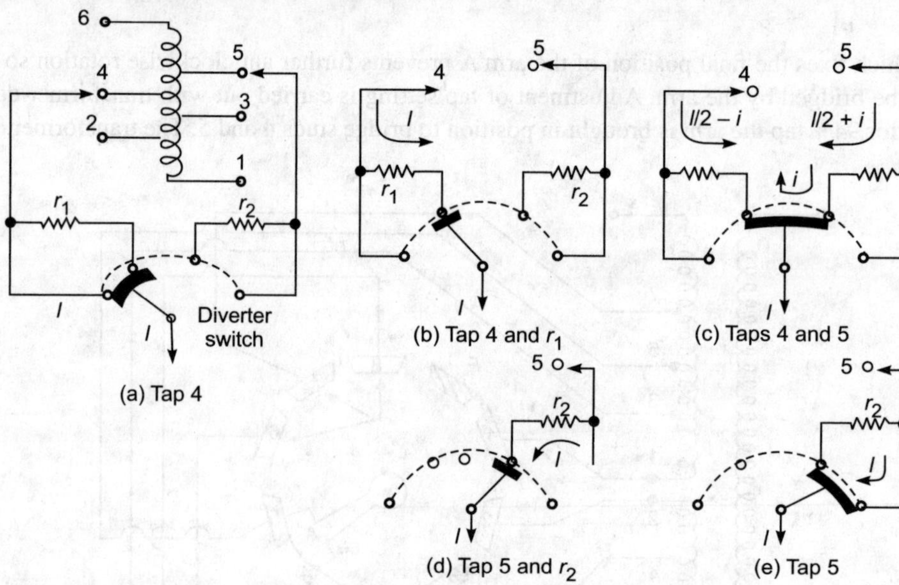

Fig. 3.72 *Simple switching sequence for on-load tap changing*

To ensure that the transition once started gets completed, an energy storage (usually a spring device) is provided which acts even if the auxiliary power supply happens to fail. In resistor-aided tap changing the current break is made easier by the fact that the short-circuit resistor causes the current to be opened to have unity power factor.

On-load tap changer control gear can be from simple push-button initiation to complex automatic control of several transformers operating in parallel. The aim is to maintain a given voltage level within a specified tolerance or to raise it with load to compensate for the transmission line voltage drop. The main components are an automatic voltage regulator, a time delay relay, and compounding elements. The time delay prevents unwanted initiation of a tap change by a small transient voltage fluctuation. It may be set for a delay upto 1 min.

At present tap changers are available for the highest insulation level of 1475 kV (peak) impulse and 630 kV power frequency voltage. Efforts are underway to develop tap changers suitable for still higher insulation levels. More compact tap changers with high reliability and performance are being made by employing vacuum switches in the diverter switch. Also, now thyristorized tap changers are available for special applications where a large number of operations are desired.

Concept Check LO 9

1. What is a three-winding transformer? ○○●
2. What is the purpose of a tertiary winding? ○○●
3. Why tertiary winding is called a stabilizing winding? ○●●
4. What is the significance of performing open-circuit test on a three-winding transformer? ○●●
5. What is the use of a tap-changing transformer? ○○●
6. State the different uses of tapping. ○○●
7. What is meant by no-load tap changing? ○○●
8. What is meant by on-load tap changing? ○○●
9. Explain the basic purpose of a tertiary winding. To what additional use can it be put? ●●●

3.18 | VOLTAGE AND CURRENT TRANSFORMERS

LO 10 Discuss special types of transformers used for different applications

These transformers are designed to meet the specific need of measurement and instrumentation systems, which accept voltages in the range of 0–120 V and currents upto 5 A. Power system voltages can be as high as 750 kV and currents upto several tens of kA. Their measurement requires accurate ratio voltage and current transformations, which is accomplished by potential and current transformers.

Potential Transformer (PT)

!Attention It must transform the input voltage accurately to output voltage both in magnitude and phase. The impedance presented by the instrument on measurement system to the transformer output terminals is called *burden*. It is mainly resistive in nature and has a large value, e.g. the impedance (practically a resistance) of a voltmeter. The circuit model of a PT is drawn in Fig. 3.73. It is the same as that of an ordinary transformer but ideally should have

$$\frac{\bar{V}_1}{\bar{V}_2} = \frac{N_1}{N_2} \angle 0°$$

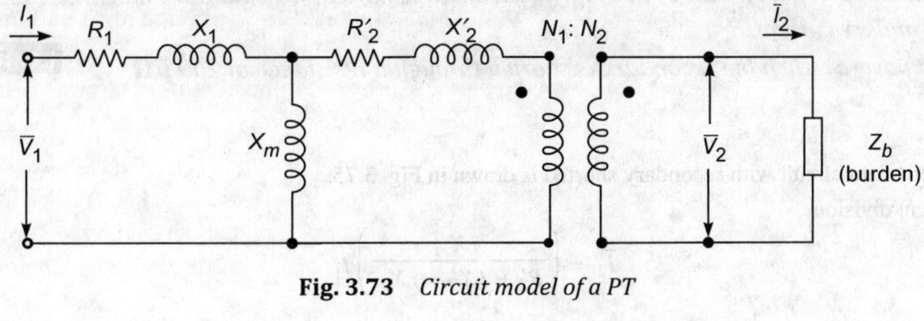

Fig. 3.73 *Circuit model of a PT*

The current drawn by the burden causes a voltage drop in $(R'_2 + j X'_2)$ and this current referred to primary plus the magnetizing current (all phasors) causes a voltage drop in $(R_1 + j X_1)$. Therefore $\bar{V}_2/\bar{V}_1$ differs from the desired value (N_1/N_2) in magnitude and phase resulting in magnitude and phase errors. The errors are to be kept within the limit defined by the precision required. In order to achieve this a PT is designed and constructed to have low leakage reactance, low loss and high magnetizing reactance (low magnetizing current).

Low reactance is achieved by interlacing primary and secondary both on core limb. High magnetizing reactance requires minimum iron path and high permeability steel. Low loss requires low-loss steel and very thin laminations.

Most important thing for low PT errors is to make the burden (Z_b) as high as feasible.

Current Transformer (CT)

It is the current ratio transformer meant for measuring large currents and provide a step down current to current measuring instruments like an ammeter. Such instruments present a short-circuit to the CT secondary. It means that burden $Z_b \approx 0$. An ideal CT current ratio is

$$\frac{\bar{I}_2}{\bar{I}_1} = \frac{N_1}{N_2}\angle 0^\circ$$

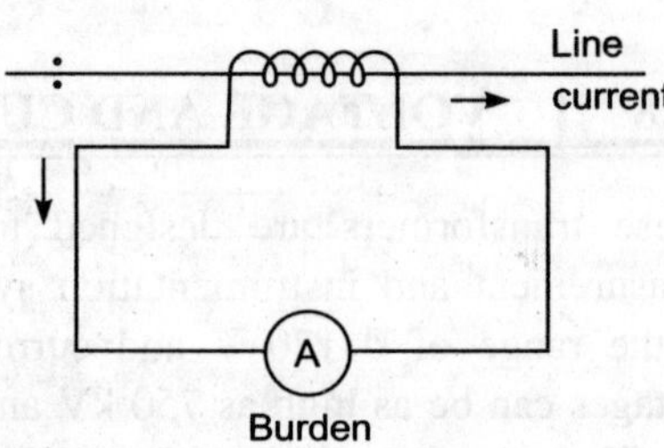

Fig. 3.74 *Current transformer for power-line current*

Causes of CT errors and their remedy are the same as for a PT discussed earlier in this section.

In power system applications CT has a single-turn primary which is the line itself as shown in Fig. 3.74. The secondary is rated 1–5 A.

The burden impedance (which in fact is practically resistive) cannot be allowed to exceed beyond a limit. Most important precaution in use of a CT is that in no case should it be open circuited (even accidently). As the primary current is independent of the secondary current, all of it acts as a magnetizing current when the secondary is opened. This results in deep saturation of the core which cannot be returned to the normal state and so the CT is no longer usable.

Example 3.27 *A 250 A/5 A, 50 Hz current transformer has the following parameters as seen on 250 A side*

$$X_1 = 505\ \mu\Omega, \quad X'_2 = 551\ \mu\Omega, \quad X_m = 256\ m\Omega$$
$$R_1 = 109\ \mu\Omega, \quad R'_2 = 102\ \mu\Omega$$

(a) The primary is fed a current of 250 A with secondary shorted. Calculate the magnitude and phase of the secondary current.

(b) Repeat part (a) when the secondary is shorted through a resistance of 200 μΩ.

LO 10

Solution

(a) The equivalent circuit with secondary shorted is drawn in Fig. 3.75.

By current division

$$\bar{I}'_2 = \left(\frac{jX_m}{R'_2 + jX'_2 + jX_m}\right) I_1$$

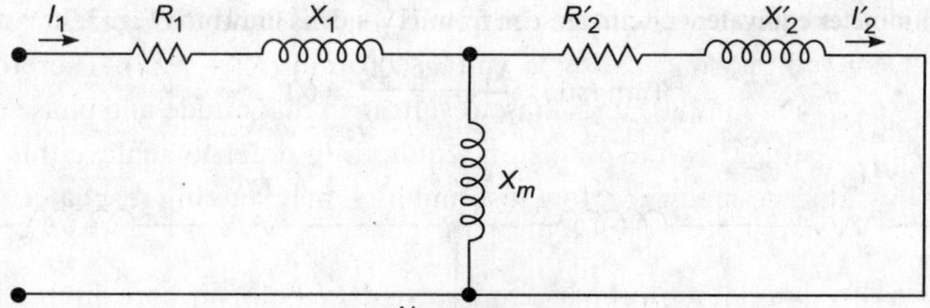

$$\frac{N_2}{N_1} = \frac{250}{5} = 50$$

Fig. 3.75

$$\bar{I}_1 = 250 \angle 0° \text{ A}$$

$$\bar{I}'_2 = \frac{j256 \times 10^3}{109 + j(551 + 256 \times 10^3)} \times 256$$

$$\bar{I}'_2 = \left(\frac{j256}{256.51\angle 89.975}\right) \times 250 \text{ A}$$

$$\bar{I}_2 = \frac{j256}{256.5\angle 89.975°} \times 5; \; I_2 = \left(\frac{N_1}{N_2}\right) I'_2 = \frac{1}{50} \times 250° = 5 \text{ A}$$

$$I_2 = 4.989 \text{ A} \quad \text{phase} = 0.025° \text{ (negligible)}$$

$$\text{Error} = \frac{5 - 4.989}{5} \times 100 = 0.22 \; \%$$

(b) $R'_b = 200\ \mu\Omega$ in series with R'_2, X'_2, $R_b = \left(\frac{1}{50}\right)^2 \times 200 = 0.08\ \mu\Omega$

$$\bar{I}'_2 = \frac{j256 \times 10^3}{(109 + 0.08) + j(551 + 256 \times 10^3)} \times 250 \text{ A}$$

$$\bar{I}'_2 = \frac{256\angle 90°}{256.551\angle 89.975°} \times 5 = 4.989 \angle 0.025°$$

No change as $R'_b = 0.08\ \mu\Omega$ is negligible

Example 3.28 *A 6000 V/100 V, 50 Hz potential transformer has the following parameters as seen from HV side.*

$R_1 = 780\ \Omega$ $\quad X_1 = 975\ \Omega$ $\quad X_m = 443\ k\Omega$

$R'_2 = 907\ \Omega$ $\quad X'_2 = 1075\ \Omega$

(a) The primary is excited at 6500 V and the secondary is left open. Calculate the secondary voltage, magnitude and phase.

(b) The secondary is loaded with 1 kΩ resistance, repeat part (a)

(c) The secondary is loaded with 1 kΩ reactance, repeat part (a)

LO 10

Solution The potentiometer equivalent circuit as seen from HV side is drawn on Fig. 3.76.

$$\text{Turn ratio, } \frac{N_1}{N_2} = \frac{6000}{1000} = 60$$

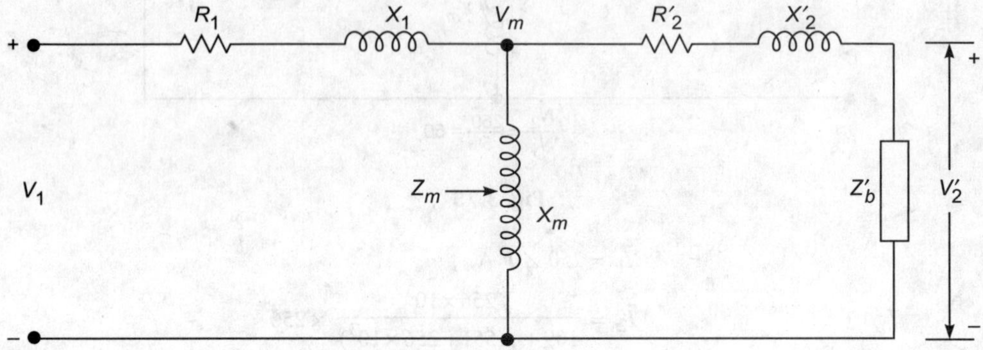

Fig. 3.76

(a) Secondary open;

$$Z_b = \infty$$
$$V_1 = 6500 \text{ V}$$
$$V'_2 = \left(\frac{jX_m}{R_1 + jX_1}\right) V_1$$
$$V'_2 = \frac{j443 \times 10^3}{780 + j443 \times 10^3} \times 6500$$
$$V'_2 = 1 \angle 0.1° \times 6500 = 6500 \angle 0.1° \text{ V}$$
$$V_2 = 6500 \times \left(\frac{N_2}{N_1}\right) = 108 \text{ V}, \angle 0.1°$$

(b)

$$Z_b = R_b = 1 \text{ kW}, R'_b = (60)^2 \times 1 = 3600 \text{ k}\Omega$$

As R'_b is far larger than R'_2 and X'_2, we can ignore R'_2, X'_2

Then

$$\bar{Z}_m = jX_m \parallel R'_b$$
$$\bar{Z}_m = \left(\frac{j443 \times 3600}{3600 + j443}\right) = 439.7 \angle 83° = 53.6 + j\,436.4 \text{ k}\Omega$$
$$(R_1 + jX_1) + \bar{Z}_m = (0.78 + j\,0.975) + (53.6 + j\,436.4)$$
$$= 54.38 + j\,473.4 = 440.77 \angle 82.9° \text{ k}\Omega$$
$$\bar{V}_m = \left[\frac{\bar{Z}_m}{(R_1 + jX_1) + \bar{Z}_m}\right] V_1$$
$$\bar{V}_m = \left[\frac{439.7 \angle 83°}{440.77 \angle 82.9°}\right] \times 6500 = 6484 \angle 0.1°$$
$$\bar{V}'_2 = \bar{V}_m = 6484 \angle 0.1° \text{ V}$$

$$V_2 = \frac{6484}{60} = 108.07 \text{ V ; phase } 0.1°$$

Exact value should be $\frac{6500}{60} = 108.33$

$$\text{Error} = \frac{108.33 - 108.07}{108.33} = 0.26\%$$

(c) $\bar{Z}_b = jX_b \; ; X_b = 1 \text{ k}\Omega$

$$\bar{Z}'_b = j\,3600 \text{ k}\Omega$$

Ignoring R'_2, X'_2 in comparison

$$\bar{Z}_m = j\,443 \parallel j\,3600 = j\frac{443 \times 3600}{443 + 3600} = j\,394.45 \text{ k}\Omega$$

$$(R_1 + jX_1) + \bar{Z}_m = (0.78 + j\,0.975) + j\,394.45 = 0.78 + j\,395.425 = 395.426 \angle 89.89°$$

$$\bar{V}'_2 = \bar{V}_m = \frac{394.45 \angle 90°}{395.426 \angle 89.89} \times 6500 = 6484 \angle 0.01°$$

$$V_2 = \frac{6484}{60} = 108.07, \text{ phase } 0.01°$$

V_2 is same as in resistive load (part (b) except for change in phase. In any case phase is almost zero.

3.19 | AUDIO-FREQUENCY TRANSFORMER

LO 10

It is used at the output stage of audio frequency electronic amplifier for matching the load to the output impedance of the power amplifier stage. Here the load is fixed but the frequency is variable over a band (audio, 20 Hz to 20 kHz), the response being the ratio V_2/V_1. A flat frequency response over the frequency band of interest is most desirable. The corresponding phase angle (angle of V_2 w.r.t. V_1) is called phase response. A small angle is acceptable.

The transformer is used in electronic circuits (control, communication, measurement etc.) for stepping up the voltage or impedance matching. They are normally small in size and have iron cores. It is essential that distortion should be as low as possible.

Equivalent Circuit

Figure 3.77 shows the exact circuit model of a transformer with frequency variable over a wide range. Here the magnetizing shunt branch is drawn between primary and secondary impedances (resistance and leakage reactance). Also represented is the shunting effect of transformer windings stray capacitance C_s. In the intermediate frequency (IF) range the shunt branch acts like an open circuit and series impedance drop is also negligibly small such that V_2/V_1 remains fixed (flat response) as in Fig. 3.77.

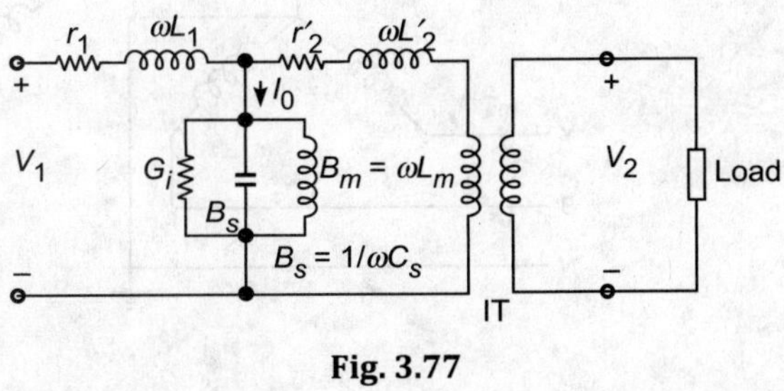

Fig. 3.77

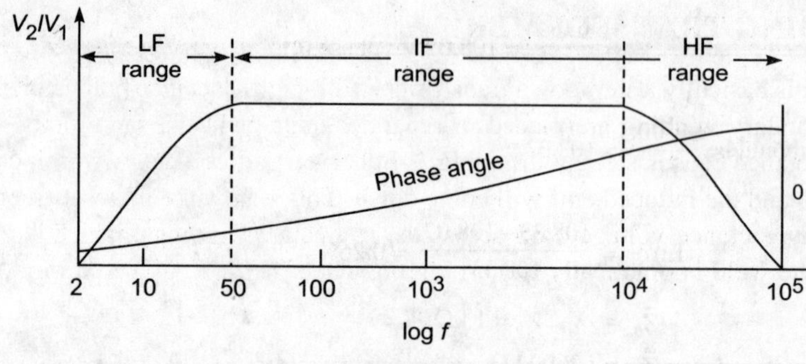

Fig. 3.78 *Frequency response (V_2/V_1 vs log f) of a transformer*

In the LF (low frequency) region the magnetizing susceptance is low and draws a large current with a consequent large voltage drop in $(r_1 + j\omega L_1)$. As a result V_2/V_1 drops sharply to zero as $B_m = 0$ (Fig. 3.78). In the HF (high frequency) region $B_s = 1/\omega C_s$ (stray capacitance susceptance) has a strong shunting effect and V_2/V_1 drops off as in Fig. 3.78, which shows the complete frequency response of a transformer on logarithmic frequency scale.

3.20 | GROUNDING TRANSFORMER

LO 10

In case the neutral of the power transformer is not available for grounding (e.g. when a Δ-Δ transformer is used), a special *Y*-Δ transformer is employed only for neutral grounding as shown in Fig. 3.79(a). Such a transformer is called a *grounding transformer* and it is a step down transformer. The star connected primaries are connected to the system and its neutral is grounded. The secondaries are in delta and generally do not supply any load but provide a closed path for triplen harmonic currents to circulate in them. Under balanced conditions the current in a grounding transformer is its own exciting current. Under fault conditions (such as LG fault) large current may flow in it. Hence a grounding transformer should be of sufficient rating to withstand the effects of LG (line to ground) faults. Transformers with 'zigzag' connection are sometimes employed as grounding transformers as shown in Fig. 3.78(b).

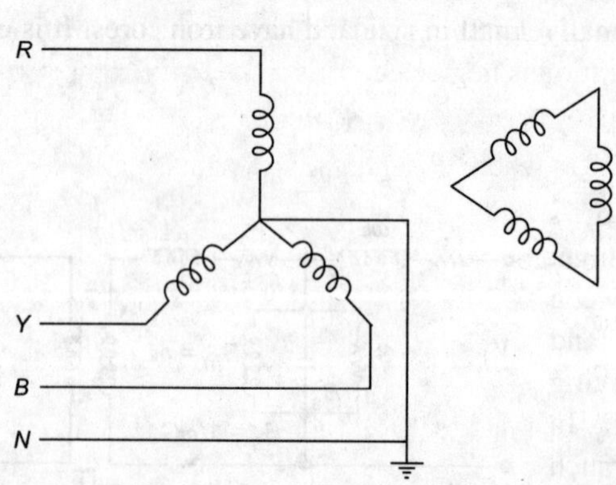

Fig. 3.79(a) *Grounded transformer connections*

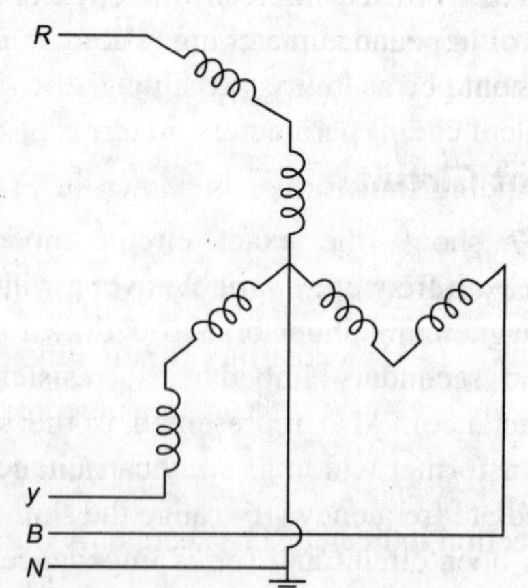

Fig. 3.79(b) *Zig-zag grounding connections*

3.21 | WELDING TRANSFORMER

LO 10

Welding transformer is basically a step-down transformer with high reactance both in primary and secondary. Its primary and secondary winding are placed in separate limbs or in the same limbs but spaced distance apart. This high reactance causes steeply drooping *V-I* characteristics. That is with increase in current, the leakage flux increase and the induced emf will come down. This is why the increase in primary or secondary current increases the reactance voltage drop across the respective windings, which is essential to limit the welding current as the weld is practically a short circuit. The schematic of a welding transformer is shown in Fig. 3.80.

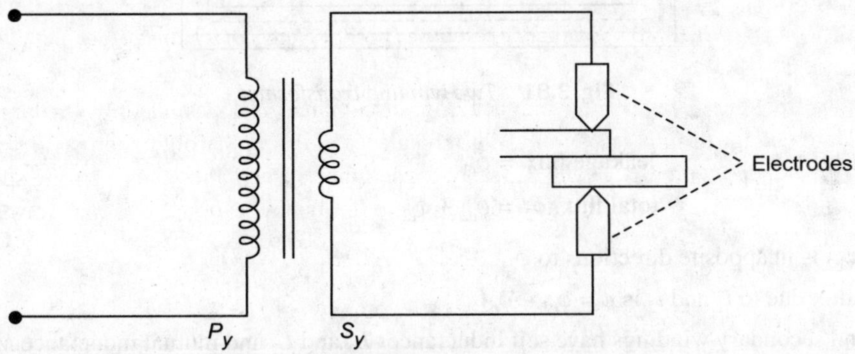

Fig. 3.80 *Welding transformer*

3.22 | TRANSFORMER AS A MAGNETICALLY COUPLED CIRCUIT

LO 10

In Sections 3.3, 3.4 and 3.5 the equivalent circuit of a transformer was developed in terms of primary and secondary resistances, leakage reactances, magnetizing shunt reactance, core loss shunt resistance and an ideal transformer. This development was through magnetizing current needed to setup core flux and emf s induced in the winding by sinusoidally varying core flux, the concept of the ideal transformer and representing core loss by an equivalent shunt resistance. The following section treats the transformer as mutually coupled circuit wherein voltages and currents are related in terms of resistances and inductances. Here the core is assumed to have constant permeability so the magnetic saturation is neglected. This model gives more physical meaning of equivalent circuit parameters, in terms of transformer magnetic field.

A two winding transformer is shown in Fig. 3.81, where R_1 and R_2 are primary and secondary winding resistances.

The primary current i_1 into the dotted terminal produces

$$\text{core flux (mutual flux)} = \phi_{c1}$$

$$\text{leakage flux} = \phi_{l1}$$

$$\text{total flux, } \phi_1 = \phi_{c1} + \phi_{l1}$$

in the direction indicated. The secondary current i_2 out of the dotted terminal produces

$$\text{core flux (mutual flux)} = \phi_{c2}$$

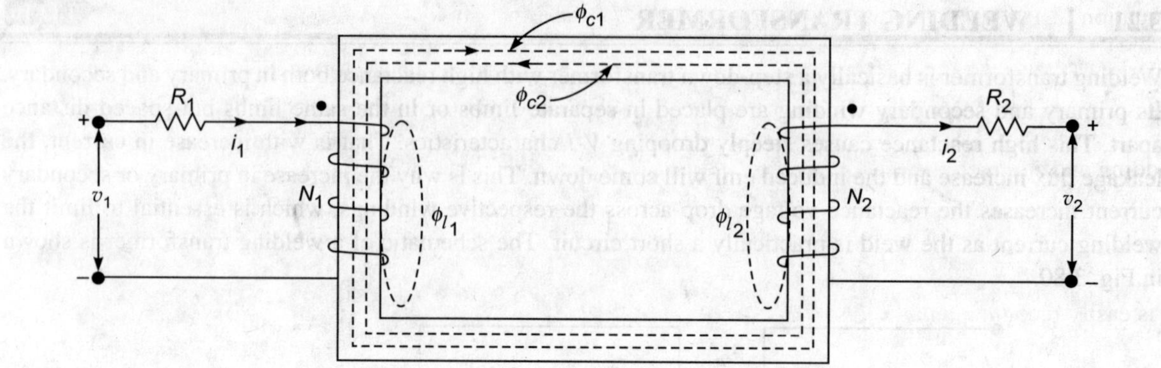

Fig. 3.81 *Two winding transformer*

$$\text{leakage flux} = \phi_{l2}$$

$$\text{total flux, } \phi_2 = \phi_{c2} + \phi_{l2}$$

The core flux ϕ_{c2} is in apposite directions to ϕ_{c1}.

Then net core flux due to i_1 and i_2 is $\phi = \phi_{c1} - \phi_{c2}$

The primary and secondary windings have self inductances L_1 and L_2 and mutual inductance M. In a bilateral circuit

$$M_{12} = M_{21} = M \quad \text{refer Section 2.4}$$

It can be easily proved as the core offers the same permeance to $(N_1 i_1)$ and $(N_2 i_2)$.

Equivalent Circuit of a Magnetically Coupled Circuit

The transformer as a coupled circuit is drawn in Fig. 3.82(a). From the basic laws the kVL equations of primary and secondary are written below.

Primary $\quad v_1 = R_1 i_1 + L_1 \dfrac{di_1}{dt} - M \dfrac{di_2}{dt}$; minus sign results from the fact that i_2 flows out of dotted terminal (3.106a)

Secondary $\quad v_2 = M \dfrac{di_1}{dt} - L_2 \dfrac{di_2}{dt} - R_2 i_2$ (3.106b)

we will now refer the secondary side quantities to the primary side.

Turn ratio, $\quad a = \dfrac{N_1}{N_2}$

then $\quad v_2' = a v_2$

$$i_2' = \frac{i_2}{a}$$

Equation (3.106(a)) is then written as

$$v_1 = R_1 i_1 + L_1 \frac{di_1}{dt} - aM \frac{d}{dt}\left(\frac{i_2}{a}\right)$$

Adding and subtracting $aM \frac{di_1}{dt}$, we get

$$v_1 = R_1 i_1 + (L_1 - aM)\ \frac{di_1}{dt} + aM \frac{d}{dt}\left(i_1 - \frac{i_2}{a}\right) \tag{3.107}$$

It is easily recognized that

$$i_1 - \frac{i_2}{a} = i_m, \text{ the core } \textit{magnetizing current} \tag{3.108}$$

Converting secondary equation to primary side

$$av_2 = aM \frac{di_1}{dt} - a^2 L_2 \frac{d}{dt}\left(\frac{i_2}{a}\right) - a^2 R_2 \left(\frac{i_2}{a}\right) \tag{3.109}$$

Adding and subtracting $aM \frac{d}{dt}\left(\frac{i_2}{a}\right)$ and reorganizing we get

$$v_2' = av_2 = aM \frac{d}{dt}\left(i_1 - \frac{i_2}{a}\right) - (a^2 L_2 - aM)\frac{d}{dt}\left(\frac{i_2}{a}\right) - a^2 R_2 \left(\frac{i_2}{a}\right) \tag{3.110}$$

From Eqs (3.107) and (3.110) the equivalent circuit is drawn in Fig. 3.82(b).

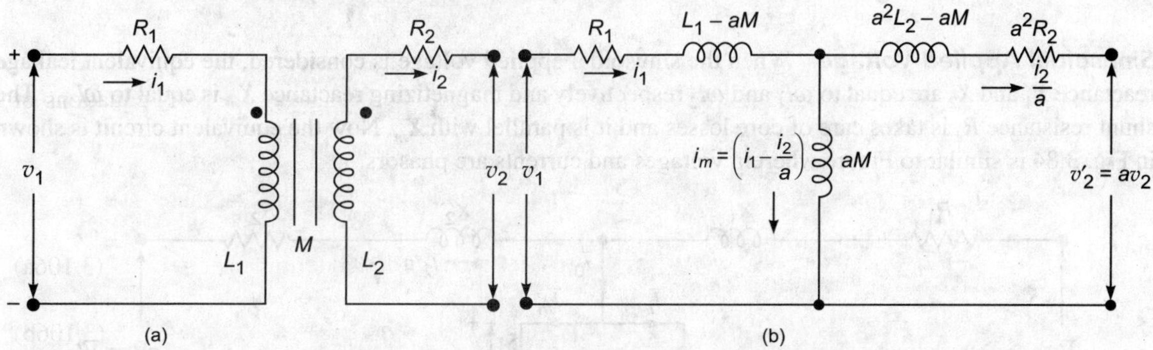

Fig. 3.82 *(a) Magnetically coupled circuit, (b) Equivalent circuit with secondary parameters referred to primary*

We now need to recognize leakage inductance. Using the basic definitions

$$L_1 - aM = \frac{N_1(\phi_{c1} + \phi_l)}{i} - \frac{N_1}{N_2} \cdot \frac{N_2 \phi_{c1}}{i_1} = \frac{N_1 \phi_{l1}}{i_1} = l_1$$

$$(a^2 L_2 - aM) = a^2 \left(L_2 - \frac{M}{a}\right) = a^2 \left[\frac{N_2(\phi_{c2} + \phi_{l2})}{i_2} - \frac{N_2}{N_1} \cdot \frac{N_1 \phi_{c2}}{i_2}\right]$$

$$= a^2\left(\frac{N_2 \phi_{l2}}{i_2}\right) = a^2 l_2$$

$$aM = \frac{N_1}{N_2} \cdot \frac{N_2 \phi_{c1}}{i_1} = \frac{N_1 \phi_{c1}}{i_1} = L_{m1}$$

Equations (3.107) and (3.110) can now be written as

$$v_1 = R_1 i_1 + l_i \frac{di_1}{dt} + L_{m1} \frac{di_m}{dt} \tag{3.111a}$$

$$av_2 = L_{m1} \frac{di_m}{dt} - a^2 l_2 \frac{d}{dt}\left(\frac{i_2}{a}\right) - a^2 R_2 \left(\frac{i_2}{a}\right) \tag{3.111b}$$

From these equations the equivalent circuit is drawn in Fig. 3.83.

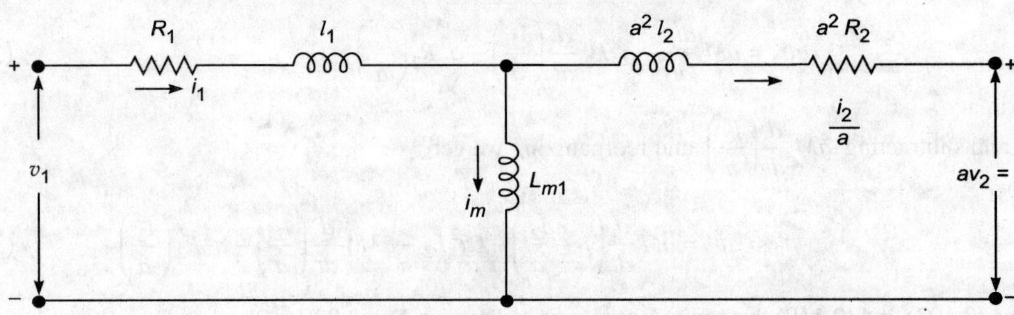

Fig. 3.83 *Conductively coupled equivalent circuit of transformer*

Sinusoidal Applied Voltage When the sinusoidal applied voltage is considered, the equivalent leakage reactance X_1 and X_2 are equal to ωl_1 and ωl_2 respectively and magnetizing reactance X_m is equal to ωL_{m1}. The shunt resistance R_c is takes care of core losses and it is parallel with X_m. Now the equivalent circuit is shown in Fig. 3.84 is similar to Figure wherein voltages and currents are phasors.

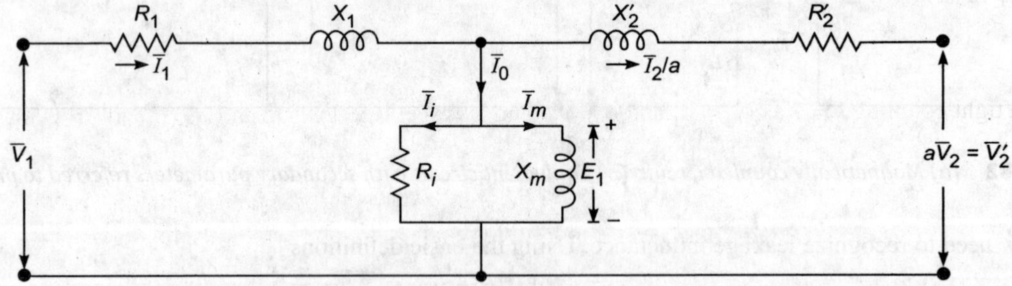

Fig. 3.84 *Equivalent circuit of a transformer with sinusoidal applied voltage*

Induced emf in primary winding, $E_1 = X_m I_m = \omega L_{m1} I_m$

It may be noted that R_i has been connected in parallel to X_m to account for core loss.

$$R_i = \frac{E_1^2}{P_i}$$

For the coupled-circuit of Fig. 3.82(a), Eqs 3.106(a) and (b) take the form

$$\left.\begin{aligned} \bar{V}_1 &= R_1\bar{I}_1 + j\omega L_1\bar{I}_1 - j\omega M\bar{I}_2 \\ \bar{V}_2 &= j\omega M\bar{I}_2 - j\omega L_2\bar{I}_2 - R_2\bar{I}_2 \end{aligned}\right\} \tag{3.112}$$

Coupled Coefficient It is a measure of leakage fluxes in a magnetically coupled-circuit. We begin by defining the coupling factor of each winding as

$$k = \frac{\text{mutual flux due to winding current}}{\text{total flux due to winding current}} = \frac{\phi_c}{\phi}$$

The factor is less than unity as $\phi = \phi_c + \phi_l$.

For primary winding

$$k_1 = \frac{\phi_{c1}}{\phi_1} = \frac{(N_2\phi_{c1})/i_1}{(N_2\phi_1)/i_1} \tag{i}$$

or

$$k_1 = \frac{M}{\left(\dfrac{N_2}{N_1}\right)\left(\dfrac{N_1\phi_1}{i_1}\right)} = \left(\frac{N_1}{N_2}\right)\frac{M}{L_1} \tag{ii}$$

Similarly,

$$k_2 = \left(\frac{N_2}{N_1}\right)\cdot\frac{M}{L_2}$$

Taking the geometric mean yields the coupling coefficient as

$$k = \sqrt{k_1 k_2} = \frac{M}{\sqrt{L_1 L_2}}$$

or

$$M = k\sqrt{L_1 L_2}\,;\, k < 1 \tag{iii}$$

For a tight coupling $k = 1$ as $\phi_{c1} = \phi_1$ and $\phi_{c2} = \phi_2$, no leakage. It follows from Eqs (i) and (ii) that

$$\left(\frac{N_1}{N_2}\right)\frac{M}{L_1} = \left(\frac{N_2}{N_1}\right)\frac{M}{L_2}$$

or

$$\frac{N_1}{N_2} = \sqrt{\frac{L_1}{L_2}}$$

Methods of Reducing Leakage Flux To reduce primary and secondary voltage drop leakage flux of both winding should be kept low which would result in high M (tight coupling). The methods for reducing leakage flux have already been discussed in Section 3.2.

Example 3.29 *A transformer has turn ratio of a = 10. The results of two open-circuit tests conducted on the transformer are given below:*

(a) The primary on application of 200 V draws 4 A with secondary open circuited which is found to have a voltage of 1950 V.

(b) The secondary on application of 2000 V draws 0.41 A with the primary open circuited.

Calculate L_1 and L_2 and coupling coefficient. What is the voltage of primary in part (b). LO 10

Solution

(a)
$$X_m = \frac{240}{4} = 50\ \Omega,\ X_m = 2\pi f L_1$$

$$L_1 = \frac{200}{2\pi \times 50} = 0.159 \text{ H}$$

$$1950 = \sqrt{2}\ \pi N_2 \phi_{max} = \sqrt{2}\ \pi \psi_{max}$$

$$\psi_{max} = \frac{1950}{\sqrt{2}\pi} = 8.78 \text{ Wb–T}$$

$$M = \frac{\psi_{max}}{i_1(\text{max})} = \frac{8.78}{\sqrt{2} \times 4} = 1.55 \text{ H}$$

(b)
$$E_1 = \sqrt{2}\ \pi f N_2 \phi_{max} = \sqrt{2}\ \pi f \psi_{max}$$

$$\frac{\psi_{max}}{i_2(\text{max})} = M, \quad \psi_{max} = \sqrt{2} \times 0.42 \times 1.55$$

$\therefore$
$$E_1 = \sqrt{2}\ \pi \times 50 \times \sqrt{2} \times 0.41 \times 1.55 = 199.6 \text{ A}$$

$$L_2 = \frac{2000}{\sqrt{2}\pi \times 50} \cdot \frac{1}{\sqrt{2} \times 0.41} = 15.53 \text{ H}$$

Coupling coefficient,
$$k = \frac{1.55}{\sqrt{0.159 \times 15.53}} = 0.986$$

Example 3.30 *A 150 kVA transformer 2400/240 V rating has the following parameters:*

$R_1 = 0.2\ \Omega$, $\quad R_2 = 2 \times 10^{-3}$

$X_1 = 0.45\ \Omega$, $\quad X_2 = 4.5 \times 10^{-3}$

$R_i = 10\ k\Omega$ $\quad X_m = 1.6\ k\Omega$ *(referred to HV)*

Calculate the leakage inductances, magnetizing inductance, mutual inductance and self-inductances. LO 10

Solution

$$a = \frac{N_1}{N_2} \approx \frac{2400}{240} = 10$$

$$X_1 = 2\pi f l_1,\ l_1 = \frac{0.45}{314} \times 10^{-3} = 0.01433 \text{ mH}$$

$$X_2 = 2\pi f\, l_2,\; l_2 = \frac{4.5\times 10^{-3}}{314} = 0.01433 \text{ mH}$$

Magnetizing inductance

$$2\pi f L_{m1} = X_m = 1.6\times 10^3$$

$$L_{m1} = 5.096 \text{ H}$$

Self inductances

$$l_1 = L_1 - L_{m1}$$

$$L_1 = 5.096 + 0.01433\times 10^{-3} = 5.096 \text{ H}$$

$$L_{m1} = aM,\; M = \frac{L_{m1}}{a} = \frac{5.096}{10} = 0.5096 \text{ H}$$

$$l_2 = L_2 - \frac{M}{a}$$

$$L_2 = l_2 + \frac{M}{a} = 0.01433\times 10^{-3} + \frac{0.5096}{10} = 0.05098 \text{ H}$$

Coupling factor $$k = \frac{M}{\sqrt{L_1 L_2}} = \frac{0.5096}{\sqrt{5.096\times 0.05098}} = 0.09998 \approx 1$$

Example 3.31 *Solve Problem 3.8 using Matlab*[†††††]*. Also calculate % voltage regulation and η at full load and 0.8 pf lagging.*

LO 10

Solution Steps for computing circuit model parameters, voltage regulation and Efficiency at full load for a Transformer using MATLAB.

Open-Circuit Test The equivalent circuit as seen on open-circuit test is given in Fig. 3.23(b).

Applied voltage = V_1 (rated) Current drawn = 10

Power input = I_0

Power input = $P_0 = P_i$ (core loss)

$$Y_0 = \frac{I_0}{V_1},\; G_i = \frac{P_0}{V_1^2}$$

$$B_m = \sqrt{Y_0^2 - G_i^2}$$

Short-Circuit Test The equivalent circuit as seen during short-circuit test is drawn in Fig. 3.23(b).

Applied voltage = V_{sc} (a fraction of rated value)

Current drawn = I_{sc} (nearly full load value)

Power input = $P_{sc} = P_c$ (copper loss)

$$Z = \frac{V_{sc}}{I_{sc}},\qquad R = \frac{P_{sc}}{(I_{sc})^2}$$

$$X = \sqrt{Z^2 - R^2}$$

[†††††] For detailed write-up on MATLAB, the reader is encouraged to read Appendix G of the authors' book "Modern Power System Analysis", 3rd ed. Tata McGraw-Hill, New Delhi, 2003.

Voltage Regulation

$$\% \text{ voltage regulation} = \frac{\text{voltage drop}}{\text{rated secondary voltage at full load and specified pf}} \times 100$$

or,

$$VR = \frac{I(R\cos\phi \pm X\sin\phi)}{V_2} \times 100; \text{ + for lagging pf; – for leading pf}$$

where I = secondary current

R = equivalent resistance referred to secondary

X = equivalent reactance referred to secondary

ϕ = power factor angle

Efficiency at full load

$$\text{Efficiency at full load} = \frac{\text{Full load output} \times 100}{\text{Full load output + Core loss + Copper loss at full load}} = \frac{P}{P + P_i + P_c} \times 100$$

MATLAB PROGRAM

```
P=50000;
V1=2200;
V2=110;
V0=110;
I0=10;
P0=400;
Y0=I0./V0
Gi=P0./(V0^2)
Bm=sqrt (Y0^2-Gi^2)
Vsc=90;
Isc=20.5;
Psc=808;
Z=Vsc./Isc
R=Psc./Isc^2)
X=sqrt (Z^2-R^2)
TR=2200/110;
Gi_HV=Gi./(TR^2)
Bm_HV=Bm./(TR^2)
R_LV=R./(TR^2)
X _LV=X./(TR^2)
I2=P./V2
```

```
pf=0.8;
Th=acos(pf)
dV=I2.*(R_LV.*cos(Th)+X_LV.*sin(Th))
VR=(dV./V2)*100
Pi=P0
Pc=Psc
EFF_Full_load=(P*100)./(P+Pi+Pc)
y0 =
   0.0909
Gi =
   0.0331
Bm =
   0.0847
Z =
   4.3902
R =
   1.9227
X =
   3.9468
Gi_HV =
   8.2645e-005
Bm_HV =
   2.1171e-004
R_LV =
   0.0048
X_LV =
   0.0099
I2 =
   454.5455
Th =
   0.6435
dV =
   4.4389
VR =
   4.0354
Pi =
   400
Pc =
   808
EFF_Full_load =
   97.6410
```

Note For manual solution, refer solved Problem 3.3 of the Authors' book [76].

Concept Check LO 10

1. What are voltage and current transformers? ○○●
2. What is meant by burden of PT? ○●●
3. How to reduce PT errors? ○●●
4. What is a current transformer? ○○●
5. What is an audio-frequency transformer? ○○●
6. What are the applications of an audio-frequency transformer? ○○●
7. What is a grounding transformer? ○○●
8. Which type of transformer is normally preferred for grounding purpose? ●●●
9. What is a welding transformer? ○○●

SUMMARY

LO 1

- A transformer is a static device comprising coupled coils (primary and secondary) wound on common magnetic core. The arrangement transfers electric energy from one coil (primary) at a particular voltage level to the other coil (secondary) at another voltage level via the magnetic flux carried by the core.
- In a transformer, all voltages and currents are sinusoidal. The device is bilateral i.e. electric energy can be made to flow in either direction with reversal of roles of the two coils.
- Two types of transformer cores are commonly employed in practice - core type and shell type. In the core type, the windings are wound around the two legs of a rectangular magnetic core, while in shell type, the windings are wound on the central leg of a three legged core.
- Transformer windings are made of solid or stranded copper or aluminium strip conductors. For electronic transformers, 'magnetic wire' is normally used as conductor.

LO 2

- *Ideal transformer*
  (a) The core is infinitely permeable and is lossless.
  (b) Both windings have no resistance and there is no leakage flux; so no voltage drop in either winding.
- Transformer ratio $a = \dfrac{E_1}{E_2} = \dfrac{N_1}{N_2} = \dfrac{I_2}{I_1}$; (Ideal transformer)
- Primary induced emf

$$E_1 = \sqrt{2}\,\pi f N_1 \phi_{max} = 4.44 f N_1 \phi_{max}$$

Correspondingly, $E_2 = E_1/a$

- When a current I_2 is drawn from secondary, the current I_1 drawn from primary comprises three compound
  I_m = magnetizing current to establish core flux. It lags by 90°
  I_i = core (iron) loss current in phase with E_1
  I_2' = current to counter secondary

$$\text{AT} = \left(\frac{N_2}{N_1}\right) I_2 = \frac{1}{a} I_2$$

Then

$$\bar{I}_1 = (\bar{I}_m + \bar{I}_i) + \bar{I}_2' = \bar{I}_0 + \bar{I}_2'$$

$\bar{I}_0 = \bar{I}_m + \bar{I}_i$ = no-load current, secondary open circuits.

- Voltage drop in a transformer is due to primary and secondary resistance and leakage reactance (a series effect).

- Impedance is transferred from one side to the other in direct square of turn-ratio. Susceptance to transfer in inverse square of turn-ratio. Thus

Secondary impedance as seen on primary side

$$Z_2' = a^2 Z_2,\ a = \frac{N_1}{N_2}$$

Similarly, $Z_1' = \frac{1}{a^2} Z_1$

- The transformer equivalent circuit as seen from any side;

Shunt branch ($G_i \parallel B_m$) draws I_m and I_i

Series branch ($R + jX$) carries load current I_2'

$$R = R_1 + a^2 R_2,\ X = X_1 + a^2 X_2$$

LO 3

- The equivalent circuit parameters are determined by two non-loading tests:

Open-circuit test – rated voltage applied on one side the other left open

Determines: G_i , B_m and core loss, P_i

Short-circuit test – one side shorted, reduced voltage applied on other to carry full-load current

Determines: R, X and full-load copper loss, P_c

- *Transformer losses*

P_i = core (iron) loss, constant at constant primary voltage

P_c = copper loss (I^2R loss) proportional to square of load current

LO 4

- Per-unit value of the impedance of a transformer is equal when referred to both primary and secondary sides.

LO 5

- *Transformer efficiency*

$$\eta = \frac{P_0\left(=V_2 I_2 \cos\theta\right)}{P_0 + P_i + P_c}$$

For η(max)

$$P_c = I_2^2 R_2(\text{eq}) = P_i$$

or $$I_2(\text{load}) = \sqrt{\frac{P_i}{R_2(\text{eq})}}$$

- Voltage registration, $VR = \frac{V_{20} - V_2}{V_2} \times 100$

$$= \frac{I(R\cos\phi \pm X \sin\phi)}{V_2} \times 100;$$

+ for laging pf

– for leading pf

LO 7

- In an auto-transformer the primary and secondary windings are electrically connected so that a part of the windings is common to the two. As a result, a part of the power is transferred conductively. It, therefore has higher efficiency and kVA compared to the corresponding two-winding transformer.
- In a 3-phase transformer, three single-phase transformers are connected in star/delta (various possible connection). Depending on labeling and phase sequence, the line voltage undergoes a phase shift of ±30 or ± 90°. The common practice is to use 30° phase shift.

LO 8

- The necessary conditions to be satisfied for parallel operation of transformers are:
  1. The transformers must be connected properly as far as their polarities are concerned so that the net voltage around the local loop is zero.
  2. The transformers must have the same voltage-ratio to avoid no-load circulating current when transformers are in parallel on both primary and secondary sides.

LO 9

- Tap-changing transformer is mainly used for system voltage control and real/reactive power control by providing appropriate tap settings.
- A three-winding transformer is the one which has three windings, namely, primary, secondary and tertiary windings which transfer voltage at three different levels.

LO 10

- Special types of transformers used for various applications are instrument transformers, grounding transformers, audio-frequency transformers, and welding transformer.

PRACTICE PROBLEMS

3.1 The emf per turn for a single-phase 2200/220- V, 50-Hz transformer is approximately 12 V. Calculate **LO 2**
(a) the number of primary and secondary turns, and
(b) the net cross-sectional area of core of a maximum flux density of 1.5 T.

3.2 A transformer has primary and secondary turns of 1250 and 125 respectively. It has core cross-section of 36 cm^2 and its rms flux density is to be limited to 1.4 T (to prevent core saturation). What maximum 50 Hz voltage can be applied on the primary side and the corresponding open-circuit secondary voltage? **LO 2**
The core has a mean length of 150 cm and its relative permeability can be assumed to be 8000. What would be the rms exciting current when the transformer's primary winding is excited at a voltage as calculated above? Also calculate the magnetizing susceptance as seen from primary and secondary sides.
If the transformer were to be excited at 60 Hz, what should be the maximum primary voltage for the core flux density limit not to be exceeded? What would be the magnetizing susceptance as seen on each side in this case?

3.3 A single-phase transformer is rated 600/200 V, 25 kVA, 50 Hz. The transformer is supplying full load on secondary side at 0.707 pf lagging. What is the load impedance? Assuming the transformer to be ideal what impedance is seen on the primary side; also the primary current and its pf. **LO 2**

3.4 A single-phase 50 Hz transformer has a voltage ratio of 230/2300 V. A capacitor rated 30 kVAR is connected on the 2300 V side. Calculate the value of this capacitor. What is the kVAR of the capacitor and the value of its capacitance as seen on 230-V side. Assume the transformer to be ideal. **LO 2**

3.5 A transformer has 200 primary and 400 secondary turns. The primary draws a current of $(44.68 + j\,2.37)$ A when the secondary supplies a load current of $(21.62 + j0)$ A. (a) Find the exciting current. (b) If the core has a permeance of 7.69×10^{-5} H/T^2, find the peak value of core flux. (c) Find the primary and secondary induced emfs if the frequency is 50 Hz. (d) Find the core loss. **LO 3**
Remark: This is a learning exercise. This is not how the exciting current is measured in a transformer because this needs differentiating of two nearly equal quantities which introduces large measurement error. As elaborated in this chapter the exciting current is measured by a no-load test.

3.6 A 23 kVA, 50 Hz, 2300/230 V transformer has primary and secondary turns of 200/20. When rated voltage is applied, calculate the mutual core flux neglecting the winding voltage drops. At full load the leakage flux linking each winding is 1% of the mutual flux. Calculate the primary and secondary leakage reactances and the total reactance as referred to either side. **LO 2**
Hint: ϕ_{l1} is caused by I_1N_1 and by I_1N_2, while the mutual flux is caused by $(I_1N_1 - I_2N_2)$. Refer Fig. 3.11.

3.7 A 100 kVA, 1100/230 V, 50-Hz transformer has an HV winding resistance of 0.1 Ω and a leakage reactance of 0.4 Ω. The LV winding has a resistance of 0.006 Ω and a leakage reactance of 0.01 Ω. Find the equivalent winding resistance, reactance and impedance referred to the HV and LV sides. Convert these to pu values. **LO 2** **LO 4**

3.8 A 50 kVA, 2200/110 V transformer when tested gave the following results: **LO 5**
OC test, measurements on the LV side: 400 W, 10 A, 110 V
SC test, measurements on the HV side; 808 W, 20.5 A, 90 V

Compute all the parameters of the equivalent circuit referred to the HV and LV sides of the transformer. Also calculate % voltage regulation and efficiency at full load and 0.8 pf lagging.

3.9 A 22/127 kV, 125 MVA transformer has primary and secondary impedances of $0.015 + j\,0.06$ pu each. Its magnetizing reactance is j 120 pu. The pu values are expressed on the base of the transformer rating. Calculate the primary and secondary impedances in ohms and also the magnetizing reactance in ohms on the LV side. LO 5

3.10 A 20 kVA, 2000/200 V, 50 Hz transformer is operated at no-load on rated voltage, the input being 150 W at 0.12 power factor. When it is operating at rated load, the voltage drops in the total leakage reactance and the total resistance are, respectively, 2 and 1 per cent of the rated voltage. Determine the input power and power factor when the transformer delivers 10 kW at 200 V at 0.8 pf lagging to a load on the LV side. LO 5

3.11 A single-phase load is fed through a 66-kV feeder whose impedance is $120 + j\,400\ \Omega$ and a 66/6.6 kV transformer whose equivalent impedance (referred to LV) is $0.4 + j\,1.5\ \Omega$. The load is 250 kW at 0.8 leading power factor and 6 kV. LO 5

(a) Compute the voltage at the sending-end of the feeder.
(b) Compute the voltage at the primary terminals of the transformer.
(c) Compute the complex power input at the sending-end of the feeder.

3.12 An audio-frequency ideal transformer is employed to couple a 60-Ω resistance load to an electric source which is represented by a constant voltage of 6 V in series with an internal resistance of 2400 Ω. LO 3

(a) Determine the turn-ratio required to ensure maximum power transfer by matching the load and source impedances (i.e. by making the 60 Ω secondary impedance to 2400 Ω when referred to the primary).
(b) Find the load current, voltage and power under the conditions of maximum power transfer.

3.13 Draw a clear phasor diagram of a transformer operating at rated values. Refer to Fig. 3.14(a) and assume $N_1/N_2 = 1.5$ and $I_1R_1 = 0.15\,E_1$, $I_2R_2 = 0.15\,V_2$, $I_1X_1 = 0.3\,E_1$, $I_2X_2 = 0.25\,V_2$, $I_i = 0.1\,I_2'$, $I_m = 0.25\,I_2'$ LO 2

Consider the load power factor to be (a) 0.8 lagging (b) 0.8 leading. Use V_2 as the reference phasor.

3.14 An ideal transformer has a primary winding of 200 turns. On the secondary side the number of turns between A and B is 600 and between B and C is 400 turns, that between A and C being 1000. The transformer supplies a resistor connected between A and C which draws 10 kW. Further, a load of $2000\ \angle 45°\ \Omega$ is connected between A and B. The primary voltage is 2 kV. Find the primary current. LO 2

3.15 A 5-kVA, 400/80-V transformer has R_{eq} (HV) = 0.25 Ω and X_{eq} (HV) = 5 Ω and a lagging load is being supplied by it resulting in the following meter readings (meters are placed on the HV side). LO 2

$I_1 = 16$ A, $V_1 = 400$ V, $P_1 = 5$ kW

For this condition calculate what a voltmeter would read if connected across the load terminals. Assume the exciting current to be zero.

3.16 A 25-kVA, 230/115-V, 50-Hz transformer has the following data LO 3

$R_1 = 0.12\ \Omega$ $R_2 = 0.04\ \Omega$

$X_1 = 0.2\ \Omega$ $X_2 = 0.05\ \Omega$

Find the transformer loading which will make the primary induced emf equal in magnitude to the primary terminal voltage when the transformer is carrying the full load current. Neglect the magnetizing current.

3.17 The resistances and leakage reactances of a 10 kVA, 50 Hz, 2200/220 V distribution transformer are as follows: LO 5

$R_1 = 4\ \Omega$ $\quad R_2 = 0.04\ \Omega$

$X_1 = 5\ \Omega$ $\quad X_2 = 0.05\ \Omega$

Each quantity is referred to its own side of the transformer. (Suffix '1' stands for HV and '2' for LV).

(a) Find the total leakage impedance referred to (i) HV side (ii) LV side.

(b) Consider the transformer to give its rated kVA at 0.8 pf lagging to a load at rated voltage. Find the HV terminal voltage and % voltage regulation.

(c) Repeat (b) for a pf of 0.8 leading.

(d) Consider the core-loss to be 80 W. Find the efficiency under the conditions of part (b). Will it be different for the conditions under part (c)?

(e) If the load in part (b) gets short-circuited, find the steady-state current in the HV lines, assuming that the voltage applied to the transformer remains unchanged.

3.18 For Problem 3.10, assume that the load power factor is varied while the load current and secondary terminal voltage are held fixed. With the help of a phasor diagram, find the load power factor for which the voltage regulation is zero. LO 5

3.19 A 20 kVA, 2000/200 V, single-phase transformer has the following parameters: LO 5

HV winding: $R_1 = 3\ \Omega$ $\quad X_1 = 5.3\ \Omega$

L V winding: $R_2 = 0.05\ \Omega$ $\quad X_2 = 0.05\ \Omega$

(a) Find the voltage regulation at (i) 0.8 pf lagging (ii) upf (iii) 0.707 pf leading.

(b) Calculate the secondary terminal voltage at (i) 0.8 pf lagging (ii) upf (iii) 0.707 pf leading when delivering full-load current with the primary voltage held fixed at 2 kV.

3.20 The approximate equivalent circuit of a 4 kVA, 200/400 V single-phase transformer, referred to the LV side, is shown in Fig. P3.20. LO 3

(a) An open-circuit test is conducted by applying 200 V to the LV side, keeping the HV side open. Calculate the power input, power factor and current drawn by the transformer.

(b) A short-circuit test is conducted by passing full-load current from the HV side keeping the LV side shorted. Calculate the voltage required to be applied to the transformer and the power input and power factor.

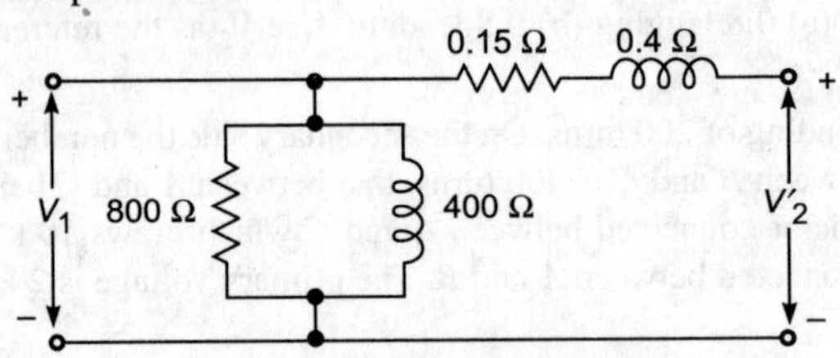

Fig. P 3.20

3.21 A 20 kV A, 2000/200 V transformer has name plate leakage impedance of 8%. What voltage must be applied on the HV side to circulate full-load current with the LV shorted? LO 3

3.22 Derive the condition for zero voltage regulation. Also show that the magnitude of maximum voltage regulation equals the pu value of equivalent leakage impedance. LO 5

3.23 The following test results were obtained for a 20 kVA, 50 Hz, 2400/240 V distribution transformer: LO 3

Open-circuit test (LV): 240 V, 1.066 A, 126.6 W

Short-circuit test (HV): 57.5 V, 8.34 A, 284 W

(a) When the transformer is operated as a step-down transformer with the output voltage equal to 240 V, supplying a load at unity power factor, determine the maximum efficiency and the unity power factor load at which it occurs.

(b) Determine the power-factor of the rated load, supplied at 240 V, such that the terminal voltage observed on reducing the load to zero is still 240 V.

3.24 In a 25 kVA, 2000/200 V transformer, the iron and copper losses are 300 and 400 W respectively. LO 3

(a) Calculate the efficiency on unity power-factor at (i) full-load (ii) half-load. LO 5

(b) Determine the load for maximum efficiency and the iron-and the copper-loss in this case.

3.25 The efficiency of a 1000 kVA, 110/220 V, 50 Hz, single-phase transformer is 98.5% at half full-load at 0.8 pf leading and 98.8% at full-load upf. LO 3

Determine: (a) iron-loss, (b) full-load copper-loss, and (c) maximum efficiency at upf. LO 5

3.26 Open and short-circuit tests performed on a 500 kVA, 6600/2300 V, 50 Hz transformer yielded the following data: LO 5

No-load loss = 3 kW

Full-load short circuit loss = 4 kW

(a) Calculate the load (kVA) at which the transformer efficiency would be maximum for a given power factor. Calculate this efficiency for a pf of 0.85.

(b) The transformer supplies the following load cycle.
12 hours, full load 0.8 pf.
12 hours, half full load 0.9 pf.
Calculate the energy efficiency of the transformer.

3.27 A transformer has its maximum efficiency of 0.98 at 20-kVA at unity power factor. During the day it is loaded as follows: LO 5

12 hours; 2 kW at power factor 0.6
6 hours; 10 kW at power factor 0.8
6 hours; 20 kW at power factor 0.9

Find the 'all-day' efficiency of the transformer.

3.28 A 20 kVA, 200/500 V, 50 Hz, single-phase transformer is connected as an auto-transformer as shown in Fig. P3.28. Determine its voltage-ratio and the kVA rating. Mark on the diagram, the magnitudes and relative directions of the currents in the winding as well as in the input and output lines when delivering the rated kVA to load. LO 7

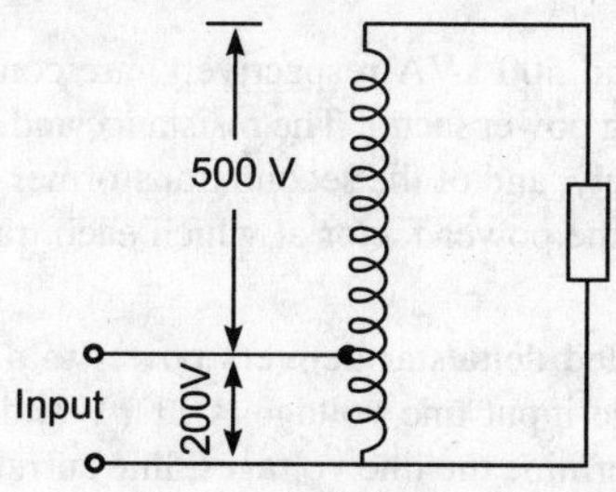

Fig. P3.28

3.29 A 400/100 V, 10 kVA, 2-winding transformer is to be employed as an autotransformer to supply a 400 V circuit from a 500 V source. When tested as a 2-winding transformer at rated load, 0.85 pf lagging, its efficiency is 0.97%. LO 7

(a) Determine its kVA rating as an autotransformer.

(b) Find its efficiency as an autotransformer.

3.30 A 20 kVA, 2000/200 V, two-winding transformer is to be used as an autotransformer, with a constant source voltage of 2000 V. At full-load of unity power factor, calculate the power output, power transformed and power conducted. If the efficiency of the two-winding transformer at 0.7 pf is 97%, find the efficiency of the autotransformer. LO 7

3.31 A 200/400 V, 20 kVA, and 50 Hz transformer is connected as an autotransformer to transform 600 V to 200 V. LO 7
(a) Determine the autotransformer ratio a'.
(b) Determine the kVA rating of the autotransformer.
(c) With a load of 20 kVA, 0.8 pf lagging connected to 200 V terminals, determine the currents in the load and the two transformer windings.

3.32 An audio frequency output transformer couples a variable frequency source of output resistance 4.5 kΩ to a load of 10 Ω. The transformer has a turn ratio of 25.4. On test the following inductance data are measured on the transformer. LO 10
(i) Inductance seen on the primary side with secondary open = 18.7 H.
(ii) Inductance seen on the primary side with secondary shorted = 0.215 H.
In terms of the frequency response calculate (a) lower corner frequency (b) upper corner frequency and (c) voltage gain and phase angle at the geometric mean of the frequencies in parts (a) and (b)
Hint: It is sufficiently accurate to assume that test
(i) yields L_m, the magnetizing inductance and test
(ii) yields the leakage inductance seen on primary side. Transformer winding resistance is ignored.

3.33 A 20 kVA, 4400/220 V transformer with an equivalent impedance of 0.01 Ω is to operate in parallel with a 15 kVA, 4400/220 V transformer with an equivalent impedance of 0.015 Ω. The two transformers are connected in parallel and made to carry a load of 25 kVA. Assume both the impedances to have the same angle. LO 8
(a) Find the individual load currents.
(b) What per cent of the rated capacity is used in each transformer?

3.34 Two single-phase transformers, rated 1000 kVA and 500 kVA respectively, are connected in parallel on both HV and LV sides. They have equal voltage ratings of 11 kV/400 V and their per unit impedances are $(0.02 + j\,0.07)$, and $(0.025 + j\,0.0875)$ Ω respectively. What is the largest value of the unity power factor load that can be delivered by the parallel combination at the rated voltage? LO 8

3.35 Two single-phase transformers rated 600 kVA and 500 kVA respectively, are connected in parallel to supply a load of 1000 kVA at 0.8 lagging power factor. The resistance and reactance of the first transformer are 3% and 6.5% respectively, and of the second transformer 1.5% and 8% respectively. Calculate the kVA loading and the power factor at which each transformer operates. LO 8

3.36 An ideal 3-phase step-down transformer, connected delta/star delivers power to a balanced 3-phase load of 120 kVA at 0.8 power factor. The input line voltage is 11 kV and the turn-ratio of the transformer, phase-to-phase is 10. Determine the line voltages, line currents, phase voltages and phase currents on both the primary and the secondary sides. LO 7

3.37 A Δ/Y connected bank of three identical 60 kVA 2000/100 V, 50 Hz transformers is fed with power through a feeder whose impedance is $0.75 + j\,0.25$ Ω per phase. The voltage at the sending-end of the feeder is held fixed at 2 kV line-to-line. The shortcircuit test when conducted on one of the transformers with its LV terminals shortcircuited gave the following results: LO 7

$V_{HV} = 40$ V $f = 50$ Hz
$I_{HV} = 35$ A $P = 800$ W

(a) Find the secondary line-to-line voltage when the bank delivers rated current to a balanced 3-phase upf load.

(b) Calculate the currents in the transformer primary and secondary windings and in the feeder wires on the occurrence of a solid 3-phase short-circuit at the secondary line terminals.

3.38 Each phase of 3-phase transformer is rated 6.6 kV/230V, 200 kVA with a series reactance of 8%. LO 7

(a) Calculate the reactance in ohm referred to HV/LV sides.

(b) The transformer is connected Y/Y. What is its 3-phase rating (voltage and kVA) and the per unit reactance.

(c) Calculate the pf of load (rated) at which voltage regulation would be maximum. If this load is fed at rated voltage on LV side, what should be the HV side line voltage?

3.39 A 2400/220 V, 300 kVA, 3-phase transformer has a core loss of 33 kW at rated voltage. Its equivalent resistance is 1.6%. Calculate the transformer efficiency at 1.8 pf at (i) full load (ii) at half load. LO 4 LO 5

What is the load at which the transformer efficiency would be maximum? Calculate its value at a pf of 0.8.

Hint: Use the pu method.

3.40 A 3-phase 50 kVA, 6.6/0.4 kV 50 Hz transformer is Δ/Y connected. It yielded the following test results: LO 5

OC Test	SC Test
$P_0 = 520$ W	$P_{SC} = 610$ W
$I_0 = 4.21$ A	$I_{SC} = 4.35$ A
$V_0 = 400$ V	$V_{SC} = 340$ V

Calculate the pu circuit parameters of the transformer. Determine its efficiency and voltage regulation at full load 0.8 pf lagging. Calculate also the maximum efficiency and the load (0.8 pf) at which it will occur.

3.41 A 6.6/0.4 kV, 100 kVA distribution transformer is connected Δ/Y. The transformer has 1.2% resistance and 5% reactance. Find the voltage regulation at full load, 0.8 pf leading. With 0.4 kV as secondary voltage (on load), what is the primary voltage? LO 5

Hint: Use pu system.

3.42 A single-phase, 50 Hz, three-Winding trans-former is rated at 2200 V on the HV side with a total of 250 turns. Of the two secondary windings, each can handle 200 kVA, one is rated at 550 V and the other at 220 V. Compute the primary current when the rated current in the 220 V winding is at upf and the rated current in the 550 V winding is 0.6 pf lagging. Neglect all leakage impedance drops and magnetizing current. LO 9

3.43 A small industrial unit draws an average load of 100 A at 0.8 lagging pf from the secondaries of its 2000/200 V, 60 kVA Y/Δ transformer bank. Find: LO 2

(a) The power consumed by the unit in kW,

(b) the total kVA used,

(c) the rated line currents available from the transformer bank,

(d) the rated transformer phase currents of the Δ-secondaries,

(e) per cent of rated load on transformers,

(f) primary line and phase currents, and

(g) the kVA rating of each individual transformer.

3.44 The HV terminals of a 3-phase bank of three single-phase transformers are connected to a 3-wire, 3-phase, 11 kV (line-to-line) system. The LV terminals are connected to a 3-wire, 3-phase load rated of 1000 kVA and 2200 V line-to-line. Specify the voltage, current and kVA ratings of each transformer (both HV and LV windings) for the following connections: LO 7

(a) HV – Y, LV – Δ (b) HV – Δ, LV – Y (c) HV – Y, LV – Y (d) HV – Δ, LV – Δ.

3.45 A 3-phase bank consisting of three single-phase 3-winding transformers (Y/Δ/Y) is employed to step-down the voltage of a 3-phase, 220 kV transmission line. The data pertaining to one of the transformers are given below: **LO 7**

Ratings
Primary 1: 20 MVA, 220 kV
Secondary 2: 10 MVA, 33 kV
Tertiary 3: 10 MVA, 11 kV
Short-circuit reactances on 10 MV A base

$$X_{12} = 0.15 \text{ pu}; \quad X_{23} = 0.1 \text{ pu}; \quad X_{13} = 0.2 \text{ pu}$$

Resistances are to be ignored. The Δ-connected secondaries supply their rated current to a balanced load at 0.85 power factor lagging, whereas the tertiaries provide the rated current to a balanced load at upf (constant resistance).

(a) Compute the primary line-to-line voltage to maintain the rated voltage at the secondary terminals.
(b) For the conditions of part (a) find the line-to-line voltage at the tertiary terminals.
(c) If the primary voltage is held fixed as in part (a), to what value will the tertiary voltage increase when the secondary load is removed?

3.46 A 500-kVA, 11/0.43-kV, 3-phase delta/star connected transformer has on rated load HV copper-loss of 2.5 kW and LV loss of 2 kW. The total leakage reactance is 0.06 pu. Find the ohmic values of the equivalent resistance and leakage reactance on the delta side. **LO 7**

3.47 Two transformers each rated 250-kVA, 11/2-kV and 50-Hz are connected in open-delta on both the primary and secondary. **LO 7**

(a) Find the load kVA that can be supplied from this transformer connection.
(b) A delta connected three-phase load of 250 kVA, 0.8 pf, 2 kV is connected to the low-voltage terminals of this open-voltage transformer. Determine the transformer currents on the 11 kV side of this connection.

3.48 Two 110-V, single-phase furnaces take loads of 500 kW and 800 kW respectively at a power factor of 0.71 lagging and are supplied from 6600 V, 3-phase mains through a Scott-connected transformer combination. Calculate the currents in the 3-phase lines, neglecting transformer losses. Draw the phasor diagram. **LO 9**

3.49 Figure P3.49 shows a Scott-connected transformer, supplied from 11 kV, 3-phase, 50 Hz mains. Secondaries, series connected as shown, supply 1000 A at a voltage of $100\sqrt{2}$ to a resistive load. The phase sequence of the 3-phase supply is *ABC*. **LO 9**

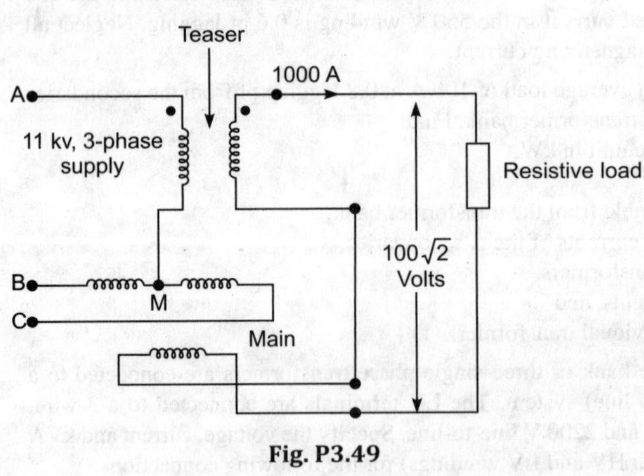

Fig. P3.49

(a) Calculate the turn-ratio of the teaser transformer.
(b) Calculate the line current I_B and its phase angle with respect to the voltage of phase A to neutral on the 3-phase side.

3.50 A 15 kVA, 2200/220 V, 50 Hz transformer gave the following test results: LO 5

OC (LV side)	V = 220 V	I = 2.72 A	P = 185 W
SC (HV side)	V = 112 V	I = 6.3 A	P = 197 W

Compute the following:
(a) Core loss
(b) Full-load copper loss
(c) Efficiency at full-load 0.85 lagging of
(d) Voltage regulation at full-load 0.8 lagging/leading pf

3.51 A transformer of rating 20 kVA, 2000/200 V has the following parameters: LO 5

R_{eq}(HV side) = 2.65 Ω
Z_{eq}(HV side) = 4.23 Ω

Core loss at rated voltage = 95 W
(a) Calculate transformer efficiency when delivering 20 kVA at 200 V at 0.8 pf lagging.
(b) What voltage must be applied on the HV side for load as in part (a).
(c) Find the percentage voltage regulation.

3.52 A 100 kVA, 11 kV/231 V transformer has HV and LV winding resistances of 8.51 Ω and 0.0038 Ω respectively. It gave the following test results: LO 5

OC (LV side)	231 V	15.2 A	1.25 kW
SC (HV side)	440 V	9 A	Not measured

Calculate
(a) Equivalent leakage reactance of the transformer
(b) Full load copper loss
(c) Efficiency at full-load and half full-load at 0.85 lagging power factor

3.53 A 100 kVA, 2200 V/220 V transformer has the following circuit parameters. LO 5

R_1 = 0.23 Ω	R_2 = 0.0023 Ω
X_1 = 1.83 Ω	X_2 = 0.013 Ω
R_i (HV side) = 5.6 kΩ	X_m(HV side) = 1.12 kΩ

The transformer is subjected to the following daily load cycle = 4 h on no load, 8 h on 1/4th full-load at 0.8 pf, 8 h on 1/2 full-load at upf, and 4 h on full-load at 0.9 pf.
Determine the all-day energy efficiency of the transformer.

3.54 A 400/200 V, 50 Hz transformer has a primary impedance of 1.2 + j 3.2 Ω and secondary impedance of 0.4 + j 1.0 Ω. A short-circuit occurs on the secondary side with 400 V applied to the primary. Calculate the primary current and its power factor. LO 2

3.55 A 50 Hz, 3-winding transformer can be considered as an ideal transformer. The primary is rated 2400 V and has 300 turns. The secondary winding is rated 240 V, 400 kVA and supplies full-load at upf. The tertiary is rated 600 V, 200 kVA and supplies full-load at 0.6 pf lagging. Determine the primary current. LO 9

3.56 An ideal transformer has 200 primary turns and 360 secondary turns, the primary being excited at 600 V. The full secondary has a resistive load of 8 kW. The secondary is also tapped at 240 turns which supplies a pure inductive load of 10 kVA. Find the primary current and its pf. LO 2

3.57 A 50 kVA, 2300 V/230 V transformer draws power of 750 W at 0.5 A at no load when 2300 V is applied to the HV side. The HV winding resistance and leakage reactance are 1.8 Ω and 4 Ω respectively. Calculate: LO 2
(a) the no load pf
(b) the primary induced emf

(c) the magnetizing current and
(d) the core loss component of current.

3.58 Two single-phase transformers operate in parallel to supply a load of $44 + j\,18.6\ \Omega$. The transformer A has a secondary emf of 600 V on open circuit with an internal impedance of $1.8 + j\,5.6\ \Omega$ referred to the secondary. The corresponding figures for transformer B are 610 V and $1.8 + j\,7.4\ \Omega$. Calculate the terminal voltage, current and power factor of each transformer. LO 8

3.59 Each phase of a 3-phase transformer is rated 6.6 kV/230 V, 200 kVA with a series reactance of 8% LO 7
(a) Calculate the reactance in ohm referred to HV/LV sides.
(b) The transformer is connected Y/Y. What is its 3-phase rating (voltage and kVA) and the per unit reactance.
(c) Calculate the pf of load (rated) at which voltage regulation would be maximum. If this load is fed at rated voltage on LV side, what should be the HV side line voltage?

Answers

3.1 (a) 183,18 (b) $0.036\ \text{m}^2$
3.2 16.7 mA; 8.44×10^{-6} ℧, 0.844×10^{-3} ℧; 2375 V, 7×10^{-6} ℧, $0.7\ 10^{-3}$ ℧
3.3 $1.6\angle 45°$, $14.4\angle 45°$; 41 2/3 A, pf = 0.707 lag
3.4 18.05 μF, 30, 1.805 μF
3.5 (a) 2.77 A (b) 51.54 mWb
(c) 2288/4576 V (d) 3.3 kW
3.6 $X_{11} = 3.27\ \Omega$, $X_{12} = 0.0327\ \Omega$, X_{12} (primary) = 6.54 Ω, X_{eq} (secondary) = 0.0654 Ω
3.7 $(0.237 + j\,0.629\ \Omega)$, $(0.0104 + j\,0.0275)\ \Omega$ $(0.019 + j\,0.052)$ pu
3.8 HV side: $g_i = 2.28 \times 10^{-4}$ ℧, $b_m = 0.828 \times 10^{-4}$ ℧
$R = 1.932\ \Omega$, $X = 3.946\ \Omega$
LV side: $g_i = 0.0331$ ℧, $b_m = 0.0848$ ℧
$R = 4.808 \times 10^{-3}\ \Omega$, $X = 9.865 \times 10^{-3}\ \Omega$
3.9 $0.058 + j\,0.232\ \Omega$, $1.0935 + j\,7.42\ \Omega$, $j\,464.6\ \Omega$
3.10 10.23 kW, 0.742 lagging
3.11 (a) 58.95 kV (b) 59.7 kV
(c) $(254.34 - j\,172.6)$ kVA, 0.827 leading
3.12 (a) $N_1/N_2 = 6.325$ (b) 7.91 mA, 0.474 V, 3.75 mW
3.14 13.02 A
3.15 70.4 V
3.16 19.57 kW, 0.79 leading
3.17 (a) (i) $8 + j\,10$ (ii) $0.08 + j\,0.1$
(b) 2256.4 V, 2.56% (c) 2201.8 V, 0.08%
(d) 97%, No (e) 171.9 A
3.18 0.9 leading
3.19 (a) (i) 6.29% (ii) 4% (iii) −0.815%
(b) (i) 187.4 V (ii) 192 V (iii) 201.6 V
3.20 (a) 50 W, 0.447 lag, 0.56 A (b) 17.1 V, 60.2 W, 0.352 lag
3.21 160 V
3.23 (a) 13.37 kVA, 98.14% (b) 0.805 leading
3.24 (a) (i) 97.27% (ii) 96.9%
(b) 21.6 kVA, 300 W, 300 W

3.25 (a) 4071 W (b) 8079 W
(c) 98.9%

3.26 (a) 86.6%, 98.4% (b) 98.27%

3.27 96.77%

3.28 28 kVA

3.29 (a) 50 kVA (b) 99.38%

3.30 220 kW, 20 kW, 200 kW, 99.7%

3.31 (a) 3 (b) 30
(c) 100 A, 50 A, 50 A

3.32 (a) 141.7 Hz (b) 50.9 kHz
(c) 2.68 kHz, 23.2×10^3, $\approx 0°$

3.33 (a) 68.2 A, 45.6 A (b) 75%, 66.7%

3.34 1400 kVA

3.35 584.2 kVA, 0.858 pf lag; 426.5 kVA at 0.705 pf lag

3.36 11 kV, 1.9 kV; 63 A, 36.4 A; 11 kV, 1.1 kV; 3.64 A, 36.5 A

3.37 (a) 164.9 V (b) 12090 A

3.38 (a) 7.05×10^{-3} Ω (LV side), 5.81 Ω (HV side)
(b) 600 kVA, 398.4 V, 11.43 kV, 0 pf lag, 12.34 kV

3.39 (i) 96.85% (ii) 96.62%, 237.2 kVA, 96.93%

3.40 $G_i = 0.0104$ pu, $B_i = 0.0574$ pu, $R_{eq} = 0.0123$ pu, $X_{eq} = 0.050$ pu, Voltage reg = 4%,
$\eta = 97.74$, η (max) = 97.47% at 91.95% of f1

3.41 –2.04%, 6.465 kV

3.42 162.6 ∠ –26.6° A

3.43 (a) 27.7 kW (b) 34.6 kVA (c) 173.2 A
(d) 100 A (e) 57.7% (f) 17.32 A
(g) 20 kVA

3.44 (a) 6.35/2.2 kV, 52.6/151.5 A, 333.3 kVA
(b) 11/1.27 kV, 30.3/262.4 A, 333.3 kVA
(c) 6.35/1.27 kV, 52.6/262.4 A, 333.3 kVA
(d) 11/2.2 kV, 30.3/151.5 A, 333.3 kVA

3.45 (a) 243.98 kV (b) 11.16 kV (c) 11.89 kV

3.46 6.5 Ω, 43.4 Ω

3.47 (a) 433 kVA (b) 13.12 A (phase and line)

3.48 123.2 A, 182.2 A, 182.2 A

3.49 (a) 95.26 (b) 3.85 A lagging V_{AN} by 45°

3.50 (a) 185 W (b) 231 W
(c) 96.8% (d) + 4.41%, – 1.94%

3.51 (a) 97.8% (b) 2041 V (c) 2.05%

3.52 (a) 45.81 Ω (b) 1.41 kW (c) 95.6%

3.53 97.2%

3.54 51.78 A, 0.363 lagging

3.55 226.7 A, $\angle -17.1°$

3.56 21.35 A, 0.625 lagging

3.57 (a) 0.625 lagging, (b) 2300 $\angle 0°$ volts
(c) 0.5 A (d) 0.33 A

3.58 $V_t = 538.6 \angle -3.2°$ V, $I_A = 6.39 \angle -19°$ A, $I_B = 5.82 \angle -33.5°$ A,

$Pf_A = 0.962$ lagging, $Pf_B = 0.863$ lagging

3.59 (a) 7.05 m Ω, 5.81 Ω

(b) 600 kVA, 398.4 V, 11.43 kV.

(c) OPF lagging, 12.34 kV.

MULTIPLE CHOICE QUESTIONS

1. The electric energy from primary coil at a particular voltage level to the secondary coil at another voltage level is transferred through — LO 1

(a) air (b) coil
(c) flux (d) None of these

2. A transformer core is laminated to reduce — LO 1

(a) hysteresis losses (b) eddy current losses
(c) copper losses (d) core losses

3. The no-load current drawn by a transformer is usually ______ percent of the full load current. — LO 2

(a) 0.2 to 0.5 (b) 2 to 5
(c) 12 to 15 (d) 20 to 30

4. The magnetic flux path in a transformer must have — LO 1

(a) high resistance (b) low resistance
(c) high reluctance (d) low reluctance

5. No-load loss on a transformer is carried out to determine — LO 3

(a) copper loss (b) magnetizing current
(c) magnetizing current and loss (d) efficiency of the transformer

6. Sumpner's test is conducted on transformers to determine — LO 3

(a) temperature (b) stray losses
(c) all-day efficiency (d) voltage regulation

7. The purpose of providing iron core in a transformer is — LO 1

(a) to provide support to windings (b) to reduce hysteresis losses
(c) to reduce eddy current losses (d) to decrease the reluctance of magnetic flux path

8. While conducting short-circuit test on a transformer, which of the following side is short-circuited? — LO 3

(a) High voltage side (b) Low voltage side
(c) Primary side (d) Secondary side

9. Which winding of a transformer has more number of turns? — LO 1

(a) High voltage winding (b) Low voltage winding
(c) Primary winding (d) Secondary winding

10. Efficiency of a power transformer is of the order of — LO 5

(a) 80 to 85% (b) 90 to 95%
(c) 96 to 99% (d) 100%

11. A common method of cooling a power transformer is — LO 1

(a) natural air cooling (b) air blast cooling
(c) oil cooling (d) Any of the above

12. In a transformer, efficiency depends upon LO 5
(a) supply frequency (b) load current
(c) power factor of load (d) Both (b) and (c)

13. Natural air cooling is used for transformers up to a rating of LO 1
(a) 3000 kVA (b) 1000 kVA
(c) 00 kVA (d) 250 kVA

14. Two transformers operating in parallel will share their load in proportion to their LO 8
(a) per unit impedances (b) leakage reactances
(c) efficiencies (d) kVA ratings

15. In a transformer, tapping are normally provided on LO 9
(a) primary side (b) secondary side
(c) low voltage side (d) high voltage side

16. Transformer ratings are usually expressed in LO 2
(a) Ampere-hour (b) Kilovolt Ampere
(c) Mega Watt (d) Kilovolt

17. Material used for the construction of a transformer core is usually LO 1
(a) wood (b) copper
(c) aluminum (d) silicon steel

18. In a transformer, the resistance between its primary and secondary windings is LO 2
(a) zero (b) 1 ohm
(c) 1000 ohms (d) infinite

19. A Buchholz relay is installed on LO 1
(a) welding transformers (b) auto-transformers
(c) air-cooled transformers (d) oil-cooled transformers

20. Star/Star connected transformers work satisfactorily when LO 7
(a) load is unbalanced (b) load is balanced
(c) on balanced as well as unbalanced loads (d) None of the above

21. The magnetizing current in a transformer is usually small because it has LO 2
(a) small air gap (b) large leakage flux
(c) laminated silicon steel core (d) fewer rotating parts

22. A transformer can have zero voltage regulation at LO 5
(a) leading power factor (b) lagging power factor
(c) unity power factor (d) zero power factor

23. Which one of the following is the main advantage of an autotransformer? LO 7
(a) Reduction in hysteresis losses (b) Reduction in eddy current losses
(c) Saving in winding material (d) Negligible copper losses

24. An ideal transformer is the one which has LO 2
(a) no losses and magnetic leakage
(b) interleaved primary and secondary windings
(c) a common core for its primary and secondary windings
(d) core of stainless steel and winding of pure copper metal

25. When a transformer is run at its rated voltage but reduced frequency, its LO 2
(a) flux density remains unaffected (b) iron losses are reduced
(c) core flux density is reduced (d) core flux density is increased

26. The core in a large power transformer is built of LO 1
(a) cast iron (b) mild steel
(c) ferrite (d) silicon steel

27. Cruciform shape is used in transformer core LO 1
   (a) to reduce core loss (b) to reduce winding copper
   (c) to provide mechanical strength (d) to reduce core reluctance
28. No load current in a transformer LO 2
   (a) lags the applied voltage by 90°
   (b) lags the applied voltage by somewhat less than 90°
   (c) leads the applied voltage by 90°
   (d) leads the applied voltage by somewhat less than 90°
29. A 200/100 V, 50 Hz transformer is to be excited at 40 Hz from the 100 V side. For the exciting current to remain the same, the applied voltage should be LO 2
   (a) 150 V (b) 125 V
   (c) 100 V (d) 80 V
30. Power input to a transformer on no load at rated voltage comprises predominantly LO 3
   (a) copper loss (b) hysteresis loss
   (c) core loss (d) eddy current loss
31. A 2/1 ratio, two-winding transformer is connected as an auto transformer. Its kVA rating as an auto transformer compared to a two-winding transformer is LO 7
   (a) same (b) 1.5 times
   (c) 2 times (d) 3 times
32. The high frequency hum in the transformer is mainly due to LO 1
   (a) laminations being not sufficiently tight (b) magnetostriction
   (c) oil of the transformer (d) tank walls
33. The efficiency of a transformer at full-load 0.85 pf lag is 95%. Its efficiency at full-load 0.85 pf lead will be LO 5
   (a) less than 95% (b) more than 95%
   (c) 95% (d) 100%
34. Under balanced load conditions, the main transformer rating in the Scott connection is greater than that of the teaser transformer by LO 7
   (a) 5% (b) 15%
   (c) 57.7% (d) 85%
35. Non-loading heat run test on transformer is performed by means of LO 3
   (a) SC test (b) OC test
   (c) half time on SC and half time on OC (d) Sumpner's test

Answers

1. (c)	**2.** (b)	**3.** (b)	**4.** (d)	**5.** (c)	**6.** (a)	**7.** (d)	**8.** (b)	**9.** (a)	**10.** (c)
11. (c)	**12.** (d)	**13.** (a)	**14.** (a)	**15.** (c)	**16.** (b)	**17.** (d)	**18.** (d)	**19.** (d)	**20.** (b)
21. (a)	**22.** (a)	**23.** (c)	**24.** (a)	**25.** (d)	**26.** (d)	**27.** (d)	**28.** (b)	**29.** (d)	**30.** (c)
31. (c)	**32.** (b)	**33.** (c)	**34.** (b)	**35.** (d)					

SAMPLE EXAMPLES SOLVED USING MATLAB

```
% This program solves the Example 3.15 of the text if we use the default values.
% The same program can be used for user defined values.
ch-menu ('WOULD U LIKE TO USE', 'DEFAULT VALUES', 'USER DEFINED VALUES');
```

```
if ch==l
RI=0.2;
R2=.002;
XI=0.45;
X2=.0045;
Ri=10000;
Xm=1600;
VP=2400;
VS=240;
kVA=150;
pf=0.8;
else
    R1=input('ENTER THE VALUE OF R1 IN OHMS                ');
    R2=input('ENTER THE VALUE OF R2 IN OHMS                ');
    X1=input('ENTER THE VALUE OF Xl IN OHMS                ');
    X2=input('ENTER THE VALUE OF X2 IN OHMS                ');
    Ri =input('ENTER THE VALUE OF Ri IN OHMS               ');
    Xm=input('ENTER THE VALUE OF Xm IN OHMS (as seen from HV side ');
    VP=input('ENTER THE PRIMARY VOLTAGE IN VOLTS           ');
    VS=input('ENTER THE SECONDARY VOLTAGE IN VOLTS         ');
    kVA=input('ENTER THE RATING IN kVA                     ');
    pf=input('ENTER THE POWER FACTOR (lagging)             ');
    end
    a=VP/VS;
    R_HV=RI+R2*a^2;
    X_HV=XI+X2*a^2;
    I2_fl=kVA*1000/VS;
    I2_flr=I2_fl/a;
    V_drop=I2_flr*(R_HV*pf+X_HV*sin(acos(pf)));
    V_reg=V_drop*100/VP
    V_1=VP+V_drop;
    P_out=kVA*pf*1000;
    P_closs=R_HV*I2_flr^2;
    P_iloss=V_1^2/Ri;
    P_L=P_close+P_iloss;
    Efficiency=100*P_out/(P_out+P_L)
    I_o=V_1/Ri-j*V_1/Xm
```

```
    I22=I2_flr*(exp(-j*acos(pf)));
    1_1=1_o + 122;
    I_P=abs (I _1)

%This program is used to plot the voltage regulation as a function of load
%for 0.8 pf lagging, 1.0 pf and 0.8 pf leading
%Default values are those used in Example 3.15 of the text
ch=menu('WOULD U LIKE TO USE', 'DEFAULT VALUES', 'USER DEFINED VALUES');
if ch==1
R1=0.2;
R2=.002;
X1=0.45;
X2=.0045;
Ri=10000;
Xm=1600;
VP=2400;
VS=240;
kVA=150;
pf=0.8;
else
    R1=input('ENTER THE VALUE OF R1 IN OHMS      ');
    R2=input('ENTER THE VALUE OF R2 IN OHMS      ');
    X1=input('ENTER THE VALUE OF X1 IN OHMS      ');
    X2=input('ENTER THE VALUE OF X2 IN OHMS      ');
    Ri=input('ENTER THE VALUE OF Ri IN OHMS      ');
    Xm=input('ENTER THE VALUE OF Xm IN OHMS (as seen HV=Sid   ');
    VP=input ('ENTER THE PRIMARY VOLTAGE IN VOLTS       ');
    VS=input('ENTER THE SECONDARY VOLTAGE IN VOLTS      ');
    kVA=input('ENTER THE RATING IN kVA ');
end
    a=VP/VS;
    R_HV=R1+R2*a^2;
    X_HV=X1+X2*a^2;
% Data generation for 0.8 pf lagging
    12=0.0;
    pf=0.8;
for j=1:26;
```

```
        I2_r=I2/a;
        V_drop=I2_r*(R_HV*pf+X_HV*sin(acos(pf)));
        V_reg=V_drop*100/VP;
        reg1(j) =V reg;
        I2=12+25
    end
    % Data generation of 1.0 pf
        I2=0.0;
        pf=1.0;
    for j=1:26;
        12_r=I2/a;
        V_drop=I2_r*(R_HV*pf_X_HV*sin(acos(pf)));
        V_reg=V_drop*100/VP;
        reg2(j)=V_reg;
        I2=I2+25
    end
    % Data generation for 0.8 pf lagging
        I2=0.0;
        pf=0.8;
    forj=l :26;
        I2_r=12/a;
        V_drop=12_r*(R_HV*pf-X_HV*sin(acos(pf)));
        V_reg=V-drop*100/VP;
        reg3(j)=V_reg;
        12=12+25
    end
        IL=0:25:625;
        plot (IL,reg1,'b-')
        hold on;
        plot(IL,reg2,'b.')
        hold on;
        plot(IL,reg3,'b*')
        legend('O.8 PF lagging','1.0 PF', '0.8 PF leading');
        hold off
```

CHAPTER 4

Principles of Electromechanical Energy Conversion

LEARNING OBJECTIVES

After studying this chapter, students will be able to

LO1 Demonstrate the electromechanical energy conversion process in a magnetic system with suitable illustrations of linear and non-linear circuits

LO2 Explain the mechanical force developed and flow of mechanical energy in an electromechanical energy conversion system for linear and non-linear magnetic circuits

LO3 Illustrate the electromechanical energy conversion process in a multiple-excited magnetic field system

LO4 Discuss the electromechanical energy conversion process through electric field in a system

4.1 | INTRODUCTION

The chief advantage of electric energy over other forms of energy is the relative ease and high efficiency with which it can be transmitted over long distances. Its main use is in the form of a transmitting link for transporting other forms of energy, e.g. mechanical, sound, light, etc. from one physical location to another. Electric energy is seldom available naturally and is rarely directly utilized. Obviously two kinds of energy conversion devices are needed—to convert one form of energy to the electric form and to convert it back to the original or any other desired form.

In this chapter, the concept of electromechanical conversion and principles involved in the conversion are briefly described. The energy conversion from mechanical to electrical or electrical to mechanical is done by the use of electromechanical devices.

The operating principle of energy conversion devices is similar, but their structural details differ depending upon their function.

Various electromechanical energy conversion devices may be categorized as follows:

- The *first category* of devices, involving small motion, processes only low-energy signals from electrical to mechanical or vice versa such as microphones, gramophone pick-ups, loud speakers and low-signal transducers.

- The *second category* consists of force or torque-producing devices with limited mechanical motion such as electromagnetic relays, moving-iron instruments, actuators, etc.
- The *third category* includes continuous energy conversion devices like motors and generators which are used for bulk energy conversion and utilization.

Electromechanical energy conversion takes place via the medium of a magnetic or electric field—the magnetic field being most suited for practical conversion devices. Because of the inertia associated with mechanically moving members, the fields must necessarily be slowly varying, i.e. *quasistatic* in nature. Electromechanical energy conversion process is basically a reversible process. The coupling between the electrical and mechanical systems of Electro-Mechanical Energy Conversion (EMEC) devices is through the magnetic or electric field. Generally the electromechanical energy conversions work on the principles of field energy as basis, because the energy storing capacity of the magnetic field is much greater (about 25,000 times) than the electric field. The advantages of using field energy as basis in the energy conversion are as follows:

- It can deal with both steady state and transient state of all types of electromechanical energy converters.
- It gives more physical insight into the operation of all EMEC devices which is indispensable.
- It leads physically to the generalized theory of electrical machines.

This chapter is mainly devoted to the understanding of the principle of electromechanical energy conversion. Simple examples will be used for illustrative purposes. In later chapters the analysis of continuous energy conversion equipment will be carried out.

Industry Relevance

Energy can neither be created nor be destroyed". We can only change its forms, using appropriate energy-conversion processes. The purpose of electro-mechanical conversion device is to change the form of energy. When it is converting mechanical input to electrical output the device is "generating". With electrical input, when mechanical output is obtained, the device is motoring.

Electro-mechanical energy-conversion finds applications in following categories of systems:

(a) transducers: Devices for obtaining signals for measurement/control,

(b) force-producing devices: Solenoid-actuators, relays, electromagnets,

(c) devices for continuous-energy-conversion: Motors/Generators

4.2 ENERGY IN MAGNETIC SYSTEM

LO 1 Demonstrate the electromechanical energy conversion process in a magnetic system with suitable illustrations of linear and non-linear circuits

Energy can be stored or retrieved from a magnetic system by means of an exciting coil connected to an electric source. Consider, for example the magnetic system of an attracted armature relay of Fig. 4.1. The resistance of the coil is shown by a series lumping outside the coil which then is regarded as an ideal loss-less coil. The coil current causes magnetic flux to be established in the magnetic circuit. It is assumed that all the flux ϕ is confined* to the iron core and therefore

*The leakage flux (which is of course small) does not take part in the energy conversion process. It can be accounted for by placing an imaginary coil in series with the ideal coil which produces exactly the flux linkages corresponding to the leakage flux. As in the case of transformers, the inductance of such a coil is referred to as the *leakage inductance*. Here the leakage inductance is assumed to be negligible.

links all the N turns creating the coil flux linkages of

$$\lambda = N\phi \tag{4.1}$$

The flux linkage causes a reaction emf of

$$e = \frac{d\lambda}{dt} \tag{4.2}$$

to appear at the coil terminals with polarity (as per Lenz's law) shown in the Fig. 4.1. The associated circuit equation is

$$v = iR + e$$

$$= iR + \frac{d\lambda}{dt} \tag{4.3}$$

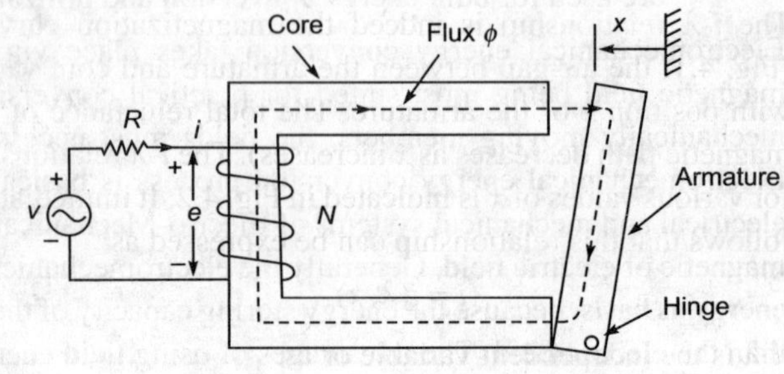

Fig. 4.1 *Attracted armature relay*

The electric energy input into the ideal coil due to the flow of current i in time dt is

$$dW_e = ei\, dt \tag{4.4}$$

Assuming for the time being that the armature is held fixed at position x, all the input energy is stored in the magnetic field. Thus

$$dW_e = ei\, dt = dW_f \tag{4.5}$$

where dW_f is the change in field energy in time dt. When the expression for e in Eq. (4.2) is substituted in Eq. (4.5), we have

$$dW_e = id\lambda = \mathcal{F}\, d\phi = dW_f \tag{4.6}$$

where $\mathcal{F} = Ni$, the magnetomotive force (mmf).

The relationship i-λ or $\mathcal{F}$-λ is a functional one corresponding to the magnetic circuit which in general is nonlinear (and is also history-dependent, i.e. it exhibits hysteresis). The energy absorbed by the field for finite change in flux linkages for flux is obtained from Eq. (4.6) as

$$\Delta W_f = \int_{\lambda_1}^{\lambda_2} i(\lambda)\, d\lambda = \int_{\phi_1}^{\phi_2} \mathcal{F}(\phi)\, d\phi \tag{4.7}$$

Conservative System

As the flux in the magnetic circuit undergoes a cycle $\phi_1 \rightarrow \phi_2 \rightarrow \phi_1$, an irrecoverable loss in energy takes place due to hysteresis and eddy-currents in the iron, assuming here that these losses are separated out and are supplied directly by the electric source. This assumption renders the ideal coil and the magnetic circuit as a *conservative system* with energy interchange between themselves so that the net energy is conserved.

Probe

What are the losses in an iron core of a magnetic circuit?

The energy absorbed by the magnetic system to establish flux ϕ (or flux linkages λ) from initial zero flux is

$$W_f = \int_0^{\lambda} i(\lambda)\, d\lambda = \int_0^{\phi} \mathcal{F}(\phi)\, d\phi \tag{4.8}$$

This then is the energy of the magnetic field with given mechanical configuration when its state corresponds to flux ϕ (or flux linkages λ).

i-λ relationship

The i-λ relationship is indeed the magnetization curve which varies with the configuration variable x (Fig. 4.1: the air-gap between the armature and core varies with position x of the armature. The total reluctance of the magnetic path decreases as x increases). The i-λ relationship for various values of x is indicated in Fig. 4.2. It immediately follows that this relationship can be expressed as

$$i = i(\lambda, x)$$

if λ is the independent variable or as

$$\lambda = \lambda(i, x)$$

if i is the independent variable.

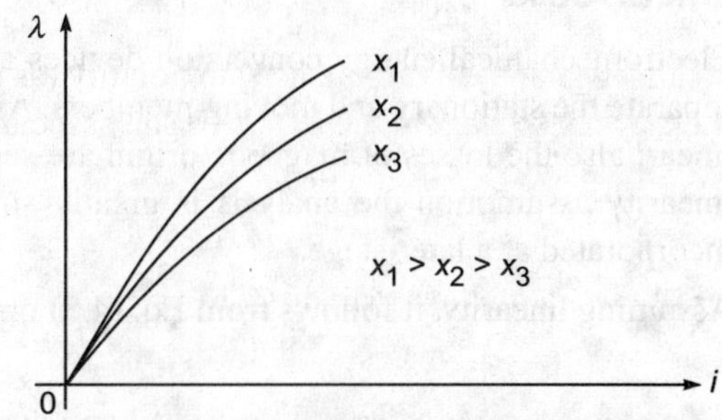

Fig. 4.2 *i-λ relationship with variable x*

Therefore, the field energy (Eq. (4.8)) is in general a function of two variables,

i.e. $$W_f = W_f(\lambda, x) \quad (4.9a)$$

or $$W_f = W_f(i, x) \quad (4.9b)$$

According to Eqs (4.9a) and (4.9b) field energy is determined by the instantaneous values of the system states (λ, x) or (i, x) and is independent of the path followed by these states to reach the present values. This means that the field energy at any instant is history independent.

A change in λ with fixed x causes electric-magnetic energy interchange governed by the circuit Eq. (4.3) and the energy Eq. (4.6). Similarly, if x is allowed to change with fixed λ, energy will interchange between the magnetic circuit and the mechanical system. The general case of such energy interchanges (electric-magnetic-mechanical) is the subject matter of Sec. 4.3.

As per Eq. (4.8) the field energy is the area between the λ-axis and i-λ. curve as shown in Fig. 4.3. A new term, *co-energy* is now defined as

$$W'_f(i, x) = i\lambda - W_f(\lambda, x) \quad (4.10)$$

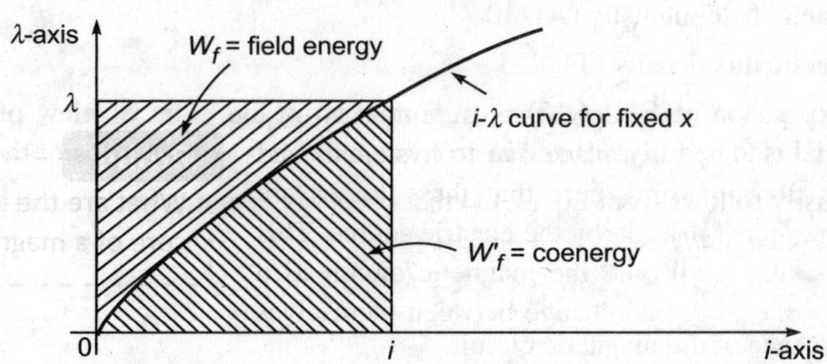

Fig. 4.3 *Field energy and coenergy*

wherein by expressing λ as $\lambda(i, x)$, the independent variables of W'_f become i and x. The coenergy on Fig. 4.3 is shown to be the complementary area of the i-λ curve. It easily follows from Fig. 4.3 that

$$W'_f = \int_0^i \lambda\, di \tag{4.11}$$

Linear Case

Electromechanical energy conversion devices are built with air-gaps in the magnetic circuit which serve to separate the stationary and moving members. As a result the i-λ relationship of the magnetic circuit is almost linear; also the losses of magnetic origin are separately accounted for by semi-empirical methods. With the linearity assumption the analysis is greatly simplified. Losses and certain nonlinear effects may then be incorporated at a later-stage.

Assuming linearity, it follows from Eq. (4.8) or Fig. 4.3 that

$$W_f = \frac{1}{2} i\lambda = \frac{1}{2} \mathcal{F}\phi = \frac{1}{2} \mathcal{R}\phi^2 \tag{4.12}$$

where, as it is known, $\mathcal{R} = \mathcal{F}/\phi$ = reluctance of the magnetic circuit. Since the coil inductance

$$L = \lambda/i$$

the field energy can be expressed as

$$W_f = \frac{1}{2}\frac{\lambda^2}{L} \tag{4.13}$$

In the linear case the inductance L is independent of i but is a function of configuration x. Thus the field energy is a special function of two independent variables λ and x, i.e.

$$W_f(\lambda, x) = \frac{1}{2}\frac{\lambda^2}{L(x)} \tag{4.14}$$

The field energy is distributed throughout the space occupied by the field. Assuming no losses and *constant permeability*, the energy density† of the field is

$$w_f = \int_0^B H dB = \frac{1}{2} HB = \frac{1}{2}\frac{B^2}{\mu} \quad \text{J/m}^3 \tag{4.15}$$

where H = magnetic field intensity (AT/m)

B = magnetic flux density (T)

The energy density expression of Eq. (4.15) is important from the point of view of design wherein the capability of the material is to be fully utilized in arriving at the gross dimensions of the device.

For the linear case it easily follows from Eq. (4.11) that coenergy is numerically equal to energy, i.e.

$$W'_f = W_f = \frac{1}{2} \lambda i = \frac{1}{2} \mathcal{F}\phi = \frac{1}{2} \mathcal{P}\mathcal{F}^2 \tag{4.16}$$

where $\mathcal{P} = \phi/\mathcal{F}$ = permeance of the magnetic circuit.

---

† If $A(\text{m}^2)$ and $l(\text{m})$ are the area and length dimensions of the field, then from Eq. (4.8)

$$w_f = \frac{W_f}{Al} = \int_0^\lambda \frac{iN}{l} d\left(\frac{\lambda}{NA}\right) = \int_0^B H\, dB$$

Also in terms of the coil inductance

$$W'_f = \int_0^i (\lambda = Li)di = \frac{1}{2}Li^2$$

or in general

$$W'_f(i, x) = \frac{1}{2}L(x)i^2 \quad (4.17)$$

The expression for coenergy density is

$$w'_f = \int_0^H B\,dH \quad (4.18a)$$

which for the linear case becomes

$$w'_f = \frac{1}{2}\mu H^2 = \frac{1}{2}\frac{B^2}{\mu} \quad (4.18b)$$

Concept Check — LO 1

1. Define field energy and co-energy. ○○●

4.3 FIELD ENERGY AND MECHANICAL FORCE

LO 2 Explain the mechanical force developed and flow of mechanical energy in an electromechanical energy conversion system for linear and non-linear magnetic circuits

Consider once again the attracted armature relay excited by an electric source as in Fig. 4.4. The field produces a mechanical force F_f in the direction indicated which drives the mechanical system (which may be composed of passive and active mechanical elements). The mechanical work done by the field when the armature moves a distance dx in positive direction is

$$dW_m = F_f\,dx \quad (4.19)$$

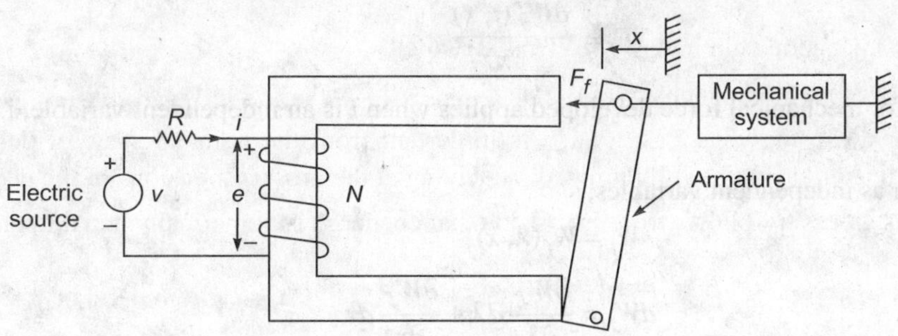

Fig. 4.4 *Production of mechanical force*

Note: Difficulty Level ⟶ ○○● — Level 1 and Level 2 Category; ○●● — Level 3 and Level 4 Category; ●●● — Level 5 and Level 6 Category

This energy is drawn from the field by virtue of change dx in field configuration. As per the principle of energy conservation

Recall Mechanical energy output = electrical energy input – increase in field energy (4.20)

or in symbolic form

$$F_f dx = i d\lambda - dW_f \tag{4.21}$$

It may be seen that $F_f dx$ is the gross mechanical output, a part of which will be lost in mechanical friction.

From Eq. (4.10)

$$W_f = i\lambda - W'_f(i, x)$$

Then

$$dW_f = d(i\lambda) - dW_f(i, x)$$

$$= i d\lambda + \lambda di - \left(\frac{\partial W'_f}{\partial i} di + \frac{\partial W'_f}{\partial x} dx \right) \tag{4.22}$$

Substituting for dW_f from Eq. (4.22) in Eq. (4.21), we have

$$F_f dx = i d\lambda - \left[i d\lambda + \lambda di - \left(\frac{\partial W'_f}{\partial i} di + \frac{\partial W'_f}{\partial x} dx \right) \right]$$

or

$$F_f dx = \left[\frac{\partial W'_f}{\partial i} - \lambda \right] di + \frac{\partial W'_f}{\partial x} dx \tag{4.23}$$

Because the incremental changes di and dx-are independent and di is not present in the left-hand side of Eq. (4.23), its coefficient on the right-hand side must be zero i.e.

$$\frac{\partial W'_f}{\partial i} - \lambda = 0$$

$$\lambda = -\frac{\partial W'_f}{\partial x} \tag{4.24}$$

It then follows from Eq. (4.23) that

$$F_f = \frac{\partial W'_f(i, x)}{\partial x} \tag{4.25}$$

This expression for mechanical force developed applies when i is an independent variable, i.e. it is a *current excited* system

If (λ, x) are taken as independent variables,

$$W_f = W_f(\lambda, x)$$

$$dW_f = \frac{\partial W_f}{\partial \lambda} d\lambda + \frac{\partial W_f}{\partial x} dx \tag{4.26}$$

Substituting Eq. (4.26) in Eq. (4.21)

$$F_f dx = i d\lambda - \frac{\partial W_f}{\partial \lambda} d\lambda - \frac{\partial W_f}{\partial x} dx$$

or $$F_f dx = -\frac{\partial W_f}{\partial x} dx + \left(i - \frac{\partial W_f}{\partial \lambda} \right) d\lambda \tag{4.27}$$

Since $d\lambda$, the independent differential, is not present on the left hand side of this equation,

$$i - \frac{\partial W_f}{\partial \lambda} = 0$$

or $$i = \frac{\partial W_f(\lambda, x)}{\partial \lambda} \tag{4.28}$$

Hence $$F_f = -\frac{\partial W_f(\lambda, x)}{\partial x} \tag{4.29}$$

In this form of expression for the mechanical force of field origin, λ is the independent variable, i.e. it is a voltage-controlled system as voltage is the derivative of λ.

In linear systems where inductances are specified it is more convenient to use coenergy for finding the force developed (Eq. (4.25)). If the system is voltage-controlled, the current can be determined by writing the necessary circuit equations (see Examples 4.11 and 4.12).

It is needless to say that the expressions of Eqs (4.25) and (4.29) for force in a translatory system will apply for torque in a rotational system with x replaced by angular rotation θ.

Direction of Mechanical Force Developed

With reference to Eq. (4.29) it immediately follows that F_f is positive (i.e. it acts in the positive reference direction of x) if $\partial W_f(\lambda, x)/\partial x$ is negative which means that stored energy of the field is reduced with increase of x while flux linkages λ are held fixed. In the particular case of Fig. 4.4 as x increases (i.e., the armature moves towards left), the field energy for fixed λ is reduced because the air-gap is reduced. It means that F_f in this case acts in the positive direction.

Important! It is therefore, concluded that the mechanical force produced by the field acts in a direction to reduce field energy or in other words the system seeks a position of minimum field energy.

Similarly, it can be concluded from Eq. (4.25) that the system seeks a position of maximum coenergy. Also in Fig. 4.4, the force acts in a direction to increase x thereby reducing the magnetic circuit reluctance and increasing the coil inductance.

Determination of Mechanical Force

Nonlinear case It was seen above that the mechanical force is given by the partial derivatives of coenergy or energy as per Eqs (4.25) and (4.29). In the general nonlinear case, the derivative must be determined numerically or graphically by assuming a small increment Δx. Thus

$$F_f \approx \left. \frac{\Delta W_f'}{\Delta x} \right|_{i=\text{cost}} \tag{4.30a}$$

or $$F_f \approx -\left. \frac{\Delta W_f}{\Delta x} \right|_{\lambda=\text{cost}} \tag{4.30b}$$

These two expressions will give slightly different numerical values of F_f because of finite Δx. Obviously F_f is the same in each case as $\Delta x \to 0$. Calculation of F_f by Eq. (4.30a) is illustrated in Ex. 4.2.

Linear case From Eq. (4.17)

$$W_f'(i, x) = \frac{1}{2} L(x) i^2$$

$$\therefore \qquad F_f = \frac{\partial W_f'}{\partial x} = \frac{1}{2} i^2 \frac{\partial L(x)}{\partial x} \tag{4.31}$$

From Eq. (4.31), it is obvious that the force acts in a direction to increase the inductance of the exciting coil, a statement already made.

Alternatively from Eq. (4.14)

$$W_f(\lambda, x) = \frac{1}{2} \frac{\lambda^2}{L(x)}$$

$$\therefore \qquad F_f = \frac{\partial W_f}{\partial x} = \frac{1}{2} \left(\frac{\lambda}{L(x)} \right)^2 \frac{\partial L(x)}{\partial x} \tag{4.32}$$

It may be seen that Eqs (4.31) and (4.32) are equivalent as $i = \lambda/L$.

Also from Eq. (4.12)

$$W_f(\phi, x) = \frac{1}{2} \mathcal{R}(x) \phi^2$$

$$\therefore \qquad F_f = -\frac{\partial W_f}{\partial x} = -\frac{1}{2} \phi^2 \frac{\partial \mathcal{R}(x)}{\partial x} \tag{4.33}$$

It must be remembered here that there is no difference between λ as independent variable or ϕ as independent variable as these are related by a constant ($\lambda = N\phi$). It follows from Eq. (4.33) that the force acts in a direction to reduce reluctance of the magnetic system, a statement that has been made already.

Another expression for F_f can be derived as below:

From Eq. (4.12)

$$W_f(\lambda, x) = \frac{1}{2} \lambda i(x)$$

$$\therefore \qquad F_f = -\frac{\partial W_f}{\partial x} = -\frac{1}{2} \frac{\lambda \partial i(x)}{\partial x} \tag{4.34}$$

Mechanical Energy

When the armature in Fig. 4.4 is allowed to move from position x_a to x_b with the coil current remaining constant at i_o, the mechanical energy output is

$$\Delta W_m = \int_{x_a}^{x_b} F_f \, dx \tag{4.35}$$

Integrating Eq. (4.25)

$$\Delta W_m = \int_{x_a}^{x_b} F_f \, dx = \Delta W_f' \text{ (}i\text{ remaining constant)}$$

$$= \text{increase in coenergy} \tag{4.36}$$

The graphical representation of Eq. (4.36) is given in Fig. 4.5(a) for the general nonlinear case while Fig. 4.5(b) gives the linear case. In each case the electrical energy input is

$$\Delta W_e = i_0(\lambda_2 - \lambda_1) \tag{4.37}$$

For the linear case, it follows from the geometry of Fig. 4.5(b) that

$$\Delta W_f = \Delta W_f' = \Delta W_m = \frac{1}{2}\, i_0(\lambda_2 - \lambda_1) = \frac{1}{2}\,\Delta W_e \tag{4.38}$$

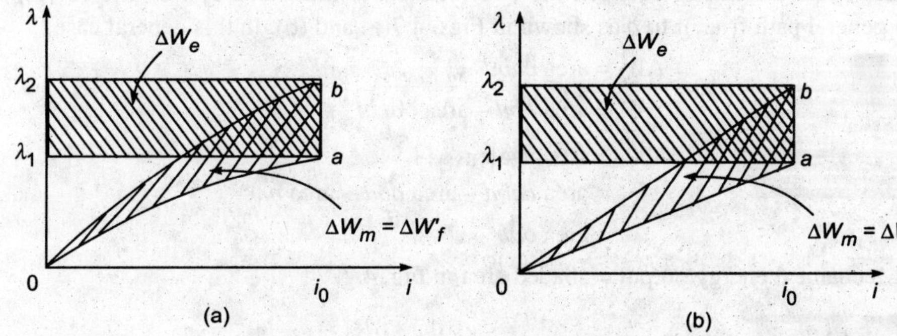

Fig. 4.5 *(a) Nonlinear case, (b) Linear case*

which means that half the electrical energy input gets stored in the field and the other half is output as mechanical energy. In this kind of operation the armature must move from position x_a to x_b infinitely slowly for the excitation coil current to remain constant.

Let now the armature in Fig. 4.4 be allowed to move from x_a and x_b with coil flux linkage λ remaining constant. Integrating Eq. (4.29),

$$\Delta W_m = \int_{x_a}^{x_b} F_f dx = -\Delta W_f (\lambda \text{ remaining constant})$$

$$= \text{decrease in field energy} \tag{4.39}$$

This is illustrated in Fig. 4.6(a) for the general nonlinear case and in Fig. 4.6(b) for the linear case. In each case

$$\Delta W_e = 0 \tag{4.40}$$

For the linear case

$$\Delta W_m = \frac{1}{2}\,\lambda_0\,(i_1 - i_2) \tag{4.41}$$

For λ to remain constant, the armature must move from x_a to x_b in zero time. Since there is no electrical input, the mechanical energy output is drawn from the field energy which reduces by an equal amount.

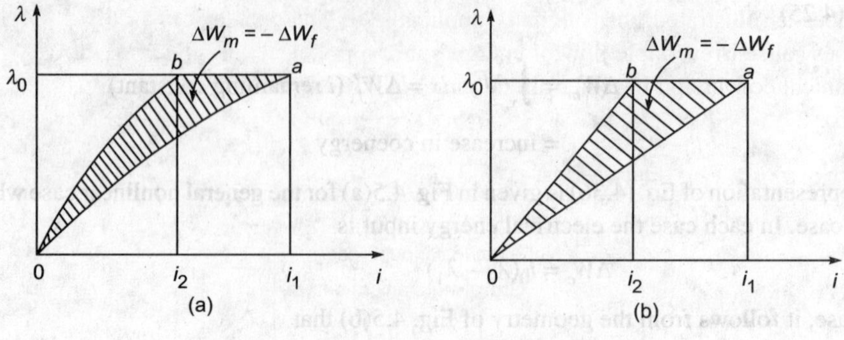

Fig. 4.6 *(a) Nonlinear case, (b) Linear case*

The actual armature movement lies between the two ideal cases illustrated above. The corresponding *i*-λ relationship is a general path from *a* to *b* as shown in Figs. 4.7(a) and (b). In this general case

$$\Delta W_e = \text{area } cabd$$

$$\Delta W_f = \text{area } obd - \text{area } oac$$

Now

$$\Delta W_m = \Delta W_e - \Delta W_f$$

$$= \text{area } cabd - \text{area } obd + \text{area } oac$$

$$= \text{area } oab$$

or Mechanical energy output = shaded area in Fig. 4.7.

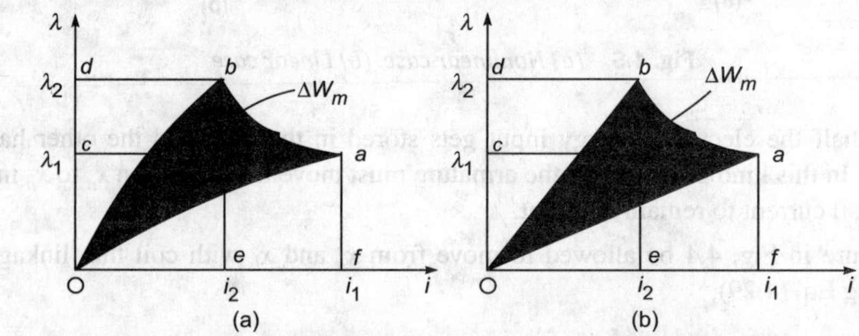

Fig. 4.7 *(a) Nonlinear case, (b) Linear case*

Since *ab* is a general movement, this area which represents the mechanical energy output has to be computed graphically or numerically.

Flow of Energy in Electromechanical Devices

!Attention Electromechanical energy conversion is a reversible process and Eqs (4.25) and (4.29) govern the production of mechanical force. In Fig. 4.4 if the armature is allowed to move on positive *x* direction under the influence of F_f, electrical energy is converted to mechanical form via the coupling field. If instead the armature is moved in the negative *x* direction under the influence of external force, mechanical energy is converted to electrical form via the coupling field. This conversion process is not restricted to translatory devices as illustrated but is equally applicable to rotatory devices (see Ex. 4.4). Electrical and mechanical losses cause irreversible flow of energy out of a practical conversion device. The flow of energy in electromechanical conversion in either direction along with irrecoverable energy losses is shown in Figs. 4.8(a) and 4.8(b).

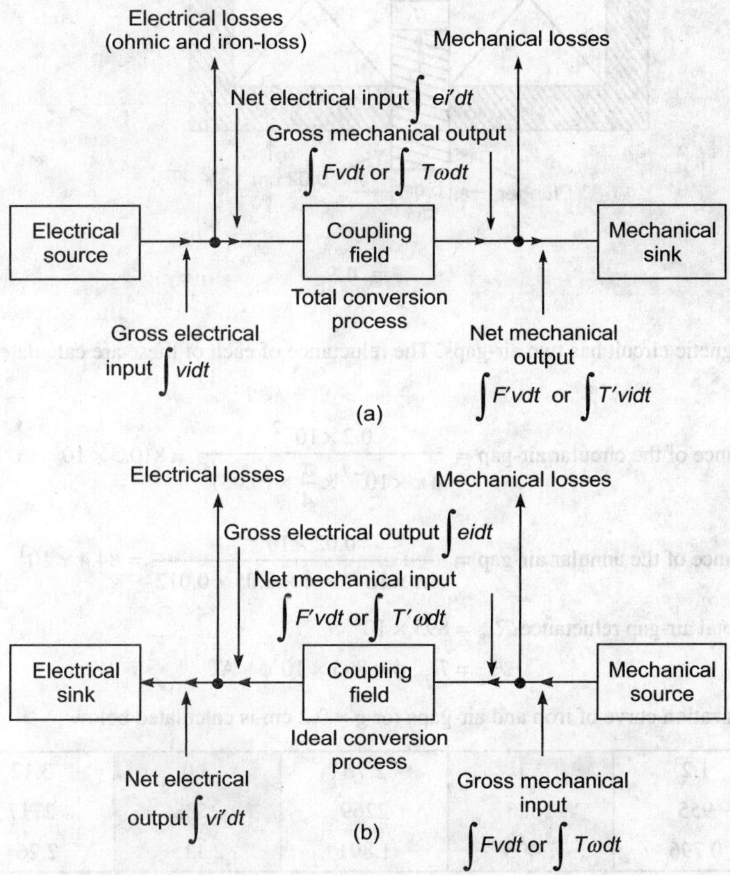

Fig. 4.8 *Flow of energy in electromechanical energy conversion via a coupling field*

Example 4.1 *Figure 4.9 shows the cross-sectional view of a cylindrical iron-clad solenoid magnet. The plunger made of iron is restricted by stops to move through a limited range. The exciting coil has 1200 turns and carries a steady current of 2.25 A. The magnetizing curve of the iron portion of the magnetic circuit is given below:*

Flux, Wb	*0.0010*	*0.00175*	*0.0023*	*0.0025*	*0.0026*	*0.00265*
MMF, AT	*60*	*120*	*210*	*300*	*390*	*510*

Calculate the magnetic field energy and coenergy for air-gap of g = 0.2 cm and g = 1 cm with exciting current of 2.25 A in each case.

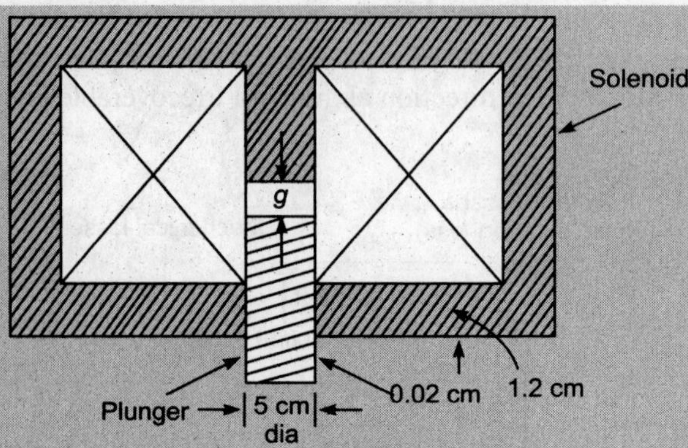

Fig. 4.9

LO 2

Solution The magnetic circuit has two air-gaps. The reluctance of each of these are calculated as follows:

Case 1: $g = 0.2$ cm

$$\text{Reluctance of the circular air-gap} = \frac{0.2 \times 10^{-2}}{4\pi \times 10^{-7} \times \dfrac{\pi}{4} \times (0.05)^2} = 810.5 \times 10^3$$

$$\text{Reluctance of the annular air-gap} = \frac{0.02 \times 10^{-2}}{4\pi \times 10^{-7} \times \pi \times 0.05 \times 0.012} = 84.4 \times 10^3$$

$$\text{Total air-gap reluctance } \mathcal{R}_{ag} = 895 \times 10^3$$

$$\mathcal{F}_{ag} = \mathcal{R}_{ag}\,\phi = 895 \times 10^3 \phi \quad \text{AT}$$

The combined magnetization curve of iron and air-gaps for $g = 0.2$ cm is calculated below:

λ (WbT)	1.2	2.1	2.76	3.0	3.12	3.18
AT	955	1686	2269	2538	2717	2882
i(A)	0.796	1.405	1.891	2.115	2.264	2.40

The i-λ curve is plotted in Fig. 4.10 from which the field coenergy found graphically is

$$\text{Area } oea = 3.73 \text{ J}$$

$$\therefore \quad \text{Energy area } oaf = 3.11 \times 2.25 - 3.73 = 3.27 \text{ J}$$

Case 2: $g = 1$ cm

$$\text{Reluctance of circular air-gap} = 4052.5 \times 10^3$$

$$\text{Reluctance of annular air-gap} = 84.4 \times 10^3$$

$$\mathcal{R}_{ag} = 4136.7 \times 10^3$$

Because of such high reluctance of air-gap, ampere-turns absorbed by iron part of the magnetic circuit can be neglected, e.g. for

$$\phi = 0.0025 \text{ Wb}$$

$$\text{AT (air-gap)} = 10342$$

$$\text{AT (iron)} = 390$$

Thus it is seen that even in the near saturation region, AT (iron) is less than 5% of the total AT required for establishing the flux.

For

$$i = 2.25 \text{ A}$$

$$\mathcal{F} = 1200 \times 2.25 = 2700 \text{ AT}$$

$$\phi = \frac{2700}{4136.7 \times 10^3} = 0.653 \times 10^{-3} \text{ Wb}$$

$$\lambda = N\phi = 1200 \times 0.653 \times 10^{-3} = 0.784 \text{ Wb-T}$$

$$\text{Field energy} = \text{coenergy} = \frac{1}{2} i\lambda$$

$$= \frac{1}{2} \times 2.25 \times 0.784 = 0.882 \text{ J}$$

Example 4.2 *In Ex. 4.1 calculate the force on the plunger for g = 0.2 cm with an exciting current of 2.25 A.*

LO 2

Solution

$$F_f = \frac{\partial W'_f(i, g)}{\partial g}$$

where g is the air-gap. Given $g = 0.2 \times 10^{-2}$ m.

Case a: Reluctance of the iron path accounted for.

This is the nonlinear case where the derivative desired can be found numerically.

Assume

$$\Delta g = -0.05 \times 10^{-2} \text{ m (a decrease)}$$

$$g + \Delta g = 0.15 \times 10^{-2} \text{ m}$$

Air-gap reluctance

$$= 810.5 \times 10^3 \times \frac{0.15}{0.2} + 84.4 \times 10^3$$

$$= 692 \times 10^3$$

For this air-gap

λ	1.2	2.1	2.76	3.0	3.12	3.18
AT	752	1331	1802	2030	2189	2344
i	0.626	1.110	1.501	1.692	1.824	1.953

This is plotted in Fig. 4.10 from which increase in co energy for $i = 2.25$ A is the area *oab*, i.e.

$$\Delta W'_f = 0.718 \text{ J}$$

$$\therefore \quad F_f = \frac{\partial W'_f}{\partial g} = \frac{0.718}{-0.05 \times 10^{-2}} = -1436 \text{ N}$$

Since a decrease in g causes an increase in W_f, the force on plunger acts in negative direction (positive direction is in the increasing direction of g). Hence the force is attractive (tending to reduce g).

It may be noted that better results will be obtained by choosing a smaller value of Δg.

Case b: The magnetization curve of iron is assumed linear (corresponding to the initial slope).

This is the linear case so that we can proceed analytically.

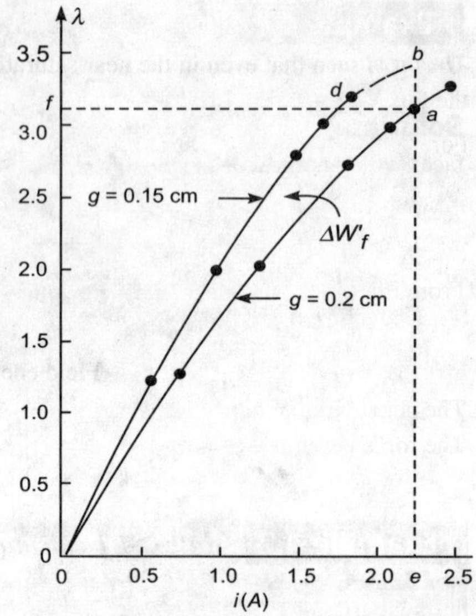

Fig. 4.10

$$\text{Total reluctance } \mathcal{R} = \frac{60}{0.001} + 84.4 \times 10^3 + \frac{g}{4\pi \times 10^{-7} \times \frac{\pi}{4}(0.05)^2}$$

$$= 144.4 \times 10^3 + 4053g \times 10^5$$

$$= 144.4\,(1 + 28.1 \times 10^2 g) \times 10^3$$

$$\phi = \frac{1200i}{144.4(1 + 28.1 \times 10^2 g) \times 10^3}\ \text{Wb}$$

$$\lambda = N\phi = \frac{(1200)^2 i}{144.4(1 + 28.1 \times 10^2 g) \times 10^3}\ \text{WbT}$$

$$W'_f(i, g) = \frac{1}{2} i\lambda = \frac{(1200)^2 i^2}{2 \times 144.4(1 + 28.1 \times 10^2 g) \times 10^3}$$

$$F_f = \frac{\partial W'_f}{\partial g} = -\frac{(1200)^2 i^2 \times 28.1 \times 10^2}{2 \times 144.4(1 + 28.1 \times 10^2 g)^2 \times 10^3}$$

For $g = 0.2 \times 10^{-2}$ m

$$F_f = -\frac{(1200)^2 \times (2.25)^2 \times 28.1 \times 10^2}{2 \times 144.4(1 + 28.1 \times 0.2)^2 \times 10^3} = -1619\ \text{N}$$

The linearity assumption causes the values of force to differ by about 13% from that obtained by the nonlinear method. The actual difference will be still less as in the nonlinear case the derivative is obtained by approximation ($\Delta g = 0.05 \times 10^2$ m). It may be noted here that the linearity assumption renders great computational saving and is hence commonly employed.

Example 4.3 *In Ex. 4.1, assume the reluctance of the iron path to be negligible. The exciting current is 2.25 A. The plunger is now allowed to move very slowly from g = 1 cm to g = 0.2 cm. Find the electrical energy input to the exciting coil and the mechanical output.*

LO 2

Solution As already calculated in Ex. 4.2, the reluctance of the magnetic path as a function of the air-gap length is

$$\mathcal{R} = (84.4 + 4053 \times 10^2 g) \times 10^{-3}$$

Flux linkages for an exciting current of 2.25 A are

$$= \frac{(1200)^2 \times 2.25}{(84.4 + 4053 \times 10^2 g)^2 \times 10^3} = \frac{38.4}{(1 + 48 \times 10^2 g)}\ \text{WbT}$$

Since the plunger moves very slowly, the exciting current remains constant at 2.25 A. Hence

$$\Delta W_e = i_0\,(\lambda_2 - \lambda_1) = 2.25 \times 38.4 \left(\frac{1}{1 + 48 \times 0.2} - \frac{1}{1 + 48 \times 1}\right) = 2.25 \times 38.4\,(0.094 - 0.020) = 6.39\ \text{J}$$

$$\Delta W_m = \frac{1}{2}\Delta W_e = 3.195\ \text{J}$$

Example 4.4 *The magnetic flux density on the surface of an iron face is 1.6 T which is a typical saturation level value for ferromagnetic material. Find the force density on the iron face.* LO 2

Solution Let the area of the iron face be $A(\text{m})^2$. Consider the field energy in the volume contained between the two faces with a normal distance x. From Eq. (4.15)

$$W_f(B, x) = \frac{1}{2}\frac{B^2 Ax}{\mu}$$

From Eq. (4.29), the mechanical force due to the field is

$$F_f = -\frac{\partial W_f(B, x)}{\partial x} = -\frac{1}{2}\frac{B^2 A}{\mu}$$

The negative sign indicates that the force acts in a direction to reduce x (i.e. it is an attractive force between the two faces). The force per unit area is

$$|F_f| = \frac{1}{2}\frac{B^2}{\mu} = \frac{1}{2}\times\frac{(1.6)^2}{4\pi\times10^{-7}} = 1.02\times10^6 \text{ N/m}^2$$

Example 4.5 *In the electromagnetic relay of Fig. 4.11 excited from a voltage source, the current and flux linkages are related as*

$$i = \lambda^2 + 2\lambda(1-x)^2 \; ; x < 1$$

Find the force on the armature as a function of λ. LO 2

Solution

$$W_f(\lambda, x) = \int_0^\lambda i d\lambda = \frac{1}{3}\lambda^3 + \lambda^2(1-x)^2$$

$$F_f = -\frac{\partial W_f}{\partial x} = 2\lambda^2(\lambda - x)^2$$

Example 4.6 *The electromagnetic relay of Fig. 4.11 is excited from a voltage source*

$$v = \sqrt{2}\, V \sin \omega t$$

Assuming the reluctance of the iron path of the magnetic circuit to be constant, find the expression for the average force on the armature, when the armature is held fixed at distance x.

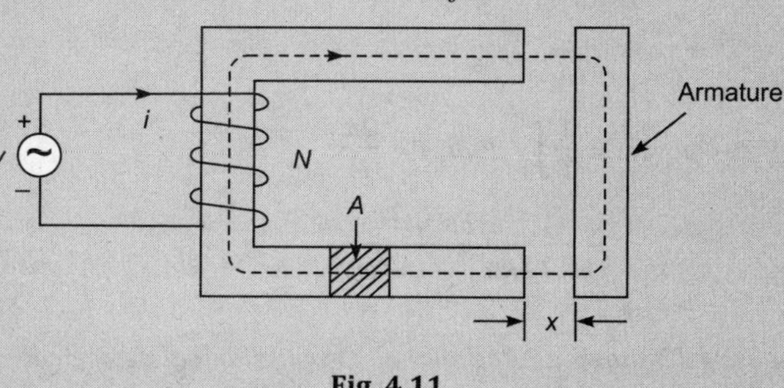

Fig. 4.11

LO 2

Solution

$$\text{Reluctance of the iron path} = a \text{ (say)}$$

$$\text{Reluctance of the air path} = \frac{2x}{\mu_0 A} = bx$$

$$\text{Total reluctance of the magnetic path, } \mathcal{R} = a + bx$$

$$W_f(\phi, x) = \frac{1}{2}\mathcal{R}(x)\phi^2 \qquad (\text{Note that } \lambda = N\phi)$$

$$F_f = -\frac{\partial W_f(\phi, x)}{\partial x} = -\frac{1}{2}\phi^2\frac{\partial \mathcal{R}}{\partial x} = -\frac{1}{2}b\phi^2 \tag{4.42}$$

Notice that $\partial\mathcal{R}/\partial x = b$ is positive, so that F_f is negative, i.e. it acts in a direction to reduce x (which means in a direction to reduce reluctance $\mathcal{R}$).

Now i and v are related by the circuit equation

$$v = \mathcal{R} + L\frac{di}{dt}$$

whose steady-state solution is

$$\bar{I} = \frac{V}{\sqrt{R^2 + \omega^2 L^2}} \angle -\tan^{-1}\frac{\omega L}{R} \tag{4.43}$$

Then

$$i = \frac{\sqrt{2}V}{\sqrt{R^2 + \omega^2 L^2}} \sin\left(\omega t - \tan^{-1}\frac{\omega L}{r}\right) \tag{4.44}$$

From Eq. (2.21) $L = N^2/\mathcal{R}$

Then
$$\phi = \frac{Ni}{\mathcal{R}} = \frac{\sqrt{2}NV}{\sqrt{(R\mathcal{R})^2 + (N^2\omega)^2}} \sin\left(\omega t - \tan^{-1}\frac{\omega N^2}{R\mathcal{R}}\right) \tag{4.45}$$

Substituting ϕ in Eq. (4.42)

$$F_f = -\frac{bN^2V^2}{(R\mathcal{R})^2 + (N^2\omega)^2} \sin^2\left(\omega t - \tan^{-1}\frac{N^2\omega}{R\mathcal{R}}\right)$$

Time-average force is then

$$F_f(av) = \frac{1}{T}\int_0^T F_f dt;\ T = \frac{2\pi}{\omega}$$

$$= -\frac{1}{2}\frac{bN^2V^2}{(R\mathcal{R})^2 + (N^2\omega)^2} \tag{4.46}$$

Example 4.7 *Figure 4.12 shows a rotational electromechanical device called the reluctance motor. It is required to determine the torque acting on the rotor as a function of current input to the exciting coil and the angle (θ) of rotor and stator overlap.*

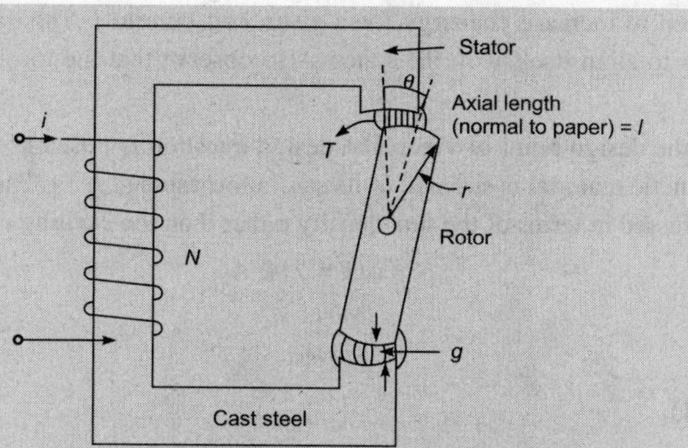

Fig. 4.12 *Elementary reluctance machine*

Obviously the torque expression from coenergy, i.e. $W_f'(i, \phi)$ must be developed. It is assumed that the cast steel magnetic path has negligible reluctance so that the reluctance encountered in the magnetic path is that due to the two annular air-gaps.

LO 2

Solution

$$\text{Air-gap area normal to flux, } A = \left(r + \frac{1}{2}g\right)\theta l$$

Total air-gap length along the flux path = 2g

$$\text{Reluctance of the air-gaps} = \frac{2g}{\mu_0\left(r + \frac{1}{2}g\right)\theta l} \tag{4.47}$$

$$\text{Flux established, } \phi = \frac{Ni\mu_0\left(r + \frac{1}{2}g\right)\theta l}{2g} \tag{4.48}$$

$$\text{Flux linkage, } \lambda = \phi N$$

$$= \frac{N^2 i\mu_0\left(r + \frac{1}{2}g\right)\theta l}{2g} \tag{4.49}$$

$$\text{Coenergy, } W_f'(i, \theta) = \frac{1}{2}\lambda i$$

$$= \frac{N^2 i^2\mu_0\left(r + \frac{1}{2}g\right)\theta l}{4g} \tag{4.50}$$

$$\text{Torque developed, } T_f = \frac{\partial W_f'(i, \theta)}{\partial \theta}$$

$$= \frac{N^2 i^2\mu_0\left(r + \frac{1}{2}g\right)l}{4g} \tag{4.51}$$

This torque acts in a direction to increase coenergy for a given coil current i. This happens (see Eq. (4.50)) when θ increases, i.e. the rotor tends to align itself with the stator. Also observe that the torque in this case is independent of angle θ.

Consider the problem from the design point of view. The design question is posed as to the maximum torque that can be developed, when the magnetic material is stressed to its saturation value (1.6 T). The torque expression of Eq. (4.51) above must therefore be expressed in terms of the flux density rather than the exciting coil current.

$$B = \phi/A = \mathcal{F}/\mathcal{R}\,A$$

$$= \frac{\mu_0 Ni}{2g} \tag{4.52}$$

Using Eq. (4.52) in Eq. (4.51),

$$T_f = \frac{B^2 gl\left(r + \frac{1}{2}g\right)}{\mu_0} \tag{4.53}$$

Since the maximum value of B is fixed from consideration of the magnetic material, any desired torque can be achieved by a suitable combination of g, r and l. The considerations in relative adjustment of these three dimensions of the device are beyond the scope of this book.

Let

$$g = 0.0025 \text{ m}$$

$$l = r = 0.025 \text{ m}$$

$$B = 1.6 \text{ T}$$

Then

$$T_f = \frac{(1.6)^2 \times 0.0025 \times 0.025\left(0.025 + \frac{1}{2} \times 0.0025\right)}{4\pi \times 10^{-7}} = 3.34 \text{ Nm}$$

In the nonoverlapping region of the rotor and stator, the field geometry is very complex and an analytical expression for torque is not possible. It is interesting to examine this problem from the point of view of the coil inductance as a function of the rotor's angular position. From Eq. (4.49)

$$L(\theta) = \lambda/i$$

$$= \frac{N^2 \mu_0 l\left(r + \frac{1}{2}g\right)\theta}{2g} \tag{4.54}$$

With constant air-gap length, in the overlapping region, the inductance increases linearly with θ acquiring a maximum value when rotor is in the vertical position in Fig. 4.12 and then decreases linearly. From Eq. (4.54), the inductance is zero at $\theta = 90°$, i.e. no overlap between rotor and stator pole faces. However, it is known that the coil inductance is not zero for $\theta = 90°$ which means that the inductance model of Eq. (4.54) is not valid in the region of low θ. In fact, the inductance has a least value in the horizontal position of the rotor and rises to a maximum value when the rotor goes to the vertical position travelling from either direction. In a practical device the region between the two pole faces of the stator is narrow (Fig. 4.13) and the rotor and stator pole faces are so shaped that the reluctance of the magnetic

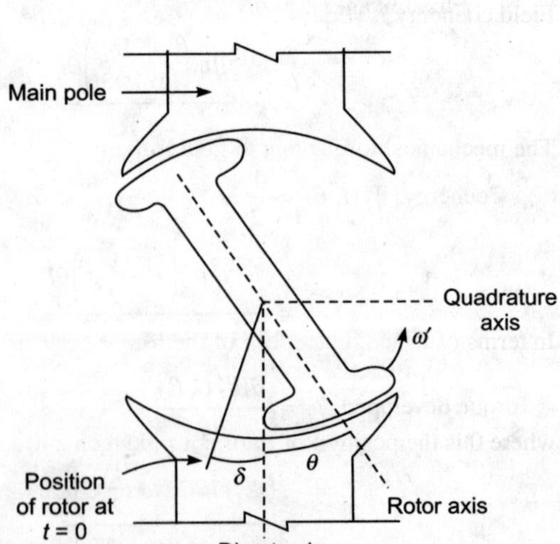

Fig. 4.13

circuit and therefore the coil inductance varies almost sinusoidally. The coil inductance has a maximum value when the rotor is aligned along the main pole axis (called the *direct axis*) and a minimum value when the rotor is at 90° to the main pole axis (called the *quadrature axis*) as shown in Fig. 4.13. It is convenient to choose the direct axis as the reference for angle θ.

It is readily seen from Fig. 4.13 that the coil inductance is a double frequency function of θ (there are two cycles of L-variation in one complete rotation of the rotor). Therefore, $L(\theta)$ can be written as:

$$L(\theta) = L_1 + L_2 \cos 2\theta \tag{4.55}$$

This variation of inductance is depicted in Fig. 4.14.

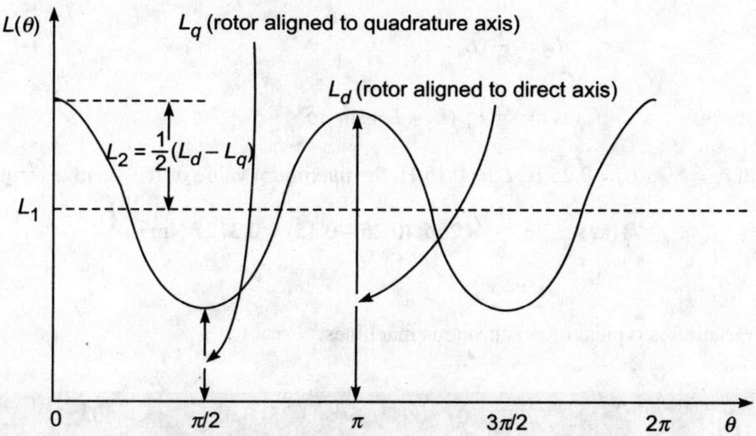

Fig. 4.14 *Variation of the coil inductance with rotor position in a reluctance machine*

Assume the excitation current to be sinusoidal,

$$i = I_m \cos \omega t \tag{4.56}$$

Field coenergy is (Eq. (4.17))

$$W'_f(i, \theta) = \frac{1}{2} L(\theta) i^2$$

The mechanical torque due to field is then

$$T_f = \frac{\partial W'_f}{\partial \theta} = \frac{1}{2} i^2 \frac{\partial L(\theta)}{\partial \theta}$$

$$= -I_m^2 L_2 \sin 2\theta \cos^2 \omega t \tag{4.57}$$

In terms of the angular speed of the rotor (ω')

$$\theta = \omega' t - \delta \tag{4.58}$$

where 0 is the position of rotor at t = 0 when current i is maximum. Then

$$T_f = -I_m^2 L_2 \sin 2(\omega' t - \delta) \sin \omega t$$

$$= -\frac{1}{2} I_m^2 L_2 \sin 2(\omega' t - \delta)(1 + \cos 2\omega t)$$

$$= -\frac{1}{2} I_m^2 L_2 \{\sin 2(\omega' t - \delta) + \sin 2(\omega' t - \delta)\cos 2\omega t\}$$

$$= -\frac{1}{2} I_m^2 L_2 \left\{\sin 2(\omega' t - \delta) + \frac{1}{2}[\sin 2(\omega' t + \omega t - \delta) + \sin 2(\omega' t - \omega t - \delta)]\right\} \tag{4.59}$$

It is observed from Eq. (4.59) that the torque is time-varying with the average value zero if of $\omega' \neq \omega$. However, when the rotor runs at speed $\omega' = \omega$, called the *synchronous speed*, the average torque is

$$T_f(\text{av}) = \frac{1}{4} I_m^2 L_2 \sin 2\delta \tag{4.60}$$

From Fig. 4.14

$$L_2 = \frac{1}{2}(L_d - L_q)$$

$$\therefore \quad T_f(\text{av}) = \frac{1}{8} I_m^2 (L_d - L_q) \sin 2\delta \tag{4.61}$$

Thus, for example, with $I_m = 5$ A, $L_d = 0.25$ H, $L_q = 0.15$ H, the maximum value of the average torque is

$$T_f(\text{av})|_{\max} = \frac{1}{8} \times 25 \times (0.25 - 0.15) = 0.3125 \text{ Nm}$$

when $\delta = 45°$

Sinusoidal torque – δ variation is typical of synchronous machines.

Concept Check — LO 2

1. What is a current-excited system? ○○●
2. What is a voltage-controlled system? ○○●
3. What is electromechanical energy conversion? ○○●
4. Why do all practical energy conversion devices make use of the magnetic field as a coupling medium rather than an electric field? ●●●
5. What are the special applications where the electric field is used as a coupling medium for electromechanical energy conversion? Also explain why electric field coupling is preferred in such applications. ○●●

4.4 | MULTIPLE-EXCITED MAGNETIC FIELD SYSTEMS

LO 3 Illustrate the electromechanical energy conversion process in a multiple-excited magnetic field system

Singly-excited devices discussed earlier, are generally employed for motion through a limited distance or rotation through a prescribed angle. Electro-mechanical transducers have the special requirement of producing an electrical signal proportional to forces or velocities or producing force proportional to electrical signal (current or voltage). Such transducers require two excitations—one excitation establishes a magnetic field of specified strength while the other excitation produces the desired signal (electrical or mechanical). Also continuous energy conversion devices—motors and generators—require multiple excitation. One continuous energy conversion device has already been studied in Ex. 4.6 which is singly-excited (reluctance motor).

Probe

What are the losses in an iron core of a magnetic circuit?

Expressions for Field Energy and Coenergy

Figure 4.15 shows a magnetic field system with two electrical excitations—one on stator and the other on rotor. The system can be described in either of the two sets of three independent variables; (λ_1, λ_2, θ) or (i_1, i_2, θ). In terms of the first set

$$T_f = -\frac{\partial W_f(\lambda_1, \lambda_2, \theta)}{\partial \theta} \tag{4.62}$$

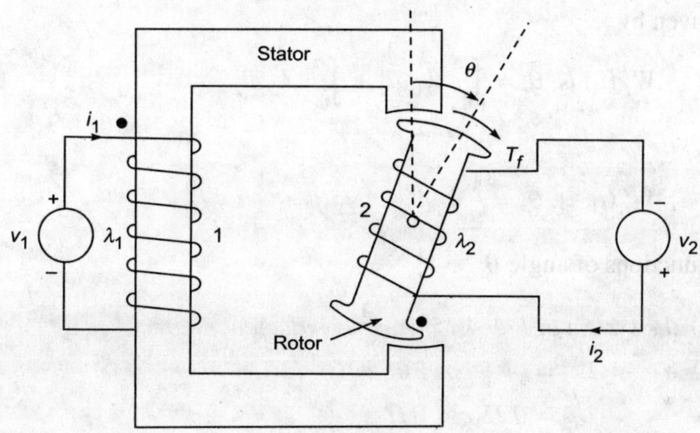

Fig. 4.15

where the field energy is given by

$$W_f(\lambda_1, \lambda_2, \theta) = \int_0^{\lambda_1} i_1 d\lambda_1 + \int_0^{\lambda_2} i_2 d\lambda_2 \tag{4.63}$$

Analogous to Eq. (4.28)

$$i_1 = \frac{\partial W_f(\lambda_1, \lambda_2, \theta)}{\partial \lambda_1}$$

$$i_2 = \frac{\partial W_f(\lambda_1, \lambda_2, \theta)}{\partial \lambda_2}$$

Assuming linearity

$$\lambda_1 = L_{11} i_1 + L_{12} i_2 \tag{4.64a}$$

$$\lambda_2 = L_{21} i_1 + L_{22} i_2; \ (L_{12} = L_{21}) \tag{4.64b}$$

Solving for i_1 and i_2 in terms of λ_1, λ_2 and substituting in Eq. (4.63) gives upon integration‡

$$W_f(\lambda_1, \lambda_2, \theta) = \frac{1}{2}\beta_{11}\lambda_1^2 + \beta_{12}\lambda_1\lambda_2 + \frac{1}{2}\beta_{22}\lambda_2^2 \tag{4.65}$$

where

$$\beta_{11} = L_{22}/(L_{11}L_{22} - L_{12}^2)$$
$$\beta_{22} = L_{11}/(L_{11}L_{22} - L_{12}^2)$$
$$\beta_{12} = \beta_{21} = -L_{12}/(L_{11}L_{22} - L_{12}^2)$$

The self- and mutual-inductance of the two exciting coils are functions of angle θ.

If currents are used to describe the system state

$$T_f = \frac{\partial W_f'(i_1, i_2, \theta)}{\partial \theta} \tag{4.66}$$

where the coenergy is given by

$$W_f'(i_1, i_2, \theta) = \int_0^{i_1} \lambda_1\, di_1 + \int_0^{i_2} \lambda_2\, di_2 \tag{4.67}$$

In the linear case

$$W_f'(i_1, i_2, \theta) = \frac{1}{2}L_{11}i_1^2 + L_{12}i_1i_2 + \frac{1}{2}L_{22}i_2^2$$

where inductances are functions of angle θ.

Example 4.8 *For the system of Fig. 4.15, various inductances are:*

$$L_{11} = (4 + \cos 2\theta) \times 10^{-3}\ H$$
$$L_{12} = 0.15 \cos \theta\ H$$
$$L_{22} = (20 + 5 \cos 2\theta)\ H$$

Find the torque developed if $i_1 = 1$ A, $i_2 = 0.02$ A.

LO 3

Solution

$$W_f'(i_1, i_2, \theta) = \frac{1}{2}(4 + \cos 2\theta) \times 10^{-3} \times i_1^2 + (0.15 \cos \theta)i_1i_2 + \frac{1}{2}(20 + 5 \cos 2\theta)i_2^2$$

---

‡
$$i_1 = \beta_{11}\lambda_{11} + \beta_{12}\lambda_2$$
$$i_2 = \beta_{21}\lambda_1 + \beta_{22}\lambda_2;\ \beta_{21} = \beta_{12}$$

$$W_f(\lambda_1, \lambda_2, \theta) = \int_0^{\lambda_1}(\beta_{11}\lambda_1 + \beta_{12}\lambda_2)d\lambda_1 + \int_0^{\lambda_2}(\beta_{12}\lambda_1 + \beta_{22}\lambda_2)d\lambda_2$$

$$= \beta_{11}\int_0^{\lambda_1}\lambda_1 d\lambda_1 + \beta_{12}\left[\int_0^{\lambda_1}\lambda_2 d\lambda_1 + \int_0^{\lambda_2}\lambda_1 d\lambda_2\right] + \beta_{22}\int_0^{\lambda_2}\lambda_2 d\lambda_2$$

$$= \beta_{11}\int_0^{\lambda_1}\lambda_1 d\lambda_1 + \beta_{12}\int_0^{\lambda_1,\lambda_2} d(\lambda_1\lambda_2) + \beta_{22}\int_0^{\lambda_2}\lambda_2 d\lambda_2$$

$$= \frac{1}{2}\beta_{11}\lambda_1^2 + \beta_{12}\lambda_1\lambda_2 + \frac{1}{2}\beta_{22}\lambda_2^2$$

$$T_f = \frac{\partial W_f'}{\partial \theta} = (\sin 2\theta) \times 10^{-3}\, i_1^2 - 0.15\, (\sin \theta) i_1 i_2 - 5(\sin 2\theta) i_2^2$$

$$= -10^{-3} \sin 2\theta - 3 \times 10^{-3} \sin \theta$$

The first term $-10^{-3} \sin 2\theta$ is the reluctance torque which arises if the self-inductances are functions of space angle θ. If L_1 and L_2 are independent of θ (the rotor and stator are round with uniform air-gap, known as the *round rotor* construction) the reluctance torque becomes zero. The second term is the torque produced by the mutual component. It is also seen that the reluctance torque is a double frequency of the space angle as compared to the second term. The negative sign indicates that the torque is restoring in nature, i.e. it opposes the displacement θ.

Example 4.9 *In the electromagnetic relay shown in Fig. 4.16*

$$L_{11} = k_1/x,\ L_{22} = k_2/x,\ L_{12} = k_3/x$$

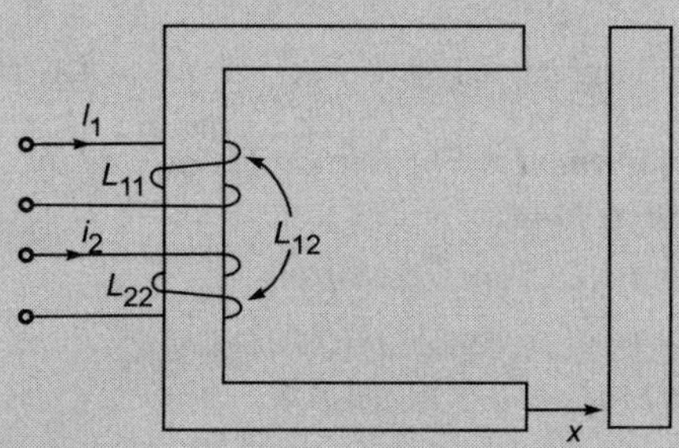

Fig. 4.16

Find the expression for the force on the armature, if

$$i_1 = I_1 \sin \omega_1 t,\ i_2 = I_2 \sin \omega_2 t$$

write an expression for the average force. For what relationship between ω_1 and ω_2, the average force is (i) maximum (ii) minimum.

LO 3

Solution

$$W_f'(i_1, i_2, x) = \frac{1}{2}\frac{k_1}{x} i_1^2 + \frac{k_2}{x} i_1 i_2 + \frac{1}{2}\frac{k_3}{x} i_2^2$$

$$F_f = \frac{\partial W_f'}{\partial x} = -\frac{1}{2}\frac{k_1}{x^2} i_1^2 - \frac{k_2}{x^2} i_1 i_2 - \frac{1}{2}\frac{k_3}{x^2} i_2^2$$

Substituting for i_1, i_2

$$F_f = -\frac{1}{2}\frac{k_1}{x^2} I_1^2 \sin^2 \omega_1 t - \frac{k_2}{x^2} I_1 I_2 \sin \omega_1 t \sin \omega_2 t - \frac{1}{2}\frac{k_3}{x^2} I_2^2 \sin^2 \omega_2 t$$

$$F_f = -\frac{1}{4}\frac{k_1^2}{x^2} I_1^2 + \frac{1}{4}\frac{k_1^2}{x^2} \cos 2\omega_1 t - \frac{1}{2}\frac{k_2}{x^2} I_1 I_2 \cos(\omega_1 - \omega_2)t$$

$$+ \frac{1}{2}\frac{k_1}{x^2} I_1 I_2 \cos(\omega_1 + \omega_2)t - \frac{1}{4}\frac{k_2}{x^2} I_2^2 - \frac{1}{4}\frac{k_3}{x^2} I_2^2 \cos 2\omega_2 t$$

Since these are mixed frequency terms

$$F_f(\text{av}) = \lim_{T\to\infty} \frac{1}{T}\int_0^T F_f(t)\,dt$$

If $\omega_1 \neq \omega_2$,

$$F_f(\text{av}) = -\frac{1}{4}\frac{k_1^2}{x^2} I_1^2 - \frac{1}{4}\frac{k_2}{x^2} I_2^2 \text{ (minimum force)}$$

If $\omega_1 = \omega_2$,

$$F_f(\text{av}) = -\frac{1}{4}\frac{k_1^2}{x^2} I_1^2 - \frac{1}{2}\frac{k_2}{x^2} I_1 I_2 - \frac{1}{4}\frac{k_2}{x^2} I_2^2 \text{ (maximum force)}$$

Example 4.10 *Two coupled coils have self- and mutual-inductance of*

$$L_{11} = 2 + \frac{1}{2x}; \quad L_{22} = 1 + \frac{1}{2x}; \quad L_{12} = L_{21} = \frac{1}{2x}\Big)$$

over a certain range of linear displacement x. The first coil is excited by a constant current of 20 A and the second by a constant current of –10 A. Find:

(a) Mechanical work done if x changes from 0.5 to 1 m.

(b) Energy supplied by each electrical source in part (a).

(c) Change in field energy in part (a).

Hence verify that the energy supplied by the sources is equal to the increase in the field energy plus the mechanical work done. LO 3

Solution Since it is the case of current excitations, the expression of coenergy will be used

$$W_f'(i_1, i_2, x) = \frac{1}{2}L_{11}i_1^2 + L_{12}i_1 i_2 + \frac{1}{2}L_{22}i_2^2$$

$$= \left(2 + \frac{1}{2x}\right) \times 200 + \frac{1}{2x} \times (-200) + \left(1 + \frac{1}{2x}\right) \times 50$$

$$= 450 + \frac{25}{x}$$

(a)

$$F_f = \frac{\partial W_f'}{\partial x} = -\frac{25}{x^2}$$

$$\Delta W_m = \int_{0.5}^{1} F_f dx = \int_{0.5}^{1} -\frac{25}{x^2} dx = -25 \text{ J}$$

(b)

$$\Delta W_{el} = \int_{\lambda_1(x=0.5)}^{\lambda_1(x=1)} i_1\, d\lambda_1 = i_1[\lambda_1(x=1) - \lambda_1(x=0.5)]$$

$$\lambda_1 = L_{11}i_1 + L_{12}i_2$$

$$= \left(2 + \frac{1}{2x}\right) \times 20 + \frac{1}{2x} \times (-10) = 40 + \frac{5}{x}$$

$$\lambda_1(x = 0.5) = 50,\ \lambda_1(x = 1) = 45$$

$\therefore$ $$\Delta W_{e1} = 20(45 - 50) = -100 \text{ J}$$

Similarly $$\Delta W_{e2} = i_2[\lambda_2(x = 1) - \lambda_2(x = 0.5)]$$

$$\lambda_2 = L_{12}i_1 + L_{22}i_2$$

$$\lambda_2 = \frac{1}{2x} \times 20 + \left(1 + \frac{1}{2x}\right) \times (-10) = -10 + \frac{5}{x}$$

$$\lambda_2(x = 0.5) = 0,\ \lambda_2\ (x = 1) = -5$$

$$\Delta W_{e2} = -10(-5) = 50 \text{ J}$$

Net electrical energy input, $\Delta W_e = \Delta W_{e1} + \Delta W_{e2} = -100 + 50 = -50$ J

(c) For calculating the change in the field energy, β's have to be obtained.

$$\beta_{11} = \frac{L_{22}}{D};\ D = L_{11}L_{22} - L_{12}^2$$

$$= \frac{2x+1}{4x+3}$$

Similarly, $$\beta_{22} = \frac{4x+1}{4x+3}$$

$$\beta_{12} = -\frac{1}{4x+3}$$

At $x = 0.5$; $$\beta_{11} = \frac{2}{5},\quad \beta_{22} = \frac{3}{5},\quad \beta_{12} = -\frac{1}{5}$$

At $x = 1$; $$\beta_{11} = \frac{3}{7},\quad \beta_{22} = \frac{5}{7},\quad \beta_{12} = -\frac{1}{7}$$

The values of λ have already been calculated at $x = 0.5$, 1 m.

As per Eq. (4.65), the field energy is given by

$$W_f = \frac{1}{2}\beta_{11}\lambda_1^2 + \beta_{12}\lambda_1\lambda_2 + \beta_{22}\lambda_2^2$$

The field energy at $x = 0.5$ m and $x = 1$ m is then calculated as

$$W_f(x = 0.5) = \frac{1}{2} \times \frac{2}{5} \times (50)^2 = 500 \text{ J}$$

$$W_f(x = 1) = \frac{1}{2} \times \frac{3}{7} \times (45)^2 - \frac{1}{7} \times 45 \times (-5) + \frac{1}{2} \times \frac{5}{7} \times (-5)^2 = 475 \text{ J}$$

Hence $$\Delta W_f = W_f(x = 1) - W_f(x = 0.5) = 475 - 500 = -25 \text{ J}$$

$$\Delta W_f + \Delta W_m = -25 - 25 = -50 = \Delta W_e \text{ (verified)}$$

Note: In the linear case with constant current excitation

$$\Delta W_f = \Delta W_f'$$

ΔW_f can be easily calculated from part (a) without the need of calculating β's. Thus

$$W_f' = 450 + \frac{25}{x}$$

$$\Delta W_f' = W_f'(x = 1) - W_f'(x = 0.5)$$

$$= 475 - 500 = -25 \text{ J}$$

Example 4.11 *Two coupled coils have self- and mutual-inductances as in Ex. 4.10. Find the expression for the time-average force of field origin at x = 0.5 m if:*

(a) both coils are connected in parallel across a voltage source of 100 cos 314t V,

(b) both coils are connected in series across a voltage source of 100 cos 314t V,

(c) coil 2 is shorted and coil 1 is connected to a voltage source of 100 cos 314t V, and

(d) both coils are connected in series and carry a current of 0.5 cos 314t A.

LO 3

Solution Though cases (a), (b) and (c) pertain to voltage excitation, the coenergy approach works out to be more convenient and will be used here.

$$W_f'(i_1, i_2, x) = \frac{1}{2}L_{11}i_1^2 + L_{12}i_1i_2 + \frac{1}{2}L_{22}i_2^2$$

$$= \frac{1}{2}\left(2+\frac{1}{2x}\right)i_1^2 + \left(1+\frac{1}{2x}\right)i_1i_2 + \frac{1}{2}\left(\frac{1}{2x}\right)i_2^2$$

$$F_f = \frac{\partial W_f'(i_1,i_2,x)}{\partial x}$$

$$= -\frac{1}{4x^2}i_1^2 - \frac{1}{2x^2}i_1i_2 - \frac{1}{4x_2^2}i_2^2$$

For $x = 0.5$ m

$$F_f = -i_1^2 - 2i_1i_2 - i_2^2$$

The force acts in a direction to decrease x.

(a) Both coils connected in parallel across the voltage source:

$$L_{11} = 2 + \left.\frac{1}{2x}\right|_{x=0.5} = 3$$

$$L_{22} = 1 + \left.\frac{1}{2x}\right|_{x=0.5} = 2$$

$$L_{12} = L_{21} = \left.\frac{1}{2x}\right|_{x=0.5} = 1$$

From Eqs (4.64a) and (4.64b)

$$v = e_1 = \frac{d\lambda_1}{dt} = 3\frac{di_1}{dt} + \frac{di_2}{dt} = 100 \cos 314t$$

$$v = e_2 = \frac{d\lambda_2}{dt} = \frac{di_1}{dt} + 2\frac{di_2}{dt} = 100 \cos 314t$$

Solving we get

$$\frac{di_1}{dt} = 20 \cos 314t$$

$$\frac{di_2}{dt} = 40 \cos 314t$$

Integrating

$$i_1 = \frac{20}{314}\sin 314t$$

$$i_2 = \frac{40}{314}\sin 314t$$

Substituting for i_1 and i_2 in the expression for F_f,

$$F_f = -\frac{1}{(314)^2}[(20)^2 + 2\times 20\times 40 + (40)^2]\sin^2 314t$$

$$= -\left(\frac{60}{314}\right)^2 \sin^2 314t$$

But $$\frac{1}{T}\int_0^T \sin^2 \omega t\, dt = \frac{1}{2}$$

$\therefore$ $$F_f(\text{av}) = -\frac{1}{2}\left(\frac{60}{314}\right)^2 = 0.0183 \text{ N}$$

(b) Both coils connected in series across the voltage source:

$$v = \frac{d\lambda_1}{dt} + \frac{d\lambda_2}{dt}$$

$$= \left(3\frac{di_1}{dt} + \frac{di_2}{dt}\right) + \left(\frac{di_1}{dt} + 2\frac{di_2}{dt}\right)$$

But $i_1 = i_2 = i$ (series connection)

$\therefore$ $$v = 7\frac{di}{dt} = 100\cos 314t$$

Integrating we get

$$i = \frac{100}{7\times 314}\sin 314t$$

Substituting in the expression for F_f,

$$F_f = -4\times\left(\frac{100}{7\times 314}\right)^2 \sin^2 \omega t$$

or $$F_f(\text{av}) = -2\left(\frac{100}{7\times 314}\right)^2 = -0.00144 \text{ N}$$

(c) Coil 2 shorted, coil 1 connected to voltage source:

$$100\cos 314t = 3\frac{di_1}{dt} + 2\frac{di_2}{dt}$$

$$0 = \frac{di_1}{dt} + 2\frac{di_2}{dt}$$

Solving we have

$$\frac{di_1}{dt} = 40\cos 314t$$

$$\frac{di_2}{dt} = 20\cos 314t$$

Upon integration we get

$$i_1 = \frac{40}{314}\sin 314t$$

$$i_2 = -\frac{20}{314}\sin 314t$$

Substituting for i_1 and i_2 in the expression for F_f,

$$F_f = -\frac{1}{314^2}[(40)^2 - 2\times 40\times 20 + (20)^2]\sin^2 314t$$

$$F_f(\text{av}) = -\frac{1}{2}\left(\frac{20}{314}\right)^2 = -0.0203 \text{ N}$$

(d) Both coils in series carrying current:

$$i = 0.5 \cos 314t$$

Substituting in the expression for F_f,

$$F_f = -(1+2+1)\times(0.5)^2\cos^2 314t$$

$$F_f(\text{av}) = -0.5 \text{ N}$$

Example 4.12 *A doubly-excited magnetic field system has coil self-and mutual-inductances of*

$$L_{11} = L_{22} = 2$$

$$L_{12} = L_{21} = \cos\theta$$

where θ is the angle between the axes of the coils.

(a) The coils are connected in parallel to a voltage source $v = V_m \sin \omega t$. Derive an expression for the instantaneous torque as a function of the angular position θ. Find therefrom the time-average torque. Evaluate for $\theta = 30°$, $v = 100 \sin 314t$.

(b) If coil 2 is shorted while coil 1 carries a current of $i_1 = I_m \sin \omega t$, derive expressions for the instantaneous and time-average torques. Compute the value of the time-average torque when $\theta = 45°$ and $i_1 = \sqrt{2}\sin 314t$.

(c) In part (b) if the rotor is allowed to move, at what value of angle will it come to rest?

LO 3

Solution

$$T_f = \frac{\partial W_f'(i_1, i_2, \theta)}{\partial \theta}$$

$$= \frac{1}{2}\left(\frac{\partial L_{11}}{\partial \theta}\right)i_1^2 + \left(\frac{\partial L_{12}}{\partial \theta}\right)i_1 i_2 + \frac{1}{2}\left(\frac{\partial L_{22}}{\partial \theta}\right)i_2^2$$

Substituting the values of inductances,

$$T_f = -(\sin\theta)\, i_1 i_2$$

From circuit equations

$$V_m \cos \omega t = 2\frac{di_1}{dt} + (\cos\theta)\frac{di_2}{dt}$$

$$V_m \cos \omega t = (\cos\theta)\frac{di_1}{dt} + 2\frac{di_2}{dt}$$

Solving these we get

$$\frac{di_1}{dt} = \frac{di_2}{dt} = \frac{V_m \sin \omega t}{(2+\cos\theta)}$$

Integrating

$$i_1 = i_2 = \frac{V_m \sin \omega t}{\omega(2+\cos\theta)}$$

Substituting in T_f,

$$T_f = -\frac{V_m^2 \sin\theta}{(2+\cos\theta)^2\omega^2}\sin^2 \omega t$$

$$T_f\,(\text{av}) = \frac{V_m^2 \sin\theta}{2(2+\cos\theta)^2\omega^2}$$

Given: $\theta = 30°,\ v = 100 \sin 314t$

$$\therefore \quad T_f\,(\text{av}) = -\frac{(100)^2 \sin 30°}{2(2+\cos 30°)^2 \times (314)^2} = -0.069 \text{ Nm}$$

(b) From circuit equations

$$0 = (\cos\theta)\frac{di_1}{dt} + 2\frac{di_2}{dt}$$

or

$$\frac{di_2}{dt} = -\frac{1}{2}(\cos\theta)\frac{di_1}{dt}$$

or

$$i_2 = -\frac{1}{2}(\cos\theta) i_1$$

Given: $i_1 = I_m \sin \omega t$

$$\therefore \quad i_2 = -\frac{1}{2} I_m (\cos\theta) \sin \omega t$$

Substituting in T_f,

$$T_f = -(\sin\theta) \times \frac{1}{2} I_m^2 (\cos\theta) \sin^2 \omega t$$

$$= -\frac{1}{2} I_m^2 (\sin\theta)(\cos\theta)\sin^2 \omega t$$

$$T_f\,(\text{av}) = -\frac{1}{8} I_m^2 (\sin 2\theta)$$

Given: $\theta = 45°,\ I_m = \sqrt{2}$

$$\therefore \quad T_f\,(\text{av}) = \frac{1}{8} \times 2 \sin 90° = 0.25 \text{ Nm}$$

(c) The average torque is zero and changes sign at $\theta = 0°, 90°, 180°$. The rotor can come to rest at any of these values of θ but the position of stable equilibrium will only be $\theta = 90°, 270°, \ldots$ (The reader should draw T_f(av) versus θ and reason out).

4.5 | FORCES/TORQUES IN SYSTEMS WITH PERMANENT MAGNETS LO 3

Method of finding forces in systems with permanent magnets is best illustrated by an example. Figure 4.17(b) shows a moving armature relay excited by a permanent magnet (PM).

The dc magnetizing curve of the permanent magnet is drawn in Fig. 4.17(a) which upon linear extrapolation at the lower B-end can be expressed as

$$B_m = \mu_R (H_m - H_c') = \mu_R H_m + B_r \tag{4.68}$$

μ_R = recoil permeability of the PM material

$= B_r/H_c'$; on linearized basis

Soft iron portion of the magnetic circuit including the armature is assumed to have $\mu = \infty$. For finding the force on the armature, it will be convenient to use Eq. (4.25) for which we will need the expression of system coenergy $W_f'(i, x)$ which must be independent of i as there is no exciting current in the system. Thus coenergy will be a function of the space variable x only i.e.,

$$W_f'(i, x) \Rightarrow W_f'(x)$$

Coenergy is given by the expression of Eq. (4.11), i.e.

$$W_f' = \int_0^i \lambda\, di$$

Permanent Magnet Excited Systems

This expression needs to be carefully interpreted for the case of permanent magnetic excited systems. The limits of integration in this expression mean from state of zero flux to a state of certain flux (but with zero current). The state of zero flux is imagined by means of a fictitious exciting coil (of N_f turns) carrying current i_f as shown in Fig. 4.17(c). The current is assumed to be adjusted to value I_f causing the core flux to reduce to zero and the original state is then reached by imaging the current (i_f) to reduce to zero. Thus

$$W_f' = \int_{I_f}^0 \lambda_f\, di_f \tag{4.69}$$

At any value of i_f, we can write

$$N_f i_f = H_m d + 2H_g x \tag{4.70}$$

Continuity of flux allows us to write

$$B_m A = B_g A$$

or

$$B_m = \mu_R H_m = B_g = \mu_0 H_g \tag{4.71}$$

Substituting for H_g in Eq. (4.70)

$$N_f i_f = H_m d + 2(x/\mu_0) B_m$$

Substituting for H_m from Eq. (4.68) and solving we get

$$B_m = \frac{\mu_R (N_f i_f - H_c' d)}{d + 2(\mu_R/\mu_0)x} \tag{4.72}$$

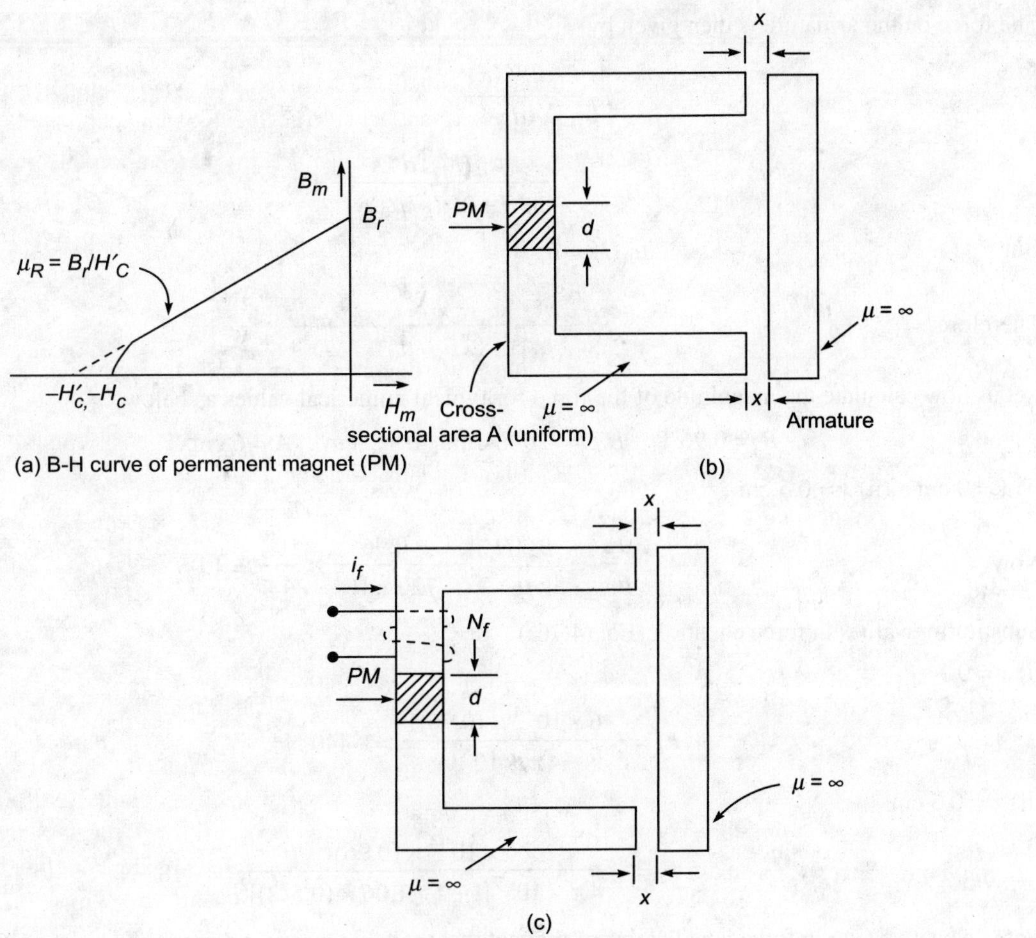

Fig. 4.17

Flux linkages of the fictitious coil are given by

$$\lambda_f = AB_m N_f = \frac{\mu_R N_f A(N_f i_f - H_c' d)}{d + 2(\mu_R/\mu_0)x} \tag{4.73}$$

Flux and flux linkages would be zero at

$$i_f = (H_c' d/N_f) \tag{4.74}$$

Substituting Eq. (4.69), we get

$$W_f'(x) = \int_{i_f}^{0} \frac{\mu_R N_f A(N_f i_f - H_c' d)}{d + 2(\mu_R/\mu_0)x} \, di_f$$

$$= \frac{\mu_R A H_c' d^2}{2[d + 2(\mu_R/\mu_0)x]} \tag{4.75}$$

The force on the armature is then given by

$$F_f = \frac{dW_f'(x)}{dx}$$

$$= \frac{(\mu_R H_c')^2 d^2 A}{\mu_0[d + 2(\mu_R/\mu_0)x]^2}$$

But $$\mu_R H_c' = B_r$$

Therefore, $$F_f = \frac{AB_r^2}{\mu_0[1 + 2(\mu_R/\mu_0)(x/d)^2]} \tag{4.76}$$

Let us now calculate the magnitude of the force for typical numerical values as below,

$$B_r = 0.96 \text{ T}, \quad H_c' = 720 \text{ kA/m}, \quad d = 2 \text{ cm}, \quad A = 6 \text{ cm}^2$$

(i) $x = 0$ cm (ii) $x = 0.5$ cm

Now $$\frac{\mu_R}{\mu_0} = \frac{B_r/H_c'}{\mu_0} = \frac{0.96}{720 \times 10^3} \times \frac{10^7}{4\pi} = 1.06$$

Substituting values in force equation (Eq. (4.76))

(i) $x = 0$

$$F_f = -\frac{6 \times 10^{-4} \times (0.96)^2}{4\pi \times 10^{-7}} = -440 \text{ N}$$

(ii) $x = 0.5$ cm

$$F_f = -\frac{6 \times 10^{-4} \times (0.96)^2}{4\pi \times 10^{-7}[1 + 2 \times 1.06 \times (0.5/2)]^2} = -188 \text{ N}$$

Note: Negative sign in force is indicative of the fact that the force acts in a direction to reduce x (air-gap).

Similar treatment could be used for mixed situation where system has both permanent magnets and exciting coils. It should be stressed here that an alternative procedure is to use the finite element method to evaluate the coenergy from the vector form of Eq. (4.18a) integrated over the volume, i.e.,

$$W_f' = \int_V \int_0^{B_0} \vec{B} \cdot d\vec{H} dV \tag{4.77}$$

To calculate force $-dW_f'/dx$ is obtained by numerical differentiation (see Example 4.2). This method is of general applicability wherever the magnetic circuit analysis cannot be carried out.

Concept Check LO 3

1. What is a singly-excited magnetic field system? ○○●
2. What is a doubly-excited magnetic field system? ○○●
3. Elaborate the statement, "In a round rotor machine (uniform air-gap) with exciting coil placed in stator slots no reluctance torque is developed". ●●●

4.6 | ENERGY CONVERSION VIA ELECTRIC FIELD

LO 4 Discuss the electromechanical energy conversion process through electric field in a system

Electromechanical energy conversion via the electric field is analogous to the magnetic field case studied earlier.

Important! Charge in the electric field is analogous to flux linkages and voltage to current in the magnetic field case.

Electric Field Energy

Figure 4.18 shows a parallel plate condenser with a fixed and a movable plate. The condenser is fed from a current source. The leakage current of the condenser is represented outside by conductance so that the condenser's electric field is *conservative*. Let us assume that the movable plate of the condenser is held fixed in position x. The electric energy input to the ideal condenser gets stored in the electric field so that

$$dW_e = v\,dq = dW_f \tag{4.78}$$

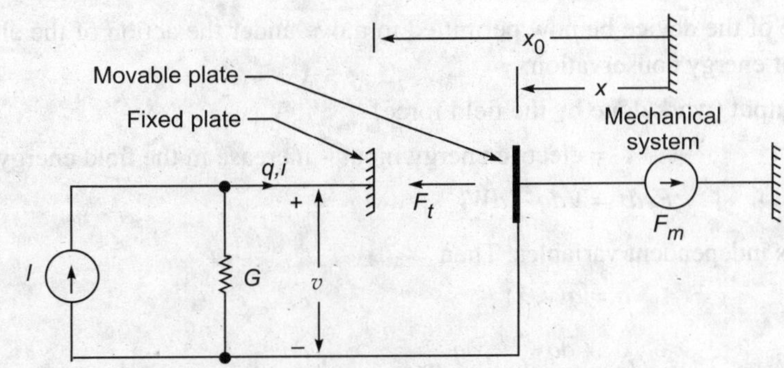

Fig. 4.18 *Electric field electromechanical device*

The total field energy is

$$W_f = \int_0^q v\,dq \tag{4.79}$$

In a condenser v and q are linearly related as

$$C = q/v = \text{capacitance of the device}$$

$$\therefore \qquad W_f = \frac{1}{2}\frac{q^2}{C} \tag{4.80}$$

The capacitance C is a function of configuration (position x of the movable plate) and can be expressed as

$$C = \frac{\varepsilon_0 A}{(x_0 - x)}$$

where A = plate area and ε_0 = permittivity of free space. Thus W_f, the field energy is a function of two independent variables q and x, i.e.

$$W_f(q, x) = \frac{1}{2}\frac{q^2}{C(x)} \tag{4.81}$$

The expression of Eq. (4.81) for the field energy immediately reveals that the electric field energy can be changed electrically by changing q or mechanically by changing x, i.e. moving the movable plate.

The field energy can also be written as

$$W_f(v, x) = \frac{1}{2} vq = \frac{1}{2} C(x)v^2 \tag{4.82}$$

The energy density in the electric field can be expressed as

$$\omega_f = \int_0^D EdD = \frac{1}{2}\frac{D^2}{\varepsilon_0} = \frac{1}{2}\varepsilon_0 E^2 \tag{4.83}$$

where E = electric field intensity or potential gradient
$= D/\varepsilon_0$
D = electric field flux density

Energy Conversion

Let the movable plate of the device be now permitted to move under the action of the electric field force F_f. As per the principle of energy conservation:

Mechanical energy output (work done by the field force)

= electric energy input – increase in the field energy

or

$$F_f dx = v\,dq - dW_f \tag{4.84}$$

Let us choose (v, x) as independent variables. Then

$$q = q(v, x)$$

$$dq = \frac{\partial q}{\partial v} dv + \frac{\partial q}{\partial x} dx \tag{4.85}$$

and

$$W_f = W_f(v, x)$$

$$dW_f = \frac{\partial W_f}{\partial v} dv + \frac{\partial W_f}{\partial x} dx \tag{4.86}$$

Substituting Eqs (4.85) and (4.86) in Eq. (4.84) and reorganizing

$$F_f dx = \left(v\frac{\partial q}{\partial x} - \frac{\partial W_f}{\partial x} \right) dx + \left(v\frac{\partial q}{\partial v} - \frac{\partial W_f}{\partial v} \right) dv \tag{4.87}$$

Since v and x are independent variables, the coefficient of dv in Eq. (4.87) must be zero. Hence

$$v\frac{\partial q}{\partial x}(v, x) - \frac{\partial W_f}{\partial x}(v, x) = 0$$

so

$$Fdx = \frac{\partial}{\partial x}(vq(v, x) - W_f(v, x)) \tag{4.88}$$

Defining coeneregy as

$$W_f'(v, x) = vq(v, x) - W_f(v, x) \tag{4.89}$$

the electric field force F_f can be written as

$$F_f = \frac{\partial W_f'(v,x)}{\partial x} \tag{4.90}$$

If instead (q, x) are taken as independent variables

$$W_f = W_f(q, x)$$

$$dW_f = \frac{\partial W_f}{\partial q} dq + \frac{\partial W_f}{\partial x} dx \tag{4.91}$$

Substituting in Eq. (4.84)

$$F_f\, dx = \left(v - \frac{\partial W_f}{\partial q} \right) dq - \frac{\partial W_f}{\partial x} dx \tag{4.92}$$

Since v and q are independent variables, the coefficient of dq in Eq. (4.92) must be zero. Hence Eq. (4.92) gives

$$F_f = -\frac{\partial W_f(q,x)}{\partial x} \tag{4.93}$$

From Eqs (4.89) and (4.79), the coenergy[§] can be written as

$$W_f'(v, x) = \int_0^v q\, dv \tag{4.94}$$

For a linear system $q = Cv$

$$W_f'(v, x) = \frac{1}{2} C(x) v^2 \tag{4.95}$$

The coenergy density is given by

$$W_f' = \frac{1}{2} \varepsilon_0 E^2 \tag{4.96}$$

Example 4.13 *Find an expression for the force per unit area between the plates of a parallel plate condenser in terms of the electric field intensity. Use both the energy and coenergy methods. Find the value of the force per unit area when $E = 3 \times 10^6$ V/m, the breakdown strength of air.* LO 3

Solution With reference to Fig. 4.18, the energy in the electric field is

$$W_f(q, x) = \frac{1}{2} \frac{q^2}{C} = \frac{1}{2} \frac{q^2 (x_0 - x)}{A \varepsilon_0}$$

---

§

$$W_f'(v, x) = vq - \int_0^q v\, dq$$

Integrating the second term by part

$$W_f'(v, x) = vq - vq + \int_0^v q\, dv = \int_0^v q\, dv$$

From Eq. (4.95)

$$F_f = -\frac{\partial W_f(q, x)}{\partial x} = \frac{1}{2}\frac{q^2}{A\varepsilon_0}$$

But

$$q = DA = \varepsilon_0 EA$$

$$F_f = \frac{1}{2}\varepsilon_0 E^2 A \quad \text{or} \quad E_f/A = \frac{1}{2}\varepsilon_0 E^2$$

$$= \frac{1}{2} \times (3 \times 10^6)^2 \times 8.85 \times 10^{-12} = 39.8 \text{ N/m}^2$$

From Eq. (4.95), the field coenergy is

$$W_f'(v, x) = \frac{1}{2}Cv^2 = \frac{1}{2}v^2\frac{A\varepsilon_0}{(x_0 - x)}$$

Now from Eq. (4.90)

$$E_f = \frac{\partial W_f'(v, x)}{\partial x} = \frac{1}{2}v^2\frac{A\varepsilon_0}{(x_0 - x)^2}$$

But

$$v = E(x_0 - x)$$

$\therefore$

$$F_f = \frac{1}{2}\varepsilon_0 E^2 A \quad \text{or} \quad F_f/A = \frac{1}{2}\varepsilon_0 E^2 \text{ (as before)}$$

It may be observed here that while the force density on the bounding surface in a magnetic field near saturation was found to be 1.02×10^6 N/m^2, it has a value[¶] of only 39.8 N/m^2 in an electric field with the electric intensity at its breakdown value. This indeed is the reason why all practical energy conversion devices make use of the magnetic field as the coupling medium rather than the electric field. Electric field devices are sometimes used as transducers.

4.7 | DYNAMICAL EQUATIONS OF ELECTROMECHANICAL SYSTEMS

LO 4

Figure 4.19 shows an electromagnetic relay whose armature is loaded with spring K, damper B, mass M and a force generator F. Figure 4.20 shows the abstracted diagram of a general electromechanical system. It is easily noticed that the electromechanical device has one electrical port and one mechanical port (one terminal of the mechanical port being the ground) through which it is connected to the electrical source on one side and mechanical load on the other side. In general there can be more than one electrical port (multiple excited system).

Let the electromechanical device has an inductance

$$L = L(x)$$

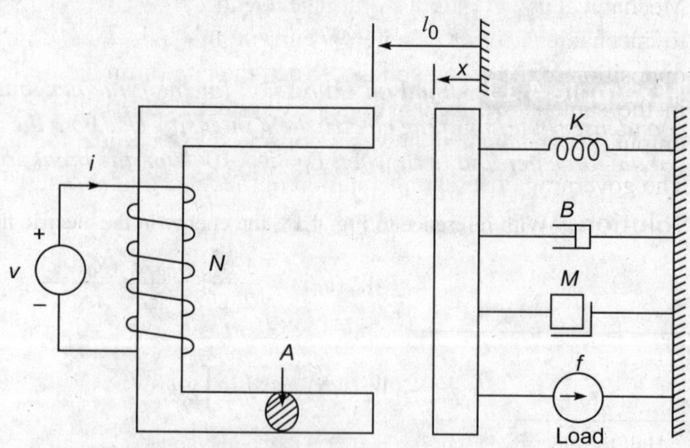

Fig. 4.19

[¶]Such a low value results from $\varepsilon_0 = 8.85 \times 10^{-12}$ compared to $\mu_0 = 4\pi \times 10^{-7}$

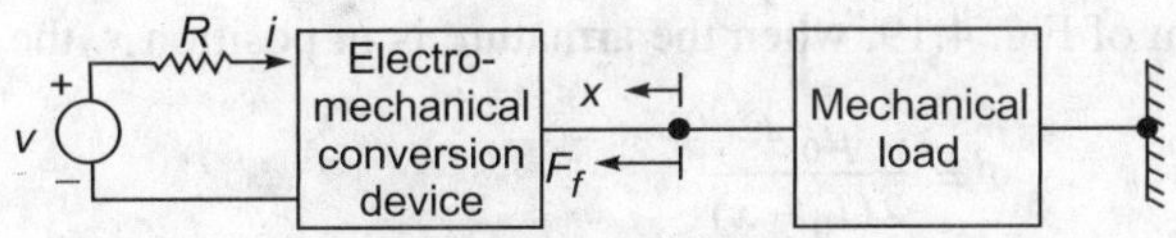

Fig. 4.20 *Abstracted electromechanical system diagram*

The governing electrical equation is

$$v = iR + \frac{d\lambda}{dt} = iR + \frac{d}{dt}[L(x)i]$$

$$= iR + \underbrace{L(x)\frac{di}{dt}}_{\text{self-inductance voltage}} + \underbrace{i\frac{dL(x)}{dx}\frac{dx}{dt}}_{\text{speed voltage}} \tag{4.97}$$

Now $\quad W'_f(i, x) = \frac{1}{2}L(x)i^2$

$$\therefore \quad F_f = \frac{\partial W'_f}{\partial x} = \frac{1}{2}i^2\frac{dL(x)}{dx} \tag{4.98}$$

The mechanical power output is given by

$$P_m = F_f\frac{dx}{dt} = \frac{1}{2}i^2 \times \frac{dL(x)}{dx} \times \frac{dx}{dt}$$

$$= \frac{1}{2}i\left(i\frac{dL(x)}{dx} \times \frac{dx}{dt}\right) \tag{4.99a}$$

$$= \frac{1}{2}\text{ current} \times \text{speed voltage}^{**} \tag{4.99b}$$

Mechanical power output results (i.e. electrical power is converted to mechanical form) when the current in the device flows in opposition to the speed voltage. When the current in the device is in the same direction as the speed voltage, electrical power is output, i.e., mechanical power is converted to electrical form.

Probe

When does a machine give mechanical power or electrical power as output?

The governing differential equation of the mechanical system is

$$F_f = M\frac{d^2x}{dt^2} + B\frac{dx}{dt} + Kx + f \tag{4.100a}$$

$$\text{or} \quad \frac{1}{2}i^2\frac{dL(x)}{dx} = M\frac{d^2x}{dt^2} + B\frac{dx}{dt} + Kx + f \tag{4.100b}$$

** Half the electrical input (ei) is stored in the magnetic field. This agrees with Eq. (4.38). In continuous energy conversion devices (electric motors and generators), however, the average energy stored in the magnetic field remains constant over a cycle of operation, so that the electrical power input (EI for a dc device where E and I are dc speed voltage and current input or $EI\cos\phi$ for an ac device where E is the rms speed voltage and I is the rms current input while $\cos\phi$ is the power factor) is fully converted to mechanical form or vice versa.

Now for the specific system of Fig. 4.19, when the armature is in position x, the self-inductance L is found below:

$$\mathcal{P} = \frac{\mu_0 A}{2(I_0 - x)}$$

$$L(x) = N^2 \mathcal{P} = \frac{\mu_0 A N^2}{2(I_0 - x)}$$

$$\frac{dL(x)}{dx} = \frac{\mu_0 A N^2}{2(l_0 - x)^2}$$

Substituting in Eqs (4.97) and (4.98), the two differential equations defining the system's dynamic behavior are obtained as:

$$v = iR + \frac{\mu_0 A N^2}{2(l_0 - x)^2} \times \frac{di}{dt} + \frac{\mu_0 A N^2 i}{2(l_0 - x)^2} \times \frac{dx}{dt} \tag{4.101}$$

$$\frac{\mu_0 A N^2 i^2}{4(l_0 - x)^2} = M\frac{d^2x}{dt^2} + B\frac{dx}{dt} + Kx + f \tag{4.102}$$

These are nonlinear differential equations which can be solved numerically on the digital computer. However, for small movement around the equilibrium point the following procedure can be adopted.

Let the equilibrium point be (V_0, I_0, X_0, f_0). At equilibrium the system is stationary and all derivatives are zero. Thus from Eqs (4.101) and (4.102), the following relationships between equilibrium values are obtained.

$$V_0 = I_0 R \tag{4.103}$$

$$\frac{\mu_0 A N^2 I_0^2}{4(l_0 - x_0)^2} = KX_0 + f_0 \tag{4.104}$$

Let the departure (small) from the equilibrium values be (v_1, i_1, x_1, f_1)

Then

$$V_0 + v_1 = (I_0 + i_1)R + \frac{\mu_0 A N^2}{2(l_0 - X_0 - x_1)^2} \cdot \frac{di_1}{dt} + \frac{\mu_0 A N^2 (I_0 + i_1)}{2(l_0 - X_0 - x_1)^2}\frac{dx_1}{dt}$$

$$\frac{\mu_0 A N^2 (I_0 + i_1)^2}{4(l_0 - X_0 - x_1)^2} = M\frac{d^2x_1}{dt^2} + B\frac{dx_1}{dt} + K(X_0 + x_1) + f_0 + f_1$$

Neglecting products of small departures and small departures compared to equilibrium values, and also cancelling out equilibrium terms as per Eqs (4.103) and (4.104),

$$v_1 = i_1 R + \frac{\mu_0 A N^2}{2(l_0 - X_0)}\frac{di_1}{dt} + \frac{\mu_0 A N^2 I_0}{2(l_0 - X_0)^2}\frac{dx_1}{dt} \tag{4.105}$$

$$\frac{2\mu_0 A N^2 I_0 i_1}{4(l_0 - X_0)^2} = M\frac{dx_1}{dt^2} + B\frac{dx_1}{dt} + Kx_1 + f_1 \tag{4.106}$$

Equations (4.105) and (4.106) are linear differential equations governing the system behavior for small incremental values around the equilibrium values (called the *operating point*). These can be easily solved analytically for dynamic or steady-state conditions.

Concept Check LO 4

1. Which field energy is suited for electromechanical energy conversion process?
2. How energy can be stored in or retrieved from a magnetic system?
3. How energy can be stored in or retrieved from an electric system?

SUMMARY

LO 1

- Electromechanical energy conversion takes place via the medium of a magnetic or electric field–the magnetic field being most suited for practical conversion devices.
- Electromagnets, relays, actuators, motors and generators are few examples of electromechanical energy conversion devices.
- Energy can be stored or retrieved from magnetic system by means of an exciting coil connected to an electric source. The field energy can be given by

  $W_f = W_f(\lambda, x)$ or $W_f = W_f(i, x)$

  λ – flux linkage

  x – air-gap between the armature and core

  i – current
- Mechanical force is given by

$$F_f = -\frac{\partial W_f(\lambda, x)}{\partial x} \quad \text{or} \quad F_f = \frac{\partial W'(\lambda, x)}{\partial x}$$

LO 2

- For doubly excited magnetic field system, field energy is given by

$$W_f = (\lambda_1, \lambda_2, \theta) = \int_0^{\lambda_1} i_1 d\lambda_1 + \int_0^{\lambda_2} i_2 d\lambda_2$$

LO 3

- Electromechanical energy conversion via the electric field is analogous to the magnetic field, charge in the electric field is analogous to flux linkages and voltage to current in the magnetic field case

$$W_f = \frac{1}{2}\frac{q^2}{C}$$

  q – charge in Coulombs

  C – capacitance of the device

PRACTICE PROBLEMS

4.1 In the electromagnetic relay of Fig. 4.11, the exciting coil has 1000 turns. The cross-sectional area of the core is A = 5 cm × 5 cm. Reluctance of the magnetic circuit may be assumed negligible. Also neglect fringing effects. **LO 1**

(a) Find the coil inductance for an air-gap of x = 1 cm. What is the field energy when the coil carries a current of 2.0 A? What is the force on the armature under these conditions?

(b) Find the mechanical energy output when the armature moves from x_e = 1 cm to x_b = 0.5 cm assuming that the coil current is maintained constant at 2.0 A.

(c) With constant coil current of 2.0 A, derive an expression for force on armature as a function of x. Find the work done by the magnetic field when x changes from x_e = 1cm to x_b = 0.5 cm from $\int_{x_e}^{x_b} F_f\, dx$. Verify the result of part (b).

(d) Find the mechanical energy output in part (b) if the flux linkages are maintained constant corresponding to a coil current of 2.0 A.

4.2 In Fig. 4.7(b) if the i-λ curve ab is assumed to be a straight line, find the expression for the mechanical energy output. If this figure pertains to the electromagnetic relay of Fig. 4.11, find the value of the mechanical energy output, given that: $i_1 = 2.0$ A, $i_2 = 1.5$ A, $x_a = 1$ cm and $x_b = 0.5$ cm. LO 1

4.3 Consider the cylindrical iron-clad solenoid magnet of Fig. 4.9. The data for magnetizing curve of the iron part of the solenoid are given in Ex. 4.1. For $g = 0.2$ cm, find the force on the plunger if λ is assumed constant corresponding to an exciting current of 2.25 A. Why does this value differ from that calculated in Ex. 4.2? LO 2

4.4 For the cylindrical iron-clad solenoid magnet of Fig. 4.9, assume that the magnetic path reluctance remains constant at a value corresponding to the linear part of the magnetization curve. LO 2

(a) Derive an expression for the force in terms of g for constant coil current 2.25 A. Calculate the value of the force for $g = 1$ and 0.2 cm.

(b) What is the electrical energy input to the system when g changes from 1 to 0.2 cm, while the coil current is maintained constant at 2.25 A.

(c) Calculate the work done on the plunger during the movement specified in part (b).

(d) With the coil current maintained constant at 2.25 A. What is the direction and magnitude of the electrical energy flow if the plunger is made to move from $g = 0.2$ to 1 cm?

4.5 Repeat part (c) of Problem 4.4 with the nonlinear magnetization curve for the iron path. Compare the two results and comment. LO 2

4.6 For the electromagnetic relay of Fig. 4.11, calculate the maximum force on armature if saturation flux density in the iron part is 1.8 T. Given: cross-sectional area of core = 5 cm × 5 cm, coil turns = 1000. LO 3

4.7 For the electromagnetic device shown in Fig. P4.7, assume the reluctance of the iron part of the magnetic circuit to be negligible. Determine the time-average force on the movable member at any fixed position of the moving member, if LO 3

(a) $i = I \cos \omega t$ (b) $v = V \cos \omega t$

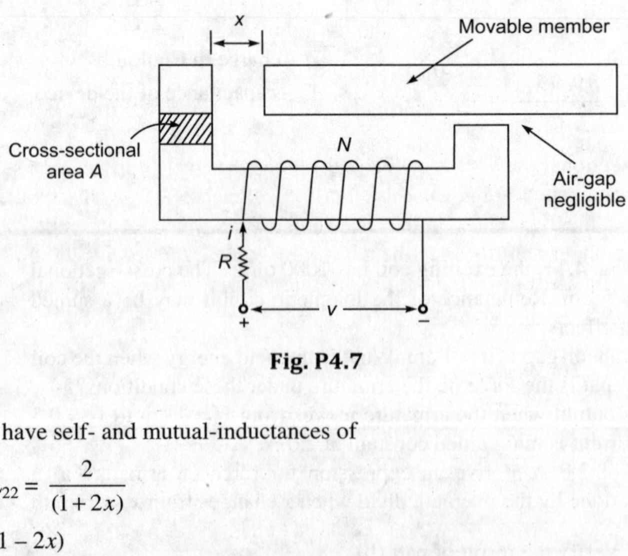

Fig. P4.7

4.8 Two coils have self- and mutual-inductances of LO 3

$$L_{11} = L_{22} = \frac{2}{(1+2x)}$$

$$L_{12} = (1 - 2x)$$

The coil resistances may be neglected.

(a) If the current I_1 is maintained constant at 5 A and I_2 at –2 A, find the mechanical work done when x increases from 0 to 0.5 m. What is the direction of the force developed?

(b) During the movement in part (a), what is the energy supplied by sources supplying I_1 and I_2?

4.9 Two coils have self- and mutual-inductances of **LO 3**

$$L_{11} = L_{22} = \frac{2}{(1+2x)}$$

$$L_{12} = \frac{1}{1+2x}$$

Calculate the time-average force and coil currents at $x = 0.5$ m if:

(a) both the coils connected in parallel across a voltage source of 100 cos 314t,
(b) coil 2 is shorted while coil 1 is connected across a voltage source of 100 cos 314t, and
(c) the two coils are connected in series across a voltage source of 100 cos 314t.

4.10 The doubly-excited magnetic field system of Fig. 4.15 has coil self- and mutual-inductances of **LO 3**

$$L_{11} = L_{22} = 2 + \cos 2\theta$$
$$L_{12} = \cos \theta$$

where θ is the angle between the axes of the coils. The coils are connected in series and carry a current of $i = \sqrt{2}\, I \sin \omega t$. Derive an expression for the time-average torque as a function of angle θ.

4.11 In the rotary device of Fig. 4.15, when the rotor is the region of $\theta = 45°$, the coil inductances can be approximated as **LO 3**

$$L_{11} = L_{12} = 2 + \frac{\pi}{2}\left(1 - \frac{\theta}{45}\right)$$

$$L_{12} = L_{21} = \frac{\pi}{2}\left(1 - \frac{\theta}{90}\right)$$

where θ is in degrees

Calculate the torque of field origin if the rotor is held in position $\theta = 45°$ with

(a) $i_1 = 5$ A, $i_2 = 0$ (b) $i_1 = 0$, $i_2 = 5$ A
(c) $i_1 = 5$ A, $i_2 = 5$ A (d) $i_1 = 5$ A, $i_2 = -5$ A
(e) Find the time-average torque if coil 1 carries a current of 5 sin 314t while coil 2 is short circuited.

4.12 Figure P4.12 shows the cross-sectional view of a cylindrical plunger magnet. The position of the plunger when the coil is unexcited is indicated by the linear dimension D. Write the differential equations describing the dynamics of the electromagnetic system. Determine the equilibrium position of the plunger and linearize the describing equation for incremental changes about the equilibrium point. Assume the iron to be infinitely permeable. **LO 3**

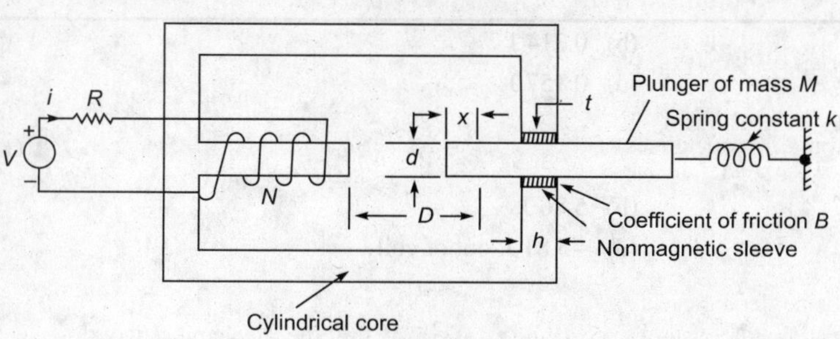

Fig. P4.12

4.13 For the electromagnet of Fig. P4.13 write the dynamical equation. Assume the cross-sectional area of each limb of the magnet to be A and that the coupling between the two coils to be tight. Iron is to be taken as infinitely permeable. LO 4

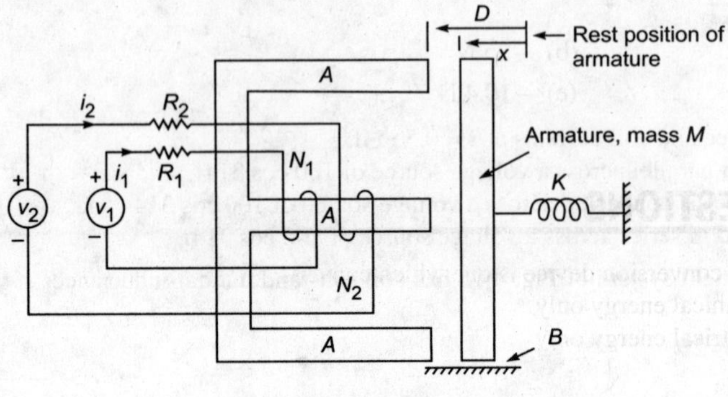

Fig. P4.13

4.14 For the electromechanical system shown in Fig. P4.14, the air-gap flux density under steady operating condition is LO 4

$$B(t) = B_m \sin \omega t$$

Find (a) the coil voltage, (b) the force of field origin as a function of time, and (c) the motion of armature as a function of time.

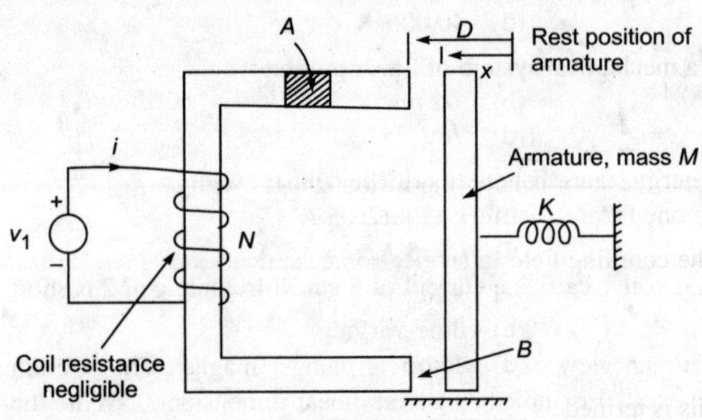

Fig. P4.14

Answers

4.1 (a) 0.157 H, –31.4 N (b) 0.314 J
(c) 0.314 J (d) 0.157 J

4.2 $\frac{1}{2}(\lambda_2 i_1 - \lambda_1 i_2)$, 0.2355 J

4.3 –1335 N

4.4 (a) –83.5 N, –1130 N (b) 5.86 J
(c) 2.91 J (d) –5.81 J, out of coil

4.6 -6.446×10^3 N

4.7 (a) $-\frac{1}{4} I^2 \left(\frac{a}{x^2}\right)$, $a = N^2 \mu_0 A$ (b) $\omega^2 \dfrac{aV^2}{4(r^2x^2 + a^2\omega^2)}$, $a = N^2 \mu_0 A$

4.8 (a) −4.5 J; for $0 < x < 0.35$ tends to decrease x and for $x > 3.5$ tends to increase x (b) −7.5 J, 3 J

4.9 (a) −0.034 N (b) −0.034 N (c) −0.034 N

4.10 $-I^2 (\sin 2\theta + \sin \theta)$

4.11 (a) −25 N (b) −25 N (c) −75 N
(d) −25 N (e) −10.4 N

MULTIPLE CHOICE QUESTIONS

1. An electro-mechanical energy conversion device is one which converts ______ LO 1
(a) electrical energy to mechanical energy only
(b) mechanical energy to electrical energy only
(c) Both of these
(d) None of these

2. What coupling field is used between the electrical and mechanical systems in an energy conversion device? LO 1
(a) Magnetic field (b) Electric field
(c) Magnetic field or Electric field (d) None of these

3. The energy storing capacity of magnetic field is about ______ times greater than that of electric field. LO 1
(a) 50,000 (b) 25,000
(c) 10,000 (d) 40,000

4. The formula for energy stored in a mechanical system of linear motion type is _____. LO 1
(a) $1/2\ Jw_r^2$ (b) $1/2\ mv^2$
(c) $1/2\ mv$ (d) Jw_r^2

5. The principle of conservation of energy states that energy can be neither created nor destroyed, but can be merely converted from one form to another. (True/False) LO 1

6. Electromagnetic radiation from the coupling field in an electromechanical energy conversion device is LO 1
(a) constant (b) time-varying
(c) negligible (d) zero

7. The emf induced by flux variations is termed as ______________. LO 2
(a) dynamically induced emf (b) statically induced emf
(c) self-induced emf (d) mutually induced emf

8. ______________ is an example of a singly-excited magnetic field system. LO 2
(a) Synchronous machine (b) dc machine
(c) Reluctance motor (d) Transducer

9. An alternator is an example of ________ . LO 2
(a) singly-excited magnetic system (b) multiply-excited magnetic system
(c) singly-excited electric system (d) multiply-excited electric system

10. Electromechanical energy conversion is a ________ process. LO 1
(a) unidirectional (b) bidirectional
(c) reversible (d) constant

Answers

1. (c) **2.** (c) **3.** (b) **4.** (b) **5.** True **6.** (c) **7.** (b) **8.** (c) **9.** (b) **10.** (c)

CHAPTER 5

Basic Concepts in Rotating Machines

LEARNING OBJECTIVES

After studying this chapter, students will be able to

- **LO1** Interpret the common features of rotating machines
- **LO2** Explain the operation of basic machine types
- **LO3** Examine on various losses that occur in machines and how improvement in characteristics of magnetic, conducting and insulating materials has brought about efficiency in machine performance
- **LO4** Analyze the need for matching characteristics of electric machines and load in ensuring steady operating behavior of the complete system

5.1 | INTRODUCTION

It was seen in Ch. 4 that electromechanical energy conversion takes place whenever a change in flux is associated with mechanical motion. Speed voltage is generated in a coil when there is relative movement between the coil and magnetic field. Alternating emf is generated if the change in flux linkage of the coil is cyclic. The *field windings* which are the primary source of flux in a machine are, therefore, arranged to produce cyclic north-south space distribution of poles. A cylindrical structure is a natural choice for such a machine. Coils which are the seats of induced emf's are several in number in practical machines and are suitably connected in series/parallels and in star/delta 3-phase connection to give the desired voltage and to supply the rated current. This arrangement is called the *armature winding*. When the armature coils carry currents they produce their own magnetic field which interacting with the magnetic field of the field winding produces electro-magnetic torque tending to align the two magnetic fields.

The field winding and armature winding are appropriately positioned on a common magnetic circuit composed of two parts—the *stator* (stationary member) and the *rotor* (rotating member). The stator is the annular portion of a cylinder in which rotates a cylindrical rotor; there being an appropriate clearance (*air-gap*) between the two. The rotor axle is carried on two bearings which are housed in two end-covers bolted on the two sides of the stator as shown in Fig. 5.1. The stator and rotor are made of high permeability magnetic material—silicon steel. Further, the member in which the flux rotates is built up of thin insulated laminations to reduce eddy-current loss.

Since *electromechanical energy conversion requires relative motion between the field and armature winding*, either of these could be placed on the stator or rotor. Because of practical convenience, field windings are normally placed on the rotor in the class of machines called the *synchronous machines*; the cross-sectional view of one such machine is shown in Fig. 5.1. The armature winding is housed in suitably shaped *slots* cut out in the stator. The field winding is supplied with dc from an external source, called the *exciter*, through a pair of *slip-rings* as shown in Fig. 5.1. The exciter is generally coupled directly to the rotor shaft of the synchronous machine.

In an *induction machine* the stator has a 3-phase winding which draws a component of current from the mains to set up a cyclic flux pattern in the air-gap which rotates at a speed corresponding to supply frequency (*synchronous speed*) and the rotor is either properly wound and the winding is short-circuited or is merely a set of copper (or aluminium) bars placed in rotor slots short-circuited at each end by means of end-rings.

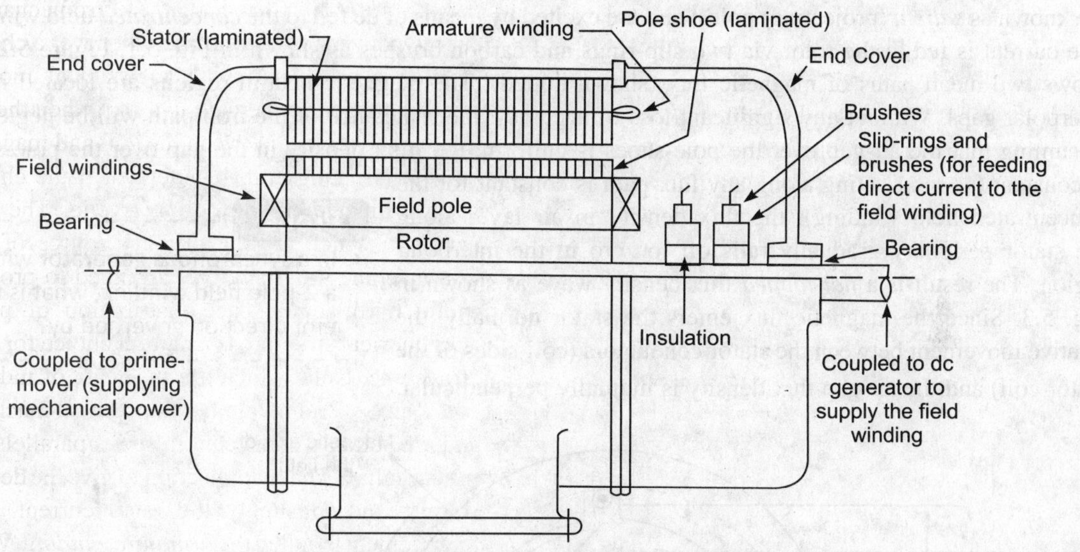

Fig. 5.1 *ac machine-synchronous type*

In *dc machines* a most convenient and practical arrangement is to generate alternating voltages and to convert these to dc form by means of a rotating mechanical rectifier. Therefore, it is necessary that the armature winding in a dc machine is on the rotor and the field winding on the stator.

In this chapter, through suitable approximations, the physical machine will be reduced to a simple mathematical model.

Industry Relevance

Electrical machines follow similar design rules and operating principles of electric motors. However their applications vary. Electrical machines are used in applications where it is required to convert electrical energy into mechanical energy such as fans, air blowers, etc. There are also used to produce electricity on a larger scale in industries or power plants. Their uses vary from household appliances, power tools and traction motors to ship propulsion, pipeline compression, and storage applications.

5.2 ELEMENTARY MACHINES

LO 1 Interpret the common features of rotating machines

Synchronous Machine

Salient Pole Rotor

Figure 5.2 shows the simplified version of an ac synchronous generator with a 2-pole field winding on the rotor and a single coil aa' on the stator. The type of rotor poles are known as *salient* (projecting) poles; and are excited by means of dc fed to the *concentrated* field winding. The current is fed to the rotor via two slip-rings and carbon brushes as shown in Fig. 5.1. Figure 5.2 also shows two mean paths of magnetic flux (shown dotted). The magnetic neutral regions are located in the interpolar gaps. Without any significant loss of accuracy, the reluctance of the iron path will be neglected. Assuming that the air-gap over the pole-shoes is uniform, the flux density in the gap over the pole-shoes is constant (as mmf acting along any flux path is constant for the concentrated field winding); the flux density in air layer along the stator periphery gradually falls off to zero in the interpolar region. The result is a *flat-topped* flux density wave as shown in Fig. 5.3. Since the magnetic flux enters the stator normally, the relative movement between the stator conductors (coil-sides of the stator coil) and the air-gap flux density is mutually perpendicular.

Probe

In a synchronous generator with a 2-pole field winding, what is the emf direction governed by?

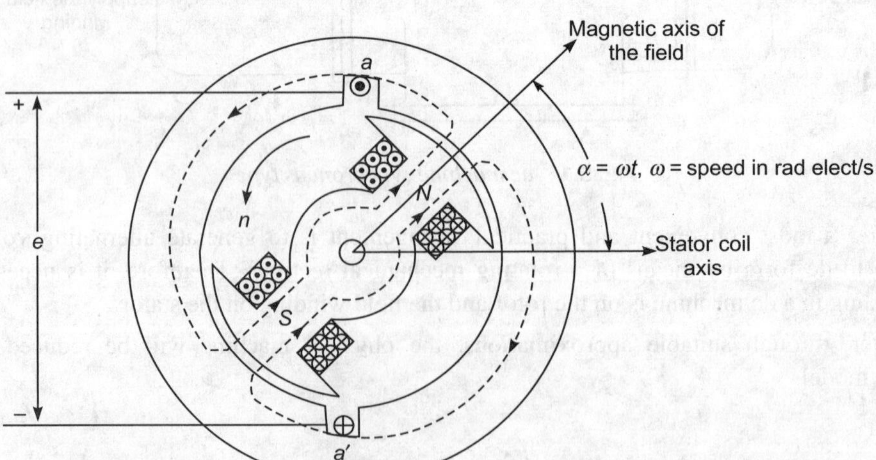

Fig. 5.2 *Elementary synchronous generator—salient-pole 2-pole rotor*

This results in the emf being induced along the stator conductors as per the *Blv* rule, with emf direction on being governed by $\vec{v} \times \vec{B}$ or the Fleming's right-hand rule. It is to be further observed that the magnetic flux density is constant along the axial length; the end effects (fringing) at either end of the cylindrical structure are neglected in this model.

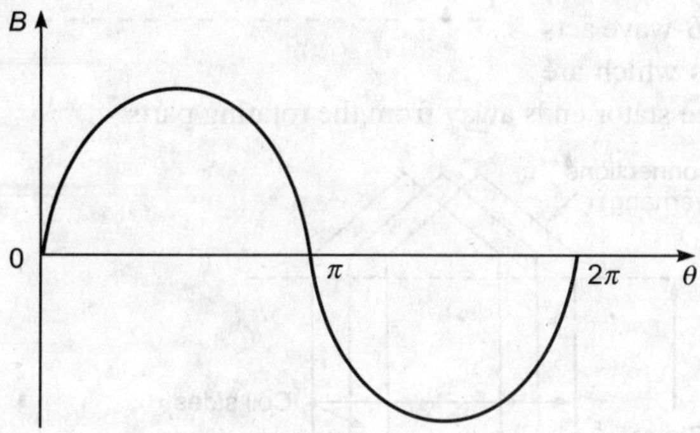

Fig. 5.3 *Flat-topped flux density wave*

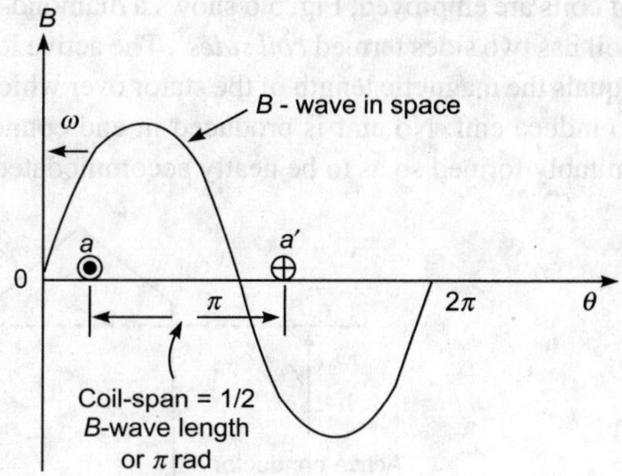

Fig. 5.4 *Sinusoidal air-gap flux density space wave*

In ac machines it is desirable for the induced emfs to be sinusoidal in waveform, therefore, flux density wave in the machine air-gap must be sinusoidal. This is achieved in the salient pole construction (with concentrated field coils) by providing nonuniform air-gap above pole-shoes; minimum air-gap in the middle of the pole-shoe progressively increasing towards outer edges. Here it will be assumed the air-gap flux density wave is *perfectly* sinusoidal as shown in Fig. 5.4.

Probe

What type of waveform is desirable for the induced emf in ac machines?

Nonsalient Pole Rotor

Another method of obtaining a sinusoidal *B*-wave in air-gap is the use of the *nonsalient* pole structure, i.e. a cylindrical rotor with uniform air-gap but with a suitably distributed field winding along the rotor periphery as shown in Fig. 5.5. In this structure as one moves away from the pole-axis, the flux paths link progressively smaller number of field ampere-turns. The ampere-turns can be so distributed as to give a nearly sinusoidal* *B*-wave in space as shown in Fig. 5.4.

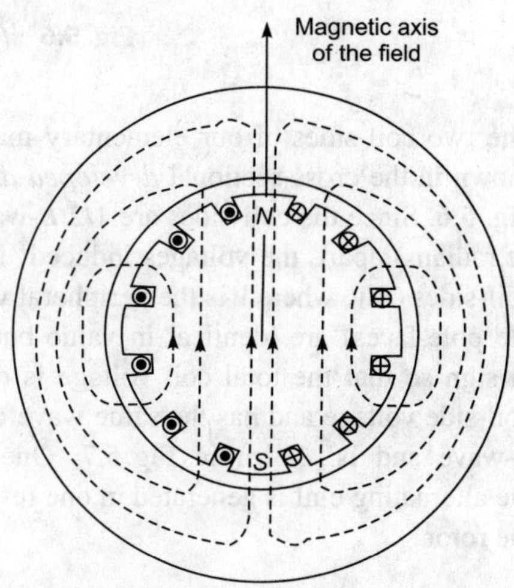

Fig. 5.5 *Nonsalient pole (cylindrical) rotor*

* Actual *B*-wave in this case will be a stepped wave, whose fundamental is being considered in our model. The high frequency harmonics corresponding to the steps are ignored as they do not affect machine performance significantly.

Diamond-shaped Elementary Coil

The armature coil in the elementary 2-pole machine of Fig. 5.2 is placed in two diametrically opposite slots notched out on the inside of the stator. The coil can have many turns. Various shapes of coils are employed; Fig. 5.6 shows a diamond-shaped coil. Each coil has two sides termed *coil sides*†. The active length of coil-sides equals the magnetic length of the stator over which the *B*-wave acts to induce emf. No emf is produced in end connections which are suitably formed so as to be neatly accommodated on the stator ends away from the rotating parts.

Probe

How is the armature coil in the elementary 2-pole machine placed?

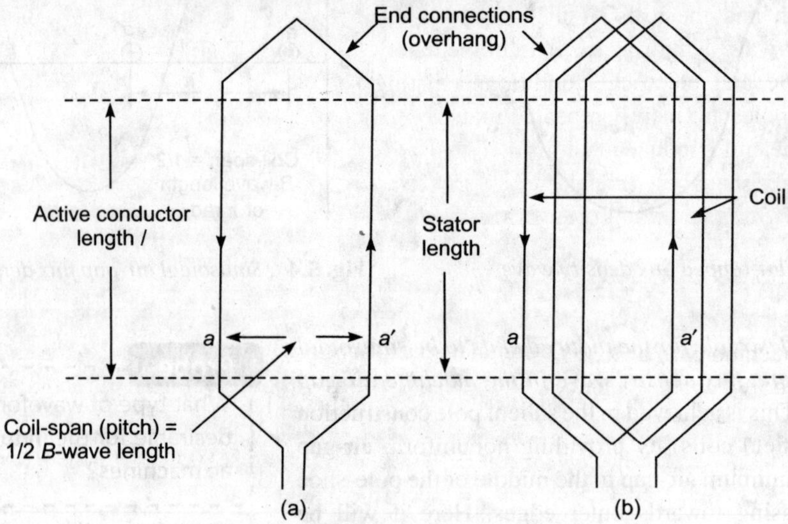

Fig. 5.6 *(a) Single-turn coil, (b) Multi-turn coil*

The two coil-sides of our elementary machine are shown in the cross-sectional *developed diagram* of Fig. 5.6. Since the coil-sides are 1/2 *B*-wave length (π radians) apart, the voltages induced‡ in the two coil-sides (*Blv*, where *v* is the peripheral velocity of the pole-faces) are identical in value but opposite in sign so that the total coil voltage is double the coil-side voltage and has the same waveform as the *B*-wave and is shown in Fig. 5.7. One cycle of the alternating emf is generated in one revolution of the rotor.

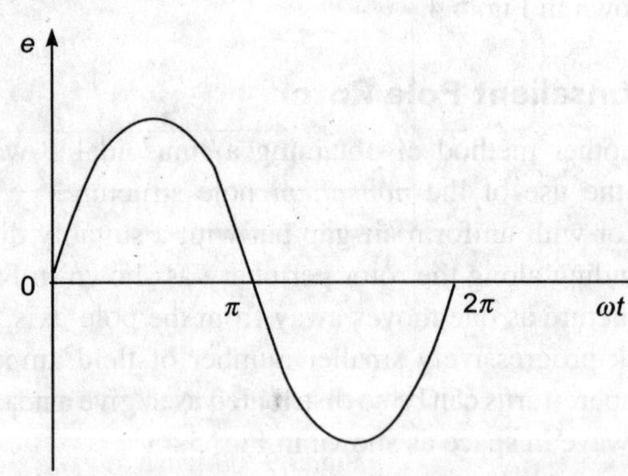

Fig. 5.7 *Coil emf in the elementary machine of Fig. 5.2 (ω = rotor speed in rad/sec)*

† A coil-side in a single-turn coil is a conductor and in a multi-turn coil it is a bundle of conductors (equivalently it can be considered as one conductor).

‡ The direction of induced emf can be established by Fleming's right-hand rule in which the motion is that of a conductor with respect to the field.

Elementary 4-pole Synchronous Generator with Full-Pitched Coils

Consider now a machine with a 4-pole structure, as shown in Fig. 5.8, the poles being alternately north and south. The flux-density space wave of this structure is drawn in Fig. 5.9. It has two complete cycles in the total angle of 2π radians which will now be referred to as the *mechanical angle*. It is obvious from Figs 5.8 and 5.9 that now two coils can be placed symmetrically–one coil under each pair of poles. Each coil has a *span* (*pitch*) of 1/2 *B*-wave length and the axes of the two coils are spaced one full *B*-wave length apart. It is immediately seen that magnetic and electrical conditions existing under one pair of poles are merely repeated under every other pole-pair. The emf's induced in coils a_1a_1' and a_2a_2' are both alternating, equal in magnitude and in time phase to each other. It is convenient to work in terms of one pole pair of a multipolar machine (number of poles must obviously be even). The total angular span of one pole pair is therefore taken as 2π and is referred to as *electrical angle* as different from the actual mechanical angle. Let

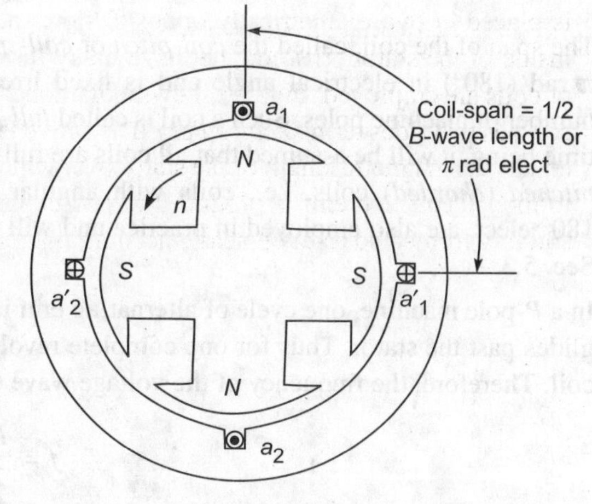

Fig. 5.8 *An elementary 4-pole synchronous generator with full-pitched coils*

Probe

Define electrical angle in a synchronous machine with a 4-pole structure.

$$\theta = \text{electrical angle}$$

$$\theta_m = \text{mechanical angle}$$

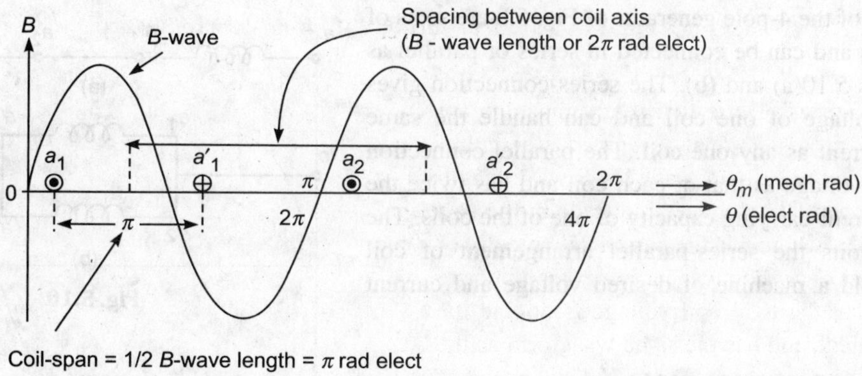

Fig. 5.9 *B-wave and the stator coils of a 4-pole synchronous generator*

These angles are both shown in Fig. 5.9 Taking the ratio of total electrical and mechanical angles of a *P*-pole machine,

$$\frac{\theta}{\theta_m} = \frac{2\pi \times (P/2)}{2\pi} = \frac{P}{2}$$

$$\therefore \qquad \theta = \left(\frac{P}{2}\right)\theta_m \qquad (5.1)$$

The span of the coil, called the *coil-pitch* or *coil-span*, which was indicated to be 1/2 *B*-wavelength, will be π rad (180°) in electrical angle and is fixed irrespective of the number of machine poles. Such a coil is called *full-pitched*. For the time being, it will be assumed that all coils are full-pitched. *Short-pitched* (*chorded*) coils, i.e., coils with angular span less than 180° elect. are also employed in practice and will be discussed in Sec. 5.3.

Probe

What are full-pitched and short-pitched coils?

In a *P*-pole machine, one cycle of alternating emf is generated in each coil as one pole-pair of the rotor poles glides past the stator. Thus for one complete revolution of the rotor, *P*/2 cycles of emf are generated in the coil. Therefore, the frequency of the voltage wave (it is a time wave) is

$$f = \frac{P}{2} \times \frac{n}{60} = \frac{nP}{120} \text{ Hz} \qquad (5.2)$$

where n = speed of the rotor in revolution per minute (rpm)

Differentiating each side of Eq. (5.1) with respect to time

$$\omega = \left(\frac{P}{2}\right)\omega_m \qquad (5.3)$$

where ω = rotor speed in electrical rad/s

ω_m = rotor speed in mechanical rad/s

Obviously $\qquad \omega = 2\pi f \quad$ rad (elect.)/s $\qquad$ (5.4)

The two coils of the 4-pole generator of Fig. 5.8 are seats of identical emfs and can be connected in series or parallel as shown in Figs 5.10(a) and (b). The series connection gives double the voltage of one coil and can handle the same maximum current as any one coil. The parallel connection has the same voltage as that of each coil and has twice the maximum current-carrying capacity of one of the coils. The designer exploits the series-parallel arrangement of coil groups to build a machine of desired voltage and current rating.

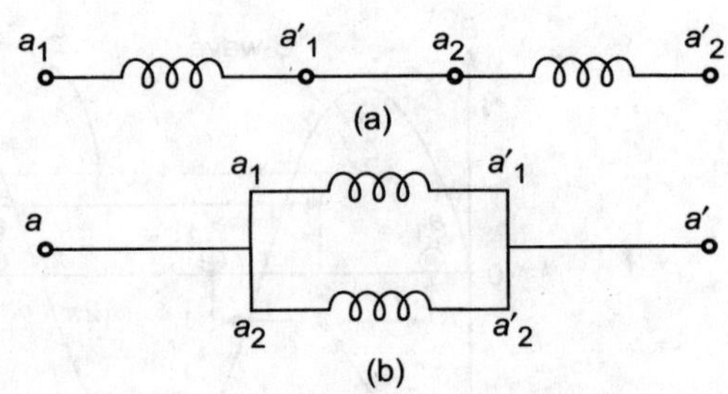

Fig. 5.10

Three-phase Generator (Alternator)

Important! Practical synchronous generators are always of the 3-phase kind because of the well-known advantages[§] of a 3-phase system. If two coils were located at two different space locations in the stator of Fig. 5.2, their emf's will have a time phase difference corresponding to their electrical space displacement. If three coils are located in the stator of the 2-pole machine of Fig. 5.2 at relative electrical spacing of

[§] All power generation and transmission systems are of 3-phase. Except for fractional-kW and for certain special purposes all motors are of 3-phase.

120° (or $2\pi/3$ rad), an elementary 3-phase machine results as is shown in Fig. 5.11(a). The corresponding 4-pole arrangement is depicted in Fig. 5.11(b) where each phase has two symmetrically placed coils corresponding to each pair of poles. The coils of each phase are series/parallel connected and the three phases of a synchronous generator are generally connected in star as shown in Figs 5.12(a) and (b).

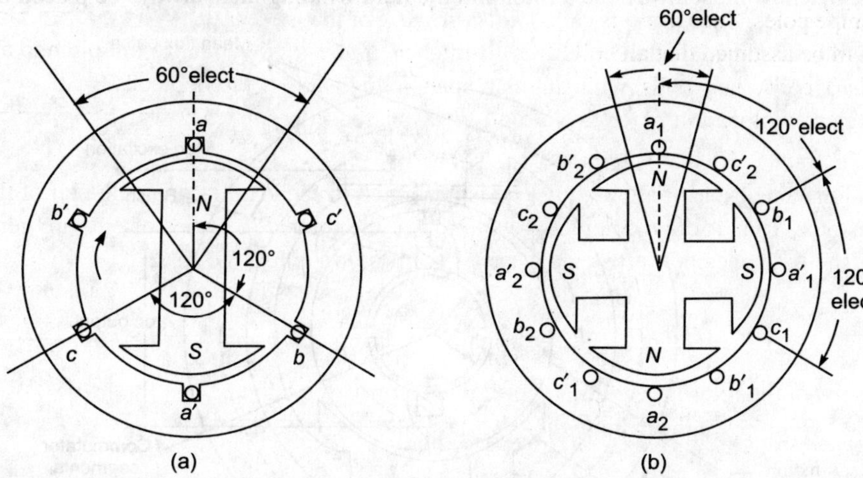

Fig. 5.11 *(a) Three-phase, 2-pole synchronous generator, (b) Three-phase, 4-pole synchronous generator*

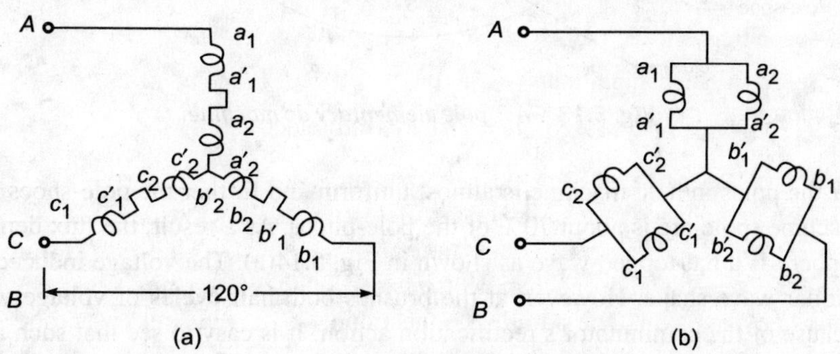

Fig. 5.12 *Star connected elementary 3-phase generator*

The process of torque production in a machine will be explained in Sec. 5.5 after gaining some familiarity with the rotating magnetic field produced by a current-carrying 3-phase winding.

The dc Machine

Figure 5.13 shows a 2-pole elementary dc machine with a single coil rotating armature. It may be seen that the field winding is stationary with salient poles whose pole-shoes occupy a major part of the pole-pitch. An alternating emf is induced in the coil due to rotation of the armature past the stator (field) poles. The two ends of the armature coil are connected to two conducting (copper) segments which cover slightly less than a semicircular arc. These segments are insulated from each other and from the shaft on which they are mounted. This arrangement is called the *commutator*. Current is collected from the commutator segments by means of copper-*carbon brushes*. The connection of the coil to the outside circuit reverse each half cycle in such a manner that the polarity of the one brush is always positive and the other negative. This indeed is the *rectification* action of the commutator-brush arrangement. As already stated, it is because of this requirement of the mechanical rectification action, to obtain direct voltage from the alternating induced emf, that the armature in a dc machine must always be a rotor and the field winding must always be placed on the stator.

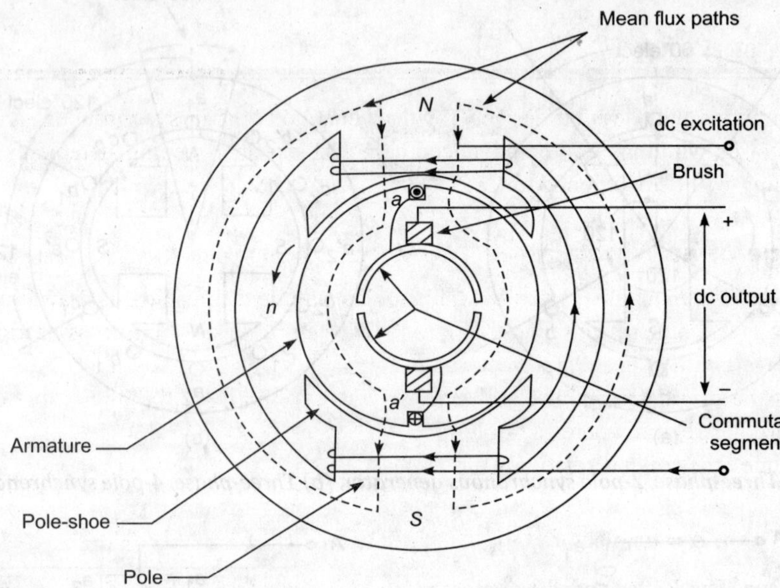

Fig. 5.13 *A 2-pole elementary dc machine*

The air-gap under the poles in a dc machine is almost uniform and further the pole-shoes are wider than in a synchronous machine (pole-arc is about 70% of the pole-pitch). As a result, the flux density in air around the armature periphery is a flat-topped wave as shown in Fig. 5.14(a). The voltage induced in the coil (full-pitched) has a similar wave shape. However, at the brushes both half cycles of voltage wave are positive, (Fig. 5.14(b)) because of the commutator's rectification action. It is easy to see that such a waveform has a higher average value for a given value of B_{peak}, which therefore explains why a dc machine is designed to have a flat-topped *B*-wave.

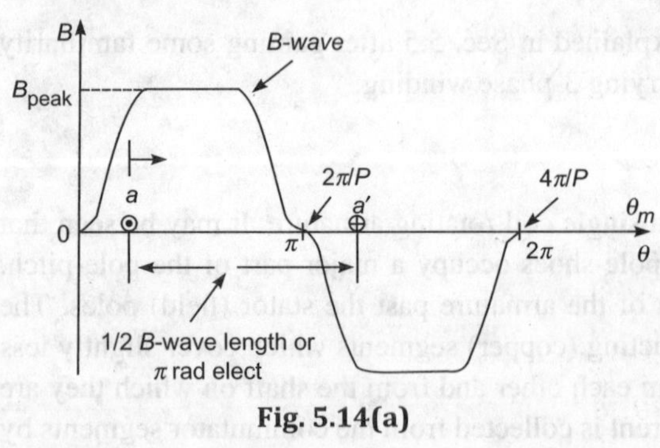

Fig. 5.14(a)

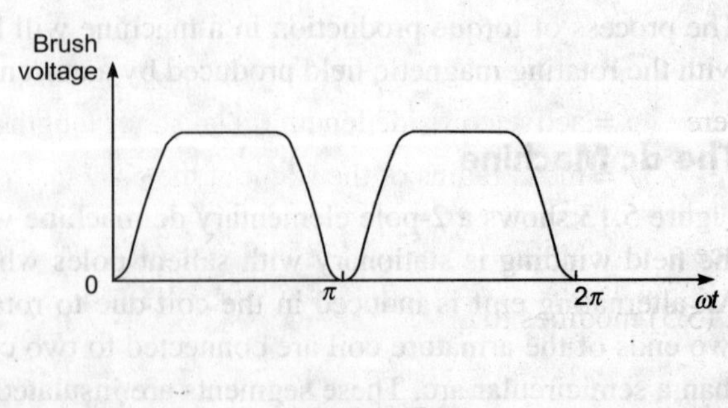

Fig. 5.14(b)

The actual dc machine armature has several coils placed in slots around the armature which are connected in the from of a *lap or wave* winding. The addition of coil emfs (waveform as in Fig. 5.14(b)) with time phase displacement results in almost constant dc voltage at brushes. The reader may add two such waveforms at 90° elect. Displacement and find that dc (average) voltage becomes double but per cent voltage variations reduce and the frequency of variations doubles.

5.3 | GENERATED EMF

LO 1

The quantitative expressions will now be derived for the generated emf in synchronous and dc machine armatures. Some idea of ac windings will be advanced here and certain sweeping statements made, but the discussion on dc winding will be postponed to Ch. 7.

Generated Voltage of ac Winding

The *B*-wave of a synchronous machine (in general multi-polar) assumed sinusoidal is drawn in Fig. 5.15 and a single full-pitched coil (coil-side space separation π rad (180°) elect.) is shown in cross-sectional from. The *B*-wave moves towards left with a speed of ω elect. rad/s or ω_m mech. rad/s. At the origin of time the coil-sides are located in the interpolar region where the pole flux links the coil. At any time *t* the coil has relatively moved by

$$\alpha = \omega t \text{ elect. rad}$$

to the right of the *B*-wave. The *B*-wave can be expressed as

$$B = B_p \sin \theta$$

$$= B_p \sin\left(\frac{P}{2}\theta_m\right)$$

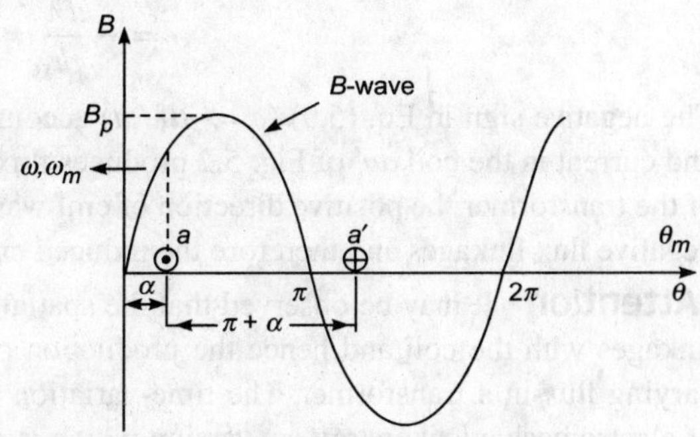

Fig. 5.15 *Relative location of the B-wave and armature coil at any time t*

where B_p = peak flux density

Since the flux is physically spread over the mechanical angle, the flux ϕ linking the coil can be computed by integrating over the mechanical angle. Thus

$$\phi = \int_{2\alpha/P}^{2(\pi+\alpha)/P} B_p \sin\left(\frac{P}{2}\theta_m\right) lr\, d\theta_m \tag{5.5}$$

where l = active coil-side length (axial stator length) and

r = mean radius of the stator at the air-gap.

Since $\theta_m = \dfrac{2}{P}\theta$

Eq. (5.5) modifies to

$$\phi = \left(\frac{2}{P}\right)\int_{\alpha}^{\pi+\alpha} B_p lr \sin\theta\, d\theta$$

$$= \left(\frac{2}{P}\right) 2B_p lr \cos \alpha$$

$$= \left(\frac{2}{P}\right) 2B_p lr \cos \omega t = \Phi \cos \omega t \quad (5.6)$$

It is, therefore, seen that the flux linking the coil varies sinusoidally and has a maximum value of

$$\Phi = \frac{4}{P} B_p lr \text{ (flux/pole)} \quad (5.7)$$

at $\alpha = \omega t = 0$, which indeed is flux/pole. The flux linkages of the coil at any time t are

$$\lambda = N\phi = N\Phi \cos \omega t \quad (5.8)$$

Probe

Derive an expression for the flux linkages of a coil.

where N = number of turns of the coil.

Hence the coil induced emf is

$$e = -\frac{d\lambda}{dt} = \omega N\Phi \sin \omega t \quad (5.9)$$

The negative sign in Eq. (5.9) ($e = -d\lambda/dt$) accounts for the fact that the assumed positive direction of emf and current in the coil *aa′* of Fig. 5.2 produces flux along the coil axis causing positive flux linkages. In case of the transformer the positive direction of emf was assumed such as to cause a current which would produce negative flux linkages and therefore the induced emf law used was $e = +d\lambda/dt$.

!Attention It may be observed that the spatial flux density wave upon rotation causes time-varying flux linkages with the coil and hence the production of emf, an effect which is produced by a fixed-axis time-varying flux in a transformer. The time-variation factor is introduced by rotation causing the phenomenon of electromechanical energy conversion which is not possible in a transformer with fixed-axis time-varying flux.

The rms value of emf induced in the coil from Eq. (5.9) is

$$E = \sqrt{2}\pi f N\Phi = 4.44 \, fN\Phi \text{ volts} \quad (5.10)$$

Probe

What is the rms value of emf induced in a coil?

which is the same result as in a transformer except for the fact that Φ here is the flux/pole.

It may be observed from Eqs (5.6) and (5.9) that the sinusoidally varying flux linking the coil (represented by the phasor $\bar{\Phi}$) leads the sinusoidally varying emf (represented by the phasor $\bar{E}$) by 90°. The phasor relationship is illustrated by the phasor diagram of Fig. 5.16. This is in contrast of the transformer case wherein the flux phasor lags the emf phasor by 90° (refer Fig. 3.4). This difference is caused by the negative sign in the induced emf of Eq. (5.9) while a positive sign was used for the transformer.

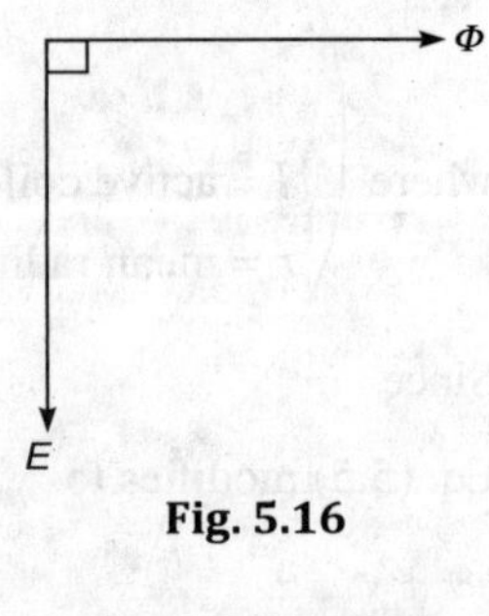

Fig. 5.16

Distributed Winding

It may be seen from Eq. (5.7) that the flux/pole is limited by the machine dimensions and the peak flux density which cannot exceed a specified value dictated by saturation characteristic of iron. Therefore, for inducing an emf of an appropriate value in a practical machine (it may be as high as $11/\sqrt{3}$ to $37/\sqrt{3}$ kV phase), a large number of coil turns are needed and it is not possible to accommodate all these in a single slot-pair. Furthermore, it may be also noticed that with one coil/pole pair/phase, i.e. one slot/pole/phase, the periphery of the stator is far from being fully utilized. It is, therefore, natural to create more slots/pole/phase (SPP) on the stator periphery. In a practical machine with S slot distributed uniformly round the stator periphery,

$$\text{SPP} = m = \frac{S}{qP} \qquad q = \text{number of phases (generally } q = 3) \tag{5.11}$$

Figure 5.17 illustrates a 2-pole, 3-phase machine with $m = 3$. The angle between adjacent slots is

$$\gamma = \frac{\pi P}{S} \text{ elect. rad} \tag{5.12}$$

The winding of phase a in the machine has three coils (11′, 22′ and 33′) which are placed in three slot-pairs distributed in space with an angular separation of γ elect. rad. The total angle $\sigma = m\gamma$ occupied by the phase winding along the armature periphery is called the *phase spread*. Such a winding is referred to as the *distributed winding*. Since the machine is always wound with identical coils, the sinusoidal emfs induced in coils 11′, 22′ and 33′ have the same rms value (E) but have a progressive time phase difference of γ because coil are uniformly distributed in space.

Important! These coils are series connected to yield the phase voltage E_a which is the phasor sum of the coil emf's as shown in Fig. 5.18. It is observed from this figure that because of distribution, the rms phase voltage is less than the algebraic sum of the rms coil voltages. This reduction ratio called the *breadth factor* (also *distribution factor*) is to be determined now, for the general case of SPP = m.

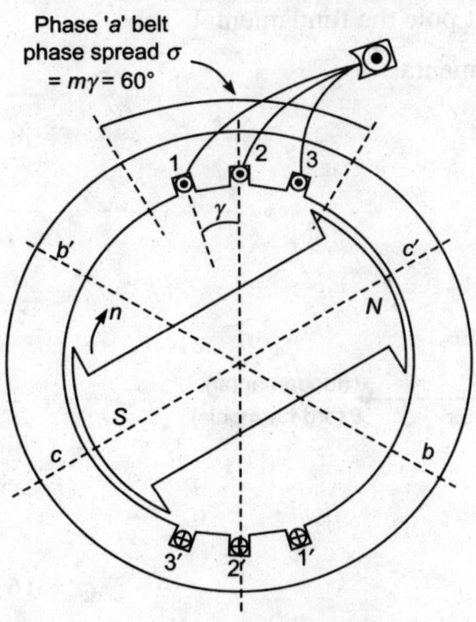

Fig. 5.17 *Synchronous generator with distributed winding*

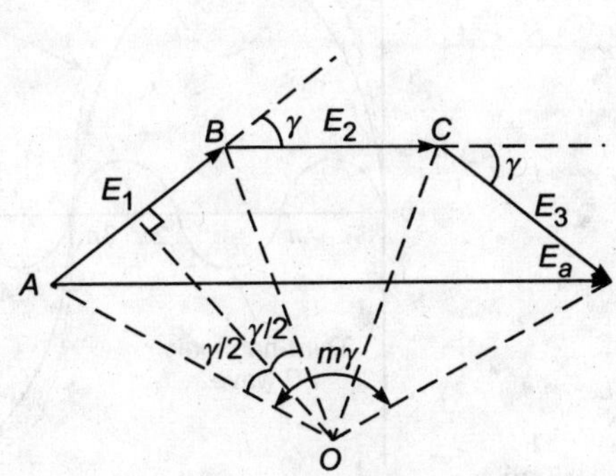

Fig. 5.18 *Phasor diagram of coil emf's to yield phase voltage*

It is easily seen that in Fig. 5.18 the coil emf phasors form sides of a regular polygon, the centre of whose circumscribing circle is constructionally located in the diagram. The phase voltage E_a is given by the resultant phasor (*AD* in Fig. 5.18). The breadth factor is then defined as

$$K_b = \frac{AD}{3AB} \quad \text{or} \quad \frac{AD}{mAB} \text{ (in general)} \tag{5.13}$$

From the geometry of Fig. 5.18

$$AB = 2\ OA \sin \gamma/2$$
$$AD = 2\ OA \sin \gamma m/2$$

Hence

$$K_b = \frac{\sin m\gamma/2}{m \sin \gamma/2} < 1 \text{ for } m > 1 \tag{5.14}$$

It is seen from Eqs (5.11) and (5.12) that

$$m\gamma = \frac{\pi}{q} \text{ or } \frac{\pi}{3}\text{; a fixed value independent of } P \text{ and } S$$

The induced phase emf for a distributed winding is obtained by multiplying Eq. (5.10) by K_b. Thus

$$E = 4.44\ K_b f N_{ph}\ \Phi \tag{5.15}$$

where N_{ph} = total turns (in *series*) per phase

Harmonic Content in the Distributed Winding

The flux density wave of a synchronous machine is never exactly sine wave. Because of odd symmetry of poles (alternately north-south), the space harmonic content of the *B*-wave comprises odd harmonics only, which induce the corresponding harmonic emf's in the winding. Figure 5.19 shows the fundamental *B*-wave, and its third-harmonic; because of somewhat flat-topped nature of the *B*-wave, its third-harmonic is of the "dipping" kind, producing a flux density dip of in the middle of the main pole. It follows from Fig. 5.19 that *n* poles of the *n*th harmonic occur in the space occupied by one of pole the fundamental. Thus,

$$\theta \text{ (}n\text{th harmonic)} = n\theta \text{ (fundamental)}$$

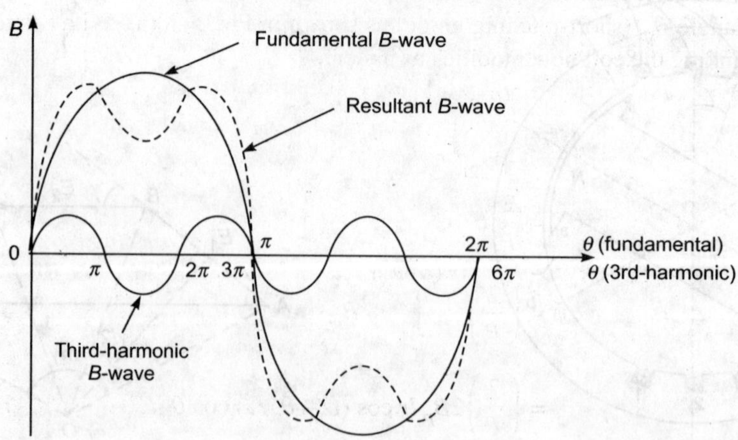

Fig. 5.19 *Fundamental and third-harmonic of the B-wave in the air-gap of a synchronous machine*

Therefore from Eq. (5.14)

$$K_b \text{ (}n\text{th harmonic)} = \frac{\sin mn\gamma/2}{m \sin n\gamma/2} \tag{5.16}$$

It can be easily shown (see Ex. 5.1) that K_b (nth harmonic) is less than K_b (fundamental) for important harmonics resulting in reduction of harmonic content of the voltage of the distributed winding–an incidental advantage.

Example 5.1 *Calculate the fundamental, third and fifth harmonic breadth factors for a stator with 36 slots wound for 3-phase, 4-poles.* LO 1 ○○●

Solution

$$m = \frac{36}{3 \times 4} = 3 \quad \text{and} \quad \gamma = \frac{180° \times 4}{36} = 20° \text{ elect.}$$

(i) $$K_b \text{ (fundamental)} = \frac{\sin \dfrac{3 \times 20°}{2}}{3 \sin \dfrac{20°}{2}} = 0.96$$

(ii) $$K_b \text{ (third harmonic)} = \frac{\sin \dfrac{3 \times 3 \times 20°}{2}}{3 \sin \dfrac{3 \times 20°}{2}} = 0.667$$

(iii) $$K_b \text{ (fifth harmonic)} = \frac{\sin \dfrac{3 \times 5 \times 20°}{2}}{3 \sin \dfrac{5 \times 20°}{2}} = 0.218$$

It is noticed that K_b for the third-harmonic is sufficiently less than the fundamental and is far less for the fifth-harmonic.

Short-pitched (Chorded) Coils

So far it was assumed that the stator coils are full-pitched (a span of π rad elect). Coils may have a span of less than the full-pitch. This arrangement offers certain advantages. Consider that the coil-span is less than the full-pitch by an elect. angle θ_{sp} (short-pitching angle) as shown in Fig. 5.20(a). With reference to Fig. 5.15, Eq. (5.5) for the flux linking the coil now modifies as under:

$$\phi = \int_{2\alpha/P}^{2(\pi+\alpha-\theta_{sp})/P} B_p \sin\left(\frac{P}{2}\theta_m\right) lr\, d\theta_m$$

Since $$\theta_m = \frac{2}{P}\theta$$

$$\phi = \left(\frac{2}{P}\right) \int_{\alpha}^{(\pi+\alpha-\theta_{sp})} B_p \sin \theta\, lr\, d\theta$$

$$= \left(\frac{2}{P}\right) 2B_p lr \cos(\alpha - \theta_{sp}/2) \cos \theta_{sp}/2$$

Note: Difficulty Level ⟶ ○○● — Level 1 and Level 2 Category; ○●● — Level 3 and Level 4 Category; ●●● — Level 5 and Level 6 Category

$$= \Phi \cos (\theta_{sp}/2) \cos (\omega t - \theta_{sp}/2) \tag{5.17}$$

The flux linking the coil and therefore the coil emf reduces by multiplicative factor of

$$K_p = \cos \theta_{sp}/2 = \textit{pitch factor} \tag{5.18}$$

The meaning of the pitch factor can be arrived at from another angle. In Fig. 5.20(a) with positive direction of coil-side emfs marked in opposite direction, the coil emf is the phasor sum of coil-side emfs, i.e.

> **Probe**
>
> What is the pitch factor in short pitched coils?

$$\bar{E}_c = \bar{E}_a + \bar{E}_{a'}$$

In the case of a full-pitch coil $\bar{E}_{a'}$ and $\bar{E}_a$ are in phase (they are π rad apart but their positive direction are marked oppositely) so that

$$E_c = 2\,E_a$$

as shown in Fig. 5.20(b).

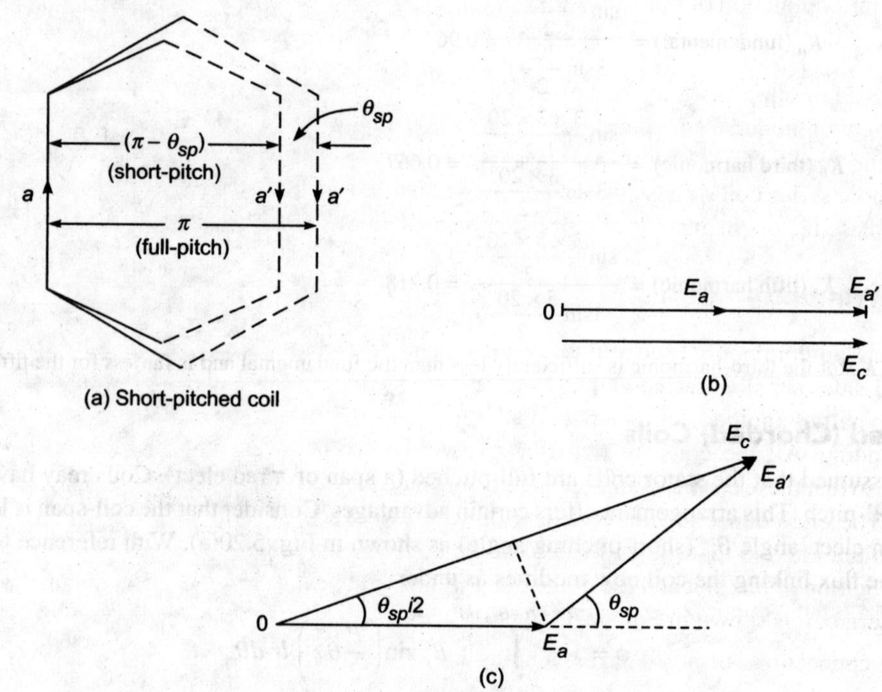

Fig. 5.20

In a coil short-pitched by θ_{sp}, $\bar{E}_{a'}$ will lead (or lag) $\bar{E}_a$ by the angle θ_{sp} depending upon direction of rotation as shown in Fig. 5.20(c). From the geometry of the phasor diagram

$$E_c = 2E_a \cos \theta_{sp}/2$$

Hence the reduction of the coil emf due to short-pitching is governed by the factor

$$K_p = \frac{2E_a \cos \theta_{sp}/2}{2E_a}$$

$$= \cos \theta_{sp}/2 \text{ (same as Eq. (5.18))}$$

Because of short-pitching the expression of Eq. (5.15) for the phase voltage modifies to

$$E = 4.44\, K_b K_p f N_{ph} \Phi \tag{5.19}$$

Let $K_w = K_b K_p =$ *winding factor*; then

$$E = 4.44\, K_w f N_{ph} \Phi \tag{5.20}$$

For the *n*th harmonic the pitch factor would be

$$K_p = \cos n\theta_{sp}/2 \tag{5.21}$$

Any particular harmonic can be eliminated by coil short-pitching by making

$$n\theta_{sp}/2 = k\frac{\pi}{2};\ k = 1, 3, \ldots \text{ (odd)}$$

For example, for elimination of the 13th harmonic

$$13\,\theta_{sp}/2 = 90° \quad \text{or} \quad \theta_{sp} \approx 14°$$

!Attention Thus short-pitching helps in elimination of any particular harmonic or in reduction of the harmonic content of the induced voltage in general. It is easily seen from Fig. 5.20(a) that short-pitched coils have shorter end connections. Thus there is a saving in copper per coil but part of this saving is wiped off by the fact that more series coils/ phase would now be needed to generate a specified phase voltage. A designer has to weigh these factors in arriving at a decision on the angle of short-pitching.

Two-layer Winding

One important and commonly used way of neatly arranging the end connection of coils in a winding is to place two coil-sides per slot. Each coil then has one coil-side in the bottom half of one slot and the other coil-side in the top half of another slot (one pole pitch away for full-pitched coils). Such a winding is known as a two-layer winding. All the coils of a two-layer winding are of similar shape so that these can be wound separately and then placed in the slots. The end connections at each end of such a coil are given an involute kink so that the coil-sides can be conveniently placed in the bottom of one slot and top of the other. A formed two-layer, multi-turn coil is shown in Fig. 5.21.

Probe

What is a two-layer winding?

The winding connections of a two-layer winding are best illustrated by means of a *developed* winding diagram which imagines that the stator has been cut and laid out flat. One phase of a 4-pole, two-layer *lap* winding is shown in Fig. 5.22. The upper layer is indicated by a continuous line and the lower layer by a dotted line and both are drawn side-by-side in the diagram. This figure illustrates the simplest case of 4-poles with 1 slot/pole/phase. The four coils could be series connected or various series-parallel connections could be employed. In Fig. 5.22 the four coils of one

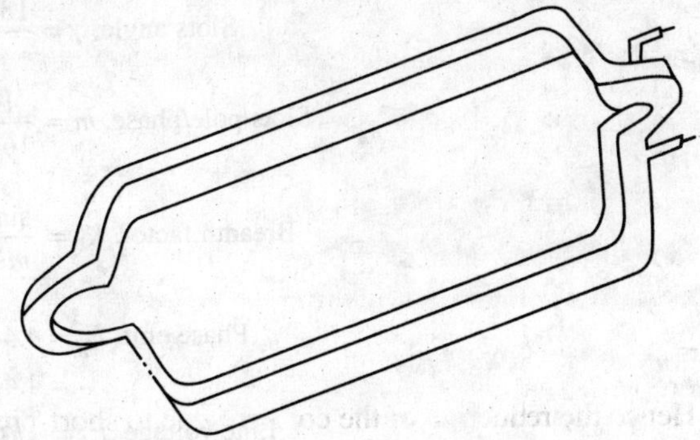

Fig. 5.21 *Formed two-layer multi-turncoil*

phase are connected in parallel in two series groups of two coils each. Obviously the phase current i_a and the conductor current i_c = are related as $i_c = i_a/2$.

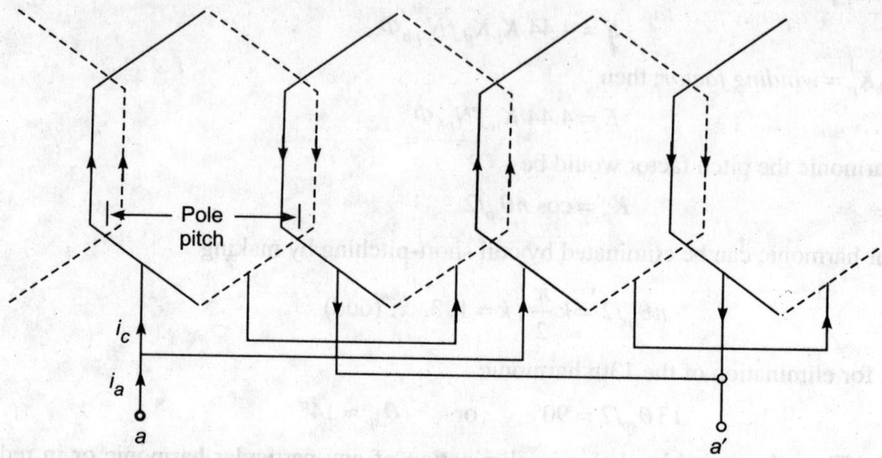

Fig. 5.22 *One phase of a 4-pole, 2-layer ac lap winding*

Example 5.2 *A 3-phase, 16-pole synchronous generator has a star-connected winding with 144 slots and 10 conductors per slot. The flux per pole is 0.04 Wb (sinusoidally distributed) and the speed is 375 rpm. Find the frequency and phase and line induced emf's. The total turns/phase may be assumed to be series connected.* LO 1

Solution

$$f = \frac{nP}{120} = \frac{375 \times 16}{120}\ 50\ \text{Hz}$$

$$\text{Total number of conductors} = 144 \times 10 = 1440$$

$$\text{Total number of turns} = \frac{1440}{2} = 720$$

$$\text{Number of turns (series)/phase, } N_{ph} = \frac{720}{3} = 240$$

$$\text{Slots angle, } \gamma = \frac{180° P}{S} = \frac{180° \times 16}{144} = 20°$$

$$\text{Slots/pole/phase, } m = \frac{144}{16 \times 3} = 3$$

$$\text{Breadth factor, } K_b = \frac{\sin m\gamma/2}{m \sin \gamma/2} = \frac{\sin \frac{3 \times 20°}{2}}{3 \sin \frac{20°}{2}} = 0.96$$

$$\text{Phase emf, } E_p = 4.44\ K_b f N_{ph}\ \text{(series)}\ \Phi$$

$$= 4.44 \times 0.96 \times 50 \times 240 \times 0.04 = 2046\ \text{V}$$

$$\text{Line voltage, } E_L = \sqrt{3}\ E_p = 3543.6\ \text{V}$$

Example 5.3 *A 3-phase, 50 Hz, star-connected alternator with 2-layer winding is running at 600 rpm. It has 12 turns/coil, 4 slots/pole/phase and a coil-pitch of 10 slots. If the flux/pole is 0.035 Wb sinusoidally distributed, find the phase and line emf's induced. Assume that the total turns/phase are series connected.* LO 1

Solution

$$P = \frac{120f}{n} = \frac{120 \times 50}{600} = 10$$

$$\text{Total slots, } S = 4 \times 3 \times 10 = 120$$

$$N_{ph} = \frac{120 \times 12}{3} = 480$$

$$\text{SPP, } m = 4$$

$$\text{Slot angle, } \gamma = \frac{180° \times 10}{120} = 15°$$

$$K_b = \frac{\sin m\gamma/2}{m \sin \gamma/2} = \frac{\sin \frac{4 \times 15°}{2}}{4 \sin \frac{15°}{2}} = 0.958$$

$$\text{Pole-pitch} = \frac{120}{10} = 12 \text{ slots}$$

$$\text{Coil-pitch} = 10 \text{ slots}$$

$$\text{Short-pitching angle, } \theta_{sp} = (12 - 10) \times 15° = 30°$$

$$K_p = \cos \theta_{sp}/2 = \cos \frac{30°}{2} = 0.966$$

$$\text{Phase emf induced, } E_p = 4.44\, K_b K_p f N_{ph} \text{ (series)} \Phi$$

$$= 4.44 \times 0.958 \times 0.966 \times 50 \times 480 \times 0.035 = 3451 \text{ V}$$

Also

$$E_L = \sqrt{3} \times 3451 = 5978 \text{ V}$$

Example 5.4 *A 2-pole, 3-phase, 50 Hz, 2300 V synchronous machine has 42 slots. Each slot has two conductors in a double layer winding. The coil pitch is 17 slots. Each phase winding has two parallel paths. Calculate the flux/pole required to generate a phase voltage of $2300/\sqrt{3}$ V.* LO 1

Solution

$$\text{SPP, } m = \frac{42}{3 \times 2} = 7$$

$$\text{Slot angle, } \gamma = \frac{2 \times 180°}{42} = 8.57°$$

$$K_b = \frac{\sin m\gamma/2}{m \sin \gamma/2} = \frac{\sin(7 \times 8.57°/2)}{7 \times \sin 8.57°/2} = 0.956$$

$$\text{Coil pitch} = 17 \text{ slots}$$

$$\text{Pole pitch} = \frac{42}{2} = 21 \text{ slots}$$

$$\text{Short pitching angle, } \theta_{sp} = (21 - 17) \times 8.57° = 34.28°$$

$$K_p = \cos 34.28°/2 = 0.956$$

$$N_{ph} \text{ (series)} = \frac{42 \times 2}{2 \times 3 \times 2} = 7$$

$$E_a = 4.44\, K_b K_p f N_{ph} \text{ (series)} \Phi$$

$$\frac{2300}{\sqrt{3}} = 4.44 \times 0.952 \times 0.956 \times 50 \times 7 \times \Phi$$

or

$$\Phi = 0.94 \text{ Wb}$$

The dc Machine

With reference to the single-coil elementary dc machine of Figs 5.13 and 5.14(a) which shows the *B*-wave of the machine relative to the elementary full-pitched coil, let

$$\Phi = \text{flux/pole}$$

Consider that the coil is lying in the interpolar region so that the full/pole (Φ) links is positively. Let it now move through one pole-pitch (π rad (elect.)) in time Δt so that its flux linkages reverse in sign. The change in flux linkages during this movement is

$$\Delta\lambda = -2N_c\Phi$$

where N_c are the coil turns. For a *P*-pole machine

$$\Delta t = \frac{2\pi}{P\omega_m} \text{ s} \tag{5.22}$$

Hence the *average* coil-induced emf is

$$E_a = -\frac{\Delta\lambda}{\Delta t} = \frac{\Phi\omega_m N_c P}{\pi} \tag{5.23}$$

Important! The dc machine armature is always wound with 2-layer winding forming a closed circuit. The coils are suitably connected to the commutator segments made of copper, insulated from each other and from the shaft and formed into a cylinder. The current from the armature winding is tapped through brushes placed on the commutator periphery, 180° elect. apart. The alternate brushes are of opposite polarity and all the brushes of one polarity are connected in parallel resulting in two armature terminals. This arrangement causes the closed armature winding to form a series-parallel circuit. The dc voltage between the brushes of opposite polarity is the sum of the average voltages of series turns between the brushes, each turn having the same average voltage. The number of parallel paths depends upon the type of armature winding *wave* and *lap* type and are

$$\begin{aligned} A &= 2; \text{ for wave winding} \\ A &= P; \text{ for lap winding} \end{aligned} \tag{5.24}$$

Let the machine be wound with *Z* conductors (a conductor is the active part of the side of a turn). Since two conductors form a turn,

Probe

Derive an equation for the armature emf of a dc machine.

$$\text{Series turns/parallel path, } N = \frac{Z}{2A}$$

Using Eq. (5.23), the armature emf which equals the parallel path emf is given by

$$E_a = \frac{\Phi \omega_m Z}{2\pi}\left(\frac{P}{A}\right) \tag{5.25}$$

But
$$\omega_m = \frac{2\pi n}{60} \text{ rad (mech.)/s}$$

Hence
$$E_a = \frac{\Phi n Z}{60}\left(\frac{P}{A}\right) \text{ V; } n = \text{speed in rpm} \tag{5.26}$$

It follows from Eq. (5.24) that

$$\frac{P}{A} = \frac{P}{2} \text{ for wave winding}$$
$$= 1 \text{ for lap winding} \tag{5.27}$$

Since P has a minimum value of 2, the wave winding will have a larger armature emf than a lap winding for the same values of Φ, n and Z.

The details of the dc machine winding, where the brushes are placed on the commutator and how the parallel paths are formed, will be taken up in Ch. 7.

The circuit schematic (circuit model) of a de generator is drawn in Fig. 5.23. As the current in the field and armature circuits are dc, the circuit inductances do not play any role. The armature circuit has a voltage E_a induced in it by rotation of armature in the flux/pole Φ as per the relationship of Eq. (5.25). When the armature feeds current to the external circuit (the load), there is a voltage drop I_aR_a in the armature circuit, where R_a is the effective resistance of the armature including the voltage drop at brush contacts. The Kirchhoff voltage law applied to the armature circuit yields

$$E_a = V_t + I_aR_a$$

where V_t is the voltage at generator terminals. This equation is usually written as

$$V_t = E_a - I_a R_a \tag{5.28}$$

Example 5.5 *A 1500 kW, 600 V, 16-pole, dc generator runs at 200 rpm. What must be the useful flux per pole, if there are 2500 lap-connected conductors and full-load copper losses are 25 kW? Also calculate the area of the pole-shoe if the average gap flux density is 0.85 T. The generator is feeding full load of 1500 kW at the terminal voltage of 600 V.* LO 1

Solution Figure 5.23 shows the circuit schematic of the dc generator

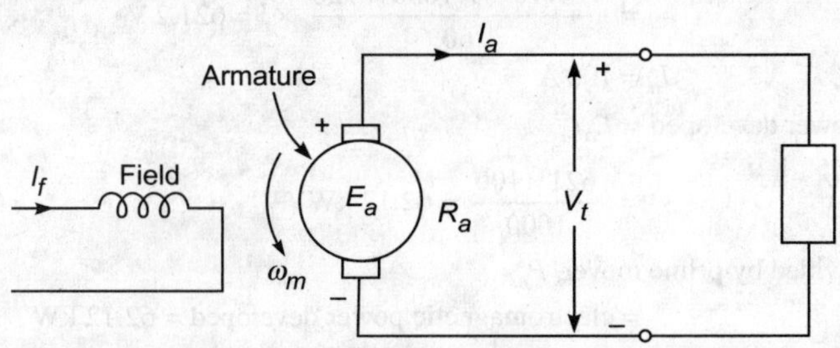

Fig. 5.23 *Schematic circuit diagram of a dc generator*

$$\text{Full-load armature current, } I_a = \frac{1500 \times 1000}{600} = 2500\text{A}$$

$$\text{Copper-loss} = I_a^2 R_a = 25 \times 1000$$

where R_a is the effective resistance of the armature circuit,

$$R_a = \frac{25 \times 1000}{2500 \times 2500} = 4 \times 10^{-3} = \Omega$$

Kirchhoff's voltage law equation for the armature circuit is

$$E_a = V_t + I_a R_a$$

$$= 600 + 2500 \times 4 \times 10^{-3} = 610 \text{ V}$$

But $$E_a = \frac{\Phi n Z}{60}\left(\frac{P}{A}\right)$$

where Φ is the useful flux/pole which links armature coil. Some of the pole flux will complete its circuit bypassing the armature—called the *leakage flux*.

Substituting the values,

$$610 = \frac{\Phi \times 200 \times 2500}{60}\left(\frac{16}{16}\right)$$

or, $$\Phi = 0.0732 \text{ Wb}$$

$$\text{Area of pole-shoe} = \frac{0.0732}{0.85} = 861 \text{ cm}^2$$

Example 5.6 *A 4-pole, lap-wound dc machine has 728 armature conductors. Its field winding is excited from a dc source to create an air-gap flux of 32 mWb/pole. The machine (generator) is run from a primemover (diesel engine) at 1600 rpm. It supplies a current of 100 A to an electric load.*

(a) Calculate the electromagnetic power developed.

(b) What is the mechanical power that is fed from the primemover to the generator?

(c) What is torque provided by the primemover?

LO 1

Solution

(a) $$E_a = \frac{\Phi n Z}{60} \times \left(\frac{P}{A}\right)$$

$$= \frac{32 \times 10^{-3} \times 1600 \times 728}{60} \times 1 = 621.2 \text{ V}$$

$$I_a = 100 \text{ A}$$

$$\text{Electromagnetic power developed} = E_a I_a$$

$$= \frac{621 \times 100}{1000} = 62.12 \text{ kW}$$

(b) Mechanical power provided by prime mover, P_m

$$= \text{electromagnetic power developed} = 62.12 \text{ kW}$$

But $$P_m = T\omega_m$$

$$\text{Primemover torque, } T = \frac{62.12 \times 100}{\left(\dfrac{2\pi \times 1600}{60}\right)} = 370.75 \text{ Nm}$$

5.4 MMF OF DISTRIBUTED AC WINDINGS

LO 1

It has been seen earlier that the armature of a practical machine has distributed winding wound for the same number of poles as the field winding. As the armature carries current, the resultant field of its current-carrying coils has the same number of poles as the field winding. Our approach will be to find the mmf space distribution of the current-carrying armature by superimposing the mmf space waves of individual coils.

MMF Space Wave of a Single Coil

A cylindrical rotor machine with small air-gap is shown in Fig. 5.24(a). The stator is imagined to be wound for two-poles with a single N-turn full-pitch coil carrying current i in the direction indicated. The figure shows some flux lines of the magnetic field set up. A north and corresponding south pole are induced on the stator periphery. The magnetic axis of the coil is from the stator north to the stator south. Each flux line radially crosses the air-gap twice, normal to the stator and rotor iron surfaces and is associated with constant mmf Ni. On the assumption that the reluctance of the iron path is negligible, half the mmf ($Ni/2$) is consumed to create flux from the rotor to stator in the air-gap and the other half is used up to establish flux from the stator to rotor in the air-gap. *Mmf and flux radially outwards from the rotor to the stator (south pole on stator) will be assumed to be positive and that from the stator to rotor as negative.*

The physical picture is more easily visualized by the developed diagram of Fig. 5.24(b) where the stator with the winding is laid down flat with rotor on the top of it. It is seen that the mmf is a rectangular space wherein mmf of + $Ni/2$ is consumed in setting flux from the rotor to stator and mmf of – $Ni/2$ is consumed in setting up flux from the stator to the rotor. It has been imagined here that the coil-sides occupy a narrow space on the stator and the mmf changes abruptly from – $Ni/2$ to + $Ni/2$ at one slot and in reverse direction at the other slot. The mmf change at any slot is

Probe

What is the mmf change at any slot of a single coil?

$$Ni = \text{ampere-conductors/slot}$$

and its sign depends upon the current direction.

The mmf space wave of a single coil being rectangular, it can be split up into its fundamental and harmonics. It easily follows from the Fourier series analysis that the fundamental of the mmf wave as shown in Fig. 5.24(b) is

$$\mathcal{F}_{a1} = \frac{4}{\pi}\left(\frac{Ni}{2}\right) \cos\theta = F_{1p} \cos\theta \tag{5.29}$$

where θ is the electrical angle measured from the magnetic axis of the coil which coincides with the positive peak of the fundamental wave.

From Eq. (5.29)

$$F_{a1}\text{ (peak)} = F_{1p} = \frac{4}{\pi}\left(\frac{Ni}{2}\right) \tag{5.30}$$

Probe

Write an expression for fundamental mmf wave of a current-carrying coil.

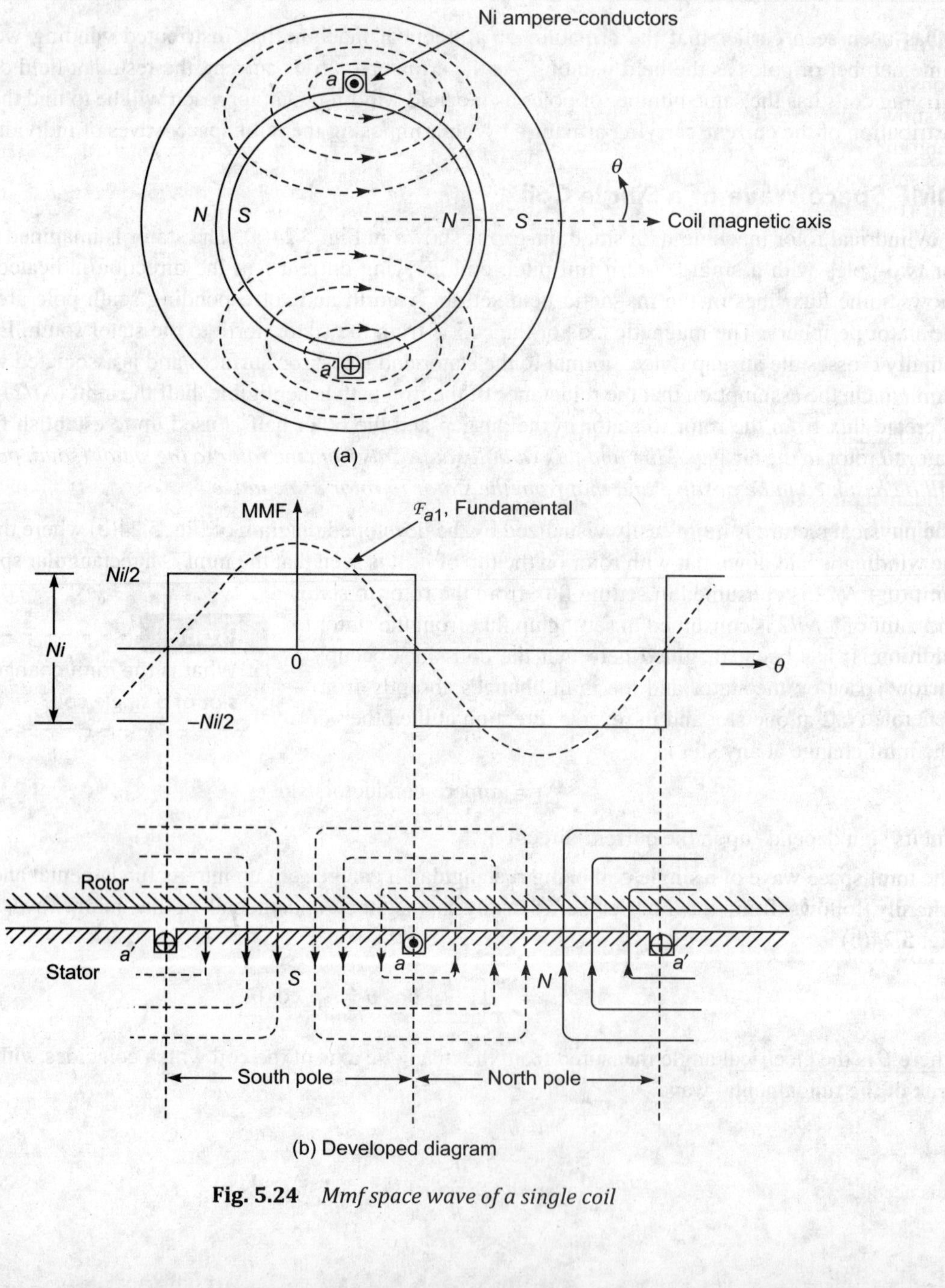

Fig. 5.24 *Mmf space wave of a single coil*

From now onwards only the fundamental mmf wave of a current-carrying coil will be considered, neglecting its associated harmonic space waves whose amplitudes are small[¶]. Furthermore, as will soon be seen, in a distributed winding, the rectangular mmf waves of individual phase group coils add up to produce a mmf wave closer to a sine wave, i.e. the harmonics tend to cancel out.

It may be noted here that with the assumption of the narrow air-gap, the mmf distribution will be the same if the coils were located in the rotor slots instead.

MMF Space Wave of One Phase of a Distributed Winding

Consider now a basic 2-pole structure with a round rotor, with 5 slots/pole/phase (SPP) and a 2-layer winding as shown in Fig. 5.25. The corresponding developed diagram is shown in Fig. 5.26(a) along with the mmf diagram which now is a stepped wave–obviously closer to a sine wave than the rectangular mmf wave of a single coil (Fig. 5.24(b)). Here since SPP is odd (5), half the ampere-conductors of the middle slot of the phase group a and a' contribute towards establishment of south pole and half towards north pole on the stator. At each slot the mmf wave has a step jump of $2N_c i_c$ ampere-conductors where N_c = coil turns (equal to conductors/layer) and i_c = conductor current.

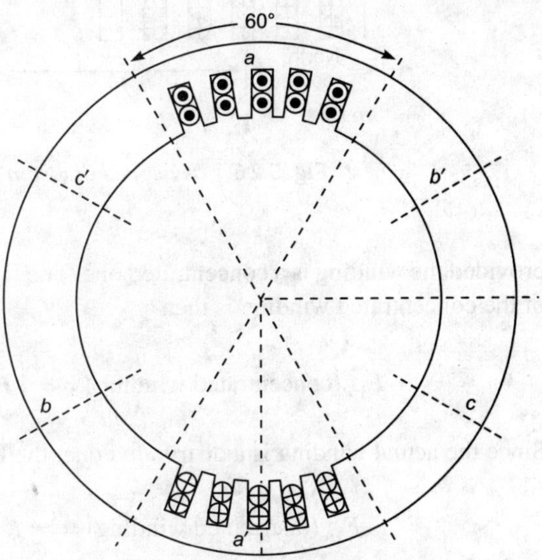

Fig. 5.25 *A 3-phase, 2-pole structure with two-layer winding*

Now F_{1p}, the peak of the fundamental of the mmf wave, has to be determined. Rather than directly finding the fundamental of the stepped wave, one can proceed by adding the fundamentals of the mmf's of individual slot-pairs (with a span of one pole-pitch). These fundamentals are progressively out of phase (space phase as different from time phase) with each other by the angle γ. This addition is easily accomplished by defining the breadth factor K_b, which will be the same as in the case of the generated emf of a coil group.

Let N_{ph} (series) = series turns per parallel path of a phase

A = number of parallel paths of a phase

Then AT/parallel path = N_{ph} (series) i_c

AT/phase = $A(N_{ph}\text{ (series)}i_c) = N_{ph}\text{ (series) } i_a$

where i_a = phase current = $A\, i_c$

It now follows that

$$\text{AT/pole/phase} = \left(\frac{N_{ph}\text{ (series)}}{P}\right) i_a$$

[¶]For a rectangular wave the normalized peaks of the harmonic waves are:

$$F_{3p} = \frac{4}{3\pi},\ F_{5p} = \frac{4}{5\pi},\ F_{7p} = \frac{4}{7\pi},\ \ldots$$

It is needless to say that there cannot be any even harmonics.

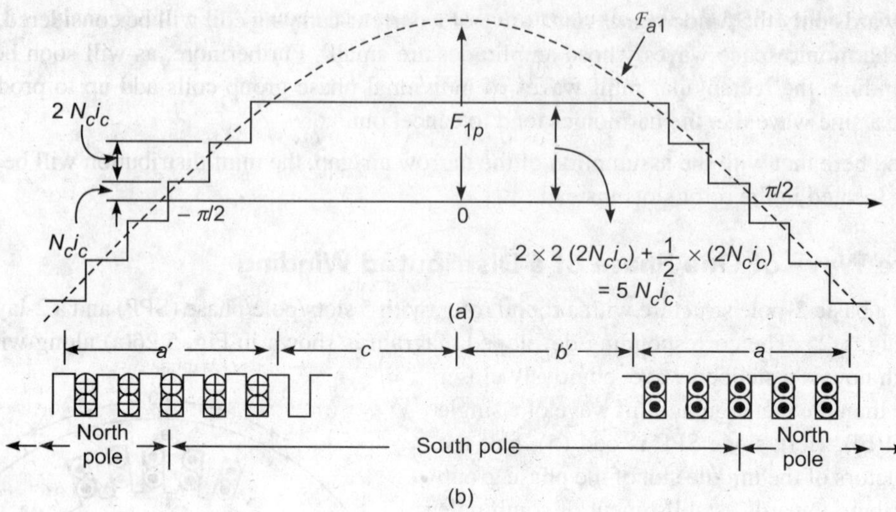

Fig. 5.26 *Developed diagram and mmf wave of the machine of Fig. 5.25*

provided the winding is a concentrated one. The fundamental peak of the concentrated winding is then

$$F_{1p}\,(\text{concentrated winding}) = \frac{4}{\pi}\left(\frac{N_{ph}\,(\text{series})}{P}\right)i_a$$

> **Probe**
> Derive the expression for the fundamental peak of concentrated and distributed winding.

Since the actual winding is a distributed one, the fundamental peak will be reduced by the factor K_b. Thus

$$F_{1p}\,(\text{distributed winding}) = \frac{4}{\pi}K_b\left(\frac{N_{ph}\,(\text{series})}{P}\right)i_a \tag{5.31}$$

Hence

$$\mathcal{F}_{a1} = \frac{4}{\pi}K_b\left(\frac{N_{ph}\,(\text{series})}{P}\right)i_a\cos\theta \tag{5.32}$$

where the pole axis is taken as the angle reference (Fig. 5.26(b)).

MMF Wave of Short-Pitched Coils

The effect on the mmf wave of short-pitched coils can be visualized by Fig. 5.27 in which the stator has two short-pitched coils (a_1a_1', a_2a_2') for phase a of a 2-pole structure**. The mmf of each coil establishes one pole. From the developed diagram of Fig. 5.27(b) it is seen that the mmf wave is rectangular but of shorter space length than a pole-pitch. The amplitude of the fundamental peak gets reduced by a factor K_p, called the *pitch*

** The arrangement of coils for 3-phase 2-pole, two-layer winding, with $S = 12$, $m = 2$ and coil-pitch = 5 (chorded by one slot) is shown in Fig. 5.28.

It is seen from Fig. 5.28 that the coil-sides in a given slot do not belong to the same phase. This is in contrast to Fig. 5.25 depicting a full-pitch winding.

factor, compared to the full-pitch rectangular mmf wave. It can be shown[††] by Fourier analysis that

$$F_{1p} = \frac{4}{\pi} F_{\text{pole}} \cos \theta_{sp}/2$$

Hence $$K_p = \cos \theta_{sp}/2 \tag{5.33}$$

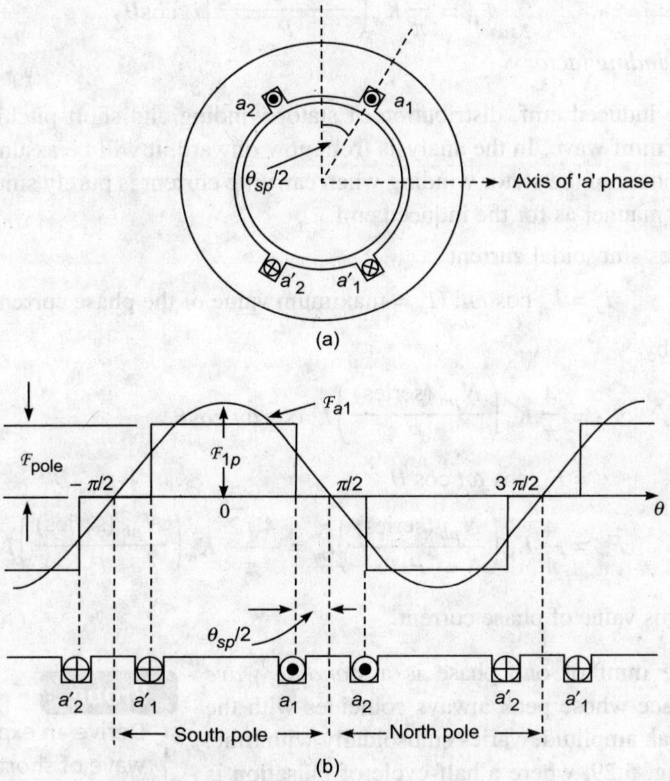

Fig. 5.27 *(a) Stator with short-pitched coil at phase 'a' (single-layer winding), (b) Effect of short-pitched coil on mmf wave*

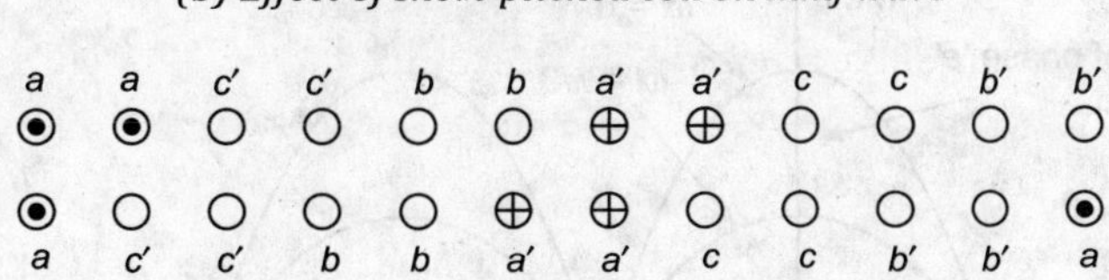

Fig. 5.28 *Two-layer short pitch winding*

[††] From the Fourier series

$$F_{1p} = \frac{2}{\pi} \int_{-\pi/2+\theta_{sp}/2}^{\pi/2-\theta_{sp}/2} F_{\text{pole}} \cos \theta \, d\theta$$

$$= \frac{4}{\pi} F_{\text{pole}} \cos \theta_{sp}/2$$

It is not surprising that the same result is obtained for K_p for the space mmf wave as for the generated emf in a short-pitched coil.

In general when the winding is both distributed and short-pitched, the fundamental space mmf of phase a as in Eq. (5.32) generalizes to

$$\mathcal{F}_{a1} = \frac{4}{\pi} K_w \left(\frac{N_{ph}\,(\text{series})}{P} \right) i_a \cos\theta \tag{5.34}$$

where $K_w = K_b K_p$ = *winding factor*

Like in the case of the induced emf, distribution of stator winding and short-pitching both help reducing harmonics in the space mmf wave. In the analysis from now onward, it will be assumed that the space mmf wave created by each phase of the stator winding when carrying current is purely sinusoidal. Also K_b and K_p are defined in the same manner as for the induced emf.

When the phase a carries sinusoidal current

$$i_a = I_m \cos \omega t;\ (I_m = \text{maximum value of the phase current}) \tag{5.35}$$

the mmf wave is given by

$$\mathcal{F}_{a1} = \frac{4}{\pi} K_w \left(\frac{N_{ph}\,(\text{series})}{P} \right) I_m \cos \omega t \cos\theta$$

$$= F_m \cos \omega t \cos \theta \tag{5.36}$$

where $$F_m = \frac{4}{\pi} K_w \left(\frac{N_{ph}\,(\text{series})}{P} \right) I_m = \frac{4\sqrt{2}}{\pi} K_w \left(\frac{N_{ph}\,(\text{series})}{P} \right) I \tag{5.37}$$

where $I = I_m/\sqrt{2}$ = rms value of phase current.

As per Eq. (5.36), the mmf of one phase is a *standing wave* (*pulsating wave*) in space whose peak always coincides with the phase axis while the peak amplitude varies sinusoidally with time. This is illustrated in Fig. 5.29, where a half-cycle of pulsation is indicated.

Probe

Derive an expression for the mmf wave of short pitched coil.

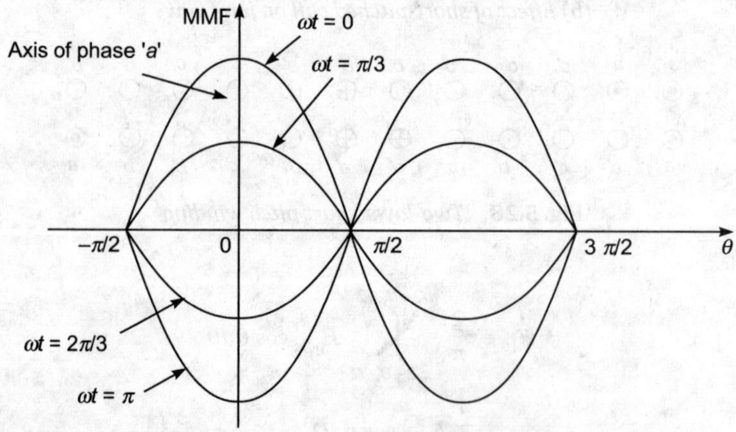

Fig. 5.29 *Standing (pulsating) mmf wave caused by sinusoidal current in phase a coils*

Current-sheet Concept

It was seen above that a distributed winding gives rise to a stepped mmf wave having a strong fundamental component which will be considered in machine modelling while all the harmonic components will be neglected (justification for the same has been advanced). Now the kind of space distribution of current on the stator (or rotor) has to be found which will produce a purely sinusoidally distributed mmf wave. An intuitive answer is that it would be a sinusoidally distributed current.

Figure 5.30 shows a sinusoidally distributed current-sheet with a peak linear density of A A/m length of the air-gap periphery. The current flows from one end of the stator (or rotor) to the other end parallel to the stator axis. With respect to the angle reference shown, the current distribution is given by

$$i = A\cos\theta \tag{5.38a}$$

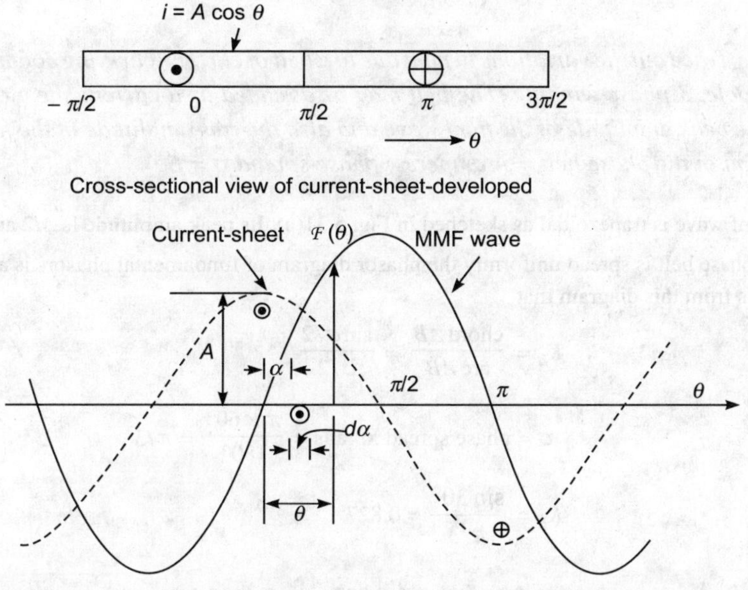

Fig. 5.30 *Sinusoidally distributed current-sheet and its associated mmf wave*

The mmf at angle θ is contributed by the current contained in angular spreads of $\pi/2$ on either side of it. The current in the differential angle $d\alpha$ at angle α is found by converting it to mechanical angle and multiplying it by mean radius. Thus the differential current is

$$(A\cos\alpha)\frac{D}{2}d\left(\frac{2}{P}\alpha\right) = \frac{AD}{P}\cos\alpha\, d\alpha$$

where D = mean air-gap diameter

The mmf at angle θ is then given by

$$\mathcal{F}(\theta) = \frac{1}{2}\left[\frac{AD}{P}\int_{-\pi/2+\theta}^{\theta}\cos\alpha\, d\alpha - \frac{AD}{2}\int_{\theta}^{\pi/2+\theta}\cos\alpha\, d\alpha\right]$$

Probe

Derive an expression for mmf at an angle θ in a current sheet.

$$= \frac{AD}{P} \sin\theta \tag{5.38b}$$

whose peak value is

$$F_{peak} = \frac{AD}{P}$$

As per Eqs (5.38a) and (5.38b), the mmf wave associated with the sinusoidally distributed current-sheet is also sinusoidal but displaced 90°(elect.) from the current-sheet. Also the peak value of the mmf wave equals half the current contained in one loop of the current-sheet, i.e.

$$\frac{1}{2}\left(\frac{2}{\pi} A \times \frac{\pi D}{P}\right) = \frac{AD}{P} = F_{peak} \tag{5.39}$$

Example 5.7 *Trace out the variations in mmf due to a belt of current-carrying conductor representing one phase of a 2-pole, 3-phase winding. The belt may be assumed as a current sheet of uniform current density. What is the peak amplitude of the mmf wave and also the rms amplitude of the fundamental wave. Given: Total current in the phase belt = A amperes; phase spread σ = 60°.* LO 1

Solution The mmf wave is trapezoidal as sketched in Fig. 5.31(a). Its peak amplitude is $A/2$ ampere-turns.

As the current in the phase belt is spread uniformly the phasor diagram of fundamental phasors is arc of 60° as shown in Fig. 5.31(b). It follows from this diagram that

$$K_b = \frac{\text{chord } AB}{\text{arc } AB} = \frac{\sin \sigma/2}{\sigma/2}$$

where σ = phase spread in rads $= \dfrac{\pi \times 60°}{180°} = \pi/3$

$$K_b = \frac{\sin 30°}{\pi/3} = 0.827$$

RMS amplitude of fundamental mmf wave of one phase $= 0.827 \dfrac{A}{2} = 0.4135\ A$

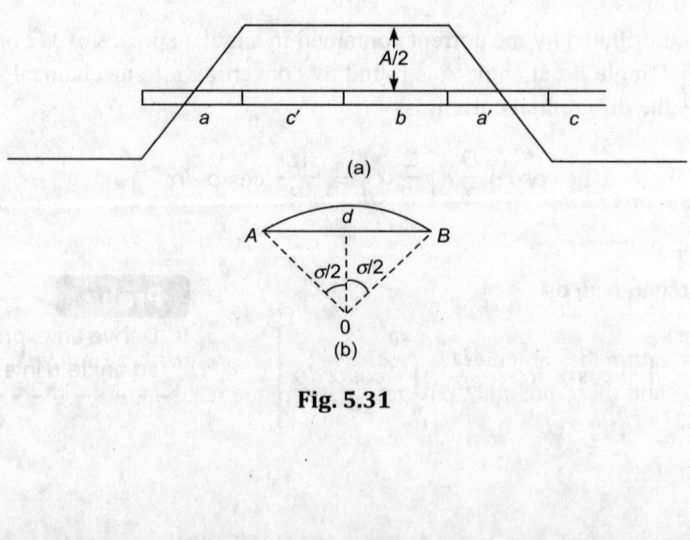

Fig. 5.31

5.5 | ROTATING MAGNETIC FIELD

LO 1

It was seen in Sec. 5.4 that the sinusoidal current in any phase of an ac winding produces a pulsating mmf wave in space whose amplitude varies sinusoidally with time. The expression for the fundamental component of this mmf is given in Eq. (5.36). The harmonics of the mmf wave are rendered inconsequential in a distributed winding and are still further reduced in amplitude if short-pitched coils are used. Therefore, the fundamental of the mmf wave will only be considered in the machine model given here.

Machine Model of Rotating Magnetic Field with 3-phases

Consider now that the three phases of an ac winding are carrying balanced alternating currents.

$$\begin{aligned} i_a &= I_m \cos \omega t \\ i_b &= I_m \cos (\omega t - 120°) \\ i_c &= I_m \cos (\omega t - 240°) \end{aligned} \tag{5.40}$$

Three pulsating mmf waves are now set up in the air-gap which have a time phase difference of 120° from each other. These mmf's are oriented in space along the magnetic axes of phases *a*, *b* and *c* as illustrated by the concentrated coil in Fig. 5.32. Since the magnetic axes are located 120° apart in space from each other, the three mmf's can be expressed mathematically as

$$\begin{aligned} \mathcal{F}_a &= F_m \cos \omega t \cos \theta \\ \mathcal{F}_b &= F_m \cos (\omega t - 120°) \cos (\theta - 120°) \\ \mathcal{F}_c &= F_m \cos (\omega t - 240°) \cos (\theta - 240°) \end{aligned} \tag{5.41}$$

wherein it has been considered that three mmf waves differ progressively in time phase by 120°, i.e. 2 π/3 rad elect. and are separated in space in space phase by 120°, i.e. $2\pi/3$ rad elect. The resulting mmf wave which is the sum of the three pulsating mmf waves is

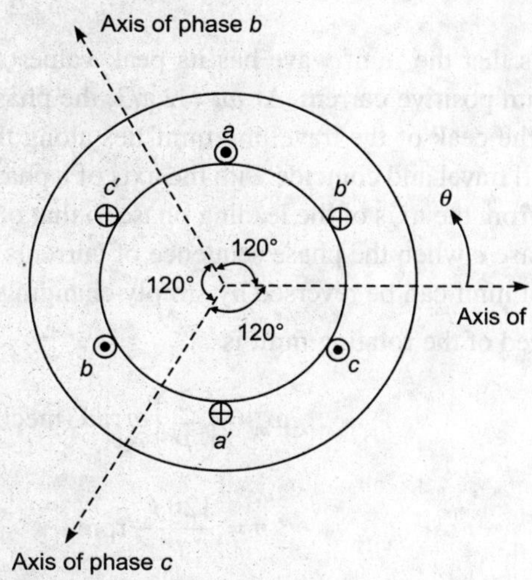

Fig. 5.32 *Relative location of the magnetic axes of three phases*

$$\mathcal{F} = \mathcal{F}_a + \mathcal{F}_b + \mathcal{F}_c$$

or $$\mathcal{F}(\theta, t) = F_m [\cos \omega t \cos \theta + \cos(\omega t - 120°) \cos(\theta - 120°) + \cos(\omega t - 240°) \cos(\theta - 240)]$$

It simplifies trigonometrically to

$$\mathcal{F}(\theta, t) = \frac{3}{2} F_m \cos(\theta - \omega t) + \frac{1}{2} F_m [\cos (\theta + \omega t) + \cos(\theta + \omega t - 240°) + \cos(\theta + \omega t - 480°)] \tag{5.42}$$

Recognizing that

$$\cos (\theta + \omega t - 480°) = \cos (\theta + \omega t - 120°)$$

it is found that the terms within the capital bracket of Eq. (5.42) represent three unit phasors with a progressive phase difference of 120° and therefore add up to zero. Hence, the resultant mmf is

$$\mathcal{F}(\theta, t) = \frac{3}{2} F_m \cos(\theta - \omega t) \tag{5.43}$$

It is found from Eq. (5.43) that the resultant mmf is distributed in both space and time. It indeed is a *travelling wave* with sinusoidal space distribution whose space phase angle changes linearly with time as ωt. It, therefore, rotates in the air-gap at a constant speed of ω rad (elect.)/s.

The peak value of the resultant mmf is

$$F_{peak} = \frac{3}{2} F_m \tag{5.44a}$$

Probe

What is the peak value of the resultant mmf in a rotating magnetic field having 3 phases?

Using the value of F_m from Eq. (5.37)

$$F_{peak} = 3 \times \frac{2\sqrt{2}}{\pi} K_w \left(\frac{N_{ph}(\text{series})}{P}\right) I \tag{5.44b}$$

At $\omega t = 0$, i.e. when the a phase current has maximum positive value, it follows from Eq. (5.43) that

$$\mathcal{F}(\theta, 0) = \frac{3}{2} F_m \cos\theta$$

It means that the mmf wave has its peak value (at $\theta = 0$) lying along the axis of phase a when it carries maximum positive current. At $\omega t = 2\pi/3$, the phase b (assumed lagging) has its positive current maximum, so that the peak of the travelling mmf lies along the axis of b phase. By the same argument the peak of the mmf will travel and coincide with the axis of c phase at $\omega t = 4\pi/3$. It is, therefore, seen that the resultant mmf travels from the axis of the leading phase to that of the lagging phase, i.e., from phase a towards phase b and then phase c when the phase sequence of currents is abc (a leads b leads c). The direction of rotation of the resultant mmf can be reversed by simply changing the phase sequence of currents.

The speed of the rotating mmf is

$$\omega_m = \left(\frac{2}{P}\right) \omega \text{ rad (mech.)/s} \quad \text{or} \quad \frac{2\pi n}{60} = \frac{2}{P} \times 2\pi f$$

or

$$n = \frac{120 f}{P} \text{ rpm} \tag{5.45}$$

This indeed is the rotor speed to induce emf of frequency f in the ac winding. Hence this is called the *synchronous* speed and now onwards would be denoted by the symbol n_s.

Conclusion

Whenever a balanced 3-phase winding with phases distributed in space so that the relative space angle between them is $2\pi/3$ rad (elect.), is fed with balanced 3-phase currents with relative time phase difference of $2\pi/3$ rad (elect.), the resultant mmf rotates in the air-gap at a speed of $\omega_s = 2\pi f$ elect. rad/s where f is the frequency of currents in Hz. The direction of rotation of the mmf is from the leading phase axis towards the lagging phase axis.

The above conclusion is easily generalized to q phases ($q > 2$), where the relative space-phase angle between the phase winding is $2\pi/q$ rad (elect.) and the relative time phase angle between currents is also $2\pi/q$ rad (elect.) For a 2-phase winding the time and space phase angles should be $\pi/2$ rad or 90° (elect.).

Physical Picture

Sinusoidally distributed mmf along space angle θ can be regarded as a *two dimensional space vector* oriented along its positive peak (towards south pole on the stator) having an amplitude equal to its peak value. For example, the mmf of a phase a in Fig. 5.33(a) is represented by the vector $\vec{F}_a$ along the positive direction of the axis of coil aa'. Sinusoidally distributed mmf's can be added by the vector method.

The mmf of phase a varies in the time and space as (Eq. (5.41))

$$F_a = F_m \cos \omega t \cos \theta \tag{5.46a}$$

It is a pulsating wave with a fixed axis (Fig. 5.29), i.e. the amplitude of the vector $\vec{F}_a$ oscillates sinusoidally with time (the exciting current is $i_a = I_m \cos \omega t$ with fixed axis (axis of phase a). Trigonometrically, we can write Eq. (5.46a) as

$$F_a = \frac{1}{2} F_m \cos (\theta - \omega t) + \frac{1}{2} F_m \cos (\theta + \omega t) \tag{5.46b}$$

or

$$\vec{F}_a = \vec{F}^+ + \vec{F}^- \tag{5.46c}$$

It may be seen from Eq. (5.48b) that a pulsating mmf can be considered as the sum of two (sinusoidally distributed) mmf's of same amplitude $(1/2)F_m$ which rotates at constant speed of ω elect. rad/sec. in opposite directions; $\vec{F}^+$ of amplitude $1/2F_m$ rotates in the positive direction of θ and $\vec{F}^-$ also of amplitude $1/2F_m$ rotates in the negative direction of θ. The resolution of a pulsating sinusoidally distributed mmf into two sinusoidally distributed rotating mmf's is illustrated in Fig. 5.33(a). As the two component mmf's rotate, their resultant ($\vec{F}_a$) oscillates sinusoidally in time with its axis remaining fixed.

The mmf's along the axes of coils bb' and cc' can be similarly resolved into two oppositely rotating mmfs of equal amplitude. Consider the instant when the phase a current is maximum positive, i.e. $+I_m (i_a = I_m \cos \omega t)$, the phase b current is at that time $-\frac{1}{2} I_m$ and is becoming less negative ($i_b = I_m \cos (\omega t - 120°)$) and the phase c current is $-\frac{1}{2} I_m$ but is becoming more negative ($i_c = I_m \cos (\omega t - 240°)$). This is also obvious from the phasor diagram of Fig. 5.33(c). Instantaneous locations of the component rotating of all three phases are shown in Fig. 5.33(b). It is immediately observed from this figure that the three $\vec{F}^+$ are coincident and are rotating in the positive direction at speed ω adding up to a positively rotating field of amplitude $3/2\, F_m$. The three $\vec{F}^-$ are location at relative angles of 120° and are rotating in the negative direction at speed ω and thereby cancel themselves out (zero resultant field). It may be, therefore, concluded that the resultant of three pulsating mmf's of maximum amplitude F_m with 120° time phase difference having axes at relative space angles of 120° is a single rotating field of strength $3/2F_m$ rotating at speed ω elect. rad/s in the positive direction.

If the reluctance of the iron path of flux is neglected, the sinusoidal rotating mmf wave creates a coincident sinusoidal rotating B-wave in the air-gap with a peak amplitude of

$$B_{\text{peak}} = \frac{3\mu_0 F_m}{2g}$$

where g is the length of the machine air-gap.

Probe

What is the expression for the peak amplitude of the rotating B-wave in the air gap created by a sinosoidal rotating mmf wave?

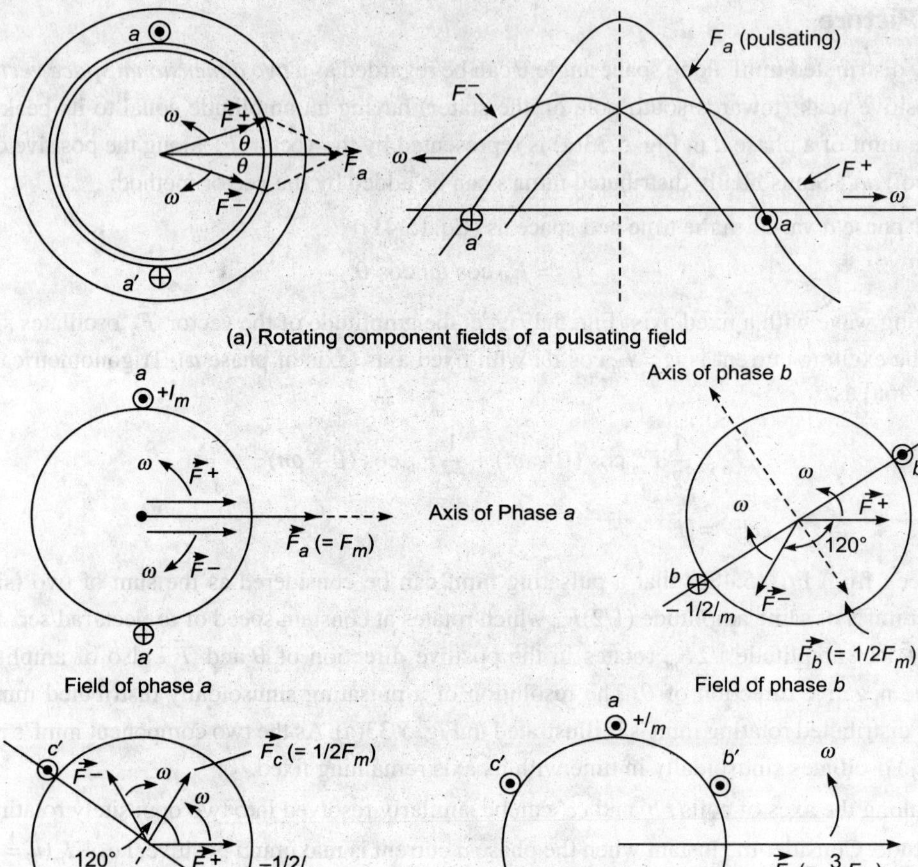

a
a'
ω
θ
θ
F_a (pulsating)
F^-
F^+
ω
(a) Rotating component fields of a pulsating field
$+I_m$
Axis of Phase a
F_a (= F_m)
Field of phase a
Axis of phase b
b'
b
120°
$-1/2I_m$
F_b (= $1/2F_m$)
Field of phase b
c'
c
F_c (= $1/2F_m$)
120°
$-1/2I_m$
Axis of phase c
Field of phase c
F_r (= $\frac{3}{2}F_m$)
$-\frac{1}{2}I_m$
Resultant field
(b) Location of component rotating fields of all the three phases and the resultant field at time t = 0
I_a
$+I_m$
I_b
I_c
(c)

Fig. 5.33

Example 5.8 *Consider a 3-phase ac machine with star connected stator (neutral not connected). It is excited by 3-phase balanced currents of rms magnitude I. If one phase is disconnected while the other two carry the same magnitude current (by adjustment of voltage across the two lines), determine the relative peak magnitude of rotating mmf waves.*

LO 1

Solution Let for single-phase rms current I, peak value of oscillating mmf be F_m. With only two phases carrying currents, the phase current will be equal and opposite.

$$F_{a1} = F_m \cos\theta \cos \omega t$$

$$F_{a2} = -F_m \cos(\theta - 120°) \cos \omega t$$

$$F_{net} = F_{a1} + F_{a2} = F_m[\cos\theta \cos \omega t - \cos(\theta - 120°) \cos \omega t]$$

$$= F_m(\cos\theta - \cos(\theta - 120°))\cos \omega t$$

$$= 2F_m[\sin(\theta - 60°)\sin(-60°)] \cos \omega t$$

$$= \sqrt{3}\, F_m \cos(\theta + 30°) \cos \omega t$$

$$F_{net} = \frac{\sqrt{3}}{2} F_m \cos(\theta + 30° - \omega t)$$

$$+ \frac{\sqrt{3}}{2} \cos(\theta + 30° + \omega t)$$

$$= \quad \vec{F}^{+} \quad + \quad \vec{F}^{-}$$

$\uparrow$ Forward $\quad$ $\uparrow$ Backward

Relative strength of rotating mmf = $\frac{\sqrt{3}}{2} F_m$

Vector diagram The vector diagram is drawn in Fig. 5.34.

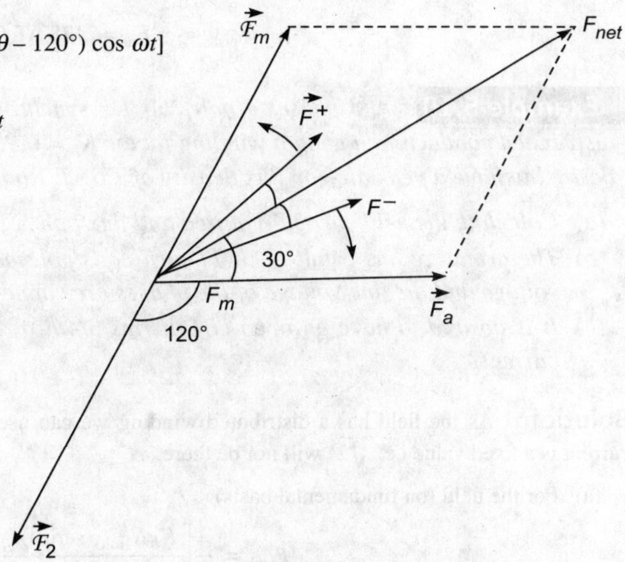

Fig. 5.34

Example 5.9 *A 3-phase, 400 kVA, 50 Hz, star-connected alternator (synchronous generator) running at 300 rpm is designed to develop 3300 V between terminals. The armature consists of 180 slots, each slot having one coilside with eight conductors. Determine the peak value of the fundamental mmf in AT/pole when the machine is delivering full-load current.*

LO 1

Solution

$$P = \frac{120 f}{n_s} = \frac{120 \times 50}{300} = 20$$

$$I_L = I_P = \frac{400 \times 1000}{\sqrt{3} \times 3300} = 70 \text{ A}$$

Maximum value of the phase current, $I_m = 70\sqrt{2}$ A

$$\text{Slot angle, } \gamma = \frac{180° \times 20}{180} = 20°$$

$$\text{SPP, } m = \frac{180}{3 \times 20} = 3$$

$$K_b = \frac{\sin m\gamma/2}{m\sin\gamma/2} = 0.96$$

$$\text{Turns/phase (series)} = \frac{180\times 8}{2\times 3} = 240$$

$$F_m = \frac{4}{\pi} K_b \left(\frac{N_{ph}(\text{series})}{P}\right) I_m$$

$$= \frac{4}{\pi} \times 0.96 \times \left(\frac{240}{20}\right) \times 70\sqrt{2} = 1452 \text{ AT /pole /phase}$$

$$F_{\text{peak}} = \frac{3}{2} F_m = 2178 \text{ AT /pole}$$

Example 5.10 *A 3-phase, 4-pole, 50 Hz synchronous machine has its rotor winding with 364 distributed conductors having a winding factor $K_b = 0.957$. The machine has an air-gap of 0.8 mm. It is desired to have a peak air-gap flux density of 1.65 T. Rotor length = 1.02 m, rotor diameter = 41 cm.*

(a) Calculate the field current required and flux/pole.

(b) The armature has 3 full-pitch 11-turn coils/pole pair per phase. Calculate the open-circuit phase voltage and the line voltage of the phases are connected in star.

(c) It is desired to have an open-circuit rms 3/4th of that found in part (b). What should be the field current?

LO 1

Solution As the field has a distributed winding we can use the result for armature winding except that the field current is a fixed value i.e. $\sqrt{2}$ will not be there.

(a) For the field (on fundamental basis)

$$\mathcal{F}_{\text{peak}} = \frac{4}{\pi}\left[\frac{K_b N_{\text{field}}(\text{series}) I_f}{P}\right]; \text{ Eq. (5.31)} \quad \text{(i)}$$

$$H_{\text{peak}} = F_{\text{peak}}/g \quad \text{(ii)}$$

$$B_{\text{peak}} = \mu_0 H_{\text{peak}} \quad \text{(iii)}$$

or

$$B_{\text{peak}} = \frac{4\mu_0}{\pi g}\left[\frac{K_b N_{\text{field}}(\text{series}) I_f}{P}\right] \quad \text{(iv)}$$

$$1.65 = \frac{4\times 4\pi\times 10^{-7}}{\pi\times 0.8\times 10^{-2}}\left[\frac{0.957\times(364/2)\times I_f}{4}\right]$$

$$I_f = \frac{1.65\pi\times 8\times 10^{-3}\times 4\times 2}{4\times 4\,\pi\times 10^{-7}\times 0.957\times 364}$$

$$= 18.95 \text{ A}$$

(b)

$$\text{Flux/pole, } \Phi = \frac{4}{P} B_{\text{peak}}\, lr \quad \text{(Eq (5.7))} \quad \text{(v)}$$

Substituting values

$$\Phi = 1.65\times 1.02\times 0.41/2 = 0.345 \text{ Wb}$$

$$N_{ph}\text{ (series)} = 3\times 11\times \frac{P}{2} = 66$$

$$\text{Slot angle } \gamma = \frac{60°}{3} = 20°,\ m = 3$$

$$\text{Breadth factor, } K_b = \frac{\sin 3 \times 10°}{3 \sin 10°} = 0.96$$

$$E_{ph} = \sqrt{2}\ \pi K_b f N_{ph} \text{ (series) } \Phi$$

Substituting values

$$E_{ph} = \sqrt{2}\ \pi \times 0.96 \times 50 \times 66 \times 0.345 = 4855 \text{ V}$$

$$E_{\text{line}} = 4853\sqrt{3} = 8409 \text{ V}$$

(c) Open-circuit voltage required is 3/4th of E_{line} as above. On linear basis

$$I_f \text{(new)} = 0.75 \times 18.95 = 14.21 \text{ A}$$

Example 5.11 *A 3-Phase 50 kW, 4-pole, 50 Hz induction motor has a winding (ac) designed for delta connection. The winding has 24 conductors per slot arranged in 60 slots. The rms value of the line current is 48A. Find the fundamental of the mmf wave of phase-A when the current is passing through its maximum value. What is the speed and peak value of the resultant mmf/pole?* LO 1

Solution

$$\gamma = \frac{180° \times 4}{60} = 12°$$

$$\text{SPP} = \frac{60}{4 \times 3} = 5$$

$$K_b = \frac{\sin \dfrac{5 \times 12°}{2}}{5 \sin \dfrac{12°}{2}} = 0.957$$

$$I_P = \frac{I_L}{\sqrt{3}} = \frac{48}{\sqrt{3}} \text{ A}$$

$$I_{P,\max} = \frac{48\sqrt{2}}{\sqrt{3}} = 39.2 \text{ A}$$

$$\text{Turns/phase} = \frac{60 \times 24}{2 \times 3} = 240$$

$$F_m \text{ (phase } a) = \frac{4}{\pi} K_b \left(\frac{N_{ph}\text{(series)}}{P} \right) I_{P,\max}$$

$$= \frac{4}{\pi} \times 0.957 \times \left(\frac{240}{4} \right) \times 39.2$$

$$= 2866 \text{ AT/pole}$$

$$F_{\text{peak}} = \frac{3}{2} F_m = 4299 \text{ AT/pole}$$

$$\text{Speed of rotation of the resultant mmf} = \frac{120 f}{P} = \frac{120 \times 50}{4} = 1500 \text{ rpm}$$

5.6 TORQUE IN ROUND ROTOR MACHINE

LO 1

Torque Produced By Stator and Rotor Windings

When the stator and rotor windings of a machine both carry currents, they produce their own magnetic fields along their respective axes which are sinusoidally distributed along the air-gap. Torque results from the tendency of these two fields to align themselves. The flux components set up by the stator and rotor currents cross the air-gap twice and complete their circuits via the stator and rotor iron. These components fields cause the appearance of north and south poles on stator and rotor surfaces; the field axes being along north-south and out of the north pole. This is illustrated in Fig. 5.35(a) for a 2-pole structure. The torque tending to align the two fields is produced only if the two fields have the same number of poles and are stationary with respect to each other. Two relatively rotating fields will produce alternating torque as they cross each other so that the average torque is zero. All varieties of electric machines (synchronous, induction and dc) are therefore devised to produce interacting fields with zero relative velocity.

Probe
When do the stator and rotor fields result in producing torque?

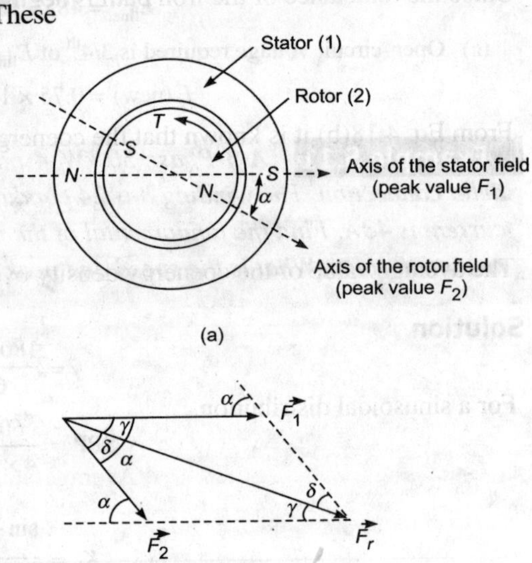

Fig. 5.35 *(a) Stator and rotor fields (b) Vector sums of stator and rotor fields*

Certain underlying assumptions are made at this stage:

1. Stator and rotor mmf's are sinusoidal space waves; this is sufficiently ensured by distributed windings.
2. Rotor is cylindrical (nonsalient pole) so that the air-gap is uniform throughout.
3. The air-gap is narrow so that flux established in it is radial (negligible tangential component) and further the flux density does not vary significantly[‡‡] along a radial path in the gap. As a result, the field intensity H along any radial path is constant in the air-gap. The mmf across the air-gap at any space point is
$$F_{\text{air-gap}} = Hg$$
where g is the radial air-gap length.
4. Reluctance of the iron path of flux is assumed negligible. As a consequence of assumptions 1–3, a sinusoidal space mmf wave produces a sinusoidal flux density wave in space in phase with it.
5. Most of the resultant flux is common to both stator and rotor windings, i.e. it is *mutual flux*. The *leakage flux*[§§] linking either winding produces the *leakage inductance* as in a transformer. These affect only the net voltages applied to the ideal machine.

Probe
What produces the leakage inductance?

[‡‡] This is so because the cylindrical area presented to gap flux does not vary appreciably with radius.

[§§] Refer Sec. 5.7.

Let F_1 F_2 be the peak values of the spatial sinusoidal mmf waves of the stator and rotor respectively as shown in Fig. 5.35(a) for a 2-pole machine; the angle between their respective positive peaks being denoted by α. As stated earlier, these mmf's can be represented as space vectors with magnitudes equal to their peak values and angles corresponding to their positive peaks. The resultant space mmf (which will also be sinusoidal being the sum of sinusoids) can be obtained by the vector summation as shown in Fig. 5.35(b). Using trigonometric relations, it easily follows that the peak value of the resultant mmf is

$$F_r^2 = F_1^2 + F_2^2 + 2F_1F_2 \cos \alpha \tag{5.47a}$$

Since the reluctance of the iron path is negligible, the peak value of the resultant field intensity is

$$H_r = \frac{F_r}{g} \tag{5.47b}$$

From Eq. 4.18(b) it is known that the coenergy density is

$$\frac{1}{2}\mu_0 H^2$$

The average value of the coenergy density over the air-gap volume is

$$\frac{1}{2}\mu_0 \text{ (average value of } H^2)$$

For a sinusoidal distribution,

$$\text{Average value of } H^2 = \frac{1}{2}H_r^2$$

$$\therefore \qquad \text{Average coenergy density} = \frac{1}{4}\mu_0 H_r^2$$

$$\text{Volume of air-gap} = \pi Dlg$$

where D is the mean diameter at air-gap and l = rotor length = stator length. The total coenergy of the field is then

$$W_f' = \frac{\pi}{4}\mu_0 H_r^2 Dlg = \frac{\pi}{4}\mu_0\left(\frac{F_r}{g}\right)^2 Dlg$$

$$= \frac{\mu_0 \pi Dl}{4g} F_r^2 \tag{5.48}$$

Substituting for F_r from Eq. (5.47a),

$$W_f' = \frac{\mu_0 \pi Dl}{4g}(F_1^2 + F_2^2 + 2F_1F_2 \cos \alpha) \tag{5.49}$$

The torque developed is given by

$$T = +\frac{\partial W_f'}{\partial \alpha} = -\frac{\mu_0 \pi Dl}{2g} F_1F_2 \sin \alpha \tag{5.50}$$

For a *P*-pole machine

$$T = -\left(\frac{P}{2}\right)\frac{\mu_0 \pi Dl}{2g} F_1 F_2 \sin \alpha \tag{5.51}$$

Interpretation From the torque expression (5.51) it is seen that the torque developed is proportional to the peak values of the stator and rotor mmfs and is proportional to the sine of the angle between the axes of the two fields. The negative sign in Eq. (5.52) indicates that the torque acts in a direction to reduce α, i.e. to align the two fields. Obviously, equal and opposite torques will act on the stator and rotor.

With reference to the geometry of Fig. 5.34(b) it is found that

$$F_1 \sin \alpha = F_r \sin \delta \tag{5.52}$$

and

$$F_2 \sin \alpha = F_r \sin \gamma \tag{5.53}$$

The torque expression of Eq. (5.51) can therefore be expressed in two more alternative forms,

$$T = -\left(\frac{P}{2}\right)\frac{\mu_0 \pi Dl}{2g} F_1 F_r \sin \gamma \tag{5.54}$$

$$T = -\left(\frac{P}{2}\right)\frac{\mu_0 \pi Dl}{2g} F_2 F_r \sin \delta \tag{5.55}$$

The torque in the expressions of Eqs (5.54) and (5.55) can be immediately imagined to be the result of the interaction of the stator and the resultant mmf's or the rotor and the resultant mmf's and is proportional to the sine of the angle between the interacting mmf's. The expression of Eq. (5.55) is generally preferred in simplified machine models.

Since $\mu_0 F_r/g = B_r$; (peak value of resultant (or air-gap) flux density) the expression of Eq. (5.55) can be written as

$$T = -\left(\frac{P}{2}\right)\frac{\pi Dl}{2} F_2 B_r \sin \delta \tag{5.56}$$

Interpretation This expression reveals a design viewpoint; the torque of electromagnetic origin is limited by the saturation value of flux density (B_r) in the magnetic material and maximum permissible mmf (F_2). The maximum value of F_2 is dictated by the maximum allowable winding current without exceeding the specified temperature rise.

An alternative and more useful form of the torque expression is in terms of the resultant flux/pole, Φ_r, rather than the peak flux density B_r Now

$$\Phi_r = B_{av} \text{ (over a pole)} \times \text{(pole area)}$$

$$= \left(\frac{2}{\pi} B_r\right)\left(\frac{\pi Dl}{P}\right) = \frac{2 B_r Dl}{P}\text{; same as Eq. (5.7)} \tag{5.57}$$

$$D = 2r$$

Substituting for B_r from Eq. (5.57) in Eq. (5.55),

$$T = -\frac{\pi}{2}\left(\frac{P}{2}\right)^2 \Phi_r F_2 \sin \delta \tag{5.58}$$

where Φ_r = *resultant (i.e. air-gap) flux/pole* due to superimposition of the stator and rotor mmf's

Equation (5.57) can be written as

$$\Phi_r = \left(\frac{2Dl}{P} \times \frac{\mu_0}{g}\right) F_r = \mathcal{P} F_r \tag{5.59}$$

where $\mathcal{P} = \left(\frac{2Dl}{P} \times \frac{\mu_0}{g}\right)$ = effective permeance per pole

Probe

Derive an expression for the resultant flux/pole due to superimposition of stator and rotor mmfs.

The effective permeance/pole relates the peak value of sinusoidal mmf wave and the flux/pole created by it. It is a constant quantity¶¶ so long as the permeability of iron is assumed infinite, otherwise it is a function of F_r (i.e. $\mathcal{P}(F_r)$) which decreases as F_r increases due to saturation of iron.

Alternative Derivation

Figure 5.36(a) shows the stator mmf wave and the corresponding air-gap flux density wave. The rotor mmf wave makes an angle of α with the stator mmf wave. The rotor current wave which is the cause of rotor mmf leads it by an angle of 90° as shown in Fig. 5.36(b). The force (and torque) is produced by the interaction of B_1-wave and A_2-wave as per the ***Bli*** rule. It may be seen that positive B_1 and positive A_2 produce negative force (opposite to the positive direction of θ). An angular element $d\theta$ of the machine located at an angle θ from the origin produces a torque

$$d\tau = -rBldi$$

Substituting values

$$dT = -\frac{1}{2} D(B_1 \cos\theta)\, l \underbrace{\left(-A_2 \sin(\theta - \alpha)\frac{D}{P} d\theta\right)}_{di}$$

where di = elemental current in differential periphery

$$\text{Differential periphery} = \frac{P}{2} d\theta_m = \frac{D}{P} d\theta$$

$$= \frac{1}{2}\frac{D^2 l}{P} B_1 A_2 \cos\theta \sin(\theta - \alpha)\, d\theta$$

The torque for one pole-pair is obtained by integrating the above equation from $\theta = 0$ to $\theta = 2\pi$. It yields

$$T = -\frac{1}{2P} \pi D^2 l B_1 A_2 \sin\alpha \tag{5.60}$$

Now

$$B_1 = \mu_0 F_1 / g$$

$$A_2 = F_2 P / D \qquad \text{(see Eq. (5.39))}$$

It then follows

$$T = -\frac{\mu_0 \pi}{2g} D l F_1 F_2 \sin\alpha \tag{5.61}$$

a result already established in Eq. (5.50).

¶¶ So long as the machine has uniform air-gap (round rotor), the permeance/pole offered by it is independent of the spatial orientation of the axis of the mmf wave. In the salient-pole construction two different permeances—one along the axis of the projecting poles and the other at 90° elect. to it—will be accounted for in Sec 8.11.

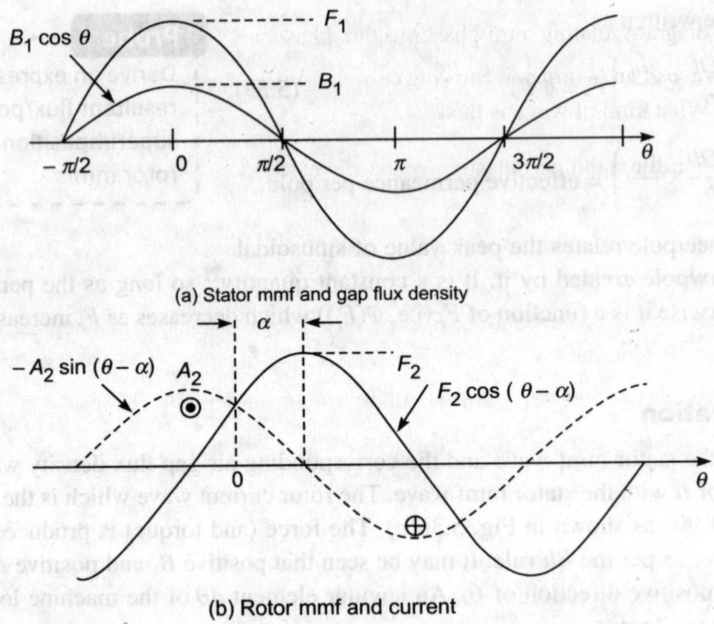

Fig. 5.36 *Interaction of air-gap flux density established by the stator mmf and rotor current wave corresponding to the rotor mmf wave*

Concept Check

LO 1

1. What measures are adopted to make the *B*-wave in a synchronous machine nearly sinusoidal? Why should the *B*-wave be sinusoidal?
2. A synchronous machine has *P* poles and generate voltage of frequency *f* Hz. Write the expressions for its speed in rad (elect.)/s, rad (mech)/s and rpm.
3. Explain the terms, coil span, coil pitch, short-pitching and cording of coils.
4. What is SPP? Write the expressions for a stator having *S* slots and *P* poles.
5. In a distributed winding why is the phase emf less than the algebraic sum of phase current in series?
6. Derive the expression for the breadth factor by means of a phasor diagram.
7. Repeat Question 6 for the pitch factor.
8. What is the purpose of using short-pitched coils in ac windings?
9. Write the expression for phasor emf in a synchronous machine. Use standard symbols and explain what each symbol stands for.
10. Taking the *B*-wave to be sinusoidal, derive the expression for flux/pole.
11. Write the expression for flux linkages of an *N*-turn coil if the *B*-wave is sinusoidal and rotating at synchronous speed ω rad/s.

12. Draw the phasor diagram relating emf phasor to flux phasor.

13. Sketch the mmf wave of an N-turn coil carrying current i. Write the expression for its fundamental if i is sinusoidal. What kind of wave is this?

14. Write the expression for standing pulsating space wave and sketch it at $\omega t = 0, \frac{\pi}{3}, \frac{2\pi}{3}$ and π rad.

15. Rotate the peak value of a rotating magnetic field and the maximum value of the fundamental of the mmf space wave of one phase.

16. State the conditions or a 3-phase winding of a stator to create a rotating magnetic field and its speed and direction of rotation.

17. State the conditions for two interacting rotating fields to create steady torque.

18. Two interacting fields $\vec{F_1}$ and $\vec{F_2}$ have a resultant field $\vec{F_r}$. Write the expression for torque developed in terms of $\vec{F_2}$ and $\vec{F_r}$. Explain the significance of the angle between them.

5.7 | OPERATION OF BASIC MACHINE TYPES

LO 2 Explain the operation of basic machine types

An elementary explanation of the torque-production process of basic machine types through the general torque expression of Eq. (5.58) will be given in this section.

Synchronous Machine

Figure 5.37 shows a synchronous machine with a round rotor. The rotor is initially stationary with fixed north-south poles created by dc excitation. Let the 3-phase winding of the stator be connected to a 3-phase supply of fixed voltage V (line) and fixed frequency f (this is known as the *infinite bus*). As a result, 3-phase currents flow in the stator winding creating a rotating magnetic field rotating at synchronous speed n_s ($=120f/P$) in the counter-clockwise direction (say). Since the rotor is stationary and cannot pick up speed instantaneously (inertia effect), the two fields move relative to each other resulting in zero average torque. As such the motor is *non-self-starting*.

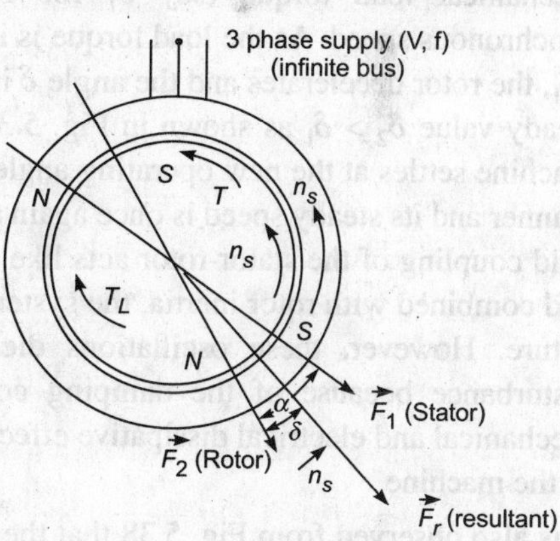

Fig. 5.37 *Torque production in synchronous machine (motoring)*

Consider now that the rotor is run by auxiliary means to a speed close to synchronous in the direction of rotation of the stator field. The two fields now have the opportunity of locking into each other or, in other words, the rotor *pulls into step* with the stator field and then on runs at exactly synchronous speed. It is easily seen from Fig. 5.37 that the electromagnetic torque developed (T) acts on the rotor in the direction of rotation of rotor and balances the load torque T_L. The mechanical power

Probe

Explain the functioning of a non-self-starting motor.

therefore flows to the load (motoring action) and, by the principle of conservation of energy, an equal amount of electrical power (plus losses in the device) are drawn from the electric supply. It is also seen from Fig. 5.37 that for a given T_L, the rotor field lags behind the stator field by an angle α or behind the resultant field by an angle δ. The torque developed by the synchronous motor is given by the expression of Eq. (5.58), i.e.

$$T = \frac{\pi}{2}\left(\frac{P}{2}\right)^2 \Phi_r F_2 \sin \delta \tag{5.62}$$

It may be seen that the negative sign has been deleted from the torque expression with the understanding that the torque acts in a direction to align the fields (it is indicated by an arrow sign in Fig. 5.37).

If the stator winding resistance and leakage reactance are assumed negligible (a fair assumption), the induced emf of the stator winding balances the terminal voltage, i.e.

$$V \approx \sqrt{3} \times 4.44\, K_w f \Phi_r N_{ph} \text{ (series)} \tag{5.63}$$

For a fixed terminal voltage, therefore, the resultant flux/pole is almost constant, independent of the shaft load. F_2, the peak of rotor mmf wave being dependent upon the rotor current (dc), is constant for fixed excitation. Equation (5.62) under conditions of constant terminal voltage and constant rotor excitation can therefore be written as

$$T = K \sin \delta \tag{5.64}$$

where δ is positive when the rotor field lags behind the resultant field and would be negative otherwise. The angle δ is known as the *torque angle* or *power angle.*

Interpretation The plot of electromagnetic torque developed by the synchronous machine is shown in Fig. 5.38. The machine operates at fixed δ for a given mechanical load torque (say δ_1 for T_{L1}) and runs at synchronous speed. As the load torque is increased to $T_{L2} > T_{L1}$, the rotor decelerates and the angle δ increases to a new steady value $\delta_2 > \delta_1$ as shown in Fig. 5.38. Of course, the machine settles at the new operating angle in an oscillatory manner and its steady speed is once again synchronous. The field coupling of the stator-rotor acts like a spring coupling and combined with rotor inertia, the system is oscillatory in nature. However, these oscillations die out after every disturbance because of the damping contributed by the mechanical and electrical dissipative effects that are present in the machine.

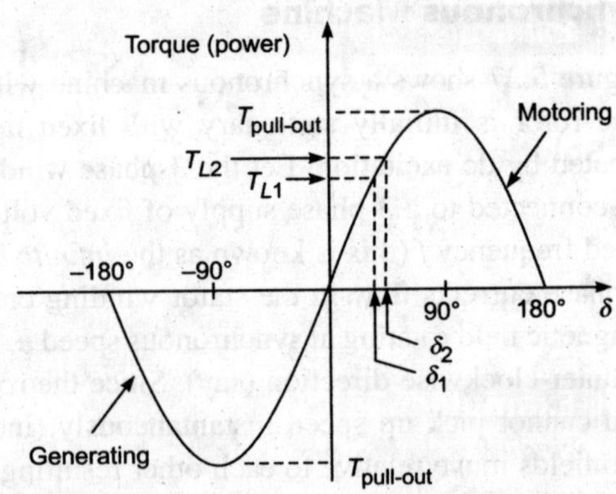

Fig. 5.38 *Torque-angle (T – δ) characteristic of synchronous machine*

It is also observed from Fig. 5.38 that the maximum torque developed by the motoring machine is at $\delta = 90°$ and is called the *pull-out torque* (or *pull-out power*); power being proportional to torque as the machine speed is synchronous, independent of load. If the motor is loaded with torque (power) more than $T_{\text{pull-out}}$, the developed torque reduces (Fig. 5.38) the rotor lag angle increases monotonically till the rotor-stator field-bond snaps, i.e. the rotor *falls out of step*. The machine will finally come to a stop and must, as a precautionary measure, be disconnected from the supply much before that. It is easily seen from the expression of Eq. (5.62) that the pull-out torque can be increased by increasing the stator terminal voltage, Φ_r increases with terminal voltage (Eq. (5.63)) and/or rotor field excitation.

With negative δ, i.e. rotor field leading the resultant field in the direction of rotation of the rotor, the electromagnetic torque as seen from Fig. 5.39 is now developed in a direction opposite to that of the rotor rotation and must be balanced by an external mechanical torque T_{PM} (provided by a prime-mover) for the rotor to run at synchronous speed maintaining the locking of the rotor and stator fields. The mechanical power now flows into the rotor and the electrical power flows out of the stator to the infinite bus. The rotor readjusts its angle of lead such that the electrical output equals the mechanical input minus losses. If the mechanical input is more than the maximum electrical power developed (corresponding to generator pull-out torque, $\delta = 90°$), the rotor accelerates and falls out of step, i.e. the synchronism between the rotor and stator fields is lost.

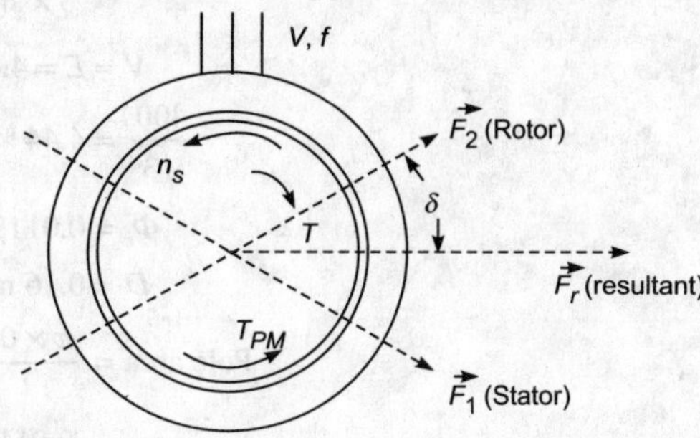

Fig. 5.39 *Torque production in synchronous machine (generating)*

To summarize, a synchronous machine has a synchronous locking between the stator and rotor fields with the rotor field lagging the resultant air-gap field in motoring operation and leading the resultant field in generating operation. The electromagnetic torque developed is a sine function of angle δ, between the rotor field and the resultant field; as a result the machine falls out of step and loses synchronism if conditions are created for δ to increase beyond ±90°. Within this range the machine operates at synchronous speed under varying load conditions. Further, the machine is non-self-starting as a motor.

Example 5.12 *In a certain electric machine $F_2 = 850$ AT, $F_1 = 400$ AT, $\alpha = 123.6°$ and $\mathcal{P}$(permeance/pole) $= 1.408 \times 10^{-4}$ Wb/AT. Calculate the value of the resultant air-gap flux/pole.* LO 2

Solution

$$F_r = (F_1^2 + F_2^2 + 2F_1F_2 \cos \alpha)^{1/2}$$

$$= [(400)^2 + (850)^2 + 2 \times 400 \times 850 \cos 123.6°]^{1/2} = 711.5 \text{ AT}$$

$$\Phi_r = \mathcal{P}F_r = 1.408 \times 10^{-4} \times 711.5 = 0.1 \text{ Wb}$$

Example 5.13 *A 50 Hz, 400 V, 4-pole cylindrical synchronous generator has 36 slots, two-layer winding with full-pitch coils of 8 turns each. The mean air-gap diameter is 0.16 m, axial length 0.12 m and a uniform air-gap of 2 mm. Calculate the value of the resultant AT/pole and the peak air-gap flux density. The machine is developing an electromagnetic torque of 60 Nm as a generator at a torque angle 26°. What should be the rotor AT/pole? What is the stator AT and the angle it makes with the resultant AT? Also find the stator current.* LO 2

Solution

$$N_{ph}(\text{series}) = \frac{36 \times 8 \times 2}{2 \times 3} = 96$$

$$\gamma = \frac{4 \times 180°}{36} = 20°$$

$$m = \frac{36}{4 \times 3} = 3$$

$$K_b = \frac{\sin(3 \times 20°/2)}{3 \times \sin(20°/2)} = 0.96$$

$$V \approx E = 4.44\, K_d f \Phi_r N_{ph}$$

$$\frac{400}{\sqrt{3}} = 4.44 \times 0.96 \times 50 \times \Phi_r \times 96$$

$$\Phi_r = 0.0113 \text{ Wb/pole}$$

$$D = 0.16 \text{ m}, l = 0.12 \text{ m}$$

$$\text{Pole area} = \frac{\pi \times 0.16 \times 0.12}{4} = 0.0151 \text{m}^2$$

$$B_r(\text{av}) = \frac{0.0113}{0.0151} = 0.749 \text{ T}$$

$$B_r(\text{peak}) = \frac{\pi}{2} \times 0.749 = 1.18 \text{ T}$$

$$F_r = \frac{g B_r(\text{peak})}{\mu_0} = \frac{2 \times 10^{-3} \times 1.18}{4\pi \times 10^{-7}}$$

$$= 1878 \text{ AT/pole (peak)}$$

$$\text{Torque developed, } T = \frac{\pi}{2}\left(\frac{P}{2}\right)\Phi_r F_2 \sin \delta$$

$$60 = \frac{\pi}{2} \times \left(\frac{4}{2}\right)^2 \times 0.0113\, F_2 \sin 26°$$

or $$F_2 = 1928 \text{ AT/pole (peak)}$$

Mmf phasor diagram is drawn in Fig. 5.40 with $\vec{F}_2$ leading $\vec{F}_r$ by $\delta = 26°$. This is the case in a generating machine. From the phasor diagram we find

$$F_1^2 = (1928)^2 + (1878)^2 - 2 \times 1928 \times 1878 \sin 26°$$

or $$F_1 = 858 \text{ AT/pole (peak)}$$

$$\cos \psi = \frac{(858)^2 + (1878)^2 - (1928)^2}{2 \times 858 \times 1878}$$

or $$\psi = 80°$$

$\vec{F}_1$ lags $\vec{F}_r$ by 80°

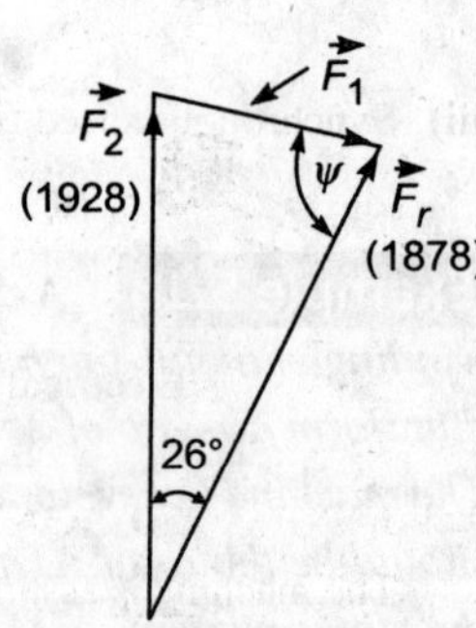

Fig. 5.40

$$F_1 = \frac{3}{2} \times \frac{4\sqrt{2}}{\pi} k_w \left(\frac{N_{ph}(\text{series})}{P}\right) I_a$$

$$K_w = K_b = 0.96; N_{ph} \text{ (series)} = 96$$

$$858 = \frac{3}{2} \times \frac{4\sqrt{2}}{\pi} \times 0.96 \times \frac{96}{4} \times I_a$$

or $$I_a = 13.8 \text{ A}$$

Example 5.14 *A 3ϕ, 4-pole, 50 Hz synchronous machine has the data as given below.*

$$K_w \text{ (rotor)} = 0.976$$
$$\text{Air-gap} = 1.5 \text{ mm (cylindrical rotor)}$$
$$\text{Mean air-gap diameter} = 29 \text{ cm}$$
$$\text{Axial length} = 35 \text{ cm}$$
$$\text{Rotor winding turns} = 746$$
$$\text{Stator winding SPP} = 4; \text{ conductor/slot} = 20$$
$$\text{Field current} = 20 \text{ A}$$

(a) The machine is being used in motoring mode with B_r = 1.6 T. Determine

(i) F_2, peak rotor AT

(ii) Maximum torque

(iii) Electrical input at maximum torque

(b) If the machine is used as generator with same field current, determine its open-circuit (no-load) voltage.

LO 2

Solution (a) **Motoring mode**

(i)
$$F_2 = \frac{4}{\pi} K_w \text{ (rotor)} \left(\frac{N_{\text{pole}}\text{(series)}}{P} \right) I_f$$
$$= \frac{4}{\pi} \times 0.976 \times \frac{746}{4} \times 20$$
$$= 4635 \text{ AT}$$

(ii)
$$B_r = 1.6 \text{ T (given)}$$

Then
$$T_{\max} = \left(\frac{P}{2}\right)\left(\frac{\pi DL}{2}\right) F_2 B_r$$
$$= \frac{4}{2} \times \frac{\pi \times 0.29 \times 0.35}{2} \times 4635 \times 1.6$$
$$= 2364$$

(iii) Synchronous speed (motor speed)

$$\omega_m = \frac{2\omega}{P} = \frac{4\pi f}{P} = \frac{4\pi \times 50}{4}$$
$$= 50\,\pi \text{ rad (mech)/sec}$$

$$\text{Electrical power input, } P_{\text{in}} = T_{\max} \times \omega_m$$
$$= 2632 \times 50\pi \text{ W}$$
$$= 371 \text{ kW}$$

The machine is assumed loss-less, so

$$P_{\text{in}} = P_{\text{in, max}} = 371 \text{ kW}$$

(b) **Generating Mode**

Open Circuit Voltage The armature is not carrying any current, i.e. $F_1 = 0$, $F_2 = 4630$ AT as calculated above.

$$\text{SPP} = \frac{S}{3P} = 4$$

$$\gamma = \frac{180°P}{S} = \frac{180}{12} = 15°$$

$$K_w = K_b = \frac{\sin 30°}{3\sin 15°/2} = 0.958$$

$$\Phi_r = \left(\frac{2Dl}{P} \times \frac{\mu_0}{g}\right) F^2$$

$$= \frac{2 \times 0.29 \times 0.35 \times 4\pi \times 10^{-7}}{4 \times 1.5 \times 10^{-3}} \times 4635$$

$$= 0.197 \text{ Wb}$$

$$N_{ph} = \frac{20 \times 4 \times 4}{2} = 160$$

$$E_{ph} = 4.44\, K_b f N_{ph}\, \Phi$$

$$= 4.44 \times 0.958 \times 50 \times 160 \times 0.197$$

$$= 6704 \text{ V}$$

$$E_{\text{line}} = 11610 \text{ V} \quad \text{or} \quad 11.61 \text{ kV}$$

Induction Machine

This machine has not been introduced so far. Consider a cylindrical rotor machine with both the stator and rotor wound for three phases and identical number of poles as shown in Fig. 5.41. Assume initially the rotor winding to be open-circuited and let the stator be connected to an infinite bus (V, f). The stator currents set up a rotating magnetic field in the air-gap which runs at synchronous speed inducing emf in the stator winding which balances the terminal voltage under the assumption that the stator resistance and leakage reactance are negligible. Also the rotating field induces emf in the rotor winding but no rotor current flows because the rotor is open-circuited. The frequency of rotor emf is of course f. Since the rotor mmf $F_2 = 0$, no torque is developed and the rotor continues to be stationary. The machine acts merely as a transformer where the stator (primary) and rotor (secondary) have emfs of the same frequency induced in them by the rotating magnetic flux rather than by a stationary time-varying flux as in an ordinary transformer.

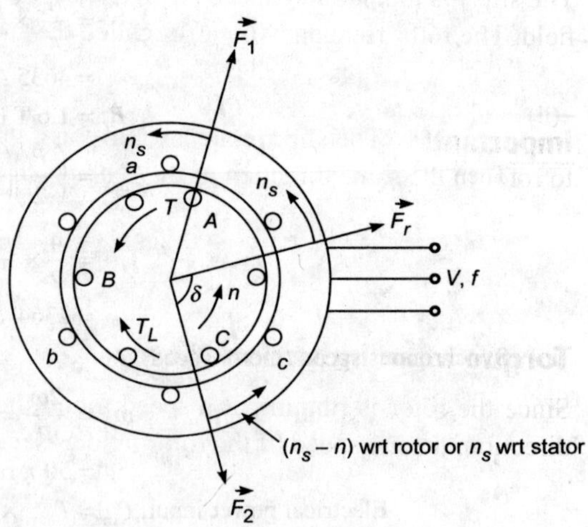

Fig. 5.41 *Illustrating the principle of induction machine*

Let the rotor be now held stationary (*blocked* from rotation) and the rotor winding be short-circuited. The rotor now carries 3-phase currents creating the mmf F_2 rotating in the same direction and with the same speed as the stator field. F_2 causes reaction currents to flow into the stator from the busbar (just as in an ordinary transformer) such that the flux/pole Φ_r of the resultant flux density wave (rotating in the air-gap at synchronous speed) induces a stator emf to just balance the terminal voltage. Obviously Φ_r must be the same as when the rotor was open-circuited. In fact, Φ_r will remain constant independent of the operating conditions created by load on the motor. The interaction of Φ_r and F_2, which are stationary with respect to each other, creates the torque tending to move the rotor in the direction of F_r or the stator field F_1. The induction motor is therefore a self-starting device as different from the synchronous motor.

Let the short-circuited rotor be now permitted to rotate. It runs in the direction of the stator field and acquires a steady speed of n. Obviously $n < n_s$, because if $n = n_s$, the relative speed between the stator field and rotor winding will be zero and therefore the induced emfs and rotor currents will be zero and hence no torque is developed. The rotor thus cannot reach the synchronous speed n_s and hence cannot exceed n_s. With the rotor running at n, the relative speed of the stator field with respect to rotor conductors is $(n_s - n)$ in the direction of n_s. The frequency of induced emfs (and currents) in the rotor is therefore

Probe

How is an induction motor different from a synchronous motor?

$$f_2 = \frac{(n_s - n)P}{120} = \left(\frac{n_s - n}{n_s}\right)\left(\frac{n_s P}{120}\right)$$

$$= sf \tag{5.65}$$

where
$$s = \frac{n_s - n}{n_s} = \textit{slip of the rotor} \tag{5.66}$$

The slip s is the per unit speed (with respect to synchronous speed) at which the rotor slips behind the stator field. The rotor frequency $f_2 = sf$ is called the *slip frequency*. From Eq. (5.66), the rotor speed is

$$n = (1 - s)n_s \tag{5.67}$$

Important! The slip frequency currents in the rotor winding produces a rotor field rotating with respect to rotor in the same direction as the stator field at a speed of

$$\frac{120sf}{P} = \frac{120(n_s - n)f}{n_s P} = (n_s - n) \tag{5.68}$$

Torque in an Induction Motor

Since the rotor is running at a speed n and the rotor field at $(n_s - n)$ with respect to the rotor in the same direction, the net speed of the rotor field as seen from the stator (ground reference) is

$$n + (n_s - n) = n_s$$

i.e., the same as the stator field. Thus the reaction field F_2 of the rotor is always stationary with respect to the stator field F_1 or the resultant field F_r (with flux Φ_r per pole). Since the rotor mmf F_2 is proportional to the rotor current I_2 and the resultant flux/pole Φ_r is fixed by terminal voltage independent of operating conditions, the induction motor torque is given by (see Eq. (5.58))

$$T = KI_2 \sin \delta \tag{5.69}$$

It is observed here that the torque is produced by the induction motor at any mechanical speed other than synchronous; such a torque is called the *asynchronous torque*.

The angle δ by which F_2 lags behind F_r, the resultant mmf needs to be known. Before proceeding to determine δ, it must be observed that shorting the rotor winding is equivalent to shorting all the winding conductors individually. As a result the rotor does not necessarily have to be properly wound; it may be constructed of conducting bars placed in the rotor slots slightly skewed and shorted by conducting *end-rings* on each side of the rotor. Such a rotor is called the *squirrel-cage* rotor; the conducting cage is separately illustrated in Fig. 5.42 The squirrel-cage rotor has a cheap and rugged construction and is adopted in a vast majority of induction motor applications. The machine with a properly wound rotor is called the *wound-rotor* induction motor and is provided with three slip-rings which provide the facility of adding external resistance in the rotor winding before shorting these. Such motors are used in on-load starting situations (see Ch. 9).

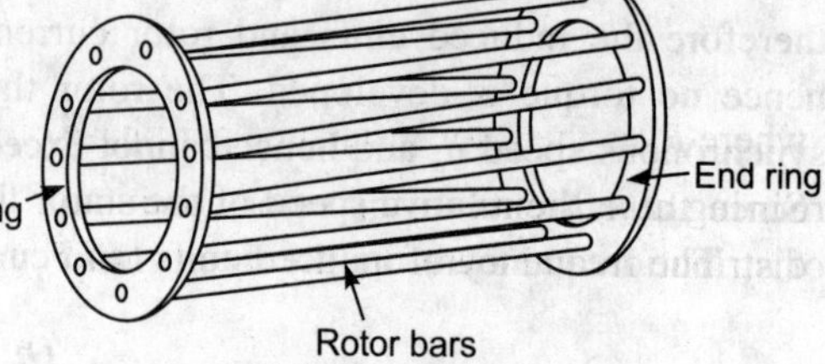

Fig. 5.42 *Squirrel-cage rotor*

Normally the full-load slip of a squirrel-cage induction motor is small 3–10%. Consequently the rotor impedance is mainly resistive, the rotor leakage reactance being proportional to $f_2 = sf$ is negligible. Furthermore, the rotor induced emf is proportional to the rotor-slip as Φ_r is fixed and rotates at speed $n_s - n = sn_s$ with respect to rotor. This results in the rotor current being very nearly in phase with the rotor emf and proportional to the rotor slip. This conclusion would obviously apply to individual rotor conductors as well.

Figure 5.43 shows the resultant flux density wave B_r gliding past the rotor conductors at speed $(n_s - n) = sn_s$ in a developed diagram. The induced currents in the shorted rotor conductors are sinusoidally distributed–the distribution moves at speed $(n_s - n)$ with respect to the rotor in synchronism with the B_r-wave. Further, because the rotor conductors are assumed resistive, i.e. currents in them are in phase with their respective emf's, the rotor current distribution is therefore in space phase with B_r-wave. The sinusoidal rotor current distribution produces a sinusoidal rotor mmf wave F_2 which lags 90° behind rotor current distribution or 90° behind B_r-wave. It is, therefore, concluded that for small values of slip, the angle δ in the induction motor is 90°. Hence,

$$T = kI_2 \text{ (for small slip)} \tag{5.70}$$

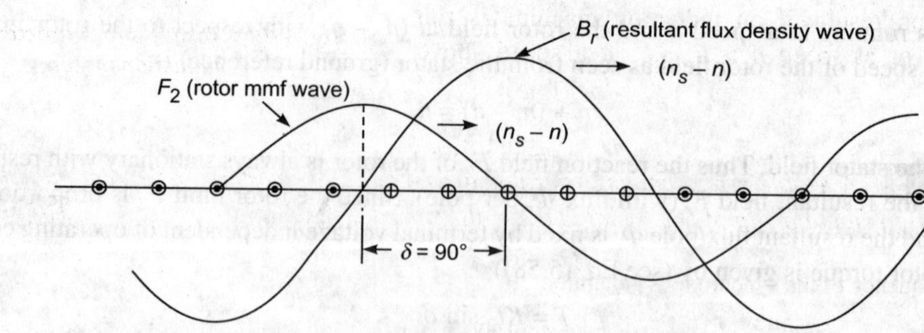

Fig. 5.43 *Resultant air-gap flux density and rotor mmf waves—resistive rotor*

Analysis Since rotor emf is linearly proportional to slip***, so is the rotor current for mainly a resistive rotor at small values of slip. Hence, the torque developed in the induction motor is a linearly increasing function of slip for small value of slip, being zero for $s = 0$, i.e. at synchronous speed.

As slip increases further, the leakage reactance of the rotor can no longer be neglected. Its value at slip s is sX_2, where X_2 is the rotor leakage reactance per phase at frequency f, i.e. when the rotor is at a *stand-still*. The rotor current now lags the rotor induced emf by

$$\theta = \tan^{-1}\frac{sX_2}{R_2}$$

where R_2 is the rotor resistance per phase.

Since the currents in the rotor conductors lag the induced emf's by angle θ, the rotor conductor current distribution and therefore the rotor mmf F_2 shifts to the left in Fig. 5.41 by an angle θ, so that

$$\delta = 90° + \theta \tag{5.71}$$

It means $\sin\delta < 1$. Further, since the rotor impedance is increasing with s, rotor current is less than proportional to slip. These two factors cause the motor torque to pass through a maximum value and then begin to decrease gradually as s is continuously increased.

Torque-slip Characteristic of an Induction Motor

The nature of the complete torque-slip characteristic of the induction motor is exhibited in Fig. 5.44. The maximum torque is known as the *break-down torque.* The motor would come to rest if loaded for short time with torque load larger than the breakedown value.

As already mentioned, the slip of an induction motor is 3–10% at full-load. Therefore, it is substantially a constant speed drive unlike the synchronous motor which runs at constant speed independent of load.

Generating action results if an induction machine is run at negative slip or at speed $n > n_s$, i.e. at a speed above synchronous.

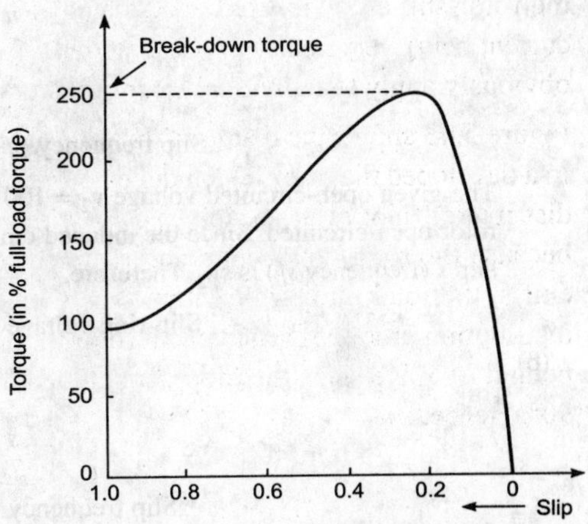

Fig. 5.44 *Torque-slip characteristic of induction motor*

Example 5.15 *A 4-pole synchronous generator driven at 1500 rpm feeds a 6-pole induction motor which is loaded to run at a slip of 5%. What is the motor speed?* LO 2

Solution Frequency of the synchronous generator,

$$f = \frac{4 \times 1500}{120} = 50 \text{ Hz}$$

*** Since Φ_r, is practically constant, independent of operating conditions, the rotor emf is proportional to the relative speed between the resultant field and the rotor. i.e.

$$(n_s - n) = sn_s$$

Synchronous speed of the induction motor,

$$n_s = \frac{120 \times 50}{6} = 1000 \text{ rpm}$$

Motor slip, $s = 0.05$

Motor speed $= (1 - s)n_s = 0.95 \times 1000 = 950$ rpm

Example 5.16 *A 6-pole, 50-Hz wound-rotor induction motor when supplied at the rated voltage and frequency with slip-rings open-circuited, developed a voltage of 100 V between any two rings. Under the same conditions its rotor is now driven by external means at*

(a) 1000 rpm opposite to the direction of rotation of stator field, and

(b) 1500 rpm in the direction of rotation of stator field.

Find the voltage available between slip-rings and its frequency in each of these cases. LO 2

Solution

Synchronous speed,

$$n_s = \frac{120 \times 50}{6} = 1000 \text{ rpm}$$

(a)

$$n = -1000 \text{ rpm}$$

$$s = \frac{n_s - n}{n_s} = \frac{1000 - (-1000)}{1000} = 2$$

Slip frequency, $sf = 2 \times 50 = 100$ Hz

The given open-circuited voltage $v_2 = 100$ V corresponds to a slip of $s = 1$, the motor being stationary with the rotor open-circuited. Since the induced emf in ac winding is proportional to frequency, the rotor induced emf at slip s (frequency sf) is sv_2. Therefore,

Slip-ring voltage $= sv_2 = 2 \times 100 = 200$ V

(b)

$$n = 1500 \text{ rpm}$$

$$s = \frac{1000 - 1500}{1000} = -0.5$$

Slip frequency $= 0.5 \times 50 = 25$ Hz

It may be seen that the negative sign has been dropped, which merely implies a reversal in the phase angle of the voltage.

Slip-ring voltage $= 0.5 \times 100 = 50$ V

It is found that the induction machine can be regarded as a *generalized transformer* with rotor voltage and frequency both being proportional to slip.

Example 5.17 *The stator of the induction motor of Ex. 5.16 is fed at the rated voltage and frequency while its slip-rings are connected to a 25-Hz supply.*

(a) Will there be a starting torque?

(b) At what speed will steady operation result?

(c) At what speed will steady operation result if the rotor is also fed with a 50-Hz supply? LO 2

Solution

(a) Under stationary conditions of the rotor, no torque will be developed as stator and rotor fields will rotate relative to each other, i.e. no starting torque.

(b) For steady operation, the stator and rotor fields must be stationary relative to each other.

Speed of the stator field (with respect to the stator surface)

$$= \frac{120 \times 50}{6} = 1000 \text{ rpm}$$

Speed of the rotor field (with respect to the rotor surface)

$$= \frac{120 \times 25}{6} = 500 \text{ rpm}$$

Steady synchronous operation will result when the rotor is run at 500 rpm in the same direction as the stator field.

(c) Speed of the rotor field (with respect to the rotor surface)

$$= 1000 \text{ rpm}$$

(i) If the rotor field rotates in the same direction as the stator, steady (synchronous) operation is only possible at zero speed. At any other speed of the two fields will have relative motion and will produce zero torque.

(ii) If the rotor field rotates opposite to the stator field, the steady (synchronous) operation will result when the rotor moves at 2000 rpm in the direction of the stator field; it is only at this speed of the rotor that the two fields are relatively stationary.

Note: The induction motor fed as above from both stator and rotor sides operates in the synchronous mode as different from the induction mode.

Example 5.18 *A 3-phase, 50 Hz induction motor runs at a speed of 576 rpm at full load.*

(a) How many poles does the motor have?

(b) What is its slip and frequency of rotor currents at full load? Also find rotor speed with respect to the rotating field.

(c) What is the motor speed at twice full-load slip?

(d) By what factor should the rotor resistance be increased for the motor to run at a speed of 528 rpm at full-load torque?

LO 2

Solution

(a) Nearest synchronous speed, $n_s = 600$ rpm

$$P = \frac{120 \times 50}{600} = 10$$

$$s = \frac{600 - 576}{600} = 0.04$$

$$f_2 = 0.04 \times 50 = 2 \text{ Hz}$$

Rotor speed with respect to the rotating field = 0.04 × 600 = 24 rpm

(c)

$$s = 2 \times 0.04 = 0.08$$

$$n = (1 - s)n_s = (1 - 0.08) \times 600 = 552 \text{ rpm}$$

(d) $$s\,(\text{new}) = \frac{600-528}{600} = 0.12$$

$$\frac{s(\text{new})}{s(\text{old})} = \frac{0.12}{0.04} = 3$$

For the same motor torques, the rotor current must remain constant. As the rotor slip becomes 3 times, the rotor induced emf increases by the same factor. Therefore, for rotor current to remain the same, its resistance must be increased 3 times.

The dc Machine

Figure 5.13 showed the essential constructional features of an elementary 2-pole dc machine. The stator has a fixed pole structure with dc excitation which means that the stator-created flux density wave acting on the rotor periphery remains fixed in space. For torque to be created, the armature (rotor) when carrying currents must produce an mmf pattern that remains fixed in space while the armature moves. After the study of the dc winding in Ch. 7 and how it is connected to commutator segments, it will be seen that the armature mmf in a dc machine is indeed fixed in space and makes an angle of 90° with the main field. As the dc machine structure is necessarily of the salient pole type, the main pole flux density wave is far from sinusoidal[†††] (see Fig. 5.14) and the armature mmf is stepped triangular (to be shown in Ch. 7). Equation (5.58) will not be used in finding the torque expression for the dc machine as this result applies to sinusoidally distributed fields.

> **Probe**
> What is the condition to be satisfied in a dc machine for torque to be created?

5.8 LINEAR MACHINES

LO 2

So far we have dealt with rotary machines which find universal use. Each of these machines can have their linear motion version. The developed diagrams that we have used extensively are linear versions of the corresponding machine which are employed for specific purposes where linear motion is the requirement like in transportation and reciprocation, machine tools, and also in limited range linear motion in robotics. In these applications, linear induction motors are used because of constructional convenience and low cost.

In rail transportation, the 'rotor' of the normal induction motor is the conducting stationary rail, which acts as short circuited conductor. The wound stator is on the moving vehicle. The details and performance characteristics of the linear induction motor shall be taken up in Ch. 9. Here we shall present the basic analysis of a linearly moving field. The mmf diagram for one phase of linear concentrated winding is the same as the developed diagram of Fig. 5.24(b) and is redrawn in Fig. 5.45 where the linear dimension is z in place of angle θ. Of course, torque would now be force.

> **Probe**
> What acts as a short circuited conductor in rail transportation?

It is seen from Fig. 5.45 that the wavelength of the fundamental of the mmf wave is β, which corresponds to 2 poles with electrical angle 2π. The fundamental mmf can be expressed as

[†††] In fact it is desired to make the B-wave as flat-topped as possible to yield high value of flux/pole for given physical dimensions.

$$F_{a1} = \frac{4}{\pi}\left(\frac{Ni}{2}\right)\cos\left(\frac{2\pi}{\beta}z\right) \tag{5.72}$$

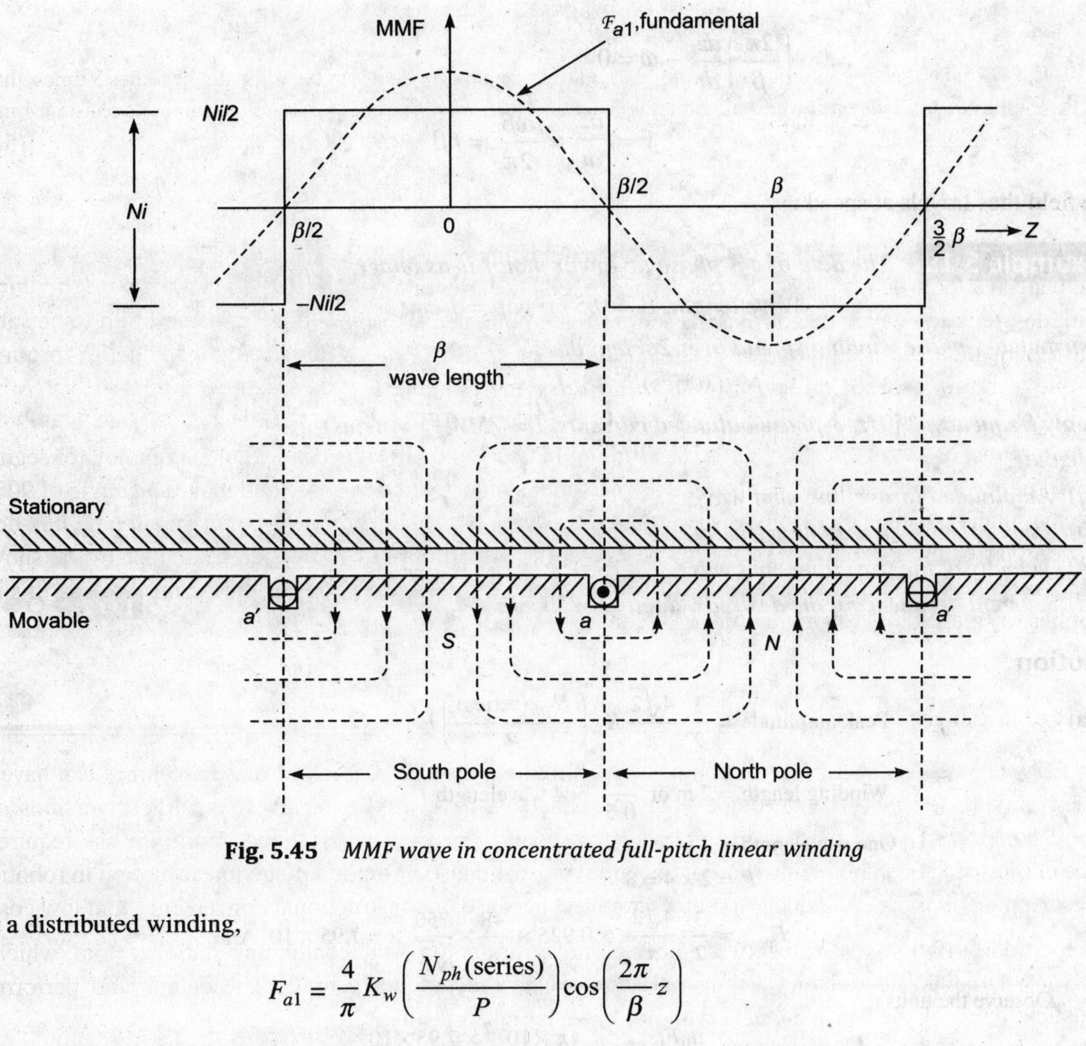

Fig. 5.45 *MMF wave in concentrated full-pitch linear winding*

For a distributed winding,

$$F_{a1} = \frac{4}{\pi}K_w\left(\frac{N_{ph}\,(\text{series})}{P}\right)\cos\left(\frac{2\pi}{\beta}z\right)$$

or
$$F_{a1} = \frac{4\sqrt{2}}{\pi}K_w\left(\frac{N_{ph}\,(\text{series})}{P}\right)I\cos\omega t\cos\left(\frac{2\pi}{\beta}z\right);\ I = \text{rms current} \tag{5.73}$$

For a 3-phase winding carrying 3-phase balanced currents at frequency $\omega = 2\pi f$, it can be shown (on the lines of rotating magnetic field) that

$$F(z, t) = \frac{3}{2}F_m\cos\left(\frac{2\pi}{z}\beta - \omega t\right) \tag{5.74}$$

where
$$F_m = \frac{4\sqrt{2}}{\pi}K_w\left(\frac{N_{ph}\,(\text{series})}{P}\right)I \tag{5.75}$$

It is observed from Eq. (5.76) that the resultant field is a travelling wave, whose speed (linear) is found as

$$\frac{2\pi}{\beta}z - \omega t = K \text{ (any constant value)}$$

$$\left(\frac{2\pi}{\beta}\right)\frac{dz}{dt} - \omega = 0$$

or
$$v = \frac{dz}{dt} = \frac{\omega\beta}{2\pi} = f\beta \tag{5.76}$$

The field thus travels at speed v.

Example 5.19 *The data of a 3-phase ac linear motor is as under:*

Wave length, β = 0.5 m; gap = 1 cm

Distributed 3-phase winding spread over 2m length

N_{ph}(series) = 48; K_w = 0.925

Supply frequency, 25 Hz, 3-phase balanced currents, $I = 750/\sqrt{2}$ A (rms)

Calculate:

(a) Amplitude of travelling mmf wave
(b) Peak value of air-gap flux density
(c) Velocity of the travelling mmf wave
(d) Current, frequency if the desired velocity is 72 km/h

LO 2

Solution

(a)
$$\text{Peak amplitude} = \frac{3}{2}\cdot\frac{4\sqrt{2}}{\pi}K_w\left(\frac{N_{ph}(\text{series})}{P}\right)I$$

$$\text{Winding length} = 2 \text{ m or } \frac{2}{0.5} = 4 \text{ wavelength}$$

$$\text{One wavelength} = 2 \text{ pole}$$

$$P = 2 \times 4 = 8$$

$$F_{\text{peak}} = \frac{3}{2}\cdot\frac{4\sqrt{2}}{\pi} \times 0.925 \times \frac{48}{8} \times \frac{750}{\sqrt{2}} \times = 7.95 \times 10^3 \text{ A/m}$$

Observe the units.

(b)
$$B_{\text{peak}} = \frac{\mu_0 F_{peak}}{g} = \frac{4\pi \times 10^{-7} \times 7.95 \times 10^3}{1 \times 10^{-2}} = 0.999 \text{ or } 1 \text{ T}$$

(c)
$$v = \frac{\omega\beta}{2\pi} = 25 \times 0.5 = 12.5 \text{ m/s}$$

(d)
$$v = \frac{72 \times 10^3}{3600} = 20$$

$$20 = f \times 0.5 \quad \text{or} \quad f = 40 \text{ Hz}$$

5.9 MAGNETIC LEAKAGE IN ROTATING MACHINES

LO 2

The leakage flux in rotating machines is that flux which links only the stator or only the rotor windings. Because of the presence of air-gap in the magnetic circuit of machines, the leakage in these is quite significant and cannot be neglected in analysis as could be done in the case of transformers. The leakage in machines falls into the following two broad categories:

1. Leakage in main poles, and
2. Leakage in armature.

Leakage in Main Poles

The main poles of a dc machine and that of a synchronous machine are excited by means of dc to produce a steady properly distributed flux density; the chief difference between the two being that while the poles of a dc machine form the stator and that of a synchronous machine are located on rotor. The *useful flux* is that flux which coming from the main poles crosses the air-gap and enters the armature. Some of the flux leaks through via two typical paths indicated in Fig. 5.46(a) without entering the armature. This then constitutes the leakage flux whose only effect is to increase the flux density in the roots of the poles without contributing to useful flux and therefore must be accounted for in the magnetic circuit design of the machine. Similar leakage takes place in the poles of a synchronous machine as shown in Fig. 5.46(b).

Probe

What is a useful flux?

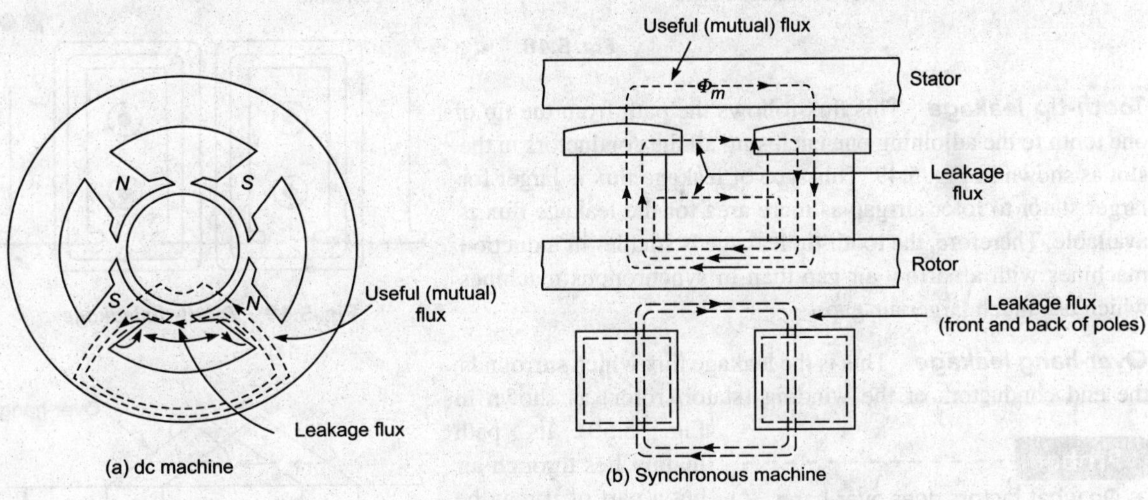

Fig. 5.46 *Leakage in main poles*

Leakage in Armature

The complexity of identifying the leakage paths in the wound armature arises from the fact that the winding is distributed and the armature surface slotted. The total leakage flux of the armature can be divided into several components identified in the following.

Slot leakage This is the flux which follows the path from tooth to tooth across the slots as shown in Fig. 5.47 and in the process links stator/rotor windings only. It is observed that the path of the slot-leakage flux is perpendicular to that of the main flux, which passes radially down the teeth, and a very small part of it straight down the slots. It is further to be noticed that a smaller amount of leakage flux links the bottom conductors in slots than the top conductors. The slot leakage is very much dependent upon the shape of slots. It is larger in semi-closed slots (Fig. 5.48(a)) used in induction machines, because of narrow (low reluctance) slot opening, compared to open slots (Fig. 5.48(b)) used in synchronous and dc machines.

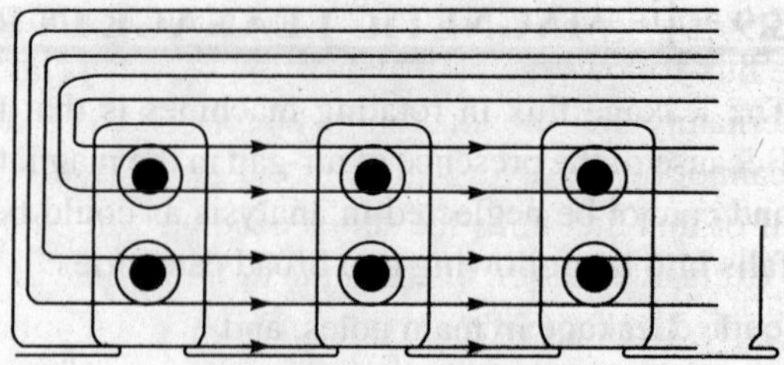

Fig. 5.47 *Slot leakage*

Probe

What is the slot leakage dependent on?

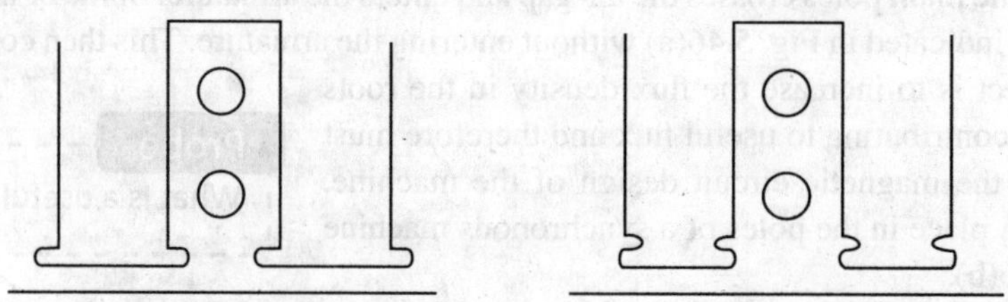

(a) Semiclosed slots (b) Open slots

Fig. 5.48

Tooth-tip leakage This flux follows the path from the tip of one tooth to the adjoining one enclosing all the conductors in the slot as shown in Fig. 5.49. This type of leakage flux is larger for larger stator to rotor air-gap as more area for the leakage flux is available. Therefore, the tooth-tip leakage is smaller in induction machines with a narrow air-gap than in synchronous machines which use much larger air-gaps.

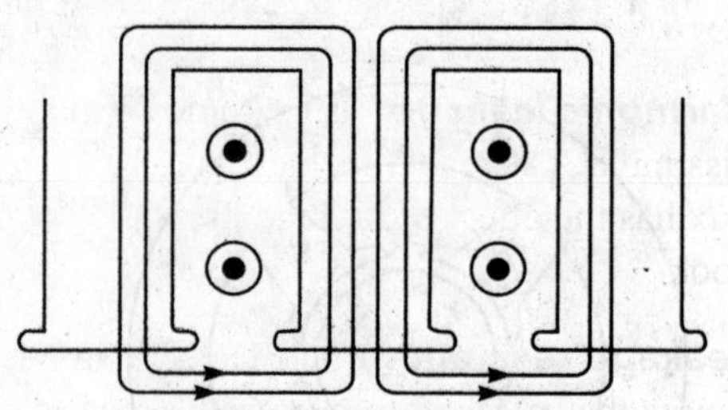

Fig. 5.49 *Tooth-tip leakage*

Over-hang leakage This is the leakage flux which surrounds the end conductors of the winding (stator/ rotor) as shown in Fig. 5.50. Its path mainly lies through air but a part of it may be located in the core-iron or the iron of end shields. The amount of this leakage depends upon the proximity of conductors and their relative location with respect to both core and end-shields. This leakage is generally small because of the large air paths involved. It is particularly insignificant in the squirrel-cage induction machine rotor which has no over-hang.

Probe

On what factors does over-hang leakage mainly depends on?

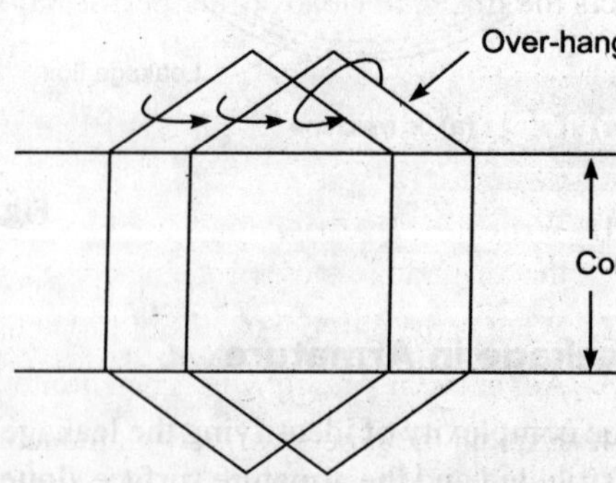

Fig. 5.50 *Over-hang leakage*

Zig-zag leakage In the case of induction machines both the stator and rotor are slotted so that some of the flux follows the path alternating between stator and rotor teeth as shown in Fig. 5.51. This flux therefore alternately links conductors in stator and rotor slots and is known as zig-zag leakage. Because of its nature it cannot be clearly assigned to either the stator or rotor windings. It is usually considered empirically that half of this flux links the stator winding while the other half links the rotor winding. This type of leakage is an exclusive feature of the induction machine and its value is a function of the percentage of the slot-pitch occupied by tooth in the rotor and stator and upon the length of the air-gap.

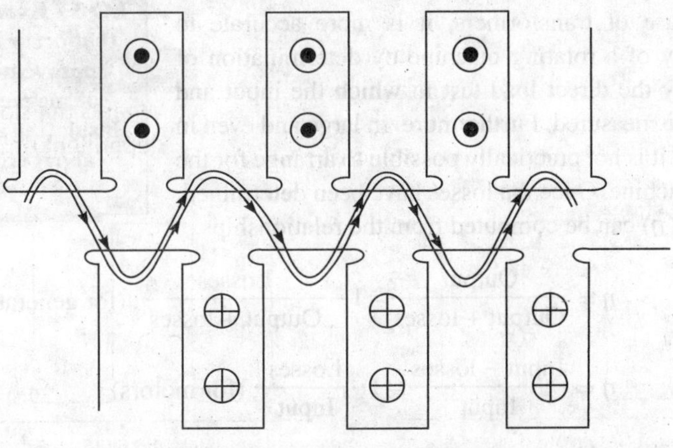

Fig. 5.51 *Zig-zag leakage*

Harmonic leakage This kind of leakage results when the winding distribution on the stator and rotor are dissimilar. The main flux then has a harmonic component not corresponding to either winding and this excess flux has the effect of leakage flux. The detailed treatment of this kind of leakage is beyond the scope of this book.

Leakage reactance The leakage flux of various kinds as enumerated above, linking one of the windings, causes that winding to possess leakage reactance which can be considered as a lumped parameter in series in the circuit model of the machine whose effect is to cause a voltage drop in the machine. Since the field windings of dc and synchronous machines carry direct current, the leakage flux linking them in no way affects the machine steady-state performance.

Concept Check LO 2

1. Write the expression for stator line voltage of a synchronous machine and show that it determines the air-gap flux/pole of the machine. ○○●
2. What is the pull-out torque of a synchronous machine and the meaning of loss of synchronism? ○○●
3. Advance the reason why a synchronous motor is not self-starting. ○●●
4. Explain the process of how an induction motor develops torque when ac supply is connected to its stator. Why it cannot develop torque at synchronous speed? ○●●

5. Define slip of an induction motor. At full-load what is the range of the value of slip.
6. What is the frequency of the rotor currents of an induction motor?
7. Why an induction motor is called asynchronous motor?

5.10 LOSSES AND EFFICIENCY

LO 3 Examine on various losses that occur in machines and how improvement in characteristics of magnetic, conducting and insulating materials has brought about efficiency in machine performance

The losses and efficiency of a transformer have been studied in Sec. 3.6. As in the case of transformers, it is more accurate to determine the efficiency of a rotating machine by determination of its losses rather than by the direct load test in which the input and output are required to be measured. Furthermore, in large and even in medium-size machines, it is not practically possible to arrange for the actual loading of the machine. Once the losses have been determined, the machine efficiency (η) can be computed from the relationships:

$$\eta = \frac{\text{Output}}{\text{Output + losses}} = 1 - \frac{\text{Losses}}{\text{Output + losses}}; \text{ (for generators)}$$

$$\eta = \frac{\text{Input} - \text{losses}}{\text{Input}} = 1 - \frac{\text{Losses}}{\text{Input}}; \text{ (for motors)}$$

The efficiency thus determined is more accurate because the error involved is only in losses, whereas in the direct method there is error in measurement of both the input and output.

The study of losses is essential for design purposes because (i) losses directly influence the economy of operation of the machine; and (ii) the rating of a machine depends on the maximum temperature that the insulation can withstand, which in turn is dictated by the heat developed in the core and conductors by the losses. Of course, the rating of a machine for a given frame size and losses can be raised by proper design of the ventilation system.

Probe

What is the importance of study of losses and why is it essential?

The process of energy conversion in rotating machines involves currents, fluxes and rotation which cause losses in conductors and ferromagnetic materials, and mechanical losses of rotation. Various losses can be classified conveniently by the tree-diagram shown in Fig. 5.52.

Constant Losses

A machine is normally designed to run at constant voltage mains and at a substantially constant speed (variable speeds are also required for certain applications). As a result, some of the losses remain nearly constant in the working range of the machine and are, therefore, named constant losses. The constant losses can be further classified as no-load core-loss and mechanical-loss.

No-load Core (Iron)-Loss

This loss consists of hysteresis and eddy-current loss caused by changing flux densities in the iron core of the machine when only the main winding is excited. The core-loss is largely confined to the armature of a dc machine, the armature of a synchronous machine and the stator of an induction machine. The frequency of flux density variation in the rotor core of the induction machine is so low (*sf*) under normal operating conditions that it has negligible core-loss.

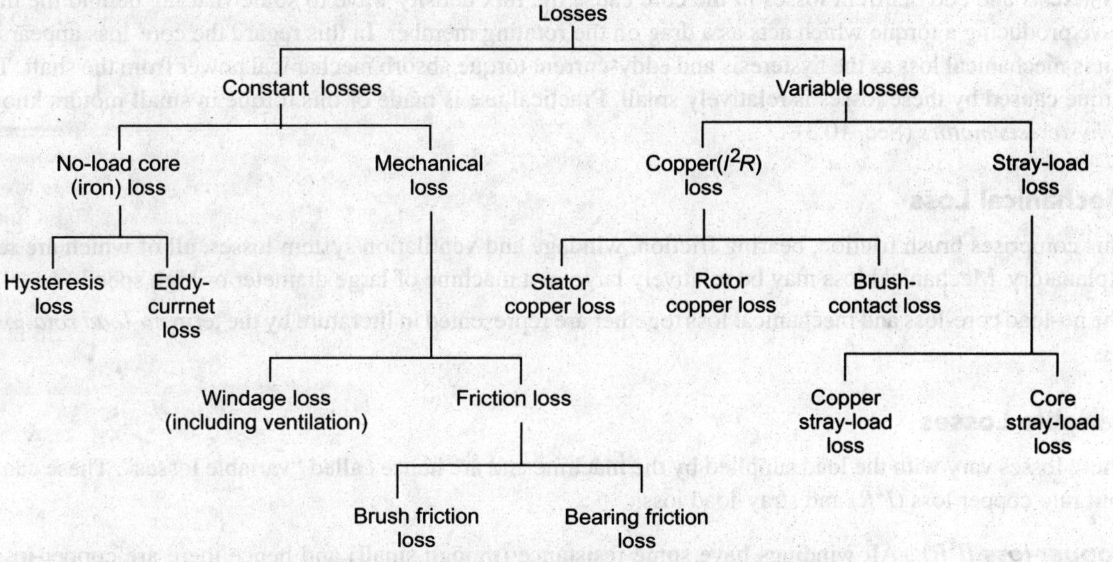

Fig. 5.52 *Classification of losses in rotating machines*

While in the case of transformers the core-loss arises because of time-variation of the flux density with the axis of flux remaining fixed; in the case of rotating machines, this loss results from both time-variation of the flux density and rotation of its axis. As a consequence the specific core-loss is larger in rotating machines than that in transformers.

Probe

What is pulsation loss?

The time- and axis-variation of the flux density in a rotating machine is illustrated by means of the cross-sectional view of a dc machine as shown in Fig. 5.53. It is easily seen from this figure that as the machine armature rotates, the flux density in the elemental volume of the core shown shaded varies cyclically in magnitude as well as in direction.

Additional hysteresis and eddy-current loss called *pulsation loss* also occurs in rotating machines on account of high-frequency flux density variations caused by slotting of the stator/rotor or both. In the case of dc and synchronous machines, the relative movement between the slotted armature and the poles causes high-frequency flux density variation in the pole-shoes because of the difference in reluctance of the flux paths corresponding to the teeth and slots. In case of induction machines where both the stator and rotor are slotted, the pulsation frequency is different in the two. In order to reduce the pulsation loss, it is a common practice to use laminated pole-shoes for dc and synchronous

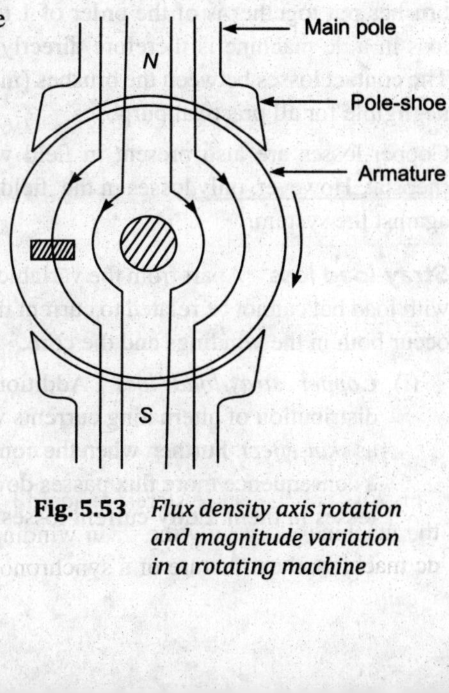

Fig. 5.53 *Flux density axis rotation and magnitude variation in a rotating machine*

machines; also for small machines of this type, the main pole itself may be built-up of laminations. Of course, much thicker laminations are used in pole-shoe than in the machine core.

Hysteresis and eddy current losses in the core cause the flux density wave to somewhat lag behind the mmf wave producing a torque which acts as a drag on the rotating member. In this regard the core-loss appears as if it is mechanical loss as the hysteresis and eddy-current torque absorb mechanical power from the shaft. The torque caused by these losses is relatively small. Practical use is made of this torque in small motors known as *hysteresis motors* (Sec. 10.3).

Mechanical Loss

This comprises brush friction, bearing friction, windage and ventilation system losses, all of which are self-explanatory. Mechanical loss may be relatively large in a machine of large diameter or high speed.

The no-load core-loss and mechanical loss together are represented in literature by the term *no-load rotational loss.*

Variable Losses

These losses vary with the load supplied by the machine and are hence called "variable losses". These can be split into copper loss (I^2R) and stray-load loss.

Copper-loss (I^2R) All windings have some resistance (though small) and hence there are copper-losses associated with current flow in them. The copper-loss can again be subdivided into the stator copper-loss, rotor copper-loss and brush-contact loss. The stator and rotor copper-losses are proportional to the current squared and are computed with the dc resistance of windings at 75 °C.

The conduction of current between the brushes (made of carbon) and the commutator of a dc machine is via short arcs in the air-gaps which are bound to exist in such a contact. As a consequence, the voltage drop at the brush contact remains practically constant with load; its value for positive and negative brushes put together is of the order of 1 to 2 V. The brush-contact loss in a dc machine is therefore directly proportional to current.

Probe

Classify copper-loss.

The contact losses between the brushes (made of copper-carbon) and slip-rings of a synchronous machine are negligible for all practical purposes.

Copper-losses are also present in field windings of synchronous and dc machines and in regulating the rheostat. However, only losses in the field winding are charged against the machine, the other being charged against the system.

Stray-load loss Apart from the variable losses mentioned above, there are some additional losses that vary with load but cannot be related to current in a simple manner. These losses are known as "stray-load loss" and occur both in the windings and the core.

(i) *Copper stray-load loss* Additional copper-loss occurs in the conductors due to nonuniform distribution of alternating currents which increase the effective resistance of conductors and is known as *skin-effect*. Further, when the conductors carry load current, the teeth of the core get saturated and as a consequence more flux passes down the slots through the copper conductors setting up eddy-current losses in them. Eddy-current losses are also present in the winding overhang.

(ii) *Core stray-load loss* Due to the flow of load current in a machine, the flux pattern in teeth and core gets distorted. The flux density decreases at one end of the flux density wave and increases at the other. Since the core-loss is almost proportional to the square of the flux density, its reduction due to a reduction in the flux density is less than the increase due to an increase in the flux density and as a consequence there is a net increase in the core-loss, predominantly in the teeth, which is known as the stray-load loss in the core.

Probe
Explain the types of stray-load loss.

Under loaded conditions, the teeth are highly saturated and as a result more flux leaks through the stator frame and end-shields causing eddy-current loss in them which, indeed, is another component of the core stray-load loss.

The stray-load loss is difficult to calculate accurately and therefore it is taken as 1 % of the output for a dc machine and 0.5% of the output for both synchronous and induction machines.

Machine Efficiency

Because of the presence of fixed and variable losses in a machine, the machine efficiency continuously increases with the load acquiring a maximum value at a particular load related to the design of the machine. Further, the full-load efficiency varies with the rating of a machine and is considerably higher for large-size machines; for example, the efficiency is close to 75% for 1 kW machine, 90% for 35 kW, 93% for 350 kW size and as high as 97% for 3500 kW. Efficiency of low-speed machines is usually lower than that of high-speed machines, the spread being 3 to 4%.

!Attention For a machine operating at a substantially constant voltage and speed, the various losses as enumerated earlier are:

(1) Constant losses[‡‡‡],

$$P_k = P_{i0} + P_{wf} \tag{5.77}$$

where P_{i0} = no-load core (iron)-loss (constant)

P_{wf} = windage and friction loss (constant)

(2) Variable losses,

$$P_v = P_c + P_{st} + P_b \tag{5.78}$$

where $P_c = 3I^2R$, the copper-loss (factor 3 will not be present in a dc machine);

R is the resistance parameter of the machine.

P_{st} = stray-load loss (copper + iron) = αI^2

(Here the stray-load loss is assumed proportional the square of the load current)

$P_b = V_bI$ = brush-contact loss (in dc machines); V_b being the brush-contact voltage drop

Hence $P_v = 3I^2 R + \alpha I^2 + V_bI = (3R + \alpha) I^2 + V_bI = K_vI^2 + VI$ (5.79)

Thus total machine losses can be expressed as a function of the current as

$$P_L = P_k + K_vI^2 + V_bI \tag{5.80}$$

Probe
Derive an expression for total machine losses.

[‡‡‡] Field copper-loss for a dc and synchronous machine is constant and can be lumped with constant losses.

Generating machine

Power output, $P_{out} = CVI$

C = constant ($\sqrt{3}$ × pf for 3-phase ac machine)

V = machine voltage (line)

$$\text{Efficiency, } \eta = \frac{P_{out}}{P_{out} + P_L}$$

$$= \frac{CVI}{CVI + P_k + K_v I^2 + V_b I} \tag{5.81a}$$

$$= \frac{CV}{(CV + V_b) + \left(\frac{P_k}{I} + K_v I\right)} \tag{5.81b}$$

The maximum efficiency is obtained at (minimum denominator in Eq. (5.81))

$$\frac{P_k}{I} = K_v I$$

or $$P_k = K_v I^2 \tag{5.82}$$

> **Probe**
> When is the maximum efficiency reached in a generating machine?

Thus the maximum efficiency is reached at a load when the losses proportional to the square of current equal the constant losses. This is the same conclusion as arrived at for a transformer (Eq. (3.59(a))).

Motoring machine

Power input, $P_{in} = CVI$

$$\text{Efficiency, } \eta = \frac{P_{in} - P_L}{P_{in}}$$

$$= \frac{CVI - P_k - K_v I^2 - V_b I}{CVI} \tag{5.83a}$$

$$= \frac{(CV - V_b) - \left(\frac{P_k}{I} + K_v I\right)}{CV} \tag{5.83b}$$

This also reaches the maximum value when

$$P_k = K_v I^2 \tag{5.84}$$

i.e. Constant losses = losses proportional to the square of current

Interpretation As per the condition of Eq. (5.82) or (5.84) for maximum efficiency, the constant losses and variable losses (proportionality constant K_v) are so proportioned by the choice of machine dimensions as to yield maximum efficiency near about the full-load. The constant losses are mainly determined by the choice of flux density and the volume of iron used and the variable losses are governed by the choice of current density and the volume of copper used. Further, the flux density used is limited to slightly saturated values and the current density is limited by the allowable temperature rise (depending upon the class of insulation). Therefore, adjusting the machine efficiency to yield the maximum value at a particular load is an exercise in proportioning iron and copper to be used in the machine.

Maximum Output

Consider for example the motoring machine. The power output is expressed as

$$P_{out} = CVI - P_k - K_v I^2 - V_b I$$

It is a fairly good assumption to neglect V_b; in fact this term is not present in ac machines. Then

$$P_{out} = CVI - P_k - K_v I^2$$

For maximum power output

$$\frac{dP_{out}}{dI} = CV - 2K_v I = 0$$

or

$$I = \frac{CV}{2K_v}$$

The maximum power output is then given by

$$P_{out}\,(\max) = CV\left(\frac{CV}{2K_v}\right) - P_K - K_v\left(\frac{CV}{2K_v}\right)^2$$

$$= \frac{(CV)^2}{4K_v} - P_k$$

> **Probe**
> Derive an expression for maximum power output.

The power input is

$$P_{in} = CVI = \frac{(CV)^2}{2K_v}$$

The efficiency at maximum power output is given by

$$\eta = \frac{(CV)^2/4K_v - P_k}{(CV)^2/2K_v}$$

Obviously this will be slightly less than 50%. This is too low a value to be acceptable for a power-delivering device. Further, under maximum output operation, the losses being almost half the input, it would be impossible to limit the temperature rise to the allowable value.

Conclusion

Thus the electromechanical power devices are never operated to deliver maximum output. In fact these are operated at a load (nearly full-load) at which the efficiency is maximum. This is in contrast to electronic devices (low power) which are usually operated to deliver maximum power output as the total power being very small, the efficiency is of secondary consideration. Further, the problem of heat (caused by losses) dissipation is not so intense as in large power rotating machines.

5.11 RATING AND LOSS DISSIPATION

LO 3

Rating

The rating of a synchronous generator is its operating voltage, frequency, speed and kVA/MVA output at a specified power factor[§§§]. In case of motors the output rating is given in kW (older practice was to specify the rated output in horse-power (1 hp = 746 W)). The dc machines are rated in terms of voltage and power output in kW. The frequency specification of ac machines is normally the standard supply frequency, i.e. 50 Hz—it is 60 Hz on the American continent. The rated voltage is specified as a standard value, viz 230 V/ 400 V/ 3.3 kV/6.6 kV/11 kV. Most manufacturers build small and medium sized motors in standard kW sizes (consult ISI: 325–1970).

Insulation which is used in intricate forms is the most vulnerable part of a machine being highly susceptible to temperature which is the major factor[¶¶¶] in determining its lifetime and therefore that of the machine. As a rule of thumb the lifetime of a machine is reduced to one-half for every 8–10 °C rise in temperature. Acceptable life expectancy of electric power equipment being 10–30 years, the highest temperature of insulation anywhere in the machine has to be limited to a value depending upon the class of insulation employed. The maximum output that a machine can supply without exceeding a specified temperature rise above the ambient (40°C as per ISI), is known as the *continuous rating* of the machine. The continuous rating, as determined by temperature rise, limits the machine losses generally assuring an acceptable value of machine efficiency. Any other specific performance figure(s) (say the breakdown torque of induction motor) must be independently met by the machine designer.

Both iron and copper being the seats of losses in a machine, the temperature distribution in the machine is quite complex. This problem is greatly simplified by assigning a single temperature to define the thermal state of the machine. The single temperature is determined by the rise in resistance of windings as measured from the machine terminals. In super-size synchronous generators, thermocouples are embedded during manufacture to track temperature of *hot-spots* predetermined by the designer through heat transfer studies.

ISI specifications classify insulation for industrial machines as Class B, Class F and Class H. Class B insulation includes mica, glass fibre, asbestos, and other similar materials with suitable binding substances. The highest temperature rise (40°C above ambient) allowed for this class of insulation is 130°C. Class F insulation also includes mica, glass fibre, and synthetic substances, but it must be able to withstand a higher temperature of 155°C. Class H can withstand a still higher temperature of 180°C and may consist of materials like silicone elastomer and combinations in various forms of mica, glass fibre, asbestos, etc. with silicone resins for bonding.

Continuous-Rated Motor

A continuous-rated motor must operate successfully ±10% variation of the rated voltage and ±5% variation of the rated frequency. The combined variation of voltage and frequency in a direction to adversely affect losses (V increases, f reduces) must not exceed 10%. It is further expected that the continuous-rated motors

Probe

At what variation of rated voltage and rated frequency should a continuous rated motor operate on?

[§§§] kVA/MVA rating and specified pf determine the mechanical power rating of the prime mover to which it is coupled. It can of course be operated at other power factors.

[¶¶¶] Other factors affecting the life of insulation are oxidation and ingress of dirt and moisture.

are built with ample safety margin so as be capable of withstanding short-time overloads of 25% with 10% reduction in voltage without excessive temperature rise.

In industrial applications certain situations require the motor to be loaded for short time periods followed by long cooling intervals. Such motors are *short-time rated* for standard periods of 5, 15, 20, 30 and 60 min. These motors are specially designed with higher flux densities in iron and higher current densities in copper. As a result they have better torque producing capability but lower thermal capacity when compared to continuous-rated motors.

Short-time rating of a continuous-rated motor is much more than its continuous rating because of the heat storage in thermal capacity of the machine during the heat transient under short-time loading. This is illustrated in Fig. 5.54. The thermal transient**** has an exponential growth (a single time constant) with a steady temperature rise of ($P_L R_T$) where P_L represents the motor loss in heat units at a particular load and R_T is the overall thermal resistance of the cooling system. It is evident from Fig. 5.54 that P_L^{st} (short-time loading loss) allowed for a specified loading period is more than P_L^c (continued loading loss) without the machine exceeding the allowable temperature rise. Hence as the loading period is reduced, the motor short-time rating increases.

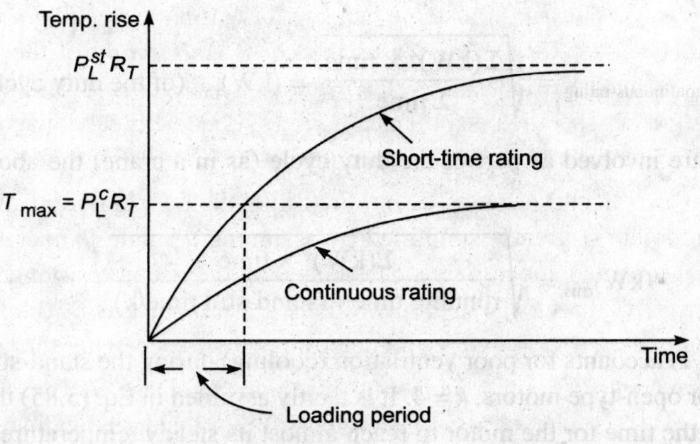

Fig. 5.54 *Continuous and short-time rating*

**** Figure 5.55 gives the simplified, lumped, analogous circuit model of the machine for heat transfer. The temperature (single temperature representing the overall thermal state of the machine) is measured with respect to ground reference (ambient). Here

P_L = motor loss in heat units

R_T = overall thermal resistance of the cooling system

C_T = thermal capacity of the machine

The differential governing the thermal transient is

$$C_T \frac{dT}{dt} + \frac{T}{R_T} = P_L$$

whose solution for initial temperature rise of 0 °C (machine starting from cold conditions) is

$$T(t) = T_s (1 - e^{-t/\tau})$$

where $\tau = R_T C_T$, the thermal time-constant

$T_s = P_L R_T$ the steady-state temperature rise

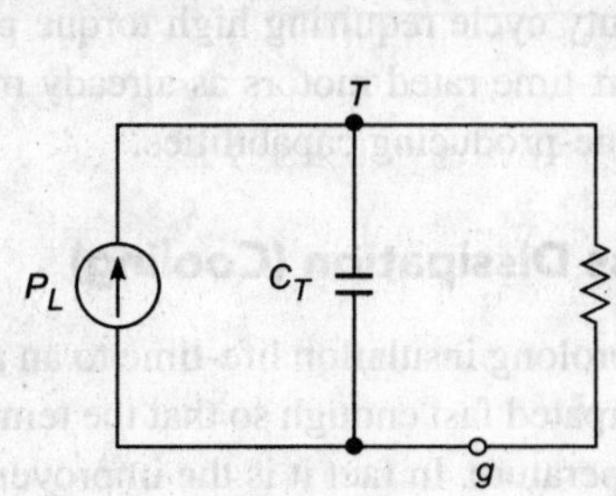

Fig. 5.55

In industrial applications of motors a typical problem is the determination of the size of a continuous-rated motor for a given duty cycle—the duty cycle of a "planer" may be visualized as a simple example; during the forward stroke the motor is on full-load while it is practically unloaded during the return stroke. A crude yet reliable method of motor selection is to assume that the motor losses are proportional to the square of loading (this overemphasises I^2R loss as compared to the constant coreloss). The average loss during a duty cycle is proportional to

Probe

What is the typical problem faced in the industrial applications of motors?

$$\left[\frac{\Sigma(\text{kW})^2 \times \text{time}}{\Sigma\,\text{time}}\right]$$

where kW = motor loading in a period of duty cycle. The continuous-rated motor which has the same loss is given by

$$(\text{kW})^2_{\text{continuous-rating}} = \left[\frac{\Sigma(\text{kW})^2 \times \text{time}}{\Sigma\,\text{time}}\right]$$

or

$$(\text{kW})_{\text{continuous-rating}} = \sqrt{\frac{\Sigma(\text{kW})^2 \times \text{time}}{\Sigma\,\text{time}}} = (\text{kW})_{\text{rms}}\ \text{(of the duty cycle)}$$

If stand-still period(s) are involved as part of the duty cycle (as in a crane) the above relationship must be modified as under

$$(\text{kW})_{\text{rms}} = \sqrt{\frac{\Sigma(\text{kW})^2 \times \text{time}}{\text{running time} + (\text{stand-still time}/k)}} \tag{5.85}$$

where the constant ($k > 1$) accounts for poor ventilation (cooling) during the stand-still period(s) where there is no forced cooling. For open-type motors, $k = 4$. It is tacitly assumed in Eq. (5.85) that the duty cycle period is sufficiently less than the time for the motor to reach almost its steady temperature rise when continuously loaded to its continuous-rating.

The errors involved in the $(\text{kW})_{\text{rms}}$-method are swamped out when the nearest higher standard rating, e.g. 90 kW motor is selected for $(\text{kW})_{\text{rms}} = 85$.

A duty cycle requiring high torque peaks cannot be satisfied by $(\text{kW})_{\text{rms}}$ choice which has a thermal basis. Short-time rated motors as already mentioned are better suited for such applications because of their better torque-producing capabilities.

Loss Dissipation (Cooling)

To prolong insulation life-time to an acceptable value, the heat generated owing to loss in a machine must be dissipated fast enough so that the temperature rise does not exceed the allowable limit for a specified ambient temperature. In fact it is the improvement in heat transfer technology that has helped in a major reduction in machine sizes for given ratings, in particular for large-size machines.

Combined conduction and forced convective cooling are the practical means of removing heat of losses from all electric machinery. Because of limited allowable temperature rise, radiation does not make any significant contribution to loss dissipation.

Radial ventilation Radial ventilation is commonly employed wherein the natural centrifugal action of the rotor may be supplemented by the rotor fan. Figure 5.56 shows the radial ventilation scheme suitable for machines up to 20 kW.

Axial ventilation Axial ventilation scheme of Fig. 5.57 is suitable for machines of moderate outputs and high speeds.

Probe

Axial ventilation is suitable for what kind of machines?

Combined radial and axial ventilation This is employed for large machines as shown in Fig. 5.58 for an induction motor.

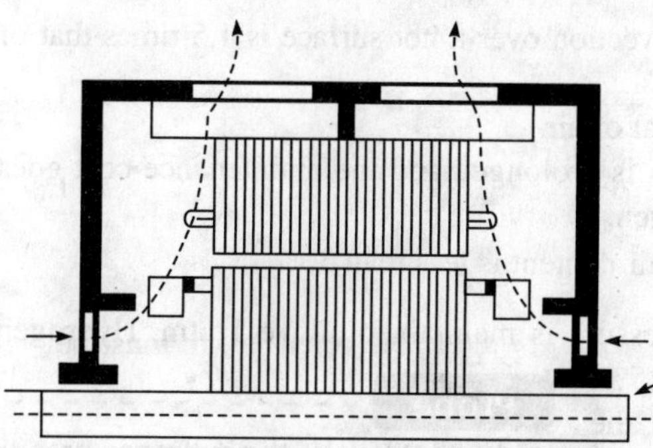

Fig. 5.56 *Radial ventilation*

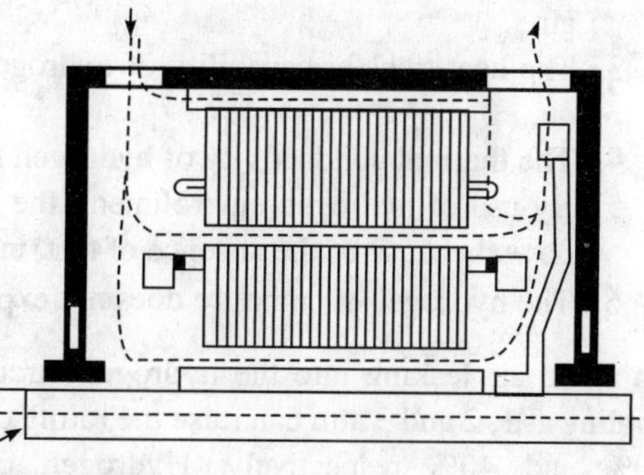

Fig. 5.57 *Axial ventilation*

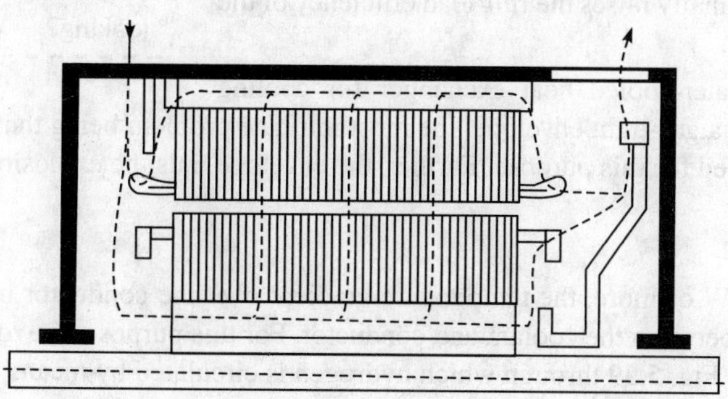

Fig. 5.58 *Combined axial and radial ventilation*

Totally-enclosed Totally enclosed machine presents a special ventilation problem as the inside of the machine has no air-connection with outside. In such machines heat is transferred to the enclosure (called *carcass*) by an internal fan and from where it is removed to the ambient by an external fan mounted on the shaft. The cooling in a totally enclosed machine cannot be as efficient as in an open-type machine.

Losses being roughly proportional to the volume of material increase as the cube of the linear dimensions, while the cooling surface increases as the square of the same. Therefore, the loss dissipation problem becomes more intense in large turbo-generators.

For large machines, which may require several tonnes of cooling air/hour, forced ventilation is used wherein air is passed through a cleaning filter before being forced into the machine for cooling purposes. A more compact scheme of securing clean cooling air is the *closed-circuit* system as employed for turbo-generators of small rating. In this system hot air is cooled by a water-cooled heat exchanger.

Hydrogen Cooling

For large turbo-generators, hydrogen is commonly used as a cooling medium in a closed circuit. The following properties of hydrogen make it most suited for this purpose.

1. Hydrogen has a density of 1/14 of that of air at the same temperature and pressure, reducing thereby windage losses and noise.
2. On an equal weight basis, the specific heat of hydrogen is 14 times that of air. Therefore, for the same temperature and pressure, the heat-storing capacity/ unit volume of hydrogen is the same as that of air.
3. The heat-transfer capability of hydrogen by forced convection over a hot surface is 1.5 times that of air.
4. The thermal conductivity of hydrogen is seven times that of air.
5. By use of hydrogen environment, the life of insulation is prolonged and the maintenance cost goes down because of the absence of dirt, moisture and oxygen.
6. The hydrogen-air mixture does not explode so long as air content is less than 30%.

To avoid air leaking into the hydrogen circuit, hydrogen pressure is maintained above 1 atm. Hydrogen cooling at 1, 2 and 3 atm can raise the rating of a machine by 15%, 30% and 40% respectively. Hydrogen cooling reduces the temperature and resistance of windings and hence the losses to be dissipated. This fact marginally raises the full-load efficiency of the machine (by about 0.5%).

Probe

At what level is the hydrogen pressure maintained at, to avoid air leaking?

The machine and its water-cooled heat exchanger for cooling hydrogen are enclosed in a gas-tight envelope; the most intricate problem being that of sealing the bearings. Oil-filled gas-seals are used for this purpose. Further, the envelope must be explosion-proof.

Direct Gas cooling

For a machine of 100 MW or more, the temperature gradient over the conductor insulation is high enough to call for direct contact between the coolant and conductor. For this purpose, the rotor conductor comprises hollow tubes as shown in Fig. 5.59 through which hydrogen is circulated by means of flexible connections.

Direct Water Cooling

Turbo-generators of the highest rating have a hydrogen-cooled stator core and a direct water-cooled stator and rotor windings. The speed of circulating water must be limited to 2.5 m/s to avoid erosion and cavitation. Figure 5.60 shows the arrangement for direct water-cooling of the rotor winding which is most desirable because of high electric loading of rotor and is mechanically most difficult. Direct water-cooling of the stator winding requires flexible water-tube connections with insulation against high voltages and low water conductivity.

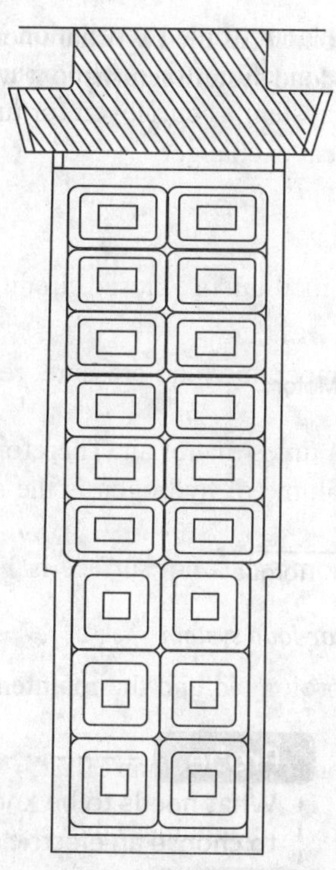

Fig. 5.59 *Direct gas cooling*

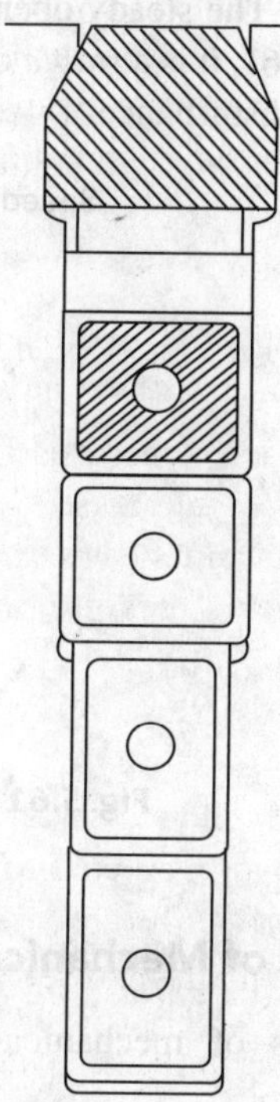

Fig. 5.60 *Direct water cooling*

Concept Check — LO 3

1. List the type of losses in electric machine. What is the nature of each loss? ○○●
2. What is the relative speed between stator and rotor rotating fields in an induction motor? ○○●
3. Sketch the torque-slip characteristic of an induction motor. Explain the nature of the low slip part of the characteristic. Locate on the characteristic the full-load torque operating point. ○●●

5.12 MATCHING CHARACTERISTICS OF ELECTRIC MACHINE AND LOAD

The machine and the load are the two components of an electro-mechanical energy-conversion system, and the machine characteristics, generally, play a predominant part in the operating behavior of the complete system.

LO 4 Analyze the need for matching characteristics of electric machines and load in ensuring steady operating behavior of the complete system

In choosing an electric motor its speed-torque characteristic is needed to be known to a fair degree of accuracy and further it has to be properly matched to the speed-torque characteristic of the mechanical load. Figure 5.61 shows the speed-torque characteristic[††††] of an induction motor with a fan-type load (load torque roughly proportional

[††††] This is the same characteristic as in Fig. 5.43 except with the ordinates reversed.

to square of speed). The steady operating point is the intersection point P of the two characteristics. As can be seen from Fig. 5.61, it is a *stable* operating point and the machine-load system returns to it when subjected to a short-duration disturbance.

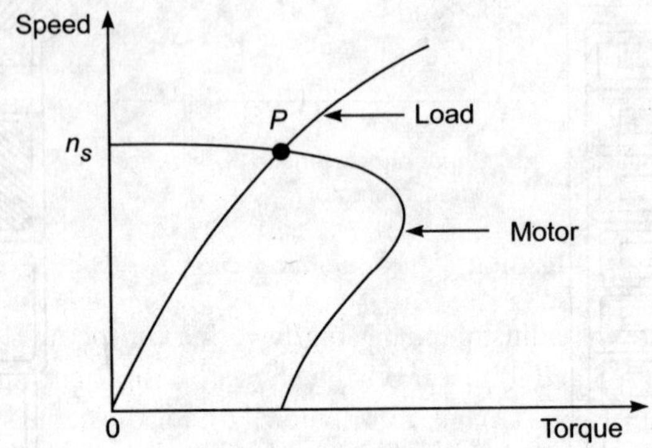

Fig. 5.61 *Steady operating point of a motor-load system*

Characteristics of Mechanical Loads

> **Probe**
> What needs to be known in order to choose an electric motor?

The characteristics of mechanical loads can be classified as below:

1. Constant-speed loads These can be of two kinds. Certain loads require approximately constant speed as the load torque varies, e.g. machine tools, hydraulic pumps, fans, etc. Certain special loads like paper mill drives require exactly a constant speed independent of the load torque.

2. Variable-speed (or constant kW) loads Certain loads, such as cranes, hoists and other traction-type drives, demand high torque at low speeds and low torque at high speed so that the kW demanded from the mains remains substantially constant. This nature is imparted to the load wherever heavy inertias are to be accelerated.

3. Adjustable-speed loads These are of a constant adjustable speed kind as in certain machine tool applications or of a variable adjustable speed kind as in cranes. The range of speed adjustment in certain drives can be highly demanding.

The motor characteristics can be classified as:

1. Constant-speed type The speed remains exactly constant independent of torque as in Fig. 5.62(a). This characteristic is possessed by the synchronous motor.

2. Shunt-type Here the motor speed drops by a few per cent from no-load to full-load as in Fig. 5.62(b). The ac induction motor (over the operating region) and dc shunt motor both possess this characteristic.

3. Series-type Here the speed rises sharply as the load torque reduces as in Fig. 5.62(c). This type of characteristic is possessed by a dc series motor ideally suitable for traction-type loads.

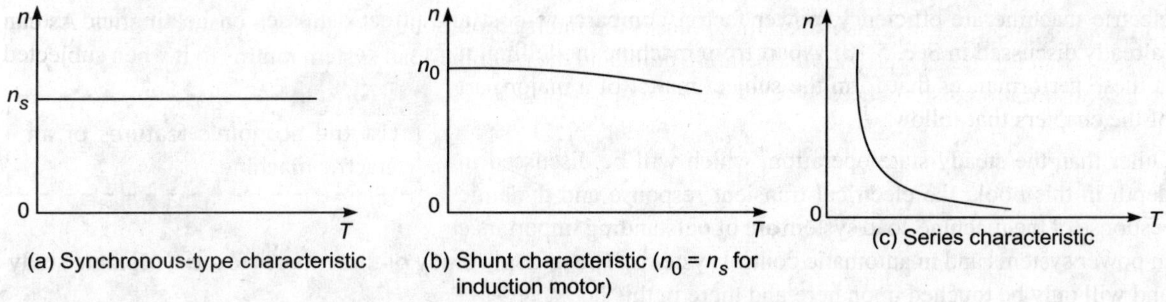

Fig. 5.62 *Types of motor characteristic*

Adjustable speed drives require the adjustment (raising/lowering) of the three motor characteristic types. It will be seen in later Chapters (7, 8 and 9) that this is much more easily accomplished in dc motors than in ac motors. Solid-state power control (Ch. 12) has contributed a lot to adjustable speed drives.

Accelerating and Decelerating Characteristics of Motor-load

The accelerating *(starting)* and decelerating *(braking)* characteristics of motor-load systems are also of equal importance in their industrial applications. The system should be capable of coming to full speed from rest and be able to be stopped in an acceptable time period. These requirements are stringent in starting *on-load* and in fast braking and reversal in certain special applications (rolling mill drives). A motor has three regions of operation—*generating, motoring* and *braking*. In generating region it returns the decelerating intertial energy back to electric main preventing the system from acquiring dangerously high speeds—as in lowering a hoist or down-the-gradient traction. In braking region the machine absorbs mechanical energy (as well as some electric energy) in form of losses in it appearing as heat. A dc motor offers excellent starting and braking characteristics, much superior to those of an ac induction motor.

Probe

Describe the regions of operations of a motor.

Similar to the case of motors, the operating point of a generator-load system is determined by the characteristics of the two as shown in Fig. 5.63 for a dc shunt generator (Sec. 7.11). Similar is the case with ac (synchronous) generators. In modern systems, generators operating in parallel feed loads spread over geographically wide areas through transmission lines. The system must meet the requirement of a substantially constant voltage as load varies over a wide range. A captive generator feeding a single motor is used in certain speed-control schemes in which the terminal voltage may be required to vary in a peculiar fashion.

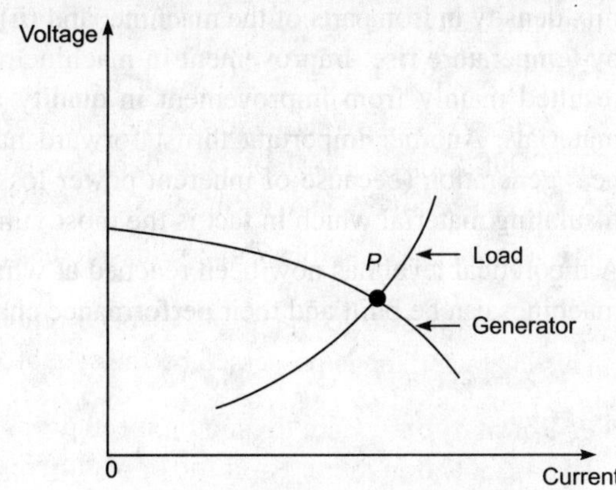

Fig. 5.63 *Steady operating point of a generator-load system*

It is therefore seen that among the features of great importance are the torque-speed characteristic of a motor and the *V-I* characteristic of a generator. Equally important can be the limits through which these characteristics can be varied. Other relevant important economic features of an electric machine are efficiency, power factor, comparative cost and effect of losses on heating and rating (already discussed in Sec. 5.10). Apart from machine modelling, it is these performances that form the subject matter of a major part of the chapters that follow.

Probe

List the economic features of an electric machine.

Other than the steady-state operation, which will be discussed in depth in this book, the electrical transient response and dynamic response of the machine-load system are of outstanding importance in power systems and in automatic control systems. The in-depth study of these topics forms a separate study and will only be touched upon here and there in this book.

Concept Check LO 4

1. Explain how an induction motor can self-start but cannot run at synchronous speed. ○●●
2. Explain why rotor induced emf is proportional to slip. ○●●
3. Distinguish between time phase difference and space phase difference. ○○●
4. State the condition of maximum efficiency of an electric machine. Compare it with that of a transformer. ○○●
5. When a conducting coil is moving past a sinusoidal *B*-wave, what is the relative position of the coil axis when the induced emf in it is (1) maximum, and (ii) zero? ○○●

5.13 RESUME

In this chapter the following common features of rotating machines have been studied—ac windings (elementary treatment), emf and mmf of ac winding, concepts of rotating magnetic field and process of torque production by two interacting fields. Differences have also been discovered, particularly in a dc machine which has a commutator, the study of whose action has been postponed to Ch. 7. Through expressions of emf and torque, it was seen that the machine capability for a given frame size is limited by (i) the saturation flux density in iron parts of the machine, and (ii) the current-carrying capability of windings which is limited by temperature rise. Improvement in machine performance and the cost per unit power have over the years resulted mainly from improvement in quality and characteristics of magnetic, conducting and insulating materials. Another important thrust forward has been in the direction of heat removal from the seats of heat generation (because of inherent power loss) so as to limit the temperature rise to that permitted in the insulating material which in fact is the most vulnerable part of a machine.

A theoretical level has now been reached at which simple mathematical models of various types of rotating machines can be built and their performance characteristics studied through these models.

SUMMARY

LO 1

- Constructional Feature – Electric Machines
  Rotating electric machines have two flux carrying parts which are made of laminated silicon steel. These two are the following:

  Stator: It is a stationary annular cylinder

  Rotor: It rotates within the stator supported by a shaft, ball bearings and *end rings* bolted to the stator. There is a narrow *air-gap* between the stator and rotor.

  Windings: There are two windings made of copper. These are placed in stator and rotor *slots* or in one of these wound on projecting poles. In *synchronous* and *dc machines* the main field is created by the *field poles* (even in number) and dc excited. The other winding which interchanges electric power with the external circuit and so carries the *load current* is called the *armature winding* and in the *seat of induced emf.*

  In a synchronous machine, the field poles are on the rotor and armature winding on the stator. It is the preferred construction and universally adopted. Excitation current is passed to the field poles through *slip-ring brush* arrangement. In a dc machine it is a must that the field poles are on the stator and the armature on the rotor. The rotor also carries a *commutator* whose segments are suitably connected to the armature windings and act to convert the alternating armature current to dc for the external connection.

  In an *induction machine* both stator and rotor are slotted and carry armature windings; rotor may carry just slot conductors shorted by end rings.

  Thus the electric machines are of two types:
  - ac Machines: Synchronous and induction
  - dc Machines
- *Field-pole types*

  Salient projecting poles, non-salient cylindrical poles

  Synchronous machine can have both types, dc machine has only salient poles
- Induced dc emf in rotating machines. It is the speed rmf. The relative motion between *B*-wave and coils, which causes change in flux linkage and emf induction
- *Mechanical and electrical angles*

$$\frac{\theta_e}{\theta_m} = \frac{2}{P};\ P = \text{number of poles}$$

- *Speed-Poles-Frequency*

$$n = \frac{120f}{P},\ \text{synchronous speed in rpm}$$

Or
$$f = \frac{np}{120}\ \text{Hz}$$

- *Armature Coils* Could be single-turn or multi-turn with two end connections.

  Coil-side – each active side of a coil
- *Coil span (pitch)* – full-pitched, angle between coil sides is π rad or 180° electrical

  – short-pitched; angle between coil sides is less than π in terms of number of slots
- *Two-layer windings*–two coil sides per slot
- Induced emf (*ac*) of a single *N* turn full-pitch coil

$$E(\text{rms}) = \sqrt{2}\ \pi f N\Phi = 4.44 f N\Phi$$

$$\Phi = \text{flux/polej}$$

Induced emf phasor lags the flux phasor by 90°

- *Distributed winding* More than one coil/phase

Slots/pole/phase, SPP $= m = \dfrac{S}{qP}$; S = slot,

q = number of phases, generally three

Slot angle, $\gamma = \dfrac{\pi P}{S}$ rad (elect.)

Phase spread, $\sigma = m\gamma$

- *Breadth factor, k_b* Because of distributed winding the phase emf is less than the algebraic sum of series turns/phase by the breadth factor

$$k_b = \frac{\sin m\gamma/2}{m \sin \gamma/2} < 1$$

Generally k_b (harmonics) < k_b (fundamental)
Therefore, distributed winding incidentally reduces the harmonic content of emf induced.

- *Short-pitched (corded) coils* The emf of a short-pitched coils is less than that for a full-pitched coil by the pitch factor

$$K_p \cos\frac{\theta_{sp}}{2} < 1;\ \theta_{sp} = \text{short-pitching angle in rad elect.}$$

- Chording of coil saves in overhang copper by proper choice of θ_{sp} any particular harmonic can be eliminated
- *Winding factor*

$$K_\omega = K_b K_p < 1$$

- General formula for induced emf phase

$$E_p = \sqrt{2}\,\pi K_w f N_{ph}\,(\text{series})\,\Phi\ V$$

It is applicable for synchronous machine (stator) and induction machine stator and rotor. For 3-phase synchronous and induction machine the windings of the three phase are laid 120° elect. apart from each other.

- *MMF of ac Winding* A single coil produces rectangular mmf wave (with north and south poles, strength $(Ni/2)$)

Fundamental emf wave

$$F_{a1} = \frac{4}{\pi}(Ni/2)\cos\theta, \quad \theta = \text{space angle elect.}$$

For sinusoidal current ($i_a = \sqrt{2}\,I\cos\omega t$)

$$F_{a1} = \sqrt{2}\,K' I\cos\omega t\cos\theta$$

$$= F_m\cos\omega t\cos\theta$$

It is a standing pulsating wave.

For a distributed winding

$$F_{a1} = \frac{4\sqrt{2}}{\pi}K_w\left(\frac{N_{ph}(\text{series})}{P}\right)I\cos\omega t\cos\theta$$

$$= F_m\cos\omega t\cos\theta$$

- *Rotating Magnetic Field* A 3-phase winding with their axis located at *120° elect space phase difference* from each other and fed with 3-phase balanced currents with a *time phase difference of 120° elect.* The *mmf*-wave rotates at synchronous speed $\omega_s = 2\pi f$ rad (elect.)/s (or $n_s = 120\,f/P$ rpm). The direction of rotation of mmf wave is from the leading phase to the lagging phase axis. The number of poles of the mmf wave is same as for which the winding is wound.
- *Torque in Round Rotor Machine* Necessary conditions for production of steady torque by two interacting magnetic fields.
  1. The two fields must be relatively stationary
  2. The two fields must have the same number of poles

Torque expression F_1, F_2 are peak values of sinusoidally distributed fields rotating at synchronous speed and F_r is the resultant field.

Torque, $T = k\,F_1 F_2 \sin\alpha$,

α = angle between F_1 and F_2

$= k\,F_r F_2 \sin\delta$;

δ = angle between F_r and F_2

It is F_r that produces the *air-gap flux, Φ_r/pole*

LO 2

- *Synchronous Machine*

Generating F_2 leads F_r by angle δ. Electromagnetic torque $T = T_{PM}$ in opposite direction to synchronous

Motoring F_2 lags F_r by angle δ. Electromagnetic torque $T = T_L$ (load torque) in opposite direction to synchronous speed. Non-self starting

Terminal voltage (equal to induced emf)

$$V(\text{line}) = \sqrt{3} \times 4.44\,K_w f\,\Phi_r N_{ph}\,(\text{series})$$

For fixed terminal voltage Φ_r is constant; same as in a transformer. Therefore

$$T = K_T \sin\delta$$

Pullout torque, $T_{max} = K_T$, $\delta = 90°$

Higher torque load causes *loss of synchronism* or the machine *pullout*.

- *Induction Machine* Two types: 1. *wound rotor*, terminals shorted externally 2. *squirrel cage* rotor, copper or aluminum bars in slots shorted by conducting end rings.

Operation When the stator is excited from 3-phase mains (V, f), the rotating field induces currents in shorted windings and their interaction produces torque. The machine is therefore *self-starting*. The steady rotor speed n must be less than n_s, the synchronous speed for induction currents in rotor and torque production. The induction motor is therefore *asynchronous motor*.

$$\text{Slip, } s = \frac{n_s - n}{n_s}$$

Rotor frequency, $f_2 = sf$

The air-gap flux Φ_r is determined by the applied voltage. The torque-slip ($T - s$) characteristic is non-linear and slip at full-load is 2—5%. So speed less than synchronous is nearly constant (shunt characteristic).

Maximum torque is the break-down above which the motor stalls.

PRACTICE PROBLEMS

5.1 Determine the breadth and pitch factors for a 3-phase winding with 2 slots per pole per phase. The coil span is 5 slot-pitches. LO 1

If the flux density wave in the air-gap consists of the fundamental and a 24% third-harmonic, calculate the percentage increase in the rms value of the phase voltage due to the harmonic.

5.2 A 50-Hz, 6-pole synchronous generator has 36 slots. It has two-layer winding with full-pitch coils of 8 turns each. The flux per pole is 0.015 Wb (sinusoidally distributed). Determine the induced emf (line-to-line) if the coils are connected to form (a) 2-phase winding (b) star-connected 3-phase winding. LO 1

5.3 The air-gap flux density distribution of a 6-pole, 50-Hz synchronous generator is LO 1

$B(\theta) = B_1 (\sin\theta + 0.3 \sin 2\theta + 0.15 \sin 5\theta)$

The total flux/pole is 0.015 Wb. Find the fundamental, third-harmonic and fifth harmonic flux/pole.

5.4 Show that the limiting value of the breadth factor for the fundamental is LO 1

$$K_b = \frac{\sin\frac{1}{2}\sigma}{\frac{1}{2}\sigma}$$

where $\sigma = m\gamma =$ phase spread

and *m, the* slots pole per phase, tend to be large.

5.5 A 50-Hz synchronous salient pole generator is driven by a hydro-electric turbine at a speed of 125 rpm. There are 576 stator slots with two conductors per slot. The air-gap diameter is 6.1 m and the stator length is 1.2 m. The sinusoidally distributed flux density has a peak value of 1.1 T. (a) Calculate the maximum rms single-phase voltage that can be produced by suitably connecting all the conductors. (b) Find the per phase emf if the conductors are connected in a balanced 3-phase winding. LO 1

5.6 Find the number of series turns required for each phase of a 3-phses, 50-Hz, 10-pole alternator with 90 slots. The winding is to be star-connected to give a line voltage of 11 kV. The flux/pole is 0.2 Wb. LO 1

5.7 A dc armature is built up of laminations having an external diameter of 80 cm and the internal diameter of 42 cm. The length of the armature is 32 cm. The flux density in the armature core is 0.85 T. The armature is wave connected with 72 slots and 3 conductors/slot. If the number of poles is 6, find the emf induced when the armature is rotated at a speed of 600 rpm. LO 1

(*Hint*: Flux passing through armature core is half the flux/pole. See Fig. 5.13.)

5.8 A 6-pole, wave-connected dc armature has 300 conductors and runs at 1200 rpm. The emf generated is 600 V. Find the useful flux/pole. LO 1

5.9 A 4-pole, dc machine has a lap-connected armature having 60 slots and eight conductors per slot. The flux per pole is 30 mWb. If the armature is rotated at 1000 rpm, find the emf available across the armature terminals. Also calculate the frequency of emf in the armature coils. LO 1

5.10 Trace out the variations in mmf due to a belt of current-carrying conductors representing one phase of a 2-pole, 3-phase winding. The belt may be assumed to be a current-sheet with uniform current density. What is the peak amplitude of the mmf wave if the total current in the belt is *A* amperes? LO 1

(*Hint*: The mmf wave is trapezoidal.)

5.11 Each phase belt of a 2-pole, 3-phase winding carrying balanced 3-phase currents can be assumed to be a current-sheet with uniform density (as in Prob. 5.10). Sketch the resultant mmf wave at $\omega t_1 = 0$, $\omega t_2 = \pi/3$ and $\omega t_3, = 2\pi/3$. LO 1 ●●●

5.12 Phase *a* of a 3-phase stator at the instant of carrying maximum current has 60 amperes-conductors in the phase belt. Sketch the mmf wave of the phase when the slots/pole/phase are 1, 2, 3, 4 and 5 respectively. Comment upon the change in shape of mmf wave with the number of slots/pole/phase. LO 1 ○●●

5.13 A 2-pole, 3-phase ac winding is housed in 18 slots, each slot having 12 conductors. Consider the time instant at which the current in phase *a* has its maximum value of 10 A. LO 1 ●●●
(a) Sketch all the 18 slots on a horizontal axis. Mark the direction of currents in the conductors occupying the slots relevant to phase *a*. Make a proportional sketch of the mmf wave of phase *a* only.
(b) Mark the maximum value of the mmf wave on the sketch.
(c) Calculate the peak value of the fundamental of the mmf wave.

5.14 A 4-pole, 50-Hz induction motor has 24 stator slots with 2-layer winding. It has 16-turn coils chorded (short-pitched) by one slot. The machine is delta-connected and connected to a 440 V, 3-phase supply. If the stator resistance and leakage reactance are assumed negligible, find the flux/pole of the rotating flux density wave. LO 1 ○●●

5.15 The induction machine of Prob. 5.14 has a stator length of 28 cm and a mean air-gap diameter of 18 cm. The machine air-gap is 1 mm. What line current will it draw when running at no-load? Assume the iron to be infinitely permeable. LO 1 ○●●

(*Hint*: At no-load the machine draws only the magnetizing current to establish the flux/pole as calculated in Prob. 5.14.)

5.16 In Ex. 5.6 what will be the peak value of resultant mmf/pole, if the winding is 3-phase and is chorded by one slot. LO 1 ○●●

5.17 A 3-phase induction motor runs at a speed of 1485 rpm at no-load and at 1350 rpm at full-load when supplied from a 50-Hz, 3-phase line. LO 1 ○●●
(a) How many poles does the motor have?
(b) What is the % slip at no-load and at full-load?
(c) What is the frequency of rotor voltages at no-load and at full-load?
(d) What is the speed at both no-load and full-load of: (i) the rotor field with respect to rotor conductors, (ii) the rotor field with respect to the stator, and (iii) the rotor field with respect to the stator field.

5.18 A 4-pole, 3-phase synchronous motor fed from 50-Hz mains is mechanically coupled to a 24-pole, 3-phase synchronous generator. At what speed will the set rotate? What is the frequency of the emf induced in the generator? LO 1 ○●●

5.19 A 20-pole synchronous generator running at 300 rpm feeds a 6-pole induction motor which is loaded to run at a slip of 5%. Find the speed at which the induction motor runs and the frequency of the currents induced in its rotor. LO 1 ○●●

5.20 A slip-ring induction motor runs at 285 rpm on full-load when connected to 50-Hz supply. Calculate: (a) the number of poles; (b) the slip; and (c) the slip for full-load torque if the total resistance of the rotor circuit is doubled. Assume the rotor leakage reactance to be negligible in the range of slips being considered. LO 2 ○○●

5.21 Consider the synchronous machine with dimensional and other data. The machine is excited with same field current $I_f = 20$ A. LO 2 ○●●

Calculate the air gap (resultant) flux density needed to generate a terminal voltage of 3 kV (line). What is the corresponding maximum torque it can develop as a motor? What would be the mechanical load for a torque angle of 25°?

5.22 Consider a synchronous machine which is used in motor mode. Its mechanical dimensions and winding particulars are: LO 2

Mechanical dimension

Air-gap length = 1.3 mm
Mean air-gap diameter = 22 cm
Axial length = 41 cm

Rotor winding

Total field series turns = 880
K_w (rotor) = 0.965

Stator winding

SSP = 3
Conductors/slot = 12

Operating conditions

Field current, I_f = 4 A
Peak air-gap flux density, B_r = 1.35 T

(a) Find peak rotor ampere-turns F_2
(b) Find open-circuit voltage.

5.23. A three-phase ac linear motor has armature winding wavelength of 30 cm. The supply currents have a frequency of 75 Hz. LO 2

(a) Calculate the linear velocity of mmf wave.
(b) The vehicle velocity of the motor is synchronous type.
(c) The vehicle velocity of the motor is induction type with a slip of 0.05.

5.24 The armature of three-phase linear motor has a winding wavelength of 25 cm and winding length of six wavelength. The 3-phase winding has 240 turns/phase with a winding factor of 0.92. For an air-gap 0.95 cm it is desired to have peak fundamental air-gap flux density of 1.25 T. Calculate the rms value of armature currents needed. LO 2

5.25 The dimensions, stator and rotor windings of a synchronous motor are: LO 2

Phases = 3, Frequency = f, No. of poles = P

Rotor length (m) = l_1
Axial length (m) = l
Air-gap length (mm) = l_g
Rotor winding turns (series) = N_1
Rotor winding factor = K_{w1}
Field current (A) = I_f
Peak air-gap flux density = B_r
Peak rotor mmf = F_2
Stator winding turns (series) = N_2
Stator short pitching angle = θ_{sp}
Peak stator mmf = F_1
Torque angle = δ

Write METLAB script to calculate

(i) Open-circuit voltage, V_{oc}
(ii) Developed torque T_{dev}
(iii) Developed power P_{dev}

5.26 For the synchronous motor of Problem 5.22, determine LO 2

(a) developed torque T_{dev} as a function of δ
(b) power developed P_{dev} as a function of δ
(c) maximum T_{dev} and P_{dev}.

5.27 The outside diameter of the rotor of an alternation is 0.74 and the axial length is 1.52 m. The machine has four poles and the flux density at the rotor surface is given by 1.12 cos θ_e where θ_e = elect. angle. LO 2

(a) Find the flux/pole (b) If the peak value of $\mathcal{F}_r$, is 18000 AT, calculate the permeance/pole.

5.28 A synchronous generator of 50 Hz with 6 poles has a flux/pole of 0.15 Wb. Each stator coil has two turns and a coil pitch of 150° elect. Calculate the coil voltage (rms). LO 2

5.29 Calculate the short-pitching angle to eliminate the fifth harmonic in the induced emf of a synchronous generator. What is the corresponding reduction in the fundamental and the thirteenth harmonic? LO 2

5.30 A 50 Hz, 8-pole, pole 3-phase synchronous generator has 48 slots. Calculate the % reduction in the fundamental, third and fifth harmonic strengths on account of distributed windings. LO 2

5.31 A synchronous generator has 12 poles and 3-phase winding placed in 144 slots; the coil span is 10 slots. Determine the distribution factor, pitch factor and winding factor. LO 2

5.32 The phase voltage of a 50 Hz synchronous generator is 3.3 kV at a field current of 10 A. Determine the open-circuit voltage at 50 Hz with a field current of 8 A. Neglect saturation. LO 2

5.33 A 50 Hz, 3-phase hydroelectric generator has a rated speed of 100 rpm. There are 540 stator slots with two conductors per slot. The air-gap dimensions are: D = 6.25 m, L = 1.16 m. The maximum flux density B_m = 1.2 T. Calculate the generated voltage/phase. LO 2

5.34 Calculate the voltage induced in the armature of a 4-pole lap-wound dc machine having 728 conductors and running at 1600 rpm. The flux/pole is 32 mWb. LO 2

If this armature carries a current of 100 A, what is the electromagnetic power and torque developed?

5.35 A 240 V dc motor takes 25 A when running at 945 rpm. The armature resistance is 0.24 Ω. Determine the no-load speed assuming negligible losses. Flux/pole is constant. LO 2

5.36 A 4-pole dc motor has a lap-connected armature with 60 slots and 8 conductors/ slot. The armature has an applied voltage of 240 V. It draws a current of 50 A when running at 960 rpm. The resistance of the armature is 0.1 Ω. Find the flux/pole that would be necessary for this operation. LO 2

5.37 In a given machine $\mathcal{F}_2$ (rotor mmf) 850 AT and $\mathcal{F}_1$ (stator mmf) 400 AT, α (included angle) = 123.6° and $\mathcal{P}$(permeance/pole) 1.408 × 10^{-4} Wb/AT. Find the value of the resultant air-gap flux/pole. LO 2

5.38 A P-pole machine has a sinusoidal field distribution as shown in Fig. P5.38. The armature carries a uniform current sheet of value J A/m causing a triangular mmf distribution as shown in the figure. LO 2

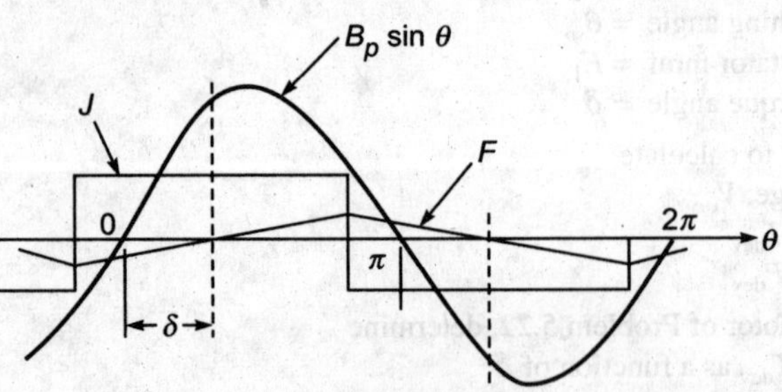

Fig. P5.38

The machine has an axial length of l and a mean air-gap diameter of D.
(a) Find the peak value of the armature mmf.
(b) Derive an expression for the electromagnetic torque developed.

5.39 A 3-phase, 50 Hz, 4-pole, 400 V wound rotor induction motor has a stator winding Δ-connected and a rotor winding Y-connected. Assume effective turn ratio stator/rotor = 2/1 (phase basis). For a rotor speed of 1440 rpm, calculate: LO 2
(a) the slip
(b) the standstill rotor induced emf/phase
(c) the rotor induced emf/phase at this speed
(d) the rotor frequency in (b) and (c)

5.40 A 50 Hz induction motor runs at 576 rpm at full load. Determine: LO 2
(a) the synchronous speed and the number of poles
(b) the frequency of rotor currents
(c) the rotor speed relative to the revolving field

5.41 A 3-phase induction motor runs at a speed of 940 rpm at full-load when supplied with power at 50 Hz, 3-phase. LO 2
(a) How many poles does the motor have?
(b) What is its slip at full-load?
(c) What is the corresponding speed of:
(i) the rotor field wrt the rotor surface
(ii) the rotor field wrt the stator
(iii) what is the rotor speed at twice full-load slip?

Answers

5.1 0.966, 0.966, 0.75%
5.2 617.78 V, 534.86 V
5.3 0.0133, 0.0013, 0.0002 Wb
5.5 (a) 27412 V (b) 13720 V
5.6 150
5.7 1787 V
5.8 0.033 Wb
5.9 240 V, $33\frac{1}{3}$ Hz
5.10 $A/2$
5.13 (b) 180 AT (c) 230 AT
5.14 0.0166 Wb
5.15 11.26 A
5.16 10.368 AT
5.17 (a) 4 (b) 1%, 10% (c) 0.5, 5 Hz
(d) (i) 15, 150 rpm (ii) 1500 rpm (iii) 0
5.18 1500 rpm, 300 Hz
5.19 950 rpm, 2.5 Hz
5.20 (a) 20 (b) 5% (c) 10%
5.27 (a) 0.63 Wb (b) 0.035 m Wb/AT
5.28 57.7 V
5.29 5%, 51.2%
5.30 9%, 66.7%, 75.5%

5.31 0.866, 0.958, 0.83
5.32 3.168 kV
5.33 11.125 kV
5.34 62.12 kW, 370.75 N m
5.35 969 rpm
5.36 31.25 m Wb
5.37 0.1 Wb
5.38 (a) πJD/2P (b) 2 k sin δ, K = JBD^{2}l/2P.
5.39 (a) 4% (b) 115.5 V
(c) 4.62 V (d) 50 Hz, 2 Hz.
5.40 (a) 10, 4 (b) 2 Hz (c) 24 rpm.
5.41 (a) 6 (b) 6%
(c) 60 rpm, 1000 rpm, 880 rpm.

MULTIPLE CHOICE QUESTIONS

1. A full-pitched coil in a 6-pole machine has a mechanical angle span of **LO 1**
(a) 30° (b) 60°
(c) 90° (d) 180°

2. To eliminate the fifth harmonic a short pitched coil should have a short-pitching angle of **LO 1**
(a) 36° (b) 18°
(c) 15° (d) 12°

3. Armature winding of a synchronous generator can be connected (i) single-phase and (ii) 3-phase. Compare the kVA rating of the generator in the two cases: **LO 1**
(a) both will have the same kVA
(b) kVA (single-phase) > kVA (3-phase)
(c) kVA (3-phase) > kVA (single-phase)
(d) armature winding cannot be connected in both the ways stated

4. Phase relationship between mmf phasor $\bar{F}$ and emf phasor $\bar{E}$ in a synchronous machine is **LO 2**
(a) $\bar{F}$ leads $\bar{E}$ by 90°
(b) $\bar{F}$ lags $\bar{E}$ by 90°
(c) $\bar{F}$ and $\bar{E}$ are in phase
(d) this angle depends upon the pf of the load

5. A full-pitched coil of *Ni* ampere-turns placed in stator slots causes a fundamental mmf wave of peak amplitude: **LO 1**
(a) $\frac{4}{\pi}(Ni)$ (b) $\frac{\pi}{4}(Ni)$
(c) $\frac{4}{\pi}\left(\frac{Ni}{2}\right)$ (d) $\frac{\pi}{4}\left(\frac{Ni}{2}\right)$

6. In a dc machine the angle between the stator and rotor fields is **LO 1**
(a) dependent upon the load (b) 45°
(c) 90° (d) 180°

7. A 4-pole 50 Hz induction motor runs at a speed of 950 rpm. The frequency of rotor currents is **LO 1**
   (a) 47.5 Hz (b) 50 Hz
   (c) 5 Hz (d) 2.5 Hz

8. If the rotor of an induction motor is assumed to be purely resistive, the angle between the resultant flux density wave and rotor mmf wave is **LO 1**
   (a) dependent upon the load (b) 180°
   (c) 90° (d) 45°

9. In a non-salient pole synchronous machine the distribution of field mmf around the air-gap is a **LO 2**
   (a) sinusoidal wave (b) rectangular wave
   (c) stepped triangular wave (d) flat topped stepped wave

10. For a cyclic load variation of a motor the rating is determined by **LO 1**
    (a) average load (b) the peak load
    (c) the rms load (d) 3/4th of the peak load

11. Practical synchronous motors are always of the _____ kind. **LO 1**
    (a) 3-phase (b) 2-phase
    (c) 6-phase (d) 1-phase

12. In ac machines, it is desirable for the induced emfs to be _____in waveform. **LO 1**
    (a) sinusoidal (b) cosine
    (c) triangular (d) rectangular

13. The _____ in a dc machine should always be a rotor. **LO 1**
    (a) armature (b) field winding
    (c) commutator (d) pole

14. Which of the following is a self-starting device? **LO 2**
    (a) Inductor motor (b) Synchronous motor
    (c) dc motor (d) Linear motor

15. In rotating machines, _____ is the flux which follows the path from tooth to tooth across the slots. **LO 1**
    (a) slot leakage (b) tooth-tip leakage
    (c) over-hang leakage (d) zig-zag leakage

Answers

1. (b)	2. (a)	3. (b)	4. (a)	5. (c)	6. (c)	7. (c)	8. (c)	9. (d)	10. (c)
11. (a)	12. (a)	13. (a)	14. (a)	15. (a)					

SAMPLE EXAMPLES SOLVED USING MATLAB

```
%This program shows the development of rotating magnetic field
% by 3-Phase windings
% enter phase sequence as abc or aca
% enter yes for is the supply balanced
x=input('Please enter phase sequence\n','s')
test='abc';
```

```
test1='yes';
y=input('Is the supply balanced\n','s')
m=strcmp(x,test);
n=strcmp(y,test1);
bmax=1;
freq=60;
t=linspaee(0,5/60,100);
th=linspaee(0,2*pi ,100);
w=2*pi*freq;
Baa=sin(w*t).*(cos(0)+j*sin(0));
if m==1 & n==1
Bbb=sin(w*t-2*pi/3).*(cos(2*pi/3)+j*sin(2*pi/3));
Bce=sin(w*t+2*pi/3)  .*(cos(-2*pi/3)+j*sin(-2*pi/3));
end
if m==0 & n==1
Bbb=sin(w*t+Z*pi/3).*(cos(2*pi/3)+j*sin(2*pi/3));
Bcc=sin(w*t-2*pi/3).*(cos(-2*pi/3)+j*sin(-2*pi/3));
end
if m==1 & n==0

Bbb=sin(w*t-pi/3).*(cos(2*pi/3)+j*sin(2*pi/3));
Bcc=sin(w*t+2*pi/3).*(cos(-2*pi/3)+j*sin(-2*pi/3));
end
if m==0 & n==0
Bbb=sin(w*t+pi/3).*(cos(2*pi/3)+j*sin(Z*pi/3));
Bcc=sin(w*t-2*pi/3).*(cos(-2*pi/3)+j*sin(-2*pi/3));
end
Bnet=(Baa+Bbb+Bcc);
Bnet1=Bnet/1.5;
alpha=linspace(0,2*pi,100);
circle=1.5*exp(j*alpha);
circlee=2*exp(j*alpha);
rcircle=0.85*exp(j*alpha);

for    ii=1:length(t)
       circl=[circle circlee];
     fill(real(circl),imag(circl),'c');
       hold on;
       plot (circl,'c');
       fill(real(rcircle),imag(rcircle),'c')
```

```
        th=angle(Bnet(ii));
        beta=linspace(th-80*pi/180,th+80*pi/180,60);
        cir=cos(beta)+j*sin(beta);
        beta=linspace(th+20*pi/180,th-20*pi/180,60);
        cirr=1.5*(cos(beta)+j*sin(beta));
          hold on;
        th1=angle(-Bnet(ii));
        beta1=linspace(th1-80*pi/180,th1+80*pi/180,60);
        cir1=cos(beta1)+j*sin(beta1);
        beta1=linspace(th1+20*pi/180,th1-20*pi/180,60);
        cirr1=1.5*(cos(beta1)+j*sin(beta1));
        zeta=180*angle(Bnet1(ii))/pi-90;
        zeta1=180*angle(Bnet1 (ii))/pi;
        cir2=[cir cirr];
        cir21=[cir1 cirr1];
        fill(real(cir2),imag(cir2),'m');
        fill(real(cir21),imag(cir21),'b');
        text(real(Bnet1(ii)),imag(Bnet1(ii),'\bfNorth-Pole','Rotation',zeta,
'color','k' ...
        ,'FontSize',10,'HorizontalAlignment','center','Vertical','bottom');
text(real(-Bnet1(ii)),imag(-Bnet1(ii)),'\bfSouth-Pole','Rotation',zeta,
'color', 'k' ...
        ,'FontSize',10,'HorizontalAlignment','center','Vertical','top');
          if m==1
        title('Phase sequence is abc & field rotates anticlockwise');
end
if m~=1
        title('Phase sequence is abc & field rotates clockwise');
end
        axis square;
        drawnow;
        hold off;
end

%%%%%%%%%%%%%%%%%%%%%%%%%%%%%%%%%%%%%%%%%%%%%%%%%%%%%%%%%%%%%%%%%%%%%%%%%%
```

CHAPTER 6

ac Armature Windings

LEARNING OBJECTIVES

After studying this chapter, students will be able to

LO1 Define armature windings and differentiate between the two types of windings

LO2 Interpret the concept of phase grouping in an ac winding

LO3 Explain the different types of windings

LO4 Discuss on tooth ripple and deduce the related expressions

6.1 | INTRODUCTION

Chapters 4 and 5 emphasized that field and armature windings are the essential features of electric machines. In this chapter we will understand the applications of different types of armature windings that are used in electrical machines. Single-layered windings are used in high voltage machines. However, double-layered windings have a lot of advantages when compared to the single-layered windings; some of these advantages are given below:

- easy to manufacture
- low cost manufacturing
- fractional slot windings can be Employed
- fractional pitch coil can be used with a double-layered winding

The double layer windings are also the most widely used class of windings. In the following sections, we will learn what are armature windings and their different types.

6.2 | ARMATURE WINDINGS

> **LO 1 Define armature windings and differentiate between the two types of windings**

The field windings are simple arrangements with concentrated coils (i.e. coils in which all the turns have the same magnetic axis). Armature windings on the other hand comprise a set of coils (single or multiturn) embedded in the slots, uniformly spaced round the armature periphery. The emfs are induced in armature coils due to relative motion between them and the *B*-wave in the air-gap established by the field windings. In an ac machine (3-phase) the armature coils are symmetrically connected to form a set of three balanced phases (equal emf magnitudes with a relative phase displacement of $2\pi/3$ rad). In a dc machine the armature coils are connected via commutator segments which are tapped by stationary brushes so as to give a constant (dc) voltage between brushes. It was also seen in Chapter 5 that when the armature winding carries current, it establishes the same number of alternating (north-south) poles for which it is wound.

> **Probe**
> How are the armature coils connected in ac and dc machines?

A coil may be of single-turn having two conductors with end connections or multiturn with two coil-sides each composed of several conductors. The active coilside (or conductor) length in which the emf is induced equals the armature length (over which the flux density is established). The *pitch of a coil* is the space angle (electrical) between its two sides and must equal an integral number of slots. The coil pitch may be full (equal to one pole pitch or 180° elect.) or short-pitch (*chorded*) coils may be used. The pitch of a coil could be expressed in terms of its angular span or in terms of slots. The slots/pole must be an integral number for a full-pitch coil.

Types of Armature Windings

Practically there are two types of windings, viz. single-layer and two-layer (or double-layer). In a single-layer winding each coil—side of a coil occupies the whole slot as shown in Fig. 6.1(a). In a double-layer winding one coil-side of a coil occupies the upper position in one slot and the second coil-side occupies the lower position in a slot displaced from the first coil-side by the coil-span as shown in Fig. 6.1(b). In a double-layer winding each slot is occupied by two coil-sides, one placed on top of the other, referred to as top and bottom coil-sides. It easily follows from Fig. 6.1 that

$$U = 2 \text{ coil-sides/slot} \tag{6.1a}$$

$$C = S/2 \text{ (single-layer winding)} \tag{6.1b}$$

$$C = S \text{ (double-layer winding)} \tag{6.1c}$$

where C = number of armature coils

S = number of armature slots

The primary difference in single-and double-layer windings is in the arrangement of the overhang. In single-layer the coils are arranged in groups and the overhang of one group of coils is made to cross the other appropriately by adjusting the size (corresponding to axial length of armature) and shape of individual coil groups. This means a variety of coils differing both in size and shape resulting in inconvenience and higher cost in production. Single-layer windings are, therefore, rarely used in modern machine practice except in small sizes. Machines are of course still found in use with single-layer windings.

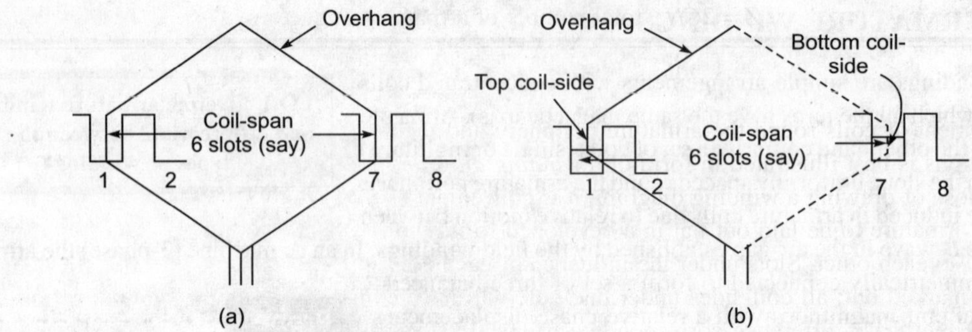

Fig. 6.1 *(a) Single-layer coil, (b) Double-layer coil (bottom coilside is shown dotted)*

In a double-layer winding all the coils are identical in shape and size (diamond shape as shown in Fig. 5.21 is employed) with two coil-sides lying in two different planes. Each slot has one coil-side entering its bottom half from one side and the other coil-side leaving its top half on the other side. A special kink at each end of the diamond shape allows neat symmetrical packing of coil overhangs and the problem of overhang crossing as in a single-layer winding is avoided. Because of identical coils, production is facilitated and results in a reduction of cost.

dc machine windings are invariably double-layered.

Probe

What type of windings is found in a dc machine?

Nomenclature

S = number of slots (must be divisible by 3 in a 3-phase ac machine)

C = number of coils

$2C$ = number of coil-sides

N_c = number of turns/coil

S/P = slots/pole, i.e. pole pitch in terms of slots (is noninteger for fractional slot winding)

$\gamma = \dfrac{180°P}{S}$ slot angle (electrical) or slot-pitch (angular displacement between midpoints of adjacent slots)

q = number of phases (generally $q = 3$)

$\beta = 2\pi/q$, time phase displacement between emfs of successive phases (generally $\beta = 2\pi/3$)

m = number of slots/pole/phase (SPP)

A = number of parallel paths

In ac machines it is possible to have only one path or more paths in parallel-phase (each phase is an open-circuit); but dc machine windings are always of closed-circuit type with two or more (even) number of parallel paths.

In large dc machines windings may be arranged with more than one coil-side in top and bottom halves of the slots. Nomenclature specific to dc machine winding is:

$$U = 2C/S, \text{ number of coil-sides/slot (even)}$$

$$Z_s = UN_c, \text{ number of conductors/slot}$$

$Z = 2CN_c$, total number of armature conductors

Remark

The arrangement of coils round the armature periphery and their interconnections is best illustrated in form of a *winding diagram.* For the purpose of drawing a winding diagram, it is convenient to imagine the armature to be laid out flat in a developed form with slots parallel to each other. Slots under the influence of each pole can then be marked out; all coil-sides under one pole will have emfs induced in the same direction with a progressive time phase difference corresponding to the slot angle, Cross-sectional developed view is also handy in illustrating the underlying ideas of ac windings. In a developed form of dc winding, the field poles are also indicated.

Probe

What is the major use of a winding diagram?

Concept Check LO 1

1. What is the significance of a winding diagram? ○○●
2. Why double layer winding is preferred? ○●●

6.3 ac WINDINGS

LO 2 Interpret the concept of phase grouping in an ac winding

ac windings are generally of a 3-phase kind because of the inherent advantages of a 3-phase machine. The armature coils must be connected to yield balanced (equal in magnitude and successive phase difference of $2\pi/3$ rad) 3-phase emfs. To begin with the slots around the armature periphery must be divided into phase-bands.

Phase Grouping

Initially a simple case will be assumed where SPP is an integral number; such winding is referred to as *integral-slot* winding. For illustrative purposes, let $m = 2$ which means 12 slots per pole pair for a 3-phase armature. Slot angle is $360°/12 = 30°$. Further let the coil-pitch be full six slots. Figure 6.2(a) shows the 12 slots numbered from left to right; six slots are under the influence of one pole with a particular direction of emfs in coil-sides and the remaining six slots are under opposite pole with opposite direction of emfs as indicated.

Probe

What is an integral-slot?

In Fig. 6.2(a) the 12 armature slots are divided into six phase-bands of two (= m) slots each having an angular spread of $\sigma = 60°$; in fact each pole is divided into three band (as many as the number of phases). If coil-sides in slots (1, 2) belong to phase band A, those in slots (5, 6) which are 120° (or four slots) away belong to phase band B and those in slots (3, 4) are 60° (or two slots) away from A which when reverse-connected would belong to phase C. (See phasor diagram of Fig. 6.2(b).) Therefore coil-sides in slots-(3, 4) are said to belong

Note: Difficulty Level ⟶ ○○● — Level 1 and Level 2 Category; ○●● — Level 3 and Level 4 Category; ●●● — Level 5 and Level 6 Category

to phase band *C′*. As a result of this arrangement the phase-band sequence is *AC′ BA′ CB′* which will repeat for each pair of poles. This arrangement of phase-band is called 60° phase grouping. The four coil-sides of each pair of coils of a phase can be connected additively in any order. For example, the order of the coil-side connection for phase *A* could be (1–8–2–7) as used in a single-layer winding with concentrated coils or it could be (1–7–2–8) in a two-layer lap winding (these are explained soon after). The phasor diagrams of all the three phases for the former kind of connection is given in Fig. 6.2(c).

The 60° phase grouping discussed above can be used for single-layer or double-layer windings. It is also possible to use a 120° phase grouping where the slots under a pole pair are divided into three phase-bands as in Fig. 6.2(d). For the example in hand there are four slots per phase-band. It is obvious that slots for return coil-side for this phase-grouping will not be available in single-layer winding. It can only be used for a double-layer winding. A phase-grouping of 120° is rarely adopted and will not be discussed any further.

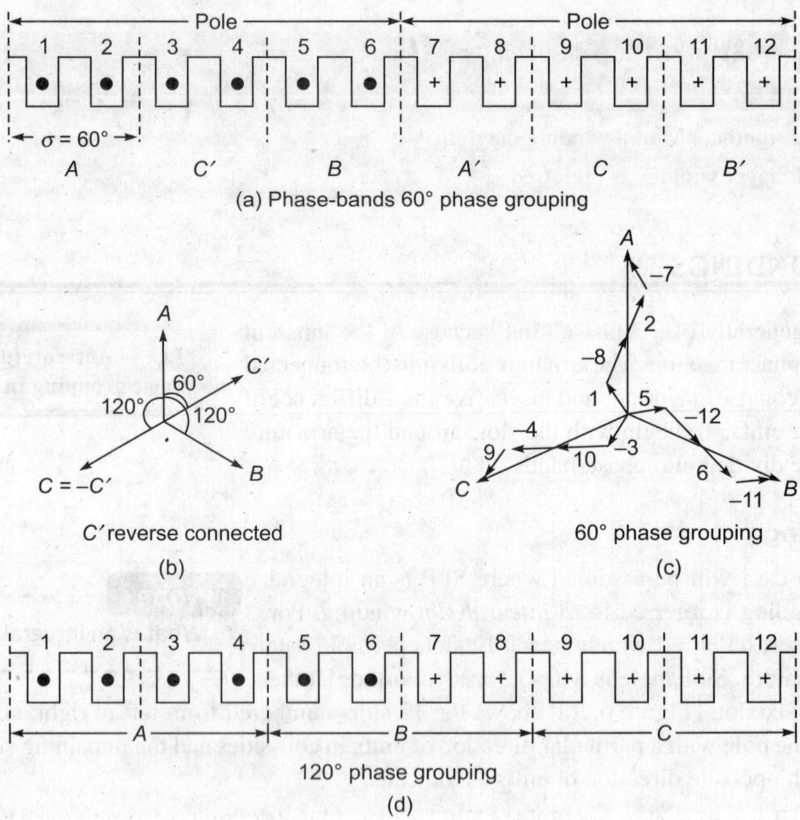

(a) Phase-bands 60° phase grouping

C′ reverse connected

(b)

60° phase grouping

(c)

120° phase grouping

(d)

Fig. 6.2

6.4 TYPES OF ARMATURE WINDINGS

Single-layer Windings

LO 3 Explain the different types of windings

Single-layer windings are not commonly used in practice except for machines of a few kW because of the disadvantages mentioned earlier. Single-layer winding may be *concentric, lap* or *wave* type. Here only the concentric type winding will be illustrated while the lap type will be explained in two-layer winding. Wave winding because of certain problems in end connections is not used in ac machines.

Concentric Windings

Concentric windings may be classified into two main categories, viz. *unbifurcated* (or half-coiled) and *bifurcated* (whole-coiled). In the former type the coils comprising a phase group in adjacent pole pitches are concentric as indicated in Fig. 6.3(a). The individual coils may have a span greater or less than a pole pitch but the average coil-span equals one pole-pitch. This kind of arrangement is provided to avoid crossing of two coils under one phase-group. In bifurcated winding, each coil group is split into two sets of concentric coils and the return coil-sides are shared with those of another group as shown in Fig. 6.3(b). It is clearly evident from the figure that this kind of arrangement is only possible when SPP is even.

It is easily seen from Figs 6.3(a) and (b) that for accommodating the windings for all the three phases, the overhang must be arranged in two or three planes. Figure 6.4(a) which corresponds to unbifurcated winding (Fig. 6.3(a)), the overhang is arranged in a continuous chain with sequence (if seen from a fixed reference) $A \uparrow B \downarrow C \uparrow A \downarrow B \ldots$, where upward and downward arrows indicate the upper and lower planes. The three plane overhang arrangement of a bifurcated winding is depicted in Fig. 6.4(b).

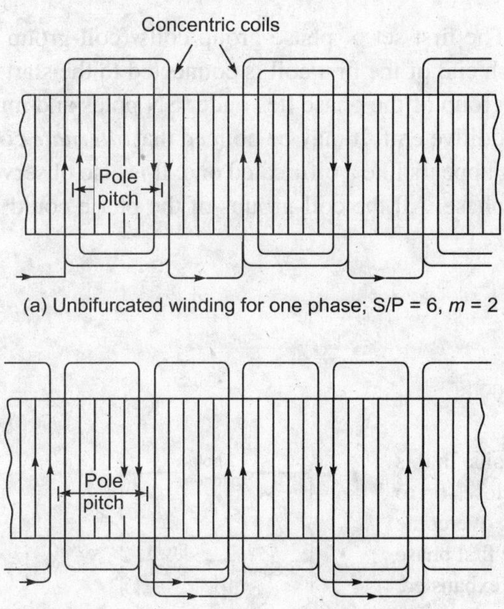

(a) Unbifurcated winding for one phase; S/P = 6, m = 2

(b) Bifurcated winding for one phase; S/P = 6, m = 2

Fig. 6.3 *Single-layer winding with concentric coils*

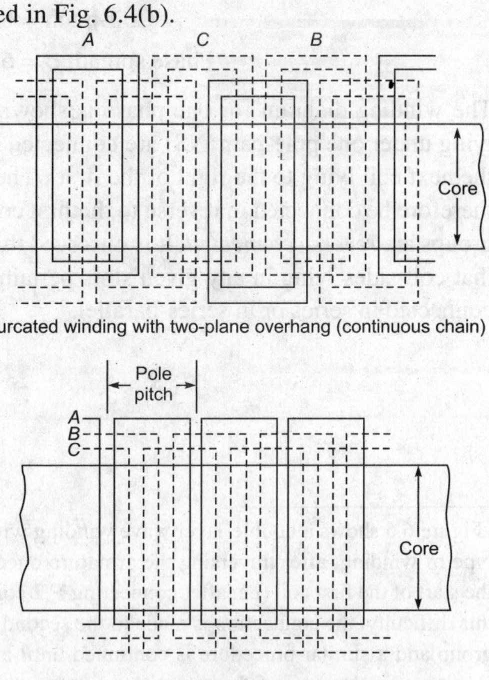

(a) Unbifurcated winding with two-plane overhang (continuous chain)

(b) Bifurcated winding with three-plane overhang

Fig. 6.4 *Arrangement of overhang in single-layer concentric winding*

Single-layer coils can be arranged in semi-closed slots (the coil is opened and pushed in slots from one side, the coil then being reformed and reconnected by buff-welding).

Chording and the use of fractional SPP is not possible in a single-layer winding. As will be seen in the next section, it is a serious drawback.

Double-layer Windings

Double-layer windings are the most widely used class of windings. Though both lap and wave types are possible, because of inherent problems of a wave winding*, it is now an accepted practice to use the lap type for double-layer ac winding. Double-layer windings fall into two main classes depending upon the value of *SPP-integral slot winding* when SPP is an integer and *fractional slot winding* when SPP is a fractional number. To meet the requirement of symmetry among phases, the number of slots/phase (*S*/3) must be a whole number.

Probe
What is the most widely used class of windings?

Probe
Depending on what quantity, is a double-layer winding classified?

Integral Slot Winding

Here SPP is an integer. This type of winding has already been illustrated in Fig. 5.22. The winding arrangement is further illustrated through an example. Let

$$m = 2 \text{ slots} \quad \text{and} \quad S/P = 6 \text{ slots}$$

$$\text{Coil-pitch} = 6 \text{ slots (full-pitch coils)}$$

$$\text{Phase spread, } \sigma = 60° \text{ elect.}$$

The winding diagram for one phase is shown in Fig. 6.6. The first set of phase-group coils (coil-group 1) lying under one pole-pair (NS) are connected in series (finish end of the first coil is connected to the start of the next coil lying to the right of the first). The second coil-group of the phase lies under SN poles and must therefore be connected in reverse to the first coil-group for additive emf. It may be noticed that *alternate* coil-groups are *reverse connected.* It is observed that the winding appears like a bifurcated one. It is also observed that coil-sides lying in any given slots pertain to the same phase. All the coil-groups of the phase could be connected in series or in series-parallel.

*Figure 6.5 shows a double-layer wave winding with single-turn coils. In this type of winding, after traversing the armature once the winding closes on to the start of the first coil (i.e. after connecting *P*/2 coils in series). To overcome this difficulty, the connection is made to the second coil-side of the first phase group and a similar procedure is continued until all the coils are exhausted. However, in the case of fractional slot windings this problem is even more complicated because, after all the turns around the armature are completed, some coils remain unconnected.

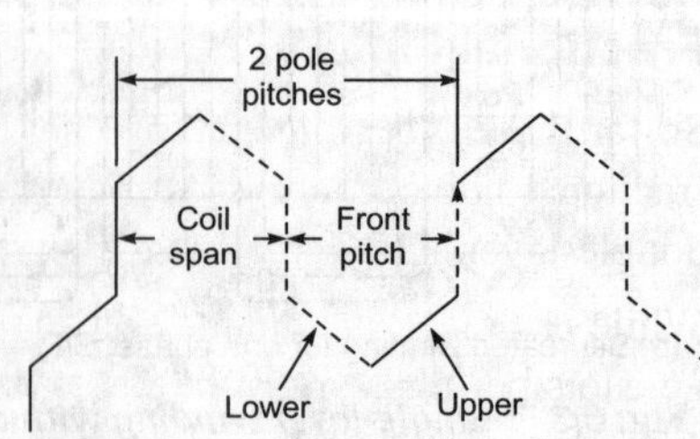

Fig. 6.5

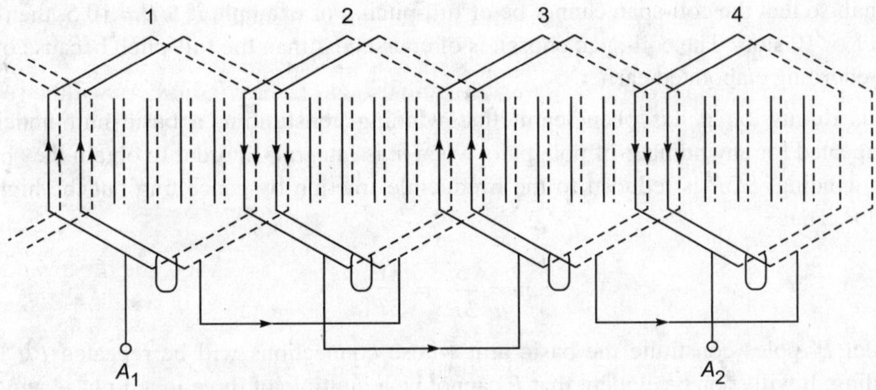

Fig. 6.6 *Double-layer lap winding; 4 poles, m = 2, σ = 60° coil-pitch = 6 slots (full-pitch)*

In practice, however, it is common to use chorded or short-pitched coil. As already mentioned in Ch. 5, this type of arrangement offers certain inherent advantages such as reduction in copper needed for end connections. Further, certain harmonics present in the emf wave are greatly suppressed. However, coil chording lead to a reduction in the emf generated (refer Eqs (5.18) and (5.20). This type of winding is best illustrated by means of a cross-sectional developed diagram. Here letters (a, b, c) refer to the coil-sides of the corresponding phase. For the example considered,

$$S/P = 9 \text{ slots} \quad \text{and} \quad m = 3$$

$$\text{Coil-span} = 8 \text{ slots (chorded by one slot)}$$

$$= 60° \text{ electrical}$$

The cross-sectional view of winding for one pole-pair is drawn in Fig. 6.7. Certain observations can be made from this figure. The top bottom coil-side phase grouping is merely displaced by one slot (equal to chording). Further, in each-group of three slots, the coil-side of two different phases are placed in one of the slots.

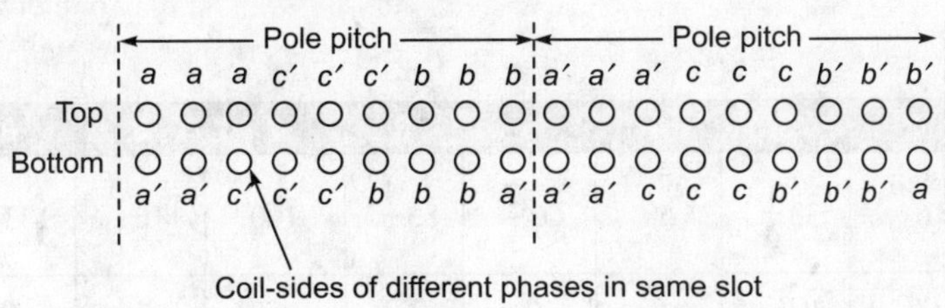

Fig. 6.7 *Cross-sectional view of two-layer winding: m = 3; σ = 60°, coil-pitch = 8 slots (chorded by one slot)*

Fractional Slot Windings

So far these types of windings have been studied in which SPP is an integer. Windings, wherein SPP is a fractional number, are known as fractional slot windings. Fractional slot winding is easily adopted with a double-layer arrangement.

While $m = S/3P$ is a noninteger in a fractional slot winding, S, the number of slots, must be divisible by 3, i.e. slots per phase must be integral in order to obtain a symmetrical 3-phase winding. The pole pitch, S/P, is also fractional, so that the coil-span cannot be of full-pitch. For example if $S/P = 10.5$, then the coil-span can be either 11 or 10 slots. The coil-span chosen is of course less than the full-pitch because of the inherent advantages of chording elaborated earlier.

In the previous discussion it has been learnt that winding constituting a basic unit under a pole-pair (N and S) is repeated for any number of pole pairs when m is integral. In order to obtain the "basic unit" for fractional slot winding, $S/3P$ is reduced to the irreducible fraction by cancelling out the highest common factor in S and P. Thus

$$m = \frac{S}{3P} = \frac{S'}{3P'} \tag{6.2}$$

Coil-sides under P' poles constitute the basic unit whose connections will be repeated P/P' times in the complete winding. It will soon be evident that P cannot be a multiple of three in a 3-phase winding.

In a double-layer winding, the phase-grouping of coil-sides for the top layer is repeated in the bottom layer with corresponding coil-sides being located one coil-span away. Therefore, all that is needed is to establish the phase-grouping of top layer of coil-sides.

Fractional slot winding is easily understood by means of an example Let

$$S = 108, \qquad P = 10$$

then
$$m = \frac{S}{3P} = \frac{108}{3 \times 10} = 3\frac{3}{5} \text{ slots}$$

Now
$$\frac{S}{3P} = \frac{108}{3 \times 10} = \frac{54}{3 \times 5} = \frac{S'}{3P'}$$

Thus the basic unit in the winding has 5 poles (note that this need not be even) covering 54 slots.

Coil Group Table

A coil-group table (Table 6.1) is now prepared on the following lines.

Table 6.1

Slot No.	1	2	3	4	5	6	7	8	9	10	11
Angle	0	$16\frac{2}{3}$	$33\frac{1}{3}$	50	$66\frac{2}{3}$	$83\frac{1}{3}$	100	$116\frac{2}{3}$	$133\frac{1}{3}$	150	$166\frac{2}{3}$
Phase	[a]	a	a	a	c'	c'	c'	c'	b	b	b
← Pole-ptich 1 →											
Slot No.	**12**	**13**	**14**	**15**	**16**	**17**	**18**	**19**	**20**	**21**	**22**
Angle	$3\frac{1}{3}$	20	$36\frac{1}{3}$	$53\frac{1}{3}$	70	$86\frac{2}{3}$	$103\frac{1}{3}$	120	$136\frac{2}{3}$	$153\frac{1}{3}$	170
Phase	a'	a'	a'	a'	c	c	c	[b']	b'	b'	b'
← Pole-Pitch 2 →											

(*Contd.*)

Slot No.	23	24	25	26	27	28	29	30	31	32	33
Angle	$6\frac{2}{3}$	$23\frac{1}{3}$	40	$56\frac{2}{3}$	$73\frac{1}{3}$	90	$106\frac{2}{3}$	$123\frac{1}{3}$	140	$156\frac{2}{3}$	$173\frac{2}{3}$
Phase	a	a	a	a	c'	c'	c'	b	b	b	b
← Pole-pitch 3 →											
Slot No.	**34**	**35**	**36**	**37**	**38**	**39**	**40**	**41**	**42**	**43**	**44**
Angle	10	$26\frac{2}{3}$	$43\frac{1}{3}$	60	$76\frac{2}{3}$	$93\frac{1}{3}$	110	$126\frac{2}{3}$	$143\frac{1}{3}$	160	$176\frac{2}{3}$
Phase	a'	a'	a'	$[c]$	c	c	c	b'	b'	b'	b'
← Pole-ptcih 4 →											
Slot No.	**45**	**46**	**47**	**48**	**49**	**50**	**51**	**52**	**53**	**54**	
Angle	$13\frac{1}{3}$	30	$46\frac{2}{3}$	$63\frac{1}{3}$	80	$96\frac{2}{3}$	$113\frac{1}{3}$	130	$146\frac{2}{3}$	$163\frac{1}{3}$	
Phase	a	a	a	c'	c'	c'	c'	b	b	b	
← Pole-pitch 5 →											

1. Calculate the slot angle,

$$\gamma = \frac{10 \times 180°}{108} = \frac{50}{3} = 16\frac{2°}{3} \quad \text{(Keep in fractional form)}$$

2. Beginning with 0°, calculate the angle for serially arranged slots in a table. Every time the angle exceeds 180°, subtract 180°, the angle of one pole-pitch. This takes care of the fact that the positive direction of emfs under adjacent poles are in opposite direction.
3. Phase-group A is located corresponding to:

$$0 \le \text{angle} < 60°$$

Phase-group C: $60° \le \text{angle} < 120°$

Phase-group B: $120° \le \text{angle} < 180°$

Phase-group ordering is $AC'\ BA'\ CB'$ …

Probe

What is the location of phase-group A?

4. The starting points of phases are located at 0°, 60° and 120°. While 0° is the start, the other two angles are located at $S'/3 = 18$ and $2S'/3 = 36$ slots away.

Probe

At what angles are the starting points of phases located at?

$$\text{Angle } S'/3 \text{ slots away} = \frac{\gamma S'}{3} = \frac{180P'}{S'} \times \frac{S'}{3} = 60\ P'$$

$$\text{Angle } 2S'/3 \text{ slots away} = \frac{2\gamma S'}{3} = \frac{180P'}{S'} \times \frac{2S'}{3} = 120\ P'$$

These angles are 60°, 120° or their multiples for any P' except when P' is multiple of 3 (= number of phase). Thus if P' (or P) is multiple of 3, balanced winding is not possible.

It is immediately seen from the last rows of Table 6.1 that there are three cycles of (4, 4, 3, 4, 3) slot distribution starting at slot no. 1(0°, phase A), slot no. $(1 + S'/3) = 19$ (120°, phase B) and slot no. $(1 + 2S'/3) = 37$ (60°, phase C). So this pattern results in a balanced winding.

Figure 6.8 shows the layout of the winding for the basic unit of five poles indicating the number of coils and their phase-groups. The dashed phase-groups are reverse connected. The start of all the three phases is made at the start of an undashed coil-group. This winding pattern would repeat for every 5-pole unit. It, therefore, represents the winding diagram for 10, 20, … pole armature.

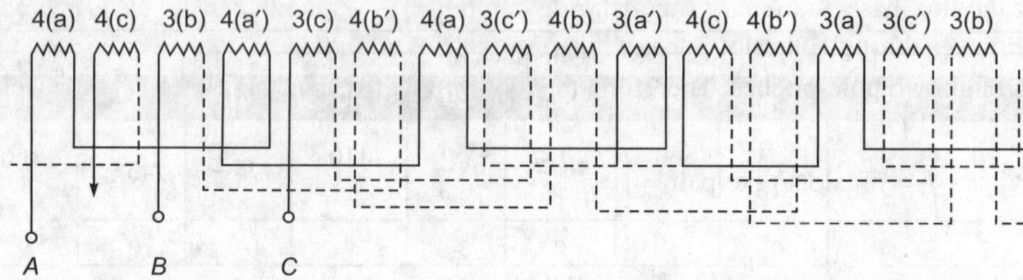

Fig. 6.8 *Layout of basic unit of five poles of fractional slot winding example*

Though the fractional slot winding may appear to be somewhat complicated, it has certain technical advantages and can be easily manufactured. The number of armature slots chosen need not necessarily be an integral multiple of the number of poles. Consequently one may choose a particular number of slots for which the notching gear is available. This results in saving in machine tools. This flexibility can be effectively used where the number of poles (machine speed) varies over a wide range as in the case of synchronous machines.

Furthermore, *the high-frequency harmonics caused by slotting are considerably reduced by the use of fractional slot winding.* As the poles move past a slotted armature, the flux/pole fluctuates (does not move) as the slot-teeth pattern facing a pole repeats. This causes induction (static) of emf harmonics of high frequency. If S/P is integral, the disposition of armature slots relative to pole simultaneously repeats (in time phase) at every pole so that the harmonics in phase-groups of a given phase are in phase. Nonintegral S/P causes the harmonics in phase-group to become out-of-phase thereby reducing their strength in the phase voltage.

The generation of harmonics of slot origin is explained in detail below for the interested reader.

Concept Check — LO 2

1. When do you use concentric winding?
2. What are the advantages of fractional slot winding over integral slot winding?
3. Compare lap and wave winding. Where each type is used and why?

6.5 TOOTH RIPPLE

LO 4 Discuss on tooth ripple and deduce the related expressions

Because of lower reluctance of the magnetic path corresponding to the teeth as compared to the slots, the flux density wave in the air-gap is rippled as shown in Fig. 6.9. Unlike the space harmonics of the B-wave which move at the same peripheral speed as the fundamental, the tufts of flux embracing the teeth move back and forth as the teeth move relatively past the pole-shoes. Since there is no net movement of B-wave ripples, no space harmonic is produced by these. However, due to cyclic variation in reluctance of the magnetic path offered to the field poles by the toothed armature, the total flux/pole contains a space stationary but time-varying component. This effect is most pronounced when the pole-arc covers an integral plus one-half slot-pitches as illustrated in Figs 6.10(a) and (b). The relative position of the armature and poles in Fig. 6.10(a) corresponds to low reluctance and maximum flux/pole and Fig. 6.10(b) corresponds to high reluctance and minimum flux. It is thus seen that the flux has a stationary time-varying component. A complete cycle of flux variation occurs when the pole moves through one slot-pitch, while one cycle of the fundamental wave is generated when the poles move through two pole-pitches. Therefore, the tooth ripple frequency in flux is

$$f_s = f \times \text{slots/pole-pair} = \left(\frac{2S}{P}\right) f$$

Probe

How is the tooth ripple frequency in flux expressed as?

where f = fundamental frequency

f_s = frequency of flux variation due to slotting

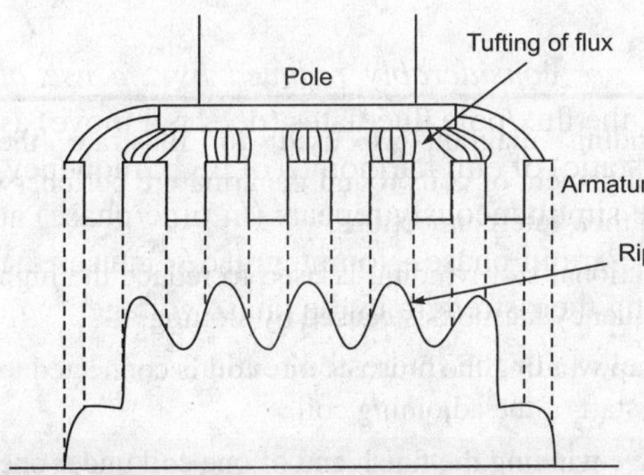

Fig. 6.9 *Tooth ripple in B-wave*

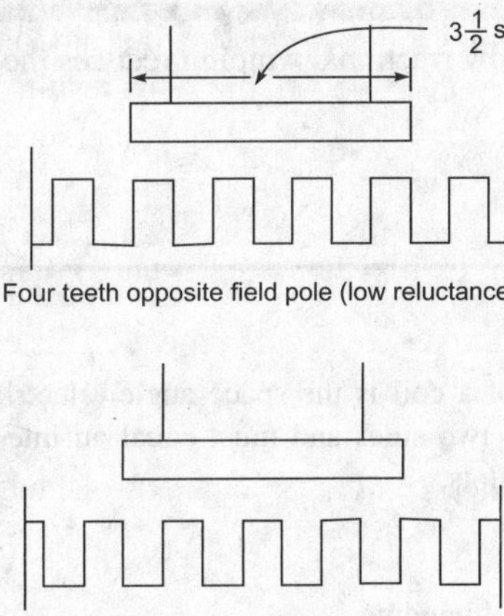

(a) Four teeth opposite field pole (low reluctance, maximum flux)

(b) Three teeth opposite field pole (high reluctance, minimum flux)

Fig. 6.10

Flux linking an armature coil can be expressed as

$$\phi = (\Phi + \phi_s \sin 2\pi f_s t) \cos 2\pi f t$$

and the emf due to this flux is

$$e = -N\frac{d\phi}{dt} = 2\pi f N \Phi \sin 2\pi f t$$

$$-\frac{1}{2} N\phi_s \,[2\pi(f_s - f)\cos 2\pi(f_s - f)$$

$$+ 2\pi(f_s + f)\cos 2\pi(f_s + f)] \quad (6.3)$$

Thus, apart from the fundamental, the two frequencies present in the emf wave owing to armature slotting are

$$(f_s \pm f) = f\left(\frac{2S}{P} \pm 1\right) \tag{6.4}$$

If $S/P = 9$, then

$$(f_s \pm f) = 50(18 \pm 1) = 950, 850 \text{ Hz}$$

These frequencies are high enough to cause interference in telecommunication in the neighbourhood of power lines.

If S/P is fractional, the space relation between the slotted armature and a given field pole is not the same as in succeeding poles as a result of which the emfs of ripple frequencies ($f_s \pm 1$) in various armature coils become out-of-phase and tend to cancel out in an interconnected set of phase coils.

Probe

What happens when the S/P value is fractional?

Concept Check LO 4

1. Explain how fractional winding reduces the emfs of ripple frequencies.

SUMMARY

LO 1

- The pitch of a coil is the space angle (electrical) between its two sides and must equal an integral number of slots.

LO 3

- Two types of winding
  - Single layer: each coil side of a coil occupies the whole slot
  - Double layer: each slot is occupied by two coil sides
- Winding diagram is used to illustrate the arrangement of coils round the armature periphery and their interconnections.
- Fractional slot winding is used to reduce the high frequency harmonics caused by slotting.
- In lap winding the finish of one coil is connected to the start of the adjoining coil.
- Wave winding the finish end of one coil under one pole pair is connected to the start of a coil under the next pole pair.

PRACTICE PROBLEMS

6.1 Draw a single-layer unbifuracated winding for a 3-phase, 4-pole machine having 24 armature slots. Assume one coil-side. Clearly show the end connection of a continuous chain arrangement is used. **LO 3**

6.2 For the same number of slot and poles as in Problem 6.1, draw a bifurcated type of winding. If the number of slots are changed from 24 to 36, is it possible to have bifurcated winding? If not, why? **LO 3**

6.3 Give a developed view of double-layer armature winding for a 3-phase machine with 6 poles and 36 slot, If full-pitched coils are used. Indicate all end connection and the start and finish of each phase. LO 3

6.4 A 3-phase machine has 4 poles and 48 armature slots. If the coils are chorded by one slot, draw the double-layer winding diagram for all three phases. Why is it that chording more than 1/3 pole-pitch is not used in practice? LO 3

6.5 The armature of a 3-phase machine with 16 poles and 180 slots is wound with fractional slot windings. Construct the winding table for one basic unit of poles. Indicate the start of each phase. For the basic unit determine the distribution of coil groups and the phase sequence. LO 3

6.6 A 3-phase, 10-pole machine has 72 slots. Construct the winding table for fractional slot winding. Draw the winding diagram with a coil-span of seven slots. LO 3

6.7 A 3-phase, 50-Hz, 10-pole machine has 120 armature slots. What harmonic frequencies will be present in the generated emf on account of slotting ? How do these affect the operation of the machine? LO 4

MULTIPLE CHOICE QUESTIONS

1. How are concentric windings classified? LO 3
(a) Unbifurcated and bifurcated (b) Single-layer and double-layer
(c) Lower and upper (d) Unfractional and fractional

2. What is the most widely used class of windings? LO 3
(a) Double-layer windings (b) Integral-slot windings
(c) Fractional slot windings (d) Concentric windings

3. High-frequency harmonics caused by slotting can be reduced by the use of _____ winding. LO 3
(a) fractional slot (b) integral-slot (c) double-layer (d) tooth ripple

4. What is used to illustrate the arrangement of coils around the armature periphery and their interconnections? LO 1
(a) Winding diagram (b) Tooth ripple
(c) Wave winding (d) Phase grouping

5. _____ windings comprise a set of coils embedded in the slots, uniformly spaced around the periphery. LO 1
(a) Armature (b) Field (c) Phase (d) Rotor

Answers

1. (a) **2.** (a) **3.** (a) **4.** (a) **5.** (a)

CHAPTER 7

dc Machines

LEARNING OBJECTIVES

After studying this chapter, students will be able to

- **LO1** Explain the constructional details, working principle and applications of dc machines and review the armature winding connections
- **LO2** Discuss the significance of EMF and torque equation of dc machines and hence deduce its equivalent circuit model
- **LO3** Review the concepts of armature reaction and commutation and discuss the need for compensating winding
- **LO4** Outline the excitation methods used for different types of dc machines
- **LO5** Analyze the operation of dc generators by its performance characteristics
- **LO6** Explain the parallel operation of dc generators
- **LO7** Analyze the performance of dc motors by its performance characteristics
- **LO8** Discuss the starting methods and speed control methods of dc motors
- **LO9** Describe the braking methods of dc motors
- **LO10** Discuss various testing methods of dc machines for predetermination of efficiency
- **LO11** Review the dc machine dynamics and constructional features of PMDC motors

7.1 | INTRODUCTION

A dc machine is constructed in many forms and for a variety of purposes, from the 3 mm stepper drawing a few μA at 1.5 V in a quartz crystal watch to the giant 75000 kW or more rolling mill motor. It is a highly versatile and flexible machine. It can satisfy the demands of load requiring high starting, accelerating and retarding torques. A dc machine is also easily adaptable for drives with a wide range of speed control and fast reversals.

dc motors are used in rolling mills, in traction and in overhead cranes. They are also employed in many control applications as actuators and as speed or position sensors. With ac being universally adopted for generation, transmission and distribution, there are almost no practical uses now of the dc machine as a power generator. Its use as a motor-generator (ac motor-dc generator) for feeding dc drives has also been replaced in modern practice by rectifier units. In certain applications dc motors act as generators for brief time periods in the "regenerative" or "dynamic braking" mode, especially in electric traction systems.

The basic principles underlying the operation and constructional features of a dc machine were discussed in Sec. 5.2 (refer to Fig. 5.13) while the emf equation was given in Eq. (5.26). It was stated there that the *field winding* (*concentrated type*) is mounted on salient-poles on the stator and the *armature winding* (*distributed type*) is wound in slots on a cylindrical rotor. Constructional features of a practical machine are brought out by half cross-sectional views of Figs 7.1 and 7.2 wherein all important machine parts are named. Both small and large industrial machines have generally the conventional *heteropolar cylindrical rotor* structure, although some unconventional homopolar machines have also been devised.

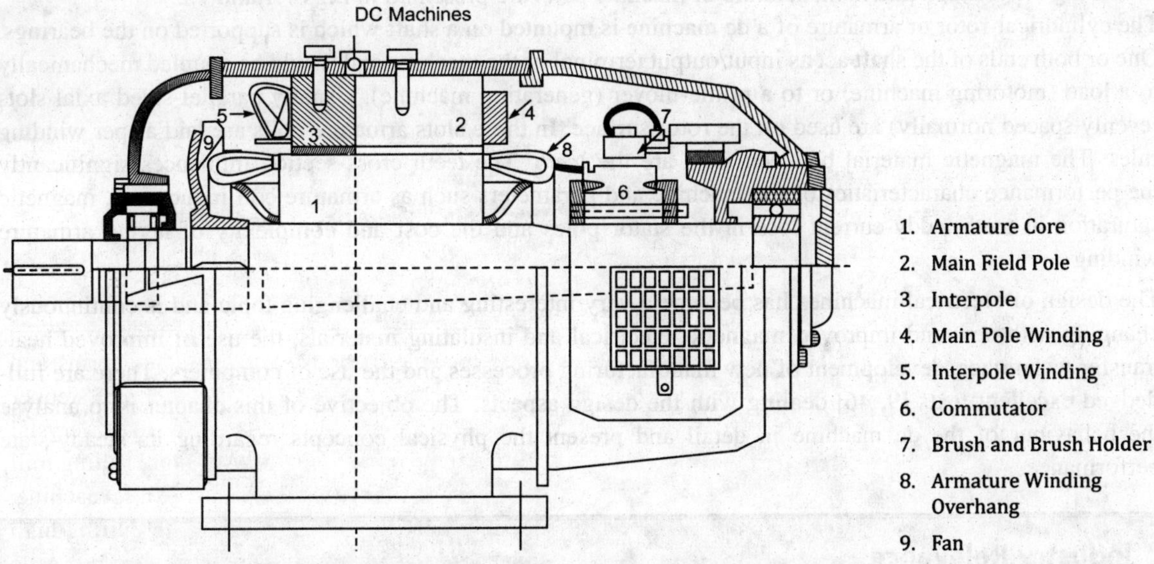

Fig. 7.1 *Sectional view of a dc machine*

The magnetic circuit of a dc machine consists of the armature magnetic material (*core*), the air-gap, *field poles* and yoke as shown in Figs 5.13 and 7.2. The yoke of a dc machine is an annular ring on the inside of which are bolted field poles and the interpoles. The *interpoles or commutation poles* are narrow poles fixed to the yoke, midway between the main field poles. Interpoles and *compensating* windings, which will be described later in this chapter in connection with *commutation* problems, are required to be excited suitably.

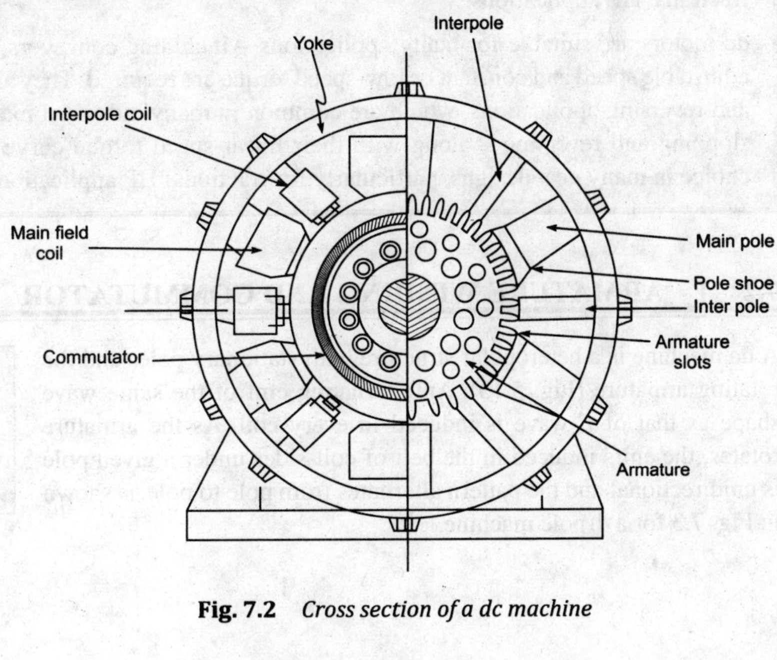

Fig. 7.2 *Cross section of a dc machine*

The use of an electric field winding, which supplies electric energy to establish a magnetic field in the magnetic circuit, results in the great diversity and a variety of performance characteristics. The armature winding is connected to the external power source through a *commutator-brush* system (see Fig. 7.1 item 6), which is a mechanical rectifying (switching) device for converting the alternating currents and induced emfs of the armature to dc form. Figure 7.4(a) shows a single commutator segment and Fig. 7.4(b) is the cross-sectional view of a built-up commutator. The longitudinal and perpendicular to the machine axis cross-sectional view of a dc machine, indicating the location and nomenclature of machine parts are presented in Figs 7.1 and 7.2.

The cylindrical-rotor or armature of a dc machine is mounted on a shaft which is supported on the bearings. One or both ends of the shaft act as input/output terminal of the machine and would be coupled mechanically to a load (motoring machine) or to a prime-mover (generating machine). Usually parallel-sided axial slots (evenly spaced normally) are used on the rotor surface. In these slots armature coils are laid as per winding rules. The magnetic material between slots are the *teeth.* The teeth cross-section influences significantly the performance characteristics of the machine and parameters such as armature coil inductance, magnetic saturation in teeth, eddy-current loss in the stator poles and the cost and complexity of laying armature winding.

The design of electrical machines has become a very interesting and challenging topic and is continuously changing with new and improved magnetic, electrical and insulating materials, the use of improved heat-transfer techniques, development of new manufacturing processes and the use of computers. There are full-fledged excellent texts [9, 46] dealing with the design aspects. The objective of this chapter is to analyse the behaviour of the dc machine in detail and present the physical concepts regarding its steady-state performance.

Industry Relevance

dc motors are often selected instead of ac motors for many reasons. dc motors and controllers are often the low-cost option when compared to inverter-duty ac motors and drives. This is especially true for fractional HP applications.

dc motors are suitable for many applications – including conveyors, turntables and others for which adjustable speed and constant or low-speed torque are required. They also work well in dynamic braking and reversing applications, which are common in many industrial machines. Their quick acceleration, stopping and reversing – along with their linear-speed torque curve – make the dc motor a popular choice in many new designs, particularly for fractional HP applications.

7.2 ARMATURE WINDING AND COMMUTATOR

LO 1 Explain the constructional details, working principle and applications of dc machines and review the armature winding connections

A dc machine is a heteropolar structure with stationary poles and the rotating armature (Fig. 5.13). An alternating emf of the same wave shape as that of *B*-wave is induced in every coil. As the armature rotates, the emfs induced in the belt of coil-sides under a given pole is unidirectional and the pattern alternates from pole to pole as shown in Fig. 7.3 for a 4-pole machine.

The coil side current pattern is the same as the emf pattern. The only difference is that while the coil-side emf reduces towards the outer side of poles, the current remains the same in all the coil-sides except for alternations from pole to pole, while the coil-side current reverses, the current exchanged with external circuit must be unidirectional and voltage must be constant and of same polarity (d.c.). This is the rectification process which is carried out by mechanical rectifier comprising commutator-brush assembly.

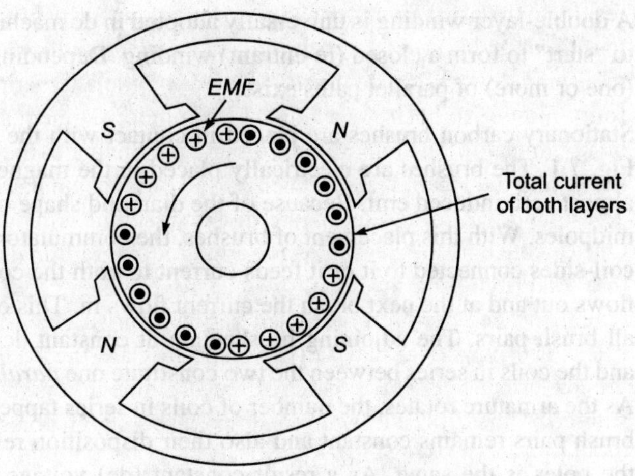

Fig. 7.3 *4-pole dc machine*

Commutator-Brush Assembly

The commutator is a cylindrical assembly of wedge-shaped copper segments (Fig. 7.4(a)) insulated from one another and the shaft by thin mica or *micanite* sheets. In high-speed machines the segments are so shaped that they can be clamped by two cast-iron *V*-shaped rings as shown in Fig. 7.4(b). Each commutator segment forms the junction between two armature coils ("finish" of one coil and "start" of the other). In large machines flat copper strips known as *risers* are used forming clip connections to armature bar conductors (Fig. 7.4(b)).

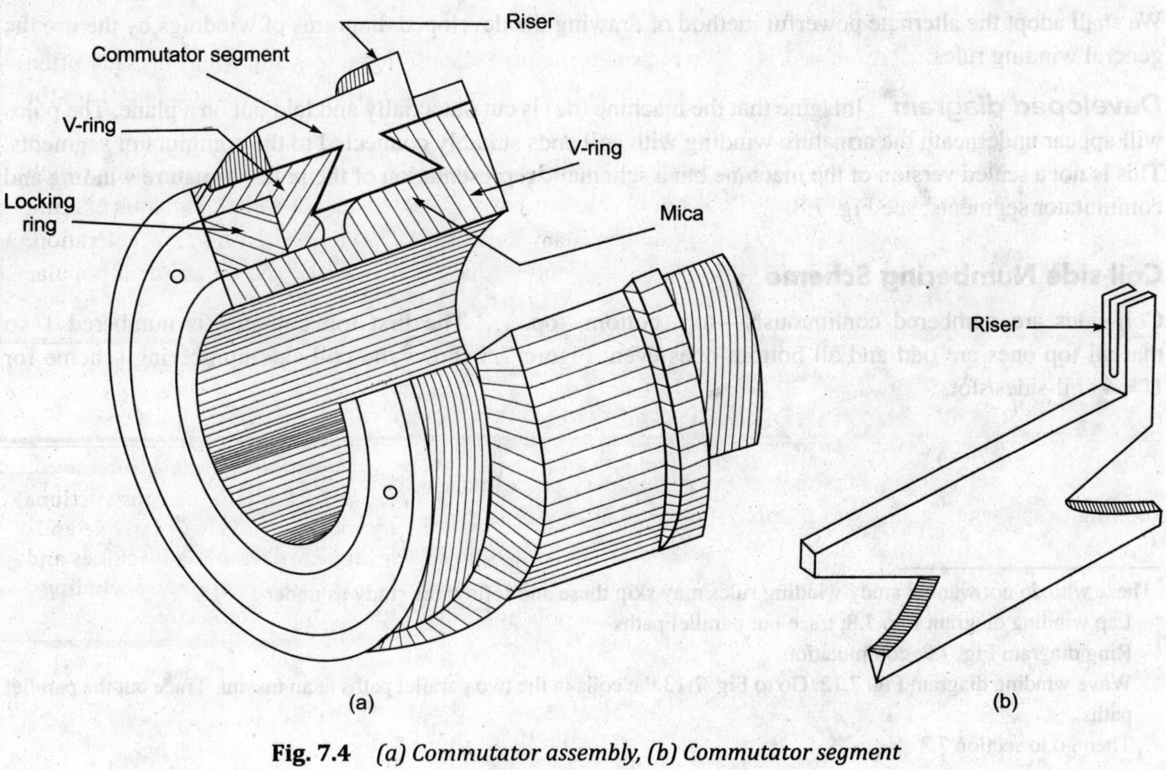

Fig. 7.4 *(a) Commutator assembly, (b) Commutator segment*

A double-layer winding is universally adopted in dc machines. The coils are continuously connected "finish" to "start" to form a closed (re-entrant) winding. Depending upon the type of connection (lap or wave), pairs (one or more) of parallel paths exist.

Stationary carbon brushes are placed in contact with the commutator under spring pressure; see item 7 of Fig. 7.1. The brushes are electrically placed in the magnetic neutral regions where the armature coils have almost zero induced emf. Because of the diamond shape of coils, the brushes are physically placed opposite midpoles. With this placement of brushes, the commutator segment contacted is either fed current from both coil-sides connected to it or it feeds current to both the coil-sides. Thus, at one brush the current constantly flows out and at the next brush the current flows in. This occurs at all brush pairs. The adjoining brushes are at constant dc voltage and the coils in series between the two constitute one *parallel path.* As the armature rotates, the number of coils in series tapped by the brush pairs remains constant and also their disposition relative to the poles is the same. As a result constant (dc) voltage appears across brush pairs. As a coil crosses the brush position, the current in it must reverse which is the commutation process.

Probe

What is the significance of commutator and brushes in a dc machine?

Armature Winding Commutator Connections*

For elementary explanation, cross-sectional view of armature winding and commutator may be employed. As only conductor cross-sections appear in the diagram, clarity does not emerge with out lengthy detailed account and general winding rules get left out.

We shall adopt the alternate powerful method of drawing the developed diagrams of windings by the use the general winding rules.

Developed diagram Imagine that the machine (dc) is cut out axially and laid out on a plane. The poles will appear underneath the armature winding with coil ends suitably connected to the commutator segments. This is not a scaled version of the machine but a schematic representation of the poles, armature winding and commutator segments; see Fig. 7.8.

Coil-side Numbering Scheme

Coil-sides are numbered continuously—top, bottom, top, …. The first top coil-side is numbered 1 so that all top ones are odd and all bottom ones even. Figure 7.5 shows the coil-side numbering scheme for $U = 4$ coil-sides/slot.

---

*These who do not want to study winding rules may skip these and follow the study as under

Lap winding diagram Fig. 7.8; trace out parallel paths

Ring diagram Fig. 7.9; commutation

Wave winding diagram Fig. 7.12. Go to Fig. 7.13 the coils in the two parallel paths at an instant. Trace out the parallel paths

Then go to section 7.3

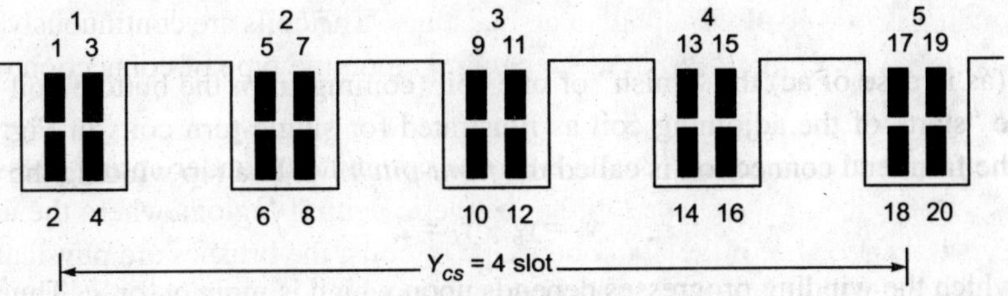

Fig. 7.5

Coil-pitch/Back-pitch

In Fig. 7.5 the coil-span Y_{cs} is assumed 4 slots (or the coil spans 4 teeth). Therefore, coil-side 1 will form a coil with coil-side 18 and 3 with 20, and so on. This is depicted in Fig. 7.6. The coil-span in terms of coil-sides is $y_{cs} = 18 - 1 = 17$ or $20 - 3 = 17$. This indeed is the distance measured in coil-sides between two coil-sides connected at the back end of armature (end away from commutator) to form a coil. It is known as back-pitch denoted as y_b. Obviously coil-span

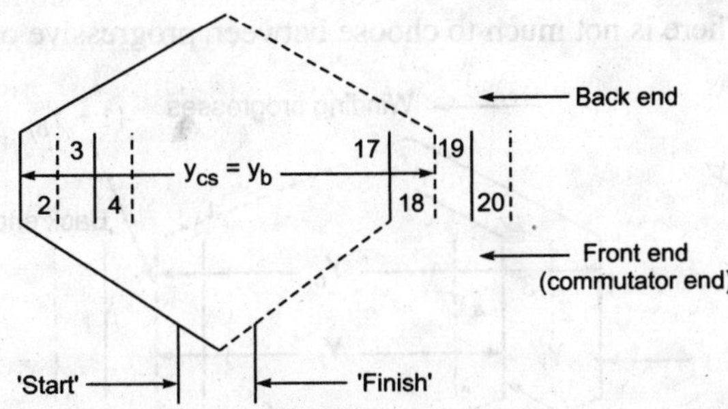

Fig. 7.6

$$Y_{cs} = y_b = 17 \text{ (odd)}$$

which is odd in this case and must always be so. It indeed equals

$$y_b = 4 \times 4 + 1 = 17$$

or in general $\quad y_b = UY_{cs} + 1$ (odd because U is even) $\quad$ (7.1)

where $\quad Y_{cs}$ = coil-span in slots

Probe

What is coil-pitch or back-pitch?

Commutator-pitch

The junction of two coils ("finish" - "start") is connected to one commutator segment. Therefore,

$$\text{Number of commutator segments} = C \text{ (number of armature coils)} \tag{7.2}$$

The number of commutator segments spanned by the two ends of a coil is called commutator-pitch, y_c. In Fig. 7.7(a), $y_c = 2 - 1 = +1$.

Coil-span

The coil-span in terms of slots is always nearly full-pitch. This ensures that coil-side voltages around the coil are additive most of the time (except when coil-sides lie near the magnetic neutral region). Thus

$$Y_{cs} = \frac{S}{P} \text{ (nearest lower integer)} \tag{7.3}$$

which means that for nonintegral S/P, the coils are short-pitched.

Lap Winding

In a lap winding (as in case of ac) the "finish" of one coil (coming from the bottom coil-side) is connected to (lapped on) the "start" of the adjoining coil as illustrated for single-turn coils in Fig. 7.7. The coil-side displacement of the front end connection is called the *front-pitch*, y_f. In a lap winding the *resultant-pitch*

$$y_r = y_b \sim y_f = 2 \tag{7.4}$$

The direction in which the winding progresses depends upon which is more, y_b or y_f. Thus

$$y_b > y_f \quad \text{(Progressive winding, Fig. 7.7(a))} \tag{7.5a}$$

$$y_b < y_f \quad \text{(Retrogressive winding, Fig. 7.7(b))} \tag{7.5b}$$

There is not much to choose between progressive or retrogressive winding; either could be adopted.

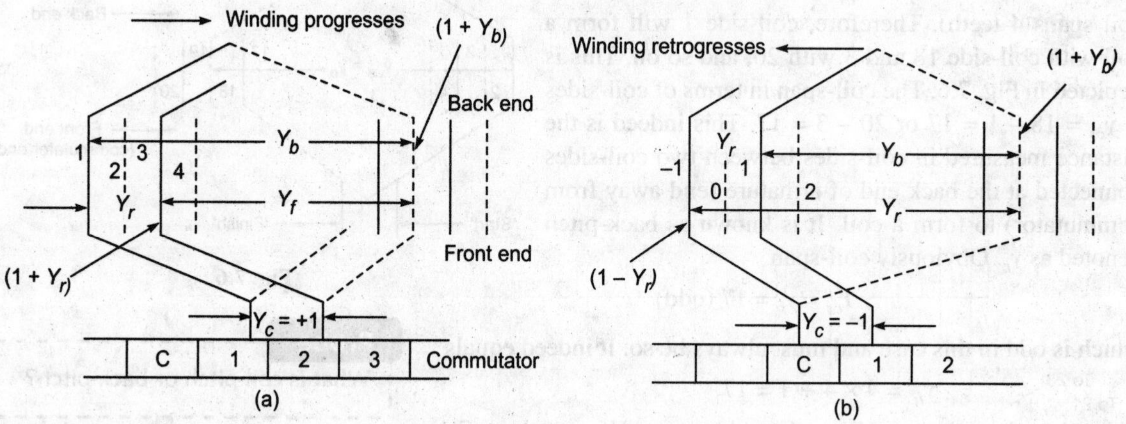

Fig. 7.7

As shown in Figs 7.7(a) and (b), the two ends of a coil are connected across adjacent commutator segments. Thus commutator-pitch,

$$y_c = \pm 1 \text{ (+1 for progressive, } -1 \text{ for retrogressive)} \tag{7.6}$$

Coils in lap winding are continuously connected as per the above rule and in the end it closes onto itself (as it must). In the process all coils have been connected.

To learn certain further aspects of lap winding—location of brushes, etc., an example is worked out.

Example 7.1 *Draw the lap-winding diagram in the developed form for a 4-pole, 12-slot armature with two coil-sides/slot. Assume single-turn coils. Indicate the number and position of brushes on the commutator. What is the number of parallel paths?* LO 1 ○○●

Solution

$$C = S = \text{number of commutator segments} = 12$$

$$Y_{cs} = \frac{12}{4} = 3 \text{ slots}$$

Note: Difficulty Level ⟶ ○○● — Level 1 and Level 2 Category; ○●● — Level 3 and Level 4 Category; ●●● — Level 5 and Level 6 Category

$$Y_b = 2Y_{cs} + 1 = 7$$

$$y_f = 7 - 2 = 5 \text{ (we choose progressive winding)}$$

This information is sufficient to draw the developed winding diagram of Fig. 7.8. The ends of coil formed by coil-sides (1-8) are connected to commutator segments 1, 2, and so on. The four poles are shown on the developed diagram. All coil-sides under one pole have emf induced in the same direction and the pattern alternates. The arrows on the coil sides indicate the direction of current flow. If the armature moves left to right, the emfs in coil sides are induced in the same direction as currents, the machine is generating, supplying power to the external circuit. On the other hand, if the armature moves right to left, the induced emfs are in opposite direction to currents, the machine is motoring; receiving power from the supply.

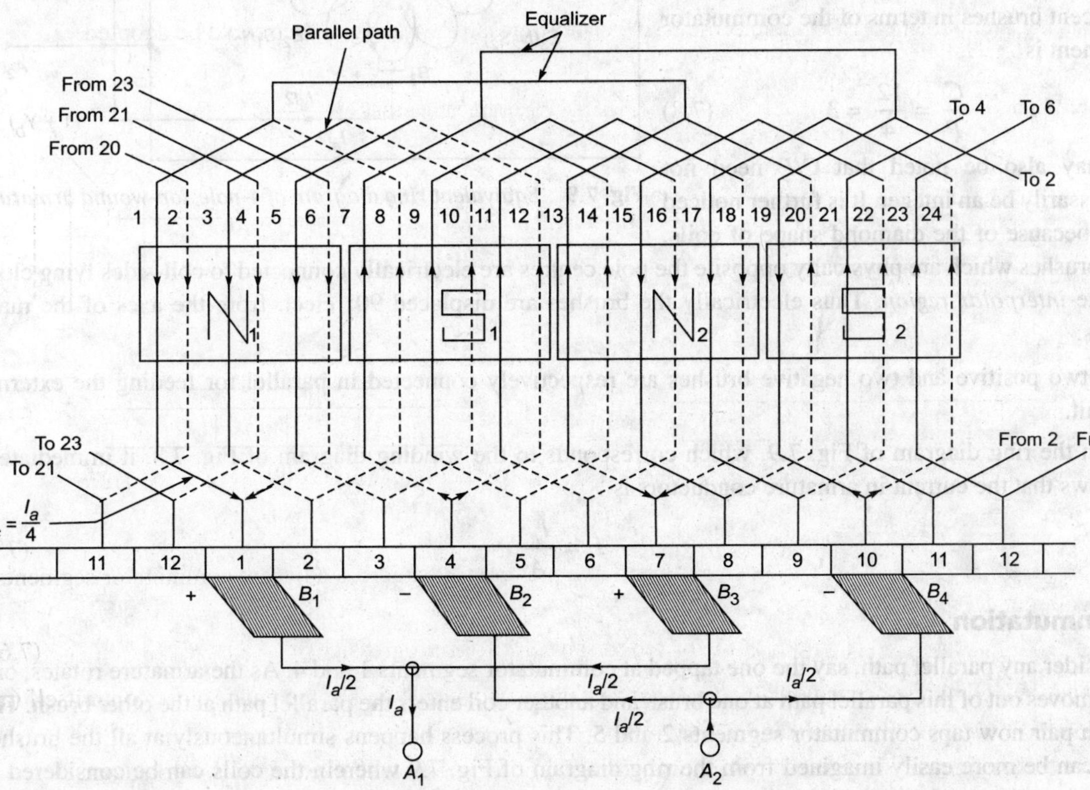

Fig. 7.8 *Lap winding for 4-poles, 12-slot armature (single-turn coils, 2 coil-sides/slot)*

Parallel Paths/Brushes

It is easily found from the winding diagram that three coils ($C/P = 12/4$) are located under one pole-pair (N_1S_1) and are series connected so that their emfs add up. This constitutes one parallel path. The complete winding can be divided into four such parallel paths lying under four different pole-pairs (N_1S_1, S_1N_2, N_2S_2, S_2N_1). It is, therefore, concluded that the number of parallel paths (A) in a lap-wound machine in general equal the number of poles (P), i.e.

$$A = P \tag{7.7}$$

The four parallel paths in the winding of Fig. 7.8 electrically form a close ring as shown in Fig. 7.9 in which the parallel path emfs and currents around the loop alternate.

The ends of parallel paths meet at commutator segments 1, 4, 7 and 10 at the instant shown. These are the locations of the brushes (equal to the number of poles) and are alternately positive and negative. It is also found that the brushes are physically located opposite the pole centres and the electrical angle between them is, therefore, 180°. The spacing between adjacent brushes in terms of the commutator segment is

$$\frac{C}{P} = \frac{12}{4} = 3 \qquad (7.8)$$

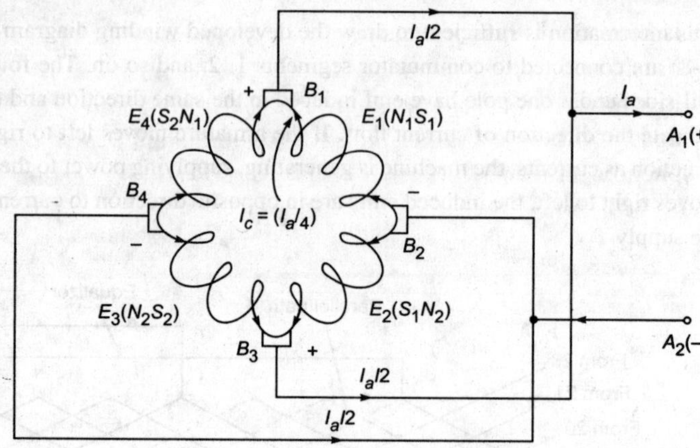

Fig. 7.9 *Equivalent ring diagram of 4-pole, lap-wound armature*

It may also be noted that C/P need not necessarily be an integer. It is further noticed that because of the diamond shape of coils, the brushes which are physically opposite the pole centres are electrically connected to coil-sides lying close to the *interpolar region.* Thus electrically the brushes are displaced 90° elect. from the axes of the main poles.

The two positive and two negative brushes are respectively connected in parallel for feeding the external circuit.

From the ring diagram of Fig. 7.9, which corresponds to the winding diagram of Fig. 7.8, it immediately follows that the current in armature conductors is

$$I_c = \frac{I_a}{A} \qquad (7.9)$$

Commutation

Consider any parallel path, say the one tapped at commutator segments 1 and 4. As the armature rotates, one coil moves out of this parallel path at one brush and another coil enters the parallel path at the other brush. The brush pair now taps commutator segments 2 and 5. This process happens simultaneously at all the brushes and can be more easily imagined from the ring diagram of Fig. 7.9 wherein the coils can be considered to rotate in a circular fashion. In this way a brush pair always taps C/P coils (in series) and as a consequence the voltage available at each brush pair is maintained constant (dc). This indeed is the *commutation action.* As a matter of fact the voltage tapped varies slightly for a brief period when the changeover of coils in parallel paths takes place. However, this voltage variation is negligible in practical machines which have a large number of coils.

Ideal Commutation

It easily follows from the winding diagram of Fig. 7.8 and from the equivalent ring diagram of Fig. 7.9 that as a coil moves out of one parallel path into another, the current in it must reverse. In Fig. 7.8 as the armature moves by one commutator segment, currents in four coils—(1, 8), (7, 14), (13, 20) and (19, 2)—must reverse.

These coils are said to undergo *current commutation*. It is also observed that during the brief period in which a coil undergoes commutation, its coil-sides are passing through the interpolar region so that negligible emf is induced in the commutating coil. At the same time, in this period the coil remains short-circuited by the brush, bridging the adjoining commutator segments to which the coil ends are connected. Ideal commutation in which the conductor current change from $+I_c$ to $-I_c$ is sketched in Fig. 7.10(a). If the current in a coil does not reverse fully at the end of commutation period, there will be sparking at the brush contact. This phenomenon and its remedy will be discussed in a Section 7.8

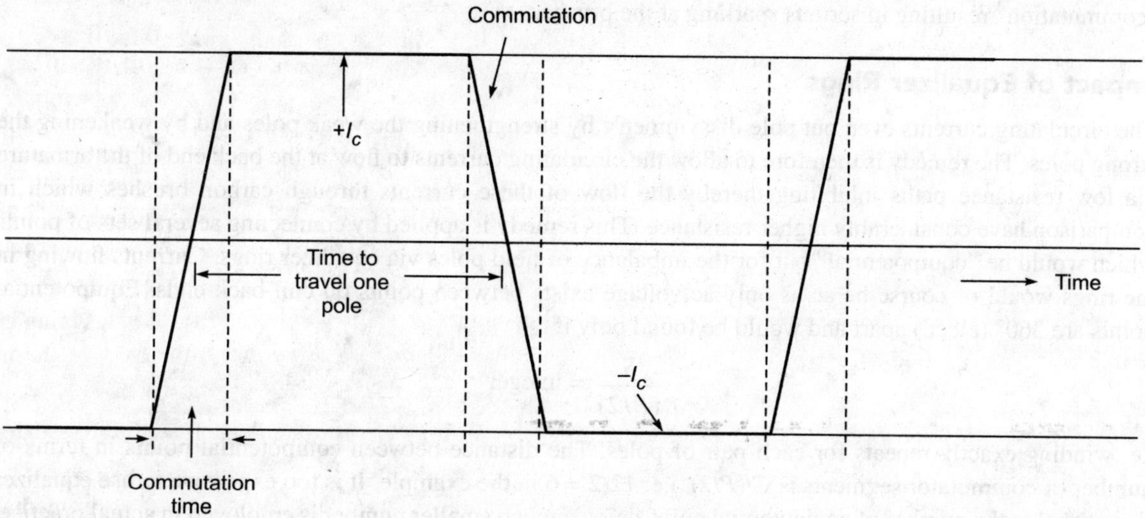

Fig. 7.10(a) *Ideal Commutation*

Symmetry Requirement

To avoid no-load circulating currents and certain consequential commutation problems, all the parallel paths must be identical so as to have the same number of coil-sides. Symmetry therefore requires that

$$\frac{2C}{P} = \frac{US}{P} = \text{integer} \tag{7.10}$$

Remark

To practically wind a dc armature the above winding rules are not needed except that the coils of the double-layer winding are to be continuously connected from "finish" to "start" till the winding closes onto itself. Further the "finish-start" junction is connected to the commutator segment physically opposite the midpoint of the coil; the ends of each coil being connected to the adjacent commutator segment. This is illustrated in Fig. 7.10(b). Of course the winding rules given above help in providing an insight to the reader into the winding and commutator action to produce dc at the brushes.

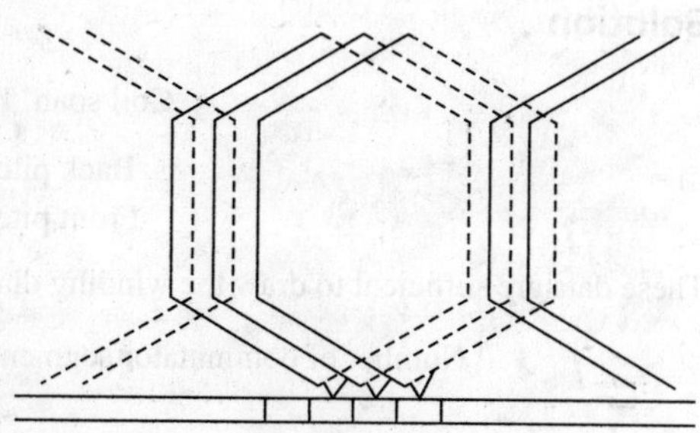

Fig. 7.10(b)

Equalizer Rings

The poles in a dc machine cannot be made identical so as to have the same value of flux/pole. Any dissymmetry among poles leads to inequality in parallel path emfs e.g. in Fig. 7.9, $E_1(N_1S_1)$, $E_2(S_1N_2)$, $E_3(N_2S_2)$ and $E_4(S_2N_1)$ will not be identical. As a result the potential of the positive and negative brush sets are no longer equal so that circulating current will flow in the armature via the brushes to equalize the brush voltage even when the armature is not feeding current to the external circuit. Apart from causing unbalanced loading around the armature periphery when the armature feeds current, the circulating currents also interfere with "commutation" resulting in serious sparking at the brushes.

Impact of Equalizer Rings

The circulating currents even out pole dissymmetry by strengthening the weak poles and by weakening the strong poles. The remedy is therefore to allow the circulating currents to flow at the back end of the armature via low resistance paths inhibiting thereby the flow of these currents through carbon brushes which in comparison have considerably higher resistance. This remedy is applied by connecting several sets of points which would be "equipotential" but for the imbalance of field poles via equalizer rings. Currents flowing in the rings would of course be ac as only ac voltage exists between points on coil back ends. Equipotential points are 360° (elect.) apart and would be found only if

$$\frac{S}{(P/2)} = \text{integer}$$

i.e. winding exactly repeats for each pair of poles. The distance between equipotential points in terms of number of commutator segments is $C/(P/2)$, i.e. 12/2 = 6 in the example. It is too expensive to use equalizer rings equal to the number of equipotential point-pairs; a much smaller number is employed in actual practice. Two equalizer rings are shown properly connected in Fig. 7.8 for the example.

While equalizer rings inhibit the flow of circulating currents via brushes, they are not a prevention for the circulating currents which do cause additional copper losses.

It will soon be seen that wave winding scheme does not have the need of equalizer rings and would naturally be preferred except in large heavy current machines.

Example 7.2 *Give the relevant details for drawing lap winding for a dc machine with 4 poles, 22 slots and 6 coil-sides/slot. What should be the spacing between brushes?* LO 1

Solution

$$\text{Coil span, } Y_{cs} = \frac{22}{4} \approx 5 \text{ slots}$$

$$\text{Back pitch} = 6 \times 5 + 1 = 31$$

$$\text{Front pitch} = 31 - 2 = 29 \text{ (progressive winding)}$$

These data are sufficient to draw the winding diagram.

$$\text{Number of commutator segments} = \frac{1}{2} US = 3 \times 22 = 66$$

$$P = A = 4$$

$$\text{Number of brushes} = 4$$

$$\text{Spacing between adjacent brushes} = \frac{66}{4} = 16\frac{1}{2} \text{ segments}$$

Wave Winding

Fact to Know In wave winding the "finish" end of one coil under one pole-pair is connected to the start of a coil under the next pole-pair as shown in Fig. 7.11.

The process is continued till all the armature coils are connected and the winding closes onto itself. Certain conditions must be fulfilled for this to happen. The winding has the appearance of a wave and hence the name. The ends of each coil spread outwards and span y_c (commutator pitch) segments. As the number of coil-sides is double the number of segments, the top coil-side of the second coil will be numbered $(1 + 2y_c)$. The numbering of other coil-sides is clear from the figure. It follows that

$$1 + 2y_c - y_f = 1 + y_b$$

or

$$y_f + y_b = 2y_c = y_r \tag{7.11}$$

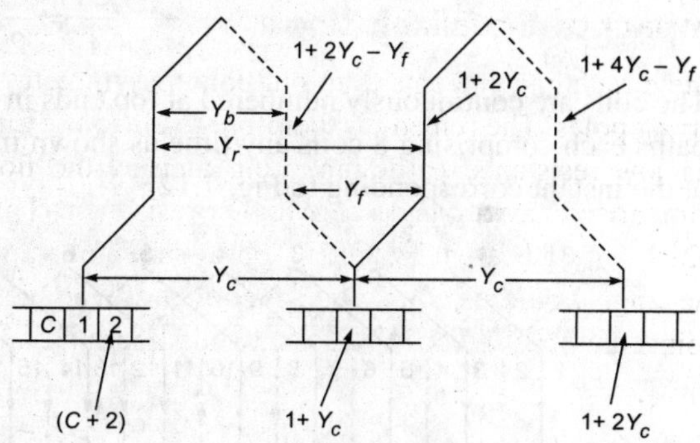

Fig. 7.11

Starting at segment 1 and after going through $P/2$ coils or y_c $(P/2)$ segments, the winding should end up in segment 2 for progressive winding or segment (C) for retrogressive winding. This means that for the winding to continue and cover all the coils before closing onto itself, i.e.

$$y_c\left(\frac{P}{2}\right) = (C \pm 1)$$

or

$$y_c = \frac{2(C \pm 1)}{P} \quad \text{(must be integer)} \tag{7.12}$$

In Eq. (7.12) the winding is progressive if + sign is used and is retrogressive otherwise.

Once y_b is known from the coil-span and y_c is determined from Eq. (7.12), the back-pitch y_f is calculated from Eq. (7.11). The winding diagram can now be drawn.

In wave winding the coils are divided into two groups—all coils with clockwise current are series connected and so are all coils with counter-clockwise current–and these two groups are in parallel because the winding is closed.

Important! Thus a wave winding has always *two parallel paths irrespective of the number of poles;* also only two brushes are required, i.e.

$$A = 2 \tag{7.13}$$

These statements are corroborated by an illustrative example below:

Example 7.3 *For a 6-pole dc armature with 16 slots having two coil-sides per slot and single-turn coils, calculate the relevant pitches for a wave winding and draw the developed winding diagram.*

LO 1

Solution

$$Y_{cs} = \frac{16}{6} \approx 2 \text{ slots (nearest lower integral value)}$$

$$y_b = 2 \times 2 + 1 = 5$$

$$C = 16$$

$$y_c = \frac{2(16 \pm 1)}{6} = 5 \text{ segments}$$

$$y_f = 2y_c - y_b = 5$$

As per the above values of various pitches, the developed diagram of the winding is drawn in Fig. 7.12. It is observed that the armature winding has two parallel paths; in Fig. 7.29)—current is going in at segment 6 and is coming out at segment 14 (coil-side 27 has negligible emf and the direction of current in it is determined by the next coil-side in series with it, i.e. 32). Only two brushes are, therefore required—one brush is opposite a north pole and the other opposite the diametrically opposite south pole. The spacing between brushes is

$$\frac{C}{A} = \frac{C}{2} = 8 \text{ segments} \tag{7.14}$$

The coils are continuously numbered at top ends in Fig. 7.12. Between the two brushes there are two parallel paths each comprising 8 coils any time as shown in Fig. 7.13 where coil of each parallel path are numbered at the instant corresponding to Fig. 7.12.

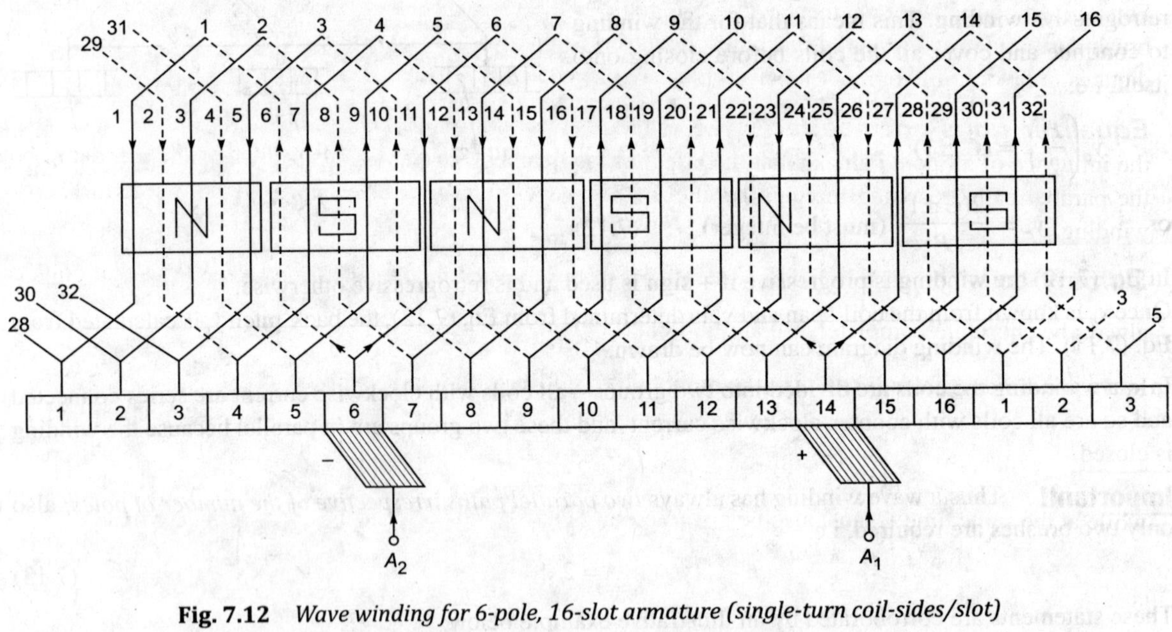

Fig. 7.12 *Wave winding for 6-pole, 16-slot armature (single-turn coil-sides/slot)*

We thus conclude that *number parallel path* in a wave winding is $A = 2$, *irrespective of number of poles.* Therefore conductor current in a wave would machine is

$$I_c = \frac{I_a}{2} \tag{7.15}$$

Spacing between the two brushes is

$$\frac{C}{A} = \frac{16}{2} = 8 \text{ segments}$$

Electrical spacing = 180°

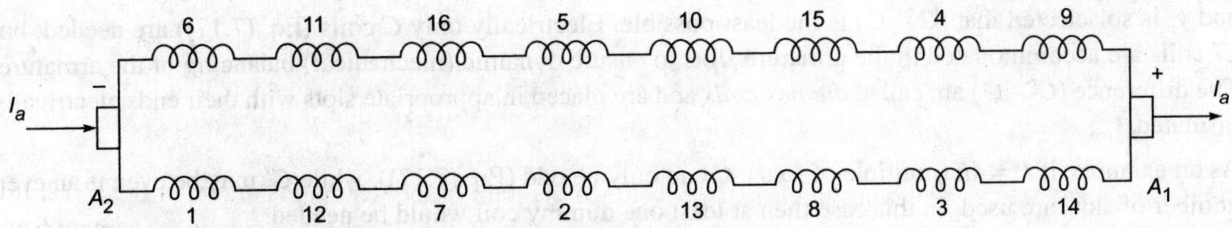

Fig. 7.13 *Parallel paths of wave winding*

In practical wave-wound machines as many brushes as number of poles are used with spacing between adjacent brushes being *C*/*P* commutator segments and the brushes are alternatively positive and negative. All positive and all negative brushes are respectively connected in parallel to feed the external circuit. This reduces the current to be carried by each brush to a value.

$$I_{\text{Brush}} = \frac{I_a}{(P/2)} \tag{7.16}$$

For a given commutator segment width this reduces the segment length required for maximum allowable brush current density. In small machines, however, economy—cost of brush gear relative to commutator—dictates in favour of two brushes only, which are placed opposite two adjoining poles.

Equalizer rings not needed The armature coils forming each of the two parallel paths are under the influence of all pole-pairs so that the effect of the magnetic circuit asymmetry is equally present in both the parallel paths resulting in equal parallel-path voltages. Thus equalizer rings are not needed in a wave winding.

Dummy coils In a lap winding $y_c = \pm 1$ irrespective of the number of armature coils so that coils can always be chosen to completely fill all the slots ($C = \frac{1}{2}US$).

In a wave winding from Eq. (7.12), the number of coils must fulfil the condition

$$C = \frac{P}{2}y_c \pm 1 \tag{7.17}$$

while at the same time *C* must also be governed by

$$C = \frac{1}{2}US \tag{7.18}$$

For a certain-design value of *P* and the choice of *S* restricted by manufacturing considerations (availability of a certain notching gear for armature stamping), the values of *C* as obtained from Eqs (7.17) and (7.18) may not be the same. In such a situation the number of coils *C*′ is dictated by

$$C' = \frac{1}{2}US$$

in such a manner that

$$C' > C$$

and y_c is so selected that $(C' - C)$ is the least possible. Electrically only C coils (Eq. (7.17)) are needed, but C' coils are accommodated in the armature slots to ensure dynamic (mechanical) balancing of the armature. The difference $(C' - C)$ are called *dummy coils* and are placed in appropriate slots with their ends electrically insulated.

As an example, if $P = 4n$ (multiple of four), C can only be odd (Eq. (7.17)), while C' may be even if an even number of slots are used. In this case then at least one dummy coil would be needed.

Example 7.4 *For a 4-pole dc armature with 28 slots and 8 coil-sides per slot, find the winding pitches and the commutator pitch for a wave winding. What is the distance between brushes in terms of commutator segments?*

LO 1

Solution The number of coils that can be accommodated in slots,

$$C = \frac{28 \times 8}{2} = 112$$

$$y_c = \frac{2}{P}(C \pm 1)$$

$$= \frac{2}{4}(112 \pm 1) = 55\frac{1}{2} \text{ or } 56\frac{1}{2}$$

Since no integral value of y_c is possible for $C = 112$, dummy coil would be present.

Now $$C' = 112$$

$$C = \frac{P}{2} y_c \pm 1$$

Choosing $$y_c = 55$$

$$C = 111 \text{ (number of coils needed electrically}$$

$$= \text{number of commutator segments)}$$

Therefore $$(112 - 111) = 1 \text{ dummy coil will be placed on the armature}$$

$$Y_{cs} = \frac{28}{4} = 7$$

$$y_b = 8 \times 7 + 1 = 57$$

$$y_f = 2y_c - y_b = 2 \times 55 - 57 = 53$$

Distance between brushes if four brushes are used = 111/4

$$= 27\frac{3}{4} \text{ segments}$$

Remark

As per Fig. 7.11, to wind a wave armature, the coils are prepared as per y_b $(= UY_{cs} + 1)$. Once the coils are ready only y_c is needed to connect the coils. Practical winders may prefer coil numbering instead of coil-side numbering–a coil beginning with a top coil-side numbered $(1 + 2y_c)$ in Fig. 7.11 is indeed a coil numbered $(1 + y_c)$.

Comparative Summary of Lap and Wave Winding

Table 7.1 Differences between lap winding and wave winding

Lap Winding	Wave Winding
1. Coil-span, $Y_{cs} = \frac{S}{P}$ (lower integer)	$Y_{cs} = \frac{S}{P}$ (lower integer)
2. Back-pitch, $y_b = UY_{cs} + 1$	$y_b = UY_{cs} + 1$
3. Commutator pitch, $y_c = \pm 1$ (+ for progressive, – for retrogressive)	$y_c = \frac{2(C \pm 1)}{P}$ (must be integral) (+ for progressive, – for retrogressive)
4. Front-pitch, $y_f = y_b \pm 2$ (+ for progressive, – for retrogressive)	$y_f = 2y_c - y_b$
5. Parallel paths, $A = P$; Conductor current, $I_c = I_a/A$	$A = 2$; $I_c = I_a/2$
6. Number of brushes $A = P$	Number of brushes = 2 (Large machine use P brushes)
7. No dummy coil needed	Dummy coil may be needed
8. Equalizer rings needed	Equalizer rings not needed

!Attention Choice between Lap and Wave Windings

Wave winding's greatest attraction is that it does not require equalizer rings† which means a less expensive machine compared to lap winding. Lap winding has the advantage of a larger number of parallel paths and lower conductor current ($I_c = I_a/A$) and is therefore adopted for low-voltage high-current machines. The use of wave winding is prohibited for armature currents exceeding 300 A because of certain commutation difficulties.

Steps for Designing Armature Winding Using MATLAB

Designing steps of a lap connected dc winding are given below:

1. Coil span $Y_{cs} = \frac{S}{P}$
2. Back-pitch, $y_b = UY_{cs} + 1$
3. Commutator pitch, $y_c = \pm 1$ (+ for progressive – for retrogressive)
4. Front-pitch, $Y_f = Y_b \pm 2$ (+ for progressive – for retrogressive)
5. Parallel paths, $A = P$
6. Conductor current, $I_c = \frac{I_a}{A}$
7. A loop program to construct winding table.
   Given in a dc machine:

$$\text{No of slots } (S) = 18$$
$$\text{Poles } (P) = 6$$
$$\text{Coil sides/slot } (U) = 2$$
$$\text{Armature current } (I_a) = 25 \text{ A}$$

† A duplex wave winding not discussed here would require equalizer rings.

MATLAB PROGRAM

```
S = 18;                                    % Winding Parameters
P = 6;
Ycs =S./P
U = 2;
Yb = U.*Ycs + 1
Yc_progressive = 1
Yc_retrogressive=-1
Yf_progressive = Yb + 2
Yf_retrogressive = Yb-2
A = P
Ia = 25;
Ic = Ia./A
for i = 1: 2 : 2 (2*S -Yb)        % Winding Table
[i i + Yb]
end
Ycs =
    3
Yb =
    7
Yc_progressive =
    1
Yc_retrogressive =
    - 1
Yf_progressive =
    9
Yf_retrogressive =
    5
A =
    6
Ic =
    4.1667
ans =
    1    8
ans =
    3    10
ans =
    5    12
ans =
    7    14
```

```
ans =
    9    16
ans =
   11    18
ans =
   13    20
ans =
   15    22
ans =
   17    24
ans =
   19    26
ans =
   21    28
ans =
   23    30
ans =
   25    32
ans =
   27    34
ans =
   29    36
```

7.3 CERTAIN OBSERVATIONS

LO 1

From this discussion in sections 7.1 and 7.2, certain observations about the dc machine are summarized as follows. These will help in visualizing the machine behaviour in this chapter.

1. Brushes in a dc machine are normally placed electrically in the interpolar regions and therefore make an angle of 90° elect. with the axes of the adjoining field poles.
2. A lap winding has $A = P$ parallel paths such that the armature current I_a divides out into A paths giving a conductor current of $I_c = I_a/A$. In the case of wave winding, $A = 2$, independent of P.
3. The brushes are alternately positive and negative (elect. angle between adjacent pair being 180° elect). Only two brushes are needed in wave winding though P brushes are commonly provided for heavy-current armatures.
4. The armature periphery is divided into "belts" (P in number) each under influence of a pole. Emfs and currents in all the conductors of a belt are unidirectional–conductors that go out of a belt due to rotation are simultaneously replaced by an equal number coming into the belt. Magnitude of conductor emfs in a belt follows the pattern (wave) of flux density in the airgap while the current in all these conductors (I_c) is the same in all the belts except that the current pattern in the belts alternate in space but remain fixed in time. This basically results from the action of the commutator.
5. If the conductor current flows in the same direction as the conductor emf, the machine outputs electrical power (and absorbs mechanical power), i.e. the machine is operating in *generating mode*. On the other hand, when the conductor current and emf oppose each other, the machine absorbs electrical power and outputs mechanical power, i.e. it operates in the *motoring mode*.

6. Barring irrecoverable losses (of both electric and magnetic origin), there is a balance between electrical and mechanical powers of the machine; the average energy stored in the magnetic field remains constant independent of the armature rotation.

Concept Check — LO 1

1. Each commutator segment is connected to how many coil ends? ○○●
2. What are interpoles, their purpose, location and excitation? Explain why of each item. ○●●
3. Compare the number of parallel paths in the lap and wave windings. ○○●
4. State the condition which determines if a dc machine is generating or motoring. ○○●
5. Write the expression relating the electrical power converted to the mechanical form in a dc motor. How are the electrical power input and mechanical power output different from these powers? ○●●
6. What is the significance of a winding diagram? ○●●
7. When do you use concentric winding? ○○●
8. What are the advantages of fractional slot winding over integral slot winding? ○●●
9. Compare lap and wave winding. Where each type is used and why? ○●●
10. Why double layer winding is preferred? ○●●
11. Explain how fractional slot winding reduces the emfs of ripple frequencies. ○●●

7.4 EMF AND TORQUE

LO 2 Discuss the significance of EMF and torque equation of dc machines and hence deduce its equivalent circuit model

It was shown in Sec. 5.2 (Fig. 5.14(a) that in a dc machine the magnetic structure is such that the flux density wave in the air-gap is flat-topped with quarter-wave symmetry so long as the armature is not carrying any current. It will be seen in Sec. 7.7 that the flux density wave gets distorted when the armature carries current (armature reaction effect destroys quarter-wave symmetry). However, this fact does not affect the constancy of emf (between brushes) and torque developed by the machine with magnitudes of each of these being determined by the flux/pole independent of the shape of the *B*-wave.

EMF Equation

As per Eq. (5.23), the average coil emf is

$$E_a^c = \frac{\Phi \omega_m N_c P}{\pi} \tag{7.19}$$

where Φ = flux/pole

ω_m = armature speed in rad/s

N_c = number of coil turns

P = number of poles

Let $\quad C_p$ = coils/parallel path

This number is fixed independent of the armature rotation; as one coil moves out of the parallel path another comes in and takes its place. Thus, the parallel path emf which equals the armature emf is given by

$$E_a = \frac{\Phi\omega_m(C_pN_c)P}{\pi} = \frac{\Phi\omega_m N_P P}{\pi} \tag{7.20}$$

Here $\quad N_p$ = turns/parallel path = $\dfrac{Z}{2A}$

where $\quad Z$ = total armature conductors

A = number of parallel paths

Hence
$$E_a = \frac{\Phi\omega_m Z}{2\pi}\left(\frac{P}{A}\right) = K_a\Phi\omega_m \tag{7.21}$$

where
$$K_a = \frac{ZP}{2\pi A}$$

or
$$E_a = \frac{\Phi n Z}{60}\left(\frac{P}{A}\right);\ \omega_m = \frac{2\pi n}{60} \tag{7.22}$$

where $\quad n$ = armature speed in rpm

E_a remains constant (dc) by virtue of fixed coils per parallel path independent of rotation of armature (commutation action). It is also observed that E_a depends upon the flux/pole and not upon the shape of the flux density wave.

Probe

What is the significance of EMF equation?

Torque Equation

Figure 7.14 shows the flux density wave in the air-gap and the conductor current distribution in the developed armature for one pole-pair. It is immediately seen that the force on conductors is unidirectional. Each conductor as it moves around with the armature experiences a force whose time variation is a replica of the *B*-wave.

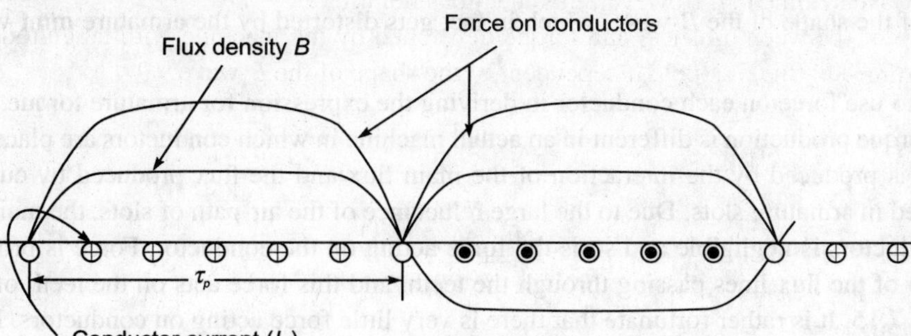

Fig. 7.14 *Torque production in dc machine*

Therefore, the average conductor force

$$f_{c,\text{av}} = B_{\text{av}}\, l I_c \tag{7.23}$$

where B_{av} = average flux density over pole

l = active conductor length

Total force

$$F = Z f_{c,\text{av}} = B_{\text{av}} I_c l Z$$

This force (and therefore torque) is constant (independent of time) because both the flux density wave and current distribution are fixed in space at all times. Now the torque developed

$$T = B_{\text{av}}\, I_c l Z r \tag{7.24}$$

where r = mean air-gap radius

The flux/pole[‡] can be expressed as

$$\Phi = B_{\text{av}}\, \tau_p l$$

where

$$\tau_p = \text{pole-pitch (Fig. 7.14)} = \frac{2\pi r}{P}$$

$$\therefore \quad \Phi = B_{\text{av}}\left(\frac{2\pi r}{P}\right) l$$

or

$$B_{\text{av}} = \frac{\Phi P}{2\pi} \times \frac{1}{rl} \tag{7.25}$$

Substituting for B_{av} in Eq. 7.24,

$$T = \frac{1}{2\pi}\, \Phi I_c Z P$$

$$= \frac{1}{2\pi}\, \Phi I_a Z \left(\frac{P}{A}\right) \text{ Nm} \tag{7.26}$$

$$= K_a \Phi I_a \text{ Nm} \tag{7.27}$$

It is, therefore, seen that the machine torque is uniform for given flux/pole and armature current. Further, it is independent of the shape of the *B*-wave, which in fact gets distorted by the armature mmf when it carries current.

It is convenient to use force on each conductor in deriving the expression for armature torque. However, the mechanism of torque production is different in an actual machine in which conductors are placed in armature slots. The force is produced by the interaction of the main flux and the flux produced by current carrying conductors placed in armature slots. Due to the large reluctance of the air-path of slots, the main flux passing through the conductors is negligible and so is the force acting on the conductor. Force is produced mainly by the distortion of the flux lines passing through the teeth, and this force acts on the teeth of the armature as shown in Fig. 7.15. It is rather fortunate that there is very little force acting on conductors. If all the force were to act on conductors, it would crush the insulation between conductors and slots.

[‡] It is to be noted that the flux/pole Φ is not dependent on the actual shape of flux density distribution in the airgap. The flux density distribution in fact gets distorted by the armature current.

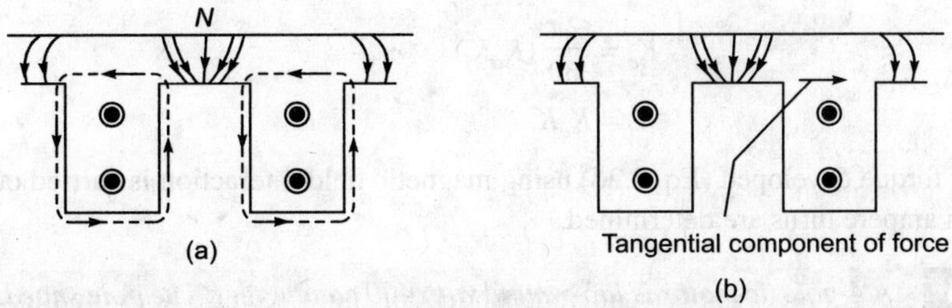

Fig. 7.15 *(a) Main flux (solid) and flux (dotted) produced by armature current-carring conductors, (b) Resultant flux*

Power Balance

Mechanical power

$$T\omega_m = K_a\,\Phi\omega_m I_a$$
$$= E_a I_a \text{ W} \tag{7.28}$$

This is nothing but a statement of energy conservation, i.e. electrical and mechanical powers must balance in a machine.

$E_a I_a$ is referred to as *electromagnetic power.* From Eq. (7.28) we get the *electromagnetic torque* as

$$T = (E_a I_a)/\omega_m$$

Sum-up

$$\text{Armature emf, } E_a = K_a\,\Phi\omega_m \tag{7.29}$$
$$\text{Electromagnetic torque, } T = K_a\,\Phi I_a \tag{7.30}$$

where $K_a = \dfrac{ZP}{2\pi A}$; machine constant

$$\omega_m = \frac{2\pi n}{60} \text{ rad/s}$$

n = speed in rpm

Power balance $$T\omega_m = E_a I_a = \text{Electromagnetic power} \tag{7.31}$$

In frequent use we may drop the suffix m in ω_m, i.e., write ω in place of ω_m.

Linear Magnetization

If the magnetic circuit of the machine is assumed linear[§]

$$\Phi = K_f I_f$$

where I_f = field current

K_f = field constant

Then $$E_a = K_a K_f I\omega_m = K_e\, I_f n \text{ V} \tag{7.32}$$

and $$T = K_a K_f I_f I_a = K_t I_f I_a \text{ Nm} \tag{7.33}$$

[§]Presence of air-gap justifies this approximation so long as iron is lightly in saturated state.

where $$K_e = \frac{2\pi}{60}(K_a K_f) \tag{7.34}$$

$$K_t = K_a K_f \tag{7.35}$$

The derivation of torque developed (Eq. 7.26) using magnetic field interaction is carried out in Sec. 7.6 after armature reaction ampere turns are determined.

Example 7.5 *A 4-pole dc motor is lap-wound with 400 conductors. The pole shoe is 20 cm long and average flux density over one-pole-pitch is 0.4 T, the armature diameter being 30 cm. Find the torque and gross mechanical power developed when the motor is drawing 25 A and running at 1500 rpm.*

LO 2

Solution

$$\text{Flux/pole} = \frac{\pi \times 30 \times 10^{-2}}{4} \times 20 \times 10^{-2} \times 0.4 = 0.0188 \text{ Wb}$$

$$\text{Induced emf} = \frac{\Phi n Z}{60}\left(\frac{P}{A}\right)$$

$$= \frac{0.0188 \times 1500 \times 400}{60} \times \left(\frac{4}{4}\right) = 188 \text{ V}$$

$$\text{Gross mechanical power developed} = E_a I_a = \frac{188 \times 25}{1000} = 4.7 \text{ kW}$$

$$\text{Torque developed} = \frac{4.7 \times 1000}{\left(\frac{2\pi \times 1500}{60}\right)} = 29.9 \text{ Nm}$$

7.5 CIRCUIT MODEL

LO 2

The parallel paths of dc machine armature are symmetrical and each has an induced emf E_a and a resistance R_p. as shown in Fig. 7.16 below for $A = 4$. Its Thevenin equivalent is drawn by the side in which

$$V_{oc} = E_a,\ R_{TH} = R_p/A = R_a \tag{7.36}$$

The armature can therefore be represented by the symbol as shown in Fig. 7.16 with E_a within circle and the series resistance R_a written by its side. We may later on skip writing R_a with the understanding that it present within.

Circuit Model of dc Machine

The armature resistance is quite small so as to limit the copper-loss to an acceptable value. Figure 7.16 also shows the field circuit of the machine and the field coil axis is placed at 90° to the brush axis as per the actual arrangement in the

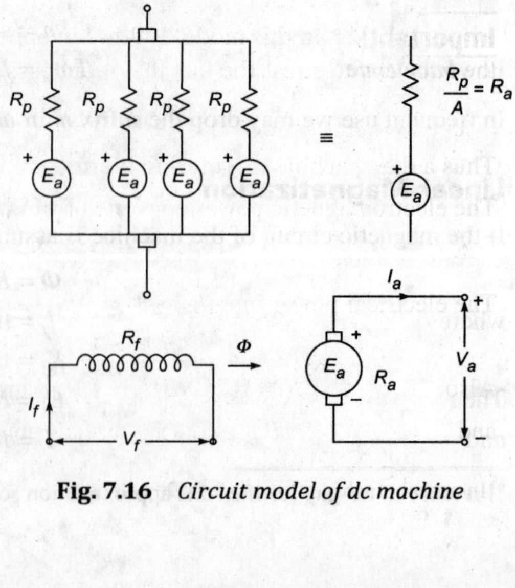

Fig. 7.16 *Circuit model of dc machine*

machine[¶]. From circuit point of view it is not necessary to rigidly follow this scheme. Since most of the time steady-state dc behaviour of the machine will be considered, the inductances of field and of armature (this is negligible any way) circuits are of no consequence and are not shown in the circuit model. The armature induced emf and machine torque are governed by the relationships of Eqs (7.29) and (7.30).

The voltage drop at brush-commutator contact is fixed (1–2 V), independent of armature current as the conduction process is mainly through numerous short arcs. However, this voltage being small is modelled as linear resistance and lumped with R_a. From now onwards it will be assumed that R_a includes the effect of brush voltage drop.

Generating Mode

Important! The machine operates in generating mode (puts out electrical power) when I_a is in the direction of induced emf E_a as in Fig. 7.17(a). For the armature circuit

$$V_t \text{ (armature terminal voltage)} = E_a - I_a R_a;\ E_a > V_t \tag{7.37}$$

Thus a dc machine is generating if its armature induced emf (E_a) is more than its terminal voltage (V_t)

The electromagnetic power converted from mechanical to electrical from is

$$E_a I_a = P_{\text{mech}}(\text{in})|_{\text{net}} = P_{\text{elect}}(\text{out})|_{\text{gross}} \tag{7.38}$$

The net electrical power output is

$$P_0 = V_t I_a \tag{7.39}$$

Also
$$E_a I_a - V_t I_a = I_a^2 R_a = \text{armature copper-loss (inclusive of brush loss)} \tag{7.40}$$

and
$$P_{\text{mech}}(\text{in})|_{\text{gross}} = \text{shaft power} = P_{\text{mech}}(\text{in})|_{\text{net}} + \text{rotational loss} \tag{7.41}$$

In this mode torque (T) of electromagnetic origin is opposite to the direction of rotation of armature, i.e., mechanical power is absorbed and a prime-mover is needed to run the machine.

The conductor emf and current are also in the same direction for generating mode as shown in the cross-sectional view of Fig. 7.17(c).

Motoring Mode

Important! In this mode, I_a flows in opposition to induced emf E_a as in Fig. 7.17(b). E_a is now known as the ***back emf*** to stress the fact that it opposes the armature emf. For the armature circuit

$$V_t \text{ (armature terminal voltage)} = E_a + I_a R_a;\ V_t > E_a \tag{7.42}$$

Thus a d.c. machine is motoring if armature terminal voltage (V_a) is more than its induced emf (E_a).

The electromagnetic power converted from mechanical to electrical from is

$$E_a I_a = P_{\text{elect}}(\text{in})|_{\text{net}} = P_{\text{mech}}(\text{out})|_{\text{gross}} \tag{7.43}$$

The electrical power input is

$$P_i = V_t I_a \tag{7.44}$$

Also
$$V_t I_a - E_a I_a = I_a^2 R_a = \text{armature copper-loss (inclusive of brush loss)} \tag{7.45}$$

and
$$P_{\text{mech}}(\text{out})|_{\text{net}} = \text{shaft power} = P_{\text{mech}}(\text{out})|_{\text{gross}} - \text{rotational loss} \tag{7.46}$$

[¶]In actual machine this angle is 90° elect.

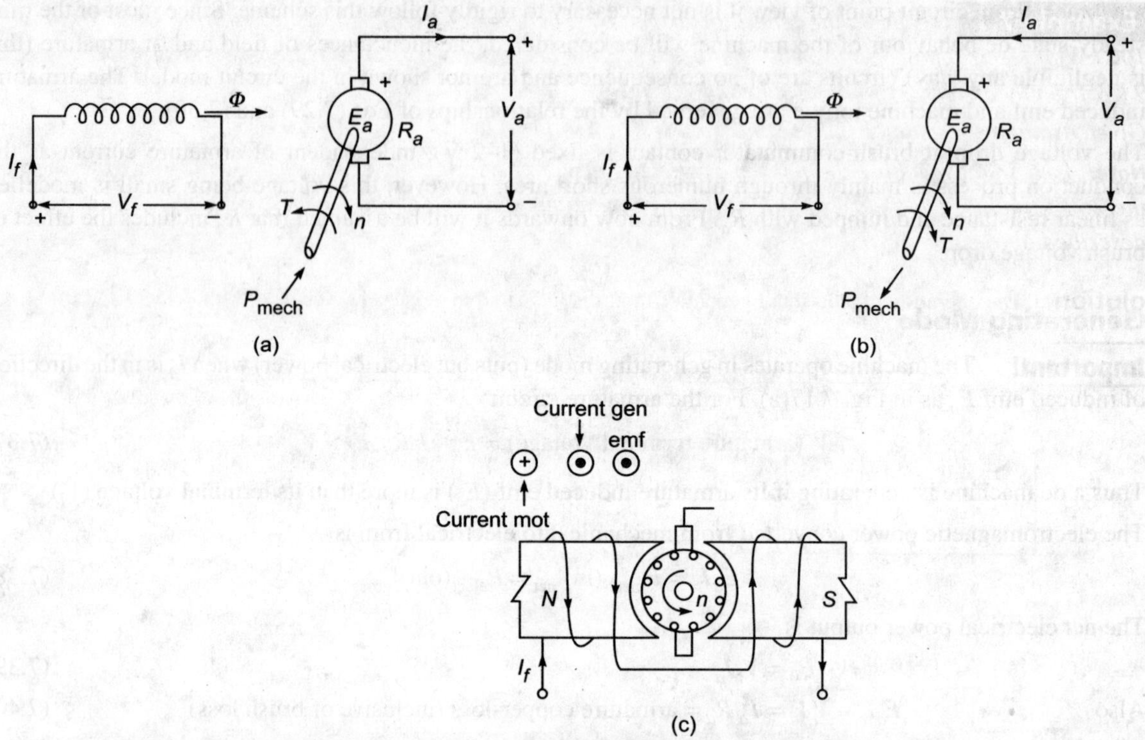

Fig. 7.17 *(a) Generating mode, (b) Motoring mode, (c) Generating/motoring modes*

In this mode torque (T) of electromagnetic origin is in the direction of armature rotation, i.e., mechanical power is put out and is and absorbed by load (mechanical).

Conductor emf and current are also in opposite directions for motoring mode as shown in Fig. 7.17(c).

Example 7.6 *A 220 V dc generator supplies 4 kW at a terminal voltage of 220 V, the armature resistance being 0.4 Ω. If the machine is now operated as a motor at the same terminal voltage with the same armature current, calculate the ratio of generator speed to motor speed. Assume that the flux/pole is made to increase by 10% as the operation is changed over from generator to motor.* LO 2

Solution From Eq. (7.22)

$$n \propto \frac{E_a}{\Phi} \tag{i}$$

As a generator

$$I_a = \frac{4 \times 1000}{220} = 18.18 \text{ A}$$

$$E_{ag} = 220 + 0.4 \times 18.18 = 227.3 \text{ V} \tag{ii}$$

As a motor

$$E_{am} = 220 - 0.4 \times 18.18 = 212.7 \text{ V} \tag{iii}$$

Also

$$\Phi_m = 1.1\, \Phi_g \tag{iv}$$

Substituting in Eq. (i)

$$\frac{n_g}{n_m} = \frac{227.3}{212.7} \times \frac{\Phi_m}{\Phi_g} = \frac{227.3}{212.7} \times 1.1 = 1.176$$

Example 7.7 *A dc shunt generator driven by a belt from an engine runs at 750 rpm while feeding 100 kW of electric power into 230 V mains. When the belt breaks it continues to run as a motor drawing 9 kW from the mains. At what speed would it run?*

Given armature resistance 0.08 Ω and field resistance 115 Ω.

Note: *In a shunt machine the field is connected across the armature and is also connected directly to the 230 V mains. The field excitation therefore remains constant as the machine operation changes as described above.*

LO 2

Solution The operation of a dc shunt generator/motor is indicated in the circuit models of Figs 7.18(a) and (b).

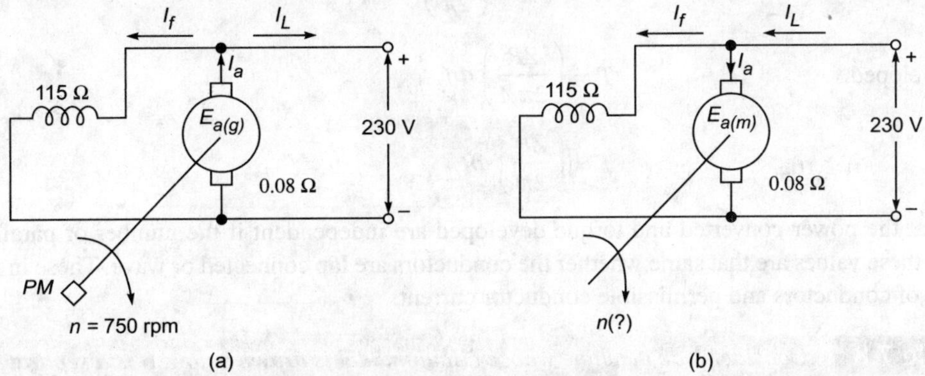

Fig. 7.18 *(a) Generating, (b) Motoring (prime-mover disconnected)*

Field current $I_f = \dfrac{230}{115} = 2$ A; remains constant in operation change-over.

Running as generator (feeding power to mains)

$$I_L \text{ (line current)} = \frac{100 \times 1000}{230} = 434.8 \text{ A}$$

$$I_f = 2 \text{ A}$$

$$I_a = I_L + I_f = 434.8 + 436.8 \text{ A}$$

$$E_a(g) = 230 + 0.08 \times 436.8 = 264.9 \text{ V}$$

$$n(g) = 750 \text{ rpm}$$

Running as motor (drawing power from mains)

$$I_L = \frac{9 \times 1000}{230} = 39.13 \text{ A}$$

$$I_f = 2 \text{ A}$$

$$I_a = I_L - I_f = 39.13 - 2 = 37.13 \text{ A}$$

$$E_a(m) = 230 - 0.08 \times 37.13 = 227 \text{ V}$$

As field current (and so flux/pole) do not change during the two kinds of operation, the induced imf (E_a) is proportional to armature speed. Hence

$$\frac{n(\text{motor})}{n(\text{generator})} = \frac{n(\text{motor})}{750} = \frac{227}{264.9}$$

$$n \text{ (motor)} = 642.7 \text{ rpm.}$$

Lap Versus Wave Winding

Consider a P pole machine having flux/pole Φ and rotating at ω_m rad/s. It has a total of Z conductors and maximum permissible conductor current is I_c. Let us derive the expression for power converted and torque developed.

$$E_a = \left(\frac{ZP}{2\pi A}\right)\Phi\omega_m$$

$$I_a \text{ (permitted)} = A\, I_c$$

$$\text{Power converted} = E_a I_a = \left(\frac{ZP}{2\pi}\right)\Phi\omega_m I_c \quad \text{(i)}$$

Torque developed,

$$T = \left(\frac{ZP}{2\pi A}\right)\Phi I_a$$

or

$$T = \left(\frac{ZP}{2\pi}\right)\Phi I_c \quad \text{(ii)}$$

We find that the power converted and torque developed are independent if the number of parallel paths. It means that these values are that same whether the conductors are lap connected or wave. These in fact depend on number of conductors and permissible conductor current.

Example 7.8 *Consider a dc machine whose circuit model is drawn below. It is a separately excited machine as its field coil (winding) is excited from a voltage source independent of the armature circuit. A 25 kW, 250 V dc machine is separately excited. The field current is held constant at a speed of 3000 rpm. The open circuit voltage is 250 V. Calculate the terminal power, electromagnetic power and torque at terminal voltage of (a) 255 V, and (b) 248 V. The armature resistance is 0.05 Ω. Speed is held constant at 3000 rpm.*

LO 2

Solution Open circuit ($I_a = 0$), Then

$$V_t = E_a = 250 \text{ V at } 3000 \text{ rpm}$$

E_a remains constant

(a) $$V_t = 255 \text{ V}$$

As $V_t > E_a$, the machine is acting as a motor

$$I_a = \frac{V_t - E_a}{R_a} = \frac{255-250}{0.05} = 100 \text{ A}$$

The current flowing into the positive terminal in opposition to E_a. Therefore

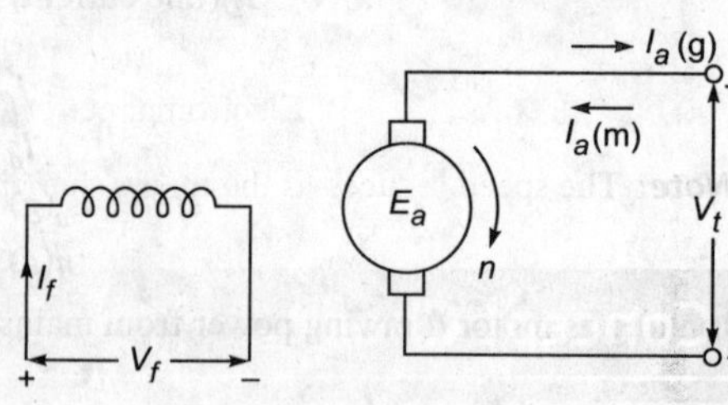

Fig. 7.18(P)

$$\text{Electromagnetic power (in)} = E_a I_a = 250 \times 100 = 25 \text{ kW} = \text{mechanical power output}$$

$$\text{Speed} = 3000 \text{ rpm or } \frac{3000 \times 2\pi}{60} = 314.16 \text{ rad/s}$$

Electromagnetic torque,

$$T = \frac{E_a I_a}{\omega_m} = \frac{25 \times 10^3}{314.16} = 79.58 \text{ Nm}$$

The torque is in the direction of rotation driving the mechanical load which absorbs the mechanical power produced by the motor.

(b) $V_t = 248$ V, $E_a > V_t$, the machine is acting as generator

$$I_a = \frac{250 - 248}{0.05} = 40\ A$$

I_a flows out of positive terminal and is in same direction as E_a.

$$\text{Electromagnetic power (out)} = E_a I_a = 248 \times 40 = 9.92 \text{ kW} = \text{mechanical power in}$$

$$\text{Electromagnetic Torque,} \quad T = \frac{9.92 \times 10^3}{314.16} = 31.58 \text{ Nm}$$

The torque is opposite direction to direction of rotation. The mechanical is absorbed by the machine. It is supplied by the prime mover.

Example 7.9 *In the machine of Example 7.8 the field is held constant with a terminal voltage of 250 V, the armature speed is found to be 2900 rpm. Is the machine motoring or generating? Calculate terminal current, terminal power and electromagnetic power.* LO 2

Solution

As $$E_a = K_a \Phi \omega$$

With Φ constant, at 2950 rpm

$$E_a = 250 \times \frac{2950}{3000} = 245.8 \text{ V}$$

$$V_t = 250 \text{ V}$$

As $V_t > E_a$, the machine is motoring

Terminal quantities

$$I_a = \frac{250 - 245.8}{0.05} = 84 \text{ A (in)}$$

$$\text{Power, } P_{in} = 250 \times 84 = 21 \text{ kW}$$

$$\text{Electromagnetic power} = E_a I_a = 245.8 \times 84 = 20.65 \text{ kW}$$

Note: The speed reduces as the motor shaft carries the load.

Concept Check — LO 2

1. Discuss that emf and torque of a dc machine depend on the flux/pole but are independent of the flux density distribution under the pole.
2. Write the expressions for the induced emf and torque of a dc machine using standard symbol. What is the machine constant? What is the value of the constant relating ω_m and n?
3. When does a dc machine operate in generating mode?
4. When does a dc machine operate in motoring mode?

7.6 ARMATURE REACTION

LO 3 Review the concepts of armature reaction and commutation and discuss the need for compensating winding

When the armature of a dc machine carries current, the distributed armature winding produces its own mmf (distributed) known as armature reaction. The machine airgap is now acted upon by the resultant mmf distribution caused by simultaneous action of the field ampere-turns (AT_f) and armature ampere-turns (AT_a). As a result the air-gap flux density gets distorted as compared to the flat-topped (trapezoidal) wave with quarter-wave symmetry when the armature did not carry any current.

Figure 7.19 shows the cross-sectional view of a 2-pole machine with single equivalent conductor in each slot (current/conductor = UN_cI_c where U = coil-sides/slot, N_c = conductors/coil-side (turns/coil) and I_c = conductor current). Two axes can be recognized—the axes of main poles called the *direct axis* (*d*-axis) and the axis at 90° to it called the *quadrature axis* (*q-axis*). Obviously the *q-axis* is the geometric neutral axis (GNA) of the machine. The brushes in a dc machine are normally located along the *q-axis*.

Armature Current Distribution

Because of commutator action, armature current distribution is as shown in Fig. 7.19 for a 2-pole machine (or Fig. 7.3 for a 4-pole machine). All the conductors on the armature periphery between adjacent brushes carry currents (of constant value, UN_cI_c) in one direction and the current distribution alternates along the periphery. This current pattern remains fixed in space independent of armature rotation. Since the brushes are placed along GNA, the stationary armature current pattern is congruent with the main poles. Also, the current pattern can be shifted by moving all the brushes simultaneously to either side; this is not a normal operation in a dc machine.

Armature Reaction Flux

Probe

What is cross-magnetizing effect?

It is easy to see from Fig. 7.19 that the axis of AT_a lies along the *q*-axis at 90° elect. to the of main poles which lies along the *d*-axis (AT_a lags behind AT_f with respect to the direction of armature rotation for the motoring mode and vice versa for the generating mode). It may be noticed that AT_a and AT_f as shown by arrows are not vectors as their space-angle distribution is non-sinusoidal (though periodic). Armature reaction with axis at 90° to the main field axis is known as cross-magnetizing mmf. Figure 7.19 also shows the flux pattern, in dotted lines, caused by armature reaction acting alone. It is immediately observed that the armature reaction flux strengthens each main pole at one end and weakens it at the other end (*crossmagnetizing* effect). If the iron in the magnetic circuit is assumed unsaturated (therefore linear), the net flux/pole remains unaffected by armature reaction though the air-gap flux density distribution gets distorted. If the main pole excitation is such that iron is in the saturated region of magnetization (this is the case in a practical machine), the increase in flux density at one end of the poles caused by armature reaction is less than the decrease at the other end, so that there is a net reduction in the flux/pole, a *demagnetizing* effect; the decrement being dependent upon the state of magnetization of iron and the amount of AT_a (i.e. the armature current).

It may be summarized here that the nature of armature reaction in a dc machine is cross-magnetizing with its axis (stationary) along the *q*-axis (at 90° elect. to the main pole axis). It causes no change in flux/pole if the iron is unsaturated but causes reduction in flux/pole (demagnetizing effect) in presence of iron saturation.

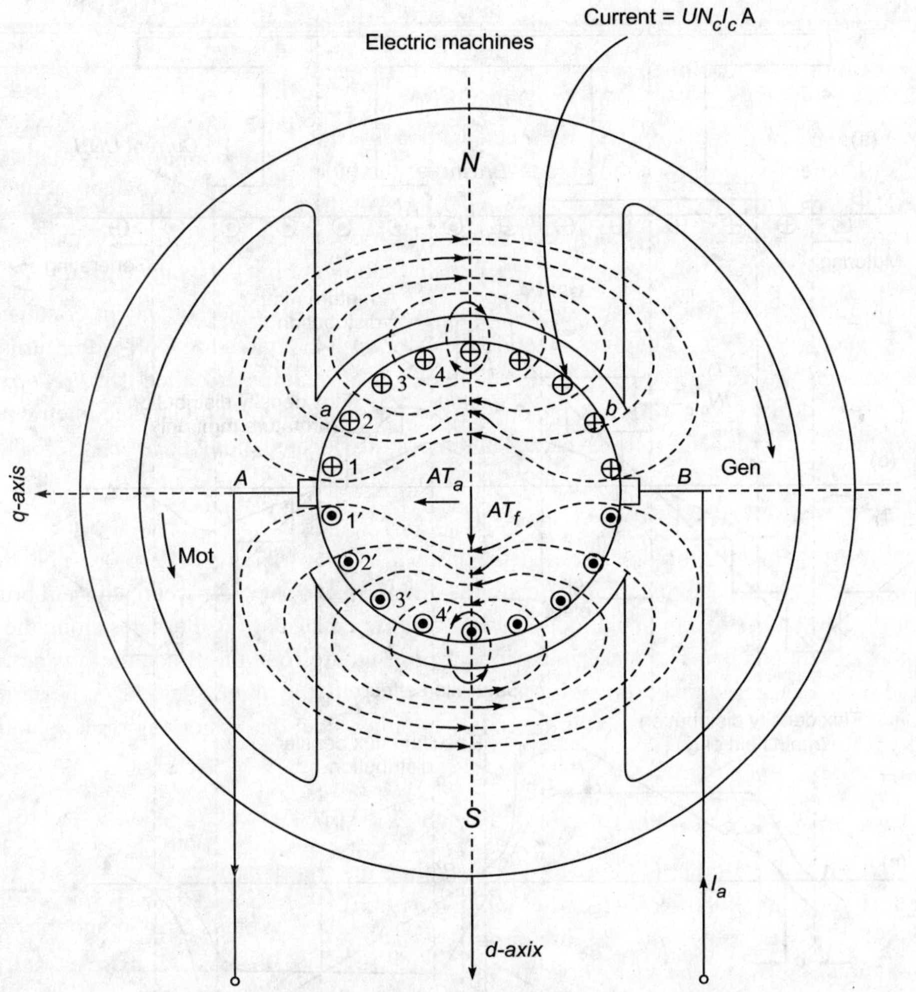

Fig. 7.19 *Armature reaction flux in dc machine*

Graphical Picture of Flux Density Distribution

For a better understanding of the interaction between the field and the armature magnetic field, consider the developed diagram of Fig. 7.20(a) for one pole-pair with brushes placed in geometrical neutral axis (GNA), which is also magnetic neutral axis (MNA) when armature is not carrying current. Using the principles evolved in Sec. 5.4, the armature mmf distribution is drawn in Fig. 7.20(b) which is a *stepped wave* with axis shift of 90° elect. from the main pole axis (*d-axis*), i.e. it is cross-mangnetizing. Each step of the wave has a height of UN_cI_c, where U = coil-sides/slot, N_c = conductors/coil-side (turns/coil) and I_c = conductor current. The stepped-wave of mmf can be well approximated as a triangular wave as shown in Fig. 7.20(b). The peak value of armature ampere-turns is obtained as follows:

$$\text{Ampere-conductors/pole} = \frac{ZI_c}{P} = \frac{ZI_a}{AP}$$

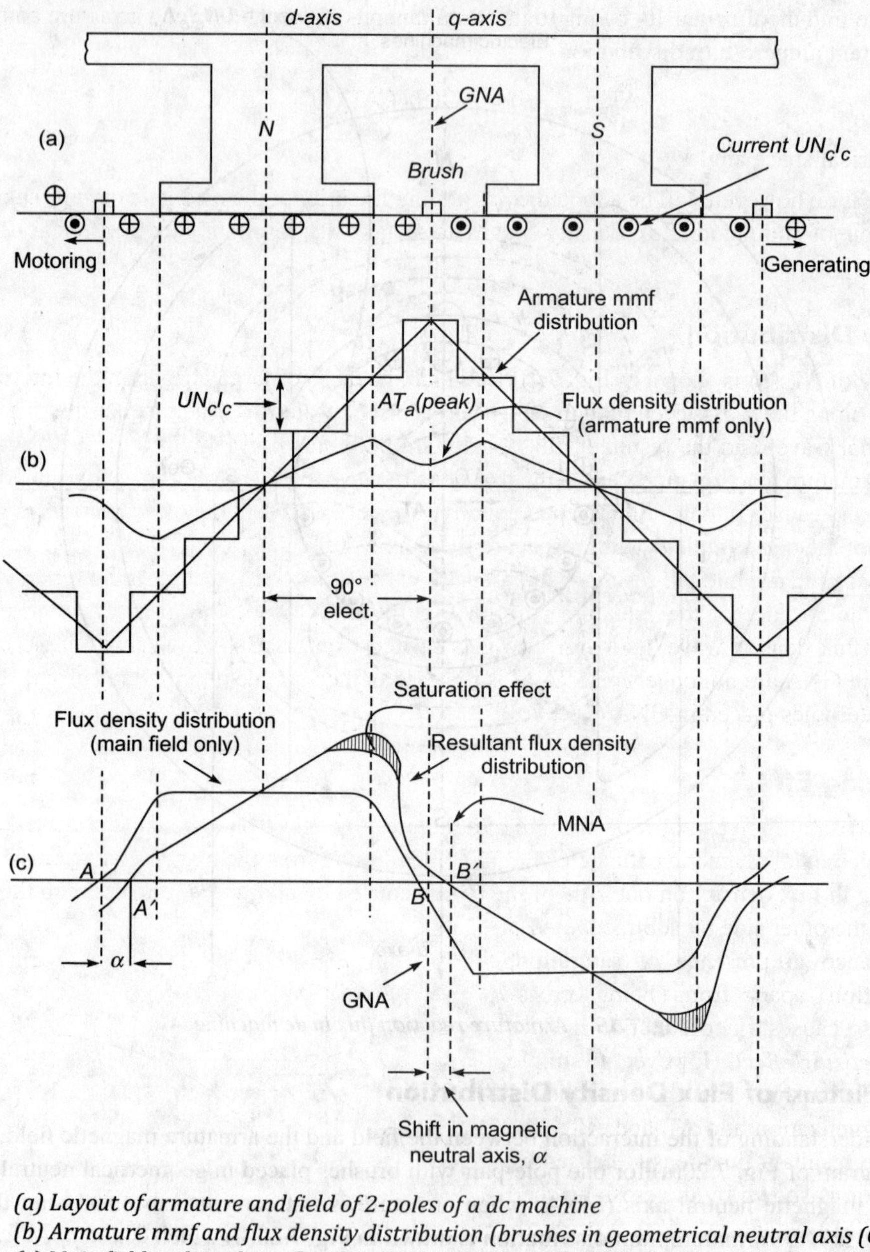

Fig. 7.20 *(a) Layout of armature and field of 2-poles of a dc machine*
(b) Armature mmf and flux density distribution (brushes in geometrical neutral axis (GNA))
(c) Main field and resultant flux density distribution—shift (α) in magnetic neutral axis (MNA)

$$\text{Ampere-turns/pole} = \frac{Zl_a}{2AP} = \text{AT}_a\text{ (peak)} \tag{7.47}$$

Also

$$\text{AT}_a\text{ (peak)} = \frac{\text{AT}_a\text{(total)}}{P}$$

The exact way to find the flux density owing to the simultaneous action of field and armature ampere-turns is to find the resultant ampere-turn distribution

$$\text{AT}_{\text{resultant}}(\theta) = \text{AT}_f(\theta) + \text{AT}_a(\theta)$$

where θ is electrical space angle.

A simpler procedure, however, will be adopted by assuming linearity of the magnetic circuit making possible the superposition of individual flux density waves to obtain the resultant flux density as illustrated in Fig. 7.20(c).

Flux Density Distribution

The flux density of $\text{AT}_a(\theta)$ is shown in Fig. 7.20(b) which, because of large air-gap in the interpolar region, has a strong dip along the *q*-axis even though AT_a(peak) is oriented along it. The flux density of the main field alone (trapezoidal wave) and the resultant flux density are both drawn in Fig. 7.20(c). It is found from this figure that the armature reaction mmf causes the flux density wave to get distored so as to be depressed in one half of the pole and causes it to be strengthened equally (linearity effect) in the other half (cross-magnetizing effect) because of the odd symmetry with respect to the *d*-axis of the flux density wave of the armature mmf. It is, therefore, seen that while the resultant flux density wave is distorted, the flux/pole remains unchanged at its value in the absence of the armature current. Figure 7.20(c) also reveals that apart from distortion of the resultant flux density wave, its MNA also gets shifted from its GNA by a small angle α so that the brushes placed in GNA are no longer in MNA as is the case in the absence of armature current. This effect is countered by interpoles placed in GNA (Sec. 7.8).

Demagnetizing Effect

The effect of iron saturation can now be brought into picture. At angle θ on either side of the *d*-axis, ampere-turn acting on elemental magnetic path are $(\text{AT}_f \pm \text{AT}_a)$. It is found from the magnetization curve of Fig. 7.21 that the increase in flux density on one side of the *d*-axis caused by additive AT_a is less than the decrease in flux density on the other side by subtractive AT_a.

As a consequence in presence of saturation, armature reaction apart from being cross-magnetizing also causes a net reduction in flux/pole, a *demagnetizing effect*. However, no simple quantitative relationship can be established between demagnetization and AT_f and AT_a. The reduction in the resultant flux density caused by saturation of iron is shown by the cross-hatched areas in Fig. 7.20(c).

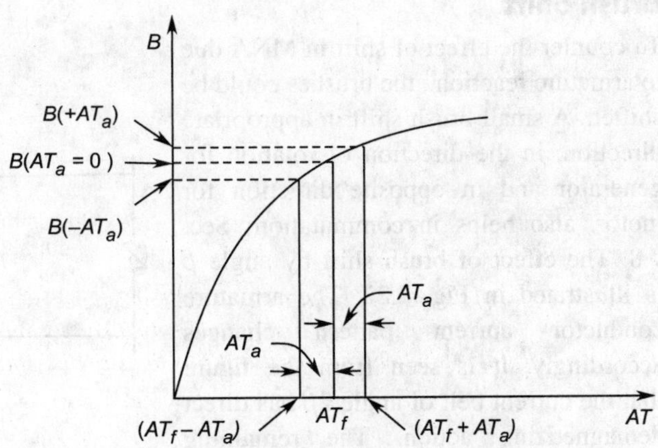

Fig. 7.21 *Magnetization curve*

Summary

To summarize, the armature reaction in a dc machine is cross-magnetizing causing distortion in the flux density wave shape and a slight shift in MNA. It also causes demagnetization because a machine is normally designed with iron slightly saturated.

Other Deleterious Effects

(i) Increase in iron loss Increase if flux density under one half of the pole and decrease on the other causes iron loss in armature teeth to increase as the loss is proportional to square of flux density.

(ii) Commutation Shift of MNA causes induced emf in coils undergoing commutation to oppose the current reversal. It is seen from Fig. 7.20 that for generating case the coil-side to the right of the brush will have ⊕ emf while it should be – emf. The same holds for the motoring case. The details will be discussed in Sec. 7.8 on commutation.

(iii) Possibility of commutator sparking Under heavy load (large armature current) and so deep distortion both coil sides of the coil passing the maximum flux density region will have much larger induced emf than the average coil emf. If the emf value exceeds 30-40 V, there may be sparkover of the commutator segment connected to the coil to the adjacent segment. The result may be flash-over of the complete commutator.

Remedies

The cross-magnetizing effect of the armature reaction can be reduced by making the main field ampere-turns larger compared to the armature ampere-turns such that the main field mmf exerts predominant control over the air-gap flux. This is achieved by:

Probe

How to reduce the effects of armature reaction?

(i) Introducing saturation in the teeth and pole-shoe.

(ii) By *chamfering* the pole-shoes which increases the air-gap at the pole tips. This method increases the reluctance to the path of main flux but its influence on the cross-flux is much greater. This is because the cross flux has to cross the air-gap twice: see Fig. 7.19.

(iii) The best yet the most expensive method is to compensate the armature reaction mmf by a compensating winding located in the pole-shoes and carrying a suitable current. This method is discussed in detail in Sec. 7.7.

Brush Shift

To counter the effect of shift in MNA due to armature reaction , the brushes could be shifted. A small brush shift in appropriate direction, in the direction of rotation for generator and in opposite direction for motor, also helps in commutation; Sec. 7.8. The effect of brush shift by angle β is illustrated in Fig. 7.22. The armature conductor current pattern changes accordingly. It is seen from the figure that the current belt of angle 2β has direct demagnetizing action. The remaining current belt of angle $(180° - 2\beta)$ is cross-magnetizing. This is illustrated by the following example.

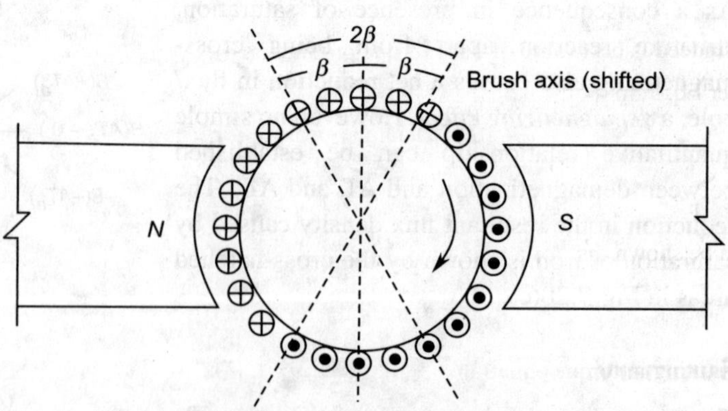

Fig. 7.22 *Effect of brush shift*

Example 7.10 *A 250 kW, 400 V, 6-pole dc generator has 720 lap wound conductors. It is given a brush lead of 2.5 angular degrees (mech). from the geometric neutral. Calculate the cross and demagnetizing turns per pole. Neglect the shunt field current.*

LO 3

Solution

$$\text{Armature current, } I_a = \frac{25 \times 10^3}{400} = 625 \text{ A}$$

Number of parallel paths = 6

$$\text{Conductor current, } I_c = \frac{625}{6} = 104.2 \text{ A}$$

$$\text{Total armature ampere-turns, AT}_a = \frac{1}{2}\left(\frac{720 \times 104.2}{6}\right)$$

$$= 6252 \text{ AT/pole}$$

With reference to Fig. 7.22, it is easily observed that in 180° elect., conductors in that belt of 2β elect. degrees are demagnetizing where β is the brush shift in electrical degrees.

$$\beta = 2.5\,(6/2) = 7.5° \text{ elect.}$$

Hence,

$$\text{cross-magnetizing ampere-turns} = 6250\left(1 - \frac{2 \times 7.5}{180}\right) = 5731 \text{ AT/pole}$$

$$\text{Demagnetizing ampere-turns} = 6250 \times \frac{2 \times 7.5}{180} = 521 \text{ AT/pole}$$

Torque Equation *(Based on magnetic field interaction)*

The magnetic field interaction torque given by Eq. (7.48) is reproduced below

$$T = (\pi/2)\,(P/2)^2\, \Phi_r F_2 \sin \delta \tag{7.48}$$

It has been shown above that the resultant flux/pole Φ in a dc machine is always oriented at 90° to the armature reaction *AT* (i.e.F_2). Thus

$$\delta = 90°(\text{fixed})$$

This indeed is best value of δ for torque production

As per Eq. 7.46

$$F_2 \text{ (triangular peak)} = \text{AT}_a \text{ (peak)}$$

$$= \frac{ZI_a}{2AP} \text{ per pole}$$

It can be shown that for a triangular periodic wave, the fundamental is $8/\pi^2$ of the peak value. Thus

$$F_2 \text{ (fundamental peak)} = \frac{8}{\pi^2}\frac{ZI_a}{2AP}$$

Substituting in torque equation

$$T = \frac{\pi}{2}\left(\frac{P}{2}\right)^2 \Phi \frac{8}{\pi^2}\frac{ZI_a}{2AP}$$

or

$$T = \frac{1}{2\pi}\Phi\left(\frac{P}{A}\right) ZI_a; \text{ same as Eq. (7.26)}$$

7.7 COMPENSATING WINDING

LO 3

It was seen in Sec. 7.6, Fig. 7.20(c) that armature reaction causes the flux density wave to be so badly distorted that when a coil is passing through the region of peak flux densities, the emf induced in it far exceeds the average coil voltage. If this emf is higher than the breakdown voltage across adjacent segments, a sparkover could result which can easily spread over and envelop the whole commutator as the environment near the commutator is always somewhat ionized and conditions are favourable for flashover. The result is complete short circuit of armature. The maximum allowable voltage between adjacent segments is 30–40 V, limiting the average voltage between them to much less than this figure. The choice of the average coil voltage determines the minimum number of commutator segments for its design.

In spite of the above safe design of the commutator there is another factor which can cause severe overvoltages to appear between commutator segments. This is the time variation of the armature reaction and its associated flux owing to sudden changes in machine load. Consider coil *aa′* of Fig. 7.23 located midway between the main poles so that the full armature flux/pole, Φ_a (shaded area), links the coil. If the load on the machine undergoes a fast change, I_a and Φ_a change accordingly resulting in statically induced emf in the coil proportional to $d\Phi_a/dt$. The voltage is over and above the dynamically induced emf in the coil. Worst conditions occur when these two emfs are additive. This happens when load is dropped from a generator or added to a motor. (The reader should verify by application of the right-hand rule and Lenz's law to coil *aa′* of Fig. 7.23.) The only way to remedy this situation is to neutralize the armature reaction ampere-turns by a *compensating winding* placed in slots cut out in pole faces such that the axis of this winding coincides with the brush axis (along which lies the axis of AT_a). For automatic neutralization of AT_a at any current, it is necessary that the compensating winding be series excited with armature current in such a direction as to oppose AT_a. The compensating winding appropriately connected is shown schematically in Figs 7.24(a) and (b).

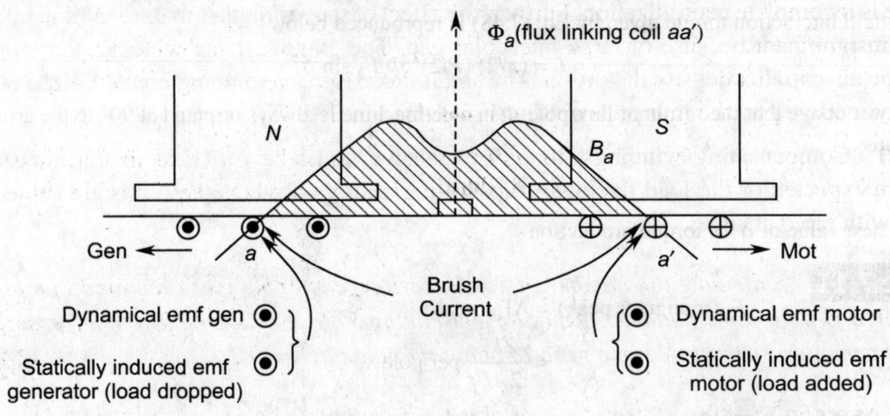

Fig. 7.23

Effect of Compensating Winding on Armature mmf

It is found from Fig. 7.24(a) that complete neutralization of the armature mmf is not possible with this arrangement, since the distributions of armature and compensating mmfs are not identical. It is customary to compensate part of the armature mmf directly under the pole shoes. The number of ampere-turns required for this purpose is

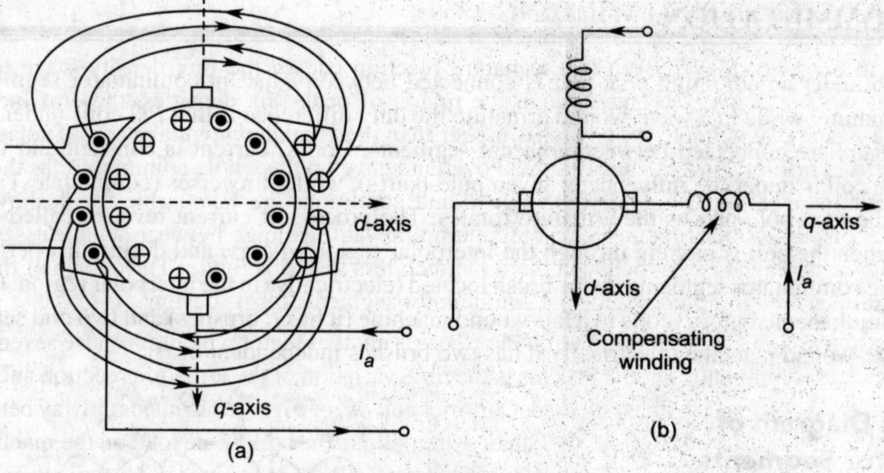

Fig. 7.24 *Compensating Winding*

$$AT_{cw}/\text{pole} = AT_a\text{ (peak)} \times \left(\frac{\text{pole arc}}{\text{pole pitch}}\right)$$

$$= \frac{I_a Z}{2AP} \times \frac{\text{pole arc}}{\text{pole pitch}} \tag{7.49a}$$

The compensating winding neutralizes the armature mmf directly under the pole while in the interpolar region, there is incomplete neutralization. Further, the effect of the resultant armature mmf in interpolar region is rendered insignificant because of large interpolar gap. The compensating winding, therefore, practically eliminates the air-gap flux density distortion. The small flux density remaining unneutralized in GNA will be appropriately modified by the interpole windings discussed in Sec. 7.8.

!Attention Compensating windings, though expensive, must be provided in machines where heavy overloads are expected or the load fluctuates rapidly, e.g. motors driving steel-mills are subjected to severe duty cycles with rapid changes.

Example 7.11 *Calculate the number of conductors on each pole piece required in a compensating winding for a 6-pole lap-wound dc armature containing 286 conductors. The compensating winding carries full armature current. Assume ratio of pole arc/ pole pitch = 0.7.*

LO 3

Solution As per Eq. (7.48)

$$AT_{cw}/\text{pole} = \frac{I_a Z}{2AP}\left(\frac{\text{pole arc}}{\text{pole pitch}}\right)$$

$$\therefore \quad N_{cw}/\text{pole} = \frac{Z}{2AP}\left(\frac{\text{pole arc}}{\text{pole pitch}}\right) = \frac{286}{2 \times 6 \times 6} \times 0.7 = 2.78$$

Compensating conductors/pole = 2 × 2.78 = 6 (nearest integer).

7.8 COMMUTATION

LO 3

One coil each under an adjoining pole-pair is connected between adjacent commutator segments in a lap-wound dc armature, while in a wave-wound armature the only difference is that $P/2$ coils under the influence of $P/2$ pole-pairs are connected between adjacent segments. Coil(s) current is constant and unidirectional so long as the coil is under the influence of given pole-pair(s), while it reverses (commutates) when the coil passes onto the next pole-pair as the armature rotates. The process of current reversal called commutation takes place when the coil is passing through the interpolar region (q-axis) and during this period the coil is shorted via the commutator segments by the brush located (electrically) in the interpolar region. Commutation takes place simultaneously for P coils in a lap-wound machine (it has P brushes) and two coil sets of $P/2$ coils each in a wave-wound machine (electrically it has two brushes independent of P).

Schematic Diagram of Commutator Segments

Figure 7.25 shows the schematic diagram of commutator segments in developed form connected to armature coils. Attention will now be focussed on coil C_c as it undergoes commutation. Various symbols used in the figure are:

I_c = coil current

$I_b = 2I_c$ = brush current

w_c = width of one commutator segment

w_m = width of mica insulation between segments

w_b = brush width
$= (w_c + w_m)$ for the case illustrated; in practice, however, $w_b = 1.5\ w_c$

v_c = peripheral speed of commutator

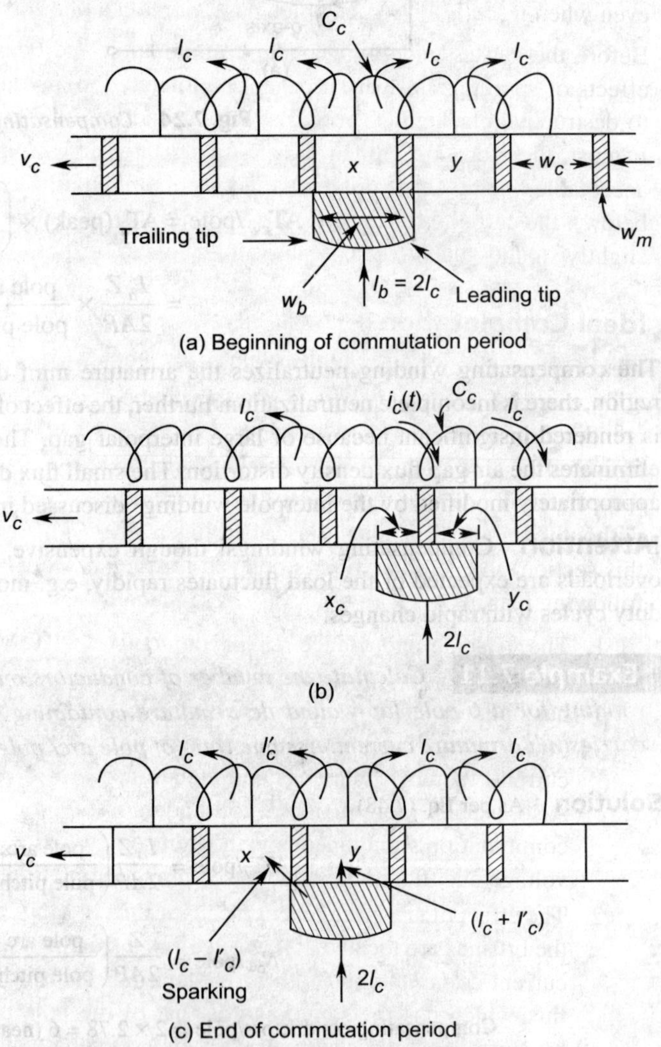

Fig. 7.25 *Commutation process*

At the instant the commutation of the coil C_c begins, the leading tip of the brush is making full contact with the segment x and is just going to make contact with segment y as shown in Fig. 7.25(a). At this instant all coils to the right of segment x carry current I_c flowing from left to right and those on the left current I_c in the opposite direction. During the period of commutation as the coil passes from right to the left of the brush, the coil current must reverse. During this period the brush short-circuits the coil via segments x and y as shown in Fig. 7.25(b). The contact width, x_c, between brush and segment x reduces linearly while the contact width, y_c, between brush and segment y increases. The coil current $i_c(t)$ during this period is changing. If at the end of the commutation period, when the trailing tip of the brush is going to break contact with segment x as shown in Fig. 7.25(c), the coil current has not reversed and acquired full value I_c but as $I_c' < I_c$, the breaking of current $(I_c - I_c')$ at the trailing brush tip takes place causing sparking. This is known as *under-commutation* (or *delayed-commutation*). It is easy to see from Fig. 7.25, that the period of commutation is given by

$$t_c = \frac{w_b - w_m}{v_c} \tag{7.49b}$$

even when $w_b > (w_c + w_m)$ in which case more than one coil undergoes commutation simultaneously.

Before the causes underlying under-commutation and consequent sparking are explained, the deleterious effects of sparking and why it cannot be tolerated to any large degree may now be studied. Sparking leads to destructive blackening, pitting and wear and eventual burning of commutator copper and brush carbon. It must, therefore, be limited to a tolerable intensity to prolong life of commutator-brush assembly to an acceptable value. As completely sparkless commutation is not possible practically (for reasons advanced below), the carbon brushes must be replaced after some time and less frequently commutator "turned" to a slightly smaller diameter to prepare a fresh clean surface.

Ideal Commutation

Ideal Commutation (also called straight-line commutation) is that in which the current of the commutating coils changes linearly from $+I_c$ to $-I_c$ in the commutation period as shown in Fig. 7.26. The figure also shows delayed commutation and the current $(I_c - I_c')$ in the spark. In a machine without commutation aids (described later in this section) the commutation is delayed for the following reasons:

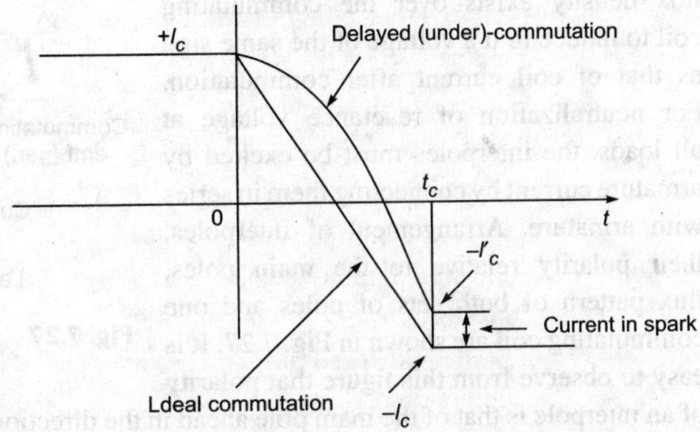

Fig. 7.26 *Straight-line commutation*

1. The leakage inductance L_c of the coil (see Sec. 5.7) undergoing commutation has induced in it *reactance voltage* L_c (di_c/dt) which opposes the change in current thereby delaying commutation. Also, usually more than one coil undergo commutation simultaneously, the induced voltage due to mutual inductance among them also tends to prevent current reversal.
2. The effect of armature reaction causes shift in MNA as shown in Fig. 7.20(c) from A, B to A', B'. Since the brushes are located at A, B(GNA's), a small voltage is induced in the commutating coil. It opposes current commutation (both for generating/motoring machine) as the commutating coil is cutting the flux which has the same sign as that of the pole being left behind. It could be partially remedied by shifting the brushes towards MNA but that causes direct demagnetization and is therefore not employed in practice.

There are two ways of achieving good commutation–close to straight-line commutation. These are resistance commutation and voltage commutation. The former is always used to give marginal support to the latter.

> **Probe**
> How can we achieve ideal commutation?

Resistance Commutation

High contact resistance between commutator segments and brushes, achieved by using carbon brushes, adds resistance to the circuit of the commutating coil thereby reducing the time-constant (L/R) of the current transient ($i_c(t)$), helping it to change faster in the desired direction. Carbon brushes are invariably used in dc machines. They also help reduce commutator wear and are themselves easily replaceable.

Voltage Commutation

To speed up the commutation process, the reactance voltage must be neutralized by injecting a suitable polarity dynamical (speed) voltage into the commutating coil. In order that this injection is restricted to commutating coils, narrow interpoles (also called *commutating poles* or *compoles)* are provided in the interpolar region. These apply a local correction to the air-gap flux density wave such that a pip of appropriate flux density exists over the commutating coil to induce in it a voltage of the same sign as that of coil current after commutation. For neutralization of reactance voltage at all loads, the interpoles must be excited by armature current by connecting them in series with armature. Arrangement of interpoles, their polarity relative to the main poles, flux pattern of both sets of poles and one commutating coil are shown in Fig. 7.27. It is easy to observe from this figure that polarity of an interpole is that of the main pole ahead in the direction of armature rotation for the generating mode and that of the main pole left behind with respect to the direction of rotation for motoring mode. The interpolar air-gap is kept larger than that of the main pole so that their magnetic circuit is linear resulting in cancellation of the reactance voltage (a linear derivative term) at all loads. Large air-gap results in greater amount of leakage flux which is accommodated by tapering the interpoles with a wider base as shown in Fig. 7.27.

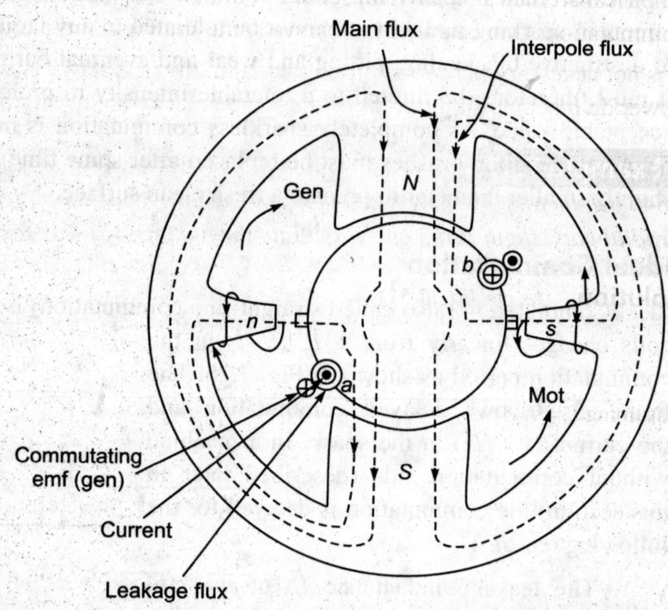

Fig. 7.27 *Interpoles—arrangement and action, 2-pole machine*

For cancellation of reactance voltage on an average basis

$$2[B_i(\text{av})l_i v_a]\, N_c = L_c \frac{di_c}{dt} = L_c\left(\frac{2I_c}{t_c}\right) \tag{7.50}$$

where B_i(av) = average flux density in interpolar air-gap

l_i = iron length of interpoles (it is less than that of main poles)

v_a = armature peripheral speed

N_c = number of turns of commutating coil.

With B_i determined from Eq. (7.50), the ampere-turns needed to cancel the armature reaction ampere-turns and then to create the necessary flux density are given by

$$AT_i = AT_a\,(\text{peak}) + \frac{B_i}{\mu_0} l_{gi} \tag{7.51}$$

where l_{gi} = air-gap of interpoles.

As relationships of Eqs (7.50) and (7.51) are based on sweeping approximation and also accurate estimate of the coil leakage inductance L_c cannot be obtained, the attainment of good commutation is more an empirical art than an analytical science. Furthermore, only good commutation can be achieved but not perfect commutation. One can always observe some sparking at the brushes of a dc machine in operation.

It is not necessary to have interpoles equal to the number of main poles and to reduce cost, especially in low-power dc machines, interpoles of the same polarity are often fitted in alternate interpolar spaces only.

Example 7.12 *A 440 V, 4-pole, 25 kW, dc generator has a wave-connected armature winding with 846 conductors. The mean flux density in the air-gap under the interpoles is 0.5 Wb/m² on full load and the radial gap length is 0.3 cm. Calculate the number of turns required on each interpole.* LO 3

Solution As per Eq. (7.51)

$$AT_i = AT_a(\text{peak}) + \frac{B_i}{\mu_0} l_{gi} = \frac{I_a Z}{2AP} + \frac{B_i}{\mu_0} l_{gi}$$

Assuming $I_a = I_{\text{line}}$

$$I_a = \frac{25 \times 10^3}{440} = 56.82 \text{ A}$$

$$\therefore \quad AT_i = \frac{56.82 \times 846}{2 \times 2 \times 4} + \frac{0.5}{4\pi \times 10^{-7}} \times 0.3 \times 10^{-2} = 4198$$

$$\therefore \quad N_i = \frac{AT_i}{I_a} = \frac{4198}{56.82} = 73.88 = 74$$

Concept Check LO 3

1. Why do we need a compensating winding to nearly over come the armature reaction and how is this winding excited and why?
2. What is demagnetizing effect?
3. Under what circumstances, compensating winding needs to be provided to reduce the effects of armature reaction?

7.9 METHODS OF EXCITATION

LO 4 Outline the excitation methods used for different types of dc machines

The performance characteristics of a dc machine are greatly influenced by the way in which the field winding is excited with direct current. There are two basic ways of exciting a dc machine.

1. Shunt field Here the field winding is provided with a large number (hundreds or even thousands) of turns of thin wire and is excited from a voltage source. The field winding, therefore, has a high resistance and carries a small current. It is usually excited in parallel with armature circuit and hence the name *shunt field winding*. Since the armature voltage of a dc machine remains substantially constant, the shunt field could be regulated by placing an external series resistance in its circuit.

2. Series field Here the field winding has a few turns of thick wire and is excited from armature current by placing it in series with armature, and therefore it is known as *series field winding*. For a given field current, control of this field is achieved by means of a *diverter,* a low resistance connected in parallel to series winding. A more practical way of a series field control is changing the number of turns of the winding by suitable *tappings* which are brought out for control purpose.

Figure 7.28 shows the physical arrangement of shunt and series field windings on one pole of a machine.

Excellent and versatile ways of controlling the shunt and series excitations are now possible by use of solid-state devices and associated control circuitry.

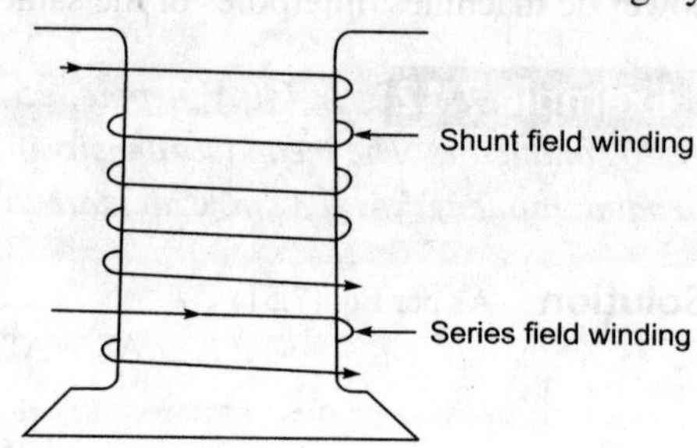

Fig. 7.28 *Shunt and series field windings*

Classification of Excitation Systems

The dc machine excitation is classified in two ways—separate excitation and self-excitation explained below.

Note: In drawing the excitation diagrams of a dc machine, the field winding will be drawn at 90° to the armature circuit. As pointed out earlier the actual spatial orientation of the magnetic fields produced by the field and armature circuits is 90° elect.

Where the excitation diagrams are to be drawn repeatedly, we may not necessarily use this convention.

Separate excitation The field is excited from a source independent of the armature circuit as shown in Fig. 7.29(a). Permanent magnet excitation fall into this category.

Shunt excitation The shunt field is excited from the armature voltage as shown in Fig. 7.29(b).

Series excitation The series field is excited from the armature current as in Fig. 7.29(c).

Remark

As in a dc generator there is no initial voltage or current the shunt field resistance or total circuit resistance in series excitation and the generator speed must meet certain condition for the generation to excite and build up voltage; to be discussed in Section 7.10.

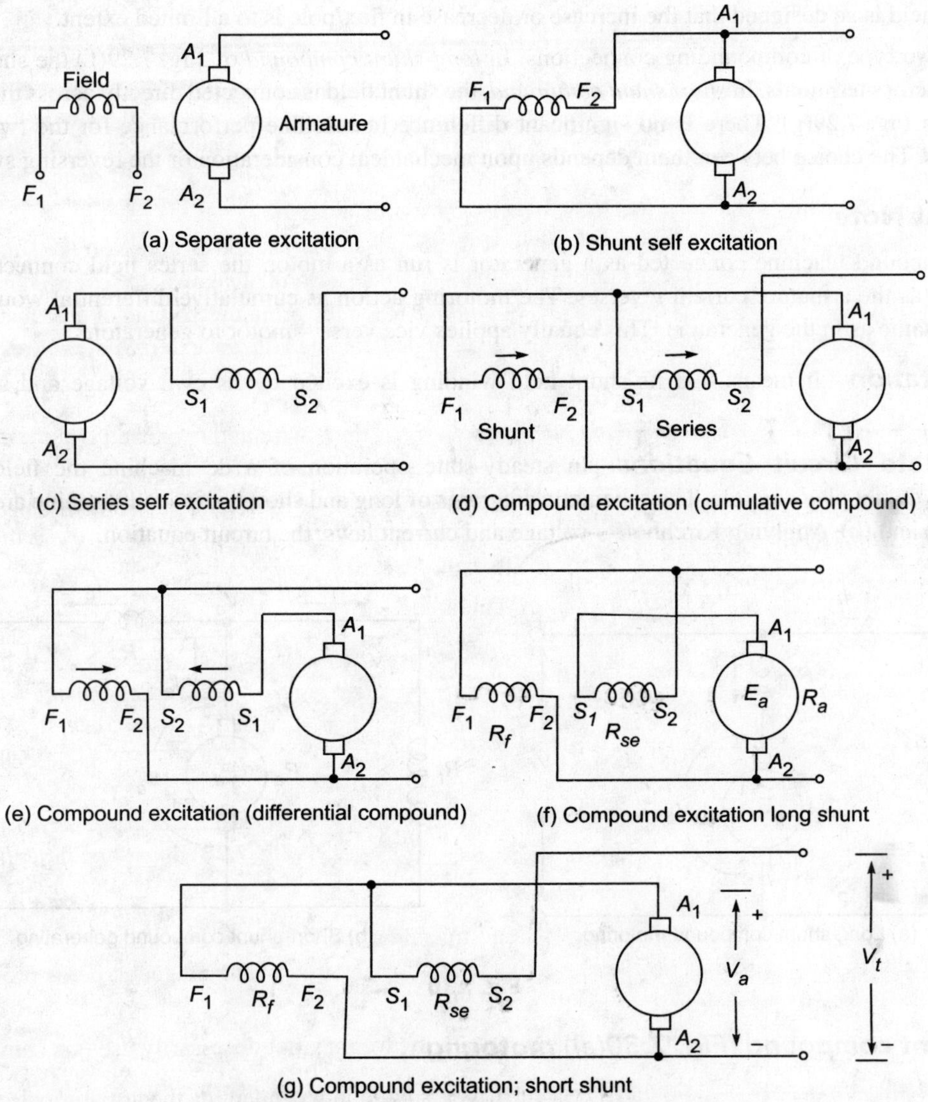

Fig. 7.29 *Methods of excitation of dc machine*

Compound Excitation

!Attention In compound excitation both shunt and series field are excited. If the two field aid each other (their ampere-turn are additive), the excited is called *cumulative compound* as shown in Fig. 7.29(d). The shunt field is much stronger than the series field. The air gap flux increases with armature current.

If the two fields oppose each other, the excitation is called *differential compound* as in Fig. 7.29(e). The air gap flux/pole decreases with armature current.

Probe

What is the difference between cumulative and differential compound?

The series field is so designed that the increase or decrease in flux/pole is to a limited extent.

There are two type of compounding connections. In *long shunt compound* of Fig. 7.29(f) the shunt field is connected across terminals. In *short shunt compound*, the shunt field is connected directly across the armature as shown in Fig. 7.29(g). There is no significant difference in machine performance for the two types of connections. The choice between them depends upon mechanical consideration or the reversing switches.

Important Note

If a dc compound machine connected as a generator is run as a motor, the series field connections must be reversed as the armature current reverses. The motoring action as cumulative/differential would then be preserved (same as in the generator). This equally applies vice versa – motor to generator.

Self–excitation It means that its shunt field winding is excited by its own voltage and series field winding.

Steady-state Circuit Equations In steady-state operation of a dc machine the field winding inductances do not play any role. The schematic diagrams of long and short compound machine are shown in Figs 7.30(a) and (b). Applying Kirchhoff's voltage and current laws, the circuit equation.

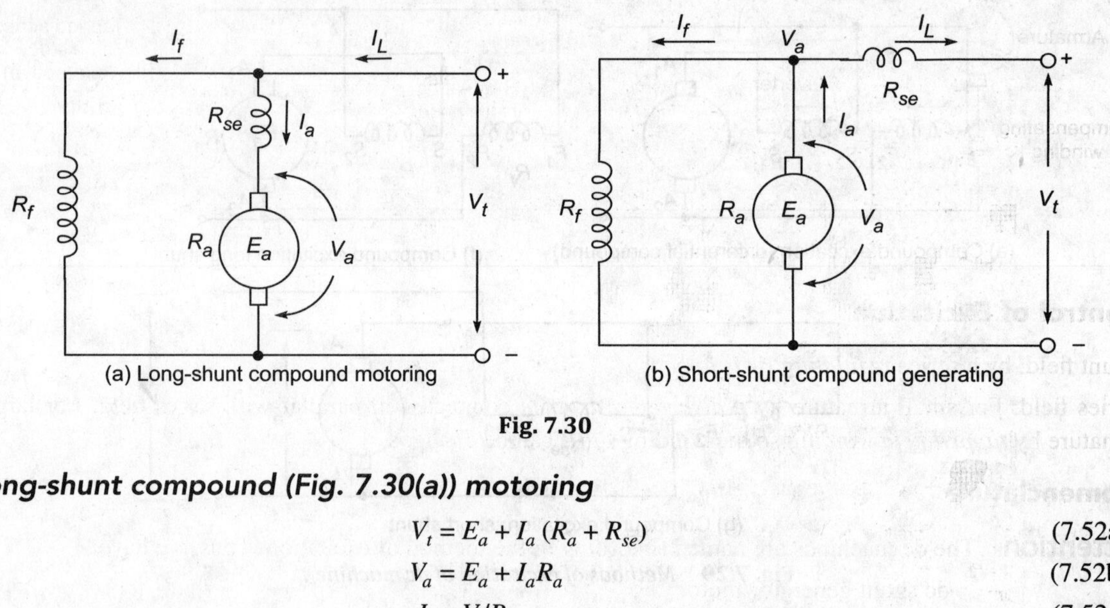

(a) Long-shunt compound motoring (b) Short-shunt compound generating

Fig. 7.30

Long-shunt compound (Fig. 7.30(a)) motoring

$$V_t = E_a + I_a (R_a + R_{se}) \tag{7.52a}$$

$$V_a = E_a + I_a R_a \tag{7.52b}$$

$$I_f = V_t/R_f \tag{7.52c}$$

$$I_L = I_a + I_f \tag{7.52d}$$

Generating

Directions of I_a and I_L reverse while I_f direction does not changes. Correspondingly, in Eqs. (7.52)(a) and (b) + sign changes to – sign.

Short-shunt compound (Fig. 7.30(b)) Though the figure is drawn for generating the equations for both operations can be directly written down

$$V_t = E_a \mp I_a R_a \mp I_L R_{se} \quad (7.53a)$$

$$V_a = E_a \mp I_a R_a \quad (7.53b)$$

$$I_f = V_a / R_f \quad (7.53c)$$

$$I_L = I_a \mp I_f \quad (7.53d)$$

where – sign is for generating and + sign for motoring.

General Connection Diagram of a Compound dc Machine

The connection diagram of a compound d.c. machine with commuting winding and compensating windings is drawn in Fig. 7.31. Total resistance in series with armature is $(R_a + R_i + R_c)$; R_i = interpole (compole) winding resistance, R_c = compensating winding resistance.

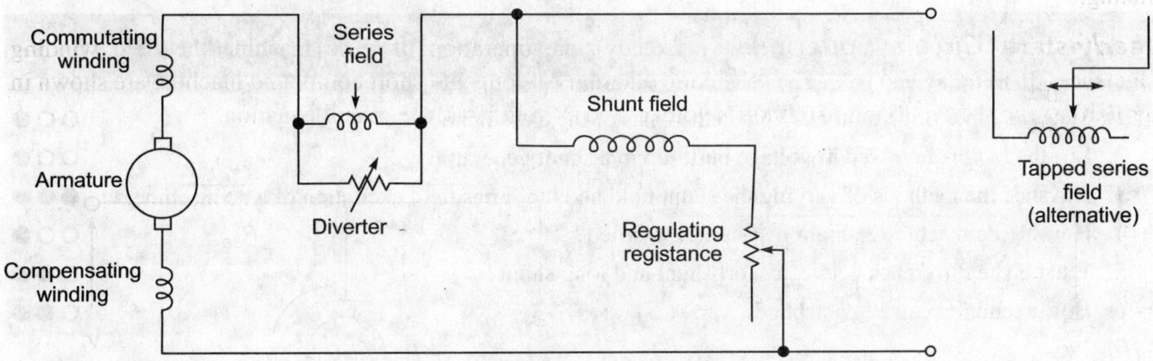

Fig. 7.31 *Compound dc machine, long shunt*

Control of Excitation

Shunt field: by a series *regulating resistance*

Series field: For small armature by a *diverter resistance* connected in parallel with series field. For large armature by *tapped field winding* so the winding turns can be changed.

Nomenclature

!Attention The dc machines are named according to the method of excitation. Thus, we have

dc shunt generator/motor

dc series generator/motor

dc compound generator/motor

As we have seen in section 7.2 a dc machine can operate as a generator or motor; but the machine designed specifically as generator/motor has certain distinguishing features.

Net Excitation

For a machine on load, the next AT excitation is

$$\text{AT (net)} = \text{AT (shunt)} \pm \text{AT (series)} - \text{AT}_d$$

$$= N_f I_f \pm N_{se} I_{se} - AT_d$$

where N_f = shunt field turns

N_{se} = series field current can be reduced by tapping

I_{se} = Series winding Current I_a or I_L; can be reduced by a diverter (a resistance in parallel to the series winding)

AT_d = demagnetizing ampere-turn caused by armature reaction

Usually the net excitation is stated in terms of equivalent shunt field current

$$I_f, \text{eq} = I_f \text{ (net)} = \text{AT (net)}/N_f$$

I_f, eq determines the flux/pole Φ whose measure is the induced emf E_a at specified speed ($E_a = K_a \Phi\omega_m$). It will be discussed in section 7.10.

Concept Check LO 4

1. State the types of dc motors. What is the basis of the classification? ○○●
2. List the factors involved in voltage build-up in a shunt generator. ○○●
3. Advance the methods of varying the shunt field and the series field excitation of a dc machine. ●●●
4. How are dc machine excitation systems classified? ○○●
5. What is the difference between short shunt and long shunt? ○●●
6. How excitation can be controlled? ○●●

7.10 OPERATING CHARACTERISTICS OF dc GENERATOR

LO 5 Analyze the operation of dc generators by its performance characteristics

In the operation of a dc generator, the four basic variable of concern are terminal voltage V_t, armature current I_a, field current I_f and the speed n. To investigate their interrelationship the generator is run (by a prime mover) at rated speed (n constant) of the remaining three variables one is held constant at a certain value and of the last two one is varied to study its relationship with the other. The relationship has to be presented graphically because of the magnetic saturation effect. Four characteristics of importance are the following:

No-load Characteristic

With $I_a = 0$ (no load) at constant n, it is the presentation of V_t (= E_a) vs I_f. This is the most important characteristic as it reveals the nature of the magnetization of the machine. It is easy to determine as the generator is on no load and so only low rated prime mover will serve the purpose. It is commonly called the *open–circuit/magnetization characteristic*.

Load Characteristic

If the plot of V_t vs I_f with I_a with held is constant at rated value and constant speed, it is indeed the magnetization characteristic on load.

External Characteristic

With I_f and n constant at present, the variation of V_t vs I_a is indeed the characteristic when the generator feeds a load (load is normally variable)

Armature Characteristic

It is the presentation of I_a vs I_f with V_t held constant (at rated value) and generator run at constant n and load varied. It reveals the armature reaction affect on the flux/pole. It is also called regulation characteristic.

The characteristics 2, 3 and 4 are to be determined by on load tests which cannot be easily conducted on a large size machine as both prime mover and load are to be fully rated.

Experimental Set-up for Determining Characteristics

The experimental set-up for determining the dc generator characteristics is shown in Fig. 7.32. The generator is run by a prime mover at rated speed. The field current can be varied form a low value to full value by a potentiometer arrangement. The load can be switched 'ON' or 'OFF' and when 'ON' can be varied.

If only no load test is to be conducted, a small motor (usually an induction motor) is used as a prime motor.

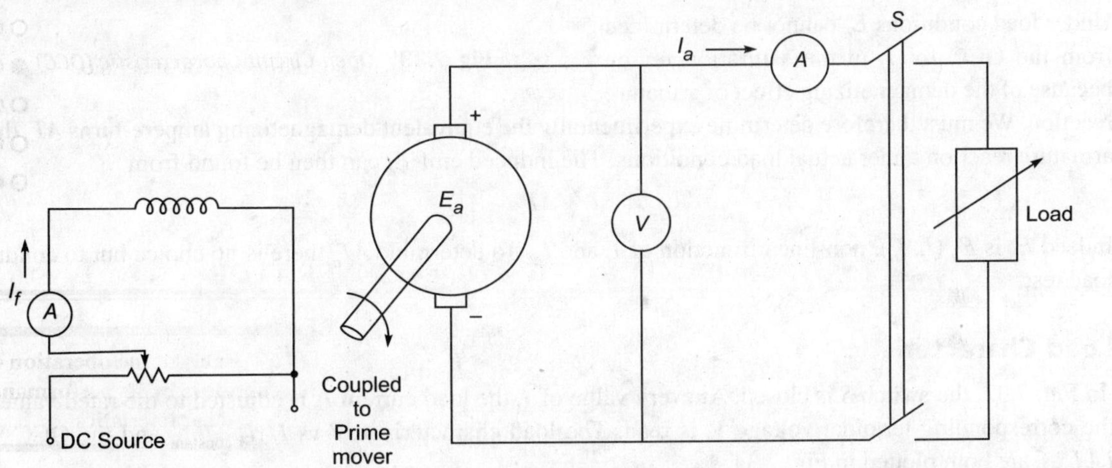

Fig. 7.32 *Experimental set up to determine dc generator characteristics–separately excited*

No Load Test–Open Circuit Test

The switch S in Fig. 7.32 is kept open. The open circuit voltage

$$V_{OC} = E_a \propto \Phi \text{ at } n = \text{constant}$$

Thus, V_{OC} is a measure of Φ. Therefore, the plot, of V_{OC} vs I_f is the *magnetization characteristic* of the machine. It is rightly called the Open Circuit Characteristic (OCC).

In conducting the OCC test, I_f must be raised gradually only in the forward direction otherwise the curve would exhibit local hysteresis loops. Further, as the machine would have been previously subjected to magnetization, a small *residual voltage* would be present with field unexcited. As will be seen presently, this is necessary for generator self-excitation.

A typical magnetization characteristic is shown in Fig. 7.33. It is known as open-circuit characteristic (OCC) because of the method by which it is determined. It exhibits all the important characteristic of the magnetization curve of iron, modified by the presence of air-gap in the magnetic circuit.

The extension of the liner portion of the magnetization curve, shown dotted in Fig. 7.33, is known as the *air-gap line* as it represents mainly the magnetic behaviour of the machine's air-gap, the iron being unsaturated in this region consumes negligible ampere-turns; in any case the effect of iron is also linear here.

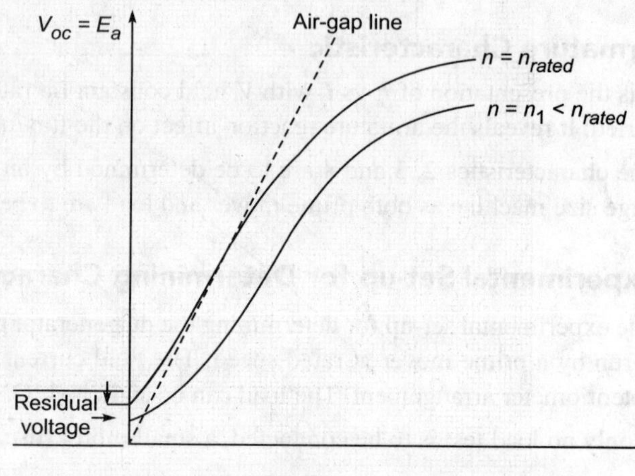

Fig. 7.33 *Open-Circuit Characteristic (OCC)*

The open-circuit characteristic at a speed other than the one at which the test is conducted is a mere proportional translation of the characteristic as shown in Fig. 7.33. This is because of the direct proportionality between E_a and n.

Under load conditions E_a cannot be determined from the OCC for I_f in the saturation region because of the demagnetizing effect of armature reaction. We must therefore determine experimentally the equivalent demagnetizing ampere-turns AT_d due to armature reaction under actual load conditions. The induced emf E_a can then be found from

$$AT_{\text{net}} = AT_f + AT_{se} - AT_d \tag{7.54}$$

Indeed E_a is E_a (I_f, I_a), non-linear function of I_f and I_a. To determine AT_d there is no choice but to conduct a load test.

Load Characteristic

In Fig. 7.32, the switch S is closed. At every value of I_f the load current I_a is adjusted to the rated value and the corresponding terminal voltage V_t is read. The load characteristic V vs $I_f \mid I_{a\ \text{constant}}$ and the OCC V vs $I_f \mid I_{a=0}$ are both plotted in Fig. 7.34.

To the load characteristic we add I_aR_a drop to get E_a induced emf with load. At $I_f = OA$, $E_a = AD + DC$, $DC = I_aR_a$. Therefore, the voltage drop caused by armature reaction is BC. In the low I_f region the magnetic circuit is unsaturated and armature reaction drop is almost zero and so OCC and E_a vs I_f merge.

At $I_f = OA$, the voltage induced with load is AC. With no load the same voltage is induced at $I_f = OF$. Therefore $AF = I_{fd}$, the demagnetizing field current equivalent of armature reaction. Thus

$$AT_d = I_{fd}\,N_f$$

We need only one point (D) on the load characterization. Further, it is sufficiently accurate to assume that

$$AT_d \propto I_a \tag{7.55}$$

Another type of load characterization is described below:

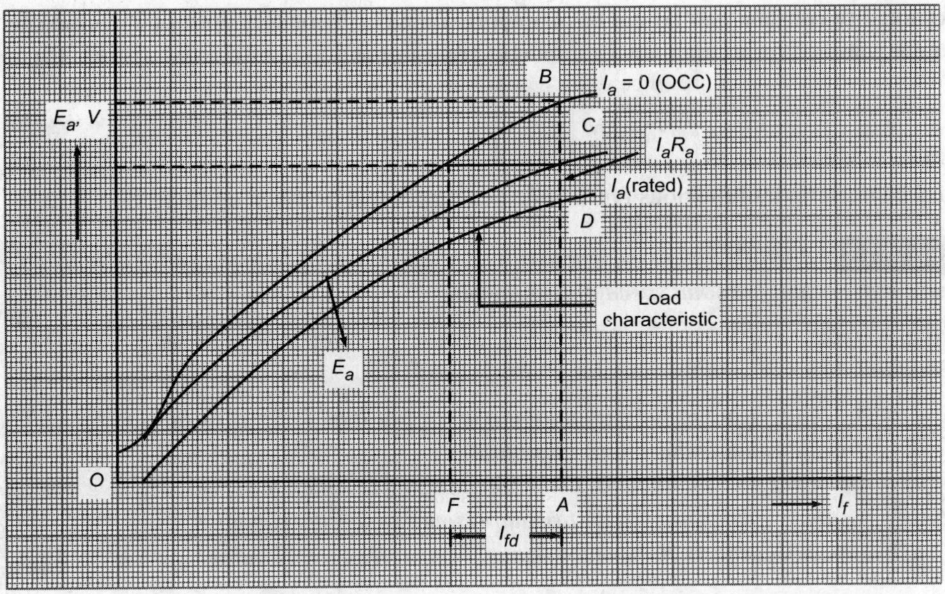

Fig. 7.34 *OCC and load characteristic*

Armature Characteristic

With switch S open in Fig. 7.32, I_f is adjusted to give V_{OC} = V (rated). The switch is then closed the load current I_a is increased and also I_f is increased so as to keep the terminal voltage constant at rated value. The plot of I_f vs I_a sketched in Fig. 7.35 is the armature characteristic.

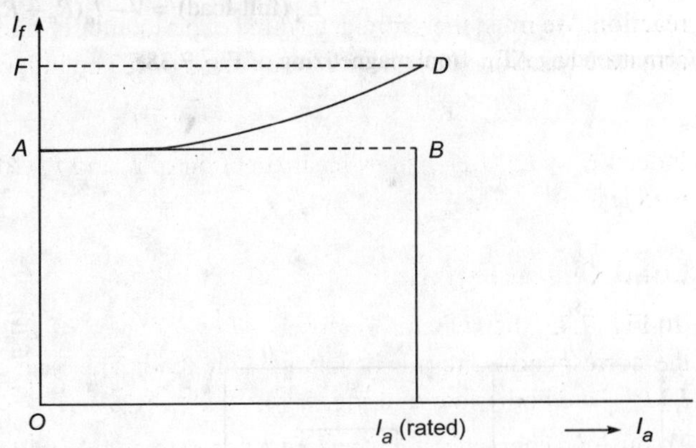

Fig. 7.35

It is seen from the figure that at low values of I_a, the increase in I_f is very small to provide for increasing I_aR_a drop. At large values of I_a there is a sharp increase in I_f to compensate for voltage drop caused by armature reaction. Thus at I_a (rated) $\Delta I_f = AF$ provides for I_aR_a drop plus armature reaction drop.

Let us examine this data on the OCC plotted in Fig. 7.36.

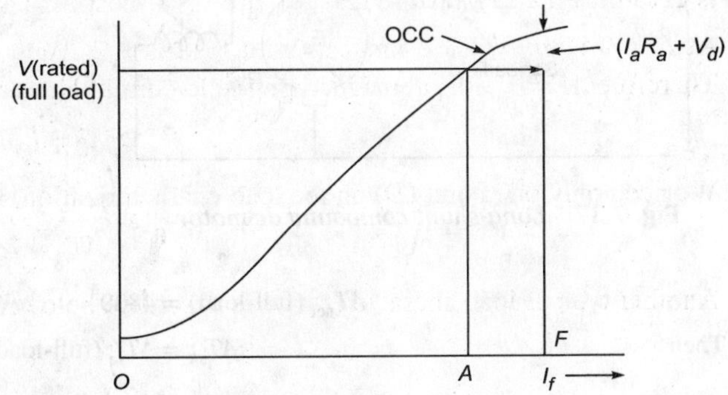

Fig. 7.36

The voltage drop $I_aR_a + V_d$ is indicated on the figure. By subtracting I_aR_a, we can then calculate the constant

$$K = \frac{V_d}{I_a\,(\text{rated})} \tag{7.56}$$

The proportionality constant can be used as an approximation for other values of I_a. From the ΔI_f we can compute the N_{se} needed for a cumulative compound generator as

$$N_{se} = N_f\left(\frac{\Delta I_f}{I_a}\right) \tag{7.57}$$

The external characteristics of various types of dc generators will be taken up in Section 7.12.

Example 7.13 *A 240-V compound (cumulative) dc motor has the following open-circuit magnetization characteristic at normal full-load speed of 850 rev/min:*

Excitation, AT/pole	*1200*	*2400*	*3600*	*4800*	*6000*
Generated emf, V;	*76*	*135*	*180*	*215*	*240*

The resistance voltage drop in the armature circuit at full-load is 25 V. At full-load the shunt and the series winding provide equal ampere-turn excitation.

Calculate the mmf per pole on no load. Estimate the value to which the speed will rise when full-load is removed, the resistance voltage drop in the armature circuit under that condition being 3 V. Ignore armature-reaction and brush-contact effects. Assume long-shunt cumulative compounding. LO 5

Solution At full load, from Fig. 7.37,

$$E_a\text{ (full-load)} = V - I_a(R_a + R_{se}) = 240 - 25 = 215\text{ V}$$

Corresponding AT_{net} from magnetizing of Fig. 7.38,

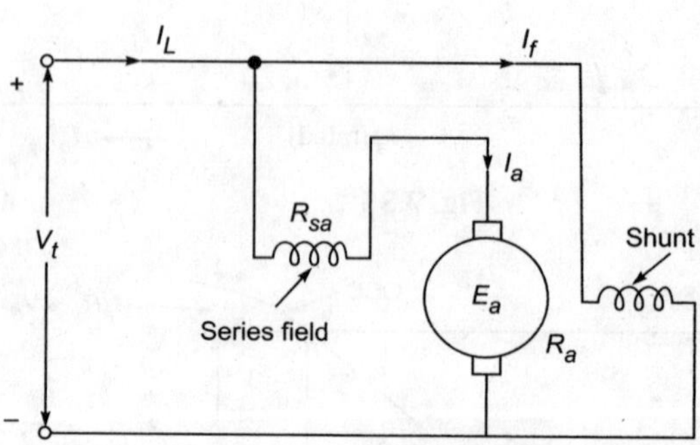

Fig. 7.37 *Long-shunt compound dc motor*

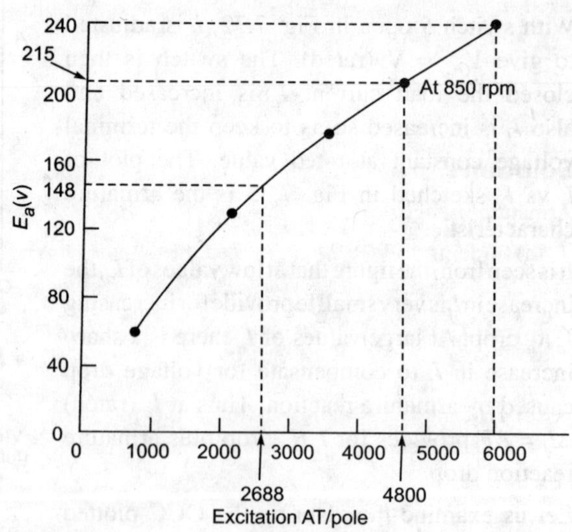

Fig. 7.38

$$AT_{net}\text{ (full-load)} = 4800$$

Then $$AT_{sh} = AT_{se}\text{ (full-load)} = 2400$$

Now $$I_a\,(\text{no-load}) = \frac{3}{25}\,I_a\,(\text{full-load});\; I_a\,(\text{no-load}) = I_f\,(\text{shunt field})$$

$$AT_{se}\,(\text{no-load}) = \frac{3}{25} \times 2400 = 288$$

$$AT_{sh} = 2400 \text{ (no change)}$$

$$AT_{net}\,(\text{no-load}) = 2400 + 288 = 2688$$

$$E_a \text{ (from the magnetizing curve)} = 148 \text{ V at 850 rpm, at } AT_{net} \text{ (no-load)}$$

$$E_a\,(\text{no-load}) = 240 - 3 = 237 \text{ V}$$

Now $$E_a \propto n \text{ at given } AT_{net}$$

Hence, $$n(\text{no-load}) = 850 \times \frac{237}{148} = 1361 \text{ rpm}$$

It will be seen that in a cumulatively compound dc motor, full-load speed is much less than no-load speed.

Example 7.14 *The following OC test data was recorded for a separately dc generator:*

$I_f(A)$	0.0	0.2	0.4	0.6	0.8	1.0	1.2	1.4
$V_{oc}(V)$	10	52	124	184	220	244	264	276

Its load test data is as under

$$I_a\,(fl) = 50 \text{ A},\ V = 240 \text{ V and } I_f = 1.4 \text{ A}$$

The armature resistance inclusive of the brush voltage drop is $R_a = 0.3$ W

Estimate at full load

(a) The internal induced emf

(b) The voltage drop caused by armature reaction

(c) The field current equivalent of armature reaction demagnetization

LO 5

Solution The OCC is plotted in Fig. 7.39. The load test point ($V = 240$ V, $I_f = 1.4$ A) for $I_a = 50$ A is located at A.

Armature voltage drop, $$I_aR_a = 50 \times 0.3 = 15 \text{ V}$$

(a) Adding 15 V to 240 V located the point B which is the internal induced emf,

$$E_a = 240 + 15 = 255 \text{ V}$$

(b) At $I_f = 1.4$ A $V_{OC} = 276$ at point C. Therefore armature reaction voltage drop,

$$V_d = BC = 276 - 255 = 21 \text{ V}$$

(c) To induce $E_a = 255$ V (point B) the field current corresponds to point D on OCC. The field current equivalent of the armature reaction demagnetization is

$$I_f = BD = 0.36 \text{ A}$$

Assuming linear relationship $$K_{ar} = \frac{I_{fd}}{I_a(fl)} = \frac{0.36}{50} = 7.2 \times 10^{-3}$$

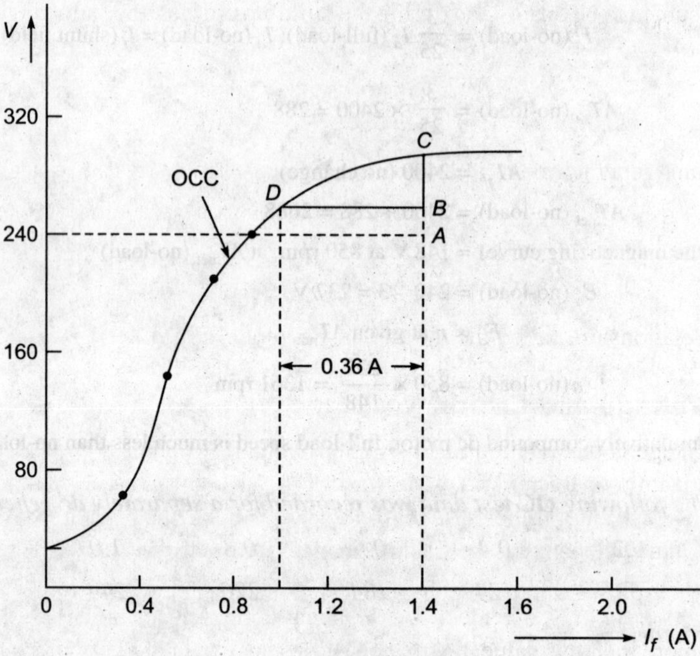

Fig. 7.39

7.11 SELF-EXCITATION

LO 5

Important! Rather than arranging a separate dc source for excitation purposes, practical generators are always excited from their own armature terminals, this method of excitation being known as self-excitation.

A self-excited generator with connection as shown in Fig. 7.40 is known as a shunt generator by virtue of the parallel connection of the field winding with the armature.

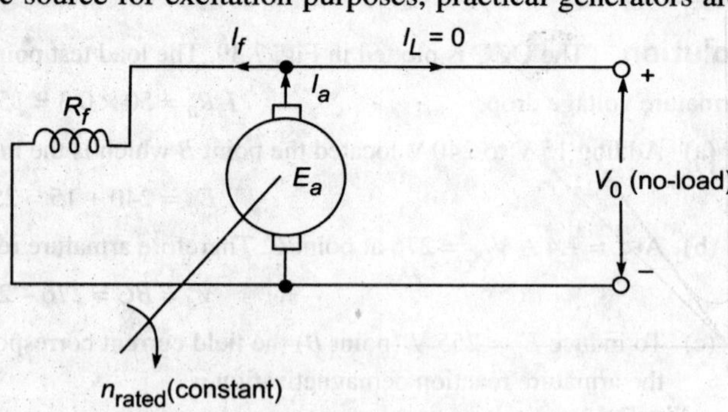

Fig. 7.40 *Self-excited generator (shunt generator)*

Voltage Build-Up

Assume that the field is introduced into the circuit after the armature has been brought to rated speed. At the instant of switching on the field, the armature voltage corresponds to a small residual value which causes a small field current to flow. If the field is connected such that this current increases the field mmf and therefore the induced emf, the machine will continuously build-up. This indeed is a positive feedback connection and the machine builds up to a final steady value only because of the saturation characteristic of the machine's magnetic circuit. Analytical study of voltage build-up is presented in Section 7.21.

Since the generator is assumed to be on no-load during the build-up process, the following circuit relationships apply (Fig. 7.40).

$$I_a = I_f \tag{7.56(a)}$$

$$V = E_a - I_f R_a \tag{7.56(b)}$$

The field current in a shunt generator being very small, the voltage drop $I_f R_a$ can be neglected so that

$$V \approx E_a(I_f) \text{ (magnetization characteristic)} \tag{7.57}$$

For the field circuit

$$V = I_f R_f \tag{7.58}$$

which is a straight line relationship, called the R_f-line, in V-I_f plot of Fig. 7.41.

No-load Terminal Voltage and Critical Resistance

The no-load terminal voltage is the solution of Eqs (7.58b) and (7.59). Thus the intersection point P of the R_f-line with the magnetization characteristic as shown in Fig. 7.41 gives the no-load terminal voltage (V_0) and the corresponding field current. Further, it is easy to visualize from Fig. 7.41 that the no-load voltage can be adjusted to a desired value by changing the field resistance. It can also be seen with reference to Fig. 7.42 that as the field resistance is increased the no-load voltage decreases. The no-load voltage is undefined for a field resistance ($R_{f3} = R_{fc}$) whose line coincides with the linear portion of the magnetization curve. With field resistance even slightly more than this value, the machine does not excite to any appreciable value and would give no-load voltage close to the residual value. The machine with this much resistance in the field fails to excite and the corresponding resistance is known as the *critical resistance* (R_{fc}).

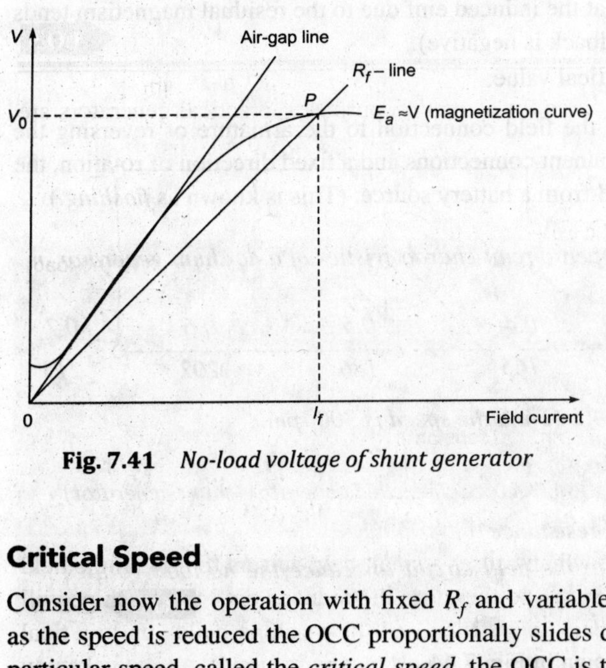

Fig. 7.41 *No-load voltage of shunt generator*

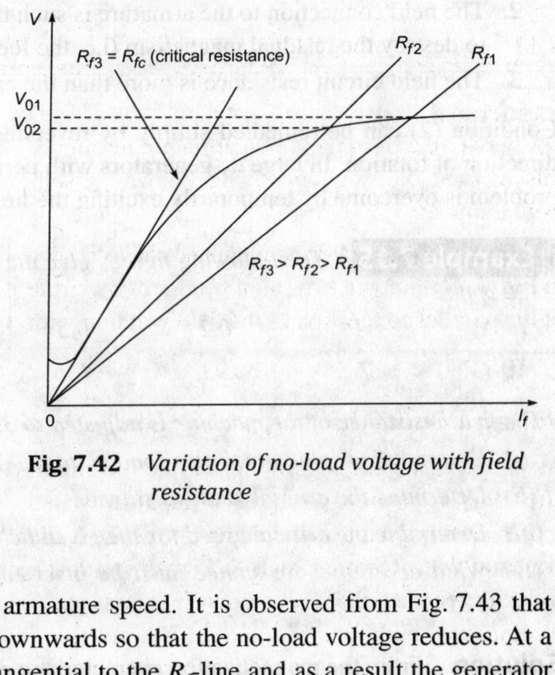

Fig. 7.42 *Variation of no-load voltage with field resistance*

Critical Speed

Consider now the operation with fixed R_f and variable armature speed. It is observed from Fig. 7.43 that as the speed is reduced the OCC proportionally slides downwards so that the no-load voltage reduces. At a particular speed, called the *critical speed,* the OCC is tangential to the R_f-line and as a result the generator would fail to excite.

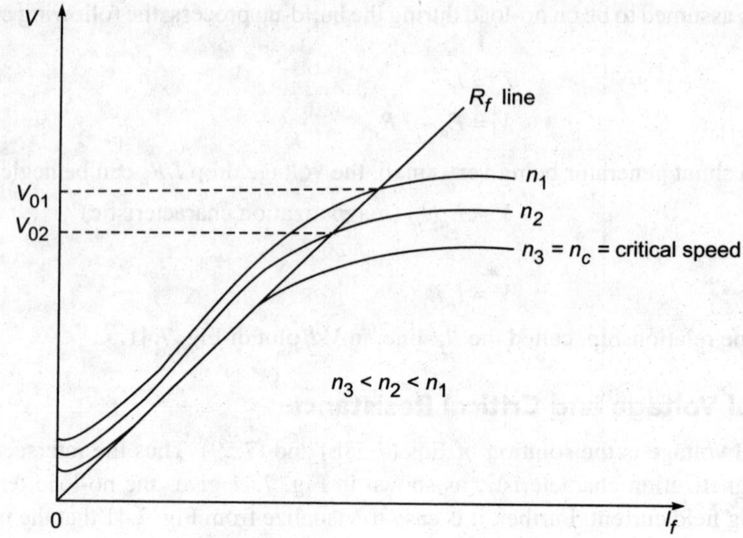

Fig. 7.43 *Effect of speed on no-load voltage*

!Attention To summarize, a dc shunt generator may fail to self-excite for any of the three reasons mentioned below:

1. Residual magnetism is absent.
2. The field connection to the armature is such that the induced emf due to the residual magnetism tends to destroy the residual magnetism (i.e. the feedback is negative).
3. The field circuit resistance is more than the critical value.

Condition (2) can be remedied simply by reversing the field connection to the armature or reversing the direction of rotation. In large dc generators with permanent connections and a fixed direction of rotation, the problem is overcome by temporarily exciting the field from a battery source. (This is known as *flashing*.)

Example 7.15 *The following figures give the open-circuit characteristics of a dc shunt generator at 300 rpm:*

I_f (A)	0	0.2	0.3	0.4	0.5	0.6	0.7
V_{oc} (V)	7.5	93	135	165	186	202	215

The field resistance of the machine is adjusted to 354.5 Ω and the speed is 300 rpm.

(i) Determine graphically the no-load voltage.

(ii) Determine the critical field resistance.

(iii) Determine the critical speed for the given field resistance.

(iv) What additional resistance must be inserted in the field circuit to reduce the no-load voltage to 175 V.

LO 5

Solution From the magnetization characteristic drawn in Fig. 7.44.

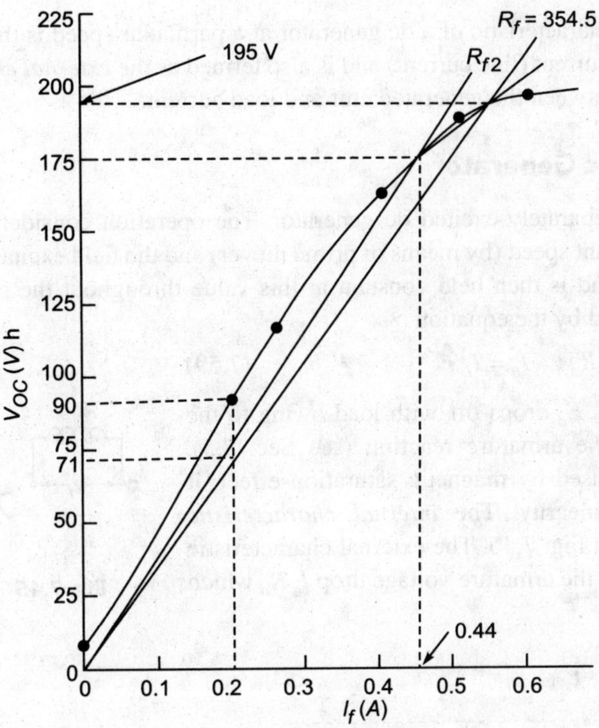

Fig. 7.44

(i) $$V\,(\text{no-load})|_{R_f=354.5\Omega} = 195\text{ V}$$

(ii) $$R_f\,(\text{critical}) = \frac{90}{0.2} = 450\ \Omega$$

(iii) $$n\,(\text{critical}) = 300 \times \frac{71}{90} = 236.7\text{ rpm}$$

(iv) $$V\,(\text{no-load}) = 175\text{ V}$$

$$R_{f2} = \frac{175}{0.44}\ 397.7\ \Omega$$

Additional resistance to be inserted in field circuit

$$= 397.7 - 354.5 = 43.2\ \Omega$$

7.12 | CHARACTERISTICS OF dc GENERATORS

LO 5

With the advent of silicon-controlled rectifiers, the importance of the dc machine as a generator has considerably reduced as SCRs can be employed to draw ac power from standard ac supply and convert it to dc; also the dc voltage can also be varied with ease. For the sake of completeness the characteristics of dc generators will be briefly discussed here, which are still found in older installations in industry (as a motor-generator set for speed control of dc motors; see Sec. 7.14).

Important! The load characteristic of a dc generator at a particular speed is the relationship between its terminal voltage and load current (line current) and is also termed as the *external characteristic*. The *internal characteristic* is the plot between the generated emf and load current.

Separately-excited dc Generator

Figure 7.45 is that of a separately-excited dc generator. The operation considered here assumes that the armature is driven at constant speed (by means of prime mover) and the field excitation (I_f) is adjusted to give rated voltage at no-load and is then held constant at this value throughout the operation considered. The armature circuit is governed by the equation

$$V = E_a - I_a R_a;\quad I_a = I_L \tag{7.59}$$

In spite of fixed excitation, E_a drops off with load owing to the demagnetizing effect of the armature reaction (see Sec. 7.5). As the voltage drop is caused by magnetic saturation effect, it increases with load nonlinearity. The *internal characteristic* ($E_a - I_L$) is shown dotted in Fig. 7.46. The external characteristic differs from the internal by the armature voltage drop $I_a R_a$ which is also shown in Fig. 7.46.

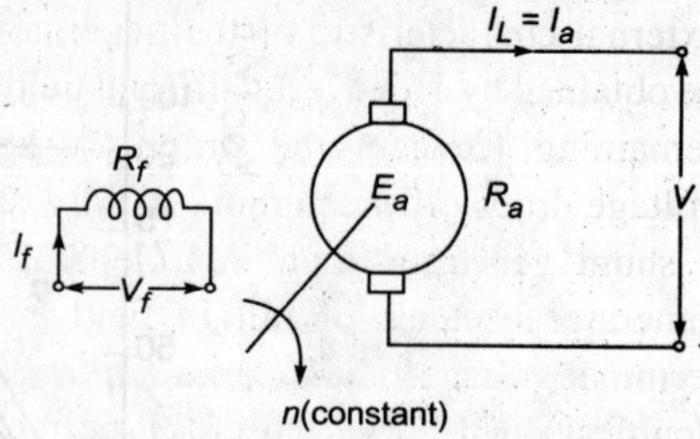

Fig. 7.45 *Separately excited dc generator*

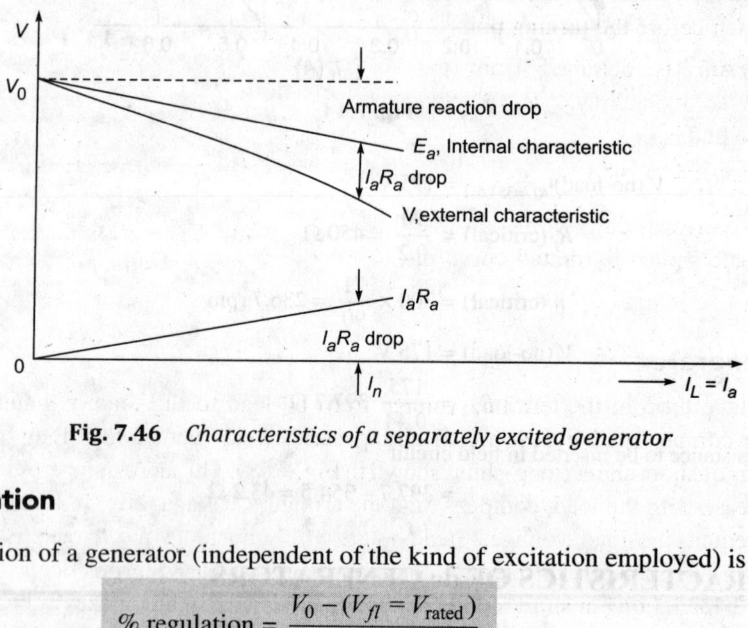

Fig. 7.46 *Characteristics of a separately excited generator*

Voltage Regulation

The voltage regulation of a generator (independent of the kind of excitation employed) is defined as

$$\%\ \text{regulation} = \frac{V_0 - (V_{fl} = V_{\text{rated}})}{V_{\text{rated}}} \tag{7.60}$$

where V_{fl} = full-load voltage = V_{rated}

V_0 = no-load voltage corresponding to rated voltage at full-load excitation remaining unchanged**

** In case a shunt field winding is provided, it would mean that the total resistance in a filed circuit remains unchanged in the operation.

Shunt Generator

A dc shunt generator is a self-excited generator. The phenomenon of voltage buildup on no-load and the conditions necessary for the same have already been discussed in Sec. 7.11. Figure 7.47 shows a shunt-connected generator. With field resistance adjusted to a certain value by means of the regulating resistance, the desired no-load voltage can be obtained (refer to Fig. 7.41). The external characteristic of the generator can then be obtained by a load test with total field resistance remaining fixed in the process. The terminal voltage drops off much more rapidly with load in a shunt generator than in a separately-excited generator because of fall in field current with terminal voltage. The *external characteristic* is a double-valued curve with a certain I_L (max) as shown in solid line in Fig. 7.48. As indicated in Fig. 7.48, the useful parts of the external characteristic is much before the turning point.

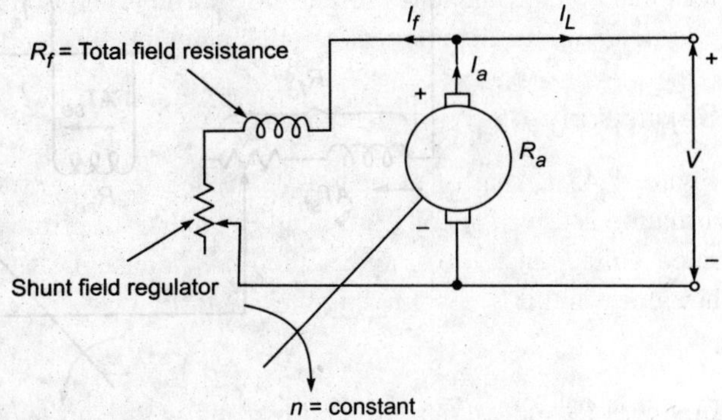

Fig. 7.47 *dc shunt generator on load*

Internal characteristic is obtained from the external characteristic by adding I_aR_a. At any point P (V_1, I_{L1}) we find

$$I_{a1} = I_{L1} + (V_1/R_f)$$

This locates the corresponding point on the internal characteristic shown by dotted curve in Fig. 7.48.

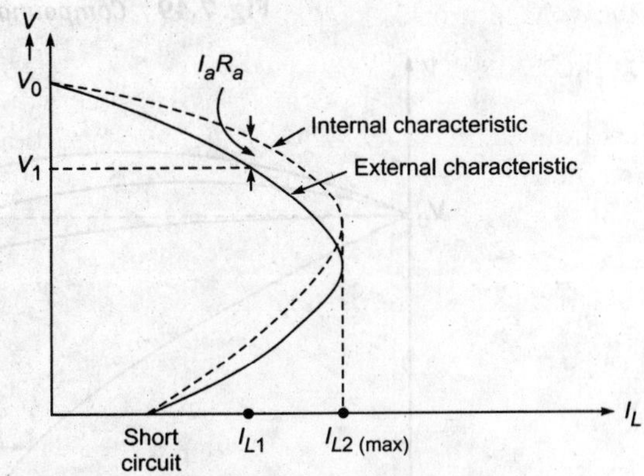

Fig. 7.48 *External Characteristic of dc shunt generator*

Compound Generator

The causes of voltage drop in the terminal voltage from no-load to full-load in a shunt generator can be partially/fully/over-compensated by use of an aiding series field (cumulative compound), which can be connected in a long- or short-shunt (long-shunt shown in Fig. 7.49). The aiding ampere-turns of the series field automatically increase with the load, compensating the armature voltage drop. In level-compound generator full-load voltage equals no-load voltage. Steady-state volt-ampere (*V-I*) characteristics of a compound generator are shown in Fig. 7.50. Differential compounding is not used in practice as the terminal voltage falls off steeply with load. Long or short-shunt connection of series winding makes only a marginal difference in the *V-I* characteristic of a compound generator. Compounding level can be adjusted by a diverter in parallel to the series field; see Section 7.9, Fig. 7.31.

Series Generator

A series generator is employed in series with a dc line to compensate for the line voltage drop which is proportional to line current.

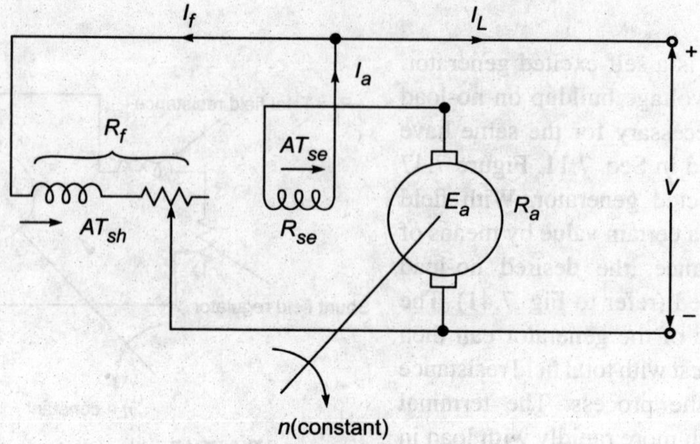

Fig. 7.49 *Compound generator (long-shunt)*

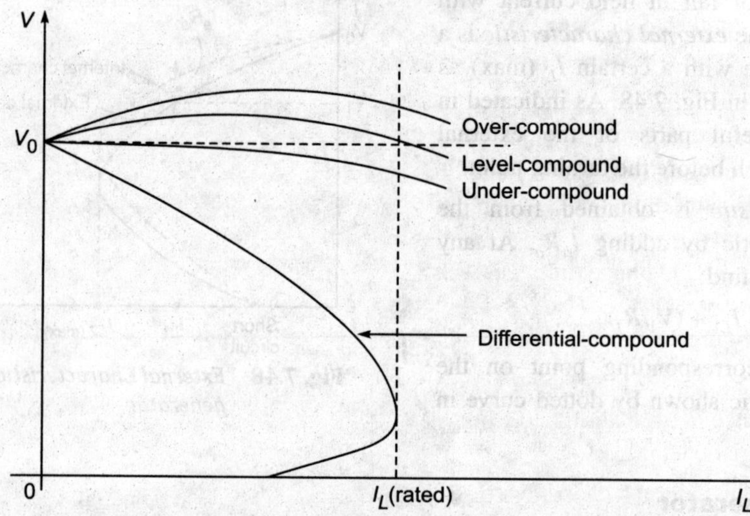

Fig. 7.50 *External characteristics of a compound generator*

Connection Diagram

A series self-excited dc generator connection has been given in Fig. 7.29(c). It is drawn in Fig. 7.51 along with a load which can be switched on. The terminal voltage is given by

$$V = E_a - I_a (R_a + R_{se}) \tag{7.61}$$

As the load current incurrent increase the excitation $N_{sc}\, I_a$ increase and E_a rises sharply but levels off with saturation and begins to drop due to armature reaction.

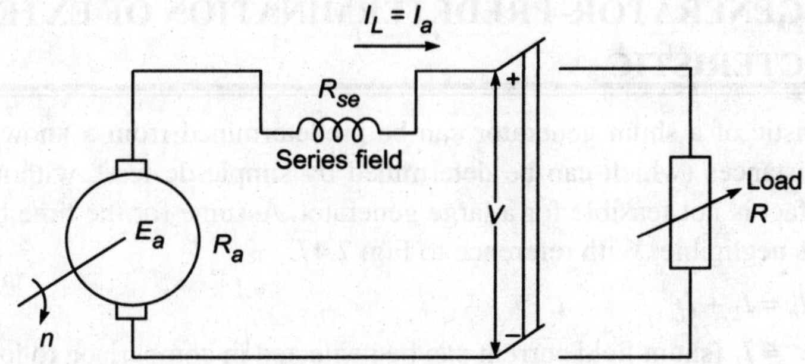

Fig. 7.51

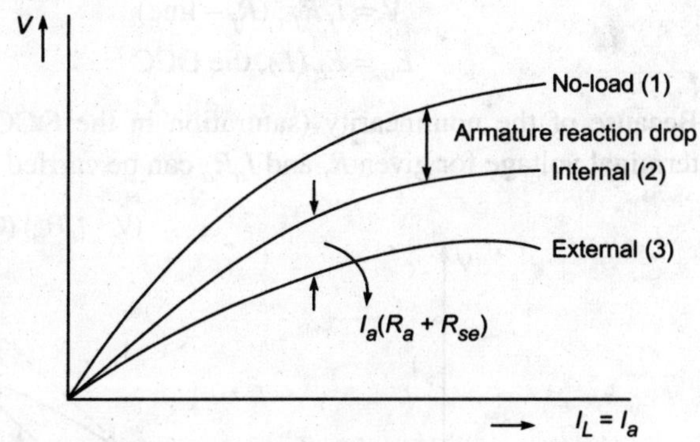

Fig. 7.52

Characteristics

No load test is conducted by exciting the series field separately from a low voltage source. It is the saturation curve V_{OC} vs $I_f = I_a$ drawn in Fig. 7.52 curve 1. *Internal characteristic*.

It is the induced emf E_a which is less than V_{OC} by the armature reaction voltage drop. The voltage drop sharply increases with saturation and E_a beings to reduce; curve 2. *External characteristic* lies below lies below curve 2 by $I_a (R_a + R_{se})$ voltage drop, curve 3.

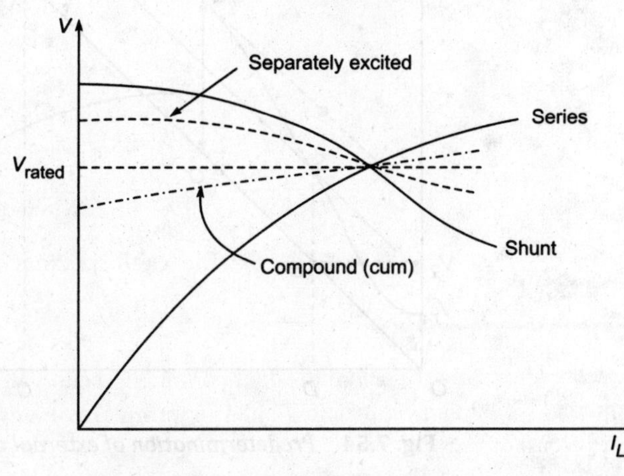

Fig. 7.53 *V-I characteristics of dc generators*

Condition for Self-excitation

The total resistance $(R_a + R_{se} + R_L)$ should be less than the critical resistance R_C as determined from the saturation characteristic; curve 1.

It is observed from the external characteristic, curve 3, that the maximum line boost is limited.

Comparison of Volt Ampere (*V-I*) Characteristics of dc Generators

The *VI* characteristic (steady state) of various types of dc generators is shown in Fig. 7.53. These and drawn with rated terminal voltage at full-load.

7.13 SHUNT GENERATOR–PREDETERMINATION OF EXTERNAL CHARACTERISTIC

LO 5

The external characteristic of a shunt generator can be predetermined from a knowledge of the OCC and armature and field resistances (which can be determined by simple dc test), without actually performing the load test which in fact is not feasible for a large generator. Assume for the time being that the armature reaction voltage drop is negligible. With reference to Fig. 7.47.

$$I_a = I_L + I_f$$

$$\approx I_L \text{ (shunt field current can be neglected in comparison to load current)} \quad (7.62)$$

$$E_a = V + I_a R_a \quad (7.63)$$

$$V = I_f R_f, \ (R_f\text{– line}) \quad (7.64)$$

$$E_a = E_a(I_f), \text{ the OCC} \quad (7.65)$$

Because of the nonlinearity (saturation in the OCC), the solution of Eqs (7.62)–(7.65) to determine the terminal voltage for given R_f and I_aR_a can be carried out graphically as shown in Fig. 7.54.

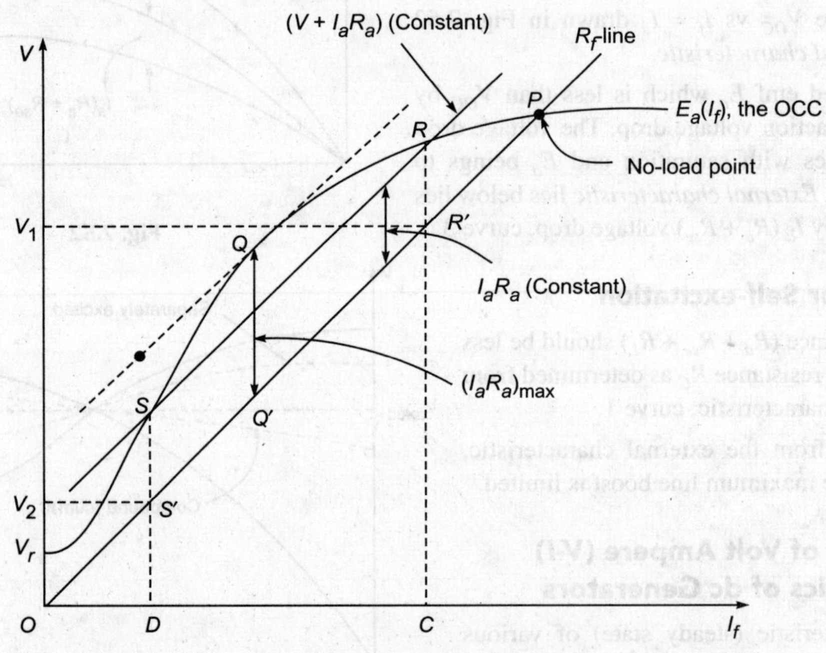

Fig. 7.54 *Predetermination of external characteristic of dc shunt generator*

Residual Voltage

For given I_aR_a, Eq. (7.64) is represented by a straight line parallel to the R_f line and above it by vertical intercept of I_aR_a. The intersection of this line with OCC gives two solution points R and S for the induced emf. For these solution points the two terminal voltage values V_1 and V_2 can be read from the R_f line corresponding to points R' and S'. Maximum possible armature current corresponds to point Q at which a line parallel to the R_f line is tangent to the OCC. Corresponding to each solution point, values of V and I_f can be read and

$I_L = I_a - I_f$, can be calculated and subsequently the external characteristic plotted as shown in Fig. 7.48. The short-circuit point on the external characteristic corresponds to $I_a R_a = V_r$, the residual voltage.

Effect of Armature Reaction

It has been shown in Section 7.10 that the *demagnetization caused by armature reaction can be quantified by equivalent field current I_{fd} which can be taken as proportional to the armature current I_a*. There are two ways in which we can account for I_{fd} in predetermining the external characteristic of a shunt generator starting from its OCC. With reference to Fig. 7.55

Shift the origin by I_{fd} to left along the I_f–axis. This is equivalent to shifting the OCC to right parallel to the I_f–axis by I_{fd} from the new origin.

We shall use the first approach and illustrate the second through an example. A typical OCC is sketched in Fig. 7.55 and the R_f line is drawn from the origin. Their intersection point P determines the no load voltage V_0.

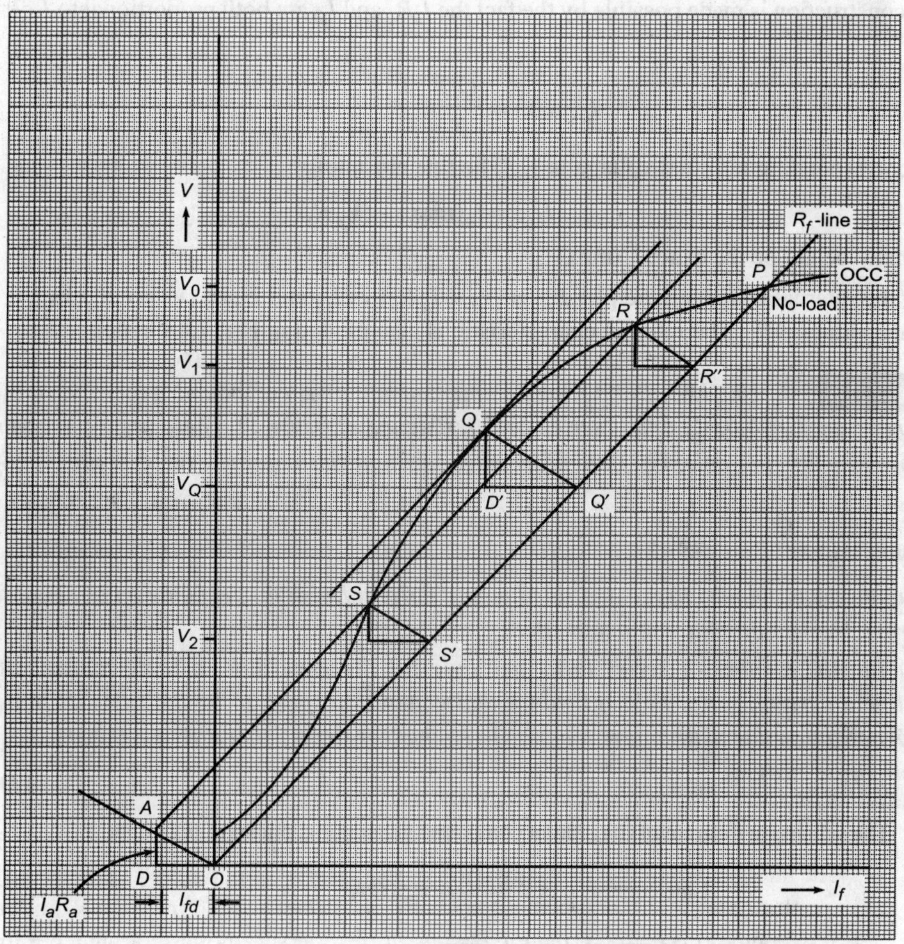

Fig. 7.55

Constructional Steps Method

⇒ Located D to the left of origin such that $OD = I_{fd}$

⇒ Locate A vertically above D such that $AD = I_aR_a$ for an assumed value of I_a (about the rated value)

⇒ From A draw a line parallel to the R_f line intersecting OCC at R and S

⇒ Shift ΔODA at R and S as shown in Fig. 7.55. This step locates R' and S' on the R_f line which yield the two terminal voltages V_1 and V_2 for the assumed value of I_a.

⇒ For V_1, V_2 calculate $I_{f1} = V_1/R_f$ and $I_{f2} = V_2/R_f$

⇒ Calculate the corresponding line currents $I_{L1} = I_a - I_{f1}$ and $I_{L2} = I_a - I_{f2}$

⇒ Plot V_1 vs I_{L1} and V_2 vs I_{L2} on the V–I_L coordinate plane

⇒ Repeat the above steps for different values of I_a

⇒ The complete plot is the *external characteristic* of double-valued shape as sketched in Fig. 7.48

⇒ To find I_a (max), draw a tangent to OCC parallel to the R_f line giving the point Q (corresponding to I_a (max)). Draw now QQ' parallel to OA and the $\Delta Q'D'Q$ similar to ΔODA. QD' then gives max I_aR_a. This construction is made possible by the fact the I_aR_a and I_{fd} are both proportional to I_a. It then follows that

$$I_a(\text{max}) = \frac{QD'}{R_a}$$

The corresponding terminal voltage is V_Q. The line current is

$$I_{LQ'} = I_a(\text{max}) - (V_Q/R_f)$$

This may be not be the exact I_L(max)

For plotting the internal characteristic plot $E_{a1} = (V_1 + I_a R_a)$ vs I_{L1} and $E_{a2} = (V_2 + I_aR_a)$ vs I_{L1}; similarly for other assumed values of I_a. The internal characteristic is also plotted in Fig. 7.48 in dotted line.

Example 7.16 *A 100 kW, 200 V, long-shunt commutative compound generator has equivalent armature resistance of 0.03 Ω and a series field resistance of 0.004 Ω. There are 1200 shunt turns per pole and 5 series turn per pole. The data of its magnetization characteristic at 1000 rpm is given below:*

I_f (A)	0	1	2.2	3.3	4.2	5.3	7.1
E_a (V)	11	33	105	150	175	200	225

(a) Calculate the terminal voltage and rated output current for shunt field current of 5 A and a speed of 950 rpm. Neglect armature reaction effect.

(b) Calculate the number of series turns per pole for the generator to level compound with same shunt field current, series field resistance and generation speed to be the same as in part (a). The demagnetization due to armature reaction in equivalent shunt field current is 0.001875 I_a where I_a is the armature current.

LO 5

Solution The schematic diagram of long-shunt compound generator is drawn in Fig. 7.56(a)

$$\text{Line current, } I_L = \frac{100 \times 10^3}{200} = 500 \text{ A}$$

Shunt field current, $I_f = 5$ A (given)

From Fig. 7.56(a)

$$I_a = 500 + 5 = 505 \text{ A}$$

$$I_{f,eq} = I_f + \frac{N_{se}}{N_f} I_{se}$$

$$= 5 + \frac{5}{1200} \times 505 = 7.1 \text{ A}$$

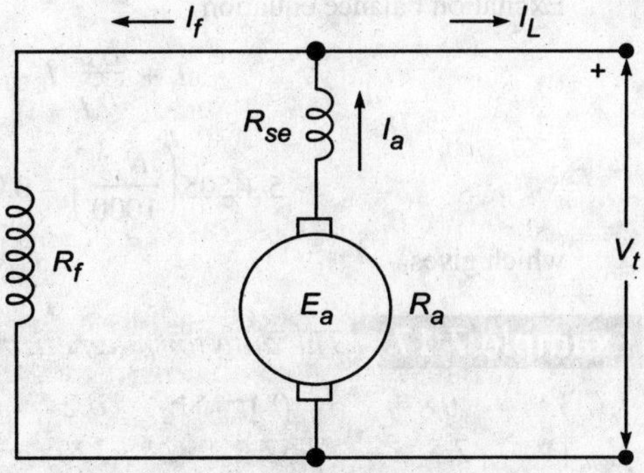

Fig. 7.56(a)

From the magnetization characteristic of Fig. 7.56(b) (as per data given)

(a) $E_a = 225$ V at 1000 rpm

$$E_a \text{ (950 rpm)} = 225 \times \frac{950}{1000} = 213.75 \text{ V}$$

From Fig. 7.56(a)

$$V_t = E_a - I_a (R_a + R_{se}) = 213.75 - 505 (0.03 + 0.004)$$

The generator is under – compounded (armature reaction ignored)

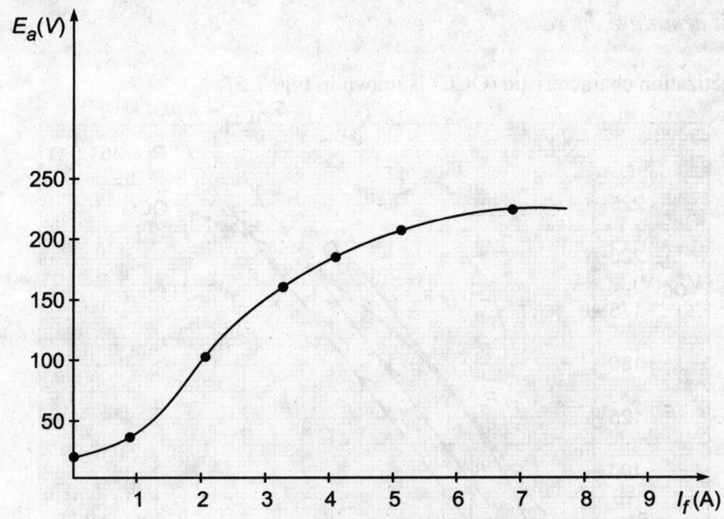

Fig. 7.56(b)

(b) $I_{fd} = 0.001875 \times 505 = 0.95$ A (demag)

For level compound

$$V_t = 200 \text{ V}$$

$$E_a = 200 + 505 (0.03 + 0.04) = 217.17 \text{ (950 rpm)}$$

$$E_a \text{ (1000 rpm)} = 217.17 \times \frac{1000}{950} = 228.6 \text{ V}$$

From the magnetization characteristic

$$I_f \text{(net)} = 7.5 \text{ A}$$

Excitation balance equation

$$I_f + \frac{N_{se}}{N_f} I_a - I_{fd} = I_f \text{ (net)}$$

$$5 + 505\left(\frac{N_{se}}{1000}\right) - 0.95 = 7.5$$

which gives $$N_{se} = 6.87 \text{ or } 7 \text{ turns}$$

Example 7.17 *The data for magnetization characteristic of a dc shunt generator is as under:*

$I_f(A)$	0	0.1	0.2	0.3	0.4	0.5	0.6	0.7
$V_{OC}(V)$	7.5	62.5	120	152.5	175	192.5	205	215

The shunt field has a resistance of 354.5 Ω and the armature resistance is 0.5 Ω.
Determine

(a) the no-load voltage
(b) the terminal voltage at an armature current of 40 A
(c) the maximum possible armature current and the corresponding terminal voltage
(d) the short circuit armature current

LO 5

Solution The magnetization characteristic (OCC) is drawn in Fig. 7.57.

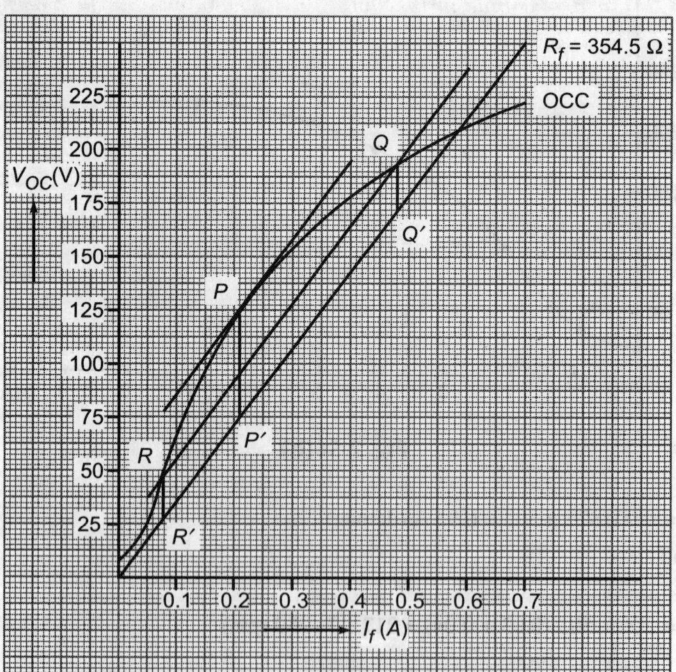

Fig. 7.57

(a) $R_f = 354.5$ Ω line is drawn from the origin. Its intersection with OCC gives no-load voltage $V_0 = 200$ V

(b) $I_a = 40$ A, $I_aR_a = 40 \times 0.5 = 20$ V

We have drawn a line parallel to the R_f line and 20 V (vertically) above it. It intersects the OCC at Q and R. The points Q' and R' vertically below on the R_f line give the voltages at I_a = 40 A, 170 V, 25 V

The generation will be operated at 170 V.

(c) Draw a line tangent to the OCC. It locates the point P from which we locate P' vertically below it on the R_f line. It is found that

$$I_a R_a \text{ (max)} = PP' = 50 \text{ V}$$

or

$$I_a \text{ (max)} = 50/0.5 = 100 \text{ A}$$

(d) Short circuit current

$$I_a(\text{sc}) = \frac{V_r}{R_a} = \frac{7.5}{0.5} = 15 \text{ A}$$

Example 7.18 *A 100 kW, 200 V, 1000-rpm dc shunt generator has R_a = 0.03 Ω. The generator is driven at rated speed and excitation is such that it gives a rated voltage of 200 V at no-load. The data for the magnetization characteristic of the generator are as follows:*

$I_f^*(A)$	0	1	2.2	3.3	4.5	7.1
EMF(V)	10	32.5	100	167.5	200	225

Determine the voltage appearing across the generator terminals when I_a = 500 A

A series field of 5 turns per pole having a total resistance of 0.005 Ω is to be added in a long-shunt compound. There are 1200 turns per pole in the shunt field. The generator is to be level-compounded so that the full-load voltage is 200 V when the resistance in the shut field circuit is adjusted to give a no load voltage of 200 V. Find the value of the series field diverter resistance to obtain the desired performance. Assume that the armature reaction AT has been compensated for. **LO 5**

Solution The magnetization characteristic is drawn in Fig. 7.58. Drawing the R_f-line corresponding to 200 V.

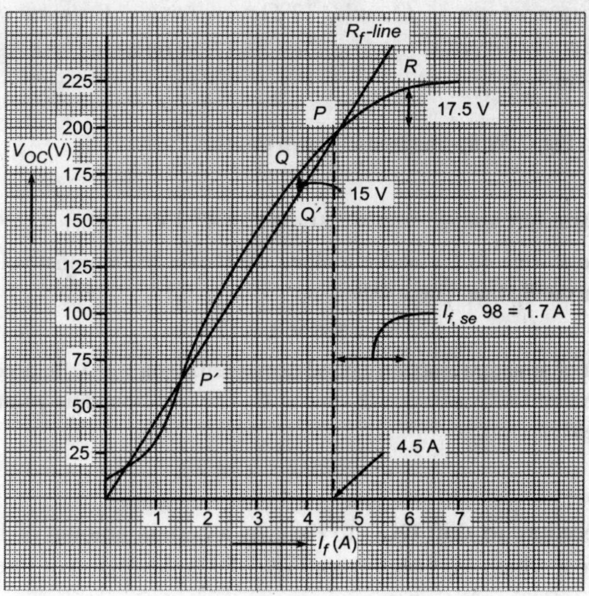

Fig. 7.58

$$R_f = \frac{200}{4.5} = 44.4\ \Omega$$
$$I_a = 500\ \text{A}$$
$$I_a R_a = 500 \times 0.03 = 15\ \text{V}$$

Drawing a line parallel to the R_f-line at 15 V (= QQ')

$$V(\text{terminal}) = 165\ \text{V (corresponding to } Q')$$

With series winding added in the long-shunt compound (cumulative):

$$\text{Armature circuit resistance} = 0.03 + 0.005 = 0.035\ \Omega$$
$$\text{Armature circuit voltage drop} = 500 \times 0.035 = 17.5\ \text{V}$$

To compensate for this voltage drop, the operating point on OCC must lie at R, i.e. (200 + 17.5) = 217.5 V. Both full-load and no-load voltage will now be 200 V. From Fig. 7.58 the series field current measured in terms of shunt of field current is

$$I_{f,\,se} = 1.75\ \text{A}$$
$$\text{AT}_{se} = I_{f,\,se} \times 1200 = 1.75 \times 1200 = 2100$$
$$I_{se} = \frac{2100}{5} = 420\ \text{A}$$
$$I_{\text{diverter}} = 500 - 420 = 80\ \text{A}$$
$$R_{\text{diverter}} = 0.005 \times \frac{420}{80} = 0.0265\ \Omega$$

Example 7.19 *The open-circuit characteristic of a shunt generator at rated speed is*

$I_f(A)$	1	2.5	5	7	9	12	15	18
$V_{OC}(V)$	22	231	400	479	539	605	642	671

The field and armature resistances are 46 Ω and 0.12 Ω respectively. Estimate the terminal voltage when the armature current is 360 A in two cases (a) Armature reaction ignored (b) 1 A, field current is needed to counteract the effect of armature reaction.

LO 5

Solution OCC is plotted in Fig. 7.59 R_f-line is also drawn thereon.

(a) $I_a = 360$ A

$$\text{Armature voltage drop} = 360 \times 0.12 = 43.2\ \text{V}$$

Drawing line parallel to R_f-line 43.2 V above it along the V axis, the terminal voltage and corresponding field current are at point B on the R_f-line. These values are

$$V_t = 566\ \text{V},\ I_f = 12\ \text{A}$$

(b) Field current equivalent of armature reaction caused demagnetization = 1 A. To account for this the OCC shifted to right by 1 A. The shifted OCC intersects R_f-line at C. The corresponding point D on the R_f-line yields the terminal voltage and field currents as

$$V_t = 520\ \text{V},\ I_f = 10.8\ \text{A}$$

Remark The empirical procedure to shift OCC to right by a certain amount of field current applies in the region around full load current which as is seen from Fig. 7.59 is the region of interest.

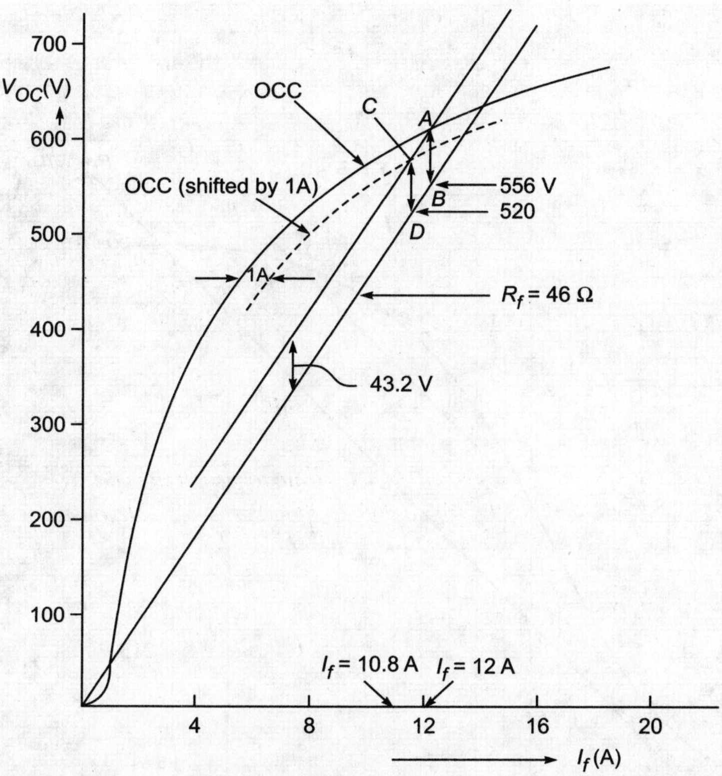

Fig. 7.59

Example 7.20 *The OCC and R_f = 167 Ω line are drawn to scale in Fig. 7.60 for a dc shunt generator with armature resistance of 0.5 Ω. The generator is driven at constant speed. It is estimated that at armature current of 40 A, the armature reaction demagnetization has field current equivalent of 0.14 A. Estimate thee acceptable terminal voltage when the armature is delivering 50 A. Also find the line current.*

LO 5

Solution Using proportionality

$$I_{fd}|_{at\ I_a = 50\ A} = 0.14 \times \frac{50}{40} = 0.175\ A$$

$$I_aR_a = 50 \times 0.5 = 25\ V$$

Choose an arbitrary point D' on the R_f line. Draw a right angle triangle $D'AD$ with $D'A$ = 0.175 A (horizontal) $\angle D'AD = 90°$, AD = 25 (vertical). Through D draw a line parallel to the R_f line interacting the OCC at R (higher voltage point) Now draw RR' parallel to DD' locating R' on the R_f line. Reading from R'

$$V\ (\text{terminal}) = 232.5\ V$$

$$I_f = 1.5\ A$$

Then
$$I_L = I_a - I_f = 50 - 1.5\ A = 48.5\ A$$

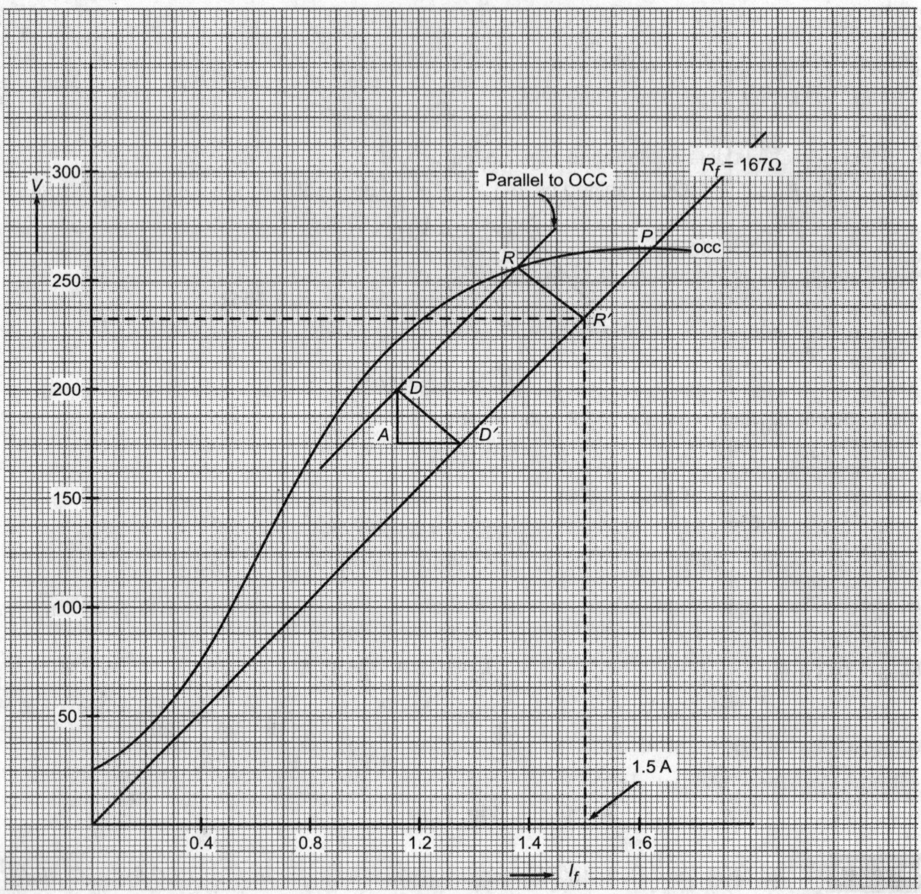

Fig. 7.60

Example 7.21 *A 20 kW, 240 V shunt generator is driven by a prime mover whose speed drops uniformly from 1190 rpm at no-load to 1150 rpm at full-load. By adding a series winding it is converted to a short shunt compound such that its voltage rises from 230 V at no-load to 240 V at I_a = 83.3 A. The resistance of the series winding is estimated to be 0.045 Ω. The armature resistance including brush voltage drop is found to be 0.12 Ω. The shunt field turns are 550/pole.*

The generator is separately excited and the following tests are conducted.

No-load Test (1190 rpm)

$V_{OC}(V)$	220	230	240	250	260	270
$I_f(A)$	1.0	1.15	1.35	1.50	1.69	2.02

Load Test (1150 rpm)

Armature terminal voltage = 240 V

Armature current = 83.3 A

Field current = 2.1 A

LO 5

Determine
(a) the demagnetizing AT/pole at armature current of 83.3 A
(b) the number of series turns required

Solution The no-load test data is plotted in Fig. 7.61(b)

(a) Load Test – 1150 rpm

$$E_a = 240 + 83.3 \times 0.12 = 250 \text{ V}$$

Looking up mag. curve we find E_a at 1190 rpm (OCC prime mover speed)

$$E_a \text{ (1190 rpm)} = 250 \times \frac{1190}{1150} = 258.7 \text{ V}$$

$$I_f = 2.1 \text{ A (given)}$$

The field current needed to induce 258.7 V on no-load

$$I_f \text{ (net)} = 1.65 \text{ A}$$

Demagnetization in terms of field current

$$I_{fd} = 2.1 - 1.65 = 0.45 \text{ A}$$

Then $$AT_d = 550 \times 0.45 = 247.5$$

(b) Series turns added – compound generator – short shunt. The connection diagram is drawn in Fig. 7.61(a). The performance required:

No-load (1190 rpm)

$$V_t = 230 \text{ V}$$

We can assume $$E_a \approx 230 \text{ V}$$

Consulting mag. curve, we find

$$I_f = 1.43 \text{ ; } AT_f = 1.43 \times 550 = 786.5$$

Then $$R_f = \frac{230}{1.43} = 161 \ \Omega \text{ remains constant}$$

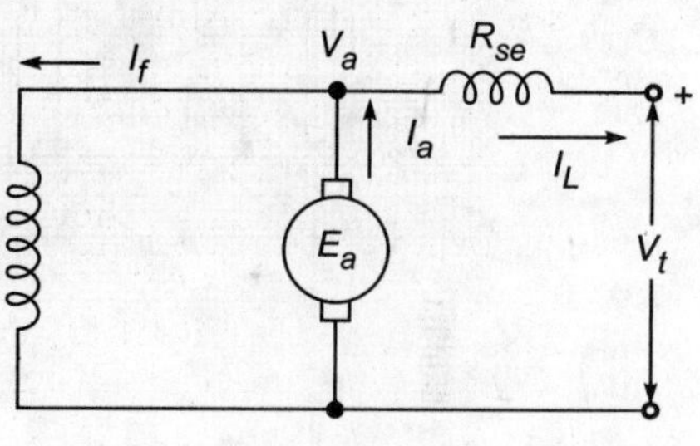

Fig. 7.61(a)

Load (1150 rpm)

$$V_t = 240 \text{ V}, I_a = 83.3 \text{ A}$$

$$I_L = I_a - I_f = 83.3 - \frac{V_a}{R_f} \text{ ; } V_a = \text{armature voltage}$$

$$V_a = 240 + 0.045\, I_L = 240 + \left(83.3 - \frac{V_a}{161}\right) \times 0.045$$

$$V_a \left(1 + \frac{0.045}{161}\right) = 240 + 83.3 \times 0.045 = 243.7$$

or $$V_a = 243.7 \text{ V}$$

Then $$E_a = 243.7 + 83.3 \times 0.12 = 253.3 \text{ V}$$

To consult mag. curve

$$E_a \text{ (1190 rpm)} = 253.3 \times \frac{1190}{1150} = 262.1 \text{ A}$$

$$I_f \text{ (needed)} = 1.675 \text{ A}$$

$$\text{AT (needed)} = 550 \times 1.675 = 921.25$$

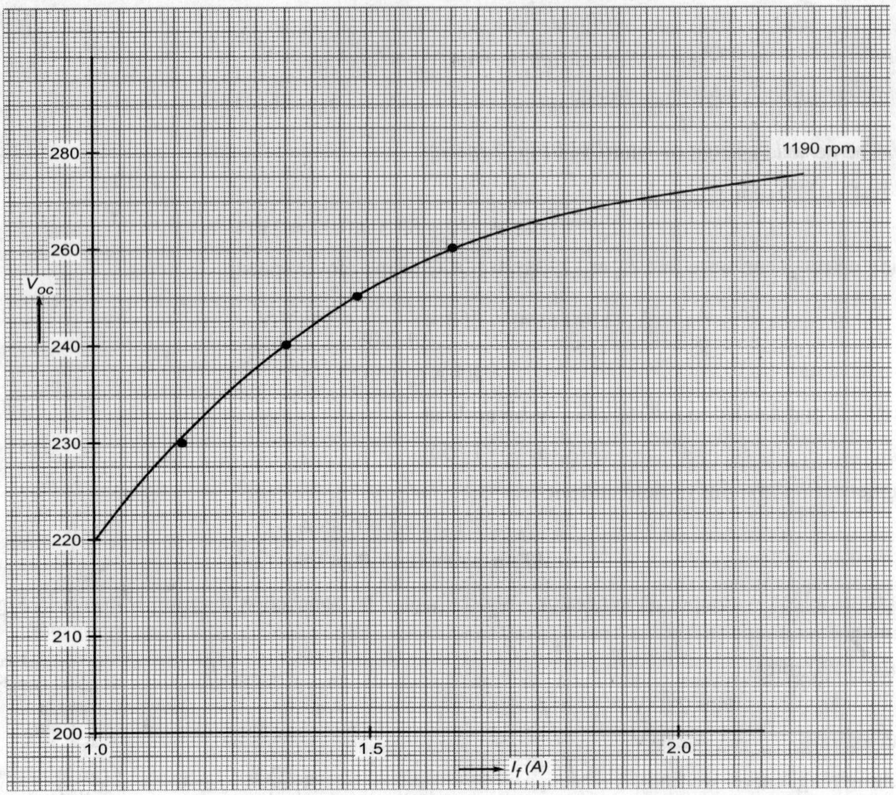

Fig. 7.61(b)

$$I_f = \frac{V_a}{R_f} = \frac{243.7}{161} = 1.51 \text{ A}$$

$$AT_f = 550 \times 1.51 = 830$$

AT balance equation

$$AT_f + AT_{se} - AT_d = AT \text{ (needed)}$$

$$830 + AT_{se} - AT_d = 925.25$$

or

$$AT_{se} = 343$$

$$I_L = 83.3 - 1.51 \approx 81.0 \text{ A}$$

$$N_{se} = \frac{343}{81.8} \approx 4 \text{ turns /pole}$$

Example 7.22 *A 50 kW, 250 V, 200 A long shunt cumulative compound dc generator has the following data:*

$$R_a \textit{ (inclusive of brush voltage drop)} = 0.05\ \Omega$$

$$\textit{Series field resistance } R_{se} = 0.01\ \Omega$$

The magnetization characteristic at 1200 rpm is drawn in Fig. 7.62(a)

LO 5

Shunt field turns, $N_f = 1000$

Series field turns, $N_{se} = 3$

The generator is run at 1150 rpm and its shunt field current is adjusted to 5.6 A. Compute the terminal voltage at rated voltage current.

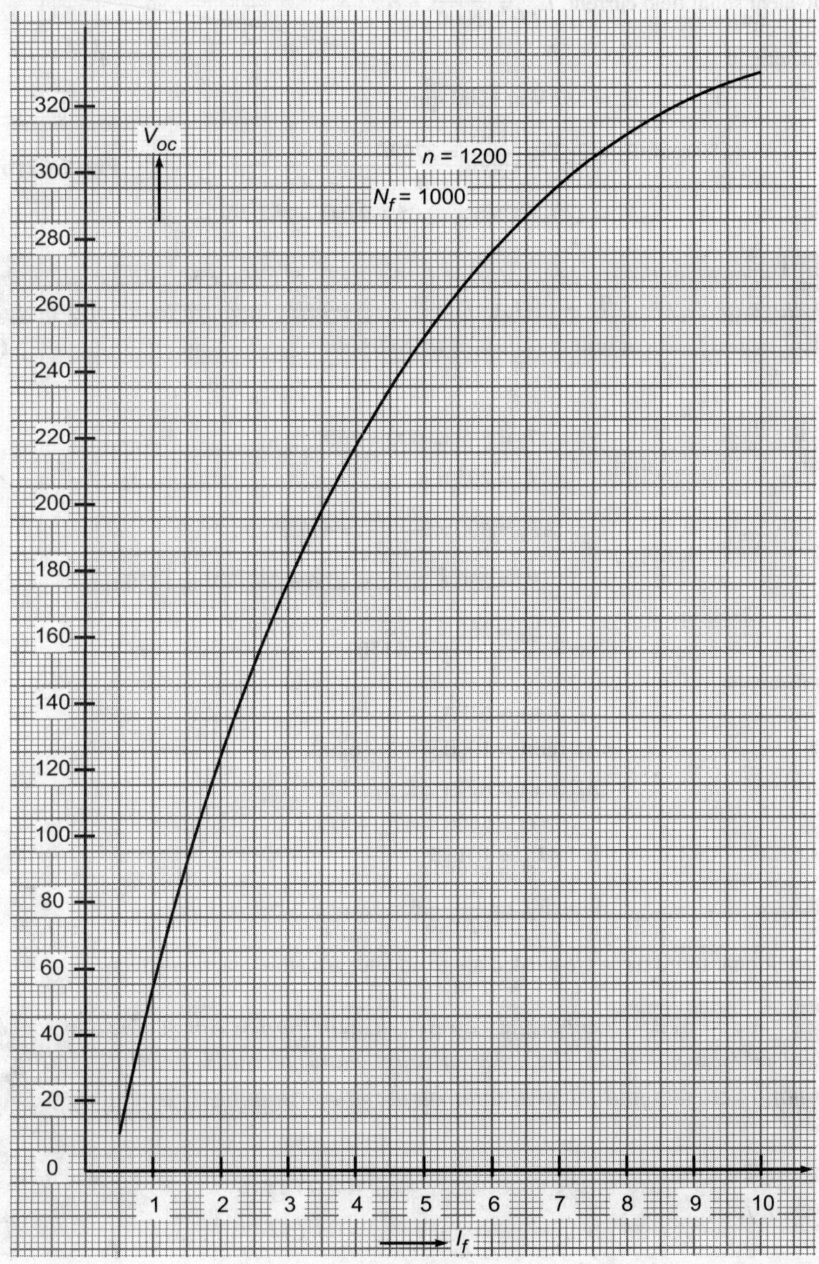

Fig. 7.62(a)

Solution The connection diagram of long shunt compound generator is drawn in Fig. 7.62(b).

$$\text{Load current, } I_L = 200 \text{ A}$$

$$I_a = 200 + 5.6 = 205.6 \text{ A}$$

Excitation ampere-turns $1000 \times 5.6 + 205.6 \times 3 = 6217$

$$\text{Equivalent shunt field current, } I_{f eq} = \frac{6217}{1000} \approx 6.22 \text{ A}$$

From the mag. characteristic

$$E_a = 282 \text{ V at 1200 rpm}$$

At $$1150 \text{ rpm, } E_a = 282 \times \frac{1150}{1200} = 270 \text{ V}$$

Terminal voltage $$V_t = 270 - 205.6 \times (0.05 + 0.01) = 257.7 \text{ V}$$

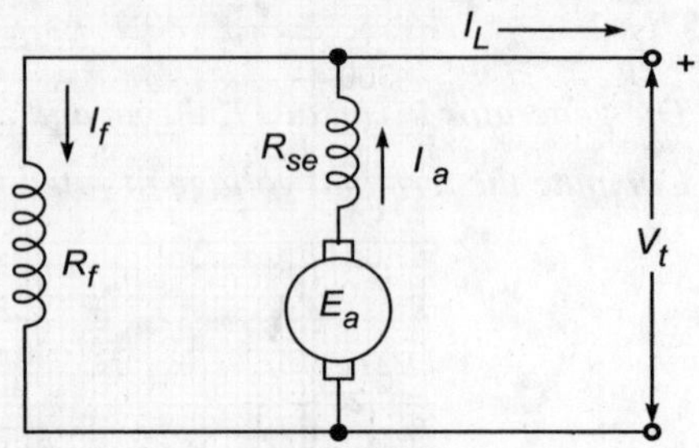

Fig. 7.62(b)

Example 7.23 *Consider the dc generator of Example 7.22. What should be the number of series for a terminal voltage or 250 A at I_L = 200 A. The demagnetizing ampere-turns of armature reaction at I_a = 200 A are 400. Shunt field current is adjusted to 5.6 A.* LO 5

Solution

$$AT_f = 5.6 \times 1000 = 5600$$

$$I_a = 205.6 \text{ A}$$

$$AT_d = 400 \times \frac{205.6}{200} = 411.2$$

$$E_a = 250 + 205.6 \times 0.06 = 262.3 \text{ V at 1150 rpm}$$

At 1200 rpm, $$E_a = 262.3 \times \frac{1200}{1150} = 273.7 \text{ V}$$

From the mag. characteristic

$$I_f \text{ (net)} = 6.2 \text{ A}$$

$$AT \text{ (net)} = 1000 \times 6.2 = 6200$$

AT balance equation $$AT \text{ (net)} = AT_f + AT_{se} - AT_d$$

$$6200 = 5600 + AT_{se} - 411.2$$

or $$AT_{se} = 1011.2$$

$$N_{se} = \frac{1011.2}{205.6} = 4.9 \text{ or 5 turns}$$

We conclude that 2 additional series turns are needed are to counter armature reaction effect.

Example 7.24 *A 20 kW, 250 V separately excited dc generator is run at constant speed of 1000 rpm has an armature resistance of 0.16 Ω. The magnetization characteristic is presented in Fig. 7.63.*

(a) At rated armature current find the generator output at constant field current of (i) 1 A, (ii) 2 A, and (iii) 2.5A.

(b) Repeat part (a) if the generator speed is decreased to 800 rpm.

Note: *Neglect armature reaction.* LO 5

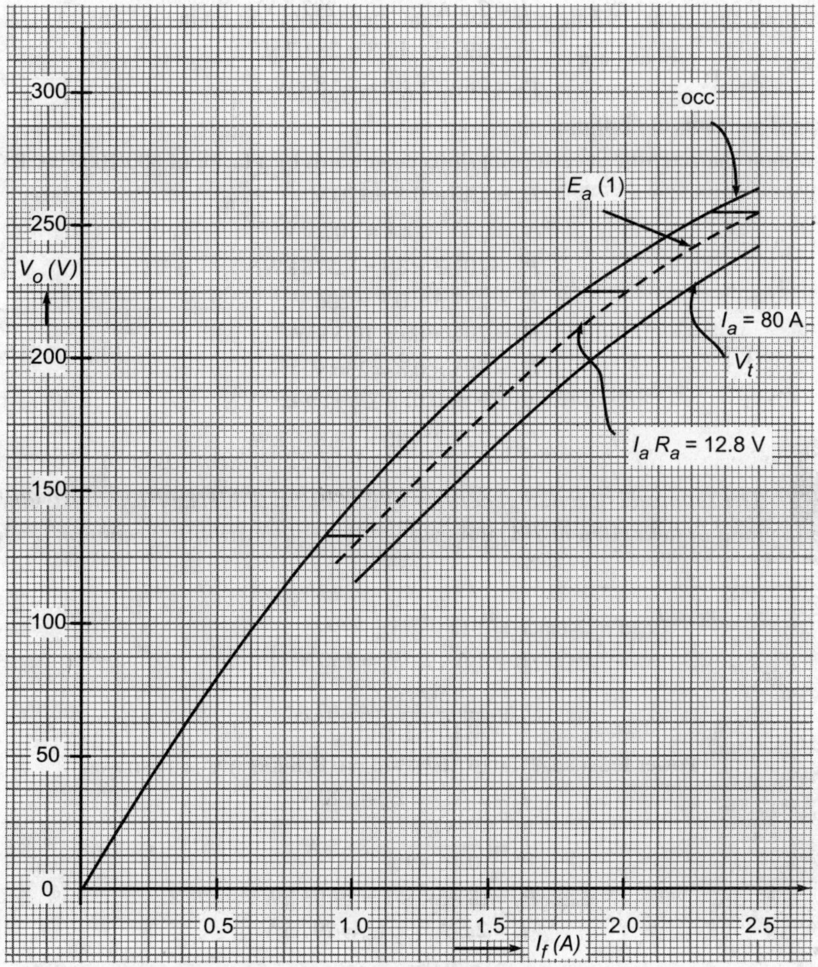

Fig. 7.63

Solution

$$\text{Rated armature current} = \frac{20 \times 10^3}{250} = 80 \text{ A}$$

(a) (i) $I_f = 1$ A From the mag. characteristic $E_a = 150$ V

$$I_a = 80 \text{ A} \quad R_a = 0.16$$

$$\text{Internal voltage drop} = 80 \times 0.16 = 12.81$$

$$V_t = 150 - 12.8 = 137.2 \text{ V}$$

$$\text{Generator output, } P_0 = 137.2 \times 80 = 10.976 \text{ kW}$$

(ii)

$$I_f = 2 \text{ A} \qquad E_a = 257.5 \text{ V}$$

$$V_t = 257.5 - 12.8 = 238.2$$

$$P_0 = 238.2 \times 80 = 19 \text{ kW}$$

(iii) $I_f = 2.5$ A $\quad E_a = 297.5$ V

$$V_t = 297.5 - 12.8 = 284.7 \text{ V}$$

$$P_0 = 284.7 \times 80 = 22.8 \text{ kW}$$

(b) Generator speed increases to 1200 rpm

The magnetization characteristic at each point shifts downwards by a factor of 800/1000 = 0.8

(i)

$$E_a = 150 \times 0.8 = 120 \text{ V}$$

$$V_t = 120 - 12.8 = 107.2 \text{ V}$$

$$P_0 = 107.2 \times 80 = 8.576 \text{ kW}$$

(ii)

$$E_a = 257.5 \times 0.8 = 206 \text{ V}$$

$$P_0 = 206 \times 80 = 16.5 \text{ kW}$$

(iii)

$$E_a = 295 \times 0.8 = 236 \text{ V}$$

$$V_t = 236 - 12.8 = 223.2 \text{ V}$$

$$P_0 = 223.2 \times 80 = 17.856 \text{ kW}$$

Example 7.25 *The separately excited generator of Example 7.24 feeds a load resistance of 3 Ω. Find the power fed to the load at field current 1 A, 2 A and 25 A. Neglect armature reaction.* LO 5

Solution

Load resistance = 3 Ω

$I_f = 1$ A $\quad E_a = 150$ V

$$I_a = \frac{150}{0.16+3} = 47.67 \text{ A}$$

$$\text{Power fed load, } P_0 = (47.67)^2 \times 3 = 6.808 \text{ kW}$$

$I_f = 2$ A $\quad E_a = 240$ V

$$I_a = \frac{257.5}{0.16+3} = 81.5 \text{ A}$$

$$P_0 = (81.5)^2 \times 3 = 19.93 \text{ kW}$$

$I_f = 2.5$ A $\quad E_a = 297.5$ V

$$I_a = \frac{297.5}{0.16+3} = 94.15 \text{ A}$$

$$P_0 = (94.15)^2 \times 3 = 26.56 \text{ kW}$$

Example 7.26 *The separately excited dc generator of Example 7.24 is run at 1000 rpm and the load test is conducted. At each field current, the load current is adjusted to 80 A and the readings of the terminal voltage taken. The data recorded is as under:* LO 5

I_f (A)	2.5	2.0	1.0
V_t (V)	240	200	120

Find the armature reaction demagnetization in terms of field current (I_{fd}) at each data point.

Solution On Fig. 7.63 the magnetization characteristic $V_t - I_f$ is now plotted on it. To this plot, we add $I_aR_a = 80 \times 0.16 = 12.8$ V (vertically). This gives us the internal characteristic E_a(i) – I_f; shown dotted.

I_{fd} is the field current difference between E_a(i) and E_a (open circuit) at given field current. I_{fd} at the three values of I_f are shown by horizontal lines between magnetization characteristic (E_a (open circuit)) and internal characteristic (E_a(i)). The data is recorded below:

I_f (A)	1	2	2.5
I_{fd} (A)	0.2	0.75	1

It is observed that I_{fd} increases as the magnetic circuit gets into saturation region.

Example 7.27 *The OCC of a dc shunt generator is drawn to scale in Fig. 7.64. The generator has an armature resistance of 0.5 Ω. The generator is run at constant speed.*

(a) Find the field resistance for a no-load voltage of 230 V and the field current.

(b) At an armature current of 80 A find the terminal voltage, field current and load power.

(c) It is required that the armature current be 80 A. What would be the terminal voltage accounting for 5% flux reduction? Also calculate load power.

LO 5

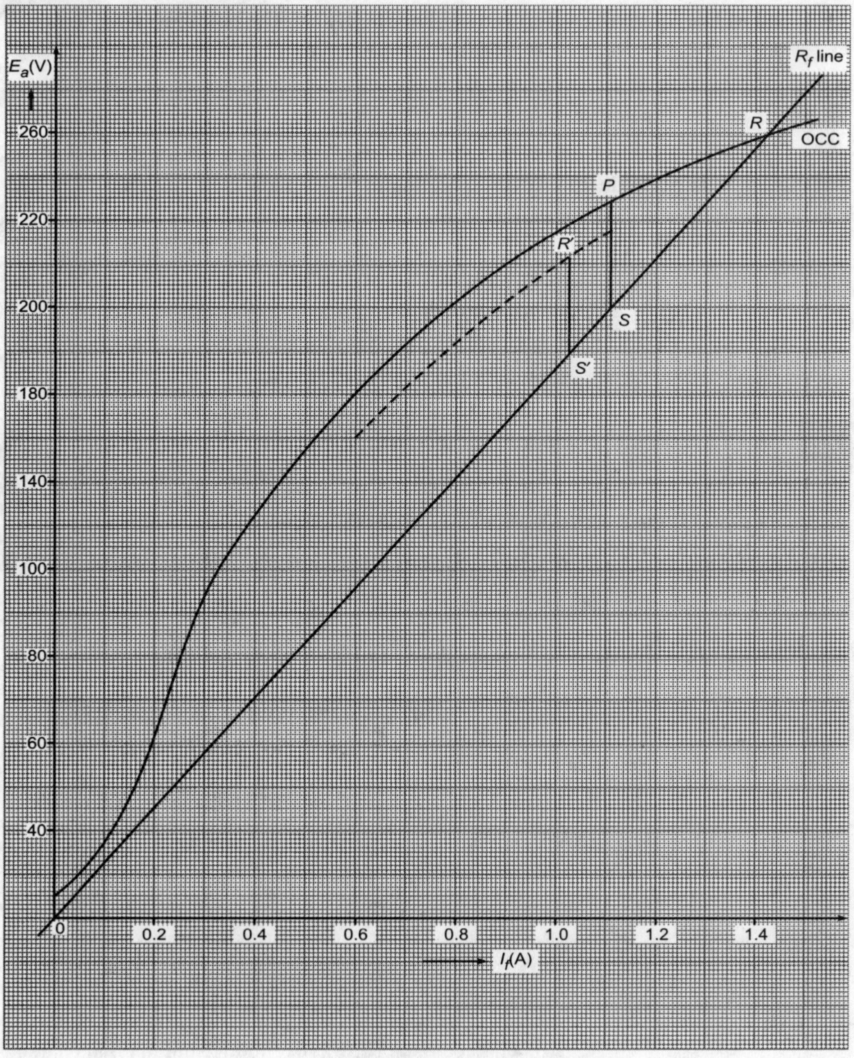

Fig. 7.64

Solution Locate 230 V, point P on the OCC. Draw the R_f line from the origin to P. We find

(a) $$I_f = 1.3 \text{ A} \quad \therefore R_f = 230/1.3 = 176.9\ \Omega$$

$$I_a = 80 \text{ A}, \quad R_a = 0.5, \quad I_aR_a = 40 \text{ V}$$

Draw a line parallel to R_f line 40 V above it, It intersects OCC at P.
The point S below it on R_f line in the terminal voltage

$$V_t = 202 \text{ V}$$

Corresponding field current is

$$I_f = 202/176.9 = 1.14 \text{ A}$$

(b) Given $$I_a = 80 \text{ A},$$

$$R_a = 0.5\ \Omega$$

$$I_aR_a = 80 \times 0.5 = 40 \text{ V}$$

Between OCC and R_f line locate (by scale) vertical line RS = 40 V. The point given the terminal voltage

$$V_t = 200 \text{ V}$$

Corresponding $$I_f = 1.07 \text{ A}$$

$$\text{Line current, } I_L = 80 - 1.07 = 78.93 \text{ A}$$

$$\text{Load power} = 200 \times 78.93 = 15.786 \text{ kW}$$

(c) Induced emf current with 5% reduction (Corresponding to 5% reduction in flux/pole) is drawn in dotted line in Fig. 7.64. Its intersection with line parallel to R_f line 20 V above it is R'. The corresponding point S on R_f line gives

$$V_t = 182 \text{ V}$$

$$I_f = 182/176.9 = 103 \text{ A}$$

$$I_L = 40 - 1.03 = 38.97 \text{ A}$$

$$\text{Load power} = 182 \times 38.397 = 7.09 \text{ kW}$$

Example 7.28 *A 75 kW, 250 V compound dc generator has the following data:*

$$R_a = 0.04\ \Omega, \quad R_{se} = 0.004\ \Omega \quad R_f = 100\ \Omega$$

Brush contact drop, V_b = 2 V (1 volt each brush)

Compare the generator induced emf when fully loaded in (i) long shunt compound, and (ii) short shunt compound

LO 5

Solution Long shunt and short shunt connection are drawn in Figs 7.65(a) and (b).

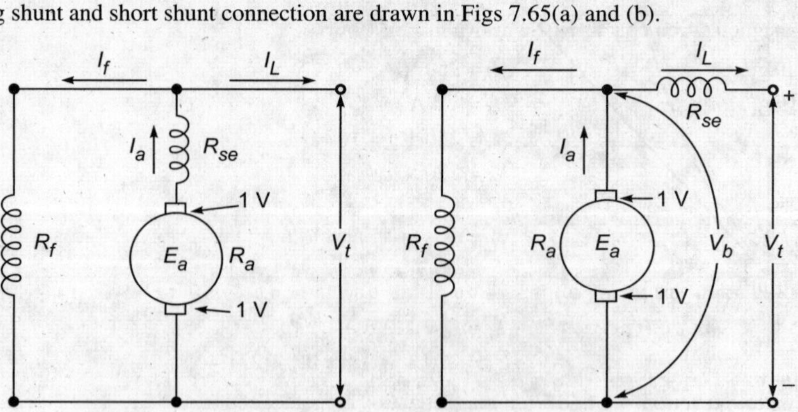

(a) Long shunt (b) Short shunt

Fig. 7.65 *Compound Generator*

At full load

$$V_t = 250 \text{ V}, I_L = \frac{75 \times 10^3}{250} = 300 \text{ A}$$

Long Shunt (LS)

$$I_f = 250/100 = 2.5 \text{ A}, \quad I_a = 300 + 2.5 = 302.5 \text{ A}$$

$$E_a = 250 + (0.04 + 0.004) \times 302.5 + 2 = 265.31 \text{ A}$$

Short Shunt (SS)

$$V_b = 250 + 0.004 \times 300 = 251.2 \text{ V}$$

$$I_f = 251.2 /100 = 2.512 \text{ A}$$

$$I_a = 300 + 2.512 = 302.\ 512 \text{ A}$$

$$E_a = 250 + 0.04 \times 302.512 + 2 = 264.1 \text{ V}$$

$$E_a \text{ (LS)} - E_a \text{ (SS)} = 265.31 - 264.1 = 1.21 \text{ V}$$

$$\text{Difference as percentage of } V_t = \frac{1.21 \times 100}{250} = 0.484\%$$

Conclusion: There is no significant difference in LS and SS.

Example 7.29 *A 25 kW, 220 V, 1600 rpm dc shunt generator with R_a = 0.1 Ω has magnetization characteristic data given below:*

I_f (A)	0.0	0.25	0.50	0.75	0.1	1.25	1.5
E_a (V)	10	90	150	190	220	243	250

(a) what would be the current and field resistance at terminal voltage of 220 V?

(b) At rated current and rated terminal voltage, find the value of field current and field resistance. Ignore the effect of armature reaction.

(c) Find the value of electromagnetic power and torque in part (b).

(d) Under load conditions in part (b), the field current is I_f = 1.25 A. Find the field current needed to counter the effect of armature reaction.

LO 5

Solution The magnetization characteristic drawn in Fig. 7.66.

(a) No-load voltage, $V_0 = 250$ V

Corresponding, $I_f = 1.5$ A

$$R_f = 250/1.5 = 167\ \Omega$$

(b) $V_t = 220$ V

$$I_L = \frac{25 \times 10^3}{220} = 113.6 \text{ A}$$

As I_f is small, we can assume

$$I_a \approx I_L = 113.6 \text{ A}$$

$$E_a = 220 + 113.6 \times 0.1 = 231.4 \text{ V}$$

Ignoring armature reaction, we find from magnetization characteristic

$$I_f = 1.1 \text{ A} \quad \text{(about 1\% of } I_L)$$

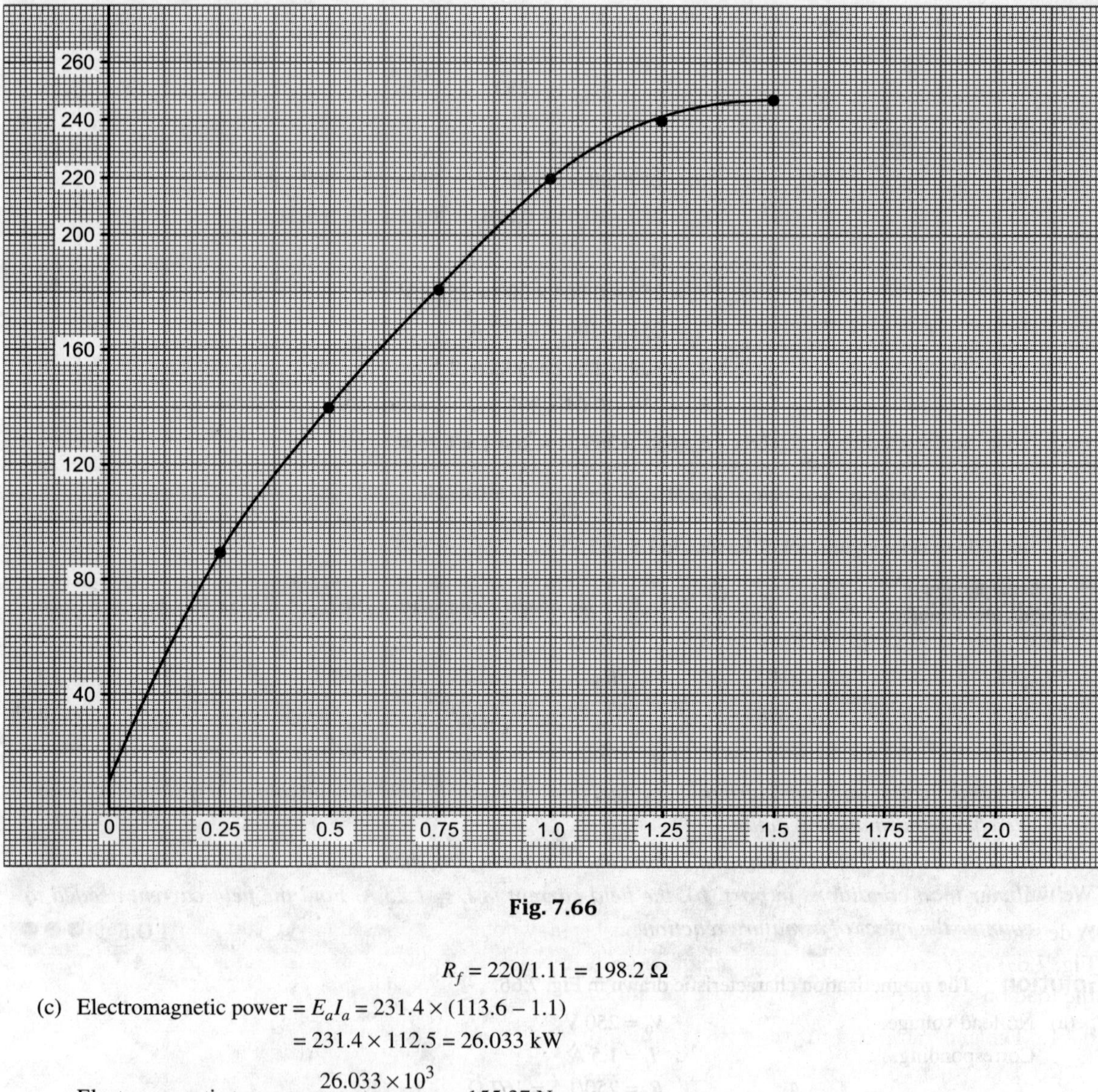

Fig. 7.66

$$R_f = 220/1.11 = 198.2\ \Omega$$

(c) Electromagnetic power $= E_a I_a = 231.4 \times (113.6 - 1.1)$

$$= 231.4 \times 112.5 = 26.033 \text{ kW}$$

$$\text{Electromagnetic torque} = \frac{26.033 \times 10^3}{1600 \times \dfrac{2\pi}{60}} = 155.37 \text{ Nm}$$

(d) Under load condition as in part (b)

$$I_f = 1.1 \text{ A, armature reaction ignored}$$

$$\text{Actual } I_f = 1.25 \text{ A}$$

I_f needed to counter effect of armature reaction $= 1.25 - 1.1 = 0.15$ A

Concept Check LO 5

1. What is the effect of magnetic saturation on the external characteristics of a dc shunt generator? ○●●
2. Explain the meaning and significance of the critical field resistance of a shunt generator. ○●●
3. What is OCC and what information does it reveal about a dc machine? At what speed is it determined? What is the air-gap line? ○●●
4. Write the basic proportionality relationships of a dc machine. What form these take for linear magnetization? ○○●
5. Sketch the external characteristic of a shunt generator and explain the reason for its special nature: part of it is two-valued. ○●●
6. List the performance characteristics of dc machines. ○●●

7.14 PARALLEL OPERATION OF dc GENERATORS

LO 6 Explain the parallel operation of dc generators

For supplying a large dc load it is desirable to use more than one generator in parallel. This arrangement provides the security that if one generator gives way, the other(s) can feed part load.

Desirable Conditions for dc Generators in Parallel

⇒ Same voltage rating
⇒ Same percentage voltage regulation
⇒ Same percentage speed regulation of the prime movers

As the generator voltage is easily adjustable (in a range) so the above conditions are not a must.

Paralleling a dc Generator to Busbars (Mains)

We will consider the case of dc shunt generators.

A dc shunt generator (G_1) is connected to the busbars (switches S, S_1 closed) and feeding the load as shown in Fig. 7.67(a). $I_1 = I_L$. Another shunt generator G_2 is to be connected to the busbars to share part load relieving the load on G_1. The stepwise procedure for this operation is described below:

1. The generator G_2 is driven by its prime mover and brought to rated speed
2. The switch S of G_2 is closed. The centre-zero voltmeter across the switch S_2 reads the voltage difference between busbars and voltage of G_2
3. The voltage of G_2 is adjusted by its shunt field regulator till the voltage across S_2 is nearby zero. The switch S_2 is then closed. G_2 now floats on the busbars without exchanging any current
4. The regulator resistance of G_2 is now reduced (increasing its field current) so that it feeds current I_2 to busbars. The regulating resistance of G_1 is now increased (reducing its field current) so that it feeds smaller current to the busbar. The adjustment process finally results in proper current sharing between the two generators such that $I_1 + I_2 = I_L$
5. In the above adjustment the bus bar voltage is maintained constant.

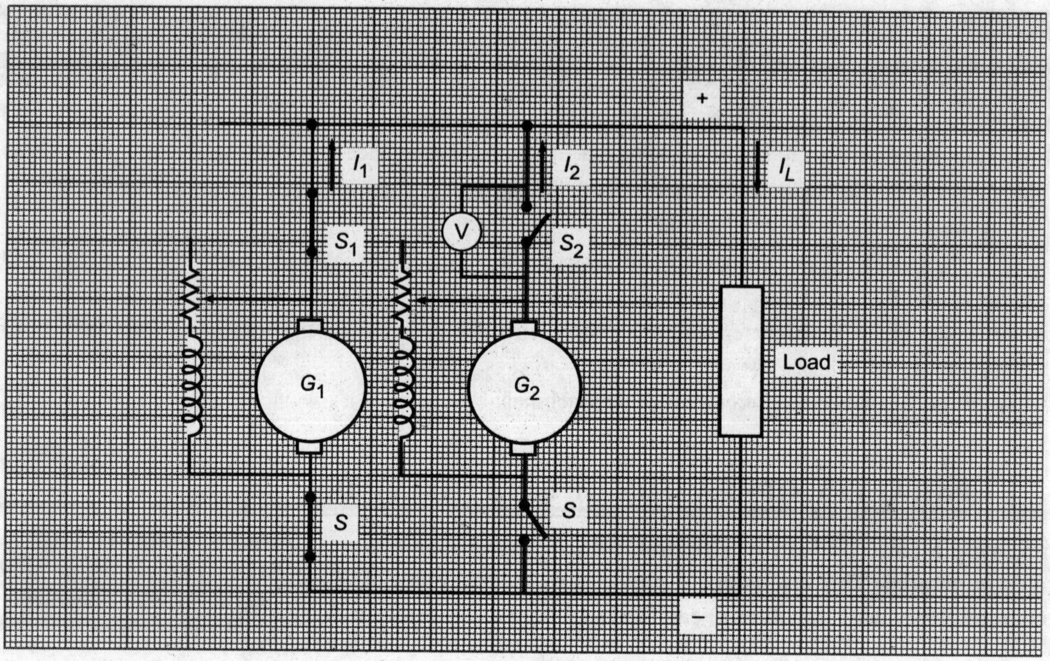

Fig. 7.67(a)

Determination of Load Sharing

The load sharing between the generators is determined by their external characteristics. Shown in Fig. 7.67(b) are the external characteristics of the two generators with their field currents adjusted for the same no-load voltage. The combined characteristic of the system is also drawn. At busbar voltage V_{bus} the total load current is supplied by the two operators such that

$$I_1 + I_2 = I_L$$

The adjustments of the field currents modifies the external characteristic changing the load sharing.

Parallel Operation of Compound Generators

Two compound generators in parallel feeding a load are sketched in Fig. 7.68. With switches S_1 and S_2 closed they are sharing the load. This system is found to be unstable because of the positive voltage feedback through series windings. We will present the qualitative arguments.

If because of any reason the emf of G_1 increases, it causes its load current I_1 to increase and correspondingly I_2 to decrease. The series excitation of G_1 increases and that of G_2 decreases and so the internal voltage of G_1 increases further and that of G_2 decreases. Consequently current I_1 fed by G_1 to load sharply increases and that of G_2 sharply decreases. Finally all the load shifts to G_1 from G_2. In fact G_2 may begin to act as a motor. All this leads to heavy overloading of G_1, an unacceptable operation.

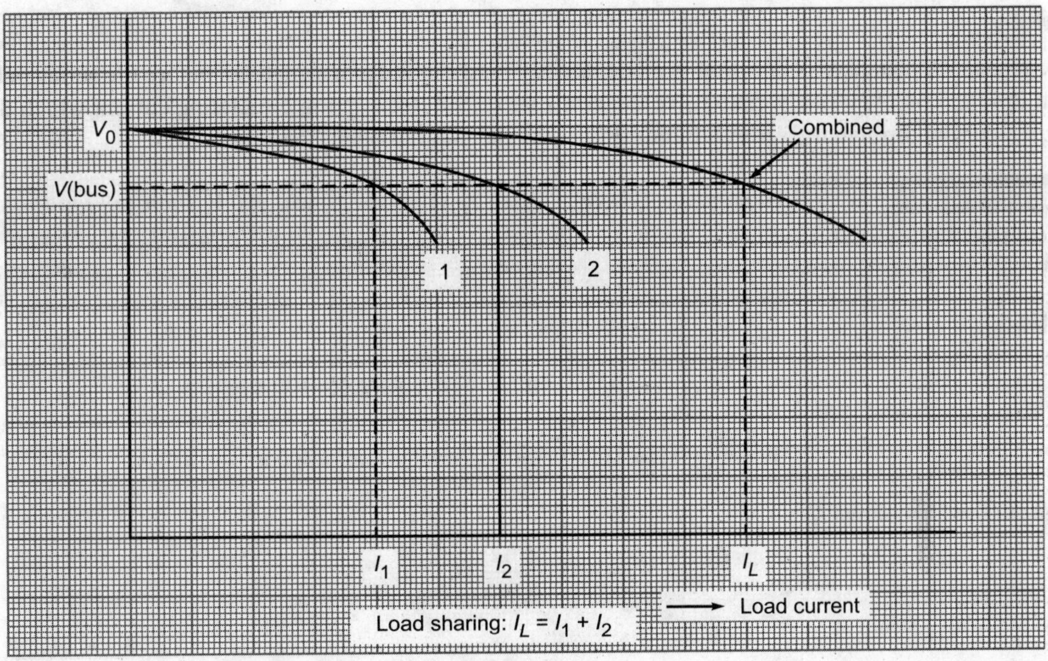

Fig. 7.67(b)

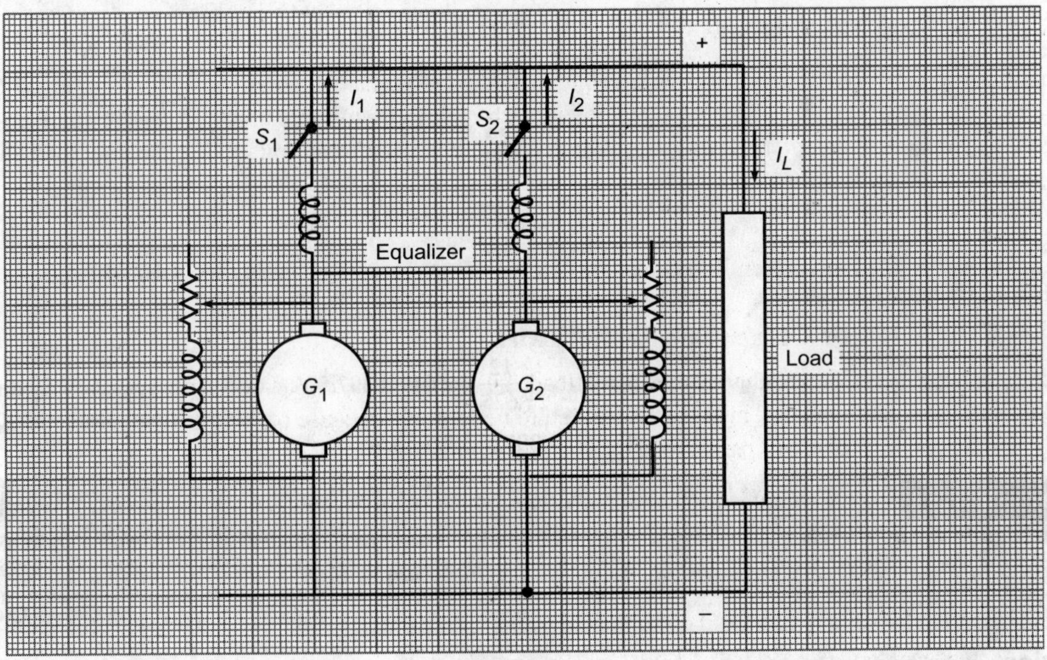

Fig. 7.68 *Compound generators in parallel S_1 and S_2 closed*

Remedy

A low-resistance equalizer connection is made directly between the two armatures before the series fields as shown in Fig. 7.68. Any emf variations of the armatures causes equalizing circulating current which do not affect the current through the series windings. Thereby the parallel operation is stabilized.

Example 7.30 *Two dc shunt generators are rated 230 kW and 150 kW, 400 V. Their full load voltage drops are 3% and 6% respectively. They are excited to no load voltages of 410 V and 420 V respectively. How will they share a load of 1000 A and the corresponding bus voltage?* LO 5

Solution

$$G_1 \quad I_{f1} = \frac{250 \times 10^3}{400} = 625 \text{ A; Full load voltage drop} = 400 \times \frac{3}{100} = 12 \text{ V}$$

$$G_2 \quad I_{f2} = \frac{150 \times 10^3}{400} = 375 \text{ A; Full load voltage drop} = 400 \times \frac{6}{100} = 24 \text{ V}$$

Voltage of G_1 at load current I_1

$$V_1 = 410 - \left(\frac{12}{625}\right) I_1 \tag{i}$$

Voltage of G_2 at load current I_2

$$V_2 = 420 - \left(\frac{24}{375}\right) I_2 \tag{ii}$$

Load on generation in parallel, $I_L = I_1 + I_2 = 1000$ A, Bus voltage = V (?)

In parallel operation

$$V_1 = V_2 = V \tag{iii}$$

From Eqs (i) and (ii)

$$410 - \left(\frac{12}{625}\right) I_1 = 420 - \left(\frac{24}{375}\right) I_2$$

$$0.064\, I_2 - 0.0192\, I_1 = 10 \tag{iv}$$

$$I_1 + I_2 = 1000 \tag{v}$$

Solving Eqs (iv) and (v), we get

$$I_1 = 649 \text{ A}, I_2 = 351 \text{ A}$$

$$\text{Bus voltage, } V = 410 - \left(\frac{12}{625}\right) \times 649 = 397.5 \text{ V}$$

$$\text{Load} = 10^3 \times 397.5 = 397.5 \text{ kW}$$

Example 7.31 *In Example 7.30, the two generators are excited to equal no-load voltages. What should be the percentage voltage drop of 150 kW generator in order that the share load is in the ratio of their ratings? What is the no load voltage for a bus voltage of 400 V and load current of 1000 A?* LO 6

Solution Let the full load voltage drop of G_2 be x volts. Then

$$V_1 = V_0 - \left(\frac{12}{625}\right) I_1 \tag{i}$$

$$V_2 = V_0 - \left(\frac{x}{375}\right) I_2 \tag{ii}$$

For parallel operation

$$V_1 = V_2 = V$$

So from Eqs (i) and (ii)

$$\left(\frac{x}{375}\right) I_2 = \left(\frac{12}{625}\right) I_1 \tag{iii}$$

We want
$$\frac{I_1}{I_2} = \frac{250}{150} = \frac{5}{3} \tag{iv}$$

Solving (iii) and (iv)

$$\frac{I_1}{I_2} = \frac{625x}{12 \times 375} = \frac{5}{3}$$

or
$$x = 12 \text{ V}$$

$$\text{Percentage voltage drop of } G_2 = \frac{12}{400} \times 100 = 3\%$$

Hence the percentage voltage drop of the two generators must be equal

$$I_1 + I_2 = 1000$$

$$\frac{I_1}{I_2} = \frac{5}{3}$$

which give
$$I_1 = 625 \text{ A}, \quad I_2 = 375 \text{ A}$$

$$V_{Bus} = V_0 - \left(\frac{12}{625}\right) \times 625 = 400 \text{ V}$$

or
$$V_0 = 412 \text{ V}$$

Concept Check — LO 6

1. What are the desirable conditions for parallel operation of dc generators? ○○●
2. How can we determine the load sharing of generators? ○●●
3. How to stabilize parallel operation of dc compound generators? ○●●

7.15 CHARACTERISTICS OF dc MOTORS

LO 7 Analyze the performance of dc motors by its performance characteristics

Important! The power of the dc motor lies in its versatility and ease with which a variety of speed-torque characteristics can be obtained, and the wide range of speed control which is possible without the need of elaborate control schemes while a high level of operating efficiency is maintained. In a dc generator the speed is fixed by the primemover and remains nearly constant throughout the operating part of the characteristics,

While the field excitation is adjusted to yield the desired terminal voltage at a given load. In a motor, on the other hand, the need is to match the speed-torque characteristic of the load and to run the load at a specified speed or speed by adjustment of the field and the armature voltage in case the speed control over a wide range is required.

Expressions for Back EMF and Torque

The fundamental emf and torque relationships of Eqs. (7.29) and (7.30) are reproduced below:

$$\text{Induced emf, } E_a = \frac{\Phi n Z}{60}\left(\frac{P}{A}\right) = K_a\,\Phi\omega \text{ V} \tag{7.66}$$

$$\text{Electromagnetic torque, } T = \frac{1}{2\pi}\,\Phi I_a Z\left(\frac{P}{A}\right) = K_a \Phi I_a \text{ Nm} \tag{7.67}$$

where
$$K_a = \frac{ZP}{2\pi \text{A}}$$

and
$$\omega = \left(\frac{2\pi}{60}\right) n = \text{speed in rad (mech)/s and } n = \text{speed in rpm} \tag{7.68}$$

In place of electromagnetic torque may also use the term developed torque with symbol T. For repeated use we may use the term torque only.

In motoring operation it is convenient to express armature speed in rpm. So we write the above relationships as

$$n = \frac{1}{K'_a}\left(\frac{E_a}{\Phi}\right);\ K'_a = \left(\frac{2\pi}{60}\right) K_a \tag{7.69}$$

$$T = K_a \Phi I_a \tag{7.70}$$

The induced emf E_a in a motor is known as *back emf* as it opposes the applied terminal voltage V_t. It is related to the terminal voltage by the armature circuit equation. For the short-compound motor (given in Fig. 7.30(b); Eq. 7.53(a))

Probe

What is significance of back emf in a dc motor?

$$E_a = V_t - I_a(R_a + R_{se}) \tag{7.71}$$

The flux/pole Φ is found from the magnetization characteristic against the equivalent excitation

$$I_{feq} = I_f \pm \left(\frac{N_{se}}{N_f}\right) I_a \tag{7.72}$$

corrected for the demagnetizing effect of armature reaction.

As in the case of dc generators depending on the type of excitation there are three types of dc motors – shunt motor, series motor and compound motor. Self-excitation has no meaning for a dc motor.

!Attention Three important operating characteristic of dc motor are

- Speed-armature current characteristic
- Torque-armature current characteristic
- Speed-torque characteristic

In this section, we shall examine the nature of the three characteristics mentioned above on quantitative-qualitative basic because of magnetic saturation and armature reaction caused demagnetization.

Shunt Motor

The connection diagram of dc shunt motor is shown in Fig. 7.69. Its operation with fixed terminal voltage and constant field current (fixed field resistance) will now be considered.

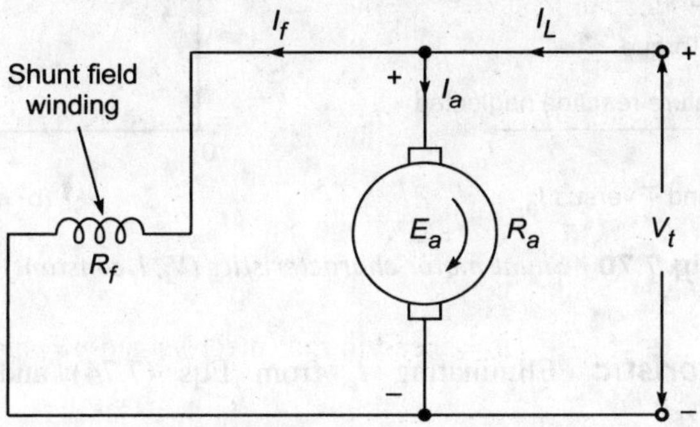

Fig. 7.69 *Shunt motor*

Speed-current characteristic The armature circuit equation is

$$E_a = V_t - I_a R_a = K'_a \Phi n \tag{7.73}$$

which gives

$$n = \frac{V_t - I_a R_a}{K'_a \Phi} \tag{7.74}$$

which indeed is the speed-current characteristic. At no load the armature current I_{ao} is quite small (2 – 5% of $I_a\,(fl)$), we will, therefore, assume $I_{ao} = 0$ for sketching the characteristic. The no load speed is then

$$n_0 = \left(\frac{V_t}{K'_a \Phi}\right) \tag{7.75}$$

Armature reaction effect ignored.

Φ remains constant. It is seen from Eq. (7.74) that speed drops linearly due to $I_a R_a$ drop. The characteristic is drawn in dotted line in Fig. 7.70(a).

Effect of armature reaction caused demagnetization.

Φ reduces with increasing I_a. It then follows from Eq. (7.74) that the speed at any I_a is higher than in the armature reaction neglected case. However, the effect of $I_a R_a$ drop predominates and the speed drop with I_a. Of course, the no load speed is the same as given by Eq. (7.77a). The characteristic is drawn in solid line in Fig. 7.70(a) which lies above the characteristic with armature reaction effect ignored.

Torque-current characteristic As per Eq. (7.70), motor torque is

$$T = K_a \Phi I_a \tag{7.76}$$

The characteristic is linear if armature reaction effect is ignored as in shown Fig. 7.70(a) by dotted line. Otherwise reduction in Φ causes the characteristic to bend downwards as shown by solid line in Fig. 7.70(a).

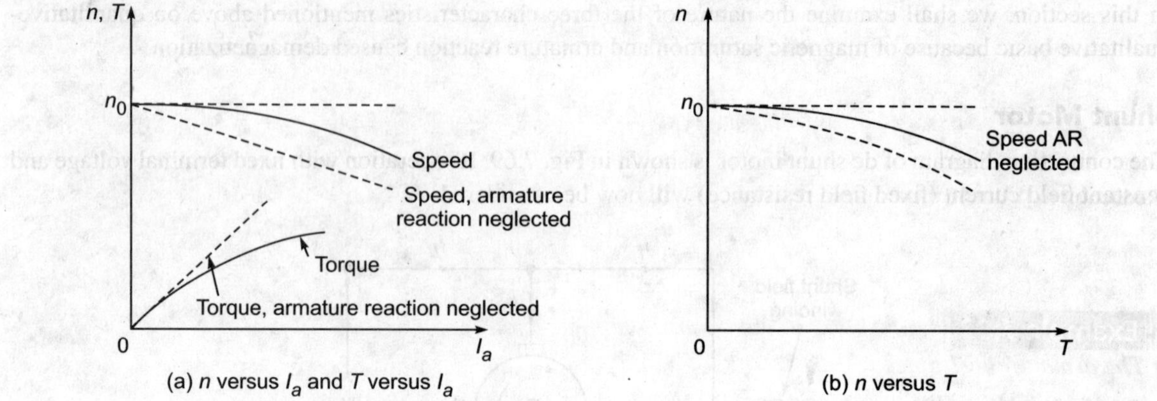

Fig. 7.70 *Shunt motor characteristics (V_t, I_f constant)*

Speed-torque characteristic Eliminating I_a from Eqs (7.74) and (7.76), the speed-torque characteristic is obtained as

$$n = \left(\frac{V_t}{K'_a \Phi}\right) - \left[\frac{R_a}{K'_a K_a \Phi^2}\right] T \tag{7.77}$$

If the armature reaction is ignored Φ remains constant. Therefore the speed drops linearly as per Eq. (7.77). The no load speed is

$$n_0 = \left(\frac{V_t}{K'_a \Phi}\right) \tag{7.77a}$$

The characteristic is drawn in dotted line in Fig. 7.70(b).

Armature reaction effect causes Φ to reduce with increasing torque (and so increasing current). The speed drops more sharply with torque than in the linear case because of Φ^2 in the denominator of the second them in Eq. (7.77). The characteristic is sketched in solid line in Fig. 7.70(b).

Sum-up The speed drops from no load to full load by a few per cent; in fact the speed remains substantially constant. Such a characteristic is known as "shunt characteristic".

Example 7.32 *An 8 kW, 230 V, 1200 rpm dc shunt motor has R_a = 0.7 Ω. The field current is adjusted until, on no load with a supply of 250 V, the motor runs at 1250 rpm and draws armature current of 1.6 A. A load torque is then applied to the motor shaft, which causes the I_a to rise to 40 A and the speed falls to 1150 rpm. Determine the reduction in the flux per pole due to the armature reaction.* LO 7

Solution Equation (7.73) gives

$$n = K'_a\left(\frac{E_a}{\Phi}\right) = K'_a\left(\frac{V - I_a R_a}{\Phi}\right)$$

or

$$\Phi = K'_a\left(\frac{V - I_a R_a}{n}\right)$$

$$\Phi(\text{no-load}) = K'_a\left(\frac{250 - 1.6 \times 0.7}{1250}\right) = 0.2\,K'_a$$

$$\Phi(\text{load}) = K'_a\left(\frac{250 - 40 \times 0.7}{1150}\right) = 0.193\,K'_a$$

Reduction in Φ due to the armature reaction

$$= \frac{0.2 - 0.193}{0.2} \times 100 = 3.5\%$$

Example 7.33 *A 20 kW, 250 V dc shunt motor has a full-load armature current of 85 A at 1100 rpm. The armature resistance is 0.18 Ω.*

Determine:

(a) the internal electromagnetic torque developed;

(b) the internal torque if the field current is suddenly reduced to 80% of its original value;

(c) The steady motor speed in part (b) assuming the load torque to have remained constant.

Assume: magnetic circuit to be linear. **LO 7**

Solution

(a)
$$E_a = 250 - 0.18 \times 85 = 234.7 \text{ V}$$
$$E_a I_a = T\omega$$
$$234.7 \times 85 = T \times \frac{2\pi \times 1100}{60}$$
or
$$T = 173.2 \text{ Nm}$$

(b) Magnetic circuit linearity is assumed i.e. $\Phi \propto I_f$. Field suddenly reduced to 0.8 of original value. I_a is assumed to remain constant for that instant. Then

$$T_1 = 0.8 \times 173.2 = 138.6 \text{ Nm}$$

(c) Under steady condition for the motor internal torque to build to the original value, new values of armature current and speed are established. These are obtained below:

$$T = K'_a K_f I_f \times 85 = K'_a K_f \times 0.8\, I_f \times I_{a1} = 173.2 \text{ Nm}$$
or
$$I_{a1} = 106.25 \text{ A}$$
$$E_{a1} = 250 - 0.18 \times 106.25 = 230.9 \text{ V}$$
and
$$234.7 = K_a' Kf\, I_f \times 1100$$
$$230.9 = K'_a\, K'_f\, 0.8\, I_f n_1$$
Dividing we get
$$\frac{230.9}{234.7} = \frac{0.8 n_1}{1100}$$
or
$$n_1 = 1353 \text{ rpm.}$$

Observe that both motor speed and armature current increase by reducing field current with constant load torque.

Example 7.34 *A dc shunt motor runs at 1200 rpm on no-load drawing 5 A from 220 V mains. Its armature and field resistances are 0.25 Ω and 110 Ω respectively. When loaded (at motor shaft), the motor draws 62 A from the mains. What would be its speed? Assume that the armature reaction demagnetizes the field to the extent of 5%.*

Also calculate the internal torque developed at no-load and on load. What is the motor shaft torque at load (this torque drives the mechanical load).

LO 7

Solution In a dc shunt motor

$$I_L = I_a + I_f$$

Shunt field current,
$$I_f = \frac{220}{110} = 2 \text{ A (constant)}$$

At no load

$$I_{a0} = 5 - 2 = 3 \text{ A}$$

$$E_{a0} = 220 - 0.25 \times 3 = 219.25 \text{ V} \qquad \text{(i)}$$

$$n_0 = 1200 \text{ rpm}$$

$$E_{a0} = K'_a \Phi n_0$$

or
$$219.25 = K'_a \Phi \times 1200$$

$$T_0 \omega_0 = E_{a0} I_{a0}$$

$$\left(\frac{2\pi \times 1200}{60}\right) T_0 = 219.25 \times 3$$

or
$$T_0 = 5.23 \text{ Nm}$$

This torque is absorbed in iron loss, and windage and friction losses of the motor (the shaft being at no load).

On load

$$I_{a1} = 62 - 2 = 60 \text{ A}$$

$$E_{a1} = 220 - 0.25 \times 60 = 205 \text{ V} \qquad \text{(ii)}$$

$$\text{flux/pole } \Phi_1 = 0.95\, \Phi$$

$$E_{a1} = K'_a \Phi_1 n_1$$

$$205 = K'_a \times 0.95\, \Phi n_1 \qquad \text{(iii)}$$

Dividing Eq. (ii) by Eq. (i)

$$\frac{0.95 n_1}{1200} = \frac{205}{219.25}$$

or
$$n_1 = 1181 \text{ rpm}$$

Drop in motor speed is only 1.6% on being loaded. This because the reduction in flux/pole due to armature reaction counters the drop in speed caused by armature resistance drop.

$$T_1\, \omega_1 = E_{a1} I_{a1}$$

$$\left(\frac{2\pi \times 1181}{60}\right) T_1 = 205 \times 60$$

or
$$T_1 = 99.45 \text{ Nm}$$

$$T_1 \text{ (shaft)} = 99.45 - 5.23$$

$$= 94.22 \text{ Nm}$$

Series Motor

The connection diagram of a series motor is drawn in Fig. 7.71. From the armature circuit equation

$$E_a = V_t - I_a (R_a + R_{se}) = K'_a \Phi n \qquad (7.78)$$

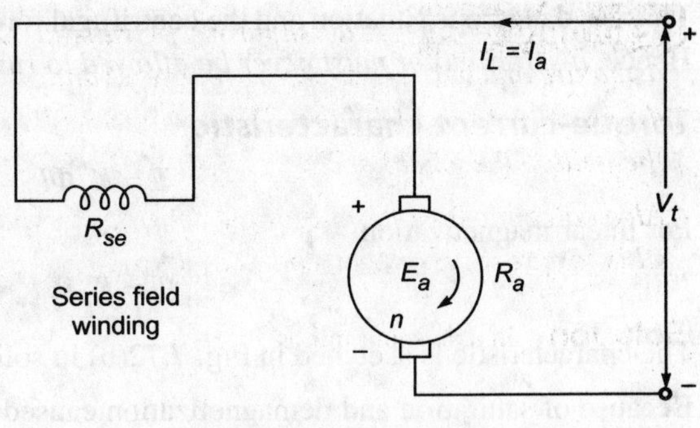

Fig. 7.71 *Series Motor*

Speed-current characteristic From Eq. (7.78), we can express motor speed as

$$n = \left(\frac{V_t}{K'_a \Phi}\right) - \left(\frac{R_a + R_{se}}{K'_a \Phi}\right) I_a \qquad (7.79)$$

This is the exact speed-current equation.

In a series motor $\Phi - I_a$ is the magnetization characteristic, which as we know has an initial linear region beyond which saturation sets in. In the linear region

$$\Phi = K_f I_a;\ K_f \text{ is constant} \qquad (7.80)$$

Substituting in Eq. (7.79), the speed can be expressed as

$$n = \frac{1}{K'_a K_f}\left[\left(\frac{V_t}{I_a}\right) - (R_a + R_{se})\right] \qquad (7.81)$$

This is a shifted rectangular hyperbola sketched in Fig. 7.72(a) in solid line. As the armature current increases rate of increase of Φ reduces. It means

$$\Phi < K_f I_a$$

As a result, the actual speed found from Eq. (7.79) is higher than based on linear assumption. The actual speed-current characteristic lies above the characteristic based on linear magnetization; the characteristic is sketched in dotted line in Fig. 7.72(a).

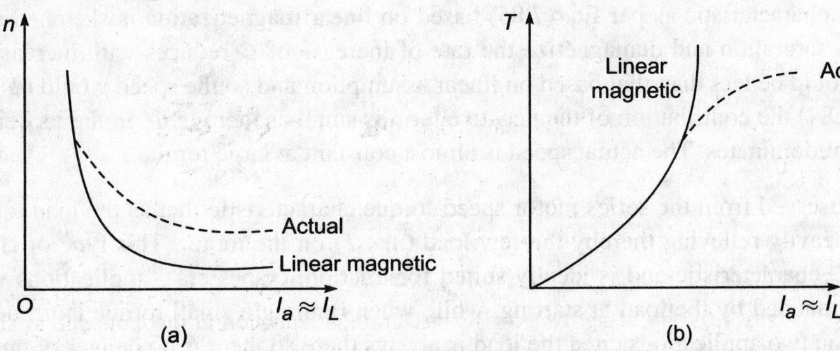

Fig. 7.72 *Series Motor Characteristics*

!Attention *Some observations*

1. As the armature current increases with load the speed comes down sharply
2. At no load

$$I_a \to 0,\ \Phi \to 0,\ n \to \infty$$

as shown from Eq. (7.79); the second term is very small because of $(R_o + R_{se})$

This is a dangerous situation and the centrifugal forces will destroy the armature and may harm the personnel. Hence, *a series motor must never be allowed to run at no load even accidently.*

Torque-current characteristic

$$T = K_a \Phi I_a \tag{7.82}$$

For linear magnetization

$$T = K_a K_f I_a^2 \text{ : a parabola} \tag{7.83}$$

The characteristic is sketched in Fig. 7.72(b) in solid line.

Because of saturation and demagnetization caused by armature reaction, the torque tends to level off in the actual characteristic shown in dotted line in Fig. 7.72(b).

Speed-torque characteristic From Eq. (7.82)

$$I_a = \frac{1}{K_a}\left(\frac{T}{\Phi}\right)$$

Substituting in Eq. (7.79) and organizing

$$n = \left(\frac{V_t}{K'_a \Phi}\right) - \left(\frac{R_a + R_{se}}{K'_a K_a \Phi^2}\right) T \tag{7.84}$$

For linear magnetization

Eliminating I_a between Eqs (7.81) and (7.83) we get

$$n = \frac{1}{K'_a K_f}\left[\frac{V_t \sqrt{K_a K_f}}{\sqrt{T}} - (R_a + R_{se})\right] \tag{7.85}$$

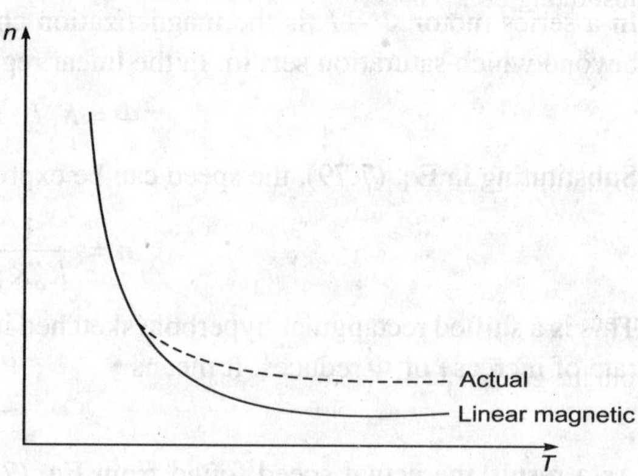

Fig. 7.73 *Speed-Torque characteristics of series motor*

The speed-torque characteristic as per Eq. (7.85) based on linear magnetization is sketched in Fig. 7.73 in solid line. Due to saturation and demagnetize the rate of increase of Φ reduces with increasing torque. At large torque, Φ would be less than that based on linear assumption and so the speed would be higher. This is because in Eq. (7.84) the contribution of the negative term is small as there is Φ^2 in the its denominator and so the first term predominates. The actual speed is almost constant at large torque.

Remark It is observed from the series motor speed-torque characteristic that as the load torque increases the speed drops heavily relieving thereby the few load ($n \times T$) on the motor. This type of characteristic is known as "series" characteristic and is ideally suited for traction, cranes etc—applications where in large accelerating is demanded by the load at starting, while when running a small torque is needed to maintain steady speed. In the two applications cited the load is always there so there is no danger of under-loading or no-loading.

Example 7.35 *The magnetization characteristic of a 4-pole de series motor may be taken as proportional to current over a part of the working range; on this basis the flux per pole is 4.5 mWb/A. The load requires a gross torque proportional to the square of the speed equal to 30 Nm at 1000 rev/min. The armature is wave-wound and has 492 active conductors. Determine the speed at which the motor will run and the current it will draw when connected to a 220 V supply, the total armature resistance of the motor being 2.0 Ω.*

LO 7

Solution Referring to Eq. (7.22)

$$E_a = \frac{\Phi n Z}{60}\left(\frac{P}{A}\right) = \frac{(4.5\times10^{-3}\times I_a)n\times492}{60}\left(\frac{4}{2}\right)$$

$$= 0.0738\ nI_a \quad \text{(i)}$$

Recalling Eq. (7.26), the torque developed

$$T = \frac{1}{2\pi}\Phi I_a Z\left(\frac{P}{A}\right) = \frac{1}{2\pi}(4.5\times10^{-3} I_a)\, I_a\times492\left(\frac{4}{2}\right)$$

$$= 0.705\ I_a^2 \quad \text{(ii)}$$

Further $$E_a = V_t - I_a(R_a + R_{se}) = 220 - 2I_a \quad \text{(iii)}$$

Substituting Eq. (i) in (iii),

$$0.0738\ nI_a = 220 - 2I_a$$

or $$I_a = \frac{220}{2 + 0.0738n} \quad \text{(iv)}$$

Substituting this expression for I_a in Eq. (ii),

$$T = 0.705\left[\frac{220}{2+0.0738n}\right]^2$$

Given: Load Torque $T_L = K_L n^2$

From the given data K_L can be evaluated as

$$K_L = \frac{30}{(1000)^2} = 3\times10^{-5}\ \text{Nm/rpm}$$

Under steady operation conditions, $T_L = T$ (developed)

or $$3\times10^{-5} n^2 = 0.705\left(\frac{220}{2+0.0738n}\right)^2$$

Solving, $$n = 662.6\ \text{rpm}$$

Substituting for n in Eq. (iv)

$$I_a = \frac{220}{2 + 0.0738\times663.2}$$

$$= 4.32\ \text{A}$$

Example 7.36 *A 250-V dc series motor with compensating winding has a total armature circuit resistance of (R_a + R_{se}) = 0.08 Ω. It is run at 1200 rpm by means of a primemover with armature circuit open and series field separately excited. This test yielded the following magnetisation data:*

I_{se}(A)	40	80	120	160	200	240	280	320	360	400
V_{OC} (V)	62	130	180	222	250	270	280	288	290	292

Sketch the speed-torque characteristic of the series motor connected to 250 V main by calculating the speed and torque values at armature currents of 75, 100, 200, 300, 400 A.

LO 7

Solution Rather than drawing the magnetisation curve, we shall use the above data by linear interpolation between the data point given.

Sample Calculation

$$I_a = I_{se} = 75 \text{ A}$$

$$E_a = 250 - 0.08 \times 75 = 244 \text{ V}$$

Using linear interpolation, at $I_{se} = 75$ A

$$E_a \text{ (1200 rpm)} = 130 - \frac{130 - 62}{40} \times 5 = 121.5 \text{ V}$$

$$n = \frac{1200}{121.5} \times 244 = 2410 \text{ rpm}$$

$$T\omega = E_a I_a$$

or

$$T = \left(\frac{60}{2\pi \times 2410}\right) \times 244 \times 75 = 72.5 \text{ Nm}$$

Computations can be carried in tabular from below:

$I_a = I_{se}$ (A)	75	100	200	300	400
E_a (V)	244	242	234	226	218
E_a (V) at 1200 rpm	121.5	155	250	283	292
n (rpm)	2410	1874	1123	958	902
T (Nm)	72.5	123	398	676	923

The speed-torque ($n - T$) characteristic is drawn in Fig. 7.74.

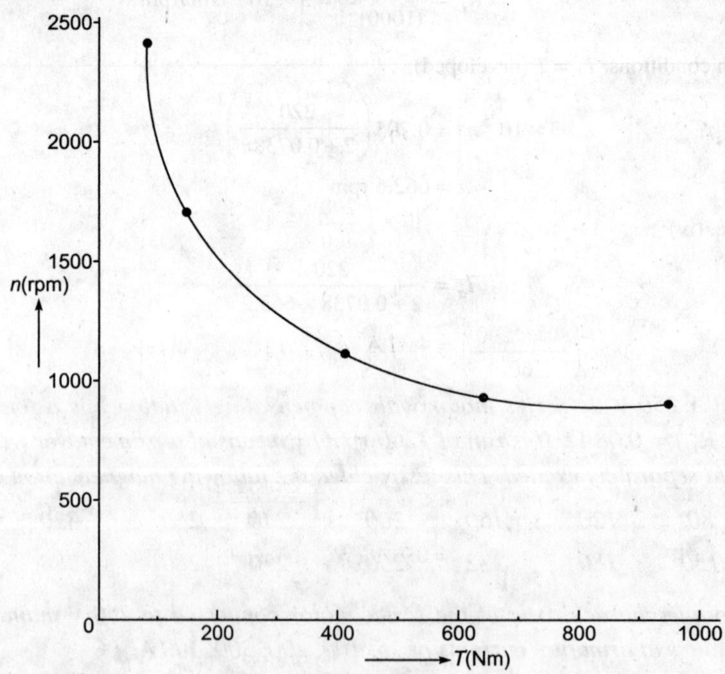

Fig. 7.74

Example 7.37 *A 220 V, 7.5 kW series motor is mechanically coupled to a fan. When running at 400 rpm the motor draws 30 A from the mains (220 V). The torque required by the fan is proportional to the square of speed. $R_a = 0.6\ \Omega$, $R_{se} = 0.4\ \Omega$. Neglect armature reaction and rotational loss. Also assume the magnetisation characteristic of the motor to be linear.*

(a) Determine the power delivered to the fan and torque developed by the motor.

(b) Calculate the external resistance to be added in series to the armature circuit to reduce the fan speed to 200 rpm. Calculate also the power delivered to the fan at this speed.

LO 7

Solution

$$P_{\text{fan}} = P_{\text{dev}} = P;\ \text{no rotational loss}$$

$$P = E_a I_a \tag{i}$$

$$T_{\text{dev}} = K'_a K_f I_{se} I_a = K'_a K_f I_a^2;\ I_{se} = I_a;\ \text{linear magnetization}$$

$$T_{\text{fan}} = K_F n^2$$

But

$$T_{\text{dev}} = T_{\text{fan}} = T$$

$\therefore$

$$I_a \propto n \tag{ii}$$

(a) Operation at 400 rpm ($R_{\text{ext}} = 0$)

$$E_a = 220 - (0.6 + 0.4) \times 30 = 190\ \text{V}$$

$$I_a = 30\ \text{A}$$

$$P = 190 \times 30 = 5.7\ \text{kW}$$

$$T\omega = E_a I_a$$

or

$$T = \frac{5700}{\dfrac{2\pi \times 400}{60}} = 136\ \text{Nm}$$

(b) Operation at 200 rpm ($R_{\text{ext}} = ?$)

$$T = 136 \times \left(\frac{200}{400}\right)^2 = 34\ \text{Nm}$$

$$I_a = 30 \times \left(\frac{200}{400}\right) = 15\ \text{A}$$

$$T\omega = E_a I_a$$

$$34 \times \left(\frac{2\pi \times 200}{60}\right) = [220 - (0.6 + 0.4 + R_{\text{ext}}) \times 15] \times 15$$

Solving we get

$$R_{\text{ext}} = 10.5\ \Omega$$

$$P = T\omega = 34 \times \left(\frac{2\pi \times 200}{60}\right)$$

$$= 0.721\ \text{kW}$$

Example 7.38 *A 180 kW, 600 V dc series motor runs at 600 rpm at full load current of 300 A. The total resistance of its armature circuit 0.105 Ω. The magnetization curve data at 500 rpm is as below. The series field excitation is provided separately.*

$V_{OC} = E_a$ (V)	440	470	500	530	560	590
Series field current (A)	250	277	312	356	406	466

Determine the starting torque developed when the starting current is limited to 500 A. Assume that the armature reaction mmf is proportional to the square of armature current. LO 7

Solution The magnetization curve at 500 rpm is drawn in Fig. 7.75.

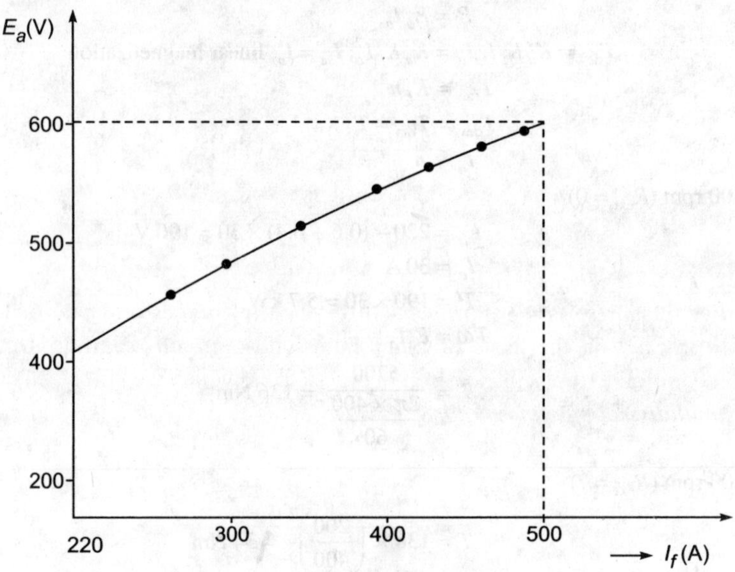

Fig. 7.75

$$\text{Full load current, } I_a = \frac{180 \times 10^3}{600} = 300 \text{ A, Speed, } n = 600 \text{ rpm}$$

Back emf

$$E_a = 600 - 300 \times 0.105 = 568.5 \text{ V}$$

To look up magnetization curve, we should find the emf at 500 rpm

$$E_a \text{ (500 rpm)} = 568.5 \times \frac{500}{600} = 473 \text{ V}$$

From the magnetization curve to induce 473 V, the series field current needed is

$$I_{se} = I_a = 282 \text{ A; we will write } I_a$$

Therefore the series field equivalent demagnetizing current is

$$I_{ad} = 300 - 282 = 18 \text{ A}$$

As I_{ad} is proportional to the square of armature current,

$$I_{ad}\text{/ampere of armature current} = \frac{18}{(300)^2} = 0.2 \times 10^{-3}\text{A/A}$$

At start

$$I_a = 500 \text{ A}$$

$$\text{Effective } I_a = 500 - I_{ad} = 500 - (500)^2 \times 0.2 \times 10^{-3} = 450 \text{ A}$$

From the magnetization curve, the corresponding induced emf is

$$E_a = 590 \text{ V at } n = 500 \text{ rpm}$$

Power balance equation

$$T\omega = E_a I_a \tag{i}$$

Therefore

$$T_{\text{start}} = \frac{E_a I_a}{\omega} \tag{ii}$$

Substituting values

$$T_{\text{start}} = \frac{590 \times 500}{500 \times \frac{2\pi}{60}} = 5634 \text{ Nm}$$

Compound Motor

It has been said earlier that there is no significant difference in the performance of long and short shun connection. We shall proceed on the basis of long shunt whose circuit diagram is given in Fig. 7.30(a). Further as differential compound motor is not used in practice, for which we will advance the reason later, we will consider only *cumulative compound motor*.

Speed equation

$$n = \frac{V_t - I_a(R_a + R_{se})}{K_a' \Phi} \tag{7.84}$$

Torque equation

$$T = K_a \Phi I_a \tag{7.85}$$

In a compound motor Φ is determined from the magnetization characteristic for the combined (additive) net excitation

$$I_f\text{(net)} = I_f + \left(\frac{N_{se}}{N_f}\right) I_a,\ I_f = \text{constant}$$

It will help matters if we could separate out the flux created by shunt and series fields but the superposition cannot be applied. To overcome this problem we consider the magnetization characteristic (Φ vs I_f) sketched in Fig. 7.76 it is found from this figure that $\Phi = \Phi_{sh} + \Phi_{se}$

where Φ_{sh} = flux of shunt excitation

Φ_{se} = flux of series excitation

Further Φ_{se} can be taken to be proportional to I_a corresponding to slope of the magnetization characteristic in the region beyond Φ_{sh}.

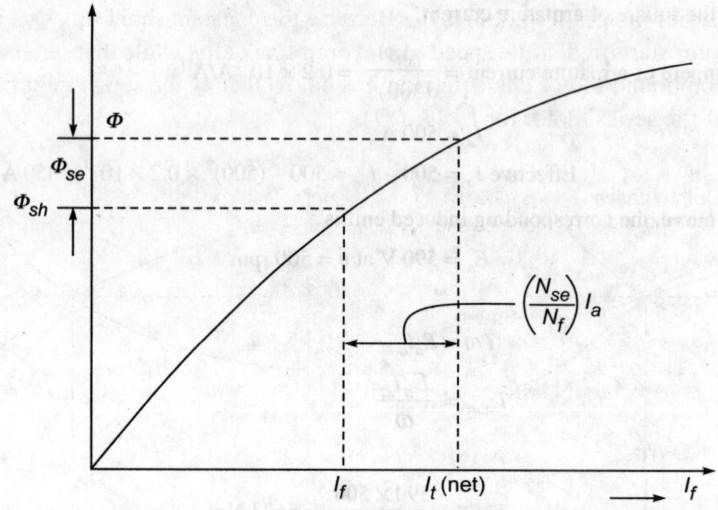

Fig. 7.76

It may be noted that the above method accounts for saturation but not demagnetizing effect of armature reaction.

Speed-current characteristic

$$n = \frac{1}{K'_a}\left[\frac{V_t - I_a(R_a + R_{se})}{\Phi_{sh} + \Phi_{se}}\right] \tag{7.86}$$

As I_a increases, numerator decreases and Φ_{se} in the denominator increases and as a result the motor speed falls much more sharply than in shunt motor. At large values of current the characteristic is similar to that of a series motor. The characteristic lies in between those of shunt and series motors, closer to one whose field is stronger near full load. The compound motor has a great advantage over series motor because it has a definite no load speed given by

$$n_0 = \frac{1}{K'_a}\left(\frac{V_t}{\Phi_{sh}}\right); \text{ shunt motor no load speed} \tag{7.87}$$

Yet its characteristic can made closer to the relieving characteristic of series motor.

A differentially compound motor has flux/pole $\Phi = (\Phi_{sh} - \Phi_{se})$. It is seen from Eq. (7.84) that on over-load Φ reduces sharply and so the motor acts like a series motor on no load. This is why the *differential compound motor is not used in practice*.

Advantages of compound motors over series/shunt motors Comparison of compound motor speed-current characteristic with shunt at the same full load speed and also the series motor is depicted in Figs. 7.77(a) and (b) in which the conclusions made above are borne out. Certain observations are made below:

1. Compound and shunt motors (Fig. 7.77(a)) Because of Φ_{se} increasing with I_a, the speed of the compound motor falls much more sharply than the shunt motor. Therefore, the $n - I_a$ characteristic of the compound motor lies above that of the shunt motor for $I_a < I_a\,(fl)$ and lies below for $I_a > I_a\,(fl)$

2. Compound and series motors (Fig. 7.77(b)) Because the constant shunt flux Φ_{sh}, $n - I_a$ characteristic of compound motor starts at definite speed n_0 and drop gradually, while that series motor drop steeply. Therefore, the compound motor characteristic lies below that of the series motor for $I_a < I_a\,(fl)$ and lies above that of the series motor for $I_a > I_a\,(fl)$

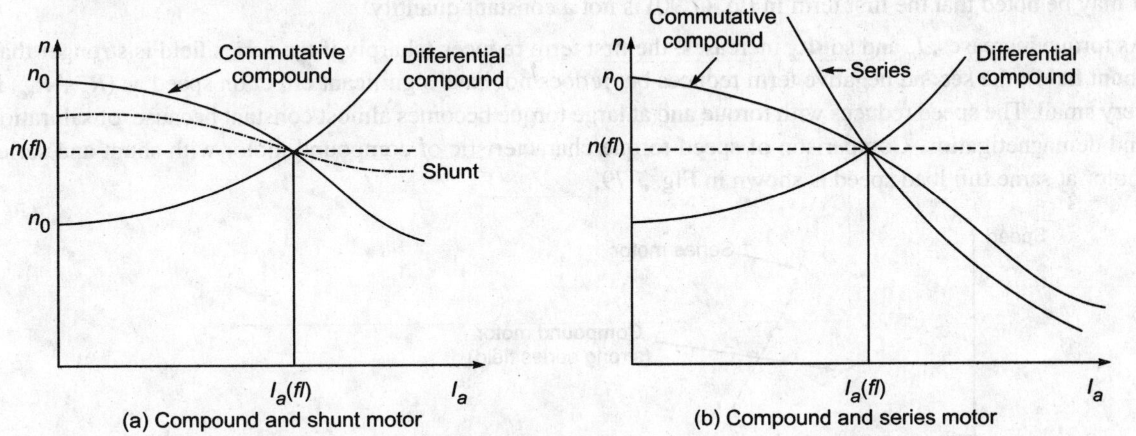

Fig. 7.77 *Comparison for same full load speed for n – I_a characteristics*

Torque–current characteristic From Eq. (7.85)

$$T = K_a(\Phi_{sh} + \Phi_{se})\, I_a \tag{7.88}$$

$$= K_a \Phi_{sh}\, I_a + K_a \Phi_{se}\, I_a$$

$$\Phi_{se} \approx K_{se}\, I_a$$

Thus

$$T = K_a \Phi_{sh}\, I_a + K_a K_{se}\, I_a^2 \tag{7.89}$$

shunt type + series type

Due to saturation and demagnetization both torque components level off. The torque-current characteristic of the compound motor as sum of shunt and series components is sketched in Fig. 7.78. This approach of dividing of Φ into Φ_{sh} and Φ_{se} is heuristic and so reveals the nature of characteristic; can not to used quantitatively. Quantitative approach (graphical) will be illustrated in examples that follow.

Incidentally the figure provides the comparison of the nature of the $(T - I_a)$ characteristics of the three kinds of dc motors.

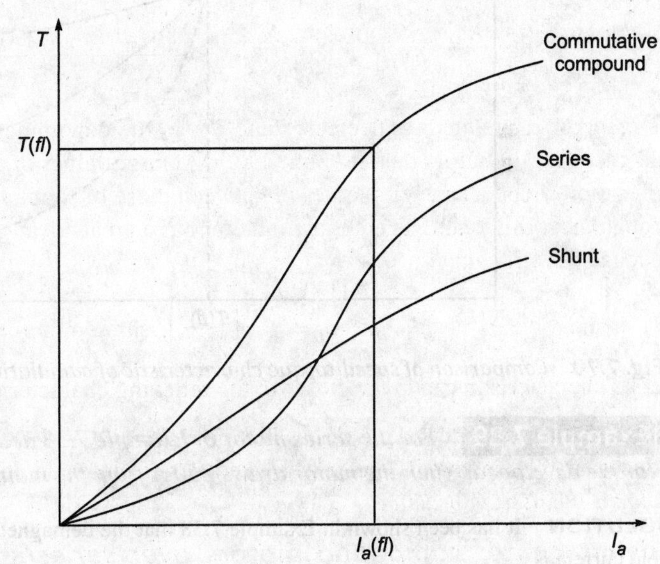

Fig. 7.78 *Torque-current characteristic of compound motor.*

Speed-torque characteristic Eliminating I_a between Eqs 7.86 and (7.88) we obtain

$$n = \frac{1}{K'_a}\,\frac{V_t}{\Phi_{sh} + \Phi_{se}} - \frac{1}{K'_a K_a}\,\frac{R_a + R_{se}}{(\Phi_{sh} + \Phi_{se})^2}\,T \tag{7.90}$$

where as before no load speed is

$$n_0 = \frac{1}{K'_a}\left(\frac{V_t}{\Phi_{sh}}\right) \tag{7.91}$$

It may be noted that the first term in Eq. (7.90) is not a constant quantity.

As torque increases, I_a and so Φ_{se} increases, the first term reduces (sharply if the series field is stronger than shunt field); the second negative term reduces but it does not have significant effect on speed as $(R_a + R_{se})$ is very small. The speed reduces with torque and at large torque becomes almost constant because of saluration and demagnetigation. Comparison of speed-torque characteristic of compound motor with shunt and series motor at same full load speed is shown in Fig. 7.79.

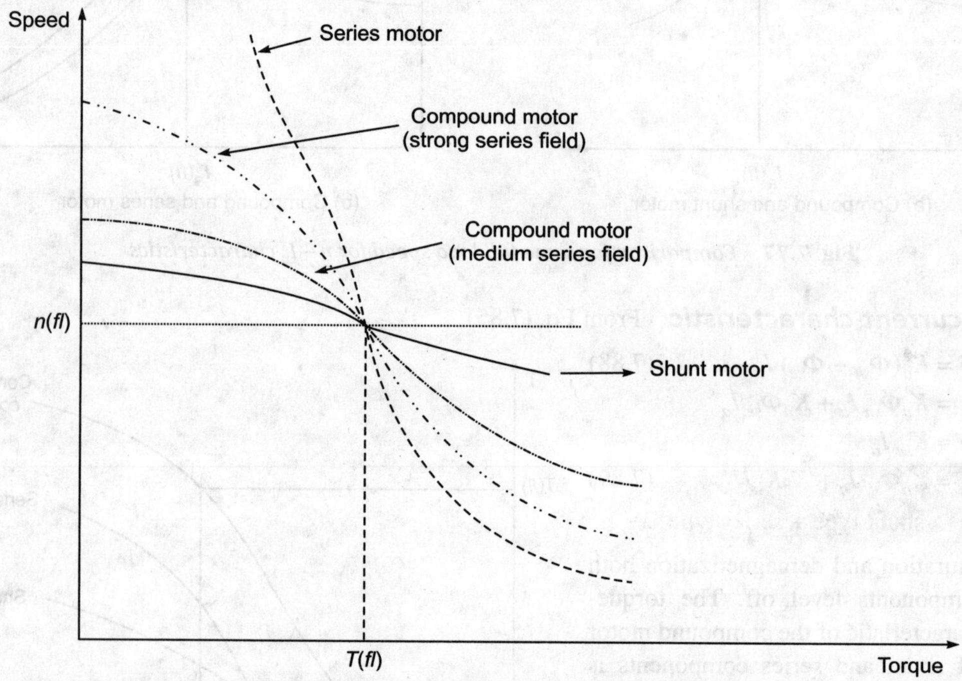

Fig. 7.79 *Comparison of speed-torque characteristic of cumulative compound motor with shunt and series motor*

Example 7.39 *For the series motor of Example 7.38 determine the speed and mechanical power and torque developed, when the motor draws 250 A from the mains.*

LO 7

SOLUTION It has been shown in Example 7.38 that the demagnetizing effect of armature reaction in terms of series field current is

$$I_{ad} = 0.2 \times 10^{-3}\, I_a^2$$

At

$$I_a = 250 \text{ A}$$

$$I_{ad} = 0.2 \times 10^{-3} \times (254)^2 = 12.5 \text{ V}$$

Therefore

$$I_a\,(\text{net}) = 250 - 12.5 = 237.5 \text{ A}$$

From the magnetization curve

$$E_a \text{ (at 500 rpm)} = 428 \text{ V}$$

$$E_a \text{ (actual)} = 600 - 250 \times 0.105 = 573.75 \text{ V}$$

Therefore, motor speed

$$n = 500 \times \frac{573.75}{428} = 670 \text{ rpm}$$

Mechanical power developed

$$P_{mech} = E_a I_a = 573.75 \times 250$$

$$= 143.44 \text{ kW}$$

$$\text{Torque developed} = \frac{143.44 \times 10^3}{670 \times \frac{2\pi}{60}} = 2044 \text{ Nm}$$

Example 7.40 *A 240 V compound (cumulative) dc motor has the following open circuit magnetization characteristic at normal full load speed of 850 rev/min:*

Excitation, AT/pole	*1200*	*2400*	*3600*	*4800*	*6000*
Generated emf, V	*76*	*135*	*180*	*215*	*240*

The resistance voltage drop in the armature circuit at full load is 25 V. At full load the shunt and the series windings provide equal ampere-turn excitation.

Calculate the mmf per pole on no load. Estimate the value to which the speed will rise when full load is removed, the resistance voltage drop in the armature circuit under that condition being 3V. Ignore armature-reaction and brush-contact effects. Assume long-shunt cumulative compounding. LO 7

Solution At full load, from Fig. 7.81,

$$E_a \text{ (full load)} = V - I_a (R_a + R_{se}) = 240 - 25 = 215 \text{ V}$$

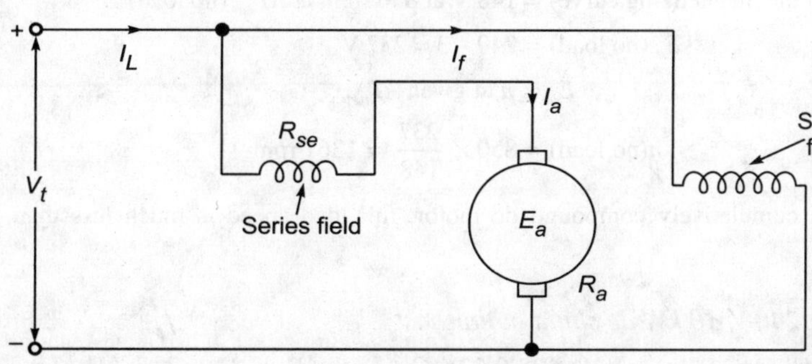

Fig. 7.80 *Long-shunt compound dc motor*

Corresponding AT_{net} from magnetizing curve of Fig. 7.81.

$$AT_{net} \text{ (full load)} = 4800$$

Then

$$AT_{sh} = AT_{se} \text{ (full load)} = 2400$$

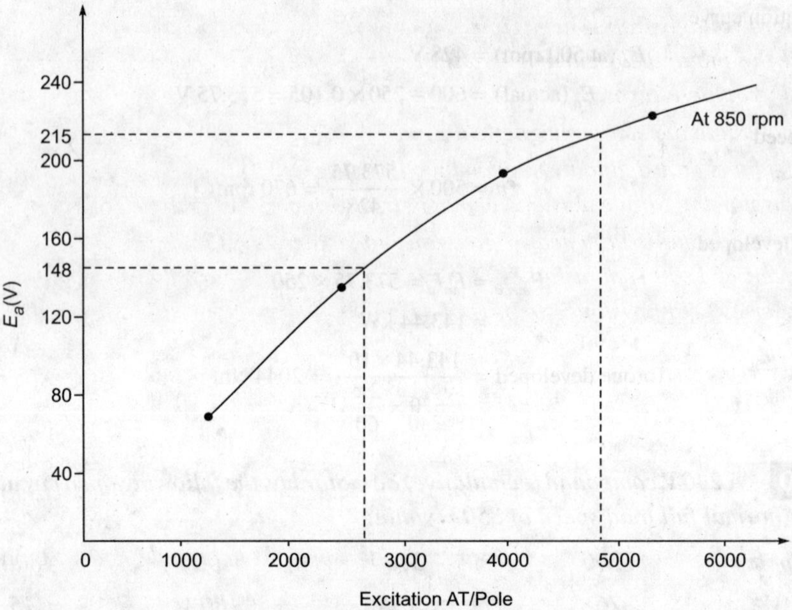

Fig. 7.81

Now $$I_a \text{ (no load)} = \frac{3}{25} I_a \text{ (full load)}; I_a \text{ (no load)} = I_f \text{ (shunt field)}$$

$$AT_{se} \text{ (no load)} = \frac{3}{25} \times 2400 = 288$$

$$AT_{sh} = 2400 \text{ (no change)}$$

$$AT_{net} \text{ (no load)} = 2400 + 288 = 2688$$

$$E_a \text{(from the magnetizing curve)} = 148 \text{ V at 850 rpm, at } AT_{net} \text{ (no load)}$$

$$E_a \text{ (no load)} = 240 - 3 = 237 \text{ V}$$

Now $$E_a \propto n \text{ at given } AT_{net}$$

Hence, $$n\text{(no load)} = 850 \times \frac{237}{148} = 1361 \text{ rpm}$$

It may be seen that in a cumulatively compound dc motor, full load speed is much less than no load speed (see Fig. 7.71).

Example 7.41 *A 240 V, 10 kW dc shunt motor has*

$$R_a = 0.18\ \Omega, \quad R(\text{inter poles}) = 0.025\ \Omega \quad \text{and} \quad R_f = 320\ \Omega \text{ (field only)},$$

$$\text{Shunt field turns} = 2000$$

LO 7

No load test conducted at 240 V by varying the field current yielded the following data:

Speed, n	*1160*	*1180*	*1220*	*1250*	*1300*
Field current, I_f(A)	*0.664*	*0.632*	*0.584*	*0.558*	*0.522*

At no load armature voltage drop is negligible.

(a) At full load the motor field current is adjusted to 0.6 A for a speed of 1218 rpm. Calculate the demagnetizing ampere-turns.

(b) In part (a) calculate the electromagnetic torque.

(c) The field current is adjusted to the maximum value. The starting armature current is limited to 75 A. Calculate the starting torque. Assume demagnetizing ampere-turns to be 165.

(d) The shunt field current is adjusted to give a no load speed of 1250 rpm. A series field is provided to give a speed of 1200 rpm at full load current. Calculate the number of turns of the series field. Assume the series field resistance to be 0.04 Ω.

Solution

Full load current,
$$I_a = \frac{10 \times 10^3}{240} = 41.7 \text{ A}$$

At no load
$$E_a \approx V_t = 240 \text{ V}$$
$$E_a = K'_a \Phi n$$

or
$$n = \frac{1}{K'_a} \frac{E_a}{\Phi}, \; \Phi = \Phi(I_f)$$
$$n = \frac{1}{K'_a} \cdot \frac{E_a}{\Phi(I_f)}; \; \Phi = \Phi(I_f)$$

Thus $n - I_f$ characteristic is the inverse of $\Phi - I_f$ characteristic scaled by $\left(\frac{E_a}{K'_a}\right)$, pulled in Fig. 7.82

(a) At full load current,
$$I_a = 41.7 \text{ A}, I_f = 0.6 \text{ A}$$
$$E_a = V_t - I_a(R_a + R_i) = 240 - (0.18 + 0.025) \times 41.7$$
$$E_a = 231.45 \text{ V}$$
$$n = 1218 \text{ rpm}$$
$$n \text{ (at } E_a = 240 \text{ V)} = 1218 \times \frac{240}{231.45} = 1262 \text{ rpm}$$

The corresponding field current required is found from the $n - I_f$ characteristic at
$$I_f = 0.548 \text{ A}$$

Actual field current required is,
$$I_f = 0.6 \text{ A}$$

Therefore, equivalent demagnetizing field current is
$$I_{fd} = 0.6 - 0.548 = 0.052 \text{ A}$$

Demagnetizing ampere-turns,
$$AT_d = 0.052 \text{ A} \times 2000 = 104$$

(b) Torque developed
$$T\omega = E_a I_a$$

or
$$T = \frac{231.45 \times 41.7}{1218 \times \frac{2\pi}{60}} = 75.7 \text{ Nm}$$

(c) Field current (max),
$$I_f = \frac{240}{320} = 0.75 \text{ A}$$
$$AT_d = 165 \text{ (given)}$$
$$I_{fd} = \frac{165}{2000} = 0.0825 \text{ A}$$

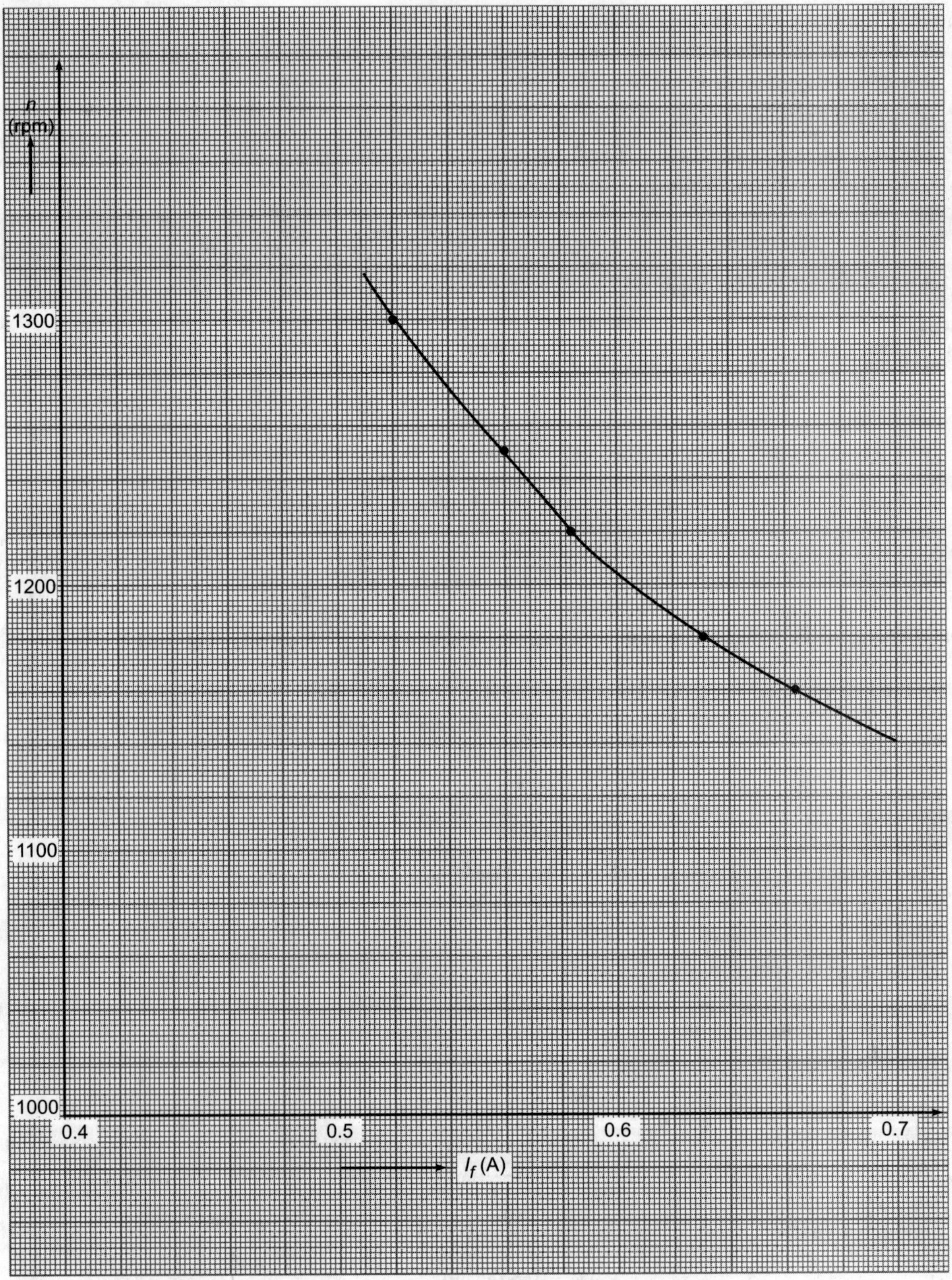

Fig. 7.82

$$I_f \text{ (net)} = 0.75 - 0.0825 = 0.6675 \text{ A}$$

At this field current at $E_a = 240$ V, the motor speed is from the $n - I_f$ plot as

$$n = 1150 \text{ rpm or } \omega = 121.27 \text{ rad/s}$$

We know that $$E_a = K_a \Phi \omega \quad \text{(i)}$$

and $$T = K_a \Phi I_a \quad \text{(ii)}$$

At $I_f = 0.75$, $AT_d = 165$, Φ has the same value in Eqs (i) and (ii).

From Eq. (i) $$K_a \Phi = \frac{E_a}{\omega} = \frac{240}{121.27} = 1.98$$

Using this value in Eq. (ii) with I_a (start) = 75 A, we get

$$T \text{ (start)} = 1.98 \times 75 = 148.5 \text{ Nm}$$

(d) No load speed, $$n_0 = 1250 \text{ rpm}$$

$$E_a = 240 \text{ V}$$

Corresponding field current, $I_f = 0.56$ A ; from $n - I_f$ characteristic of Fig. 7.82.

At full load speed specified, $$n = 1200 \text{ rpm}$$

$$R_{se} = 0.04\ \Omega$$

$$\text{Total resistance in armature circuit} = 0.18 + 0.25 + 0.04 = 0.245\ \Omega$$

$$E_a = 240 - 0.245 \times 41.7 = 229.8 \text{ V}$$

At, $$E_a = 240 \text{ corresponding speed} = 1200 \times \frac{240}{229.8} = 1150 \text{ rpm}$$

From $n - I_f$ characteristic

$$I_f \text{ (net)} = 0.684 \text{ A}$$

Equivalent I_f to be produced by the series field = 0.684 – 0.56 = 0.124 A

We then field $$N_{se} = \frac{0.124 \times 2000}{41.7} = 5.95 \text{ or } 6 \text{ turns}$$

Example 7.42 *A shunt motor draws a full load armature current of 56.5 A from 230 V supply and runs at 1350 rpm. It has an armature-circuit resistance of 0.15 Ω and shunt field turns of 1250. Its OCC data at 1350 rpm is given below:*

V_{OC} (V)	180	200	220	240	250
I_f (A)	1.18	1.4	1.8	2.4	2.84

(a) Determine the shunt field current of the motor at no load speed of 1350 rpm.

(b) Determine the demagnetizing effect of armature reaction in ampere-turns/pole at full load current.

(c) By adding series-field turns and connecting the motor in long shunt compound, it is required to have a speed of 1230 rpm when drawing an armature current of 56.5 A from 230 V supply. Assume the series field resistance to be 0.033 Ω. Determine the series field turns/pole.

(d) In part (c) the series field turns provided are 25/pole with series field resistance of 0.025 Ω. Determine the speed at which the motor would run when connected to 230 V Supply drawing armature current of 56.5 A.

LO 7

Solution The shunt motor connections are drawn in Fig. 7.84 No regulating resistance in the shunt, so the shunt field current remains constant. The magnetization characteristic from the OCC data is plotted in Fig. 7.85.

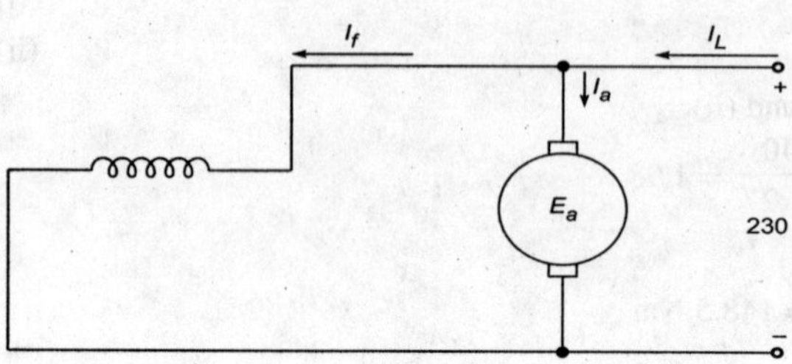

Fig. 7.83

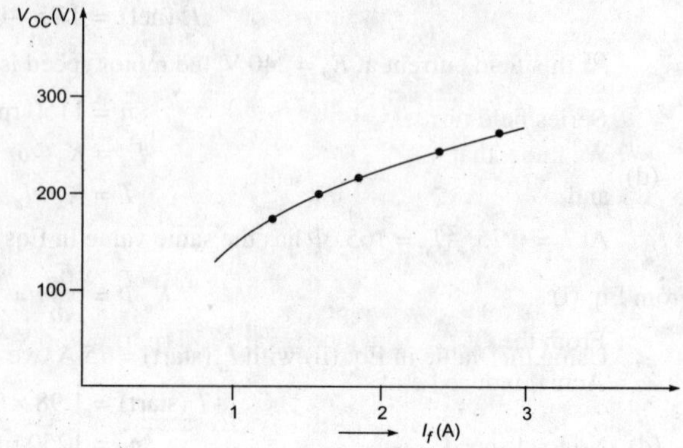

Fig. 7.84

(a) At no load $\qquad n_0 = 1350$ rpm

The no load armature voltage drop $I_{ao}R_a$ can be neglected. So

$$E_a = V_t = 230 \text{ V}$$

We find from the OCC at $\qquad E_a = 230$ V

$$I_f = 1.08 \text{ A}$$

The shunt field current remains constant throughout as the field has fixed resistance.

(b) Full load $\qquad$ Speed $n = 1350$ rpm

Armature current, $\qquad I_a = 56.5$ A

Therefore $\qquad E_a = 230 - 56.5 \times 0.15 = 221.5$ V

At $\qquad E_a = 221.5$ V (speed 1350) it is found from the OCC that

$$I_f \text{ (net)} = 1.8 \text{ A}$$

Actual $\qquad I_f = 2.08$ A

The difference is the demagnetizing effect of armature reaction in terms of field current. So

$$I_{fd} = 2.08 - 1.8 = 0.28 \text{ A}$$

Demagnetizing ampere-turns, $\qquad AT_d = 0.28 \times 1200 = 336$

(c) Long shunt compound connections are drawn in Fig. 7.85. Shunt field current is same as calculated in part (a)

$$I_f = 2.08 \text{ A}$$

At full load $\qquad E_a = 230 - 56.5\,(0.15 + 0.033)$

$$= 219.7 \text{ V}$$

Speed required, $\qquad n = 1230$ rpm

To look up OCC, we find E_a at 1350 rpm

$$E_a(1350 \text{ rpm}) = 219.7 \times \frac{1350}{1230} = 241.1 \text{ V}$$

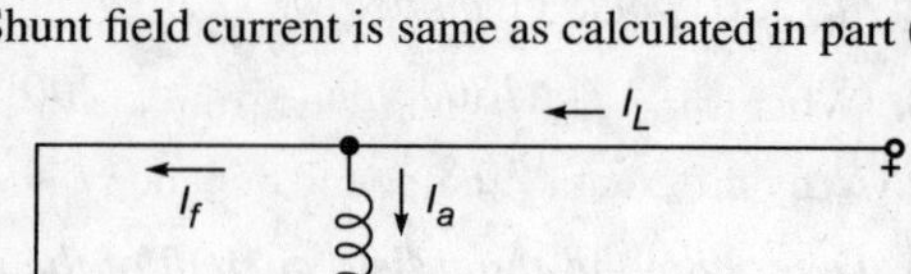

Fig. 7.85

To induce this emf

$$I_f\text{(net)} = 2.41 \text{ A} \quad \text{or} \quad 2.41 \times 1200 = 2982 \text{ AT}$$

Actual shunt field current

$$I_f = 1.08 \text{ A} \quad \text{or} \quad 1296 \text{ AT}$$

AT balance equation

$$AT_{net} = AT_{sh} - AT_d + AT_{se}$$

$$2892 = 1296 - 336 + AT_{se}$$

or $$AT_{se} = 1932$$

Series field turns, $$N_{se} = \frac{1932}{65.5} = 34/\text{pole}$$

(d) $$AT_{se} = 25 \times 56.5 = 1412.5$$

$$AT_{net} = 1296 - 334 + 1412.5 = 2374.5$$

$$I_f\ (\text{net}) = \frac{2374.5}{1200} = 1.98\ \text{A}$$

From the OCC $$E_a(\text{at 1350 rpm}) = 226\ \text{V}$$

Actual induced emf

$$E_a\ (\text{actual}) = 230 - 56.5\ (0.15 + 0.025) = 221.1\ \text{V}$$

We then find

$$\text{Speed, } n = 1350 \times \frac{221.1}{226} = 1320\ \text{rpm}$$

Concept Check — LO 7

1. How can one choose between a short shunt and long shunt cumulatively compound dc motor? ○●●
2. Explain how the back emf of a motor causes the development of mechanical power. ●●●
3. Sketch the speed-torque characteristic of a shunt motor at fixed field current. Explain the nature of the characteristic through relevant fundamental relationships of the dc machine. ○●●
4. Sketch the speed-torque characteristic of a dc series motor and advance the underlying reasoning for the nature of the characteristic based on fundamental relationships of the dc machine. ●●●
5. Explain through sketch and derivations the speed-torque characteristic of a differentially compound dc motor. ○●●
6. Assuming magnetic linearity derive the expression for speed-torque characteristic of a dc series motor using suitable symbols. ○●●
7. Compare the speed torque characteristics of a series and commutative compound motor. Why does the compound motor have a definite no-load speed? ●●●
8. What is meant by shunt-characteristic of a dc shunt motor? ○○●
9. Why differential compound dc motor is not used in practice? ○●●

7.16 | STARTING OF dc MOTORS

LO 8 Discuss the starting methods and speed control methods of dc motors

At the time of starting ($n = 0$), the induced emf of a motor is zero such that the current drawn from rated voltage supply would be

$$I_a^s = \frac{V}{R_a} \tag{7.92}$$

for a shunt motor. The series field resistance would be included in the denominator for a series motor. For large motors the armature resistance may be 0.01 pu or less; even for lower kW motors it ranges from 0.0625 to 0.125.

Important! Thus in full-voltage starting of a dc motor, the armature current can be several times (about 100 times for large motors) the rated value. For several reasons mentioned below such a large current cannot be allowed to flow in a motor even for the short starting period.

(i) It would cause intolerably heavy sparking at the brushes which may destroy the commutator and brush-gear.
(ii) Sudden development of large torque causes mechanical shock to the shaft, reducing its life.
(iii) Such heavy current cannot be generally permitted to be drawn from the source of supply.

Methods of Starting

For the above-mentioned reasons all motors, except for small and fractional-kW motors, must be started with external resistance included in armature circuit to limit the starting current safe values. Where variable-voltage dc supply is available, e.g. in the Ward-Leanard control scheme, this can be used for motor starting and no starting resistance would be needed.

One point in favour of direct starting must be mentioned here. Since the motor torque with direct start is much higher, the motor starts much more quickly. As a consequence the Joule input per start is much less than that with resistance start. This would be helpful in repeated starting–saving energy and causing less temperature rise.

!Attention Maximum allowable starting current is not more than 1.5 to 2.0 times the rated value.

These values are safe and yet at the same time permit a high starting torque for quick acceleration of the motor. Obviously the dc motor can start on load. Where a variable-voltage supply is available for speed control, starting poses no problem at all.

Shunt and Compound Motor Starters

In shunt and compound motors starting the shunt field should be switched on with full starting resistance in armature circuit. A short time delay in this position allows the field current to build up to the steady value of the inductive field transients. Also all the regulating resistance in the field circuit must also be cut out before starting for the field current to be maximum as $T_{start} \propto I_f$.

We shall illustrate shunt motor starters, but these are applicable for compound motors as well. There are two types of shunt motor starters:

⇒ Three-point starter – employed where motor field current can be varied in a narrow range and so does the motor speed

⇒ Four-point starter – motor field current can vary over a wide range and so does the motor speed

Three-point starter The connection diagram of a three-point shunt motor starter is shown in Fig. 7.86. The starter terminals to be connected to the motor are *A* (armature); *F* (field) and *L* (line). The starting resistance is arranged in steps between conducting raised studs. As the starting handle is rotated about its fulcrum, it moves from one stud to the next, one resistance step is cut out, and it gets added to the field circuit. There is a short time wait at each stud for the motor to build up speed. This arrangement ensures a high average starting torque.

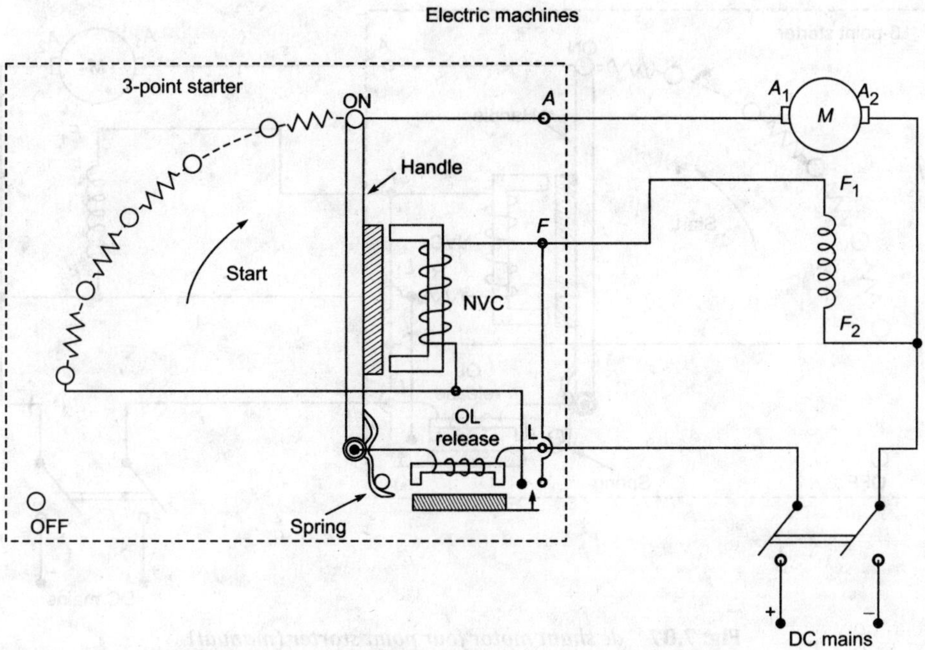

Fig. 7.86 *dc shunt motor there point starter (manual)*

At start the handle is brought to stud one. The line voltage gets applied to the armature with full starting resistance in series with armature and to the field with NVC in series. Thus the motor starts with maximum torque. As it pick up speed the handle is moved from stud to stud to the 'ON' position shown in Fig. 7.86. The starting resistance has been fully cut out and is now included in the field circuit; being small it makes little difference in the field current. The resistance of NVC is small and forms part of the field resistance. The voltage across the armature is the line voltage. The handle is held in this position by the electromagnet excited by the field current flowing through NVC.

Important! Two protections are incorporated in the starter.

1. NVC (no volt coil): In case of failure of field current (due to accidental or otherwise open circuiting), this coil releases the handle (held electromagnetically), which goes back to the OFF position under the spring action.
2. OL (over-load) release: The contact of this relay at armature current above a certain value (over load/ short circuit) closes the NVC ends, again bringing the handle to OFF position.

In the three-point starter if the field regulator is used to reduce the field current to low values for high motor speed NVC may release the handle causing the motor to shut down where such variation of field current is desired a four-point starter is used.

Four-point starter To overcome the problem caused when the field current is low, NVC is connected across the two lines, one line connected to *F* terminal through the starter and other directly to the second line from another *L* terminal of the starter. To limit the NVC current a protective resistance *R* is connected in series with it. The starter diagram is drawn in Fig. 7.87. It now has four terminals = *A F LL*. The rest operation remains the same.

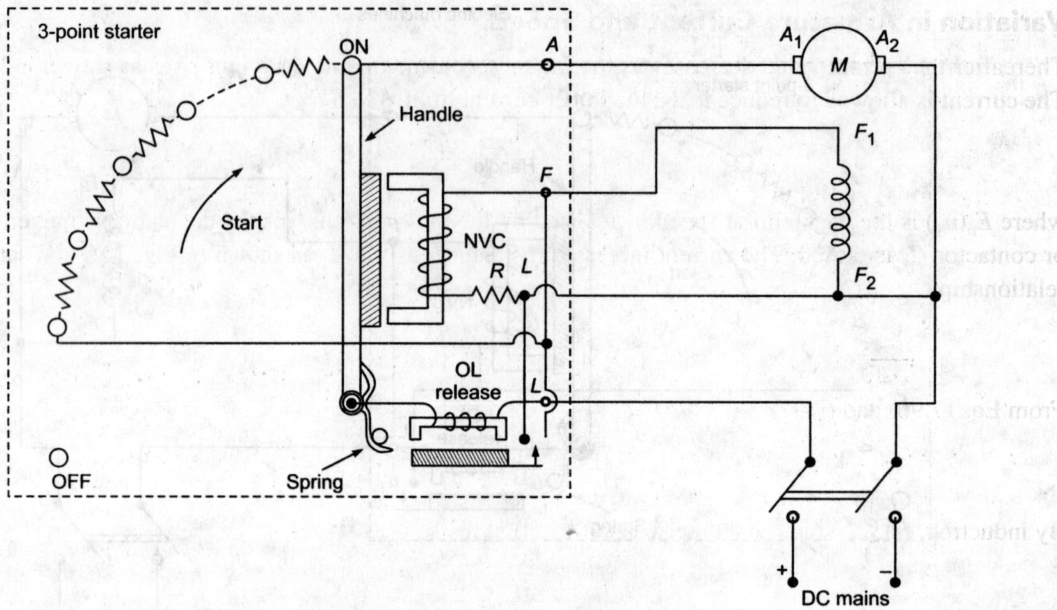

Fig. 7.87 *dc shunt motor four point starter (manual)*

Automatic Starters

The modern practice is to use a push-button type automatic starter in industries. Automatic starters carry out essentially the same functions as the manual ones with electromagnetic relays that short out sections of the robust metallic starting resistors either in a predetermined time sequence or when the armature current has dropped to a preset value. Such an arrangement is shown in Fig. 7.88. Most automatic starters embody extra control and safety features.

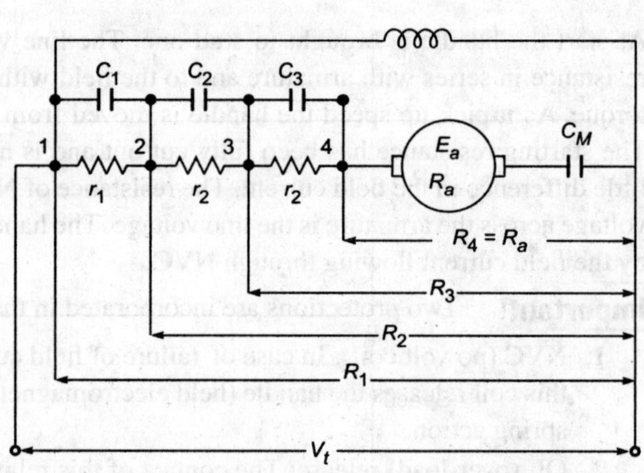

Fig. 7.88 *Automatic Shunt motor starting*

Starter Step Calculation for dc Shunt Motor

From Fig. 7.88 it is seen that the instant the starter is moved to stud 1 or conductor[††] C_M is closed, the current in the circuit reaches a value I_1, designated as the upper current limit, given by

$$I_1 = \frac{V}{R_1} \tag{7.93}$$

[††]Resistance arranged in a circular arc with studs on which a conductor arm moves are employed in a manual starter. Contactors as shown in Fig. 7.88 are used in automatic starters.

Variation in Armature Current and Speed

Thereafter the current value decreases as the motor speeds up and its back emf rises as shown in Fig. 7.89. The current is allowed to reduce to I_2, the lower current limit, given by

$$I_2 = \frac{V - E_a(n_1)}{R_1} \tag{7.94}$$

where $E_a(n_1)$ is the back emf at speed n_1 reached by the motor. At this instant the starter is moved to stud 2 or contactor C_1 is closed. The current increases instantaneously to I_1 as shown in Fig. 7.89 and satisfies the relationship

$$I_1 = \frac{V - E_a(n_1)}{R_2} \tag{7.95}$$

From Eqs (7.96) and (7.97)

$$\frac{I_1}{I_2} = \frac{R_1}{R_2} \tag{7.96a}$$

By induction, for a k-stud ($(k - 1)$ sections of external resistance) starter,

$$\frac{I_1}{I_2} = \frac{R_1}{R_2} = \frac{R_2}{R_3} = \cdots = \frac{R_{k-1}}{R_k} \tag{7.96b}$$

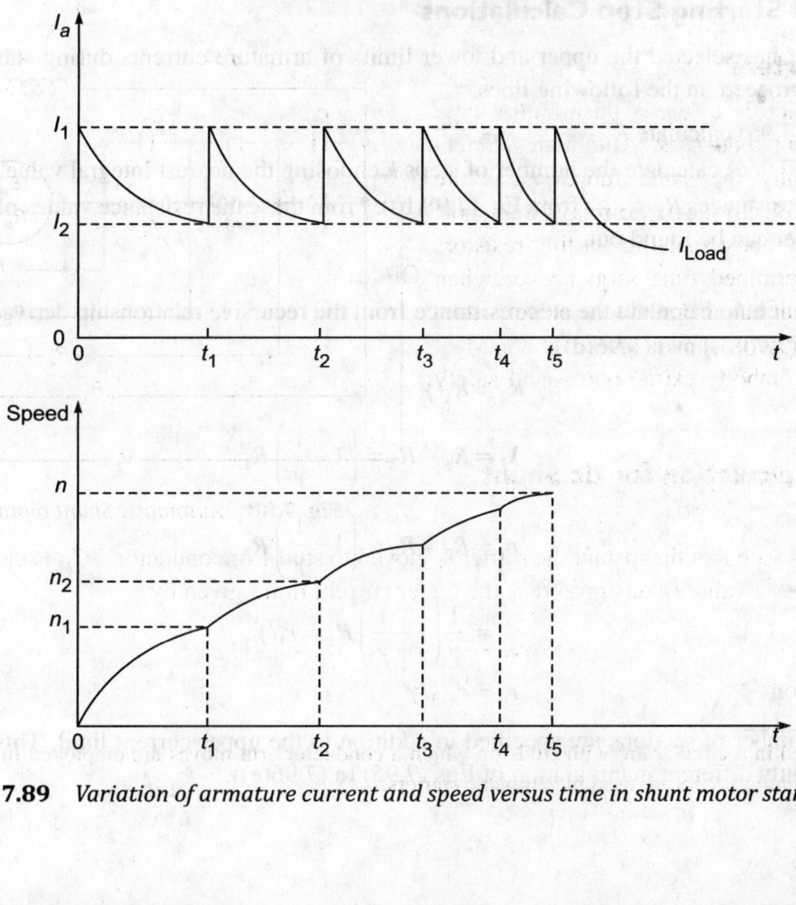

Fig. 7.89 *Variation of armature current and speed versus time in shunt motor starting*

It immediately follows from Eq. (7.98b) that

$$\frac{R_1}{R_2}\cdot\frac{R_2}{R_3}\cdots\frac{R_{k-1}}{R_k}=\left(\frac{I_1}{I_2}\right)^{k-1}=\gamma^{k-1}$$

where
$$\gamma=\frac{I_1}{I_2}=\frac{\text{upper current limit}}{\text{lower current limit}} \tag{7.96c}$$

Hence
$$\frac{R_1}{R_k}=\gamma^{k-1} \tag{7.96d}$$

From Fig. 7.89 it is obvious that

$$R_k = r_k = R_a \text{ (armature resistance)}$$

$\therefore$
$$\frac{R_1}{R_a}=\gamma^{k-1} \tag{7.96e}$$

Figure 7.89 shows the plot of the armature current and speed versus time as the starter resistance is cut out in steps. During waiting time at each step, the current falls and the speed rises exponentially according to a single dominant time-constant (Sec. 7.21). This is the reason why the waiting time at each step progressively reduces as is easily evident from Fig. 7.89.

Procedure for Starting Step Calculations

Once the designer has selected the upper and lower limits of armature currents during starting, starter step calculations can proceed on the following lines:

(i) From Eq. (7.95) calculate R_1.

(ii) From Eq. (7.96e) calculate the number of steps k choosing the nearest integral value.

(iii) Calculate resistances $R_1, R_2 \ldots$ from Eq. (7.98(b)). From these the resistance values of various sections of the starter can be found out.

OR

It is convenient to calculate the step resistance from the recursive relationship derived below

From Eqs (7.96(b)) and (7.96(d))

$$R_2 = R_1/\gamma$$

Then
$$r_1 = R_1 - R_2 = \left(1-\frac{1}{\gamma}\right)R_1 \tag{7.97}$$

Next
$$r_2 = R_2 - R_3 = \left(1-\frac{1}{\gamma}\right)R_2$$

$$=\frac{1}{\gamma}\left(1-\frac{1}{\gamma}\right)R_1 = r_1/\gamma$$

By induction
$$r_n = r_{n-1}/\gamma \tag{7.98}$$

Sometimes the number of sections are specified in addition to the upper current limit. This problem can be handled by a slightly different manipulation of Eqs (7.95) to (7.96(e)).

In the series motor the flux/pole changes with the armature current making the starter step resistance calculation somewhat more involved. These are not dealt with in this book.

Example 7.43 *A starter is required for a 220-V shunt motor. The maximum allowable current is 55 A and the minimum current is about 35 A. Find the number of sections of starter resistance required and the resistance of each section. The armature resistance of the motor is 0.4 Ω.* LO 8

Solution

$$I_1 = 55\text{ A}; \quad I_2 = 35\text{ A}$$

$$\gamma = \frac{I_1}{I_2} = 1.57$$

From Eq. (7.95)

$$R_1 = \frac{V_1}{I_1} = \frac{200}{55} \; 4\,\Omega$$

From Eq. (7.98(e))

$$\gamma^{n-1} = \frac{R_1}{R_a} = \frac{4}{0.4} = 10$$

or

$$n = 6.1$$

For an integral choice of $n = 6$,

$$\gamma = 1.585, \quad 1/\gamma = 0.631$$

Using Eq. (7.98(b)) the values of the resistances are obtained as

$$R_1 = 4\,\Omega$$

$$r_1 = (1 - 0.631) \times 4 = 1.476\,\Omega$$

$$r_2 = 1.476 \times 0.631 = 0.931\,\Omega$$

Proceeding on the above lines

$$r_3 = 0.587\,\Omega, \quad r_4 = 0.370\,\Omega, \quad r_5 = 0.235\,\Omega$$

Example 7.44 *A 25 kW, 230 V has an armature resistance of 0.12 Ω and a field resistance of 120 Ω. Its speed at full load is 2000 rpm. The field current may be neglected.*

(a) For a four step (resistance steps) calculate the resistances of the steps if the allowable maximum armature current is 1.5 times the full load current. What is the lower current limit?

(b) Calculate the motor speed at each stud when the current reaches the lower limit and starting handle is to be moved to the next stud. LO 8

Solution

$$I_a\,(fl) = \frac{25 \times 10^3}{230} = 108.7\text{ A}$$

(a) Four resistance steps require five studs ($k = 5$) upper current limit,

$$I_1 = 108.7 \times 1.5 = 163\text{ A}$$

$$R_1 = \frac{230}{I_1} = \frac{230}{163} = 1.41\,\Omega$$

$$R_a = 0.12\,\Omega\text{ (given)}$$

$$\frac{R_1}{R_a} = \gamma^{k-1}$$

$$\frac{1.41}{0.12} = \gamma^4 \text{ or } \gamma = \left(\frac{1.41}{0.12}\right)^{1/4}$$

$$\gamma = 1.851$$

$$\frac{I_1}{I_2} = \gamma = 1.851, \; 1/\gamma = 0.54$$

or $$I_2 \text{ (lower limit)} = \frac{1.63}{1.85} = 88.1 \text{ A}$$

Resistance steps $$r_1 = (1 - 0.54) \times 1.41 = 0.6486 \; \Omega = 0.649 \; \Omega$$

$$r_2 = 0.6486 \times 0.54 = 0.3502 \; \Omega = 0.350 \; \Omega$$

$$r_3 = 0.1891 \; \Omega = 0.189 \; \Omega$$

$$r_4 = 0.1021 \; \Omega = 0.102 \; \Omega$$

(b) At full load

$$I_a(fl) = 103.7 \text{ A}$$

$$E_a = 230 - 103.7 \times 0.12 = 217.6 \text{ V}$$

Speed n = 2000 rpm

$E_a = K'_a \Phi n = K''_a n$; filed current is constant, armature reaction effect ignored.

or $$217.6 = K''_a \times 2000 \text{ or } K''_a = 0.1088$$

At stud 1 when armature current reduces to I_2 = 88.1 A

$$E_a(1) = V_t - I_2 \times R_1 = 230 - 88.1 \times 1.41 = 105.8 \text{ V}$$

$\therefore$ $$n_1 = \frac{105.8}{0.1088} = 971 \text{ rpm}$$

At stud 2

$$E_a(2) = 230 - 88.1 \times (1.41 - 0.649) = 230 - 88.1 \times 0.761$$

$$= 162.96 \text{ V}$$

$$n_2 = \frac{162.96}{0.1088} = 1498 \text{ rpm}$$

At stud 3

$$E_a(3) = 230 - 88.1 \times (0.761 - 0.35) = 230 - 88.1 \times 0.411$$

$$= 193.8 \text{ V}$$

$$n_3 = \frac{193.8}{0.1088} = 1778 \text{ rpm}$$

At stud 4

$$E_a(4) = 230 - 88.1 \times (0.411 - 0.189) = 230 - 88.1 \times 0.222$$

$$= 210.4 \text{ V}$$

$$n_4 = \frac{210.4}{0.1088} = 1934 \text{ rpm}$$

At stud 5 (ON)

The armature current instantaneously rises to I_1 = 163 A. The speed increases from 1934 rpm to 2000 rpm and armature current falls to 108.7 A to match the full load on the motor of the load is any different the armature current and speed adjust accordingly.

Note: To determine the times at which speeds $n_1, \ldots, n_4$ are reaches needs electromagnetic study; Section 7.21.

The Example 7.42 is also solved, by MATLAB.

```
clc
clear
Pop=25*1000;
Vt=230;
Ra=0.12;
rf=120;
Nfl=2000;
Iafl=Pop /Vt;
Iamax=1.5*Iafl;
k=5;
I1=Iamax;
R1=Vt/ I1;
r=(R1/Ra)^(l/(k-l));
I2 = I1/r ;
for i=l:(k-l)
   if i==l
         rest(i)=(l-(l/r) )*Rl
   else
         rest(i)=rest(i-l)*(l/r)
   end
end
```

Answer:
rest =
   0.6488
rest =
   0.6488   0.3504
rest =
   0.6488   0.3504   0.1892
rest =
   0.6488   0.3504   0.1892   0.1022

MATLAB code for Example 7.44 is given below:

```
clc
clear
Pop=25*1000;              % Power Output
Vt=230;                   % Terminal Voltage
Ra=0.12;                  % Armature resistance
rf=120;     % Field Resistance
Nfl=2000;                 % Full load speed
Iafl=Pop/Vt; % Full load armature current
Iamax=1.5*Iafl;           % Max armature current
```

```
k=5;                          % No of studs
I1=Iamax;
R1=Vt/I1;
r=(R1/Ra)^(1/(k-1));
I2=I1/r;
for i=1 : (k-1)
   if i==1
       rest (i)=(1-(1/r))*R1;  % rest=resistance of the studs
   else
       rest (i)=rest (i-1)*(1/r);
   end
end
% part (b)
Iaf1 = 103.7;
Ea=Vt-Iaf1*Ra; % back EMF
Ka=Ea/Nf1;
for i=1: (k-1)
   if i==1
       Eastud (1)=Vt-I2*R1;
       Rnew=R1
   else
       Eastud (i)=Vt- I2*(Rnew-rest(i-1));
       Rnew=Rnew-rest(i-1)
   end
end
N=Eastud/Ka;
display (Renw)
```

Answer:

Rnew=1.4107

Rnew=0.7618

Rnew=0.4114

Rnew=0.2222

Rnew=0.2222

7.17 | SPEED CONTROL OF dc MOTORS

LO 8

The dc motors are in general much more adaptable speed drives than ac motors which are associated with a constant-speed rotating field. Indeed one of the primary reasons for the strong competitive position of dc motors in modem industrial drives is the wide range of speeds afforded.

Methods of Speed Control

From Eq. (7.69)

$$n = \frac{1}{K'_a}\frac{E_a}{\Phi} = \frac{1}{K'_a}\left(\frac{V_t - I_a R_a}{\Phi}\right) \tag{7.99}$$ ‡‡

Since the armature drop is small, it can be neglected.

$$\therefore \quad n \approx \frac{1}{K'_a}\frac{V_t}{\Phi} \tag{7.100}$$

This equation gives us two methods of effecting speed changes, i.e. the variation of field excitation, I_f and that of terminal voltage, V_t. The first method causes a change in the flux per pole, Φ and is known as the field control and the second method is known as the armature control.

Base Speed It is the speed at which the motor runs at rated terminal voltage and rated field current. It is indeed the name plate speed of the motor.

Speed Regulation

$$\% \text{ speed regulation} = \frac{n_o - n_{fl}}{n_{fl}} \times 100$$

n_o = no load speed

n_{fl} = full load speed i.e. *rated speed.*

Field Control

For fixed terminal voltage, from Eq. (7.102)

$$\frac{n_2}{n_1} = \frac{\Phi_1}{\Phi_2} \tag{7.101}$$

which for linear magnetization implies

$$\frac{n_2}{n_1} = \frac{\Phi_1}{\Phi_2} = \frac{I_{f1}}{I_{f2}} \tag{7.102}$$

Limitations of field control method Certain limitations of the field control method are:

1. Speeds lower than the rated speed cannot be obtained because the field cannot be made any stronger; it can only be weakened.
2. Since the speed is inversely proportional to the flux/pole while the torque is directly proportional to it for a given armature current, it can cope with *constant kW drives* only where the load torque falls with speed.
3. For motors requiring a wide range of speed control, the field ampere-turns are much smaller than the armature ampere-turns at high speeds causing extreme distortion of the flux density in the air-gap. This leads to unstable operating conditions or poor commutation. Compensating winding can be used to increase the speed range which can be 2 to 1 for large motors, 4 to 1 for medium sized ones and 8 to 1 for small motors. Even then the field control is restricted to small motors.

‡‡ In case a series field is also provided, the armature drop would be $I_a(R_a + R_{se})$.

4. This control method is not suited to applications needing speed reversal; since the only way to reverse speed is to disconnect the motor from the source and reverse the field/armature polarity. The field circuit being highly inductive, it is normally the armature which is reversed.

Shunt Motor

Figure 7.90(a) illustrates the field control for shunt motors; the control being achieved by means of a rheostat (regulator) in the field circuit. Reproducing Eq. (7.77) here for convenience:

$$n = \frac{V_t}{K_a'\Phi} - \left(\frac{R_a}{K_a'K_a\Phi^2}\right)T \qquad (7.103)$$

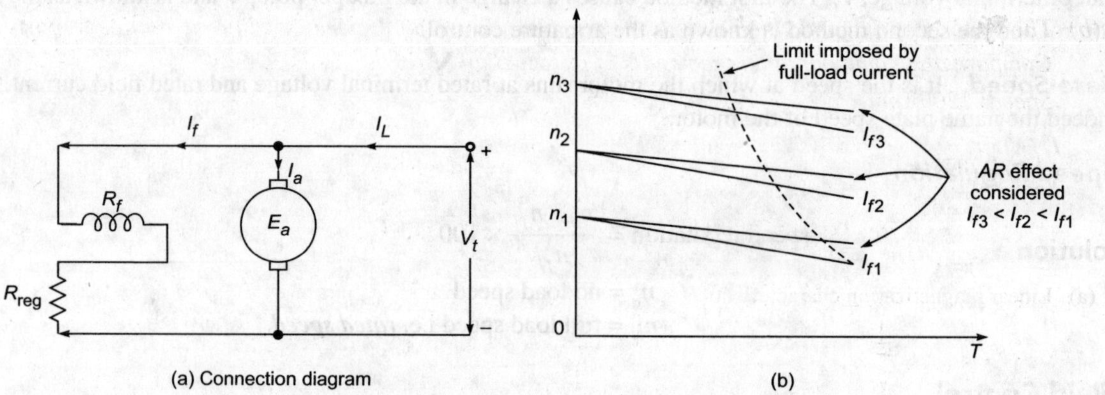

Fig. 7.90 *Field control for shunt motor*

The speed-torque characteristic which has a small linear drop due to the second term (R_a effect) and translates upwards as the field is weakened due to the armature reaction is shown in Fig. 7.90(b). The demagnetizing effect of the armature reaction causes the characteristics to somewhat bend upwards with increasing torque (increasing load current). The working range of the speed-torque characteristic reduces with increasing speed in order for the armature current not to exceed the full-load value with a weakening field.

Example 7.45 *A 400-V dc shunt motor takes a current of 5.6 A on no-load and 68.3 A on full-load. Armature reaction weakens the field by 3%. Calculate the ratio of full-load speed to no-load speed. Given $R_a = 0.18\ \Omega$, brush voltage drop = 2 V, $R_f = 200\ \Omega$.*

LO 8

Solution

$$I_f = \frac{400}{200} = 2\text{ A}$$

No-load

$$I_{a0} = 5.6 - 2 = 3.6\text{ A}$$

$$E_{a0} = 400 - 0.18 \times 3.6 - 2 = 397.4\text{ V}$$

Full-load

$$I_a(fl) = 68.3 - 2 = 66.3\text{ A}$$

$$E_a(fl) = 400 - 0.18 \times 66.3 - 2 = 386.1\text{ V}$$

$$\frac{n(fl)}{n(nl)} = \frac{386.1}{397.4} \times \frac{1}{0.97} = 1$$

Observation Because of field weakening caused by armature reaction, full-load speed is equal to no-load speed.

Example 7.46 *A 750 kW, 250 V, 1200 rpm dc shunt motor has armature and field resistances of 0.03 Ω and 41.67 Ω respectively. The motor is provided with compensating winding to cancel out the armature reaction. Iron and rotational losses can be ignored. The motor is carrying constant torque load drawing 126 A and running at 1105 rpm.*

(a) Assuming magnetic linearity, what would be the line current and motor speed if the field current is reduced to 5 A. Also calculate the load torque.

(b) The motor has magnetization data as below. Calculate motor speed and line current as in part (a). Compare and comment upon the results of parts (a) and (b)

Speed: 1200 rpm

I_f (A)	4	5	6	7
V_{OC} (V)	217	250	267	280

LO 8

Solution

(a) Linear magnetization characteristic

$$I_{f1} = \frac{250}{41.67} = 6 \text{ A}$$

$$I_{a1} = 126 - 6 = 120 \text{ A}$$

$$E_{a1} = 250 - 0.03 \times 120 = 246.4 \text{ V}$$

$$n_1 = 1105 \text{ rpm}, \omega_1 = 115.7 \text{ rad/s}$$

Now we know

$$E_a = K_a \Phi \omega = K''_a I_f \omega \quad \text{(i)}$$

$$T = K_a \Phi I_a = K''_a I_f I_a \quad \text{(ii)}$$

From Eq. (i)

$$E_{a1} = 246.4 = K''_a \times 6 \times 115.7$$

or

$$K''_a = 0.355$$

$$T = 0.355 \times 6 \times 120 = 255.6 \text{ Nm (constant)}$$

Field current reduced

$$I_{f2} = 5 \text{ A}$$

$$T \text{ (constant)} = K''_a I_{f1} I_{a1} = K''_a I_{f2} I_{a2}$$

or

$$I_{a2} = 120 \times \frac{6}{5} = 144 \text{ A}$$

$$I_{L2} = 144 + 2 = 146 \text{ A}$$

From Eq. (i)

$$E_{a2} = 250 - 0.03 \times 144 = 245.68 \text{ V}$$

$$245.68 = 0.355 \times 5 \times \omega_2$$

or

$$\omega_2 = 138.4 \text{ rad/s or } 1322 \text{ rpm}$$

(b) Rather than drawing the magnetisation curve, we shall linearly interpolate between data points.

$$I_{f1} = 6\text{ A} \Rightarrow V_{OC} = 267\text{ V at 1200 rpm or 125.7 rad/s}$$

$$267 = K_a\,\Phi_1 \times 125.7$$

or
$$K_a\Phi_1 = 2.124$$

$$I_{f2} = 5 \Rightarrow V_{OC} = 250\text{ V at 1200 rpm or 125.7 rad/s}$$

or
$$250 = K_a\Phi_2 \times 125.7$$

$$K_a\Phi_2 = \frac{250}{125.7} = 1.989$$

$$T(\text{constant}) = K_a\Phi_2 I_{a2} = K_a\Phi_1 I_{a1}$$

or
$$I_{a2} = I_{a1}\left(\frac{K_a\Phi_1}{K_a\Phi_2}\right)$$

$$= 120 \times \frac{2.124}{1.989} = 128.1\text{ A}$$

$$I_{L2} = 128.1 + 2 = 130.1\text{ A}$$

$$E_{a2} = 250 - 0.03 \times 128.1 = 246.16\text{ V}$$

$$246.16 = K_a\Phi_2\omega_2$$

or
$$\omega_2 = \frac{246.16}{1.989} = 123.76\text{ rad /s or 1182 rpm}$$

Comparison of results Speed as calculated by linear assumption is $\frac{1322-1182}{1182} \times 100 = 11.8\%$ higher than obtained by consideration of the actual magnetisation characteristic. So the linearization is not very good for speed estimation. As armature voltage drop is very small the armature induced emf is nearly the same in both cases (245.68 V and 246.16 V). Therefore, the speed is mainly governed by the flux/pole. On linear basis the flux reduces by a factor of 5/6 = 0.833 while the actual reduction (obtained from the magnetization characteristic) is 250/267 = 0.936. This is why the actual speed is lower than that calculated on linear basis.

Example 7.47 *A 250-V dc shunt motor has $R_f = 150\ \Omega$ and $R_a = 0.6\ \Omega$. The motor operates on no-load with a full field flux at its base speed of 1000 rpm with $I_a = 5$ A. If the machine drives a load requiring a torque of 100 Nm, calculate armature current and speed of motor. If the motor is required to develop 10 kW at 1200 rpm, what is the required value of the external series resistance in the field circuit? Neglect saturation and armature reaction.* **LO 8**

Solution Assuming linear magnetization characteristic

$$E_a = K_a\Phi\omega_m = K\omega_m \tag{i}$$

$$T = K_a\Phi I_a = KI_a \tag{ii}$$

At no load

$$250 - 5 \times 0.6 = K \times \frac{2\pi \times 1000}{60}$$

or
$$K = 2.36$$

When driving a load of 100 N m,

$$I_a = \frac{T}{K} = \frac{100}{2.36} = 42.4\text{ A}$$

Now $$\omega_m = \frac{E_a}{K} = \frac{250 - 42.4 \times 0.6}{2.36} = 95.15 \text{ rad/s}$$

∴ $$n = \frac{60}{2\pi}\omega_n = 909 \text{ rpm}$$

Given: Output = 10 kW at 1200 rpm

Assuming linear magnetization, Eqs (i) and (ii) can be written as

$$E_a = K' I_f \omega_m \quad \text{(iii)}$$

$$T = K' I_f I_a \quad \text{(iv)}$$

From the data of no-load operation

$$I_f = \frac{250}{150} = 1.67 \text{ A}$$

∴ $$K' = \frac{K}{I_f} = \frac{2.36}{1.67} = 1.413$$

Now $$(250 - 0.6 I'_a) I'_a = 10 \times 1000$$

Solving $I'_a = 44.8$ A (the higher value is rejected)

Substituting values in Eq. (iii)

$$250 - 0.6 \times 44.8 = 1.413 \times I'_f \times \frac{2\pi \times 1200}{60}$$

or $$I'_f = 1.257 \text{ A}$$

∴ $$R_f \text{ (total)} = \frac{250}{1.257} = 199\ \Omega$$

Hence $$R_f \text{ (external)} = 199 - 150 = 49\ \Omega$$

Series Motor

Speed control is achieved here by adjusting the field ampere-turns. There are three ways of changing them:

1. Diverter field control A diverter resistor is connected across the field winding as shown in Fig. 7.91. By varying R_d the field current and hence the field ampere-turns can be reduced. From the circuit, it is obvious that

$$I_{se} = I_a\left(\frac{R_d}{R_{se} + R_d}\right) = K_d I_a \quad (7.104)$$

where $$K_d = \frac{R_d}{R_{se} + R_d} = \frac{1}{R_{se}/R_d + 1}$$

For linear magnetization

$$\Phi = K_d K_f I_a \quad (7.105)$$

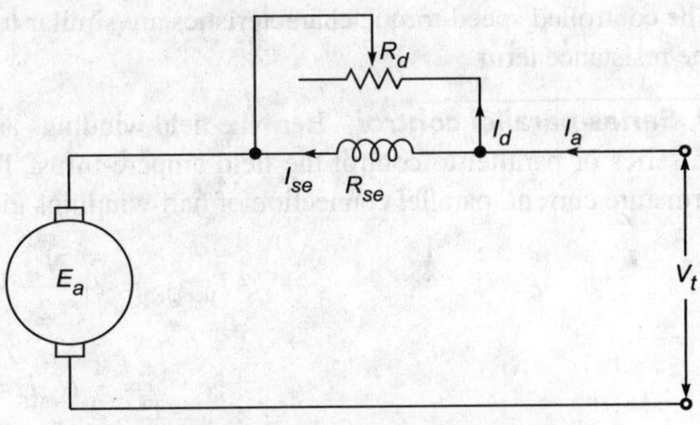

Fig. 7.91 *Diverter resistor control circuit*

Hence Eq. (7.79) can be rewritten as

$$n = \frac{1}{K'_a K_f K_d}\left[\frac{V_t\sqrt{K_a K_f K_d}}{\sqrt{K_d}\sqrt{T}} - \{R_a + (R_{se} \| R_d)\}\right] \quad (7.106)$$

From Eq. (7.106), speed-torque characteristics for decreasing values of K_d (decreasing R_d) are plotted in Fig. 7.92. These can be corrected for saturation and armature reaction effects.

One precaution to be taken in this method in order to avoid oscillations in speed initiated by load changes is to use an inductively wound diverter resistor.

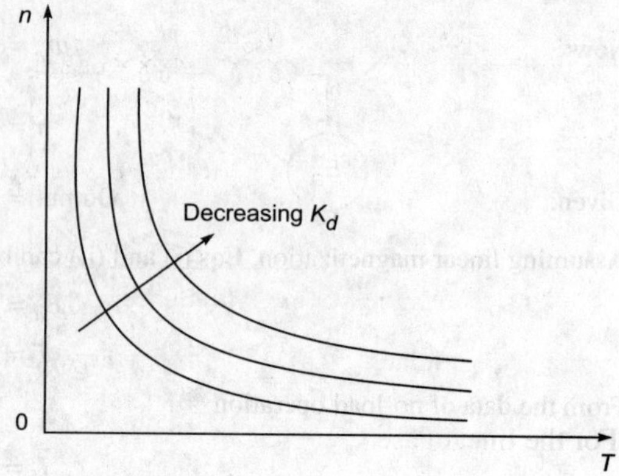

Fig. 7.92 *Effect of diverter parameter on speed-torque characteristics*

2. Tapped-field-control Here the field ampere-turns are adjusted in steps by varying the number of turns included in the circuit. The circuit is shown in Fig. 7.93, from which the following relations are obtained:

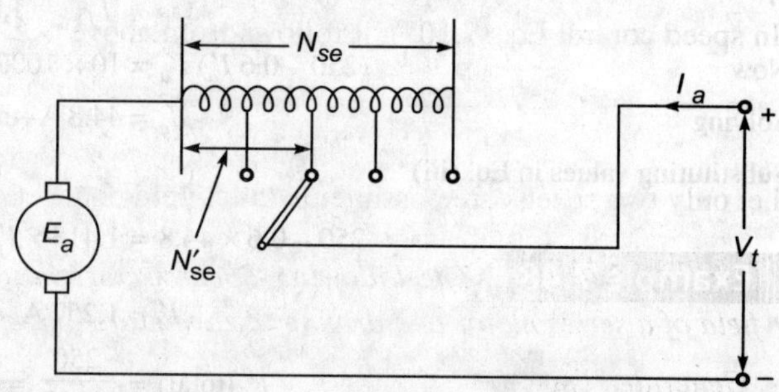

Fig. 7.93 *Tapped-field control*

$$I_{se}\text{ (effective)} = \frac{N'_{se}}{N_{se}} I_a = K_{se} I_a$$

where N'_{se} = tapped field turn with resistance

$R'_{se} = K_{se} R_{se}$.

Now $\Phi = K_f K_{se} I_a$, $T = K_a K_f K_{se} I_a^2$, linear case

substituting Φ in Eq. (7.79 and eliminating I_a with R_{se} replaced by $K_{se} R_{se}$)

$$n = \frac{1}{K'_a K_f K_{se}}\left[\frac{V_t\sqrt{K_a K_f K_{se}}}{\sqrt{T}} - (R_a + K_{se} R_{se})\right] \quad (7.107)$$

The controlled speed-torque characteristics are similar to those of Fig. 7.92 except for the marginal effect of the resistance term.

3. Series-parallel control Here the field windings are divided into two equal halves and then connected in series or parallel to control the field ampere-turns. The circuits shown in Figs 7.95(a) and (b). For any armature current, parallel connection of half windings gives

$$AT_{\text{parallel}} = 2\left(\frac{N_{se}}{2} \times \frac{I_a}{2}\right)$$

$$= \frac{1}{2} N_{se} I_a = \frac{1}{2} AT_{\text{series}}$$

and

$$R'_{se} = \left(\frac{R_{se}}{2} \| \frac{R_{se}}{2}\right) = \frac{1}{4} R_{se} \quad (7.108)$$

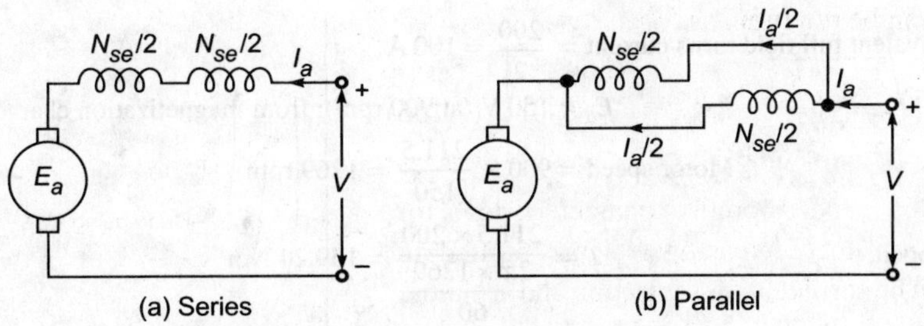

Fig. 7.94 *Series-parallel control*

For the linear case

$$\Phi = \frac{1}{2} K_f I_a$$

In speed control Eq. (7.107), it follows from above

$$K_{se} = 1 \text{ or } \frac{1}{2} \tag{7.109}$$

i.e. only two speeds are possible; parallel field connection gives the higher speed.

Example 7.48 *The following open-circuit characteristic was obtained by separately exciting the field of a series motor and driving the armature at a speed of 900 rev/min.*

Generated emf (V)	*0*	*78*	*150*	*192*	*220*
Field current (A)	*0*	*50*	*100*	*150*	*200*

The armature and series-field resistances are 0.035 and 0.015 Ω respectively. Determine the speed and torque of the motor for a current of 200 A when fed from a 220 V supply and operated with (a) the full field-winding, and (b) the field turns reduced by half. (c) diverter of resistance 0.03 Ω. LO 8

Solution

$$\text{Total armature resistance} = 0.035 + 0.015 = 0.05\ \Omega$$

$$E_a = 220 - 200 \times 0.05 = 210\text{ V}$$

(a) Full field winding:

At 200 A current, 900 rpm, induced emf = 220 V (from magnetization characteristics)

$$\text{Motor speed} = 900 \times \frac{210}{220} = 859.1\text{ rpm}$$

Torque developed,
$$T = \frac{E_a I_a}{\omega} = \frac{210 \times 100}{\frac{2\pi \times 859.1}{60}} = 159.24\text{ Nm.}$$

(b) Field winding terms reduced to half;

$$R_{se} = \frac{0.015}{2} = 0.0075\ \Omega$$

$$R\text{ (total)} = 0.035 + 0.0075 = 0.\ 0425\ \Omega$$

$$E_a = 220 - 200 \times 0.0425 = 211.5\text{ V}$$

$$\text{Equivalent full field turns current} = \frac{200}{2} = 100 \text{ A}$$

$$E_a = 150 \text{ V (at 900 rpm); from magnetization characteristic}$$

$$\text{Motor speed} = 900 \times \frac{211.5}{150} = 1269 \text{ rpm}$$

Torque developed,
$$T = \frac{211.5 \times 200}{\dfrac{2\pi \times 1269}{60}} = 159.24 \text{ Nm}$$

(c) Diverter across series field

$$R_a = 0.03, R_{se} = 0.15$$

$$K_d = \frac{1}{R_{se}/R_d + 1} = \frac{1}{1/2 + 1} = \frac{2}{3}$$

$$I_{se} = K_d I_a = \frac{2}{3} \times 200 = 133.3 \text{ A}$$

From the open-circuit characteristic (by interpolation) at $I_{se} = 133.3$ A

$$E_a = 150 + \frac{192 - 150}{150 - 100} \times (133.3 - 100)$$

or
$$E_a = 178 \text{ V at 900 rpm}$$

$$R_{se} \parallel R_d = \frac{2}{3} \times 0.015 = 0.1\ \Omega$$

$$E_a = 220 - 200\,(0.035 + 0.01) = 211 \text{ V}$$

$$\text{Motor speed} = 900 \times \frac{211}{178} = 1067 \text{ rpm}$$

$$\text{Torque developed} = \frac{211 \times 200}{\dfrac{2\pi}{60} \times 1067} = 377.87 \text{ Nm.}$$

Example 7.49 *A 4-pole series-wound fan motor draws an armature current of 50 A, when running at 2000 rpm on a 230 V dc supply with four field coil connected in series. The four field coils are now connected in two parallel groups of two coils in series. Assuming the flux/pole to be proportional to the exciting current and load torque proportional to the square of speed, find the new speed and armature current. Neglect losses. Given: armature resistance = 0.2 Ω, resistance of each field coil = 0.05 Ω.* **LO 8**

Solution

$$E_a = K_E I_{se} n \tag{i}$$

$$T_L = K_L n^2 = T_M = K_M I_{se} I_a \tag{ii}$$

Field coils in series

$$I_{se} = I_a$$

$$R_{se} = 4 \times 0.05 = 0.2\ \Omega, \quad R_a = 0.2\ \Omega$$

$$R_A = 0.2 + 0.2 = 0.4\ \Omega,$$

$$I_{a1} = 50 \text{ A}$$

$$E_{a1} = 230 - 0.4 \times 50 = 210 \text{ V}$$

$$210 = K_E \times 50 \times 2000 \tag{iii}$$

and
$$K_L \times (2000)^2 = K_M \times (50)^2 \tag{iv}$$

Field coils in two series groups in parallel

$$I_{se}\text{ (effective)} = I_a/2$$

$$R_{se} = \frac{2 \times 0.05}{2} = 0.05\ \Omega$$

$$R_A = 0.2 + 0.05 = 0.25\ \Omega$$

$$E_{a2} = 230 - 0.25\, I_{a2} = K_E \times (I_{a2}/2)\, n_2 \quad \text{(v)}$$

and

$$K_L n_2^2 = K_M \times (I_{a2}/2) \times I_{a2} \quad \text{(vi)}$$

From Eqs (iv) and (vi)

$$\frac{n_2^2}{(2000)^2} = \frac{I_{a2}^2}{2 \times (50)^2}$$

or

$$n_2 = 28.3\, I_{a2} \quad \text{(vii)}$$

From Eqs (iii) and (v)

$$\frac{230 - 0.25\, I_{a2}}{210} = \frac{I_{a2} n_2}{2 \times 50 \times 2000}$$

Substituting for n_2 from Eq. (vii) and simplifying

$$I_{a2}^2 + 8.4\, I_{a2} - 7740 = 0$$

or

$$I_{a2} = 83.9\text{ A; negative value is rejected}$$

$$n_2 = 28.3 \times 83.9 = 2374\text{ rpm}$$

Armature Control

Important! The main requirement of this control scheme is a *variable voltage supply* to the armature whose current rating must be somewhat larger than that of the motor. It is superior to the field control scheme in three respects, outlined below:

(i) It provides a *constant-torque drive.* In the shunt motor case by keeping the field current at maximum value full motor torque can be obtained at full-load armature current at all speeds.

(ii) Since the main field ampere-turns are maintained at a large value, flux density distortion caused by armature reaction is limited.

(iii) Unlike field control scheme, speed reversal can be easily implemented here.

There are three main types of armature control schemes. These are discussed below:

Rheostatic Control

Series armature-resistance control Here the applied armature voltage is varied by placing an adjustable resistance R_e in series with the armature as shown in Fig. 7.95 along with the speed-torque characteristics.

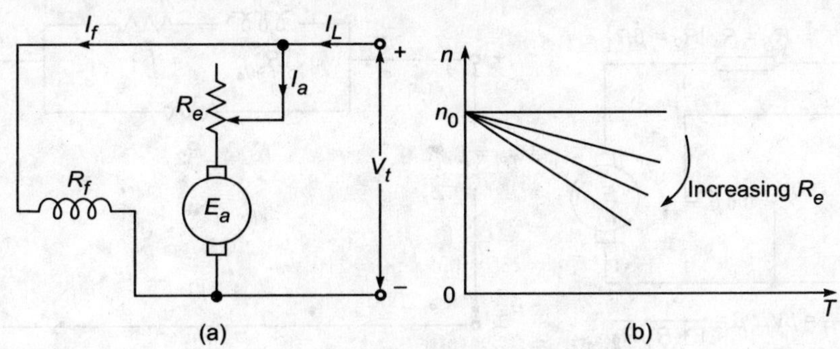

Fig. 7.95 *Series armature resistance control and speed-torque characteristics*

Some of the limitations of the rheostatic control method are enumerated below:

(i) Only speeds below the rated value can be obtained. This can be shown by using Eq. (7.100) (armature resistance is negligible) that

$$n_1 = \frac{1}{K_a'}\frac{V_t}{\Phi}; \quad n_2 = \frac{1}{K_a'}\frac{V_t - I_a R_e}{\Phi} \tag{7.110}$$

which gives

$$\frac{n_1 - n_2}{n_1} = \frac{I_a R_e}{V_t} \tag{7.111}$$

(ii) Range of speeds available is limited because efficiency is reduced drastically for large speed reductions. By definition of armature efficiency

$$\eta \approx \frac{(V_t - I_a R_e) I_a}{V_t I_a} = 1 - \frac{I_a R_e}{V_t} = \frac{n_2}{n_1} \tag{7.112}$$

(iii) The speed regulation of the method is poor as for a fixed value of series armature resistance, the speed varies directly with load, being dependent upon the resistance voltage drop.

In general, rheostatic control is economically feasible only for very small motors (fractional kW) or for short-time, intermittent show-downs for medium-sized motors.

Shunted armature control It is a variation of the rheostatic control. The principle of voltage division is used to reduce the voltage across the armature as shown for a shunt motor in Fig. 7.96(a). The Thevenin equivalent circuit as seen from the armature terminals is drawn in Fig. 7.96(b). The no-load armature speed is governed by V_{TH}, which can be independently adjusted by the ratio R_2/R_1. The series resistance (Thevenin resistance) is $R_e = R_1 \| R_2 = \beta R_1$ is very small so the control circuit gives better speed regulation compared to the circuit arrangement of Fig. 7.96(a). $V_{TH} = \beta V_r$

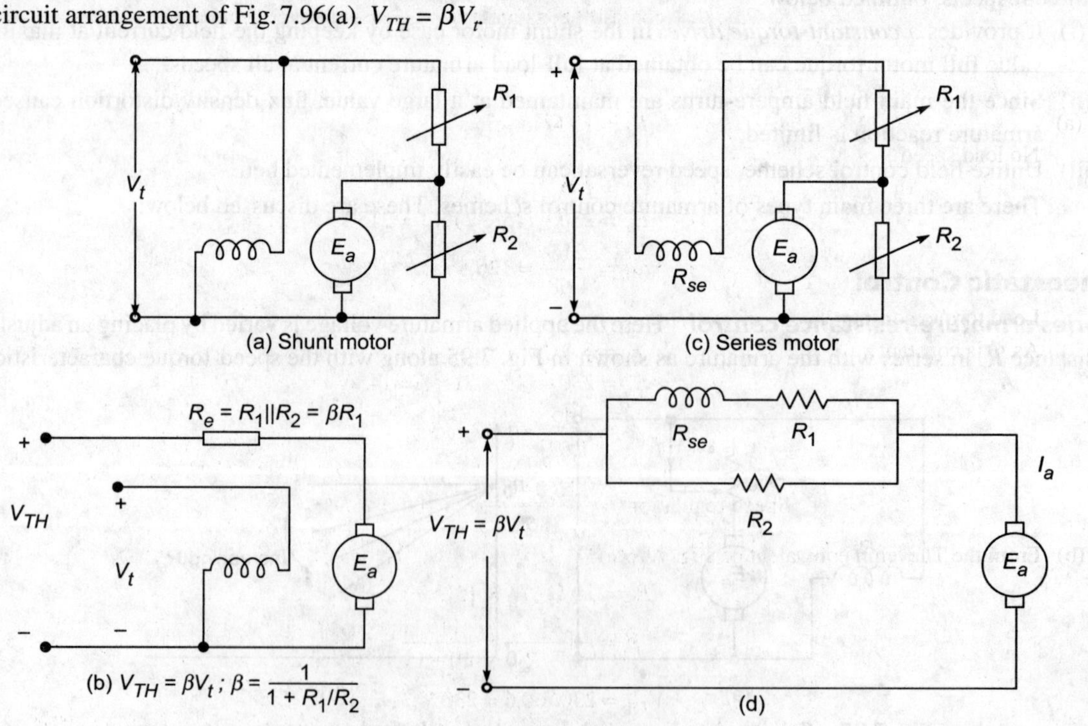

Fig. 7.96 *Shunted armature speed control*

The potential divider circuit for series motor speed control is shown in Fig. 7.96(c) and its Thevenin equivalent is drawn in Fig. 7.96(d).

On approximate basis neglecting voltage drop in $(\beta R_1 + R_a)$,

$$E_a \approx \beta V_t = K'_a \Phi n$$

or
$$n = \frac{1}{K'_a} \frac{\beta V_t}{\Phi}; \; \Phi(I_{se}) = \Phi(\beta I_a)$$

At light load through Φ is quite small the reduced value of βV_t result in *finite motor speed* unlike a series motor operating at rated V_t.

Example 7.50 *A 230 V dc shunt motor having armature resistance of 2 Ω draws an armature current of 5 A to drive a constant torque load at 1250 rpm. At no load it draws a current of 1 A.*

(a) *A resistance of 15 Ω is added in series to the armature. Find the motor speed with load torque as above. Also determine the speed regulation.*

(b) *A resistance of 15 Ω is shunted across the armature and 10 Ω in series with the supply line (as in Fig. 7.96(a). Calculate the load speed and speed regulation.*

(c) *Compare the power wasted in external resistance (s) in parts (a) and (b).*

Rotational loss torque is negligible. The armature reaction effect is to be ignored. **LO 8**

Solution

$$E_a = 230 - 2 \times 5 = 220 \text{ V}$$

$$n = 1250 \text{ rpm}, \; \omega = 130.9 \text{ rad/s}$$

$$E_a = K_a \Phi \omega; \; \Phi \text{ is constant; constant shunt field current, no armature reaction effect}$$

$$220 = K_a \Phi \times 130.9$$

or
$$K_a \Phi = 1.68$$

(a)
$$R_e = 15 \; \Omega \text{ in series}$$

No load speed

$$I_{ao} = 1 \text{ A}$$

$$E_a = 230 - (15 + 2) \times 1 = 213 \text{ V}$$

$$\omega_0 = \frac{213}{1.68} = 126.8 \text{ rad/s}$$

Load torque constant

As Φ is constant

$$I_a = 5 \text{ A}$$

$$E_a = 230 - (15 + 2) \times 5 = 145 \text{ V}$$

$$\omega = \frac{145}{1.68} = 86.3 \text{ rad/s}$$

$$\text{Speed regulation} = \frac{126.8 - 86.3}{86.3} \times 100 = 46.9\%$$

(b) From the Thevenin equivalent of Fig. 7.96(c)

$$R_1 = 10 \; \Omega, \; R_2 = 15$$

$$\beta = \frac{R_2}{R_1 + R_2} = \frac{15}{10 + 15} = 0.6$$

$$V_{TH} = 230 \times 0.6 = 230 \text{ V}$$

$$R_{TH} = \beta R_1 = 0.6 \times 10 = 6\ \Omega$$

No load speed

$$I_{ao} = 1\text{ A}$$
$$E_a = V_{TH} - (R_{TH} + R_a)\, I_{ao}$$
$$= 230 - (6 + 2) \times 1 = 130\text{ V}$$
$$\omega_0 = \frac{130}{1.68} = 77.38\text{ rad/s}$$

On load

$$I_a = 5\text{ A}$$
$$E_a = 138 - (6 + 2) \times 5 = 98\text{ V}$$
$$\omega = \frac{98}{1.68} = 58.33\text{ rad/s}$$
$$\text{Speed regulation} = \frac{77.38 - 58.33}{58.33} \times 100 = 32.6\%$$

Observation Speed regulation is much better (less) in shunted armature control than in rheostatic control.

(c) Power loss

(i) Rheostatic control

$$P_e = (5)^2 \times 15 = 375\text{ W}$$

(ii) Shunted armature control

$$V_a(\text{across armature}) = 98 + 2 \times 5 = 108\text{ V}$$
$$P(15\ \Omega) = \frac{(108)^2}{15} = 777.6\text{ W}$$
$$I(15\ \Omega) = \frac{108}{15} = 7.24\text{ A}$$
$$I(10\ \Omega) = 7.2 + 5 = 12.2\text{ A}$$
$$P(10\ \Omega) = (12.2)^2 \times 10 = 1488.4\text{ W}$$

Then

$$P_e = 777.6 + 1488.4 = 2266\text{ W}$$

Observation External power loss is far larger in shunted armature control than in rheostatic control. In fact it is much larger than power of the motor (230 × 5 = 1150 W) being controlled.

Remark Shunted armature control is employed for very small motors where speed regulation requirement is stringent.

Example 7.51 *A dc shunt motor is connected to a constant voltage source and is driving a constant torque load. Show that if $E_a > 0.5\ V_t$ increasing the resultant flux reduces the speed and if $E_a < 0.5\ V_t$, increasing the resultant flux increases the speed. The back emf E_a is changed by a series resistance in the armature circuit.*

LO 8

Solution

Let

$$R_t = R_a + R_e$$
$$E_a = V_t - I_a R_t > 1/2\ V_t \qquad \text{(i)}$$
$$\omega = \frac{E_a}{K_a \Phi} = \frac{V_t}{K_a \Phi} - \frac{I_a R_1}{K_a \Phi} \qquad \text{(ii)}$$

As $I_a = \dfrac{T}{K_a \Phi}$, Eq. (i) is converted to the form with Φ as the only variable (T is constant).

Thus $$\omega = \frac{V_t}{K_a\Phi} - \frac{T R_1}{K_a^2\Phi^2} \tag{iii}$$

For ω to decrease with Φ, $\dfrac{d\omega}{d\Phi}$ should be negative. It then follows from Eq. (ii)

$$\frac{d\omega}{d\Phi} = -\frac{V_t}{K_a\Phi^2} + \frac{2T R_t}{K_a^2\Phi^3} < 0 \tag{iv}$$

Substituting $T = K_a\Phi I_a$, we get

$$-\frac{V_t}{K_a\Phi^2} + \frac{2K_a\Phi I_a R_t}{K_a^2\Phi^3} < 0$$

or $$-V_t + 2\,I_a R_t < 0$$

or $$V_t - 2I_a R_t > 0$$

But $I_a R_t = V_t - E_a$, so

$$V_t - 2V_t + 2E_a > 0$$

or $$E_a > 0.5\,V_t$$

For speed to increase by increasing flux

$$\frac{d\omega}{d\Phi} > 0$$

By increasing the inequality sign in Eq. (iv) it follows:

$$E_a < 0.5\,V_t$$

Series-parallel control Here two identical motors are coupled together mechanically to a common load. Two speeds at constant torque are possible in this method—one by connecting the motors armatures in series and the other by connecting them in parallel as shown in Fig. 7.97. When connected in series, the terminal voltage across each motor is $\dfrac{V_t}{2}$ whereas when they are connected in parallel it is V_t. Thus armature control of speed is achieved; speed (series): speed (parallel) ::1:2.

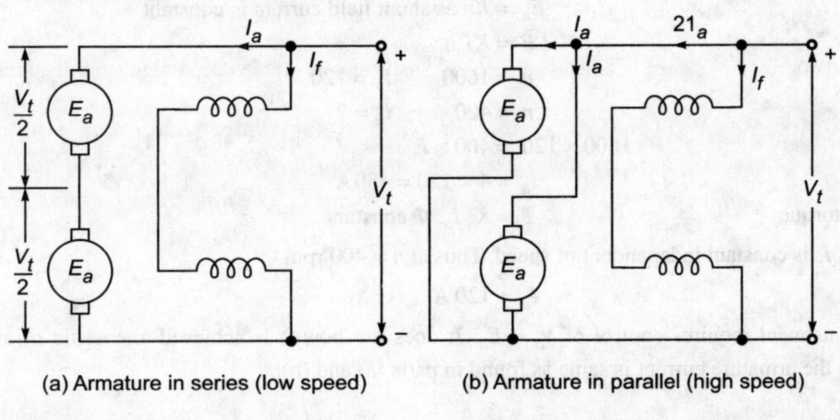

Fig. 7.97 *Series-parallel speed control (shunt-motors); case of constant load torque is illustrated; speed ratio 1:2*

Figure 7.98(a) and (b) gives the connections for series-parallel speed control of two identical series motors.

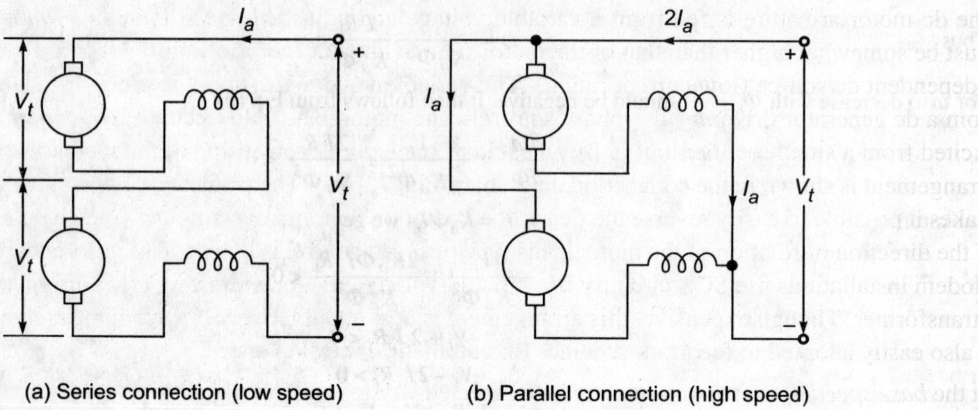

Fig. 7.98 *Series-parallel speed control of series motors; case of constant load torque is illustrated; speed ratio 1:2*

This method is superior to the rheostatic control insofar as efficiency is concerned. It is, however, limited to two speed steps. The method is commonly employed for speed control of series traction motors.

Example 7.52 *A dc shunt motor has speed control range of 1600 rpm to 400 rpm by rheostatic control. All losses and armature reaction effect may be neglected.*

(a) The motor drives a constant power load. It has a speed of 1600 rpm drawing 120 A armature current. What would be the armature current at 400 rpm?

(b) Repeat part (a) if the load is constant torque.

(c) Repeat parts (a) and (b) if speed is controlled by armature voltage.

LO 8

Solution All losses neglected means $R_a = 0$

(a) Constant power

$$P = E_a I_a$$

$$E_a = Kn \text{ as shunt field current is constant}$$

$$\therefore \quad P = KI_a n$$

$$n = 1600 \qquad I_a = 120$$

$$n = 400 \qquad I_a = ?$$

$$1600 \times 120 = 400 \times I_a$$

$$I_a = 4 \times 120 = 480\,\text{A}$$

(b) Constant torque

$$T = K_a I_a, \ \Phi \text{ constant}$$

Therefore I_a is constant independent of speed. Thus at $n = 400$ rpm

$$I_a = 120\,\text{A}$$

(c) Speed adjustment requires control of $V_a \approx E_a$. It does not how it is achieved–rheostatic or armature voltage. Therefore the armature current is same as found in parts (a) and (b).

Important! Ward Leonard Speed Control

It is combined armature and field control and is therefore, operationally the most efficient method of speed control with a wide range.

The dc motor armature is fed from a variable voltage and adjustable polarity supply whose current rating must be somewhat higher than that of the motor. The field (shunt) of the motor is separately excited from an independent dc source (low current rating). The variable voltage dc supply in older installations is obtained from a dc generator driven by a 3-phase squirrel-cage motor. The field circuit of the generator is separately excited from a small rectifier unit or by an excitor coupled to an extension of the motor shaft. The complete arrangement is shown in the connection diagram of Fig. 7.99(a). The connection of the potentiometer (Pot 1) makes it possible to easily reverse the generator excitation thereby reversing the voltage polarity for reversal of the direction of rotation of the motor. This type of speed control is known as Ward-Leonard speed control. Modem installations use SCR circuitry for variable-voltage dc supply drawing power from ac mains through a transformer. Though expensive, this arrangement is neat and relatively free from maintenance problems. It is also easily adopted to feedback schemes for automatic control of speed.

At the *base speed* n_b the motor armature is fed at rated voltage and its field current is adjusted to the maximum value, i.e. the field is excited at rated voltage. Reducing the armature voltage provides a *constant-torque speed control* where the speed can be reduced below the base value, while the motor has full torque capability (as I_f = max and I_a can have rated value). For obtaining speeds above n_b, the field is gradually weakned maintaining armature voltage at rated value. The motor torque therefore reduces as its speed increases which corresponds is to *constant-kW (or hp) drive*. The kind of control over torque-speed characteristic achieved is illustrated in Fig. 7.99(b) where the nature of power-speed characteristic is also revealed.

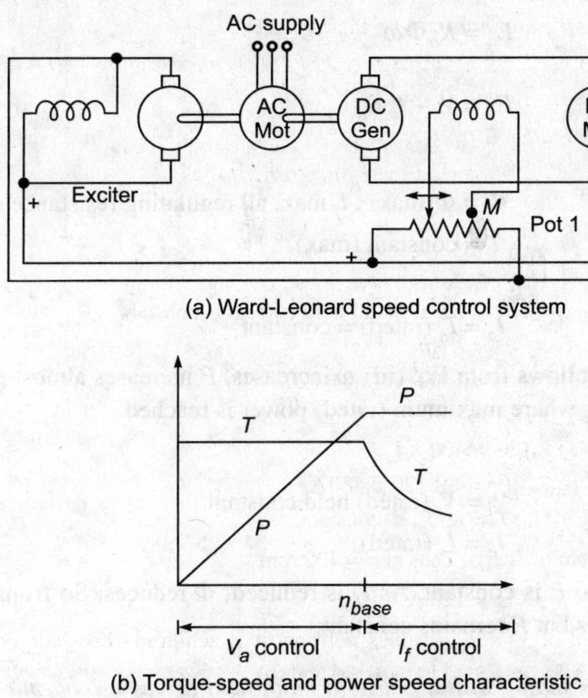

Fig. 7.99

!Attention Some of the attractive features of the Ward-Leonard system are listed below in addition to the advantages mentioned for armature control in general:

(i) The absence of an external resistance considerably improves the efficiency at all speeds. Another feature which enhances the efficiency is that when the generator emf becomes less than the back emf of the motor, electrical power flows back from motor to generator, is converted to mechanical form and is returned to the mains via the driving ac motor. The latter aspect makes it an ideal choice if frequent starting, stopping and reversals are required.

(ii) No special starting gear is required. As the generator induced voltage is gradually raised from zero, the motor starts up smoothly.

(iii) Speed reversal is smoothly carried out.

Explanation through Fundamental Relationships

Fundamental relationships are reproduced below:

Electromagnetic power

$$P = E_a I_a = T\omega \quad \text{(i)}$$

Electromagnetic torque

$$T = K_a \Phi I_a \quad \text{(ii)}$$

Back emf

$$E_a = K_a \Phi \omega \quad \text{(iii)}$$

Armature circuit equation

$$E_a = V_t - I_a R_a \quad \text{(iv)}$$

Constant torque operation

$$\Phi = \Phi(\text{max}) \text{ ; } I_f \text{ max, all regulating resistance cut out}$$

$$T = \text{constant (max)}$$

From Eq. (ii)

$$I_a = I_a \text{ (rated)} = \text{constant}$$

As V_t is increased, E_a increases. It follows from Eq. (iii) ω increases, P increases almost linearly (I_aR_a drop ignored). At $V_t = V_t$ (rated), $n = n_{\text{base}}$ where maximum (rated) power is reached.

Constant power operation

$$V_t = V_t \text{ (rated) held constant}$$

$$I_a = I_a \text{ (rated)}$$

From Eq. (iv), E_a is constant and so P is constant. As I_f is reduced, Φ reduces. So from Eq. (iii) speed ω increases and from Eq. (ii) T reduces but P remains constant.

Example 7.53 *A 200-V shunt motor with a constant main field drives a load, the torque of which varies at the square of the speed. When running at 600 rpm, it takes 30 A. Find the speed at which it will run and the current it will draw, if a 20-Ω resistor is connected in series with armature. Neglect motor losses.*

LO 8

Solution Armature resistance is assumed negligible. Further field current is ignored in comparison to armature current, i.e.,

$$I_L = I_a$$

As per the data given

$$200 = K_e \times 600 \quad \text{(i)}$$

$$T = K_t \times 30 = K_L \times (600)^2 \quad \text{(ii)}$$

With a 20-Ω resistor added in the armature circuit

$$(200 - 20I_a) = K_e \times n$$

$$K_t I_a = K_L n^2 \quad \text{(iv)}$$

Dividing Eq. (iii) by (i) and (iv) by (ii)

$$\frac{200 - 20I_a}{200} = \frac{n}{600} \quad \text{(v)}$$

$$\frac{I_a}{30} = \frac{n^2}{(600)^2} \quad \text{(vi)}$$

Solving

$$n = 260.5 \text{ rpm}$$

$$I_a = 5.66 \text{ A}$$

Example 7.54 *A 400 V series motor has a total armature resistance of 0.25 Ω. When running at 1200 rpm it draws a current of 25 A. When a regulating resistance of 2.75 Ω is included in the armature circuit, it draws current of 15 A. Find the speed and ratio of the two mechanical outputs. Assume that the flux with 15 A is 70% of that with 25 A.*

LO 8

Solution

$$E_a = K'_a \Phi n$$

$$400 - 0.25 \times 25 = K'_a \Phi_1 \times 1200 \quad \text{(i)}$$

$$400 - (2.75 + 0.25) \times 15 = K'_a \Phi_2 \times n_2 \quad \text{(ii)}$$

Dividing Eq. (ii) by (i)

$$\frac{355}{393.75} = \frac{n_2}{1200} \times \frac{\Phi_2}{\Phi_1} = \frac{n_2}{1200} \times 0.7$$

which gives

$$n_2 = 1545.6 \text{ rpm}$$

Ratio of mechanical outputs,

$$\frac{P_{02}}{P_{01}} = \frac{355 \times 15}{393.75 \times 25} = 0.541$$

Example 7.55 *A dc shunt motor is driving a centrifugal pump whose load torque varies as square of speed. The pump speed is controlled by varying the armature voltage of the motor with the field current remaining constant. At full load with an armature voltage of 500 V, the armature current is 128 A. Calculate the armature voltage required to reduce the speed to $1/\sqrt{2}$ of its original value. R_a = 0.28 Ω. Ignore the effect of armature reaction and loss torque (reduction in torque output on account of rotation losses).*

LO 8

Solution

$$I_{a1} = 128 \text{ A}$$

$$E_{a1} = 500 - 0.28 \times 128 = 464.2 \text{ V}$$

As the field current remains constant

$$T = K_T n_1^2 \propto 128; \text{ field current constant} \quad \text{(i)}$$

Speed is to be reduced to $n_2 = n_1/\sqrt{2}$. Then

$$K_T (n_1/\sqrt{2})^2 \propto I_{a2} \quad \text{(ii)}$$

From Eqs (i) and (ii)

$$\frac{I_{a2}}{128} = \frac{1}{2} \quad \text{or} \quad I_{a2} = 64 \text{ A}$$

$$\text{New applied armature voltage} = V_{t2}$$

$$(V_{t2} - 0.28\, I_{a2}) \propto \frac{n_1}{\sqrt{2}} \quad \text{(iii)}$$

$$464.2 \propto n_1 \quad \text{(iv)}$$

From Eqs (iii) and (iv) we get

$$\frac{V_{t2} - 0.28 \times 128}{464.12} = \frac{1}{\sqrt{2}}$$

or $$V_{t2} = 346.1 \text{ V}$$

Example 7.56 *The Ward Leonard speed control system of Fig. 7.99 uses two identical machines of rating 230 V, 4.5 kW, 1500 rpm. The generator is driven at a constant speed of 1500 rpm, $R_a = 0.5\ \Omega$ each machine.*

The magnetization characteristic data obtained at 1500 rpm is as under

I_f (A)	0.0	0.2	0.3	0.4	0.5	0.6	0.7	0.8	1.0	1.2
Voc(V)	45	110	148	175	195	212	223	230	241	251

Neglect the effect of armature reaction.

(a) The motor field current is held constant at 0.8 A. To obtain a motor speed at range of 300-1500 rpm with power (mechanical) output of 4.5 kW, determine the range of the generator field current.

(b) The generator field current is kept constant at 1 A, while the motor field current is reduced to 0.2 A. Determine the motor current and speed for a power output of 4.5 kW.

LO 8

Solution OCC will not be drawn. Instead interpolation will be used in the regions between the given data points.

(a)
$$I_{fm} = 0.8 \text{ A} \Rightarrow E_{am} = 230 \text{ V at 1500 rpm}$$

$$n_m = 300 - 1500 \text{ rpm (range)}$$

(i)
$$n_m = 300 \text{ rpm}$$

$$E_{am} = \frac{230 \times 300}{1500} = 46 \text{ V}$$

$$P_{mot} = 4500 = 46\, I_a$$

or $$I_a = 97.8 \text{ A}$$

$$E_{ag} = 46 + 2 \times 0.5 \times 97.8 = 143.8 \text{ V}$$

From the magnetisation characteristic we get

$$I_f = 0.3 - \frac{0.1}{(148-110)} \times (148 - 143.8) = 0.29 \text{ A}$$

(ii) $$n_m = 1500 \text{ rpm} \quad \text{and} \quad E_{am} = 230 \text{ V}$$

$$P_{mot} = 230 \times I_a = 4500$$

or $$I_a = \frac{4500}{230} = 19.6 \text{ A}$$

$$E_{ag} = 230 + 2 \times 0.5 \times 19.6 = 249.6 \text{ V}$$

From the magnetisation characteristic

$$I_f = 1.2 - \frac{0.2}{(241-230)} \times (251 - 249.6) = 1.18 \text{ A}$$

Hence range of I_f is 0.29 – 1.18 A

(b) $$I_{fg} = 1 \text{ A} \Rightarrow E_{ag} = 241 \text{ A, 1500 rpm (constant)}$$

$$\left(\frac{241 - E_{am}}{2 \times 0.5}\right) \times E_{am} = 4500$$

or $$E_{am}^2 - 241\, E_{am} + 4500 = 0$$

or $$E_{am} = 220.5 \text{ V; lower value is rejected}$$

$$I_{fm} = 0.2 \text{ A} \Rightarrow E_{am} = 110 \text{ V at 1500 rpm}$$

$$n_m = 1500 \times \left(\frac{220.5}{110}\right) = 3007 \text{ rpm}$$

$$I_o = \frac{241 - 220.5}{2 \times 0.5} = 20.5 \text{ A}$$

Concept Check — LO 8

1. Based on emf and torque equation compare and contrast the two methods of speed control of dc motor.
2. Discuss the method of speed control of a dc series motor.
3. How is a shunt motor started? Why it should not be started direct on line?
4. Why large current must not be allowed to flow in a dc motor even for a short duration?
5. Mention the starting methods for dc shunt/compound motors?
6. Explain the speed control methods adopted in dc series motor.

7.18 | BRAKING OF dc MOTORS

LO 9 Describe the braking methods of dc motors

Important! Controlled slowing or stopping of a motor and its driven load is as important as starting in many applications (e.g. cranes, traction on a slope to avoid excessive speed, etc.). Braking methods based on friction, electromechanical action, eddy-currents, etc. are independent of the motor but sometimes electric braking is better justified owing to its greater economy and absence of brake wear. The dc motor is still being widely used for traction purposes. One of the main reasons for this is its excellent braking characteristics and ability of smooth transition from the motor to the generator mode and vice versa. During the braking period, the motor is operated as a generator and the kinetic or gravitational potential energy (cranes or hoists) is dissipated in resistors (plugging) or returned to the supply (regenerative braking).

There are three methods of electrical braking:

(i) *plugging* or *counter-current,* (ii) *dynamic* or *rheostatic, and* (iii) *regenerative.* These will be discussed here briefly. The dynamics of the braking problem is discussed in Ref. [9].

Plugging

This involves the sudden reversal of the connections of either the field or armature[§§] winding during motor operation. A strong braking torque is achieved by maintaining the supply voltage to the armature with connections reversed (Fig. 7.100). The effective armature voltage ($E_a + V_t$) is initially $\approx 2V_t$ so that a limiting braking resistor (may be a starting resistor) must be brought into the circuit. The kinetic energy of the moving system is dissipated in the armature and braking resistances.

Electrical braking of any variety becomes less effective as speed decreases with a consequent decrease in the braking torque. This is because the braking torque

$$T_b = \frac{P_b\,(\text{breaking power})}{n}$$

$$= [E_a^2/R_b]/n$$

$$= \frac{(nK_a)^2/R_b}{n}$$

$$= n(K_a^2/R_b)$$

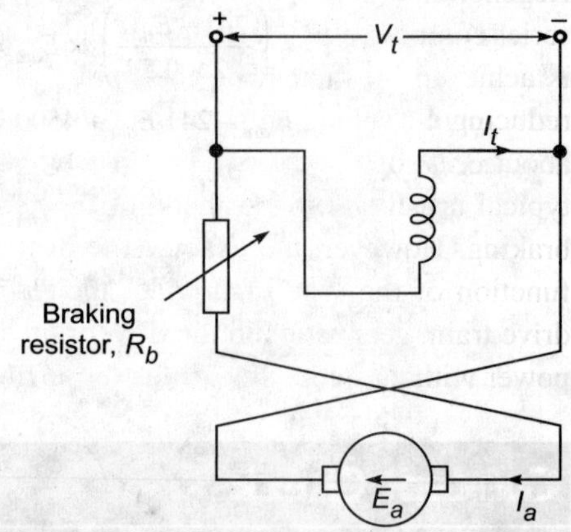

Fig. 7.100 *Plugging connections for a shunt motor*

The supply must be switched off close to zero speed (unless the intention is to run the motor in the reverse direction), using a current or speed directional relay and applying back-up mechanical or hydraulic brakes to bring the motor to a halt. The large initial current and the resultant high mechanical stress restrict the application of plugging to small motors only.

Dynamic Braking

The armature is disconnected from the supply and then a braking resistor R_b is immediately connected across it (Fig. 7.101). The motor acts as a generator, driven by the stored kinetic energy dissipating power in R_b. This is a simple method of bringing a motor nearly to a standstill. The braking time is a function of the system inertia, load torque and motor rating. The field circuit is left connected to the supply. The only danger is that if the supply

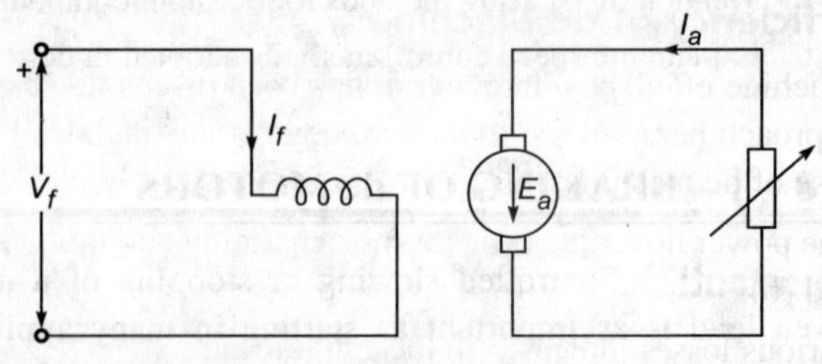

Fig. 7.101 *Dynamic braking, shunt motor*

[§§] Because of the problem of interrupting highly inductive field current and the time needed for the field current to build up in opposite direction, it is a common practice to reverse armature connections.

fails, braking also fails. If the field is left connected across the armature, then initially the braking torque is the same but starts falling sharply with speed, and the problem arises once the speed falls below the critical value for self-excitation. For a series motor, it is necessary for braking to reverse either the field or the armature winding connections for build-up of the armature emf. The value of R_b should be such that $(R_b + R_a + R_{se})$ is less than the critical resistance for the speed at which the braking is commenced.

Regenerative Braking

Good to Know In this method most of the braking energy is returned to the supply and is used specially where the duty cycle requires the braking or slowing of the machine more frequently and is most useful in holding a descending load of high potential energy at a constant speed. The condition for regeneration is that the rotational emf is more than the applied voltage so that the current is reversed and the mode of operation changes from motoring to generating.

Regeneration is possible with a shunt and separately excited motors and with compound motors with weak series compounding. Series motors need a reversal of either the field or the armature connections. Regeneration is achieved by increasing the field current or armature speed or reducing the supply voltage. It has been shown in literature [9] that about 35% of the energy put into an automotive vehicle during typical urban traction is theoretically recoverable by regenerative braking. However, the exact value of the recoverable energy is a function of the type of driving, the terrain, the efficiency of the drive train, gear ratios in the drive/train, etc. The method needs a supply capable of accepting the generated power without undue rise of the terminal voltage.

Probe

Why dc motor is preferred for electric train?

Concept Check LO 9

1. Explain the different methods of braking of dc motors. ○●●
2. What is plugging? ○○●

7.19 | EFFICIENCY AND TESTING

LO 10 Discuss various testing methods of dc machines for predetermination of efficiency

Efficiency of dc Machine

Machine efficiency, in general, has been discussed in Sec. 5.10. The approach here will be to apply the general principles for the specific case of dc machines.

The power flow diagrams for the generating and motoring modes of a dc machine are shown in Figs. 7.102(a) and (b).

Various losses indicated in these figures are:

P_{wf} = winding and friction in loss

P_i = total core loss

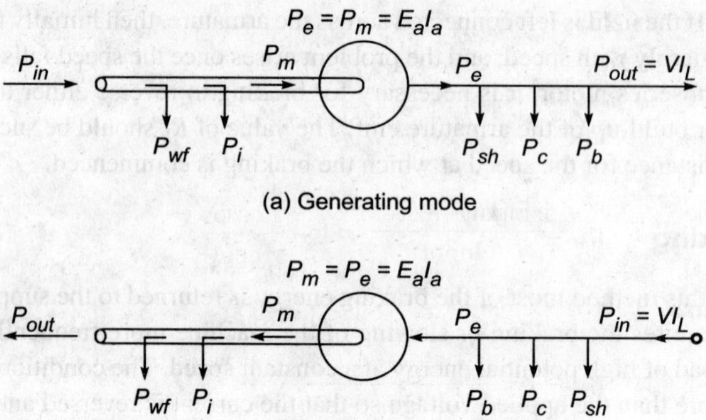

(a) Generating mode

(a) Motoring mode

Fig. 7.102 *dc machine power flow diagrams (stray iron-loss is included in P_i and stray copper-loss in P_c)*

$$= P_{io} + \text{(stray load iron loss)}$$

(This break up is possible only for shunt machine)

where

$$P_{io} = \text{no load core loss} \tag{7.113}$$

We define rotational loss as

$$P_{rot} = P_{io} + P_{\omega f} = \text{rotational loss} \tag{7.114}$$

P_{sh} = shunt field loss in shunt and compound machine

P_c = armature copper losses including loss in series winding and stray load copper

P_b = brush contact losses

We will combine these losses as

Constant loss, $$P_k = (P_{io} + P_{wf}) + P_{sh} \tag{7.115}$$

Variable loss, $$P_v = (R_a + R_{se})\, I_a^2 + P_{st}^l \tag{7.116}$$

where P_{st}^l = total stray load loss (iron plus copper)

$\approx$ proportional to square of armature current

We can than write

$$P_v = K_v I_a^2$$

The brush contact loss P_b will be treated separately as it proportional to I_a.

The expressions of dc machine efficiencies are derived below:

$$\text{Machine efficiency, } \eta = \frac{\text{Output}}{\text{Input}}$$

For generating machine

$$\eta_G = \frac{\text{Output}}{\text{Output + losses}} = 1 - \frac{\text{Losses}}{\text{Output + losses}}$$

In form of symbols in Fig. 7.102(a)

$$\eta_G = 1 - \frac{P_k + K_v I_a^2 + V_b I_a}{VI_L + P_k + K_v I_a^2 + V_b I_a} \tag{7.117}$$

For motoring machine

$$\eta_M = \frac{\text{Input} - \text{losses}}{\text{Input}} = 1 - \frac{\text{Losses}}{\text{Input}}$$

In terms of symbols of Fig. 7.102(b)

$$\eta_M = 1 - \frac{P_k + K_v I_a^2 + V_b I_a}{VI_L} \tag{7.118}$$

It is known from Eqs. (5.82) and (5.84) that the maximum efficiency occurs when

$$\text{Variable loss} = \text{constant loss}$$

or
$$K_v I_a^2 = P_k \tag{7.119}$$

or
$$I_a = \sqrt{\frac{P_k}{K_v}}$$

The ratio $I_a/I_a(fl)$ can be adjusted in machine design by apportioning iron and copper content of the machine.

7.20 TESTING OF DC MACHINES

LO 10

Swinburne's Test

Important! There is a wide variety of non-loading tests that could be performed on dc machines. Swinburne's test and Hopkinson's test are the most important and actually conducted in practice on shunt motors. For obvious reasons the non-loading test cannot be conducted on a series motor.

This is a no-load test and hence cannot be performed on a series motor. Figure 7.103 gives the connections for the test. The motor is run at no-load at rated speed by adjusting the field current to a rated value for accurate determination of no-load loss (P_{i0} + P_{wf}). The machine would run at higher than rated speed with a rated armature voltage. Therefore a series in the armature circuit is employed to reduce voltage applied to the motor armature such that it runs at rated speed.

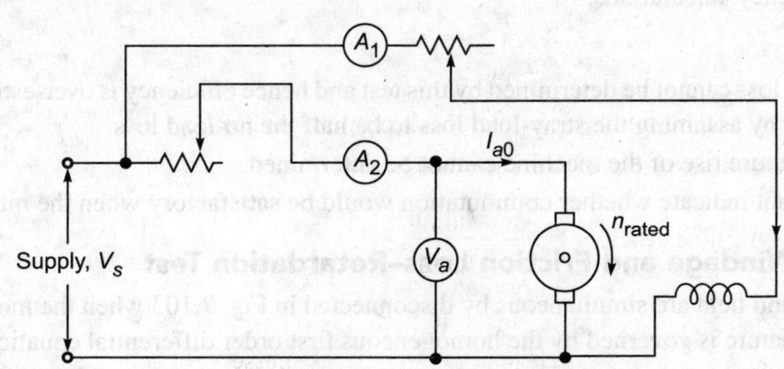

Fig. 7.103 *Swinburne's test*

Constant loss In Fig. 7.103

$$\text{Motor input, } V_a I_{a0} = P_{i0} + P_{wf} + I_{a0}^2 R_a$$

$$\text{Rotational loss} = P_{i0} + P_{wf} = V_a I_{a0} - I_{a0}^2 R_a \tag{7.120}$$

Now
$$P_{sh} = \text{shunt field loss}$$
$$= I_f^2 R_f = V_f I_f \tag{7.121}$$

Hence
$$P_k = \text{constant loss}$$
$$= (V_a I_{a0} - I_{a0}^2 R_a) + I_f^2 R_f \tag{7.122}$$

Variable loss The armature resistance (inclusive of brush contact drop assumed approximately linear) is measured by a dc test by passing a rated armature current from a battery supply. Then

$$P_v = I_a^2 R_a$$

The stray load-loss can be neglected or estimated as 1% of rated output at full load.

Total loss

$$P_L = P_k + P_v$$
$$= (V_a I_{a0} - I_{a0}^2 R_a) + I_f^2 R_f + I_a^2 R_a \tag{7.123}$$

Efficiency can now be calculated at any load current on the following lines:

Generator

$$I_a = I_L + I_f$$

Then
$$\eta_G = 1 - \frac{P_L}{V_t I_L + P_L} \tag{7.124}$$

Motor

$$I_a = I_L - I_f$$

Then
$$\eta_M = \left(1 - \frac{P_L}{V_t I_L}\right) \tag{7.125}$$

Note: Since the resistances (R_a and R_f) are measured cold, temperature correction must be applied to these before using in efficiency calculations.

Disadvantages

(i) The stray-load loss cannot be determined by this test and hence efficiency is over-estimated. Correction can be applied by assuming the stray-load loss to be half the no-load loss.

(ii) Steady temperature rise of the machine cannot be determined.

(iii) The test does not indicate whether commutation would be satisfactory when the machine is loaded .

Separating out Windage and Friction Loss–Retardation Test

If both the armature and field are simultaneous by disconnected in Fig. 7.103 when the motor was running at steady speed, the armature is governed by the homogeneous first order differential equation.

$$J\frac{d\omega}{dt} + f\omega = 0, \text{ motor torque is given} \tag{i}$$

where J = moment of inertia of the armature in Kg-m$_2$, f = windage and friction coefficient in Nm/rad/s

The natural solution of this equation is

$$\omega(t) = Ae^{st} \quad \text{(ii)}$$

Substituting in Eq. (i), we get

$$JAs\, e^{st} + fAe^{st} = 0$$

or

$$(Js + f) = 0$$

or

$$s = -f/J \quad \text{(iii)}$$

Substituting in Eq. (ii)

$$\omega(t) = Ae^{-(f/J)t}$$

Let

$$J/f = \tau, \text{ time constant}$$

then

$$\omega(t) = Ae^{-t/T} \quad \text{(iv)}$$

At $t = 0$, switch-off time $\omega(0) = \omega_0$, then $A = \omega_0$

Therefore at any time t,

$$\omega(t) = \omega_0\, e^{-t/\tau} \quad \text{(v)}$$

where τ is known as the *time constant*. At $t = \tau$, the speed will be

$$\omega(t = \tau) = \omega_0\, e^{-1} = 0.368\, \omega_0$$

Thus at $t = \tau$, the speed reduces to 36.8% of the initial value. This result can be used to determine $\tau = J/f$.

or initial slope of $\omega(t)$

$$\left.\frac{d\omega}{dt}\right|_{t=0} = -(\omega_0/t)\, e^{-t/\tau}\Big|_{t=0} = -(\omega_o \tau) \quad \text{(vi)}$$

Tangential line to $\omega(t)$ at $t = 0$, intersects the t-axis at τ.

This is the alternative way to determine τ

The Retardation Test

The motor is run to rated speed (or any high speed) and the supply is switched-off. As the motor decelerates (retards), several speed-time readings are taken, by a speedometer and watch with seconds hand. Initial readings are taken at small time intervals and the time interval is increased as the motor slows down. The readings (ω vs t) are plotted as shown in adjoining figure. From the graph we find the time T at which the speed reduces to 36.8% of the initial value. Now

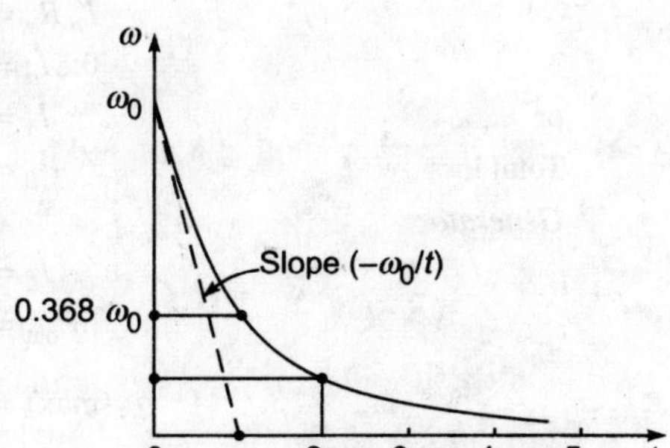

ω vs t plot

$$T = J/f \text{ seconds}$$

The moment of inertia is estimated from the measured dimensions and estimated density of the armature, commutator, fan and axle. Therefore, we find

$$f = J/T, \text{ Nm/rad/s}$$

The windage and friction loss at any speed is then

$$P_{\omega f} = f\,\omega \text{ W} \quad \text{(vi)}$$

Example 7.57 *A 10 kW, 250 V, dc shunt motor with an armature resistance of 0.8 Ω and a field resistance of 275 Ω takes 3.91 A, when running light at rated voltage and rated speed.*

(a) What conclusions can you draw from the above data regarding machine losses?

(b) Calculate the machine efficiency as a generator when delivering an output of 10 kW at rated voltage and speed and as a motor drawing an input of 10 kW. What assumption if any do you have to make in this computation?

LO 10

(c) Determine the maximum efficiencies of the machine when generating and when motoring.

Solution

(a) Shunt field loss,

$$P_{sh} = \frac{(250)^2}{275} = 227.3 \text{ W}$$

$$\text{Rotational loss} = P_{rot} = 250 \times 3.91 - (3.91)^2 \times 0.8 = 965 \text{ W}$$

(b) *Generator*

$$I_L = \frac{10 \times 10^3}{250} = 40 \text{ A}; \; I_f = \frac{250}{275} = 0.91 \text{ A}$$

$$I_a = 40 + 0.91 = 40.91 \text{ A}$$

$$P_L = 965 + 227.3 + (40.91)^2 \times 0.8 = 2.53 \text{ kW}$$

$$\eta_G = 1 - \frac{2.53}{10 + 2.53} = 79.8\%$$

Motor

$$I_a = I_L - I_f = 40 - 0.91 = 39.1 \text{ A}$$

$$P_L = 965 + 227.3 + (39)^2 \times 0.8 = 415 \text{ kW}$$

$$\eta_M = 1 - \frac{2.415}{10} = 75.85\%$$

Assumption: Stray-load loss has been neglected.

(c) The condition for maximum efficiency is

$$I_a^2 R_a = P_{rot} + P_{sh}$$

$$0.8 I_a = 965 + 227.3 = 1192.3 \text{ W}$$

or

$$I_a = 38.6 \text{ A}$$

Total loss

$$P_L = 2 \times 1192.3 = 2384.6 \text{ W}$$

Probe

What is the condition of maximum efficiency?

Generator

$$I_L = I_a - I_f = 38.6 - 0.91 = 37.69 \text{ A}$$

$$P_{\text{out}} = 250 \times 37.69 = 9422.5 \text{ W}$$

$$\eta_G \text{ (max)} = I - \frac{2384.6}{9422.5 + 2384.6} = 79.8\%$$

Motor

$$I_L = I_a + I_f = 38.6 + 0.91 = 39.51 \text{ A}$$

$$P_{in} = 250 \times 39.51 = 9877.5 \text{ W}$$

$$\eta_M \text{ (max)} = 1 - \frac{2384.6}{9877.5} = 75.85\%$$

The MATLAB program for the Example 7.57 is shown below.

```
clc
clear
Pop=10*1000; Vt=250; Ra=0.8; Rf=275; Ia=3.91;
%% part (a)
Psh=Vt^2/Rf;
Prot=Vt*Ia-Ia^2*Ra;
%% part (b)
%% generator
I1=Pop/Vt;
If=Vt/Rf;
Ia=I1 + If;
Ploss=Prot+Psh+Ia^2*Ra;
Eff_gen=(1-Ploss/(Ploss+Pop))*100
%% motor
Ia=I1-If;
Ploss=Prot+Psh+Ia^2*Ra;
Eff_rnotor=(1-Ploss/(Pop))*100
%% part (c)
Ia=sqrt((Prot+Psh)/Ra);
Ploss_tot=2*(Prot+Psh)
%% generator
I1=Ia-If;
Pout=Vt*I1;
Eff_gen_rnax=(1-Ploss_tot/(Ploss_tot+Pout))*100
% motor
I1=Ia+If;
Pin = Vt*I1;
Eff_motor_max=(1-Ploss_tot/Pin)*100
```

Answer:

Eff_gen = 79.7996

Eff_ motor = 75.8498

EfC_gen_max = 79.8048

Eff_motor_max = 75.8585

Example 7.58 *A 50-kW, 250 V, 1200 rpm dc shunt motor when tested on no-load at 250 V draw an armature current of 13.2 A, while its speed is 1215 rpm. Upon conducting other tests it is found that $R_a = 0.06\ \Omega$ and $R_f = 50\ \Omega$ while V_b (brush voltage drop) = 2 V.*

Calculate the motor efficiency at a shaft load of 50 kW at rated voltage with a speed of 1195 rpm. Assume that the stray load loss is 1% of the output.

LO 10

What would be the load for the motor to have maximum efficiency and what would be its value?

Solution *No-load test*

$$\text{Armature input} = 250 \times 13.2 = 3300 \text{ W}$$

$$P_{rot} = P_{i0} + P_{fw} = 3300 - 0.06 \times (13.2)^2 - 2 \times 13.2$$

$$= 3263 \text{ W}$$

As speed varies very little from no-load to full-load P_{rot} almost remains constant.

On-load

$$P_{out} = 50 \text{ kW (at shaft)}$$

Let the armature current be I_a. We write the power balance equation.

$$250\, I_a - 0.06\, I_a^2 - 2\, I_a - 3263 - \underbrace{0.01 \times 50 \times 10^3}_{P_{st}} = 50 \times 10^3$$

or $$0.06\, I_a^2 - 248\, I_a + 53763 = 0$$

Solving we get

$$I_a = 229.6 \text{ A}$$

$$P_{\text{in}} = 250 \times 229.6 + \frac{(250)^2}{50} = 58650 \text{ W}$$

$$\eta = \frac{50000}{58650} \times 100 = 85.25\%$$

We shall assume that P_{st} remains mainly constant in the range of load, we are investigating. Then

$$\eta = \frac{250 I_a - 0.06 I_a^2 - 2 I_a - 3763 (P_{out} + P_{st})}{250 I_a + 250 \times 5}$$

For maximum efficiency $$\frac{d\eta}{dI_a} = 0$$

Solving we get $$I_a = 284 \text{ A}$$

Substituting this value of I_a in η expression, we get

$$\eta_{max} = 86.36\%$$

Example 7.59 *A 600 V dc shunt motor drives a 60 kW load at 900 rpm. The shunt field resistance is 100 Ω and the armature resistance is 0.16 Ω. If the motor efficiency at the load is 85%, determine*

(a) the rotational loss

(b) the no load armature current and speed. Also find speed regulation

(c) the armature current for electromagnetic torque of 600 Nm

LO 10

Solution

$$P_{out} = 60 \text{ kW}$$

$$P_L = \left(\frac{1}{\eta} - 1\right) P_{out} = \left(\frac{1}{0.85} - 1\right) \times 60$$

$$= 10.59 \text{ W}$$

$$P_{in} = 60 + 10.59 = 70.59 \text{ kW}$$

$$I_L = \frac{70.59 \times 10^3}{600} = 117.65 \text{ A}$$

$$I_f = \frac{600}{100} = 6 \text{ A}$$

$$I_a = 117.65 - 6 = 111.65 \text{ A}$$

$$E_a = 600 - 111.65 \times 0.16 = 582.14 \text{ V}$$

$$n = 900 \text{ rpm (given)}$$

(a)
$$P_L = I_a^2 R_a + P_{rot} + P_{sh}$$

$$10.59 \times 10^3 = (111.65)^2 \times 0.16 + P_{rot} + 600 \times 6$$

or
$$P_{rot} = 4995 \text{ W or } 4.995 \text{ kW}$$

(b) No load

Armature resistance loss can be ignored

$$\text{Rotational loss} = P_{rot} = 4995 \text{ W (loss in nearly independent of speed)}$$

$$\text{Input power, } P_0 \approx P_{rot} = 4995 \text{ W}$$

$$I_{ao} = \frac{4995}{600} = 8.325 \text{ A}$$

At no load $I_{ao}R_a$ drop can be neglected. Therefore

$$E_{ao} \approx 600 \text{ V}$$

$$n_0 = 900 \times \frac{600}{582.14} = 927.6 \text{ rpm}$$

$$\text{Speed regulation} = \frac{927.6 - 900}{900} \times 100 = 3.07\%$$

(c)
$$E_a = K_a \Phi \omega$$

Substituting full load values

$$582.14 = K_a \Phi \times 900 \times \frac{2\pi}{60}$$

or
$$K_a \Phi = 6.177 \text{ (constant as } \Phi \text{ is constant)}$$

Electromagnetic torque
$$T = K_a \Phi I_a$$

$$600 = 6.177 \, I_a$$

or
$$I_a = 97.13 \text{ A}$$

Example 7.60 *A dc shunt motor rated 10 kW connected to 250 V supply is loaded to draws 35 A armature current running at a speed of 1250 rpm. Given $R_a = 0.5\ \Omega$*

(a) Determine the load torque if the rotational loss is 500 Ω.

(b) Determine the motor efficiency if the shunt field resistance is 250 Ω.

(c) Determine the armature current for the motor efficiency to be maximum and its value. What is the corresponding load torque and speed?

LO 10

Solution

(a)
$$E_a = 250 - 0.5 \times 35 = 232.5 \text{ V}$$

Electromagnetic power,
$$P_e = E_a I_a = P_{out} \text{ (gross)}$$

$$P_{out}\,(\text{gross}) = 232.5 \times 35 = 8137.5 \text{ W}$$

$$P_{rot} = 500 \text{ W}$$

$$P_{out}\,(\text{net}) = 8137.5 - 500 = 7637.5 \text{ W}$$

Speed $$\omega = \frac{2\pi}{60} \times 1250 = 130.9 \text{ rad/s}$$

Load torque, $$T_L = \frac{7637.5}{130.9} = 58.35 \text{ Nm}$$

(b) $$I_f = \frac{250}{250} = \text{A}$$

$$I_L = 35 + 1 = 36 \text{ A}$$

$$P_{\text{in}} = 250 \times 36 = 9000 \text{ W}$$

Efficiency, $$\eta = \frac{P_{out}(\text{net})}{P_{in}} = \frac{7637.5}{9000} = 84\ 86\%$$

(c) Constant loss, $$P_c = P_{rot} + P_{sh}\,;\ P_{sh} \text{ is neglected}$$

or $$P_k = 500 + 250 \times 1 = 750 \text{ W}$$

For maximum efficiency $$I_a^2 R_a = P_k$$

or $$I_a = \sqrt{\frac{750}{0.5}} = 38.73 \text{ A}$$

$$\text{Total loss} = 2 \times 750 = 1500 \text{ W}$$

$$I_L = 38.73 + 1 = 39.73 \text{ A}$$

$$P_{in} = 250 \times 39.73 \text{ A} = 9932.5 \text{ W}$$

$$\eta_{max} = 1 - \frac{1500}{9932.5} = 84.9\%$$

Speed $$E_a = 250 - 0.5 \times 38.73 = 230.64 \text{ V}$$

$$n = 1250 \times \frac{230.64}{232.5} = 1240 \text{ rpm (marginally different)}$$

$$\omega = \frac{2\pi}{60} \times 1240 = 129.85 \text{ rad/s}$$

$$P_{out}\,(\text{gross}) = 230.64 \times 38.73 = 8932.7 \text{ W}$$

$$P_{out}\,(\text{net}) = 8932.7 - 500 = 8432.7 \text{ W}$$

$$T_L = \frac{8432.7}{129.85} = 84.94 \text{ Nm}$$

Example 7.61 *A 250 V, 25 kW shunt motor has maximum efficiency of 89% at shaft load of 20 kW and speed of 850 rpm. The field resistance is 125 Ω. Calculate the rotational loss and armature resistance. What will be the efficiency, line current and speed at an armature current of 100 A?* **LO 10**

Solution Total loss, $$P_L = \frac{P(\text{shaft})}{\eta} - P(\text{shaft}) = \left(\frac{1}{\eta} - 1\right) P(\text{shaft})$$

$$P_L = \left(\frac{1}{0.89} - 1\right) \times 20 = 2.472 \text{ kW or } 2472 \text{ W}$$

Power input, $$P_{in} = 20 + 2.472 = 22.472 \text{ kW}$$

Line current, $$I_L = \frac{22.472}{250} \times 10^3 = 89.89 \text{ A}$$

$$I_f = \frac{250}{125} = 2\text{ A}$$

$$I_a = 89.89 - 2 = 87.89\text{ A}$$

At maximum efficiency

$$I_a^2 R_a = P_{rot} + P_{sh} = 2472/2 = 1236\text{ W}$$

$$R_a = \frac{1236}{(89.89)^2} = 0.153\ \Omega$$

$$P_{sh} = 250 \times 2 = 500\text{ W}$$

$$\therefore \quad P_{rot} = 1236 - 500 = 736\text{ W}$$

$$E_a = 250 - 89.89 \times 0.153 = 236.25\text{ V}$$

At speed, $n = 850$ rpm

Now shaft load is raised till armature current becomes $I_a = 100$ A

Armature copper loss, $P_c = (100)^2 \times 0.153 = 1530$ W

$$P_L = P_c + P_{rot} + P_{sh}$$

$$= 1530 + 1236 = 2766\text{ W}$$

$$P_{in} = 250 \times I_L = 250 \times (100 + 2) = 25.5\text{ kW}$$

$$\eta = \left(1 - \frac{2.766}{25.5}\right) \times 100 = 89.15$$

$$E_a = 250 - 100 \times 0.153 = 234.7\text{ V}$$

$$n \propto E_a,\ \text{constant } I_f$$

$$\therefore \quad n = 850 \times \frac{234.7}{236.25} = 844.4\text{ rpm}$$

Hopkinson's Test

Important! This is a regenerative test in which two identical dc shunt machines are coupled mechanically and tested simultaneously. One of the machines is made to act as a motor driving the other as a generator which supplies electric power to motor. The set therefore draws only loss-power from the mains while the individual machines can be fully loaded. (Compare this test with Sumpner's test on two identical transformers).

Figure 7.104 shows the connection diagram for Hopkinson's test. One of the machines of the set is started as a motor (starter connections are not shown in the figure) and brought to speed. The two machines are made parallel by means of switch *S* after checking that similar polarities of the machine are connected across the switch. If this is the case, the voltage across the switch can be almost reduced to zero by adjustment of the field currents of the machines. Otherwise the polarities of either one of the armatures or one of the fields must be reversed and the set restarted. The switch *S* is closed after checking that the voltage across it is negligible so that heavy circulating current will not flow in the local loop of armatures on closing the switch.

The speed of the set and electric loading of the machines can be adjusted by means of rheostats placed in the two field circuits. The cause-effect relationship to load variation is given below:

$$I_{fg} \uparrow \rightarrow E_{ag} \uparrow \rightarrow E_{ag} > E_{am} \rightarrow I_{ag} \uparrow,\ I_{am} \uparrow$$

$$I_{fm} \downarrow \rightarrow n \uparrow \rightarrow E_{ag} > E_{am} \rightarrow I_{ag} \uparrow,\ I_{am} \uparrow.$$

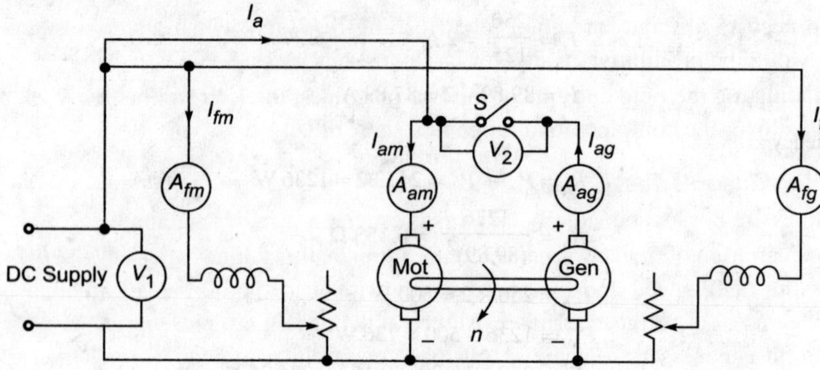

Fig. 7.104 *Hopkinson's test*

Computation of losses and efficiencies

Current drawn from supply. $I_a = I_{am} - I_{ag}$; motor draws larger current as it runs the generator (7.126)

$$\text{Total input to armature circuit} = V_t I_a$$

$$= \text{total armature losses since set output is zero} \quad (7.127)$$

Hence

$$\text{Total stray loss} = V_t I_a - I_{am}^2 R_{am} - I_{ag}^2 R_{ag} \quad (7.128)$$

= [(Windage and friction loss) + (no-load iron-loss) + (stray-load loss)] of both machines

Since the generating machine excitation is more than that of the motoring machine, their no-load iron (P_{i0}) and stray-load loss (P_{st}^l) are not quite equal. As there is no way of separating these in this test and the difference in any case is small, these are regarded as equal in the two machines. Hence

$$P_{stray} \text{ (each machine)} = \frac{1}{2}[V_t I_a - I_{am}^2 R_{am} - I_{ag}^2 R_{ag}] \quad (7.129)$$

$$\text{Motor field copper-loss} = V_t I_{fm}$$

$$\text{Generator field copper-loss} = V_t I_{fg}$$

Therefore,

$$P_{Lm} = (P_{stray} + V_t I_{fm}) + I_{am}^2 R_{am} \quad (7.130)$$

and

$$P_{Lg} = (P_{stray} + V_t I_{fg}) + I_{ag}^2 R_{ag} \quad (7.131)$$

Now

$$\eta_M = 1 - \frac{P_{Lm}}{P_{in,m}} \quad (7.132)$$

and

$$\eta_G = 1 - \frac{P_{Lg}}{P_{out,g} + P_{Lg}} \quad (7.133)$$

Advantages of Hopkinson's Test

(i) The two machines are tested under loaded conditions so that stray-load losses are accounted for.

(ii) Since it is a regenerative test, the power drawn from the mains is only that needed to supply losses. The test is, therefore, economical for long duration test like a "heat run".

(iii) There is no need to arrange for actual load (loading resistors) which apart from the cost of energy consumed, would be prohibitive in size for large-size machines.

(iv) By merely adjusting the field currents of the two machines, the load can be easily changed and a load test conducted over the complete load range in a short time.

Drawbacks of Hopkinson's Test

(i) Both machines are not loaded equally and this is crucial in smaller machines.

(ii) Since a large variation of field currents is required for small machines, the full-load set speed is usually higher than the rated speed and the speed varies with load. The full load in small machines is not obtained by cutting out all the external resistance of the generator field. Sufficient reduction in the motor field current is necessary to achieve full-load conditions resulting in speeds greater than the rated value.

(iii) There is no way of separating the iron-losses of the two machines, which are different because of different excitations.

Thus the test is better suited for large machines.

Example 7.62 *The following test results were obtained while Hopkinson's test was performed on two similar dc shunt machines:*

Supply voltage = 250 V

Field current of motor = 2 A

Field current of generator = 2.5 A

Armature current of generator = 60 A

Current taken by the two armatures from supply = 15 A

Resistance of each armature circuit = 0.2 Ω

Calculate the efficiency of the motor and generator under these conditions of load. **LO 10**

Solution

$$I_{am} = I_{ag} + I_a = 60 + 15 = 75 \text{ A}$$

Using Eq. (7:128)

$$P_{stray} = \frac{1}{2}[V_t I_a - I_{am}^2 R_{am} - I_{ag}^2 R_{ag}]$$

$$= \frac{1}{2}[250 \times 15 - 75^2 \times 0.2 - 60^2 \times 0.2] = 952.5 \text{ W}$$

$$P_{in,m} = V_t I_{am} + V_t I_{fm} = 250 \times 75 + 250 \times 2 = 19250 \text{ W}$$

Recalling Eq. (7.129)

$$P_{Lm} = (P_{stray} + V_t I_{fm}) + I_{am}^2 R_{am}$$

$$= (952.5 + 250 \times 2) + 75^2 \times 0.2 = 2577.5 \text{ W}$$

The motor efficiency is given by

$$\eta_M = 1 - \frac{P_{Lm}}{P_{in,m}} = 1 - \frac{2577.5}{19250} = 86.66\%$$

Using Eq. 7.130

$$P_{Lg} = (P_{stray} + V_t I_{fg}) + I_{ag}^2 R_{ag}$$
$$= (952.5 + 250 \times 2.5) + 60^2 \times 0.2 = 2297.5 \text{ W}$$
$$P_{out, g} = V_t I_{ag} = 250 \times 60 = 15000 \text{ W}$$
$$\eta_G = 1 - \frac{P_{Lg}}{P_{out,g} + P_{Lg}} = 1 - \frac{2297.5}{15000 + 2297.5} = 86.7\%$$

Field's Test: Two Identical Series Motors

Regenerative test on two identical series motors is not feasible because of instability of such an operation and the possibility of run-away speed. Therefore, there is no alternative but to conduct a *loading test*. In Field's test the two motors are mechanically coupled with the motoring machine driving the generator which feeds the electrical load. The connection diagram of the Field's test is shown in Fig. 7.105. It is observed that

- The generator field is connected in series with the motor circuit. The generator is thus separately excited and its excitation is identical to that of motor at all loads. This ensures that the iron-loss of both the machines are always equal.
- The load is connected to the generator directly to the load without any intervening switch such that the load cannot be switch off accidentally.
- The supply voltage is adjusted to account for the voltage drop in the generator field such that $V_m = V_{rated}$

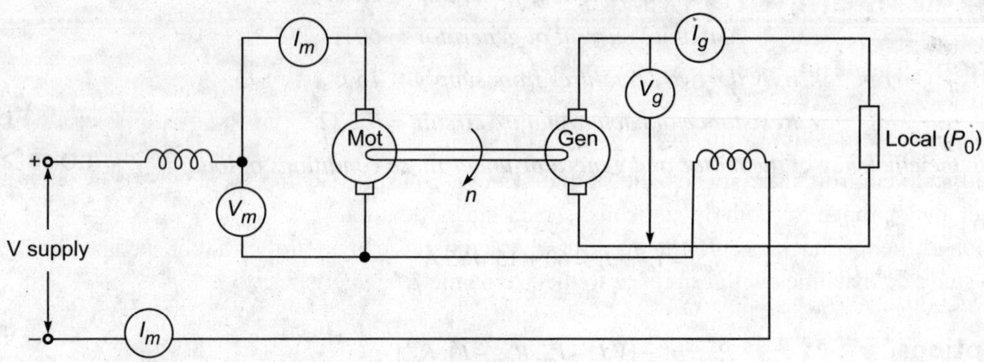

Fig. 7.105

Under load conditions:

Input to the set,	$P_i = V_m I_m$	
Output of the set,	$P_0 = V_g I_g$	
Total loss,	$P_L = V_m I_m - V_g I_g$	(i)
Total copper loss,	$P_c = (R_{am} + 2R_{se}) I_m^2$	(ii)
Total rotational loss,	$P_{rot}(\text{total}) = P_i - P_0 - P_c$	(iii)
Rotational loss of each machine	$P_{rot} = P_{rot}(\text{total})/2$	(iv)

Efficiencies

$$\eta_M = 1 - \frac{P_{rot} + (R_{am} + R_{se})I_m^2}{V_m I_m} \quad \text{(v)}$$

$$\eta_G = 1 - \frac{P_{rot} + (R_{ag}I_g^2 + R_{se}I_m^2)}{V_g I_g + [P_{rot} + (R_{ag}I_g^2 + R_{se}I_m^2)]} \quad \text{(vi)}$$

Note: Series traction motors are normally available in pairs because of the speed control needs. Being a load test even though the load is electrical it can be conducted on small motors only.

Concept Check — LO 10

1. Enumerate and classify the losses in a dc shunt motor.
2. How to determine the load current of a dc shunt motor at which the motor efficiency is maximum?
3. What are the advantages of Hopkinson's test over Swinburne's test and what are its limits?
4. By what test on dc reparalely excited generator would you determine the armature reaction equivalent field current at rated armature current.

7.21 dc MACHINE DYNAMICS

LO 11 Review the dc machine dynamics and constructional features of PMDC motors

!Attention The dc machines are quite versatile and are capable of giving a variety of V-A and speed-torque characteristics by suitable combinations of various field windings.

With solid-state controls their speeds and outputs can be controlled easily over a wide range for both dynamic and steady-state operation. By addition of the feedback circuit, the machine characteristics can be further modified. The aim of this section is to study dc machines with reference to their dynamic characteristics.

Assumptions

For illustration, let us consider the separately excited dc machine shown schematically in Fig. 7.106. For ease of analysis, the following assumptions are made:

(i) The axis of armature mmf is fixed in space, along the q-axis.

(ii) The demagnetizing effect of armature reaction is neglected.

(iii) Magnetic circuit is assumed linear (no hysteresis and saturation). As a result all inductances (which came into play in dynamic analysis) are regarded as constant.

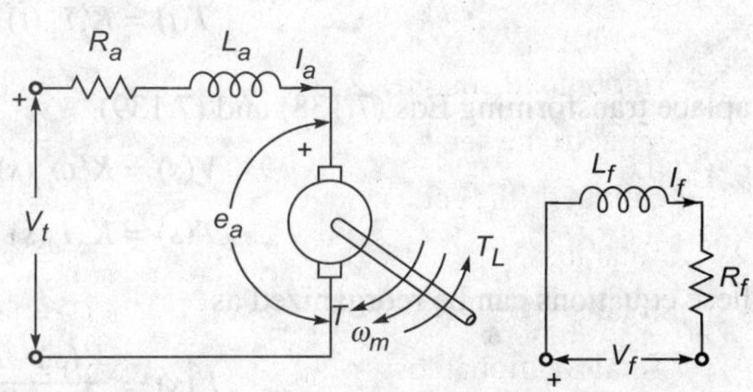

Fig. 7.106 *Schematic representation of a separately excited dc motor for dynamic analysis*

The two inductance parameters appearing in Fig. 7.106 are defined below:

L_a = armature self-inductance caused by armature flux; this is quite small[¶¶] and may be neglected without causing serious error in dynamic analysis

L_f = self-inductance of field winding; it is quite large for shunt field and must be accounted for

Mutual inductance (between field and armature) = 0; because the two are in space quadrature.

Further for dynamic analysis it is convenient to use speed in rad/s rather than rpm.

Applying Kirchhoff's law to the armature circuit,

$$V_t = e_a(t) + R_a i_a(t) + L_a \frac{d}{dt} i_a(t) \tag{7.134}$$

where $$e_a(t) = K_e i_f(t)\omega_m;\ K_e = \text{constant}\ (\phi(t) \propto i_f(t)) \tag{7.135}$$

Similarly for the field circuit,

$$v_f(t) = R_f i_f(t) + L_f \frac{d}{dt} i_f(t) \tag{7.136}$$

For motoring operation, the dynamic equation for the mechanical system is

$$T(t) = K_t i_f(t) i_a(t) = J \frac{d}{dt}\omega_m(t) + D\omega_m(t) + T_L(t) \tag{7.137}$$

where J = moment of inertia of motor and load in Nms^2

D = viscous damping coefficient representing rotational torque loss, Nm rad/s

Energy storage is associated with the magnetic fields produced by i_f and i_a and with the kinetic energy of the rotating parts. The above equations are a set of *nonlinear*[***] (because of products $i_f(t)\,\omega_m$ and $i_f(t)i_a(t)$) *state equations* with *state variables* i_f, i_a and ω_m. The solution has to be obtained numerically.

Transfer Functions and Block Diagrams

In the simple linear case of motor response to changes in armature voltage, it is assumed that the field voltage is constant and steady-state is existing on the field circuit, i.e. I_f = constant. Equations (7.134), (7.136) and (7.137) now become linear as given below

$$v(t) = K'_e\omega_m(t) + R_a i_a(t) + L_a \frac{d}{dt} i_a(t) \tag{7.138}$$

$$T(t) = K'_t i_a(t) = J \frac{d}{dt}\omega_m(t) + D\omega_m(t) + T_L(t) \tag{7.139}$$

Laplace transforming Eqs (7.138) and (7.139)

$$V(s) = K'_e\omega_m(s) + (R_a + sL_a)\, I_a(s) \tag{7.140}$$

$$T(s) = K'_t I_a(s) = (sJ + D)\omega_m(s) + T_L(s) \tag{7.141}$$

These equations can be reorganized as

$$I_a(s) = \frac{V(s) - K'_e\omega_m(s)}{(R_a + sL_a)}$$

[¶¶] The armature mmf is directed along the low permeance *q*-axis.

[***] This is inspite of the fact that the magnetic circuit has been regarded as linear.

$$= [V(s) - K'_e\omega_m(s)] \times \frac{1/R_a}{(1+s\tau_a)} \tag{7.142}$$

where $\tau_a = L_a/R_a$ = armature circuit time-constant

Also
$$\omega_m(s) = [T(s) - T_L(s)] \times \frac{1/D}{(1+s\tau_m)} \tag{7.143}$$

where $\tau_m = J/D$ = mechanical time-constant

$$T(s) = K'_t I_a(s) \tag{7.144}$$

From Eqs (7.142) – (7.144), the block diagram of the motor can be drawn as in Fig. 7.107. It is a second-order feedback system with an oscillatory response in general. It is reduced to simple first-order system, if L_a and therefore τ_a is neglected

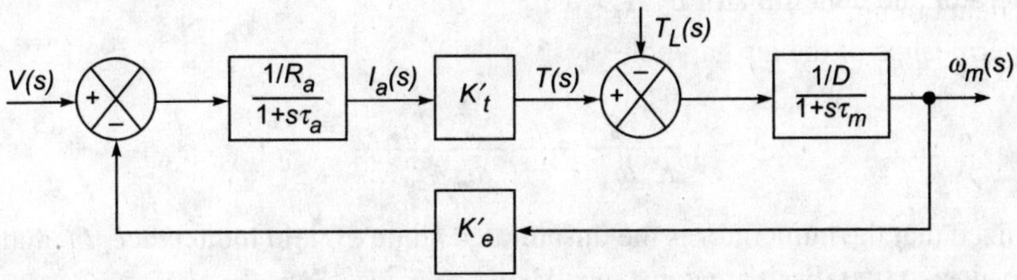

Fig. 7.107 *Block diagram of separately-excited dc motor; inputs V(s) and $\omega_m(s)$*

Shunt Generator Voltage Build-up

The qualitative explanation for the voltage build-up process in a shunt generator has already been advanced in Sec. 7.11. Here the mathematical treatment of this problem will be given, which in fact boils down to the solution of a nonlinear differential equation.

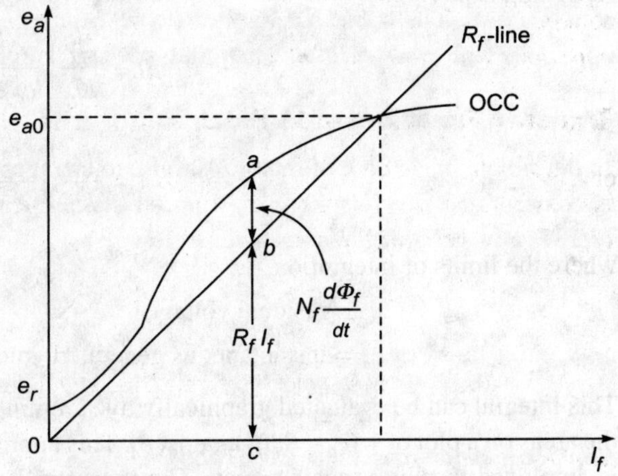

Fig. 7.108 *Magnetization curve and R_f-line*

Referring to Fig. 7.108 it is seen that for any field current the intercept *ab*, between the OCC and the R_f-line gives the voltage drop caused by the rate of change of Φ_f and the intercept *bc* gives the drop in the field resistance. The two together balance out the generated emf e_a (neglecting $i_f R_a$, the armature drop). Thus

$$N_f \frac{d\Phi_f}{dt} = e_a - R_f i_f \tag{7.145}$$

where Φ_f = field flux/pole

N_f = number of turns of field winding

The field flux Φ_f is greater than the direct axis air-gap flux Φ_d because of leakage.

Taking this into account

$$\Phi_f = \sigma \Phi_d \tag{7.146}$$

Here σ is known as the *coefficient of dispersion.*

Recalling Eq. (7.3),

$$\Phi_d = \frac{e_a}{K_a \omega_m} \tag{7.147}$$

Substituting Eqs (7.146) and (7.147) in Eq. (7.148),

$$\frac{N_f \sigma}{K_a \omega_m} \cdot \frac{de_a}{dt} = e_a - i_f R_f \tag{7.148}$$

Multiplying numerator and denominator by $N_f P_{ag}$

where P_{ag} is the permeance of the air-gap/pole

$$\frac{N_f \sigma}{K_a \omega_m} = \frac{N_f^2 \sigma \mathcal{P}_{ag}}{K_a \omega_m \mathcal{P}_{ag} N_f}$$

It is easily recognized that the numerator is the unsaturated value of field inductance, L_f, and the denominator is the slope of the air-gap line. Both are constants. Hence,

$$\frac{L_f}{K_g} \frac{de_a}{dt} = e_a - R_f i_f \tag{7.149}$$

Rewriting Eq. (7.145)

$$dt = \frac{L_f}{K_g} \left(\frac{de_a}{e_a - R_f i_f} \right)$$

or

$$t = \frac{L_f}{K_g} \int_{e_r}^{e_a} \frac{de_a}{e_a - R_f i_f} \tag{7.150}$$

where the limits of integration,

e_r = residual voltage

e_a = instantaneous generated voltage

This integral can be evaluated graphically by summing up the areas on a plot of $1/(e_a - R_f i_f)$ against e_a. This approach is employed to plot e_a against time. The theoretical time needed for the generated emf to attain the no-load value, e_{a0} would be infinite; hence in practice the time needed to reach 0.95 e_{a0} is taken as the time needed to reach e_{a0}. The variation of e_a with time is plotted in Fig. 7.109.

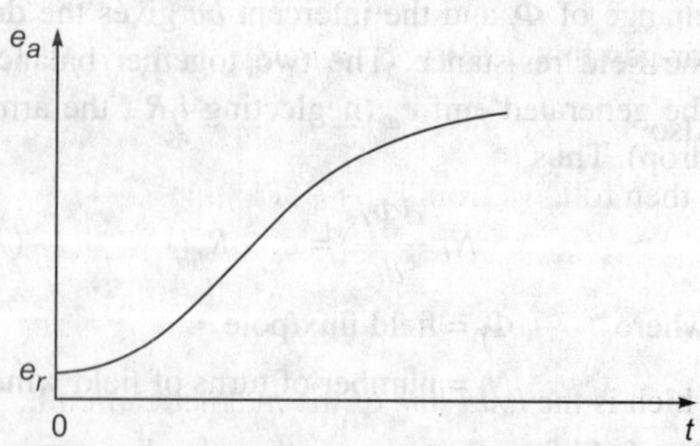

Fig. 7.109 *Voltage build-up of a shunt generator*

The response is rather sluggish since only small voltage differences (= $e_a - R_f i_f$) contribute to the flux build-up (Φ_f).

7.22 | PERMANENT MAGNET dc (PMDC) MOTORS

LO 11

As has been discussed in Chapter 2, a number of new permanent magnet materials—ceramics, and rare earth magnetic materials-have become available commercially. These materials have high residual flux as well as high coercivity. Smaller fractional and sub-tractional hp dc motors are now constructed with PM poles. As no field windings are needed, so no field current and continuous field loss.

Important! As a result PMDC motors are smaller in size than the corresponding rated field wound type motors, this fact partially off-sets the high cost of permanent magnets. Obviously, these motors offer shunt type characteristic and can only be armature controlled.

The risk of permanent magnetism getting destroyed by armature reaction (at starting/reversing or heavy overloads) has been greatly reduced by the new PM materials.

Constructional Features

The stator is an annular cylindrical shell of magnetic material on the inside of which are bonded fractional cylindrical permanent magnets (usually two poles) as shown in the cross-sectional view of Fig. 7.110. The magnetics are radially magnetized as shown by arrows. The rotor is laminated magnetic material with slotted structure in which the winding is placed whose coil ends suitably connected to the commutator (usual construction).

Magnetic Circuit

It is seen from Fig. 7.110 that the flux crosses the airgap length (l_g) twice and the thickness of the permanent magnet t_m twice. The iron path in the shell and rotor teeth and core being highly permeable ($\mu \to \infty$) can be assumed to consume any mmf. The mmf balance equation is then

$$2l_g H_g + 2t_m H_m = 0; \text{no external mmf is applied}$$

or $$l_g H_g + t_m H_m = 0$$

or $$H_g = -\left(\frac{t_m}{l_g}\right) H_m \tag{7.151}$$

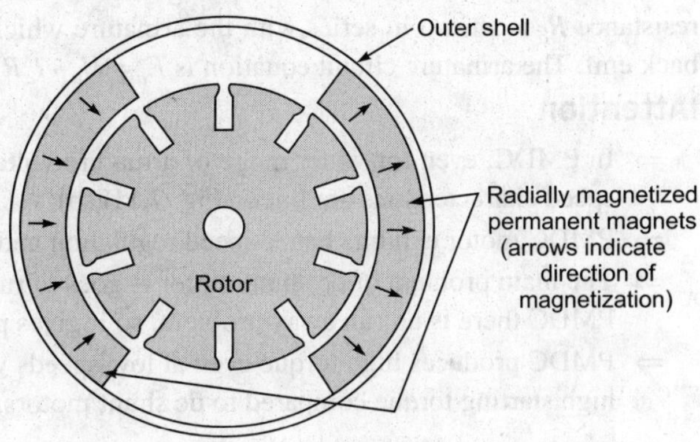

Fig. 7.110 *Cross-section view of permanent-magnet motor*

Air-gap flux density, $B_g = \mu_0 H_g$ (7.152)

Also $$B_g = B_m$$

it then follows from Eq. (7.151) that

$$B_g = -\mu_0 \left(\frac{t_m}{l_g}\right) H_m = B_m \tag{7.153}$$

which is the *load line* of the magnetic circuit.

The dc magnetization characteristics of various PM materials is presented in Fig. 2.20. Obvious choice of PMDC motor is *neodymium-iron-boron* which has high coercivity and high retentivity. Its characteristic is almost a straight line which can be expressed as

$$B_m = -\left(\frac{1.25}{940\times10^3\times4\pi\times10^{-7}}\right)\mu_0 H_m + 1.25$$

or $$B_m = -1.06\,\mu_0 H_m + 1.25 \tag{7.154}$$

The solution of Eqs. (7.153) and (7.154) yields $B_m = B_g$ from which we can find the flux/poles as

$$\Phi = B_g A_g$$

Unlike a normal dc motor Φ is a constant quantity

Emf and torque equations

$$E_a = K_a \Phi \omega$$

$$T = K_a \Phi I_a$$

As Φ is constant

$$E_a = K_m \omega \tag{7.155}$$

$$T = K_m I_a \tag{7.156}$$

where K_m = motor torque constant

Electromagnetic power, $P_e = E_a I_a$

Circuit model As there is no electrically excited field and the permanent magnetic creates a constant flux/pole, the circuit of a permanent magnet motor is as drawn in Fig. 7.111, where the armature resistance R_a is shown in series with the armature which has induced back emf. The armature circuit equation is $E_a = V_t - I_a R_a$.

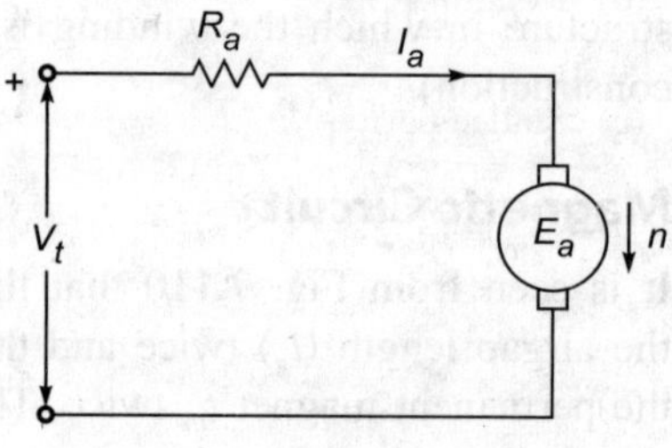

Fig. 7.111 *Circuit model*

!Attention

⇒ In PMDC, even for wider range of armature voltage the torque speed characteristics are linear, Fig. 7.112(b)

⇒ PMDC motor exhibits better speed regulation and efficiency than dc shunt motor.

⇒ The main problem of dc shunt motor is goes to run away when the field terminals are opened. But in PMDC there is no run away problem, so it gives practical benefit to the industry applications.

⇒ PMDC produces high torque even at low speeds which is shown in Fig. 7.112(a) and also it produces high starting torque compared to dc shunt motors.

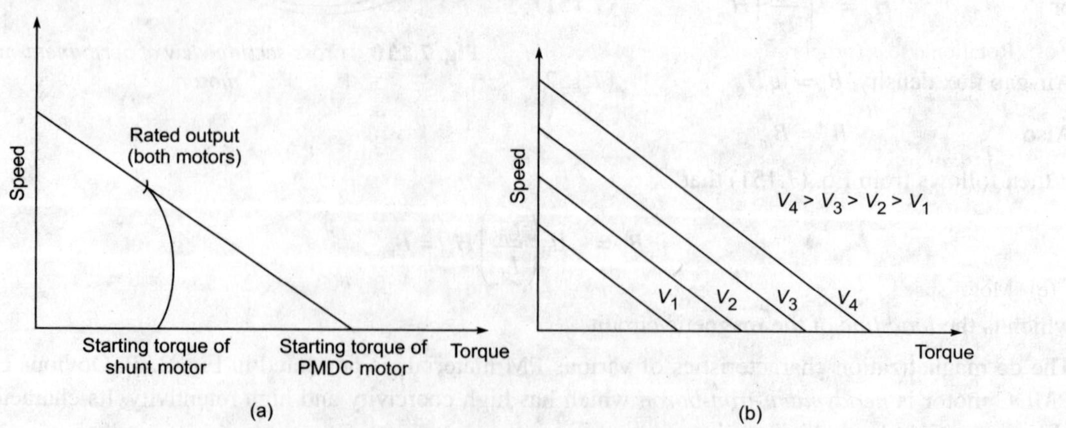

Fig. 7.112 *Speed torque characteristics of PMDC motors*

Example 7.63 *A PMDC motor has an armature resistance of 4.2 Ω. When 6 V supply is connected to the motor it runs at a speed 12,125 rpm drawing a current of 14.5 mA on no-load*

(a) Calculate its torque constant

(b) What is the value of rotational loss?

With an applied voltage of 6 V,

(c) calculate the stalled torque and stalled current of the motor (motor shaft held stationary)

(d) at a gross output of 1.6 W, calculate the armature current and efficiency. Assume that the rotational loss varies as square of speed.

(e) calculate the motor output at a speed of 10,250 rpm and the efficiency.

LO 11

Solution

No load $V_t = 6$ V, $I_{ao} = 14.5$ mA, $n = 12125$ rpm or $\omega = 1269.7$ rad/s

(a) $E_a = 6 - 14.5 \times 10^3 \times 4.2 = 5.939$ V

$$5.939 = K_m \omega = K_m \times 1269.7$$

or $K_m = 4.677 \times 10^{-3}$

(b) Rotational loss, $P_{rot} = E_a I_a$; there is no load

$$= 5.939 \times 14.5 \times 10^{-3} = 0.0861 \text{ W}$$

(c) Stalled current

$$\omega = 0 \text{ so } E_a = 0$$

$$I_a(\text{stall}) = \frac{6}{4.2} = 1.4285 \text{ A}$$

$$\text{Torque (stall)} = K_m I_a(\text{stall}) = 4.677 \times 10^{-3} \times 1.428 = 6.67 \text{ m Nm}$$

(d) $P_{out}(\text{gross}) = 1.6 \text{ W} = E_a I_a$

$$(6 - 4.2\, I_a)\, I_a = 1.6$$

$$4.2\, I_a^2 - 6 I_a + 1.6 = 0$$

Solving we find $I_a = 0.354$ A, 1.074 A

Thus

$$I_a = 0.354 \text{ A; higher value rejected}$$

$$E_a = 6 - 0.854 \times 4.2 = 4.513 \text{ V} = K_m \omega$$

$$\omega = \frac{4.513 \times 10^3}{4.677} = 965 \text{ rad/s}$$

Rotational loss (proportional to square of speed)

$$P_{rot} = 0.0861 \times \left(\frac{965}{1269.7}\right)^2 = 0.05 \text{ W}$$

$$P_{out}(\text{net}) = P_{out}(\text{gross}) - P_{rot} = 1.6 - 0.05 = 1.55 \text{ W}$$

Power input, $P_i = V_t I_a = 6 \times 0.354 = 2121$ W

$$\eta = \frac{1.55}{2.124} \times 100 = 73\%$$

(e) Motor speed, $n = 10250$ rpm or $\omega = 1073.4$ rad/s

$$E_a = K_m^{\omega} = 4.513 \times 10^{-3} \times 1073.4 = 4.844 \text{ V}$$

$$I_a = \frac{6 - 4.844}{4.2} = 0.275\, A$$

$$P_{out}(\text{gross}) = P_e = E_a I_a = 4.844 \times 0.275 = 1.332 \text{ W}$$

$$P_{rot} = 0.0861 \times \left(\frac{1073.4}{1269.7}\right)^2 = 0.0615 \text{ W}$$

$$P_{out} \text{ (net)} = 1.332 - 0.0615 = 1.27 \text{ W}$$

$$P_{in} = 6 \times 0.275 = 1.65 \text{ W}$$

$$\eta = \frac{1.27}{1.65} \times 100 = 77\%$$

7.23 | dc MACHINE APPLICATIONS

LO 11

Whenever the application of any machine is considered, its operating characteristics along with its economic and technical viability as compared to its competitors are the essential criteria. For a dc machine, of course, the main attraction lies in its flexibility, versatility and ease of control. This explains why in spite of its rather heavy initial investment it still retains its charm in strong competitive industrial applications. In the world, today, around 25% of the motors manufactured are dc motors.

With the advent of various power electronic devices, there is no doubt that the importance of a dc generator has gone down. Now, for ac to dc transformation, the dc generator as part of an ac-to-dc motor-generator set has to compete with SCR rectifiers and various other types of controlled power electronic devices which usually are less costly, compact, relatively noise-free in operation and need minimum maintenance, but suffer from the disadvantages of having poor power factor, harmonic generation, poor braking, etc. However, some of the important applications of a dc generator include—dynamometers, welding, cross-field generators for closed-loop control systems, tachogenerators, etc. Separately-excited generators are still in use for a wide output-voltage control such as in the Ward-Leonard system of speed control.

In dc series motor the starting torque is very high, up to five times the full-load torque. It may be interesting to note that the maximum torque in a dc motor is limited by commutation and not, as with other motors, by heating. Speed regulation of a dc series motor can be varied widely. For drives requiring a very high starting torque, such as hoists, cranes, bridges, battery-powered vehicles and traction-type loads, the series dc motor is the obvious choice. Speed control is by armature resistance control. Its closest rival is the wound-rotor induction motor with a rotor resistance control. But ultimately the availability and economics of a dc power is the deciding factor rather than the motor characteristics.

Compound motor characteristics depend naturally upon the degree of compounding. Shunt field of course restricts the no-load speed to a safe value. Its main competitor is the squirrel-cage high-slip induction motor.

!Attention A compound motor has a considerably higher starting torque compared to a shunt motor and possesses, a drooping speed-load characteristic. Compound dc motors are used for pulsating loads needing flywheel action, plunger pumps, shears, conveyors, crushers, bending rolls, punch presses, hoists, rolling mill, planing and milling machines, etc.

A dc shunt motor has a medium starting torque. Speed regulation is about 5–15%. It is used essentially for constant speed applications requiring medium starting torques, such as centrifugal pumps, fans, blowers, conveyors, machine tools, printing presses, etc. Owing to the relative simplicity, cheapness and ruggedness of the squirrel cage induction motor, the shunt motor is less preferred for constant-speed drives except at low-speeds. At low speeds, dc shunt motors are comparable with synchronous motors. The outstanding feature of a dc shunt motor however is its superb wide range flexible speed control above and below the base speed using solid-state controlled rectifiers (discussed in Ch. 12).

In general, whenever a decision is to be made for a choice of a suitable motor for a given application, it is necessary to make specific, analytic, economic, and technical comparison of all practical choices. Finally, it should be mentioned to the credit of a dc machine that it still remains most versatile, flexible, easily controllable energy conversion device whose demand and need would continue to be felt in industries in future for various applications discussed above.

Concept Check — LO 11

1. How dc machines can be controlled for dynamic and steady-state operation? ○●●
2. What state variables are involved in dynamic study? ○○●
3. What are the special constructional features of PMDC motors? ○○●
4. What are the advantages of PMDC motors? ○○●

SUMMARY

LO 1

- The main advantage of dc machines lies in their flexibility, versatility and high degree of control. The disadvantages are complexity associated with armature winding and commutator/brush system, more maintenance and less reliability.
- Winding types
  Lap winding – number of parallel paths $A = P$
  number of brushes = P, equalizer rings needed = P
  Wave winding – number of parallel paths = 2, independent of number of poles
  number of brushes, 2 needed but P used in practice
  no equalizer rings needed
- Emf Equation
  $$E_a = K_a \Phi \omega_m = \left(\frac{2\pi}{60}\right) K_a \Phi n \text{ V}$$
  where $K_a = \left(\frac{ZP}{2\pi A}\right)$, machine constant
- Torque Equation
  $T = K_a \Phi I_a$; K_a same as in emf equation
  T = electromagnetic torque developed
- Electromagnetic power – power converted
  $$T\omega_m = E_a I_a$$

LO 2

- Generating machine
  $$V_t = E_a - I_a R_a \,;\, V_t < E_a$$
  R_a = armature resistance, very small 0.01 pu
  There is constant voltage drop at brush about 1 to 2 V. I_a flow out of positive terminal.
- Motoring machine
  $$V_t = E_a + I_a R_a;\, V_t > E_a$$
  I_a flows into positive terminal
- Machine axes
  d-axis – along the axis of main poles
  q-axis – along the magnetic neutral axis at 90° electrical to d-axis

LO 3

- Armature reaction, AT_a along magnetic neutral axis (q-axis) at 90° to d-axis
  Nature – *cross magnetizing*, weakens one side of poles and strengthens the other side of poles. Φ remains constant in linear region of magnetizing. In saturation region of main poles Φ decreases, a *demagnetizing effect*.
- Armature reaction creates distortion in the flux density, shift in MNA, increased iron loss, commutation problem and commutator sparking.
- Compensating winding; provided in the pole shoes to counter armature reaction thereby mitigating the harmful effects of armature reaction
- Commutation – as an armature coil moves out the influence of one pole pair to the next pole pair, the current in it must reverse. The reactance *emf* $\left[\frac{di(\text{coil})}{dt}\right]$ opposes the change.

Interpoles – to aid current reversal in commutating coil narrow poles are placed in magnetic neutral region with polarity such that the speed emf induce in the coil opposes reactance emf.

LO 4

- *Excitation of machine poles* Separate excitation from independent source, shunt excitation (from armature voltage), series excitation (from armature current), compound excitation (combined); it could be cumulative compound (series excitation aids shunt excitation) or differential compound (series excitation opposes shunt excitation, not used in practice)
- *Machine types as per method of excitation*
  Generator: shunt, series, compound
  Motor : shunt, series, compound
  Compound connection – long shunt and short shunt
- OCC (Open Circuit Characteristic), V_{OC} (= E_a) *vs* I_f. Machine is run as separately excited generator at constant speed. This indeed is the *magnetization characteristic*.

LO 5

- Separately excited dc generators have the advantage of permitting a wide range of output voltages. But self-excited machines may produce unstable voltages at lower output voltages, where the field resistance line becomes essentially tangent to the magnetization curve.

LO 7

- Cumulative compounded generators may produce a substantially flat voltage characteristics or one which rises with load, where as shunt or separately excited generators may produce a drooping voltage characteristics.

LO 8

- *Shunt Motor*

$$n = \frac{1}{K'_o}\frac{V_t - I_aR_a}{\Phi}, \; T = K_a \Phi I_a;$$

$$K'_a = \left(\frac{2\pi}{60}\right)K_a$$

The speed n decreases slightly as load torque increases and so I_aR_a drop increases. This is typical slightly drooping shunt characteristic. However, armature reaction reduces Φ due to saturation effect. This counters drop in speed.

Speed Control

$$n = \frac{1}{K'_o}\frac{V_t}{\Phi}, \; I_aR_a \text{ ignored}$$

Field Control As shunt field current is reduced, Φ reduces and n increases and torque developed reduces. Constant-kW control.

Armature Control As V_t is increased with Φ_{max}, n increases. The machine develops constant torque. Constant – T control

- *Series Motor* As the series field is excited by armature current

$$n = \left(\frac{1}{K'_o K_f}\right)\frac{V_t}{I_a},$$

$$T = K_a\,\Phi I_a = (K_aK_f)I_a^2, \text{ on linear basis}$$

The speed reduces sharply with load torque (I_a increases). This is typical series characteristics. To be noted that at no-load I_a becomes zero and speed tends to infinite. The series motor should never be light loaded. Ideal for traction-type loads.

Speed Control

Field Control Φ is controlled by tapping series field turns or by a diverter resistance across the series field.

Armature Control V_t is controlled in two steps by series-parallel connection of two mechanically coupled identical series motors.

- *Compound Motor (cumulative)* Speed-torque characteristic lies in between shunt and series characteristic depending upon compounding. The motor has a definite no-load speed determined by the shunt field.
- *Motor starting* At start E_a = 0, direct starting current is unacceptably high. A series resistance is included in motor armature circuit and is cutout in steps as the motor speeds up.
- *Motor efficiency* Constant loss, P_k = core loss, windage and friction loss, shunt field loss
  Variable loss, armature copper loss $I_a^2 R_a$
  At η_{max}
  Variable loss = Constant loss

$$I_a^2R_a = P_k \quad \text{or} \quad I_a = \sqrt{\frac{P_k}{R_a}}$$

Relationships to remember

$$n \propto \frac{E_a}{\Phi}, \; T \propto \Phi I_a$$

If $\Phi \propto I_f$ (linear magnetization)

$$n \propto \frac{E_a}{I_f},\ T \propto I_f I_a,\quad T \propto I_a^2 \text{ (for series motor)}$$

Approximation

On no-load, $E_a \approx V_t$

This approximation may be used even on load where less accuracy is acceptable.

LO 9

Braking of dc motors is process of controlled slowing or stopping of a dc motor and its driven load and is highly useful in many applications like cranes, tractions, etc to avoid excessive speeds.

LO 10

Swinburne's and Hopkinson's test called as non-loading tests is commonly used in dc shunt motors for predetermination of efficiency.

LO 11

With solid-state controls the speeds and output of dc machines can be controlled easily over a wide range for both dynamic and steady-state operation. By addition of the feedback circuit, the machine characteristics can be further modified.

PRACTICE PROBLEMS

7.1 A compensated dc machine has 20000 AT/pole. The ratio of the pole arc to pole pitch is 0.8. The interpolar air-gap length and flux density are respectively 1.2 cm and 0.3 T. For rated I_a = 1000 A, calculate the compensating winding AT per pole and the number of turns on each interpole. LO 3

7.2 The no-load saturation curve for a generator operating at 1800 rpm is given by the following data LO 5

E_g (V)	8	40	74	113	152	213	234	248	266	278
I_f (A)	0	0.5	1.0	1.5	2.5	3.0	3.5	4.0	5.0	6.0

(a) Plot the no-load saturation curve for 1500 rpm.

(b) Calculate the generated voltage when the generator is operating on no-load with a field current of 4.6 A and at a speed of 1000 rpm.

(c) What is the field current required to generate 120 V on no-load when the generator is operating at 900 rpm?

(d) This machine is operated as a shunt generator at 1800 rpm with a field current of 4.6 A. What is the no-load voltage when the generator is operating at 1500 rpm?

7.3 The accompanying data are given for the saturation curve of an 80-kW, 220-V, 1200-rpm shunt generator, the data being for 1200 rpm: LO 5

I_f (A)	0	0.4	0.8	1.2	1.6	2.0	2.5	3.2	4.0	4.5	5.0	5.5
E_g (V)	10	38	66	96	128	157	188	222	248	259	267	275

(a) The shunt field resistance is adjusted to 50 Ω and the terminal voltage is found to be 250 V, at a certain load at 1200 rpm. Find the load supplied by the generator and the induced emf. Assume that the flux is reduced by 4% due to armature reaction. Armature resistance is 0.1 Ω.

(b) For the same field resistance and an armature current of 250 A obtain the values of E_g, V_t and I_f.

7.4 Find the resistance of the load which takes a power of 5 kW from a shunt generator whose external characteristic is given by the equation $V_t = (250 - 0.5I_L)$ LO 5

7.5 A dc shunt generator rated 175 kW, 400 V, 1800 rpm is provided with compensating winding. Its magnetization data at 1800 rpm is given below: LO 5

I_f (A)	1	2	3	4	5	6	7	8	9
V_{OC} (V)	100	200	300	370	415	440	765	475	480

Other data of the generator are

$R_a = 0.05\ \Omega$, $R_f = 20\ \Omega$, $R_{ext} = 0$ to 300

when R_{ext} = regulating resistance in the shunt field.

The generator connected as shunt is run by a prime-mover at 1600 rpm.
(a) Find the value of R_{ext} for the no-load voltage to be 400 V.
(b) A load resistance of 1 Ω is connected across the generator terminals. What should be the value of R_{ext} for the load voltage to be 400 V.
(c) What would be the generator terminal voltage when the load resistance is disconnected?

7.6 In a 110 V compound generator, the resistance of the armature, shunt and series windings are 0.06, 25 and 0.05 Ω respectively, The load consists of 200 lamps each rated at 55 W, 100 V. Find the emf and armature current, when the machine is connected for (a) long shunt (b) short shunt (c) How will the ampere-turns of the series windings be changed, if in (a) a diverter of resistance 0.1 Ω is connected across the series field? Ignore armature reaction and brush voltage drop. LO 5

7.7 A dc shunt generator has the following open-circuit characteristic when separately excited: LO 5

Field current, A	0.2	0.4	0.6	0.8	1.0	1.4	2.0
Emf, V	80	135	178	198	210	228	246

The shunt winding has 1000 turns per pole and a total resistance of 240 Ω. Find the turns per pole of a series winding that will be needed to make the terminal voltage the same at 50 A output as on no-load. The resistance of the armature winding, including the series compounding winding, can be assumed to be 0.36 Ω and constant. Ignore armature reaction.

7.8 A dc compound generator has a shunt field winding of 3600 turns per pole and a series field winding of 20 turns per pole. Its open-circuit magnetization characteristic when it is separately excited by its shunt field winding and driven at its no-load rated speed is given below: LO 5

AT/pole	3120	4680	6240	7800	9360
Emf (V)	289	361	410	446	475

The full-load armature current is 100 A and the ohmic drop in the armature circuit for this current is 20 V including brush drop. At no load the ohmic drop may be ignored and the terminal voltage is 415 V. The fall in speed from no-load to full-load is 8%; the shunt field circuit is connected across the output terminals of the machine and its resistance is kept constant.
Determine the terminal voltage and power output for the full-load armature current of 100 A. Neglect the effects of armature reaction.

7.9 A 250 kW, 6-pole, dc compound generator is required to give 500 V on no-load and 550 V on full-load. The armature is lap-connected and has 1080 conductors; the total resistance of the armature circuit is 0.037 Ω. The open-circuit characteristic for the machine at rated speed is given by: LO 5

Armature voltage, V	500	535	560	580
Field ampere-turns/pole	6000	7000	8000	9000

The field ampere-turns per pole to compensate for armature reaction are 10% of the armature ampere-turns per pole. The shunt field winding is connected across the output terminals and has a resistance of 85 Ω. Determine the required number of series turns per pole.

7.10 A 10 kW, 250 V shunt motor has an armature resistance of 0.5 Ω and a field resistance of 200 Ω. At no load and rated voltage, the speed is 1200 rpm and the armature current is 3 A. At full load and rated voltage, the line current is 47 A and because of armature reaction, the flux is 4% less than its no-load value. LO 2
(a) What is the full-load speed?
(b) What is the developed torque at full load?

7.11 A 75 kW, 250 V, dc shunt motor has the magnetisation data at 1200 rpm as below: LO 7

I_f(A)	1	2	3	4	5	6	7	8
E_a(V)	70	130	183	207	250	274	292	307

No-load test on the motor conducted at 1100 rpm yielded rotational loss (no-load iron loss + windage and frictional loss) as 2200 W. Stray load loss can be taken as 1% of output. Given: $R_a = 0.025$ Ω

(a) Determine the motor speed and mechanical (net) output at an armature current of 300 A. Assume that armature reaction causes 5% voltage reduction in induced emf.
(b) Shunt field turns/pole = 1200. Two series turns are added in cumulative compound. Assuming R_{se} to be negligible, calculate speed and power output.

7.12 A 200 V shunt motor has $R_a = 0, 1\ \Omega$, $R_f = 240\ \Omega$ and rotational loss 236 W. On full-load, the line current is 9.8 A with the motor running at 1450 rpm. Determine: (a) the mechanical power developed (b) the power output (c) the load torque, and (d) the full-load efficiency. LO 7

7.13 A 220 V unsaturated shunt motor has an armature resistance (including brushes and interpoles) of 0.04 Ω and the field resistance of 100 Ω. Find (a) the value of resistance to be added to the field circuit to increase the speed from 1200 to 1600 rpm, when the supply current is 200 A, (b) With the field resistance as in (a), find the speed when the supply current is 120 A. If the machine is run as a generator to give 200 A at 220 V, find (c) the field current at 1300 rpm, and (d) the speed when the field current is 2 A. LO 7

7.14 Derive the standard torque equation of a dc motor, from first principles. A 4-pole series motor has 944 wave-connected armature conductors. At a certain load the flux per pole is 34.6 mWb and the total mechanical power developed is 4 kW. Calculate the line current taken by the motor and the speed at which it will run with an applied voltage of 500 V. The total motor resistance is 3 Ω. LO 7

7.15 The following data pertain to 250 V dc series motor: LO 8

$Z = 180,\ \dfrac{P}{A} = 1$

Flux/pole = 3.75 m Wb/field amp

Total armature circuit resistance = 1 Ω

The motor is coupled to centrifugal pump whose load torque is

$T_L = 10^{-4}\, n^2$ Nm

where n = speed in rpm.

Calculate the current drawn by the motor and the speed at which it will run.

7.16 A dc shunt motor is being operated from 300 V mains. Its no-load speed is 1200 rpm. When fully loaded, it delivers a torque of 400 Nm and its speed drops to 1100 rpm. Find its speed and power output when delivering the same torque; if operated with an armature voltage of 600 V. Excitation is assumed unchanged, i.e. the motor field is still excited at 300 V. State any assumption you are required to make. LO 2 LO 4

7.17 A 50 kW, 230 V dc shunt motor has an armature resistance of 0.1 Ω and a field resistance of 200 Ω. It runs on no-load at a speed of 1400 rpm, drawing a current of 10 A from the mains. When delivering a certain load, the motor draws a current of 200 A from the mains. Find the speed at which it will run at this load and the torque developed. Assume that the armature reaction causes a reduction in the flux/pole of 4% of its no-load value. LO 8

7.18 A 250 V dc series motor has the following OCC at 1200 rpm: LO 7

I_f(A)	5	10	15	20	25	30
V_{OC}(V)	100	175	220	240	260	275

$R_a = 0.3\ \Omega$ and series field resistance is 0.3 Ω. Find the speed of the machine when (a) $I_a = 25$ A and (b) the developed torque is 40 Nm.

7.19 A 15 kW, 250 V, 1200 rpm shunt motor has 4 poles, 4 parallel armature paths, and 900 armature conductors; $R_a = 0.2\ \Omega$. At rated speed and rated output the armature current is 75 A and $I_f = 1.5$ A. Calculate (a) flux/pole, (b) the torque developed, (c) rotational losses, (d) η, (e) the shaft load, and (f) if the shaft load remains fixed, but the field flux is reduced to 70% of its value by field control, determine the new operating speed. LO 7

7.20 A 115 kW, 600 V dc series wound railway track motor has a combined, field and armature resistance (including brushes) of 0.155 Ω. The full-load current at rated voltage and speed is 216 A. The magnetization curve at 500 rpm is as follows. LO 7

EMF (V)	375	400	425	450	475
I_f (A)	188	216	250	290	333

(a) Neglecting armature reaction, calculate the speed in rpm at the rated current and voltage.
(b) Calculate the full-load internal (developed) torque.
(c) If the starting current is to be restricted to 290 A, calculate the external resistance to be added in the motor circuit and the starting torque.

7.21 A 100 kW, 600 V, 600 rpm dc series wound railway motor has a combined field and armature resistance (including brushes) of 0.155 Ω. The full-load current at the rated voltage and speed is 206 A. The magnetization curve at 400 rpm is as follows: LO 7

I_f (A)	188	206	216	250	290	333
EMF (V)	375	390	400	425	450	475

(a) Determine the armature reaction in equivalent demagnetizing field current at 206 A.
(b) Calculate the internal (developed) torque at the full-load current.
(c) Assuming demagnetizing armature reaction mmf proportional to (I_a^2). determine the internal starting torque at the starting current of 350 A.

7.22 A 3 kW series motor runs normally at 800 rpm on a 240 V supply, taking 16 A; the field coils are all connected in series. Estimate the speed and current taken by the motor if the coils are reconnected in two parallel groups of two in series. The load torque increases as the square of the speed. Assume that the flux is directly proportional to the current and ignore losses. LO 7

7.23 A 20 kW, 500 V shunt motor has an efficiency of 90% at full load. The armature copper-loss is 40% of the full-load loss. The field resistance is 250 Ω. Calculate the resistance values of a 4-section starter suitable for this motor in the following two cases: LO 8

Case 1: Starting current ≤ $2I_{fl}$

Case 2: Starting current (min) = 120% I_{fl}.

7.24 A starter is to be designed for a 10 kW, 250 V shunt motor. The armature resistance is 0.15 Ω. This motor is to be started with a resistance in the armature circuit so that during the starting period the armature current does not exceed 200% of the rated value or fall below the rated value. LO 8

That is, the machine is to start with 200% of armature current and as soon as the current falls to the rated value, sufficient series resistance is to be cut out to restore current to 200% (or less in the last step).

(a) Calculate the total resistance of the starter.
(b) Also calculate the resistance to be cut out in each step in the starting operation.

7.25 A dc motor drives a 100 kW generator having an efficiency of 87%. LO 7

(a) What should be the kW rating of the motor?
(b) If the overall efficiency of the motor generator set is 74%, what is the efficiency of the motor?
(c) Also calculate the losses in each machine.

7.26 A 600 V dc motor drives a 60 kW load at 900 rpm. The shunt field resistance is 100 Ω and the armature resistance is 0.16 Ω. If the motor efficiency is 85%, determine: LO 7

(a) the speed at no-load and the speed regulation.
(b) the rotational losses.

7.27 Enumerate the principal losses that occur in a dc generator, and where appropriate, state the general form of the physical law upon which each loss depends. Calculate the efficiency of a self-excited dc shunt generator from the following data; Rating: 10 kW, 250 V, 1000 rpm. LO 7

Armature resistance = 0.35 Ω

Voltage drop at brushes = 2 V
Windage and friction losses = 150 W
Iron-loss at 250 V = 180 W

Open circuit characteristic:

EMF (V):	11	140	227	285	300	312
Field current (A)	0	1.0	1.5	2.0	2.2	2.4

7.28 A 60 kW, 252 V shunt motor takes 16 A when running light at 1440 rpm. The resistance of armature and field is 0.2 and 125 Ω respectively when hot. (a) Estimate the efficiency of the motor when taking 152 A. (b) Also estimate the efficiency if working as a generator and delivering a load current of 152 A at 250 V. **LO 10**

7.29 A 200 V shunt motor takes 10 A when running on no-load. At higher loads the brush drop is 2 V and at light loads it is negligible. The stray-load loss at a line current of 100 A is 50% of the no-load loss. Calculate the efficiency at a line current of 100 A if armature and field resistances are 0.2 and 100 Ω respectively. **LO 10**

7.30 Hopkinson's test on two machines gave the following results for full load; line voltage 230 V, line current excluding field current 50 A; motor armature current 380 A; field currents 5 and 4.2 A. Calculate the efficiency of each machine. The armature resistance of each machine = 0.02 Ω. State the assumptions made. **LO 10**

7.31 Calculate the efficiency of a 500 V shunt motor when taking 700 A from the following data taken when the motor was hot: motor stationary, voltage drop in the armature winding 15 V, when the armature current was 510 A; field current 9 A at normal voltage. Motor running at normal speed unloaded; the armature current was 22.5 A, when the applied voltage was 550 V; allow 2 V for brush contact drop and 1 % of the rated output of 400 kW for stray-load losses. **LO 10**

7.32 A 480 V, 20 kW shunt motor took 2.5 A, when running light. Taking the armature resistance to be 0.6 Ω, field resistance to be 800 Ω and brush drop 2 V, find the full-load efficiency. **LO 10**

7.33 The open-circuit characteristic for a dc generator at 1200 rpm is: **LO 5**

Field current (A)	0	2	4	6	8
Armature voltage (V)	13	140	230	285	320

Estimate the critical field resistance at this speed. Keeping the field resistance fixed at this speed, the generator is run at 1600 rpm. What would be the no-load voltage?
The generator is now loaded to an armature current of 500 A. Estimate the terminal voltage and line current (at 1600 rpm). Neglect armature reaction effects.
Given: Armature resistance = 0.05 Ω,
Brush voltage drop = 2 V

7.34 The open-circuit voltage of a dc shunt generator is 130 V. When loaded the voltage drops to 124 V. Determine the load current. **LO 2**

Given: $R_a = 0.03\ \Omega$, $R_f = 20\ \Omega$

7.35 A 220 V dc shunt motor runs at 1400 rpm on no-load drawing an armature current of 2.4 A from the supply. Calculate the motor speed for an armature current of 60 A. Assume that the flux/pole reduces by 4% armature reaction. It is given that $R_a = 0.24\ \Omega$. **LO 7**

7.36 A dc shunt machine has an open-circuit voltage of 220 V at 1200 rpm. The armature resistance is 0.2 Ω and field resistance is 110 Ω. Calculate its speed as a motor when drawing a line current of 60 A from 230 V mains. Assume that armature reaction demagnetizes the field to the extent of 5%. **LO 8**

7.37 The full-load current in the armature of a shunt motor is 100 A, the line voltage being 400 V, the resistance of the armature circuit 0.2 Ω and the speed 600 rpm. What will be the speed and line current if the total torque on the motor is reduced to 60% of its full-load value and a resistance of 2 Ω is included in the armature circuit, the field strength remaining unaltered. **LO 2**

7.38 230 V dc is shunt motor has an armature resistance of 0.25 Ω. What resistance must be added in series with the armature circuit to limit the starting current 90 A? With this starting resistance in circuit, what would be the back emf when the armature current decreases to 30 A? LO 8

7.39 For the hoist drive system shown in Fig. P.7.39. LO 7

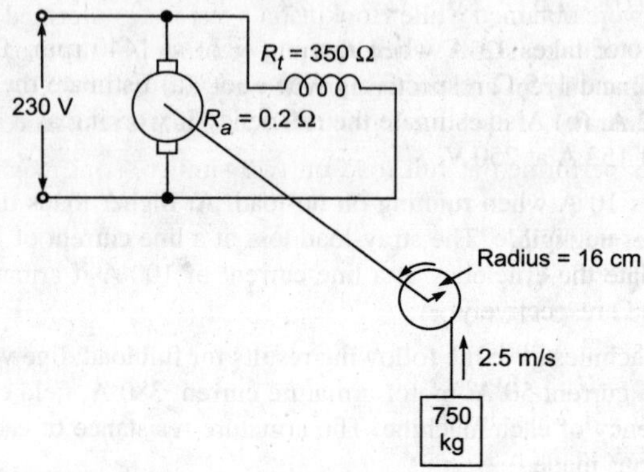

Fig. P.7.39

(a) Calculate the minimum size of the motor.
(b) If the line voltage drops to 180 V, what would be the hoisting speed?

7.40 A 400 V dc shunt motor takes a current of 5.6 A on no-load and 68.3 A on full-load. The load current weakens the field by 3%. Calculate the ratio of full-load speed to no-load speed. Given: $R_a = 0.18\ \Omega$, brush voltage drop = 2 V, $R_f = 200\ \Omega$. LO 8

7.41 What resistance must be added in series with the armature of the shunt motor of Prob. 7.40 to reduce the speed to 50% of its no-load value with gross torque remaining constant. The shunt field resistance also remains unchanged. LO 8

7.42 A 200 V dc series motor yielded the following operational data: LO 7

Speed (rpm)	640	475	400
Current (A)	20	30	40

Find the speed at which the motor will run when connected to 200 V mains with a series resistance of 2 Ω while drawing 35 A. Armature circuit resistance = 1.2 Ω.

7.43 A series motor takes 50 A at 400 V while hoisting a load at 8 m/s. The total series resistance of the armature circuit is 0.46 Ω. What resistance must be added in series with the armature circuit to slow the hoisting speed down to 6 m/s? Assume magnetic linearity. What other assumption you need to make. LO 7

7.44 The magnetization curve of a 4-pole dc series motor with a 2-circuit winding was obtained by separately exciting the field with currents of the values given below, and loading the armature to take the same current. the speed being maintained constant at 800 rpm: LO 7

Current (A)	10	20	30	40	50	60	70	80
Voltage between brushes (V)	160	295	375	425	460	485	505	525

Determine the speed and torque for this motor for currents of 10 A, 50 A and 80 A for a line voltage of 600 V, having been given that the total number of conductors is 660, the resistance of the armature circuit is 0.30 Ω and that of the field coils is 0.25 Ω.

7.45 A dc series motor has the following magnetization characteristic at 800 rpm: LO 7

Field current (A)	0	40	80	120	160	200
OC voltage (V)	40	390	680	910	1080	1220

Armature resistance = 0.2 Ω

Field resistance = 0.2 Ω
Brush voltage drop = 2 V (total)
The machine is connected to 750 V mains. If a diverter of resistance 0.2 Ω is connected across the field windings, calculate (a) the speeds, and (b) the torques developed at armature currents of 80 A and 120 A respectively.

7.46 The following test results were obtained while Hopkinson's test was performed on two similar dc Shunt Machines: supply voltage is 250 V, field current of generator and motor was 6 A and 5 A respectively and the line current and motor current were 50 A and 400 A. Calculate the efficiency of motor and generator of each having 0.0150 Ω armature resistance. **LO 10**

7.47 The Hopkinson's test was performed at full load on two similar shunt machines. The test results are: **LO 10**

Line voltage = 110 V
Line current = 48 A,
Armative current = 230 A (motor)
Field currents = 3 A and 5 A
Armature resistance = 0.035 Ω (each)
Brush contact drop = 1 V per brush

Calculate the efficiency of both the machines.

7.48 At no load, a dc shunt motor takes 5 A from 400 V supply. Armature and field resistance of that shunt motor is 0.5 Ω, 200 Ω. Find the efficiency of the motor when it takes 60 A line current. **LO 10**

7.49 The following test results are obtained while Swinburne's test is performed on dc shunt motor with input voltage is 200 V. At no load the input power is 1.1 kW, the current is 5.5 A, and the speed is 1150 rpm. Calculate the efficiency at 50 A load. Given: armature resistance is 0.6 Ω and shunt field resistance is 110 Ω. **LO 10**

7.50 A 250 V dc shunt motor takes 5 A at no load. The armature resistance including brush contact resistance is 0.2 Ω, field resistance is 250 Ω. Calculate the output and efficiency when input current is 20 A. **LO 10**

7.51 Determine the minimum speed at which the motor can hold the load by means of regenerative braking. The following parameters are: input voltage is 250 V, armature resistance is 0.1 Ω with full load condition. It has an emf of 240 V at a speed of 1200 rpm. The motor is driving an overhauling load with a torque of 180 Nm. **LO 9**

7.52 A series motor has an armature resistance of 0.7 Ω and field resistance of 0.3 Ω. It takes a current of 15 A from a 200 V supply and runs at 800 rpm. Find the speed at which it will run when a 5 Ω resistance is added in series to the motor and it takes the same current at the supply voltage. **LO 2**

7.53 Find the efficiency of a long shunt compound generator rated at 250 kW, 230 V when supplying 76% of rated load at rated voltage. The resistance of armature and series field are 0.009 Ω and 0.003 Ω. The shunt field current is 13 A. When the machine is run as a motor at no load the armature current is 25 A at rated voltage? **LO 10**

7.54 A PMDC mator connected to 48 V supply runs on no-load at 2000 rmp drawing 1.2 A. It is found to have an armature resistance of 1.02 Ω. **LO 11**
(a) Calculate torque constant and no-load relational loss
(b) It is now connected to 50 V supply and runs at 2200 rpm.
Calculate its shaft power output and electromagnetic torque

7.55 A separately excited dc generator has OCC data at 1200 rmp as below. **LO 5**

V_{OC}	10	54	160	196	248	284	300	360
I_f (A)	0	1	2	3	4	5	6	7

At $I_f = 6$ A when loaded at $I_a = 50$ A, it has a terminal voltage of 230 V at speed of 1200 rpm.

Given: $R_a = 0.5\ \Omega$.

(a) Determine the armature voltage drop due to the demagnetizing effect of armature reaction

(b) Assuming that this voltage drop in proportional to the square of armature current, determine the field current for a terminal voltage of 230 V at armature current of $I_a = 40$ A

(c) In part (a) determine the field current equivalent of demagnetization

7.56 A 260 V dc shunt mator has armature resistance of $R_a = 0.4\ \Omega$ and field resistance of $R_f = 50\ \Omega$. It is drawing full load armature current of $I_a = 50$ A. Which has a demagnetizing field current equivalent of 0.8 A. Calculate the motor speed. The mator's OCC data at 1200 rpm is as given in Problem 7.55. **LO 7**

7.57 Design a suitable double-layer lap winding for a 6-pole armature with 18 slots and two coil-sides per slot. Give the values of front-pitch, back-pitch and commutator pitch. Draw the developed diagram and show the positions of brushes. **LO 1**

7.58 Design a suitable double-layer wave winding for a 4-pole, 13-slot dc armature with two coil-sides per slot. Give values of pitches, Draw the developed diagram and show the positions of brushes. **LO 1**

7.59 A 6-pole, dc armature with 36 slot and 2 coil-sides per slot is to be wave-wound (double-layer). Calculated the back-pitch, front-pitch and commutator-pitch. How many dummy coils, if any, are required? Draw the developed diagram of the winding and show the locations of brushes and the distance between them in terms of commutator segments. If (a) 2 brushes are used, and (b) 6 brushes are used. **LO 1**

7.60 A 25 kW dc generator has a 6-pole lap-connected armature of 312 conductors. How would you change the connections to the commutator to form two armature circuits and what effect would this change have on the voltage, current and output of the machine? **LO 3**

Answers

7.1 $N_t = 23$

7.2 (b) 144.4 V (c) 3.7 A (d) 195 V

7.3 (a) 14.5 kW (b) 224 V, 50 V; 249 V, 75 V; 4.48 A, 1 A

7.4 11.98 Ω

7.5 (a) 44 Ω (b) 23.3 Ω (c) 425 V

7.6 (a) 121.5 V (b) 104.4 A, 120.28 V (c) 66.6%

7.7 6 turns **7.8** 403 V, 39.6 kW **7.9** 5 turns

7.10 (a) 1142 rpm (b) 86.8 N m

7.11 (a) 1123 rpm, 69.85 kW (b) 1086 rpm, no change in power

7.12 (a) 1.786 kW (b) 1.55 kW (c) 16.03 N m (d) 79.1%

7.13 (a) 33.3 W (b) 1624 rpm (c) 2.185 A (d) 1419 rpm

7.14 8.43 A, 436 rpm **7.15** 24.7 A, 810 rpm **7.16** 2300 rpm, 96.35 kW

7.17 1337 rpm, 298.4 N m

7.18 (a) 1085 rpm (b) 1174 rpm

7.19 (a) 0.013 Wb (b) 78.4% (c) 2.56 kW (d) 78.4% (e) 15 kW (f) 2177 rpm

7.20 (a) 708.1 rpm (b) 1650 N m (c) 1.91 Ω, 2,492 N m

7.21 (a) 13 A (b) 1862 N m (c) 3885 N m

7.22 951 rpm, 26.9 A

7.23 Case 1: 2.94 Ω, 1.63 Ω, 0.9 Ω, 0.5 Ω; Case 2: 2.7 Ω, 1.52 Ω, 0.85 Ω, 0.48 Ω

7.24 (a) 2.975 Ω (b) 1.56 Ω, 0.78 Ω, 0.39 Ω, 0.15 Ω

7.25 (a) 115 kW (b) 85% (c) 20.3 kW

7.26 (a) 927.6 rpm, 3.1%, (b) 5 kW **7.27** 86.96%

7.28 (i) 77.73% (ii) 81.36%

7.29 74.48% **7.30** 91.99%, $\eta_G = 92.05\%$

7.31 89.7% **7.32** 88.6% **7.33** 330 V, 494.8 A

7.34 193.8 A **7.35** 1379 rpm **7.36** 1137 rpm

7.37 423 rpm **7.38** 153.3 V

7.39 (a) 18.4 kW (b) 1.96 m/sec.

7.40 1 **7.41** 2.92 Ω **7.42** 242.8 rpm

7.43 1.885 Ω

7.44 (a) 3029 rpm, 1.7 N m (b) 1029 rpm, 24.14 N m (c) 888 rpm, 43.48 N m.

7.45 (a) 852 rpm, 649 N m (b) 626 rpm, 1303 N m.

7.46 92.31%, 92.13% **7.47** 87.06%, 87.85% **7.48** 79.67%

7.49 75.16% **7.50** 66.12% **7.52** 475 rpm

7.53 91.7%

7.54 (a) 0.223 V/rad/s, 56 W (b) 114.8 W, 0.792 N m

7.55 (a) 20 V (b) ≈ 5.5 A (c) 0.9 A.

7.56 960 rpm **7.57** (a) 5, 7, 6 (b) 1-13-25, 7-19-31

7.58 7, 7 **7.59** $y_b = 13$, $y_f = 11$, $y_c = 12$, one dummy coil

7.60 $E_{wave}/E_{lap} = 2.98$, $I_{a,wave}/I_{a,lap} = 1/3$, $P_{wave}/P_{lap} = 0.993$

MULTIPLE CHOICE QUESTIONS

1. Why is the armature of a dc machine made of silicon steel stampings? LO 1 ○○●
(a) To reduce hysteresis loss
(b) To reduce eddy current loss
(c) For the ease with which the slots can be created
(d) To achieve high permeability.

2. Slot wedges in a dc machine are made of LO 1 ○○●
(a) mild steel (b) silicon steel

3. The armature reaction *AT* in a dc machine LO 5 ○●●
(a) are in the same direction as the main poles
(b) are in direct opposition to the main poles
(c) make an angle of 90° with the main pole axis
(d) make an angle with the main pole axis which is load dependent.

4. A dc series motor has linear magnetization and negligible armature resistance. The motor speed is LO 7 ○●●
(a) directly proportional to $\sqrt{T}$; T = load torque (b) inversely proportion a to $\sqrt{T}$
(c) directly proportional to T (d) inversely proportional to T

5. The armature reaction mmf in a dc machine is LO 3 ○○●
(a) sinusoidal in shape (b) trapezoidal in shape
(c) rectangular in shape (d) triangular in shape.

6. What losses occur in the teeth of a dc machine armature? LO 1 ●●●
(a) Hysteresis loss only (b) Eddy current loss only
(c) Both hysteresis and Eddy current loss (d) No losses

7. The process of current commutation in a dc machine is opposed by the LO 3
(a) emf induced in the commutating coil because of the inter-pole flux
(b) reactance emf
(c) coil resistance
(d) brush resistance

8. In a level compound generator the terminal voltage at half full-load is LO 5
(a) the same as the full load voltage (b) the same as no load voltage
(c) is more than the no-load voltage (d) is less than the no-load voltage

9. Field control of a dc shunt motor gives LO 8
(a) constant torque drive (b) constant kW drive
(c) constant speed drive (d) variable load speed drive

10. In a series-parallel field control of a dc series motor with fixed armature current LO 8
(a) such connections are not used in practice (b) both series and parallel connections give the same speed
(c) series connection gives higher speed (d) parallel connection gives higher speed

11. If the magnetic circuit of dc machine is in saturation region, the armature reaction LO 3
(a) does not affect the flux/pole (b) increases flux/pole.
(c) decreases flux/pole (d) affects the flux/pole only when armature current in very small.

12. In Hopkinson's test LO 10
(a) iron loss in both machines are equal
(b) iron loss in motoring machine is more than in generating machine
(c) iron loss in generating machine is more than in motoring machine
(d) only stray load iron loss is equal in both machine.

13. In a dc motor electromagnetic torque is LO 2
(a) ω_m/E_aI_a, in the direction of ω_m (b) ω_m/E_aI_a, opposite to the direction of ω_m
(c) E_aI_a/ω_m, opposite to the direction of ω_m (d) E_aI_a/ω_m, in the direction of ω_m.

14. If field current is decreased in a shunt dc motor, the speed of the motor ______. LO 8
(a) remains same (b) increases
(c) decreases (d) None of these

15. When is the mechanical power developed by a dc series motor maximum? LO 1
(a) When back emf is equal to half the applied voltage
(b) When back emf is equal to the applied voltage
(c) When back emf is zero
(d) None of these

16. Ward-Leonard control is basically a ________ control method. LO 8
(a) armature resistance control (b) field control
(c) armature voltage control (d) field diverter control

17. Which of the following windings is necessary in case of all dc machines? LO 1
(a) Lap winding (b) Wave winding
(c) Closed winding (d) Open type winding

18. Dummy coils are used in dc machines for ______. LO 1
(a) increasing efficiency (b) reducing armature reaction
(c) mechanical balancing (d) All of these

19. Function of _____ is to collect current from the commutator and supply it to the external load. LO 1
(a) field magnet (b) armature
(c) brushes (d) yoke

20. A motor converts ____________. LO 2
(a) mechanical energy into electrical energy (b) chemical energy into electrical energy

(c) electrical energy into Mechanical energy (d) electrical energy into chemical energy

21. The mechanical losses are about _____ % of full load losses. LO 10
(a) 0 to 10 (b) 10 to 20
(c) 20 to 30 (d) 30 to 40

22. Which motor cannot be started on no-load? LO 10
(a) Shunt motor (b) Series motor
(c) Cumulative compound motor (d) Differential compound motor

23. Field weakening control method is used for ___________. LO 8
(a) Above rated speed (b) Below rated speed
(c) Both A and B (d) None of these

24. Which of the following represents the rotating losses of machine? LO 10
(a) Eddy current losses (b) Hysteresis losses
(c) Friction and windage losses (d) All of these

25. Swinburne test is conducted under which of the following conditions? LO 10
(a) No-load (b) Full-load
(c) Half-load (d) Any of these

26. Dynamic braking can be used for______. LO 9
(a) series motors (b) shunt motors
(c) compound motors (d) All of these

27. Which of following windings is connected in series with armature winding of a dc machine? LO 2
(a) Auxiliary winding
(b) Compensating winding
(c) Both compensating and commutating pole winding
(d) Commutating pole winding

28. With increase in speed of a dc motor, LO 7
(a) back emf increases but line current falls
(b) back emf falls and line current increases
(c) both back emf as well as line current increase
(d) both back emf as well as line current fall

29. The output power of an electric motor is taken from the _____________. LO 1
(a) field (b) armature
(c) coupling mounted to shaft (d) motor frame

30. Commutator in a dc generator is used for __________. LO 1
(a) collecting of current (b) increasing efficiency
(c) reducing losses (d) converting ac armature current into dc

31. The effect of _________ on main field flux is an armature reaction. LO 8
(a) armature mmf (b) armature current
(c) armature flux (d) All of these

32. In a dc machine, torque depends on which of the following? LO 8
(a) Flux (b) Armature current
(c) Both A and B (d) Speed

33. Which of the following generators are used in arc welding? LO 2
(a) Shunt generators (b) Series generators
(c) Cumulative compound generators (d) Differential compound generators

34. An essential condition for connecting two dc generators in parallel is that LO 6
(a) terminal voltage should be same (b) polarities should be same
(c) rating of generators should be same (d) Both A and B

35. How is a dc machine preferred for battery charging? LO 11

(a) As dc series generator (b) As dc shunt generator
(c) As compound dc generator (d) None of these

Answers

1. (d)	**2.** (b)	**3.** (c)	**4.** (b)	**5.** (d)	**6.** (b)	**7.** (b)	**8.** (b)	**9.** (b)	**10.** (d)
11. (c)	**12.** (a)	**13.** (c)	**14.** (b)	**15.** (a)	**16.** (c)	**17.** (c)	**18.** (c)	**19.** (c)	**20.** (c)
21. (b)	**22.** (b)	**23.** (a)	**24.** (c)	**25.** (a)	**26.** (d)	**27.** (c)	**28.** (a)	**29.** (c)	**30.** (d)
31. (c)	**32.** (c)	**33.** (d)	**34.** (c)	**35.** (b)					

MATLAB PROGRAM FOR NO-LOAD SATURATION CURVES FOR A DC GENERATOR AT DIFFERENT RPMS

```
Eg1=[8 40 74 113 152 213 234 248 266 278]
If=[0 0.5 1 1.5 2 3 3.5 4 5 6]
plot (If,Eg1)
hold
Eg2=(1500/1800) * Eg1
plot (If,Eg2)
Eg3=(1000/1800)*Eg1
plot(If,Eg3)
legend (1800 rpm', 1500 rpm', '1000 rpm')
```

Plot is shown in the below figure.

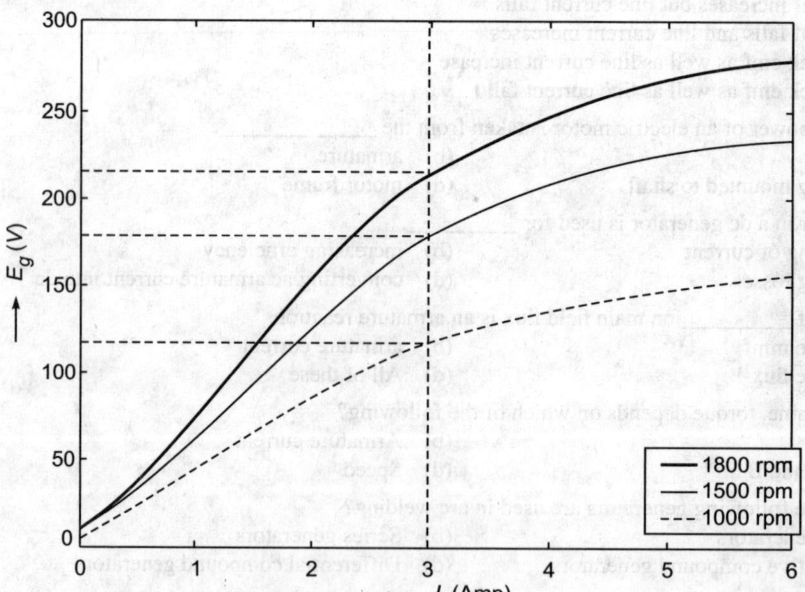

SAMPLE EXAMPLES SOLVED USING MATLAB

```
% This program solves Example 7.3 of the text
% given data are
NG=750;
PI=I00;
V=230;
P2=9;
Ra=.08;
Rf=115;
IF=V/Rf;
IL1=PI *1000/V;
IA1=IF+IL1 ;
EAI=V+Ra*IA1 ;
% EAI calculated in the text seems incorrect
IL2=P2*I000/V;
IA2=IL2- IF;
EA2=V-Ra*IA2;
NM=EA2*NG/EA1;
fprintf('Speed of the machine in motoring mode is %8.2f  RPM\n',NM)
% This program solves Example 7.15 of the text
% the data are
No=1200;
ILo=5;
V=220;
Ra=0.25;
Rf=110;
ILl=62;
% D is indicative of demagnetisation
D=.05;
% calculations for no load case
IF=V/Rf;
IAo=ILo-IF;
EAo=V-Ra*IAo;
To=EAo*IAo*60/(2*pi*No);
% KP is equal to Ka prime multiplied by No
KP=EAo/No
% calculations for on load case
IAl=ILl-IF
EAl=V-Ra*IAI;
Nl=EAl/((I-D)*KP);
TI=EAl*IAl*60/(2*pi*NI);
Tl_shaft=Tl-To;
fprintf('Machine Speed on load is is %8.2f RPM \n',Nl)
fprintf('Machine developed torque on no load is %8.2f Nm \n',To)
fprintf('Machine developed torque on load is %8.2f Nm \n', Tl)
fprintf('Machine shafttorque on load is %8.2f Nm \n', Tl_shaft)
```

CHAPTER 8 Synchronous Machines

LEARNING OBJECTIVES

After studying this chapter, students will be able to

- **LO1** Develop basic modeling by flux distribution and equivalent circuit diagram
- **LO2** Determine synchronous reactance, voltage regulation, and efficiency, by developing open circuit and short circuit characteristic, and by other methods (MMF/Potier/ASA)
- **LO3** Understand the concept of synchronous machine connected and staying synchronised to an infinite bus, and effect of armature reaction, change of excitation, and load, on machine performance
- **LO4** Examine the two reaction models incorporating saliency (calculation of direct axis and quadrature axis reactance) and power flow through synchronous machine
- **LO5** Analyze various starting methods and their necessity for synchronous motor, and various parameters that limit the operation of synchronous generator
- **LO6** Explain the parallel operation and transient analysis of synchronous machine
- **LO7** Outline the basic operating principle of single phase synchronous generator and brushless dc motor

8.1 | INTRODUCTION

A synchronous machine is one of the important types of electric machines; in fact all generating machines at power stations are of synchronous kind and are known as *synchronous generators* or *alternators*.

The structure and certain operational features of the synchronous machine have already been explained in Ch. 5, while the 3-phase ac windings used in the stator of the synchronous machine have been further elaborated in Ch. 6. In this chapter the basic model used for performance analysis of the synchronous machine and some of the operational features peculiar to this type of machine will be discussed. The methods of obtaining model parameters and active and reactive power transfer characteristics will also be discussed at length.

It has been seen in Ch. 5 that essentially two types of constructions are employed in synchronous machines–one of these is known as the *cylindrical-rotor* type and the other is called the *salient-pole* type. The cylindrical-rotor machine has its rotor in cylindrical form with dc field windings (distributed type) embedded in the rotor slots. This type of construction provides greater mechanical strength and permits more accurate dynamic balancing. It is particularly adopted for use in high-speed turbo-generators wherein a relatively long but small diameter rotor is used to limit the centrifugal forces developed. Two or at most 4-pole machines use this type of construction.

The second type of synchronous machine, known as the salient-pole type, has its rotor poles projecting out from the rotor core. This type of construction is used for low-speed hydroelectric generators and the large number of poles necessary are accommodated in projecting form on a rotor of large diameter but small length. This construction is almost universally adopted for synchronous motors.

Because of the distinguishing constructional features explained above, the cylindrical-rotor machine has uniform air-gap, so that the permeance offered to the mmf acting on the magnetic circuit is independent of the angle between the axis of the mmf and that of the rotor poles. On the other hand, in the salient-pole construction, the permeance offered to the mmf varies considerably with the angle between the mmf axis and that of the rotor poles. It is on this account that the cylindrical-rotor machine is simpler to model and analyse, compared to the salient-pole type. The modelling and analysis of both these types will be presented in this chapter.

It has been seen in Ch. 5 that the essential feature that distinguishes the synchronous machine from the other types of electric machines is the synchronous link between the rotor and stator rotating fields. As a result there is a fixed relationship between the rotor speed and the frequency of emfs and currents on the stator, which is reproduced below:

$$\text{Frequency } f = \frac{Pn_s}{120} \text{ Hz} \tag{8.1}$$

where P = number of poles

n_s = speed of rotor in rpm (called *synchronous speed)*

$\omega_s = 2\pi f$ = synchronous speed in rad/s.

A synchronous generator when supplying isolated load acts as a voltage source whose frequency is determined by its primemover speed as per Eq. (8.1). Synchronous generators are usually run in parallel connected through long distance transmission lines. The system (called power system) is so designed as to maintain synchronism in spite of electrical or mechanical disturbances. Such an interconnected system offers the advantages of continuity of supply and economies in capital and operating costs. Synchronous motors find use in industry wherever constant-speed operation is desired. Another advantage of the synchronous motor is that its power factor can be controlled simply by variation of its field current. This is the reason why in most large industrial installations a part of the load is usually handled by synchronous motors which are operated at a leading power factor so as to yield an overall high power factor for the complete installation.

Industry Relevance

In this type of machine, rotating speed of rotor is proportional to the frequency of armature induced e.m.f./current. Due to the fact that these machines can be made to supply reactive power to the system, constant speed from no load to full load, higher efficiency & capability to produce bulk electrical energy, makes it very attractive to industrial application both as generator as well as motor.

8.2 | BASIC SYNCHRONOUS MACHINE MODEL

LO 1 Develop basic modeling by flux distribution and equivalent circuit diagram

As stated above the cylindrical-rotor synchronous machine offers constant permeance to mmf waves irrespective of the mechanical position of the rotor and is, therefore, simpler to model. Figure 8.1 shows the cross-sectional view of a 2-pole* cylindrical-rotor synchronous machine.

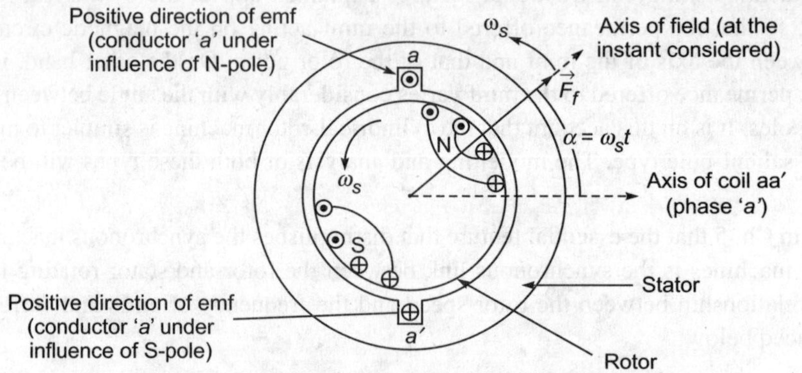

Fig. 8.1 *Cylindrical-rotor synchronous machine*

The rotor has distributed windings which produce an approximately sinusoidally distributed mmf wave in space rotating at synchronous speed ω_s rad (elect.)/s (n_s rpm) along with the rotor. This mmf wave is represented by the *space vector* $\vec{F}_f$ in the diagram and which at the instant shown makes an angle $\alpha = \omega_s t$ with the axis of coil *aa′* on the stator (coil *aa′* represents the phase *a*). The *peak value* of the vector $\vec{F}_f$ is F_f. As a consequence of the uniform air-gap, the mmf wave F_f produces sinusoidally distributed flux density wave B_f, in space phase with it. Figure 8.2 shows the developed diagram depicting the space phase relationship between B_f and F_f waves. As the rotor rotates (at synchronous speed ω_s), the B_f wave causes sinusoidally varying flux ϕ to link with the coil *aa′*. The maximum value of this flux is Φ_f, the *flux per pole*. Considering the time reference when $\vec{F}_f$ lies along the axis of coil *aa′*, this is a cosinusoidal variation, i.e.

$$\phi = \Phi_f \cos \omega_s t; \ \omega_s = 2\pi f \tag{8.2}$$

It is, therefore, seen that the flux linking the coil *aa′* is a sinusoidal time variation and can be represented as the *time phasor* $\bar{\Phi}_f$. It will be referred to as *flux phasor*.

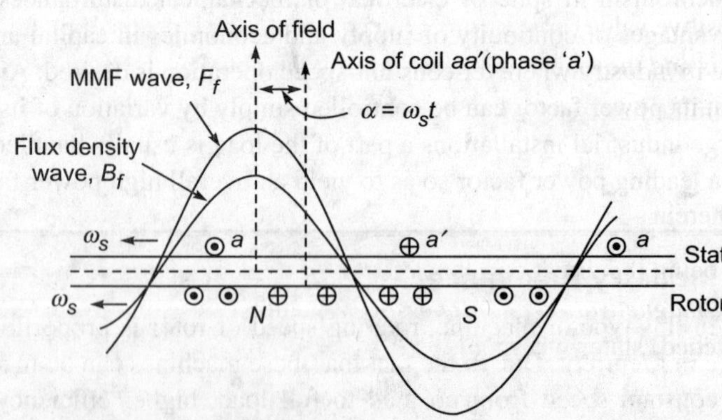

Fig. 8.2 *Developed diagram showing mmf, air-gap flux density and coil aa′ of stator*

Consider now the space vector $\vec{F}_f$ as seen from the axis of coil *aa′* on the stator. As $\vec{F}_f$ rotates at synchronous speed, it appears

*Multipolar structure is merely a cyclic repetition of the 2-pole structure in terms of the electrical angle.

to be sinusoidally time-varying at $\omega_s = 2\pi f$ elect. rad/s as is evident from the developed diagram of Fig. 8.2. Furthermore, when the maximum positive value of F_f space wave is directed along the axis of coil aa', the flux linkage of the coil has maximum positive value. It may, therefore, be considered that the rotating space vector $\vec{F}_f$ as seen from the stator is a time phasor $\bar{F}_f$ which is in phase with the flux phasor $\bar{\Phi}_f$ as shown in Fig. 8.3. The magnitude relationship between $\bar{\Phi}_f$ and $\bar{F}_f$ will be governed by the magnetization curve; this will be linear if the iron is assumed to be infinitely permeable in which case

$$\Phi_f = \mathcal{P} F_f$$

where $\mathcal{P}$ = permeance per pole (see Eq. (5.59)).

emf Equation

The emf induced in the coil aa' of N turns is given by the Faraday's law,

$$e_{af} = -N\frac{d\lambda}{dt}, \quad \lambda = \text{flux linkage of one coil}$$

$$= -N\frac{d}{dt}(\Phi_f \cos \omega_s t)$$

$$= N\omega_s \Phi_f \sin \omega_s t \qquad (8.3)$$

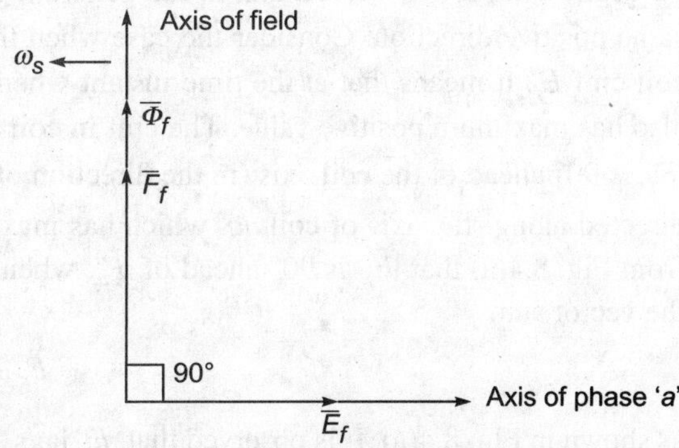

Fig. 8.3

The positive direction of the emf is indicated on the coil aa'. This is also verified by the flux cutting rule when conductor a lies under the influence of N-pole of the rotor and conductor a' simultaneously lies under the influence of the S-pole. It immediately follows from Eqs (8.2) and (8.3) that the emf e_{af} represented as time phasor $\bar{E}_f$ lags behind the mmf phasor $\bar{F}_f$ and the flux phasor $\bar{\Phi}_f$ by 90° as shown in Fig. 8.3. This figure also shows the relative location of the field axis and the axis of phase a wherein the axis of phase a is 90° behind the rotor field axis. This is indicative of the fact that the field vector $\vec{F}_f$ lies 90° ahead of the axis of coil aa' in Fig. 8.1 when the emf in coil aa' has maximum positive value (projection of the phasor $\bar{E}_f$ on the axis of phase a). The rotor field axis is known as the *direct-axis* and the axis at 90° elect. from it is known as the *quadrature-axis*.

It immediately follows from Eq. (8.3) that the rms value of the emf induced in coil aa' is

$$E_f = \sqrt{2}\,\pi f N\Phi_f \qquad (8.4)$$

wherein $\Phi_f = \Phi_f(F_f)$ or $\Phi_f(I_f)$ flux/pole (8.5)

I_f being the direct current in the rotor field. Equation (8.5) between the flux/pole and field current is indeed the *magnetization characteristic*. Equation (8.4) suitably modified for a distributed (and also possibly short-pitched) stator winding is

$$E_f = \sqrt{2}\,\pi K_w f N_{ph}\,\Phi_f \qquad \text{(see Eq. (5.20))} \qquad (8.6)$$

It easily follows that the emfs produced in the other phases of the stator would progressively differ in time phase by 120°.

The conclusions drawn from the above discussion are indeed general and are reproduced below.

Important! Whenever the magnetic structure of a cylindrical rotor synchronous machine is subjected to rotating mmf vector, it is seen as an mmf phasor from the stator with its flux phasor in phase with it, while the phasor representing the phase emf induced lags behind both these phasors by 90° (see Fig. 8.3). It was shown in Sec. 5.5 that when a 3-phase stator supplies a balanced load, it sets up its own mmf vector $\vec{F}_{ar}$ (sinusoidally distributed in space), called the *armature reaction,* rotating at synchronous speed in the same direction as the rotor.

The magnetic circuit is now subjected to two rotating mmf vectors $\vec{F}_f$ and $\vec{F}_{ar}$, both rotating at synchronous speed with a certain angle between them. The objective here is to establish what determines this angle.

Unity Power Factor

To begin with, it is observed that in the generating operation of the machine, the emf and current have the same positive direction. Consider the case when the current I_a supplied by the coil *aa′* is in phase with the coil emf E_f. It means that at the time instant when the emf is maximum positive in the coil *aa′,* its current also has maximum positive value. The emf in coil *aa′* will be maximum positive when the field mmf vector $\vec{F}_f$ is 90° ahead of the coil axis (in the direction of rotation) as shown in Fig. 8.4(a). Simultaneously, $\vec{F}_{ar}$ is directed along† the axis of coil *aa′* which has maximum positive current at this instant. It is therefore, seen from Fig. 8.4(a) that $\vec{F}_f$ is 90° ahead of $\vec{F}_{ar}$ when $\bar{E}_f$ and $\bar{I}_a$ are in phase. The resultant mmf vector $\vec{F}_r$ is the vector sum

$$\vec{F}_r = \vec{F}_f + \vec{F}_{ar} \tag{8.7}$$

as shown in Fig. 8.4(a). It is observed that $\vec{F}_r$ lags $\vec{F}_f$ by angle δ. The corresponding phasor diagram, drawn as per the general conclusions stated earlier, is shown in Fig. 8.4(b) wherein the phasor equation corresponding to the vector Eq. (8.7) is

$$\bar{F}_r = \bar{F}_f + \bar{F}_{ar} \;(\bar{F}_r \text{ lags } \bar{F}_f \text{ by angle } \delta) \tag{8.8}$$

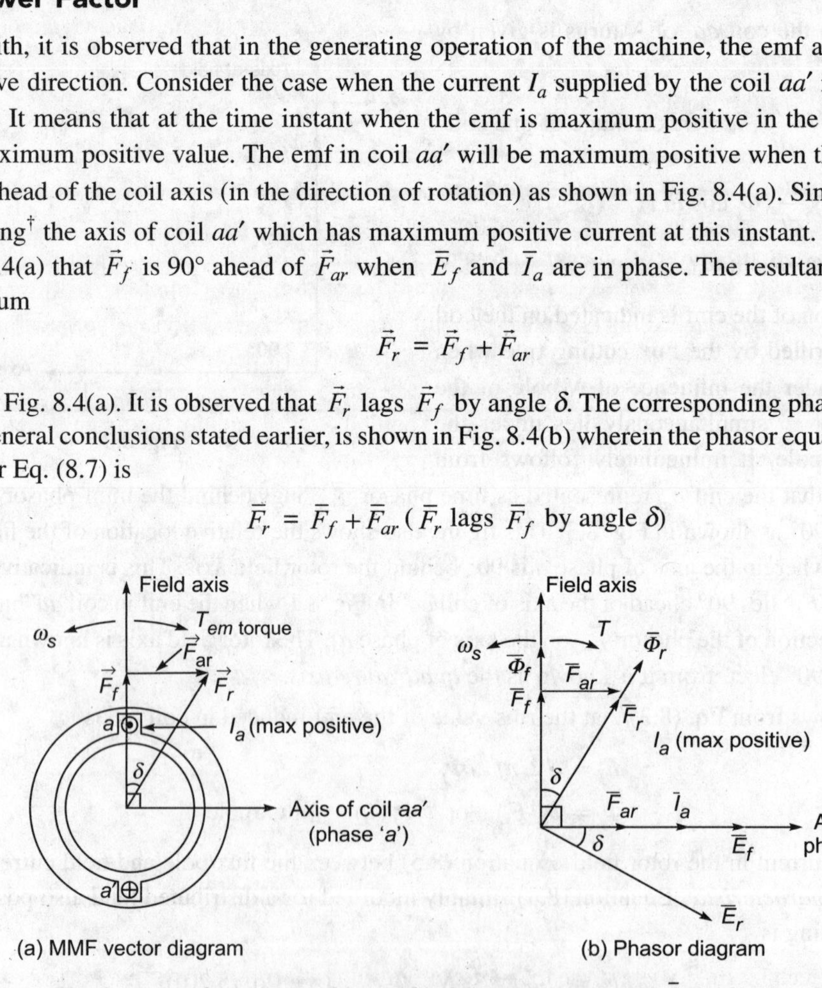

Fig. 8.4 *Synchronous machine on load (generating action), $\bar{I}_a$ in phase with $\bar{E}_f$*

† It was shown in Sec. 5.5 that the mmf vector of stator carrying balanced 3-phase currents is directed along the axis of coil *aa′* when phase *a* carries maximum positive current.

!Attention It is observed from the phasor diagram that $\bar{I}_a$ and $\bar{F}_{ar}$ are in phase; this is logical[‡] because $\bar{F}_{ar}$ is produced by $\bar{I}_a$ and is proportional to it. The resultant mmf vector $\bar{F}_r$ produces the resultant air-gap flux phasor $\vec{\Phi}_r$ which in turn induces the emf $\bar{E}_r$ in phase *a* lagging 90° behind the phasor $\bar{F}_r$. The phase emf $\bar{E}_r$ induced in the machine, called *air-gap emf,* lags by angle δ behind $\bar{E}_f$ (emf induced by $\vec{F}_f$ acting alone) which is the same angle by which $\vec{F}_r$ lags behind $\vec{F}_f$ in the vector diagram of Fig. 8.4(a). $\bar{E}_f$ is known as the *excitation emf.*

Lagging Power Factor

In case $\bar{I}_a$ lags $\bar{E}_f$ by an angle ψ, the positive current maximum in the coil *aa′* will occur angle ψ later, so that $\vec{F}_f$ now lies (90° + ψ) ahead of $\vec{F}_{ar}$ as shown in the vector diagram of Fig. 8.5(a). The corresponding phasor diagram is shown in Fig. 8.5(b) wherein $\bar{F}_f$ leads $\bar{F}_{ar}$ by (90° + ψ). The phase angle between $\bar{E}_r$ and $\bar{I}_a$ indicated by θ is the *power factor angle* provided it is assumed that the armature has zero resistance and leakage reactance so that the machine terminal voltage

$$\bar{V}_t = \bar{E}_r$$

It is seen from Figs 8.5(a) and (b) that the field poles lie an angle δ ahead of the resultant mmf (or resultant flux) wave. The electromagnetic torque developed in the machine tries to align the field poles with the resultant field and is, therefore, in a direction as shown in Fig. 8.5(a) as well as in Fig. 8.5(b). It is immediately seen that the torque on the field poles is in opposite direction to that of rotation which means that mechanical power is absorbed by the machine. This is consistent with the assumed condition of the generating action (positive current in the direction of positive emf). It may, therefore, be concluded that in generating action, the field poles are driven ahead of the resultant flux wave by an angle δ as a consequence of the forward acting torque of the primemover. Also, the field poles are dragged behind by the resultant flux from which results the conversion of mechanical energy into electrical form.

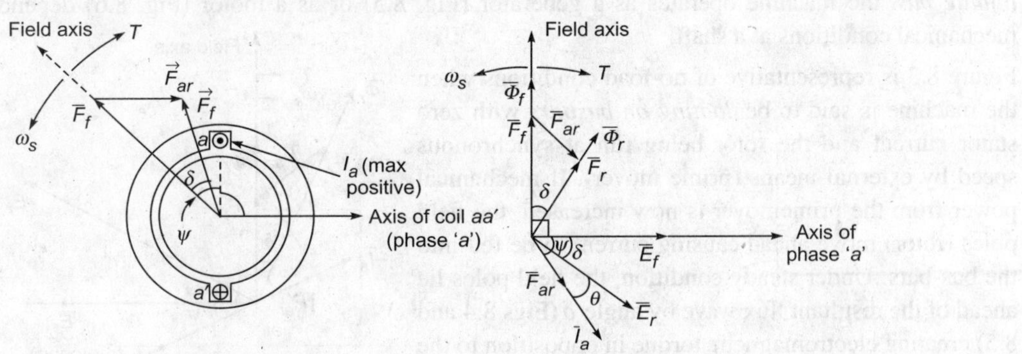

Fig. 8.5 *Synchronous machine on load (generating action), $\bar{I}_a$ lags $\bar{E}_f$ by angle ψ*

As already shown in Sec. 5.6 Eq. (5.58), if the magnetic circuit is assumed linear, the magnitude of the torque is given by

[‡] An observer on stator from the axis of phase *a* observes simultaneous occurrence of maximum positive value of F_{ar} and I_a.

$$T = \frac{\pi}{2}\left(\frac{P}{2}\right)^2 \Phi_r F_f \sin\delta \tag{8.9}$$

where Φ_r is the resultant flux/pole, F_f the peak value of field ampere-turns and δ is the angle by which $\vec{F}_f$ leads $\vec{F}_r$. It is also observed from Fig. 8.5(b) that $\bar{E}_f$ also leads E_r by the same angle δ.

When F_f and Φ_r are held constant in magnitude, the machine meets the changing requirements of load torque by adjustment of the angle δ which is known as the *torque* (*power*[§]) *angle.*

The torque expression of Eq. (8.9) can also be written in terms of voltages as

$$T = KE_r E_f \sin\delta \tag{8.10}$$

where E_r = emf induced in the machine under loaded condition; called *air-gap emf.*

E_f = emf induced by the field mmf F_f acting alone, i.e. the machine is on no-load with same F_f (or rotor field current) as on-load; called *excitation emf*.

Motoring Mode

In motoring action of the synchronous machine, the positive current flows opposite to the induced emf. Since the phasor diagrams above have been drawn with the convention of generating current (i.e. current in the direction of emf), the armature reaction phasor $\bar{F}_{ar}$ will now be located by phase reversing the motoring current for consistency of convention. Accordingly the phasor diagram for motoring action is drawn in Fig. 8.6. It is immediately observed from this figure that $\bar{F}_f$ and $\bar{E}_f$ now lag $\bar{F}_r$ and $\bar{E}_r$ respectively by angle δ. The torque of electromagnetic origin therefore acts on the field poles in the direction of rotation so that the mechanical power is output meaning thereby motoring action.

If the terminal voltage $\bar{V}_t = \bar{E}_r$ and its frequency is held constant by an external 3-phase source, called *infinite bus*, the machine operates as a generator (Fig. 8.5) or as a motor (Fig. 8.6) depending upon the mechanical conditions at a shaft.

Figure 8.3 is representative of no-load conditions when the machine is said to be *floating on busbars* with zero stator current and the rotor being run at synchronous speed by external means (prime mover). If mechanical power from the primemover is now increased, the field poles (rotor) move ahead causing current to be fed into the bus-bars. Under steady condition, the field poles lie ahead of the resultant flux wave by angle δ (Figs 8.4 and 8.5) creating electromagnetic torque in opposition to the direction of rotor rotation; the value of δ corresponds to the balance of torques (or power, $P = T\omega_s$). The electrical power output is

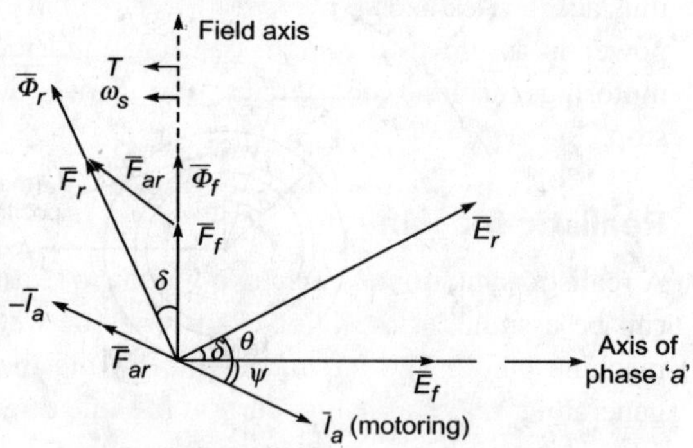

Fig. 8.6 *Synchronous machine on load (motoring action)*

$$3V_t\,(= E_r)I_a \cos\theta \quad \text{(for 3 phases)}$$

which balances the mechanical power input from the primemover because there are no losses in the stator

[§]Torque and power are proportional to each other as the machine runs at synchronous speed (ω_s) under steady operating conditions.

(resistance is assumed to be zero). Here I_a is the generating current taken to be positive in the direction of the machine's induced emf E_r.

If instead, the shaft is mechanically loaded from the no-load condition (Fig. 8.3), the field poles lag behind the resultant flux wave as in Fig. 8.6 creating electromagnetic torque in the direction of rotation thereby outputting mechanical power; the electrical input power being 3 V_t (= E_r)I_a cos θ, where I_a is the motoring current taken as positive in opposition to the positive direction of E_r.

Torque (Power)-Angle Characteristics

The torque (power)-angle ($T - \delta$) relationship of Eq. (8.10) for fixed E_f and E_r (= V_t) is drawn in Fig. 8.7 with δ taken as positive for generating action and negative for motoring action. The operating points as the generator and motor are indicated by g and m on this curve corresponding to the specific condition of loading.

!Attention The characteristic exhibits, both in generating and motoring operation, a maximum torque at δ = 90°, called the *pull-out torque,* beyond which the synchronous link between field poles and the resultant flux wave is severed and the machine falls *out-of-step* (or *loses synchronism*). The average developed machine torque now becomes zero and so the average electric power fed by the generating machine to the bus-bars, (infinite) or average electric power drawn by the motoring machine from the bus bars reduces to zero. The generating machine thus accelerates and so overspeeds (primemover power is assumed to remain constant) and the motoring machine decelerates and comes to stop.

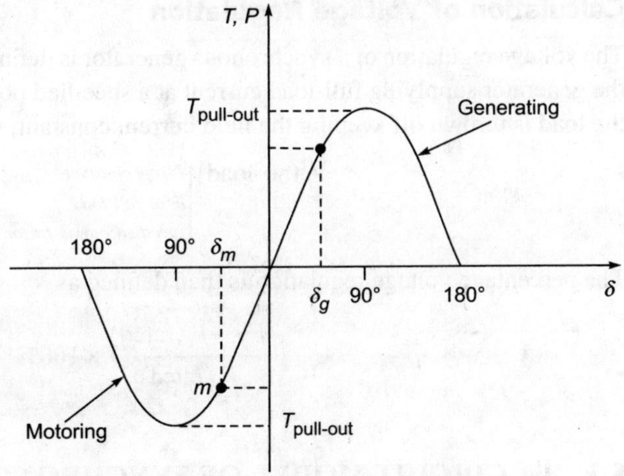

Fig. 8.7 *T-δ characteristic of synchronous machine*

Realistic Machine

A realistic synchronous machine will have resistance R_a and leakage reactance X_l per armature phase which can be assumed to be lumped in series between the terminal voltage V_t and the air-gap emf E_r for each machine phase. The circuit diagram of the machine on a per phase basis is drawn in Fig. 8.8(a) for the generating mode (armature current in same direction as E_r). From Fig. 8.8(a)

$$\bar{V}_t = \bar{E}_r - \bar{I}_a (R_a + jX_l) \text{ (generating mode)} \tag{8.11}$$

and from Fig. 8.8(b)

$$\bar{V}_t = \bar{E}_r + \bar{I}_a (R_a + jX_l) \text{ (motoring mode)} \tag{8.12}$$

The resistance of a synchronous machine armature is usually very small and can generally be neglected in performance analysis (see Sec. 8.3).

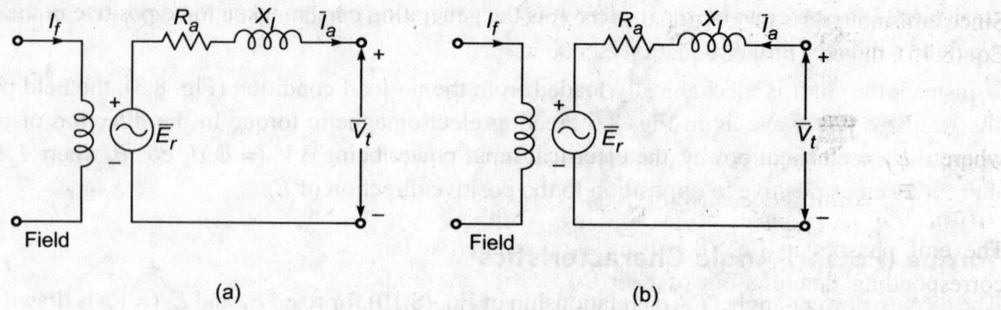

Fig. 8.8 *(a) Generating mode, (b) Motoring mode*

Calculation of Voltage Regulation

The voltage regulation of a synchronous generator is defined on lines similar to that of a transformer. Consider the generator supplying full-load current at a specified power factor and rated terminal voltage, V_t (rated). As the load is thrown off keeping the field current constant, the terminal voltage rises to

$$V_t\,(\text{no-load})\Big|_{\substack{I_f \text{ kept constant} \\ \text{as at full-load} \\ \text{specified power factor}}} = E_f, \text{ the excitation emf}$$

The percentage voltage regulation is then defined as

$$\left.\frac{E_f - V_t(\text{rated})}{V_t(\text{rated})} \times 100\%\right|_{\text{At specified power factor}} \tag{8.13}$$

8.3 | CIRCUIT MODEL OF SYNCHRONOUS MACHINE

LO 1

By assuming linearity of the magnetic circuit, it is possible to obtain simple circuit model of the synchronous machine. The validity of this assumption stems from the fact that air-gap is the predominant component of the magnetic circuit of the machine. Approximate nonlinear analysis is the subject matter of Sec. 8.4.

As per Eq. (8.7), the resultant mmf phasor is given by

$$\bar{F}_r = \bar{F}_f + \bar{F}_{ar} \tag{8.14}$$

The resultant flux Φ_r and the air-gap emf $\bar{E}_r$ must in general be obtained from $\bar{F}_r$. However, the assumption of linear magnetic circuit ($\Phi = \mathcal{P}F$; $\mathcal{P}$ is constant permeance[¶]), allows one to find the resultant flux by the principle of superposition as

$$\bar{\Phi}_r = \bar{\Phi}_f + \bar{\Phi}_{ar} \tag{8.15}$$

where $\bar{\Phi}_f$ = flux component produced by $\bar{F}_f$ acting alone

$\bar{\Phi}_{ar}$ = flux component produced by $\bar{F}_{ar}$ acting alone.

[¶]In cylindrical-rotor machine being considered here, $\mathcal{P}$ is independent of the direction along which $\vec{F}$ is directed because of a uniform air-gap.

Since emf induced is proportional to the flux/pole (Eq. (8.6)) and lags behind it by 90°, from the flux phasor Eq. (8.15), the emf phasor equation can be written as

$$\bar{E}_r = \bar{E}_f + \bar{E}_{ar} \tag{8.16}$$

where $\bar{E}_f$ = emf induced by $\bar{\Phi}_f$ acting alone (excitation emf)

$\bar{E}_{ar}$ = emf induced by $\bar{\Phi}_{ar}$ acting alone.

The emf phasors in Eq. (8.16) are proportional to the corresponding flux phasors of Eq. (8.15) with the emf phasors lagging the respective flux phasors by 90°. The phasor Eqs (8.15) and (8.16) are represented by the phasor diagram of Fig. 8.9. In this figure the flux phasor triangle and emf phasor triangle are similar to each other with the emf phasor triangle being rotated anticlockwise from the flux phasor triangle by 90°.

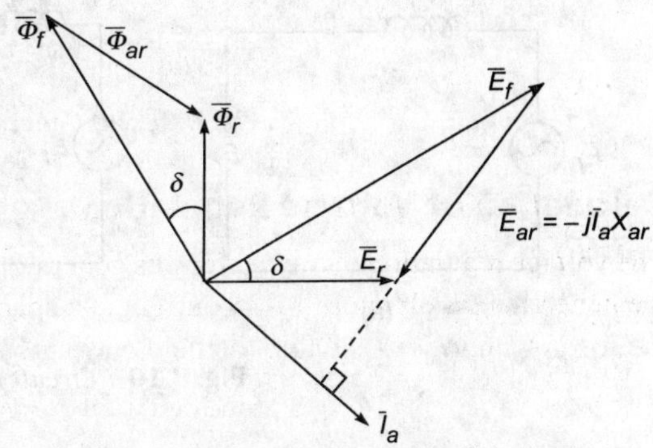

Fig. 8.9 *Phasor diagram of flux and emf components in synchronous machine*

As $\bar{\Phi}_{ar}$ is in phase with $\bar{I}_a$ (generating machine) and is proportional to it, the emf $\bar{E}_{ar}$ is proportional** to $\bar{I}_a$ and lags behind $\bar{\Phi}_{ar}$ by 90°, i.e.

$$\bar{E}_{ar} = -j\bar{I}_a X_{ar} \tag{8.17}$$

where X_{ar}, the constant of proportionality, is indeed an inductive reactance. Thus Eq. (8.16) can be written as

$$\bar{E}_r = \bar{E}_f - j\bar{I}_a X_{ar} \tag{8.18}$$

Corresponding to Eq. (8.18), Fig. 8.10(a) gives the *per phase* circuit model of the synchronous machine.

!Attention Comparing Eqs (8.15), (8.16) and (8.18), it is concluded that the reactance X_{ar} equivalently replaces the effect of the armature reaction flux. If X_{ar} is known for a machine, one can work in terms of voltages and currents and need not represent fluxes on the phasor diagram.

The effect of armature resistance R_a and leakage reactance X_l are embodied in the phasor equation as (see Eq. (8.11))

$$\bar{V}_t = \bar{E}_r - \bar{I}_a(R_a + jX_l) \tag{8.19}$$

Important! From Eq. (8.19) and Fig. 8.10(a) follows the complete circuit model of Fig. 8.10(b) which can be reduced to the simpler form of Fig. 8.10(c) by combining series reactances and by allowing the identity of $\bar{E}_r$ to be lost. In the circuit model of Fig. 8.10(c), the total reactance

** $\bar{E}_{ar} = -jK_a\bar{\Phi}_{ar}$; K_a = emf constant of armature winding

$= -jK_a\mathcal{P}\bar{F}_{ar}$; $\mathcal{P}$ = permeance/pole

$= -jK_a\mathcal{P}K_{ar}\bar{I}_a$; K_{ar} = constant of armature winding (see Eq. (5.44b))

$= -jX_{ar}\bar{I}_a$

where $X_{ar} = K_a\mathcal{P}K_{ar}$

X_{ar} will be more for a machine with higher permeance, i.e., smaller air-gap,

$$X_s = X_{ar} + X_l \text{ (per phase)} \tag{8.20}$$

is known as the *synchronous reactance* of the machine, while

$$Z_s = \sqrt{R_a^2 + X_s^2} \tag{8.21}$$

is the *synchronous impedance* of the machine.

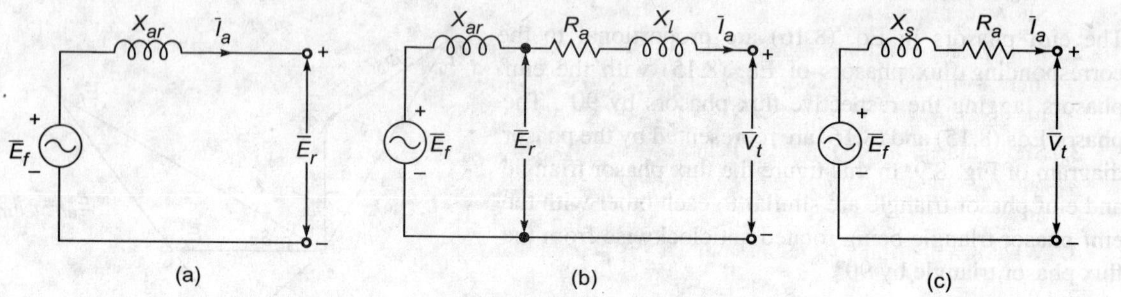

Fig. 8.10 *Circuit model of synchronous machine*

Recall The synchronous reactance takes into account the flux produced by the flow of balanced 3-phase currents in the stator as well as the leakage flux.

The *excitation emf, E_f,* accounts for the flux produced by the rotor field (dc excited). The magnitude of the excitation emf can be controlled by the dc field current (I_f) called the *excitation current*. If the load on the machine is thrown off, E_f appears at the machine terminals which then is the *open-circuit* voltage of the machine. With reference to Fig. 8.10(c) E_f is also called *voltage behind synchronous impedance or reactance* (as R_a can be neglected).

It must be remembered here that the synchronous impedance model of the synchronous machine is based on the linearity assumption and will hold for the unsaturated region of machine operation and is valid only for the cylindrical-rotor machine.

In the above method, we have converted mmfs to emfs based on magnetic linearity assumption this method is also known as the ***emf method***.

Practical Range of Synchronous Impedance

Expressed in the pu system, the synchronous reactance of synchronous machines lies in a narrow range of values. From practical data, it is observed that the armature resistance (R_a) is usually of the order of 0.01 pu, i.e. the voltage drop in the armature resistance at the rated armature current is about 1% of the rated voltage. The leakage reactance value ranges from 0.1 to 0.2 pu and the synchronous reactance ($X_s = X_{ar} + X_l$) is of the order of 1.0 to 2 pu. It is, therefore, seen that the armature resistance of a synchronous machine is so low that it can be neglected for all practical purposes except in the computation of losses, temperature rise and efficiency. It may be noted here that R_a must be small in order to minimize the I^2R loss and limit the temperature rise of the machine, and X_s should be large to limit the maximum current that may flow under fault (short-circuit) conditions. However, the modern practice is to design[††] synchronous machines with a

[††] The designer can achieve the desired value of synchronous reactance by adjusting air-gap of the machine (see footnote p. 450).

medium range of values for synchronous reactance as quick-acting[‡‡] circuit breakers are now available to disconnect a machine from the faulted line.

Concept Check LO 1

1. A synchronous generator is supplying zero power factor (i) lagging and (ii) leading current. Show that the terminal voltage V_t and the excitation emf E_f are in phase. ○○●
2. In a synchronous motor drawing leading current at pf = cos θ. Draw the phasor diagram and find there from the phase and magnitude relationships between V_t and E_f. Here phase relationship means lag/lead and magnitude relationship means greater than/less than. ○○●
3. Show that in a generating synchronous machine the phase relationship (lag/lead) between V_t and E_f is independent of the power factor (lag/lead). Draw the phasor diagrams to discover your answer. ○●●
4. What is a distributed winding and what is distribution factor? ○●●
5. What are the factors affecting the synchronous generator terminal voltage? ○○●

8.4 | DETERMINATION OF THE SYNCHRONOUS REACTANCE

LO 2 Determine synchronous reactance, voltage regulation, and efficiency, by developing open circuit and short circuit characteristic, and by other methods (MMF/Potier/ASA)

With the assumption of a linear magnetic circuit, the circuit model (per phase) of a synchronous machine is as given in Fig. 8.10(c). If R_a is neglected, it then follows that

$$\bar{E}_f = \bar{V}_t + j\bar{I}_a X_s \tag{8.22}$$

It is immediately seen from Eq. (8.22) that for a given field current under short-circuit condition ($I_a = I_{SC}$, $V_t = 0$),

$$X_s = \frac{E_f}{I_{SC}} \tag{8.23}$$

But $E_f = V_{OC}$ (open-circuit voltage, i.e. $I_a = 0$ with the same field current). Then with the linearity assumption

$$X_s = \left.\frac{V_{OC}}{I_{SC}}\right|_{I_f=\text{const}} \tag{8.24}$$

where V_{OC} = open circuit voltage and I_{SC} = short-circuit current on a per phase basis with the same field current.

Since the magnetization characteristic of the machine is nonlinear, it is necessary to determine the complete open-circuit characteristic (OCC) of the machine ($V_{OC} - I_f$ relationship). However, it will soon be shown that it is sufficient to determine one point on the short-circuit characteristic (SCC) of the machine ($I_{SC} - I_f$ relationship). as it is linear in the range of interest (for I_{SC} up to 150% of the rated current).

[‡‡] The typical circuit breaker opening time is 1–1.5 cycles.

Note: Difficulty Level ⟶ ○○● — Level 1 and Level 2 Category; ○●● — Level 3 and Level 4 Category; ●●● — Level 5 and Level 6 Category

Open-circuit Characteristic (OCC)

In this test the machine is run by a primemover at synchronous speed n_s to generate voltage at the rated frequency, while the armature terminals are open-circuited as in Fig. 8.11 with switch S open. The readings of the open-circuit line-to-line armature voltage, $V_{OC} = \sqrt{3}\,E_f$, are taken for various values of I_f, the rotor field current. It may be noted that I_f is representative of the net mmf/pole acting on the magnetic circuit of the machine. These data are plotted as OCC in Fig. 8.12 which indeed is the magnetization characteristic, i.e. the relation between the space fundamental component of the air-gap flux and the net mmf/pole acting on the magnetic circuit (space harmonics are assumed negligible).

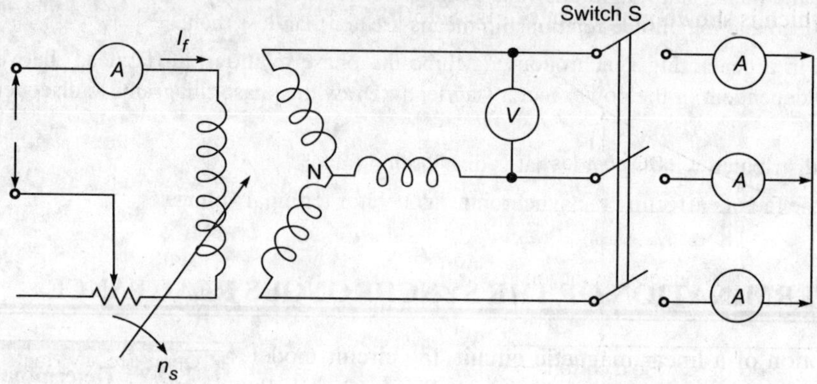

Fig. 8.11 *Schematic diagram for open-circuit and short-circuit tests*

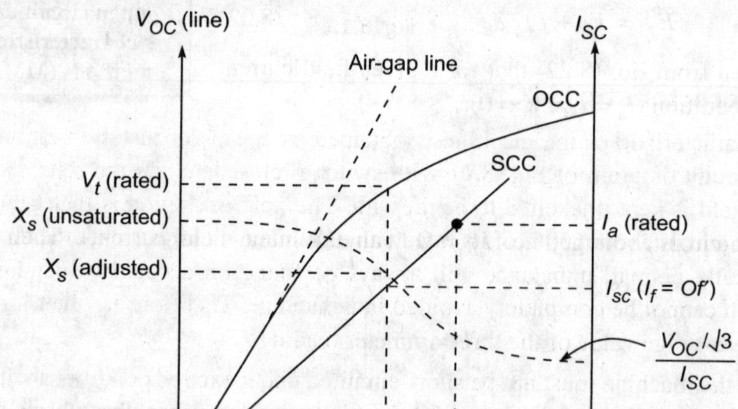

Fig. 8.12 *Open-circuit and short-circuit characteristics*

Important! The OCC exhibits the saturation phenomenon of the iron in machine. At low values of I_f when iron is in the unsaturated state, the OCC is almost linear and the mmf applied is mainly consumed in establishing flux in the air-gap, the reluctance of the iron path being almost negligible. The straight-line part of the OCC, if extended as shown dotted in Fig. 8.12, is called the *air-gap line* and would indeed be the OCC if iron did not get saturated.

Open Circuit Loss

The loss in the open-circuit method conducted as above comprises no load (OC) core loss and windage and friction loss. The power corresponding to these losses is drawn from the prime-mover running the machine, which can be measured by a dynmometer or torque meter. The plot of P_{OC} versus field current is shown in Fig. 8.13. The winding and friction loss being constant gets separated out by reducing the field current to zero. The OC core loss arises from hysteresis and eddy-current losses which very as 1.6th power of open circuit voltage which is shown in the plot of Fig. 8.13(b).

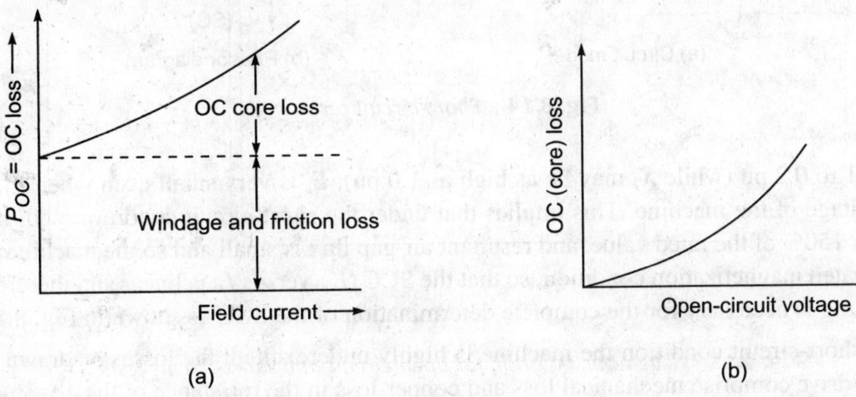

Fig. 8.13

Short-circuit Characteristic (SCC)

The short-circuit characteristic of the machine is obtained by means of the short-circuit test conducted as per the schematic circuit diagram of Fig. 8.11 with switch *S* closed. While the rotor is run at synchronous speed n_s, the rotor field is kept unexcited to begin with. The field excitation is then gradually increased till the armature current equals about 150% of its rated value. While the current in all the three ammeters should be identical, practically a small unbalance will always be found on account of winding and field current dissymmetries which cannot be completely avoided in a machine. Therefore I_{SC} the short-circuit current per phase is taken as the average value of the three ammeter readings.

It is to be noted that the machine must not be short-circuited under excited conditions with a near about rated voltage. This can give rise to intolerably large transient currents in the machine.

The circuit model of the machine under short-circuit conditions is given in Fig. 8.14(a) and the corresponding phasor diagram in Fig. 8.14(b) wherein $R_a = 0$. Since the armature circuit is assumed purely inductive, the short-circuit current lags the air-gap voltage $\bar{E}_r$ by 90° so that the armature reaction mmf phasor $\bar{F}_{ar}$ is in direct opposition to $\bar{F}_f$, i.e, the armature reaction is fully *demagnetizing* in effect.

The air-gap voltage needed to circulate the short-circuit current in the armature is given by

$$E_r = I_a\,(\text{SC})\,X_l \tag{8.25}$$

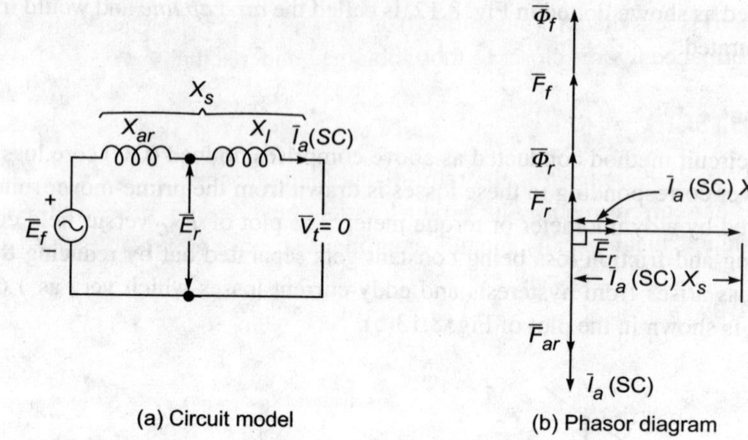

Fig. 8.14 *Short-circuit test*

As X_l is about 0.1 to 0.2 pu (while X_s may be as high as 1.0 pu), E_r is very small even when E_f has a value close to rated voltage of the machine. This implies that under the short-circuit condition with the armature current as high as 150% of the rated value, and resultant air-gap flux is small and so the machine is operating under the unsaturated magnetization condition, so that the SCC (I_{SC} versus I_f) is linear and therefore only one short-circuit reading is necessary for the complete determination of the SCC as shown in Fig. 8.12.

Since under the short-circuit condition the machine is highly underexcited, the losses as drawn in from the mechanical shaft drive comprise mechanical loss and copper-loss in the resistance of the armature, the iron-loss being negligible.

The unsaturated synchronous reactance can be obtained from the OCC and SCC of Fig. 8.12 as

$$X_s\,(\text{unsaturated}) = \left.\frac{V_{OC}/\sqrt{3}}{I_{SC}}\right|_{I_f\,\text{const}} \tag{8.26}$$

where I_f corresponds to the unsaturated magnetic region or V_{OC} value corresponding to the air-gap line could be used.

Since a synchronous machine under operating conditions works in a somewhat saturated region of the magnetization characteristic, the performance of the machine as calculated from X_s, defined above, will differ considerably from the actual value effective during normal operation. To account for the fact that the machine actually operates in the saturated region, it is a must to resort to the nonlinear analysis (Sec. 8.4) or use a heuristic technique of adjusting X_s to a suitable value.

If X_s, as defined in Eq. (8.26), is plotted for various values of the field current, the chain-dotted curve of Fig. 8.12 will be obtained. Initially in the unsaturated region the value of the synchronous reactance remains constant at X_s (unsaturated) and then drops off sharply because of saturation of the OCC.

The value of X_s corresponding to the field current, which gives rated voltage on open-circuit, is defined as

$$X_s\,(\text{adjusted}) = \left.\frac{V_t(\text{rated})/\sqrt{3}}{I_{SC}}\right|_{\substack{I_f\text{ corresponding to}\\ V_t(\text{rated})\text{ on OCC}}} ;\ I_f = Of' \tag{8.27}$$

The value of X_s (adjusted) is less than X_s (unsaturated) as shown in Fig. 8.12. Obviously X_s (adjusted) would yield the machine performance figures closer to those obtained under actual operation.

Short Circuit Ratio (SCR)

The short-circuit ratio (SCR) is defined as the ratio of the field current required to produce rated voltage on open-circuit to the field current required to produce rated armature current with the armature terminals shorted while the machine is mechanically run at synchronous speed. From the OCC and SCC as shown in Fig. 8.15,

$$\text{SCR} = \frac{of'}{of''} \tag{8.28}$$

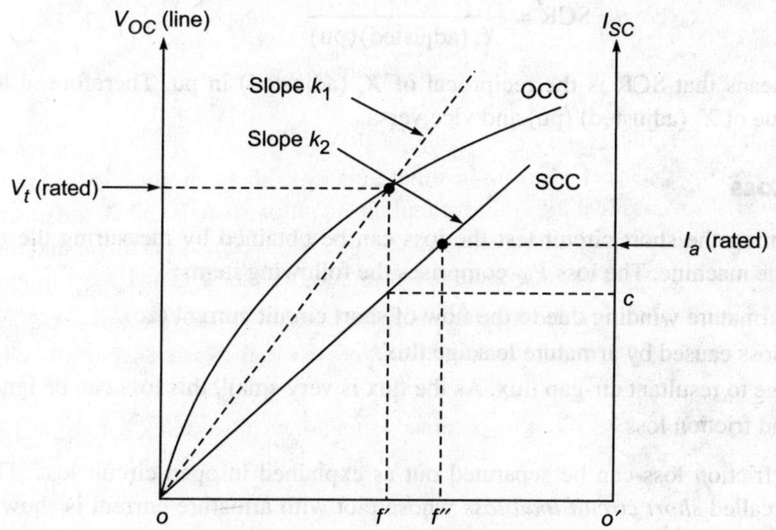

Fig. 8.15 *Short-circuit ratio (SCR)*

as per the definition. It is further noted from this figure that

$$X_s\,(\text{adjusted}) = \frac{V_t(\text{rated})/\sqrt{3}}{o'c} \tag{8.29}$$

and slopes of OCC and SCC are

$$k_1 = \frac{\text{OC voltage}}{\text{field current}} = \frac{V_t(\text{rated})}{of'} \tag{8.30}$$

$$k_2 = \frac{\text{SC current}}{\text{field current}} = \frac{I_a(\text{rated})}{of''} = \frac{o'c}{of'} \tag{8.31}$$

Substituting Eqs (8.30), and (8.31) in Eq. (8.28),

$$\text{SCR} = \frac{of'}{of''}$$

$$= \frac{V_t(\text{rated})}{k_1} \cdot \frac{k_2}{I_a(\text{rated})}$$

$$= \frac{o'c}{V_t(\text{rated})} \cdot \frac{V_t(\text{rated})}{I_a(\text{rated})}$$

$$= \frac{o'c}{V_t(\text{rated})/\sqrt{3}} \cdot \frac{V_t(\text{rated})/\sqrt{3}}{I_a(\text{rated})}$$

$$= \frac{1}{X_s(\text{adjusted})} \cdot X_{\text{Base}}$$

$$\text{SCR} = \frac{1}{X_s(\text{adjusted})(\text{pu})} \tag{8.32}$$

Equation (8.32) means that SCR is the reciprocal of X_s (adjusted) in pu. Therefore, a low value of SCR implies a large value of X_s (adjusted) (pu) and vice versa.

Short Circuit Loss

Important! During the short circuit test the loss can be obtained by measuring the mechanical power required to drive the machine. The loss P_{SC} comprises the following items:

1. I^2R loss in armature winding due to the flow of short circuit current (ac).
2. Local core loss caused by armature leakage flux.
3. Core loss due to resultant air-gap flux. As the flux is very small, this loss can be ignored.
4. Windage and friction loss.

The windage and friction loss can be separated out as explained in open circuit loss. The remaining loss (items 1 and 2) is called *short circuit load loss* whose plot with armature current is shown in Fig. 8.16. By measuring armature dc resistance correcting it for ac and the armature temperature during SC test, dc I^2R loss can then be subtracted leaving behind the stray load loss – sum of load core loss and loss due to additional conductor resistance offered to alternating current. The stray load loss is also plotted in Fig. 8.16.

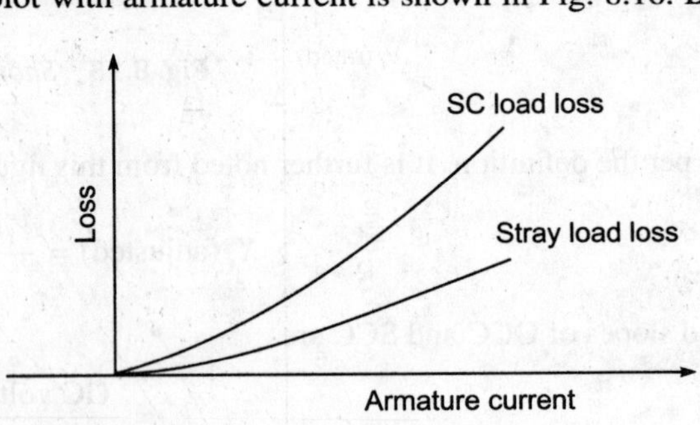

Fig. 8.16

The armature resistance as calculated from armature current and short circuit load loss is called *effective armature resistance.* It can be calculated at rated current and then assumed to remain constant. Thus

$$R_a(\text{eff}) = \frac{\text{short circuit load loss (per phase)}}{(\text{short circuit armature current})^2} \tag{8.33}$$

Example 8.1 *Draw the open-circuit and short-circuit characteristics using the data given below for a 150 MW, 13 kV, 0.85 pf, 50 Hz synchronous generator.*

Open-circuit characteristic

$I_f(A)$	200	450	600	850	1200
V_{OC} (line) (kV)	4	8.7	10.8	13.3	15.4

Short-circuit characteristic

$$I_f = 750\text{ A}, \quad I_{SC} = 8000\text{ A}$$

(a) Determine the unsaturated synchronous reactance of the machine.

(b) Determine the saturated synchronous reactance of the machine.

(c) Convert the value of reactances in part (a) and (b) in their pu value.

(d) What is the short circuit ratio of this machine?

(e) The machine supplies full-load at a pf of 0.85 lagging and a terminal voltage of 13 kV.

Case I – Find the excitation emf and voltage regulation using the synchronous reactance (linear circuit model).

Case II – Find again the excitation emf using saturated synchronous reactance and then find the field current needed to supply the specified load. With field current remaining constant, find the OC voltage and therefrom calculate the voltage regulation of the machine.

(f) Compare the results of part (e) cases I and II.

(g) For the load as in part (e) draw the phasor diagram showing voltages and current. Calculate the angle between the terminal voltage and excitation emf.

LO 2

Solution The OCC and SCC as per the data are drawn in Fig. 8.17.

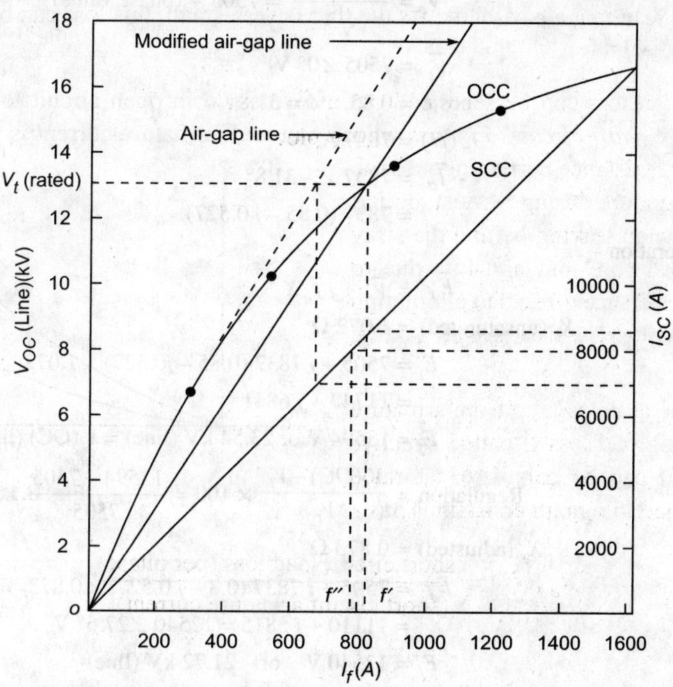

Fig. 8.17

Note: Values have been read from Fig. 8.17 drawn on enlarged scale

(a) Corresponding to V_{OC} = 13 kV on the air-gap line, I_{SC} = 7000 A for the same field current.

$$\therefore \quad X_s \text{ (unsaturated)} = \frac{13 \times 1000}{\sqrt{3} \times 7000} = 1.072\ \Omega$$

(b) Corresponding to V_{OC} = 13 kV on the OCC, I_{SC} = 8600 A for the same field current.

$$\therefore \quad X_s \text{ (adjusted)} = \frac{13 \times 1000}{\sqrt{3} \times 8600} = 0.873\ \Omega$$

$$I_a \text{ (rated)} = \frac{150 \times 10^6}{\sqrt{3} \times 0.85 \times 13 \times 10^3} = 7837 \text{ A}$$

$$\text{Base ohms} = \frac{13 \times 1000}{\sqrt{3} \times 7837} = 0.958\ \Omega$$

(c)
$$X_s \text{ (unsaturated) (pu)} = \frac{1.072}{0.958} = 1.12$$

$$X_s \text{ (adjusted) (pu)} = \frac{0.873}{0.958} = 0.911 \text{ pu}$$

(d)
I_f corresponding to $V_{OC} = V_t$ (rated) is of' = 810 A

I_f corresponding to $I_{SC} = I_a$ (rated) is of'' = 750 A

$$\therefore \quad \text{SCR} = \frac{810}{750} = 1.08 = \frac{1}{X_s\text{(adjusted)(pu)}}$$

(e)
$$V_t = \frac{13 \times 1000}{\sqrt{3}} = 7505 \text{ V (phase value)}$$

$$\bar{V}_t = 7505 \angle 0° \text{ V}$$

$$\cos\phi = 0.85, \quad \phi = 31.8°$$

$$I_a\,(fl) = 7837 \text{ A}$$

$$\bar{I}_a = 7837 \angle -31.8°$$

$$= 7837\,(0.85 - j\,0.527)$$

For generating operation

$$\bar{E}_f = \bar{V}_t + j\,\bar{I}_a X_s$$

Case I: X_s (unsaturated) = 1.072 Ω

$$\therefore \quad E_f = 7505 + j\,7837\,(0.85 - j0.527) \times 1.072$$

$$= 11747 + j\,6841$$

$$E_f = 13594 \text{ V or } 23.54 \text{ kV (line)} = V_t\text{(OC) (linear model)}$$

$$\text{Regulation} = \frac{V_t(OC) - V_t}{V_t} \times 100 = \frac{13594 - 7505}{7505} \times 100 = 81\%$$

Case II: X_s (adjusted) = 0.873 Ω

$$\therefore \quad \bar{E}_f = 7505 + j\,7837\,(0.8 - j\,0.527) \times 0.873$$

$$= 11110 + j\,5815 = 12540 \angle 27.6° \text{ V}$$

or
$$E_f = 12540 \text{ V} \quad \text{or} \quad 21.72 \text{ kV (line)}$$

The field current needed for this E_f is found from the modified air-gap line. This cannot be found from the OCC where E_f will correspond to highly saturated region. In fact the machine under lagging load operates in lightly saturated region as the load current is demagnetizing. So on empirical basis field current is found from the *modified air-gap line (line joining origin to V_t (rated) in OCC)*. From this line use find.

$$I_f = \frac{810}{13} \times 21.72 = 1353 \text{ A}$$

For this value of field current when the machine is open circuited

$$V_t\,(\text{OC}) = 16.3 \text{ kV}$$

$$\text{Voltage regulation} = \frac{16.3-13}{13} \times 100 = 25.4\%$$

Comment: The voltage regulation as obtained by X_s (adjusted) as found from the modified air-gap line is far more realistic than that obtained by use *of* X_s (unsaturated).

(f) The phasor diagram is drawn in Fig. 8.18. It is immediately seen that E_f leads V_t by $\delta = 27.2°$.

Observation The approximation used in transformer (Eq. (3.60)) in calculating voltage drop cannot be used in a synchronous machine as X_s is much larger than in a transformer where X_{eq} is 0.05-0.08 pu. So phasor calculation must be carried out.

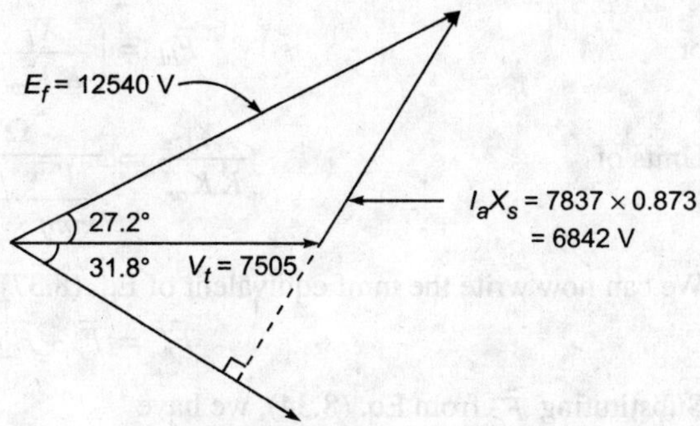

Fig. 8.18

8.5 MMF METHOD

LO 2

In the synchronous reactance method of finding voltage regulation, we linearly converted all mmfs into emfs and then heuristically adjusted the synchronous reactance to account for saturation. In the mmf we will follow the reverse procedure.

Reproducing Eqs (8.14) and (8.19), V_t and I_a are rated valued throughout,

$$\bar{F}_r = \bar{F}_f + \bar{F}_{ar} \tag{8.34}$$

$$\bar{V}_t = \bar{E}_r - j\bar{I}_a X_l - \bar{I}_a R_a \tag{8.35}$$

or

$$\bar{V}_t = \bar{E}'_r - \bar{I}_a R_a \tag{8.36}$$

where

$$\bar{E}'_r = \bar{E}_r - j\bar{I}_a X_l;\ \bar{E}'_r \text{ is not air-gap emf} \tag{8.37}$$

In the synchronous machine, induced emf is found from associated mmf through OCC and lags it 90°, i.e.,

$$\bar{E} \xleftrightarrow{OCC} -j\bar{F} \tag{8.38}$$

If we assume magnetic linearity (that is operation on the air-gap line), we can write

$$\bar{E} = -jK\bar{F};\ K = \text{constant, units}\left(\frac{V}{mmf}\right) \tag{8.39}$$

This result helps to convert the leakage reactance voltage to equivalent mmf. Thus

$$-j\bar{I}_a X_l = -jK\bar{F}_{al} \tag{8.40}$$

where $\bar{F}_{al}$ is equivalent mmf.

$$\bar{F}_{ar} = K_{ar}\bar{I}_a, \; K_{ar} = \text{armature constant, units} \left(\frac{mmf}{A}\right) \tag{8.41}$$

Substituting $\bar{I}_a$ from Eq. (8.41) in Eq. (8.40), we get

$$\frac{\bar{F}_{ar}}{K_{ar}} X_l = K\bar{F}_{al}$$

or

$$\bar{F}_{al} = \left(\frac{X_l}{K K_{ar}}\right)\bar{F}_{ar} \tag{8.42}$$

Units of

$$\frac{X_l}{K K_{ar}} = \frac{\Omega}{\frac{V}{mmf} \cdot \frac{mmf}{A}} = \text{constant dimensionless}$$

We can now write the mmf equivalent of Eq. (8.37) using Eq. (8.39)

$$\bar{F}'_r = \bar{F}_r + \bar{F}_{al}\,; -jK \text{ cancels out} \tag{8.43}$$

Substituting $\bar{F}_r$ from Eq. (8.34), we have

$$\bar{F}'_r = \bar{F}_f + \left(\bar{F}_{ar} + \bar{F}_{al}\right) \tag{8.44}$$

in which $\bar{F}_{ar}$ and $\bar{F}_{al}$ are in phase with $\bar{I}_a$

From $\bar{F}'_r$ we can find $\bar{E}'_r$ from the OCC

For clarity of understanding let us draw the phasor diagram corresponding to the phasor Eqs (8.36) and (8.44) where θ = pf angle. If I_aR_a is ignored, $\beta = \theta$

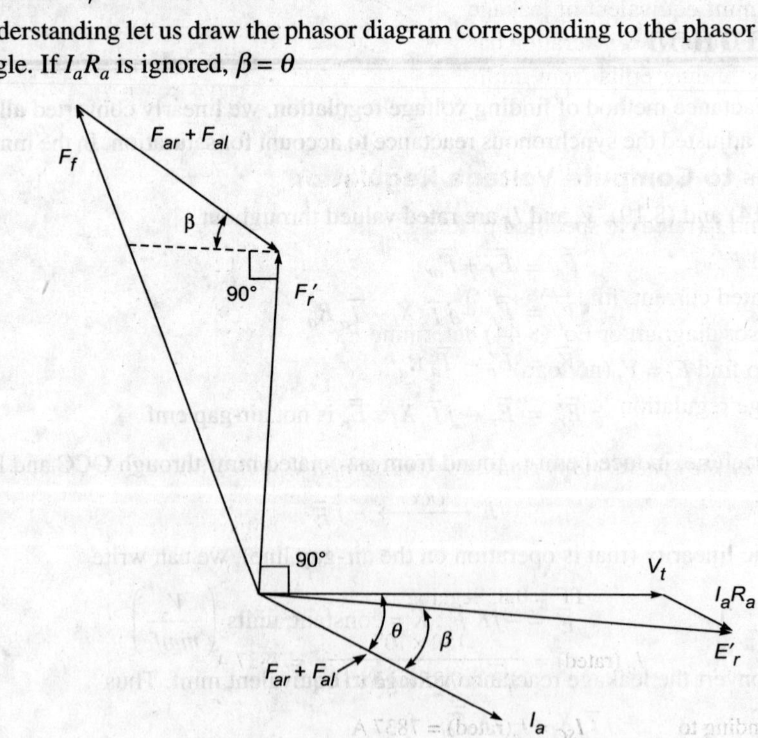

Fig. 8.19 *Phase diagram of mmf method*

Important Note

mmfs are measured in units of field current

$$\frac{mmf}{N_f} = I_f\,,\ N_f = \text{number of effective field turns which need not be known}$$

To determine $(F_{ar} + F_{al})$

Under short circuit at rated current, the field excitation is consumed in balancing armature reaction mmf and the balance induces emf to balance I_{SC}(rated) X_l voltage drop.

Thus

$$I_f\big|_{at\ rated\ SC\ current} = I_{f,ar} + I_{f,al}$$

With $\bar{F}_r'$ and $\left(\bar{F}_{ar} + \bar{F}_{al}\right)$ known as above $\bar{F}_f$ can be found by drawing the mmf phasor diagram to scale. Computationally it is convenient to use trigonometric relationship derived below.

The mmf phasor diagram is redrawn in Fig. 8.20 in convenient orientation. It easily follows that

$$F_f = \sqrt{(AB + BC\sin\beta)^2 + (BC\cos\beta)^2} \quad (8.45)$$

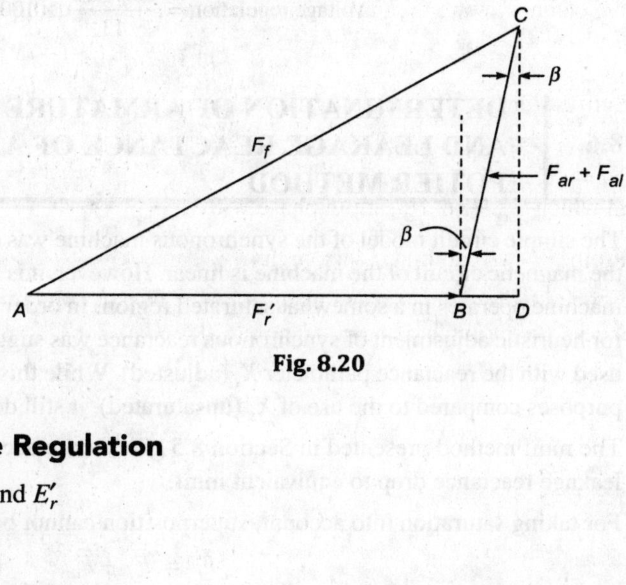

Fig. 8.20

Observation

In this (mmf) method linear assumption has only been made in finding mmf equivalent of leakage reactance voltage drop. Otherwise, saturation has been accounted for by finding emfs from mmfs and vice versa.

Important! Steps to Compute Voltage Regulation

1. For V_t (rated) and I_a(rated) at specified pf find E_r'
2. From OCC find F_r'
3. From SCC at rated current, find $(F_{ar} + F_{al})$
4. From mmf phasor diagram or Eq. (8.44) determine F_f
5. Consult OCC to find $E_f = V_t$ (no load)
6. Compute voltage regulation

Example 8.2 *For the synchronous generator of Example 8.1, determine its voltage regulation by the mmf method.* LO 2

Solution

$$\text{PF} = 0.85 \text{ lagging}$$

$$I_a\ (\text{rated}) = \frac{150 \times 10^6}{\sqrt{3} \times 0.85 \times 13 \times 10^3} = 7837\ \text{A}$$

From Fig. 8.17 corresponding to $I_{SC} = I_a(\text{rated}) = 7837$ A

$$I_f = 750 \text{ A} = F_{ar} + F_{al}$$

At rated current, 0.85 pf lag

$$E_r' = \bar{V}_t = 13 \text{ kV (line-to-line) ; } I_a R_a = 0.$$

From the OCC corresponding mmf is

$$I_f = 810 = F_r' \text{; in units (A) of field current}$$

$$\beta = \theta = \cos^{-1} 0.85 = 31.8°$$

From the phasor diagram of Fig. 8.20 and Eq. 8.44

$$F_f = \sqrt{(810 + 750 \sin 31.8°)^2 + (750 \cos 31.8°)^2} = 1505 \text{ A} = I_f$$

At I_f = 1500 A from the OCC, we find

$$E_f = 16.3 \text{ kV (line)}$$

$$V_t \text{(no load)} = E_f = 16.3 \text{ kV (line)}$$

$$\text{Voltage regulation} = \frac{16.3 - 13}{13} \times 100 = 25.4\%$$

8.6 DETERMINATION OF ARMATURE REACTION AMPERE-TURNS AND LEAKAGE REACTANCE OF A SYNCHRONOUS MACHINE—POTIER METHOD

LO 2

The simple circuit model of the synchronous machine was obtained in Sec. 8.3 by making the assumption that the magnetic circuit of the machine is linear. However, it is known that, under normal operating conditions, the machine operates in a somewhat saturated region. In order to take account of magnetic saturation a procedure for heuristic adjustment of synchronous reactance was suggested so that the simple circuit model could still be used with the reactance parameter X_s (adjusted). While this does give more accurate results for many practical purposes compared to the use of X_s (unsaturated), it still does not fully account for magnetic saturation.

The mmf method presented in Section 8.5 gives more accurate results but it also uses linearity in converting leakage reactance drop to equivalent mmf.

For taking saturation into account, superposition cannot be applied and, therefore, mmf phasor equation,

$$\bar{F}_r = \bar{F}_f + \bar{F}_{ar} \qquad \text{(see also Fig. 8.5 (b))}$$

must be used.

The induced emfs E_f and E_r corresponding to F_f and F_r can be found from the OCC (which takes magnetic saturation into account). The problem, however, is to determine F_{ar} for a given armature current I_a, i.e. the armature reaction constant

$$K_{ar} = \frac{F_{ar}}{I_a} \tag{8.46}$$

This proportionality constant[§§] must be found experimentally for a given machine.

Knowing the magnetization characteristic of the machine and the proportionality constant K_{ar}, the phasor diagram of Fig. 8.5(b) can be constructed for given operating conditions (say, for generating mode) wherein

[§§] This proportionality constant is given by Eq. (5.44b) provided the design constants of the machine, K_w and N_{ph} (series) are known. Even then a designer would be interested to determine the experimental value of this proportionality constant to verify his design calculations.

no approximation of neglecting the nonlinearity need be made. The next step then is to find the terminal voltage of the machine which immediately follows from the phasor equation

$$\bar{V}_t = \bar{E}_r - \bar{I}_a (R_a + jX_l) \text{ (generating mode)} \tag{8.47}$$

The armature resistance, R_a, can be easily measured under dc conditions and duly corrected to its ac value and operating temperature; it can even be altogether neglected without any significant loss of accuracy of analysis since its value is only about 0.01 pu as stated already. However, X_l, the leakage reactance of the machine must be determined. This can be calculated from the design parameters of the machine (provided these are known) but only to a low degree of accuracy. Therefore, it is necessary to obtain its actual value by experimental methods.

Accurate analysis of the synchronous machine performance can be carried out under any operating conditions provided K_{ar} (Eq. (8.34)) and X_l, the leakage reactance are known. Because of the nonlinearity of iron, there are a variety of experimental methods of determining these quantities to high but varying degrees of accuracy. One of the well-known methods called the *Potier method* is described here.

Steps to Find Synchronous Reactance by Potier Method

In the Potier method, tests are conducted to determine the following two characteristics with the machines running at synchronous speed.

1. **Open-circuit characteristic (OCC)** as described in Sec. 8.3, with reference to Fig. 8.12. This is redrawn in Fig. 8.21.

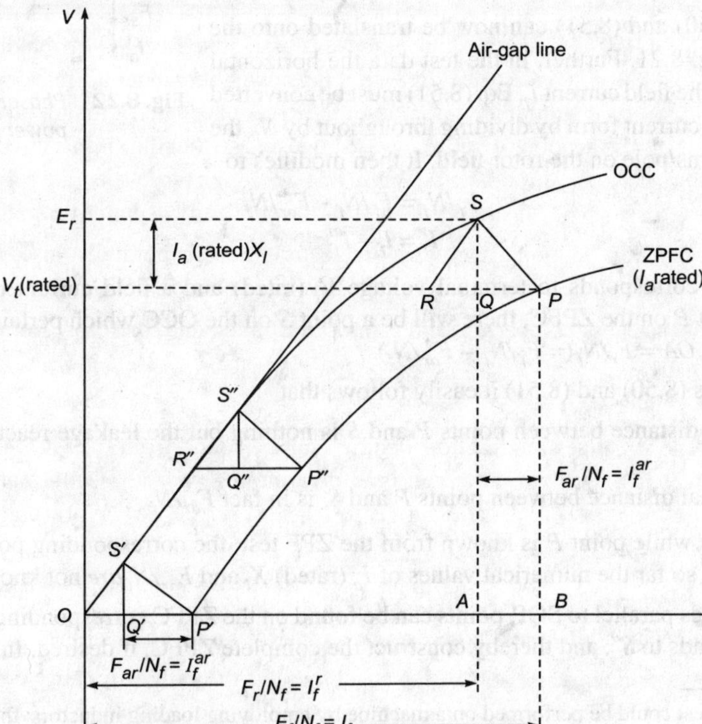

Fig. 8.21 *Graphical features of the Potier method*

2. **Zero power factor (lagging) characteristic (ZPFC)** This is conducted by loading the machine as a generator with pure inductive load[¶¶] (balanced 3-phase) which is adjusted to draw rated current from the machine while the field current is adjusted to give various values of terminal voltage. Figure 8.21 also shows the ZPFC. However, there is no need for conducting this test fully to determine the ZPFC. All one needs is two points on this characteristic—*P* corresponding to a field current which gives the rated terminal voltage while the ZPF load is adjusted to draw rated current, and the point *P'* which corresponds to the short-circuit conditions on the machine ($V_t = 0$) with the field current adjusted to give rated armature current. Since the armature resistance is of negligible order, the short-circuit current lags behind the resultant induced emf E_r by almost 90°, V_t being equal to zero. Therefore, *P'* constitutes a point on the ZPFC. It will soon be shown that the complete ZPFC, if required, can be constructed from the knowledge of the points *P* and *P'*.

Figure 8.22 gives the phasor diagram under conditions of zero power factor (lagging) load with the armature resistance neglected. It is seen from this figure that the mmf and voltage phasor equations

$$\bar{V}_t = \bar{E}_r - j\bar{I}_a X_l \tag{8.48}$$

and

$$\bar{F}_r = \bar{F}_f + \bar{F}_{ar} \tag{8.49}$$

These reduce to simple algebraic equations

$$V_t = E_r - I_a X_l \tag{8.50}$$

and

$$F_r = F_f - F_{ar} \tag{8.51}$$

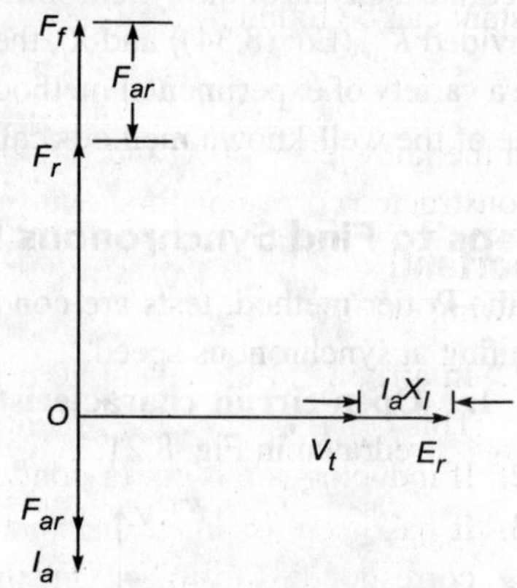

Fig. 8.22 *Phasor diagram for zero power factor (lagging) load*

The algebraic Eqs (8.50) and (8.51) can now be translated onto the OCC and ZPFC of Fig. 8.21. Further, in the test data the horizontal axis of Fig. 8.21 being the field current I_f, Eq. (8.51) must be converted into its equivalent field current form by dividing throughout by N_f, the effective number of turns/pole on the rotor field. It then modifies to

$$F_r/N_f = F_f/N_f - F_{ar}/N_f \tag{8.52a}$$

or

$$I_f^r = I_f - I_f^{ar} \tag{8.52b}$$

Point *P* on the ZPFC corresponds to terminal voltage V_t (rated) and a field current of $OB = F_f/N_f = I_f$. Corresponding to point *P* on the ZPFC, there will be a point *S* on the OCC which pertains to the emf E_r and the resultant excitation $OA = F_r/N_f (= F_f/N_f - F_{ar}/N_f)$.

From Fig. 8.21 and Eqs (8.50) and (8.51) it easily follows that

(i) *SQ*, the vertical distance between points *P* and *S* is nothing but the leakage reactance drop I_a (rated) X_l.

(ii) *QP*, the horizontal distance between points *P* and *S*, is in fact F_{ar}/N_f.

It is observed here that while point *P* is known from the ZPF test, the corresponding point *S* on the OCC is not yet known because so far the numerical values of I_a (rated) X_l and F_{ar}/N_f *are* not known.

By constructing triangles parallel to SQP, points can be found on the ZPFC corresponding to the points on the OCC, e.g. *P''* corresponds to *S''*, and thereby construct the complete ZPFC, if desired. In the reverse process,

[¶¶] The zero power factor test could be performed on a machine by employing loading inductors; these constitute the load of nearly but not exactly zero of and also are impractical for large machines. The ZPF test at rated voltage only could also be conducted by synchronizing it to the mains and regulating its excitation to yield zero pf operation (see Sec. 8.10).

corresponding to P', the short-circuit point, S' can be located on the OCC by drawing $S'P'$ parallel to SP. Obviously

$$S'Q' = I_a \text{ (rated) } X_l \text{ and } Q'P' = F_{ar}/N_f$$

Since the initial part of the OCC is almost linear, OS' is part of air-gap line. Therefore going back to point P and by taking horizontal distance $RP = OP'$ and drawing RS parallel to OS', the desired point S can be located on the OCC corresponding to the known point P on the ZPFC.

Once point S on the OCC has been located, the following can be measured to scale:

$$SQ = I_a \text{ (rated) } X_l \quad \text{and} \quad QP = F_{ar}/N_f = I_f^{ar} \tag{8.53}$$

from which X_l and F_{ar} can be calculated. Since I_f^{ar} corresponds to I_a (rated), the armature mmf proportionality constant can be found as

$$K'_{ar} = I_f^{ar}/I_a \text{ (rated)}$$

With the knowledge of X_l and K'_{ar} for the stator windings, the complete phasor diagram of the machine can be constructed corresponding to any operating conditions, generating or motoring.

Important! It must be observed here that the Potier method though elegant is not exact because of the following explicit and implicit assumptions made therein:

1. In arriving at the algebraic Eqs (8.50) and (8.51)/(8.52), the armature resistance has been neglected. This being a very valid assumption, introduces no error of any significance.
2. If inductors are used for conducting the ZPF test, the power factor is somewhat different from zero.
3. It has been assumed that in Fig. 8.21, $S'Q' = SQ = I_a$ (rated) X_l which means that in the ZPF test corresponding to point P and the short-circuit test corresponding to point P', the leakage reactance of the machine is assumed to remain unchanged. This is not altogether correct because the machine excitation under short-circuit conditions is OP' while it is OA for point P (this point on the ZPFC corresponds to rated terminal voltage and rated armature current). Since $OA >> OP'$, point P corresponds to saturated conditions on the machine with a larger leakage flux and hence a larger value of leakage reactance contrary to the assumption made.

Better methods [48] are available in literature which attempt to overcome assumption 3 above but these are beyond the scope of this book.

Construction Procedure—Potier Method (Fig. 8.23)

Important! From the SC test $I_{SC} = I_a$ (rated) which locates P' on I_f– axis such that $OP' = I_f^{SC}$

- From the ZPF test at V_t (rated) and I_a (rated) locate the point P such that $OB = I_f^{zpf}$
- From P draw a line parallel to I_f – axis and locate R such that $RP = OP'$
- From R draw line parallel to the air-gap line which locates point S on OCC.
- From S draw SQ perpendicular to RP
- $PQ = I_f^{ar}$; $X_l = \dfrac{SQ \text{ (to scale)}}{\sqrt{3}\, I_a \text{(rated)}}$; OC voltage is line voltage

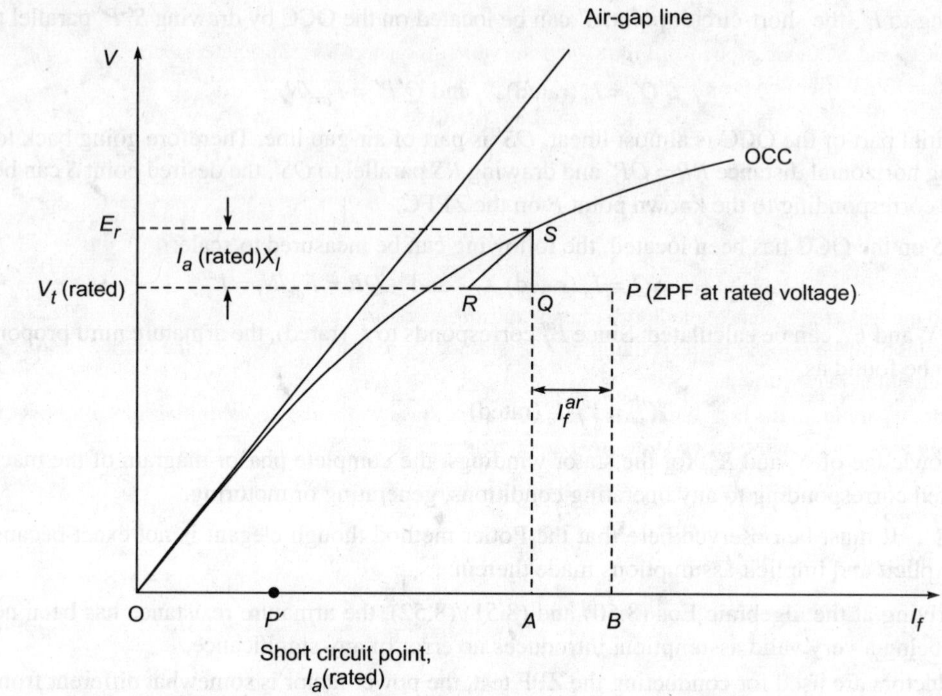

Fig. 8.23 *Construction procedure Potier method*

Example 8.3 *The following data were obtained for the OCC of a 10 MVA, 13 kV, 3-phase, 50 Hz, star-connected synchronous generator:*

I_f (A)	50	75	100	125	150	162.5	200	250	300
V_{OC} (line) (kV)	6.2	8.7	10.5	11.6	12.8	13.7	14.2	15.2	15.9

An excitation of 100 A causes the full-load current to flow during the short-circuit test. The excitation required to give the rated current at zero pf and rated voltage is 290 A.

(a) Calculate the adjusted synchronous reactance of the machine.

(b) Calculate the leakage reactance of the machine assuming the resistance to be negligible.

(c) Determine the excitation required when the machine supplies full-load at 0.8 pf lagging by using the leakage reactance and drawing the mmf phasor diagram. What is the voltage regulation of the machine?

Also calculate the voltage regulation for this loading using the adjusted synchronous reactance. Compare and comment upon the two results.

LO 2

Solution

$$I_a \text{ (rated)} = \frac{10 \times 10^6}{\sqrt{3} \times 13 \times 10^3} = 444 \text{ A}$$

(a) The OCC and SCC are plotted in Fig. 8.24 from which the short-circuit armature current corresponding to V_t (rated) = 13 kV (line) on the OCC is

$$I_{SC} = 688 \text{ A}$$

$$\therefore \quad X_s \text{ (adjusted)} = \frac{V_t\text{(rated)}}{\sqrt{3} \times I_{SC}\text{(corresponding to } V_t\text{(rated))}}$$

$$= \frac{13 \times 1000}{\sqrt{3} \times 688} = 10.9 \ \Omega$$

(b) To find the leakage reactance, the Potier triangle must be constructed. Figure 8.25 shows the plot of the OCC and the location of the point P corresponding to ZPF at rated current and voltage and the point P' corresponding to short-circuit at rated current.

At P draw parallel to the horizontal axis $PR = OP'$. At R draw RS parallel to the initial slope of the OCC thereby locating the point S on the OCC. Draw SQ perpendicular to RP. Then

$$SQ = \sqrt{3}\, I_a \text{ (rated) } X_l = 1200 \text{ V}$$

$$\therefore \quad X_l = \frac{1200}{\sqrt{3} \times 444} = 1.56 \ \Omega$$

(c) From Fig. 8.22, the armature reaction in the equivalent field current is

$$I_f^{ar} = QP = 90 \text{ A}$$

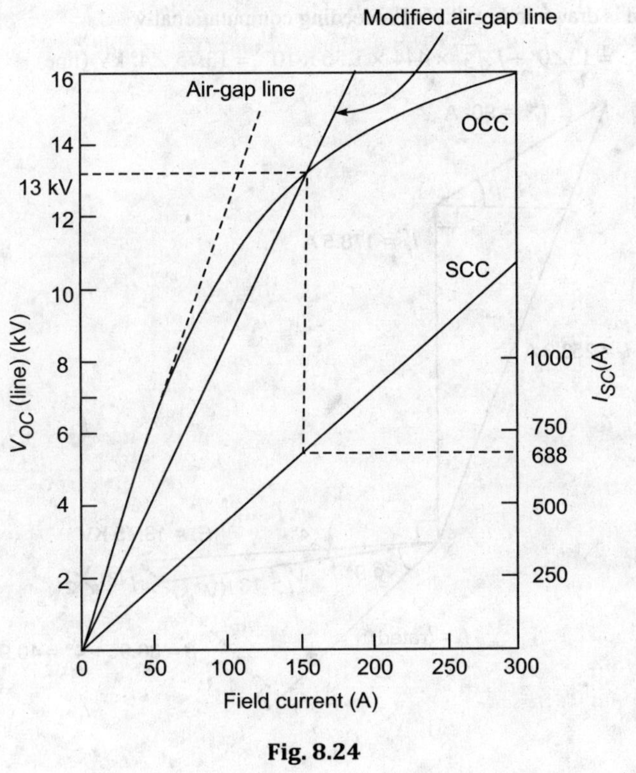

Fig. 8.24

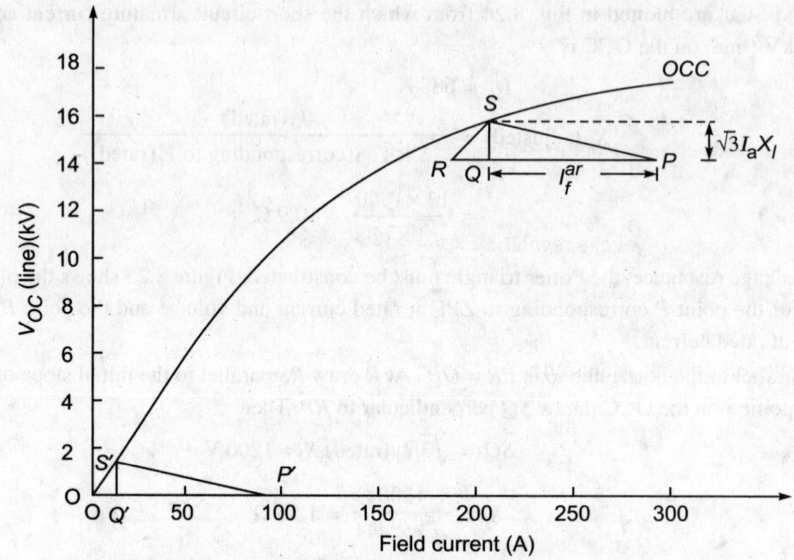

Fig. 8.25

The phasor diagram is drawn in Fig. 8.26. Proceeding computationally

$$\bar{E}_r = 13\angle 0^\circ + j\sqrt{3} \times 444 \times 1.56 \times 10^{-3} = 13.75 \angle 4^\circ \text{ kV (line)}$$

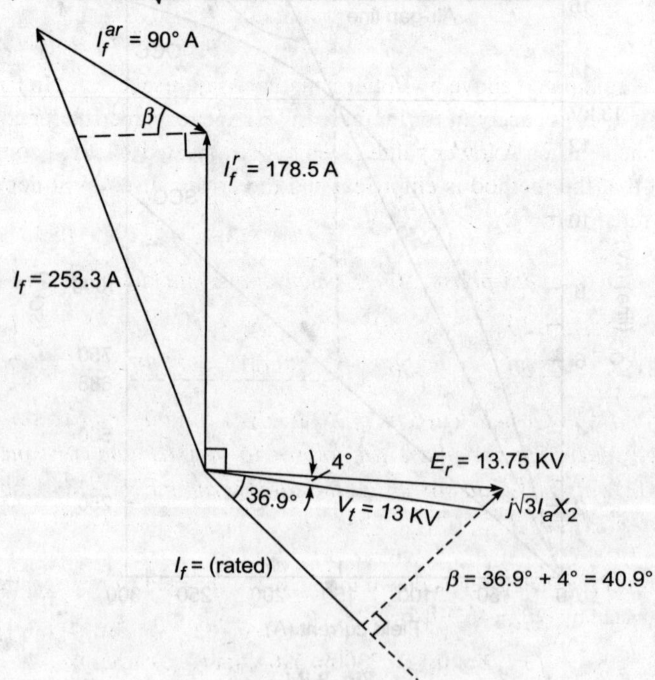

Fig. 8.26

From the OCC corresponding to $E_r = 13.75$ kV

$$I_f^r = 185 \text{ A leads } E_r \text{ by } 90°$$

From the phasor diagram

$$I_f = [(185 + 90 \sin 40.9°)^2 + (90 \cos 40.9°)^2]^{1/2} = 253 \text{ A}$$

corresponding to which the value of E_f from the OCC is

$$E_f = 15.2 \text{ kV (line)}$$

$$\therefore \quad \text{Voltage regulation} = \frac{E_f - V_t}{V_t} \times 100$$

$$= \frac{15.2 - 13}{13} \times 100 = 16.9\%$$

Voltage regulation using X_s (adjusted):

$$\bar{E}_f = \bar{V}_t + j\sqrt{3}\,\bar{I}_a \text{ (rated) } X_s \text{ (adjusted)}$$

$$= 13000 + j\sqrt{3} \times 444\,(0.8 - j\,0.6) \times 10.9$$

$$= 18029 + j\,6706$$

or

$$E_f = 19.24 \text{ kV (line)}$$

$$I_f \text{ (from modified air-gap line)} = \frac{150}{13} \times 19.24 = 222 \text{ A}$$

$$V_t\,(OC)|_{I_f = 222\text{ A}} = 14.8 \text{ kV}$$

$$\therefore \quad \text{Voltage regulation} = \frac{14.8 - 13}{13} \times 100 = 13.85\%$$

Comment

The voltage regulation as calculated above by Potier's method is quite accurate. In fact this value is somewhat lower than the actual as it does not account for increase in leakage reactance under condition of load. This fact has already been mentioned earlier. A lower value given by X_s (adjusted) method compared to X_s (unadjusted) is explained by the fact that the method is empirical and the error caused by it depends upon the saturation level of the machine at rated voltage.

Example 8.4 *The OCC of a 3-phase, 50 Hz synchronous machine is given by the following data:*

I_f (A)	15	30	50	75	90	120	160
V_{OC} (line) (V)	600	1200	2000	2900	3300	3700	4000

Under short-circuit conditions a field current of 40-A gives the full-load stator current. The armature resistance and leakage reactance per phase are known to be 0.01 and 0.12 pu. When the machine is operating as a motor drawing full-load current at the rated terminal voltage of 3.3 kV and 0.8 pf leading, calculate the field current required.

LO 2

Solution The OCC as per the data given is drawn in Fig. 8.27. The field current required to circulate the full-load short-circuit current is indicated by $OP' = 40$ A. Now

$$\sqrt{3}\, I_a X_l = 0.12 \times 3300 = 396 \text{ V (at rated current)}$$

Corresponding to 396 V on OCC, a point S' is marked and a vertical line $S'Q'$ drawn. $Q'P'$ now corresponds to $F_{ar}/N_f = I_f^{ar} = 30$ A.

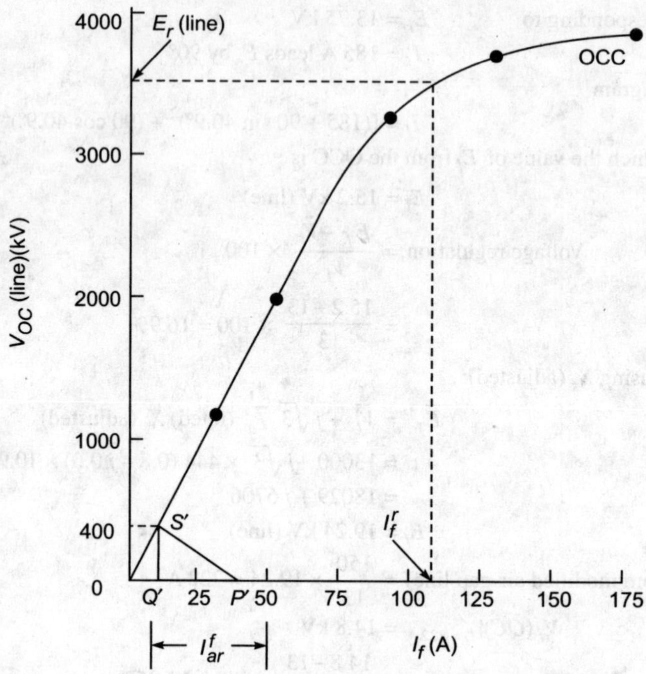

Fig. 8.27

Given that $R_a = 0.01$ pu, the voltage drop in resistance is so small that it can be neglected.

$$V_t = \frac{3.3}{\sqrt{3}} \times 1000 = 1905 \text{ (phase)}$$

$$I_a X_l = \frac{396}{\sqrt{3}} = 229 \text{ V}$$

The machine is operating as a motor at 0.8 pf leading. Therefore as per Eq. (8.12) and Fig. 8.8(b)

$$\bar{E}_r = \bar{V}_t - jI_a X_l \text{ ; } V_t \text{ (phase)} = \frac{3300}{\sqrt{3}} = 1905 \text{ V}$$

$$\bar{E}_r = 1905 \angle 0^\circ - j\,229\,(0.8 + j0.6) = 2042 - j\,183.2 = 2050 \angle -5.1^\circ \text{ V or } 3551 \text{ V (line)}$$

From the OCC for $E_r = 3551$ V, we get

$$I_f^r = 108 \text{ A}$$

The phasor diagram is drawn in Fig. 8.28. I_f^r leads E_r by 90° and I_f^{ar} is drawn anti-parallel to I_a (motoring operation) Angle $\beta = 36.9^\circ + 5.1^\circ = 42^\circ$. The construction lines are shown dotted. From the mf phasor diagram geometry

$$I_f = [(I_f + I_f^{ar} \sin \beta)^2 + (I_f^{ar} \cos \beta)^2]^{1/2}$$

or $$I_f = [(108 + 30 \sin 42^\circ)^2 + (30 \cos 42^\circ)^2]^{1/2}$$

or $$I_f = 125 \text{ A}$$

From the OCC at $I_f = 125$ A

$$E_f = 3760 \text{ V (line)}$$

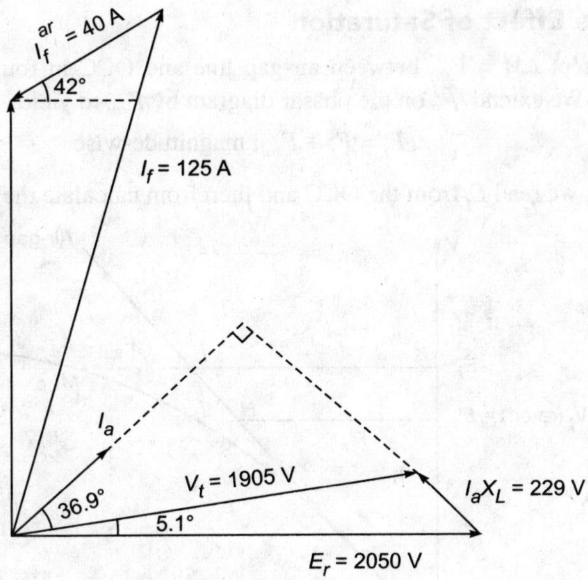

Fig. 8.28

8.7 | ASA (AMERICAN STANDARDS ASSOCIATION) METHOD (LATEST)

LO 2

ASA method of finding voltage regulation is heuristic modification of the mmf method which requires the value of X_l. Therefore, OCC and ZPFC tests are conducted and X_l is determined from the Potier triangle by the method of Section 8.6.

The OCC is sketched in Fig. 8.29 and the short circuit point is located on the I_f– axis.

The basic kVL equations of the mmf method are:

$$\bar{E}_r' = \bar{V}_t + \bar{I}_a R_a \tag{8.54}$$

or

$$\bar{E}_r' = \bar{V}_t \,;\; \bar{I}_a R_a \text{ ignored; } \beta = \theta$$

$$\bar{E}_r = \bar{V}_t + j\bar{I}_a X_l \tag{8.55}$$

where we know X_l.

Modification

In stead of finding $\bar{F}_r'$ from the OCC against $E_r' = V_t$ (rated). F_r' is read on the air-gap line as shown in Fig. 8.29(a). If means saturation has been ignored.

Now draw the mmf phasar diagram of Fig. 8.29(b). with $\beta = \theta$, the *pf* angle. This determines

$$\bar{F}_f' = \bar{F}_r' + \left(\bar{F}_{ar} + \bar{F}_{al}\right); \text{ saturation ignored}$$

Incorporating the Effect of Saturation

Against E_r, the intersept $LM = F_{sat}$ between air-gap line and OCC in found out. Heuristically this is the saturation correction. We extend $\bar{F}'_f$ on the phasar diagram by F_{sat} to yield

$$F_f = F'_f + F_{sat}\text{; magnitude-wise}$$

Having determined E_f we read E_f from the OCC and therefrom calculate the voltage regulation.

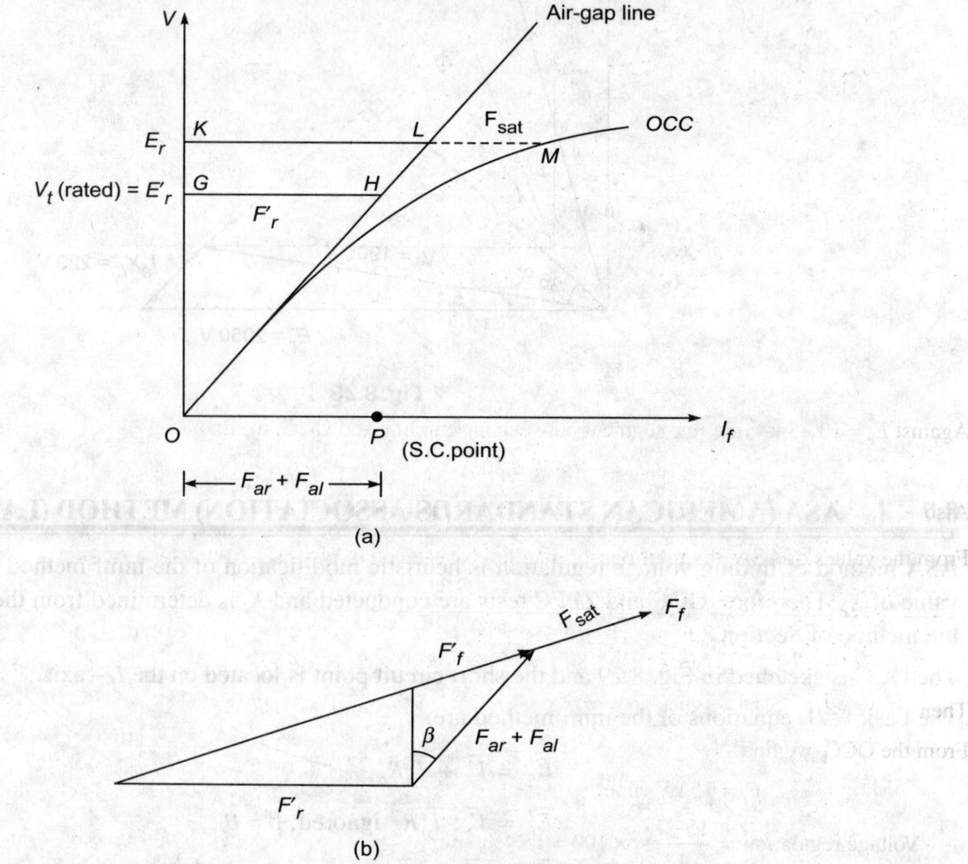

Fig. 8.29 *Determination of voltage regulation by ASA (latest) method*

Example 8.5 *For Example 8.4 calculate the voltage regulation by the ASA method.* LO 2

Solution For the data of Example 8.4 the OCC and air-gap are drawn in Fig. 8.30. Ignoring I_aR_a drop

$$E'_r = V_t \text{ (rated)} = 13 \text{ kV (line)}$$

and $$E_r = \bar{V}_L + j I_a \textit{(rated)} X_l = 13.75 \text{ kV } \textit{(line)}\text{; calculated in Example 8.4}$$

From the air-gap against $V_t = 13$ kV, we find

$$F'_r = 100 \text{ A}, \quad \beta = 36.9° + 4° = 40.9°$$

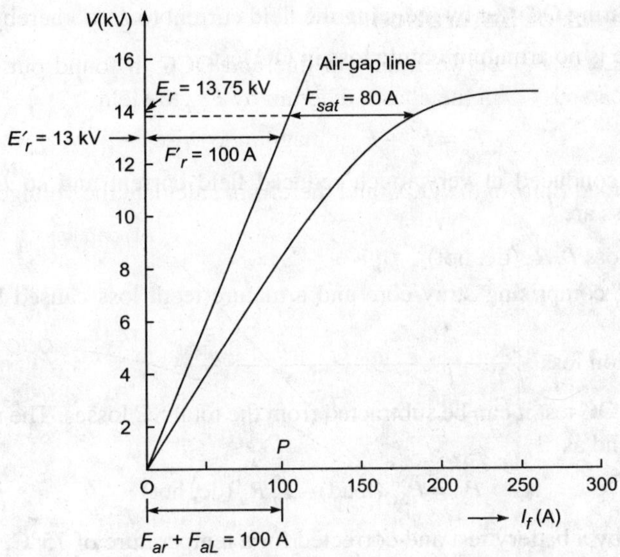

Fig. 8.30

Against $E_r = 13.75$ kV, the line segment between air-gap line and OCC, we find

$$F_{\text{sat}} = 80 \text{ A}$$

Also $$OP' = F_{ar} + F_{al} = 100 \text{ A}$$

From the values of ASA, the mmf phasor diagram is sketched in Fig. 8.31 from which we find

$$F_f' = [(100 + 100 \sin 40.9°)^2 + (100 \cos 40.9°)^2]^{1/2} = 182 \text{ A}$$

Then $$F_f = 182 + 80 = 262 \text{ A}$$

From the OCC, we find

$$E_f = 15 \text{ kV (line)}$$

$$\text{Voltage regulation} = \frac{15-13}{13} \times 100 = 15.4\%$$

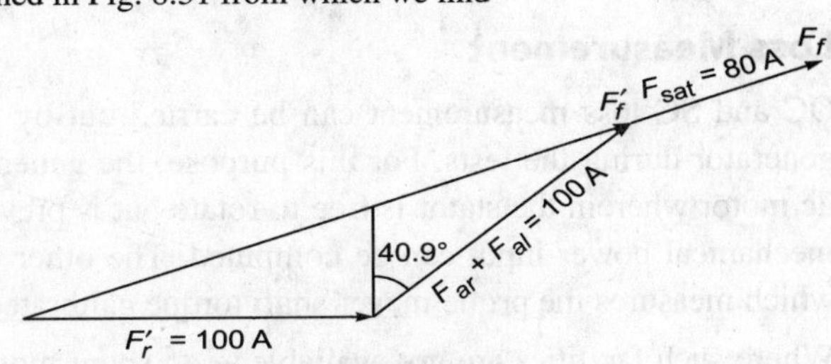

Fig. 8.31

8.8 EFFICIENCY OF SYNCHRONOUS MACHINES

LO 2

The losses in a synchronous machine have been dealt with elaborately in Section 8.4 while presenting OC and SC test. These are summarized below:

OC Test

As it is no load test, the open circuit loss P_{OC} comprises two losses.

Recall (i) Core loss, P(core, OC) (ii) windage and friction loss, P_{wf}. P(core, OC) is proportional to square of V_{OC} and P_{wf} is constant (machine speed is synchronous)

P_{wf} gets separately out during OC test by reducing the field current to zero, thereby making P(core) = 0. It is to be noted that there is no armature ohmic loss in OC test.

SC Test

Recall As the test is conduced at very much reduced field current and so P(core) is negligible. The components of the P_{SC} loss are

1. Armature copper loss $I_a^2 R_a$ (dc, hot)
2. Stray load loss, P_{st} comprising stray core and armature teeth loss caused by leakage flux and stray copper loss
3. Windage and friction loss

As P_{wf} is known from the OC test it can be subtracted from the total SC losses. The remaining loss is P_{sc}(load) loss. The stray loss is found as

$$P_{st} = P_{SC}(\text{load}) - I_a^2 R_a \text{ (dc, hot)}$$

R_a (dc) can be measured by a battery test and corrected for a temperature of 75°C.

Synchronous machine losses from OC and SC tests as separated out above are

P (core, OC)

P_{wf} constant

$I_a^2 R_a$ (dc, hot), computed

P_{st}

Loss Measurement

OC and SC loss measurement can be carried out by measuring the mechanical input to the synchronous generator during the tests. For this purpose, the generator is run at synchronous speed by a dynamometer dc motor wherein the stator is free to rotate but is prevented by spring balances from the readings of which mechanical power input can be computed. The other convenient and accurate method is the torque meter which measures the prime mover shaft torque calibrated in torque units.

Where such facilities are not available, a dc shunt motor of rating somewhat more than estimated machine losses. Before coupling the dc motor to the synchronous machine the Swinburne's test is performed on it at synchronous speed to determine the rotational loss of the motor and also its armature resistances is measured. The dc motor is now-coupled to the synchronous generator and the set driven at synchronous speed. The motor terminal voltage and armature current is measured during the test.

The OC/SC loss is found as

$$V_t I_a - I_a^2 R_a - P_{\text{rot}} = P_{OC}/P_{SC}$$

The efficiency calculation are illustrated by a comprehensive example.

Example 8.6 *A 60 kVA, 400 V, 50 Hz synchronous generator is tested for by means OC and SC tests whose data are given below:*

	Field current	OC voltage (line)	SC current (line)	SC loss (3 phase)
OC	2.85 A	400 V		
SC (at 75°C)	1.21 A	—	108 A	3.95 kW

The OC loss data are plotted in Fig. 8.32.

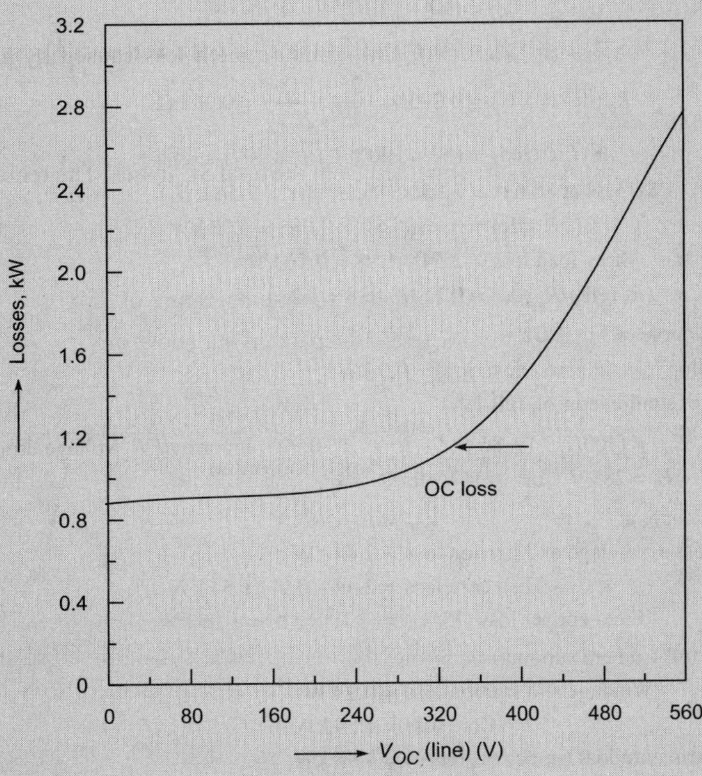

Fig. 8.32

The armature (star connected) has dc resistance/phase at 25°C of 0.075 Ω. The machine is operating at full-load 0.8 pf lagging at rated terminal voltage with a field current of 3.1 A. The field resistance is 110 Ω at 75°C.

Calculate:

(a) Effective armature resistance and synchronous reactance (saturated).

(b) Full-load stray load loss.

(c) Ratio of effective armature resistance/dc resistance.

(d) Various category of losses at full-load (75°C)

(e) Full-load efficiency at 75°C

LO 2

Solution

(a) From SC test data

$$3 \times (108)^2 \times R_a\,(\text{eff}) = 3.95 \times 100$$

or $$R_a\,(\text{eff}) = 0.113\ \Omega$$

As SC current is linear function of field current

$$I_{SC}\,(I_f = 2.85\text{ A}) = 108 \times 2.85/1.21 = 254\text{ A}$$

$$Z_s = (400/\sqrt{3})/254 = 0.91$$

$$X_s = [(0.91)^2 - (0.113)^2]^{1/2} = 0.903\ \Omega\ \text{(saturated)}$$

(b) $$R_a\,(\text{dc})\,(75°\text{C}) = 0.075 \times \left(\frac{75+273}{25+273}\right) = 0.088\ \Omega$$

$$I_a\,(\text{rated}) = (60 \times 1000)/(\sqrt{3} \times 400) = 86.6\text{ A}$$

$$\text{SC loss at 86.6 A} = 3.95 \times (86.6/108)^2 = 2.54\text{ kW}$$

$$3\,I_a^2\,R_a\,(\text{dc}) = 3 \times (86.6)^2 \times 0.088 = 1.98\text{ kW}$$

$$\text{Stray load loss} = 2.54 - 1.98 = 0.56\text{ kW}$$

(c) $$R_a\,(\text{eff})/R_a\,(\text{dc}) = 0.113/0.088 = 1.284$$

(d) From the OC loss curve of Fig. 8.32

Windage and friction loss (at zero excitation) = 0.9 kW

We will now find excitation emf on full-load

$$\bar{E}_f = (400/\sqrt{3}) + 86.6\,\angle -36.9° \times j\,0.903;\ \text{Ignoring } I_aR_a \text{ voltage drop}$$

or $$E_f = 285\text{ V} \quad\text{or}\quad 494\text{ V (line)}$$

From the OC loss curve at $V_{OC} = E_f = 494$ V

Core loss + windage and friction loss = 2.44 kW

Then core loss = 2.44 – 0.9 = 1.54 kW

Field copper loss (75°C) = $(3.1)^2 \times 110 = 1.06$ kW

Various losses on full load are summarised below:

Windage and friction loss = 0.9 kW

Core loss = 1.54 kW

Armature loss (in dc resistance) = 1.98 kW

Stray load loss = 0.56 kW

Field copper loss = 1.06 kW

Total loss = 6.04 kW

Note: Losses in exciter*** and field rheostat are not accounted against the machine.

(e) Output = 60 × 0.8 = 48 kW

Input = Output + losses = 48 + 6.04 = 54.04 kW

Efficiency, $$\eta = 1 - \frac{6.04}{48 + 6.04} = 88.8\%$$

*** In a practical arrangement field will be supplied from a dc exciter (coupled to machine shaft) through a rheostat for adjusting the field current This field current (3.1 A) with field resistance of 110 Ω requires voltage at field terminal of 341 V. So exciter voltage must be 400 V. The difference is dropped in the field rheostat.

Concept Check LO 2

1. A synchronous motor with terminal voltage V_t is drawing zero pf current. Write the phasor expression for the excitation emf, E_f. Is E_f more or less than V_t, magnitude-wise. What is the value of the power angle?
2. Under what conditions does the voltage regulation of a synchronous generator become negative?
3. Explain why the SCC is linear.
4. Distinguish between X_s (unsaturated) and X_s (saturated). Which one should be used for higher accuracy in predicting the voltage regulation of a synchronous generator.
5. The synchronous reactance of an alternator is not constant over the entire operating range. What is this so? What value would you use?
6. State the assumptions made in the Potier method and the effect they have on the accuracy of voltage regulation as calculated by this method.

8.9 NATURE OF ARMATURE REACTION

LO 3 Understand the concept of synchronous machine connected and staying synchronized to an infinite bus, and effect of armature reaction, change of excitation, and load, on machine performance

The nature of the armature reaction is dependent on the power factor at which the machine is operating mode-generating/motoring. For simplicity of explanation, it will be assumed here that the armature resistance and leakage reactance are negligible so that

$$\bar{V}_t = \bar{E}_r$$

Figure 8.33 shows the phasor diagrams, with component fluxes (i.e. the magnetic circuit is assumed linear) indicated therein for a generating machine for (a) zero power factor lagging (Fig. 8.33(a)), (b) zero power factor leading (Fig. 8.33(b)) and (c) for unity power factor (Fig. 8.33(c)). The following observations are immediately made from these phasor diagrams.

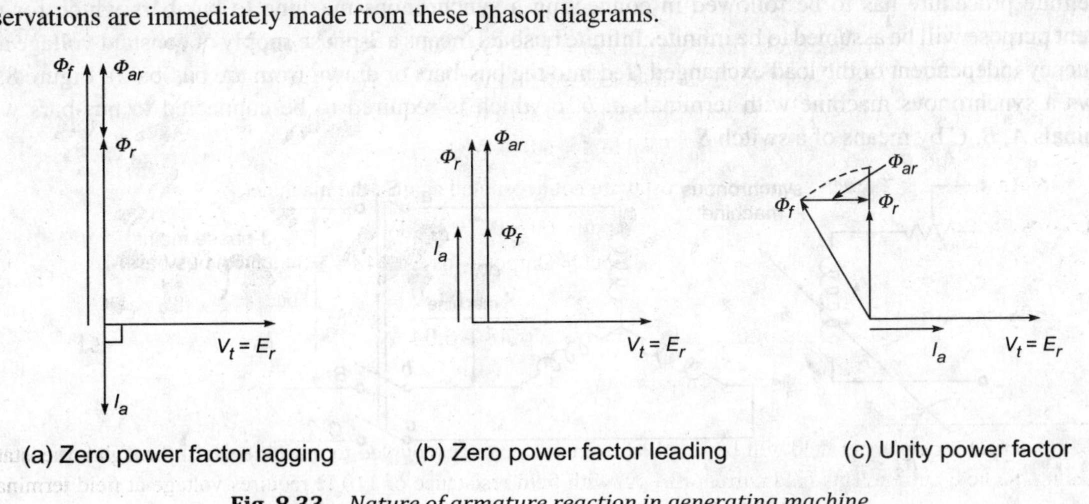

(a) Zero power factor lagging (b) Zero power factor leading (c) Unity power factor

Fig. 8.33 *Nature of armature reaction in generating machine*

Important!

1. Armature reaction is *demagnetizing* (Φ_{ar} opposes Φ_f) when a generating machine supplies zero power factor lagging current.
2. Armature reaction is *magnetizing* (Φ_{ar} aids Φ_r) when a generating machine supplies zero power factor leading current.
3. Armature reaction is mostly *cross-magnetizing* (i.e. at 90° to Φ_r) though it has a small demagnetizing component (see dotted curve in Fig. 8.33(c), when a generating machine supplies unity power factor current.

Important! From the above discussion the following more general conclusions regarding a synchronous machine in *generating mode* can be drawn:

(i) When the machine supplies a lagging power factor current, the armature reaction has both demagnetizing and cross-magnetizing components.

(ii) When the machine supplies leading power factor current, the armature reaction has both magnetizing and cross-magnetizing components.

Since in a motoring machine the armature reaction mmf and flux are in phase opposition to the armature current, the nature of the armature reaction is just the reverse of what is stated above for the generating machine.

Important! The corresponding conclusions for the *motoring machine* are stated below:

(i) When the machine draws a lagging power factor current, the armature reaction has both magnetizing and cross-magnetizing components.

(ii) When the machine draws leading power factor current, the armature reaction has both demagnetizing and cross-magnetizing components.

The effect of the above conclusions on machine operation will be seen in further detail in Sec. 8.11.

8.10 | SYNCHRONIZING TO INFINITE BUS-BARS

LO 3

A definite procedure has to be followed in connecting a synchronous machine to bus-bars which for the present purpose will be assumed to be infinite. Infinite busbars means a 3-phase supply of constant voltage and frequency independent of the load exchanged (fed into the bus-bars or drawn from the bus-bars). Figure 8.34 shows a synchronous machine with terminals *a*, *b*, c which is required to be connected to bus-bars with terminals *A*, *B*, *C* by means of a switch *S*.

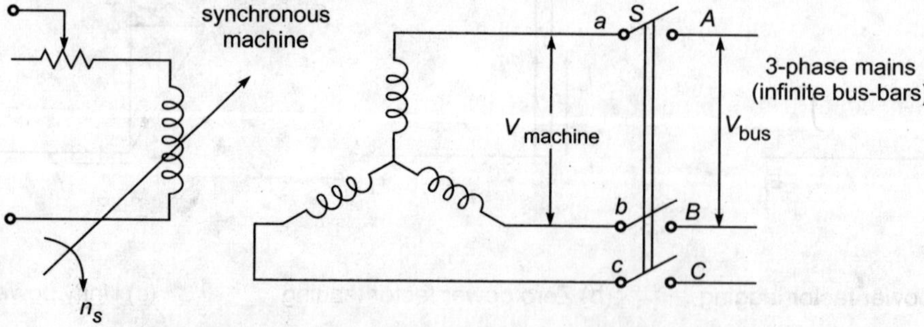

Fig. 8.34 *Connecting synchronous machine to infinite bus-bars*

Mandatory Conditions to be Satisfied for Synchronising Two Machines

The machine is run as a generator with its terminals so arranged that its phase sequence is the same as that of the bus-bars. The machine speed and field current are adjusted so as to satisfy the following conditions:

(i) The machine terminal voltage must be nearly equal to the bus-bars voltage.
(ii) The machine frequency is nearly equal to the bus-bars frequency, i.e. the machine speed is close to synchronous speed.

Determination of Instant of Synchronising

After the above conditions are satisfied the instant of switching on (*synchronising*) must be determined such that the two voltages are almost co-phasal (the acceptable phase difference is of the order of 5°). This instant is determined with the help of the method described below.

Figure 8.35 shows the phasor diagram for phase voltages (line-to-neutral) for the machine and bus-bars. As the two frequencies are not exactly equal, the machine phasors are rotating slowly with respect to the bus-bar phasors at $2\pi\Delta f$ rad/s, where Δf is the difference in the two frequencies. At the instant when the two sets of phasors are coincident (cophasal), the voltage

$$V_{aA} = 0$$
$$V_{bC} = V_{cB} \quad (8.54)$$

!Attention The condition can be easily determined by connecting three lamps—one across *aA*, the other across *bC* and the third across *cB* (in order to use standard-voltage lamps it may be necessary to employ potential transformers). The rms values of voltages V_{aA}, V_{bC} and V_{cB} oscillate at the difference frequency Δ_f so that each lamp is alternately dark and bright. At the instant of synchronization, as per the condition (8.54) stated above, the lamp across *aA* is dark while the other two lamps are equally bright. It is at this instant that switch *S* is closed. Instead of using lamps, generating stations use an instrument called *synchroscope.*

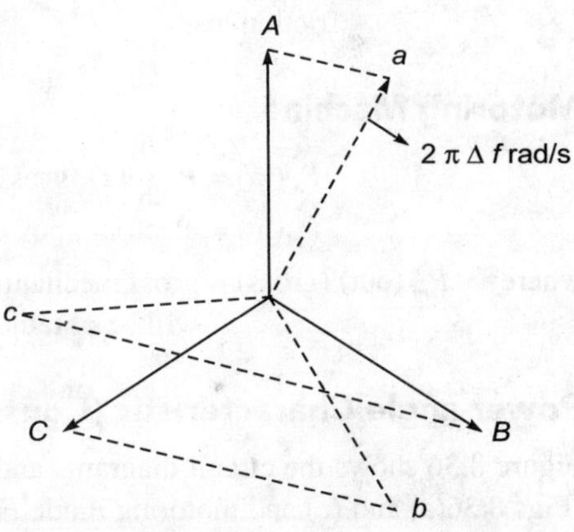

Fig. 8.35 *Determination of the synchronizing instant*

Once switch *S* is closed the stator and rotor fields of the machine lock into each other (synchronize) and the machine then onwards runs at synchronous speed. The real power exchange with the mains will now be governed by the loading conditions on the shaft while the reactive power exchange will be determined by the field excitation.

As a generator is coupled to a primemover it is easy to follow the above procedure to connect it to the bus-bars. The same procedure has to be followed for a *synchronous motor* which must be run initially by an auxiliary device (may be a small dc/induction motor) and then synchronized to the bus-bars. It may be pointed out here that the synchronous motor is non-self-starting. If, for example, switch *S* is closed in Fig. 8.35 with the rotor stationary, the stator and rotor fields will be moving relative to each other at synchronous speed so as to develop alternating torque with zero average value and as a result the motor would not start. Synchronous motors are made self-starting by providing short-circuited bars on the rotor which produce induction torque for starting (see Sec. 5.6).

8.11 | OPERATING CHARACTERISTICS

LO 3

The operating characteristics of a synchronous machine are examined here under conditions of variable load and variable excitation. One of these quantities will be assumed to be held constant at a time while the other will be allowed to vary over a wide range. Further, here too the armature resistance will be assumed negligible. This does not significantly change the operating characteristic of the machine but leads to easier understanding of the machine operation. The more general case of the machine with armature resistance accounted for will be discussed in Sec. 8.8. By virtue of negligible resistance assumption, the electrical power at the machine terminals and the mechanical power at its shaft are simply related as follows:

Generating Machine

$$P_e \text{ (out)} = P_m \text{ (in) (net)}$$

where P_e (out) = electrical power output of the machine (electrical power developed)

P_m (in) = net mechanical power input to the machine after deducting iron-loss and windage and friction loss

Motoring Machine

$$P_e \text{ (in)} = P_m \text{ (out) (gross mechanical power developed)}$$

P_e (in) = electrical power input to the machine

where P_m (out) (gross) = gross mechanical power output of the machine; the net mechanical power output will be obtained by deducting iron-loss and windage and friction loss

Power-angle Characteristic (Constant Excitation Variable Load)

Figure 8.36 shows the circuit diagrams and phasor diagrams of a synchronous machine in generating mode (Figs 8.36(a) and (c) and motoring mode Figs 8.36 (b) and (d)). The machine is assumed to be connected to infinite bus-bars of voltage V_t. It is easily observed from the phasor diagrams that in generating mode, the excitation emf E_f leads V_t by angle δ, while it lags V_t in the motoring mode. It follows from the phasor triangle *OMP* Figs 8.36(c) and (d) that

$$\frac{E_f}{\sin(90 \pm \phi)} = \frac{I_a X_s}{\sin \delta}; \; (90° + \phi), \text{ generating} \quad (90° - \phi), \text{ motoring}$$

or
$$I_a \cos \phi = \frac{E_f}{X_s} \sin \delta \tag{8.55}$$

where ϕ is the power factor angle.

Multiplying both sides of Eq. (8.55) by V_t

$$V_t I_a \cos \phi = \frac{V_t E_f}{X_s} \sin \delta$$

or
$$P_e = \frac{V_t E_f}{X_s} \sin \delta \tag{8.56}$$

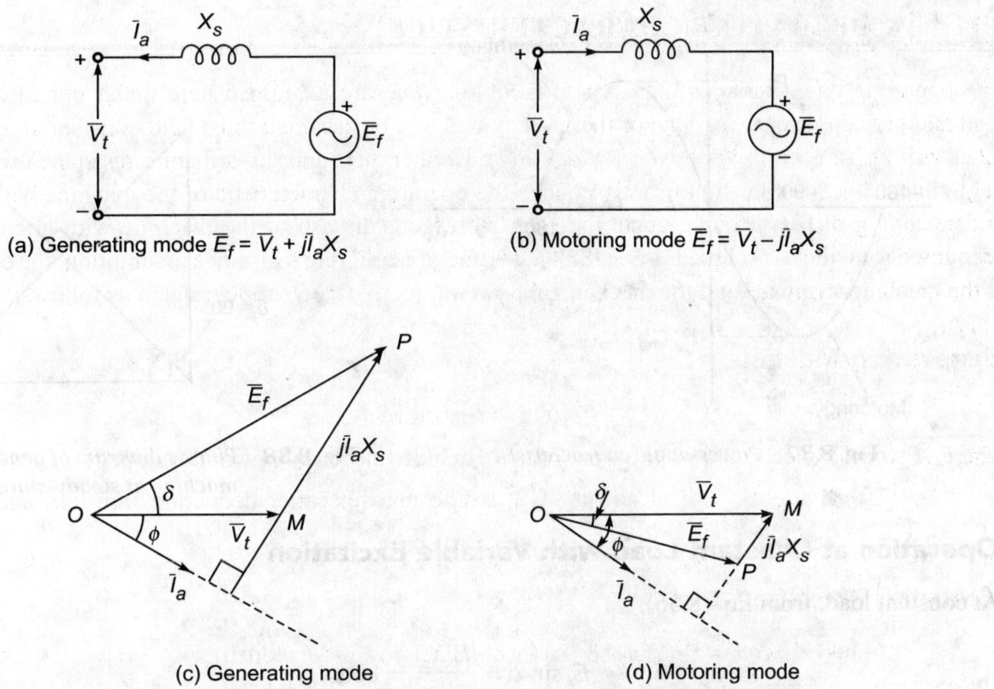

Fig. 8.36 *Synchronous machine operation (generating/motoring mode)*

where $P_e = V_t I_a \cos \phi$ = electrical power (per phase) exchanged with the bus-bars

δ = Angle between E_f and V_t and is called the *power angle*††† of the machine (δ has opposite sign for generating/motoring modes).

The relationship of Eq. (8.56) is known as the *power-angle characteristic* of the machine and is plotted in Fig. 8.37 for given V_t and E_f. The maximum power

$$P_{e,\max} = \frac{V_t E_f}{X_s} \tag{8.57}$$

occurs at $\delta = 90°$ beyond which the machine falls out of step (loses synchronism). The machine can be taken up to $P_{e,\max}$ only by gradually increasing the load. This is known as the *steady-state stability limit* of the machine. The machine is normally operated at δ much less‡‡‡ than 90°. The phasor diagram of a generating machine under condition of $P_{e,\max}$ is drawn in Fig. 8.38. Obviously I_a will be several times larger than the rated machine current in this condition.

††† The angle δ in Eq. (8.56) is between V_t and E_f while in Fig. 8.9 it is the angle between E_r and E_f. The difference between these two angles is due to the fact that now the leakage reactance of the machine is being accounted for.

‡‡‡ This is to prevent the machine from going into an unstable region (where it will fall out of step) during transient power swings. This topic concerns transient stability of the machine and is discussed in books on power systems [7].

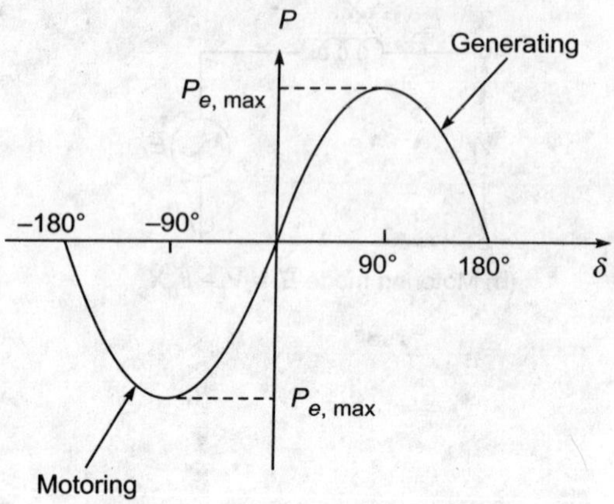

Fig. 8.37 *Power-angle characteristic*

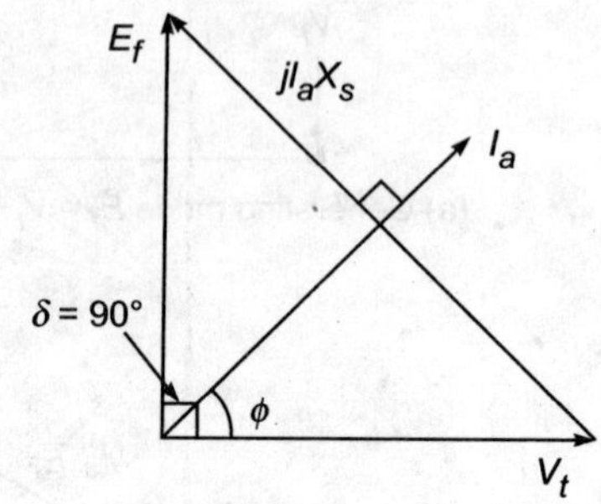

Fig. 8.38 *Phasor diagram of generating machine at steady-state stability limit*

Operation at Constant Load with Variable Excitation

At constant load, from Eq. (8.56)

$$E_f \sin \delta = \frac{P_e X_s}{V_t} = \text{const} \tag{8.58}$$

Also
$$V_t I_a \cos \phi = P_e = \text{const}$$

or
$$I_a \cos \phi = \frac{P_e}{V_t} = \text{const} \tag{8.59}$$

It is therefore, observed that at constant load, as the excitation emf E_f is varied (by varying field current I_f), the power angle δ varies such that $E_f \sin \delta$ remains constant. The machine bahaviour is depicted by the phasor diagrams of Figs 8.39(a) and (b). As E_f varies, the tip of phasor $\bar{E}_f$ moves on a line parallel to $\bar{V}_t$ and at distance $E_f \sin \delta = P_e X_s / V_t$ from it. Since $I_a \cos \phi$ = constant, the projection of the current phasor on V_t must remain constant, i.e. the tip of the current phasor traces a line perpendicular to V_t at distance $I_a \cos \phi = P_e / V_t$ from the origin. The current phasor $\bar{I}_a$ is always located at 90° to phasor $\bar{I}_a X_s$ (phasor joining tips to $\bar{E}_f$ and $\bar{V}_t$ in the direction of $\bar{E}_f$). The effect of varying excitation (E_f) on machine operating characteristics is brought out by Figs 8.39(a) and (b).

Normal excitation: At this excitation the machine operation meets the condition $E_f \cos \delta = V_t$ at which the machine power factor is unity.

Over excitation: $E_f \cos \delta > V_t$

Under excitation: $E_f \cos < V_t$

The following conclusions[§§§] are drawn from the phasor diagrams of Figs 8.39(a) and (b).

[§§§] These conclusions are corroborated by the nature of the armature reaction discussed in Sec. 8.7. For example, when a generator is overexcited it supplies lagging current which has a demagnetizing effect so that the air-gap emf E_r matches the applied voltage. Similarly an overexcited motor draws a leading current which has demagnetizing effect.

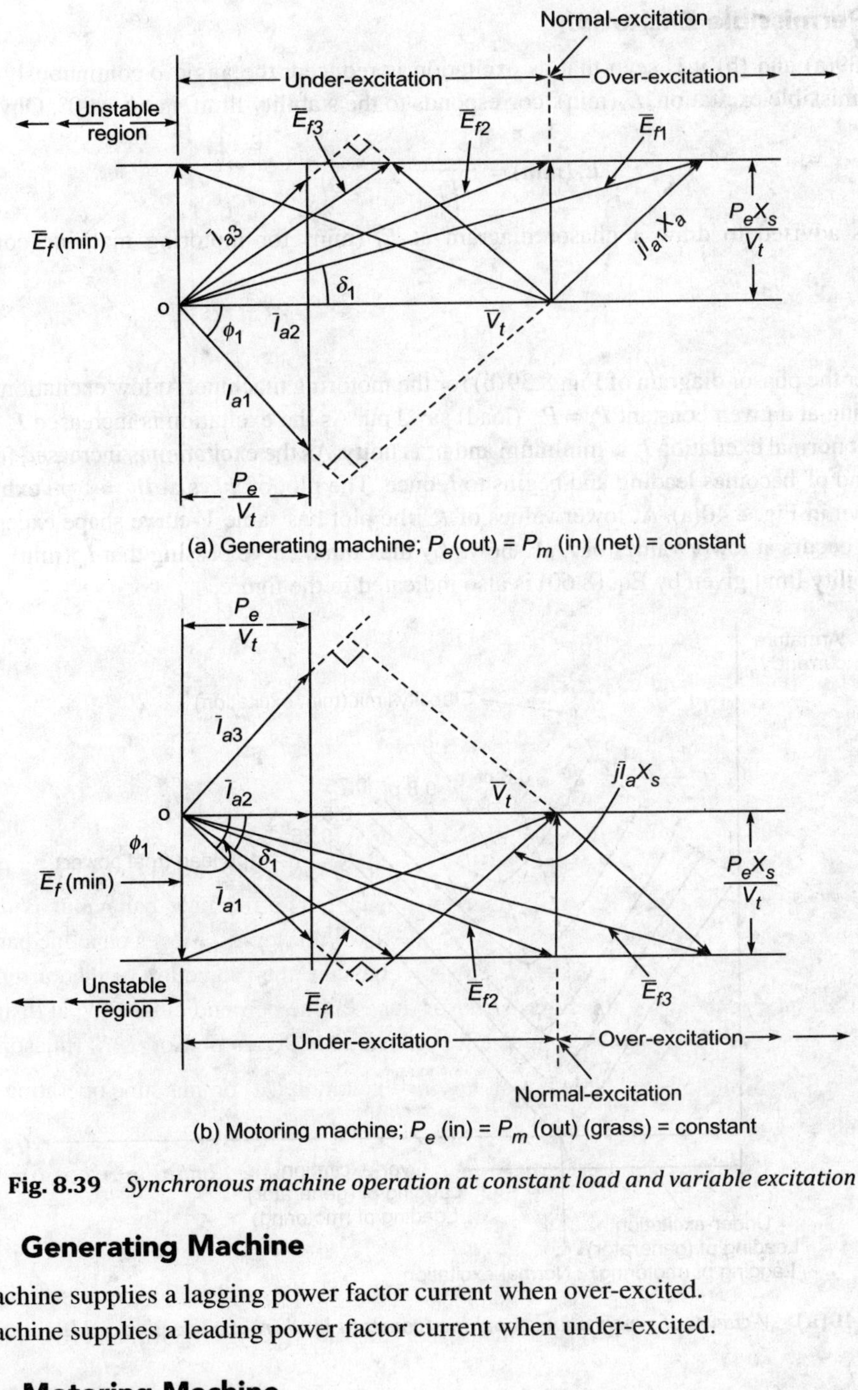

Fig. 8.39 *Synchronous machine operation at constant load and variable excitation*

Important! Generating Machine

1. The machine supplies a lagging power factor current when over-excited.
2. The machine supplies a leading power factor current when under-excited.

Important! Motoring Machine

1. The machine draws a leading power factor current when over-excited.
2. The machine draws a lagging power factor current when under-excited.

Minimum Permissible Excitation

From Figs 8.39(a) and (b) it is seen that as excitation is reduced, the angle δ continuously increases. The minimum permissible excitation, E_f(min), corresponds to the stability limit, i.e. $\delta = 90°$. Obviously

$$E_f\,(\text{min}) = \frac{P_e X_s}{V_t} \tag{8.60}$$

The reader is advised to draw a phasor diagram at E_f (min) for motoring machine corresponding to Fig. 8.37.

V– Curves

Let us consider the phasor diagram of Fig. 8.39(b) for the motoring machine. At low excitation, I_a is large and pf is low lagging at a given constant $P_e = P_m$ (load) say 1 pu. As the excitation is increased I_a reduces and pf increases till at normal excitation I_a is minimum and pf is unity. As the excitation is increased further I_a begins to increases and pf becomes leading and begins to reduce. The plot of I_a vs at P_m = 1 pu exhibits a *V*-curve nature as shown in Fig. 8.40(a). At lower values of P_m the plot has same *V*-curve shape except that I_a (min) is smaller and occurs at lower values of I_f as shown by the dotted curve passing that I_a (min). The minimum excitation stability limit given by Eq. (8.60) is also indicated in the figure.

!Attention

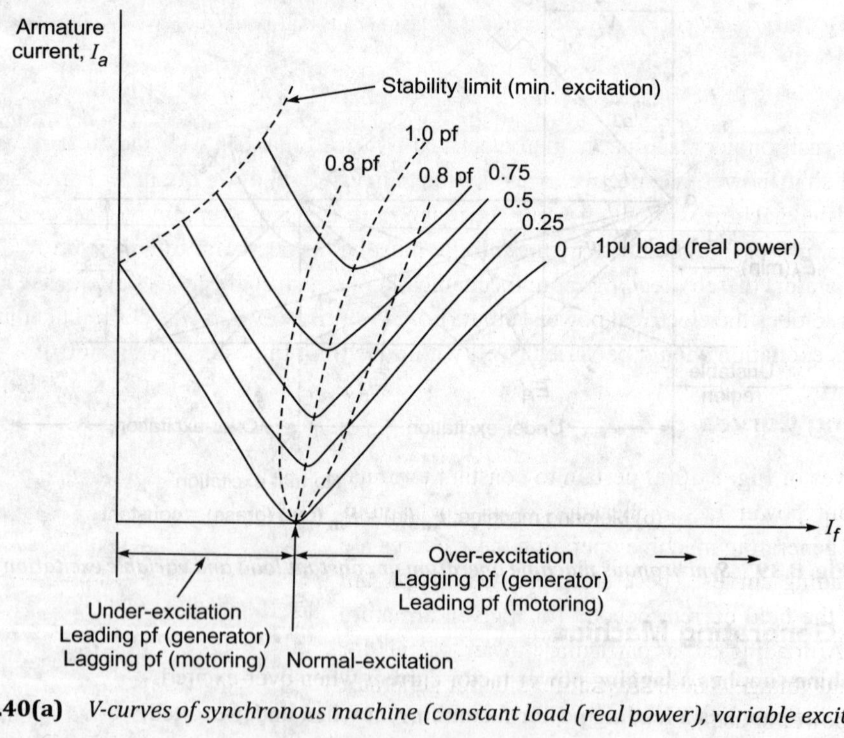

Fig. 8.40(a) *V-curves of synchronous machine (constant load (real power), variable excitation)*

It is easily seen that pf vs I_f curves for fixed P_m are inverse of I_a vs I_f plot. Thus these are *inverted V-Curves* each having maximum value of unity (pf) also shown in Fig. 8.40(b).

The generator V-curves and inverted V-curves can be found to have the same form as for the motor except for reversal of lagging and leading pf regions as shown in Figs 8.40(a) and (b).

Important!

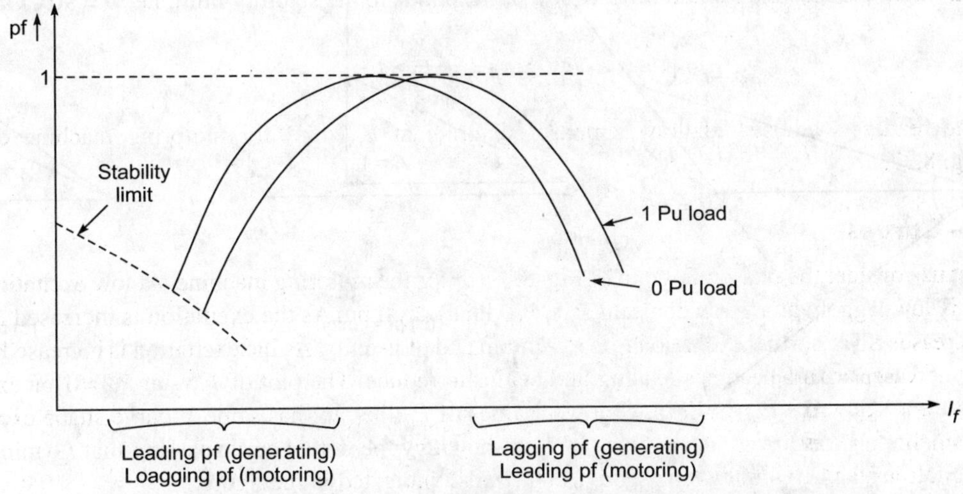

Fig. 8.40(b) *Inverted V-curves of synchronous machine (constant load (real power), variable excitation)*

Observation

Recall In a synchronous machine the real electrical power exchanged with the bus-bars is controlled by the mechanical shaft power irrespective of excitation. The excitation, on the other hand, governs only the power factor of the machine without affecting the real power flow. For example, in a generator if it is desired to feed more real power into the bus-bars the throttle must be opened admitting more steam into the turbine (coupled to generator) thereby feeding more mechanical power into shaft. As a consequence the power angle δ increases and so does the electrical power output (Eq. 8.56)). However, if it is desired to adjust the machine power factor, its excitation should be varied (well within the limit imposed by Eq. (8.60)).

Compounding Curves

The dotted curves of Fig. 8.40(a) pertain to constant terminal voltage, constant power factor operation of a synchronous machine. For a generating machine operation these curves are called compounding curves. These are presented once again in Fig. 8.41 as the field current needed for a given armature current or kVA loading at a particular power factor for constant terminal voltage. These are useful guide for generator operation in a power house.

While the shape of the compounding curves can be visualize from the dotted curves of Fig. 8.40(a), these are better understood by means of the phasor diagrams of Fig. 8.42.

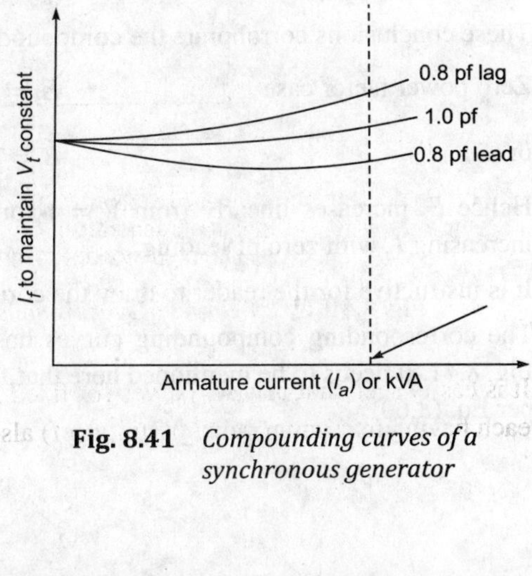

Fig. 8.41 *Compounding curves of a synchronous generator*

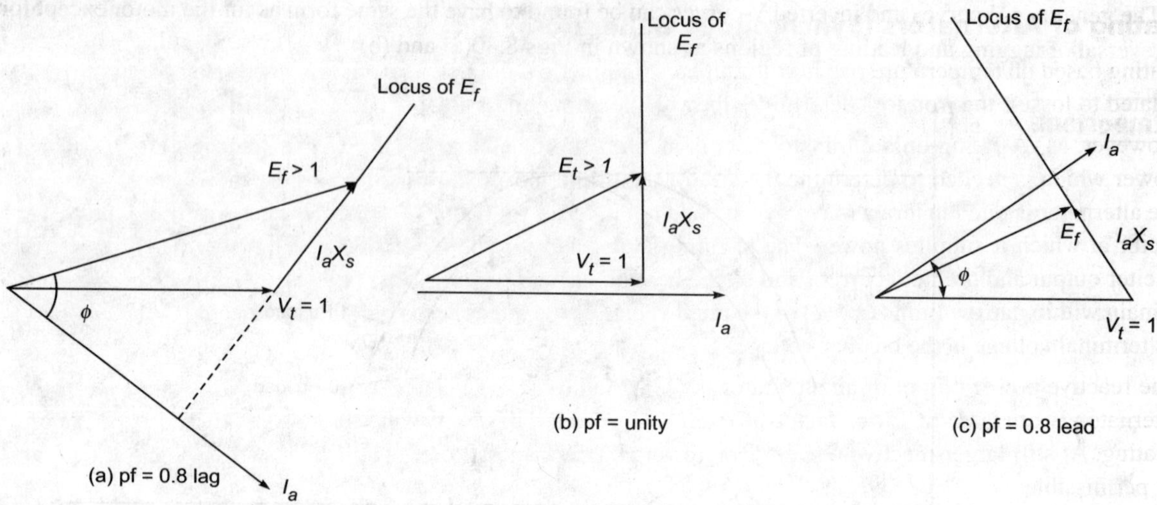

Fig. 8.42

These phasor diagrams are drawn to determine excitation emf E_f and so the field current I_f to maintain constant terminal voltage V_t = 1.0(say) for specified power factor with increasing armature current, I_a. The three phasor diagrams pertain to lagging (0.8), unity and leading (0.8) power factors.

Important! The following conclusions are drawn

(a) 0.8 pf lagging

$E_f > 1$; E_f and so I_f required increases with increasing I_a as shown by the locus of E_f tip.

(b) unity pf

$E_f > 1$ but less than E_f (lagging pf), also increases at a slower rate. Therefore I_f needed is less than lagging pf case and has to be increased.

(c) 0.8 pf lagging

$E_f < V_t$; it is seen from the locus the E_f reduces goes through a minimum and then continues to increase. Therefore, I_f should be adjusted accordingly.

These conclusions corroborate the compounding curves of Fig. 8.42.

Zero power factor case $\quad E_f = V_t - j\bar{I}_a X_s, \quad \bar{I}_a = I_a \angle \mp 90° = \mp jI_a$

or $\quad E_f = V_t \pm I_a X_s$, a scalar equation

Hence E_f increases linearly from $V_t = 1$ for increasing I_a with zero pf lagging but decreases linearly for increasing I_a with zero pf leading.

It is instructive for the reader to draw the corresponding phasor diagram.

The corresponding compounding curves has no practical significance for a motor and are not drawn in Fig. 8.41. It needs to be mentioned here that for zero pf leading load E_f cannot be reduced below E_f (min) as per Eq. (8.60).

Rating of Alternators (Synchronous Generators)

Rating based on temperature rise is volt-amperes (in practical units of kVA, MVA). As the temperature rise is related to losses, the iron loss determines the voltage rating and $I_a^2 R_a$ loss determines the current rating.

However, MVA rating unlike transformers is incomplete for alternators it gives no information on the real power which is needed to determine the size of the prime mover (turbine) to drive the alternator. Therefore, the alternator is rated in terms MW capacity and not in terms of MVA. The other alternator rating is the power factor at which it supplies power. The pf rating is normally in the range of 0.8 to 0.9 lagging. It limits the exciter output and the field current and so the heating of field winding. Of course, the terminal voltage must remain within narrow limits (± 5%) of the rated value. As the alternator is normally connected to the bus-bars, its terminal voltage is the bus-bar voltage.

The reactive power output of an alternator is $Q = P \tan (\cos^{-1} \text{pf})$. The reactive power flow increases if the alternator is operated at lower factor at rated real power. The reactive power flow is limited by armature heating. At still larger reactive power flow (lower pf), much larger field current is needed which would not be permissible.

To Sum Up

Alternator ratings are V_t (kV line), MW and pf = 8.0 – 0.9 lagging (if unspecified it should be taken as lagging because at lagging pf alternator requires larger field current)

For a *unit system* – boiler-turbine, alternator and step-up transformer form one generator unit.

$$\text{Turbine rating} = \text{MW rating of alternator plus over-load margin}$$

$$\text{Transformer rating, MVA} = \frac{\text{MW rating of alternator}}{\text{pf rating}}$$

Over Excited Synchronous Motor/Synchronous Condenser

Important! It has been seen above that a synchronous motor under over-excited condition operates at a leading power factor. Synchronous motors are therefore employed in large power installation for overall high power factor of the installation.

At no-load with losses assumed negligible, a synchronous motor operates at

$$\delta = 0 \qquad \text{(see Eq. (8.56))}$$

which means that E_f and V_t are in phase. It is seen from the phasor diagram of Figs 8.43(a) and (b), that the machine (motor) draws zero power factor leading current

$$I_a = \frac{E_f - V_t}{X_s} \qquad (E_f > V_t,\ \text{over-excited})$$

and draws zero power factor lagging current

$$I_a = \frac{V_t - E_f}{X_s} \qquad (E_f < V_t,\ \text{under-excited})$$

Thus a synchronous motor at no-load behaves as a variable condenser or inductor by simply varying its excitation. The machine operated under such a condition (motor on no-load or light load) is known as a synchronous condenser and finds application in large integrated power systems for improving the power factor under heavy-load conditions and for deproving the power factor under light-load conditions, thereby controlling the voltage profile of the power system within reasonable limits.

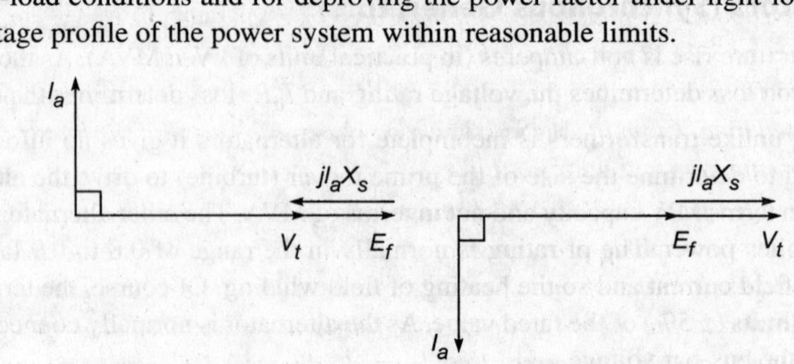

Fig. 8.43 *Synchronous condenser*

Dual-purpose Synchronous Motor

Synchronous motor is used in an industry/factory for serving two purposes. It drives a constant speed mechanical load such as a large pump, a dc generator, etc. and at the same time it also corrects an otherwise low lagging pf of the electrical load such as induction motors and fluorescent tubes. Such a synchronous motor serving dual-purpose is called dual-purpose synchronous motor.

Example 8.7 *A, 3300 V, delta-connected motor has a synchronous reactance per phase (delta) of 18 Ω. It operates at a leading power factor of 0.707 when drawing 800 kW from the mains. Calculate its excitation emf.*

LO 3

Solution On equivalent-star basis,

$$V_t = 3300/\sqrt{3} = 1905 \text{ V}$$

$$X_s = 18/3 = 6\ \Omega$$

$$I_a = \frac{800 \times 1000}{\sqrt{3} \times 3300 \times 0.707} = 198 \text{ A}$$

$$I_a X_a = 198 \times 6 = 1188 \text{ V}$$

$$\cos\phi = 0.707 \text{ or } \phi = 45° \text{ (leading)}$$

The phasor diagram is drawn in Fig. 8.44 from which E_f can be measured if the diagram is drawn to scale, or directly by calculating from the geometry of the phasor diagram.

$$\angle MPQ = 45°$$

$$MQ = PQ = 1188/\sqrt{2} = 840$$

$$E_f = OP = \sqrt{(1905 + 840)^2 + (840)^2} = 2871$$

or 4972 V (line)

It may be seen that the motor is operating over-excited.

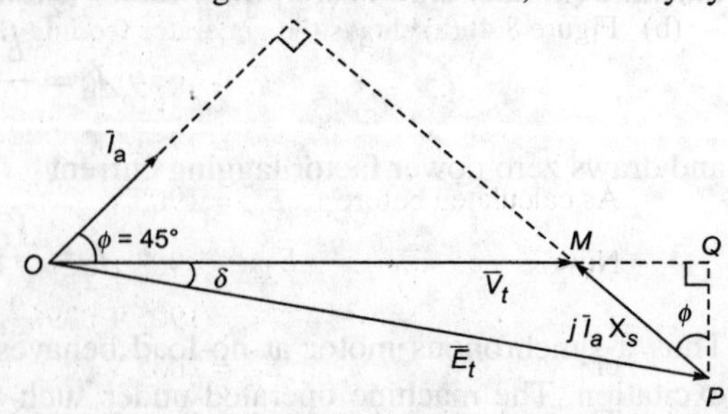

Fig. 8.44

Example 8.8 *A 1000 kW, 3-phase, star-connected, 3.3 kV, 24-pole, 50 Hz synchronous motor has a synchronous reactance of 3.24 Ω per phase; the resistance being negligible.*

(a) The motor is fed from infinite bus-bars at 3.3 kV. Its field excitation is adjusted to result in unity pf operation at rated load. Compute the maximum power and torque that the motor can deliver with its excitation remaining constant at this value.

(b) The motor is now fed from a 1200 kVA, 3-phase star-connected, 3.3 kV, 2-pole, 50 Hz synchronous generator with a synchronous reactance of 4.55 Ω per phase, the resistance being negligible. Compute the field excitations of motor and generator when the set is operating at rated terminal voltage at unity pf and the motor is delivering full-load power. The field excitations of both the machines remaining constant, the motor load is gradually raised. Compute the maximum power and torque that the motor can deliver. Also compute the terminal voltage when the motor is delivering maximum power.

LO 3

Solution

(a) The operation of motor at infinite bus-bars is shown in Fig. 8.45.

$$V_t = 3300/\sqrt{3} = 1905 \text{ V}$$

$$I_a = \frac{1000 \times 1000}{\sqrt{3} \times 3300 \times 1} = 175 \text{ A}$$

$$\cos\phi = 1, \quad \phi = 0°$$

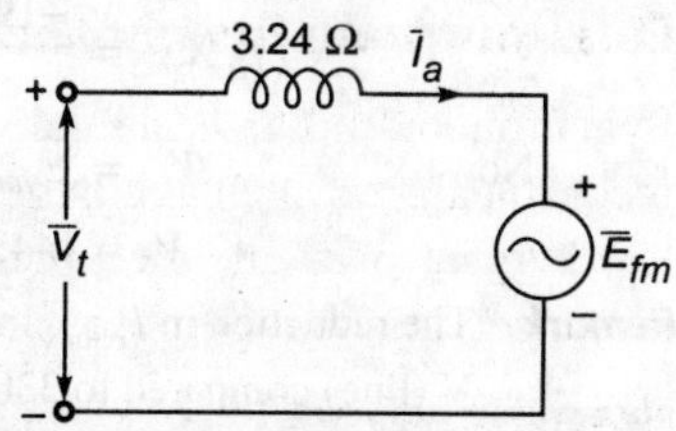

Fig. 8.45

Taking the terminal voltage as reference,

$$\bar{V}_t = 1905\angle 0° \text{ V}$$

$$\bar{I}_a = 175\angle 0° \text{ A}$$

Then the excitation emf is computed as

$$\bar{E}_{fm} = 1905\angle 0° - j\,175\angle 0° \times 3.24 = 1905 - j\,567$$

which gives $E_{fm} = 1987$ V

Excitation remaining fixed, the maximum power delivered by the motor is

$$P_{e,\max} = P_{m,\max} \text{ (gross)} = 3 \times \frac{V_t E_f}{X_{sm}} = 3 \times \frac{1905 \times 1987}{3.24 \times 1000} = 3505 \text{ kW (3-phase)}$$

$$\omega_{sm} = \frac{120 \times 50 \times 2\pi}{24 \times 60} = 26.18 \text{ rad/s}$$

$$\therefore \quad T_{\max} = \frac{3505 \times 1000}{26.18} = 133.9 \times 10^3 \text{ N m}$$

(b) Figure 8.46(a) shows the generator feeding the motor. At rated terminal voltage, unity pf, full-load operation

$$\bar{V}_t = 1905\angle 0°$$

$$\bar{I}_a = 175\angle 0°$$

As calculated before $E_{fm} = 1905$ V

Now $\bar{E}_{fg} = 1905\angle 0° + j\,175\angle 0° \times 4.55$

$= 1905 + j\,796$

or $E_{fg} = 2065$ V

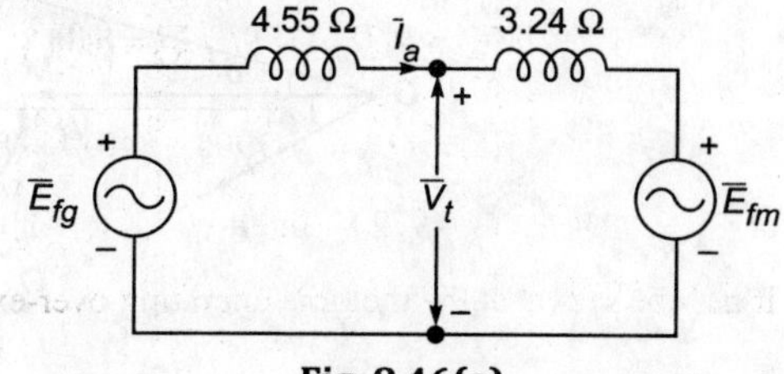

Fig. 8.46(a)

The total series reactance $X = X_{sm} + X_{sg} = 3.24 + 4.55 = 7.79\ \Omega$

The maximum power output delivered by the motor is

$$P_{e,\max} = P_{m,\max}\ (\text{gross}) = 3 \times \frac{E_{fg} \times E_{fm}}{X} = 3 \times \frac{2065 \times 1987}{7.79 \times 1000} = 1580 \text{ kW (3-phase)}$$

$$T_{\max} = \frac{1580 \times 1000}{26.18} = 60.35 \times 10^3 \text{ Nm}$$

The phasor diagram under condition of maximum power output is drawn in Fig. 8.46(b). For convenience, choosing the motor excitation as reference,

At maximum power

$$\text{Output } \delta = 90^\circ$$

$$\bar{E}_{fm} = 1987 \angle 0^\circ \text{ V}$$

$$\bar{E}_{fg} = 2065 \angle 90^\circ \text{ V}$$

Then
$$\bar{I}_a = \frac{\bar{E}_{fg} - \bar{E}_{fm}}{jX} = \frac{2065\angle 90^\circ - 1987\angle 0^\circ}{j7.79}$$

$$j\bar{I}_a X_{sm} = \frac{-1987 + j2065}{j7.79} \times j\,3.24 = -826.4 + j\,858.9$$

Now
$$\bar{V}_t = \bar{E}_{fm} + j\bar{I}_a X_{sm} = 1987 - 826.4 + j\,858.9 = 1160.6 + j\,858.9$$

or
$$V_t = 1443.85 \text{ or } 2500 \text{ V (line)}$$

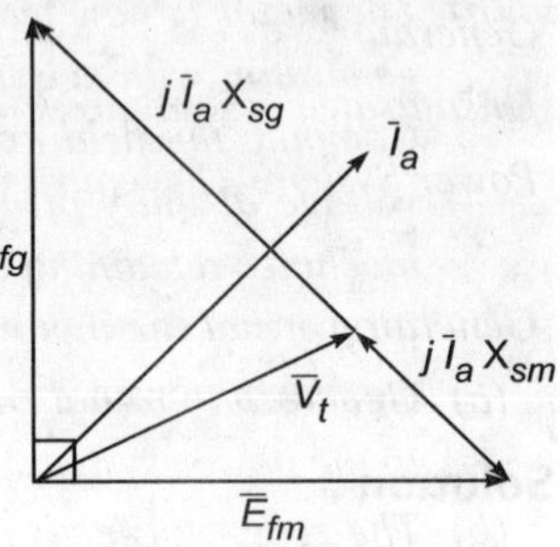

Fig. 8.46(b)

Remark The reduction in $P_{e,\max}$ in case (b) compared to case (a) is explained by the fact that V_t in this case is only 2500 V (line) compared to 3300 V in case (a).

Consideration of Armature Resistance

Figures 8.47(a) and (b) show the circuit model of synchronous machine for generating motoring modes of operation with due consideration of armature resistance. Operational analysis can be carried out by means of

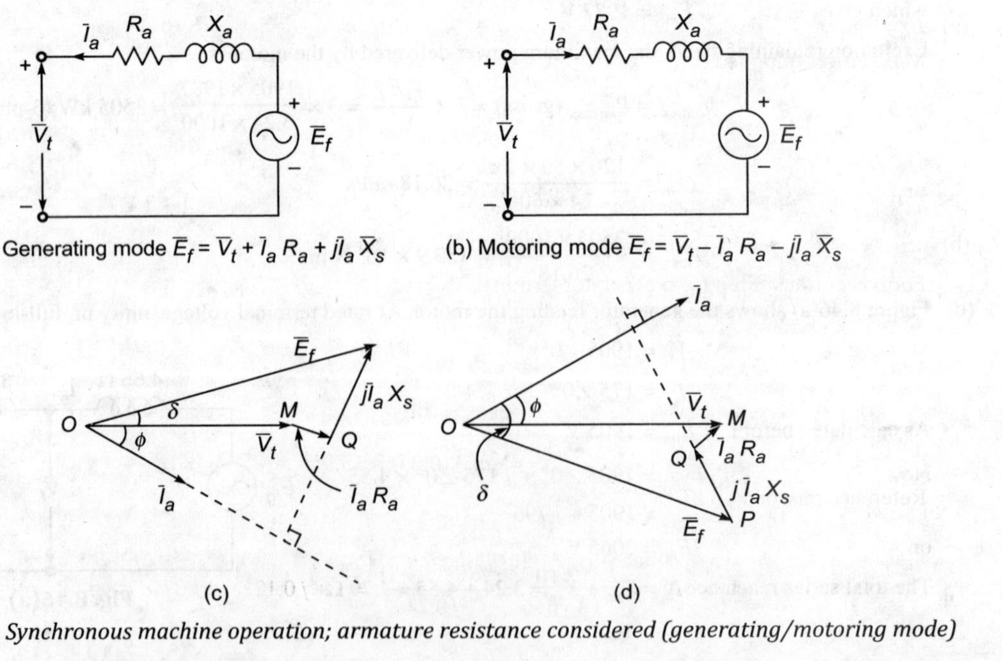

Fig. 8.47 *Synchronous machine operation; armature resistance considered (generating/motoring mode)*

the phasor equations for Figs 8.47(a) and (b) respectively or by the phasor diagrams of Figs 8.47(c) and (d). For the sake of clarity $I_a R_a$ voltage drop is shown larger, i.e. out-of-proportion. (The armature resistance has to be accounted for in efficiency calculations).

Example 8.9 *A synchronous generator feeds power to a power system. The generator and power system data are:*

Generator: *100 MVA, 11 kV*

Unsaturated synchronous reactance $X_s = 1.3$ *pu*

Power System: Thevenin's equivalent as seen from the generator terminals is

$$V_{TH} = 1 \text{ pu}, \quad X_{TH} = 0.24 \text{ pu} \quad \text{(on generator base)}$$

Generator open circuit voltage 11 kV at a field current of $I_f = 256$ *A*

(a) *Generator internal emf* E_f *is adjusted to 1 pu. What is the maximum power that the generator supplies to the power system?*

(b) *The generator feeds power* $P_e = 1$ *pu to the power system at generator terminal voltage* $V_t = 1$ *pu. Calculate the power angle* δ *of the generator and the corresponding field current* I_f.
Plot V_t *as the load is varied from a to 0.8 pu.* E_f *held constant at 1 pu. Use MATLAB.*

(c) *The generator is fitted with automatic voltage regulator, which is set for* $V_t = 1$ *pu. Load is now varied.*
Plot I_f *versus* P_e. *Use MATLAB.*

LO 3

Solution The equivalent circuit model of the generator feeding the power system is drawn in Fig. 8.48(a).

Bases
$$(\text{MVA})_B = 100$$
$$(\text{kV})_B = 11$$
$$\text{Power (MW)}_B = 100$$

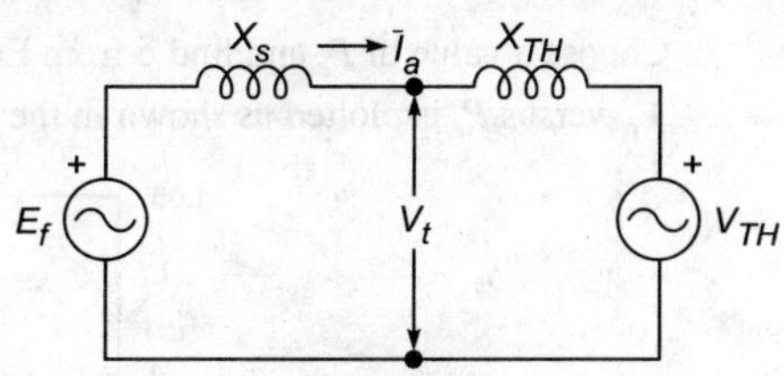

Fig. 8.48(a)

(a) $E_f = 1$ pu, $P_e = 1$ pu, specified $V_{TH} = 1$ pu

Max. Power Supplied,

$$P_{e,\max} = \frac{E_f V_{th}}{X_s + X_{TH}}$$

or
$$P_{e,\max} = \frac{1}{1.3 + 0.24} = 0.649 \text{ pu or } 64.9 \text{ MW}$$

(b) $P_e = 1$ pu, $V_t = 1$ pu; specified $V_{TH} = 1$ pu

For power transferred from generator terminals to load

$$P_e = \frac{V_t V_{TH}}{X_{TH}} \sin \delta_1$$

$$1 = \frac{1}{0.24} \sin \delta_1$$

$$\delta_1 = \sin^{-1}(0.24) = 13.9°$$

Reference phasor $V_{TH} \angle 0°$

$$\bar{V}_t = e^{j13.9°}$$

$$\bar{I}_a = \frac{V_t - V_{TH}}{jX_{TH}} = \frac{(e^{j13.9°} - 1)}{j0.24} = 1 + j\,0.122 = 1.007 \angle 7° \text{ pu}$$

To find E_f and I_f

$$\bar{E}_f = \bar{V}_{TH} + j(X_s + X_{TH})\bar{I}_a = (1 + j0) + j(1.54)(1 + j0.122) = 1.74\, e^{j62.2°}$$

$$|\bar{E}_f| = 1.741 \text{ pu} = 19.15 \text{ kV}$$

$$\delta = 62.2°$$

Slope of air-gap line, $\dfrac{V_{oc}}{I_f} = \dfrac{11}{256}$ V/A

For $F_f = 19.15$ kV

$$I_f = \frac{256}{11} \times 19.15 = 445.7 \text{ A}$$

(b) $E_f = 1$ pu (held constant), P_e varied

Let $\bar{E}_f = E_f e^{j\delta}$

$$P_e = \frac{E_f V_{TH}}{X_s + X_{TH}}$$

$$P_e = \frac{1}{1.54} e^{j\delta} \quad \text{(i)}$$

$$\bar{I}_a = \frac{E_f e^{j\delta} - V_{TH}\angle 0°}{j(X_s + X_{TH})} = \frac{e^{j\delta} - 1}{j1.54} \quad \text{(ii)}$$

We then get

$$\bar{V}_t = \bar{V}_{TH} + jX_{TH}\bar{I}_a$$

$$\bar{V}_t = 1 + \frac{0.24}{1.54}(e^{j\delta} - 1) \quad \text{(iii)}$$

Choose a value of P_e and find δ from Eq. (i). Substitute in Eq. (ii) and find the magnitude V_t from Eq. (iii). $\bar{V}_t$ versus P_e is plotted as shown in the Fig. 8.48(b) by Using MATLAB program given below.

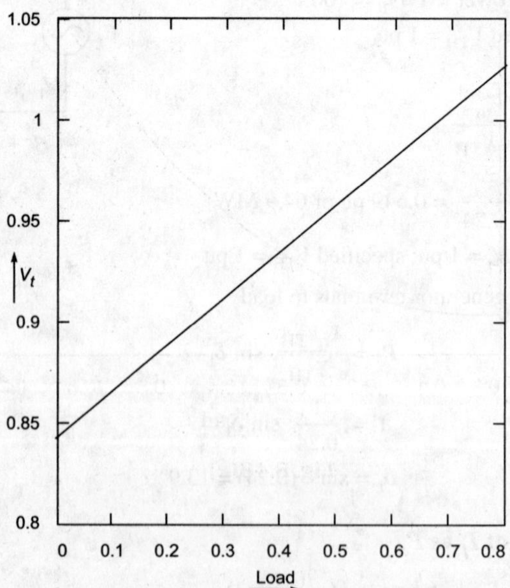

Fig. 8.48(b)

$$P_e = 0 : 0.01 : 0.8$$
$$V_t = 1 + (0.24 / 1.54).*(1.54 * P_e - 1);$$

Plot (P_e, V_t)

(c) $$V_1 = 1 \text{ pu (controlled at this value)}$$

P_e varied

$$P_e = \frac{V_t V_{TH}}{X_{TH}} \sin \delta_1 = \frac{1}{0.24} \sin \delta_1$$
$$\bar{V}_t = e^{j\delta_1}$$
$$\bar{I}_a = \frac{e^{j\delta_1} - 1}{j0.24}$$
$$\bar{E}_f = \bar{V}_{TH} + j1.54\ \bar{I}_a = 1 + (j1.54)\frac{(e^{j\delta_1} - 1)}{j0.24} = 1 + \frac{1.54}{0.24}(e^{j\delta 1} - 1)$$
$$\bar{I}_f = \frac{256}{11}\bar{E}_f = \frac{256}{11}\left[1 + \frac{1.54}{0.24}(3^{j\delta_1} - 1)\right]$$

For I_f vs P_e plot MATLAB program is given as follows and its resultant plot is shown in the Fig. 8.48(c).

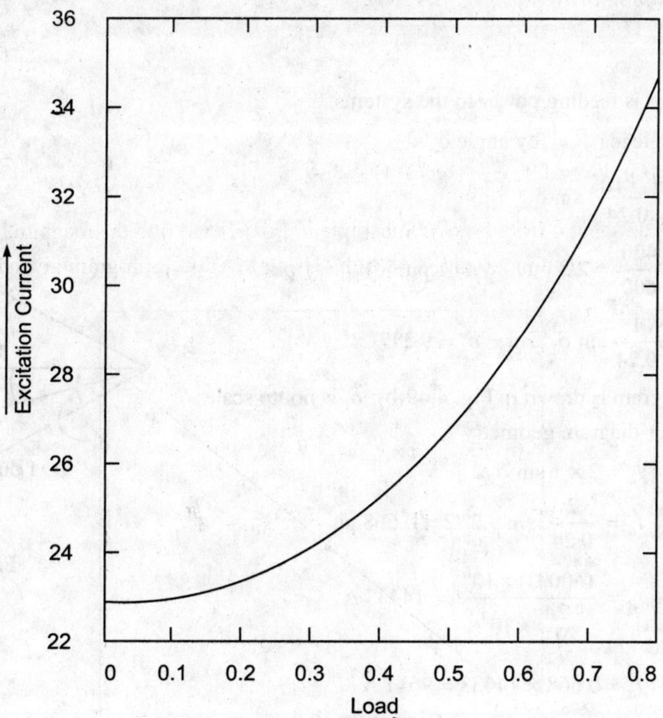

Fig. 8.48(c)

MATLAB Program to Plot I_f vs P_e

```
Pe=0: 0.01:0.8
d1=asin(0.24*Pe)
```

```
Ef=1+(1.54/0.24).*(exp(i*d1)-1)
If=(256/11).*Ef
plot(Pe,abs(If))
```

Example 8.10 *A 4 pole, 50 Hz, 24 kV, 600 MVA synchronous generator with a synchronous reactance of 1.8 pu is synchronized to a power system which can be represented by a Thevenin voltage of 24 kV in series with Thevenin reactance of 0.24 pu on generator base. The generator voltage regulator adjusts the field current to maintain its terminal voltage and 24 kV independent of load.*

(a) The generator prime mover power is adjusted so that it feeds 400 MV.

(i) Draw the phasor diagram under this operating condition

(ii) Calculate the generator current and its power factor

(b) Repeat part (a) when the generator feeds 600 MW

$(MVA)_B = 600$, $(kV)_B = 24$, $(MW)_B = 600$

The system diagram is drawn in Fig. 8.49(a).

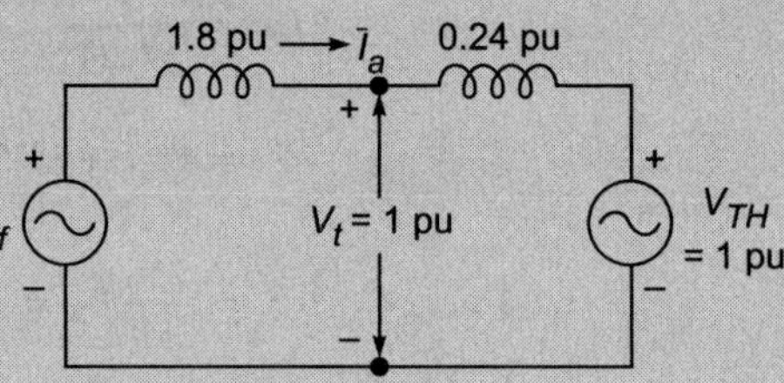

Fig. 8.49(a)

LO 3

Solution

(a) (i) As the generator is feeding power to the systems

$\bar{V}_t$ leads $\bar{V}_{TH}$ by angle δ

$$P_e = \frac{V_t V_{TH}}{0.24} \sin \delta$$

$$P_e = \frac{400}{600} = 2/3 \text{ pu}, \quad V_t = 1 \text{ pu}, \quad V_{TH} = 1 \text{ pu}$$

$$\frac{2}{3} = \frac{1}{0.24} \sin \delta_1 \quad \text{or} \quad \delta_1 = 9.2°$$

The phasor diagram is drawn in Fig. 4.49(b); δ_1 is not to scale

(ii) From the phasor-diagram geometry

$$0.24\, I_a = 2 \times 1 \sin \delta_1/2$$

or $$I_a = \frac{2}{0.24} \sin 9.2°/2 = 0.668 \text{ pu}$$

$$(I_a)_B = \frac{(600/3) \times 10^6}{\frac{24}{\sqrt{3}} \times 10^3} = 14433 \text{ A}$$

$$I_a = 0.668 \times 14433 = 9641 \text{ A}$$

Phase angle, $$\phi = \frac{\delta}{2} = 4.6° \text{ lag}$$

Power factor = cos 4.6° = 0.9968

Check $$P_e = \sqrt{3} \text{ kV } I_a \times 10^{-3} \cos \phi$$

$$= \sqrt{3} \times 24 \times 9.641 \times 0.9968 = 399.5 \approx 400 \text{ MW}$$

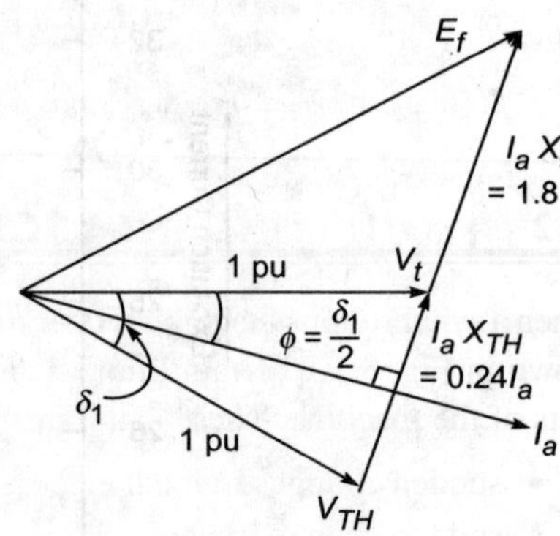

Fig. 8.49(b)

Observation $$P_e = \frac{V_t V_{TH}}{X_{TH}} \sin \delta_1 = V_t I_a \cos \phi \text{ (in pu)}$$

$$\bar{E}_f = \bar{V}_t + j\bar{I}_a X_s$$

$$= 1 \angle 0^\circ + j\,0.668 \angle -4.6^\circ \times 1.8 = 1 + 1.2 \angle 85.4^\circ = 1.622 \angle 47.4^\circ$$

or $$E_f = 1.622 \quad \text{or} \quad 38.9 \text{ kV}, \delta_2 = 47.4^\circ$$

(b) $$P_e = 600 \text{ MW or } P_e = 1 \text{ pu}$$

$$1 = \frac{1}{0.24} \sin \delta_1, \delta_1 = 13.89^\circ$$

$$I_a = \frac{2}{0.24} \sin \delta_1/2 = 1 \text{ pu}$$

or $$I_a = 14433 \times 1 = 14433 \text{ A}$$

Phase angle, $$\phi = 13.89^\circ/2 = 6.945^\circ \text{ lag}$$

Power factor = 0.993 lag

$$\bar{E}_f = 1\angle 0^\circ + j \times 1\angle -6.95^\circ \times 1.8 = 1 + 1.8 \angle 83.05^\circ = 2.157 \angle 55.6^\circ$$

$$E_f = 2.157 \text{ pu}$$

or $$E_f = 2.157 \times 24 = 31.77 \text{ kV}, \delta_2 = 55.6^\circ$$

Let us find power factor from E_f to V_{TH}

$$P_e = \frac{E_f V_{TH}}{X_s + X_{TH}} \sin(\delta_1 + \delta_2)$$

or $$P_e = \frac{2.157 \times 1}{1.8 + 0.24} \sin(13.89^\circ + 55.6^\circ) = 0.99 \text{ pu}$$

It should be 1 pu. The difference is due to rounding off errors.

8.12 | HUNTING IN SYNCHRONOUS MACHINES

LO 3

When a synchronous machine is operating at steady load indicated by point Q (P_0, δ_0) on the $P-\delta$ characteristic shown in Fig. 8.50, certain limited amplitude disturbances are bound to occur on electrical and mechanical parts of the machine. These disturbances are:

- sudden changes in load; electrical, mechanical
- sudden change in field current
- presence of harmonic variations in load (electrical/mechanical) and also in prime mover torque

As is seen from Fig. 8.50 a change in rotor angle $\Delta\delta$ caused by a disturbance produces synchronizing power (and associated synchronizing torque) given by

$$P_s = [P(\max) \cos \delta]\, \Delta\delta$$

which is proportional to $\Delta\delta$, a spring like action. The synchronizing torque acts on the drive machine or machine-mechanical load inertia to counter the change $\Delta\delta$. This synchronous link-inertia system is set into oscillations of δ and P_e about (P_0, δ_0) as shown in Fig. 8.50. Physically these oscillations are parasitic variations in rotor speed about the synchronous speed and the same frequency variation of shaft torque and electrical power (synchronizing power) exchange with bus-bars. The δ-variations are integral of the speed variations.

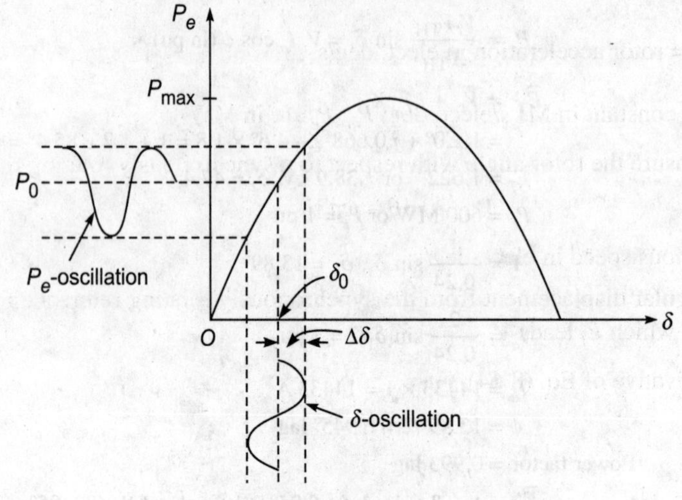

Fig. 8.50

Analytical Analysis Under Steady State Condition

As the system friction is quite small these oscillations are slow to decay out. These get again triggered by another disturbance. These oscillations must be damped out fast enough as we shall see after the analytical study which follows under steady state conditions at constant (synchronous) speed

$$P_e = P_m \tag{8.61}$$

where P_e = electrical power output of the machine

P_m = mechanical power input to the machine

the machine losses having been assumed negligible. These power flows[¶¶¶] are indicated in Fig. 8.51. If this power balance is disturbed, say for example, the electrical power output slightly decreases, the machine undergoes electromechanical dynamics wherein

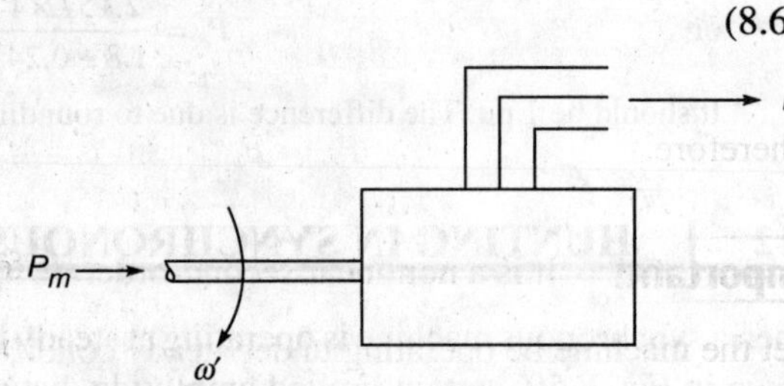

Fig. 8.51 *Power flows in synchronous machine (generator)*

$$P_m - P_e = P_a = \text{accelerating power}$$

In the absence of losses this power is absorbed in accelerating the rotor inertia, i.e.

$$P_m - P_e = \frac{1}{2} I\alpha' \tag{8.62}$$

where I = moment of inertia of rotor (all rotating members) in kg m^2

α' = rotor acceleration in rad (mech)/s^2

If powers are expressed in MW, Eq. (8.62) can be written as

$$P_m - P_e = M\alpha = M\frac{d^2\theta}{dt^2} \tag{8.63}$$

¶¶¶ Power flows will reverse in a motoring machine.

where $\alpha = \dfrac{d^2\theta}{dt^2}$ = rotor acceleration in elect. deg/s

M = inertia constant in MJ s/elect. deg; P_m, P_e are in MW

It is convenient to measure the rotor angle with respect to a synchronously rotating reference frame. Let

$$\delta = \theta - \omega_s t \tag{8.64}$$

where ω_s = synchronous speed in elect. deg/s

δ = rotor angular displacement from the synchronously rotating reference frame; this in fact is the angle by which E_f leads V_t

Taking the second derivative of Eq. (8.64),

$$\frac{d^2\delta}{dt^2} = \frac{d^2\theta}{dt^2} \tag{8.65}$$

Equation (8.62) therefore takes the form

$$P_m - P_e = M\frac{d^2\delta}{dt^2} \tag{8.66}$$

It has already been shown that P_e is governed by the power-angle characteristic of the machine, i.e.

$$P_e = \frac{V_t E_f}{X_s}\sin\delta = P_{e,\max}\sin\delta \tag{8.67}$$

Therefore,

$$P_m - P_{e,\max}\sin\delta = M\frac{d^2\delta}{dt^2} \tag{8.68}$$

Important! It is a nonlinear second-order differential equation known as the *swing equation.*

Let the machine be operating under steady conditions of

$$P_{mo} = P_{eo} = P_{e,\max}\sin\delta_0 \tag{8.69}$$

With the mechanical input power remaining constant at P_{mo}, let the angle δ be disturbed from δ_0 to $\delta_0 + \Delta\delta$. Then

$$P_e = P_{e,\max}\sin(\delta_0 + \Delta\delta)$$

$$\approx P_{e,\max}\sin\delta_0 + \left(\frac{\partial P_e}{\partial\delta}\right)_0\Delta\delta$$

or

$$P_{eo} + \Delta P_e = P_{e,\max}\sin\delta_0 + \left(\frac{\partial P_e}{\partial\delta}\right)_0\Delta\delta$$

or

$$\Delta P_e = \left(\frac{\partial P_e}{\partial\delta}\right)_0\Delta\delta \text{ (incremental machine model)} \tag{8.70}$$

Recall where $\left(\dfrac{\partial P_e}{\partial \delta}\right)_0$ = slope of the power-angle characteristic at the operating point. $\left(\dfrac{\partial P_e}{\partial \delta}\right)_0$ is known as the *synchronizing power coefficient* or *stiffness***** of the machine.

Substituting in Eq. (8.68)

$$-\Delta P_e = -\left(\frac{\partial P_e}{\partial \delta}\right)_0 \Delta\delta = M\frac{d^2(\delta_0 + \Delta\delta)}{dt^2}$$

or

$$M\frac{d^2\Delta\delta}{dt^2} + \left(\frac{\partial P_e}{\partial \delta}\right)_0 \Delta\delta = 0$$

or

$$\left[Mp^2 + \left(\frac{\partial P_e}{\partial \delta}\right)_0\right]\Delta\delta = 0;\ p = \text{differential operator} \tag{8.71}$$

The incremental change $\Delta\delta = \delta - \delta_0$ is governed by the linear second-order differential Eq. (8.71) whose characteristic roots are

$$p = \pm\left[\frac{-(\partial P_e/\partial\delta)_0}{M}\right]^{1/2} \tag{8.72}$$

Now

$$\left(\frac{\partial P_e}{\partial \delta}\right)_0 = P_{e,\max}\cos\delta_0 \tag{8.73a}$$

From Eq. (8.71), $(\partial P_e/\partial\delta)_0 > 0$ for $\delta < 90°$, i.e. in the stable†††† operating zone of the machine, the roots of the characteristic Eq. (8.71) are a complex conjugate pair indicating sinusoidal oscillatory behaviour about the operating point (P_{e0}, δ_0).

Oscillatory behaviour of the synchronous machine about the operating point has its origin in machine stiffness (or the synchronous link between rotor and stator fields which has a spring-type action). This oscillatory behaviour known as *hunting* is highly undesirable—the electrical power fed to the mains and shaft torque oscillate, which in turn causes shaft fatigue. The frequency of oscillation is given by

$$\omega_0 = \left[\frac{(\partial P_e/\partial\delta)_0}{M}\right]^{1/2} \tag{8.73b}$$

If the machine stiffness is excessive, the frequency of the shaft torque variation is higher and the shaft fatigue occurs faster. To minimize hunting to a tolerable level the oscillations must be damped out by introducing a *damping term* (proportional to $d\delta/dt$) in the swing equation (Eq. (8.68)). In fact a small amount of damping, contributed by system losses (both mechanical and electrical), is always present in the machine but this is insufficient to kill hunting. Additional damping must therefore be introduced in the machine.

**** Because of the reluctance power term, a salient-pole machine is stiffer than a cylindrical-rotor machine.

†††† If $\delta_0 > 90°$, $(P_e/\partial\delta)_0 < 0$, so that Eq. (8.66) yields two real roots—one positive and the other negative. Therefore, $\Delta\delta$ will increase without bound—an unstable system as is already known.

Damper Winding

Additional damping is provided in the salient pole synchronous machine by means of damper bars located in the main poles of the machine and short-circuited through round rings at both ends as shown in Fig. 8.99. As the rotor oscillates, the damper bars have a relative movement with respect to the air-gap flux pattern which causes induction of emfs and flow of currents in these bars. The torque created by the bar currents as per Lenz's law always opposes the relative motion. This is how a positive damping term is brought into play so that the oscillatory motion of the rotor about the operating point is considerably reduced in amplitude and the rotor quickly returns to the steady position. These short-circuited bars are known as *damper winding* or *ammortisseur winding*. These act like a squirrel cage induction motor (refer Sec 5.6) thereby providing a starting torque for the motor which otherwise being of synchronous kind is not self-starting. Therefore, the damper winding serves the dual purpose.

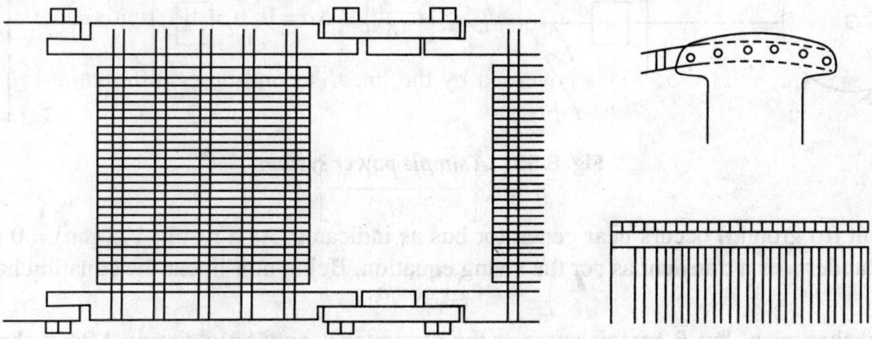

Fig. 8.52 *Damper windings for salient-pole machine*

Analytical Study under Transient Condition

In the hunting of the synchronous machine discussed above, we examine the oscillations about the steady operating point. We shall now consider the machines dynamic response to sudden change in accelerating power caused by change in electric power (P_e) fed to the busbars. As it is a fast transient it can be assumed that during this period the primemover input (P_m) remains constant.

Recalling the swing equation (Eq. [8.68]), we rewrite

$$P_e - P_{e,\max} \sin \delta = M \frac{d^2 \delta}{dt^2} = P_a; \text{ swing equation} \tag{8.74a}$$

where P_a is the accelerating power

Let a synchronous generator be operating at steady value of $\delta = \delta_0$ such that

$$P_m = P_{e,\max}\ \delta_0 = 0, \text{ i.e., } P_a = 0 \tag{8.74b}$$

where $$P_{e,\max} = \frac{E_f V}{X}$$

$P_{e,\max}$ can change suddenly due to change in V and X caused by changes in the system being fed by the generator. This will result in variations in δ governed by the non-linear swing equation (Eq. 8.74(a)) The δ dynamics can end in two conditions:

1. The machine returns to steady operation at a new value of $\delta = \delta_0$ (new) and operates stably.
2. δ may increase without bound and the generator falls out of step (loses synchronism). This is the unstable transient.

For illustration consider a simple power system wherein a generator is feeding an infinite bus through a transmission line with circuit breaker at each end as shown in Fig. 8.53.

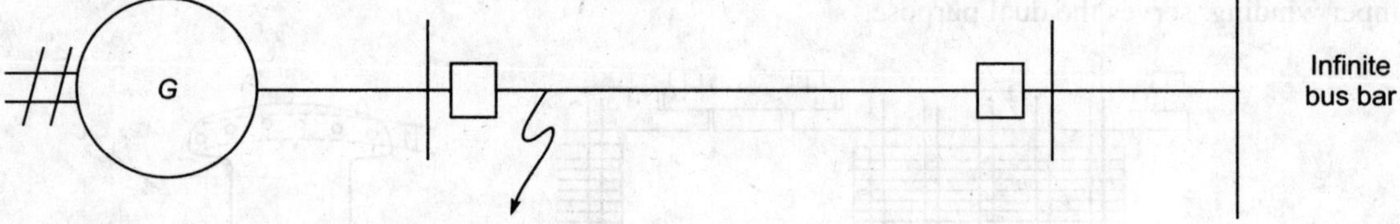

Fig. 8.53 *A simple power system*

An arcing fault (to ground) occurs near generator bus as indicated. As a result, V_b (gen) = 0 and so $P_e = 0$. The machine undergoes a transient as per the swing equation. Being non-linear the equation has to be *solved numerically*.

The circuit breaker opens but P_e remains zero as the generator is now open circuit. After a short wait during which the fault vanishes, the circuit breaker closes automatically. The system is now healthy with $P_e = P_{e,\max}$ δ.

During this kind of circuit switching δ varies as shown in Fig. 8.51. The following conclusions can be drawn from this figure.

⇒ δ increases without bound; the system is *unstable*

⇒ δ goes through a maximum and then decreases. It then decays in oscillatory manner settling at steady value. The system is thus stable having undergone the transient operation.

The stability will depend upon the initial value δ_0. For large value of δ_0 there is a stronger possibility of system instability. As the steady-state limit of δ is 90° (some what less for salient pole machine), the operating value of δ should be much less than 90°, usually about 20° to 30° in view of the transients that are likely to occur in the system.

Condition for Transient Stability

Important! It is concluded from Fig. 8.54 that the system is stable if $\frac{d\delta}{dt} = 0$ at some short time. On the other hand the system is unstable if $\frac{d\delta}{dt} > 0$ for sufficiently long time (usually about 1 second). The swing equation need to be solved for this period of time.

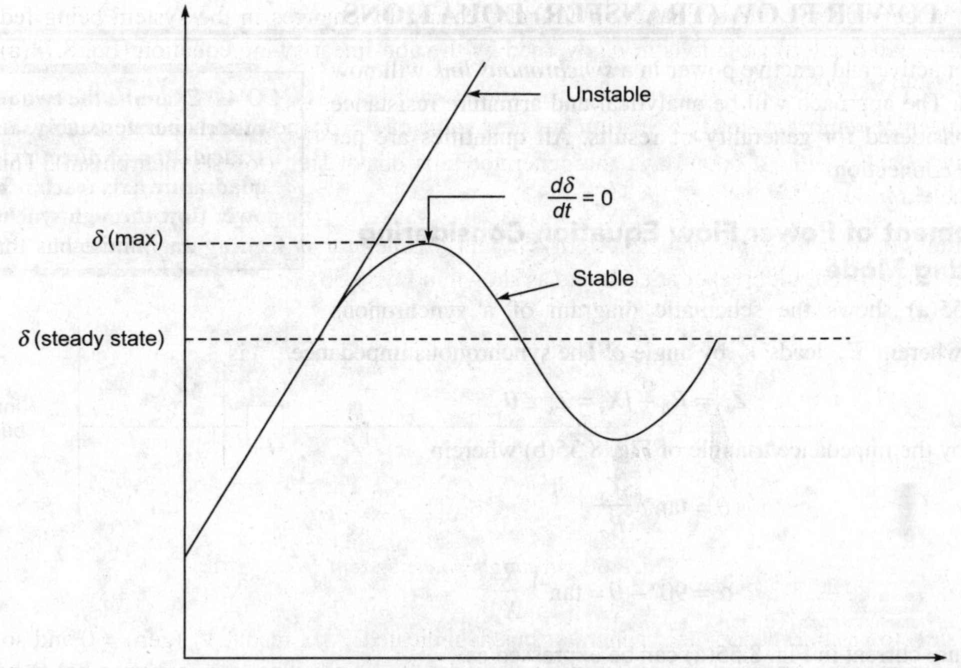

Fig. 8.54 *Plot of δ vs t for stable and unstable systems*

Concept Check — LO 3

1. In a generating synchronous machine connected to infinite bus the mechanical power input is maintained constant, while its field current is increased from a low to a high value. Draw the phasor diagram to show how the magnitude of the armature current will change. Make a sketch of I_a vs I_f.
2. In what manner does an synchronous motor adjust itself to an increasing shaft load?
3. How can the speed of an synchronous motor be varied?
4. What is a damper winding? What is the function of it and where it is located?
5. In what operating condition is a synchronous motor referred to as a synchronous condenser? How is this condition achieved?
6. Briefly describe the phenomenon of "hunting" in a synchronous motor. How is it rectified?
7. Explain briefly the process of synchronizing a synchronous motor to the bus-bars. What conditions determine the instant of synchronization?
8. What is meant by the statement that a synchronous machine is 'floating' on the bus-bars?
9. Elaborate the statement that an unloaded synchronous motor can be made to act as a capacitor or as an inductor.

8.13 | POWER FLOW (TRANSFER) EQUATIONS

LO 4 Examine the two reaction models incorporating saliency (calculation of direct axis and quadrature axis reactance) and power flow through synchronous machine

The flow of active and reactive power in a *synchronous link* will now be studied. The approach will be analytical and armature resistance will be considered for generality of results. All quantities are per phase, star connection.

Development of Power Flow Equation Considering Generating Mode

Figure 8.55(a) shows the schematic diagram of a synchronous generator wherein $\bar{E}_f$ leads $\bar{V}_t$ by angle δ. The synchronous impedance[‡‡‡‡] is

$$\bar{Z}_s = R_a + jX_s = Z_s \angle\theta \tag{8.75}$$

as shown by the impedance triangle of Fig. 8.55(b) wherein

$$\theta = \tan^{-1}\frac{X_s}{R_a} \tag{8.76a}$$

and
$$\alpha = 90° - \theta = \tan^{-1}\frac{R_a}{X_s} \tag{8.76b}$$

The armature current in Fig. 8.55(a) can be expressed as

$$\bar{I}_a = \frac{E_f\angle\delta - V_t\angle 0°}{Z_s\angle\theta} \tag{8.77}$$

The complex power output is

$$\bar{S}_e = P_e + jQ_e = V_t\angle 0°\ \bar{I}_a^* \tag{8.78}$$

where

P_e = *active power*, W

Q_e = *reactive power, vars*, positive for lagging pf and negative for leading pf

$$\text{Power factor} = \cos\tan^{-1}\frac{Q}{P}$$

Substituting for I_a from Eq. (8.77) in Eq. (8.78),

$$P_e + jQ_e = V_t\angle 0\left(\frac{E_f\angle\delta - V_t\angle 0}{Z_s\angle\theta}\right)^*$$

$$= \frac{V_tE_f}{Z_s}\angle(\theta-\delta) - \frac{V_t^2}{Z_s}\angle\theta \tag{8.79}$$

Equating the real and imaginary parts of Eq. (8.79), the following expressions for real and reactive power output are obtained as

[‡‡‡‡] If a line is present as a part of the synchronous link, the line resistance and reactance will be included in R_a and X_s respectively.

$$P_e(\text{out}) = -\frac{V_t^2}{Z_s}\cos\theta + \frac{V_t E_f}{Z_s}\cos(\theta - \delta) \tag{8.80a}$$

$$Q_e(\text{out}) = -\frac{V_t^2}{Z_s}\sin\theta + \frac{V_t E_f}{Z_s}\sin(\theta - \delta) \tag{8.80b}$$

The net mechanical power input to the machine is given by

$$P_m(\text{in}) = P'_e = Re\left[\bar{S}'_e = E_f\angle\delta\left(\frac{E_f\angle\delta - V_t\angle 0}{Z_s\angle\theta}\right)^*\right];\ Re = \text{real of}$$

$$= \frac{E_f^2}{Z_s}\cos\theta - \frac{V_t E_f}{Z_s}\cos(\delta + \theta) \tag{8.81}$$

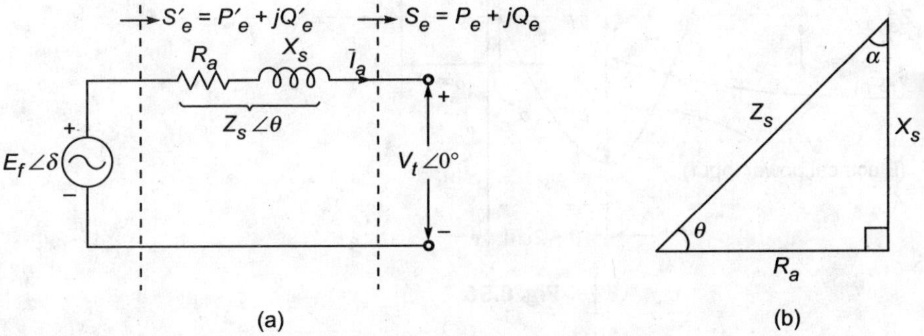

Fig. 8.55 *Impedance triangle*

It is the mechanical power converted to electric power. It is convenient to express the above results in terms of angle α defined in the impedance triangle of Fig. 8.55(b) (Eq. 8.76b). Equations (8.80a), (8.80b) and (8.81) then modify as below

$$P_e(\text{out}) = -\frac{V_t^2}{Z_s^2}R_a + \frac{V_t E_f}{Z_s}\sin(\delta + \alpha) \tag{8.82a}$$

$$Q_e(\text{out}) = -\frac{V_t^2}{Z_s^2}X_s + \frac{V_t E_f}{Z_s}\cos(\delta + \alpha) \tag{8.82b}$$

$$P_m(\text{in}) = \frac{E_f^2}{Z_s^2}R_a + \frac{V_t E_f}{Z_s}\sin(\delta - \alpha) \tag{8.83}$$

$$P_m(\text{in}) - P_e(\text{out}) = I_a^2 R_a$$

as the only loss is in resistance. This result can be proved by substituting P_m (in) and P_e (out) from Eq. (8.81) and (8.82a) followed by several steps of manipulation and reference to the phasor diagram of Fig. 8.47.

Development and Analysis of Real Electrical Output Power (P_e) vs Angle Plotting

The real electrical power output, P_e, as per Eq. (8.82a) is plotted in Fig. 8.56 from which it is observed that its maximum value is

$$P_e(\text{out})|_{\max} = -\frac{V_t^2 R_a}{Z_s^2} + \frac{V_t E_f}{Z_s}, \text{ at } \delta + \alpha = 90° \tag{8.84}$$

occurring at $\delta = \theta$, which defines the limit of steady-state stability. The machine will fall out of step for angle $\delta > \theta$. Of course, θ will be 90° if resistance is negligible in which case the stability limit will be at $\delta = 90°$ as already explained in Sec. 8.10.

!Attention

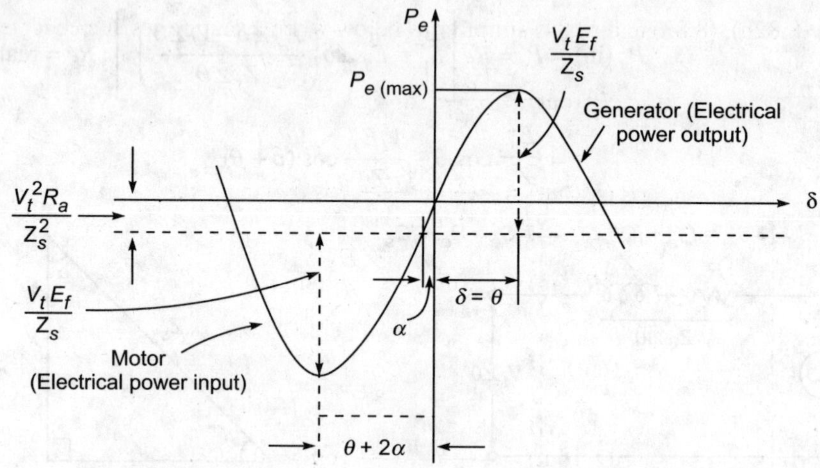

Fig. 8.56

At maximum electrical power output, the corresponding reactive power output is found from Eq. (8.82b) by substituting $\delta = \theta$

$$Q_e\,(\text{out})|\,at\ P_e\,(\text{out}),_{\max} = -\frac{V_t^2}{Z_s^2} X_s + \frac{V_t E_f}{Z_s} \cos(\delta + \alpha)$$

or

$$Q_e\,(\text{out}) = -\frac{V_t^2}{Z_s^2} X_s; \quad \text{as} \quad \delta + \alpha = 90°$$

As the Q_e (out) is negative, it means the generator var output is leading i.e. the generator operates at leading pf.

We can find $Q_e\,(\text{out})_{\max}$ from Eq. (8.82 b). It occurs at $\delta + \alpha = 180°$ or $\delta = 180° - \alpha$. Its value is

$$Q_e\,(\text{out})_{\max} = -\frac{V_t^2}{Z_s^2} X_s - \frac{V_t E_f}{Z_s}$$

Corresponding

$$P_e\,(\text{out}) = -\frac{V_t^2}{Z_s^2} R_a\text{; insignificant value as } R_a \text{ is very small}$$

As $Q_e(\text{out})_{\max}$ is negative, the generator operates at very low, leading pf (close to 90º).

These result have no significance for generator operation as that not how a generator is operated.

It follows from Eq. (8.83) that

$$P_m\,(\text{in})|_{\max} = \frac{E_f^2 R_a}{Z_s^2} + \frac{V_t E_f}{Z_s};\ \text{at } \delta = 90° + \alpha = \theta + 2\alpha \tag{8.85}$$

Since the angle δ in Eq. (8.85) is more than θ, the maximum mechanical power input (net) operation for a generator lies in the unstable region.

Equations (8.82a), (8.82b), (8.83) and (8.84) simplify as below when armature resistance is neglected

$$P_e\,(\text{out}) = \frac{V_t E_f}{X_s}\sin\delta \tag{8.86a}$$

$$Q_e\,(\text{out}) = -\frac{V_t^2}{X_s} + \frac{V_t E_f}{X_s}\cos\delta \tag{8.86b}$$

$$P_m\,(\text{in}) = P_e\,(\text{out}) = \frac{V_t E_f}{X_s}\sin\delta \tag{8.86c}$$

$$P_e\,(\text{out})|_{\max} = P_m\,(\text{in})|_{\max} = \frac{V_t E_f}{X_s};\quad \text{at } \delta = 90° \tag{8.86d}$$

For unity pf $\quad Q_e\,(\text{out}) = -\dfrac{V_t^2}{X_s} + \dfrac{V_t E_f}{X_s}\cos\delta = 0$

Recall or $\quad E_f \cos\delta = V_t$; normal excitation (generator)

For lagging pf $\quad Q_e\,(\text{out}) > 0$

or $\quad E_f \cos\delta > V_t$; over-excited (generator)

For leading pf $\quad Q_e\,(\text{out}) < 0$

or $\quad E_f \cos\delta < V_t$; under-excited (generator)

These results can be elaborated with the phasor diagram of Fig. 8.39(a).

Development of Power Flow Equation Considering Motoring Mode

Figure 8.57 shows the operation of the synchronous machine as a motor[§§§§]. Here the angle δ by which E_f lags V_t is defined as positive. Also the direction of power flow is now into the machine while the mechanical power flows out at the machine.

It then follows from Eqs (8.82a), (8.82b) and (8.83) by changing the sign of δ that for motoring operation

$$P_e\,(\text{in}) = -P_e\,(\text{out})$$

$$= \frac{V_t^2}{Z_s^2} R_a + \frac{V_t E_f}{Z_s}\sin(\delta - \alpha) \tag{8.87a}$$

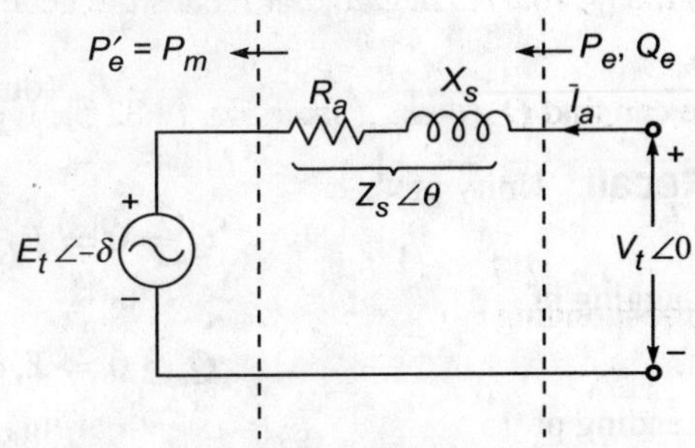

Fig. 8.57

[§§§§] The motoring operation can be analyzed by the generator power flow equation except that δ will have a negative sign and active and reactive powers will be negative of those obtained from Eqs (8.82a); (8.82b) and (8.83).

$$Q_e\,(\text{in}) = -\,Q_e\,(\text{out})$$

$$= \frac{V_t^2}{Z_s^2} X_s - \frac{V_t E_f}{Z_s} \cos(\delta - \alpha) \tag{8.87b}$$

$$P_m\,(\text{out, gross}) = -P_m\,(\text{in})$$

$$= -\frac{E_f^2}{Z_s^2} R_a + \frac{V_t E_f}{Z_s} \sin(\delta + \alpha) \tag{8.88}$$

The maximum mechanical power output from Eq. (8.88) is given by

$$P_m\,(\text{out})|_{\max} = -\frac{E_f^2}{Z_s^2} R_a + \frac{V_t E_f}{Z_s} \tag{8.89}$$

It occurs at $\delta = \theta$ which defines the limit of steady-state stability. It is easily seen from Eq. (8.73a) that the maximum electrical power input occurs at $\delta = \theta + 2\alpha$ which lies outside the stability limit. The reader should compare these results with that of the generator.

$$Q_e\,(\text{in})_{\max} = \frac{V_t^2}{Z_s^2} X_s + \frac{V_t E_f}{Z_s}$$

Occurs at $(\delta - \alpha) = 180^\circ$ or $\delta = 180^\circ - \alpha$

Corresponding $P_e\,(\text{in}) = \dfrac{V_t^2}{Z_s^2} R_a$: very small

As Q_e is positive and P_e is very small, the motor operate a low lagging pf (close to 90º). The motor acts as an inductor.

For the case of negligible resistance

$$P_e\,(\text{in}) = \frac{V_t E_f}{X_s} \sin\delta \tag{8.90a}$$

$$Q_e\,(\text{in}) = \frac{V_t^2}{X_s} - \frac{V_t E_f}{X_s} \cos\delta \tag{8.90b}$$

$$P_m\,(\text{out}) = P_e\,(\text{in}) = \frac{V_t E_f \sin\delta}{X_s} \tag{8.90c}$$

Recall Unity pf

$$Q_e = 0 \Rightarrow E_f \cos\delta = V_t \text{; normal excitation (motor)}$$

Lagging pf

$$Q_e > 0 \Rightarrow E_f \cos\delta < V_t \text{; under-excited (motor)}$$

Leading pf

$$Q_e < 0 \Rightarrow E_f \cos\delta > V_t \text{; over-excited (motor)}$$

These results can be verified from the phasor diagram of Fig. 8.39(b).

Conditions for Power Factor, R_a Accounted

These could be arrived at by the sign Q_e from Eq. (8.64b) for generator and from Eq. (8.81b) for motor. However, simple from of these conditions are found from the phasor diagrams of Fig. 8.47. These phasor diagrams are redrawn in Fig. 8.58(a) for generator with lagging power factor and in Fig. 8.58(b) for motor with leading power factor with some projections shown in dotted line.

Generator

$$AF = E_f \cos \delta$$

$$AE = V_t + I_a R_a \cos \phi$$

From the phasor diagram geometry

$$AF > AE$$

or $$E_f \cos \delta > (V_t + I_a R_a \cos \phi)$$

or $$E_f \cos \delta - I_a R_a \cos \phi > V_t \text{: over-excited, lagging pf} \quad \text{(i)}$$

Recall Other two conditions are

$$E_f \cos \delta - I_a R_a \cos \phi = V_t\text{; normal excitation, unity pf} \quad \text{(ii)}$$

$$E_f \cos \delta - I_a R_a \cos \phi < V_t \text{: under-excitation, leading pf} \quad \text{(iii)}$$

Motor The pf condition that follow similarly from Fig. 8.58(b) are

$$E_f \cos \delta + I_a R_a \cos \phi > V_t\text{; over-excited, leading pf} \quad \text{(i)}$$

$$E_f \cos \delta + I_a R_a \cos \phi = V_t\text{; normal excitation, unity pf} \quad \text{(ii)}$$

$$E_f \cos \delta + I_a R_a \cos \phi < V_t\text{; under-excitation, lagging pf} \quad \text{(iii)}$$

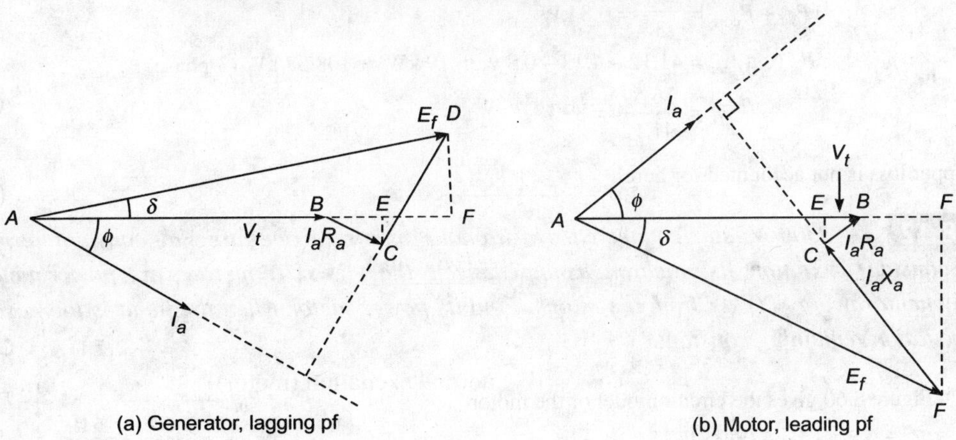

Fig. 8.58 *Determination of power factor conditions*

Example 8.11 *A 400 V, 3-phase, delta-connected synchronous motor has an excitation emf of 600 V and synchronous impedance per phase of 0.3 + j6 Ω. Calculate the net power output, efficiency, line current and power factor when the machine is developing maximum mechanical power (gross). Windage, friction and core losses may be assumed to be 2.4 kW.* LO 4

Solution

$$Z_s \text{ (eq. star)} = \frac{1}{3}(0.3 + j6) = 0.1 + j2 = 2\angle 87.14°$$

For maximum mechanical power output

$$\delta = \theta = 87.14° \ (E_f \text{ lags } V_t)$$

$$V_t = 400/\sqrt{3} = 230.9 \text{ V}$$

$$E_f = 600/\sqrt{3} = 346.4 \text{ V}$$

From the phasor diagram of Fig. 8.59

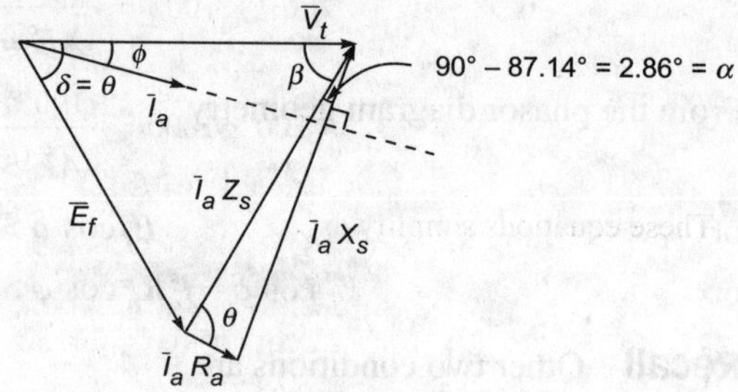

Fig. 8.59

$$I_a Z_s = \sqrt{V_t^2 + E_f^2 - 2V_t E_f \cos\theta}$$

$$I_a Z_s = \sqrt{(230.9)^2 + (346.4)^2 - 2 \times 230.9 \times 346.4 \times \cos 87.14°}$$

$$= 406.6 \text{ V}$$

$$\therefore \quad I_a = 203.3 \text{ A}$$

Further $\cos\beta = \dfrac{(230.9)^2 + (406.6)^2 - (346.4)^2}{2 \times 230.9 \times 406.6}$

or $\quad \beta = 58.31°$

From the geometry of the phasor diagram

$$\phi = 90° - 58.31° - 2.86° = +28.8°$$

$$\text{pf} = \cos\phi = 0.876 \text{ lag}$$

$$P_e \text{ (in)} = 230.9 \times 203.3 \times 0.876 = 41.12 \text{ kW}$$

$$I_a^2 R_a = (203.3)^2 \times 0.1 = 4.13 \text{ kW, (stray load loss is included as } R_a \text{ is effective value)}$$

$$P_{wf} + P_{\text{core}} = \frac{2.4}{3} = 0.8 \text{ kW}$$

$$P_m \text{ (out)}|_{\text{net}} = 41.12 - 4.13 - 0.8 = 36.19 \text{ kW or } 108.57 \text{ kW (3-phase)}$$

$$\eta = \frac{36.59}{41.12} = 89\%$$

Note: Field copper loss is not accounted for here.

Example 8.12 *A 3300 V, star-connected synchronous motor is operating at constant terminal voltage and constant excitation. Its synchronous impedance is 0.8 + j5 Ω. It operates at a power factor of 0.8 leading when drawing 800 kW from the mains. Find its power factor when the input is increased to 1200 kW, excitation remaining constant.*

LO 4

Solution

Figure 8.60 gives the circuit model of the motor

$$Z_s = 0.8 + j5 = 5.06\angle 81°; \qquad \alpha = 9°$$

$$V_t = 3300/\sqrt{3} = 1905 \text{ V}$$

$$P_e \text{ (in)} = 800/3 = 266.7 \text{ kW} \qquad \text{(per phase)}$$

$$\cos\phi = 0.8 \text{ leading}$$

$$\therefore \quad Q_e \text{ (in)} = -P_e \text{ (in)} \tan\phi = -200 \text{ kVAR}$$

Note that Q_e (in) is negative because the power factor is leading.

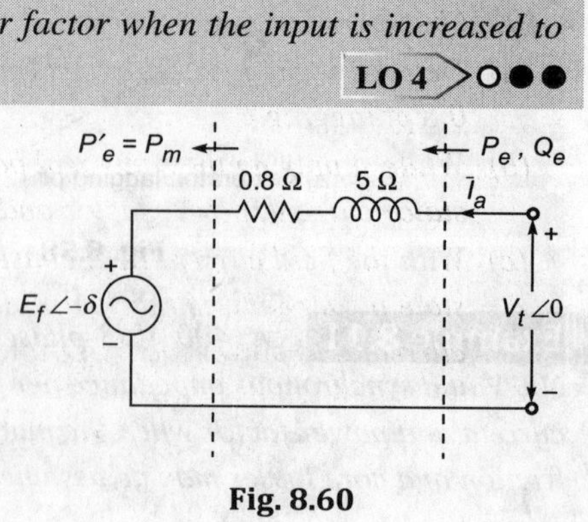

Fig. 8.60

From Eqs (8.87a) and (8.87b)

$$P_e \text{ (in)} = \frac{V_t^2}{Z_s^2} R_a + \frac{V_t E_f}{Z_s} \sin(\delta - \alpha) \qquad \text{(i)}$$

$$Q_e \text{ (in)} = \frac{V_t^2}{Z_s^2} X_s - \frac{E_f V_t}{Z_s} \cos(\delta - \alpha) \qquad \text{(ii)}$$

Substituting the values

$$266.7 \times 1000 = \frac{(1905)^2 \times 0.8}{(5.06)^2} + \frac{(1905) E_f}{5.06} \sin(\delta - \alpha)$$

$$-200 \times 1000 = \frac{(1905)^2 \times 5}{(5.06)^2} - \frac{(1905) E_f}{5.06} \cos(\delta - \alpha)$$

These equations simplify as

$$E_f \sin(\delta - \alpha) = 407.2$$

$$E_f \cos(\delta - \alpha) = 2413.6$$

from which E_f is obtained as

$$E_f = 2447.7 \text{ V}$$

Under new operating conditions

$$P_e \text{ (in)} = 1200/3 = 400 \text{ kW}$$

Substituting in (i),

$$400 \times 1000 = \frac{(1905)^2 \times 0.8}{(5.06)^2} + \frac{(1905) \times (2447.7)}{5.06} \sin(\delta - \alpha)$$

or $\quad \sin(\delta - \alpha) = 0.31 \quad \text{or} \quad \delta - \alpha = 18°$

Now Eq. (ii) is used to find Q_e for $(\delta - \alpha) = 18°$

$$Q_e = \frac{(1905)^2 \times 5}{(5.06)^2} - \frac{1905 \times 2447.7}{5.06} \cos 18° = -167.7 \text{ kVAR}$$

$$\text{Power factor} = \cos(\tan^{-1} Q_e/P_e) = \cos(\tan^{-1} 167.7/400) = 0.92 \text{ leading}$$

Example 8.13 *A three phase 10 kVA, 400 V, 4-pole, 50 Hz star connected synchronous machine has synchronous reactance of 16 Ω and negligible resistance. The machine is operating as generator on 400 V bus-bars (assumed infinite).*

(a) Determine the excitation emf (phase) and torque angle when the machine is delivering rated kVA at 0.8 pf lagging.

(b) While supplying the same real power as in part (a), the machine excitation is raised by 20%. Find the stator current, power factor and torque angle.

(c) With the field current held constant as in part (a), the power (real) load is increased till the steady-state power limit is reached. Calculate the maximum power and kVAR delivered and also the stator current and power factor. Draw the phasor diagram under these conditions. LO 4

Solution The circuit equivalent of the machine is drawn in Fig. 8.61.

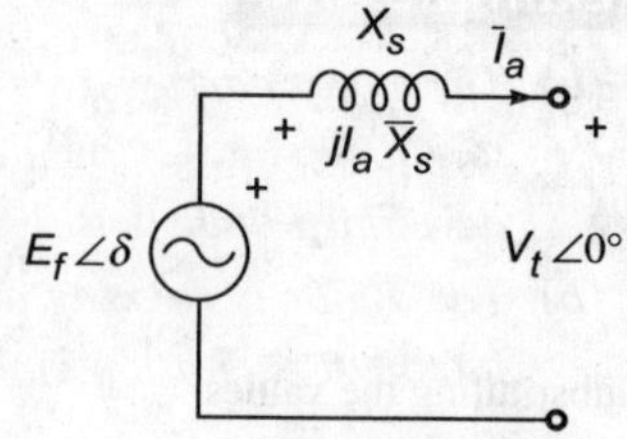

Fig. 8.61

(a) $$I_a = \frac{10 \times 10^3}{\sqrt{3} \times 400} = 14.43 \text{ A}$$

pf angle, $$\phi = \cos^{-1} 0.8 = 36.9° \text{ lag}$$

$$\bar{I}_a = 14.43 \angle -36.9°$$

$$V_t = \frac{400}{\sqrt{3}} = 231 \text{ V}$$

From the circuit equivalent $$\bar{E}_f = 231\angle 0° + j\,14.43 \angle -36.9° \times 16$$

$$= 231 + 231\angle 53.1° = 369.7 + j\,184.7$$

or $$\bar{E}_f = 413.3\angle 26.5°$$

Torque angle, $\delta = 26.5°$, E_f leads V_t (generating action)

(b) Power supplied (source), $P_e = 10 \times 0.8 = 8$ kW (3 phase)

E_f (20% more) $= 413.3 \times 1.2 = 496$ V

$$P_e = \frac{E_f V_t}{X_s} \sin \delta$$

$$\frac{8 \times 10^3}{3} = \frac{496 \times 231}{16} \sin \delta$$

Torque angle, $$\delta = 21.9°$$

From the circuit equivalent

$$\bar{I}_a = \frac{E_f\angle\delta - V_t\angle 0°}{jX_s} = \frac{496\angle 21.9° - 231}{j16}$$

$$= \frac{229 + j185}{j16} = 11.6 - j\,14.3 = 18.4\angle -50.9°$$

$I_a = 18.4$ A, pf $= \cos 50.9° = 0.63$ lagging

(c) $E_f = 413$ V; field current same as in part (a)

$$P_e\,(\text{max}) = \frac{E_f V_t}{X_s};\ \delta = 90°$$

$$= \frac{413 \times 231}{16} \times 10^{-3} = 5.96 \text{ kW/phase or } 17.38 \text{ kW, 3-phase}$$

$$\bar{I}_a = \frac{413\angle 90° - 231}{j16} = 25.8 + j\,14.43$$

$$= 29.56\angle 29.2° \text{ A}$$

$I_a = 29.56$ A, pf $= \cos 29.2° = 0.873$ leading

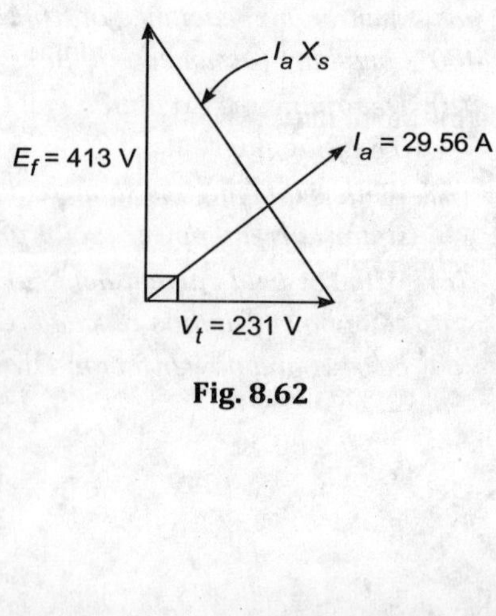

Fig. 8.62

The phasor diagram is drawn in Fig. 8.62.

kVAR delivered (negative)

$$\frac{Q_e}{P_e} = \tan - 29.2$$

or $$Q_e = 8 \times 0.559 = -4.47 \text{ kVAR}$$

Example 8.14 *The synchronous machine of Example 8.13 is acting as a motor.*

(a) The motor carries a shaft load of 8 kW and its rotational loss is 0.5 kW. Its excitation emf is adjusted to 750 V (line). Calculate its armature current, power factor and power angle. Also calculate the developed and shaft torques.

(b) The motor is running at no load and its losses can be neglected. Calculate its armature and power factor at excitation emf (line) of (i) 600 V, and (ii) 300 V. Calculate also the kVAR drawn in each case.

(c) The motor is on no load (losses to be ignored). What should be its excitation for it to draw a leading kVAR of 6? Draw the phasor diagram.

LO 4

Solution

(a)

$$P_m(\text{gross}) = 8 + 0.5 = 8.5 \text{ kW}$$

$$P_e(\text{in}) = P_m(\text{gross}), \text{ as } R_a = 0$$

$$E_f = \frac{750}{\sqrt{3}} = 433 \text{ V}, \quad V_t = 231 \text{ V (data Example 8.13)}$$

$$P_e(\text{in}) = \frac{E_f V_t}{X_s} \sin \delta$$

$$\frac{8.5 \times 10^3}{3} = \frac{433 \times 231}{16} \sin \delta$$

$$\delta = 27^\circ, E_f \text{ lags } V_t$$

$$\bar{E}_f = 433\angle -27^\circ \text{ V}, \ \bar{V}_t = 231\angle 0^\circ \text{ V}$$

In a motor current I_a will flow in Fig. 8.57.

$$\bar{I}_a = \frac{231\angle 0^\circ - 433\angle -27^\circ}{j16} = 15.6\angle 38^\circ \text{ A}$$

$$I_a = 15.6 \text{ A}, \quad \text{pf} = \cos 38.1^\circ = 0.787 \text{ leading}$$

Synchronous speed, $$n_s = \frac{120 \times 50}{4} = 1500 \text{ rpm or } 157.1 \text{ rad/s}$$

$$\text{Torque developed} = \frac{8.5 \times 10^3}{157.1} = 54.1 \text{ Nm}$$

$$\text{Shaft torque} = \frac{8 \times 10^3}{157.1} = 50.9 \text{ Nm}$$

(b) Motor running no load, no loss

$$\delta = 0$$

(i) $$E_f = \frac{600}{\sqrt{3}} = 346.4 \text{ V}, \quad V_t = 231 \text{ V}$$

$$\bar{I}_a = \frac{231 - 346.4}{j16} = j\,7.213 \text{ A leading } V_t \text{ by } 90^\circ$$

kVAR drawn $$3\,\bar{V}_t \bar{I}_a^* = 3 \times 231 \times (-j7.213) \times 10^{-3}$$

$$= -j5; \text{ minus sign is for leading kVAR}$$

The motor acts as a capacitor

$$C = \frac{7.213}{231} \times \frac{1}{2\pi \times 50} = 99.4\ \mu\text{F}$$

(ii) $$E_f = 300/\sqrt{3} = 173.2\text{ V},\quad V_t = 231\text{ V}$$

$$\bar{I}_a = \frac{231 - 173.2}{j16} = j\,3.61\text{ A lagging } V_t \text{ by } 90°$$

kVA drawn $$3\,\bar{V}_t \bar{I}_a^* = 3 \times 231 \times (j\,3.61) \times 10^{-3}$$

$$= j\,2.5;\ \text{plus sign means for lagging kVA}$$

The motor acts as inductor

$$L = \frac{231}{2.5} \cdot \frac{1}{2\pi \times 50} = 294.1\text{ mH}$$

(c) $$3\,\bar{V}_t \bar{I}_a^* \times 10^{-3} = -j\,6,\ V_t = 231$$

$$I_a^* = -j\frac{2000}{231} = -j\,8.66$$

$$\bar{I}_a = j\,8.66\ A\ (\text{leading } V_t \text{ by } 90°)$$

$$\bar{E}_f = \bar{V}_t - j\,\bar{I}_a X_s = 231 - j\,(j\,8.66) \times 16 = (231 + 138.6)\angle 0°$$

$$\bar{E}_f = 369.6\angle 0°\text{V},\ E_f = 640\text{ V (line)}$$

The phasor diagram is drawn in Fig. 8.63.

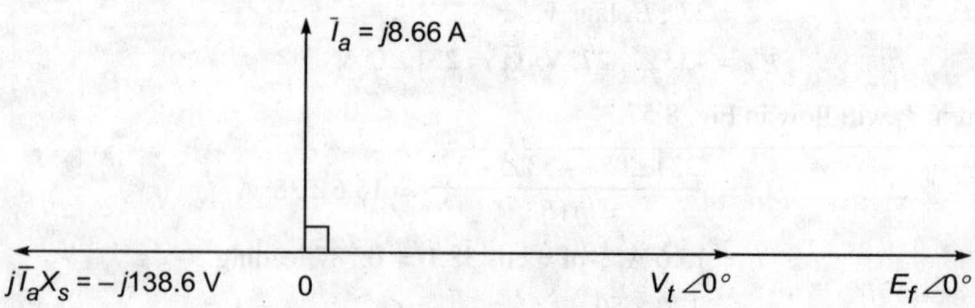

Fig. 8.63 *Phasor diagram*

Example 8.15 *A 6-pole, 3-phase, 4 MVA, 50 Hz, star connected synchronous motor is supplied from 6.6 kV bus-bars. It has a synchronous reactance of 4.8 Ω:*

(a) The motor is operating at a power angle of 20º at rated current. Find the excitation emf if the power factor is lagging leading.

(b) In part (a) find the mechanical power developed and power factor in each case. LO 4

Solution

$$V_t = \frac{6.6}{\sqrt{3}} = 3.81\text{ kV}$$

$$I_a(\text{rated}) = \frac{4 \times 10^3}{\sqrt{3} \times 6.6} = 350\text{ A}$$

$$I_a(\text{rated})\,X_s = 350 \times 4.8$$

$$= 1680\text{ V or } 1.68\text{ kV}$$

(a) In the motor

E_f lags V_t by $\delta = 20°$

$$\bar{E}_f = V_t\angle 0° - j\bar{I}_a X_s$$

The phasor diagram is drawn in Fig. 8.64.

As $I_a X_s$ = 1.68 kV constant, there are two possible solutions. From the geometry of triangle *ABD*

$$V_t^2 + E_f^2 - 2\,V_t E_f \cos\delta = (I_a X_s)^2$$

Substituting values

$$(3.81)^2 + E_f^2 - 2 \times 3.81\,E_f \cos 20° = (1.68)^2$$

or $$E_f^2 - 7.16\,E_f + 11.69 = 0$$

Its solutions are E_{f1} = 2.52 kV, E_{f2} = 4.64 kV

or E_{f1} = 4.36 kV (line), E_{f2} = 8.04 kV (line)

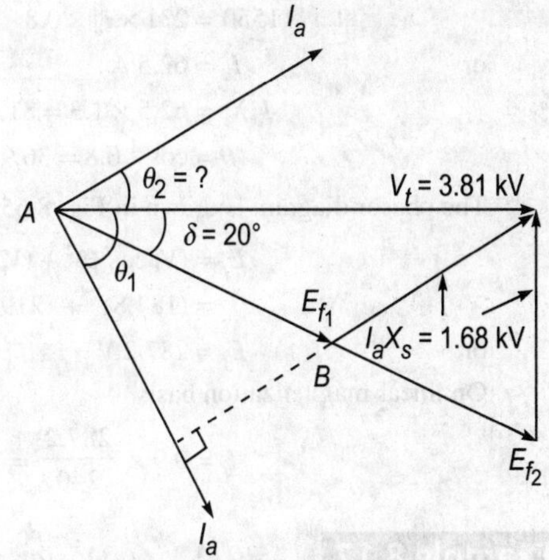

Fig. 8.64 *Phasor diagram*

(b) E_{f1} = 2.52 kV ; motor under-excited case, pf lagging

$$P_m = 3 \times \left(\frac{3.81 \times 2.52}{4.8}\right) \sin 20° = 2.052 \text{ MW}$$

$$\sqrt{3} \times 6.6 \times 350 \cos\theta_1 = 2.052 \times 10^3$$

$$pf_1 = \cos\theta_1 = 0.513 \text{ lagging}$$

$$E_{f2} = 4.64 \text{ kV ; motor over-excited pf leading}$$

$$P_m = 3 \times \left(\frac{3.81 \times 4.64}{4.8}\right) \sin 20° = 3.779 \text{ MW}$$

$$\sqrt{3} \times 6.6 \times 350 \cos\theta_2 = 3.779 \times 10^3$$

$$pf_2 = \cos\theta_2 = 0.944 \text{ leading; } \theta_2 = 19.2°$$

Example 8.16 *A synchronous motor is drawing 50 A from 400 V, three-phase supply at unity pf with a field current of 0.9 A. The synchronous reactance of the motor is 1.3 Ω.*

(a) Find the power angle.

(b) With the mechanical load remaining constant, find the value of the field current which would result in 0.8 leading power factor. Assume linear magnetization.

LO 4

Solution We proceed on per basis, star connection

(a) $$V_t = \frac{400}{\sqrt{3}} = 231\angle 0 \text{ A}$$

$$\text{pf = unity, pf angle, } \theta = 0°,\ \bar{I}_a = 50\angle 0°$$

$$\bar{E}_f = \bar{V}_t - j\bar{I}_a X_s = 231 - j\,50 \times 1.3 = 231 - j\,65 = 240\angle -15.7° \text{ V}$$

Power angle, $\delta = 15.7°$, E_f lags V_t (motor)

(b) Mechanical load in part (a)

$$P_m = V_t I_a \cos\theta = 231 \times 50 \times 1 = 11550 \text{ W}$$

It remains constant. As there is no ohmic loss

$$P_e = P_m = V_t I_a \cos\theta$$

$$11550 = 231 \times I_a \times 0.8$$

or $$I_a = 62.5 \text{ A}$$

$$I_a X_s = 62.5 \times 1.3 = 81.25 \text{ V}$$

$$\theta = \cos^{-1} 0.8 = 36.9° \text{ leading}$$

The phasor diagram is drawn in Fig. 8.65.

$$E_f^2 = (V_t \cos \theta)^2 + (V_t \sin \theta + I_a X_s)^2$$

$$= (184.8)^2 + (219.85)^2$$

or $$E_f = 287.2 \text{ V}$$

On linear magnetization basis

$$I_f = 0.9 \times \frac{287.2}{240} = 1.077 \text{ A}$$

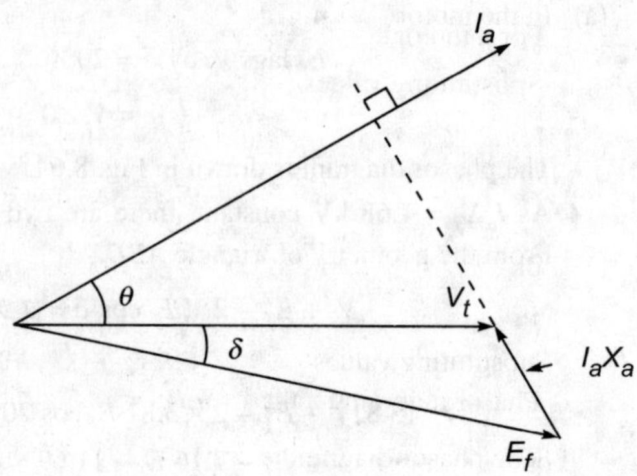

Fig. 8.65

Example 8.17 *A 40 kVA, 600 V star-connected synchronous motor has armature effective resistance of 0.8 Ω and synchronous reactance of 8 Ω It has stray loss of 2 kW.*

The motor is operating at 600 V bus-bar while supplying a shaft load of 30 kW, it is drawing rated current at leading pf.

(a) Calculate the motor efficiency.

(b) What is its excitation emf and power angle?

(c) With this excitation calculate the maximum power output (gross) and corresponding net output and the power angle.

LO 4

Solution

(a) Shaft load, $P_m(\text{net}) = 30 \text{ kW}$

Stray loss, $P_{st} = 2 \text{ kW}$

Mechanical power developed, $P_m(\text{dev}) = 30 + 2 = 32 \text{ kW}$

Armature current, $I_a = I_a(\text{rated}) = \dfrac{40 \times 10^3}{\sqrt{3} \times 600} = 38.5 \text{ A}$

Ohmic loss, $3 I_a^2 R_a = 3 \times (38.5)^2 \times 0.8 \times 10^{-3} = 3.557 \text{ kW}$

Electrical power input, $P_e(\text{in}) = 32 + 3.557 = 35.557 \text{ kW}$

Efficiency, $\eta = 1 - \dfrac{2 + 3.557}{35.557} = 84.4\%$

(b)

$$P_e(\text{in}) = \sqrt{3}\, V_t I_a \cos \theta$$

$$35.557 \times 10^3 = \sqrt{3} \times 600 \times 38.5 \cos \phi$$

$$\text{pf, } \cos \phi = 0.889 \text{ leading}$$

$$\phi = 27.3° \text{ leading}$$

$$\bar{I}_a = 38.5\angle 27.3° \text{ A}$$

Terminal voltage, $\bar{V}_t = \dfrac{600}{\sqrt{3}} \angle 0° = 346.4\angle 0° \text{ V}$

$$\bar{Z}_s = 0.8 + j\,8 = 8.04\angle 84.3° \,\Omega,\ \theta = 84.3°$$

For a motor $\quad \bar{E}_f = \bar{V}_t - \bar{I}_a \bar{Z}_a$

Substituting values

$$\bar{E}_f = 346.4\angle 0^\circ - 38.5\angle 27.3^\circ \times 8.04\angle 84.3^\circ$$

$$\bar{E}_f = 346.4 - 309.54\angle 111.6^\circ$$

or $\quad \bar{E}_f = 460.4 - j\,287.8^\circ = 542.8\angle -32^\circ$

$E_f = 542.8$ V or 940 V line; over-excited

$\delta = 32^\circ$, E_f lag V_t; motoring action

Phasor diagram method $\quad I_aR_a = 38.5 \times 0.8 = 30.8$ V, $I_aX_s = 38.5 \times 8 = 308$ V

The phasor diagram is drawn in Fig. 8.66.

$$AD = V_t \cos\phi - I_aR_a = 346.4 \times 0.889 - 30.8 = 277.1 \text{ V}$$

$$CD = V_t \sin\phi + I_aX_a = 158.62 + 308 = 466.6 \text{ V}$$

$$E_f = \sqrt{(AD)^2 + (CD)^2}$$

Substituting value, we get

$$E_f = 542.7 \text{ V}$$

$$\delta + \phi = 59.3^\circ$$

$$\delta = 59.3^\circ - 27.3^\circ = 32^\circ$$

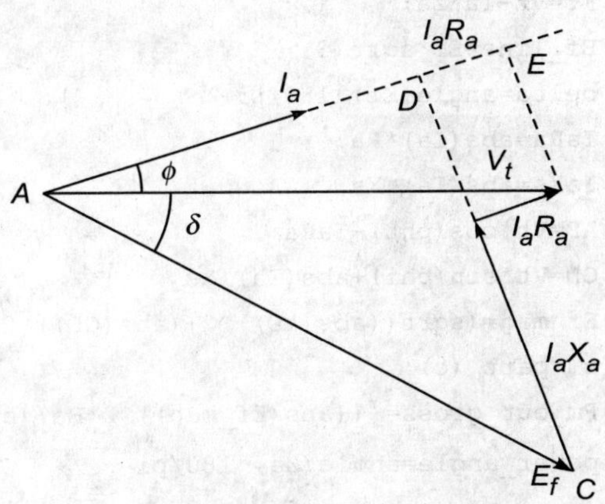

Fig. 8.66

(c) For $\quad E_f = 542.8$ V

Maximum power output (gross)

From Eq. (8.75)

$$P_m(\text{out, gross})|_{\max} = -\frac{E_f^2}{Z_s^2}R_a + \frac{V_tE_f}{Z_s}$$

Substituting values

$$P_m(\max) = -\frac{(542.8)^2}{(8.04)^2} \times 0.8 + \frac{346.4 \times 542.8}{(8.04)} = 197.40 \text{ kW}$$

$$P_m(\text{max, net}) = 197.4 - 2 = 195.4 \text{ kW}$$

Power angle, $\quad \delta = \theta = 84.3^\circ$

The Example 8.17 is also solved by MATLAB.

```
clc
clear
j=sqrt(-1);
Sop=40*1000; Vt=600; Ra=0.8; Xs=8;
%% Part (a)
Pst=2*1000; Pm_net=30*1000;
Pm_dev=Pst+Pm_net;
Ia=Sop/(sqrt(3)*Vt);
Poh=3*Ia^2*Ra;
```

```
Pin=Pm_dev+Poh;
Eff=(1-(Poh+Pst)/Pin)*100
%% part (b)
cos_phi=Pin/(sqrt(3)*Vt*Ia);
phi=acos(cos_phi);
phi_deg=phi*180/pi
Ia=Ia*(cos(phi)+sin(phi)*j);
Vt=Vt/sqrt(3);
Za=Ra+Xs*j;
Ef=Vt-Ia*Za;
Ef_line=Ef*sqrt(3)
delta=angle(Ef)*180/pi
IaRa=abs(Ia)*Ra;
IaXs=abs(Ia)*Xs;
AD=Vt*cos(phi)-IaRa
CD=Vt*sin(phi)+abs(Ia)*Xs
Ef_mag=(sqrt((abs(AD))^2+(abs(CD))^2))
%% part (c)
Pm_out_gross=-((abs(Ef_mag))^2*Ra/(abs(Za))^2)+(Vt*abs(Ef_mag)/abs(Za))
power_angle=angle(Za)*180/pi
```

Answer:

Eff = 84.3750

phi_deg = 27.2660

Ef_line = 7.9692e+002 –4.9851e+002i

delta = –32.0276

AD = 277.1281

CD = 466.6186

Ef_mag = 542.7088

Pm_out_gross = 1.9738e+004

power_angle = 84.2894

Example 8.18 *A 5 kVA, 400 V, 50 Hz synchronous generator having synchronous reactance of 25 Ω is driven by a dc motor and is delivering 4 kW, to a 400 V mains at 0.8 pf lagging. The field current of the dc motor is gradually increased till it begins to act as a generator delivering power to d.c. mains, while the synchronous machine acts as a motor drawing 4 kW from the ac mains. What is the total change in the power angle of the synchronous machine in this process? The field current of the synchronous machine is kept constant throughout. Neglect all losses.*

LO 4

Solution As generator

$$P_e = \sqrt{3}\ \times 400\, I_a \times 0.8 = 4 \times 10^3$$

or

$$I_a = 7.22 \text{ A}$$

$$\phi = \cos^{-1} 0.8 = 36.9° \text{ lag}$$

$$\bar{I}_a = 7.22\angle -36.9°$$

$$\bar{V}_t = \frac{400}{\sqrt{3}} \angle 0° = 231\angle 0° \text{ V}$$

$$\bar{E}_f = 231\angle 0° + j\,7.22\angle -36.9° \times 25 = 231 + 180.5\angle 53.1°$$

$$= 339.4 + j\,144.3 = 368.8\angle 23°$$

$$E_f = 368.6 \text{ V}, \quad \delta = +23° \quad E_f \text{ leads } V_t$$

As motor

Excitation emf (constant), $E_f = 368.8$

$$P_e\,(\text{in}) = \frac{V_t E_f}{X_s} \sin\delta$$

$$\frac{1}{3} \times 4000 = \frac{231 \times 368.8}{23} \sin\delta$$

$$\delta = 23°, \quad E_f \text{ lags } V_t$$

Total change in power angle = 23° + 23° = 46°

Example 8.19 *A three phase hydroelectric alternator is rated 110 MW, 0.8 pf lagging, 13.6 kV delta connected, 100 rpm. Determine the*

(a) number of poles

(b) MVA rating

(c) prime mover rating if the full load generator efficiency is 97.1% (field loss left out)

(d) output torque of the prime mover

LO 4

Solution

(a) $P = \dfrac{120 f}{n_s} = \dfrac{120 \times 50}{100} = 60$

(b) $(\text{MVA})_{\text{rating}} = \dfrac{110}{0.8} = 137.5$

(c) $(\text{MW})_{\text{turbine}} = \dfrac{110}{0.971} = 113.3$

(d) $T_{PM}\,(\text{shaft}) = 113.3 \times 1000 \times \dfrac{60}{2\pi \times 100} = 10.82 \times 10^2$ Nm

Example 8.20 *The per phase circuit equivalent of a synchronous generator feeding a synchronous motor through a transmission line is drawn in Fig. 8.67.*

Each machine is rated 10 kVA, 400 V, 50 Hz. The motor is driving a load of 8 kW. The field currents of the two machines are so adjusted that the motor terminal voltage is 400 V and its pf is 0.8 leading.

(a) Determine the magnitude of E_g and E_m and their relative angle.

(b) Minimum value of E_m for the machine to remain in synchronism.

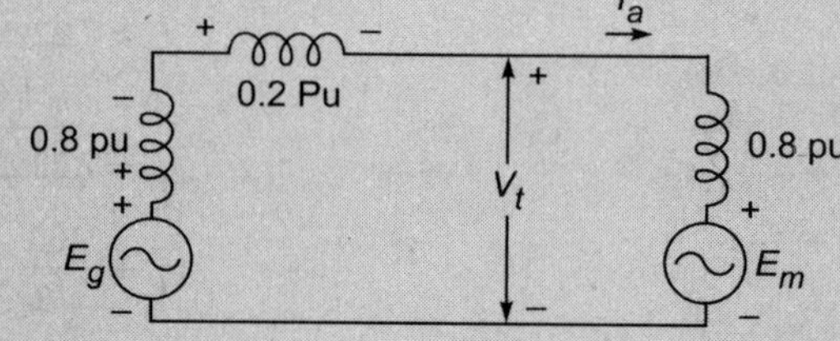

Fig. 8.67

LO 4

Solution

$$V_{\text{Base}} = 400 \text{ V}, (\text{kVA})_{\text{Base}} = 10, (\text{MW})_{\text{Base}} = 10$$

Motor load,
$$P_m = \frac{8}{10} = 0.8 \text{ pu}$$

$$V_t = 1\angle 0° \text{ pu}$$

As there is no loss

$$P_e\,(\text{in}) = P_m = 0.8 \text{ pu}$$
$$P_e\,(\text{in}) = V_t I_a \cos\theta;\ \text{pf} = \cos\theta = 0.8 \text{ leading}$$
$$0.8 = 1 \times I_a \times 0.8$$
$$I_a = 1 \text{ pu} \qquad \bar{I}_a = 1\angle 36.9°$$

Motor

$$\bar{E}_m = 1\angle 0 - j1\angle 36.9° \times 0.8 = 1 + 0.8\angle -53.1°$$
$$= 1.48 - j\,0.64 = 1.612\angle -23.4°$$
$$E_m = 1.612 \text{ pu} \quad \text{or} \quad 644.8 \text{ V (line)}$$
$$\delta_m = -23.4°$$

Generator

$$E_g = 1\angle 0° + j1\angle 36.9° \times (0.8 + 0.2)$$
$$= 1 + 1\angle 126.9° = 0.4 + j\,0.8$$
$$= 0.894\angle 63.4°$$
$$E_g = 0.894 \text{ pu} \quad \text{or} \quad 357.6 \text{ V (line)}, \quad \delta_g = 63.4°$$

Relative angle,
$$\delta_{gm} = 63.4° + 23.4° = 86.8°$$

(a) Solution by phasor diagram

The phasor diagram is drawn in Fig. 8.68. From the geometry of the diagram

$$AE = V_t \cos\theta = 0.8$$
$$BE = V_t \sin\theta = 0.6$$
$$BC = 1 \times 0.8 = 0.8$$
$$BD = 1 \times (0.8 + 0.2) = 1$$

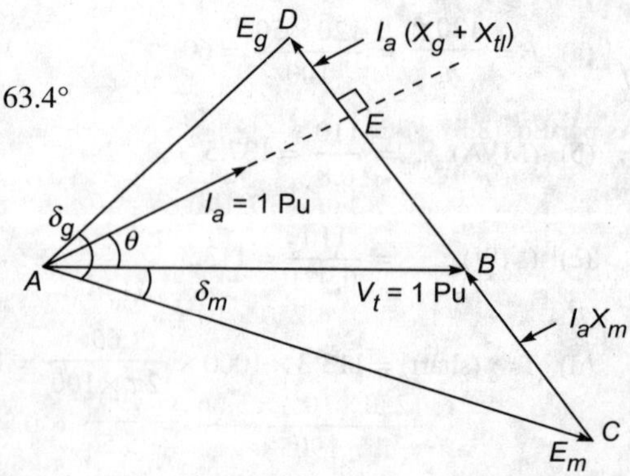

Fig. 8.68

$$CE = BC + BE = 0.8 + 0.6 = 1.4$$

$$E_m = \sqrt{(AE)^2 + (CE)^2}$$

$$= \sqrt{(0.8)^2 + (1.4)^2} = 1.612 \text{ pu}$$

$$\tan(\theta + \delta_m) = \frac{CE}{AE} = \frac{1.4}{0.8} = 1.75$$

$$\theta + \delta_m = 60°, \ \delta_m = 60° - 36.9° = 23.4°$$

$$DE = BD - BE = 1 - 0.6 = 0.4$$

$$E_g = \sqrt{(AE)^2 + (DE)^2}$$

$$= \sqrt{(0.8)^2 + (0.4)^2} = 0.894 \text{ pu}$$

$$\tan(\delta_g - \theta) = \frac{DE}{AE} = \frac{0.4}{0.8} = 0.5$$

$$\delta_g - \theta = 26.6° \quad \delta_g = 63.5°$$

(b)

$$P_e = \frac{V_t E_m}{0.8} \sin \delta_m$$

For maximum $E_{m,}$

$$\delta_m = 90°, \text{ limit of stability}$$

$$E_m(\text{min}) = \frac{0.8 \times 0.8}{1} = 0.64 \text{ pu} \quad \text{or} \quad 256 \text{ V}$$

Example 8.21 *A 3.3 kV, 50 Hz star connected synchronous motor has a synchronous impedance of (0.8 + j 55) Ω. It is synchronized to 3.3 kV main from which it is drawing 750 kW at an excitation emf of 4.27 kV (line). Determine the armature current, power factor and power angle. Also find the mechanical power developed. If the stray load loss is 30 kW, find the efficiency.*

LO 4

Solution Converting to per phase

$$V_t = \frac{3.3}{\sqrt{3}} = 1905 \text{ V}, \quad E_f = \frac{4.27}{\sqrt{3}} = 2465 \text{ V}$$

Power input to motor,

$$P_i(\text{in}) = \frac{750}{3} = 250 \text{ kW}$$

$$\bar{Z}_s = 0.8 + j\,5.5 = 5.56 \angle 81.7° \ \Omega$$

$$Z_s = 5.56 \ \Omega, \quad \theta = 81.7°, \quad \alpha = 90° - 81.7° = 8.3°$$

As per Eq. (8.87 a)

$$P_e(\text{in}) = \frac{V_t^2}{Z_s^2} R_a + \frac{V_t E_f}{Z_s} \sin(\delta - \alpha)$$

$$250 \times 10^3 = \frac{(1905)^2}{(5.56)^2} \times 0.8 + \frac{1905 \times 2465}{5.56} \sin(\delta - 8.3°)$$

$$\frac{(250 \times 10^3) \times 5.06}{1905} = \frac{1905}{5.56} \times 0.89 + 2465 \sin(\delta - 8.3°)$$

Power angle, $\delta = 18.95°$

$$\bar{I}_a \bar{Z}_s = V_t \angle 0 - E_f \angle -\delta$$

or
$$\bar{I}_a \bar{Z}_s = 1905 - 2465 \angle -18.95°$$
$$= -42.64 + j\,800.5$$
$$= 907 \angle 118°$$
$$\bar{I}_a = \frac{907\angle 118°}{5.56\angle 81.7°} = 1631\angle 36.3° \text{ A}$$
$$I_a = 163.1 \text{ A}, \quad \text{pf} = \cos 36.3° = 0.806 \text{ leading}$$
$$\bar{E}_f = 2465\angle -18.95° \quad \text{or} \quad 4270\angle -18.95° \text{ V (line)}$$

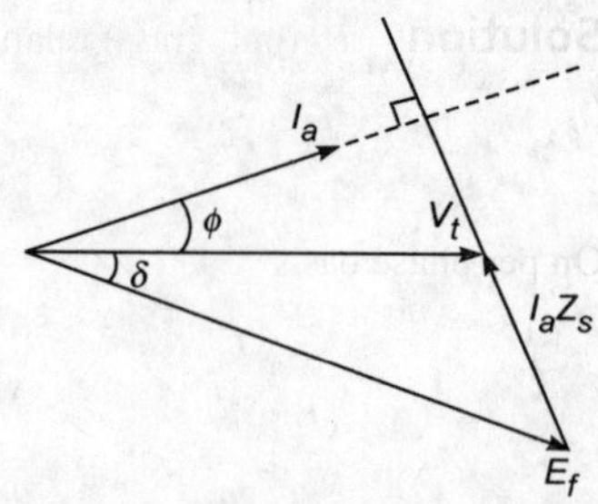

Fig. 8.69(a)

$\bar{I}_a$ leads $\bar{E}_f$ by (18.95° + 36.3°) = 55.25°. The phasor diagram is drawn in Fig. 8.69(a).

Mechanical power developed
$$P_m(\text{dev}) = 3E_f I_a \cos(\delta + \phi)$$
$$P_m(\text{dev}) = \sqrt{3} \times 4270 \times 163.1 \cos(55.25°) \times 10^{-3} = 687.5 \text{ kW}$$
$$P_{st} = 30 \text{ kW}$$
$$P_m(\text{net}) = 687.5 - 30 = 657.5 \text{ kW}$$
$$\eta = \frac{657.5}{750} \times 100 = 87.7\%$$

Alternative The phasor method

We want to find out current and pf.

$$\bar{I}_a \bar{Z}_s = \bar{V}_t - \bar{E}_f \qquad \text{(i)}$$

We convert it to current form

$$\bar{I}_a = \frac{\bar{V}_t}{\bar{Z}_s} - \frac{\bar{E}_f}{\bar{Z}_s} = \bar{I}_t - \bar{I}_e$$
$$\bar{I}_a = \left(\frac{1905}{5.56}\right)\angle -81.7° - \left(\frac{2465}{5.56}\right)\angle ?$$
$$\bar{I}_a = 342.6\angle -81.7° - 443\angle ? \qquad \text{(ii)}$$

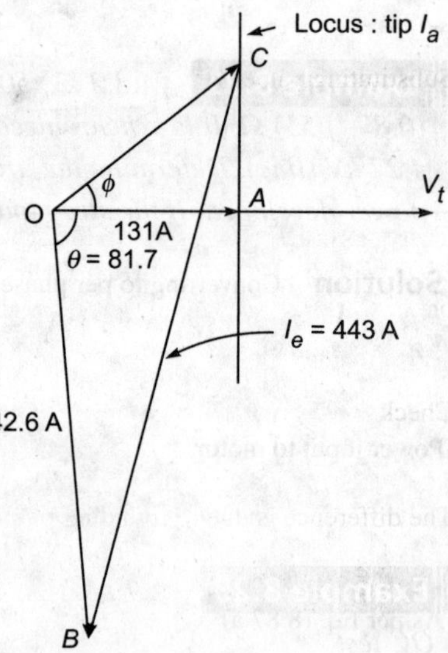

Fig. 8.69(b)

As power input is known
$$P_e(\text{in}) = \sqrt{3} \times 3.3\, I_a \cos\phi = 750 \text{ kW}$$
or
$$I_a \cos\phi = 131 \text{ A (real power component, in phase with } V_t)$$

The phasor diagram is drawn to scale in Fig. 8.69(b) in following steps.

$$OA = I_a \cos\phi = 131 \text{ A along } V_t \text{ (in phase)}$$

Draw the locus of I_a phasor tip at 90° to V_t

Draw $\quad OB = I_t = 342.6$ A lagging V_t by $\theta = 81.7°$

From B draw an arc of radius $I_e = 443$ A to locate point C on the current locus. We can now measure to scale

$$I_a = 162 \text{ A}, \ \phi = 36°, \ \cos\phi = 0.809 \text{ as found by the quantitative method.}$$

Example 8.22 *The synchronous motor of Example 8.21 when excited to an emf of 4.27 kV (line) develops mechanical power of 600 kW. Find the armature current, power faction and power input.*

LO 4

Solution P_m(out, grass) relationship of Eq. (8.88) is reproduced below

$$P_m(\text{dev}) = P_m(\text{out, gross}) = -\frac{E_f^2}{Z_s^2} R_a + \frac{E_f V_t}{Z_s} (\delta + \alpha) \quad \text{(i)}$$

On per phase basis

$$V_t = 1905 \text{ V}, \quad E_f = 2465 \text{ V}$$

$$\bar{Z}_s = 5.56 \angle 81.7°; \quad Z_s = 5.56\ \Omega, \quad \theta = 81.7°, \quad \alpha = 90° - \theta = 8.3°$$

$$P_m(\text{dev}) = \frac{600}{3} = 200 \text{ kW}$$

Substituting values in Eq. (i)

$$200 \times 10^3 = -\frac{(2465)^2}{(5.56)^2} \times 0.8 + \frac{2465 \times 1905}{5.56} \sin(\delta + \alpha)$$

$$\frac{200 \times 10^3 \times 5.56}{2465} = -\frac{2465}{5.56} \times 0.8 + 1905 \sin(\delta + \alpha)$$

$$\sin(\delta + \alpha) = 0.423$$

$$\delta + \alpha = 25°, \quad \delta = 25° - 8.3° = 16.7°$$

In a motor $\bar{E}_f$ lags $\bar{V}_t$ by δ. As per basic armature circuit equation

$$\bar{E}_f = \bar{V}_t - \bar{I}_a \bar{Z}_s$$

or

$$\bar{I}_a \bar{Z}_s = \bar{V}_t - \bar{E}_f, \ \bar{V}_t \text{ reference phasor}$$

Substituting values

$$\bar{I}_a \bar{Z}_s = 1905\angle 0° - 2465\angle -16.7° = -456 + j\,708.3$$

$$= 842.4 \angle 122° \ V$$

$$\bar{I}_a = \frac{842.4\angle 122.8}{5.56\angle 81.7°} = 151.5\angle 41.1° \text{ A}$$

or

$$I_a = 151.5 \text{ A}, \quad \phi = 41.1°, \quad \text{pf} = 0.753 \text{ leading}$$

$$P_e(\text{in}) = \sqrt{3} \times 3.3 \times 151.5 \times 0.753 = 652 \text{ kW}$$

Check

$$\text{Ohmic loss} = P_e(\text{in}) - P_m(\text{dev}) = 652 - 600 = 52 \text{ kW}$$

$$3\, I_a^2 R_a = 3 \times (151.5) \times 0.8 = 55 \text{ kW}$$

The difference is due to rounding of small angles.

Example 8.23 *A 10 MV A, 13.8 kV, 50 Hz synchronous generator yields the following test data:*

OC test $I_f = 842$ *A at rated voltage*

SC test $I_f = 226$ *A at rated armature current*

(a) Calculate its pu adjusted synchronous reactance OC test revealed that the armature resistance per phase is 0. 75 Ω.

(b) The generator is operating at power output of 8.75 MW, 0.9 pf lagging. Calculate

(i) its field current and reactive power output

(ii) the rotor power angle and reactive power output if the field current is adjusted to 842 A while the net shaft power supplied by the prime mover remains constant

LO 4

Solution

(a)
$$X_s(\text{adjusted}) = \left.\frac{V_{\text{rated}}/\sqrt{3}}{I_{SC}}\right|_{\text{At } I_f \text{ corresponding to } V_{OC}=V_{\text{rated}}}$$

Rated armature current

$$\sqrt{3}\, V_{\text{rated}}\, I_a(\text{rated}) = 10 \text{ MVA}$$

$$I_a(\text{rated}) = \frac{10\times 10^3}{\sqrt{3}\times 13.8} = 418.4 \text{ A}$$

At
$$I_f = 842 \text{ A},\quad I_{SC} = \frac{418.4}{226}\times 842 = 1558.8 \text{ A}$$

The
$$X_s\ (\text{adjusted}) = \frac{13.8\times 10^3/\sqrt{3}}{1558.8} = 5\ \Omega$$

Base values
$$(\text{MVA})_B = 10,\ (\text{kV})_B = 13.8$$

$$X_s\ (\text{pu}) = 5\times \frac{10}{(13.8)^2} = 0.2625$$

$$R_a = 0.75\ \Omega$$

$$\bar{Z}_s = 0.75 + j\,5 = 5.06\ \angle 81.5°\ \Omega$$

$$Z_s = 5.06\ \Omega,\quad \theta = 81.5°,\quad \alpha = 8.5°$$

(b) (i)
$$\text{pf} = 0.9 \text{ lagging};\ \phi = \cos^{-1} 0.9 = 25.84°$$

$$P_e = 8.75 \text{ MW}$$

$$\tan\phi = \frac{Q_e}{P_e}$$

$$Q_e = P_e \tan\phi = 8.75 \tan 25.84°$$
$$= 4.24 \text{ MVA, positive as it a lagging.}$$

To find the field current, we need excitation emf. On per phase basis

$$V_t = 13.8/\sqrt{3} = 7968 \text{ V}$$

$$\frac{P_e}{3} = V_t I_a \cos\phi$$

$$\frac{8.75\times 10^6}{3} = 7968\, I_a \times 0.9$$

or
$$I_a = 406.7 \text{ A},\quad \bar{I}_a = 406.7\ \angle -25.8°$$

$$\bar{I}_a \bar{Z}_s = 4067 \times 5.06\ \angle 81.5° - 25.8°$$
$$= 2058\ \angle 55.7° \text{ V}$$

Generator equation
$$\bar{E}_f = \bar{V}_t + \bar{I}_a \bar{Z}_s$$

or
$$\bar{E}_f = 7968\angle 0° + 2058\angle 55.7°$$
$$= 9128 + j1700 = 9285\angle 10.5° \text{ V}$$

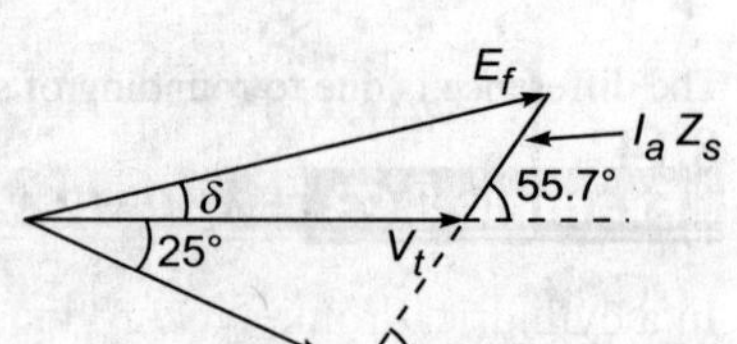

Fig. 8.70

The phasor diagram is drawn in Fig. 8.70

$$E_f = 9285 \text{ or } 16.04 \text{ kV (line)}$$

From the modified air-gap line

$$I_f = \frac{842}{13.8}\times 16.04 = 978.7 \text{ A}$$

Ohmic loss

$$3\,I_a^2\,R_a = 3 \times (406.7)^2 \times 0.75$$
$$= 0.372 \text{ MW}$$
$$P_m(\text{in}) = 8.75 + 0.372 = 9.122 \text{ MW}$$

(ii) From Eq. (8.83)

$$P_m(\text{in}) = \frac{E_f^2}{Z_s^2} R_a + \frac{V_t E_f}{Z_s} \sin(\delta - \alpha) \tag{i}$$

Field current adjusted to

$$I_f = 842 \text{ A}$$
$$E_f = V_{OC} = V_t \text{ (rated)} = 7968 \text{ V} = 7.968 \text{ kV}$$
$$P_m(\text{in}) = 9.122/3 = 3.04 \text{ MW per phase}$$

Substituting in Eq. (i)

$$3.04 = \frac{(7968)^2}{(5.06)^2} \times 0.75 + \frac{(7.968)^2}{(5.06)} \sin(\delta - \alpha)$$

$$118 = \frac{(7968)^2}{(5.06)^2} \sin(\delta - \alpha)$$

$$\sin(\delta - \alpha) = 0.094$$
$$\delta - \alpha = 5.4°$$
$$\delta = 5.4° + 8.5° = 13.9°$$

Reactive power output (Eq. 8.82b)

$$Q = -\frac{V_t^2}{Z_s^2} X_s + \frac{V_t E_f}{Z_s} \cos(\delta + \alpha)$$

or

$$Q = -\frac{(7.968)^2}{(5.06)^2} \times 5 + \frac{(7.968)^2}{(5.06)} \cos(13.9° + 8.5°)$$

or

$$Q = -0.82 \text{ MVAR/phase}$$

or

$$Q = 2.46 \text{ MVAR, leading}$$

8.14 SALIENT-POLE SYNCHRONOUS MACHINE TWO-REACTION MODEL

LO 4

In a cylindrical rotor synchronous machine the flux established by a mmf wave is independent of the spatial position of the wave axis with respect to the field pole axis. On the other hand, in a salient-pole machine as shown in the cross-sectional view of Fig. 8.71 the permeance offered to a mmf wave is highest when it is aligned with the field pole axis (called the *direct-axis* or *d-axis*) and is lowest when it is oriented at 90° to the field pole axis (called the *quadrature axis* or *q-axis*). Though the field winding in a salient-pole is of concentrated type, the *B*-wave produced by it is nearly sinusoidal because of the shaping of pole-shoes (the air-gap is least in the centre of the poles and progressively increases on moving away from the centre). Equivalently, the F_f wave can be imagined to be sinusoidally distributed and acting on a uniform air-gap. It can, therefore, be represented as space vector $\vec{F}_f$. As the rotor rotates, $\vec{F}_f$ is always oriented along the *d*-axis and is presented with the d-axis permeance. However in case of armature reaction the permeance presented to it is far higher when it is oriented along the d-axis than when it is oriented along the q-*axis*.

Figure 8.71 shows the relative spatial location of $\vec{F}_f$ and $\vec{F}_{ar}$ at the time instant when current (generating) in phase a is maximum positive and is lagging $\bar{E}_f$ (excitation emf due to $\vec{F}_f$) by angle ψ (refer to Fig. 8.5(a)). As angle ψ varies, the permeance offered to $\vec{F}_{ar}$, the armature reaction mmf, varies because of a change in its spatial position relative to the d-axis. Consequently $\vec{F}_{ar}$ produces $\bar{\Phi}_{ar}$ (armature reaction flux/pole) whose magnitude varies with angle ψ (which has been seen earlier to be related to the load power factor). So long as the magnetic circuit is assumed linear (i.e. superposition holds), this difficulty can be overcome by dividing $\vec{F}_{ar}$ into vectors $\vec{F}_d$ along the d-axis and $\vec{F}_q$ along the q-axis as shown by dotted lines in Fig. 8.71.

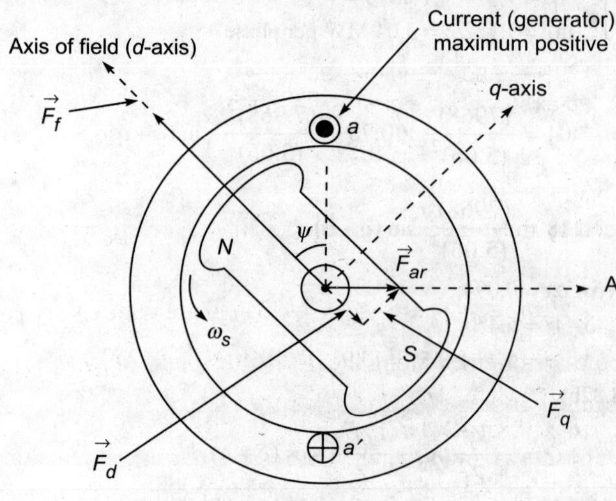

Fig. 8.71 *Salient-pole synchronous machine*

Figure 8.72 shows the phasor diagram corresponding to the vector diagram of Fig. 8.70. Here the d-axis is along $\bar{F}_f$ and 90° behind it is the q-axis along $\bar{E}_f$. The components of $\bar{I}_a$ and $\bar{F}_{ar}$ along d-and q-axis are shown in the figure from which it is easily observed, that $\bar{F}_d$ is produced by $\bar{I}_d$, the d-axis component of $\bar{I}_a$ at 90° to $\bar{E}_f$ and $\bar{F}_q$ is produced by $\bar{I}_q$, the q-axis component of $\bar{I}_a$, in phase with $\bar{E}_f$.

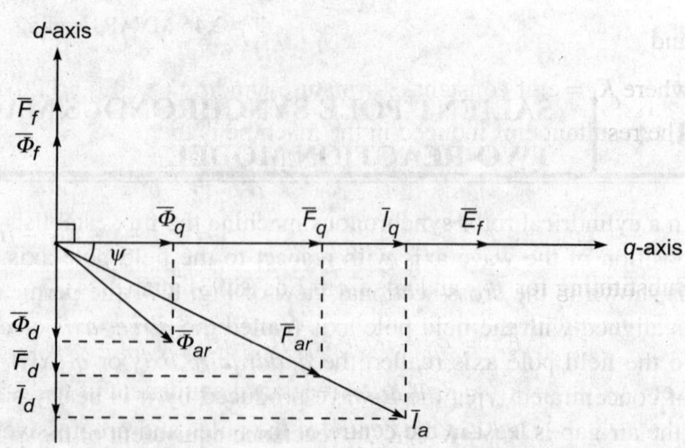

Fig. 8.72

Figure 8.73 shows the relative locations of F_d, F_q and field poles and the B-waves produced by these. It is easily seen that the B-waves contain strong third-harmonic space waves.

For reasons advanced in Ch. 5, one can proceed with the analysis on the basis of space fundamentals of B_d and B_q while neglecting harmonics.¶¶¶¶

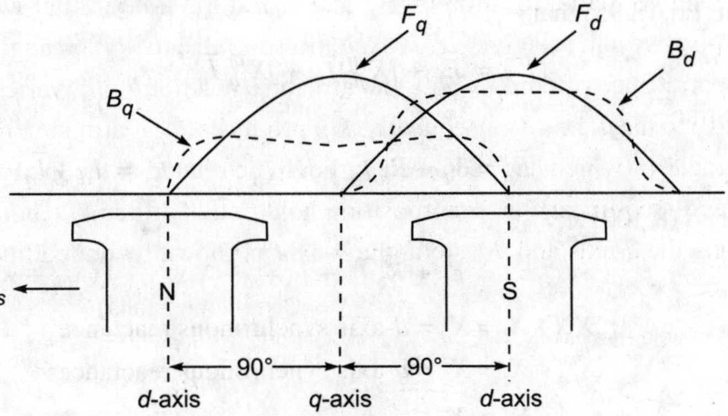

Fig. 8.73

The flux components/pole produced by the *d*- and *q*-axis components of armature reaction mmf are

$$\Phi_d = \mathcal{P}_d F_d = \mathcal{P}_d K_{ar} I_d \quad \text{(in phase with } I_d\text{)} \tag{8.91}$$

and

$$\Phi_q = \mathcal{P}_q F_q = \mathcal{P}_q K_{ar} I_q \quad \text{(in phase with } I_q\text{)} \tag{8.92}$$

where $\mathcal{P}_d$, $\mathcal{P}_q$ = permeance of pole-arc oriented along the *d*-axis/the *q*-axis ($\mathcal{P}_d > \mathcal{P}_q$).

K_{ar} = constant of armature winding (see Eq. (5.44b))

The flux phasors $\bar{\Phi}_d$ and $\bar{\Phi}_q$ are also drawn in Fig. 8.58. The resultant armature reaction flux phasor $\bar{\Phi}_{ar}$ is now no longer in phase with $\bar{F}_{ar}$ or $\bar{I}_a$ because $\mathcal{P}_d > \mathcal{P}_q$ in Eqs (8.91) and (8.92). In fact $\bar{\Phi}_{ar}$ lags or leads $\bar{I}_a$ depending upon the relative magnitudes of the *d*-axis, *q*-axis permeances.

The emfs induced by Φ_d and Φ_q are given by

$$\bar{E}_d = -jK_e\bar{\Phi}_d \tag{8.93}$$

and

$$\bar{E}_q = -K_e\bar{\Phi}_q \tag{8.94}$$

where K_e = emf constant of armature winding (see Eq. (5.20))

The resultant emf induced in the machine is then

$$\bar{E}_r = \bar{E}_f + \bar{E}_d + \bar{E}_q$$

$$= \bar{E}_f - jK_e\bar{\Phi}_d - jK_e\bar{\Phi}_q$$

Substituting for $\bar{\Phi}_d$ and $\bar{\Phi}_q$ from Eqs (8.93) and (8.94),

$$\bar{E}_r = \bar{E}_f - jK_e\mathcal{P}_d K_{ar}\bar{I}_d - jK_e\mathcal{P}_q K_{ar}\bar{I}_q \tag{8.95}$$

Let

$$X_d^{ar} = K_e\mathcal{P}_d K_{ar} = \text{reactance equivalent of the } d\text{-axis component of armature reaction} \tag{8.96}$$

$$X_q^{ar} = K_e\mathcal{P}_q K_{ar} = \text{reactance equivalent of the } q\text{-axis component of armature reaction} \tag{8.97}$$

¶¶¶¶Third-harmonic *B*-waves induce third-harmonic emfs in phases but no third-harmonic voltage appears between lines.

Obviously

$$X_d^{ar} > X_q^{ar} \qquad (\because \mathcal{P}_d > \mathcal{P}_q) \tag{8.98}$$

It then follows[*****] from Eq. (8.95) that

$$\bar{E}_f = \bar{E}_r + jX_d^{ar}\bar{I}_d + jX_q^{ar}\bar{I}_q \tag{8.99}$$

Also for a realistic machine

$$\bar{E}_r = \bar{V}_t + R_a\bar{I}_a + jX_l\bar{I}_a; \qquad \bar{I}_a = \bar{I}_d + \bar{I}_q \tag{8.100}$$

Combining Eqs (8.99) and (8.100)

$$\bar{E}_f = \bar{V}_t + R_a\bar{I}_a + j(X_d^{ar} + X_l)\bar{I}_d + j(X_q^{ar} + X_l)\bar{I}_q \tag{8.101}$$

Define $X_d^{ar} + X_l = X_d$ = d-axis synchronous reactance

$X_q^{ar} + X_l = X_q$ = q-axis synchronous reactance

It is easily seen that

$$X_d > X_q \tag{8.102}$$

Equation (8.101) can now be written as

$$\bar{E}_f = \bar{V}_t + R_a\bar{I}_a + jX_dI_a + jX_q\bar{I}_q \tag{8.103}$$

The phasor diagram depicting currents and voltages as per Eq. (8.103) is drawn in Fig. 8.70 in which δ is the angle between the excitation emf E_f and the terminal voltage V_t.

Drawing and Analysis of Phasor Diagram

In the phasor diagram of Fig. 8.74, various angles are

ψ is angle by which $\bar{I}_a$ lags $\bar{E}_f$ taken as positive, leading is negative

ϕ is the phase angle, taken as positive when $\bar{I}_a$ lags $\bar{V}_t$, leading is negative

δ is the power (or torque) angle taken as positive, generating $\bar{E}_f$ leads ∇_t by δ, motoring $\bar{E}_f$ lags $\bar{V}_t$ by δ

In the phasor diagram of Fig. 8.74, the angle $\psi = \phi + \delta$ is not known for a given V_t, I_a and ϕ. The location of E_f being unknown, I_d and I_q cannot be found which are needed to draw the phasor diagram. This difficulty is overcome by establishing certain geometric relationships for the phasor diagrams which are drawn in Fig. 8.75(a) for the generating machine and in Fig. 8.75(b) for the motoring machine.

In Fig. 8.75(a) AC is drawn at 90° to the current phasor $\bar{I}_a$ and CB is drawn at 90° to $\bar{E}_f$. Now

$$I_d = I_a \sin\psi \tag{8.104}$$

$$I_q = I_a \cos\psi \tag{8.105}$$

$$\therefore \quad I_a = I_q/\cos\psi \tag{8.106}$$

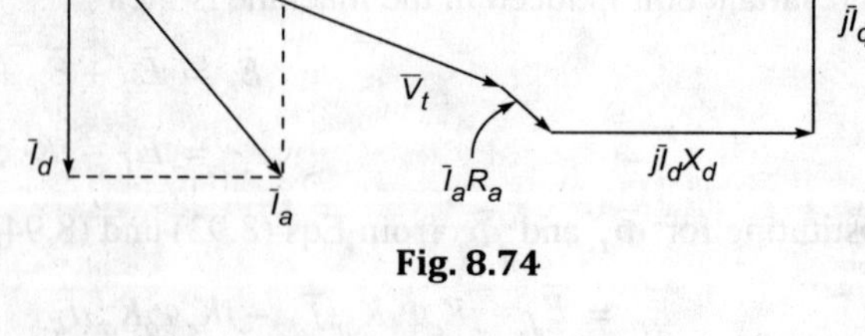

Fig. 8.74

[*****] In a cylindrical-rotor machine

$$X_d = X_q = X_{ar} \qquad (\because \mathcal{P}_d = \mathcal{P}_q \text{ independent of spatial direction of } \bar{E}_{ar})$$

so that $\bar{E}_f = \bar{E}_r + jX_{ar}(\bar{I}_d + \bar{I}_q) = \bar{E}_r + jX_{ar}\bar{I}_a$

In ΔABC

$$\cos \psi = \frac{BC}{AC} = \frac{I_q X_q}{AC}$$

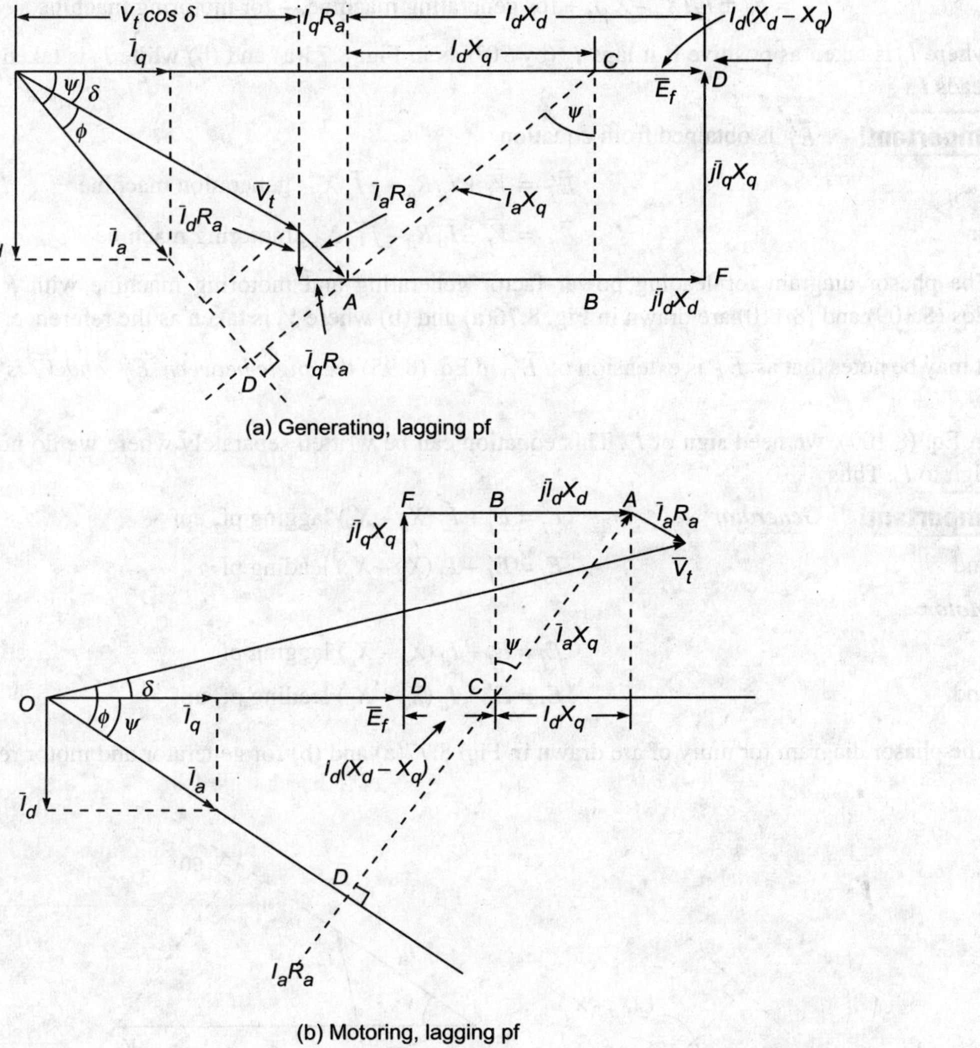

(a) Generating, lagging pf

(b) Motoring, lagging pf

Fig. 8.75 *Geometric relationships for phasor diagram of salient-pole machine*

or $$AC = \frac{I_q X_q}{\cos \psi} = I_a X_q \tag{8.107}$$

It is easily seen that $$AB = I_a X_q \sin \psi = I_d X_q$$

and $$CD = BF = I_d (X_d - X_q) \tag{8.108}$$

The phasor $\bar{E}_f$ can then be obtained by extending OC by + CD for generation machine and – CD for motoring machine (Fig. 8.75b), where CD is given by Eq. (8.108). Let us indicate the phasor OC by $\bar{E}'_f$.
The general result in terms of magnitude is

$$\bar{E}_f = E'_f \pm I_d (X_d - X_q);\ + \text{ for generating machine, } - \text{ for motoring machine} \tag{8.109}$$

where I_d is taken as positive if it lags I_q (by 90º) as in Fig. 8.75(a) and (b) while I_d is taken as negative if it leads I_q.

Important! $\bar{E}'_f$ is obtained from equation

$$\bar{E}'_f = \bar{V}_t + \bar{I}_a R_a + j\bar{I}_a X_q \text{ ; generation machine} \tag{8.110a}$$

or

$$\bar{E}'_f = \bar{V}_t - \bar{I}_a R_a - j\bar{I}_a X_q \text{ ; motoring machine} \tag{8.110b}$$

The phasor diagram for leading power factor generating and motoring machine with $R_a = 0$ based on Eqs (8.109) and (8.110) are drawn in Fig. 8.76(a) and (b) where V_t is taken as the reference.

It may be notes that as $\bar{E}_f$ is extension of $\bar{E}'_f$ in Eq. (8.95) *the angle between $\bar{E}'_f$ and $\bar{V}_t$ is the power angle δ.*

In Eq. (8.109), we need sign of I_d. This equation can be written separately where we do not need to attach sign to I_d. Thus

Important! *Generator* $E_f = E'_f + I_d (X_d - X_q)$ lagging pf, upf

and $E_f = E'_f - I_d (X_d - X_q)$ leading pf

Motor

$E_f = E'_f + I_d (X_d - X_q)$ lagging pf

and $E_f = E'_f - I_d (X_d - X_q)$ leading pf, upf

The phasor diagram for unity pf are drawn in Fig. 8.77(a) and (b) for generator and motor respectively.

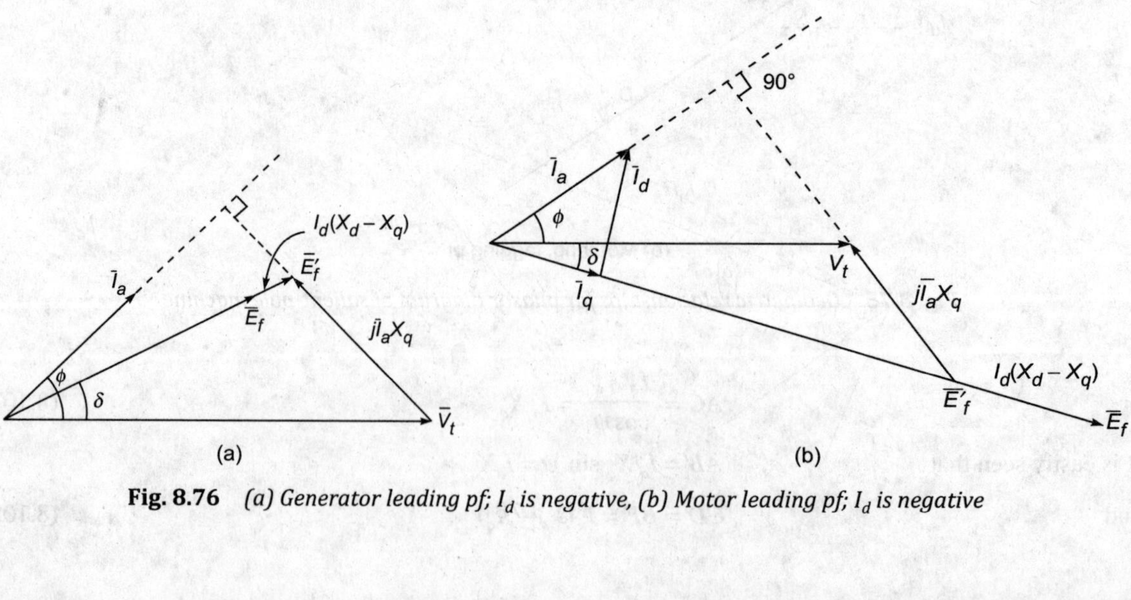

Fig. 8.76 *(a) Generator leading pf; I_d is negative, (b) Motor leading pf; I_d is negative*

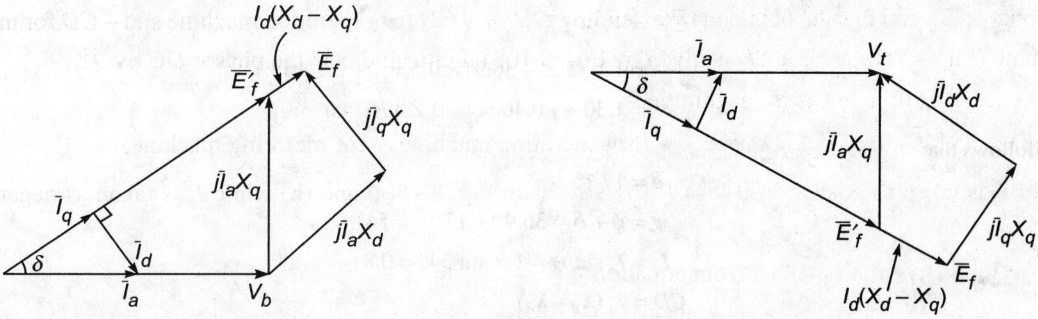

(a) Generator, upf (a) Motor, upf

Fig. 8.77

Example 8.24 *A synchronous generator has X_d (direct-axis synchronous reactance) = 0.8 pu and X_q (quadrature-axis synchronous reactance) = 0.5 pu. It is supplying full-load at rated voltage at 0.8 lagging pf. Draw the phasor diagram and calculate the excitation emf.*

Once again calculate excitation emf by ignoring X_q and assuming X_s (synchronous reactance like in a round rotor machine) = X_d. Comment upon the results. Also justify the use of this approximation from the phasor diagram.

Also calculate angle δ (power angle) with and without X_q. **LO 4**

Solution The phasor diagram in terms of symbols is drawn in Fig. 8.78.

$$\bar{V}_t = 1\angle 0° \text{ pu}, I_a = 1 \text{ pu } 0.8 \text{ lagging},$$
$$\bar{I}_a = 1\angle -36.9° \text{ pu}$$

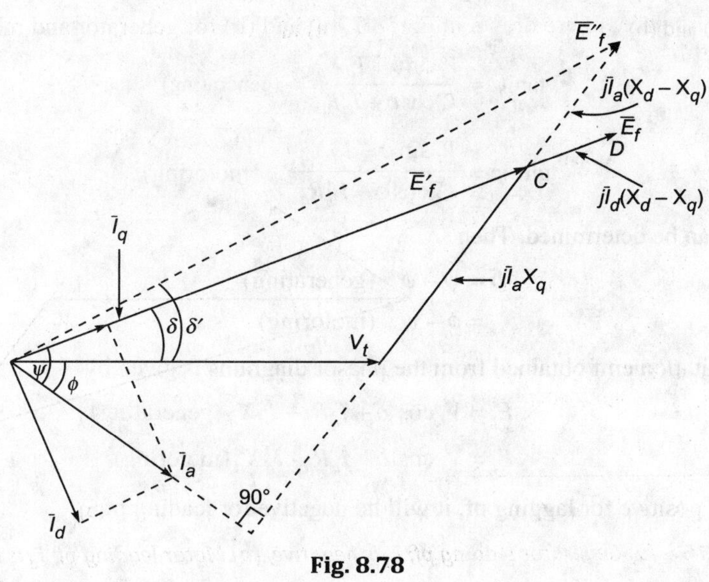

Fig. 8.78

$$\bar{E}'_f = \bar{V}_t + j\bar{I}_a X_q$$
$$= 1 + j1 \angle -36.9° \times 0.5 = 1 + 0.5\angle 53.1°$$
$$= 1.30 + j\,0.40 = 1.30 \angle 17.1° \text{ pu}$$

It then follows that

$$\delta = 17.1°$$
$$\psi = \phi + \delta = 36.9° + 17.1° = 54°$$
$$I_d = I_a \sin \psi = 1 \times \sin 54° = 0.81$$
$$CD = I_d (X_d - X_q)$$
$$= 0.81\,(0.8 - 0.5) = 0.243$$
$$\bar{E}_f = \bar{E}'_f + CD = 1.36 + 0.243 = 1.60 \text{ pu}$$

Now neglecting X_q and assuming $X_s = X_d$ the phasor diagram is indicated in dotted lines in Fig. 8.78. Here

$$\bar{E}''_f = \bar{V}_t + j\bar{I}_a X_d$$
$$= 1 + j\,1 \angle -36.9° \times 0.8$$
$$= 1.48 + j\,0.64 = 1.61 \angle 23.4°$$
$$E''_f = 1.61 \text{ pu}, \ \delta' = 23.4°$$

The error caused in excitation emf by the approximation is 1/160 or 0.625% which is negligible. It is also observed from the phasor diagram that $E_f < E''_f$

However error in δ is significant; 23.4° in place of 17.1°.

Some Useful Relationships

Instead of the above approach relationships given below derived from the phasor diagrams of Figs 8.75(a) and (b) can be used for computing power angle and excitation emf.

In ΔODC in Figs 8.74(a) and (b)

$$\tan \psi = \frac{V_t \sin \phi + I_a X_q}{V_t \cos \phi + I_a R_a} \quad \text{(generating)} \tag{8.111a}$$

and

$$\tan \psi = \frac{V_t \sin \phi - I_a X_q}{V_t \cos \phi - I_a R_a} \quad \text{(motoring)} \tag{8.111b}$$

from which angle ψ can be determined. Then

$$\delta = \psi - \phi \quad \text{(generating)} \tag{8.112a}$$
$$= \phi - \psi \quad \text{(motoring)} \tag{8.112b}$$

The magnitude of excitation emf obtained from the phasor diagrams is given by

$$E_f = V_t \cos \delta + I_q R_a + I_d X_d \ \text{(generating)} \tag{8.113a}$$
$$E_f = V_t \cos \delta - I_q R_a - I_d X_d \ \text{(motoring)} \tag{8.113b}$$

Note: Since ϕ is taken positive for lagging pf, it will he negative for leading pf.

Power-angle Characteristic and its Analysis based on Two-Reaction Theory

Figure 8.79 shows the *one-line diagram* of a salient-pole synchronous machine connected to infinite bus-bars of voltage V_b through a line of series reactance $X_{ext.}$ (per phase). The total *d*- and *q*-axis reactances are then

$$X_d = X_{dg} + X_e$$
$$X_q = X_{qg} + X_e$$

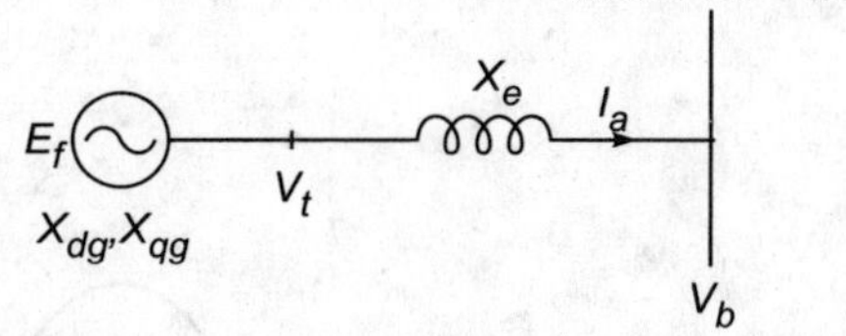

Fig. 8.79

The resistances of the machine armature and line are assumed negligible.

Figure 8.80 gives the phasor diagram when the machine is generating. It is easy to see from this figure that the real power delivered to the bus-bars is

$$P_e = I_d V_b \sin\delta + I_q V_b \cos\delta \quad (8.114)$$

Now $$I_d = \frac{E_f - V_b\cos\delta}{X_d} \quad (8.115)$$

and $$I_q = \frac{V_b \sin\delta}{X_q} \quad (8.116)$$

Substituting in Eq. (8.114),

$$P_e = \frac{E_f V_b}{X_d}\sin\delta + \underbrace{V_b^2 \frac{X_d - X_q}{2X_d X_q}\sin 2\delta}_{\text{Reluctance power}} \quad (8.117)$$

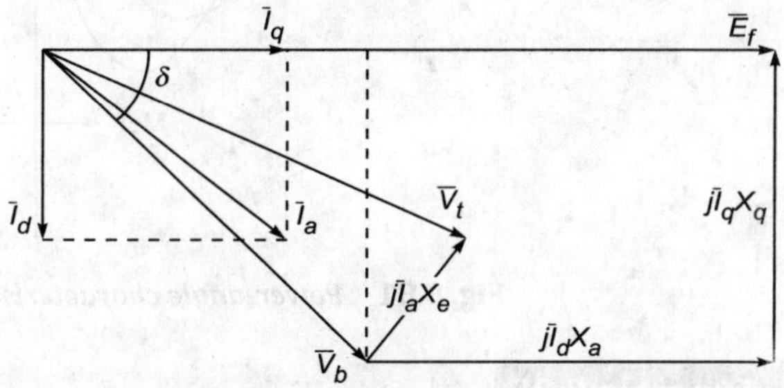

Fig. 8.80

!Attention Equation (8.117) gives the expression for the electrical power output of a salient-pole generator. The same expression would give the electrical power input of a motoring machine wherein E_f lags V_b by angle δ.

The second term in Eq. (8.117) compared to Eq. (8.42) of a cylindrical motor arises on account of saliency and is known as the *reluctance power (torque)*. The reluctance power varies as sin 2δ with a maximum value at $\delta = 45°$. It is to be further observed that this term is independent of field excitation and would be present even if the field is unexcited††††† (also refer to Sec. 4.3). A synchronous motor with salient poles but no field winding is known as the *reluctance motor.* It is used for low-power, constant-speed applications where special arrangements for dc excitation would be cumbersome.

The power angle plots of both the terms of Eq. (8.117) along with the form of the resultant power-angle curve are shown in Fig. 8.81. It is immediately observed that $P_{e,max} = P_{pull\text{-}out}$ now occurs at $\delta < 90°$ (usually at δ about 70°) and further its magnitude is larger than that for a cylindrical machine with same V_b, E_f *and* $X_s = X_d$ on account of the reluctance power term.

††††† Which means that a salient pole synchronous machine can stay synchronized to mains with its field unexcited so long as the electrical loading does not exceed

$$V_b^2 \frac{X_d - X_q}{2X_d X_q}$$

Important!

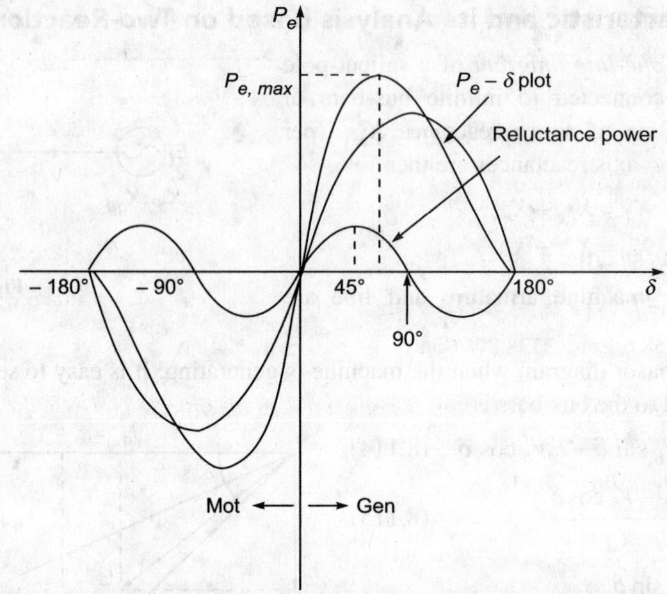

Fig. 8.81 *Power-angle characteristic of salient-pole synchronous machine*

Example 8.25 *A 1500 kW, 3-phase, star-connected, 3.3 kV synchronous motor has reactances of X_d = 4.01 and X_q = 2.88 Ω/phase. All losses may be neglected.*

Calculate the excitation emf when the motor is supplying rated load at unity pf. Also calculate the maximum mechanical power that the motor can supply with excitation held fixed at this value. LO 4

Solution Using trigonometric identity $\cos 2\delta = 2\cos^2\delta - 1$ the solution gives

$$V_t = 3300/\sqrt{3} = 1905 \text{ V}$$

$$\cos\phi = 1, \quad \sin\phi = 0, \quad \phi = 0°$$

$$I_a = \frac{1500 \times 1000}{\sqrt{3 \times 3300 \times 1}} = 262.4 \text{ A}$$

From Eq. (8.111b)

$$\tan\psi = \frac{1905 \times 0 - 262.4 \times 2.88}{1905} = -0.397$$

$$\psi = -21.6°$$

$$\delta = \phi - \psi = 0 - (-21.6°) = 21.6°$$

$$I_d = I_a \sin\psi = 262.4 \sin(-21.6°) = -96.6 \text{ A}$$

$$I_q = I_a \cos\psi = 262.4 \cos(-21.6°) = 244 \text{ A}$$

From Eq. (8.113b)

$$E_f = V_t \cos\delta - I_d X_d$$

$$= 1905 \cos 21.6° + 96.6 \times 4.01$$

$$= 2158.6 \text{ V} \quad \text{or} \quad 3738.6 \text{ V (line)}$$

Alternative The phasor diagram is drawn in Fig. 8.82.

$$V_t = 1905 \text{ V}, \quad I_a = 262.4 \text{ A}$$
$$I_a X_q = 262.4 \times 2.88 = 755.7 \text{ V}$$
$$E'_f = [(1905)^2 + (755.7)^2]^{1/2} = 2049.4 \text{ V}$$

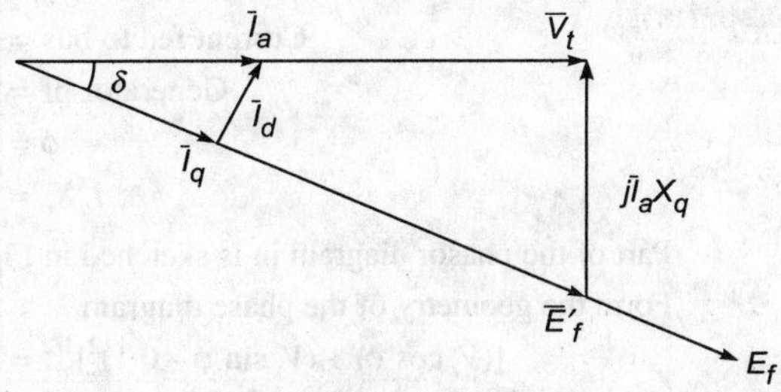

Fig. 8.82

$\bar{I}_d$ leads $\bar{I}_q$, so in a motor from Eq. (8.109)

$$I_d = -I_a \sin \delta = -96.6 \text{ A}, \ \bar{I}_d \text{ leads } \bar{I}_q$$
$$I_d (X_d - X_q) = -96.6\,(4.01 - 2.88) = -109 \text{ V}$$

Therefore $\quad E_f = E'_f - I_d (X_d - X_q) = 2049.4 + 109.2$

or $\quad E_f = 2158.6 \quad \text{or} \quad 3738.7 \text{ V (line)}$

Example 8.26 *A synchronous generator is connected to an infinite bus of voltage 1 pu through an external reactance. The reactance values are as under.*

Saturated reactance of generator

$$X_d = 1.48 \text{ pu} \quad X_q = 1.24 \text{ pu}$$

External reactance $\quad X_e = 0.1$ *pu*

The generator is feeding rated MVA to the bus at 0.9 pf lagging measured at generator terminals.

(a) Draw the complete phasor diagram showing bus voltage (V_b), generator terminal voltage (V_t), armature current (I_a) and power angle δ with respect to bus voltage.

(b) Calculate the generator terminal voltage, excitation emf and power angle. LO 4

Solution The one-line diagram of the system is drawn in Fig. 8.83(a)

$$X_d(\text{total}) = X_{dt} = X_d + X_e = 1.48 + 0.1 = 1.58 \text{ pu}$$
$$X_q(\text{total}) = X_{qt} = X_q + X_e = 1.24 + 0.1 = 1.34 \text{ pu}$$

(a) The phasor diagram is drawn in Fig. 8.83(b).

ϕ = pf angle between $\bar{V}_t$ and $\bar{I}_a$, generator terminals

ψ = angle between $\bar{E}_f$ and $\bar{I}_a$

δ = power angle between $\bar{E}_f$ and $\bar{V}_b$. $\bar{E}_f$ leads $\bar{V}_b$ as generator is feeding bus

ϕ' = pf angle between $\bar{V}_b$ and $\bar{I}_a$ at bus

(b) $\quad$ MVA (fed to bus) = 1 pu

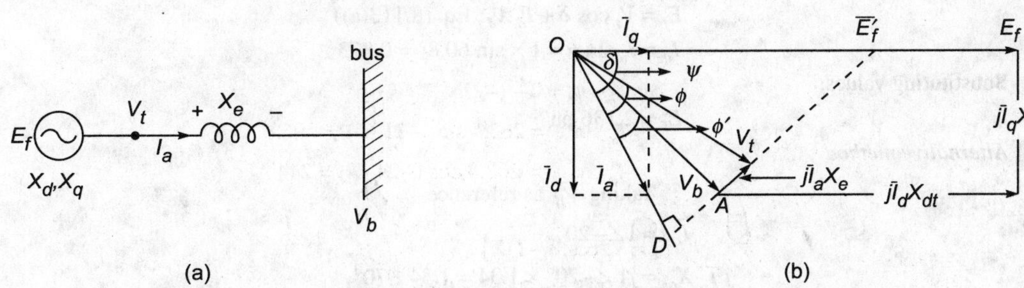

Fig. 8.83 *(a) The system, (b) Phasor diagram sketch (not to scale)*

Bus voltage $V_b = 1$ pu

Current fed to bus = generator current, $I_a = 1$ pu

Generator pf = 0.9 lagging

$$\phi = \cos^{-1} 0.9 = 25.8° \ (I_a \text{ lags } V_t)$$

$$I_a X_e = 1 \times 0.1 = 0.1 \text{ pu}$$

Part of the phasor diagram in is sketched in Fig. 8.84 (not to scale)

Form the geometry of the phase diagram

$$[(V_t \cos \phi) + (V_t \sin \phi - 0.1)^2]^{1/2} = V_b = 1$$

$$V_b = 1, \phi = 25.8°$$

Solving we find,

$$V_t = 1.04 \text{ V}, \phi' = 20°,$$

Second solution

$$V_t = 0.796 < V_b = 1 \text{ is rejected}$$

$$V_t = 1.04 \text{ pu}$$

$$\phi' = 20° \ (I_a \text{ lags } V_b)$$

$$\tan \psi = \frac{V_b \sin \phi' + I_a X_{qt}}{V_t \cos \phi'} = \frac{\sin 20° + 1 \times 1.34}{\cos 20°}, \quad \text{Eq. (8.111(a))}$$

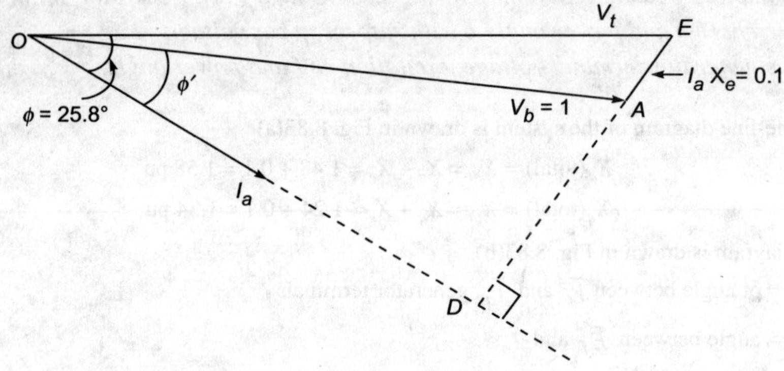

Fig. 8.84

or $\tan \psi = 1.79 \quad \psi = 60.8°$

$$\psi = \delta + \phi', \quad \delta = 60.8° - 20° = 40.8°$$

$$E_f = V_t \cos \delta + I_d X_d; \text{ Eq. (8.113(a))}$$

$$I_d = I_a \sin \psi = 1 \times \sin 60.8° = 0.873$$

Substituting values,

$$E_f = 2.136 \text{ pu}$$

Alternative method

Taking $\bar{V}_b$ as reference

$$\bar{I}_a = 1\angle{-20°}$$

$$j \bar{I}_a X_{qt} = j1 \angle{-20°} \times 1.34 = 1.34 \angle 70°$$

$$\bar{E}'_f = \bar{V}_b + j\bar{I}_a X_{qt} = 1 + 1.34\angle 70°$$
$$= 1.458 + j\,1.259 = 1.926\angle 40.8°$$
$$E'_f = 1.926 \qquad \delta = 40.8°$$
$$\psi = 40.8° + 20° = 60.8°$$
$$I_d = 1 \sin\psi = \sin 60.8° = 0.873$$

I_d lags I_q, so I_d is positive

Thus
$$E_f = E'_f + I_d(X_d - X_q) = 1.926 + 0.873\,(1.58 - 1.34)$$
$$E_f = 2.135 \text{ pu}$$

Example 8.27 *A salient pole synchronous motor having $X_d = 1.02$ pu and $X_q = 0.68$ pu is synchronized to infinite bus-bars. Its excitation is gradually reduced to zero. What maximum pu power it can deliver without losing synchronization?*

Under this operating condition calculate the pu armature current and pu reactive power it draws from the bus-bars. There are no losses. LO 4

Solution In a motor E_f lags V_t by angle δ and I_a is in phase with E_f. Therefore I_q lags V_t by δ. In the condition under consideration E_f is zero. Accordingly the phasor diagram of the motor is drawn in Fig. 8.85.

As there is no loss

$$P_m(\text{out}) = P_e(\text{in})$$

From the phasor diagram

$$P_m = I_q V_t \cos\delta - I_d V_t \sin\delta \qquad \text{(i)}$$

where $I_d V_t \sin\delta$ is negative as I_d is in phase opposition to $V_t \sin\delta$.

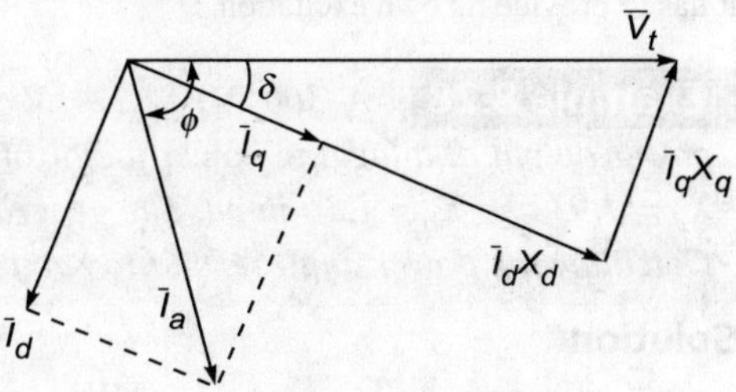

Fig. 8.85 *Phasor diagram motor with $E_f = 0$; not to scale*

Now
$$V_t \cos\delta = I_d X_d \quad \text{or} \quad I_d = \frac{V_t \cos\delta}{X_d} \qquad \text{(ii)}$$

and
$$V_t \sin\delta = I_q X_q \quad \text{or} \quad I_q = \frac{V_t \sin\delta}{X_q} \qquad \text{(iii)}$$

Substituting these values in Eq. (i), we have

$$P_m = \frac{V_t^2 \sin\delta\cos\delta}{X_q} - \frac{V_t^2 \sin\delta\cos\delta}{X_d}$$

Simplifying
$$P_m = V_t^2\left(\frac{X_d - X_q}{2X_d X_q}\right)\sin 2\delta \qquad \text{(iv)}$$

This is the *reluctance power* of a motor.

The above result follows from Eq. (8.117) by reversing the sign of δ and P_e.

P_m is maximum at $2\delta = 90°$ or $\delta = 45°$ and is given by

$$P_m(\max) = V_t^2\left(\frac{X_d - X_q}{2X_d X_q}\right) \qquad \text{(v)}$$

Substituting values

$$P_m(\max) = 1^2 \times \left(\frac{1.02 - 0.68}{2 \times 1.02 \times 0.68}\right) = 0.245 \text{ pu} \qquad \text{(v)}$$

We find that the reluctance power is much less than one pu, i.e, the motor rating.

Armature current

$$I_d = \frac{V_t \cos\delta}{X_d} = \frac{1 \times \cos 45°}{1.02} = 0.698 \text{ pu}$$

$$I_q = \frac{V_t \cos\delta}{X_q} = \frac{1 \times \sin 45°}{0.68} = 1.034 \text{ pu}$$

Observe that $I_q > I_a$

$$I_a = \sqrt{I_d^2 + I_q^2} = \sqrt{(0.693)^2 + (1.02)^2} = 1.233 \text{ pu}$$

Reactive power From the phasor diagram

$$Q_e = I_d V_t \cos\delta + I_q V_t \sin\delta \tag{vi}$$

Substituting values

$$Q_e = 0.693 \times 1 \times \cos 45° + 1.034 \times 1 \times \sin 45°$$

$$= 1.221 \text{ pu lagging VARs}$$

$$\text{Power factor} = \cos \tan^{-1}\frac{1.221}{0.245} = 0.197 \text{ lagging}$$

Observation A synchronous motor when used as a reluctance motor (no field excitation) operates at a low lagging pf as it has to provide its own excitation.

Example 8.28 *A 300 MVA, 22 kV, three phase salient-pole generator is operating at 250 MW power output at a lagging power factor of 0.85 synchronized to 22 kV bus. The generator reactances are X_d = 1.93 and X_q = 1.16 in pu. The generator gives rated open circuit voltage at a field current of 338 A. Calculate the power angle, excitation emf and the field current.*

LO 4

Solution

$$(\text{MVA})_B = 300 \quad (\text{kV})_B = 22$$

$$\text{Power output, } P_e = \frac{250}{300} = 0.833 \text{ pu}$$

$$P_e = V_t \times I_a \cos\phi$$

$$0.833 = 1 \times I_a \times 0.85$$

$$I_a = 0.98, \quad \phi = 31.8° \text{ lag}$$

$$\bar{I}_a = 0.98\angle{-31.8°}$$

$$\bar{E}'_f = \bar{V}_t + j\bar{I}_a X_q$$

$$= 1 + j\,0.98\angle{-31.8°} \times 1.16$$

$$= 1 + 1.1368\angle 58.2°$$

$$\bar{E}'_f = 1.91, \quad \delta = 28.4°$$

$$\psi = \phi + \delta = 31.8° + 28.4° = 60.2°$$

$$I_d = I_a \sin\psi = 0.98 \sin 60.2° = 0.85$$

$$I_d(X_d - X_q) = 0.85\,(1.93 - 1.16) = 0.654$$

$$E_f = E'_f + I_d(X_d - X_q) = 1.91 + 0.654 = 2.564 \text{ pu} \quad \text{or} \quad 56.4 \text{ kV}$$

The reader is advised to sketch the phasor diagram.

From the modified air-gap line

$$I_f = \frac{338}{1} \times 2.564 = 866.6 \text{ A}$$

Reactive Power Flow, Generator

The phasor diagram of a salient-pole synchronous generator at a lagging pf is redrawn in Fig. 8.86.

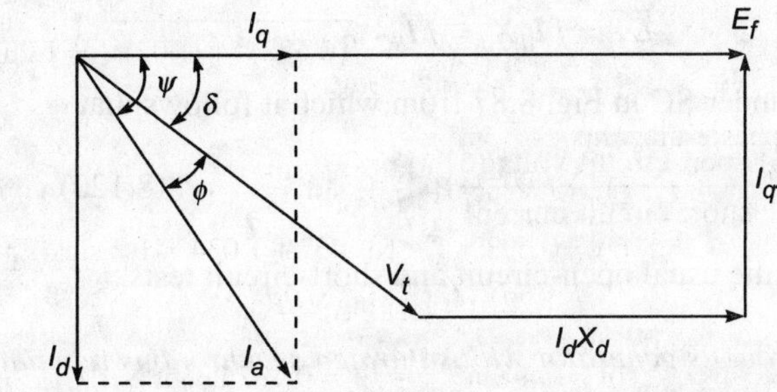

Fig. 8.86

Recall Reactive power components

$$+I_dV_t \cos\delta \text{ as } I_d \text{ lags } V_t \cos\delta \text{ by } 90°$$

$$-I_qV_t \sin\delta \text{ as } I_q \text{ leads } V_t \sin\delta \text{ by } 90°$$

Therefore

$$Q_e = I_dV_t \cos\delta - I_qV_t \sin\delta \tag{8.118}$$

From the figure

$$I_d = \frac{E_f - V_t\cos\delta}{X_d}$$

$$I_q = \frac{E_f - V_t\sin\delta}{X_q}$$

Substituting these values in Eq. (8.117), we get

$$P_e = \frac{E_fV_t}{X_d}\sin\delta + V_t^2\left(\frac{1}{X_q} - \frac{1}{X_d}\right)\sin 2\delta \tag{8.119}$$

For a cylindrical rotor generator $\quad X_d = X_q = X_s$

Eq. (8.119) reduces to

$$P_e = \frac{E_fV_t}{X_s}\sin\delta\text{; as in Eq. (8.56)}$$

It is straight forward to find the expression for Q_e(max) (Eq. (8.86b)) but it results in negative value, the generator supplies leading VARs. That is not now how a generator is operated.

Short Circuit Test

From Fig. 8.82 ($R_a = 0$)

$$V_t \sin\delta = I_q X_q$$

Under short-circuit conditions (at reduced excitation), $V_t = 0$, therefore $I_q = 0$

Hence $$\bar{I}_{SC} = \bar{I}_d + \bar{I}_q = \bar{I}_d$$

$$\bar{E}_f = j\,I_d X_d = j\,I_{SC}\;X_d$$

The phasor diagram is drawn under SC in Fig. 8.87 from which it follows that

$$X_d = \frac{E_f}{I_{SC}} = \frac{\text{open-circuit voltage}}{\text{short circuit current}}\bigg|_{\text{At } I_f} \tag{8.120}$$

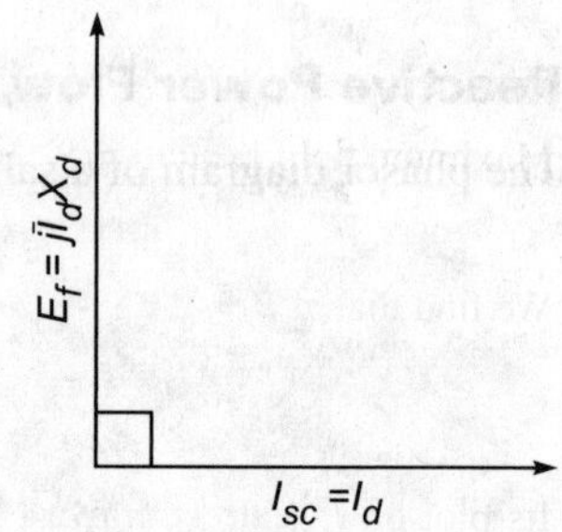

Fig. 8.87

Thus X_d can be determined by the usual open-circuit and short-circuit tests.

Example 8.29 *A synchronous generator with saturated synchronous reactance $X_d = 0.71$ pu and $X_q = 0.58$ pu connected to an infinite bus with 1 pu voltage through external reactance $X_e = 0.08$ pu. It is supplying only reactive power to the bus.*

Find the maximum and minimum pu field excitation if the armature current is not to exceed rated value.
Note: *1 pu field excitation is the field current needed to produce rated terminal voltage on no load.* LO 4

Solution Refer to one line diagram of the system drawn in Fig. 8.83(a).

$$X_d(\text{net}) = X_d + X_e = 0.71 + 0.08 = 0.79 \text{ pu}$$
$$X_q(\text{net}) = X_q + X_e = 0.58 + 0.08 = 0.66 \text{ pu}$$

For convenience we will write net values as X_d and X_q.

As the generator is supplying only reactive power, $P_e = 0$ and so $\delta = 0$ and $\phi = \pm 90°$

The phasor diagram for lagging pf is drawn in Fig. 8.88(a).

$$\bar{I}_q = 0, \quad \bar{I}_d = \bar{I}_a$$

For $$I_a = 1 \text{ pu (max limit)}$$

$$E_f = V_t + I_d X_d, \text{ scalar equation}$$

or $$E_f = 1 + 1 \times 0.79 = 1.79 \text{ pu}$$

Maximum field excitation $$I_f(\text{max}) = 1.79 \text{ pu}$$

The phasor diagram for leading pf is drawn in Fig. 8.88(b).

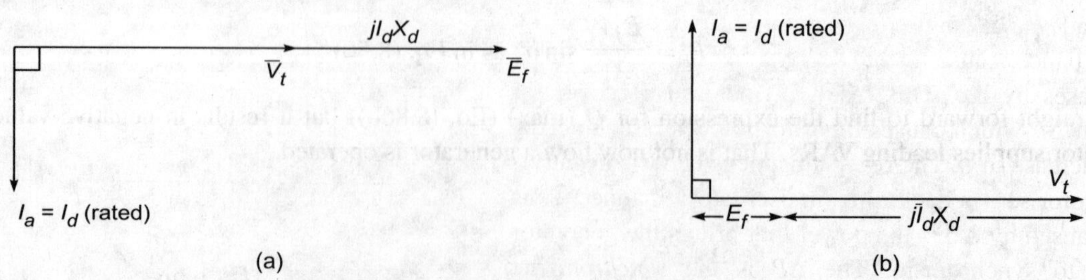

Fig. 8.88 *(a) 90° lagging pf, (b) 90° leading pf*

$$I_q = 0, \quad \bar{I}_a = \bar{I}_d$$

$$I_d = I_a = 1 \text{ pu (max limit)}$$

$$E_f = V_t + I_d X_d, \text{ scalar equation}$$

or

$$E_f = 1 - 1 \times 0.79 = 0.21 \text{ pu}$$

Maximum field excitation

$$I_f(\text{min}) = 0.21 \text{ pu}$$

We find that at $P_e = 0$

$$I_a = \left| \frac{E_f - V_t}{X_d} \right|$$

Its plot on I_f scale is drawn in Fig. 8.89. It is a *V*-curve at zero real power.

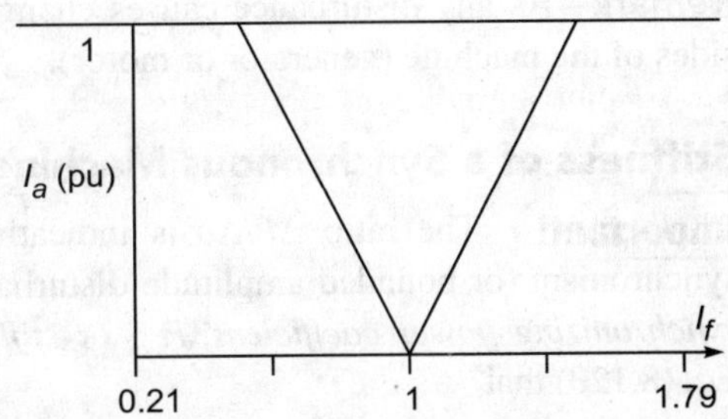

Fig. 8.89

8.15 STAYING IN SYNCHRONIZM – THE SYNCHRONIZING POWER (TORQUE)

LO 4

How a synchronous machine is synchronized to the bus-bars and the loaded electrically as a generator or mechanically as a motor has been explained in Section 8.10. The machine runs at synchronous speed (rotor and stator fields are locked together) at any load well within its maximum steady load capability. For proper operation, it is essential that the machine stays in synchronism for limited amplitude electrical/mechanical disturbances.

We will advance arguments for cylindrical rotor synchronous generator which can then be extended for a motor and also for salient-pole machines.

The power-angle characteristic of a cylindrical generator is drawn in Fig. 8.90 wherein

$$P_e = \left(\frac{E_f V_t}{X_s} \right) \sin \delta \tag{8.121}$$

$$P_e = P(\text{max}) \sin\delta$$

$$P(\text{max}) = \frac{E_f V_t}{X_s}$$

These are per phase values.

!Attention Consider that the generator is operating at load P_0 at angle δ_0. Let an electrical transitory disturbance cause rotor declaration and rotor angle increases by $\Delta\delta$. This result in increase of generator output ΔP which causes the rotor to declare countering the increase in δ. The reverse happens if $\Delta\delta$ is negative. The rotor settles back to δ_0 in oscillatory manner. Thus ΔP caused by $\Delta\delta$ is the power that brings the generator back to synchronism. The ΔP is the *synchronizing power.*

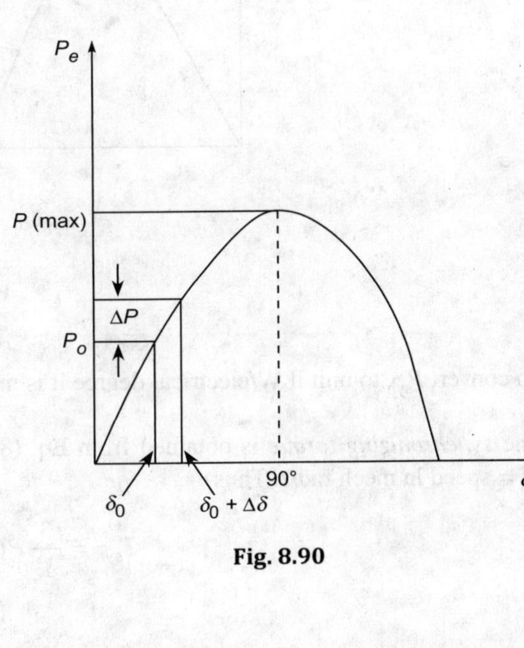

Fig. 8.90

The above arguments equally apply for a synchronous motor for disturbances of mechanical load. The reader is advised to write out the arguments.

Remark As any disturbance causes change $\Delta\delta$, the disturbance appears both electrical and mechanical sides of the machine (generator or motor).

Stiffness of a Synchronous Machine

Important! The ratio $\Delta P/\Delta\delta$ is indicative of the capability of the synchronous machine to stay in synchronism for bounded amplitude disturbances; electrical or mechanical. In the limit $dP/d\delta$ is known as *synchronizing power coefficient*, P_{syn} or *stiffness* of the electromechanical coupling. It then follows from Eq. (8.120) that

$$P_{syn} = \frac{dP}{d\delta} = \left(\frac{E_f V_t}{X_s}\right) \cos\delta \quad \text{W/elect. rad} \tag{8.122}$$

We have dropped the suffix *e* as it applies to both generating and motoring operation of the machine.

The plot of $P - \delta$ with $P_{syn} - \delta$ super imposed on it is drawn in Fig. 8.91. It is seen that $P_{syn} = P(\max)$ at $\delta = 0$ (no load) and then reduces. It is zero at $\delta = 90°$, which is the steady-state stability limit beyond which P_{syn} becomes negative.

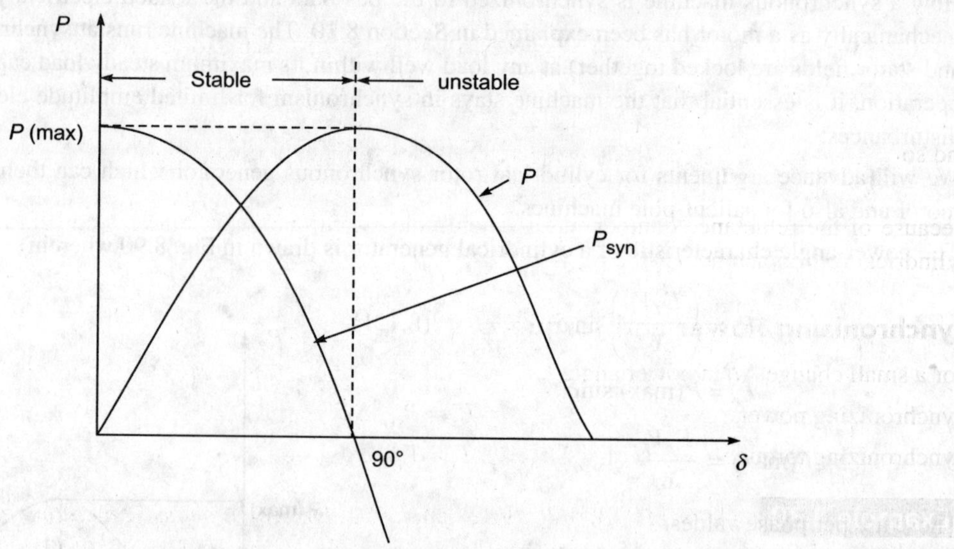

Fig. 8.91

To convert P_{syn} to unit if W/electrical degree it is multiplied by $\left(\frac{\pi}{180}\right)$.

The *synchronizing torque* is obtained from Eq. (8.122) by multiplying it by 3 (phases) and dividing it by n'_s = speed in mech rad/s. Thus

$$T_{syn} = \frac{3}{n'_s} P(\max) \cos\delta \quad \text{Nm /mech rad} \tag{8.123}$$

where $$n'_s = (2\pi f)\frac{2}{P} = \frac{4\pi f}{P} \quad \text{mech rad/s}$$

The units of T_{syn} are changed to Nm/mech degree by multiplying Eq. (8.123) by $\left(\frac{\pi}{180}\right)$.

We find from Eq. (8.122) that the machine stiffness increases with E_f (over excitation) and with reduction of X_s (a design factor; if air-gap is increased).

Important! Excessive stiffness of the electromagnetic link has two detrimental effects.

1. In a motor with no disturbances on power supply the electromagnetic stiffness almost prevents the changes in motor speed. Any sudden large load torque has a shock effect on the shaft and coupling which may damage the shaft leading to shaft fracture on repeated occurrences.
2. The frequency of oscillation of electromechanical coupling increases with stiffness. This results in shaft fatigue and consequent fracture over a period of time.

For a given machine, X_s is fixed, so it should not be operated for long periods of time with extreme over excitation.

In a salient-pole synchronous machine taking derivative of Eq. (8.117), the synchronizing coefficient is

$$P_{syn} = \frac{E_f V_t}{X_d}\cos\delta + V_t^2\left(\frac{X_d - X_q}{X_d X_q}\right)\cos 2\delta \quad \text{W/elect rad.} \tag{8.124a}$$

and so

$$T_{syn} = \frac{3}{n'_s}\left[\left(\frac{E_f V_t}{X_d}\right)\cos\delta + V_t^2\left(\frac{X_d - X_q}{X_d X_q}\right)\cos 2\delta\right] \tag{8.124b}$$

Because of the reluctance component, a salient-pole synchronous machine is comparatively stiffer than a cylindrical rotor machine.

Synchronizing Power and Torque

For a small change $\Delta\delta$ in power angle

Synchronizing power, $$P_s = P_{syn}\,\Delta\delta \tag{8.125}$$

Synchronizing torque, $$T_s = T_{syn}\,\Delta\delta \tag{8.126}$$

Example 8.30 *A 6 MV A, 11 kV, 8 pole, 50 Hz synchronous generator having saturated synchronous reactance of 0.5 pu is synchronized to 11 kV bus. Calculate its synchronizing power and torque coefficient per degree mechanical shift of rotor angle at (a) no load, and (b) at full load 0.8 lagging pf.*

LO 4

Solution

$$V_t = \frac{11\times10^3}{\sqrt{3}} = 6351 \text{ V}$$

$$I_a \text{ (rated)} = \frac{6\times10^3}{\sqrt{3}\times11\times10^3} = 314.9 \text{ A}$$

$$\text{Base ohms} = \frac{6351}{314.9} = 20.16\ \Omega$$

$$X_s = 0.5 \times 20.16 = 10\ \Omega$$

$$P_e = 3 \times \frac{E_f V_t}{X_s} \sin\delta \quad \text{W/elect rad} \qquad \text{(i)}$$

$$\text{Mechanical degree} = \text{elect rad} \times \frac{2}{P} \times \frac{180}{\pi}$$

$$= \text{electrical rad} \times \frac{45}{\pi}; P = \delta \qquad \text{(ii)}$$

$$P_{syn} = \frac{\pi}{15}\left(\frac{E_f V_t}{X_s}\right)\cos\delta \quad \text{W/mech degree} \qquad \text{(iii)}$$

$$n'_s = \frac{120 f}{P} \times \frac{2\pi}{60} = \frac{120 \times 50 \times 2\pi}{8 \times 60} = 25\pi \text{ rad mech/s}$$

$$T_{syn} = \frac{P_{syn}}{25\pi}$$

$$T_{syn} = \frac{1}{375}\left(\frac{E_f V_t}{X_d}\right)\cos\delta \quad \text{Nm/mech degree} \qquad \text{(iv)}$$

(a) No load

$$E_f = V_t = 6351 \text{ V}, \quad \delta = 0$$

$$P_{syn} = \frac{\pi}{15} \times \frac{(6351)^2}{10} \times 10^{-3} = 844.8 \quad \text{kW/degree mech}$$

$$T_{syn} = \frac{1}{375} \times \frac{(6351)^2}{10} = 10756 \quad \text{Nm/degree mech}$$

(b)

$$I_a(f_l) = 314.9 \text{ A} \quad \text{pf} = 0.8 \text{ lag} \quad \phi = 36.9° \text{ lag}$$

$$\bar{I}_a = 314.9\angle -36.9°$$

$$\bar{V}_t = 6351\angle 0° \text{ V}$$

$$\bar{E}_f = 6351\angle 0° + j\,314.9 \times 10\angle -36.9°$$

$$= 8242 + j\,2518 = 8618\angle 17°$$

or

$$E_f = 8618, \quad \delta = 17°$$

$$P_{syn} = \frac{\pi}{15} \times \frac{8618 \times 6351}{10} \cos 17° \times 10^{-3} = 1096 \quad \text{kW/rad mech}$$

$$T_{syn} = \frac{1}{375} \times \frac{8618 \times 6351}{10} \cos 17° = 13958 \quad \text{Nm /degree mech}$$

Example 8.31 *A 12 pole, 50 Hz synchronous motor has saturated X_d = 0.8 pu and X_q = 0.5 pu. It is supplying full load at rated voltage of 0.8 pf leading. Draw the phasor diagram and calculate the excitation emf in pu. Calculate its pu synchronizing power per degree elect. and pu synchronizing torque per degree mech.*

LO 4

Solution The phasor diagram is drawn in Fig. 8.92.

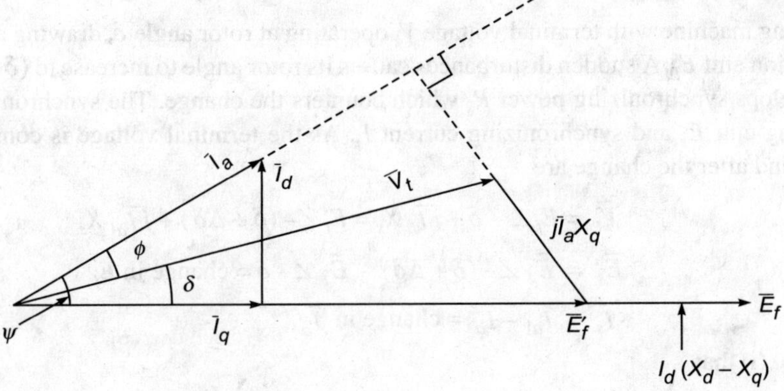

Fig. 8.92

$$\bar{V}_t = 1\angle 0^\circ,\ \bar{I}_a = 1\angle 36.9^\circ,\ \cos^{-1} 0.8 = 36.9^\circ \text{ leading}$$

$$\bar{E}'_f = \bar{V}_t - j\,\bar{I}_a X_q = 1 - j\,1 \times 0.5\angle 36.9^\circ = 1 + 0.5\angle -53.1^\circ$$

$$\bar{E}'_f = 1.3 - j\,0.4 = 1.36\angle -17^\circ$$

$$\bar{E}'_f = 1.36,\quad \delta = 17^\circ,\quad E_f \text{ lags } V_t$$

$$\psi = \phi + \delta = 36.9^\circ + 17^\circ = 53.9^\circ$$

$$\bar{I}_d = I_a \sin\psi = 1 \sin 53.9^\circ = 0.808$$

As it lead E_f by angle ϕ sign of I_a is negative in Eq. (8.109), therefore

$$E_f = E'_f + I_d\,(X_d - X_q) = 1.36 + 0.808(0.8 - 0.5)$$

$$E_f = 1.6 \text{ pu},\ \delta = 17^\circ$$

$$P_{\text{syn}} = \frac{E_f V_t}{X_d} \cos\delta + V_t^2 \left(\frac{X_d - X_q}{X_d X_q}\right) \cos 2\delta \text{ pu/elect.rad}$$

In pu system multiplier 3 is not needed

Substituting values

$$P_{\text{syn}} = \frac{1.6 \times 1}{0.8} \cos 17^\circ + 1^2 \left(\frac{0.8 - 0.5}{0.8 \times 0.5}\right) \cos\,(2 \times 17^\circ)$$

$$= 1.913 + 0.622 = 2.53 \text{ pu/elect. rad}$$

or

$$P_{\text{syn}} = \frac{\pi}{180} \times 2.53 = 0.044 \text{ pu/degree elect.}$$

$$\theta_m = \left(\frac{2}{P}\right) \theta_e = \frac{1}{6}\,\theta_e$$

$$n'_s = \frac{120 f}{P} \times \frac{2\pi}{60} = \frac{120 \times 50}{12} \times \frac{2\pi}{60} = \frac{\pi}{16.67} \text{ elect rad/s}$$

$$T_{\text{syn}} = \frac{16.67}{\pi} \times 6 \times 0.044 = 1.4 \text{ Nm (pu)/degree mech.}$$

Synchronizing EMF and Current (Incremental Variation)

Consider a motoring machine with terminal voltage V_t operating at rotor angle δ, drawing armature current I_a and having excitation emf E_f. A sudden disturbances causes its rotor angle to increase to $(\delta + \Delta\delta)$. As we have seen above it develops synchronizing power P_s which counters the change. The synchronizing power arises from synchronizing emf E_s and synchronizing current I_s. As the terminal voltage is constant the armature equations before and after the change are

$$\bar{V}_t = E_f \angle -\delta + j\bar{I}_a X_s = E_f \angle -(\delta + \Delta\delta) + j\bar{I}_{a1} X_s \quad \text{(i)}$$

$$\bar{E}_s = \bar{E}_f \angle -(\delta + \Delta\delta) - \bar{E}_f \angle -\delta = \text{change in } E_f \quad \text{(ii)}$$

$$\bar{I}_s = \bar{I}_{a1} - \bar{I}_a = \text{change in } \bar{I}_a$$

It follows from Eq. (i) that

$$E_f \angle -(\delta + \Delta\delta) - E_f \angle -\delta = -j(\bar{I}_{a1} - \bar{I}_a) X_s$$

$$E_s = -j\bar{I}_s X_s \quad \text{(iii)}$$

where $\bar{E}_s$ = synchronizing emf and $\bar{I}_s$ = synchronizing current.

Synchronizing emf

$$\bar{E}_s = E_f \angle -(\delta + \Delta\delta) - E_f \angle -\delta$$

$$= E_f \angle -\delta \, [1\angle -\Delta\delta - 1] \quad \text{(iv)}$$

$$1 \angle -\Delta\delta - 1 = \cos\Delta\delta - j\sin\Delta\delta - 1 = [-(1 - \cos\Delta\delta) - j\sin\Delta\delta]$$

$$\text{Magnitude} = [(1 - \cos\Delta\delta)^2 + \sin^2\Delta\delta]^{1/2} = 2\,(1 - \cos\Delta\delta)$$

$$= 2\left[1 - \left(1 - 2\sin^2\frac{\Delta\delta}{2}\right)\right]^{1/2} = 2\sin\frac{\Delta\delta}{2}$$

$$\text{Angle} = \tan^{-1}\frac{-\sin\Delta\delta}{-(1-\cos\Delta\delta)} = \tan^{-1}\frac{-2\sin\frac{\Delta\delta}{2}\cos\frac{\Delta\delta}{2}}{-2\sin^2\frac{\Delta\delta}{2}} = \tan^{-1}\frac{-\cos\frac{\Delta\delta}{2}}{-\sin\frac{\Delta\delta}{2}}$$

$$= -90^\circ - \frac{\Delta\delta}{2}$$

From Eq. (iv)

$$\bar{E}_s = 2E_f \sin\frac{\Delta\delta}{2} \angle -\delta - 90^\circ - \frac{\Delta\delta}{2} \quad \text{(v)}$$

Synchronizing current

$$\bar{I}_s = \frac{\bar{E}_s}{-jX_s} = \frac{2E_f}{X_s}\sin\frac{\Delta\delta}{2} \angle -\left(\delta + \frac{\Delta\delta}{2}\right) \quad \text{(iv)}$$

Alternative

Let us evaluate $(1\angle -\Delta\delta - 1)$ by the phasor method. The phasor diagram is drawn in Fig. 8.93.

$$(1\angle -\Delta\delta - 1) = \overline{AB} = 2\sin\frac{\Delta\delta}{2} \angle -\left(90 + \frac{\Delta\delta}{2}\right)$$

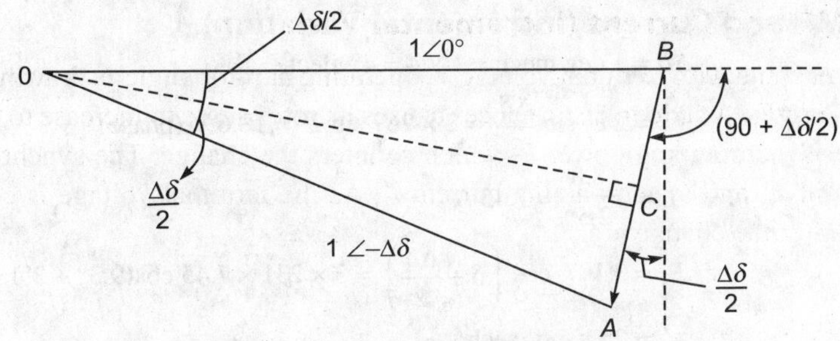

Fig. 8.93

Magnitude-wise

$$E_s = 2E_f \sin\frac{\Delta\delta}{2} \qquad I_s = \frac{E_s}{X_s} = 2\left(\frac{E_f}{X_s}\right)\sin\frac{\Delta\delta}{2}$$

$$\bar{E}_s \text{ lags } \bar{E}_f \text{ by } 90^\circ \qquad \bar{I}_s \text{ lag } \bar{V}_t \text{ by angle } \left(\delta + \frac{\Delta\delta}{2}\right)$$

Synchronizing power

$$P_s = V_t I_s \cos\left(\delta + \frac{\delta}{2}\right); \text{ per phase}$$

or
$$P_s = 2\left(\frac{V_t E_f}{X_s}\right)\sin\frac{\Delta\delta}{2}\cos\left(\delta + \frac{\Delta\delta}{2}\right) \tag{8.127}$$

For small variation of δ

Then
$$P_s = \left[\left(\frac{V_t E_f}{X_s}\right)\cos\delta\right]\Delta\delta = P_{\text{syn}}\,\Delta\delta \tag{8.128}$$

a result already obtained.

Example 8.32 *A 400 V, 8 pole, 50 Hz synchronous motor has a shaft load of 12 kW, 0.8 leading pf. Its saturated synchronous resistance X_s = 2.5 Ω. A mechanical load disturbance causes the rotor angle to slip by 1 mech. degree. Calculate the synchronizing current, power and torque.*

LO 4

Solution Load 12 kW, 0.8 pf leading

$$I_a = \frac{12\times 10^3}{\sqrt{3}\times 400\times 0.8} = 21.65\text{ A}, \quad \phi = 36.9^\circ \text{ leading}$$

$$V_t = 400/\sqrt{3} = 231\text{ V}, \quad X_s = 2.5\ \Omega$$

$$E_f = 231\angle 0^\circ - j\times 21.65\angle 36.9^\circ \times 2.5$$

$$= 231 + 54.125\angle -53.1^\circ = 263.5 + j\,43.28$$

$$= 267\angle -9.3^\circ\text{ V}$$

$$E_f = 267\text{ V}, \quad \delta = 9.3^\circ \text{ elect.}$$

Mechanical disturbance

$$\Delta\delta = 1 \text{ deg mech} = 1 \times 4 = 4° \text{ elect.}$$

$$E_s = 2E_f \sin\frac{\Delta\delta}{2} = 2 \times 267 \sin 2° = 18.6 \text{ V (phase)}$$

$$I_s = \frac{18.6}{2.5} = 7.45 \text{ A}$$

$$P_s = 3\, V_t I_s \cos\left(\delta + \frac{\Delta\delta}{2}\right) = 3 \times 231 \times 7.45 \cos(9.3° + 2°) = 5.163 \text{ kW}$$

$$n_s' = 25\,\pi \text{ rad mech/s}$$

$$T_s = \frac{5163}{25\pi} = 65.74 \text{ Nm}$$

Important! We will see from Example 8.33 that saliency effect can be ignored without causing serious errors. However under condition of low excitation saliency must be taken into account (8.117).

8.16 DETERMINATION OF X_D AND X_Q—SLIP TEST

LO 4

Direct and quadrature-axis reactances of a salient-pole synchronous machine can be estimated by means of a test known as the slip test. The machine armature is connected to a 3-phase supply whose voltage is much less than the rated voltage of the machine, while the rotor is run at speed close to synchronous with the field winding left open circuited (unexcited) as shown in Fig. 8.94(a). Since the excitation emf is zero, heavy currents would be drawn by the armature if connected to the rated voltage supply.

The currents drawn by the armature set up an mmf wave rotating at synchronous speed as shown in Fig. 8.94(b). Since the rotor is being run at a speed close to synchronous, the stator mmf moves slowly past the field poles at slip speed $(n_s \sim n)$. When the stator mmf is aligned with the *d*-axis (field poles), flux Φ_d/ pole is set up so that effective reactance offered by the machine is X_d. Similarly when the stator mmf aligns with the *q*-axis, the flux set up is Φ_q/pole and the machine reactance is X_q. The current drawn by the armature therefore varies cyclically at twice the slip frequency as shown by the current waveform drawn in Fig. 8.94(c)—the rms current is minimum when machine reactance is X_d and is maximum when it is X_q. Because of cyclic current variations and consequent voltage drop in the impedance of supply lines (behind the mains), the voltage at machine terminals also varies cyclically and has a minimum value at maximum current and maximum value at minimum current as shown by the voltage waveform of Fig. 8.94(c). The machine reactances can be found as

$$X_d = \frac{V_t(\text{at } I_a(\text{min})(\text{line})}{\sqrt{3}\, I_a\,(\text{min})} \tag{8.129}$$

and

$$X_q = \frac{V_t(\text{at } I_a(\text{max})(\text{line})}{\sqrt{3}\, I_a\,(\text{max})} \tag{8.130}$$

The phenomenon of armature current going through maximum and minimum values during the slip test is also easily seen from Eqs. (8.115) and (8.116) with $E_f = 0$

$$I_d = \frac{V_b}{X_d}\cos\delta, \qquad I_q = \frac{V_b}{X_q}\sin\delta$$

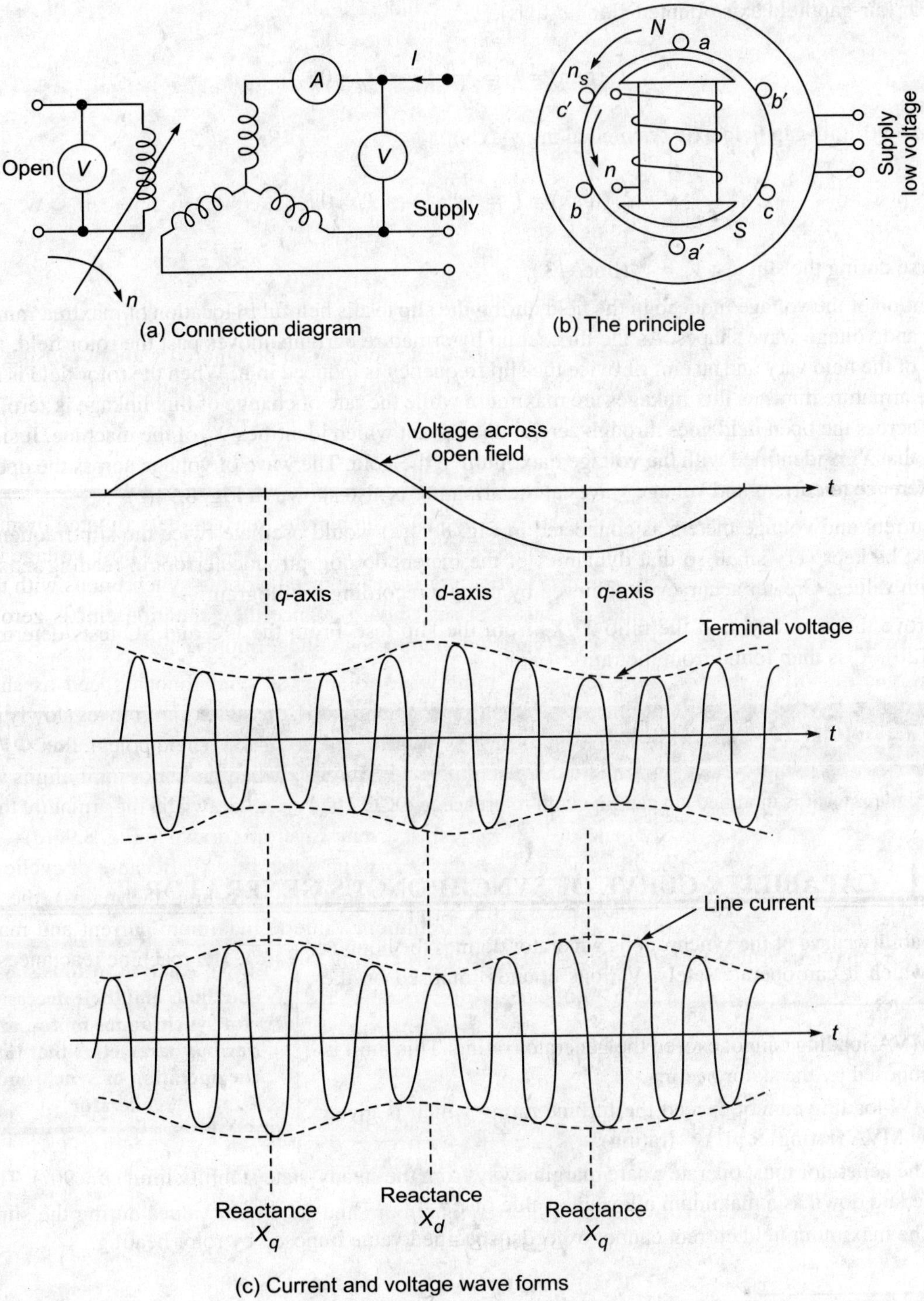

Fig. 8.94 *Slip test*

At $\delta = 0°$ (air-gap field axis oriented along d-axis),

$$I_a\,(\min) = I_d = \frac{V_b}{X_d}; \qquad I_q = 0$$

and at $\delta = 90°$ (air-gap field axis oriented along q-axis)

$$I_a\,(\max) = I_q = \frac{V_b}{X_q}; \qquad I_d = 0$$

Of course during the slip test $V_b = V_t\,(\text{line}/\sqrt{3})$.

Observation of the voltage induced in the field during the slip test is helpful in location of maxima/minima on current and voltage wave shapes. As the flux set up by armature currents moves past the rotor field, the flux linkage of the field vary and an emf of twice the slip frequency is induced in it. When the rotor field is aligned with the armature mmf, its flux linkages are maximum while the rate of change of flux linkage is zero, i.e. the voltage across the open field goes through zero at this instant which identifies X_d of the machine. It similarly follows that X_q is identified with the voltage maximum in the field. The wave of voltage across the open-field with reference to current and voltage waves at the armature is also shown in Fig. 8.94(c).

Since current and voltage meters as connected in Fig. 8.94(a) would oscillate twice the slip frequency, the slip must be kept very small so that dynamics of the meters do not introduce errors in reading maximum/minimum values. Greater accuracy is achieved by using a recording oscillogram.

To improve the accuracy, find the ratio X_q/X_d from the slip test. From the OC and SC tests determine X_d (Eq. 8.120), X_q is then found from the ratio.

Concept Check LO 4

1. Explain what is modified air-gap line with reference to OCC.

8.17 CAPABILITY CURVE OF SYNCHRONOUS GENERATOR

LO 5 Analyse various starting methods and their necessity for synchronous motor, and various parameters that limit the operation of synchronous generator

The capability curve of the synchronous generator defines the bounds within which it can operate safely. Various bounds imposed on the machine are:

1. MVA-loading cannot exceed the generator rating. This limit is imposed by the stator heating.
2. MW-loading cannot exceed the turbine rating which is given by MVA (rating) × pf[‡‡‡‡‡] (rating).
3. The generator must operate a safe margin away from the steady-state stability limit ($\delta = 90°$). This can be laid down as a maximum allowable value of δ.
4. The maximum field current cannot exceed a specified value imposed by rotor heating.

[‡‡‡‡‡] The name plate rating of a synchronous generator specifies this power factor. Of course, the machine can operate at any pf as long as MVA and MW do not exceed the ratings.

Development and Analysis of Capability Curve

To draw the capability curve of the synchronous generator, its phasor diagram is used which is redrawn in Fig. 8.95, armature resistance is neglected. After multiplying voltage magnitude of each voltage phasor by $(3\ V_t/X_s)$, the phasor diagram is redrawn in Fig. 8.96. It is immediately recognized that OMN is the complex power triangle (in 3-phase values) wherein

$$OM = S(\text{VA}); \quad NM = P(\text{W}); \quad ON = Q(\text{V AR})$$

Obviously Q is positive for lagging power factor, ϕ being the angle of OM from the *P-axis*. A mere scale change will convert these values to the units of MVA, MW and MVAR.

Recall

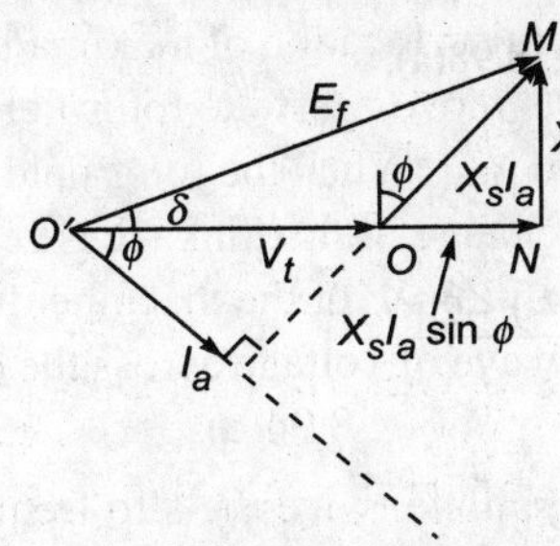

Fig. 8.95

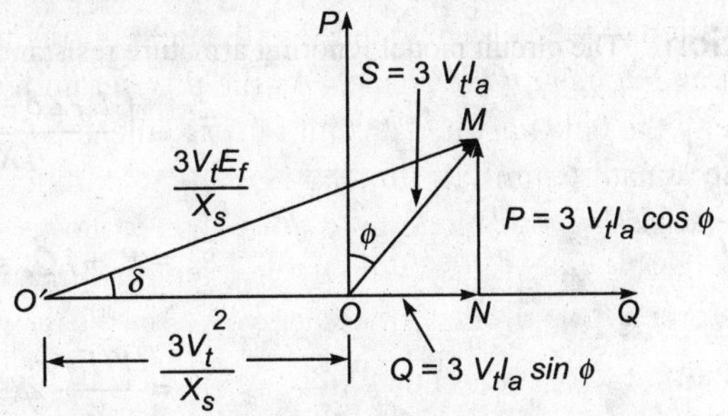

Fig. 8.96

Constant S operation will lie on a circle centred at O and radius OM. Constant P operation will lie on a line parallel to $0'Q$ -axis. Constant-excitation (F_f) operation will lie on a circle centred at O' of radius $OM\ (3V_t\ E_f/X_s)$. Constant-pf operation will lie on a radial line through O.

Now with specified upper limits of S, P and E_f (field current), the boundaries of the capability curve can be drawn in Fig. 8.97. The limit on the left side is specified by δ (max), the safe operating value from the point of view of transient stability.

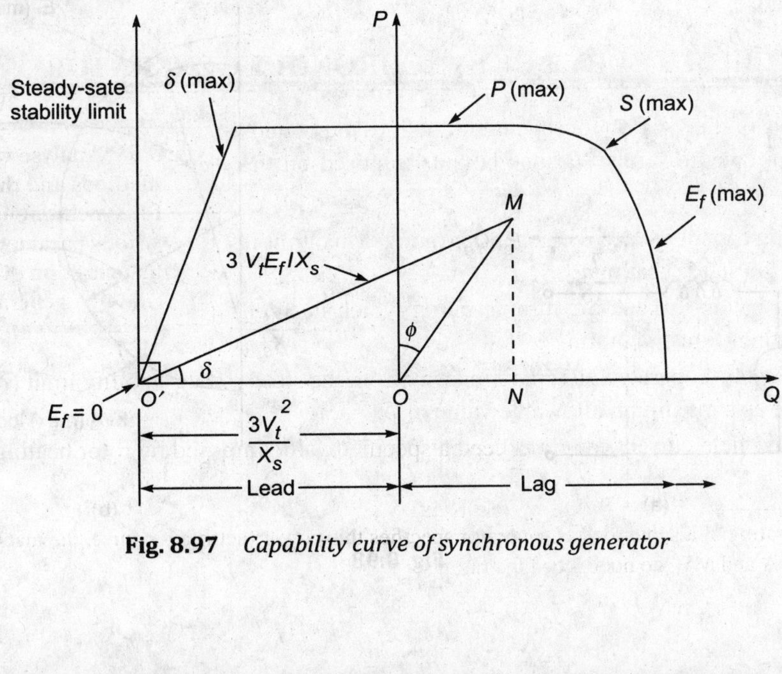

Fig. 8.97 *Capability curve of synchronous generator*

Since the minimum excitation operation corresponds to $\delta = 90°$, the machine operation is at a safe limit from E_f (min) by specifying δ (max).

The capability curve of the synchronous motor can be similarly drawn and will lie in the lower half of the PQ-plane and "lag" and "lead" regions will interchange.

Example 8.33 *Referring to the circuit model of Fig. 8.55(a) and neglecting armature resistance, derive an expression for real and reactive power outputs of a synchronous generator. Draw there from the loci of constant kVA (max) and constant field current (max). Assume the machine to operate at constant (rated) terminal voltage.*

LO 5

Solution The circuit model ignoring armature resistance is redrawn in Fig. 8.98(a).

$$\bar{I}_a = \left(\frac{E_f\angle\delta - V_t\angle 0°}{jX_s}\right)$$

$$\bar{S}_e = P_e + jQ_e = V_t\angle 0°\ \bar{I}_a^* = V_t\left(\frac{E_f\angle\delta - V_t}{jX_s}\right)^*$$

$$= \frac{V_tE_f}{X_s}\angle(90° - \delta) - \frac{V_t^2}{X_s}\angle 90°$$

It then follows that

$$S_e^2 = (V_tI_a)^2 = P_e^2 + Q_e^2 \quad \text{(i)}$$

$$P_e = Re(\bar{S}_e) = \frac{V_tE_f}{X_s}\sin\delta \quad \text{(ii)}$$

$$Q_e = Im(\bar{S}_e) = -\frac{V_t^2}{X_s} + \frac{V_tE_f}{X_s}\cos\delta \quad \text{(iii)}$$

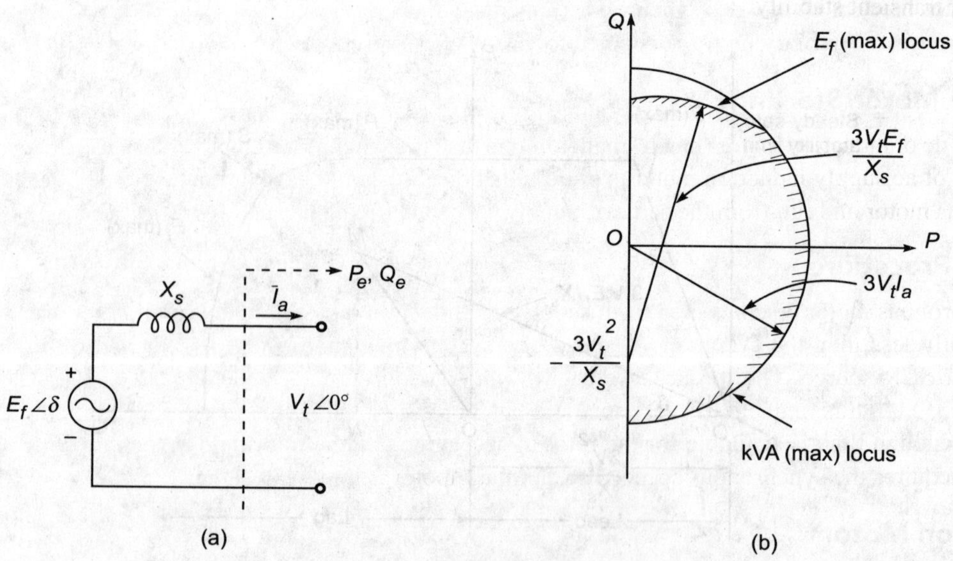

Fig. 8.98

$$\left(Q_e + \frac{V_t^2}{X_s}\right) = \frac{V_t E_f}{X_s} \cos \delta \tag{iv}$$

Squaring Eqs (ii) and (iv) and adding we get

$$S_e^2 = P_e^2 + \left(Q_e + \frac{V_t^2}{X_s}\right)^2 = \left(\frac{V_t E_f}{X_s}\right)^2 \tag{v}$$

In terms of 3 phase quantities Eqs (i) and (v) become

$$S^2 = (3\, V_t I_a)^2 = P^2 + Q^2 \tag{vi}$$

$$P^2 + \left(Q + \frac{3V_t^2}{X_s}\right)^2 = \left(\frac{3V_t E_f}{X_s}\right)^2 \tag{vii}$$

It is seen from Eq. (vi) that constant–S locus is a circle in PQ plane with radius $3V_tI_a$ centred at origin; while Eq. (vii) tells us that constant E_f-locus is a circle of radius $\frac{3V_tE_f}{X_s}$ centred at $\left(P = 0, Q = -\frac{V_t^2}{X_s}\right)$. Of course V_t is assumed constant. These circles are drawn in Fig. 8.98(b) for kVA (max) and E_f (max). These two specify the armature heating limit (I_a (max)) and field heating limit (E_f (max)) respectively as indicated in the figure.

8.18 STARTING OF SYNCHRONOUS MOTORS

LO 5

The starting of a synchronous motor has been discussed briefly in Section 8.10 on synchronization.

!Attention With the rotor field switched on if the stator is connected to the mains, the stator field rotates with respect to the rotor field producing alternating torque with zero average value. The motor, therefore, would not start. It has been stated in Section 5.6 that the condition for an electric machine to produce steady torque is the stator and rotor fields must be relatively stationary. The synchronous motor must therefore be run as a generator brought up close to synchronous speed and then synchronized to main and loaded as a motor. For this purpose an auxiliary motor is coupled to the synchronous motor.

Auxiliary Motor Starting

It is a small dc or induction motor (much smaller in size than the synchronous motor). Because of the universal availability of ac supply induction motor is preferred choice. It should have the same number of poles as the synchronous motor and run from the same ac supply, i.e., same frequency and so same synchronous speed.

Starting Procedure

With synchronous motor disconnected from supply, the induction motor is switched on. As it reaches steady speed slightly less than the synchronous speed, the synchronous motor field is switched on to dc supply – exciter or SCR source. The torque developed by interaction of stator and rotor fields alternates at slip frequency (sf). As this frequency is very small (about 2 cycles/s), during the forward torque half cycle (pull-in torque), enough time is available for the rotor to accelerate and the rotor field to lock up with stator; the rotor thus acquires the synchronous speed. The auxiliary motor is now switched off.

Synduction Motor

A synchronous motor starting on induction principle by means of damper winding is called a synduction motor. The damper windings act like the squirrel-cage rotor producing the starting torque; see Section 5.6 Fig. 5.43.

In the starting operation of a synduction motor the field is kept shorted while the stator is switched on to 3-phase ac supply. As the motor reaches close to synchronous speed, the field is energized from dc supply. The rotor now gets synchronized automatically as explained in auxiliary motor starting.

It is essential to keep the field shorted at start otherwise in the initial part of the starting time when the slip is close to unity, high voltage would be induced in the field winding it has normally large number of turns which can damage it. In fact to avoid high current in the field, it is shorted through resistance several times the field resistance. The field shorted through high resistance adds to the motor starting torque.

The above methods of starting synchronous motors can be employed when the starting torque requirement is low; may be no-load starting. The machine is loaded after it has synchronized.

Starting against High Torque

As it has been mentioned in Section 5.7 that wound rotor slip-ring induction motor provides a high starting torque by adding external resistance to rotor winding (3-phase) through slip rings. This will be elaborated in Chapter 9 on Induction Machine (see Fig. 9.3). A high starting torque synchronous motor is combination of synchronous and slip-ring induction motor into one machine. The damping winding is made in form of 3-phase winding with connection brought out through slip-ring. On starting high resistance is included.

Concept Check — LO 5

1. Why is it desirable to short circuit the dc field when a synchronous motor is started?

8.19 PARALLEL OPERATION OF SYNCHRONOUS GENERATORS

LO 6 Explain the parallel operation and transient analysis of synchronous machine

The operation considered in Sec. 8.19 is that of a synchronous machine connected to infinite bus-bars. Here the parallel operation of two finite size synchronous generators will be considered, which is the way large practical size generators are used. In a power system the generators are connected to the nodes of a grid composed of a network of transformers and transmission lines. A national level grid[§§§§§] may comprise even hundreds of generators and hundreds of kilometers of the transmission line. The grid formation is dictated by reasons of reliability (continuity of supply) and by investment and operating economics of the power plant.

Figure 8.99 shows two synchronous generators along with their primemovers to be operated in parallel. After the two generators are brought to their respective synchronous speeds and their field currents adjusted to give nearly equal terminal voltages, switch S is closed in accordance with the synchronizing procedure described in Sec. 8.10. Active and reactive powers, supplied to the common load by each generator, are controlled respectively by their primemover throttles and field currents as was observed in Sec. 8.11.

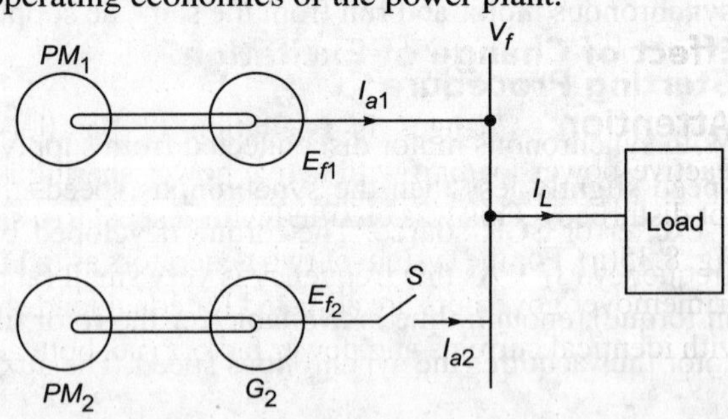

Fig. 8.99 *Parallel operation of synchronous generators*

§§§§§ India is working towards the formation of a national grid. This will result from interconnection of ever-expanding regional grids.

Adjustment of Load Sharing

The active power-sharing between paralleled generators is dependent upon the *droop* of the frequency (speed)-power characteristics of the primemovers and their governors. These characteristics are nearly linear for small changes in the operating (rated) frequency and power as shown in Fig. 8.100. These characteristics can be slided up or down by adjustment of the *set points* of their governors. Drooping¶¶¶¶¶ frequency-power characteristics are essential for proper loading between generators and the adjustment thereof.

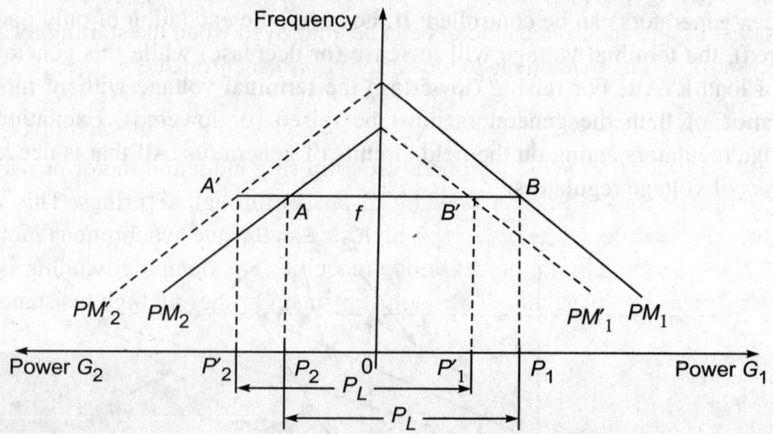

Fig. 8.100 *Primemovers' frequency (speed)-power characteristics and load sharing*

For the solid-line characteristics shown in Fig. 8.101, load $P_L = AB$ at frequency f (rated) load is shared as P_1 and P_2 such that $P_1 + P_2 = P_L$. In order to increase the load on G_2 and to correspondingly reduce the load on G_1, the frequency-power characteristic of G_2 must be raised by adjustment of the governor setting, and to keep the frequency constant the characteristic of G_1 must be simultaneously lowered. It is seen from the figure that with this adjustment, $P_L = P_1' + P_2'$ where $(P_1 - P_1') = (P_2' - P_2) = \Delta P$, the load amount which is transferred from G_1 to G_2 by adjustment of governors. It also follows that if the governor setting of only one of the primemovers is adjusted, the system frequency would change. During the process of governor setting adjustment, the system undergoes load-frequency transient which would soon die out (provided the governors are properly damped) and steady load-frequency conditions established with new load sharing.

Effect of Change of Excitation

!Attention Changes in excitation of parallel-operating generators affect the terminal voltage and reactive-power loading with active power sharing remaining unchanged (primemover governor settings are not disturbed). This was clarified with respect to a single machine connected to infinite bus-bars in Sec. 8.11, Fig. 8.35(a). For the system of two generators as in Fig. 8.99, assume that the two generators are identical, their primemover governors are adjusted for equal load sharing and their excitations are equal; so that they operate with identical currents and power factors (i.e. both active and reactive powers are equally shared). The phasor

---

¶¶¶¶¶ Speed governors belong to the class of regulating systems which are type zero, As the machine is loaded, the frequency (speed) droops to generate the error signal with respect to the reference frequency. This error signal is employed to open the throttle for regulating the turbine speed. Adjustment of the governor setting alters the reference frequency.

diagram under these operating conditions, identical for each generator, is shown in Fig. 8.101 in thick line. For the given total load and equal sharing of real load, let the excitation of G_1 be now increased and that of G_2 simultaneously reduced in such a manner as to keep the terminal voltage unchanged. Each generator, therefore, behaves as if it is connected to infinite bus-bars, i.e., $E_f \sin \delta$ will remain constant. With active power on each generator fixed (at half the total), the projection of generator (each) current on V_t (active current component) will also remain constant and as a result the power factor of G_1 goes down and that of G_2 improves (figure is drawn for upf). Thus G_1 supplies more kVAR to load and G_2 supplies an equal amount less. This is how reactive-load sharing between generators can be controlled. If, however, the excitation of only one of the generators is raised (or lowered), the terminal voltage will increase (or decrease) while this generator shares a larger (or smaller) part of load kVAR. For raising (lowering) the terminal voltage without modifying the kVAR sharing, the excitation of both the generators must be raised (or lowered). Excitation is automatically controlled by voltage regulators acting on the field circuits of generators. All that is needed to be done is to change the set-points of voltage regulators.

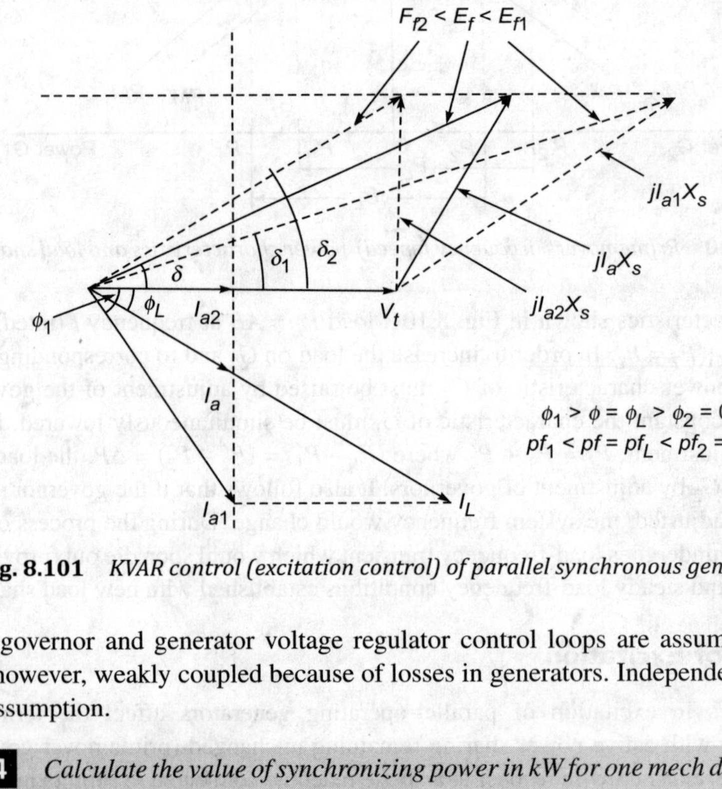

Fig. 8.101 *KVAR control (excitation control) of parallel synchronous generators*

The primemover governor and generator voltage regulator control loops are assumed independent in the above. They are, however, weakly coupled because of losses in generators. Independence of control loops is a good working assumption.

Example 8.34 *Calculate the value of synchronizing power in kW for one mech degree of displacement for a 3 ϕ, 20000 KVA, 6600 V, 50 Hz, 12 pole machine having X_s = 25% and negligible resistance.*

LO 6

Solution

$$\text{emf per phase} = \frac{6600}{\sqrt{3}} = 3810.6 \text{ volts}$$

$$\text{rotor displacement} = \delta = 1 \times \frac{P}{2} = 1 \times \frac{12}{2} = 6° \text{ elect}$$

$$I = \frac{2000 \times 10^3}{6600\sqrt{3}} = 0.577 \text{ kA} = 577.3 \text{ A}$$

$$P_{sy} = \frac{\delta E^2}{X_s} = \frac{6 \times \pi \times (3810.6)^2}{180 \times 1.65} = 921.1 \text{ kW}$$

Example 8.35 *Two 6600 V, star connected alternators in parallel supply the following loads*

400 kW at 0 pf lag - I

400 kW at 0.85 pf lag - II

300 kW at 0.8 pf lag - III

800 kW at 0.7 pf lag - IV

One alternator supplies I_a = 100 A at 0.9 pf lag. Determine output current and power factor of other machine.

LO 6

Solution

Phase angle of I load $\phi_1 = \cos^{-1}(1) = 0°$ (lag)

Phase angle of II load $\phi_2 = \cos^{-1}(0.85) = 31.78°$ (lag)

Phase angle of III load $\phi_3 = \cos^{-1}(0.80) = 36.86°$ (lag)

Phase angle of IV load $\phi_4 = \cos^{-1}(0.7) = 45.57°$ (lag)

Total real load = 400 + 400 + 300 + 800 = 1900 kW

Reactive power of I load $Q_1 = P_1 \tan \phi_1 = 400 \tan 0° = 0$

Reactive power of II load $Q_2 = P_2 \tan \phi_2 = 400 \tan (31.78) = 247.6$ kVAR

Reactive power of III load $Q_3 = P_3 \tan \phi_3 = 300 \tan (36.86) = 224.91$ kVAR

Reactive power of IV load $Q_4 = P_4 \tan \phi_4 = 800 \tan (45.57) = 816$ kVAR

Total reactive power $Q = Q_1 + Q_2 + Q_3 + Q_4 = 1288.51$ kVAR

Output of one machine $P_A = \sqrt{3} \times 6600 \times 100 \times 0.90 = 1028.8$ kW

Output of second machine $P_B = 1900 - 1028.8 = 871.2$ kW

Reactive power of I machine

$$Q_A = P_A \tan(\cos^{-1}(0.9)) = 1028.8 \tan (25.84)$$

$$= 1028.8 \times 0.485 = 498.96 \text{ kVAR}$$

Reactive power of II machine

$$Q_B = Q - Q_A = 1288.51 - 498.96 = 789.55 \text{ kVAR}$$

$$\phi = \tan^{-1}\left(\frac{Q_B}{P_B}\right) = \tan^{-1}\left(\frac{789.55}{871.2}\right) = \tan^{-1}(0.906) = 42.176$$

$$\cos \phi_B = 0.741 \text{ (lag)}$$

$$I_B = \frac{871.2 \times 10^3}{\sqrt{3} \times 600 \times 0.741} = 102.85 \text{ A}$$

Example 8.36 *Two similar 400 V, 3 phase alternators share equal kW power delivered to a balanced three phase, 50 kW, 0.8 pf lag load. If the power factor of one machine is 0.9 lag, find the power factor and current supplied by the other machine.*

LO 6

Solution

Total load $P = 50$ kW

$$\cos\phi = 0.8;\ \phi = \cos^{-1}(0.8) = 36.87°, \text{ lag}$$

Reactive power of load $Q = P\tan\phi = 50\tan 36.87° = 37.5$ kVAR

Real power supplied by machine 1, $P_1 = \dfrac{P}{2} = \dfrac{50}{2} = 25$ kW

Phase angle of machine 1 $\phi_1 = \cos^{-1}(0.9) = 25.84°$, lag

Reactive power supplied by machine 1 $Q_1 = P_1\tan\phi_1 = 25\tan(25.84°) = 12.10$ RVAR

Real power supplied by machine 2 $P_2 = \dfrac{P}{2} = \dfrac{50}{2} = 25$ kW

Reactive power supplied by machine 2 $Q_2 = Q - Q_1 = 37.5 - 12.10 = 25.391$ RVAR

Phase angle of machine 2 $Q_2 = \tan^{-1}\dfrac{Q_2}{P_2} = \tan{-1}\dfrac{25.391}{25}$

$$\phi_2 = \tan^{-1}(1.015) = 45.44°$$

power factor of machine 2 $\cos\phi_2 = \cos(45.44°) = 0.701$, lag

Current supplied by machine 2 $I_2 = \dfrac{P}{\sqrt{3}V_L\cos\phi_2} = \dfrac{25\times 1000}{\sqrt{3}\times 400\times 0.701} = \dfrac{25000}{485.652} = 51.4$ A

8.20 | SHORT-CIRCUIT TRANSIENT IN SYNCHRONOUS MACHINE LO 6

The short-circuit transient in a synchronous machine being electrical in nature is much faster than the electromechanical dynamics discussed in Sec. 8.13. It will, therefore, be assumed that by the time the major features of the short-circuit phenomenon are over, the rotor speed remains constant at the synchronous value.

The short-circuit transient in a synchronous machine is a highly complex phenomenon as a number of coupled circuits are involved and further their self-and mutual-inductances are functions of the angle and therefore of time. The detailed mathematical model of this phenomenon is beyond the scope of this book. The treatment here will be qualitative based on physical reasoning.

Physical Aspect of Transient Phenomenon

To understand the physical reasoning, which will be advanced soon, let us examine the physical picture of the transient phenomenon in an inductor switched onto a source of sinusoidal voltage as shown in Fig. 8.102(a). The resistance associated with the inductance is assumed to be negligible (as is the case in a synchronous machine) and would at the first instant be neglected. While a simple analytical approach would immediately give the quantitative answer to this problem, physical reasoning will however be followed. Figure 8.102(b) shows the waveforms of voltage, current and flux linkages in the circuit under steady-state conditions wherein the current and flux linkages lag the voltage by 90°. At the instant switch S is closed, the steady-state conditions demand that circuit current and associated flux linkages instantaneously acquire a specific value, i_s ($t = 0$). Since the flux linking the circuit has associated stored energy which cannot change

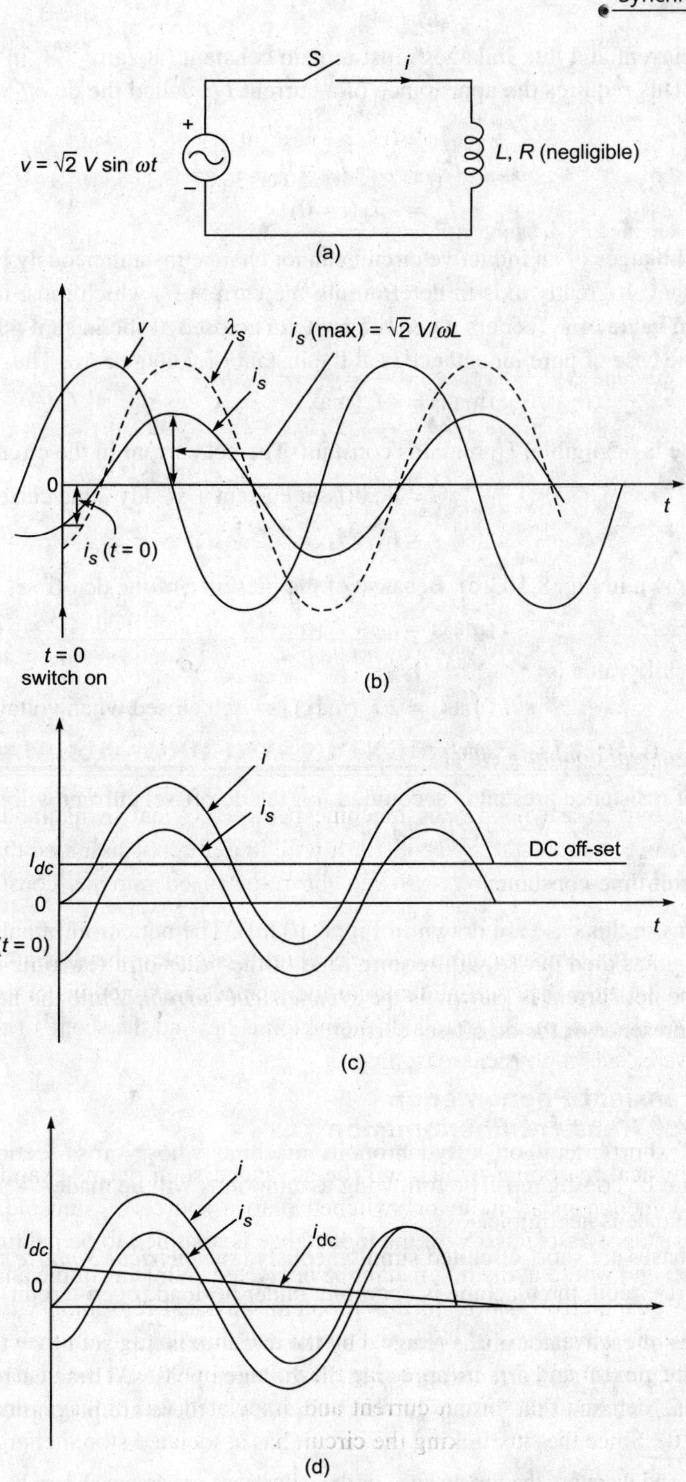

Fig. 8.102 *Switching transient in inductance*

instantaneously, the current and flux linkages must remain constant (at zero****** in the present case) at the instant of switching. This requires the appearance of a current I_{dc}, called the dc *off-set current* in the circuit such that

$$i_s(t = 0) + I_{dc} = 0$$

or

$$I_{dc} = -\, i_s\,(t = 0) \tag{8.131}$$

The fact that the flux linkages of an inductive circuit cannot change instantaneously is known as the *theorem of constant flux linkages.* It greatly aids in determining the current I_{dc} which must immediately flow in the circuit. The maximum value of I_{dc} occurs when the switch is closed at the instant where the i_s demanded is maximum which in the case of pure inductance is at the instant of voltage zero. Thus

$$I_{dc}\,(\max) = -\,i_s\,(\max) \tag{8.132a}$$

If the circuit resistance is negligible, I_{dc} remains constant. The net current in the circuit is

$$i = \text{dc off-set current} + \text{steady-state current}$$

$$= I_{dc} + i_s$$

whose waveform is drawn in Fig. 8.102(c). Because of the presence of the dc off-set current

Important!

$$i\,(\max) = I_{dc} + i_s\,(\max) \tag{8.132b}$$

whose maximum possible value is

$$i\,(\max) = 2\,i_s\,(\max)\ \text{(switch closed when voltage is zero)} \tag{8.133}$$

This is known as the *current doubling effect.*

If the small amount of resistance present is accounted for, the dc off-set current will decay as

$$i_{dc} = I_{dc} e^{-t/\pi} \tag{8.134}$$

when $\tau = R/L$ = circuit time-constant

The current waveforms in this case are drawn in Fig. 8.102(d). The net current finally settles to the steady-state current i_s after i_{dc} has died out (it will require time of the order of a few time-constants). The steady-state component of the net current is known as the *symmetrical current,* while the net current initially lacks symmetry due to the presence of the dc off-set current.

Examining the Transient Phenomenon

Now the problem of short-circuit on a synchronous machine whose cross-sectional view is shown in Fig. 8.102(e) will again be considered. The following assumptions will be made:

(i) Armature resistance is negligible.
(ii) All the three phases are short-circuited simultaneously (*symmetrical* 3-*phase short-circuit*).
(iii) Before the short-circuit, the machine is operating under no-load (open-circuit) condition.

Since the flux linkages of each stator phase (caused by the direct-axis flux set up by the field current) cannot change instantaneously, dc off-set currents appear in all the three phases. These currents are proportional to the flux linkages of each phase at the instant of short-circuit, i.e. these are proportional to the cosine of the

******The initial current and therefore the flux linkage of the inductance are assumed zero here. These could, however, have a finite value.

angle between the phase axis and d-axis. For example, if the d-axis is oriented along the a phase axis at the instant of short-circuit, the dc off-set current in phase a has a certain positive value while negative dc off-set currents of half this value would appear in phases b and c.

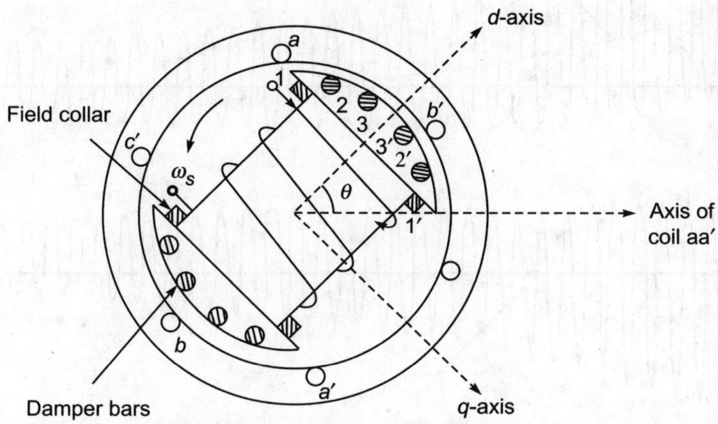

Fig. 8.102(e) *Cross-sectional view of salient-pole synchronous machine*

For short circuit at this instant the wave forms of short circuit currents in the three phases are shown in Fig. 8.103 which also indicates the dc off-set currents in dotted line. If the dc off-set currents are removed from the short circuit currents, we are left with the symmetrical short circuit current which is the same in all the three phases but for a phase difference of 120°.

The symmetrical short-circuit current (in each phase) is 90° lagging current, constituting the d-axis current, the q-axis current being zero (see Fig. 8.103). It establishes a demagnetizing armature reaction along the d-axis. By the theorem of constant flux linkages the flux Φ_f linking the field winding and damper winding (in Fig. 8.103 damper bars[††††††] form coils as 11′, 22′ and 33′ whose axis is along the d-axis) must be maintained by appearance of induced (unidirectional) currents on the field winding (over and above the normal excitation current) and damper winding. As a result the air-gap emf, E_r initially equals $E_f = V_{OC}$ (no-load voltage). The short-circuit current is then $(E_r = E_f)/X_l$, a very large value. The phasor diagram under these conditions is drawn in Fig. 8.104.

The induced currents in the damper and field windings decay at rates determined by their respective time-constants. The damper winding comprising a few thick bars has a much lower time-constant than that for the field winding and its induced current is the first to vanish. As the induced current of the damper winding decays, the net d-axis flux and therefore the air-gap emf (E_r) reduces and the symmetrical short-circuit current decays accordingly as shown in Fig. 8.105. This initial period of decay of the short-circuit current is called the *subtrasient* period in which the current decay is governed mainly[‡‡‡‡‡‡] by the damper winding time-constant.

[††††††] The axis of the damper winding AT is along the d-axis for currents induced in it because of time rate of change of the d-axis flux. The damping effect of damper winding is caused by the relative spatial movement between them and d-axis air-gap flux; the axis of the corresponding induced AT is along the q-axis of the machine. It was the latter induction which was considered in Sec. 8.17.

[‡‡‡‡‡‡] Some of the decay of the short-circuit current is contributed by the slowly-decaying field current.

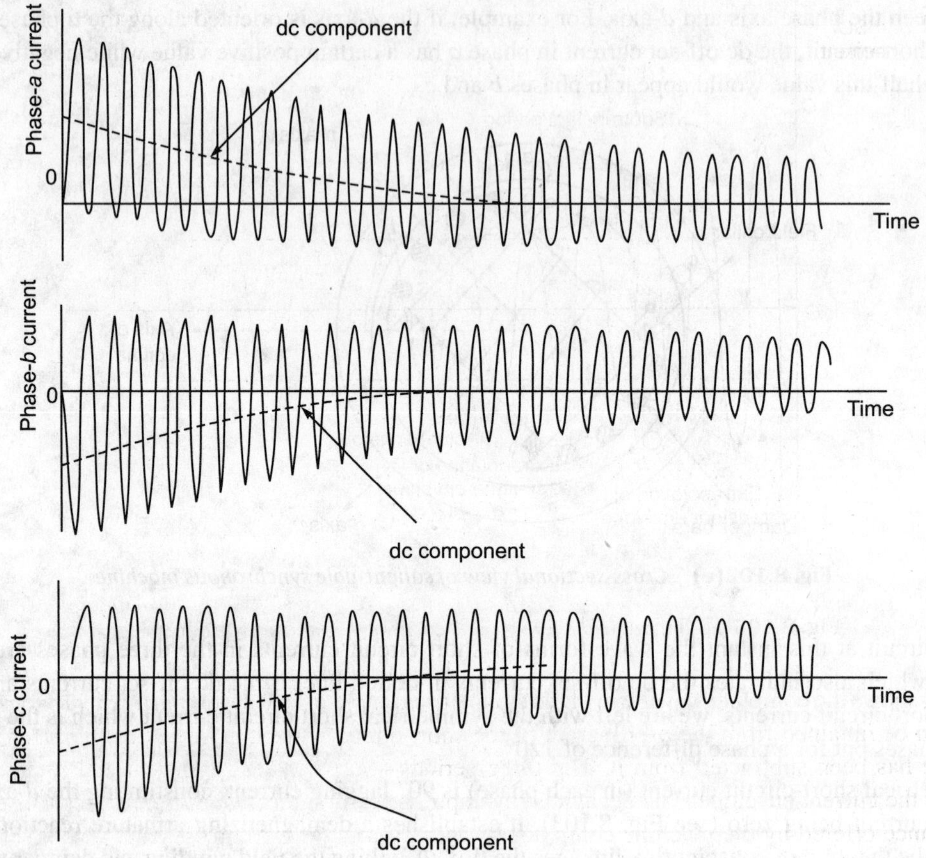

Fig. 8.103 *Short circuit current wave forms in the three phases of a synchronous generator*

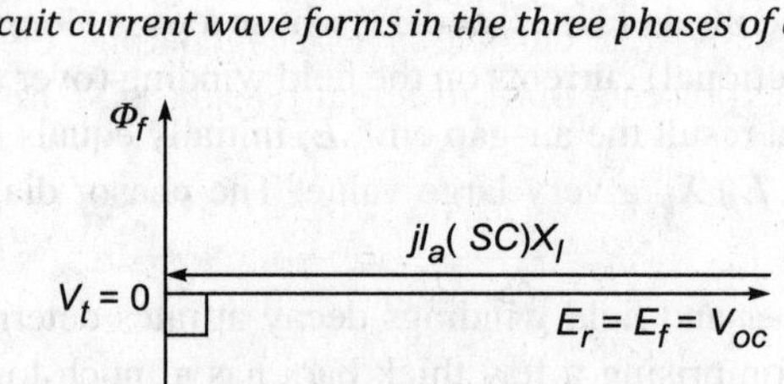

Fig. 8.104 *Phasor diagram upon initiation of 3-phase fault (Armature reaction flux is cancelled by flux set up by the additional field current and the damper winding current)*

As time progresses the induced field current continues to decay governed by its own time-constant. As a consequence the net *d*-axis flux continues to decay and so does the symmetrical short-circuit current till the steady-state short-circuit current (E_f/X_d) is established after the induced current in the field winding has died out. This period of the short-circuit transient is called the *transient* period as indicated in Fig. 8.105.

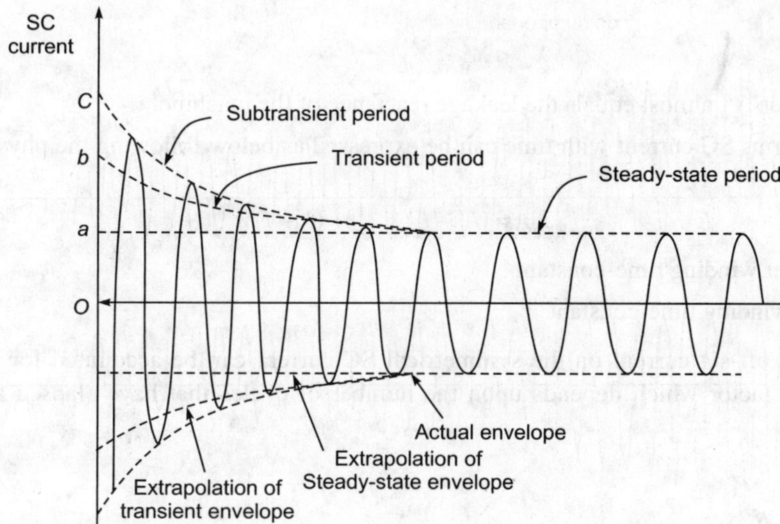

Fig. 8.105 *Symmetrical short circuit current in synchronous generator*

Figure 8.104 shows the complete waveform of the symmetrical short-circuit current in a synchronous machine. This plot can be obtained from the oscillogram of the short-circuit current in one of the phases after the dc off-set value has been subtracted from it. The three periods—substransient, transient and steady-state–are indicated on the current envelope. The decaying envelope is clearly indicative of the fact that the equivalent *d*-axis reactance offered by the machine continuously increases as time progresses and finally settles to the steady value X_d when the armature reaction demagnetizing effect becomes fully effective. Extrapolation of the subtransient, transient and steady-state current envelopes identifies the ordinates *Oa, Ob,* and *Oc* on the current coordinate. The machine presents three different reactances, during the short circuit, as defined below:

Subtransient reactance,
$$X''_d = \frac{E_f}{Oc/\sqrt{2}} = \frac{E_f}{I''} \tag{8.135}$$

Transient reactance,
$$X'_d = \frac{E_f}{Ob/\sqrt{2}} = \frac{E_f}{I'} \tag{8.136}$$

Steady-state reactance,
$$X_d = \frac{E_f}{Oa/\sqrt{2}} = \frac{E_f}{I} \tag{8.137}$$

where E_f = excitation emf (open-circuit voltage) (rms phase value)

I'' = subtransient SC current (rms)

I' = transient SC current (rms)

I = steady SC current (rms)

Obviously

$$X''_d < X'_d < X_d \tag{8.138}$$

In fact as stated earlier, X''_d almost equals the leakage reactance of the machine.

The variation of the rms SC current with time can be expressed as below following the physical arguments presented earlier

$$I_{SC} = (I'' - I')\, e^{-t/\tau_{dw}} + (I' - I)\, e^{-t/\tau_f} + I \tag{8.139}$$

where τ_{dw} = damper winding time-constant

τ_f = field winding time-constant

The effect of the dc off-set current on the symmetrical SC current can be accounted for by means of a suitable multiplying factor which depends upon the number of cycles that have elapsed after the short-circuit.

Certain Details

In the above physical picture certain finer points were purposely left out. The dc off-set stator currents establish a fixed axis flux in air-gap which gradually decays. This flux induces fundamental frequency currents in the field and damper winding which contribute a *d*-axis oscillating field. This oscillating field can be split into two fields[§§§§§§] rotating at synchronous speed in opposite directions with respect to the rotor. One of these components is stationary with respect to the stator and reacts back on the dc off-set currents. The other component travels with respect to the stator at twice the synchronous speed and therefore induces the second harmonic in it.

The plot of the field current after the short-circuit on the stator exhibiting the dc and fundamental frequency transient current is shown in Fig. 8.106.

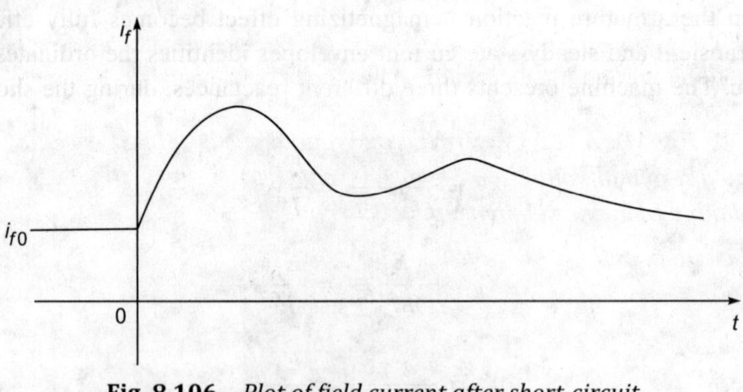

Fig. 8.106 *Plot of field current after short-circuit*

Short-Circuit under Loaded Conditions

Short-circuit analysis of a loaded synchronous machine is quite involved and is beyond the scope of this book. The method of calculating short-circuit currents during the subtransient and transient periods will be presented here, without proof. The circuit models of the machine to be used in computing subtransient

[§§§§§§] See Sec. 10.2 on single-phase induction motor.

and transient currents are given in Figs 8.107(a) and (b) wherein in place of excitation emf, voltages behind subtransient and transient reactances are used. These are given as:

$$\text{Voltage behind subtransient reactance, } E''_f = V_0 + jX''_d I_0 \tag{8.140}$$

$$\text{Voltage behind subtransient reactance, } E'_f = V_0 + jX'_d I_0 \tag{8.141}$$

where V_0 is the machine terminal voltage and I_0 is the machine current prior to occurrence of the fault.

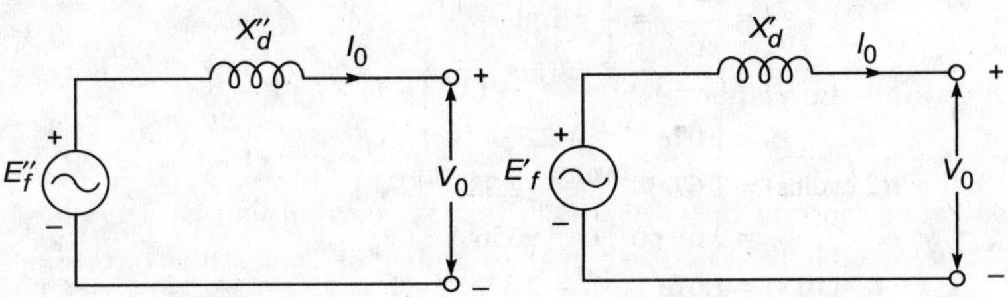

(a) Circuit model for computing subtransient current

(b) Circuit model for computing transient current

Fig. 8.107

The subtransient and transient currents during short circuit are given by

$$I'' = \frac{E''_f}{X''_d} \tag{8.142}$$

and

$$I' = \frac{E'_f}{X'_d} \tag{8.143}$$

Of course the steady-state short circuit current is given by

$$I = \frac{E_f}{X_d} \tag{8.144}$$

where E_f is the excitation emf and X_d, the steady-state d-axis reactance.

Example 8.37 *A 100-MV A, 22-kV, 50-Hz synchronous generator is operating open circuited and is excited to give rated terminal voltage. A 3-phase (symmetrical) short circuit develops at its terminals. Neglecting dc and double frequency components of current.* **LO 6**

(a) find the initial current, and

(b) find the current at the end of two cycles and at the end of 10s.

Given:

Base 100 MVA

$$X_d = 1.0 \text{ pu}, X'_d = 0.3 \text{ pu}, X''_d = 0.2 \text{ pu}$$

$$\tau_{dw} = 0.03 \text{ s}, \tau_f = 1 \text{ s}$$

Solution

(a) Initial current, $$I'' = \frac{E_f}{X''_d} = \frac{1}{0.2} = 5 \text{ pu}$$

$$I_{\text{Base}} = \frac{100 \times 1000}{\sqrt{3} \times 22} = 2624 \text{ A}$$

Then $$I'' = 5 \times 2624 = 13120 \text{ A}$$

(b) $$I' = \frac{E_f}{X'_d} = \frac{1}{0.3} = 3.33 \text{ pu}$$

$$I = \frac{E_f}{X_d} = \frac{1}{1} = 1 \text{ pu}$$

$$I_{SC}(t) = (5 - 3.33)e^{-t/0.03} + (3.33 - 1)^{e-t/1} + 1$$

$$= 1.07e^{-t/0.03} + 2.33e^{-t} + 1$$

$$t(2 \text{ cycles}) = 1.67e^{-0.04/0.03} + 2.33e^{-0.04} + 1$$

$$= 3.68 \text{ pu} \quad \text{or} \quad 9656 \text{ A}$$

$$I_{SC}(10 \text{ s}) = 1.67e^{-10/0.03} + 2.32e^{-10} + 1$$

$$= 1.0001 \text{ pu} \quad \text{or} \quad 2624 \text{ A (steady-state is practically reached)}$$

8.21 | SINGLE-PHASE SYNCHRONOUS GENERATORS

LO 7 Outline the basic operating principle of single phase synchronous generator and brushless dc motor

Certain applications, usually restricted to less than 10 kVA are better served by a single-phase synchronous generator. Examples are emergency, domestic/office supply, portable power for construction tools etc. Because of simplicity of distribution wiring, these loads are better served by a single-phase arrangement.

Designing Aspect

Single-phase generator stator winding can be arranged in two ways.

- A 12-lead three-phase winding connected to operate single-phase.
- Specialised single-phase wound stator.

In the first arrangement several connections are possible but all have to be derated compared to normal three-phase connection. Two connections are indicated in Fig. 8.108 and are compared to three-phase series star in terms of voltage and pu kVA rating.

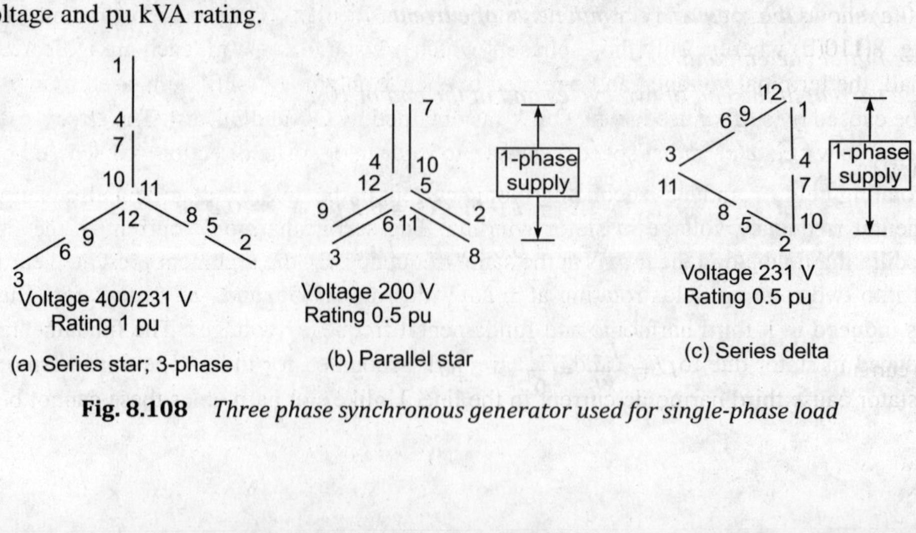

Fig. 8.108 *Three phase synchronous generator used for single-phase load*

Field Flux Distribution and its Analysis

In a single-phase generator, while rotor is similar to that of a three-phase, stator is wound single phase with short-pitched coils (short pitch γ (= SPP) slots). Thus one third of stator slots are left unwound and winding has to be single layer. Obviously the rating of the machine is 2/3rd that of three-phase wound generator. While the rotor has salient poles and produces a field distributed sinusoidally in space (because of shaping of pole faces), the stator which is single phase produces an oscillating field along a fixed axis. The expression for the fundamental component of stator field is as given in Eq. (5.36).

$$\mathcal{F}_1 = F_m \cos \omega t \cos \theta; \ \omega = \text{frequency in rad/s} \quad (8.145)$$

$$\theta = \text{space angle in rad (elect)}$$

where $F_m = \dfrac{4\sqrt{2}}{\pi} K_w \left(\dfrac{N_{ph}}{P}\right) I \cos \omega t \cos \theta; \quad I = \text{rms stator current}$

Equation (8.145) can be trigonometrically split as

$$\mathcal{F}_1 = (1/2)F_m \cos(\omega t - \theta) + (1/2\, F_m) \cos(\omega t + \theta); \quad \omega = \text{synchronous speed rad (elect)/s}$$

$$= \mathcal{F}_{1f} + \mathcal{F}_{1b} \quad (8.146)$$

when $\mathcal{F}_{1f}$ is a sinusoidally distributed field rotating in positive direction of θ (forward rotating) and $\mathcal{F}_{1b}$ is a backward rotating field. Both these fields rotate at synchronous speed as shown in Fig. 8.109 where fields are represented as rotating vectors $\vec{F}_{1f}$ and $\vec{F}_{1b}$ while the resultant field is an oscillating vector $\vec{F}_1$ field in space.

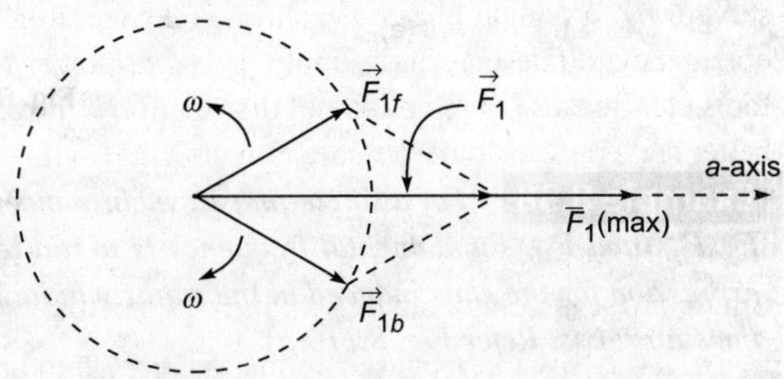

Fig. 8.109 *Stator mmf of single-phase synchronous generator*

Operating Principle

!Attention The forward rotating stator field locks into the rotor field and both move together producing electromagnetic torque and causing conversion of energy from mechanical to electrical.

Figure 8.110(a) shows the rotor and stator fields at the instant I_1 (max). The corresponding phasor diagram is drawn in Fig. 8.110(b) wherein only those phasors which are stationary w.r.t. each are indicated. Assuming X_q to be small, the terminal voltage can be related to excitation emf E_e (suffix *e* is used to avoid confusion that would be caused by suffix *f* used so far) by X_d as obtained by OC and SC test. The effects caused by $\vec{F}_{1b}$ which rotates at ω w.r.t. stator and at speed 2 ω w.r.t. to rotor in opposite direction is discussed below.

Backward rotating field F_{1b} induces second harmonic current in the field winding (and also damper winding) and fundamental frequency voltage in stator winding. The second harmonic current in the field winding causes an oscillating field along the *d*-axis at the same frequency. By the argument presented earlier this field can be split into two rotating fields rotating at $\pm 2\omega$ w.r.t. rotor or 3ω and $-\omega$ w.r.t. stator. Thus the stator winding has induced in it third harmonic and fundamental frequency voltages. The fundamental frequency voltages induced in stator due to $\vec{F}_{1f}$ and $\vec{F}_{1b}$ are both accounted for in X_d. The third harmonic voltages induced in stator cause third harmonic current in the line. Unlike slot harmonics these cannot be eliminated (or attenuated) by chording. But their amplitude is reduced by self-inductance of the field winding, eddy currents in rotor and second harmonic currents in the damper bars.

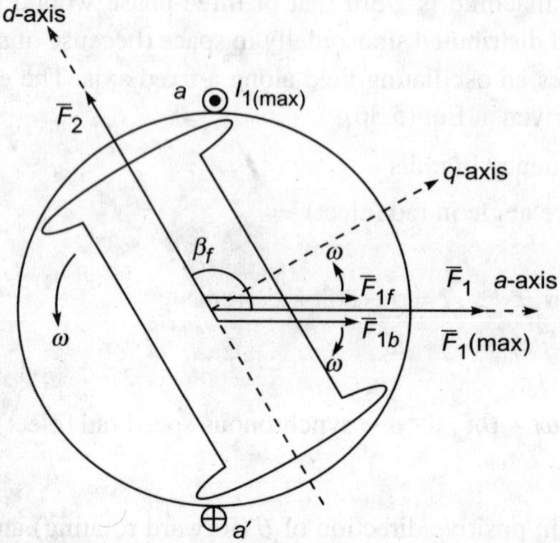

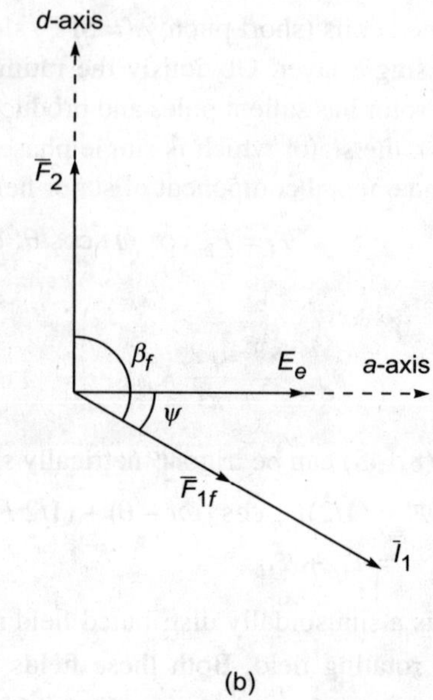

Fig. 8.110

Example 8.38 *For a single-phase synchronous generator the peak value of rotor and stator mmfs are F_2, F_{1f}, and F_{1b}. Fundamental frequency is ω rad(elect)/s. Using other appropriate symbols, derive an expression for the emfs induced in the stator winding by d-axis flux. Take $t = 0$ when F_2 is directed along the stator axis. Refer Fig. 8.110(a).*

LO 7

Solution At any time the rotor has rotated by an angle $\theta = \omega t$. The angle between $\vec{F}_2$ and $\vec{F}_{1f}$ remains fixed at β_f independent of t. Let us find the angle between $\vec{F}_2$ and $\vec{F}_{1b}$. At time $t = 0$, $\vec{F}_{1b}$ makes an angle of β_f from $\vec{F}_2$. $\vec{F}_{1b}$ and $\vec{F}_2$ are rotating in opposite direction at speed ω, so the angle between them at time t is

$$\beta_f + 2\gamma = 2\omega t + \beta_f$$

The net mmf along d-axis (axis of F_2) is in terms of peak values is

$$F_d = F_2 + F_{1f} \cos \beta_f + \underbrace{F_{1b} \cos (2\omega t + \beta_f)}_{\text{Oscillating component}}$$

Direct axis, flux/pole $\qquad \phi = \mathcal{P}_d\,[F_2 + F_{1f} \cos \beta_1 + F_{1b} \cos (2\omega t + \beta_f)]$

Angle between ϕ and stator winding axis = ωt

Flux linkage of stator coil

$$\lambda = \phi N_1 K_w \cos \omega t$$
$$= \mathcal{P}_d N_1 K_w [F_2 + F_{1f} \cos \beta_f + F_{1b} \cos (2\omega t + \beta_f)] \cos \omega t$$

where $\qquad N_1 K_w$ = effective stator turns

Reorganizing

$$= \mathcal{P}_d N_1 K_w[(F_2 + F_1 \cos \beta_f) \cos \omega t + (1/2)F_{1b} \cos (3\,\omega t + \beta_f) + (1/2)F_{1b} \cos (\omega t + \beta_f)]$$

Induced emf, $e_r = d\lambda/dt$

$$= \mathcal{P}_d N_1 K_w[F_2 + F_{1f} \cos \beta_f) \sin \omega t + (3/2)\, F_{1b} \sin (3\,\omega t + \beta_f) + (1/2)F_{1b} \sin (\omega t + \beta_f)]$$

Leaving out third harmonic induced emf

$$e_r = \sqrt{2}\, E_e \sin \omega t + \sqrt{2}\, E_{1f} \cos \beta_f \sin \omega t + \sqrt{2}\, E_{1b} \sin \omega t$$

or

$$E_r\,(\text{rms}) = \underbrace{E_e}_{\text{Excitation emf}} \underbrace{+ E_{1f} \cos \beta_f + E_{1b}}_{\substack{\text{Accounted for in } X_d \\ \text{assuming } X_q = 0}}$$

This phasor diagram is drawn in Fig. 8.110(b).

Example 8.39 *A 1000 kVA, 6.6 kV, 50 Hz, Y-connected synchronous generator has a no-load voltage of 11.4 kV at a certain field current. The generator gives rated terminal voltage at full load 0.75 lagging power factor at the same field current. Calculate:*

(a) The synchronous reactance (armature resistance being negligible);

(b) the voltage regulation;

(c) the torque angle;

(d) the electrical power developed; and

(e) the voltage and kVA rating, if the generator is reconnected in delta.

LO 7

Solution

$$I_a\,(fl) = \frac{1000}{\sqrt{3} \times 6.6 \times 1000} = 0.0875 \text{ kA}$$

$$\phi = -\cos^{-1} 0.75 = -41.4°$$

$$\bar{V}_t = 6.6/\sqrt{3} \angle 0° = 3.81 \angle 0° \text{ kV}$$

$$E_f = V_{oc} = 11.4/\sqrt{3} = 6.58 \text{ kV}$$

$$\bar{E}_f = 6.58 \angle \delta \text{ kV}$$

Refer to circuit diagram of Fig. 8.111 from which we can write

$$6.58 \angle \delta° = 3.81 \angle 0° + j\,X_s\, 0.0875 \angle -41.4° \quad \text{(i)}$$

Equating real and imaginary parts of this equation yields

$$6.58 \cos \delta = 3.81 + 0.058\, X_s \quad \text{(ii)}$$

$$6.58 \sin \delta = 0.0656\, X_s \quad \text{(ii)}$$

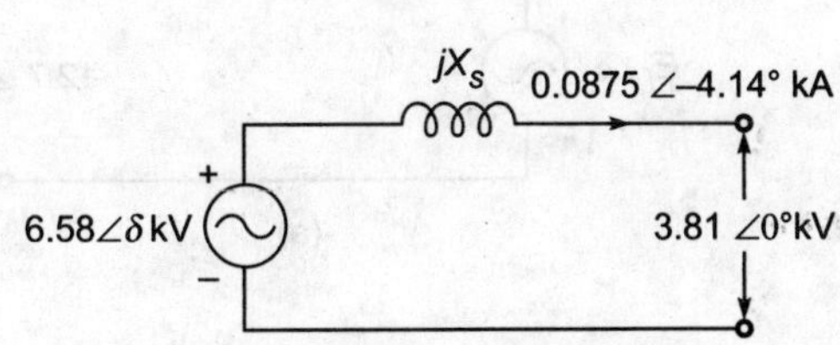

Fig. 8.111

Eliminating X_s between Eqs (ii) and (iii), we get

$$6.58 \cos \delta - 5.82 \sin \delta = 3.81 \quad \text{(iv)}$$

or $\cos (\delta + 41.4°) = 0.434 \Rightarrow \delta = 22.8°$

(a) From Eq. (iii) using the value of δ as obtained above

$$X_s = \frac{6.58 \sin 22.8°}{0.0656} = 38.9\ \Omega$$

(b) $$\text{Voltage regulation} = \frac{11.4 - 6.6}{6.6} \times 100 = 72.7\%$$

(c) Torque (power) angle $\delta = 22.8°$

(d) Electrical power developed $= 3\, E_f I_a \cos(\delta - \phi)$

$$= 3 \times 6.58 \times 0.0875 \cos(22.4° + 41.4°) = 763 \text{ kW}$$

Also directly

$$\text{electrical power developed} = 3\, V_a\, I_a \cos\phi = 3 \times 3.81 \times 0.0875 \times 0.75 = 750 \text{ kW}$$

This checks the result as there are no losses in armature resistance ($R_a \approx 0$)

(e) Machine reconnected in delta

$$\text{Voltage rating} = 6.6/\sqrt{3} = 3.81 \text{ kV}$$

$$\text{Current rating} = 0.0875\sqrt{3} \text{ kA}$$

$$\text{kVA rating} = \sqrt{3} \times 3.81 \times 0.0875\sqrt{3} \times 1000 = 1000 \text{ kVA (same as in star connection)}$$

Example 8.40 *A 22 kV, 3-phase, star-connected turbo-alternator with a synchronous impedance of j 1.2 Ω/phase is delivering 230 MW at UPF to 22 kV grid. With the turbine power remaining constant, the alternator excitation is increased by 30%. Determine machine current and power factor based upon linearity assumption.*

At the new excitation, the turbine power is now increased till the machine delivers 275 MW. Calculate the new current and power factor.

LO 7

Solution

230 MW at UPF

$$I_a = \frac{230}{\sqrt{3} \times 22 \times 1} = 6.04 \text{ kA}, \ \bar{I}_a = 6.04 \angle 0° \text{ kA}$$

$$\bar{V}_t = (22/\sqrt{3}) \angle 0° = 12.7 \angle 0° \text{ kV}$$

With reference to Fig. 8.112(a)

$$\bar{E}_f = 12.7 \angle 0° + j\, 1.2 \times 6.04 \angle 0° = 12.7 + j\, 7.25$$

or

$$E_f = 14.62 \text{ kV}$$

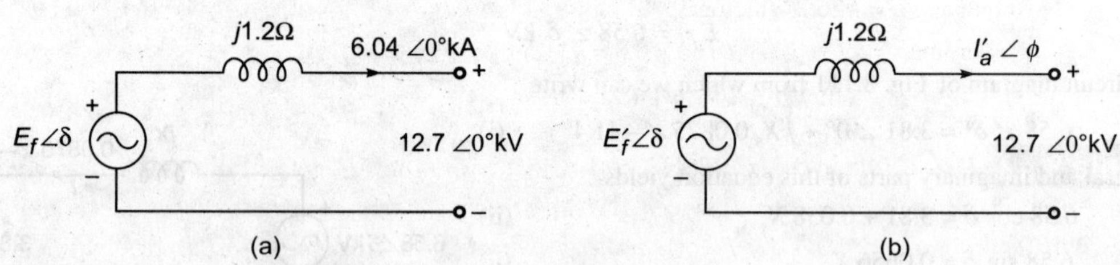

Fig. 8.112

Excitation increased by 30%; turbine power input constant at 230 MW, refer Fig. 8.112(b)

$$E'_f = 14.62 \times 1.3 = 19 \text{ kV}$$

$$\frac{230}{3} = \frac{19 \times 12.7}{1.2} \sin\delta'$$

or

$$\delta' = 22.4°$$

$$\bar{I}'_a = \frac{19\angle 22.4° - 12.7\angle 0°}{j1.2} = 7.275\angle -33.9° \text{ kA}$$

$$I_a = 7.275 \text{ kA}$$

$$p_f = \cos 33.9° = 0.83 \text{ lagging}$$

Excitation constant at new value; turbine power increased to 275 MW

$$\frac{275}{3} = \frac{19 \times 12.7}{1.2} \sin \delta''$$

or

$$\delta'' = 27.1°$$

$$\bar{I}''_a = \frac{19\angle 27.1 - 12.7\angle 0°}{j1.2} = 8.03\angle -25.9° \text{ kA}$$

$$I_a = 8.03 \text{ kA, pf} = \cos 25.9° = 0.9 \text{ lagging}$$

Example 8.41 *The full-load current of a 3.3 kVA, star-connected synchronous motor is 160 A at 0.8 pf lagging. The resistance and synchronous reactance of the motor are 0.8 Ω and 5.5 Ω per phase respectively. Calculate the excitation emf, torque angle, efficiency and shaft output of the motor. Assume the mechanical stray load loss to be 30 kW.*

LO 7

Solution Refer Fig. 8.113.

$$X_s = 0.8 + j\,5.5 = 5.56\ \angle 81.7° \ \Omega$$

$$\bar{E}_f = 1.905 - 5.56\ \angle 81.7° \times 160\ \angle -36.9° \times 10^{-3}$$

$$= 1.42\angle -26.2°$$

or

$$E_f = 1.42 \text{ kV (phase) or } 2.46 \text{ kV (line)}$$

$$P_{\text{mech}}\text{ (dev)} = 3 \times 1.42 \times 160 \cos(-36.9° + 26.2°)$$

$$= 670 \text{ kW}$$

$$\text{Shaft output} = 670 - 30 = 640 \text{ kW}$$

$$\text{Torque angle} = -26.2° \text{ (motor)}$$

$$\text{Power input} = \sqrt{3} \times 3.3 \times 160 \times 0.8 = 731.5 \text{ kW}$$

$$\eta = 640/731.5 = 87.5\%$$

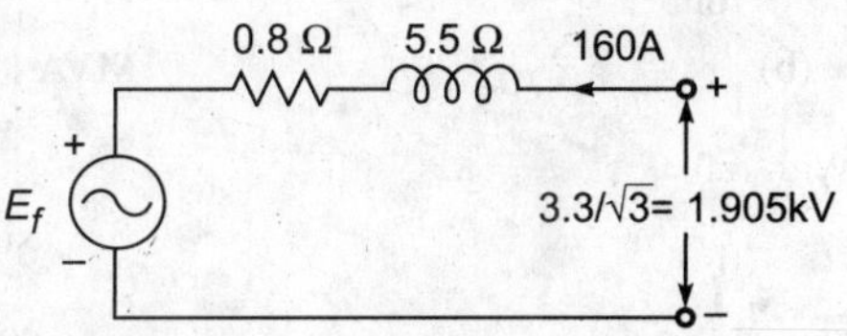

Fig. 8.113

Example 8.42 *A 4-pole, 50 Hz, 22 kV, 500 MV A synchronous generator having a synchronous reactance of 1.57 pu is feeding into a power system, which can be represented by a 22 kV infinite bus in series with a reactance of 0.4 Ω. The generator excitation is continually adjusted (by means of an automatic voltage regulator) so as to maintain a terminal voltage of 22 kV independent of the load on the generator.*

(a) Draw the phasor diagram, when the generator is feeding 250 MVA into the power system. Calculate the generator current, its power factor and real power fed by it. What is the excitation emf of the generator?

(b) Repeat part (a) when the generator load is 500 MVA.

LO 7

Solution The circuit diagram of the system is drawn in Fig. 8.114(a).

$$I_a\text{ (rated)} = \frac{500}{\sqrt{3 \times 22}} = 13.12 \text{ kA}$$

$$V_t\text{ (rated)} = 22/\sqrt{3} = 12.7 \text{ kV}$$

$$Z_b = 12.7/13.12 = 0.968\ \Omega$$

$$(MVA)_B = 500,\ (MW)_B = 500$$

$$X_{sg} = 1.57 \text{ pu (given)}$$

$$X_b = 0.4/0.968 = 0.413 \text{ pu}$$

(a)

$$\text{Load} = 250 \text{ MVA} \quad \text{or} \quad 0.5 \text{ pu}$$

$$V_t = 1 \text{ pu}, \qquad I_a = 0.5 \text{ pu}$$

$$V_b = 1 \text{pu}$$

$$X_b I_a = 0.413 \times 0.5 = 0.207 \text{ pu}$$

The phasor diagram is drawn in Fig. 8.114(b)

$$\sin\phi = \frac{(0.207/2)}{1} \quad \text{as } \overline{I}_a \text{ is at 90° to } \overline{I}_a X_b$$

or

$$\phi = 5.9°$$

$$pf = \cos\phi = 0.995 \text{ lagging}$$

$$P_e = 0.5 \times 0.995 = 0.4975 \text{ pu} \Rightarrow 124.4 \text{ MW}$$

$$\overline{E}_g = 1\angle 0° + j\,1.57 \times 0.5\ \angle -5.9° = 1.333\angle 35.9°$$

or

$$E_g = 1.333 \times 22 = 29.32 \text{ kV (line)}$$

(b)

$$\text{MVA load} = 1 \text{ pu},\ I_a = 1 \text{ pu}$$

$$X_b I_a = 0.413 \times 1 = 0.413 \text{ pu}$$

$$\sin\phi = \frac{(0.143/2)}{1} \Rightarrow \phi = 11.9°$$

$$\cos\phi = 0.98 \text{ lagging}$$

$$P_e = 1 \times 1 \times 0.98 = 0.98 \text{ pu} \quad \text{or} \quad 490 \text{ MW}$$

$$\overline{E}_g = 1\angle 0° + j\,1.57 \times 1\ \angle -11.9° = 1.324 + j\,1.536$$

or

$$E_g = 2.028 \text{ pu} \quad \text{or} \quad 44.62 \text{ kV (line)}$$

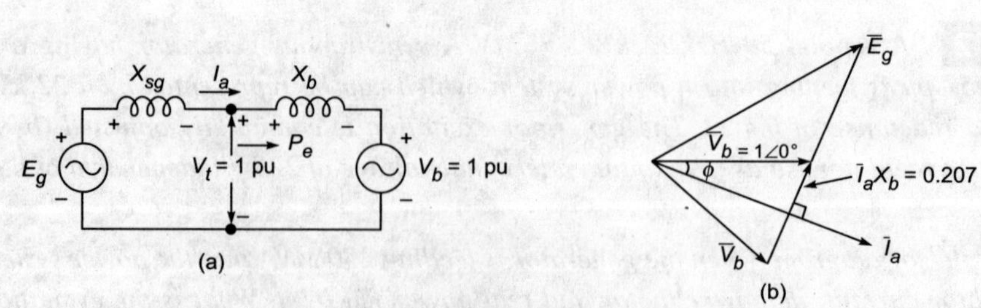

Fig. 8.114

Example 8.43 *A 25 MVA, 13 kV, 50 Hz synchronous machine has a short-circuit ratio of 0.52. For rated induced voltage on no-load, it requires a field current of 250 A.*

(a) Calculate the adjusted (saturated) synchronous reactance of the machine. What is its pu value?

(b) The machine is connected to 13-kV infinite mains and is running as a motor on no-load.

(i) Its field current is adjusted to 200 A. Calculate its power factor angle and torque angle. Ignore machine losses. Draw the phasor diagram indicating terminal voltage, excitation emf and armature current.

(ii) Does the machine act like a capacitor or an inductor to the 13 kV system? Calculate the equivalent capacitor/inductor value.

(iii) Repeat part (i) for a field current of 300 A.

LO 7

Solution

(a) Refer Fig. 8.15

$$SCR = of'/of''$$

$$of' = 250 \text{ A} \Rightarrow of'' = 250/0.52 = 481 \text{ A}$$

This is the field current needed to produce short-circuit current equal to rated value.

$$I_a \text{ (rated)} = 25/(\sqrt{3} \times 13) \times 10^3 = 1110 \text{ A}$$

$$I_{SC} \text{ (at } I_f = 250 \text{ A)} = 1110 \times 250/481 = 577 \text{ A}$$

$$X_s \text{ (adjusted)} = \frac{13 \times 1000/\sqrt{3}}{577} = 13\ \Omega$$

$$X \text{ (Base)} = \frac{(13 \times 1000)/\sqrt{3}}{1110} = 6.76\ \Omega$$

$$X_s \text{ (adjusted)} = 13/6.76 = 1.92 \text{ pu} = 1/\text{SCR (checks)}$$

(b) The circuit model of the machine is drawn in Fig. 8.115(a)

(i) $I_f = 200$ A

From the modified air-gap line

$$E_f = 13 \times (200/250) = 10.45 \text{ kV (line) or } 6.0 \text{ kV (phase)}$$

$$I_a = \frac{V_t - E_f}{X_s} = \frac{7.51 \times 6.0}{13} \times 1000 = 116 \text{ A}$$

$$\delta = 0;\ \text{no load; no losses}$$

$$pf = \cos 90^\circ = 0;\ \text{lagging (as } V_t > E_f)$$

The phasor diagram is drawn in Fig. 8.115(b)

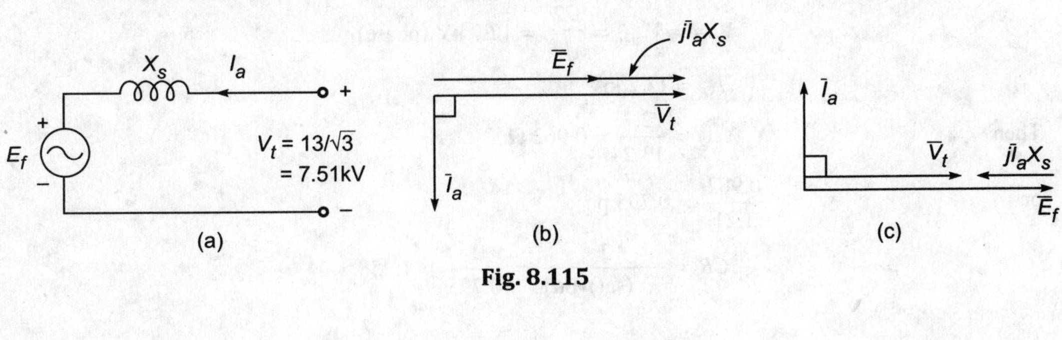

Fig. 8.115

The machine appears as an inductor

$$\omega L = \frac{13 \times 1000}{\sqrt{3} \times 116} = 2\,\pi \times 50\,L$$

or $$L = 0.206 \text{ H}$$

(ii) $$I_f = 300 \text{ A}$$

$$E_f = \frac{13 \times 300}{250} = 15.6 \text{ kV (line) or } 9.0 \text{ kV (phase)}$$

$$I_a = \frac{9-6}{13} = 231 \text{ A}$$

$$\delta = 0; \text{ no load, no losses}$$

$$pf = 90^\circ = 0; \text{ leading (as } V_t < E_f)$$

The machine appears like a capacitor.

$$\frac{1}{\omega C} = \frac{13 \times 1000}{231} = \frac{1}{2\pi \times 50 C}$$

or $$C = 56.6\ \mu\text{F}$$

Example 8.44 *A 400 MVA, 22 kV synchronous generator is tested for OCC and SCC. The following data are obtained from these characteristics extrapolated where needed.*

$$I_f = 1120\,A, \quad V_{OC} = 22\,kV, \quad I_{SC} = 13.2\,kA$$

At I_f corresponding to $I_{SC} = I_{rated}$ it is found that V_{OC} (air-gap line) = 24.2 kV (line).

(a) Determine X_s (saturated) in ohms and in pu.

(b) Determine SCR.

(c) Determine X_s (unsaturated) and I_f corresponding to V_{OC} = 22.4 kV on air gap line.

(d) The generator when operating on no-load rated terminal voltage, find the value of the generator current if it is short circuited.

LO 7

Solution

Base $$(\text{MVA})_B = 400, (\text{kV})_B = 22 \text{ (line)} \quad \text{or} \quad 12.7 \text{ (phase)}$$

$$I_B = \frac{400}{22\sqrt{3}} = 10.49 \text{ kA}$$

$$(\text{Ohm})_B = \frac{22\sqrt{3}}{10.49} = 1.21\ \Omega$$

(a) $$I_f = 1120 \text{ A}$$

$$V_{OC} = V_{\text{rated}} = \frac{22}{\sqrt{3}} = 12.7 \text{ kV (phase)}$$

$$I_{SC} = 13.2 \text{ kA}$$

Then $$X_s\text{ (sat)} = \frac{12.7}{13.2} = 0.962\ \Omega$$

or $$\frac{0.962}{1.21} = 0.795 \text{ pu}$$

(b) $$SCR = \frac{1}{X_s(\text{sat})(\text{pu})} = \frac{1}{0.795} = 1.258$$

(c)

$$I_{SC} = I_{\text{rated}} = 10.49 \text{ kA}$$

$$V_{OC} \text{ (air gap line)} = 24.2 \text{ kV (line)}$$

$$= 13.97 \text{ kV (phase)}$$

$$X_s \text{ (unsat)} = \frac{13.97}{10.49} = 1.33\ \Omega = \frac{1.33}{1.21} = 1.1 \text{ pu}$$

For finding I_f

$$\text{SCR} = \frac{of'}{of''} \qquad \text{(Fig. 8.15)}$$

$$1.258 = \frac{of'}{1120}$$

or

$$I_f = of' = 1120 \times 1.258 = 1409 \text{ A}$$

Example 8.45 *A 6.6 kV, Y-connected, 3-phase, synchronous motor operates at constant voltage and excitation. Its synchronous impedance is 2 + j 20 Ω/phase. The motor operates at 0.8 leading power factor while drawing 800 kW from the mains. Find the motor power factor when it is loaded to draw increased power of 1200 kW.*

LO 7

Solution

$$\bar{V}_t = 6.6/\sqrt{3}\ \angle 0° = 3.81\ \angle 0° \text{ kV}$$

$$\bar{I}_a = I_a \angle 36.9°$$

$$I_a = \frac{800}{\sqrt{3} \times 6.6 \times 0.8} = 87.5 \text{ A}$$

$$\bar{Z}_s = 2 + j\,20 = 20.1\ \angle 84.3°\ \Omega, \ \cos 84.3° = 0.1$$

$$\bar{E}_f = 3.81 - 20.1\ \angle 84.3° \times 0.0875\ \angle 36.9° = 4.724 - j\,1.504$$

or

$$E_f = 4.96 \text{ kV (phase)}$$

Power input increases to 1200 kW; no change in excitation

$$\bar{I}_a = \frac{V_t \angle 0° - E_f \angle -\delta}{Z_s \angle \theta} = \frac{V_t}{Z_s} \angle -\theta - \frac{E_f}{Z_s} \angle -(\delta + \theta)$$

$$P_e \text{ (in)} = 3\ Re\ [V_t \angle 0°\ \bar{I}_a^*]$$

$$= 3 \left[\frac{V_t^2}{Z_s} \cos\theta - \frac{V_t E_f}{Z_s} \cos(\delta + \theta) \right]$$

Substituting values

$$\frac{1200}{1000 \times 3} = \frac{(3.81)^2 \times 0.1}{20.1} - \frac{3.81 \times 4.96}{20.1} \cos(\delta + 84.3°)$$

Solving we get

$$\delta = 26.1°$$

$$\bar{I}_a = \frac{3.81 - 4.96 \angle -26.1°}{20.1 \angle 84.3} = 113 \angle 22.1°$$

$$pf = \cos 22.1° = 0.9265 \text{ leading}$$

Example 8.46 *A 440 V, 50 Hz, Δ-connected synchronous generator has a direct-axis reactance of 0.12 Ω and a quadrature-axis reactance of 0.075 Ω/phase; the armature resistance being negligible. The generator is supplying 1000 A at 0.8 lagging pf.*

(a) Find the excitation emf neglecting saliency and assuming $X_s = X_d$.
(b) Find the excitation emf accounting for saliency.

Compare and comment on the results of parts (a) and (b). LO 7

Solution On equivalent star basis

$$X_d = 0.12/3 = 0.04\ \Omega, \quad X_q = 0.075/3 = 0.025\ \Omega$$

(a) Saliency ignores;
$$X_s = X_d$$
$$\bar{E}_f = (440/\sqrt{3})\,\angle 0° + j\,0.04 \times 1000\,\angle -36.9°$$
$$= 279.8\,\angle 6.6°$$

or
$$E_f = 279.8 \text{ V} \quad \text{or} \quad 484.6 \text{ V (line)}$$

(b)
$$I_a = 1000 \text{ A}, \ \phi = +36.90° \text{ (lagging)}$$
$$\tan\psi = \frac{V_t \sin\phi + I_a X_q}{V_t \cos\phi + I_a R_a}$$
$$= \frac{254 \times 0.6 + 1000 \times 0.025}{254 \times 0.8}; V_t = 440/\sqrt{3} = 254 \text{ V}$$
$$= 0.873$$

or
$$\psi = 41.1°$$
$$\delta = \psi - \phi = 41.1° - 36.9° = 4.2°$$
$$E_f = V_t \cos\delta + I_d X_d; \quad I_d = I_a \sin\psi = 1000 \sin 4.2°$$
$$= 254 \cos 4.2° + 73.2 \times 0.04$$
$$= 256.3 \text{ V} \quad \text{or} \quad 444 \text{ V (line)}$$

Example 8.47 *A 1500 kVA, star-connected, 6.6-kV salient-pole synchronous motor has $X_d = 23.2\ \Omega$ and $X_q = 14.5\ \Omega$/phase respectively; armature resistance being negligible.*
Calculate the excitation emf when the motor is supplying rated load at 0.8 leading pf.
What maximum load the motor can supply without loss of synchronism, if the excitation is cut off? What will be the value of torque angle under this condition. LO 7

Solution

$$\phi = -\cos^{-1} 0.8 = -36.9° \text{ (leading)}$$
$$I_a = \frac{1500}{\sqrt{3} \times 6.6} = 131 \text{ A}; \quad V_t = \frac{6.6}{\sqrt{3}} = 3.81 \text{ kV}$$
$$\tan\psi = \frac{V_t \sin\phi - I_a X_q}{V_t \cos\phi + I_a R_a}$$
$$= \frac{-3.81 \times 0.6 - 0.131 \times 14.5}{3.81 \times 0.8 + 0} = -1.373$$

or
$$\psi = -53.9°$$
$$\delta = \phi - \psi = -36.9° - (-53.9°) = 17°$$

$$E_f = V_t \cos \delta - I_d X_d; \qquad I_d = I_a \sin \psi = 0.131 \times \sin - 53.9° = - 0.106$$

$$= 3.81 \cos 17° + 0.106 \times 23.2$$

$$= 6.1 \text{ kV} \quad \text{or} \quad 10.57 \text{ kV (line)}$$

With excitation cut off, output is only reluctance power

$$P_e = V_t^2 \left(\frac{X_d - X_q}{2 X_d X_q} \right) \sin 2\delta$$

$$= (6.6)^2 \times \frac{23.2 - 14.5}{2 \times 23.2 \times 14.5} = 5\ 63 \text{ kW}$$

Example 8.48 *Figure 8.115 shows speed (frequency)-load characteristics of two generators supplying in parallel a load of 2.8 MW at 0.8 pf lagging:*

(a) At what frequency is the system operating and what is the load supplied by each generator?

(b) If the load is now increased by 1 MW, what will be the frequency and the load sharing?

(c) In part (b) which should be the set point of G_2, for the system frequency to be 50 Hz? What would be the load sharing now?

LO 7

Solution The speed-load characteristics of the two generators are drawn in Fig. 8.116 on the sides of the load axis. These being linear can the represented by the equation of a line as

$$y = mx + c$$

where

x = generator load, it is taken positive on both sides

y = system frequency, same for both generators

c = set frequency (on frequency axis)

m = slope, Hz/MW

On the figure

AB = total load, $(x_1 + x_2)$

$AC = x_1$, load supplied by G1

$BC = x_2$, load supplied by G2

OC = system frequency

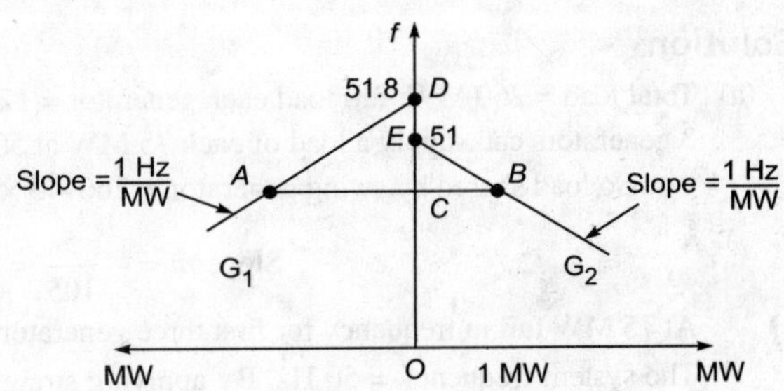

Fig. 8.116

(a) Substituting values.

G_1 $\quad y = - x_1 + 51.8$ (i) $\quad m = - 1$ Hz/MW

G_2 $\quad y = - x_2 + 51$ (ii) $\quad m = - 1$ Hz/MW

It follows

$$- x_1 + 51.8 = - x_2 + 51$$

or $\quad (x_1 - x_2) = 0.8$ (iii)

Solving $\quad x_1 + x_2 = 2.8$ MW (total load, AB)

Load supplied by G_1: $\quad x_1 = 1.8$ MW

Load supplied by G_2: $\quad x_2 = 1$ MW

System frequency. $\quad f = y = - 1.8 + 51.8 = 50$ Hz

(b) Load increased by 1 MW

$$x_1 + x_2 = 2.8 + 1 = 3.8 \text{ MW} \qquad \text{(iv)}$$

Solving Eqs (iii) and (iv)

Load G_1: $x_1 = 2.3$ MW; increase $2.3 - 1.8 = 0.5$ MW

Load G_2: $x_2 = 1.5$ MW; increase $1.5 - 1 = 0.5$ MW

Note: Additional load equally divided as the two characteristics have the same slope

system frequency $f = y = -2.3 + 51.8 = 49.5$ Hz

(c) The G_1 can supply the load of 1.8 MW at the frequency of 50 Hz. The remaining power must be shared by G_2. So the new load of G_2 is 2 MW. For the new load, the set point of G_2 must be

$$50 = -1 \times 2 + \text{set frequency}$$

$$\text{Set frequency} = 52 \text{ Hz.}$$

Example 8.49 *A generating station comprises four 125 MVA, 22 kV, 0.84 pf lagging synchronous generators with a frequency drop of 5 Hz from no-load to full. At a frequency of 50 Hz, three generators supply a steady load of 75 MW each while the balance is shared by the fourth generator (called swing generator)*

(a) For a total load of 260 MW at 50 Hz, find the no-load frequency setting of the generators.

(b) With no change in governor setting as in part (a), find the system frequency if the system load rises to 310 MW.

(c) Find the no-load frequency of the swing generator for the system frequency to be restored to 50 Hz for the load in part (b).

(d) If the swing generator trips off what will be system frequency

LO 7

Solution

(a) Total load = 260 MW; full load each generator = $125 \times 0.84 = 105$ MW

3 generators can supply a load of each 75 MW at 50 Hz.

So load shared by swing generator = $260 - 3 \times 75 = 260 - 225 = 35$ MW

$$\text{Slope } m = -\frac{5}{105} = .0476 \text{ Hz/MW} = m$$

At 75 MW fall in frequency for first three generators = $75 \times .0476 = 3.57$ Hz

The system frequency = 50 Hz. By applying straight line equation $y = mx+c$; y = system frequency, m = slope, x = load share and c = set frequency.

So set frequency of G_1, G_2 and G_3 $c = y - mx = 50 + 3.57 = 53.57$ Hz = 53.6 Hz

For 35 MW load supplied from swing generator, the set frequency = $50 - 35 \times (-0.0476) = 51.7$ Hz

(b) Since all the four generator are having same slope, the load will be shared equally.

$$\therefore \quad \text{New load sharing of each } (G_1, G_2 \text{ and } G_3) = 75 + \frac{50}{4} = 87.5 \text{ MW}$$

$$\text{So new system frequency} = 53.6 + (-0.0476) \times 87.5 = 49.43 \text{ Hz}$$

(c) If the system frequency is 50 Hz, then the three generators can supply only 75 MW each. So the remaining power is shared by swing generator

$$\text{New load of swing generator} = 310 - 3 \times 75 = 310 - 225 = 85 \text{ MW}$$

$$\text{So set frequency of swing generator} = 50 - (-0.0476) \times 85 = 54.04 \text{ Hz}$$

(d) The new system frequency after the swing generator trips off.

$$\text{New load sharing} = \frac{310}{3} = 103.33 \text{ MW}$$

$$\text{New system frequency} = 53.6 + (-0.0476) \times 103.33 = 48.68 \text{ Hz}$$

8.22 | BRUSHLESS DC MOTORS

LO 7

The term brushless dc motors is applied to many configurations of ac synchronous motors in which semiconductor control is used to control stator currents such that maximum torque is obtained at a given speed. In a conventional motor the mechanical contactor, the commutator, maintains 90° elect degrees space displacement between the rotor and stator magnetic fields to provide for the required torque.

Theoretically, the stator and rotor functions of a machine can be inverted, putting the field system on the rotor. There is no advantage to be gained if conventional commutation is used, as the commutator sections are fixed and the brush gear must rotate at the speed of the rotor field. Solid-state switching by transistors or thyristors, triggered by position sensors can, however, replace the brush gear by fully electronic commutation, endowing small machines with a valuable control facility. In this method, each phase of stator winding is energized sequentially by a power transistor (or thyristor) by means of a signal from position sensor placed on the rotor. Because of rotor position feedback triggering of thyristors/transistors, the stator and rotor field always remain in synchronism as the frequency of triggering automatically adjusts to motor speed. The length of on-time of the transistors determines motor torque magnitude. Thus by means of electronic circuitry brushless motors can be controlled for both constant and variable torque operation.

Advantages of Brushless dc Motor

The brushless dc motors, while being generally more expensive for the same kW rating, than commutator and brush motors possess certain advantages over conventional motors.

Important! These are:

(a) They require little or no maintenance.

(b) They have a much longer operating life.

(c) There is no risk of explosion or possibility of RF radiation due to arcing.

(d) They produce no brush or commutator particles or gases as by -products of operation.

(e) They are capable of operation submersed in fluids, combustible gases and may even be hermetically sealed.

(f) They are generally more efficient than brush-type dc servomotors or conventional dc motors.

(g) They provide a more rapid response and a fairly linear output torque vs input current characteristic, which lends itself to servo applications.

Schematic Representation and Operating Principle

The schematic diagram of a brushless dc motor is shown in Fig. 8.117. It also shows the three phases of the stator (armature) and rotor with *d*- and *q*-axes indicated therein. The stator is connected to a variable voltage current source through an inductor and an inverter[¶¶¶¶¶¶] comprising six SCRs (S_1 to S_6). In place of SCRs, power transistor or FETs could be used according to power rating of the motor. Diodes are connected across SCRs to protect these from the *L(di/dt)* voltage induced in the armature coil undergoing commutation. Position sensors placed on the rotor provide signal to the sensor decoders and gate drivers which cause the SCRs to be fired in sequence so as to be in synchronism with the rotor's mechanical position. The stator and rotor fields thus get locked into each other and remain in synchronism at any rotor speed.

¶¶¶¶¶¶ Inverter operation will be explained in details in Chapter 12. For the time being an SCR can be considered as a controlled switch. It goes 'ON' upon receiving 'triggering' pulse and as another SCR is triggered, the circuit facilitates the current to commute to the newly fired SCR. For conducting sequence of SCR pairs see Fig. 8.117.

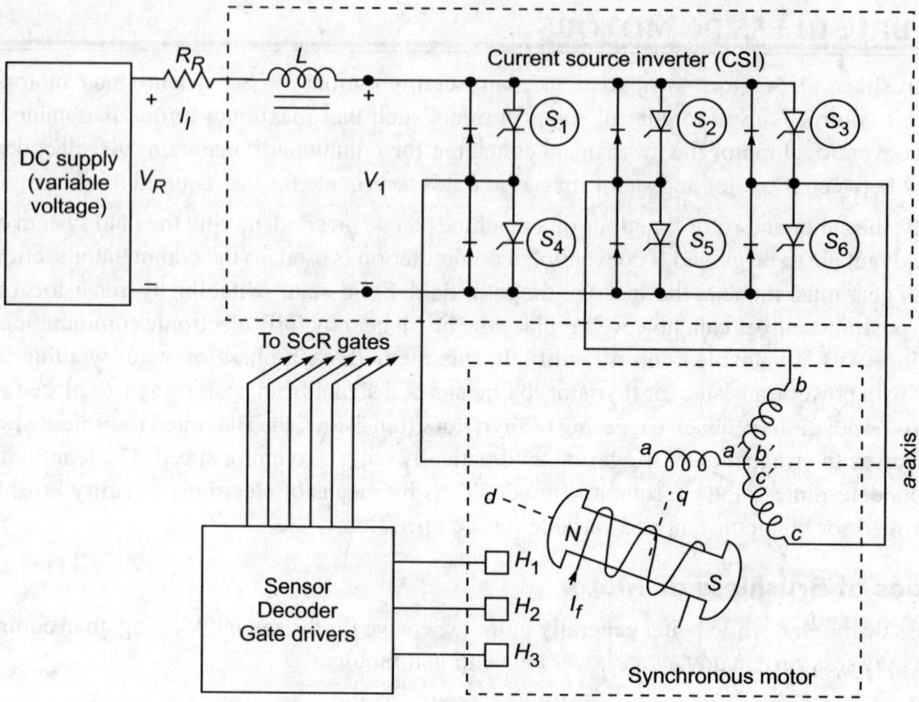

Fig. 8.117 *Brushless dc motor; schematic and operation*

Ideal phase currents are pulses of ± *I* lasting 120° elect each half and displaced from each other 120° elect phase to phase as shown in Fig. 8.118. Actual current wave forms differ from the ideal rectangular current waves by gradual rises and falls. Such an inverter where ac current flows in form of constant current pulses is known as Current Source Inverter (CSI).

Sequence of inverter firing as shown in Fig. 8.118 immediately follows from the phase current wave forms. For this sequence of SCR firing 120° or 60° elect spaced sensor codes are generated by means of light sensitive or Hall effect sensors. Figure 8.118 illustrates the case of 120° spacing wherein these light sensitive sensors are shown fixed 120° apart receiving light from a fixed light source. The rotor carries a commutating disc with 180° cut-out so that as it rotates with the rotor the light sensors receive light for 180° and are dark for 180°. Sensors produce logic '1' while receiving light and logic '0' when dark. It is easily seen that the three sensors (fixed) and commutating disc (rotating with rotor) produce sensor code sequence as given in Fig. 8.118 from which electronic circuitry generates gating pulses for firing SCRs in the sequence as indicated in the figure.

The relative position of the commutating disc can be adjusted w.r.t. the rotor poles (i.e. w.r.t. *d*- and *q*-axis). For the instantaneous rotor position (with discs fixed as indicated) it is seen that the sensor code is just going to change from 101 to 100. The phase *'a'* is in the middle of its current pulse when the current is commuting from phase *b* to c. At this instant the resultant stator field $\vec{F}_1$ is oriented along the *a*-axis as shown in Fig. 8.119 (motoring current's positive direction is opposite to the positive direction of induced emf); check in phase *'a'* by applying Fleming's right hand rule. North pole on stator is thus oriented along the *a*-axis. This north pole pushes the rotor north to create motoring torque (angle between rotor and stator N-poles is β).

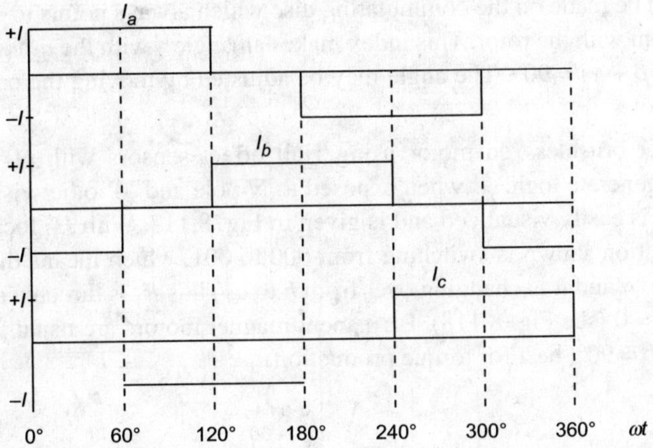

Sensor codes 120° elect spacing	1	1	1	0	0	0	H_1
	0	0	1	1	1	0	H_2
	1	0	0	0	1	1	H_3
60° elect spacing	0	0	0	1	1	1	H_1
	0	0	1	1	1	0	H_2
	0	1	1	1	0	0	H_3
SCRs conducting	1,5	1,6	2,6	2,4	3,4	3,5	

Fig. 8.118 *Current wave forms, SCRs conducting sequence and sensor codes*

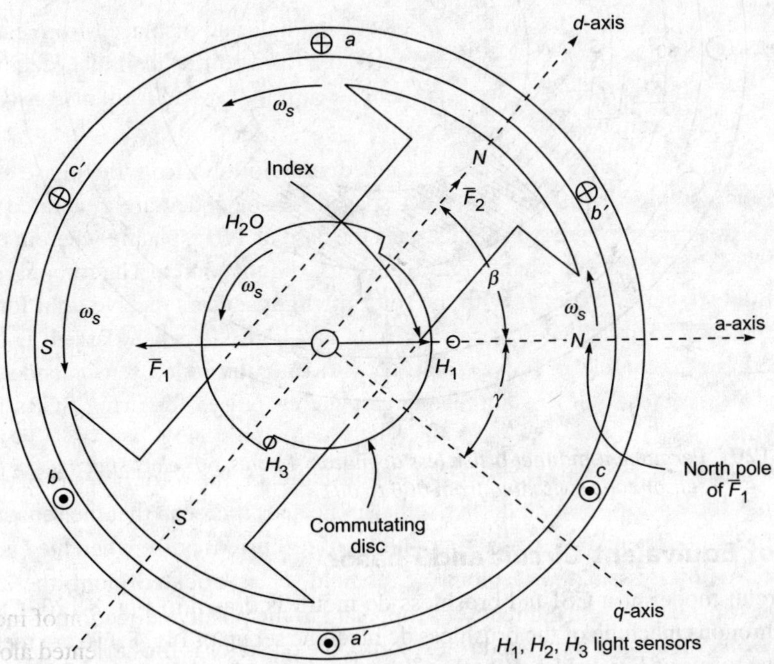

Fig. 8.119 *Brushless dc motor arrangement of sensors; 120° elect sensor code switching from 101 to 100*

An index marker can be made on the commutating disc which always points to stator north as the stator field rotates in synchronism with the rotor. This index makes an angle γ with the q-axis of the rotor (lagging d-axis by 90°). Obviously $(\beta + \gamma) = 90°$. The angle may be adjusted by moving the commutating disc on the shaft related to the rotor poles.

A permanent magnet brushless dc motor using Hall effect sensors with 60° elect spacing is shown in Fig. 8.120. Sensors generate logic' 1' when exposed to N-pole and '0' otherwise. The sensor code sequence for this arrangement is easily visualized and is given in Fig. 8.117. With H_3 located along a-axis, the sensor code at the rotor position shown is switching from 000 to 001, which means the current is in the middle of conduction for phase 'a'and it is changing over from b to c. Thus F_1 is directed along a-axis or stator N-pole is along q-axis i.e. $\gamma = 0$ (see Fig. 8.118). Permanent magnet motors are usually adjusted for this value of γ (this corresponds to $\beta = 90°$, best for torque production).

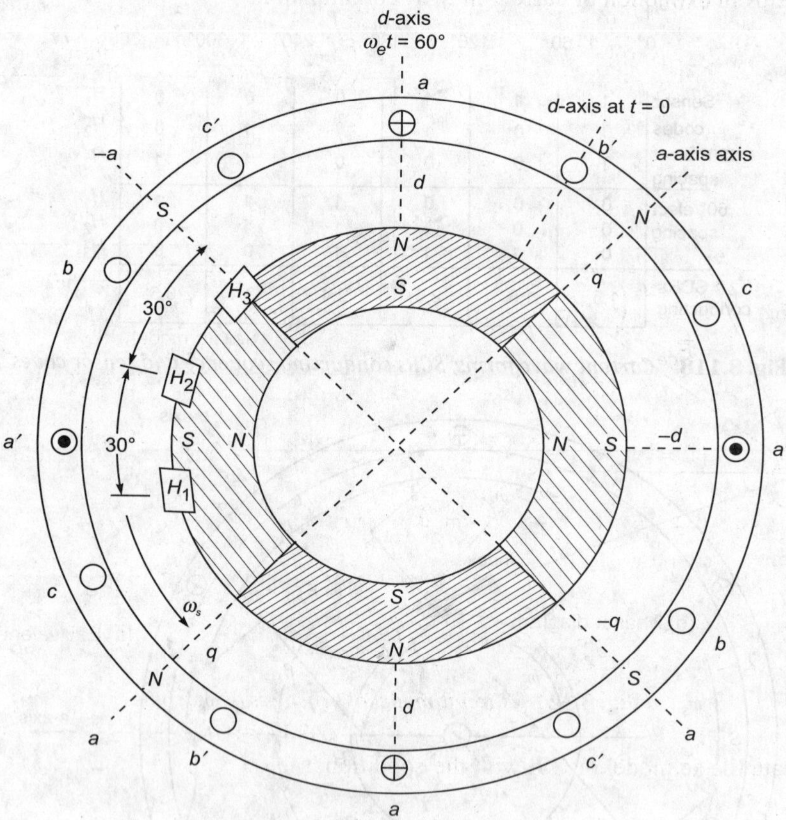

Fig. 8.120 *Permanent magnet brush less dc motor; 4 poles, 60° elect spacing, $\gamma = 0$, sensor code switching from 000 to 001*

Development of Equivalent Circuit and Phasor

Novotny-Abbas circuit model of a CSI fed brushless dc motor is drawn in Fig. 8.120. Currents (balanced) flowing in the synchronous machine of the brushless dc machine set up of Fig. 8.116 are rectangular pulses ac as shown in the wave forms of Fig. 8.118. Actual currents are somewhat rounded pulses. Our analysis will be based on the fundamental ac current and harmonic currents will be ignored. These produce space harmonic air-gap fields which being nonstationary w.r.t. the rotor field produce net zero torque. Let

$$I_I = \text{current fed to the inverter by dc source}$$

Then $\quad I_m \text{ (rms phase current)} = \text{fundamental current}$

$$= (\sqrt{6/\pi})I_I \text{; can be shown by Fourier series}$$

In Fig. 8.121(a) the synchronous machine model is the usual one characterized by direct-axis synchronous reactance X_d and excitation emf E_f where X_d and E_f both vary directly with speed which governs the frequency of operation. E_f magnitude is of course related to the rotor field current by the magnetization characteristic. The corresponding phasor diagram is drawn in Fig. 8.121(b) where I_m, the phase current, is drawn leading V (ac output voltage/terminal voltage of synchronous machine) by angle ϕ. Leading current operation is performed as it helps in extinction of current in SCR commutation.

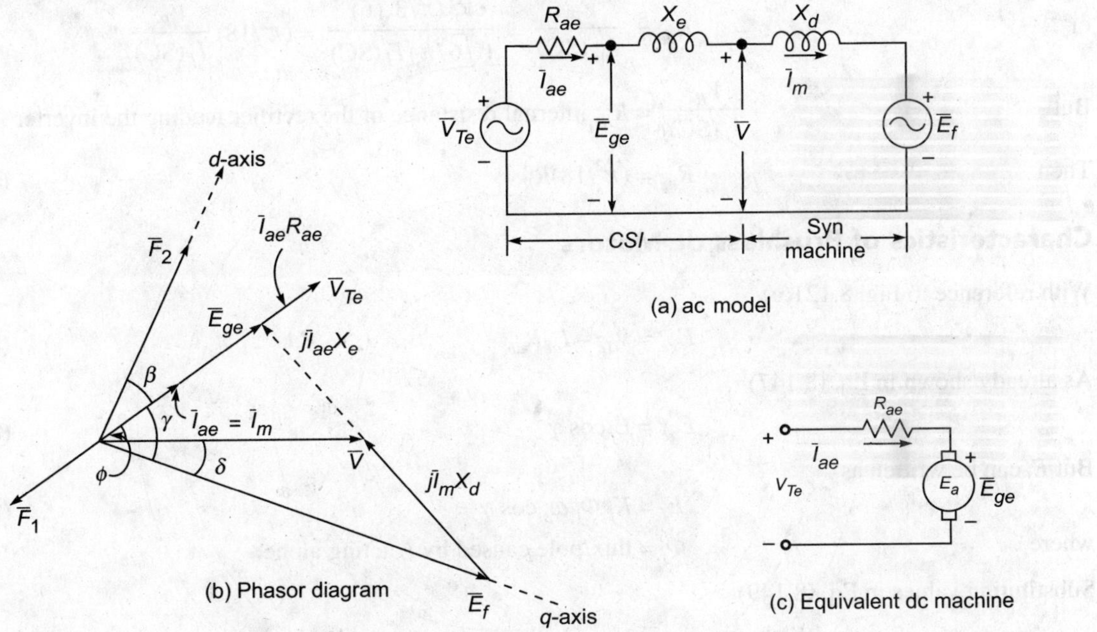

Fig. 8.121 *Circuit model of brushless dc machine*

We shall now create the ac model of CSI with the conditions:

(i) $\bar{I}_{ae} = \bar{I}_m$

(ii) Circuit parameter of model is resistance R_{ae}.

(iii) $\bar{E}_{ge}$ and $\bar{I}_{ae}$ are in phase.

These conditions will assure that this part of the model indeed represents the **equivalent dc machine.**

It is also seen from Fig. 8.119 that the phase angle between $\bar{I}_m = \bar{I}_{ae}$ and E_f is indeed angle γ. Now equating the converted power per phase of synchronous machine to that of equivalent dc machine we have

$$I_m E_f \cos \gamma = I_{ae} E_{ge}; \qquad I_m = I_{ae}$$

or
$$E_{ag} = E_f \cos \gamma \tag{8.147}$$

Translating this result to the phasor diagram (Fig. 8.120(b)), it becomes clear that V and E_{ge} are related by a fictitious reactance X_e as shown in the circuit of Fig. 8.120(a).

We shall now obtain the relationship for V_{Te} and R_{ae}. Neglecting inverter losses

$$V_R I_I = 3V_{Te} I_m = 3V_{Te} (\sqrt{6}/\pi) I_I$$

or
$$V_{Te} = V_R (\pi/3\sqrt{6})$$

Imagining a short circuit at E_{ge} i.e. $E_{ge} = 0$, we have

$$I_{ae}\,(\text{SC}) = \frac{V_{Te}}{R_{ae}}$$

or
$$R_{ae} = \frac{V_{Te}}{I_{ae}(\text{SC})} = \frac{V_R(\pi/3\sqrt{6})}{(\sqrt{6}/\pi)\, I_I(\text{SC})} = (\pi^2/18)\frac{V_R}{I_I(\text{SC})}$$

But
$$\frac{V_R}{I_I(\text{SC})} = R_R;\ \text{internal resistance of the rectifier feeding the inverter}$$

Then
$$R_{ae} = (\pi^2/18) R_R \tag{8.148}$$

Characteristics of Brushless dc Motors

With reference to Fig. 8.121(c)

$$E_{ge} = V_{Te} - I_{ae} R_{ae} \tag{8.149}$$

As already shown in Eq. (8.147)

$$E_{ge} = E_f \cos \gamma \tag{8.150}$$

But E_f can be written as

$$E_f = K_f \Phi_f \omega_s \cos \gamma \tag{8.151}$$

where Φ_f = flux/pole caused by I_f acting alone.

Substituting values in Eq. (8.149)

$$K_f \Phi_f \omega_s \cos \gamma = V_{Te} - I_{ae} R_{ae}$$

or
$$\omega_s = \frac{V_{Te} - I_{ae} R_{ae}}{K_f \Phi_f \cos \gamma} \tag{8.152}$$

Development of Torque Equitation Exerted by the Motor

Except for the effect of cos γ, this equation is the same as in conventional dc machine.

The torque developed is given by

$$T = \frac{P_{\text{out}}}{\omega_s} = \frac{3V I_{ae} \cos \phi}{\omega_s};$$

$$= \frac{3 I_{ae} E_{ge}}{\omega_s}; \qquad V\cos \phi = E_{ge} \tag{8.153}$$

Using Eq. (8.151), we get

$$T = 3K_f \Phi_f I_{ae} \cos \gamma \tag{8.154}$$

If the magnetisation characteristic is assumed liner

$$E_f = K'_f I_f \omega_s \cos \gamma \tag{8.155}$$

The speed and torque equation then are

$$\omega_s = \frac{V_{Te} - I_{ae} R_{ae}}{K'_f I_f \cos \gamma} \tag{8.156}$$

$$T = 3K'_f I_f I_{ae} \cos \gamma \tag{8.157}$$

Important! In a synchronous motor as the field current is reduced its pf becomes more lagging. But in a bruhsless dc motor (which is a synchronous motor with rotor position feedback) the decrease of field current I_f causes increase in speed as per Eq. (8.156) like in a conventional dc motor. This can be qualitatively explained by the reasoning that follows. With reference to Fig. 8.121 as I_f is decreased, E_f *and* so E_{ge} reduce and as R_{ae} is very small, this causes a disproportionate increase in $I_{ae} = I_m$. The result is rotor acceleration. Increased rotor speed counters E_f reduction and I_m increases as the voltage drop $I_m X_d$ increases with increase in frequency of operation. The result is a steady operation at a new and higher speed at a less leading or even lagging pf.

Operating Principle of PM Brushless dc Machine

Small size brushless dc machines are usually PM kinds. In such a motor rotor mmf F_2 remains fixed and also the angle γ in these machines is set to zero which mean that F_2 and F_1 (armature mmf) are displaced by an angle of 90° (best for torque developed). Further, the phase winding resistance R_1 is not negligible so must be added to R_{ae} in the dc model. The phasor diagram for $\gamma = 0$ is drawn in Fig. 8.122 wherein the following observations are made:

1. X_c = capacitive reactance = X_d (in magnitude).
2. *PF* angle is lagging.

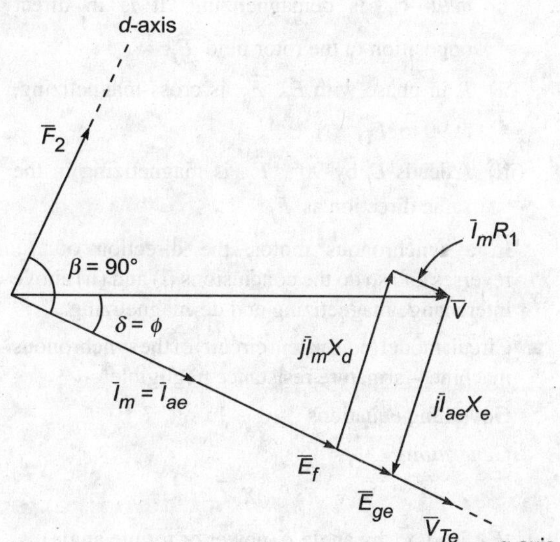

Fig. 8.122 *Phasor diagram of PM brushless motor*

The relationship of Eqs (8.152) and (8.154) for speed and torque apply except that Eq. (8.124) now modifies as

$$E_{ge} = V_{Te} - I_{ae} (R_{ae} + R_1) \tag{8.158}$$

SUMMARY

LO 1

- Synchronous machine – 3-phase balanced stator currents produce a field rotating at synchronous speed (corresponding to frequency of stator currents and number of stator poles). Rotor with dc excited poles, same number as stator poles, when rotating in same direction as stator field and same speed lock into stator field (synchronizes) producing a torque proportional to sine of the electrical angle between resultant air-gap flux (poles) and rotor poles.
- Synchronous speed, $n_s = \dfrac{120f}{P}$ rpm, $\omega_s = \dfrac{4\pi f}{P}$ rad (mech)/s; in general $n_s = \dfrac{60}{2\pi}\omega_s$
- Excitation emf (E_f) is the emf induced in the armature winding due to field current (I_f) only. At no-load, $E_f = V_t$ (terminal voltage)
- Synchronous generator ($\bar{I}_a$ in the direction of $\bar{E}_f$)
  (i) When I_a lags E_f by 90°, the *armature reaction field* $\vec{F}_a$ is demagnetizing. It is in direct opposition to the rotor field $\vec{F}_f$.
  (ii) I_a in phase with E_f, $\vec{F}_a$ is cross-magnetizing, at 90 to $\vec{F}_f$.
  (iii) I_a leads E_f by 90°, $\vec{F}_a$ is magnetizing in the same direction as $\vec{F}_f$.

  In a synchronous motor, the direction of $\bar{I}_a$ reverses and so do the conclusions (i) and (ii) above interchange magnetizing and de-magnetizing.
- Circuit model (equivalent circuit) of the synchronous machine – armature resistance negligible

  Governing equations

  Generating

  $$\bar{V}_t = \bar{E}_f - jX_s\bar{I}_a$$

  $\bar{E}_f$ lead $\bar{V}_t$ by angle δ (power or torque angle)

  Motoring

  $$\bar{V}_t = \bar{E}_f + jX_s\bar{I}_a$$

  $\bar{E}_f$ lead $\bar{V}_t$ by angle δ

LO 2

- The armature reaction effects can be simulated by an armature reactance X_a in series with I_a. Further, X_a when combined with armature leakage reactance is called the *synchronous reactance* ($X_s = X_a + X_l$) of the machine. Its value is of the order of 0.5 to 1 pu; far larger than in a transformer.
- To find the characteristics, the synchronous machine is run as generator at synchronous speed n_s by a prime mover.
- OCC – Open Circuit Characteristic – plot of V_{OC} (line) vs I_f. It is the magnetizatic characteristic of the machine. A line from the origin tangential to the linear part of OCC is called the *air-gap line*.
- SCC – Short Circuit Characteristic – plot of short circuit current I_{SC} vs I_f; it is linear as I_f is very small.

  By measuring the mechanical input to generator, we find machine losses – I^2R loss, local core loss and windage and friction loss. Windage and friction loss is found for OCC with field unexcited. The remaining loss is short circuit load loss.

  $$R_a\,(\text{eff}) = \frac{\text{short circuit load loss/phase}}{(\text{short circuit armature current})^2}$$
- Determination of synchronous reactance from OCC and SCC

  $$X_s\,(\text{unsaturated}) = \frac{V_{OC}/\sqrt{3}}{I_{SC}} \text{ at any point on air-gap line}$$

  $$X_s\,(\text{adjusted}) = \frac{V_{OC}/\sqrt{3}}{I_{SC}},\ I_f \text{ corresponding to } V_t \text{ (rated) on OCC}$$
- Voltage regulation

  $$= \left.\frac{V_t(\text{no load})\big|_{I_f \text{ same as at full load}} - V_t(\text{rated})}{V_t(\text{rated})}\right|_{\text{specified pf}}$$
- ZPFC (Zero pf Characteristic) The synchronous generator is loaded with purely inductive load

(zero *pf* lagging). The load is adjusted such that rated current is drawn from generator, while the field current is adjusted to give various terminal voltages. The plot of V_t vs I_f is the ZPFC.

- By the Potier Triangle method, the armature reaction ampere-turns (AT_a) and leakage reactance of the machine can be determined from OCC and ZPFC. The voltage regulation can then be found more accurately than by the synchronous reactance method.
- The voltage regulation can be found accurately by the MMF method and ASA method.

LO 3

- Fixed load variable excitation

  I_f controls the power factor

  Normal excitation – unity power factor

  Over-excitation – leading *pf* in motoring mode
  lagging *pf* in generating mode

  Under-excitation – lagging *pf* in motoring mode
  Leading *pf* in generating mode

- Hunting – as electrical power exchange of a synchronous machine with bus bar is governed by sin δ, the machine electromagnetic torque $T \propto$ sin δ or $T \propto \Delta\delta$ for small changes. This is a spring-like action which combined with rotor inertia causes the rotor to oscillate whenever there is electrical or mechanical disturbance. These oscillations appear both on electrical and mechanical parts of the machine. It results in shaft fatigue and unacceptable voltage oscillation. This phenomenon is called hunting and must be prevented.

  Damper winding called amortisseur windings in form of short circuited copper bars are placed in rotor pole faces which provides damping to quickly damp and the oscillation.

  Synduction motor The damping winding at start produce an induction torque which makes the synchronous motor self-starting.

LO 4

- Operating characteristic

$$P_e = \frac{V_t E_f}{X_s} \sin\delta; \textit{ power-angle characteristic}$$

where $P_e = V_t I_a \cos\theta$; $\cos\theta = pf$

$$P_e\,(\max) = \frac{V_t E_f}{X_s}$$

If $P_e > P_e$ (max), the machine loses synchronism or motor falls out of step

- Salient-pole synchronous generator/motor – Two reaction theory

  Armature reaction mmf $\bar{F}_a$ is divided into two components. One along *d*-axis (high permeance) and other along *q*-axis (low permeance). It leads to the concept of direct axis reactance X_d and quadrature axis reactance X_q ($X_d > X_q$). Accordingly, $\bar{I}_a$ is divided into two components, I_q in phase with $\bar{E}_f$ (excitation emf) and $\bar{I}_d$ in quadrature to $\bar{E}_f$. The phasor equation of the generator is then

$$\bar{E}_f = \bar{V}_t + \bar{I}_a R_a + jX_d \bar{I}_d + jX_q \bar{I}_q$$

- Angle between $\bar{E}_f$ and $\bar{I}_a$ is given by

$$\tan\psi = \frac{V_t \sin\phi + I_a X_q}{V_t \cos\phi + I_d X_q} \text{ (generating)}$$

$$\tan\psi = \frac{V_t \sin\phi - I_a X_q}{V_t \cos\phi - I_a R_a} \text{ (motoring)}$$

ϕ = *pf* angle, angle between $\bar{V}_t$ and $\bar{I}_a$
= positive for lagging, negative for leading

$\delta = \psi - \phi$ (generating)

$\delta = \phi - \psi$ (motoring)

- Excitation emf is found scalar equation

$$E_f = V_t \cos\delta + I_q R_a + I_d X_d \text{ (generating)}$$

$$E_f = V_t \cos\delta - I_q R_a - I_d X_d \text{ (motoring)}$$

- Power angle characteristic

$$P_e = \frac{E_f V_b}{X_d} \sin\delta + \underbrace{V_b^2 \frac{(X_d - X_q)}{2X_d X_q} \sin 2\delta}_{\text{Reluctance power}}$$

LO 5

- Starting – synchronous motor is non-self starting, started by an auxiliary motor (induction motor) and then *synchronized* to mains. Before loading, it is floating on the mains; drawing or delivering almost zero current.

PRACTICE PROBLEMS

8.1 The open- and short-circuit tests data on a 3-phase, 1 MVA, 3.6 kV, star-connected synchronous generator is given below: LO 2

I_f (A)	60	70	80	90	100	110
V_{OC} (line) (V)	2560	3000	3360	3600	3800	3960
SC (A)	180					

Find:

(a) The unsaturated synchronous reactance.
(b) The adjusted synchronous reactance.
(c) The short-circuit ratio.
(d) The excitation voltage needed to give rated voltage at full-load, 0.8 lagging pf. Use adjusted synchronous reactance.
(e) Voltage regulation for the load specified in part (d).

8.2 A 3-phase 2.5 MVA, 6.6 kV synchronous generator gave the following data for OCC at synchronous speed: LO 2

I_f (A)	16	20	25	32	45
V_{OC} (line) (V)	4400	5500	6600	7700	8800

With the armature short-circuited and full-load current following, the field current is 18 A. When the machine is supplying full-load current at zero pf at rated voltage, the field current is 45 A.

Determine the leakage reactance in Ω per phase and the full-load armature reactance in terms of equivalent field amperes. Find also the field current and voltage regulation when the machine is supplying full-load at 0.8 pf lagging at rated voltage. Neglect armature resistance.

8.3 In Prob. 8.1 the armature leakage reactance is estimated to be 0.15 pu. Solve for part (d) by drawing the mmf phasor diagram. LO 2

8.4 A 1 MVA, 11 kV, 3-phase star-connected synchronous machine has the following OCC test data: LO 2

I_f (A)	50	110	140	180
V_{OC} (line) (V)	7000	12500	13750	15000

The short-circuit test yielded full-load current at a field current of 40 A. The ZPF test yielded full-load current at rated terminal voltage for a field current of 150 A. The armature resistance is negligible.

Calculate the field current needed for the machine to draw full-load 0.8 pf leading current when operated as the motor connected to an 11 kV supply.

8.5 A synchronous motor is drawing power from a large system (assumed infinite) with its field current adjusted so that the armature current has unity power factor. Assume the armature resistance and leakage reactance to be negligible. LO 3

With the load torque remaining constant, the field current of the motor is raised by 10%. Discuss by means of mmf phasor diagrams the changes that occur in the power output, magnitude and phase angle of the armature current and torque angle.

If, instead of the field current, the load torque is increased by 10%, what changes will occur?

8.6 Discuss the problem posed in Prob. 8.5 by using the circuit model of the motor. LO 3

8.7 The full-load torque angle of a synchronous motor at rated voltage and frequency is 30° elect. The stator resistance is negligible. How would the torque angle be affected by the following changes? LO 3

(a) The load torque and terminal voltage remaining constant, the excitation and frequency are raised by 10%.

(b) The load power and terminal voltage remaining constant, the excitation and frequency are reduced by 10%.
(c) The load torque and excitation remaining constant, the terminal voltage and frequency are raised by 10%.
(d) The load power and excitation remaining constant, the terminal voltage and frequency are reduced by 10%.

8.8 A 1000 kVA, 3-phase, 11 kV, star-connected synchronous motor has negligible resistance and a synchronous reactance of 35 Ω per phase. LO 3

(a) What is the excitation emf of the motor if the power angle is 10° and the motor takes rated current at (i) lagging power factor, and (ii) leading power factor.
(b) What is the mechanical power developed and the power factor in part (a)?
(c) At what power angle will this motor operate if it develops an output of 500 kW at the rated line voltage and with an excitation emf of 10 kV (line)? What is the corresponding power factor?
(d) What is the minimum excitation at which the motor can deliver 500 kW at the rated line voltage without losing synchronism?

8.9 A 1000 kVA, 6.6 kV, 3-phase star-connected synchronous generator has a synchronous reactance of 25 Ω per phase. It supplies full-load current at 0.8 lagging pf and a rated terminal voltage. Compute the terminal voltage for the same excitation when the generator supplies full-load current at 0.8 leading pf. LO 3

8.10 A 750 kW, 11 kV, 3-phase, star-connected synchronous motor has a synchronous reactance of 35 Ω/phase and negligible resistance. Determine the excitation emf per phase when the motor is operating on full-load at 0.8 pf leading. Its efficiency under this condition is 93%. LO 3

8.11 A synchronous generator having a synchronous reactance of 1.0 pu is connected to infinite bus-bars of 1.0 pu voltage through two parallel lines each of 0.5 pu reactance. LO 3

(a) Calculate the generator excitation, terminal voltage and power output when it delivers rated current (1.0 pu) at unity power factor at its terminals. What active and reactive power are delivered to the infinite bus-bars?
(b) Calculate the generator excitation and terminal voltage when the generator is delivering zero active power and 0.5 pu lagging reactive power to the infinite bus-bars.
(c) With one line disconnected, can the generator deliver the same active power to the infinite bus-bars at the same excitation as in part (a)? Explain.

8.12 Consider the synchronous generator-motor set of Ex. 8.7 whose data are repeated below. LO 3

Generator: 1200 kVA, 3-phase, 3.3 kV, 2-pole, 50 Hz star-connected, X_s = 4.55 Ω/ph
Motor: 1000 kW, 3-phase, 3.3 kV, 24-pole, 50 Hz, star-connected, X_s = 3.24 Ω/ph

(a) The set is operating at rated terminal voltage and frequency with the motor drawing 800 kW at upf Compute the excitation emfs of both the machines.
With the excitation emfs held fixed at these values, what maximum torque can the motor supply? Also determine the armature current, terminal, voltage and power factor under this condition.
(b) The motor shaft load is now gradually increased while the field currents of both the generator and motor are continuously adjusted so as to maintain the rated terminal voltage and upf operation. What maximum torque can the motor now deliver without losing synchronism?

8.13 A 2500 V, 3-phase, star-connected motor has a synchronous reactance of 5 Ω per phase. The motor input is 1000 kW at rated voltage and an excitation emf of 3600 V (line). Calculate the line current and power factor. LO 3

8.14 Repeat Prob. 8.13 considering a motor resistance per phase of 0.1 Ω. LO 3

8.15 A 20 MVA, 11 kV, 3-phase, delta-connected synchronous motor has a synchronous impedance of 15 Ω/phase. Windage, friction and iron losses amount to 1200 kW. LO 3

(a) Find the value of the unity power factor current drawn by the motor at a shaft load of 15 MW. What is the excitation emf under this condition?

(b) If the excitation emf is adjusted to 15.5 kV (line) and the shaft load is adjusted so that the motor draws upf current, find the motor output (net).

8.16 A 600 V, 6-pole, 3-phase, 50 Hz, star-connected synchronous motor has a resistance and synchronous reactance of 0.4 Ω and 7 Ω respectively. It takes a current of 15 A at upf when operating with a certain field current. With the field current remaining constant, the load torque is increased until the motor draws a current of 50 A. Find the torque (gross) developed and the new power factor. LO 3

8.17 A 500 V, 3-phase, mesh-connected motor has an excitation emf of 600 V. The motor synchronous impedance is (0.4 + j5) Ω while the windage, friction and iron losses are 1200 W. What maximum power output can it deliver? What is the corresponding line current, pf and motor efficiency? LO 3

8.18 A 3-phase synchronous generator has a direct-axis synchronous reactance of 0.8 pu and a quadrature-axis synchronous reactance of 0.5 pu. The generator is supplying full-load at 0.8 lagging pf at 1.0 pu terminal voltage. Calculate the power angle and the no-load voltage if excitation remains unchanged. LO 3

8.19 A 3.5 MVA, slow-speed, 3-phase synchronous generator rated at 6.6 kV has 32 poles. Its direct- and quadrature-axis synchronous reactances as measured by the slip test are 9.6 and 6 Ω respectively. Neglecting armature resistance, determine the regulation and the excitation emf needed to maintain 6.6 kV at the terminals when supplying a load of 2.5 MW at 0.8 pf lagging. What maximum power can the generator supply at the rated terminal voltage, if the field becomes open-circuited? LO 3

8.20 A salient-pole synchronous motor has X_d = 0.85 pu and X_q = 0.55 pu. It is connected to bus-bars of 1.0 pu voltage, while its excitation is adjusted to 1.2 pu. Calculate the maximum power output, the motor can supply without loss of synchronism. LO 3

Compute the minimum pu excitation that is necessary for the machine to stay in synchronism while supplying the full-load torque (i.e. 1.0 pu power).

8.21 From the phasor diagram of the salient-pole synchronous machine given in Fig. 8.73 prove that LO 4

$$\tan \delta = \frac{I_a X_q \cos\phi - I_a R_a \sin\phi}{V_t - I_q X_q \sin\phi - I_a R_a \cos\phi}$$

8.22 Two star-connected generators are connected in parallel and supply a balanced load of 1500 kVA at 11 kV line voltage and 0.8 lagging power factor. The synchronous reactances of the two machines respectively are: 35 Ω and 40 Ω. The primemover governors of the two machines are adjusted so as to equally share the power load. The phase current in one machine is 43 A, at a lagging power factor. Calculate: LO 6

(a) the phase current in the second machine,

(b) the induced emf of each machine, and

(c) the power factor at which each machine operates.

8.23 Calculate the synchronizing coefficient (in kW and Nm per mechanical degree) at full-load for a 1000 kVA, 0.8 pf (lag), 6.6 kV, 8-pole, star-connected cylindrical rotor generator of negligible resistance and synchronous reactance of 0.8 pu. LO 3

8.24 A 3-phase hydroelectric synchronous generator is read to be 110 MW, 0.8 pf lagging, 6-kV, Y- connected, 50 Hz, 100-rpm. Determine: LO 3

(a) the number of poles
(b) the kVA rating
(c) the prime-mover rating if the full-load generator efficiency is 97.1 % (leave out field loss).
(d) the output torque of the prime-mover.

8.25 1000 kV A, 50 Hz, 2300 V, 3-phase synchronous generator gave the following test data: LO 2

Field current (A)	40	80	100	120	140	170	240
V_{OC} (Line)(V)	1000	1900	2200	2450	2600	2750	3000
I_{SC} (A)	2000						

(a) Find the field current required to deliver rated kVA at 0.8 lagging pf at rated terminal voltage.
(b) The OC voltage at this field current.
(c) The maximum kVAR that the machine can deliver as a synchronous condenser at rated voltage, if the rotor heating limits the field current to 240 A.

8.26 A 3-phase synchronous generator feeds into a 22 kV grid. It has a synchronous reactance of 8 Ω/ phase and is delivering 12 MW and 6 MVAR to the system. Determine: LO 2
(a) the phase angle of the current
(b) the power angle
(c) the generated emf.

8.27 A 6.6 kV, 3-phase synchronous machine has an open circuit characteristic given by: LO 2

Field current (A)	60	80	100	120	140	160	180
Armature emf (kV) (line)	5.3	6.2	6.8	7.2	7.5	7.7	7.9

In short-circuit a field current of 80 A gave an armature current of 360 A. Determine the saturated synchronous reactance.
When developing 400 kW of mechanical power as a motor calculate the field current for pfs of 0.8 lagging, unity and 0.8 leading.

8.28 A 200 kV A, 3.3 kV, 50 Hz three-phase synchronous generator is star-connected. The effective armature resistance is 5 Ω/phase and the synchronous reactance is 29.2 Ω/phase. At full-load calculate the voltage regulation for the following power factors: LO 1
(a) 0.707 leading
(b) unity
(c) 0.707 lagging

8.29 A 3-phase, 4-pole star-connected synchronous motor has a resistance of 0.25 Ω/phase and a synchronous reactance of j 2.5 Ω/phase. The field is excited such that the open-circuit voltage of the machine is 25 kV. The motor is synchronized to 22 kV mains. Calculate the maximum load on the motor (including rotational loss) before it could lose synchronism. What is the corresponding current and power factor? LO 3

8.30 A 6-pole, 3-phase, 50 Hz synchronous motor is supplied from 6.6 kV busbars. Its open-circuit voltage is 3.3 kV/phase. The per phase resistance and synchronous reactance are 0.6 Ω and 4.8 Ω respectively. Calculate the current, power factor and torque developed, when the excitation emf lags the busbar voltage by 15°, 25°, and 35° (elect). LO 3

8.31 A 25 kVA, 400 V, 3-phase synchronous generator delivers rated kVA at rated voltage at 0.8 pf lagging. The per phase (star basis) armature resistance and synchronous reactance respectively are 0.66 Ω and 7.1 Ω. The field winding is supplied 10.6 A at 110 V. The friction and winding loss is estimated to be 480 W and iron loss as 580 W. Calculate: LO 2
(a) the full-load efficiency.
(b) the terminal voltage when the load is thrown off.

8.32 A 15 kW. 400 V, 3-phase, star-connected synchronous motor has a synchronous impedance of 0.4 + j4 Ω. Find the voltage to which the motor should be excited to give a full-load output at LO 4

0.866 leading pf. Assume an armature efficiency of 93%. Also calculate the mechanical power developed.

8.33 A 3-phase, star-connected synchronous generator is rated at 1200 kVA, 1 1 kV. On short-circuit a field current of 55 A gives full-load current. The OC voltage with the same excitation is 1580 V/phase. Calculate the voltage regulation at (a) 0.8 lagging and (b) 0.8 leading pf. Neglect armature resistance. **LO 1**

8.34 A 1500 kW. 3-phase, star-connected 2300 V, 50 Hz synchronous motor has a synchronous reactance of 2 Ω/phase. The motor is supplied from a 3-phase, star-connected, 2300V, 1750 kVA turbo generator whose synchronous reactance is 2.8 Ω/phase. When the motor is drawing full-load power, at upf, calculate: (a) the induced emf of the generator, and (b) emf of the motor. What maximum power can flow from the generator to the motor With machine excitations held fixed? **LO 4**

8.35 A 13.8 kV 1250 kVA 3-pahse, star-connected synchronous generator has a resistance of 2.1 Ω/phase. Data for Its OCC and ZPFC characteristics is given below: **LO 2**

Field current (A)	40	50	110	140	180
Open circuit volts (line)	7.28	8.78	15.68	17.25	18.82
Zero pf volts (line)	0	1.88	10.66	13.17	15 68

Find the voltage regulation of the generator for full-load 0.8 pf lagging.

8.36 A 440 V, 50 Hz, Y-connected salient-pole synchronous generator has a direct-axis reactance of 0.12 Ω and a quadrature-axis reactance of 0.075 Ω per phase, the armature resistance being negligible. The generator is supplying 1000 A at 0.8 lagging pf. **LO 4**

(a) Find the excitation emf, neglecting saliency and assuming $X_s = X_d$.

(b) Find the excitation emf accounting for saliency.

8.37 Figure P.8.37 shows two generators supplying in parallel a load of 2.8 MW at 0.8 pf lagging: **LO 6**

(a) At what frequency is the system operating and what is the load supplied by each generator?

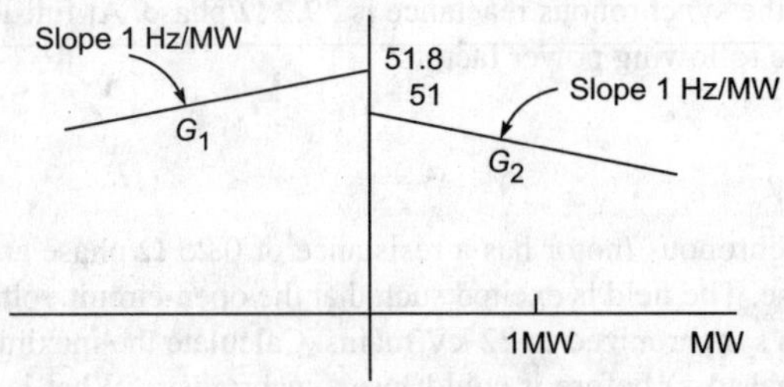

Fig. P.8.37

(b) If the load is now increased by 1 MW, what will be the new frequency and the load sharing?

(c) In part (b) which should be the set point of G_2 for the system frequency to be 50 Hz? What would be the load sharing now?

8.38 A generating station comprises four 125 kVA, 22 kV, 0.84 pf lagging synchronous generators With a frequency drop of 5 Hz from no-load to full-load, at a frequency of 50 Hz, three generators supply a steady load of 75 MW each while the balance is shared by the fourth generator (called *swing generator*). **LO 7**

(a) For a total load of 260 MW at 50 Hz, find the no-load frequency setting of the generators.

(b) With no change in governor setting as in part (a), find the system frequency if the system load rises to 310 MW.

(c) Find the no-load frequency of the swing generator for the system frequency to be restored to 50 Hz for the load in part (b).

(d) Find the system frequency if the swing generator trips off with the load as in part (b), the governor setting remaining unchanged as in part (a).

8.39 An 11 kV, 3-phase alternative has X_d and X_q of 0.6 pu and 0.1 pu respectively and negligible armature resistance. It is delivaring rated kVA at 0.8 pf lag. Determine its generated emf. LO 4

8.40 Compute the distribution factor and coil span factor for a 3 phase winding with 4 slots per pole per phase and with coil span of 10 slot pitch. LO 1

8.41 A 3.3 kV, 3 phase star connected synchronous generator has full load current of 100 A. Under shot circuit condition, it takes 5 A field current to produce full load short circuit current. The open circuit voltage is 900 V (line to line). Determine synchronous reactance per phase and voltage regulation for 0.8 pf lagging. Assume armature resistance was 0.9 ohms/phase. LO 4

8.42 Two similar 6.6 kV synchronous generators supply a total load of 1000 kW at 0.8 pf lagging so that power supplied by each machine is same. Determine the load current of second synchronous generator and power factor of each machine, when excitation of first synchronous generator is decreased so that load current reduces to 100 A. LO 6

8.43 A 3-phase, 2.5 MVA, 6.6 kV synchronous generator gave the following test. LO 2

OCC

I_f(A)	16	20	25	32	45
V_{OC} (line) (V)	4400	5500	6600	7700	8800

SC test

I_f = 18A for rated armature current.

ZPF test

I_f = 20 A for rated armature current at rated voltage.

Determine the field current and voltage regulation when the generator is supplying rated current at 0.8 pf lagging and rated voltage, $R_a = 0$

Use the following methods and compare the results and draw conclusions:

(a) Unsaturated synchronous reactance
(b) Saturated synchyranous reactance
(c) mmf method
(d) ASA method

Answers

8.1 (a) 8.21 Ω (b) 7.7 Ω (c) 1.64 (d) 5173 V (line), 43.7%

8.2 2.11 Ω, 15 A, 30.3%

8.3 150 A (read by extrapolation of OCC)

8.4 22.65 Ω, 122.5 A

8.7 (a) 33.36° (b) 30° (c) 33.36° (d) 30°

8.8 (a) (i) 4.77 kV; (ii) 7.74 kV (b) (i) 451 kW, 0.451 lag; (ii) 731.6 kW, 0.73 lead (c) 9.2°, 0.5 lag (d) 1.59 kV (line)

8.9 11.15 kV (line)

8.10 13.18 kV (line)

8.11 (a) 0.968, 0.968, –0.25 (b) 1.625, 1.125 (c) 0.926

8.12 (a) 3479 V (line), 3392 V (line), 1514.9 kW, 57.86×10^3 N m, 1416.5 V, 0.99 lead (b) 2857.5 kW, 109.15×10^3 N m

8.13 237.8 A, 0.97 lead

8.14 233.7 A

8.15 (a) 0.85 kA (upf), 13.23 kV (line) (b) 22.8 MW

8.16 420.2 N m, 0.9 lag

8.17 160.8 kW, 258.5 A, 0.842 lag, 85.5%
8.18 17.1°, 1.6 pu
8.19 50.9%, 0.454 MW
8.20 1.53 pu, 0.707 pu
8.21 $\tan\delta = \dfrac{I_a X_q \cos\phi \pm I_a r_a \sin\phi}{V_t \pm I_a x_q \sin\phi \pm I_a r_a \cos\phi}$; for ϕ lagging and + for geh and – for motoring case.
8.22 36.2 A, 0.87 lag; 12.92 kV (line, 12.43 kV (line))
8.23 7388 kW/mech deg, 94.1 × 10^3 N m/mech deg
8.24 (a) 60 (b) 137.5 kVA (c) 113.3 kW (d) 10.82 kNm.
8.25 (a) 228 A (b) 2503 V (c) 2183 kVAr lagging
8.26 (a) – 26.56° (b) 10.2° (c) 24.56 kV.
8.27 115 A, 76 A
8.28 (a) –29.7% (b) 21.7% (c) 47.93%
8.29 291 MW, 0.813 lagging
8.30 218 A, 0.87 lagging, 2.1 MW; 334 A, 0.92 lagging, 3.3 MW
8.31 (a) 90% (b) 773.5 V
8.32 490 V, 19.31 kW
8.33 (a) 11.64% (b) 7.94%
8.34 (a) 2935 V (b) 2645 V, 1617 kW
8.35 15.9%
8.36 (a) 484.6 V (b) 444 V
8.37 (a) 50 Hz (b) 49.5 Hz (c) 52 Hz
8.38 (a) 53.6 Hz, 51.7 Hz (b) 48.84 Hz (c) 48.68 Hz.
8.39 1.387
8.40 0.956, 0.991
8.41 5.12 ohm, 21.34%
8.42 54.65 A, 0.96, 0.98
8.43 (a) 21.5 A, 53% (b) 22 A, 57% (c) 38.5 A, 25.83% (d) 41 A, 28.3%

MULTIPLE CHOICE QUESTIONS

8.1 In a synchronous machine, the induced emf phasor: LO 1
(a) leads the flux phasor by 90º
(b) is in phase with the flux phasor
(c) lags behind the flux phasor by 90
(d) is in phase opposition to the flux phasor

8.2 Synchronous motor speed is controlled by varying: LO 5
(a) field excitation
(b) supply voltage
(c) supply frequency only
(d) both supply voltage and frequency

8.3 In a generating synchronous machine carrying load (usual symbols are used): LO 1
(a) E_f leads V_t by angle δ
(b) E_f lags V_t by angle δ
(c) E_f and V_t are in phase
(d) E_f and V_t are in phase opposition

8.4 Potier's method uses OCC and ZPFC to yield information about: LO 2
(a) synchronous reactance
(b) leakage reactance only
(c) field current equivalent of armature reaction only
(d) leakage reactance and field current equivalent of armature reaction

8.5 The maximum electrical power output of a synchronous generator is (usual symbols are used): LO 4

(a) $\dfrac{V_t E_f}{X_s}$ (b) $\dfrac{V_t^2}{X_s}$ (c) $\dfrac{E_f^2}{X_s}$ (d) $\dfrac{X_s}{V_t E_f}$

8.6 Synchronous generator voltage obtained by the synchronous impedance method is: LO 2

(a) higher than actual as it does not account for magnetic saturation.
(b) lower than actual as it does not account for magnetic saturation.
(c) nearly accurate as it accounts for magnetic saturation.
(d) nearly accurate as the generator is normally operated in the unsaturated region of magnetization.

8.7 A synchronous motor is running from busbars, and has a torque angle of $\delta = 15°$. The bus voltage and frequency are reduced by 10% while field current and load torque are held constant (all losses are negligible). The new torque angle will: LO 3

(a) remain the same (b) increase slightly
(c) decrease slightly (d) the motor will lose synchronism

8.8 A synchronous motor is operated from a bus voltage of 1.0 pu and is drawing 1.0 pu zero power factor leading current. Its synchronous reactance is 0.5 pu. The excitation emf of the motor will be: LO 3

(a) 2.0 (b) 1.5 (c) 1.0 (d) 0.5

8.9 It the excitation of a salient pole motor is reduced to zero: LO 3

(a) it will lose synchronism
(b) it will remain synchronized
(c) it will remain synchronized provided the load is less than a certain value
(d) it will remain synchronized provided it is operating at no load.

8.10 The armature current upon symmetrical 3-phase short-circuit of a synchronous machine (armature resistance is negligible): LO 4

(a) Constitutes q-axis current only (b) Constitutes d-axis current only
(c) has both d-axis q-axis components (d) short circuit current cannot be divided into d- and q-axis components.

Answers

1. (c) **2.** (d) **3.** (a) **4.** (d) **5.** (a) **6.** (a) **7.** (a) **8.** (d) **9.** (c) **10.** (b)

SAMPLE EXAMPLES SOLVED USING MATLAB

```
% This program solves Example 8.16 of the text
% VP is taken as reference
P=230;
PNEW=275;
VL=22;
VP=VL/sqrt(3);
XS=1.2;
IP=P/(VL*sqrt(3));
EF=VP+j*XS*IP;
EF_mag=abs(EF);
EFP=1.3*EF_mag;
% calculations for part 1
```

```
DELP=asin(P*XS/(3*EFP*VP));
IPP=(EFP*exp(j*DELP)-VP)/(j*XS);
IPP_mag=abs(I PP);
PF=cos(angle(IPP));
fprintf('Part 1 results\n')
fprintf('Machine current is %8.2f kA \n', IPP_mag)
fprintf('Power factor is %8.2f \n', PF)
% calculations for part 2
DELPP=asin(PNEW*XS/(3*EFP*VP));
IPP=(EFP*exp(j*DELPP)-VP)/(j*XS);
IPP mag=abs(IPP);
PF=cos(angle(Ipp))
fprintf('Part 2 results\n')
fprintf('Machine current is %8.2f kA \n', IPP_mag)
fprintf('Power factor is %8.2f \n', PF);
% This program solves Example 8.21 of the text
%_mag refers to magnitude of the corresponding phasor
% for exemple IP _mag is the magnitude of IP phasor
P=800;
PNEW=1200;
VL=6.6;
ZS=2+j*20;
PF=0.8;
TH=angle(ZS)
PHI=acos(PF);
% VP is taken as reference
VP=VL/sqrt(3);
IP_mag=P/sqrt(3)*VL*PF 1000);
IP=IP mag*exp(j*PHI);
EF=VP-(ZS*IP);
% EF is the phasor
EF_mag=abs(EF);
% a,b,c are required for solving expo Pe for angle ANG
a=PNEW/(1000*3);
b=VP*VP*cos(TH)/abs(ZS);
c=(abs(ZS))/(VP*EF_mag);
d=c*(b-a);
ANG=acos(d);
DEL=ANG-TH;
IPNEW=(VP-EF_mag*exp (j*DEL))/ZS;
IPNEW_mag=abs(IPNEW);
PFNEW=cos(angle(IPNEW));
fprintf ('POWER FACTOR OF THE MOTOR UNDER INCREASED POWER CONDITIONS: %f',PFNEW)
```

CHAPTER 9

Induction Machine

LEARNING OBJECTIVES

After studying this chapter, students will be able to

LO1 Explain the construction and field flux distribution of an induction motor and its equivalent circuit model development

LO2 Understand the nature of induction motor performance over a complete range of slips by circuit model parameters determination, circle diagram, and concept of transfer/conversion of power through a machine

LO3 Examine the effect of harmonics and different speed control methods of an induction motor, with various complications occurring at the start and ways to overcome them

LO4 Analyze induction motor dynamics, various constructions (deep-bar/double) of cage rotor, and several types (classes) of standard designs of squirrel-cage motors available to support different industrial applications

LO5 Define the basic operating principle of induction generator, inverted induction motor, high efficiency induction motor, and linear induction motor

9.1 INTRODUCTION

The induction machine is an important class of electric machines which finds wide applicability as a motor in industry and in its single-phase form in several domestic applications. More than 85% of industrial motors in use today are in fact induction motors. It is substantially a constant-speed motor with a shunt characteristic; a few per cent speed drop from no-load to full-load. It is a singly-fed motor (stator-fed), unlike the synchronous motor which requires ac supply on the stator side and dc excitation on the rotor. The torque developed in this motor has its origin in current induction in the rotor which is only possible at non-synchronous speed; hence the name *asynchronous* machine. Torque in a synchronous machine on the other hand, is developed only at synchronous speed when the "locking" of the two fields takes place. Therefore, the induction motor is not plagued by the stability problem inherent in the synchronous motor. Since it is a singly-fed machine, it draws its excitation current* from the mains to set up the rotating field in the air-gap which is essential for its operation. As a consequence it inherently has a power factor less than unity which usually must be corrected by means of shunt capacitors at motor terminals. There is no simple and inexpensive

*Because of the air-gap the excitation current is far larger in an induction motor than in a transformer for the same VA rating.

method of controlling the induction motor speed as is possible in a dc shunt motor. A wide range of speed control is possible only by expensive circuitry using silicon-controlled rectifiers (SCRs). It still finds stiff competition from the de shunt motor in such applications.

Industry Relevance

In this type of motor, without any electrical connection to rotor conductors, current and subsequently torque gets produced in the rotor through electromagnetic induction from the rotating magnetic field provided by three phase winding in stator. Due to simple & rugged construction with minimum maintenance requirement, this self-starting motor finds its usefulness in almost seventy percent of the motor used nowadays in industrial application.

9.2 | CONSTRUCTION

LO 1 Explain the construction and field flux distribution of an induction motor and its equivalent circuit model development

The stator of an induction motor is similar to that of a synchronous machine and is wound for three phases, modern practice being to use the two-layer winding. Figure 9.1 shows the stator of an induction motor. Two types of constructions are employed for the rotor—*wound-rotor* and *squirrel-cage rotor*. The rotor core is of laminated construction with slots suitably punched in for accommodating the rotor winding/rotor bars. The punched laminations are stacked and fitted directly onto a shaft in the case of small machines, while in the case of large machines a stack of annular punchings of a suitable cross-sectional area are fitted onto a spider-web arrangement on the shaft.

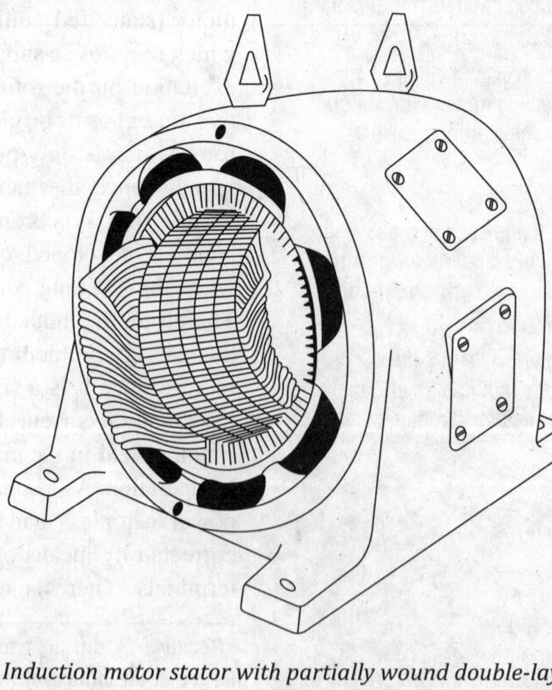

Fig. 9.1 *Induction motor stator with partially wound double-layer winding*

Wound Rotor

The winding of a wound-rotor is polyphase with coils placed in the slots of the rotor core. It is similar to that of the stator except that the number of slots is smaller and fewer turns per phase of a heavier conductor are used. The rotor is wound and connected in star with three leads brought out of the machine via slip-rings placed on the shaft. The slip-rings are tapped by means of copper-carbon brushes. Wound-rotor construction is generally employed for large size machines to be used where the starting torque requirements are stringent. Figure 9.2(a) shows the view of a slip-ring rotor. External resistance can be included in the rotor circuit through slip-rings for reducing the starting current and simultaneously improving the starting torque.

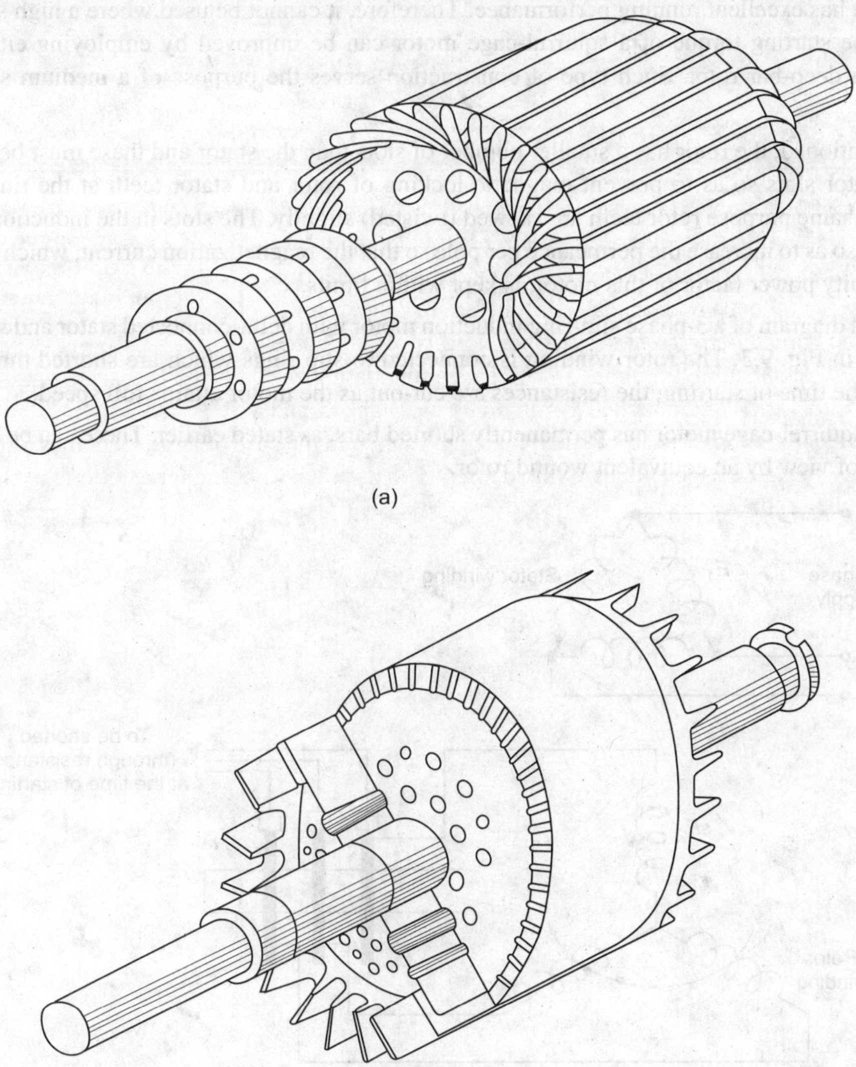

(a)

(b)

Fig. 9.2 *(a) Wound rotor for induction motor (b) Squirrel-cage rotor showing cast aluminium bars and end rings*

Squirrel-Cage Rotor

The squirrel-cage rotor has solid bars of conducting material placed in rotor slots and shorted through end-rings on each side. Figure 9.2(b) shows the view of the squirrel-cage rotor of a small-size motor. In large machines alloyed copper bars are driven in the slots and are brazed onto copper end-rings. Rotors up to 50 cm diameter usually have diecast aluminium bars wherein the end-rings are also cast simultaneously with the same material by using a suitable mould. This is an economical process and is generally employed in mass production of small size induction motors.

The rotor circuit of a squirrel-cage machine cannot be tempered with and the machine has a low starting torque, while it has excellent running performance. Therefore, it cannot be used where a high starting torque is required. The starting torque of a squirrel-cage motor can be improved by employing either a double-cage rotor or a deep-bar rotor. Such type of construction serves the purpose of a medium starting torque requirement.

As already mentioned, the rotor has a smaller number of slots than the stator and these must be a nonintegral multiple of stator slots so as to prevent magnetic locking of rotor and stator teeth at the time of starting. Further for the same purpose rotor teeth are skewed (twisted) slightly. The slots in the induction machine are semi-enclosed so as to increase the permeance per pole so that the magnetization current, which is responsible for less than unity power factor of this motor, is kept within limits.

The connection diagram of a 3-phase slip-ring induction motor with delta-connected stator and star-connected rotor is drawn in Fig. 9.3. The rotor winding is connected to slip rings which are shorted through external resistances at the time of starting; the resistances are cut-out as the motor attains full speed.

The rotor of a squirrel-cage motor has permanently shorted bars, as stated earlier. These can be replaced from a circuit point of view by an equivalent wound rotor.

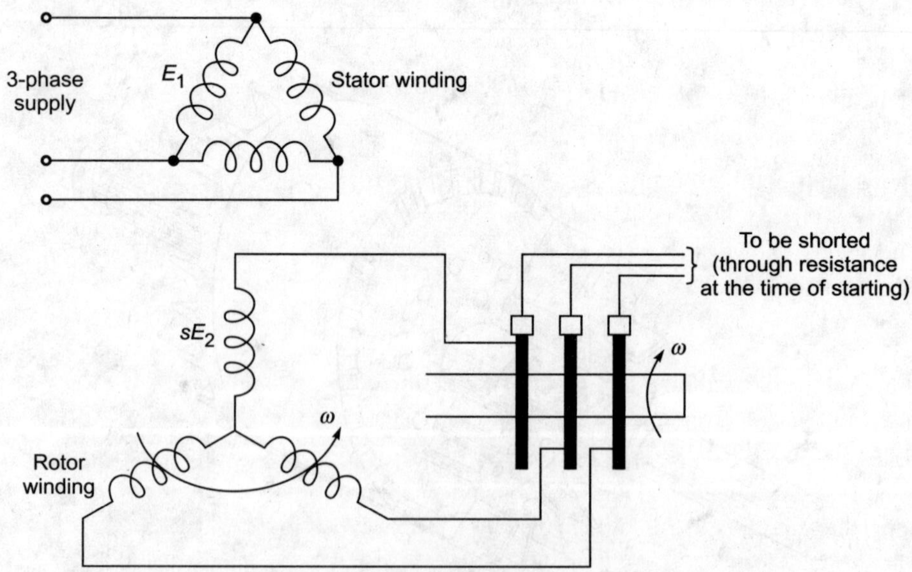

Fig. 9.3 *3-phase, slip-ring induction motor—connection diagram*

9.3 FLUX AND MMF WAVES IN INDUCTION MOTOR—PRINCIPLE OF OPERATION

LO 1

The principle of operation of the induction motor has already been discussed in Sec. 5.7. It will now be revised here and further elaborated leading to the circuit model of the motor from which performance calculations can be easily carried out.

Establishment of emf Equation

Figure 9.4 shows the cross-sectional view of an induction motor. The stator is fed from a 3-phase supply of voltage *V*/phase and frequency *f* Hz. The rotor is wound† 3-phase for as many poles as the stator and is short-circuited. It is assumed that the stator resistance and leakage reactance are both negligible so that

$$V = E_1 = \pi\sqrt{2}\, k_{w1} N_{ph1}\ \text{(series)}\ f\Phi_r \tag{9.1}$$

where
E_1 = stator induced emf/phase
k_{w1} = stator winding factor
N_{ph1} (series) = stator series turns/phase
Φ_r = resultant air-gap flux/pole

It is seen from Eq. (9.1) that irrespective of the load conditions existing on the rotor, Φ_r the flux/pole established in the air-gap is constant, related to the applied voltage in view of the assumption made. The mmf vector $\vec{F}_r$ with associated flux density vector $\vec{B}_r$ which is responsible for production of Φ_r rotates at synchronous speed as it is associated with balanced 3-phase currents drawn by the stator. The relative speed between $\vec{B}_r$ and the rotor causes induction of a current pattern in the shorted rotor. The torque produced by interaction of $\vec{B}_r$ and the rotor currents would by Lenz's law tend to move the rotor in the direction of rotation of $\vec{B}_r$ so as to reduce the relative speed. The motor is thus *self-starting* and the rotor acquires a steady speed $n < n_s$ depending upon the shaft load. It may be noted that no torque is produced at $n = n_s$ because the relative speed between $\vec{B}_r$ and rotor being zero, no currents are induced in the rotor.

Figure 9.4 shows the relative location of vectors $\vec{F}_r$, $\vec{B}_r$ (air-gap mmf and flux density), $\vec{F}_2$ (rotor mmf) wherein $\vec{F}_r$ leads $\vec{F}_2$ by angle $\delta = 90° + \theta_2$ (motoring action), θ_2 is the angle by which rotor currents lag rotor emfs. The angle θ_2 however, is very small as rotor reactance is far smaller than rotor resistance (*Reader should at this point reread Sec.* 5.6 *sub section Induction Machine)*. The stator mmf vector is then given by

$$\vec{F}_1 = \vec{F}_r - \vec{F}_2$$

is located on the vector diagram. At the instant at which the diagram is drawn, the stator and rotor phase *a* (shown as single coil) currents are maximum. The reader may verify the location of vectors from the phase *a* currents applying the right hand rule.

It must be noted that the sign convention in an induction motor is same as that of a transformer. The stator current is in opposite direction to that of induced emf (like transformer primary), while the rotor current is in same direction as the induced emf (like transformer secondary).

† The rotor can be wound for any number of phases but the same number of poles as the stator. In practice it is wound 3-phase.

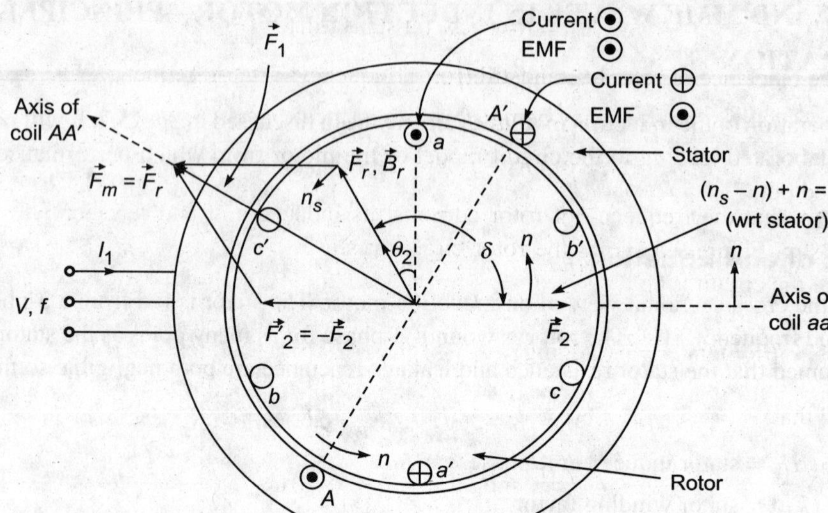

Fig. 9.4 *MMF and flux waves in induction motor*

Slip and Frequency of Rotor Currents

With reference to Fig. 9.4, it is easily observed that $\vec{B}_r$ moves at speed $(n_s - n)$ with respect to rotor conductors (in the direction of B_r). This is known as *slip speed.* The slip is defined as

$$s = \frac{\text{slip speed}}{\text{synchronous speed}} = \frac{n_s - n}{n_s} = \left(1 - \frac{n}{n_s}\right) \tag{9.2}$$

Obviously $s = 1$ for $n = 0$, i.e. for the stationary rotor and $s = 0$ for $n = n_s$, i.e. for the rotor running at synchronous speed.

The frequency of currents induced in the rotor is

$$\begin{aligned} f_2 &= \frac{(n_s - n)P}{120} \\ &= \left(\frac{n_s - n}{n_s}\right) \times \left(\frac{n_s P}{120}\right) \\ &= sf \end{aligned} \tag{9.3}$$

The normal full-load slip of the induction motor is of the order of 2% – 8%, so that the frequency of the rotor currents is as low as 1 – 4 Hz.

The per phase rotor emf at $s = 1$ (*standstill* rotor) is given by

$$E_2 = \pi\sqrt{2}\, k_{w2} N_{ph2} \text{ (series) } f \Phi_r \tag{9.4}$$

At any slip *s*, the rotor frequency being *sf*, the rotor induced emf changes to sE_2. Consider now the impedance of the rotor circuit

$$\bar{Z}_2 = R_2 + j\,X_2 \text{ (at standstill)}$$

where X_2 = leakage reactance of rotor at standstill (rotor frequency = stator frequency, f)

When the rotor runs at slip s, its frequency being sf, its impedance changes to

$$\bar{Z}_2 = R_2 + js\,X_2 \tag{9.5}$$

It is, therefore, seen that the frequency of rotor currents, its induced emf and reactance all vary in direct proportion to the slip. Figure 9.5 shows the rotor circuit at slip s. The phase angle of the circuit is

$$\theta_2 = \tan^{-1}\frac{sX_2}{R_2} \text{ (lagging)} \tag{9.6}$$

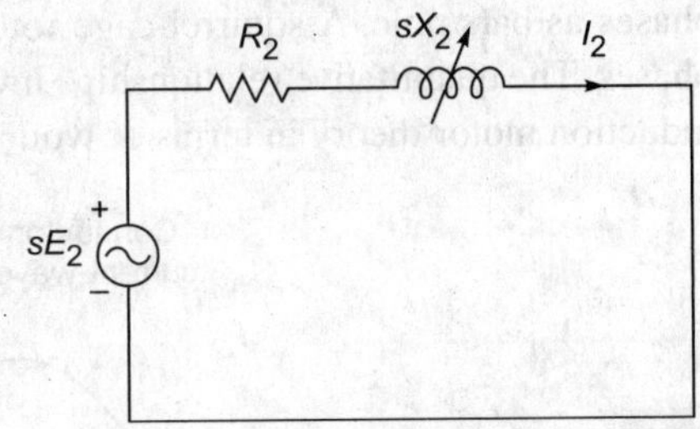

Fig. 9.5 *Rotor circuit (frequency sf)*

It is also observed that

$$\frac{E_1}{E_2} = \frac{k_{w1}N_{ph1}}{k_{w2}N_{ph2}} = \frac{N_{e1}}{N_{e2}} = a \tag{9.7}$$

where N_{e1}, N_{e2} = effective stator and rotor turns/phase

Rotor MMF and Torque Production

In Fig. 9.4 as the resultant flux density vector $\vec{B}_r$ rotates at speed $(n_s - n)$ with respect to rotor, maximum positive emf is induced in the rotor coil aa' (indicated by dot in conductor a and cross in conductor a') when $\vec{B}_r$ lies 90° ahead of the axis of the coil. Since the current in the rotor lags the emf by θ_2, the current in coil aa' will be maximum positive when $\vec{B}_r$ has moved further ahead by angle θ_2. It is at this instant of time that the rotor mmf vector $\vec{F}_2$ will lie along the axis of coil aa'. It is, therefore, seen that $\vec{B}_r$ (or $\vec{F}_r$) lies at an angle $\delta = (90 + \theta_2)$ ahead of $\vec{F}_2$. Further, $\vec{F}_2$ caused by the rotor currents of frequency $f_2 = sf$ rotates with respect to the rotor conductor at speed $(n_s - n)$ and at speed n_s with respect to the stator as the rotor itself is moving in the same direction at speed n with respect to the stator. Thus $\vec{F}_r$ and $\vec{F}_2$ both move at synchronous speed n_s with respect to the stator and are stationary relative to each other with $\vec{F}_r$ lying ahead of $\vec{F}_2$ by angle $(90° + \theta_2)$. The interaction of the rotor field and the resultant field as per Eq. (5.58) creates a torque

$$T = \frac{\pi}{2}\left(\frac{P}{2}\right)^2 \Phi_r F_2 \sin(90° + \theta_2)$$

$$= \frac{\pi}{2}\left(\frac{P}{2}\right)^2 \Phi_r F_2 \cos\theta_2 \tag{9.8}$$

in the direction of rotation of $\vec{F}_r$.

Consider now the case of the squirrel-cage rotor with conductors spread uniformly around the rotor periphery. The rotor reaction mmf F_2 is better visualized from the developed diagram of Fig. 9.6 wherein the rotor is imagined to be stationary and the B_r-wave moving with respect to it at slip speed $(n_s - n)$. Let the rotor reactance be considered negligible so that the conductor (shorted) currents are in-phase with the conductor emfs. The conductor current pattern is, therefore, sinusoidally distributed and is in space phase with B_r-wave and moves synchronously with it. The rotor mmf-wave is a stepped-sinusoidal with the same number of poles as the B_r-wave moving synchronously with it. Its fundamental (F_2) shown in Fig. 9.6 lags B_r-wave by 90°. If the rotor reactance is now brought into picture, the conductor current-wave and, therefore, the rotor mmf-wave would lag behind by angle θ_2, Thus the angle between the B_r-wave and F_2-wave would be $(90° + \theta_2)$, the same as in the wound rotor. A squirrel-cage rotor, therefore, inductively reacts in the same way as a wound rotor except that the number of phases is not obvious—one can consider it to have as many phases as bars/pole. A squirrel-cage rotor can always be replaced by an equivalent wound rotor with three phases. The quantitative relationships involved are beyond the scope of this book. Further examination of the induction motor theory in terms of wound rotor only will be undertaken.

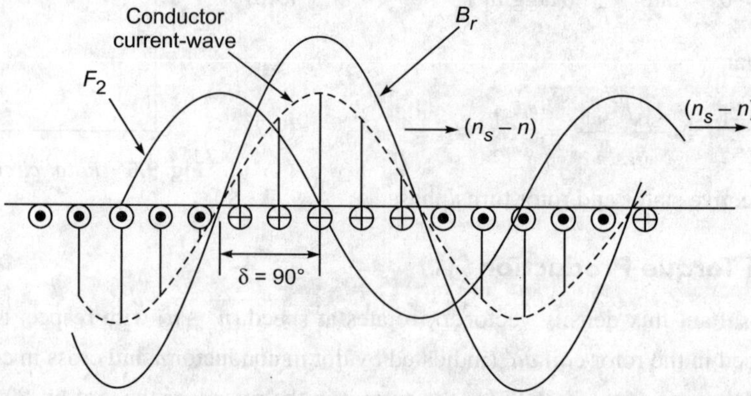

Fig. 9.6 *Rotor mmf in squirrel-cage rotor assuming negligible rotor reactance*

It is seen from Eq. (9.8) that a low-reactance rotor (low $\theta_2 = \tan^{-1} jsX_2/R_2$) will generate a larger torque for given Φ_r, F_2 and s. A squirrel-cage motor is superior in this respect as compared to a wound-rotor motor as the cage rotor has lower reactance since it does not have winding overhang.

One very important observation that can be made here is that while the rotor currents have a frequency sf, the mmf (F_2) caused by them runs at synchronous speed with respect to the stator. In other words, the reaction of rotor currents corresponds to the stator frequency (f) currents flowing on an equivalent stationary cylindrical structure placed inside the stator in place of the rotor. Or, to put in another way, the rotor currents as seen from the stator have frequency f but have the same rms value.

The stator mmf vector $\vec{F}_1$ is located on Fig. 9.4 from the vector equation

$$\vec{F}_1 = \vec{F}_r - \vec{F}_2 \tag{9.9}$$

Further, $\vec{F}_1$ can be divided into components as

$$\vec{F}_1 = \vec{F}_m + \vec{F}_2' \tag{9.10}$$

where $\vec{F}_2'$ is in opposition to $\vec{F}_2$ and equal in magnitude and

$$\vec{F}_m = \vec{F}_r \tag{9.11}$$

The stator current‡ which causes $\vec{F_1}$ can, corresponding to vector Eq. (9.10), be divided into components

$$\bar{I}_1 = \bar{I}_m + \bar{I}'_2 \tag{9.12}$$

!Attention Here $\bar{I}_m$ can be recognized as the *magnetizing current* which causes the resultant mmf F_r and the resultant flux/pole, Φ_r, while I'_2 is that component of the stator current which balances the reaction F_2 of the rotor current I_2.

Figure 9.4 also shows the relative location of stator coil AA' and the positive direction of current in it. This instantaneous vector picture holds when I_2 has maximum positive value. For F'_2 to cancel F_2, the stator current component which balances the rotor mmf must be in phase with the rotor current as seen from the stator. In terms of magnitudes

$$F'_2 = F_2$$

Using the result of Eq. (5.37)

$$\frac{4\sqrt{2}}{\pi} K_{w1} \left(\frac{N_{ph1}}{P}\right) I'_2 = \frac{4\sqrt{2}}{\pi} K_{w2} \left(\frac{N_{ph2}}{P}\right) I_2$$

or

$$I'_2 = \frac{N_{e2}}{N_{e1}} I_2 = \frac{1}{a} I_2 \tag{9.13}$$

F'_2 is oppositely directed to F_2, for them to cancel out while I'_2 and I_2 must obey the proportionality of Eq. (9.13) and must be in phase.

Further, by reference to Fig. 9.4, it can easily be seen that in the stator the positive direction of emf E_1 opposes the positive direction of I_1, while in the rotor the positive direction of I_2 is in the positive direction of sE_2. This is analogous to the transformer case.

With the direction of positive current in the stator coil AA' marked as in Fig. 9.4 and the direction of the coil axis indicated, the law of induction which will give positive emf in opposition to current is

$$e = +\frac{d\lambda}{dt}$$

(The reader should verify this.) This has the same sign as used in the transformer case so that the flux phasor Φ_r and magnetizing current which creates it lags E_1 by 90°. In the circuit model I_m would therefore be drawn by the magnetizing reactance X_m across E_1.

Remark The reader is reminded here that all the vectors (fields) are stationary with respect to one another and are rotating at n_s with respect to the stator while the rotor is rotating at n with respect to the stator.

9.4 DEVELOPMENT OF CIRCUIT MODEL (EQUIVALENT CIRCUIT) LO 1

Note: As we shall be mostly dealing with phasor magnitudes the superbar on phasor symbols would only be used for phasor equations.

In Sec. 9.3 the behaviour of the induction machine was studied in terms of the basic field phenomenon. The attempt here was purposely focussed on the transformer analogy of induction motor.

‡Current is always meant to be understood as the phase current of a balanced set of 3-phase currents.

Recall Certain facts established so far are summarized below:

1. $$\frac{E_1}{E_2} = a; \quad \frac{I_2'}{I_2} = \frac{1}{a}$$

   where E_2 = standstill rotor emf.

   Further I_2' flows into the positive terminal of E_1 and I_2 flows out of the positive terminal of E_2. Also, I_2 as seen from the stator is the current of stator frequency f and is in phase with I_2', the component of current drawn by the stator to balance the rotor mmf F_2.
2. Like in a transformer, the magnetizing current component I_m of the stator current lags the stator induced emf E_1 by 90°.
3. The induction motor is not merely a transformer which changes voltage and current levels. It in fact behaves like a generalized transformer in which the frequency is also transformed in proportion to slip such that the rotor induced emf is sE_2 and rotor reactance is sX_2.

The circuit model of the induction motor can now be drawn on a per phase basis as in Fig. 9.7(a) wherein the series elements (lumped) of the stator resistance and leakage reactance have been included in the model. The transformer linking the stator and rotor circuits is an *ideal generalized transformer* in which the standstill rotor voltage E_2 and rotor current I_2 are linked to the stator quantities via transformation ratio *a,* while the frequency parameter appears in the rotor circuit through the slip *s*, a mechanical parameter. The mechanical power output appears at the shaft indicated in the figure.

Reflection of Rotor Circuit Parameters to Stator

The rotor circuit can be referred to the stator side by a two-step process—modifying the rotor circuit so that the turn-ratio becomes unity and then carrying out a frequency transformation resulting in an equivalent rotor circuit at the stator frequency.

By multiplying the rotor voltage by *a* and the rotor current by 1/*a*, the rotor impedance gets modified to

$$\bar{Z}_2' = a^2R_2 + jsa^2X_2$$

or
$$\bar{Z}_2' = R_2' + jsX_2'; \; R_2' = a^2R_2, \; X_2' = a^2X_2 \tag{9.14}$$

In this transformation power remains *invariant.* The rotor circuit, after carrying out this step, is drawn in Fig. 9.7(b). This reduces the rotor to an *equivalent rotor* having a unity turn-ratio with the stator.

From the equivalent rotor circuit of Fig. 9.7(b)

$$\bar{I}_2' = \frac{s\bar{E}_1}{R_2' + jsX_2'} \tag{9.15a}$$

Dividing both numerator and denominator by *s*

$$\bar{I}_2' = \frac{\bar{E}_1}{R_2'/s + jX_2'} \tag{9.15b}$$

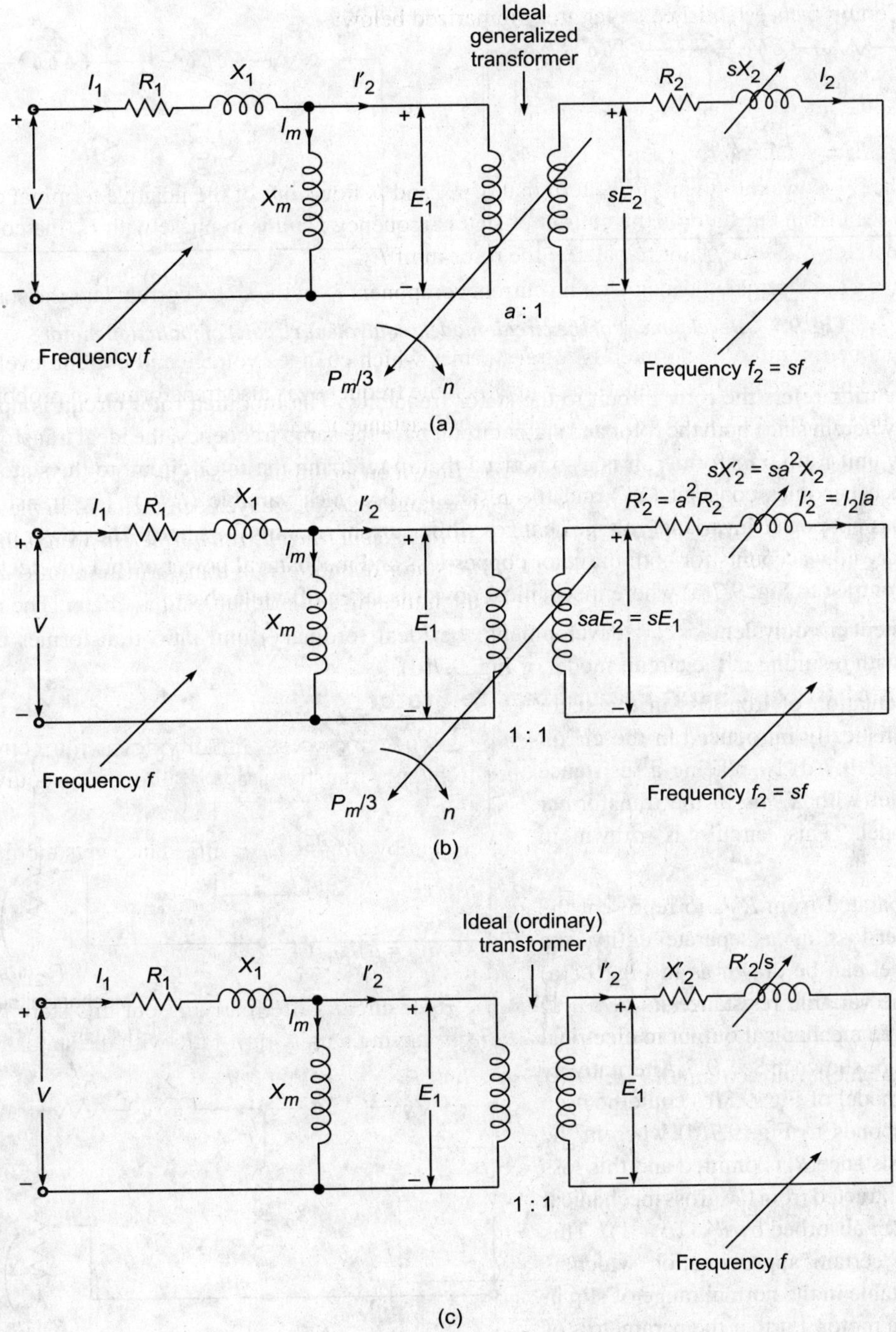
Ideal generalized transformer
I_1 R_1 X_1 I'_2 I_m X_m V E_1
R_2 sX_2 I_2 sE_2
$a : 1$
Frequency f
$P_m/3$ n
Frequency $f_2 = sf$
(a)
$sX'_2 = sa^2X_2$
$R'_2 = a^2R_2$ $I'_2 = I_2/a$
$saE_2 = sE_1$
$1 : 1$
Frequency f
$P_m/3$ n
Frequency $f_2 = sf$
(b)
Ideal (ordinary) transformer
I_1 R_1 X_1 I'_2 I_m X_m V E_1
I'_2 X'_2 R'_2/s E_1
$1 : 1$
Frequency f
(c)

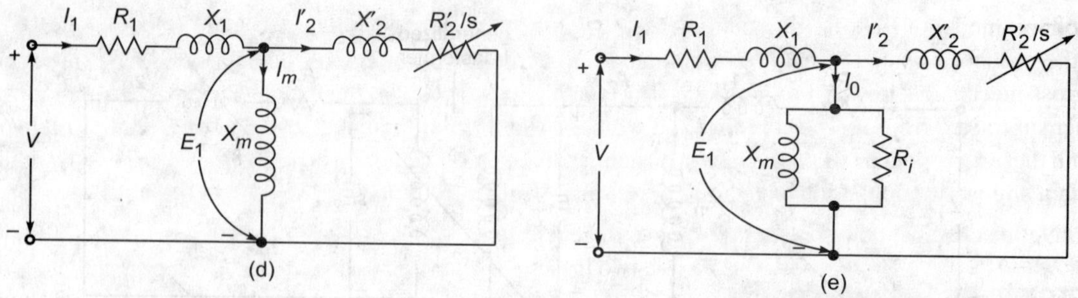

Fig. 9.7 *Development of the circuit model (equivalent circuit) of induction motor*

This simple trick refers the rotor circuit to the stator frequency. The modified rotor circuit is now drawn in Fig. 9.7(c) wherein since both the rotor and stator circuit have the same frequency, the ideal transformer is now a stationary unit-ratio transformer. It is also noticed that in referring the rotor circuit to the stator frequency the reactance becomes constant (X_2') and the resistance becomes variable (R_2'/s). The transformation of Eq. (9.15b) is not power-invariant (voltage changes while current remains the same). The power transferred to the secondary now accounts for both the rotor copper-loss and mechanical power output (in electrical form). This is in contrast to Fig. 9.7(a) where mechanical power is taken off via a shaft.

As a last step in equivalent circuit development, the ideal (ordinary) unit-ratio transformer can now be dispensed with resulting in the circuit model of Fig. 9.7(d).

The representation of iron-loss in the stator can be heuristically introduced in the circuit model of Fig. 9.7(d) by placing a resistance R_i in parallel with X_m as in the transformer circuit model. This circuit is drawn in Fig. 9.7(e).

If R_2' is separated from R_2'/s to represent the rotor copper-loss as a separate entity, the circuit model can be drawn as in Fig. 9.8(a) in which the variable resistance R_2' $(1/s - 1)$ represents the mechanical output in electrical form as will be shown in Sec. 9.5. Alternatively the circuit model of Fig. 9.8(b) could be used (this corresponds to Fig. 9.7(d) wherein the iron loss resistance R_i is omitted and this loss would be subtracted from the gross mechanical output (power absorbed by R_2' $(1/s - 1)$). This amounts to certain approximation which is quite acceptable in the normal range of slip in an induction motor. Further the parameters of this circuit (which does not need the value of R_i) can be easily obtained by two non-loading tests to be described in Section 9.6.

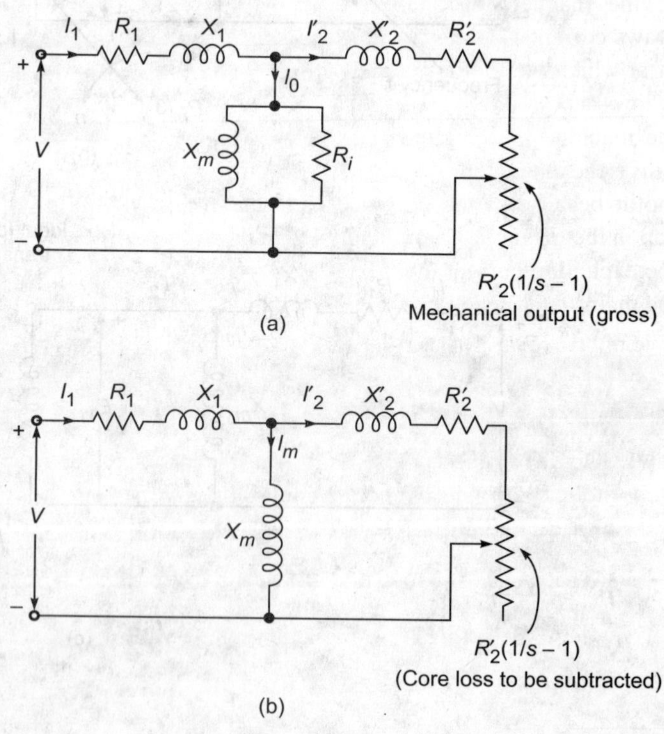

Fig. 9.8

The circuit model of Fig. 9.8(b) would be used for most of the discussion that follows. It may be noted here that the power dissipated in R'_2 (1/s – 1) includes the core loss, which must be subtracted from it to obtain the gross mechanical power. For getting net mechanical power output, the windage and friction loss must be further subtracted from it. The core loss and windage and friction loss together are lumped as *rotational loss* as both these losses occur when the motor is running. The rotational loss in an induction motor is substantially constant at constant applied voltage and motor speed varies very little from no-load to full load.

Note: Net mechanical power = shaft power

Approximate Circuit Model

An approximate circuit model of an induction motor, which results in considerable computational ease in analysis, is obtained by shifting the shunt branch in Fig. 9.8(a) to the input terminals as shown in Fig. 9.9. This step is not so readily justified as in a transformer owing to the relative magnitude of the exciting current (also referred to as the magnetizing current) which, because of the presence of the air-gap, may be as large as 30 – 50% of the full-load current. Further, the primary leakage reactance is also necessarily higher in an induction motor compared to a transformer and so ignoring the voltage drop in primary reactance is not quite justified. It is, therefore, pointed out here that the results obtained by this model are considerably less accurate than that obtained from the models of Fig. 9.8(a) and (b).

The parameters of the induction motor models as presented above are obtained from no-load and blocked-rotor tests which will be taken up in Section 9.6.

It is easily seen from Fig. 9.9 that because of the magnetizing shunt branch which draws current I_0 at almost 90° lagging, the pf at which the motor operates at full-load is low—about 0.8. At light load (small I'_2) the machine power factor is much lower. This is the inherent problem of the induction motor because of the presence of the air-gap in the magnetic circuit and the fact that the excitation current must be drawn from the mains (stator side).

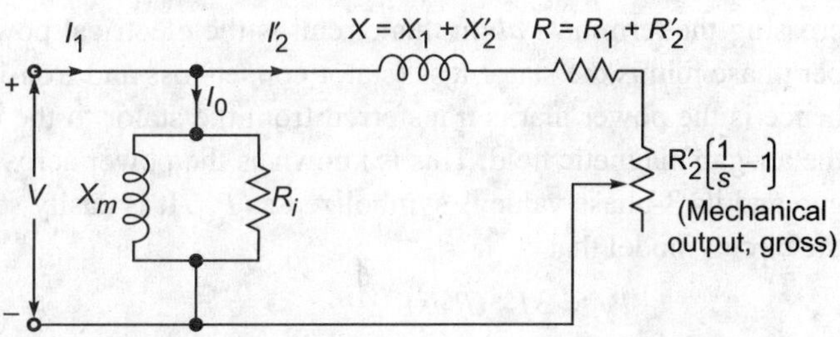

Fig. 9.9 *Approximate circuit model of induction motor*

Concept Check LO 1

1. Give a brief account of squirrel-cage induction motor. Explain qualitatively as to how it develops torque and the nature of its torque-slip characteristic. Why is it called asynchronous motor? ○●●
2. What is the effective turn-ratio of an induction motor? ○○●
3. What is standstill rotor emf and what is its frequency? How does the emf magnitude and frequency vary with speed? ○○●
4. Explain what is meant by standstill reactance of induction motor rotor. How does it vary with speed? ○●●

Note: Difficulty Level ⟶ ○○● — Level 1 and Level 2 Category; ○●● — Level 3 and Level 4 Category; ●●● — Level 5 and Level 6 Category

5. The stator of a slip-ring induction motor with slip-ring terminals open-circuited has a stator excited from 3-phase source. The rotor is run by a prime mover. What will be the frequency of rotor induced emf at the following speeds?
   (a) Half synchronous speed in the same direction as the air-gap field (AGF)
   (b) Half synchronous speed in opposite to AGF
   (c) At synchronous speed in opposite direction to AGF
6. What is meant by the excitation current of an induction motor? Draw its phasor diagram with applied voltages as the reference phasor showing its components. Which is the larger component and why?
7. What is the difference between excitation current and no-load current?
8. Draw the phasor diagram of an induction motor showing applied voltage, magnetizing, coreloss, load current and the line current. Label each component.
9. Write the expression for the resistance in the circuit model, the loss in which is equivalent to the mechanical power developed.
10. Compare and contrast the squirrel-cage and slip-ring induction motors.

9.5 POWER ACROSS AIR-GAP, TORQUE AND POWER OUTPUT

LO 2 Understand the nature of induction motor performance over a complete range of slips by circuit model parameters determination, circle diagram, and concept of transfer/conversion of power through a machine

The circuit model of Fig. 9.7(e) is redrawn in Fig. 9.10. The power crossing the terminals *ab* in this circuit is the electrical power input per phase minus the stator loss (stator copper-loss and iron-loss) and hence is the power that is transferred from the stator to the rotor via the air-gap magnetic field. This is known as the power across the air-gap and its 3-phase value is symbolized as P_G. It is easily seen from the circuit model that

$$P_G = 3I_2'^2 (R_2'/s)$$

$$= \frac{3I_2'^2 R_2'}{s} = \frac{3I_2^2 R_2}{s} \qquad (9.16)$$

or **Important!** Power across air-gap

$$= \frac{\text{rotor copper - loss}}{\text{slip}}$$

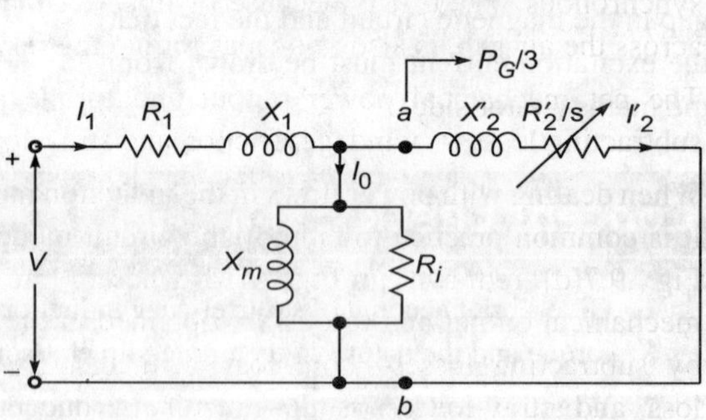

Fig. 9.10

It also follows from Eq. (9.16) that

Rotor copper loss, $P_{cr} = 3I_2'^2 R_2' = sP_G \qquad (9.17)$

Substracting the rotor copper loss from P_G gives the mechanical power output (gross), i.e.

$$P_m = P_G - 3I_2'^2 R_2'$$

$$= 3I_2'^2 R_2' \left(\frac{1}{s} - 1\right) = (1 - s) P_G \qquad (9.18)$$

This means that the gross mechanical power output is three times (3-phase) the electrical power absorbed in resistance $R_2'\left(\dfrac{1}{s}-1\right)$. Figure 9.10 can therefore be drawn as in Fig. 9.11 where R_2'/s is represented as

$$R_2'/s = R_2' + \underbrace{R_2'\left(\frac{1}{s}-1\right)}_{\text{Load resistance}}$$

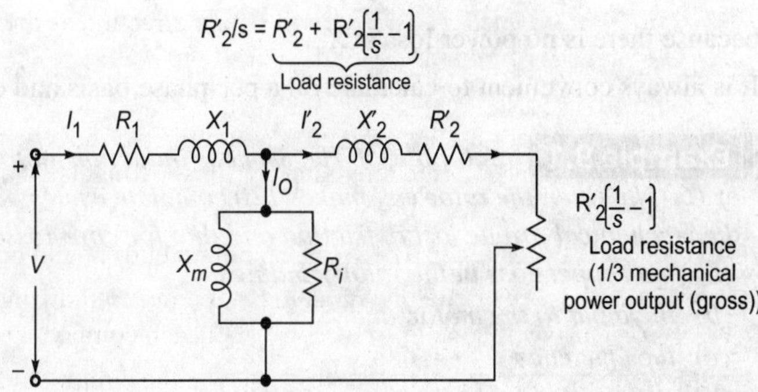

Fig. 9.11

It is noticed from Eq. (9.18) that the mechanical power output is a fraction $(1 - s)$ of the total power delivered to the rotor, while as per Eq. (9.17) a fraction s of it is dissipated as the rotor copper-loss. It is then evident that high-slip operation of the induction motor would be highly inefficient. Induction motors are, therefore, designed to operate at low slip (2–8%) at full-load.

Rotor speed is

$$\omega = (1 - s)\,\omega_s \qquad \text{rad (mech.)}/s$$

The electromagnetic torque developed is then given by

$$(1 - s)\omega_s T = P_m = (1 - s)\,P_G$$

or

$$T = \frac{P_G}{\omega_s} = \frac{3I_2'^2\,(R_2'/s)}{\omega_s}\ \text{Nm},\ I_2'^2R_2' = I_2^2R_2 \tag{9.19}$$

This is an interesting and significant result according to which torque is obtained from the power across the air-gap by dividing it with synchronous speed in rad/s as if this power was transferred at synchronous speed. It is because of this fact that P_G, the power across the air-gap, is also known as torque in *synchronous watts.*

> **Probe**
>
> Establish a relation among the power across air gap (P_G), Rotor copper loss (P_{cr}), and Mechanical power output (P_m).

The net mechanical power output and torque are obtained by subtracting losses—windage, friction and stray-load loss.

When dealing with power flows in the induction motor, it is common practice to employ the circuit model of Fig. 9.7(d) redrawn in Fig. 9.12 wherein the net mechanical output and torque are obtained in the end by subtracting losses—core loss, windage, friction loss and stray-load loss. The error introduced is negligible and the simplification is worthwhile.

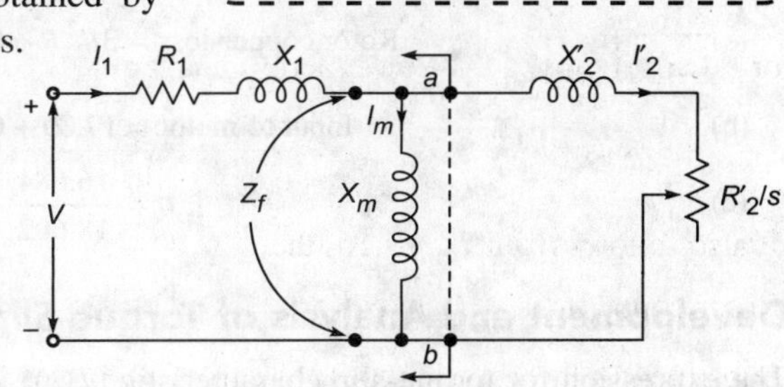

Fig. 9.12

Alternative/Computational Procedure

A convenient computational procedure for a given slip is to calculate

$$\bar{Z}_f = j\,X_m \parallel (R_2'/s + j\,X_2') = R_f + jX_f$$

in the circuit model of Fig. 9.12. Then

$$P_G = \frac{3I_2'^2 R_2'}{s} = 3I_1^2 R_f, \quad T = \frac{3I_1^2 R_f}{\omega_s} \tag{9.20}$$

because there is no power loss in X_m.

It is always convenient to calculate on a per-phase basis and convert to 3-phase values in the end.

Example 9.1 *A 6-pole, 50 Hz, 3-phase induction motor running on full load develops a useful torque of 160 Nm when the rotor emf makes 120 complete cycles per minute. Calculate the shaft power output. If the mechanical torque lost in friction and that for core-loss is 10 Nm. Compute*

(a) the copper-loss in the rotor windings,

(b) the input to the motor, and

(c) the efficiency

The total stator loss is given to be 800 W.

LO 2

Solution

$$f_2 = sf = \frac{120}{60} = 2 \text{ Hz}$$

$$\therefore \quad s = \frac{2}{50} = 0.04 \text{ or } 4\%$$

$$n_s = 1000 \text{ rpm}$$

$$n = (1 - 0.04) \times 1000 = 960 \text{ rpm}$$

$$\omega = \frac{960 \times 2\pi}{60} = 100.53 \text{ rad/s}$$

$$\text{Shaft power output} = 160 \times 100.53 = 16.085 \text{ kW}$$

Mechanical power developed, $\quad P_m = (160 + 10) \times 100.53 = 17.09$ kW

Note that torque of rotational loss is added to shaft power

(a) $$P_m = 3I_2'^2 R_2'\left(\frac{1}{s} - 1\right)$$

$$\text{Rotor copper-loss} = 3I_2'^2 R_2 = P_m\left(\frac{s}{1-s}\right) = 17.090 \times \frac{0.04}{1-0.04} = 712 \text{ W}$$

(b) $$\text{Input to motor} = 17.09 + 0.712 + 0.8 = 18.602 \text{ kW}$$

(c) $$\eta = \frac{16.084}{18.602} = 86.47\%$$

Development and Analysis of Torque-Slip Characteristic

The expression for torque-slip characteristic ($T(s)$) is easily obtained by finding the Thevenin equivalent of the circuit to the left of *ab* in Fig. 9.12.

$$\bar{Z}_{TH} = (R_1 + jX_1) \parallel jX_m = R_{TH} + jX_{TH}$$

$$V_{TH} = V\left[\frac{jX_m}{R_1 + j(X_1 + X_m)}\right]$$

The circuit then reduces to Fig. 9.13 in which it is convenient to take V_{TH} as the reference voltage.

From Fig. 9.13

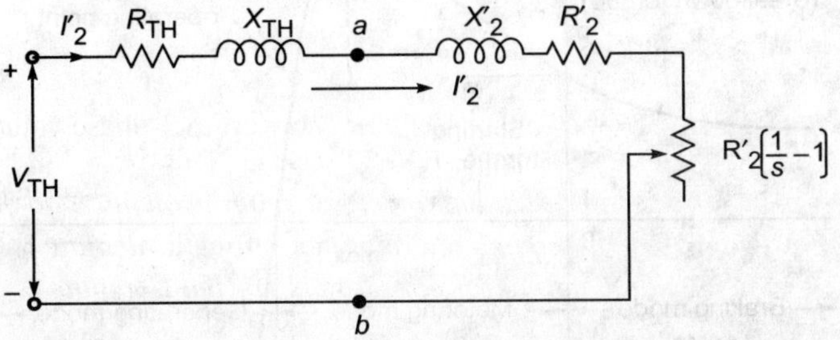

Fig. 9.13 *Thevenin equivalent of induction motor circuit model*

$$I_2'^2 = \frac{V_{TH}^2}{(R_{TH} + R_2'/s)^2 + (X_{TH} + X_2')^2} \tag{9.21}$$

$$T = \frac{3}{\omega_s} \cdot I_2'^2 (R_2'/s)$$

$$= \frac{3}{\omega_s} \cdot \frac{V_{TH}^2 (R_2'/s)}{(R_{TH} + R_2'/s)^2 + (X_{TH} + X_2')^2} \tag{9.22}$$

Equation (9.22) is the expression for torque[§] developed as a function of voltage and slip. For a given value of slip, torque is proportional to the square of voltage. The torque-slip characteristic at fixed (rated) voltage is plotted in Fig. 9.14.

!Attention Certain features of the torque-slip characteristic are listed below:

1. Motoring mode: $0 \le s \le 1$ For this range of slip, the load resistance in the circuit model of Fig. 9.13 is positive, i.e. mechanical power is output or torque developed is in the direction in which the rotor rotates. Also:

(a) Torque is zero at $s = 0$, as expected by qualitative reasoning advanced in Sec. 9.3.

(b) The torque has a maximum value, called the *breakdown torque,* (T_{BD}) at slip $s_{max,T}$. The motor would decelerate to a halt if it is loaded with more than the breakdown torque.

(c) At $s = 1$, i.e. when the rotor is stationary, the torque corresponds to the starting torque, T_s. In a normally designed motor T_s is much less than T_{BD}.

(d) The normal operating point is located well below T_{BD}. The full-load slip is usually 2% to 8%.

(e) The torque-slip characteristic from no-load to somewhat beyond full-load is almost linear.

[§] In terms of the rotor circuit (Fig. 9.7(a))

$$P_G = \frac{3I_2^2 R_2}{s} = \frac{3(sE_2)^2 (R_2/s)}{R_2^2 + (sX_2)^2} = \frac{3E_2^2 (R_2/s)}{(R_2/s)^2 + X_2^2}$$

$$\therefore \quad T = \frac{3}{\omega_s} \cdot \frac{E_2^2 (R_2/s)}{(R_2/s)^2 + X_2^2}$$

Important!

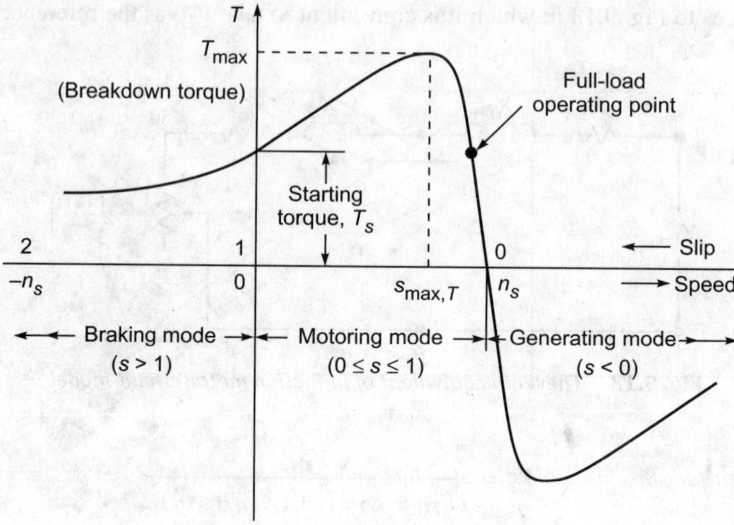

Fig. 9.14 *Torque-slip characteristics*

2. Generating mode: s < 0 Negative slip implies rotor running at super-synchronous speed ($n > n_s$). The load resistance is negative in the circuit model of Fig. 9.13 which means that mechanical power must be put in while electrical power is put out at the machine terminals.

3. Braking mode: s > 1 The motor runs in opposite direction to the rotating field (i.e. n is negative), absorbing mechanical power (braking action) which is dissipated as heat in the rotor copper.

Realization of Maximum (Breakdown) Torque that can be Developed by an Induction Motor

While maximum torque and the slip at which it occurs can be obtained by differentiating the expression of Eq. (9.22), the condition for maximum torque can be more easily obtained from the *maximum power transfer theorem* of circuit theory. As per Eq. (9.19), the torque is maximum when $I_2'^2(R_2'/s)$ is maximum, i.e. maximum power is absorbed by R'_2/s in Fig. 9.13. This condition is given as

$$\frac{R_2'}{s_{\max,T}} = \sqrt{R_{TH}^2 + (X_{TH} + X_2')^2} \quad \text{(matching of impedance magnitudes)}$$

$$s_{\max,T} = \frac{R_2'}{\sqrt{R_{TH}^2 + (X_{TH} + X_2')^2}} \tag{9.23}$$

Substituting in Eq. (9.22) and simplifying

$$T_{\max} = \frac{3}{\omega_s} \cdot \frac{0.5V_{TH}^2}{R_{TH} + \sqrt{R_{TH} + (X_{TH} + X_2')^2}} \tag{9.24}$$

It is immediately observed that the maximum torque is independent of the rotor resistance (R_2'), while the slip at which it occurs is directly proportional to it.

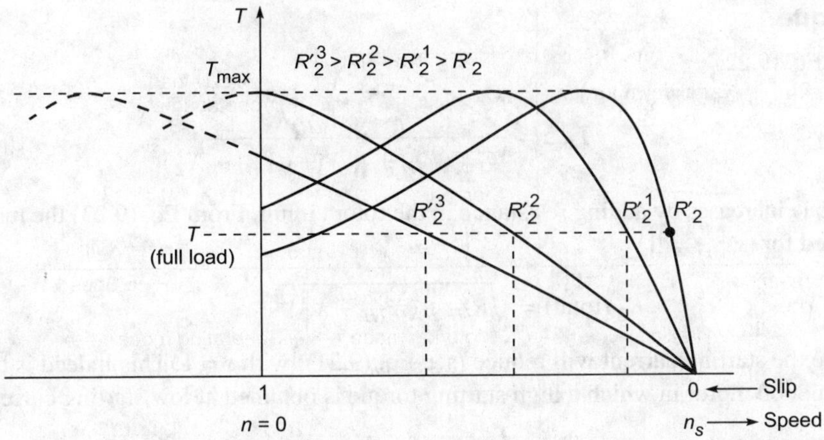

Fig. 9.15(a) *Torque-slip characteristics of induction motor with increasing values of rotor resistance*

The torque-slip characteristic of a slip-ring induction motor can be easily modified by adding external resistance as shown in Fig. 9.15(a) by four such characteristics with a progressively increasing resistance in the rotor circuit. It may be seen that as per Eq. (9.24), the maximum torque remains unchanged while as per Eq. (9.23) the slip at maximum torque proportionally increases as resistance is added to the rotor circuit.

Figure 9.15(b) indicates the *T-s* profile at various supply voltages. Speed can also be controlled in this way by changing the stator voltage. It may be noted that the torque developed in an induction motor in proportional to the square of the terminal voltage.

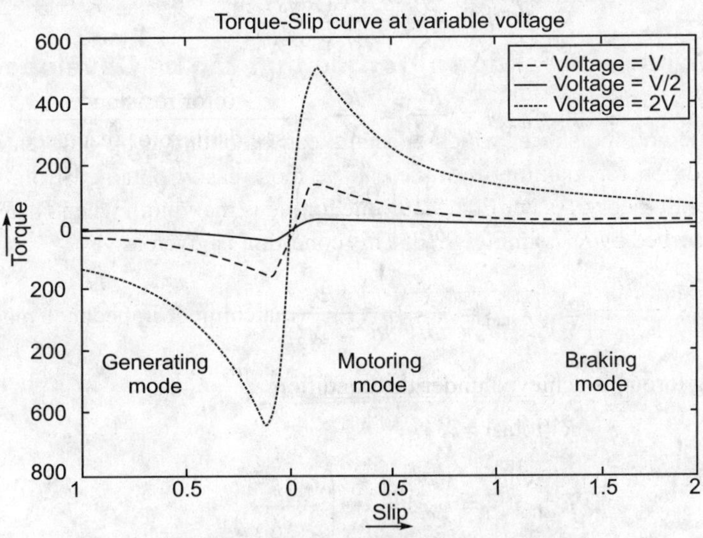

Fig. 9.15(b) *Torque-Slip curve at variable voltage*

Starting Torque

Letting $s = 1$ in Eq. (9.22)

$$T_{\text{start}} = \frac{3}{\omega_s} \cdot \frac{V_{TH}^2 R_2'}{(R_{TH} + R_2')^2 + (X_{TH} + X_2')^2} \tag{9.25}$$

The starting torque increases by adding resistance in the rotor circuit. From Eq. (9.23) the maximum starting torque is achieved for ($s_{\max,T} = 1$)

$$R_2' \text{(total)} = \sqrt{R_{TH}^2 + (X_{TH} + X_2')^2} \tag{9.26}$$

At the same time the starting current will reduce (see Eq. (9.21)) with $s = 1$. This indeed is the advantage of the slip-ring induction motor in which a high starting torque is obtained at low starting current.

An Approximation

Sometimes for getting a feel (rough answer) of the operational characteristic, it is convenient to assume the stator impedance to be negligible which leads to (see Fig. 9.13)

$$R_{TH} = 0, \quad X_{TH} = 0$$

in the Thevenin equivalent circuit of Fig. 9.13 so that $V_{TH} = V$. It then follows from Eqs (9.21) to (9.26) that

$$I_2' = \frac{V}{\sqrt{(R_2'/s)^2 + X_2'^2}} \tag{9.27}$$

$$T = \frac{3}{\omega_s} \cdot \frac{V^2 (R_2'/s)}{(R_2'/s)^2 + X_2'^2} \tag{9.28}$$

$$s_{\max,T} = \frac{R_2'}{X_2'} = \frac{R_2}{X_2} = \frac{\text{rotor resistance}}{\text{standstill rotor reactance}} \tag{9.29}$$

$$T_{\max} = \frac{3}{\omega_s} \cdot \left(\frac{0.5V^2}{X_2'} \right) \tag{9.30}$$

and

$$T_{\text{start}} = \frac{3}{\omega_s} \cdot \frac{V^2 R_2'}{R_2'^2 + X_2'^2} \tag{9.31}$$

The maximum starting torque is achieved under the condition

$$R_2'\text{(total)} = X_2' \tag{9.32}$$

and

$$T_{\text{start(max)}} = T_{\max} = \frac{3}{\omega_s} \left(\frac{0.5V^2}{X_2'} \right) \tag{9.33}$$

Note The above relationships are expressed in rator quantities (I_2, R_2, X_2) by changing V to (V/a).

Some Approximate Relationships at Low Slip

Around the rated (full-load) speed, the slip of the induction motor is so small that

$$R_2'/s >> X_2'$$

so that X_2' can be altogether neglected in a simplified analysis. Equations (9.27) and (9.28) then simplify to

$$I_2' = \frac{sV}{R_2'} \tag{9.34}$$

and

$$T = \frac{3}{\omega_s} \cdot \frac{sV^2}{R_2'} \tag{9.35}$$

It is immediately observed from Eq. (9.35) that the torque-slip relationship is nearly linear in the region of low slip which explains the linear shape of the characteristic as shown in Fig. 9.14.

Maximum Power Output

Since the speed of the induction motor reduces with load, maximum mechanical power output does not correspond to the speed (slip) at which maximum torque is developed. For maximum mechanical power output, the condition is obtained from Fig. 9.13

$$R_s'\left(\frac{1}{s} - 1\right) = \sqrt{(R_{TH} + R_2')^2 + (X_{TH} + X_2')^2} \tag{9.36}$$

The maximum power output can then be found corresponding to the slip defined by Eq. (9.36). However, this condition corresponds to very low efficiency and very large current and is well beyond the normal operating region of the motor.

Limitation of the Circuit Model

The values of the circuit model parameters must be determined under conditions closely approximating the operating condition for which the model is to be used. Circuit model parameters valid for the normal operating condition would give erroneous results when used for abnormal values of slip. At starting, the motor draws many times the rated current resulting in saturation of core and consequent increase of the stator and rotor leakage reactances. Also the rotor frequency being high (same as stator), the rotor conductors present a higher resistance. Therefore parameters good for normal operating conditions would give a pessimistic results for starting current (smaller than actual value) and starting torque higher than actual value.

Example 9.2 *A 6-pole, 50 Hz, 3-phase induction motor has a rotor resistance of 0.25 Ω per phase and a maximum torque of 10 Nm at 875 rpm. Calculate (a) the torque when the slip is 5%, and (b) the resistance to be added to the rotor circuit to obtain 60% of the maximum torque at starting. Explain why two values are obtained for this resistance. Which value will be used? The stator impedance is assumed to be negligible.*

LO 2

Solution

$$n_s = \frac{120 \times 50}{6} = 1000 \text{ rpm}; \quad \omega_s = \frac{2\pi \times 1000}{60} = 104.7 \text{ rad/s}$$

$$s_{\max,T} = \frac{1000 - 875}{1000} = 0.125$$

$$s_{\max,T} = \frac{R_2'}{X_2'} = \frac{R_2}{X_2}$$

$$X_2 = \frac{R_2}{s_{\max,T}} = \frac{0.25}{0.125} = 2\,\Omega$$

$$T_{\max} = \frac{3}{\omega_s} \cdot \left(\frac{0.5V^2}{X_2'} \right)$$

or $$10 = \frac{3}{104.7} \cdot \left(\frac{0.5V^2}{X_2'} \right) = \frac{3}{104.7} \cdot \left(\frac{0.5V^2/a^2}{X_2} \right)$$

or $$V' = V/a = \sqrt{1396}\ V$$

(a) $$T = \frac{3}{\omega_s} \cdot \frac{V_2^2 (R_s'/s)}{(R_2'/s)^2 + X_2'^2} = \frac{3}{\omega_s} \cdot \frac{V'^2 (R_2/s)}{(R_2/s)^2 + X_2^2}$$

$$s = 0.05$$

$$T = \frac{3}{104.7} \cdot \frac{1396 \times 0.25/0.05}{(0.25/0.05)^2 + (2)^2} = 6.9 \text{ Nm}$$

(b) From Eqs (9.30) and (9.31)

$$\frac{T_{\text{start}}}{T_{\max}} = \frac{(R_2 + R_{\text{ext}})}{(R_2 + R_{\text{ext}})^2 + X_2^2} \cdot \frac{X_2}{0.5}$$

$$0.6 = \left(\frac{R_t}{R_t^2 + 4} \right) \frac{2}{0.5}$$

where $$R_t = R_2 + R_{\text{ext}}$$

$$R_t^2 - 6.67\,R_t + 4 = 0$$

$$R_t = 0.667,\ 6.0\ \Omega$$

$$R_{\text{ext}} = R_t - r_2 = 0.417,\ 5.75\ \Omega$$

The former value will be used, i.e.

$$R_{\text{ext}} = 0.417\ \Omega$$

It may be noted that $R_{\text{ext}} = 5.75\ \Omega$ will correspond to $T_{\max}$ lying in the region. $s > 1$ as shown in Fig. 9.16.

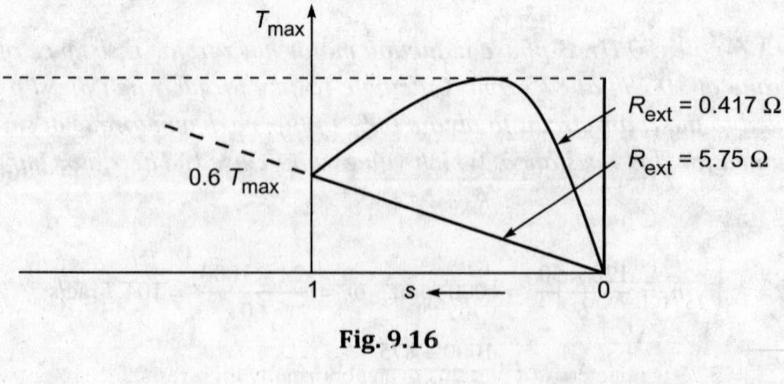

Fig. 9.16

Example 9.3 *A 3-phase induction motor has a starting torque of 100% and a maximum torque of 200% of the full-load torque.*

Find:

(a) slip at maximum torque,

(b) full-load slip, and

(c) rotor current at starting in pu of full-load rotor current.

Neglect the stator impedance.

LO 2

Solution

$$T_{fl} = \frac{3}{\omega_s} \cdot \frac{V^2 (R_s'/s_{fl})}{(R_2'/s_{fl})^2 + (X_2')^2} \quad \text{(i)}$$

$$T_s = \frac{3}{\omega_s} \cdot \frac{V^2 R_2'}{(R_2'^2 + X_2'^2)} \quad \text{(ii)}$$

$$T_{max} = \frac{3}{\omega_s} \cdot \frac{0.5V^2}{X_2'} \quad \text{(iii)}$$

(a) Dividing Eq. (iii) by Eq. (ii)

$$\frac{T_{max}}{T_s} = \frac{0.5(R_2' + X_2'^2}{X_2' R_2'} \quad \text{(iv)}$$

Let $$R_2' = kX_2' \quad \text{(v)}$$

Then $$\frac{2}{1} = \frac{0.5(1+k^2)X_2'^2}{kX_2'^2}$$

or $$k^2 - 4k + 1 = 0$$

or $$k = 0.268 = s_{max,T}^{¶} \quad \text{(vi)}$$

(b) Dividing Eq. (iii) by Eq. (i)

$$\frac{T_{max}}{T_{fl}} = \frac{0.5[(R_2'/s_{fl})^2 + (X_2')^2]}{X_2' R_2'} s_{fl}$$

$$2 = \frac{0.5[(k/s_{fl})^2 + 1]}{k} s_{fl}$$

$$s_{fl}^2 - 1.072 s_{fl} + 0.072 = 0$$

$$s_{ft} = 0.072, \ 1 \text{ (rejected)}$$

(c) $$I_2'^2(\text{start}) = \frac{V^2}{(R_2'^2)^2 + X'^2}$$

$$I_2'^2(\text{full-load}) = \frac{V^2}{(R_2'/s_{fl})^2 + X_2'^2}$$

$$\left(\frac{I_2'(\text{start})}{I_2'(\text{full-load})}\right)^2 = \frac{(R_2'/s_{fl})^2 + X_2'^2}{R_2'^2 + X_2'^2} = \frac{(k/s_{fl})^2 + 1}{k^2 + 1} = \frac{(0.268/0.072)^2 + 1}{(0.268)^2 + 1}$$

¶The second value of $s_{max,T} = 3.73$ is rejected as it pertains to an abnormally high-resistance rotor (which would be highly inefficient).

$$\frac{I_2'\text{(start)}}{I_2'\text{(full-load)}} = \sqrt{13.86} = 3.72$$

9.6 | TESTS TO DETERMINE CIRCUIT-MODEL PARAMETERS

LO 2

As the circuit model of an induction motor is similar to that of a transformer, the parameters of the model can be obtained by means of nonloading tests as in the case of the transformer—no-load test (corresponding to the OC test on the transformer) and the blocked-rotor test (corresponding to the SC test on transformer).

The No-Load (NL) Test

In this test the motor is run on no-load at rated voltage and frequency. The applied voltage and current and power input to motor are measured by the metering as per Fig. 9.17.

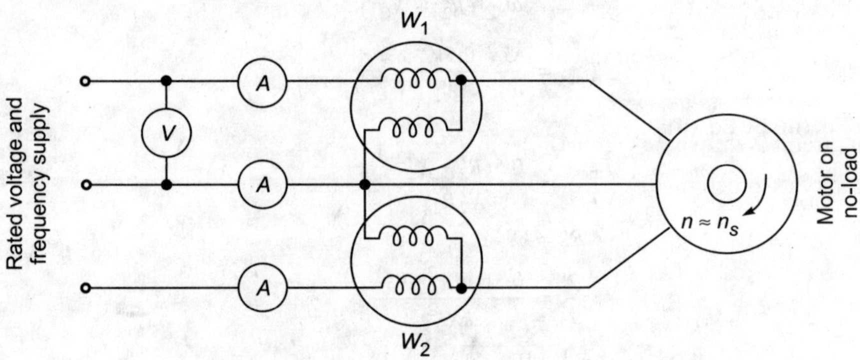

Fig. 9.17 *Connection diagram for conducting no-load test on induction motor*

Let the meter readings be

Power input = P_0 (3-phase)

Current = I_0 (average of the three meter readings)

Voltage = V_0 (line-to-line rated voltage)

Important! Power input at no-load (P_0) provides losses only as the shaft output is zero. These losses comprise,

$$P_0 \text{ (no-load loss)} = P_{c1} \text{ (stator copper loss)} + [P_i \text{ (iron/core loss)} + P_{wf} \text{ (windage and friction loss)} = \text{Rotational loss}]$$

wherein core loss occurs only in the stator as the slip is extremely low (of the order of 0.001) and so the frequency of rotor current is as low as 0.05 Hz.

The magnitude of no-load current in an induction motor is about 30–40% of full-load current because of the air-gap. So the stator copper loss at no-load needs to be accounted for. This can be estimated by measuring dc stator resistance and correcting to ac value (50 Hz) and corrected for temperature (°C).

The mechanical power developed corresponds to P_{wf} only and so, as already mentioned above the slip is very low and the output resistance

$$R_2'\,(1/s_0 - 1) = \text{very large}$$

Also $R_2'/s_0 >> X_2'$ and so X_2' can be ignored. The corresponding no-load circuit model is drawn in Fig. 9.18(a) wherein R_2'/s_0 appears in parallel to R_i. By combining the parallel shunt resistances, the final circuit at no-load is as given in Fig. 9.18(b). Here R_{iwf} accounts for rotational loss, i.e., core loss and windage and friction loss. Magnitude-wise $R_{iwf} >> X_m$.

R_1, the stator resistance, is found by dc testing of the stator winding and correcting the value to ac operation (at 50 Hz). X_1, the stator leakage reactance, will be found from the blocked-rotor test which follows. We can then find X_m and R_{iwf} from the no-load (NL) test data. By simplification of the circuit of Fig. 9.18(b), we get

$$R_0 = R_1 + \frac{X_m^2/R_{ifw}}{1+(X_m/R_{iwf})^2} \tag{9.37}$$

$$X_0 = X_1 + \frac{X_m}{1+(X_m/R_{iwf})^2} \tag{9.38}$$

The equivalent circuit is drawn in Fig. 9.18(c).

It can be justifiably assumed that $(X_m/R_{iwf})^2 = 0$, so we get from the above equations

$$R_{iwf} = \frac{X_m^2}{R_0 - R_1} \tag{9.39}$$

and

$$X_m = X_0 - X_1 \tag{9.40}$$

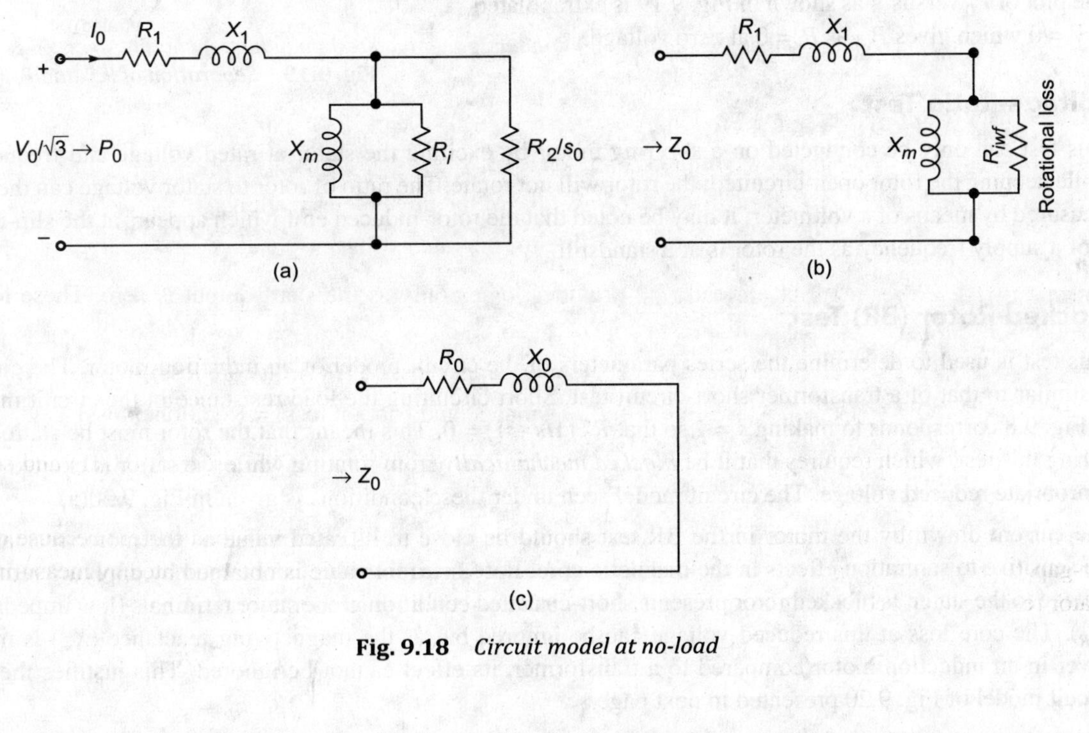

Fig. 9.18 *Circuit model at no-load*

From the NL test data (V_0, I_0, P_0) we can find from the circuit of Fig. 9.18(c).

$$Z_0 = \frac{V_0/\sqrt{3}}{I_0}$$

$$R_0 = \frac{P_0/3}{I_0^2} \quad (9.41)$$

$$X_0 = (Z_0^2 - R_0^2)^{1/2} \quad (9.42)$$

Probe

When an induction motor works at no-load, what are the losses that occur? Also give their significance.

By substituting the values of R_0 and X_0 in Eqs (9.39) and (9.40) respectively, we obtain X_m and R_{iwf}. R_i, the stator core loss resistance can be found out if the additional test of separating core loss from the windage and friction loss, as described below, is carried out.

Separation of Fixed Losses

!Attention The separation of these two losses can be carried out by the no-load test conducted from the variable-voltage, rated-frequency supply. As the voltage is reduced below the rated value, the core-loss decreases almost as the square of voltage. Since the slip does not increase significantly, the windage and friction loss remains almost constant. The voltage is continuously reduced till the machine slip suddenly begins to increase and the motor tends to stall. At no-load this happens at a sufficiently reduced voltage. The plot of P_0 versus V as shown in Fig. 9.19 is extrapolated to $V = 0$ which gives P_{wf} as $P_i = 0$ at zero voltage.

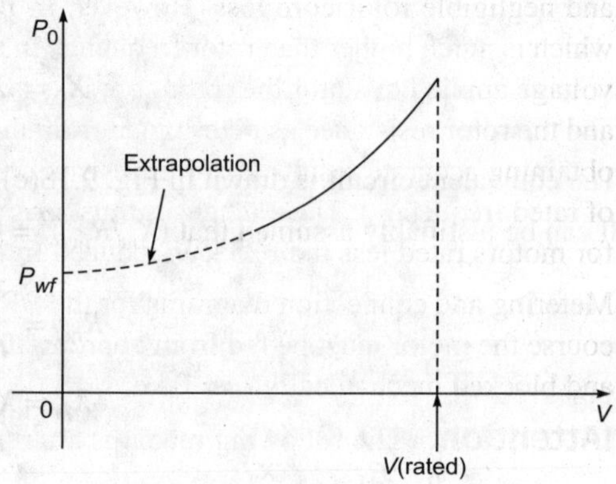

Fig. 9.19 *Separation of P_{wf} and P_i*

Voltage-Ratio Test

This test can only be conducted on a slip-ring motor by exciting the stator at rated voltage and frequency while keeping the rotor open-circuited; the rotor will not rotate. The ratio of rotor to stator voltage can then be measured by means of a voltmeter; it may be noted that the rotor-induced emf which appears at the slip-rings is of a supply frequency as the rotor is at a standstill.

Blocked-Rotor (BR) Test

This test is used to determine the series parameters of the circuit model of an induction motor. The circuit is similar to that of a transformer short-circuit test. Short circuiting the load resistance in the circuit model of Fig. 9.8 corresponds to making $s = 1$ so that R_2' $(1/s - 1) = 0$. This means that the rotor must be stationary during this test, which requires that it be *blocked mechanically* from rotating while the stator is excited with appropriate reduced voltage. The circuit model seen under these conditions is given in Fig. 9.20(a).

The current drawn by the motor in the BR test should be close to its rated value as the motor reactances are sensitive to saturation effects in the magnetic core. Rated current value is obtained by applying reduced voltage to the stator as blocked rotor presents short-circuited condition at the stator terminals (low impedance Z_{BR}). The core loss at this reduced voltage can be ignored but as the magnetizing reactance (X_m) is much lower in an induction motor compared to a transformer, its effect cannot be ignored. This justifies the BR circuit model of Fig. 9.20 presented in next page.

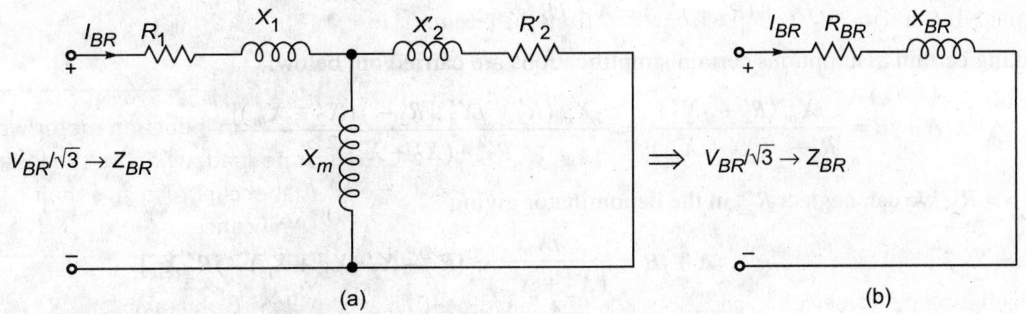

Fig. 9.20 *Circuit model in blocked-rotor test*

In normal operating range of an induction motor the slip is low (2 – 8%). This means low rotor frequency and negligible rotor core loss. However, in BR test the rotor frequency is the same as the stator frequency which is much higher than rotor frequency in normal operation (it is almost negligible). Though with reduced voltage applied to stator the rotor core loss is small, the higher rotor frequency would affect the value of R_{BR} and the rotor resistance as determined from the test it will be smaller. (see the last para of this section). So for obtaining accurate results for rotor resistance, the BR test needs to be conducted at reduced frequency (25% of rated frequency). The reactances thus obtained are then scaled up to the rated frequency (50 Hz). However, for motors rated less than 25 kW, reduced frequency test is not warranted.

Metering and connection diagrams for the BR test are the same as in the connection diagram of Fig. 9.17. Of course the motor must be fed from appropriate low voltage (variable) source of frequency as discussed above and blocked mechanically ($n = 0$ i.e. $s = 1$).

!Attention The following readings are recorded during this test:

V_{BR} = stator voltage (line-to-line)

I_{BR} = stator current (average of three ammeter readings)

P_{BR} = power fed into stator; this mainly constitutes the copper loss in stator and rotor. At reduced voltage core loss (even in stator) is negligible.

From these test readings we can compute

$$Z_{BR} = \frac{V_{BR}/\sqrt{3}}{I_{BR}} \tag{9.43}$$

$$R_{BR} = \frac{P_{BR}/3}{I_{BR}^2} \tag{9.44}$$

$$X_{BR} = (Z_{BR}^2 - R_{BR}^2)^{1/2} \tag{9.45}$$

These values constitute the series equivalent of the BR test (Fig. 9.20(b)).

We, however, need to determine the circuit model parameters R_2', X_1 X_2', while R_1 is known from the dc test. From the motor circuit in BR test as given in Fig. 9.20(b) we can write

$$\bar{Z}_{BR} = R_{BR} + jX_{BR}$$

$$= (R_1 + jX_1) + jX_m \parallel (R_2' + jX_2') \tag{9.46}$$

Let $$jX_m \parallel (R_2' + jX_2') = A + jB$$

By making certain assumptions certain simplifications are carried out below:

$$A + jB = \frac{jX_m (R_2' + jX_2')}{R_2' + j(X_2' + X_m)} = \frac{jX_m (R_2' + jX_2')[R_2' - j(X_2' + X_m)]}{R_2'^2 + (X_2' + X_m)^2} \tag{9.47}$$

As $X_m >> R_2'$, we can neglect $R_2'^2$ in the denominator giving

$$A + jB = \frac{jX_m}{(X_2' + X_m)^2}[R_2'^2 + X_2'(X_2' + X_m) - jR_2' X_m]$$

But $$R_2'^2 + X_2'(X_2' + X_m) = R_2'^2 + X_2'^2(1 + X_m/X_2')$$

$$= X_2'^2\left[1 + \frac{X_m}{X_2'} + \left(\frac{R_2'}{X_2'}\right)^2\right]$$

$$\approx X_2'(X_2' + X_m);\ \text{as} \left(\frac{R_2'}{X_2'}\right)^2 << 1$$

Then $$A + jB = R_2'\left(\frac{X_m}{X_2' + X_m}\right)^2 + j\left(\frac{X_2' X_m}{X_2' + X_m}\right)$$

Equation (9.46) then takes the form

$$\bar{Z}_{BR} = R_{BR} + j\,X_{BR}$$

$$= (R_1 + jX_1) + \left[R_2'\left(\frac{X_m}{X_2' + X_m}\right)^2 + j(X_2' \parallel X_m)\right] \tag{9.48}$$

The following results can then be written down with the knowledge that

$$X_2' \parallel X_m = X_2' \text{ as } X_m >> X_2' \tag{9.49}$$

Then $$X_1 + X_2' = X_{BR}$$

$$R_2' = (R_{BR} - R_1)\left(\frac{X_m + X_2'}{X_m}\right)^2 \approx (R_{BR} - R_1) \text{ as } X_m >> X_2' \tag{9.50}$$

If $(X_m + X_2') > 10\,R_2'$, which is usually the case, then Eq. (9.50) for R_2' causes an error of less than 1%. At this stage we need to separate out X_1 and X_2', which is not possible by the data of this (BR) test. Generally it is fairly accurate to assume that

$$X_1 = X_2' = X_{BR}/2 \tag{9.51}$$

If the BR test is conducted at rated frequency, two factors affect the value of R_2' as observed above. Firstly the rotor (winding) resistance increases as the frequency of rotor currents is the same as the rated frequency, while under normal operating conditions it is very small; few hertz or so. Secondly the frequency of rotor flux alterations is also at rated frequency. The rotor core then presents an effective resistance in parallel to R_2' thereby reducing effective R_2' as measured. These two effects tend to cancel each other out. So no reduced frequency testing is needed for small size motors (less than 25 kW).

Implementing No Load and Blocked Rotor Test in Developing Equivalent Circuit Model

With the circuit parameters obtained as above, circuit model in two alternative forms are drawn in Figs 9.21(a) and (b).

(1) Circuit Model of Fig. 9.21(a) Here R_{iwf} represents core loss and windage and friction loss. Approximation lies in the fact that actual windage and friction is taken off the shaft in mechanical form. A more exact circuit would be possible if these two losses are separated out by the test described above. Usually such accuracy is not needed.

It may be noted here that power absorbed by the output resistance $R_2'(1/s - 1)$ is the net mechanical output.

(2) IEE Circuit Model - Fig. 9.21(b) Here R_{iwf} is removed so that only shunt reactance is the magnetizing reactance X_m. Core loss and windage and friction loss must now be subtracted from the mechanical output (power in $R_2'(1/s - 1)$). This is a different type of approximation and is found *more accurate* than the approximation made in model of Fig. 9.21(a). This circuit is simpler to analyze and therefore, as mentioned already, is commonly adopted.

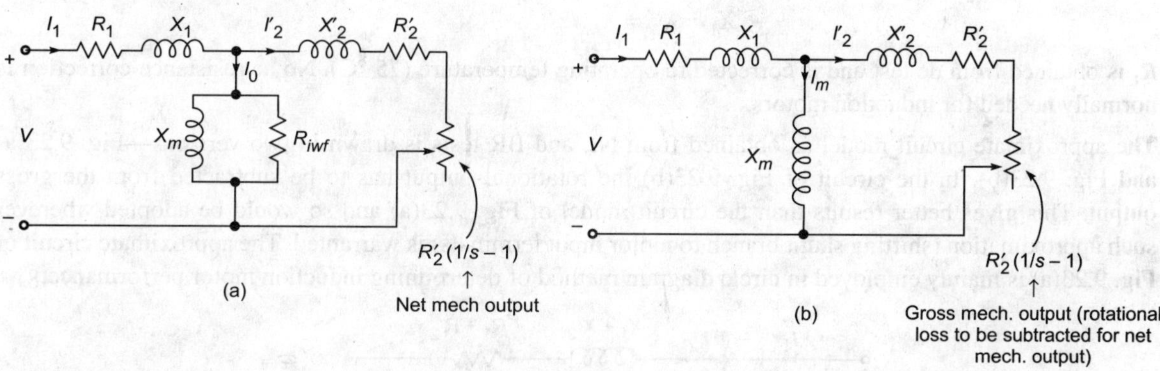

Fig. 9.21

Approximate Circuit Model

The approximate circuit model has been given in Fig. 9.9. With reference to this circuit, the circuit model seen during NL test is as given in Fig. 9.22(a) where the rotor is regarded as open circuit. The circuit model during BR test is as given in Fig. 9.22(b) wherein the shunt branch is ignored.

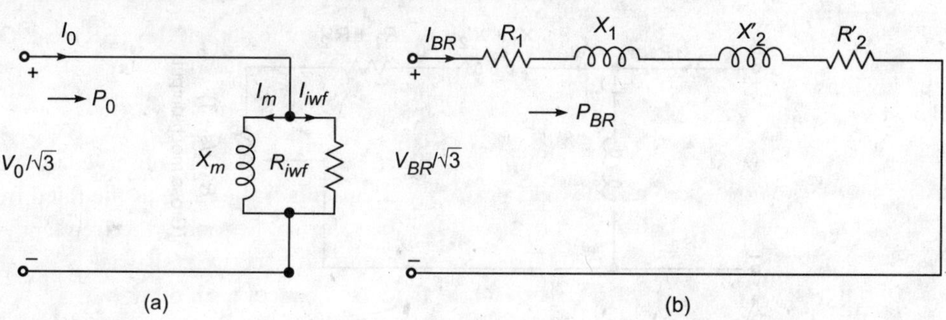

Fig. 9.22

Recall NL test

$$R_{iwf} = \frac{(V_0/\sqrt{3})^2}{(P_r/3)} = \frac{V_0^2}{P_r}\,;\; P_r = P_0 - 3I_0^2 R_1$$

$$= \text{stator core loss + windage and friction loss } (\textit{rotational loss})$$

$$I_{iwf} = \frac{V_0/\sqrt{3}}{R_{iwf}};\quad I_m = (I_0^2 - I_{iwf}^2)^{1/2}$$

$$X_m = \frac{V_0/\sqrt{3}}{I_m}$$

BR test

$$Z_{BR} = \frac{V_{BR}/\sqrt{3}}{I_{BR}} \text{ and } R_{BR} = \frac{P_{BR}/3}{I_{BR}^2} = R_1 + R_2'$$

$$X_{BR} = (Z_{BR}^2 - R_{BR}^2)^{1/2} = X_1 + X_2'$$

with usual assumption

$$X_1 = X_2' = X_{BR}/2$$

$\boldsymbol{R_1}$ is obtained from dc test and is corrected to operating temperature (75 °C). No ac resistance correction is normally needed for induction motors.

The approximate circuit model as obtained from NL and BR tests is drawn in two versions—Fig. 9.23(a) and Fig. 9.23(b). In the circuit of Fig. 9.23(b) the rotational output has to be subtracted from the gross output. This gives better results than the circuit model of Fig. 9.23(a) and so would be adopted wherever such approximation (shifting shunt branch to motor input terminals) is warranted. The approximate circuit of Fig. 9.23(a) is mainly employed in circle diagram method of determining induction motor performance.

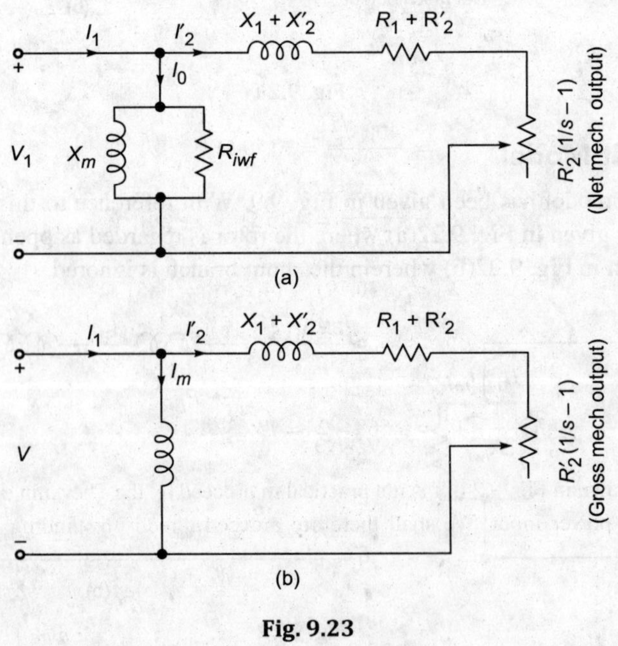

Fig. 9.23

Example 9.4 *A 400 V, 6-pole, 3-phase, 50 Hz star-connected induction motor running light at rated voltage takes 7.5 A with a power input of 700 W. With the rotor locked and 150 V applied to the stator, the input current is 35 A and power input is 4000 W; the stator resistance/phase being 0.55 Ω under these conditions. The standstill reactances of the stator and rotor as seen on the stator side are estimated to be in the ratio of 1: 0.5. When the motor is operating at a slip of 4%, calculate the stator current, its pf, net mechanical output and torque. What is motor efficiency under these conditions of operation?*

Also calculate the motor performance as above by using the parameters of the approximate circuit model. Compare the results.

LO 2

Solution BR test

$$Z_{BR} = \frac{150/\sqrt{3}}{35} = 2.47\ \Omega$$

$$R_{BR} = \frac{4000 \times 3}{(35)^2} = 1.09\ \Omega$$

$$X_{BR} = \sqrt{(2.47)^2 - (1.09)^2} = 2.22\ \Omega$$

$$X_1 + X_2' = 2.22$$

$$X_1 + 0.5\,X_1 = 2.22$$

or

$$X_1 = 1.48\ \Omega, \quad X_2' = 0.74\ \Omega$$

$$R_2' = (R_{BR} - R_1)\left(\frac{X_m + X_2'}{X_m}\right)^2$$

$$X_m = 29.03\ \Omega \text{ (obtained from NL test below)}$$

$$R_2' = (1.09 - 0.55)\left(\frac{29.03 + 0.74}{29.03}\right)^2 = 0.568\ \Omega$$

NL test

$$Z_0 = \frac{400/\sqrt{3}}{7.5} = 30.79\ \Omega$$

$$R_0 = \frac{700/3}{(7.5)^2} = 4.15\ \Omega$$

$$X_0 = \sqrt{(30.79)^2 - (4.15)^2} = 30.51\ \Omega$$

$$X_m = X_0 - X_1 = 30.51 - 1.48 = 29.03\ \Omega$$

Output resistance

$$R_2'\left(\frac{1}{s} - 1\right) = 0.568 \times 24 = 13.63\ \Omega$$

The exact circuit model is drawn in Fig. 9.24. It is not practical to proceed by the Thevenin equivalent method for finding stator, its power factor and power input. We shall therefore proceed directly by finding $\bar{Z}_{in}$ in the circuit model of Fig. 9.24.

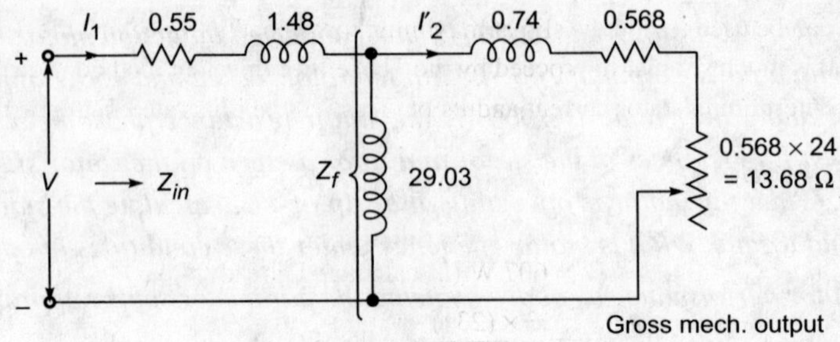

Fig. 9.24

$$\bar{Z}_f = j\,X_m \parallel (R_2'/s + jX_2')$$
$$= \text{parallel combination of rotor impedance and magnetizing reactance}$$
$$= j\,29.03 \parallel (14.2 + j\,0.74)$$
$$= 10.98 + j\,5.96 = R_f + j\,X_f$$
$$R_f = 10.98\ \Omega$$
$$\bar{Z}_{in} = (0.55 + j\,1.48) + (10.98 + j\,5.98)$$
$$= 11.53 + j\,7.44 = 13.72\ \angle 32.8°\ \Omega$$
$$I_1 = \frac{231}{13.72} = 16.84\ \text{A}$$
$$pf = \cos 32.8° = 0.84 \text{ lagging}$$
$$\text{Power input, } P_{in} = \sqrt{3} \times 400 \times 16.84 \times 0.84 = 9.80\ \text{kW}$$

As the magnetizing reactance is nondissipative, power in R_f will constitute the power across air-gap. Thus

$$P_G = 3I_1^2\,R_f\ (= 3I_2'^2\,R_2'/s) = 3 \times (16.84)^2 \times 10.98 = 9.34\ \text{kW}$$
$$\text{Gross mechanical output} = (1 - s)\,P_G = (1 - 0.04) \times 9.34 = 8.97\ \text{kW}$$
$$\text{Rotational loss} = 700 - 3 \times (7.5)^2 \times 0.55 = 607\ \text{W}$$

Net mechanical output, $$P_{\text{out}} = 8.97 - 0.607 = 8.36\ \text{kW}$$

$$\omega_s = \frac{1000 \times 2\pi}{60} = 104.72\ \text{rad/s}$$
$$\text{Torque (net)} = \frac{8360}{104.72(1 - 0.04)} = 83.16\ \text{Nm}$$

Efficiency, $$\eta = \frac{8.36}{9.80} \times 100 = 85.31\%$$

Enumeration of losses

$$\text{Rotational loss} = 607\ \text{W}\ \ (\text{core + windage + friction})$$
$$\text{Stator copper loss} = 3 \times (16.8)^2 \times 0.55 = 468\ \text{W}$$
$$\text{Rotor copper loss} = sP_G = 0.04 \times 9340 = 374\ \text{W}$$
$$\text{Total loss} = 607 + 468 + 374 = 1449\ \text{W}$$

Check: $$P_{\text{in}} - P_{\text{out}} = 9800 - 8360 = 1440\ \text{W}$$

Note: The above method can be used to find the slip at maximum torque and its value, but the derivation to be carried out would be quite complex. It is much simpler to proceed by the Thevenin equivalent method. Equally the Thevenin method becomes cumbersome in determining stator current and its pf. This will be illustrated in the two examples that follow.

Approximate Circuit Model

NL test

Rotational loss, $P_r = 607$ W (as calculated already)

$$R_{iwf} = \frac{3 \times (231)^2}{607} = 264\ \Omega$$

$$I_{iwf} = \frac{231}{264} = 0.875$$

$$I_m = \sqrt{(7.5)^2 - (0.875)^2} = 7.45$$

$$X_m = \frac{231}{7.45} = 31\ \Omega$$

BR test

$$Z_{BR} = \frac{150/\sqrt{3}}{35} = 2.47\ \Omega$$

$$R_{BR} = \frac{4000/3}{(35)^2} = 1.09\ \Omega$$

$$X_1 + X_2' = \sqrt{(7.5)^2 - (1.09)^2} = 2.22\ \Omega$$

or $X_1 = 1.48\ \Omega, \quad X_2' = 0.74\ \Omega$

The approximate circuit model is drawn in Fig. 9.25. From this figure

$$\bar{Z}\text{ (total)} = (0.55 + 0.54 + 12.96) + j\,2.22 = 14.22\ \angle 9°$$

$$I_2' = \frac{231}{14.22\angle 9°} = 16.24\angle{-9°} = 16.04 - j\,2.54$$

$$\text{Mechanical power output (gross)} = 3 \times (16.24)^2 \times 12.96 = 10.25\text{ kW}$$

$$\text{Mechanical power output (net)} = 10.25 - 0.607\text{ (rot loss)} = 9.64\text{ kW}$$

$$\text{Torque (net)} = \frac{9640}{104.72(1 - 0.04)} = 96\text{ Nm}$$

Stator current,

$$\bar{I}_1 = \bar{I}_m + \bar{I}_2'$$

$$= -j\,7.45 + (16.04 - j\,2.54)$$

$$= 16.04 - j\,10 = 18.90\ \angle -31.9°\text{ A}$$

$$I_1 = 18.90\text{ A},\ pf = \cos 31.9° = 0.85\text{ lagging}$$

$$\text{Power input} = \sqrt{3}\ 400 \times 18.9 \times 0.85 = 11.13\text{ kW}$$

$$\eta = \frac{9.64}{11.13} \times 100 = 86.61\%$$

Comparison The results as obtained from the approximate model are about 10–12% higher than those obtained from the IEEE model (the two efficiencies are of course quite close to each other). The approximate circuit leads to considerable ease in computation and visualization of the motor performance.

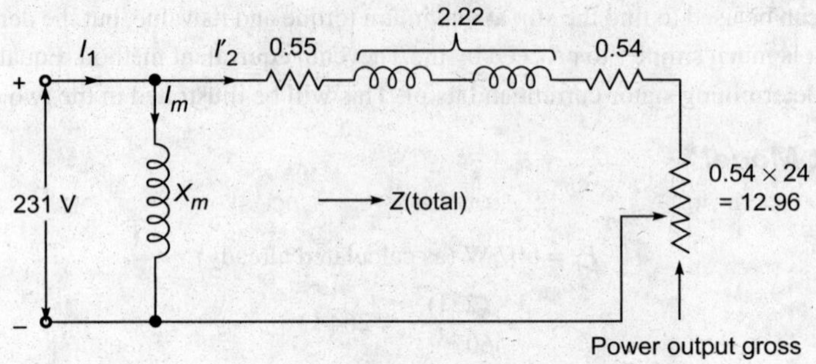

Fig. 9.25

The Example 9.4 first method is also solved by MATLAB.

```
%% example 9.4
clc
clear
j=sqrt(-1);
Vt=400; P=6; f=50; Inl=7.5; Pnl=700;
%% blocked rotor test results
Vbr=150;
Ibr=35;
Pinbr=4000;
R1=0.55
k=1/0.5;
s=0.04;
Zbr=Vbr/(sqrt(3)*Ibr);
Rbr=Pinbr/(3*(Ibr)^2);
Xbr=sqrt(Zbr^2-Rbr^2);
X1=Xbr/(1+0.5)
X2=Xbr-X1
%% no load test results
Zo=Vt/(sqrt(3)*Inl);
Ro=Pnl/(3*Inl^2);
Xo=sqrt(Zo^2-Ro^2);
Xm=Xo-X1
R2=(Rbr-R1)*((Xm+X2)/Xm)^2
Zf=1/((1/(j*Xm))+(1/((R2/s)+j*X2)));
Rf=real(Zf);
Xf=imag(Zf);
Zin=R1+j*X1+Zf;
I1=Vt/(sqrt(3)*Zin);
Pin=sqrt(3)*Vt*abs(I1)*cos(angle(I1))
```

```
Pg=3*abs(I1)*abs(I1)*Rf
Pm=(1-s)*Pg
Prot=Pnl-3*Inl*Inl*R1
Pout=Pm-Prot
ws=1000*2*pi/60;
Tnet=Pout/((1-s)*ws)
eff=Pout*100/Pin
```

Answer:

R1 = 0.5500

X1 = 1.4814

X2 = 0.7407

Xm = 29.0299

R2 = 0.5663

Pin = 9.8102e+003

Pg = 9.3422e+003

Pm = 8.9685e+003

Prot = 607.1875

Pout = 8.3613e+003

Tnet = 83.1717

eff = 85.2307

Example 9.5 *The following results were obtained on a 3-phase, 75 kW, 3.3 kV, 6-pole, 50 Hz squirrel-cage induction motor.*

NL test: *Rated frequency, 50 Hz*

$V_0 = 3.3$ *kV (line),* $I_0 = 5$ *A,* $P_0 = 2500$ *W*

BR test: *Frequency 15 Hz*

$V_{BR} = 400$ *V (line),* $I_{BR} = 27$ *A,* $P_{BR} = 15000$ *W*

DC test on stator Resistance/phase = 3.75 Ω

(a) Determine the parameters of the circuit model (exact version).

(b) Find the parameters of Thevenin equivalent as seen from the rotor circuit.

(c) Calculate the maximum torque and the slip at which it occurs.

(d) For a slip of 4%, calculate the stator current, its pf and motor efficiency.

Calculate also internal motor torque

LO 2

Solution NL test

$$Z_0 = \frac{3300/\sqrt{3}}{5} = 381\ \Omega$$

$$R_0 = \frac{2500/3}{(5)^2} = 33.33\ \Omega$$

$$X_0 = \sqrt{(381)^2 - (33.33)^2} = 379.6\ \Omega$$

$$X_m = X_0 - X_1 = ?$$

BR test

$$Z_{BR} = \frac{400/\sqrt{3}}{27} = 8.55\ \Omega$$

$$R_{BR} = \frac{15000/3}{(27)^2} = 6.86\ \Omega$$

$$X_{BR} = \sqrt{(8.55)^2 - (6.86)^2} = 5.10\ \Omega$$

$$X_1 = X_2' = X_{BR}/2 = 2.55\ \Omega \text{ (at 15 Hz)}; 8.50\ \Omega \text{ (at 50 Hz)}$$

$$\therefore \quad X_m = 379.6 - 8.50 = 371\ \Omega$$

$$R_2' = (R_{BR} - R_1)\left[\frac{X_m + X_2'}{X_m}\right]^2 = (6.86 - 3.75) \times \left[\frac{371 + 8.50}{371}\right]^2 = 3.25\ \Omega$$

(a) The circuit model is drawn in Fig. 9.26(a).

(b)

$$\bar{V}_{TH} = (3300/\sqrt{3})\angle 0° \times \frac{j371}{3.75\ j(371 + 8.5)} = 1862\ \angle 0.6°\ \text{V}$$

$$\bar{Z}_{TH} = \frac{j371 \times (3.75 + j8.5)}{3.75 + j(371 + 8.5)} = 9.08\ \angle 66.8° = 3.58 + j\,8.35 = R_{TH} + j\,X_{TH}$$

The corresponding circuit is drawn in Fig. 9.26(b)

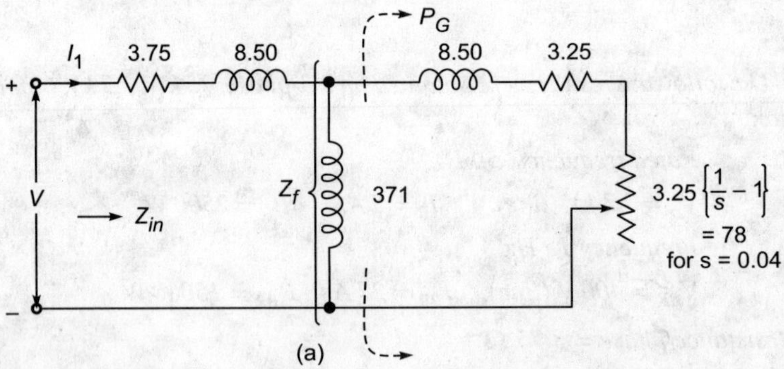

(a)

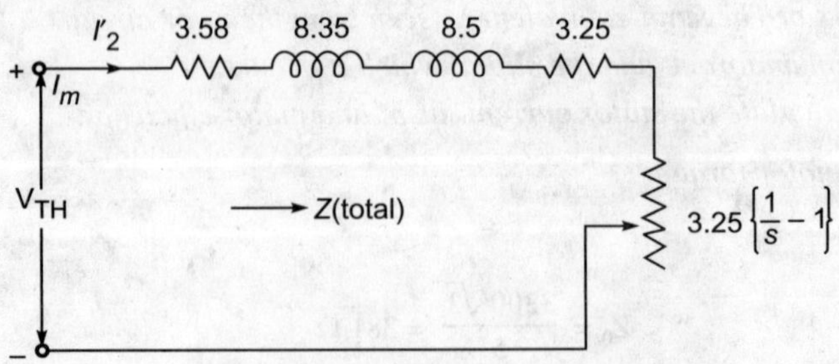

(b)

Fig. 9.26

(c) For maximum torque to be developed

$$\frac{3.25}{s_{\max,T}} = \sqrt{(3.58)^2 + (8.35 + 8.50)^2}$$

or $$s_{\max,T} = \frac{3.25}{17.22} = 0.189$$

or $$R_2'/s_{\max,T} = \frac{3.25}{0.189} = 17.2\ \Omega$$

$$n_s = 1000 \text{ rpm}, \omega_s = 104.72$$

$$\bar{Z}_{\text{total}} = (3.58 + 17.22) + j16.85 = 20.80 + j\,16.85 = 26.77\ \angle 39°\ \Omega$$

$$I_2' = \frac{1862}{26.77} = 69.56 \text{ A}$$

$$T_{\max} = \frac{3 \times (69.56)^2 \times 17.2}{104.72} = 2.387 \text{ Nm}$$

(d) As the stator current and its pf are required, we shall proceed by using the circuit of Fig. 9.26(a).

$$\bar{Z}_f = \frac{j371(81.25 + j8.50)}{81.25 + j379.50} = 78.11\ \angle 18.1°$$

$$= 74.24 + j\,24.26 = R_f + j\,X_f$$

$$\bar{Z}_{in} = (3.75 + j\,8.5) + (74.26 + j\,24.26) = 84.60\ \angle 22.8°$$

$$I_1 = \frac{3300/\sqrt{3}}{84.60} = 22.52 \text{ A}$$

$$pf = \cos 22.8° = 0.922 \text{ lagging}$$

$$\text{Gross mechanical output} = (1 - s)\,P_G = (1 - 0.04) \times 3 \times (22.52)^2 \times 74.24 = 108.43 \text{ kW}$$

$$\text{Rotational loss} = 2.500 - (5)^2 \times 3.75 = 2.41 \text{ kW}$$

$$\text{Motor input} = \sqrt{3} \times 3.3 \times 22.52 \times 0.922 = 118.22 \text{ kW}$$

$$\text{Net mechanical output} = 108.42 - 2.41 = 106.1 \text{ kW}$$

$$\eta = \frac{106.21}{118.42} = 89.1\%$$

$$\text{Internal torque developed} = (108.43 \times 1000)/[104.72\,(1 - 0.04)] = 1078.6 \text{ Nm}$$

Example 9.6 *A 400 V, 1450 rpm, 50 Hz wound-rotor induction motor has the following circuit model parameters.*

$$R_1 = 0.3\ \Omega \qquad R_2' = 0.25\ \Omega$$

$$X_1 = X_2' = 0.6\ \Omega \qquad X_m = 35\ \Omega$$

$$\text{Rotational loss} = 1500 \text{ W}$$

(a) Calculate the starting torque and current when the motor is started direct on full voltage.

(b) Calculate the full-load current, pf and torque (net).

Also find internal efficiency and overall efficiency. Internal efficiency is defined as

$$P_{out}\,(\text{gross})/P_{in};\ P_{out}\,(\text{gross}) = (1 - s)\,P_G$$

(c) Find the slip for maximum torque and the value of the maximum torque. LO 2

Solution The motor circuit model is drawn in Fig. 9.27(a) and its Thevenin equivalent version is drawn in Fig. 9.27(b). Thevenin equivalent quantities are calculated below.

$$\bar{Z}_{TH} = \frac{j35(0.3 + j0.6)}{0.3 + j35.60} = 0.29 + j\,0\;59\ \Omega$$

$$\bar{V}_{TH} = \frac{231\angle 0° \times j35}{0.3 + j35.60} = 227\ \angle 0.5°\ \text{V}$$

$$n_s = 1500 \text{ rpm} \qquad \omega_s = 157.1 \text{ rad/s}$$

(a) Starting on full voltage ($s = 1$)

From the circuit of Fig. 9.27(a)

$$\bar{Z}_f = \frac{j35(0.25 + j0.6)}{0.25 + j35.6} = 0.24 + j\,0.59\ \Omega$$

$$R_f = 0.24\ \Omega$$

$$\bar{Z}_{in} = (0.3 + j\,0.6) + (0.24 + j\,0.59) = 1.31\ \angle 65.6°\ \Omega$$

$$I_1(\text{start}) = \frac{231}{1.31} = 176.3 \text{ A}$$

$$T(\text{start}) = \frac{3I_1^2 R_f}{\omega_s}$$

$$\frac{3 \times (176.3)^2 \times 0.24}{157.1} = 142.4 \text{ Nm}$$

(b)

$$s_{fl} = \frac{1500 - 1450}{1500} = 1/30$$

$$R_2'/s_{fl} = 0.25 \times 30 = 7.5\ \Omega$$

From the circuit of Fig. 9.27(a)

$$\bar{Z}_f = \frac{j35(75 + j0.6)}{7.5 + j35.6}$$

$$= 6.94 + j\,2.05\ \Omega$$

$$R_f = 6.94\ \Omega$$

$$\bar{Z}_{in} = (0.3 + j\,0.6) + (6.94 + j\,2.05)$$

$$= 7.71\ \angle 20.1°\ \Omega$$

$$I_1 = \frac{231}{7.71} = 30 \text{ A}$$

$$p_f = \cos 20.1° = 0.94 \text{ lagging}$$

$$P_G = 3 \times (30)^2 \times 6.94 = 18.74 \text{ kW}$$

$$\text{Power output (gross)} = (1 - s)\,P_G$$

$$(1 - 1/30) \times 18.74 = 18.12 \text{ kW}$$

$$\text{Rotational loss} = 1.5 \text{ kW}$$

$$\text{Power output (net)} = 18.12 - 1.5 = 16.62 \text{ kW}$$

$$\text{Torque (net)} = \frac{16.62 \times 1000}{157.1(1 - 1/30)} = 109.4 \text{ Nm}$$

$$\text{Power input} = \sqrt{3} \times 400 \times 30 \times 0.94 = 19.54 \text{ kW}$$

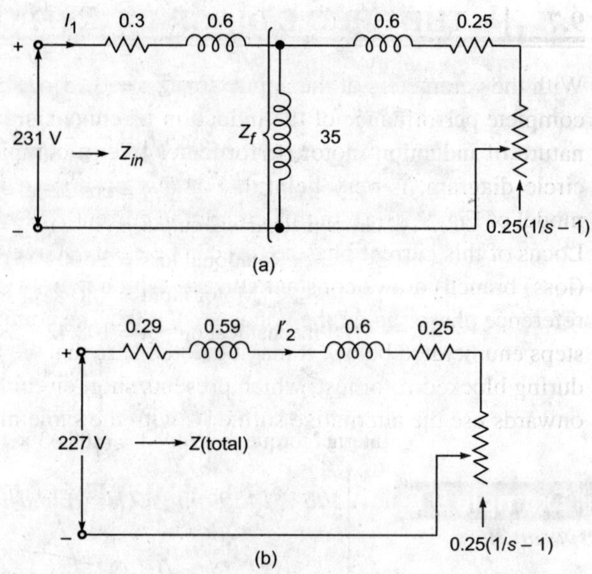

Fig. 9.27

$$\eta = \frac{16.62}{19.54} \times 100 = 85.06\%$$

$$\text{Internal efficiency} = \frac{P_G(1-s)}{P_{in}} = \frac{18.74(1-1/30)}{19.54} = 92.71\%$$

(c) We shall use the Thevenin equivalent circuit of Fig. 9.27(b). For maximum torque

$$\frac{0.25}{s_{\max,T}} = \sqrt{(0.29)^2 + (0.59 + 0.6)^2}$$

or

$$s_{\max,T} = 0.204$$

$$R_2'/s_{\max,T} = 0.25/0.204 = 1.225$$

$$\bar{Z}\,(\text{total}) = (0.29 + j\,0.59) + (1.225 + j\,0.6) = 1.93\ \angle 38.1°$$

$$I_2' = \frac{227}{1.93} = 117.6$$

$$T_{\max} = 3 \times \frac{(117.6)^2 \times 1.225}{157.1} = 323.5 \text{ Nm}$$

9.7 THE CIRCLE DIAGRAM (APPROXIMATE)

LO 2

With the parameters of the approximate circuit model obtained from the no-load and blocked-rotor tests, the complete performance of the induction machine can be calculated by varying the slip. To help visualize the nature of induction motor performance over a complete range of slips, a graphical locus approach, called circle diagram, is very helpful. For this purpose it is acceptable practice to use the approximate circuit model of Fig. 9.23(a). In this circuit the current I_2' flows in a series inductive circuit with variable resistance. Locus of this current phasor (tip) can be easily shown to be a circle. The shunt branch (magnetizing and P_{iwf} (loss) branch) draws constant current, which merely shifts the circle from the origin of phasor diagram; the reference phasor being the voltage $\angle V = 0°$. The complete circle diagram is drawn in Fig. 9.28 following the steps enumerated below. It may be noted here that we have earlier used the suffix *BR* for quantities measured during blocked-rotor test, which presents short circuit conditions in the circle model, i.e. $s = 1$. We shall now onwards use the alternative suffix *SC* with the same meaning as BR.

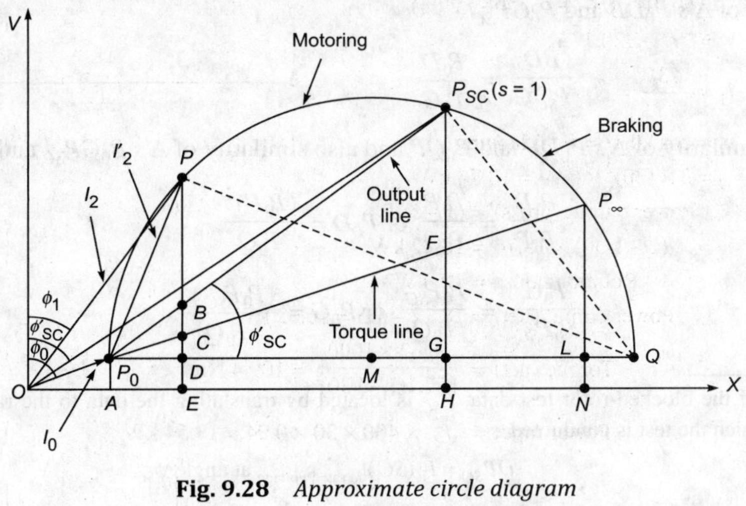

Fig. 9.28 *Approximate circle diagram*

Steps to be Followed while Developing Approximate Circle Diagram

(i) Locate the voltage-axis along which the voltage phasor $\bar{V}$ as represented by *OV*. The projection of a current phasor on the *V*-axis is the active current component while the reactive current component is the current projection on *OX*-axis (drawn lagging the *V*-axis by 90°). These current components represent real and reactive powers when multiplied by *V*, the voltage/phase.

(ii) Locate OP_0 at angle ϕ_0 (no-load pf angle) from the phasor $\bar{V}$ to represent the no-load current $\bar{I}_0$. Then

No-load loss, $P_0 = OP_0 \cos \phi_0 = P_0A$ (to scale of voltage *V*)

(iii) Draw

$$P_0Q = \frac{V}{X_1 + X_2'}$$

at 90° to the *V*-axis. This indeed is the current I_2' if all resistance in this part of the circuit could be reduced to zero.

P_0Q is the diameter of the circle of the $\bar{I}_2'$-locus. The circle is drawn with centre *M* at the midpoint of P_0Q.

(iv) The short-circuit point P_{SC} is located on the circle corresponding to $s = 1$.

$$I_2'\,(\text{SC}) = P_0P_{SC} = \frac{V'}{\sqrt{(R_1 + R_2')^2 + (X_1 + X_2')^2}}$$

$$\phi'_{SC} = \tan^{-1}\left(\frac{X_1 + X_2'}{R_1 + R_2'}\right)$$

The circle can also be drawn** by locating P_{SC} without the need of determining P_0Q.

At any operating point *P* on the circle

$OP = I_1$, input current

$OP_0 = I_0$, no-load current

$P_0P = I_2'$, rotor current as seen from stator.

From the similarity of Δ's P_0DB and P_0GP_{SC}

$$\frac{BD}{P_{SC}G} = \frac{P_0D}{P_0G} \tag{9.52}$$

Further, from the similarity of Δ's P_0DP and P_0QP and also similarity of Δ's P_0GP_{SC} and P_0QP_{SC}.

$$\frac{P_0D}{P_0P} = \frac{P_0P}{P_0Q} \text{ or } P_0D = \frac{(P_0P)^2}{P_0Q} \tag{9.53}$$

and

$$\frac{P_0G}{P_0P_{SC}} = \frac{P_0P_{SC}}{P_0Q} \text{ or } P_0G = \frac{(P_0P_{SC})^2}{P_0Q} \tag{9.54}$$

** Directly in terms of the blocked-rotor test data, P_{SC} is located by translating the data to the rated voltage from the reduced voltage at which the test is conducted.

$$OP_{SC} = I_1\,(SC)|_{\text{at rated voltage}} \text{ at angle } \phi_{SC}$$

Substituting Eqs (9.52) and (9.53) in Eq. (9.54),

$$\frac{BD}{P_{SC}G} = \frac{(P_0P)^2}{(P_0P_{SC})} = \frac{I_2'^2}{I_{2SC}'^2} \tag{9.55}$$

Multiplying the numerator and denominator of the right-hand side of Eq. (9.55) by $(R_1 + R_2')$,

$$\frac{BD}{P_{SC}G} = \frac{I_2'^2\,(R_1 + R_2')}{I_{2SC}^2\,(R_1 + R_2')} = \frac{\text{copper-loss on load}}{\text{copper-loss on } SC} \tag{9.56}$$

Since the power drawn on short-circuit[††] is

$$P_{SC}H = GH + P_{SC}H = P_0 + P_c$$

$$\therefore \quad P_{SC}G = P_c, \text{ copper-loss}^{\ddagger\ddagger} \text{ (stator + rotor)}$$

It therefore follows from Eq. (9.56) that

$$BD = I_2'^2(R_1 + R_2') = \text{copper-loss on load}$$

Now divide $P_{SC}G$ at F such that

$$\frac{P_{sc}F}{FG} = \frac{R_2'}{R_1}$$

From similarity of appropriate triangles it then easily follows that

$$BC = \text{rotor copper-loss (on-load)}$$

$$CD = \text{stator copper-loss (on-load)}$$

If P_0F is produced to meet the circle at P_∞, then

$$P_\infty L = \text{stator copper-loss, while corresponding rotor copper-loss is zero.}$$

This is only possible when

$$\frac{R_2'}{s} = 0, \quad \text{or} \quad s = \infty$$

Thus P_∞ point corresponds to infinite slip.

At load point P,

$$PE = PB + (BC + CD) + DE$$

or

Input power, $\quad P_i = P_m + (P_{cr} + P_{cs}) + P_0$

where $\quad P_m = PB = \text{net mechanical power output}$

$\quad (P_{wf}$ having been accounted for in $P_0)$

Therefore, the mechanical power output is the vertical intercept between the circle and the line P_0P_{SC}, called the *output line*.

Now

$$P_G\text{(air-gap power)} = P_m + P_{cr}$$

$$= PB + BC = PC = \text{torque in synchronous watts}$$

†† As already discussed P_0 remains constant as the slip increases to unity, where it is wholly constituted of iron-loss.

‡‡ With the approximation already made that P_c (stator) $= I_2'^2 R_1$.

Therefore, the vertical intercept between the circle and line P_0P_∞, called the *torque line* is the torque in synchronous watts. Further,

$$\text{Slip, } s = \frac{P_{cr}}{P_G} = \frac{BC}{PC}$$

By observing the circle diagram certain generalized conclusion can easily be drawn about the induction motor performance. These are:

(i) The motor has a very low pf (lagging) on no-load which improves to a certain maximum value (when *OP* is tangential to the circle) and then begins to reduce. The induction motor inherently has a lagging power factor (around 0.8) over which no control is possible unlike the synchronous motor[§§]. This is because, being a singly-fed device, it draws a large zero pf lagging current to set up the resultant air-gap flux/pole.

(ii) The mechanical power output exhibits a maximum; it can be located by drawing a line tangential to the circle and parallel to the output line.

(iii) The torque also exhibits a maximum, located at a point different from that corresponding to the maximum power. It can be located by drawing a line tangential to the circle and parallel to the torque line.

(iv) It is obvious from (ii) above that the efficiency-load plot will also exhibit a maximum. This is the inherent characteristic of any device with a constant loss component and a variable loss component proportional to the square of the load (refer Sec. 5.8).

(v) The input current increases continuously.

Typical plots of current, torque (net), pf, slip and efficiency versus pu power output for the induction motor are shown in Fig. 9.29. These plots are shown in a solid line up to 1.0 pu output (full-load) and beyond this are drawn dotted.

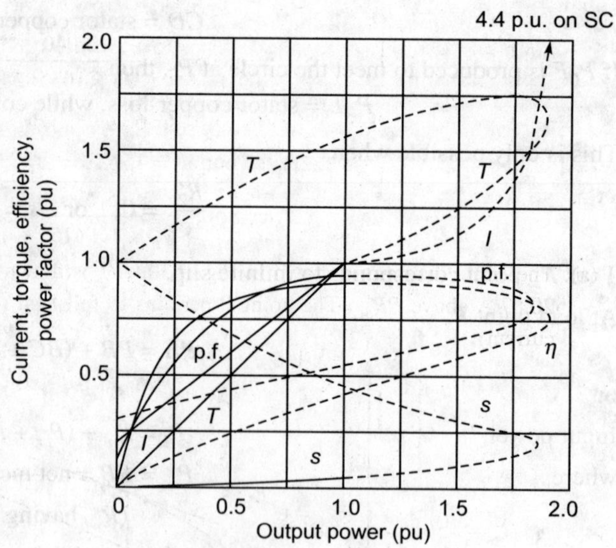

Fig. 9.29 *Typical performance curves of the induction motor*

Induction Generator

Important! As brought out by Fig. 9.14 (torque-slip characteristic) that the induction machine acts as a generator when run as super-synchronous speed ($s < 0$) by a primemover. The circle diagram for generator operation is the lower part of the circle (centre *M*) not shown in Fig. 9.28. Induction generator operation will be taken up in Section 9.14.

[§§] The resultant air-gap flux/pole in a synchronous machine can be independently adjusted by the dc excitation thereby allowing control over the machine power factor.

Example 9.7 *A 15 kW, 415 V, 4-pole, 50 Hz delta-connected motor gave the following results on test (voltages and currents are in line values):*

No-load test	*415 V*	*10.5 A*	*1,510 W*
Blocked-rotor test	*105 V*	*28 A*	*2,040 W*

Using the approximate circuit model, determine:

(a) the line current and power factor for rated output,

(b) the maximum torque, and

(c) the starting torque and line current if the motor is started with the stator star-connected.

Assume that the stator and rotor copper losses are equal at standstill.

Hint Part (a) is best attempted by means of a circle diagram. For proceeding computationally from the circuit model, we have to compute the complete output-slip curve and then read the slip for rated output.

LO 2

Solution

$$OP_0 = 10.5 \text{ A}$$

$$\cos \phi_0 = \frac{1{,}510}{\sqrt{3} \times 415 \times 10.5} = 0.2$$

or $$\phi_0 = 78.5°$$

$$OP_{SC} = \frac{28 \times 415}{105} = 110.7 \text{ A (at 415 V)}$$

$$\cos \phi_{SC} = \frac{2{,}040}{\sqrt{3} \times 105 \times 28} = 0.4$$

$$\phi_{SC} = 66.4°$$

$$\text{Rated output} = 15 \text{ kW} = \sqrt{3}\, V(I \cos \phi)_{\text{rated}}$$

$$(I \cos \phi)_{\text{rated}} = \frac{15{,}000}{\sqrt{3} \times 415} = 20.87 \text{ A}$$

(a) The circle diagram is drawn in Fig. 9.30. PP' is drawn parallel to the output line P_0P_{SC} at a vertical distance 20.87 A above PP_{SC}. The point P pertains to full load ($P'$is the second but unacceptable solution—too large a current):

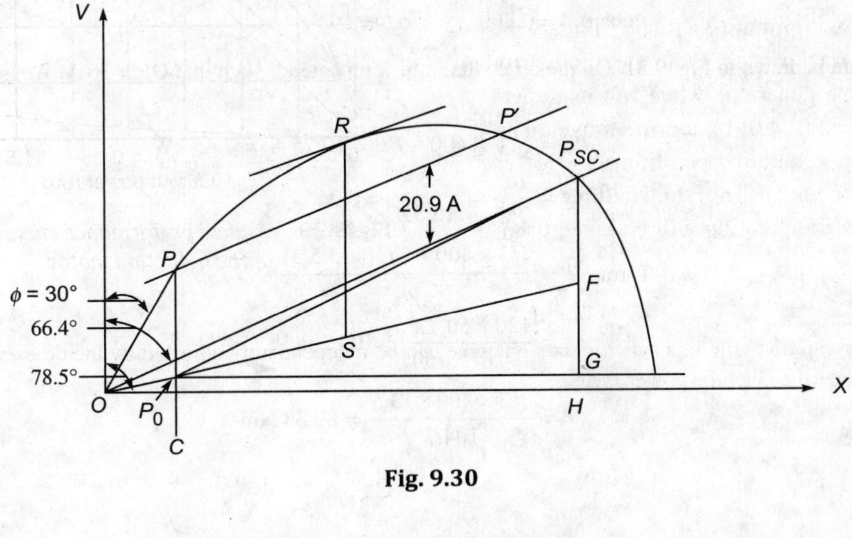

Fig. 9.30

Then,

$$I_1 = OP = 30.5 \text{ A}$$

$$\text{pf} = \cos\phi = \cos 30° = 0.866 \text{ lagging}$$

(b) Torque line P_0F is drawn by locating F as the midpoint of $P_{SC}G$ (equal stator and rotor, losses). The maximum torque point is located by drawing a tangent to the circle parallel to P_0F. The maximum torque is given as

$$RS = 44 \text{ A}$$

$$T_{\max} = 3 \times \frac{145}{\sqrt{3}} \times 44 = 31{,}626 \text{ syn watts}$$

$$n_s = 1{,}500 \text{ rpm}, \quad \omega_s = 157.1 \text{ rad/s}$$

$$T_{\max} = \frac{31{,}626}{157.1} = 201.3 \text{ Nm}$$

(c) If the motor is started to delta (this is the connection for which test data are given)

$$I_s(\text{delta}) = OP_{SC} = 110.7 \text{ A}$$

$$P_{SC}F = 22 \text{ A}$$

$$T_s(\text{delta}) = \frac{(\sqrt{3} \times 415 \times 22)}{157.1} = 100.7 \text{ Nm}$$

Star connection

$$I_s(\text{star}) = \frac{100.7}{3} = 33.6 \text{ A}$$

$$T_s(\text{star}) = \frac{100.7}{3} = 33.57 \text{ Nm}$$

Example 9.8 *A 400 V, 3-phase, 6-pole, 50 Hz induction motor gave the following test results:*

No-load	*400 V*	*8 V*	*0.16 power factor*
Blocked-rotor	*200 V*	*39 A*	*0.36 power factor*

Determine the mechanical output, torque and slip when the motor draws a current of 30 A from the mains. Assume the stator and rotor copper losses to be equal. Use circle diagram method. LO 2

Solution

$$\cos\phi_0 = 0.16; \quad \phi_0 = 80.8°$$

$$\cos\phi_{SC} = 0.36; \quad \phi_{SC} = 69°$$

The circle diagram is drawn in Fig. 9.31. On the circle diagram we located P such that OP = 30 A. We get the following results.

$$P_m = \sqrt{3} \times 400 \times PB(= 10.75 \text{ A}) = 7.45 \text{ kW}$$

$$\text{Slip } s = \frac{BC}{PC} = \frac{3.5\,A}{13.5\,A} = 0.26$$

$$\text{Torque } T = \frac{\sqrt{3} \times 400 \times PC(=13.5\,A)}{\omega_s}$$

and

$$\omega_s = \frac{120 \times 50}{6}\,\frac{2\pi}{60} = 104.7 \text{ rad/s}$$

$$T = \frac{\sqrt{3} \times 400 \times 13.5}{104.7} = 89.33 \text{ Nm}$$

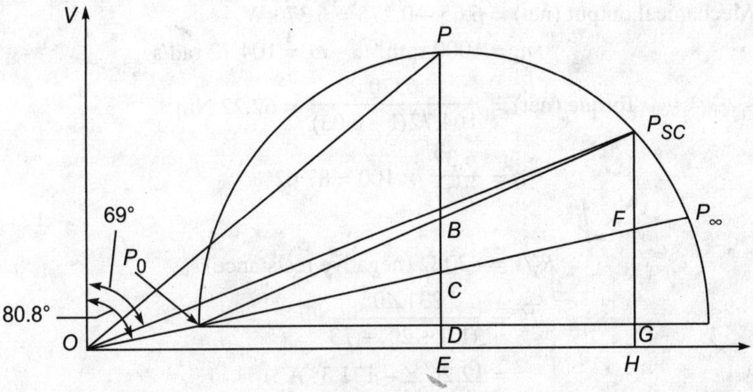

Fig. 9.31

Example 9.9 *A 3-phase, 400 V, 6-pole, 19 kW induction motor has the following parameters of its approximate circuit model.*

$$R_1 = 1.4\ \Omega \qquad R_2' = 0.6\ \Omega$$
$$X_1 = 2\ \Omega \qquad X_2' = 1\ \Omega$$
$$X_m = 50\ \Omega$$

The rotational loss is 275 W. For a slip of 0.03, – 0.03, and 1.2, determine:

(a) the line current, pf and power input.

(b) the shaft torque, and mechanical output.

(c) the efficiency.

LO 2

Solution The approximate circuit is drawn in Fig. 9.32.

(i) Slip = 0.03

(a)

$$R_2'/s = 0.6/0.03 = 20\ \Omega$$

$$\bar{I}_2' = \frac{231\angle 0°}{(1.4+20)+j3}$$
$$= 10.69\angle -8°\ \text{A}$$
$$= 10.58 - j\,1.49\ \text{A}$$

$$\bar{I}_m = \frac{231\angle 0°}{50\angle 90°} = -j\,4.62\ \text{A}$$

$$\bar{I}_1 = \bar{I}_m + \bar{I}_2'$$
$$= 10.58 - j\,1.49 - j\,4.62$$
$$= 10.58 - j\,6.11 = 12.22\angle -30°\ \text{A}$$

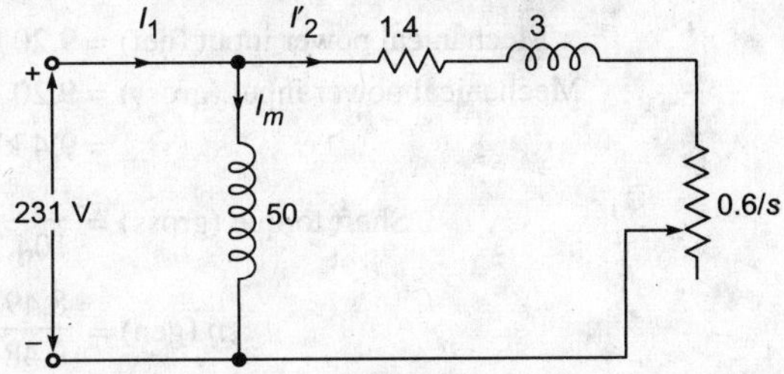

Fig. 9.32

$$I_1 = 12.22\ \text{A} \quad pf = \cos 30° = 0.866\ \text{lagging}$$

$$\text{Power input} = \sqrt{3} \times 400 \times 12.22 \times 0.866 = 7.33\ \text{kW}$$

(b)

$$P_G = \frac{3 \times (10.69)^2 \times 0.6}{0.03} = 6.86\ \text{kW}$$

$$\text{Mechanical output (gross)} = (1 - 0.03) \times 6.86 = 6.65\ \text{kW}$$

$$\text{Rotational loss} = 0.275\ \text{kW}$$

$$\text{Mechanical output (net)} = 6.65 - 0.275 = 6.37 \text{ kW}$$

$$n_s = 1000 \text{ rpm}, \qquad \omega_s = 104.72 \text{ rad/s}$$

$$\text{Torque (net)} = \frac{6370}{104.72(1-0.03)} = 62.22 \text{ Nm}$$

(c)

$$\eta = \frac{6.37}{7.27} \times 100 = 87.62\%$$

(ii) Slip = –0.03

(a)

$$R_2'/s = -20\ \Omega \text{ (negative resistance)}$$

$$\bar{I}_2' = \frac{231\angle 0°}{(1.4-20)+j3}$$

$$= 12.26 \angle -171.3° \text{ A}$$

$$= -12.12 - j\,1.85 \text{ A}$$

$$\bar{I}_m = -j\,4.62$$

$$\bar{I}_1 = (-j\,4.62) + (-12.12 - j\,1.85)$$

$$= -12.12 - j\,6.47 = 13.73 \angle{-151.9°} \text{ A}$$

$$\bar{I}_1 \text{ (out)} = -\bar{I}_1 = 13.73 \angle 28.1° \text{ A } (\textit{machine is generating})$$

$$I_1\text{(out)} = 13.73\text{A, pf} = \cos 28.1° = 0.882 \text{ leading}$$

$$\text{Power output (elect)} = \sqrt{3} \times 400 \times 13.73 \times 0.882$$

$$= 8.39 \text{ kW}$$

$$\text{Mechanical power output} = (1-s)\left(\frac{3I_2'^2 R_2'}{s}\right)$$

$$= 1.03 \times \frac{3 \times (12.26)^2 \times 0.6}{-0.03}$$

$$= -9.20 \text{ kW}$$

$$\text{Mechanical power input (net)} = 9.20 \text{ kW}$$

$$\text{Mechanical power input (gross)} = 9.20 + 0.275 \text{ (rotational loss)}$$

$$= 9.4 \text{ kW}$$

$$\text{Shaft torque (gross)} = \frac{9480}{104.72(1+0.03)} = 87.9 \text{ Nm}$$

$$\eta \text{ (gen)} = \frac{8.49}{9.48} \times 100 = 88.5\%$$

(iii) slip = 1.2

(a)

$$R_2'/s = 0.6/1.2 = 0.5\ \Omega$$

$$\bar{I}_2' = \frac{231\angle 0°}{(1.4+0.5)+j3}$$

$$= 65.1 \angle -57.7° \text{ A}$$

$$= 34.79 - j\,55.02 \text{ A}$$

$$\bar{I}_m = -j\,4.62$$

$$\bar{I}_1 = (-j\,4.62) + (34.79 + j\,55.02)$$

$$= 34.79 - j\,59.64 \text{ A}$$

$$= 69.05 \angle{-59.7°} \text{ A}$$

$$I_1 = 69.05 \text{ A}, \qquad \text{pf} = \cos 59.7° = 0.505 \text{ lagging}$$

$$\text{Power input (elect)} = \sqrt{3} \times 400 \times 69.05 \times 0.505 = 24.16 \text{ kW}$$

(b)
$$P_G = 3 \times \frac{(65.1)^2 \times 0.5}{1.2} = 5.30 \text{ kW}$$

$$\text{Mechanical power output} = (1 - s)\, P_G = (1 - 1.2) \times 5.30 = -1.06 \text{ kW}$$

or Mechanical power absorbed (net) = 1.06 kW

Rotational loss can be ignored as motor speed is

$$\eta = 1500 \times (1 - 1.2) = -200 \text{ rpm} \quad \text{or} \quad -20.94 \text{ rad/s}$$

Note: Motor runs in opposite direction of the air-gap field—absorbing mechanical power (braking action)

$$\text{Torque developed} = \frac{-1060}{-20.94} = 50.62 \text{ Nm}$$

This torque acts in direction opposite to that of the rotating field.

Total power dissipated (in the motor)

$$= 24.16 \text{ (elect)} + 1.06 \text{ (mech)}$$
$$= 25.22 \text{ kW}$$

The power to be dissipated by the motor (mostly in windings) is larger than the motor rating. This can be permitted to happen only for a short period. Further the braking power is quite low. It can be seen from Fig. 9.14 that the braking torque would reduce as the speed increases further (in negative direction).

Note: A problem with MATLAB solution is included towards the end of this chapter.

Concept Check LO 2

1. What is meant by the torque in synchronous watts? Write its expression in terms of circuit model quantities. Therefrom find the torque developed. ●●●
2. Show that the maximum torque occurs at a slip $s = \dfrac{X_2}{R_2}$ and further show that T_{max} is independent of s. ○○●
3. Draw the $T - s$ characteristic of an induction motor. Indicate the region where the characteristic is nearly linear. ○○●
4. Show that the motor can operate stably at $s_{max,T}$. Use perturbation technique. ○●●
5. Neglect the stator impedance and show that the maximum power output (developed power) occurs at slip $s = \dfrac{R_2 + \sqrt{R_2^2 + X_2^2}}{X_2}$ ●●●

   Hint: In the circuit model of Fig. 9.13 use maximum power transfer theorem. The magnitude of fixed impedance should match $\left(\dfrac{1}{s} - 1\right) R_2$.
6. Neglecting stator impedance derive the expression for the starting torque of an induction motor. Show that it increases with rotor resistance. At what resistance value it reaches the maximum. Resistance added to the rotor of a slip-ring induction motor. ○●●

7. The power input on no-load running of induction motor is consumed in what losses?
8. The power input in blocked rotor test at reduced voltage, rated current is consumed in what losses.
9. No-load test determines what parameters of the circuit model of induction motor.
10. Blocked-rotor test determines which parameters of the circuit model of induction motor?
11. Upon reducing the load on an induction motor, why does its pf come down?

9.8 STARTING

LO 3 Examine the effect of harmonics and different speed control methods of an induction motor, with various complications occurring at the start and ways to overcome them

At the time of starting the motor slip being unity, the load resistance with reference to the approximate circuit model of Fig. 9.33 is

$$R_2'\left(\frac{1}{s}-1\right)\Big|_{s=1} = 0$$

meaning thereby that short-circuit conditions prevail. Therefore, the motor current at starting can be as large as five to six times the full-load current. In comparison the exciting current in the shunt branch of the circuit model can be neglected reducing the circuit to that of Fig. 9.33.

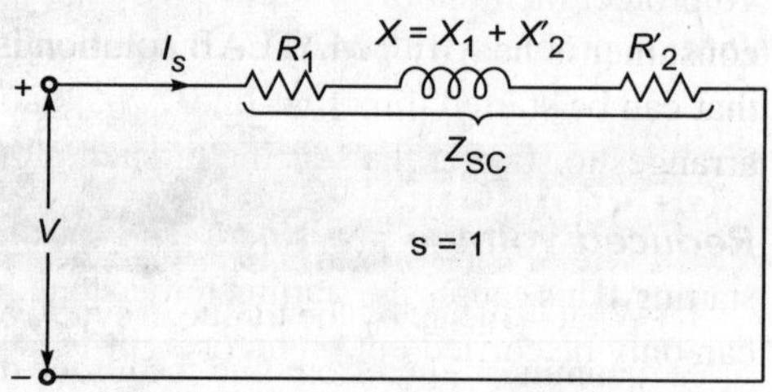

Fig. 9.33 *Approximate circuit model for starting conditions*

Now

$$\text{Starting torque, } T_s = \frac{3}{\omega_s}\cdot I_s^2\, R_2' \tag{9.57}$$

Assuming for simplicity that as a rough approximation

$$I_{fl} \approx I'_{2,fl} \tag{9.58}$$

i.e., the magnetizing current is neglected even under full-load conditions. Then

$$\text{Full-load torque, } T_{fl} = \frac{3}{\omega_s}\cdot I_{fl}^2\, R_2'/s_{fl} \tag{9.59}$$

where s_{fl} = full-load slip (2 to 8%)

The starting torque expressed as ratio of the full-load torque¶¶ is

$$\frac{T_s}{T_{fl}} = \left(\frac{I_s}{I_{fl}}\right)^2 s_{fl} \tag{9.60}$$

¶¶Two assumptions are imbedded in Eq. (9.60). These are:

1. Rotor resistance remains constant. The actual rotor resistance will vary with the frequency of the rotor current—a few hertz at full-load to 50-Hz at starting.
2. Magnetizing current is ignored. While this is a good approximation for I_s, it is a poor one for I_{fl} (Eq. (9.59)). If instead of stator current in (9.60), rotor current is used, assumption (2) above is eliminated.

Various Starting Arrangements Incorporated to Overcome the Problem of High Current at Starting

(a) Starting of Squirrel-Cage Motors

Direct starting When a squirrel-cage motor is started 'direct-on-line,'

$$I_s = I_{SC} = V/Z_{SC}$$

i.e. starting current equals the short-circuit (blocked-rotor) current. Let

$$I_{SC} = 5\, I_{fl}$$

and

$$s_{fl} = 0.04$$

Then from Eq. (9.60)

$$\frac{T_s}{T_{fl}} = (5)^2 \times 0.04 = 1$$

This means that when the starting current is as large as five times the full-load current, the starting torque just equals the full-load torque. With such a large starting current, the motor must accelerate and reach normal speed quickly otherwise overheating may damage the motor. The load on the motor at the time of starting must, therefore, be very light or preferably the motor must be on no-load.

To protect their supply systems, the electrical utilities have regulations against short-time current peaks; the consumer is heavily penalized through suitable tariff for such peaks. Therefore, it is only small-size motors that can be started direct on-line. A bulk consumer can start motors up to 10 kW direct-on-line as long as he arranges to stagger the starting of such motors.

Reduced voltage starting The starting current can be reduced to a tolerable level by reduced-voltage starting. This causes the starting torque to reduce heavily as it is proportional to the square of voltage. Such starting can only be carried out on no or light-load. Various methods of reduced voltage starting are discussed below.

Stator-impedance starting Inclusion of resistors or inductors in the three lines feeding the stator of the induction motor reduces the stator terminal voltage to $x\,V$ of the rated voltage V. The initial starting current is then

$$I_s = xI_{SC}$$

Substituting in Eq. (9.60)

$$\frac{T_s}{T_{fl}} = x^2\left(\frac{I_{SC}}{I_{fl}}\right)^2 s_{fl} \tag{9.61}$$

Thus, while the starting current reduces by a fraction x of the rated-voltage starting current (I_{SC}), the starting torque is reduced by a fraction x^2 of that obtainable with direct switching. This method can be used for small motors such as those driving centrifugal pumps; but star-delta starting (given later) is cheaper with better starting torque.

Autotransformer starting Reduced voltage for starting can be obtained from three autotransformers connected in star as shown in the schematic diagram of Fig. 9.34. If the voltage is reduced to a fraction x of the rated voltage V, the motor starting current (initial) is

$$I_s = xI_{SC} \tag{9.62}$$

where I_{SC} = starting current (line) with full-voltage

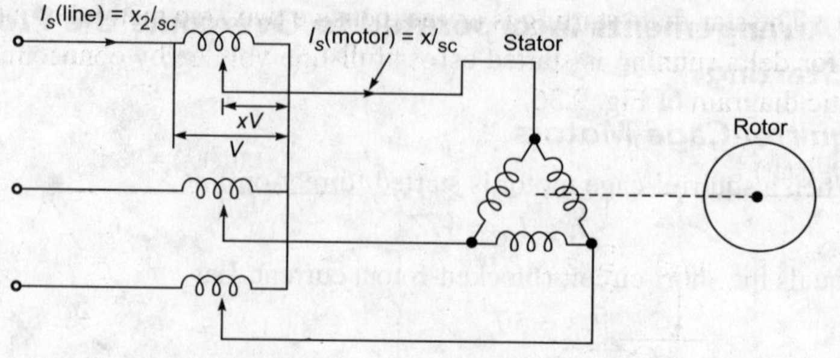

Fig. 9.34 *Autotransformer starting*

The current drawn from the supply is

$$I_s(\text{line}) = x(xI_{SC}) = x^2 I_{SC} \tag{9.63}$$

The starting/full-load torque ratio is

$$\frac{T_s}{T_{fl}} = x^2 \left(\frac{I_{SC}}{I_{fl}}\right)^2 s_{fl} \tag{9.64}$$

It is found that while the starting torque is reduced to a fraction x^2 of that obtainable by direct starting, the starting line current is also reduced by the same fraction. Compared to stator-impedance starting, the line current reduces further by a fraction x while torque remains the same. The autotransformer starting is much superior to the stator-impedance starting. Further, smooth starting and high acceleration are possible by gradually raising the voltage to the full line value.

After starting, the autotransformer is cut as shown in the wiring diagram of Fig. 9.35. It is to be observed that the autotransformer can be short-time rated. The use of an autotransformer is an expensive way of induction motor starting and is warranted for large motors only.

Probe

Establish a comparative study between Autotransformer starting and Star-delta starting.

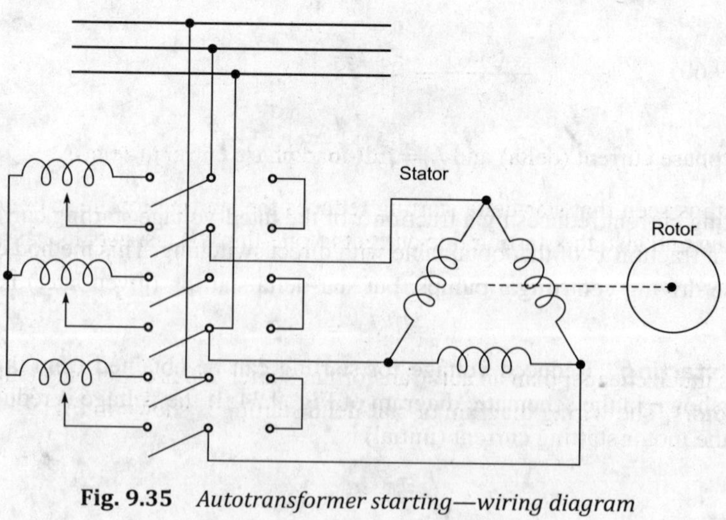

Fig. 9.35 *Autotransformer starting—wiring diagram*

Star-delta starting The star-delta starting is an inexpensive two-step method of induction motor starting. The motor designed for delta running is started across full-line voltage by connecting the phases in star as shown in the schematic diagram of Fig. 9.36.

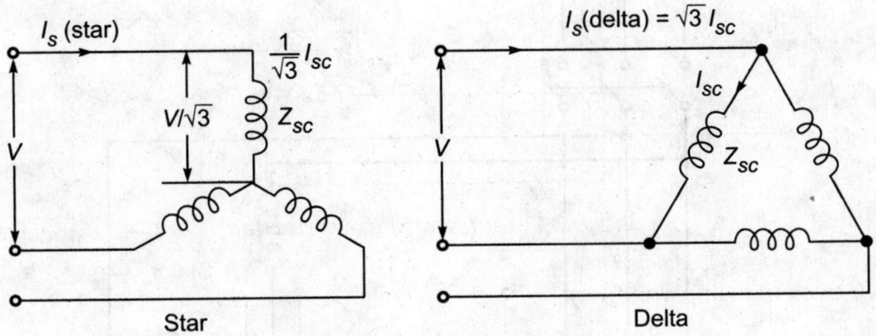

Fig. 9.36 *Star-delta starting*

In direct delta starting:

$$\text{Starting phase current, } I_{SC} = V/Z_{SC}; \quad V = \text{line voltage} \tag{9.65}$$

$$\text{Starting line current, } I_s \text{ (delta)} = \sqrt{3}\, I_{SC} \tag{9.66}$$

In star starting:

Starting line (phase) current, I_s (star)

$$= \frac{V/\sqrt{3}}{Z_{SC}} = \frac{1}{\sqrt{3}} I_{SC} \tag{9.67}$$

$$\therefore \quad \frac{I_s \text{ (star)}}{I_s \text{ (delta)}} = \frac{1}{3} \tag{9.68}$$

Using Eq. (9.60)

$$\frac{T_s \text{ (star)}}{T_{fl}} = \frac{1}{3}\left(\frac{I_{SC}}{I_{fl}}\right)^2 s_{fl} \tag{9.69}$$

where I_{SC} = starting phase current (delta) and I_{fl} = full-load phase current (delta).

!Attention It is thus seen that star-delta starting reduces the starting torque to one-third that obtainable by direct-delta starting and also the starting line current to one-third. It just acts like autotransformer starting with

$$x = 1/\sqrt{3} = 0.58$$

A star-delta starter is much cheaper than an autotransformer starter and is commonly employed for both small and medium-size motors. The wiring diagram of star-delta starting is shown in Fig. 9.37.

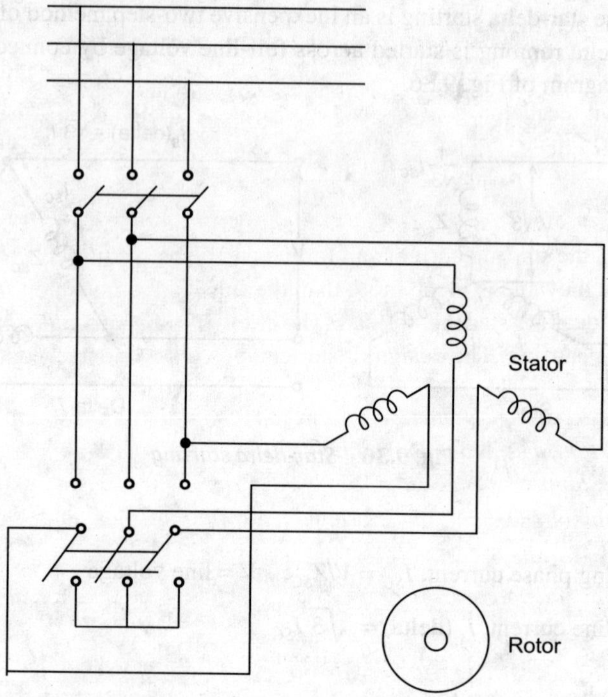

Fig. 9.37 *Wiring diagram of star-delta starting*

(b) Starting of Slip-ring Motors (Rotor-resistance Starting)

It has already been discussed that in slip-ring motors the starting current is reduced and the starting torque simultaneously increased by adding an external resistance in the rotor circuit. Figure 9.38 shows the motor circuit (magnetizing branch neglected) with external resistance is added in the rotor circuit. For simplicity, the stator impedance is also neglected. Here

R'_{ext} = external resistance in each phase of the rotor circuit as seen from the stator

$$= \left(\frac{N_{e1}}{N_{e2}}\right)^2 R_{ext}$$

where R_{ext} = actual external resistance in the rotor circuit

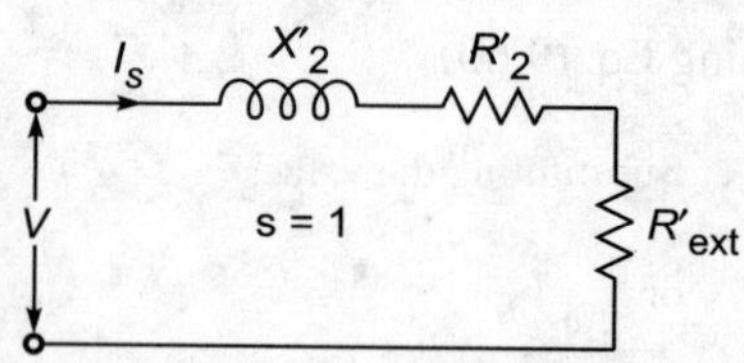

Fig. 9.38 *Starting with external resistance in the rotor circuit*

Now

$$I_s = \frac{V}{\sqrt{(R'_2 + R'_{ext})^2 + X'^2_2}} \tag{9.70}$$

$$T_s = \frac{3}{\omega_s} \cdot \frac{V^2 (R'_2 + R'_{ext})}{(R'_2 + R'_{ext})^2 + X'^2_2} \tag{9.71}$$

Maximum starting torque is achieved at

$$R_2' + R_{ext}' = X_2' \tag{9.72}$$

which equals the breakdown torque or

$$T_s(\max) = \frac{3}{\omega_s}\left(\frac{0.5V^2}{X_2'}\right) = T_{\text{breakdown}} \tag{9.73}$$

By a suitable choice of R_{ext}, the starting current and torque can be both adjusted to desirable levels. Since the maximum starting torque achievable is much more than the full-load torque, the use of slip-ring motors with rotor-resistance starting is ideal for starting on load. The external rotor resistance is arranged in steps which are gradually cut out during starting. The design of starter steps is beyond the scope of this book.

Example 9.10 *A squirrel-cage induction motor has a slip of 4% at full load. Its starting current is five times the full-load current. The stator impedance and magnetizing current may be neglected; the rotor resistance is assumed constant.*

(a) Calculate the maximum torque and the slip at which it would occur.

(b) Calculate the starting torque.

Express torques in pu of the full-load torque.

LO 3

Solution

(a)

$$I_s^2 = \frac{V^2}{R_2'^2 + X_2'^2}; \ (s = 1) \tag{i}$$

$$I_{fl}^2 = \frac{V^2}{(R_2'^2/s_{fl})^2 + X_2'^2} \tag{ii}$$

Dividing Eq. (i) by (ii)

$$\left(\frac{I_s}{I_{fl}}\right)^2 = \frac{(R_2'/s_{fl})^2 + X_2'^2}{R_2'^2 + X_2'^2} = \frac{s_{\max,T}^2 + s_{fl}^2}{s_{fl}^2\,(s_{\max,T}^2 + 1)} \tag{iii}$$

Substituting the values

$$25 = \frac{s_{\max,T}^2 + (0.04)^2}{(0.04)^2\,(s_{\max,T}^2 + 1)}$$

or

$$s_{\max,T} = 0.2 \quad \text{or} \quad 20\%$$

$$T_{\max} = \frac{3}{\omega_s}\cdot\frac{0.5V^2}{X_2'^2} \tag{iv}$$

$$T_{fl} = \frac{3}{\omega_s}\cdot\frac{V^2\,(R_2'^2/s_{fl})}{(R_2'^2/s_{fl})^2 + X_2'^2} \tag{v}$$

Dividing Eq. (iv) by (v)

$$\frac{T_{\max}}{T_{fl}} = 0.5 \times \frac{R_2'^2 + s_{fl}^2 X_2'^2}{R_2' X_2' s_{fl}} = 0.5 \times \frac{s_{\max,T}^2 + s_{fl}^2}{s_{\max,T}\, s_{fl}}$$

$$= 0.5 \times \frac{(0.2)^2 + (0.04)^2}{0.2 \times 0.04} = 2.6$$

or

$$T_{\max} = 2.6 \text{ pu}$$

(b) As per Eq. (9.60)

$$\frac{T_s}{T_{fl}} = \left(\frac{I_s}{I_{fl}}\right)^2 s_{fl} = (5)^2 \times 0.04 = 1$$

or $$T_s = 1 \text{ pu}$$

Example 9.11 *A 150 kW, 3000 V, 50 Hz, 6-pole star-connected induction motor has a star-connected slip-ring rotor with a transformation ratio of 3.6 (stator/rotor). The rotor resistance is 0.1 Ω/phase and its per phase leakage inductance is 3.61 mH. The stator impedance may be neglected. Find (a) the starting current and torque on rated voltage with short-circuited slip-rings, and (b) the necessary external resistance to reduce the rated-voltage starting current to 30 A and the corresponding starting torque.*

LO 3

Solution

$$R_2 = 0.1\ \Omega; \qquad X_2 = 314 \times 3.61 \times 10^{-3} = 1.13\ \Omega$$

$$R_2' = (3.6)^2 \times 0.1 = 1.3\ \Omega; \qquad X_2' = (3.6)^2 \times 1.13 = 14.7\ \Omega$$

(a)
$$I_s = \frac{3000/\sqrt{3}}{\sqrt{(1.3)^2 + (14.7)^2}} = 117.4 \text{ A}$$

$$T_s = \frac{3}{\omega_s} \cdot \frac{V^2 R_2'^2}{R_2'^2 + X_2'^2}$$

$$\omega_s = \frac{2\pi \times 1000}{60} = 104.7 \text{ rad/s}$$

$$T_s = \frac{3}{104.7} \cdot \frac{(3000/\sqrt{3})^2 \times 1.3}{(1.3)^2 + (14.7)^2} = 513.1 \text{ Nm}$$

(b)
$$I_s = \frac{3000/\sqrt{3}}{\sqrt{(1.3 + R_{ext}')^2 + (14.7)^2}} = 30 \text{ A}$$

$$R_{\text{ext}}' = 54.33$$

$$R_{\text{ext}} = 54.53/(3.6)^2 = 4.21\ \Omega$$

$$T_s = \frac{3}{104.7} \cdot \frac{(3000/\sqrt{3})^2 \times (1.3 + 54.33)}{(1.3 + 54.33)^2 + (14.7)^2} = 1440 \text{ Nm}$$

Remark By adding an external resistance in the rotor, starting current reduces by a factor of 117.4/30 = 3.91 while the starting torque increases by 1440/513.1 = 2.8.

Example 9.12 *A small squirrel-cage induction motor has a starting current of six times the full-load current and a full-load slip of 0.05. Find in pu of full-load values, the current (line) and starting torque with the following methods of starting ((a) to (d)).*

(a) direct switching.

(b) stator-resistance starting with motor current limited to 2 pu,

(c) autotransformer starting with motor current limited to 2 pu, and

(d) star-delta starting.

(e) what autotransformer ratio would give 1 pu starting torque?

LO 3

Solution

(a) Direct switching

$$I_s = 6 \text{ pu}$$
$$T_s = (6)^2 \times 0.05 = 1.8$$

(b) Stator-resistance starting

$$I_s = 2 \text{ pu (limited to)}$$
$$T_s = (2)^2 \times 0.05 = 0.2 \text{ pu} \quad \text{(Eq. (9.60))}$$

(c) Autotransformer starting

$$x = 2/6 = 1/3$$
$$I_s \text{ (motor)} = 2 \text{ pu}$$
$$I_s \text{ (line)} = \frac{1}{3} \times 2 \text{ pu} = 0.67 \text{ pu}$$
$$T_s = (2)^2 \times 0.05 = 0.2 \text{ pu} \quad \text{(Eq. (9.60))}$$

(d) Star-delta starting

$$I_s = \frac{1}{3} \times 6 = 2 \text{ pu} \quad \text{(Eq. (9.68))}$$
$$T_s = \frac{1}{3} \times (6)^2 \times 0.05 = 0.6 \text{ pu} \quad \text{(Eq. (9.69))}$$

(e) Autotransformer starting

$$T_s = x^2(6)^2 \times 0.05 = 1.0 \text{ pu} \quad \text{(Eq. (9.64))}$$
$$x = 0.745 \; (\approx 75\% \text{ tap})$$

9.9 | COGGING AND CRAWLING

LO 3

Cogging

A squirrel-cage rotor may exhibit a peculiar behaviour in starting for certain relationships between the number of poles and the stator and rotor slots. With the number of stator slots S_1 equal to or an integral multiple of rotor slots S_2, the variation of reluctance as a function of space will be quite pronounced resulting in strong alignment forces at the instant of starting. These forces may create an aligning torque stronger than the accelerating torque with consequent failure of the motor to start. This phenomenon is known as *cogging*. Such combination of stator and rotor slots must, therefore, be avoided in machine design.

Crawling

Certain combinations of S_1 and S_2 cause accentuation of certain space harmonics of the mmf wave, e.g. fifth and seventh harmonics which correspond to poles five and seven times that of the fundamental. Since the space-phase difference between fundamental poles of the winding phase is (0°, 120°, 240°), this (space-phase) difference is (0°, 240°, 120°) for the fifth harmonic poles and (0°, 120°, 240°) for the seventh. Hence the fifth harmonic poles rotate backwards with synchronous speed of $n_s/5$ and the seventh harmonic poles rotate forward at $n_s/7$.

!Attention These harmonic mmfs produce their own asynchronous (induction) torques of the same general torque-slip shape as that of the fundamental. Figure 9.36 shows the superimposition of the fundamental, fifth and seventh harmonic torque-slip curves. A marked saddle effect is observed with *stable* region of operation (negative torque-slip slope) around l/7th normal motor speed ($s = 6/7$). In Fig. 9.39 the load torque curve intersects the motor torque curve at the point M resulting in stable operation. This phenomenon is known as *crawling* (running stably at low speed).

Important!

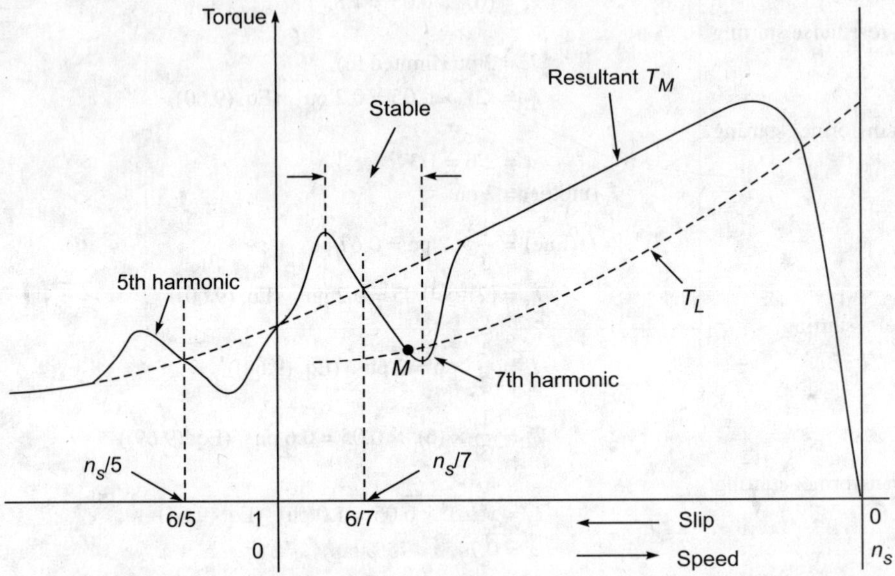

Fig. 9.39 *Torque-slip characteristic of a 3-phase induction motor showing the effect of harmonic asynchronous (induction) torques*

Certain slot combinations, e.g. $S_1 = 24$ and $S_2 = 18$ cause the stator mmf to possess a reversed 11th and a forward 13th harmonic mmf while the rotor has a reversed 13th and a forward 15th. The stator 13th harmonic mmf rotates at speed $+ n_s/13$ with respect to the stator and the rotor mmf of the 13th harmonic rotates at $-(n_s - n)/13$ with respect to the rotor when the rotor is running at speed n. These two mmf's lock into each other to produce a synchronous torque when

$$\frac{n_s}{13} = n - \left(\frac{n_s - n}{13}\right)$$

or

$$n = \frac{n_s}{7}$$

Thus there is a discontinuity at $n_s/7$ in the torque-slip characteristic produced by not the seventh but the 13th harmonic as shown in Fig. 9.40.

Arrangements to Minimize the Effect of Cogging and Crawling

Cogging and crawling are much less prominent in slip-ring induction machines as these possess higher starting torques. The induction harmonic torque cannot be avoided, but can be reduced by making a proper choice of coil-span and by *skewing*. The synchronous harmonic torques can be avoided totally by a proper combination of stator and rotor slots.

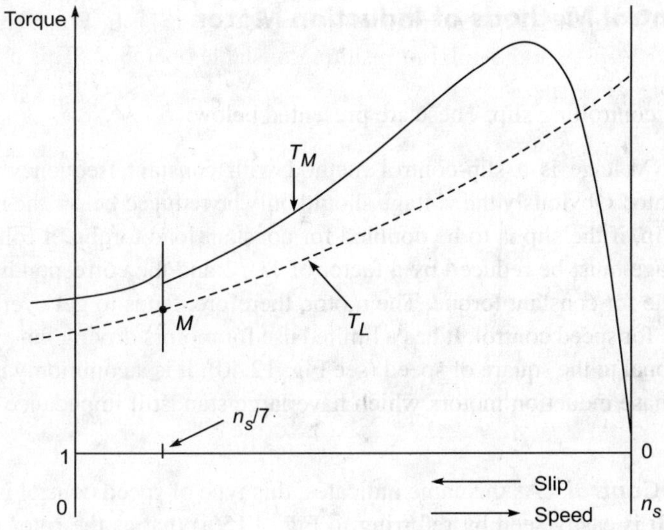

Fig. 9.40 *Synchronous harmonic torque in induction motor*

Skewing

Important! The rotor teeth are given a slight twist as shown in Fig. 9.41. As a result, each rotor tooth is opposite several stator teeth, thereby mitigating the effect of locking of stator and rotor teeth and so eliminating cogging. Further mmf harmonics get phase shifted along the rotor length reducing the effective harmonic torques.

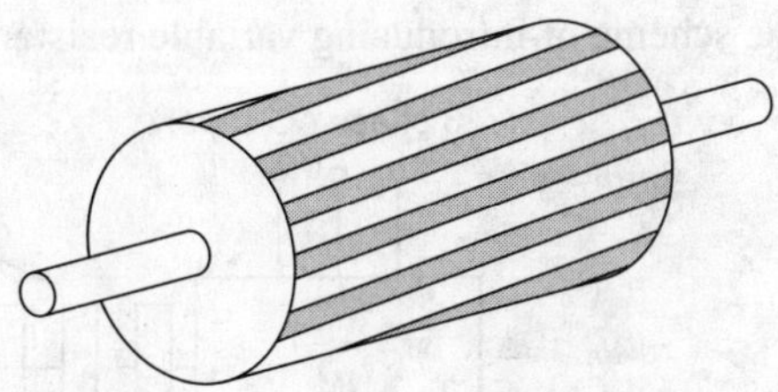

Fig. 9.41 *Rotor with skewed teeth*

9.10 SPEED CONTROL

LO 3

Parameters Effecting Speed

Stepless control of speed of induction motors cannot be carried out as efficiently and inexpensively as for dc motors. Various methods of controlling the speed of the induction motor can be visualized by consideration of the speed equation

$$n = (1 - s)\, n_s \tag{9.74}$$

It is seen from this equation that there are two basic ways of speed control, namely (i) slip-control for fixed synchronous speed, and (ii) control of synchronous speed. Since

Recall

$$n_s = \frac{120\, f}{P} \tag{9.75}$$

there are two ways to control synchronous speed—control of supply frequency and control of stator poles. The latter method gives a step control as poles can be changed in multiple of two. Pole-changing is carried out in a squirrel-cage motor only and that too for two steps.

Various Speed Control Methods of Induction Motor

(a) Silp Control

There are three ways of controlling slip. These are presented below.

(i) Voltage Control Voltage is a slip-control method with constant frequency variable-voltage begin supplied to the motor stator. Obviously the voltage should only be reduced below the rated value. For a motor operating at full-load slip, if the slip is to be doubled for constant load torque, it followers from Eqs (9.34) and (9.35) that the voltage must be reduced by a factor of $1/\sqrt{2}$ and the corresponding current (I'_2) rises to $\sqrt{2}$ of the full-load value for constant torque. The motor, therefore, tends to get overheated.*** The method, therefore, is not suitable for speed control. It has a limited use for motors driving fan-type loads whose torque requirement is proportional to the square of speed (see Fig. 12.40). It is a commonly used method for ceiling fans driven by single-phase induction motors which have large standstill impedance limiting the current by the stator.

(ii) Rotor-Resistance Control As the name indicates, this type of speed control is only possible for slip-ring induction motors. It is easily seen by referring to Fig. 9.15(a) that as the rotor resistance is increased, the motor slip increases(speed falls) for a fixed load torque. The stator current varies to a limited extent as the effect of change in slip and rotor resistance tend to cancel out (refer to Eq. (9.43)) for small values of slip. The input power, however, increases. This provides for power lost in rotor additional resistance.

The scheme of introducing variable resistance in the rotor circuit is shown in Fig. 9.42(a)

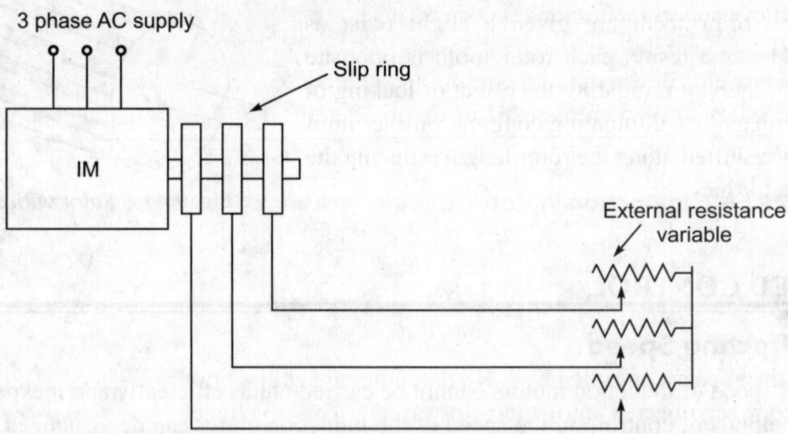

Fig. 9.42(a) *Slip control by external variable resistance*

As the slip control is achieved by additional power loss in the rotor circuit, it is more conveniently carried out by a controlled rectifier and a single fixed resistance as in Fig. 9.42(b). The additional power loss in the external resistance in rotor circuit, causes the efficiency to decrease sharply. This method of speed control as such is, therefore, adopted for a narrow speed range and usually for a short-time operation.

*** This is the reason why induction motors tend to overheat when the supply voltage falls by a large percentage of the rated motor voltage.

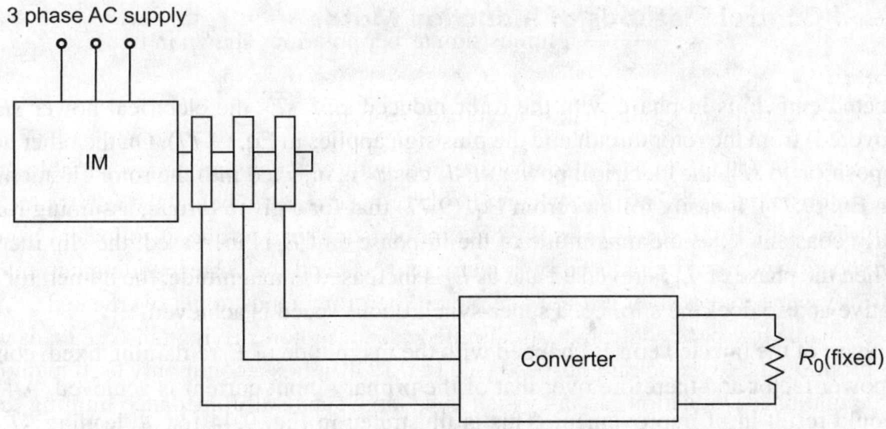

Fig. 9.42(b) *Silp control of wound-rotor motor with converter*

The efficiency of this type of speed-control scheme can be improved by returning the power out of the rotor mechanically to the rotor shaft or electrically to the mains. The first scheme can be implemented by a rectifier and dc motor coupled to the rotor shaft. The second scheme requires a frequency converter which converts variable-frequency power to a fixed (supply) frequency for feeding the electrical power back to the mains. These schemes can be devised by SCR circuitry (see Figs 12.42 and 12.43).

It is possible to achieve supersynchronous speeds by injecting power at appropriate frequency into the rotor by means of an adjustable frequency source. It may be seen that this is just the reverse of adding rotor resistance in which power is drawn out of the rotor (and wasted in external resistance). A large range of speed control both above and below synchronous is made possible by including frequency-converting equipment in the rotor circuit.

(iii) Slip-frequency EMF Injunction in Rotor Circuit It has been shown in Eq. 9.19 the torque developed in an inductor motor is

$$T = \frac{3 I_2^2 R_2}{s\omega_s} \text{ Nm} \tag{9.76}$$

For a given torque the slipe is controlled by $I_2^2 R_2$, the rotor copper loss or, in more general terms, the electrical power removed from (or injected into) the rotor circuit. This is efficiently achieved by injecting a slip-frequency emf into the rotor circuit as shown in Fig. 3.43. The torque-slip relation of Eq. (9.76) now generalizes to

$$T = \frac{3 I_2^2 R_2 \pm E_i I_2 \cos\phi_2}{s\omega_s} \tag{9.77}$$

where ϕ_2 is phases angle between $\bar{E}_i$ and $\bar{I}_2$

From the rotor circuit of Fig. 9.43

$$s\bar{E}_2 \pm \bar{E}_i = \bar{I}_2 \bar{Z}_2$$

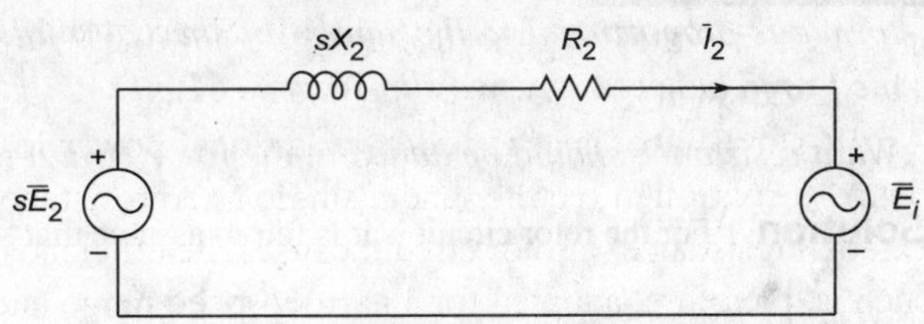

Fig. 9.43 *Slip control by injected slip-frequency emf into rotor circuit*

$$\bar{I}_2 = \frac{s\bar{E}_2 - \bar{E}_i}{\bar{Z}_2}; \text{ minus sign as per polarities shown in figure} \tag{9.78}$$

When the injected emf $\bar{E}_i$ is in-phase with the rotor induced emf $s\bar{E}_2$ the electrical power $sE_2I_2 \cos \phi_2$ is removed (recovered) from the rotor circuit and the plus sign applies in Eq. (9.77). On the other hand, when E_i is in phase opposition to sE_2, the electrical power $sE_2I_2 \cos \phi_2$ is injected into the rotor circuit and the minus sign applies in Eq. (9.77). It easily follows from Eq. (9.77) that for a given torque, assuming rotor current I_2 to remain nearly constant[†††], as the magnitude of the in-phase emf E_i is increased, the slip increases (speed decreases). When the phase of $\bar{E}_i$ is reversed and as E_i is increased in magnitude, the numerator in Eq. (9.77) becomes negative and so does the slip, i.e. a super-synchronous speed is achieved.

If instead the phase of the injected emf is changed with the magnitude of E_i remaining fixed, control over the rotor current power factor and therefore over that of the primary input current is achieved. A leading phase angle of E_i would result in pf improvement. This is illustrated in Fig. 9.44 for $\bar{E}_i$ leading $s\bar{E}_2$ by 90°. To a rough approximation the component of E_i in-phase (or in phase opposition) controls motor speed and the quadrature component control the motor power factor.

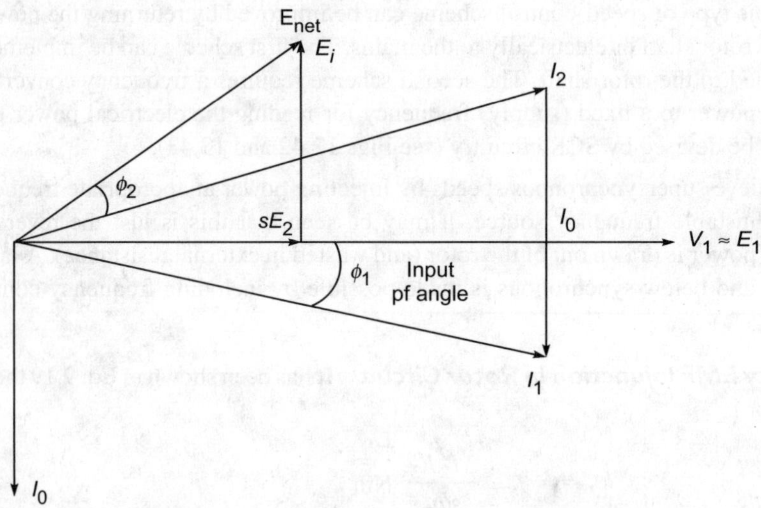

Fig. 9.44 *power factor control by rotor injection*

Example 9.13 *A 3 phase 50 Hz, 12 pole, 200 kW slip-ring induction motor drives a fan whose torque is proportional to the square of speed. At full load, the motor slip is 0.045. The rotor resistance measured between any two slip-rings is 61 mΩ.*

What resistance should be added in the rotor circuit to reduce the fan speed to 450 rpm? LO 3

Solution For the rotor circuit is it is fair to assume that

$$R_2/s >> X_2$$

So we will neglect X_2

$$R_2 = 61/2 = 30.5 \text{ m}\Omega$$

[†††]For large value of slip. The rotor power factor ($\cos \phi_2$) decreases and so current for given torque increases

$$\omega_s = \frac{120\,f}{P} \times \frac{2\pi}{60} = \frac{4\pi\,f}{P} = \frac{4\pi \times 50}{12} = 52.36 \text{ rad/s}$$

$$\omega = (1 - s)\,\omega_s = (1 - 0.045) \times 52.36 = 50 \text{ rad/s}$$

At full load $\quad 200 \times 10^3 = \omega T \text{ (fan)} = 50\,T$

or $\quad T \text{ (fan)} = 4000 \text{ Nm}$

Rotor current $\quad T = \dfrac{3}{\omega_s}\,\dfrac{I_2^2 R_2}{p}$

$$4000 = \frac{3}{52.36} \times \frac{I_2^2 \times 30.5 \times 10^{-3}}{0.045}$$

or $\quad I_2 = 321 \text{ A}$

Rotor circuit voltage (stand still)

$$E_2 = I_2\,(R_2/s) = 321 \times \frac{30.5 \times 10^{-3}}{0.045} = 217.6 \text{ V}$$

The fan speed is to be reduced to

$$n = 450 \text{ rpm} \quad \text{or} \quad \frac{450 \times 2\pi}{60} = 47.12 \text{ rad/s}$$

$$\text{Slip, } s = \frac{500 - 450}{500} = 0.1$$

$$I_2 \text{ (new)} = \frac{217.6}{R_t/s} = \frac{21.76}{R_t}, \; R_t = \text{total rotor resistance}$$

$$T \text{ (new)} = 4000 \times \left(\frac{47.12}{50}\right)^2 = 3552 \text{ Nm}$$

$$T \text{ (new)} = \frac{3}{\omega_s}\,\frac{I_2^2\,(n\omega)\,R_t}{s\,(n\omega)}$$

$$3552 = \frac{3}{52.36} \times \frac{(21.76)^2\,R_t}{R_t^2 \times 0.1}$$

From which we find $\quad R_t = 76.4 \text{ m}\Omega$

$$R_{\text{ext}} = 76.4 - 30.5 = 45.9 \text{ m}\Omega \text{ (to be added)}$$

Example 9.14 *A 440 V, 50 Hz, 4-pole 3-phase, delta-connected motor has a leakage impedance of (0.3 + j 5.5 + 0.25/s) Ω/phase (delta phase) referred to the stator. The stator to rotor voltage ratio is 2.5. Determine the external resistance to be inserted in each star phase of the rotor winding such that the motor develops a gross torque of 150 Nm at a speed of 1250 rpm.*

LO 3

Solution

$$\omega_s = \frac{2\pi \times 1500}{60} = 157.1 \text{ rad/s}$$

$$s = \frac{1500 - 1250}{1500} = 0.167$$

Total leakage impedance referred to stator (equivalent star basis)

$$= \frac{1}{3}[(0.3 + 0.25/s) + j\,5.5] = \left(0.1 + \frac{0.083}{s} + j1.83\right) \Omega\text{/phase (star)}$$

From Eq. (9.22) with

$$R'_{2t} = R'_2 + R'_{ext}; \quad R'_2 = 0.083\ \Omega,\ R'_{2t} = R'_2 \text{ (total)}$$

$$T = \frac{3}{\omega_s} \cdot \frac{V^2 (R'_{2t}/s)}{(0.1 + R'_{2t}/s)^2 + (X_1 + X'_2)^2}$$

Substituting values

$$150 = \frac{3}{157.1} \cdot \frac{(440/\sqrt{3})^2 (R'_{2t}/00.167)}{(0.1 + R'_{2t}/0.167)^2 + (1.83)^2}$$

$$R'_{2t} - 1.34\, R'_{2t} + 0.093 = 0$$

$$R'_{2t} = 1.27\ \Omega;\ 0.073\ \Omega$$

(The second value is less than $R'_2 = 0.083\ \Omega$)

$$R'_{ext} = 1.27 - 0.083\ \Omega$$

$$R_{ext}(\text{rotor}) = \frac{1.19}{(2.5)^2} = 1.19\ \Omega\text{/phase}$$

Example 9.15 *A 3-phase, 25 kW, 400 V, 50 Hz, 8-pole induction motor has rotor resistance of 0.08 Ω and standstill reactance of 0.4 Ω. The effective stator/rotor turn ratio is 2.5/1. The motor is to drive a constant-torque load of 250 Nm. Neglect stator impedance.*

(a) Calculate the minimum resistance to be added in rotor circuit for the motor to start up on load.

(b) At what speed would the motor run, if the added rotor resistance is (i) left in the circuit, and (ii) subsequently short circuited.

LO 3

Solution Motor circuit seen on rotor side is shown in Fig. 9.45; stator impedance having been neglected.

$$I_2 = \frac{V_1/a}{\sqrt{(R_2/s)^2 + X_2^2}}$$

$$T = \frac{3}{\omega_s} \cdot I_2^2 R_2/s$$

$$= \frac{3}{\omega_s} \cdot \frac{(V/a)^2 (R_2/s)}{(R_2/s)^2 + X_2^2} \quad \text{(i)}$$

Fig. 9.45

$$T(\text{start}) = \frac{3}{\omega_s} \cdot \frac{(V/a)^2 R_2}{R_2^2 + X_2^2};\ s = 1 \quad \text{(ii)}$$

Let external resistance added to rotor circuit be R_2 (ext). Then

$$R_2(\text{total}) = R_{2t} = R_2 + R_2(\text{ext}) \quad \text{(iii)}$$

Then

$$T(\text{start}) = \frac{3}{\omega_s} \cdot \frac{(V/a)^2 R_{2t}}{R_{2t}^2 + X_2^2} \quad \text{(iv)}$$

(a) $a = 2.5, \quad X_2 = 0.4\ \Omega, \quad R_2 = 0.08\ \Omega$

T(start) = T(load) = 250 Nm; This is minimum starting torque.

Actual starting must be sufficiently more than this.

$$V = 400/\sqrt{3} = 231\text{ V}$$

$$n_s = 750\text{ rpm} \quad \text{or} \quad \omega_s = 78.54\text{ rad/s}$$

Substituting values in Eq. (ii)

$$250 = \left(\frac{3}{78.54}\right) \cdot \frac{(231/2.5)^2 R_{2t}}{R_{2t}^2 + (0.4)^2} \tag{v}$$

or

$$R_{2t}^2 - 1.304\, R_{2t} + 0.16 = 0$$

or

$$R_{2t} = 0.137\ \Omega,\ 1.167\ \Omega$$

The T-s characteristics with these two values of R_{2t} are drawn in Fig. 9.46. It is easy to see that with $R_{2t} = 1.167\ \Omega$ the motor will not start as motor torque reduces ($T_{motor} < T_{load}$) for $s < 1$.

So we select $R_{2t} = 0.137\ \Omega$

$\Rightarrow \quad R_2(\text{ext}) = 0.137 - 0.08 = 0.057\ \Omega$

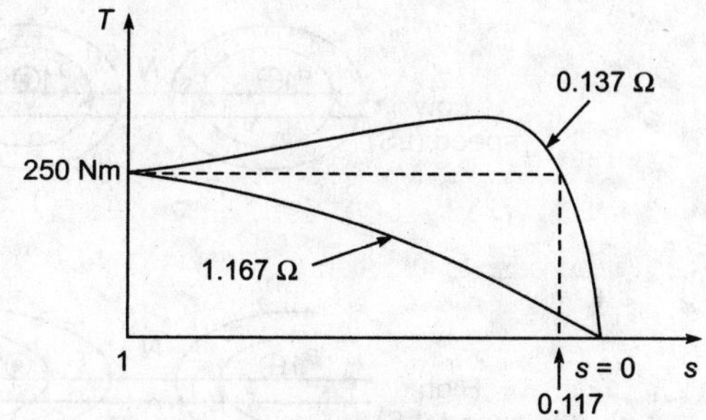

Fig. 9.46

(b)(i) $R_{2t} = 0.137\ \Omega$; External resistance included in circuit.

Substituting values in Eq. (i)

$$250 = \left(\frac{3}{78.54}\right) \frac{(231/2.5)^2 (R_{2t}/s)}{(R_{2t}/s)^2 + (0.4)^2}$$

This equation has the same solution as Eq. (v). Thus

$$R_{2t}/s = 0.137,\ 0.167$$

With $R_{2t} = 0.137\ \Omega$, we get

$$s = 1, 0.137/1.167$$

$$= 0.117 \text{ (as shown in Fig. 9.42)}$$

Motor speed, $n = 750\,(1 - 0.117) = 662$ rpm

(ii) With external resistance cut out

$$250 = \left(\frac{3}{78.54}\right) \frac{(231/2.5)^2 (R_2/s)}{(R_2/s)^2 + (0.4)^2}$$

The solution would yield as before

$$R_2/s = 0.137,\ R_2/s = 1.167$$

With $R_2 = 0.08$, we now get

$$s = 0.584,\ 0.067$$

The solution points are indicated in the T-s characteristic drawn in Fig. 9.47. The motor will run at

$$s = 0.067 \Rightarrow 700\text{ rpm}$$

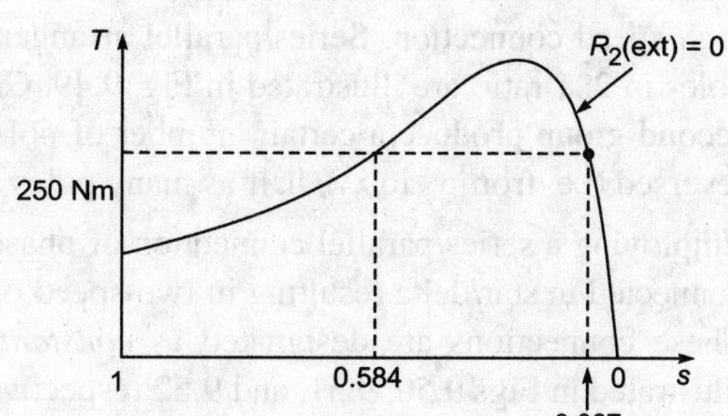

Fig. 9.47

The point $s = 0.584$ on T-s characteristic is unstable as the torque-speed slope here is positive and the motor will speed up beyond this.

(b) Pole Changing Method

The number of stator poles can be changed in the ratio of 2 : 1 by simple changes in coil connections. Either of the two synchronous speeds can be selected. In a pole-changing motor, the rotor must be squirrel-cage to avoid complications caused by reconnection of the rotor winding if a wound rotor is used. A squirrel-cage rotor automatically reacts to create the same number of poles as the stator. If two independent sets of stator windings are employed, each arranged for pole changing, as many as four synchronous speeds can be obtained.

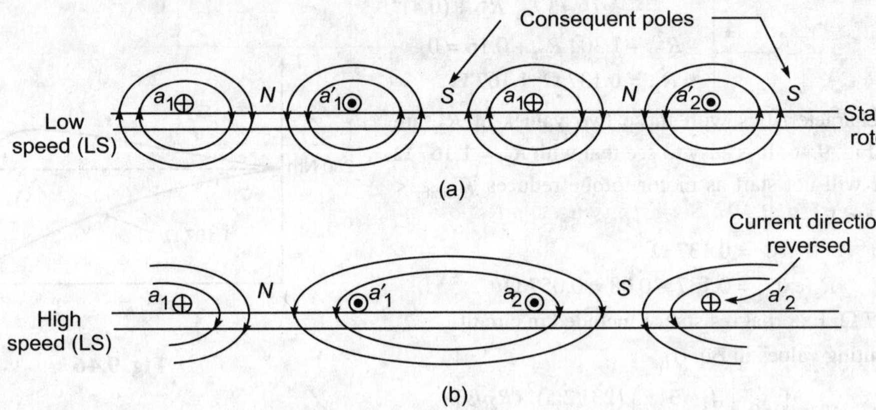

Fig. 9.48 *Principle of pole changing*

Figure 9.48(a) shows two coils a_1a_1' and a_2a_2' pertaining to phase a of the stator winding—several coils which may be short-pitched would normally be used for each phase. If the coils are connected to carry equal currents, both clockwise (or anticlockwise) as seen from top. Each coil forms a north pole and *consequent* south poles are created in the intervening space between coils. This is the lowspeed connection—a larger number of poles equal twice the number of coils. If the two coils are reconnected by means of a *controller* so that the current in one of the coils is reversed as shown in Fig. 9.48(b), the number of poles is now halved—high-speed connection. Series/parallel arrangements of reconnecting coil groups to change the number of poles in 2: 1 ratio are illustrated in Fig. 9.49. Currents flowing from x_1 to y_1 in one group and x_2 to y_2 in the second group produce a certain number of poles (case of consequent poles). If the current in one group is reversed (i.e. from y_2 to x_2), half as many poles are produced).

Employing a series/parallel connection of phase groups of individual phases, the phases can themselves be connected in star/delta resulting in two-speed operation with the three types of torque-speed characteristics. These connections are designated as *constant-torque, constant-horsepower* and *variable-torque* and are illustrated in Figs 9.50, 9.51, and 9.52 respectively. In each of these figures, let

V = line-to-line voltage

I = allowable conductor current

η_l, η_h = respective efficiencies of LS and HS connections

$\cos \phi_l$, $\cos \phi_h$ = respective power factors of LS and HS connections.

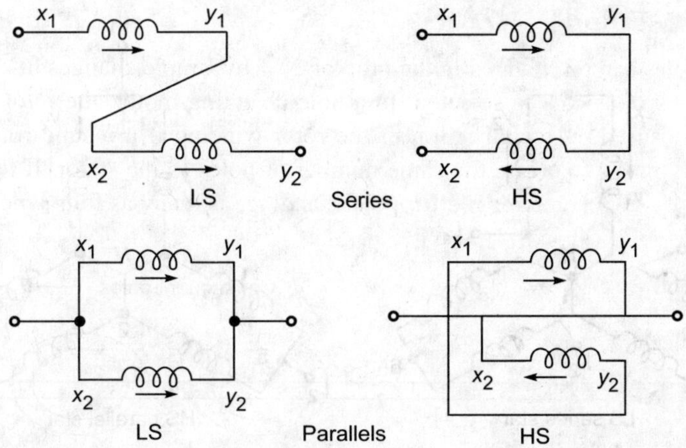

Fig. 9.49 *Series/parallel connections of phase groups for pole changing*

Important!

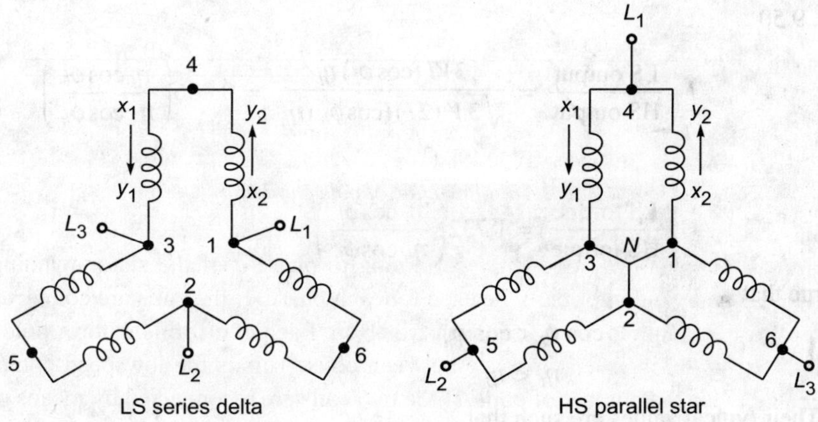

Fig. 9.50

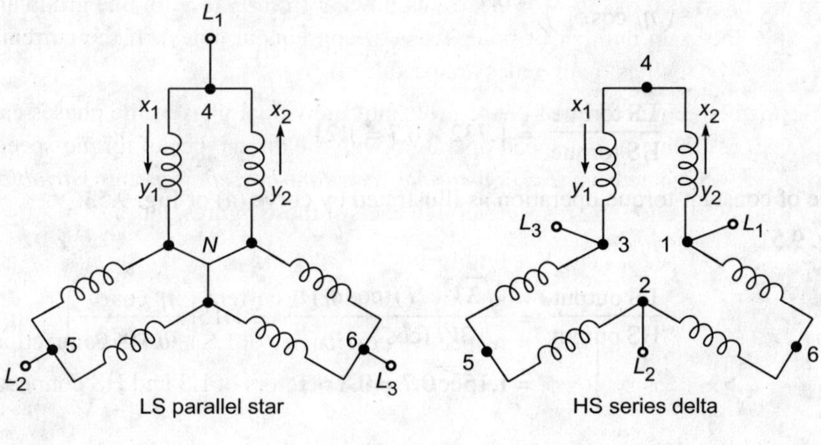

Fig. 9.51

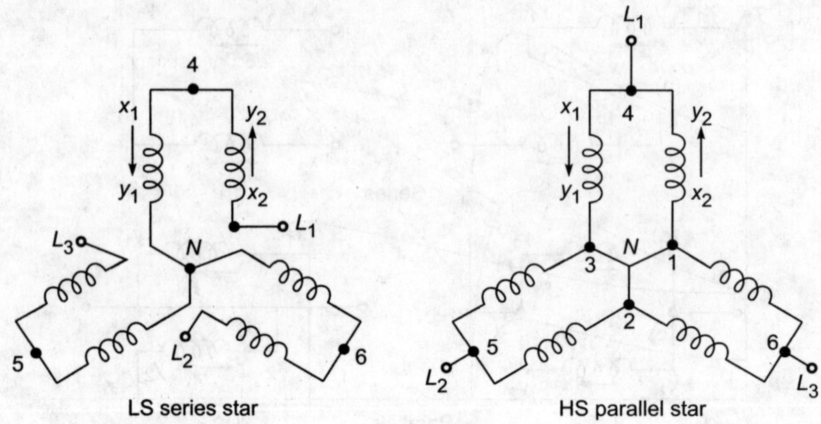

Fig. 9.52

Referring to Fig. 9.50

$$\frac{\text{LS output}}{\text{HS output}} = \frac{3VI(\cos\phi_l)\eta_l}{\sqrt{3}V(2I)(\cos\phi_h)\eta_h} = 0.866\left(\frac{\eta_l\cos\phi_l}{\eta_h\cos\phi_h}\right)$$

Hence

$$\frac{\text{LS torque}}{\text{HS torque}} = 1.732\left(\frac{\eta_l\cos\phi_l}{\eta_h\cos\phi_h}\right)$$

It is in general true that

$$\cos\phi_l < \cos\phi_h$$

$$\eta_l < \eta_h$$

Important! Their typical values are such that

$$\left(\frac{\eta_l\cos\phi_l}{\eta_h\cos\phi_h}\right) = 0.7$$

Therefore,

$$\frac{\text{LS torque}}{\text{HS torque}} = 1.732 \times 0.7 = 1.21$$

This is indicative of constant-torque operation as illustrated by curve (a) of Fig. 9.53.

Referring to Fig. 9.51

$$\frac{\text{LS output}}{\text{HS output}} = \frac{\sqrt{3}V(2I)(\cos\phi_l)\eta_l}{3VI(\cos\phi_h)\eta_h} = 1.15\left(\frac{\eta_l\cos\phi_l}{\eta_h\cos\phi_h}\right)$$

$$= 1.15 \times 0.7 = 0.8 \approx 1$$

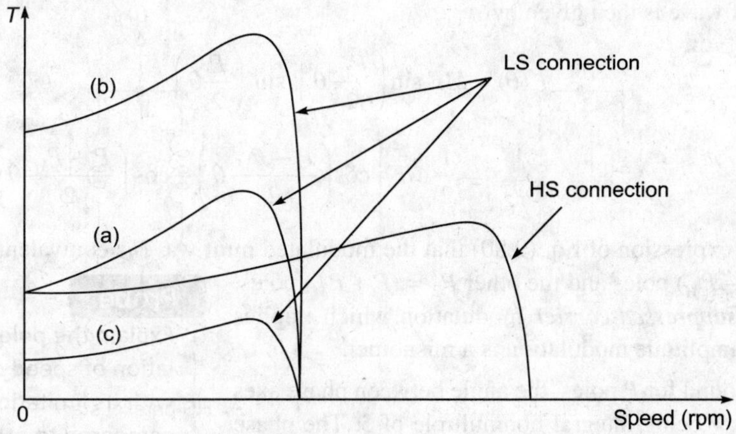

Fig. 9.53 *Torque-speed characteristics of various LS/HS connections*

and $$\frac{\text{LS torque}}{\text{HS torque}} = 2 \times 0.8 = 1.6$$

This is indicative of constant-horsepower operation as illustrated by the curve (b) of Fig. 9.53.

Similarly referring to Fig. 9.52,

$$\frac{\text{LS output}}{\text{HS output}} = \frac{\sqrt{3}VI\,(\cos\phi_l)\,\eta_t}{\sqrt{3}\,V(2I)(\cos\phi_h)\,\eta_h} = 0.5 \times 0.7 = 0.35$$

and $$\frac{\text{LS torque}}{\text{HS torque}} = 2 \times 0.35 = 0.7$$

This is the variable-torque connection illustrated by curve (c) of Fig. 9.53.

Pole-Amplitude Modulation

This method of pole-changing has the advantage that the ratio of the two speeds need not be necessarily two-to-one as in the method of consequent poles.

Consider the sinusoidally distributed mmf wave of one phase of the stator

$$\mathcal{F}(\theta) = F \sin\left(\frac{P}{2}\theta\right) \tag{9.78}$$

where P is the number of poles and θ the mechanical angle.

Also remember that F varies sinusoidally with time as it is caused by a sinusoidally-varying phase current. Let this mmf wave be modulated by another sinusoidal mmf wave of P_M poles

$$\mathcal{F}_M(\theta) = M \sin\left(\frac{P_M}{2}\theta\right) \tag{9.79}$$

The modulated mmf wave is then given by

$$\mathcal{F}'(\theta) = MF \sin\left(\frac{P_M}{2}\theta\right) \sin\left(\frac{P}{2}\theta\right)$$

$$= \frac{1}{2} MF \left[\cos\left(\frac{P-P_M}{2}\theta\right) - \cos\left(\frac{P+P_M}{2}\theta\right)\right] \tag{9.80}$$

It is found from the expression of Eq. (9.80) that the modulated mmf wave is equivalent to two mmf waves, one having $P_1 = (P - P_M)$ poles and the other $P_2 = (P + P_M)$ poles. This indeed is the *suppressed-carrier* modulation which implies that the name pole-amplitude modulation is a misnomer.

Probe

Explain the pole amplitude modulation of speed control technique with its limitations and benefits compared to other methods.

Since the stator is wound for P poles, the angle between phase axes is $r(2\pi/3)$ elect. rad; r being integral nonmultiple of 3. The phase axes angle for modulated poles is then

$$\left(\frac{P \mp P_M}{P}\right) r \left(\frac{2\pi}{3}\right) = (1 \mp P_M/P)\, r \left(\frac{2\pi}{3}\right)$$

where r = integer, nonmultiple of 3.

To suppress one of the poles, say P_1, the angle between its phase axes must be multiple of 2π, i.e.

$$(1 - P_M/P) r \left(\frac{2\pi}{3}\right) = n(2\pi);\ n = \text{integer}$$

or

$$P_M/P = \left(1 - \frac{3n}{r}\right) \tag{9.81}$$

The angle between phase axes for P_2 poles is then

$$(1 + P_M/P) r \left(\frac{2\pi}{3}\right) = \left(2 - \frac{3n}{r}\right) r \left(\frac{2\pi}{3}\right)$$

$$= 2r\left(\frac{2\pi}{3}\right) - n(2\pi)$$

$$= 2r\left(\frac{2\pi}{3}\right) \tag{9.82}$$

Since r is selected to be an integer nonmultiple of 3, the angle in Eq. (9.82) is integral multiple of $(2\pi/3)$ and hence the rotating field corresponds only to $P_2 = (P + P_M)$ poles. Similarly P_2 poles could be suppressed and P_1 poles developed.

Sample Pole Modulation for an Arbitrary Selected IM

For example, let $P = 8$. Then from Eq. (9.81) if the values $n = 1$ and $r = 4$ are chosen

$$P_M = \frac{1}{4} P = 2$$

Hence
$$P_1 = P - P_M = 6 \text{ poles (suppressed)}$$
$$P_2 = P + P_M = 10 \text{ poles}$$

In practice the modulating mmf used is rather crude; a rectangular wave of amplitude $M = 1$. This modulation is simply achieved by dividing the complete winding of each phase into P_M groups and reverse connecting alternate groups.

Figure 9.54 shows the eight poles of one phase for the example in hand. It also indicates the 2-pole ($P_M = 2$) rectangular modulating function symmetrically located with respect to $P = 8$ poles. The value of $M = -1$ means reverse connecting pole groups 5, 6, 7 and 8 resulting in sign-reversed dotted poles. On this figure, $P_1 = 6$ poles are identified by circled numbers but $P_2 = 10$ poles are not evident. These can only be visualized from Eq. 9.54 wherein the modulating function is the fundamental of the rectangular space wave. The harmonic components of the rectangular wave create their own modulated poles but these can be ignored. It has already shown above that in this example $P_1 = 6$ poles get suppressed resulting in $P_2 = 10$-pole field, when the pole groups 5, 6, 7 and 8 are reverse connected. The motor can, therefore, run corresponding to 8 (original) and 10 (modulated) poles.

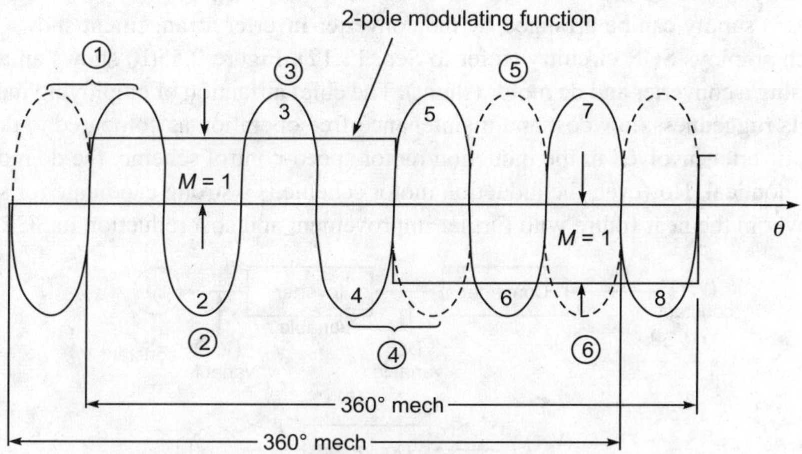

Fig. 9.54 *8-pole winding with 2-pole rectangular modulation*

Pole-amplitude modulation technique permits two speeds to be obtained which need not be in the ratio 2 : 1. It also allows great reduction in size and cost of the machine.

Example 9.16 *A 50 Hz, induction motor wound for pole-amplitude modulation has 20 initial poles and the modulating function has 8 poles. At what two speeds will the motor run?* LO 3

Solution

$$P = 20, \quad P_M = 8$$
$$P_1 = 20 - 8 = 12$$
$$P_2 = 20 + 8 = 28$$

From Eq. (9.81)
$$P_M/P = \frac{8}{20} = \frac{2}{5} = \left(1 - \frac{3n}{r}\right); \quad n = 1, \quad r = 5$$

Hence P_1 poles are suppressed

$$\text{Speed (20 poles)} = \frac{20 \times 50}{20} = 300 \text{ rpm}$$

$$\text{Speed (28 poles)} = \frac{20 \times 50}{28} = 214.286 \text{ rpm}$$

Frequency Control

The synchronous speed of the induction motor can be controlled in a stepless way over a wide range by changing the supply frequency. As per Eq. (9.1) the resultant air-gap flux per pole is given by

$$\Phi_r = \frac{1}{4.44\, K_{wl}\, N_{phl}} \cdot \left(\frac{V}{f}\right) \tag{9.83}$$

Therefore, in order to avoid saturation in stator and rotor cores which would cause sharp increase in magnetization current, the flux Φ_r must be kept constant as f is varied. To achieve this, it follows from Eq. (9.83) that when f is varied, V must also be varied such that (V/f) remains constant. Variable (V, f) supply from constant (V, f) supply can be arranged by the converter-inverter arrangement shown schematically in Fig. 9.55(a) which employs SCR circuitry (refer to Sec. 11.12). Figure 9.55(b) shows an alternative speed-control scheme using a converter and dc motor (shunt). The chief attraction of employing induction motor for speed control is its ruggedness, low cost and maintenance-free operation as compared to dc motor. Because of the cost of the inverter involved in the induction motor speed-control scheme, the dc motor scheme as of today is more economical. However, the induction motor scheme is a strong candidate for speed control and is likely to take over in the near future with further improvement and cost reduction in SCR technology.

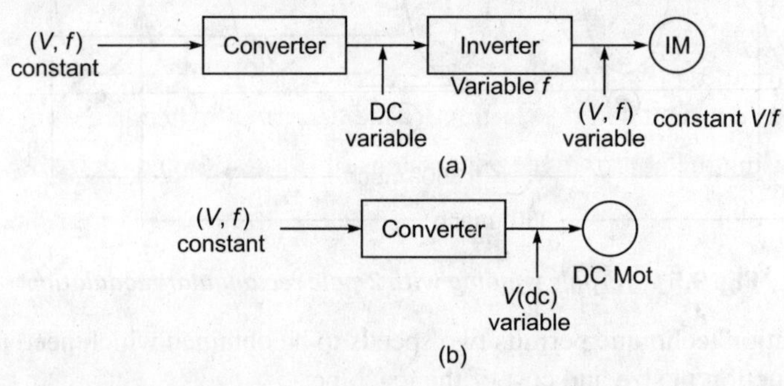

Fig. 9.55 *Comparative schemes for speed control using induction and dc motors*

Torque-Slip Characteristic—Variable *f*, Constant *V/f*

V_0 = nominal voltage

f_0 = nominal frequency

X'_{20} = nominal rotor standstill reactance referred to stator (at frequency f_0)

Then at any frequency f,

$$V = \left(\frac{f}{f_0}\right) V_0$$

$$X_2' = \left(\frac{f}{f_0}\right) X_{20}'$$

$$\omega_s = \left(\frac{f}{f_0}\right) \omega_{s0}$$

Neglecting the stator impedance, from Eqs (9.29) – (9.31)

$$s_{\max,T} = \left(\frac{f_0}{f}\right)\left(\frac{R_2}{X_{20}'}\right) = \left(\frac{f_0}{f}\right) s_{\max,T,0} \tag{9.84}$$

$$T_{\max,T} = \frac{3}{\left(\frac{f}{f_0}\right)\omega_{s0}} \cdot \left[\frac{0.5\left(\frac{f}{f_0}\right)^2 V_0^2}{\left(\frac{f}{f_0}\right) X_{20}'}\right]$$

$$= \frac{3}{\omega_{s0}} \cdot \frac{0.5 V_0^2}{X_{20}'} = \text{constant, independent of } f \tag{9.85}$$

and

$$T_{\text{start}} = \frac{3}{\left(\frac{f}{f_0}\right)\omega_{s0}} \cdot \frac{\left(\frac{f}{f_0}\right)^2 V_0^2 R_2'}{R_2'^2 + \left(\frac{f}{f_0}\right)^2 X_{20}'} \tag{9.86}$$

It is easily concluded from Eqs (9.84)-(9.86) that slip at maximum torque decreases with f increasing, the maximum torque remains constant and the starting torque decreases with f increasing.

For low values of slip, using the approximate expressions of Eqs (9.34) and (9.35) one can write

$$I_2' = \frac{s\left(\frac{f}{f_0}\right) V_0}{R_2'} \tag{9.87}$$

and

$$T = \frac{3}{\left(\frac{f}{f_0}\right)\omega_{s0}} \cdot \frac{s\left(\frac{f}{f_0}\right)^2 V_0^2}{R_2'} = \frac{3}{\omega_{s0}} \cdot \left(\frac{f}{f_0}\right)\left(\frac{s V_0^2}{R_2'}\right) \tag{9.88}$$

It is seen from Eqs (9.87) and (9.88) that current and torque both increase with f at a given slip.

From the above conclusions the torque-slip characteristics of the motor can be sketched at frequencies above and below the nominal as shown in Fig. 9.56. Figure 12.38 may be referred to for torque-speed characteristics.

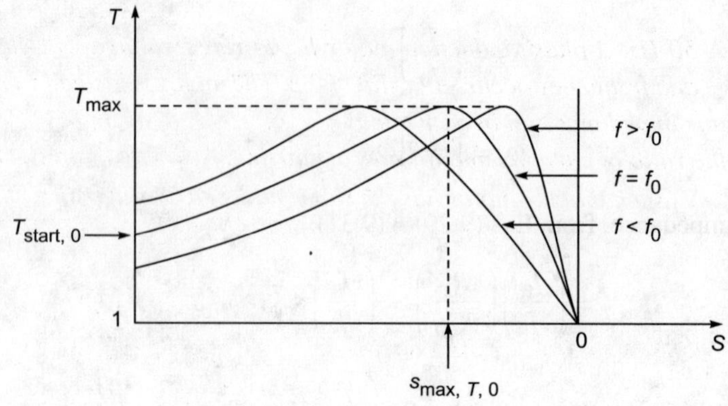

Fig. 9.56 *Torque-slip characteristics; variable f, constant (V/f)*

!Attention The torque-speed characteristics of (V/f) control of induction motor are presented in Fig. 9.57. For speeds less than ω_0 (corresponding to base frequency f_0) (V/f) is kept constant, so that the maximum motor torque remains constant and the motor can drive a constant torque load as indicated in the figure. For speeds higher than the base speed, V needed to keep (V/f) constant is more than the rated value which cannot be provided by the inverter (Fig. 9.55). So (V/f) in this speed region is allowed to reduce and the motor torque reduces proportional to the $(V/f)^2$ (this corresponds to Φ^2_{max}). This region of speed control can then drive constant-kW loads (torque demand reduces with speed). This method of speed control thus produces overall torque-speed characteristic just like that of Ward Leonard speed control of dc shunt motor.

With rapid strides made in power electronic devices and circuits combined with the ruggedness and higher efficiency of an induction motor, motors with this type of drive are replacing dc motors in several applications.

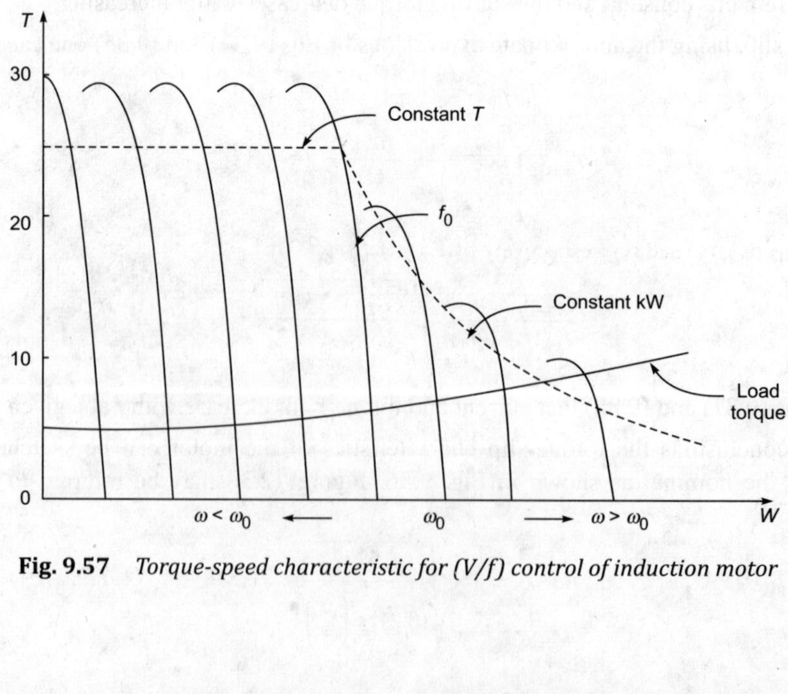

Fig. 9.57 *Torque-speed characteristic for (V/f) control of induction motor*

Example 9.17 *A 50 Hz, 3-phase induction motor has a rated voltage V_1. The motor's breakdown torque at rated voltage and frequency occurs at a slip of 0.2. The motor is instead run from a 60 Hz supply of voltage V_2. The stator impedance can be neglected.*

(a) If $V_2 = V_1$, find the ratio of currents and torques at starting. Also find the ratio of maximum torques.

(b) Find the ratio V_2/V_1 such that the motor has the same values of starting current and torque at 50 and 60 Hz.

LO 3

Solution

(a)
$$s_{\max,T} = \frac{R_2'}{X_2'} = 0.2$$

where X_2' = standstill 50 Hz rotor reactance

At rated voltage (V_1) and frequency (50 Hz)

$$I_s(1) = I_2'(1) = \frac{V_1}{\sqrt{R_2'^2 + X_2'^2}} \tag{i}$$

$$T_s(1) = \frac{V_1^2 R_2'}{R_2'^2 + X_2'^2} \tag{ii}$$

$$T_{\max}(1) = \frac{3}{\omega_s} \cdot \frac{0.5 V_1^2}{X_2'} \tag{iii}$$

At voltage V_2 and frequency 60 Hz.

$$I_2(2) = I_2'(2) = \frac{V_2}{\sqrt{R_2'^2 + \left(\frac{6}{5}\right)^2 X_2'^2}} \tag{iv}$$

$$T_s(2) = \frac{V_2^2 R_2'}{R_2'^2 + \left(\frac{6}{5}\right)^2 X_2'^2} \tag{v}$$

$$T_{\max}(2) = \frac{3}{\left(\frac{6}{5}\right)\omega_s} \frac{0.5 V_2^2}{\left(\frac{6}{5}\right) X_2'} \tag{vi}$$

Dividing Eqs (iv), (v) and (vi) respectively by Eqs (i), (ii) and (iii)

$$\frac{I_s(2)}{I_s(1)} = \frac{V_2}{V_1} \cdot \sqrt{\frac{R_2'^2 + X_2'^2}{R_2'^2 + \left(\frac{6}{5}\right)^2 X_2'^2}}$$

$$= \frac{V_2}{V_1} \cdot \sqrt{\frac{s_{\max,T}^2 + 1}{s_{\max,T}^2 + \left(\frac{6}{5}\right)^2}}$$

$$= 1 \times \sqrt{\frac{(0.2)^2 + 1}{(0.2)^2 + \left(\frac{6}{5}\right)^2}} = 0.838$$

$$\frac{T_s(2)}{T_s(1)} = \frac{V_2^2}{V_1^2} \cdot \frac{s_{\max,T}^2 + 1}{s_{\max,T}^2 + \left(\frac{6}{5}\right)^2}$$

$$= 1 \times \frac{(0.2)^2 + 1}{(0.2)^2 + \left(\frac{6}{5}\right)^2} = 0.703$$

$$\frac{T_{\max}(2)}{T_{\max}(1)} = \frac{V_2^2}{V_1^2}\left(\frac{5}{6}\right)^2 = 0.694$$

(b)

$$\frac{V_2^2}{V_1^2} \cdot \sqrt{\frac{(0.2)^2 + 1}{(0.2)^2 + \left(\frac{6}{5}\right)^2}} = 1$$

$$\frac{V_2}{V_1} = 1.19$$

This ratio will also give equal staring torques.

Concept Check — LO 3

1. Show that in star/delta starting of squirrel-cage induction motor the starting current and torque are reduced by a factor of 1/3 compared to DOL starting. ○○●
2. Elaborate the statement "rotor resistance starting of slip-ring induction motor reduces starting current and increases starting torque". ○○●
3. Why is DOL starting current very high but the starting torque is still low? Why DOL is not permitted in starting even though the short duration current cannot harm the motor? ○●●
4. What methods are used in starting squirrel-cage induction motor? Which method is used in what size of motor? Which is the most common method and what is its superiority? ○●●
5. Compare the speed control features of induction motor with dc shunt motor. ○●●
6. From no-load to full-load, what is the type speed-load characteristic of induction motor? ●●●

9.11 | DEEP-BAR/DOUBLE-CAGE ROTOR

Implement of Deep-Bar Conductors or Double-Cage Rotor in Squirrel-Cage Motor

LO 4 Analyze induction motor dynamics, various constructions (deep-bar/double) of cage rotor, and several types (classes) of standard designs of squirrel-cage motors available to support different industrial applications

The chief advantage of the slip-ring induction motor compared to the squirrel-cage one lies in the fact that while its rotor is designed with low resistance to give good running performance (high efficiency, low slip, etc.), excellent starting characteristic (low starting current, high starting torque, etc.) is simply achieved by adding an external resistance in the rotor circuit at the time of starting which is then cut out gradually while the rotor reaches normal speed. The rotor circuit in the squirrel-cage motor, however, cannot be tampered with so that while its resistance is designed to give excellent running performance, it has high starting current and low starting torque which is further impaired by a reduced-voltage starting employed to limit the starting current (starting torque is proportional to the square of the voltage applied to the motor stator). The attractive qualities of low-cost, ruggedness and maintenance-free operation of the squirrel-cage motor has impelled designers to find ways of improving its starting characteristic without sacrificing heavily its excellent running performance. The fact that the rotor currents are of stator frequency (50 Hz) at the time of starting while this frequency reduces to $f_2 = sf$ (may be as low as 2.5 Hz) under running condition, can be exploited as it causes automatic variation of rotor resistance from a high value at starting (50-Hz resistance) to a low value under running (about 2.5-Hz resistance). This phenomenon is basically the skin and proximity effect which occurs in any conductor carrying alternating current. For the conductor cross-sectional shape (round or square) normally employed for rotor bars, this variation is not prominent enough to give low starting current and high starting torque. To enhance the variation in the effective (ac) resistance of rotor bars, deep-bar conductors or double-cage rotor are employed.

Deep-Bar Rotor

In this type of construction bars of narrow width are laid down in deep semi-enclosed slots as shown in Fig. 9.58. The magnetic leakage flux pattern set up by the bar current is indicated in dotted lines in the figure. The rotor bar can be imagined to be composed of elementary strips in parallel—topmost and bottom-most strips are shown in the figure. It is easily seen that a much larger flux links the bottom elementary strip compared to the top elementary strip. As a consequence, the starting reactance (50-Hz reactance) for the bottom strip is much larger than that of the top strip. It then follows that the current in the top strip is much more than the current in the bottom strip and further the top-strip current somewhat leads the bottom-strip current because of its lower reactance. Similar arguments when applied to other elementary strips would reveal that the current is unevenly distributed over the bar cross-section with the current density progressively increasing while moving upwards from the bottom strip. Nonuniform current distribution causes greater ohmic loss meaning thereby that the effective bar resistance becomes much more than its dc resistance As the rotor speeds up to a value close to synchronous, the frequency of rotor currents ($f_2 = sf$) becomes very low. The reactances of various elementary strips at this low frequency become almost equal and the current density over the conductor cross-section becomes nearly uniform so that it offers a resistance almost equal to its dc value. By choice of bar cross-sectional dimensions, it is possible to obtain a starting rotor resistance (50-Hz resistance) to be many times the running rotor resistance (almost dc value). A deep-bar rotor, therefore, has a low starting current and a high starting torque and a low running resistance which means that it can satisfactorily meet both the desired starting and running performances. Since the net rotor reactance at standstill is somewhat higher than in a normal bar design, the breakdown torque is somewhat lower. The torque-slip characteristic of a deep-bar rotor compared to a normal-bar design with low and high rotor resistance is illustrated in Fig. 9.59.

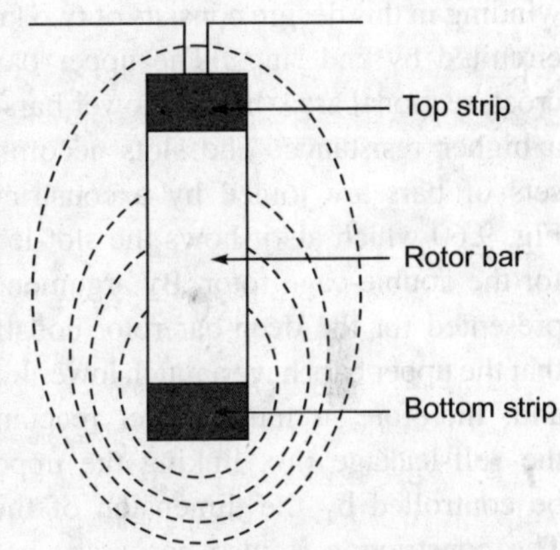

Fig. 9.58 *Deep-bar rotor and leakage flux pattern*

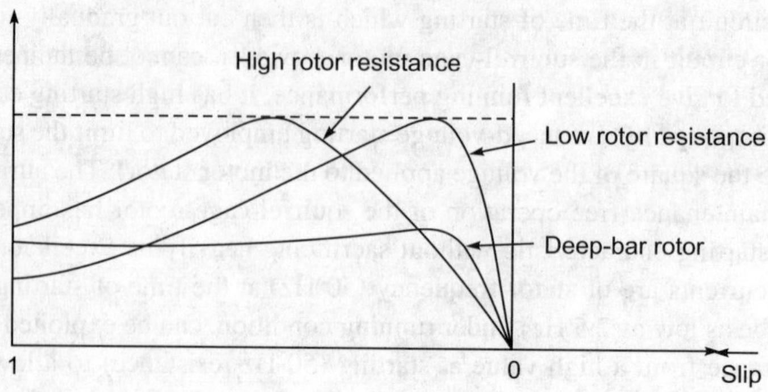

Fig. 9.59

Double-Cage Rotor

A rotor design, which though more expensive gives still better starting, and running performances than the deep-bar design, is the double-cage rotor. The squirrel-cage winding in this design consists of two layers of bars short-circuited by end rings. The upper bars have a smaller cross-sectional area than the lower bars and consequently a higher resistance. The slots accommodating the two sets of bars are joined by a constriction as shown in Fig. 9.60 which also shows the slot-leakage flux pattern for the double-cage rotor. By arguments similar to those presented for the deep-bar rotor construction, it is seen that the upper bars have a much lower leakage flux linkage and, therefore, a much lower reactance. Furthermore, the self-leakage flux linking the upper/lower bars can be controlled by the dimension of the air-constriction. The constriction is also necessary because of the fact that in its absence the main flux will return via the iron path between the two slots thereby "missing" the inner bars which then would not contribute to torque development.

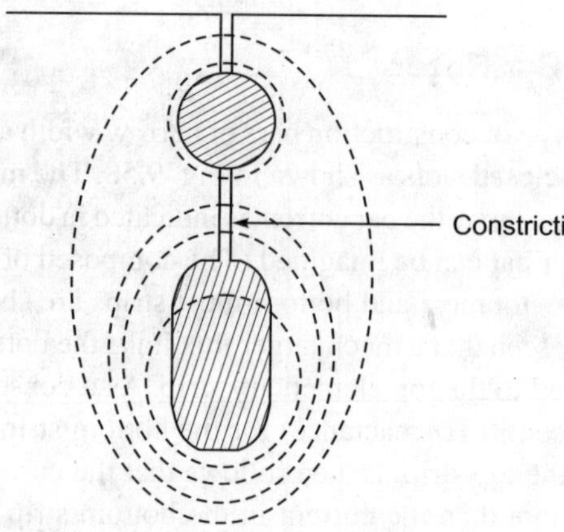

Fig. 9.60 *Slot-leakage flux pattern for double-cage rotor*

It is then found that the outer cage has high resistance and low reactance while the inner cage has low resistance and high reactance. Therefore, in the starting, the current is mainly confined to the outer cage with a consequent decrease in starting current and an increase in starting torque. Under running condition the reactance difference between the two cages evens out because of low frequency of rotor currents such that these act to conduct current in inverse proportion to their dc resistances and as a group present a low-resistance rotor to the air-gap flux giving an excellent running performance. Since the starting current is mainly confined to the upper cage, this design is susceptible to frequent starting which would cause overheating and burning out of the upper high-resistance cage.

Another type of a double-cage rotor construction in which the two cages are placed in staggered slots is shown in Fig. 9.61 along with its slot-leakage flux pattern.

The two cages though somewhat coupled magnetically can be treated as independent for simplified yet fairly accurate analysis. The approximate circuit model of the double-cage rotor on this assumption is given in Fig. 9.62. The resultant torque-slip characteristic as obtained by summation of the torque developed by the inner and outer cages is shown in Fig. 9.63.

Probe

Describe the designing aspect and advantage of a deep-cage rotor.

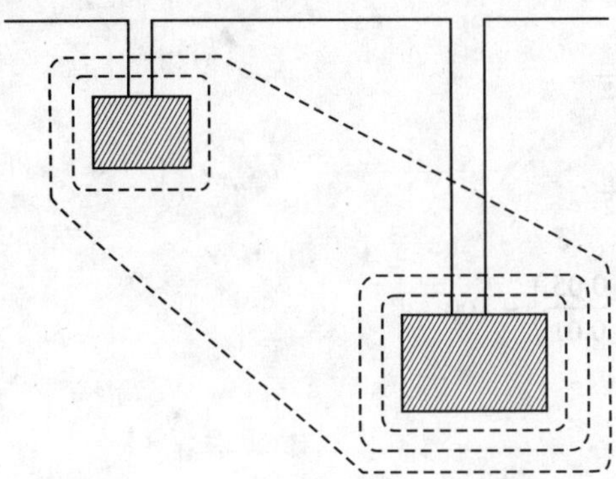

Fig. 9.61

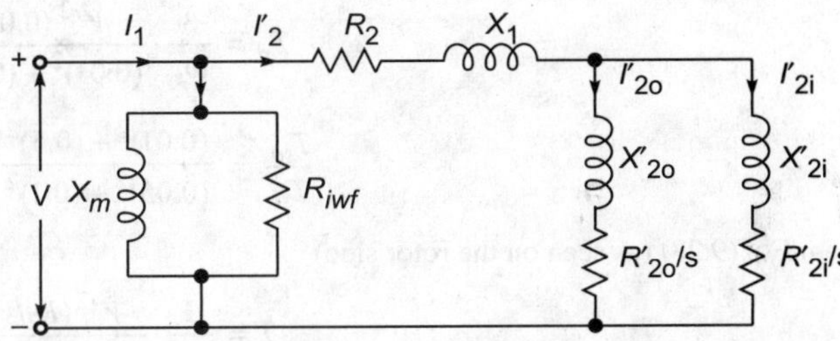

Fig. 9.62 *Approximate circuit model of double-cage induction motor*

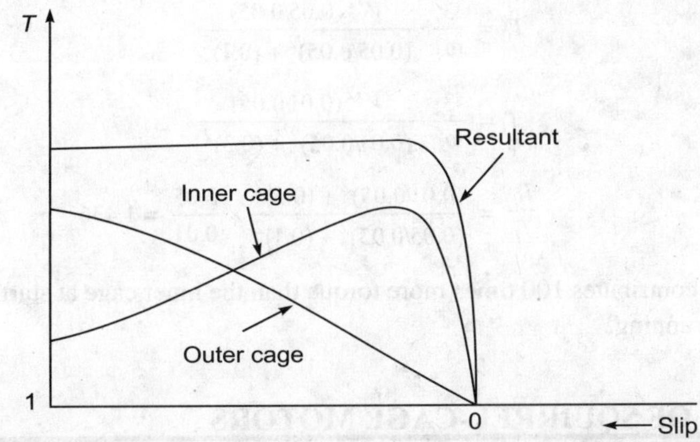

Fig. 9.63 *Torque-slip characteristic—double-cage induction motor*

It will be evident to the reader by now that for low starting torque requirement, which is the case with a vast majority of induction motor applications, the low-cost ordinary squirrel-cage construction is employed. A deep-bar rotor construction is adopted for higher starting torque applications and double-cage construction for still higher starting torque needs. For large-size motors with stringent starting torque needs, the most expensive slip-ring construction is used.

Example 9.18 *The impedances at standstill of the inner and outer cages of a double-cage rotor are (0.01 + j 0.5) Ω and (0.05 + j 0.1) Ω respectively. The stator impedance may be assumed to be negligible. Calculate the ratio of the torques due to the two cages (i) at starting, and (ii) when running with a slip of 5%.*

LO 4

Solution From Eq. (9.31) (as seen on the rotor side)

$$T_s = \frac{3}{\omega_s} \cdot \frac{V'^2 R_2^2}{R_2^2 + X_2^2}$$

where V' = rotor induced emf

Substituting values

$$T_{so} = \frac{3}{\omega_s} \cdot \frac{V'^2 (0.05)}{(0.05)^2 + (0.1)^2}$$

$$T_{si} = \frac{3}{\omega_s} \cdot \frac{V'^2 (0.01)}{(0.01)^2 + (0.5)^2}$$

$$\therefore \quad \frac{T_{so}}{T_{si}} = \frac{(0.01)^2 + (0.5)^2}{(0.05)^2 + (0.1)^2} \times \left(\frac{0.05}{0.01}\right) = 100$$

From Eq. (9.28) (as seen on the rotor side)

$$T = \frac{3}{\omega_s} \cdot \frac{V'^2 (R_2/s)}{(R_2/s)^2 + X_2^2}$$

Substituting values

$$T_0 = \frac{3}{\omega_s} \cdot \frac{V'^2 (0.05/0.05)}{(0.05/0.05)^2 + (0.1)^2}$$

$$T_i = \frac{3}{\omega_s} \cdot \frac{V'^2 (0.01/0.05)}{(0.01/0.05)^2 + (0.3)^2}$$

$$\therefore \quad \frac{T_0}{T_i} = \frac{(0.01/0.05)^2 + (0.5)^2}{(0.05/0.05)^2 + (0.1)^2} \times \frac{0.05}{0.01} = 1.436$$

Remark The outer cage contributes 100 times more torque than the inner cage at starting while it contributes only 1.436 times during running.

9.12 | CLASSES OF SQUIRREL-CAGE MOTORS

LO 4

To cater to the different starting and running requirements of a various industrial applications, several standard designs of squirrel-cage motors are available in the market. The torque-speed characteristics of the most common designs are shown in Fig. 9.64. The effective resistance of the rotor cage circuit is the most important design variable in these motors.

Class A Motors

They have normal starting torque, high starting current, and low operating slip. (0.005–0.015) with low rotor circuit resistance, they work efficiently with a low slip at full load. They are used where the load torque is low at start (e.g. fan, pump load) so that full speed is achieved quickly, thus avoiding the problem of overheating while starting.

Class B Motors

Class B motors have normal starting torque, low starting current, and low operating slip. The starting current is reduced by designing for relatively high leakage reactance by using either deep-bar rotors or double-cage rotors.

These motors have similar values of full-load slip and efficiency and are good general purpose motors used for various industrial applications such as constant speed drives (e.g. fans, pumps, blowers).

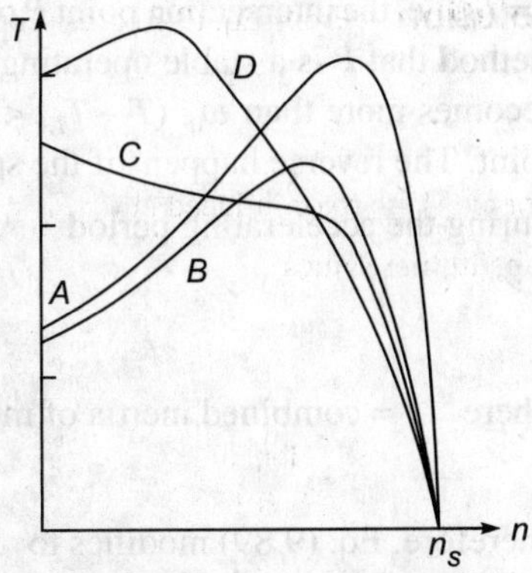

Fig. 9.64 *Torque-speed characteristics for different classes of induction motors*

Class C Motors

They have high starting torque and low starting current. A double cage rotor is used with higher rotor resistance. Class C motors are normally used for driving compressors, conveyors, crushers, etc.

Class D Motors

They have high starting torque, low starting current, and high operating slip. The torque-slip characteristics is similar to that of a wound-rotor motor with some external resistance in the rotor circuit. The full load slip is high making running efficiency low. These motors are used for driving intermittent loads requiring rapid acceleration and high impact loads such as punch presses or shears. For impact loads a flywheel is fitted to the system to provide kinetic energy during the impact.

Probe

Analyse the difference between torque-speed characteristics of various classes of squirrel cage IM based on their applications.

9.13 | INDUCTION MACHINE DYNAMICS: ACCELERATION TIME — LO 4

In general, the mechanical time-constant for any machine is much larger than the electrical time constant. Therefore, the dynamic analysis can be simplified by neglecting the electrical transient without any loss of accuracy of results.

Significance of Accelerating Time to Reach Steady Speed (Slip)

The circuit model of induction motor of Fig. 9.8(b) holds for constant slip but would also apply for slowly varying slip as is generally the case in motor starting. Figure 9.65 shows the typical torque-slip (speed) characteristic of a motor and also the load torque as a function of slip (speed). Each point on $(T_L - s)$ characteristic represents the torque (frictional) demanded by the load and motor when running at steady speed. The motor would start only if $T > T_L$ and would reach a steady operating speed of ω_0 which corresponds to

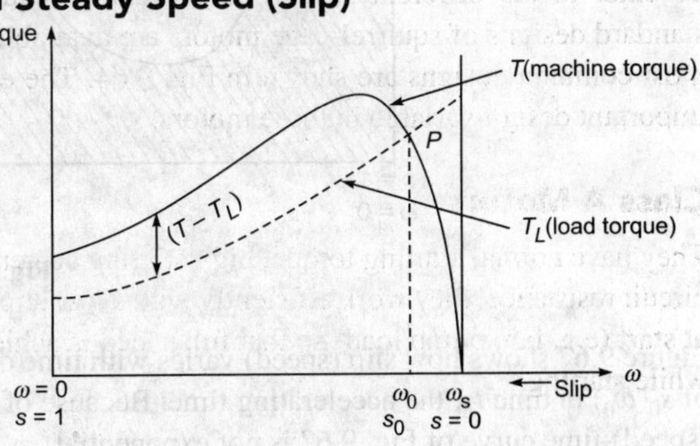

Fig. 9.65

$T = T_L$, i.e. the intersection point P of the two torque-speed characteristics. It can be checked by the perturbation method that P is a stable operating point for the load-speed characteristic shown. If for any reason the speed becomes more than ω_0, $(T - T_L) < 0$ the machine-load combination decelerates and returns to the operating point. The reverse happens if the speed decreases below ω_0.

During the accelerating period

$$T - T_L = J\frac{d\omega}{dt} \tag{9.89}$$

where J = combined inertia of motor and load. Now

$$\omega = (1 - s)\omega_s \tag{9.90}$$

Therefore, Eq. (9.89) modifies to

$$T - T_L = -J\omega_s\frac{ds}{dt} \tag{9.91}$$

Integrating

$$t\Big|_0^{t_A} = -\frac{1}{J\omega_s}\int_{s=1}^{s=s_0}\frac{1}{T - T_L}\,ds \tag{9.92}$$

Since, the term $1/(T - T_L)$ is nonlinear, the integration in Eq. (9.92) must be carried out graphically (or numerically) as shown in Fig. 9.66 for the case when $s_1 = 1$ and $s_2 = s_0$. Since $1/(T - T_L)$ becomes ∞ at s_0, the practical integration is carried out only up to 90 or 95% of s_0 depending upon the desired accuracy.

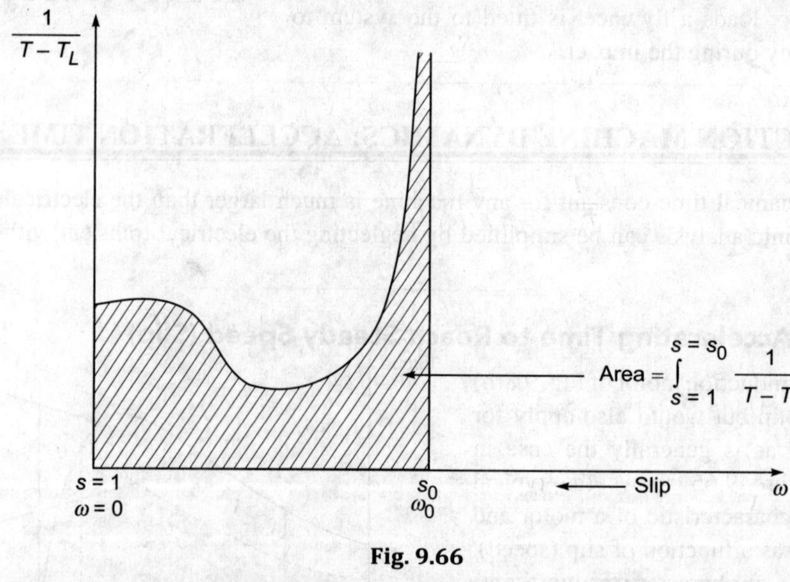

Fig. 9.66

Figure 9.67 shows how slip (speed) varies with time during the acceleration period reaching the steady value of $s_0(\omega_0)$ in time t_A, the accelerating time. Because of the nonlinearity of $(T - T_L)$ as function of slip, the slip (speed)-time curve of Fig. 9.67 is not exponential.

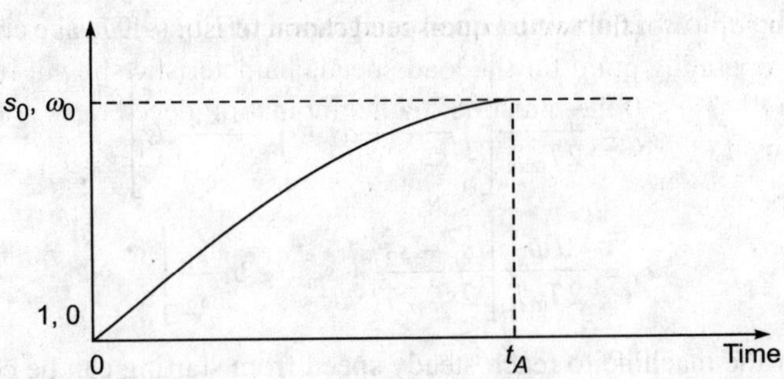

Fig. 9.67

Acceleration Time to Reach Steady Speed from Starting at No-Load

In this particular case it is assumed that the machine and load friction torque $T_L = 0$.

Assuming stator losses to be negligible (i.e $R_1 = 0$), the motor torque as obtained from Eq. (9.22) is

$$T = \frac{3}{\omega_s} \cdot \frac{V_{TH}\,(R_2'/s)}{(R_2'/s)^2 + (X_1 + X_2')^2} \tag{9.93}$$

Also from Eq. (9.24)

$$T_{\max} = \frac{3}{\omega_s} \cdot \frac{0.5\,V_{TH}^2}{(X_1 + X_2')} \tag{9.94}$$

at a slip of (Eq. (9.23))

$$s_{\max,T} = \frac{R_2'}{X_1 + X_2'} \tag{9.95}$$

From Eqs (9.93) and (9.94),

$$\frac{T}{T_{\max}} = \frac{2(X_1 + X_2')\,(R_2'/s)}{(R_2'/s)^2 + (X_1 + X_2')^2} = \frac{2}{\left[\dfrac{(R_2'/s)}{X_1 + X_2'} + \dfrac{X_1 + X_2'}{(R_2'/s)}\right]} \tag{9.96}$$

Substituting Eq. (9.94) in (9.95)

$$\frac{T}{T_{\max}} = \frac{2}{\dfrac{s_{\max,T}}{s} + \dfrac{s}{s_{\max,T}}} \tag{9.97}$$

Since T_L is assumed to be zero, the motor torque itself is the accelerating torque, i.e.

$$T = \frac{2T_{\max}}{s/s_{\max,T} + s_{\max,T}/s} = J\frac{d\omega}{dt} = -J\omega_s\frac{ds}{dt} \tag{9.98}$$

The time t_A to go from slip s_1 to s_2 is obtained upon integration of Eq. (9.97) as

$$t_A = \frac{-J\omega_s}{2T_{max}}\left[\int_{s_1}^{s_2} \frac{s}{s_{max,T}}\,ds + \int_{s_1}^{s_2} \frac{s_{max,T}}{s}\,ds\right]$$

$$t_A = \frac{J\omega_s}{2T_{max}}\left[\frac{s_1^2 - s_2^2}{2s_{max,T}} + s_{max,T}\ \ln\frac{s_1}{s_2}\right] \tag{9.99}$$

The acceleration time for the machine to reach steady speed from starting can be computed from Eq. (9.98) with $s_1 = 1$ and $s_2 = s$, i.e.

$$t_A = \frac{J\omega_s}{2T_{max}}\left[\frac{1 - s^2}{2s_{max,T}} + s_{max,T}\ \ln\frac{1}{s}\right] \tag{9.100}$$

Optimum $s_{max,T}$ for Minimum Acceleration Time

To find the optimum value of $s_{max,T}$ for the machine to have minimum acceleration time to reach s_2 from s_1, Eq. (9,98) must be differentiated with respect to $s_{max,T}$ and equated to zero, This gives

$$(s_{max,T})_{opt} = \sqrt{\frac{(s_1 - s_2)^2}{2\ln(s_1/s_2)}} \tag{9.101}$$

For minimum acceleration time for the machine to reach any slip s from start, the optimum value of $s_{max,T}$ is given by Eq. (9.100) with $s_1 = 1$ and $s_2 = s$. Then

$$(s_{max,T})_{opt} = \sqrt{\frac{(1 - s)^2}{2\ln(1/s)}} \tag{9.102}$$

and

$$t_A(\text{min}) = \frac{J\omega_s}{2T_{max}}\left[\frac{1 - s^2}{2(s_{max,T})_{opt}} + (s_{max,T})_{opt}\ \ln\frac{1}{s}\right] \tag{9.103}$$

Further, to enable us to compute the optimum value of the rotor resistance to accelerate the machine to slip s_2 from s_1, Eq. (9.101) is substituted in Eq. (9.95) giving

$$(R_2')_{opt} = (X_1 + X_2')\left[\frac{s_1^2 - s_2^2}{2\ln(s_1/s_2)}\right] \tag{9.104}$$

$$= (X_1 + X_2')\,(s_{max,T})_{opt} \tag{9.105}$$

Example 9.19 *A 3-phase, 415 V, 6-pole, 50 Hz, star-connected slip-ring induction motor has a total stator and rotor reactance of 1.5 Ω referred to the stator. The machine drives pure inertia load; the moment of inertia of the rotor and load being 11 kg m^2. Direct on-line starting is used and the rotor circuit resistance is adjusted so that the load is brought to 0.96 of the synchronous speed from rest in shortest possible time. Neglecting stator losses, compute the acceleration time and the value of the rotor resistance referred to the stator.*

LO 4

Solution

$$s = 1 - 0.96 = 0.04$$

Substituting in Eq. (9.101),

$$(s_{\max,T})_{\text{opt}} = \left[\frac{1-(0.04)^2}{2\ln(1/0.04)}\right]^{1/2}$$

$$= 0.394$$

From Eqs (9.105) and (9.103)

$$(R_2')_{\text{opt}} = 1.5 \times 0.394 = 0.591\ \Omega$$

$$t_A(\min) = \frac{J\omega_s}{2T_{\max}}\left[\frac{1-s^2}{2(s_{\max,T})_{\text{opt}}} + (s_{\max,T})_{\text{opt}}\ln\frac{1}{s}\right]$$

From Eq. (9.94), assuming $V_{TH} = V$

$$T_{\max} = \frac{3}{2\pi \times 1000}\cdot\frac{0.5(415/\sqrt{3})^2}{1.5} = 548.2$$

$$= \frac{11(2\pi \times 1000/60)}{2 \times 548.2}\left[\frac{1-(0.04)^2}{2 \times 0.394} + 0.394 \times \ln\left(\frac{1}{0.04}\right)\right]$$

$$= 2.66\ \text{s}$$

Computation Procedure to be Adopted on MATLAB

For the approximate circuit model. For given slip (s) compute stator current, pf, torque (net) and efficiency (η).

No-load Test Measured P_0, I_0, V_0 (rated value)

$$R_1 = \text{dc resistance of stator (corrected to ac)}$$

$$R_{iwf} = \frac{V_0^2}{p_r};\ P_r = P_0 - 3I_0^2 R_1 \tag{i}$$

(Rotational loss = stator core loss + winding and friction loss)

$$I_{iwf} = \frac{V_0/\sqrt{3}}{R_{iwf}} = \frac{P_r}{\sqrt{3}V_0}$$

$$I_m = (I_0^2 - I_{iwf}^2)^{1/2} = \left(I_0^2 - \frac{P_r^2}{3V_0^2}\right)^{1/2} \tag{ii}$$

$$X_m = \frac{V_0/\sqrt{3}}{I_m} = \frac{V_0/\sqrt{3}}{\left(I_0^2 - \dfrac{P_r^2}{3V_0^2}\right)^{1/2}} \tag{iii}$$

Needed for circuit model R_1, X_m.

Blocked rotor test Measured V_{BR}, I_{BR}, P_{BR}

$$Z_{BR} = \frac{V_{BR}/\sqrt{3}}{I_{BR}}$$

$$R_{BR} = \frac{P_{BR}/3}{I_{BR}^2} = R_1 + R_2'$$

$$R_2' = R_{BR} - R_1 = \frac{P_{BR}/3}{(I_{BR})^2} - R_1 \quad \text{(iv)}$$

$$X_{BR} = (Z_{BR}^2 - R_{BR}^2)^{1/2} = \left[\frac{V_{BR}^2}{3I_{BR}^2} - \frac{P_{BR}^2}{9I_{BR}^4}\right]^{1/2} \quad \text{(v)}$$

$$X_1 = X_2' = \frac{X_{BR}}{2} = \frac{1}{2}\left[\frac{V_{BR}^2}{3I_{BR}^2} - \frac{P_{BR}^2}{9I_{BR}^4}\right]^{1/2} \quad \text{(vi)}$$

Needed for circuit model R_2', X_2'.

Procedure Performance

Circuit model

$$V = \frac{V_1}{\sqrt{3}} \text{ per phase}$$

$$\bar{Z}_f = \left(\frac{R_2'}{s} + jX_2'\right) \| jX_m + R_f + jX_f \quad \text{(vii)}$$

$$\bar{Z}_{in} = (R_1 + R_f) + j(X_1 + X_f) = R_{in} + jX_{in} \quad \text{(viii)}$$

$$Z_{in} = (R_{in}^2 + X_{in}^2)^{1/2}$$

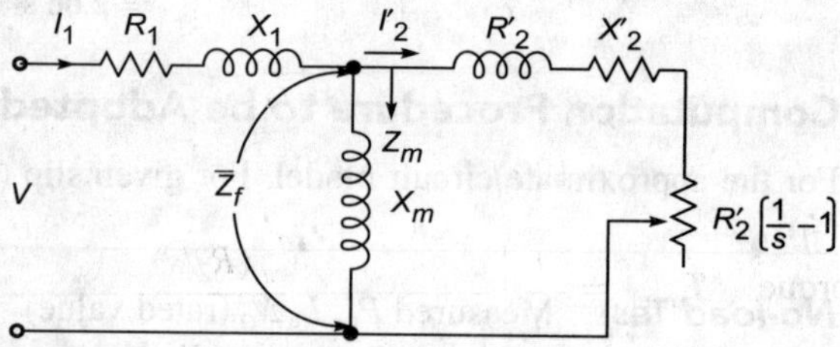

Fig. 9.68(a)

(a) *Stator current*

$$I_1 = \frac{V_1 \angle 0°}{\sqrt{3}\, Z_{in}}, \; pf \text{ angle } \phi = \tan^{-1}\frac{X_{in}}{R_{in}} \quad \text{(ix)}$$

(b) *Power factor*

$$pf = \cos\phi, \text{ lagging} \quad \text{(x)}$$

(c) *Efficiency*

Power input, $P_{in} = 3VI_1 \cos\phi$ (xi)

Power across air-gap, $P_G = 3I_1^2 R_f$ (xii)

$$\omega_s = \frac{120 f}{P}$$

$$P_m \text{ (gross)} = (1 - s)\, P_G \quad \text{(xiii)}$$

$$P_m \text{ (net)} = P_m \text{ (gross) } P_r \quad \text{(xiv)}$$

$$\eta = \frac{P_m\,(\text{net}) \times 100}{P_{in}} \times 100 \qquad \text{(xv)}$$

(d) *Torque*

$$\omega_s = 120\,f/P$$

$$\omega = (1 - s)\,\omega_s$$

$$T(\text{net}) = \frac{P_G - P_r}{(1 - s)\,\omega_s}$$

$$s = \text{given} \qquad \text{(xvi)}$$

The above procedure is now illustrated by an example.

Steps for Calculating Torque and Plotting Torque Versus Slip Using MATLAB by Thevenin's Equivalent Circuit

Induction-motor equivalent circuit simplified by Thevenin's theorem is shown in Fig. 9.68(b).

where $$Z_{1eq} = \frac{j X_m (R_1 + j X_1)}{R_1 + j X_m + X_1} = R_{1eq} + jX_{1eq}$$

$$V_{1eq} = \frac{V_1 (j X_m)}{R_1 + j (X_1 + X_m)}$$

$$I_2 = \frac{V_{1eq}}{Z_{1eq} + j X_2 + R_2/s}$$

$$\omega_s = 120\,f/P$$

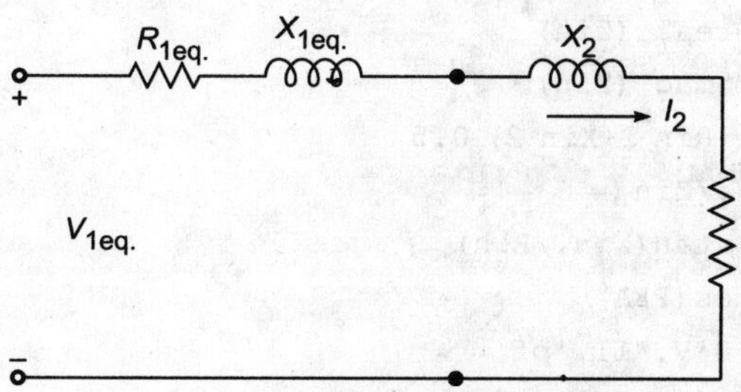

Fig. 9.68(b)

Torque $$T_{\text{mech}} = \frac{1}{\omega_s}\left[\frac{3V_{1eq}^2\,(R_2/s)}{\left(R_{1eq} + \frac{R_2}{s}\right)^2 + (X_{1eq} + X_2)^2}\right]$$

MATLAB PROGRAM

```
>>P0=467;
I0=6.8;
V0=400;
f =50;
P=8;
r1=0.68;
Pr=P0-3*I0^ 2.*r1
Riwf=V0^ 2/Pr
Iiwf=(V0/1.732)/Riwf
Im=(10^ 2-Iiwf ^2)^0.5
Xm=(V0/1.732)/Im
Vbr=180;
```

```
Ibr=17;
Pbr=1200/3;
Zbr=(Vbr/1.732)/Ibr
Rbr=Pbr/(1br^ 2)
R2=Rbr-r1
Xbr=(Zbr^ 2-Rbr^ 2)^0.5
X1=Xbr/2
X2=Xbr/2
V=V0/1.732
s=.05;
Zf=(R2/s+i*X2).*(i*Xm)./(R2/s+i*X2+i*Xm)
Rf=real(Zf)
Xf=imag(Zf)
Zin=(r1+Rf)+i*(X2+Xf)
Rin=real (Zin)
Xin=imag (Zin)
Zin=(Rin^2+Xin^2)^0.5
I1=V./Zin
PFA=atan(Xin./Rin)
pf=cos(PFA)
Pin=3*V.*I1.*pf
Pg=3*I1^2.*Rf
Ws=120*f./P
Pmgross=(1-s).*Pg
Pmnet=Pmgross-Pr
W=(1-s).*Ws
Tnet=Pmnet./W
Eff=Pmnet./Pi n
Pr =
   372.6704
Riwf =
   429.3338
Iiwf =
   0.5379
Im =
   6.7787
Xm =
   34.0695
Zbr =
   6.1133
Rbr =
```

```
  1.3841
R2 =
  0.7041
Xbr =
  5.9546
X1 =
  2.9773
X2 =
  2.9773
V =
  230.9469
Zf =
  10.4058 + 6.6933i
Rf =
  10.4058
Xf =
  6.6933
Zin =
  11.0858 + 9.6706i
Ri n =
  11.0858
Xin =
  9.6706
Zin =
  14.7111
I1 =
  15.6988
PFA =
  0.7173
pf =
  0.7536
Pin =
  8. 1964e+003
Pg =
  7.6937e+003
Ws =
  750
Pmgross =
  7.3090e+003
Pmnet =
  6.9363e+003
W
```

```
    712.5000
Tnet =
    9.7352
Eff =
    0.8463
>>
```

Example 9.20 *The results of the no-load and blocked-rotor tests on a 3-phase, Y-connected 10 kW, 400 V, 17 A, 50 Hz, 8 pole induction motor with a squirrel-cage rotor are given below.*

No-load test: *Line to line voltage = 400 V*
Total input power = 467 W
Line current = 6.8 A

Blocked rotor test: *Line to line voltage = 180 V*
Total input power = 1200 W
Line current = 17 A

The dc resistance of the stator measured immediately after the blocked rotor test is found to have an average value of 0.68 ohm/phase.

Calculate the parameters of the circuit model of the induction motor. Draw IEEE circuit model.

Using MATLAB calculate and plot against motor speed or slip the following variables:

(i) Torque (net)
(ii) Stator current
(iii) Power factor
(iv) Efficiency

LO 4

Solution

No-Load Test

$$Y_0 = \frac{6.8}{400/\sqrt{3}} = 0.0294 \ ℧$$

$$G_{iwf} = \frac{467 - 3(6.8)^2 \times 0.68}{(400)^2} = 2.33 \times 10^{-3} \ ℧$$

$$B_m = \sqrt{Y_0^2 - G_{iwf}^2} = 2.93 \times 10^{-2} \ ℧$$

$$X_m = 34.12 \ \Omega$$

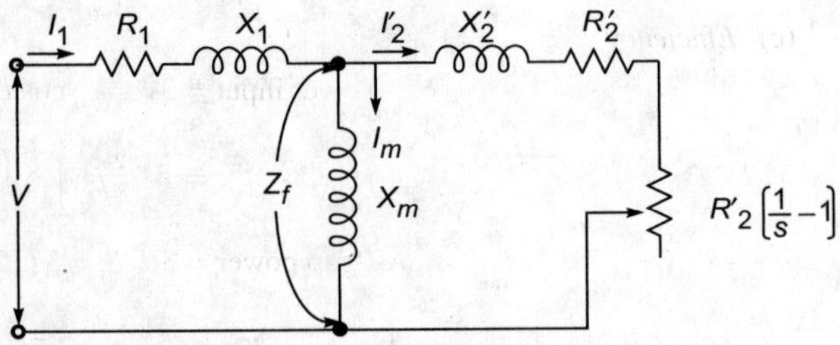

Fig. 9.68(c)

The IEEE circuit model is given in the Fig. 9.68 (c).

Blocked-Rotor Test

$$Z = \frac{180/\sqrt{3}}{17} = 6.113 \ \Omega$$

$$R = \frac{1200/3}{(17)^2} = 1.384 \ \Omega$$

$$X = \sqrt{Z^2 - R^2} = 5.95 \ \Omega$$

$$R = R_1 + R_2'$$

$$R_2' = 1.384 - 0.68 = 0.704\ \Omega$$

$$X_1 = X_2' = \frac{X}{2} = 2.975\ \Omega$$

Performance Calculation

$$s = 0.05$$

$$\bar{Z}_f = \left(\frac{R_2'}{s} + jX_2\right) \| (jX_m)$$

$$= \left(\frac{0.704}{0.05} + j2.975\right) \| (j34.12)$$

$$= (14.08 + j\,2.975) \| (j\,34.12)$$

$$= \frac{14.39\angle 11.9^\circ \times 34.12\angle 90^\circ}{39.677\angle 69.2^\circ}$$

$$= 12.375\ \angle 32.7^\circ$$

$$= 10.414 + j\,6.685$$

$$\bar{Z}_{(\text{total})} = 0.68 + j\,2.975 + 10.414 + j\,6.685$$

$$\simeq 14.7\ \angle 41.3^\circ\ \Omega$$

$$|\bar{Z}|_{(\text{total})} = 14.7\ \Omega$$

(a) *Stator Current*

$$I_1 = \frac{400}{\sqrt{3} \times 14.7} = 15.69\text{ A}$$

(b) *Power Factor*

$$pf = \cos(14.3^\circ) = 0.751\ (\text{lagging})$$

(c) *Efficiency*

$$\text{Power input} = 3V_{ph}\,I_{ph}\cos\phi$$

$$= 3\left(\frac{400}{\sqrt{3}}\right) \times 15.69 \times 0.751 = 8163.4\text{ W}$$

$$\text{Air-gap power} = 3I_1^2\,R_f = 3\,(15.69)^2 \times 10.374 = 7661.5\text{ W}$$

$$\omega_s = \frac{120\,f}{P} = 750\text{ rpm}$$

$$P_m\ (\text{gross}) = (1 - s)P_g = 0.95 \times 7661.5 = 7278.4\text{ W}$$

$$P_m\ (\text{net}) = P_m\ (\text{gross}) - P_r$$

where Pm (net) = Pm (gross) – Pr

$$P_r = P_0 - 3I_0^2\,R_1 = 372\text{ W}$$

(d) *Torque*

$$\omega = \omega_s\,(1 - s) = 750 \times 0.95 = 712\text{ rpm}$$

$$T_{(\text{net})} = \frac{P_{m(net)}}{\omega} = \frac{6906.4}{712} = 9.7\text{ Nm}$$

$$\text{Efficiency} = \frac{P_{m(net)}}{P_{\text{input}}} = \frac{6906.4}{8163.4} = 0.846$$

$$\% \text{ Efficiency} = 84.6\%$$

Torque Expression for Variable Slip

From the IEEE circuit model shown in the Fig. 9.68(d)

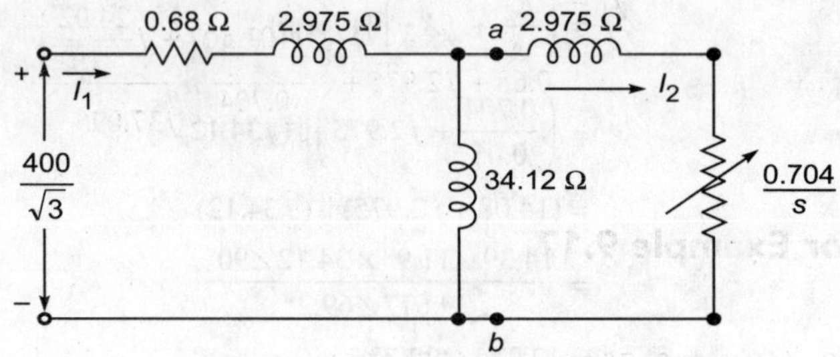

Fig. 9.68(d)

$$V_{TH} = \frac{230.9 \times 34.12}{|0.68 + j(2.975 + 34.12)|} \text{ (Thevenin's voltage across } ab \text{ terminals)}$$

$$V_{TH} = 212.35 \text{ V}$$

Torque,

$$T = \frac{3V_{TH}^2\left(\frac{R_2'}{s}\right)}{\omega_s\left[\left(R_1 + \frac{R_2'}{s}\right)^2 + (X_1 + X_2')^2\right]} = \frac{3}{78.54}\,\frac{(212.35)^2 \times \frac{0.704}{s}}{\left[\left(0.68 + \frac{0.704}{s}\right)^2 + (5.95)^2\right]}$$

$$T = \frac{1212.57/s}{\left(0.68 + \frac{0.704}{s}\right)^2 + 35.4025}$$

This expression will be used later to plot T-s curve using MATLAB.

Stator current expression for the variable slip

$$I_1 = \frac{V\angle 0^\circ}{\bar{Z}_{\text{Total}}}$$

$$\bar{Z}_f = \frac{jX_m\left(\frac{R_2'}{s} + jX_2'\right)}{\frac{R_2'}{s} + jX_2' + jX_m}$$

$$= \frac{(j34.12)\left(\frac{0.704}{s} + j2.975\right)}{\left(\frac{0.704}{s} + j37.095\right)}$$

$$= \frac{-101.507 + j\dfrac{24.02}{s}}{\dfrac{0.704}{s} + j37.095}$$

$$\bar{Z}_{\text{Total}} = (0.68 + j\,2.9785) + \bar{Z}_f$$

$$I_1 = \frac{400\,\angle 0^\circ}{0.68 + j\,2.975 + \dfrac{\left(-101.507 + j\dfrac{24.02}{s}\right)}{\dfrac{0.704}{s} + j37.095}}$$

MATLAB Program for Example 9.17

```
s=0.0001: 0.01:1
Zf =(-101.507+i*24.02./s)./((0.704./s)+i*37.095)
Zt=0.68+i*2.975+Zf
I1=(400/1.732)./Zt
Rf=real(Zf)
X=imag(I1)
R=real(I1)
PHI=atan(X./R)
pf=cos(PHI)
figure(1)
Plot(s,pf)
Pin=3*(400/1.732).*abs(I1).*pf
Pg=3*abs(I1).*abs(I1).*Rf
Pmgross=(1-s).*Pg
Pmnet=Pmgross-0.03726704
Eff=Pmnet./Pin
figure(2)
plot(s,Eff)
figure(3)
plot(s,abs (I1))
s=-1:.0001:2
y=(0.68+0.704./s).*(0.68+0.704./s)+35.4025
T=(1212.57./s)./y
figure(4)
plat(s,T)
```

Plots are shown in Fig. 9.68 (e)–(h)

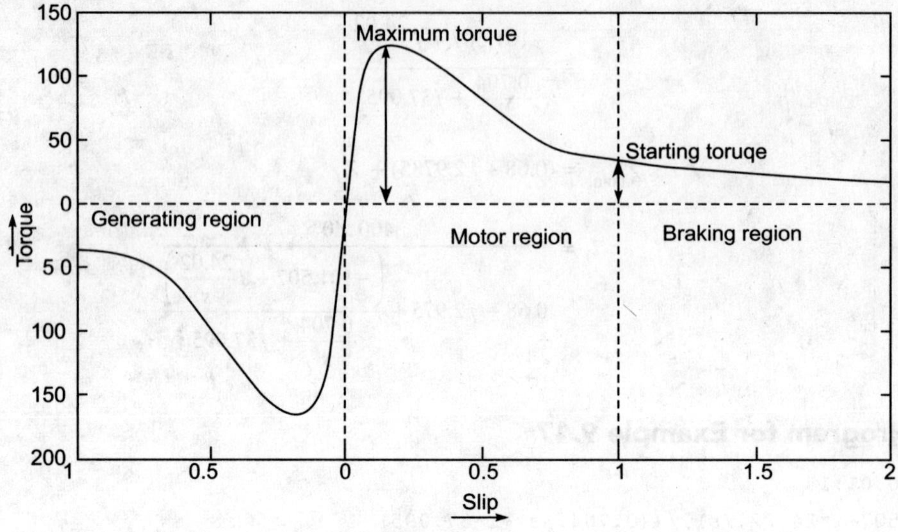

Fig. 9.68(e)

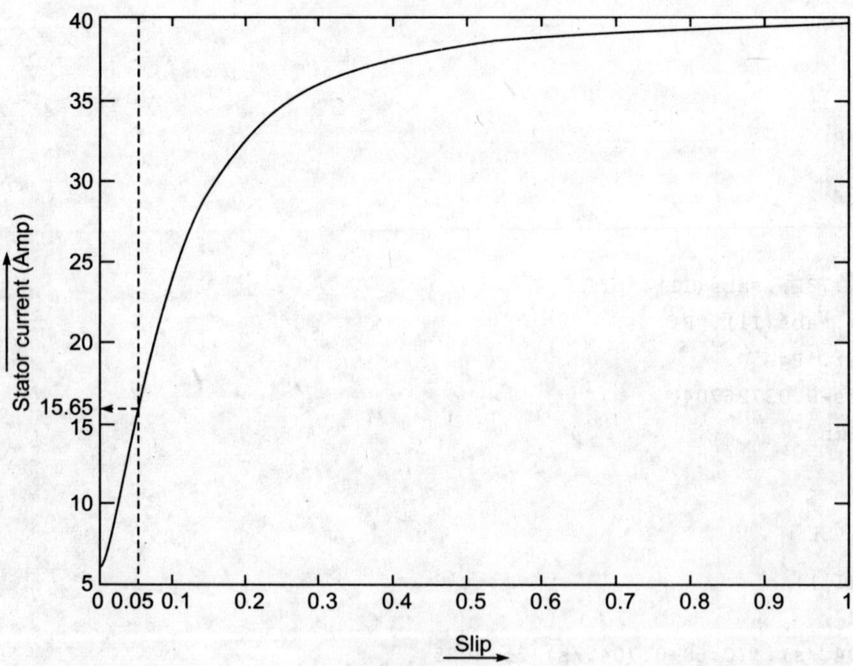

Fig. 9.68(f)

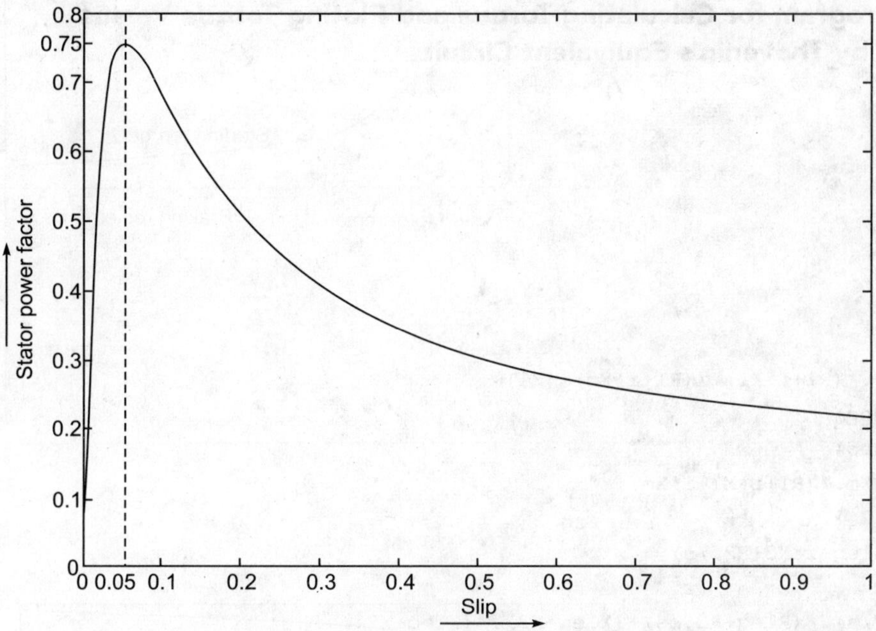

Fig. 9.68(g)

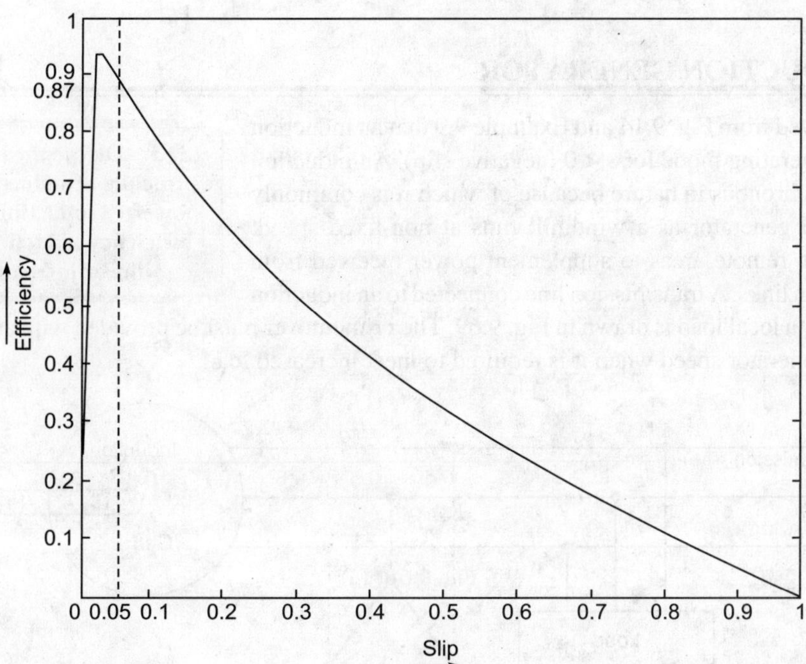

Fig. 9.68(h)

MATLAB Program for Calculating Torque and Ploting Torque Versus Frequency by Thevenin's Equivalent Circuit

```
R1=0.68;
X1=2.975;
Xm=34.12;
X2=2.975;
R2=0.704;
V1=400/1.732;
f=50;
P=8;
Z1eq=(Xm*i).*(R1+i*X1)./(R1+i*Xm+i*X1)
R1eq=real(Zeq)
X1eq=imag(Zeq)
V1eq=V1.*i*Xm./(R1+i*X1+i*Xm)
s=-1:0.001:2
I2=V1eq./(Z1eq+i*X2+R2./s)
ws=120*f./P
Z=(R1eq+R2./s).*(R1eq+R2./s)+(X1eq+X2).*(X1eq+X2)
Tmech=(1/ws).*(3.*V1eq.*V1eq.*(R2./s))./Z
plot(s,Tmech)
```

9.14 | INDUCTION GENERATOR

LO 5 Define the basic operating principle of induction generator, inverted induction motor, high efficiency induction motor, and linear induction motor

It has been observed from Fig. 9.14 and Example 9.9 that an induction machine is in generating mode for $s < 0$ (negative slip). An induction generator is asynchronous in nature because of which it is commonly used as windmill generator as a windmill runs at non-fixed speed. These are used in remote areas to supplement power received from weak transmission links. A transmission line connected to an induction generator feeding a local load is drawn in Fig. 9.69. The primemover must be provided with automatic control to increase the generator speed when it is required to meet increased load.

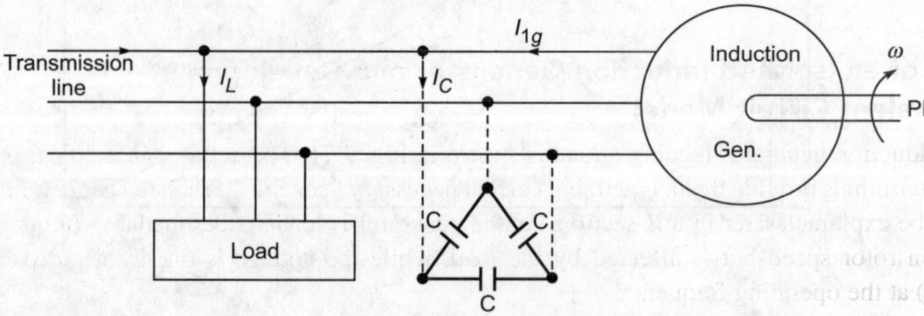

Fig. 9.69

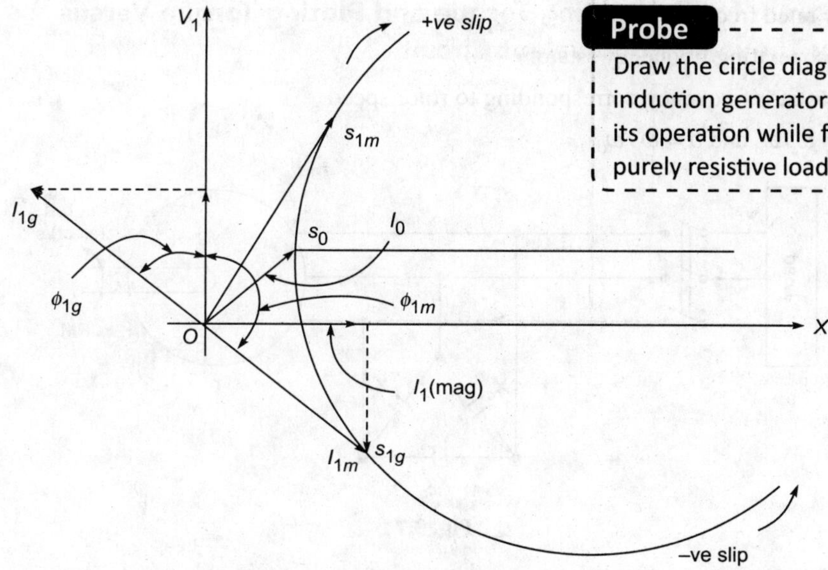

Fig. 9.70

Probe

Draw the circle diagram of an induction generator and explain its operation while feeding a purely resistive load.

Figure 9.70 shows the circle diagram of an induction machine extended to the generating region i.e., below the *OX* co-ordinate, which is the negative slip region. At slip = s_{1g}, the motor draws current I_{1m} which lags the applied voltage V_1 by $\Phi_{1m} > 90°$. This means negative pf (cos Φ_{1m}) or that the electric power flows out of the machine resulting in generating operation. The generating current fed to the line is then

$$I_{1g} = -I_{1m}$$

with a leading pf cos Φ_{1m}. This alternately means that the machine draws a 90°-lagging current component to provide its magnetizing current need. The transmission line has then to feed the lagging current component of the load as well as the magnetizing current of the induction generator. This places a severe lagging VARs load on the already weak lines. This burden must be relieved by connecting balanced shunt capacitors (in delta) across the induction generator terminals. These draw leading current or equivalently feed lagging magnetizing current of the generator.

It is to be observed here that the operating frequency of the system of Fig. 9.69 is fixed by the line frequency.

Operation of an Isolated Induction Generator and Development of its Equivalent Circuit Model

An isolated induction generator feeding a load is shown in Fig. 9.71. The delta-connected capacitors across the generator terminals provide the magnetizing current necessary to excite the isolated generator. The voltage build-up will be explained later in this section. As the generator is loaded, the operating frequency depends primarily upon rotor speed but is affected by the load, while the voltage is mainly decided by capacitor reactance (X_C) at the operating frequency.

Let ω_0 = rated frequency

ω_s = operating frequency (stator)

ω_r = stator frequency corresponding to rotor speed

$a = \omega_s/\omega_0$ and $b = \omega_r/\omega_0$

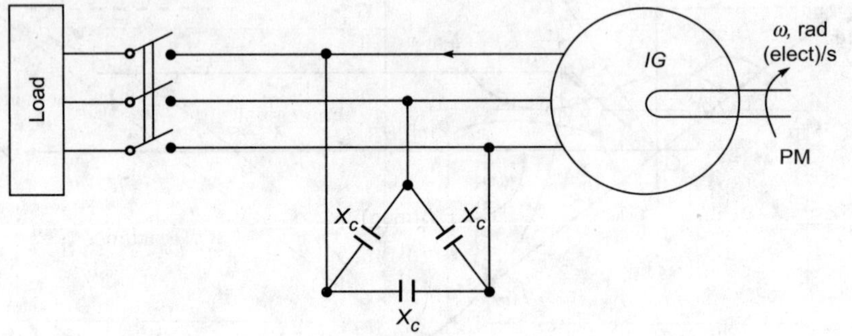

Fig. 9.71

The machine slip (which should be negative) can be expressed as

$$s = (\omega_s - \omega_r)/a\omega_0$$
$$= (a\omega_0 - b\omega_0)/a\omega_0 = (a - b)/a;\ b < a$$

Recall Assuming that the machine reactances, excitation reactances and machine induced emf correspond to the rated frequency (ω_0), the per phase equivalent circuit of the system at operating frequency $\omega_s = a\omega_0$ is drawn in Fig. 9.72(a) with load considered as purely resistive. Dividing throughout by '*a*', the circuit is reduced to the rated (fixed) frequency and is drawn in Fig. 9.72(b).

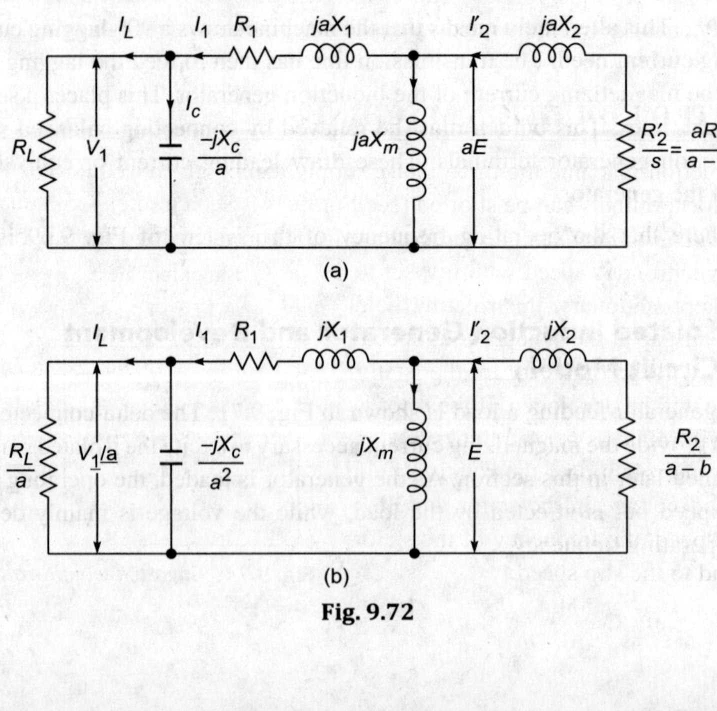

Fig. 9.72

Voltage Build Up Process at No-Load

With reference to Fig. 9.71, consider that the generator is run at synchronous speed at no-load (switch off) with Δ-connected capacitors hanging on its terminals. The rotor circuit behaves as open and so the circuit model is as drawn in Fig. 9.73(a).

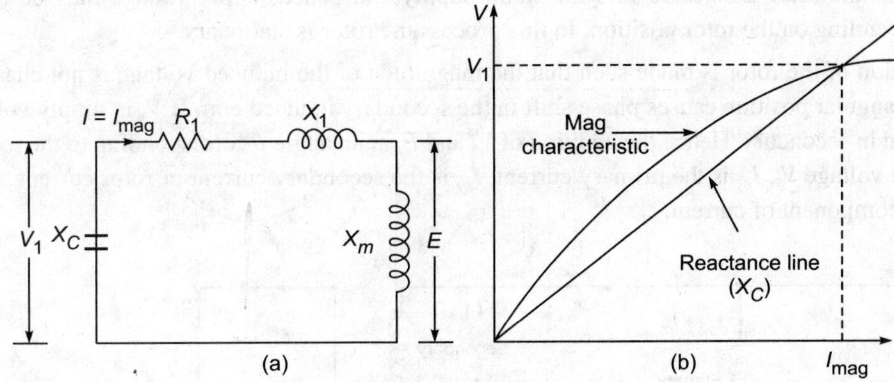

Fig. 9.73

The current fed into the motor is $I_1 = I_{mag}$, the magnetizing current. If the voltage drop in stator impedance, R_1, X_1, is ignored, we have

$$V_1 \approx E \text{ (induced emf)}$$

$$= E(I_{mag}), \text{ the magnetization characteristic}$$

Also $$V_1 = X_C I_{mag}; \text{ the reactance line on Fig. 9.73(b)}$$

Like in a dc shunt generator, the magnetization characteristic and reactance line are drawn in Fig. 9.73(b). Their intersection gives the no-load voltage. This is the case at rated frequency. V_1 will vary slightly as the actual stator frequency changes from the rated value. This variation may be ignored.

9.15 | INVERTED INDUCTION MACHINE

LO 5

In a wound-rotor induction machine the three-phase supply can be given to the rotor windings through the slip rings and the stator terminals can be shorted (see Fig. 9.74). Such a rotor-fed induction machine is also called an *inverted induction machine.* The three-phase rotor current will generate a rotating field in the air gap, rotating at the synchronous speed with respect to rotor. If the rotor is kept stationary, this rotating field will also rotate in the air gap at the synchronous speed. Voltage and current will be induced in the stator windings and a torque will be developed. If the rotor is allowed to move, it will rotate as per the Lenzs law, in opposite direction to the rotation of the rotating field decreasing the induced voltage in the stator winding. Thus, at a particular speed, the frequency of the stator circuit will correspond to the slip speed.

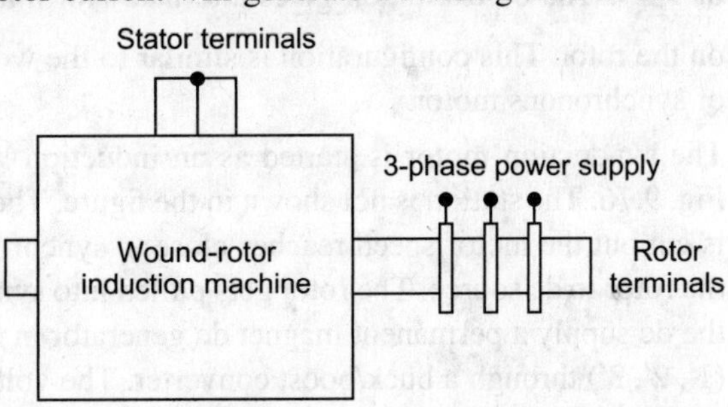

Fig. 9.74 *Inverted wound-rotor induction machine*

Induction Regulator

The Induction Regulator is used in the applications requiring constant voltage under variable load conditions and variable ac voltage supply for speed control. The basic principle of operation is the three phase supply is supplied to stator and produces revolving magnetic flux which induces emf in the rotor windings. This induced emf in the rotor is injected into the stator supply and hence supply voltage may be increased or decreased depending on the rotor position. In this process the rotor is stationary.

The construction of the rotor is made such that the magnitude of the induced voltage is not changed but by changing the angular position causes phase shift in the secondary induced emf. If V_s is supply voltage and E_r is emf induced in secondary. Hence the addition of V_s and E_r at an angle θ corresponding to the rotor position to give output voltage V_o. I_{py} is the primary current, I_{sy} is the secondary current or rotor current and I_m is the magnetizing component of current.

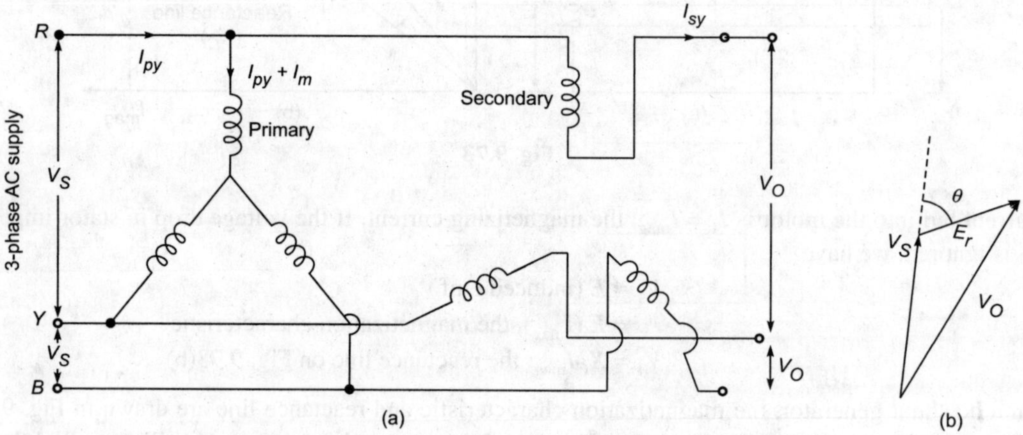

Fig. 9.75 (a) *Three phase induction regulator,* (b) *Phasor diagram*

Synchronous Induction Motor

It is slip ring induction motor which can be operated as synchronous motor when the rotor excited with a dc source. One phase of slip ring carries a current I_a and the other two phases carry a current of $-\frac{I_a}{2}$. The dc excitation gives alternate North and South poles on the rotor. This configuration is similar to the working principle of synchronous motor.

> **Probe**
>
> Explain briefly the working principle of synchronous induction motor.

The synduction motor is started as an induction with external resistances in the rotor circuit as shown in Fig. 9.76. The station is not shown in the figure. The throw-over switch is in position 1, 2, 3. As the resistance is cut out the motor speed reaches close to synchronous, the switch is thrown to position 1, 2, 3 connecting the rotor to dc source. The rotor gets pulled into synchronism and theorem runs as a synchronous motion. For the dc supply a permanent magnet dc generator in coupled to the motor and connected to attached terminals (1′, 2′, 3′) through a buck/boost converter. The voltage of the converter can be varied to control the *pf* of the motor.

The synduction motor provides a high starting torque and low starting current and *pf* control when running as synchronous motor.

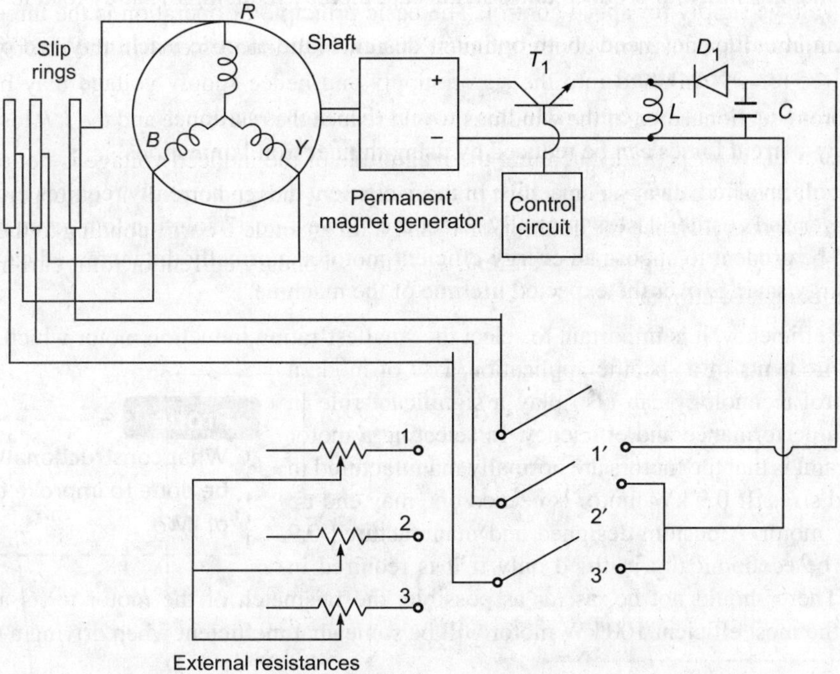

Fig 9.76 *Synchronous Induction motor with brushless excitation*

9.16 | HIGH EFFICIENCY INDUCTION MOTORS

LO 5

Necessity of High Efficiency Induction Motors

With ever-increasing energy cost the life-time operating cost of an induction motor can be traded against a high efficiency and high capital cost induction motors. With rising demand for high efficiency or energy efficient induction motors, designers and manufacturers are stepping up their production of such motors.

Arrangements to be Employed to Make an IM More Efficient

Some of the important techniques, that are employed to construct a higher efficiency induction motor compared to the standard design, are listed below:

- Reduced current density in copper of stator winding resulting in reduced copper loss but increased *copper weight and so cost.*
- Reduced flux density in air-gap by increasing stator and rotor core length. This leads to reduced *magnetic saturation and core loss.*
- Larger stator steel volume improving heat transfer out of motor and so reduced operating temperature. *Rotor fans are designed with refined aerodynamic fin shapes to reduce windage loss.*
- Use of high grade electrical steel with low hysteresis loss.

- Use of very high resistivity steel and very thin gauge laminations with consequent reduction in eddy-*current loss*.
- Rotor is fine machined to produce uniform air-gap, thereby reducing the stray load loss.

Achieving maximum efficiency needs both optimum design of the electric machinery and proper matching of machine and desired application.

Increasing the cross-sectional area of the windings would reduce the resistance and the I^2R losses. For a given flux density eddy-current losses can be reduced by using thinner iron laminations.

There is a trade-off involved always. A machine of more efficient design normally requires more material and therefore is bigger and costlier. Users normally select the "lowest-cost" solution to a given requirement. Of course it would be prudent to choose an energy-efficient motor as normally the increased capital cost would be offset by energy savings over the expected lifetime of the machine.

To optimize the efficiency, it is important to select the smallest-rating induction motor which can adequately satisfy the requirements of a specific application. Use of modern solid-state control technology can also play a significant role in optimizing both performance and efficiency. In selecting a motor the main constraint is that the motors are normally manufactured in certain standard sizes. If 0.9 kW motor is needed one may end up buying a 1 kW motor. A custom-designed and manufactured 0.9 kW motor can be economically justified only if it is required in large number. There should not be, as far as possible, the mismatch of the motor to its application. For example, even the most efficient 100 kW motor will be somewhat inefficient when driving a 60 kW load.

Probe

What constructional changes can be done to improve the efficiency of IM?

9.17 | LINEAR INDUCTION MOTOR (LIM)

LO 5

The basic principle and operation have already been explained in Section 5.8. The expression for linear velocity was derived there. Linear induction motor to produce linear motion has been devised in many different configurations. We shall mainly discuss the configuration most commonly adopted for traction purposes.

Construction and Operation

LIM is a developed version of cylindrical induction motor. If the elementary induction motor of Fig. 9.77(a) is cut axially and spread out flat as in Fig. 9.77(b) it corresponds to LIM. Of course suitable magnetic circuit must be provided (this will be illustrated later in Fig. 9.77(b)). The secondary conductor now appears in form of a sheet. For obvious reasons the air gap is much larger than in a normal induction motor as the secondary conductor contributes towards the air gap. A typical air gap in LIM is of the order of 25 mm while in a cylindrical motor it is about 1 mm.

The air-gap field now moves linearly in the direction (acb) (assumed phase sequence abc). The induced currents in the secondary sheet interacting with this field produce thrust on it and therefore its motion in the direction of the movement of the field as indicated in Fig. 9.77(b). Unlike a cylindrical motor the field in an LIM has end discontinuities. As the field reaches the back end (in relation to direction of motion) it must reduce to zero while a fresh field must build up at the front end. This rise and decay of field causes statically induced current in the secondary sheet which cannot contribute any thrust and only adds to ohmic loss of the motor.

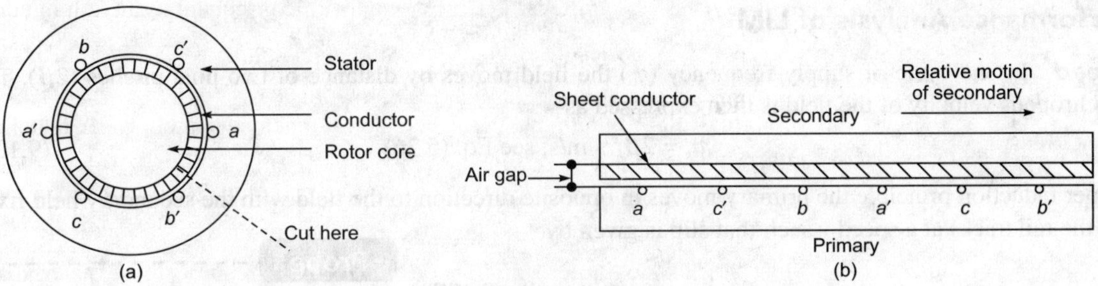

Fig. 9.77 *Linear version of a squirrel-cage rotary induction motor*

Either the primary or secondary could be made mobile. The stationary member must of course be continuous throughout the length of intended travel.

Magnetic attraction between members can be balanced out in a rotary machine but cannot be so balanced in a flat machine unless double-sided construction is adopted as illustrated in Fig. 9.78(a) where there is double layer three-phase winding on either side of the secondary. This construction is commonly adopted for traction wherein the finite length primaries are carried on the vehicle while the secondary takes the form of a continuous conducting (aluminium is used) rail as shown in Fig. 9.78(b).

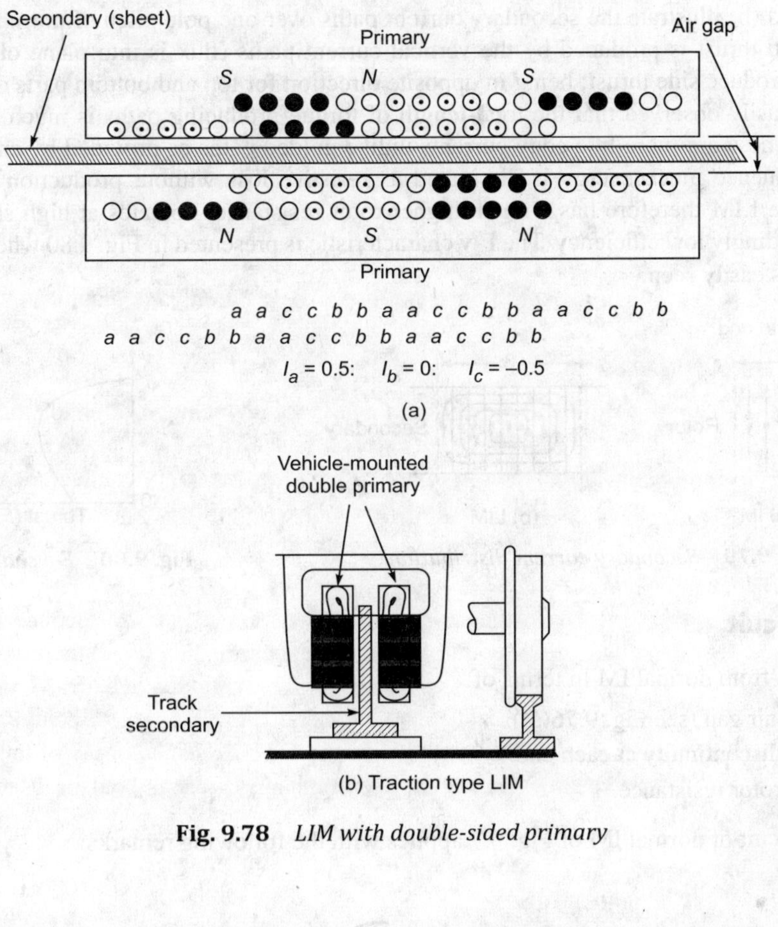

Fig. 9.78 *LIM with double-sided primary*

Performance Analysis of LIM

Speed In one cycle of supply frequency (f) the field moves by distance of two pole pitches (2β). The synchronous velocity of the field is then expressed as

$$v_s = 2\beta f \text{ m/s; see Eq. (5.76)} \tag{9.106}$$

As per induction principle the primary moves in opposite direction to the field with the secondary held fixed (on the rail track) at a speed v such that slip is given by

$$s = \frac{v_s - v}{v_s} \tag{9.107}$$

Probe

How can we vary the linear movement speed of LIM?

As per Eq. (7.106) the synchronous speed is not governed by the number of poles and any desired linear speed can be obtained by adjusting pole pitch. For example, for $\beta = 1\text{m}$

$$v_s = 2 \times 50 = 100 \text{ m/s} \quad \text{or} \quad 180 \text{ km/s}$$

which means that traction speeds are easily achieved. It may be observed here that the motor may have even or odd number of poles.

Secondary Resistance and F-v Characteristic

In Fig. 9.79(a) and (b) illustrate the secondary current paths over one pole pitch of normal induction motor and LIM. Forward thrust is produced by the vertical current paths (flux is into plane of paper) while the horizontal paths produce side thrust; being in opposite direction for top and bottom parts of secondary these cancel out. It is easily observed that the total length of torque producing paths is much more in a normal IM than in LIM (these are more like eddy currents unlike guided paths as provided by end rings in normal IM). As a consequence much energy is dissipated as resistive loss without production of corresponding mechanical power. LIM therefore has a much higher rotor resistance, operates at high slip at given thrust and has correspondingly low efficiency. The F-v characteristic is presented in Fig. 9.80 where high resistance secondary effect is easily seen.

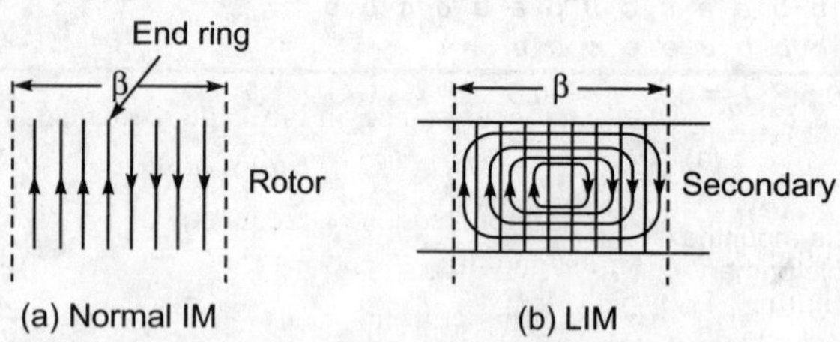

Fig. 9.79 *Secondary current distribution*

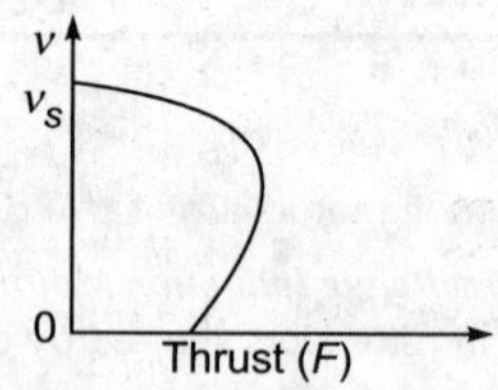

Fig. 9.80 *F-v characteristics*

Equivalent Circuit

LIM distinguishes from normal IM in terms of:

- it has large air gap (see Fig. 9.76(a)).
- it has field discontinuity at each end.
- it has high rotor resistance.

The equivalent circuit of normal IM of Fig. 9.7 applies with the following remarks:

- Large air-gap means low magnetizing reactance (therefore high magnetizing current and low pf)
- High secondary leakage inductance because of secondary current distribution (Fig. 9.79(b)), where much of the current does not link the whole of the air-gap flux.
- High rotor resistance (as explained already).

Basic Operating Principle of Magnetic Levitation Motor

In this motor, the rotor is suspended in the magnetic field and the gravitational force is overcome by electromagnetic force. The rotor will float over a guideway using the basic principles of magnets. The three main components of the system are:

1. large electrical power source
2. metal coils lining a guide way or track
3. large guidance magnets attached to the underside of the train

Here, the electric current supplied to the coils in the guideway walls is constantly alternating to change the polarity of the magnetized coils so that the magnetic field in front of the rotor is to pull forward, while the magnetic field behind the rotor adds more forward force. The guidance magnets embedded in the rotor body keep it stable during run. Once the rotor is levitated, power is supplied to the coils within the guide way walls to create a unique system of magnetic fields that pull and push the rotor along the guideway.

Concept Check LO 4

1. Show that at super-synchronous speed the induction machine acts as a generator. Write the expression for P_G in which direction does it flow? How to find the net mechanical power input and net electrical power output.

SUMMARY

LO 1

- Induction motors are of two types:

  Squirrel-cage Induction Motor – Copper/aluminum bars in rotor slots shorted by end rings. Aluminum is commonly used being of low cost.

  Wound-rotor or slip-ring Induction Motor The rotor has 3-phase windings with connections brought out through three slip-rings, the winding is shorted externally, also external resistances can be included at the time of starting; expensive, used only where high-starting torque is must.
- Results and statements made here are on per phase basis, powers – active and reactive on 3-phase basis. Winding connection will be assumed star (or equivalent star) except where specified otherwise.
- Stator resistance and leakage reactance ignored

  $$V_1 \approx E_1 = \sqrt{2}\ \pi K\omega_1 N_{ph1}(series)\, f\, \Phi_r\ V$$

  f= stator frequency (frequency of V_1), Φ_r= resultant air-gap flux
- Exciting current

  $$\bar{I}_0 = \bar{I}_m + \bar{I}_i$$

  I_m = magnetizing current; 90° lagging

  I_i = core loss current; in-phase

  Magnitude-wise $I_m >> I_i$

  PF of I_0 is very low, phase angle slightly less than 90°
- Rotor standstill emf E_2, frequency f

  $$\frac{E_1}{E_2} = \text{a (ratio of effective turns)}$$

- At speed n (slip, s)

  Rotor induced emf $= sE_2$

  Rotor frequency $f_2 = sf$

- $\dfrac{I'_2}{I_2} = \dfrac{1}{a}$; I'_2 stator current to counter I_2

- The net stator current

$$\bar{I}_1 = \bar{I}_0 + \bar{I}'_2 = \bar{I}_0 + \left(\frac{1}{a}\right)\bar{I}_2$$

  I_0 is almost 40% of I_1 (full-load)

- *Power Factor* Because of large I_0 with phase angle slightly less than 90°, the pf of the line current is of the order of 0.8 to 0.85. At light load, I'_2 reduces and so does the power factor. Therefore, the induction motor should not be run at light load for a long period of time.

- Rotor standstill reactance $= X_2$

  Rotor circuit impedance

$$\bar{Z}_2 = R_2 + jsX_2$$

LO 2

- Power across air-gap

  P_G = Gross mechanical power output + rotor copper loss

  $= P_m + 3I_2^2 R_2$

- Rotor resistance equivalent of mechanical power output

$$\left(\frac{1}{s} - 1\right)R_2$$

  which means $\dfrac{P_m}{3} = \left(\dfrac{1}{s} - 1\right)I_2^2 R_2$,

  Note $I'^2_2 R'_2 = I_2^2 R_2$

- $P_G = 3\dfrac{I'^2_2 R_2}{s} = 3\dfrac{I_2^2 R_2}{s}$

  $P_m = (1 - s) P_G$

  $T = \dfrac{P_G}{\omega_s}$, ω_s = synchronous speed in rad (mech)/s

  P_G is known as torque in *synchronous watts*

- $T = \dfrac{3}{\omega_s} \cdot \dfrac{V_1^2 (R'_2/s)}{(R'_2/s)^2 + X'^2_2} = \dfrac{3}{\omega_s} \cdot \dfrac{(V_1/a)^2 (R_s/s)}{(R_2/s)^2 + X_2^2}$

  Stator impedance ignored

- Torque-slip characteristic; see Fig. 9.14

  Motoring $0 \le s \le 1$

  Generating $s < 0$

  Breaking $s > 1$

- Determination of circuit model parameters

  Quantities measured in test: voltage, current and power

  No-load test

  Conducted at rated voltage

  Determines

  Some of core loss and windage and friction loss, X_m, R_{iwf}

  Blocked rotor test

  Conducted at reduced voltage, full-load current

  Determines

  Full-load copper loss

$$R = R_1 + R'_2,\ X = X_1 + X'_2$$
$$(R'_2 = a^2 R_2,\ X'_2 = a^2 X_2)$$

  Important Note In the computation of performance based the circuit model as determined by the above two tests, the windage and friction loss is accounted for in R_{iwf}. Therefore the mechanical power output $P_m = 3\left(\dfrac{1}{s} - 1\right)I_2^2 R_2$ is the net mechanical power output called the *shaft power*.

LO 3

- $T_{max} = T_{breakdown}\ (T_{BD})$

  For $T > T_{BD}$; motor stalls

- Resistance added in rotor circuit – slip-ring induction motor only. T_{BD} no change, slip at given torque increase, motor speed reduces, T(start) increases, I(start) reduces

- At T_{max}

$$R_2 = s_{maxT} X_2$$

$$T_{max} = T_{BD} = \frac{3}{\omega_s}\left(\frac{V_1^2}{2X'_2}\right) = \frac{3}{\omega_s}\left(\frac{(V_1/a)^2}{2X_2}\right)$$

  Stator impedance ignored

- Starting – stator impedance ignored

$$I_2\text{(start)} = \frac{V_1}{\sqrt{R'^2_2 + X'^2_2}} = \frac{V_1/a}{\sqrt{R_2^2 + X_2^2}}$$

$$T(\text{start}) = \frac{3}{\omega_s}\left[\frac{V_1^2 R_2}{R_2^2 + X_2^2}\right] = \frac{3}{\omega_s}\left[\frac{(V_1/a)^2 R_2}{R_2^2 + X_2^2}\right]$$

- *Methods of starting*

Squirrel-cage motor DOL (direct-on-line) starting not permitted for motors 5 kW and above, as the motor current is 5 to 6 times full-load current; the power supply companies do not allow such heavy short-time currents to be drawn.

- Reduced voltage start
- Series resistance starting – can be used for fractional – kW motors only
- Star/delta starting
  Start in star, run in delta
  Starting current and torque both reduce by a factor of 1/3.
- Auto transformer starting
  Expensive, used for very large motor
  Both starting current and torque reduce by a factor of x^2, x = voltage reduction factor

- Speed control – slip control, frequency control
  - *slip control* Reduced voltages for very small motors, inefficient
    Rotor resistance control for slip-ring induction motor, reduce efficiency drastically. Not suitable
  - *frequency control* In varying frequency (V/f) must be maintained constant for constant air-gap flux
    Requires expensive full rated thyristor convertor/inverter equipment

PRACTICE PROBLEMS

9.1 A 4-pole wound-rotor induction motor is used as a frequency changer. The stator is connected to a 50 Hz, 3-phase supply. The load is connected to the rotor slip rings. What are the possible speeds at which the rotor can supply power to this load at 25 Hz? What would be the ratio of voltages at load terminals at these speeds? Assume the stator and rotor impedances to be negligible. **LO 2**

9.2 A 6-pole, 50 Hz, 3-phase induction motor running on full-load develops a useful torque of 160 Nm and the rotor emf is observed to make 120 cycles/min. Calculate the net mechanical power developed. If the torque loss in windage and friction is 12 Nm, find the copper-loss in rotor windings, the input to the motor and efficiency. Given: stator losses 800 W (inclusive of windage and friction loss). **LO 3**

9.3 A 12-pole, 3-phase, 50 Hz, delta-connected induction motor draws 280 A and 110 kW under blocked-rotor test at rated voltage and frequency. Find the starting torque when switched on direct to the supply. Assume the stator and rotor copper losses to be equal under the blocked-rotor test. What would be the starting torque if the motor is started by connecting the phase windings in star. (Try this part after studying Sec. 9.8). **LO 3**

9.4 A 3.3 kV, 20-pole, 50 Hz, 3-phase, star-connected induction motor has a slip-ring rotor of resistance 0.025 Ω and standstill reactance of 0.28 Ω per phase. The motor has a speed of 294 rpm when full-load torque is applied. Compute (a) slip at maximum torque, and (b) the ratio of maximum to full-load torque. Neglect stator impedance. **LO 2**

9.5 An 8-pole, 3-phase, 50 Hz induction motor is running at a speed of 710 rpm with an input power of 35 kW. The stator copper-loss at this operating condition is known to be 1200 W while the rotational losses are 600 W. Find (a) the rotor copper-loss, (b) the gross torque developed, (c) the gross mechanical power developed, and (d) the net torque and mechanical power output. **LO 2**

9.6 A 7.5 kW, 440 V, 3-phase, star-connected, 50 Hz, 4-pole squirrel-cage induction motor develops full-load torque at a slip of 5% when operated at rated voltage and frequency. Rotational losses (core, windage and friction) are to be neglected. Motor impedance data are as follows: **LO 2**

$R_1 = 1.32\ \Omega$
$X_1 = X_2' = 1.46\ \Omega$
$X_m = 22.7\ \Omega$

Determine the maximum motor torque at rated voltage and the slip at which it will occur. Also calculate the starting torque.

9.7 The motor of Prob. 9.6 is fed through a feeder from 440 V, 50 Hz mains. The feeder has impedance of $(1.8 + j\,1.2)\ \Omega$/phase. Find the maximum torque that the motor can deliver and the corresponding slip, stator current and terminal voltage. LO 2

9.8 A 400 V, 3-phase, stator-connected induction motor gave the following test results: LO 2

No-load	400 V	8.5 A	1100 W
Blocked-rotor	180 V	45 A	5700 W

Determine the ohmic values of the components in the circuit model and calculate the line current and power factor when the motor is operating at 5% slip. The stator resistance per phase is $0.5\ \Omega$ and the standstill leakage reactance of the rotor winding referred to the stator is equal to that of the stator winding.

9.9 A 15 kW, 415 V, 4-pole, 50 Hz delta connected motor gave the following results on test (voltages and currents are in line values) LO 2

No-load test	415 V	10.5 A	1510 W
Blocked-rotor test	105 V	28 A	2040 W

Using the approximate circuit model, determine
(a) the line current and power factor for rated output,
(b) the maximum torque, and
(c) the starting torque and line current if the motor is started with the stator star-connected.
Assume that the stator and rotor copper losses are equal at standstill.
Hint: Part (a) is best attempted by means of a circle diagram. For proceeding computationally from the circuit model, the complete output-slip curve has to be computed and then the slip for the rated output read.

9.10 A 400 V, 3-phase, 6-pole, 50 Hz induction motor gave the following test results: LO 2

No-load	400 V, 8 A, 0.16 power factor
Blocked-rotor	200 V, 39 A, 0.36 power factor

Determine the mechanical output, torque and slip when the motor draws a current of 30 A from the mains. Assume the stator and rotor copper losses to be equal.

9.11 A 4-pole, 3-phase, 400 V, 50 Hz induction motor has the following parameters of its circuit model (referred to the stator side on an equivalent-star basis): LO 2

$R_1 = 1.2\ \Omega$; $X_1 = 1.16\ \Omega$
$R_2' = 0.4\ \Omega$; $X_2' = 1.16\ \Omega$
$X_m = 35\ \Omega$

Rotational losses are 800 W.
(a) For a speed of 1440 rpm, calculate the input current, power factor, net mechanical power output, torque and efficiency.
(b) Calculate the maximum torque and the slip at which it occurs.

9.12 A 3-phase, 3.3 kV, 50 Hz, 10-pole, star-connected induction motor has a no-load magnetizing current of 45 A and a core-loss of 35 kW. The stator and referred rotor standstill leakage impedances are respectively $(0.2 + j\,1.8)$ and $(0.45 + j\,1.8)\ \Omega$/phase. The motor is supplied from 3.3-kV mains through a line of reactance $0.5\ \Omega$/phase. Use approximate circuit model. LO 2
(a) The motor is running at 0.03 slip. Estimate the gross torque, stator current and power factor. Assume voltage at motor terminals to be 3.3 kV.
(b) Calculate the starting torque and current when the motor is switched on direct to line with voltage at far end of the line being 3.3 kV.

9.13 A 6-pole, 440 V, 3-phase, 50 Hz induction motor has the following parameters of its circuit model (referred to the stator on an equivalent-star basis). LO 2

$R_1 = 0.0\ \Omega$ (stator copper-loss negligible);
$X_1 = 0.7\ \Omega$
$R_2' = 0.3\ \Omega$; $X_2' = 0.7\Omega$
$X_m = 35\ \Omega$
Rotational loss = 750 W

Calculate the net mechanical power output, stator current and power factor when the motor runs at a speed of 950 rpm.

9.14 A 75 kW, 440 V, 3-phase, 6-pole, 50 Hz, wound-rotor induction motor has a full-load slip of 0.04 and the slip at maximum torque of 0.2 when operating at rated voltage and frequency with rotor winding short-circuited at the slip-rings. Assume the stator resistance and rotational losses to be negligible. Find: LO 2

(a) maximum torque,
(b) starting torque, and
(c) full-load rotor copper-loss.

The rotor resistance is now doubled by adding an external series resistance. Determine:

(d) slip at full-load,
(e) full-load torque, and
(f) slip at maximum torque.

9.15 A 3-phase induction motor has a 4-pole, star-connected stator winding and runs on 50 Hz with 400 V between lines. The rotor resistance and standstill reactance per phase are 0.4 Ω and 3.6 Ω respectively. The effective ratio of rotor to stator turns is 0.67. Calculate (a) the gross torque at 4% slip; (b) the gross mechanical power at 4% slip; (c) maximum torque, (d) speed at maximum torque; and (e) maximum mechanical power output (gross). Neglect stator impedance. LO 2

9.16 A 30 kW, 440 V, 50 Hz, 3-phase, 10-pole, delta-connected squirrel-cage induction motor has the following parameters referred to a stator phase: LO 2

$R_1 = 0.54\ \Omega \qquad R_2' = 0.81\ \Omega$
$X_1 + X_2' = 6.48\ \Omega$
$R_1 = 414\ \Omega \qquad X_m = 48.6\ \Omega$

Calculate the machine performance (input current, power factor, mechanical output (gross), torque developed (gross)) for the following conditions:

(a) as a motor at a slip of 0.025,
(b) as a generator at a slip of – 0.025, and
(c) as a brake at a slip of 2.0.

9.17 The following test results were obtained on a 7.5 kW, 400 V, 4-pole, 50 Hz, delta-connected induction motor with a stator resistance of 2.1 Ω/phase: LO 2

No-load	400 V,	5.5 A,	410 Ω
Rotor blocked	140 V,	20 A,	1550 W

Estimate the braking torque developed when the motor, running with a slip of 0.05, has two of its supply terminals suddenly interchanged.

9.18 A 3-phase, wound-rotor induction motor has a star-connected rotor winding with a rotor resistance of 0.12 Ω/phase. With the slip-rings shorted, the motor develops a rated torque at a slip of 0.04 and a line current of 100 A. What external resistance must be inserted in each rotor phase to limit the starting current to 100 A? What pu torque will be developed with rotor-resistance starting? LO 2

9.19 In Prob. 9.18 what external resistance must be inserted per rotor phase to develop full-load torque at three-fourths the synchronous speed with a line current of 100 A? LO 2

9.20 A 4-pole, 50 Hz, 3-phase induction motor has a rotor resistance of 4.5 Ω/phase and a standstill reactance of 8.5 Ω/phase. With no external resistance in the rotor circuit, the starting torque of the motor is 85 Nm. LO 2

(a) What is the rotor voltage at standstill?
(b) What would be the starting torque if 3 Ω resistance were added in each rotor phase?
(c) Neglecting stator voltage drop, what would be the induced rotor voltage and the torque at a slip of 0.03?

9.21 Calculate the ratio of transformation of an autotransformer starter for a 25 kW, 400 V, 3-phase induction motor if the starting torque is to be 75% of full-load torque. Assume the slip at full-load to be 3.5% and short-circuit current to be six times full-load current. Ignore magnetizing current of the transformer and of the motor. LO 3

9.22 With reference to the circuit model of Fig. 9.13 (as reduced by Thevenin theorem) show that LO 2

$$\frac{T}{T_{max}} = \frac{1\sqrt{K^2+1}}{1+\dfrac{1}{2}\sqrt{K^2+1}\left(\dfrac{s}{s_{max,T}}+\dfrac{s_{max,T}}{s}\right)}$$

where $K = \dfrac{X_1 + X_2'}{R_1}$

9.23 A 3-phase, 50 Hz, 75 kW induction motor develops its rated power at a rotor slip of 2%. The maximum torque is 250% of rated torque (i.e., the torque developed at rated power). The motor has a K-ratio (defined in Prob. 9.22) of $K = 4.33$. Find LO 2

(a) slip ($s_{max,T}$) at maximum torque,
(b) rotor current referred to stator at maximum torque,
(c) starting torque, and
(d) starting current.

The answers to parts (b), (c) and (d) should be expressed in terms of current and torque at full-load speed.

9.24 A 3-phase induction motor is wound for P poles. If the modulation poles are P_M, obtain the general condition to suppress $P_2 = (P + P_M)$ poles. Under this condition show that the angle between the phase axes for $P_1 = (P - P_M)$ poles is $2r\,(2\pi/3)$, where r = integer non-multiple of 3. If $P = 10$, find P_M and P_1. LO 3

9.25 The two cages of a 3-phase, 50-Hz, 4-pole, delta-connected induction motor have respective standstill leakage impedances of $(2 + j8)$ and $(9 + j2)$ Ω/phase. Estimate the gross-torque developed LO 2

(a) at standstill, the effective rotor voltage being 230 V/phase, and
(b) at 1450 rpm when the effective rotor voltage is 400 V/phase. What is the gross starting torque if a star-delta starter is used? Rotor quantities given are all referred to the stator; the stator impedance is negligible.

9.26 A 3-phase, 50 Hz, 4-pole, 400 V, wound rotor induction motor has a Δ-connected stator winding and Y-connected rotor winding. There are 80% as many rotor conductors as stator conductors. For a speed of 1425 rpm, calculate LO 2

(a) the slip,
(b) the rotor induced emf between the two slip rings, and
(c) the rotor frequency

9.27 A squirrel-cage induction motor is rated 25 kW, 440 V, 3-phase, 50 Hz. On full-load it draws 28.7 kW with line current 50 A and runs at 720 rpm. Calculate LO 2

(a) the slip,

(b) the power factor, and
(c) the efficiency.

9.28 A 3-phase, 400 V, 6-pole, 50 Hz induction motor develops mechanical power of 20 kW at 985 rpm. Calculate: LO 2
(a) the rotor copper loss,
(b) the total input power, and
(c) rotor frequency.
The stator losses are equal to 1800 W. Neglect mechanical loss.

9.29 A 400 V, 5 kW, 50 Hz induction motor runs at 1445 rpm at full-load. The rotational losses are 285 W. If the maximum torque occurs at 900 rpm. Calculate its value. LO 2

9.30 The rotor of a 6-pole, 50 Hz slip ring induction motor has a resistance of 0.25 Ω/phase and runs at 960 rpm. Calculate the external resistance/phase to be added to lower the speed to 800 rpm with load torque reducing to 3/4th of the previous value. LO 2

9.31 For the motor of Prob P9.27 LO 2
(a) calculate the starting torque when rated voltage is applied to the stator.
(b) calculate the slip at which the motor develops maximum torque and the value of this torque.
(c) what is the output torque and power in part (b)?

9.32 A 5 kW, 400 V, 50 Hz, 4-pole induction motor gave the following lest data: LO 2
No-load test: $V_0 = 400$ V, $P_0 = 350$ W, $I_0 = 3.1$ A
Blocked rotor test: $V_{SC} = 52$ V, $P_{SC} = 440$ W, $I_{SC} = 7.6$ A, 24 V,
dc when applied between the two stator terminals causes a current of 7.6 A to flow.
Calculate the motor efficiency at rated voltage at a slip of 4%.

9.33 A 3-phase, 20 kW, 600 V, 50 Hz, 6-pole, *Y*-connected squirrel-cage induction motor has the following parameters/phase referred to the stator: LO 1

$R_1 = 0.937$ W $R_2' = 0.7\ \Omega$
X (equivalent) $= 3.42\ \Omega$
$X_m = 72.9\ \Omega$

The rotational and core losses equal 545 W. For a slip of 3.5% find:
(a) the line current and power factor
(b) the mechanical output and shaft torque
(c) the efficiency.

9.34 A 7.5 kW, 400 V, 4-pole induction motor gave the following test results: LO 2
No-load test $V_0 = 400$ V $P_0 = 330$ W $I_0 = 3.52$ A
Blocked-rotor test $V_{SC} = 110$ V $P_{SC} = 615$ W $I_{SC} = 13$ A
The effective ac resistance between the stator terminals is 2.2 Ω and the full-load slip is 4%.
Determine:
(a) the parameters of the per phase circuit model.
(b) the stator current and its pf when the motor is delivering full-load.
(c) the efficiency in part (b).

9.35 A 30 kW, 440 V squirrel-cage induction motor has a starting torque of 182 Nm and a full-load torque of 135 Nm. The starting current of the motor is 207 A when rated voltage is applied. LO 2
Determine:
(a) the starting torque when the line voltage is reduced to 254 V.
(b) the voltage that must be applied for the motor to develop a starting torque equal to the full-load torque.
(c) the starting current in parts (a) and (b).
(d) the starting voltage to limit the starting current to 40 A, and the corresponding starting torque.

9.36 A 400 V, 4-pole, 7.5 kW, 50 Hz, 3-phase induction motor develops its full-load torque at a slip of 4%. The per phase circuit parameters of the machine are **LO 2**

$R_1 = 1.08\ \Omega \qquad R'_2 = ?$

$X_1 = 1.41\ \Omega \qquad x'_2 = 1.41\ \Omega$

Mechanical, core and stray losses may be neglected.

(a) Find the rotor resistance (as referred to stator)

(b) Find the maximum torque, slip at maximum torque and the corresponding rotor speed.

9.37 A 3-phase, 440 V, 4-pole 50 Hz induction motor has a star-connected stator and rotor. The rotor resistance and standstill reactance/phase are 0.22 Ω and 1.2 Ω respectively; the stator to rotor turn ratio being 1.3. The full-load slip is 4%. Calculate the full-load torque and power developed. Find also the maximum torque and the corresponding speed. **LO 2**

9.38 A 3-phase, 3.3 kV, 6-pole wound rotor induction motor has the following test data: **LO 2**

No-load test	3.3 kV	18.5 A	15.1 kW
Blocked-rotor test	730 V	61 A	3.5 kW

The resistance of the stator winding is 1.6 Ω and the rotational loss is 6.2 kW. Calculate the circuit model parameters (rotational loss not to be accounted in R_i core loss resistance). Assume $X_1/X'_2 = R_1/R'_2$ Calculate

(a) the slip at maximum developed torque

(b) the maximum developed torque and the corresponding shaft torque

(c) the starting torque at half the rated voltage

Note: Do not approximate the circuit model.

9.39 A 6-pole, 50 Hz induction motor has a rotor resistance of 0.25 Ω and a maximum torque of 180 Nm while it runs at 860 rpm. Calculate: **LO 2**

(a) the torque at 4.5% slip

(b) the resistance to be added to the rotor circuit to obtain the maximum torque at starting.

9.40 At rated voltage the blocked rotor current of an induction motor is five times its full-load current and full-load slip is 4%. Estimate its starting torque as a percentage of full-load torque when it is started by means of (a) a star-delta starter, and (b) by an autotransformer with 50% tapping. **LO 3**

9.41 A squirrel-cage induction motor has a full-load slip of 4% and a blocked-rotor current of six times the full-load current. Find the percentage of tapping of the autotransformer starter to give full-load torque on starting and the line current as a percentage of full-load current. **LO 3**

9.42 A 440 V, 22 kW, 50 Hz, 8-pole induction motor has its rotor and stator winding star-connected. The effective stator to rotor turn ratio is 2.5/1. The parameters of its circuit model are **LO 2**

$R_1 = 0.4\ \Omega \qquad R_2 = 0.07\ \Omega$

$X_1 = 1.03\ \Omega \qquad X_2 = 0.18\ \Omega$

$R_i = 127.4\ \Omega \qquad X_m = 25.9\ \Omega$

Turn ratio, $a = 2.4$

(includes rotational loss)

Neglecting any change in mechanical losses due to changes in speed, calculate the added rotor resistance required for the motor to run up to the speed 675 rpm for a constant load torque of 300 Nm. At what speed would the motor run if the added rotor resistance is:

(a) left in the circuit

(b) subsequently shorted out. Also compare the motor efficiency under these two conditions.

9.43 A 40 kW. 400 V, 3-phase, 6-pole. 50 Hz wound rotor induction motor develops a maximum torque of 2.75 times full-load torque at a slip of 0.18 when operating at rated voltage and frequency with slip rings short-circuited. Stator resistance and rotational losses may be ignored. Determine: **LO 2**

(a) the full-load slip.

(b) the full-load rotor copper loss.
(c) the starting torque at half the rated voltage.
The rotor circuit resistance is now doubled by adding an external resistance through the slip rings. Determine:
(d) the developed torque at full-load current.
(e) the slip in part (d).

9.44 Determine the slip at maximum torque and ratio of maximum to full load torque for a 3 phase star connected 6.6 kV, 20 pole, 50 Hz induction motor has rotor resistance of 0.12 Ω and standstill reactance of 1.12 Ω. The motor speed at full load is 292.5 rpm. LO 2

9.45 Compute the full load copper losses per phase and total mechanical power developed for the following specifications: 3 phase, 50 kW induction motor operating at 3% slip. Assume the stator losses are neglected. LO 2

9.46 A 3 phase, 5 hp (3.7 kW), 50 Hz, 4 pole star connected induction motor has the following test results: LO 2

No load	200 V	350 W	5 A
Short circuit	100 V	1700 W	26 A

Draw the circle diagram for full load condition, the line current, power factor and maximum torque in terms of full load torque. Rotor copper loss at standstill is half the total copper loss.

Answers

9.1 750 rpm, 2250 rpm; unity
9.2 16.08 kW, 720 W, 18.81 kW, 85.5%
9.3 1051 N m
9.4 0.09, 2.34
9.5 1.79 kW, 4304 N m, 32 kW, 31.4 kW, 422 N m
9.6 126.6 N m, 0.3, 79.3 N m
9.7 66.9 N m, 0.195, 29.26 A, 330.3 V
9.8 $x_m = 27.55$ Ω, $r_i = 161.3$ Ω, $r'_2 = 0.438$ Ω, $x_1 = x'_2 = 1.055$ Ω; 26.5 A, 0.882 lagging
9.9 (a) 30.5 A, 0.866 lagging (b) 201.3 N m (c) 36.9 A, 33.57 N m
9.10 7.45 kW, 89.33 N m, 0.26
9.11 (a) 21.1 A, 0.9 lagging, 11.91 kW, 78.98 N m, 90.5% (b) 128.5 N m, 0.155
9.12 (a) 10,660 N m, 144.4 A, 0.863 lagging (b) 4527 N m, 459 A
9.13 27.21 kW, 41.8 A, 0.923 lagging
9.14 (a) 1940 N m (b) 746.2 N m (c) 3.125 kW
(d) 0.083 (e) 781.2 N m (f) 0.4
9.15 (a) 40.51 N m (b) 6.11 kW (c) 63.49 N m
(d) 1335 rpm (e) 8.934 kW
9.16 (a) 31.35 A, 077 lagging, 16.28 kW, 265.7 N m (b) 33.22 A, 0.747 lagging, 25 kW (in), – 388 N m
(c) 132.27 A, 0.141 lagging, 5.5 kW (in), 87.49 N m (in the direction of field but opposite to the direction in which the rotor is rotating)
9.17 19.65 N m
9.18 2.4 Ω, 1 pu
9.19 0.525 Ω
9.20 (a) 523.9 V (line) (b) 102 N m (c) 15.7 V (line), 11.6 N m
9.21 0.77
9.23 (a) 0.09 (b) $3.08I'_{2fl}$ (c) $0.458\ T_{fl}$
(d) $22.9I'_{2fl}$
9.24 $P_M = 2$, $P_1 = 8$
9.25 136.7 N m, 61.4 N m, 136.7 N m

9.26 (a) 5% (b) 8.54 V (c) 2.5 Hz
9.27 (a) 0.04 (b) 0.753 lagging (c) 87.1%
9.28 (a) 304.6 W (b) 2.211 kW (c) 0.75 Hz
9.29 192.2 N m
9.30 1.42 Ω
9.31 (a) 24.68 N m (b) 0.656, 102 N m (c) 94.3 N m, 3.396 kW
9.32 85.2%
9.33 (a) 17.73 A, 0.909 lagging (b) 147.4 N m (c) 88.9%
9.34 (a) $R_1 = 11\ \Omega$, $R_2^1 = 1.113\ \Omega$, $X = 4.732\ \Omega$, $X_m = 66.43\ \Omega$, $R_i = 476.2\ \Omega$
(b) 9.53 A, 0.86 lagging (c) 86.9%
9.35 (a) 45 N m (b) 379 V
(c) 119.5 A, 178.3 A (d) 6.8 N m
9.36 (a) 0.71 Ω (b) 0.504, 744 rpm, 116 N m
9.37 126.5 N m, 0.18
9.38 (a) 0.43 (b) 4.546×10^3 N m, 4.442×10^3 N m (c) 590 N m
9.39 (a) 32.9 N m (b) 1.536 Ω
9.40 33%, 25%
9.41 82%
9.42 (a) no change, 57.2% (b) 704 rpm, 92.4%
9.43 (a) 0.034 (b) 2.694 kW
(c) 1009 N m (d) 0.068
9.44 (a) 0.107 (b) 2.2595
9.45 2 kW, 43.2 kW

MULTIPLE CHOICE QUESTIONS

9.1 In an induction motor the stator mmf comprises: LO 1
(a) mmf equal to rotor mmf
(b) mmf required to cancel rotor mmf
(c) vector sum of magnetizing mmf and component to cancel rotor mmf vector
(d) magnetizing mmf only

9.2 Rotor impedance seen from the stator is (usual symbols are used): LO 1
(a) $R_2' + j\,s\,X_2'$ (b) $R_2 + j\,sX_2$ (c) $R_2/s + j\,X_2$ (d) $R_2'/s + j\,X_2'$

9.3 At low slip the torque-slip characteristic is LO 2
(a) $T \propto \dfrac{1}{s^2}$ (b) $T \propto s^2$ (c) $T \propto \dfrac{1}{s}$ (d) $T \propto s$

9.4 For maximum starting torque in an induction motor LO 2
(a) $R_2 = 0.5X_2$ (b) $R_2 = X_2$ (c) $R_2 = 2X_2$ (d) $R_2 = 0$

9.5 The starting current of an induction motor is five times the full load current while the full load slip is 4%. The ratio of starting torque to full load torque is LO 2
(a) 0.6 (b) 0.8 (c) 1.0 (d) 1.2

9.6 In stator impedance starting of a squirrel-cage induction motor, the stator current is reduced by a factor x compared to direct on-line starting. The starting torque is reduced by the factor (compared to direct-on-line starting) LO 3
(a) x (b) x^2 (c) $1/x$ (d) $1/x^2$

9.7 For controlling the speed of an induction motor the frequency of supply is increased by 10%. For magnetizing current to remain the same, the supply voltage must LO 3

(a) be reduced by 10%
(b) remain constant
(c) be increased by 10%
(d) be reduced or increased by 20%

9.8 The speed of an induction motor is controlled by varying supply frequency keeping *V/f* constant: LO 3

(a) Breakdown torque and magnetizing current would both remain constant.
(b) Breakdown torque would remain constant but magnetizing current would increase.
(c) Breakdown torque would decrease but magnetizing current would remain constant.
(d) Breakdown torque and magnetizing current would both decrease.

9.9 The power input to an induction motor is 40 kW when it is running at 5% slip. The stator resistance and core loss are assumed negligible. The torque developed in synchronous watts is LO 2

(a) 42 kW (b) 40 kW (c) 38 kW (d) 2 kW

9.10 A squirrel cage induction motor having a rated slip of 2% on full load has a starting torque of 0.5 full load torque. The starting current is LO 2

(a) equal to full load current
(b) twice full load current
(c) four times full load current
(d) five times full load current

Answers

1. (d) **2.** (a) **3.** (d) **4.** (b) **5.** (b) **6.** (a) **7.** (a) **8.** (a) **9.** (c) **10.** (d)

SAMPLE EXAMPLES SOLVED USING MATLAB

```
%A program to find the various quantities in an induction
%machine for given circuit parameters and nameplate values function general
%**********************************************************
%                               user input data
%********************************************************
%kVa=input('Enter the KVA rating of the machine                ');
r1=input('Enter the stator ckt resistance of the machine       ');
x1=input('Enter the stator ckt reactance of the machine        ');
r2=input('Enter the rotor ckt resistance of the machine        ');
x2=input('Enter the rotor ckt reactance of the machine         ');
Xm=input('Enter the value of xm of the machine                 ');
p=input('Enter the no. of poles                                ');
f=input('Enter the frequency of the supply in Hz               ');
V=input('Enter the value of the supply voltage in volts        ');
nr=input('Enter the speed of the rotor in rpm                  ');
prot=i nput (' Enter the rot. losses in Watts                  ');
%Bmax=input('Enter the maximum value of the magnetic field     ');
xm=Xm;
p=p*2 ;
%*******************************************************
%                                  calculations
```

```
%*********************************************
v=V/sqrt(3);
ns=120*f/p;
ws=2*pi*ns/60;
s=(ns-nr)/ns;
vth=xm*v/(x1+xm);
rth=r1*xm/(x1+xm);
xth=x1*xm/(x1+xm);
sq=sqrt(rth^2+(xth+x2)^2);
z2=r2./s+j*x2;
zf=j*(z2*xm)/(z2+j*xm);
b=real (zf);
rf=real  (zf);
xf=imag(zf) ;
zi=(r1+rf)+j*(x1+xf);
i1=v/zi;
pf=cos(atan(imag(zi)/real  (zi)));
a=abs(i1);
pin=3*v*a*pf;
scl=3*a*a*r1;
pg=3*a*a*b;
rcl=s*pg;
dmp=(1-s)*pg;
psh=dmp-prot;
hp=psh/746;
tsh=psh.*60/(2*pi*nr);
eff=psh*100/pin;
T=(3*vth*vth*r2./s)/(ws*((rth+r2./s).*(rth+r2/s) + ...
  (xth+x2)*(xth+x2)));
Tst=(3*vth*vth*r2)/(ws*((rth+r2)*(rth+r2)  + (xth+x2)  *...
  (xth+x2)));
Tmax=(3*vth*vth)/((2*ws)*(sq+rth));
smax=r2/sq;
for i=1:15
     ans=menu('Which value do u want?', 'Slip',...
        'Power Factor', 'Input Power  (Watts)',...
        'Stator Cu Loss  (Watts)', 'Air Gap Power(Watts)',...
        'Rotor Cu Loss  (Watts)', 'Dev. Mech Power  (Watts),...
        'Shaft Power(Watts)','Shaft Torque(N-m)',...
        'Efficiency(%)','EM Torque  (N-m)',...
        'Starting Torque(N-m)', 'Max. Torque  (N-m)',...
        'Slip at Max Torque'...
        ,'Quit')
```

```
        if ans==1
          s
        elseif      ans==2
          pf
        elseif      ans==3
          pin
        elseif      ans==4
          scl
        elseif      ans==5
          pg
        elseif      ans==6
          rcl
        elseif      ans==7
          dmp
        elseif      ans==8
          psh
        elseif      ans==9
          tsh
        elseif      ans==10
          eff
        elseif      ans==11
          T
        elseif      ans==12
          Tst
        elseif      ans==13
          Tmax
        elseif      ans==14
          smax
        elseif      ans==15
          break
        elseif      ans==20
        end
      end
%A program to plot the torque speed characteristics
%of an induction machine for different values of rotor
%resistances
function torque speed
s=l inespace(1,.001,100);
ch=menu('WOULD YOU LIKE TO HAVE', 'DEFAULT VALUES',...
    'USER INPUT VALUES');
%***********************************************
%                        GIVEN DATA
%***********************************************
```

```
if ch==1
     Zl=.3+.6*j;
     z2=.25+.6*j;
     Zm=35*j;
     v=400;
     f=50;
     p=4;
else
     Zl=input('ENTER THE VALUE OF Zl            ');
     z2=input('ENTER THE VALUE OF Z2            ');
     Zm=input('ENTER THE VALUE OF Zm            ');
     v=input('ENTER THE VALUE OF VOLTAGE        ');
     f=input('ENTER THE VALUE OF FREQUENCY      ');
     p=input('ENTER THE NUMBER OF POLES         ');
end
Z2=real (z2)./s+j*imag(z2);
V=v/sqrt(3);
%********************************************
%                            TORQUE CALCULATIONS
%********************************************
rl=real(Zl);
r2=real (Z2);
r22=real (Z2);
xl=imag(Zl) ;
x2=imag(Z2);
xm=imag(Zm);
ns=120*f/p;
ws=2*pi*f;
vth=xm*V/(xl+xm);
rth=rl*xm/(xl+xm);
xth=xl*xm/(xl+xm);
test=l;
figure('Position',[l 29 800 534])
k=l;
g=l;
while test==1
for i=1:length(s)
    Te(i) =3*vth^2*r2(i)/(ws*s(i)*((rth+r2(i)/s(i))^2+. ...
       (xth+x2(i)) ^2));
end
Ns=linspaee(0,ns,100);
if g==1
      plot (Ns, Te, 'k-')
```

```
      g=g+1;
elseif g==2
      plot(Ns,Te, 'm-')
      g=g+1;
else
      plot(Ns,Te, 'c-')
end
hold on
axis([0 max (Ns) -inf inf]);
xlabel('nsync')
ylabel('Tind')
xlabel('\bfROTOR SPEED (rpm)')
ylabel('\bfTORQUE PRODUCED (N-m)');
title ('\bfTORQUE SPEED CHARACTERISTICS OF AN INDUCTION MOTOR');
[tmax nn]=max(Te);
Tmax=num2str(tmax);
Ntmax=num2str(Ns(nn));
%text(Ns(nn),tmax/2.['\bfTmax=' ,Tmax], 'color','k')
%text(Ns(nn),tmax/2.2.['\bfAt Speed=',Ntmax],'color','k')
choice=menu('WOULD U LIKE TO ADD ROTOR RESISTANCE',...
           'YES,' 'NO');
if choice== 1
      Rr=input('ENTER VALUE OF ADDITIONAL ROTOR RESISIANCE');
      r2=r22+Rr;
else
      test=0;
   end
end
% This program calculates induction motor performance
% (nm,I,pf,Pin,Pscl,Pag,Pdev,Prcl,Psh,Tsh,Tin,
% efficiency) based on equivalent circuit.
% Reads equivalet circuit parameters from User input data
%
%%%%%%%%%%%%%%%%%%%%%%%%%%%%%%%%%%%%%%%%%%%%%%%%%%%%%%%%%%%%%%%%%%%%%%%

%%%%%%%%%%%%%%%%%%%%%%%%%%%%%%%%%%%%%%%%%%%%%%%%%%%%%%%%%%%%%%%%%%%%%%%
%                              User input data
%%%%%%%%%%%%%%%%%%%%%%%%%%%%%%%%%%%%%%%%%%%%%%%%%%%%%%%%%%%%%%%%%%%%%%%

clear;clc;
%kva=input('Enter the kva rating of the m/c;       ');
r1=input('Enter the stator ckt resistance:');
x1=input('Enter the stator ckt reactance: ');
```

```
r2=input('Enter the rotor ckt resistance: ');
x2=input('Enter the rotor ckt reactance:   ');
xm=input('Enter the magnetising reactance:');
p= input('Enter the no. of pole pairs:  ');
f=i nput('Enter the frequency in Hz:  ');
v=input('Enter the voltage rating of the m/c:  ');
Prot=input('Enter the rotational losses:  ');
s=input('Enter the slip:  ');
Pcore=input('Enter the core losses:  ');
%%%%%%%%%%%%%%%%%%%%%%%%%%%%%%%%%%%%%%%%%%%%%%%%%%%%%%%%%%%%%%%%%%%%%%%%%%
%                    Calculation for different parameters
%
%%%%%%%%%%%%%%%%%%%%%%%%%%%%%%%%%%%%%%%%%%%%%%%%%%%%%%%%%%%%%%%%%%%%%%%%%%

P=p*2;
ns=120*f/P;
wS=2*pi*ns/60;
nm= (1-s) *ns;
zf=((r2/s)+j*x2)*j*xm)/((r2/s)+j*x2+j*xm);
zb=(((r2/(1.95))+(j*x2))*j*xm)/((r2/(1.95))+(j*x2)+(j*xm));
rf=real (zf);
xf=imag(zf);
rb=real(zb);
I=v/(rl+j*xl+.5*zf+.5*zb);
pf=cos(angle(I));
Pin=v*I*pf;
a=abs(I) ;
Pscl=a*a*rl;
Pagf=a*a*.5*rf;
Pagb=a*a*.5*rb;
Pag=Pagf-Pagb;
Pdev=(1-s)*Pag;
Prclf=s*Pagf;
Prclb=(2-s)*Pagb;
Prcl=Prclf+Prclb;
Psh=Pdev-Prot-Pcore;
Tind=Pag/ws;
Tsh=Psh/(2*pi*nm/60);
Eff=Psh/Pin*100;
%%%%%%%%%%%%%%%%%%%%%%%%%%%%%%%%%%%%%%%%%%%%%%%%%%%%%%%%%%%%%%%%%%%%%%%%%%
%          Select the output parameters to display on command window
%
%%%%%%%%%%%%%%%%%%%%%%%%%%%%%%%%%%%%%%%%%%%%%%%%%%%%%%%%%%%%%%%%%%%%%%%%%%
```

```
for i=1:15
    ans=menu('Which value do you want?','slip'...
    ,'synchronous speed (rpm)',' speed (rpm)',...
    'stator current in amperes'. 'power factor',...
    'input power (watts) ', 'stator cu loss(watts)',...
    'air gap power (watts)',...
    'developed mech. power (watts)',...
    'rotor cu losses (watts)','shaft power',...
    'torque induced (N-m)',' shaft torque',...
    'efficiency','quit');
  if ans==1;
    s
  elseif ans==2;
    ns
  elseif ans==3;
    nm
  elseif ans==4;
    I=abs(I)
  elseif ans==5;
    pf
  elseif ans==6;
    Pin=abs(Pin)
  elseif ans==7;
    Pscl=abs(Pscl)
  elseif ans==8;
    Pag=abs(Pag)
  elseif ans==9;
    Pdev
  elseif ans==10;
    Prcl
  elseif ans== 11;
    Psh
  elseif ans==12;
    Tind
  elseif ans==13;
    Tsh
  elseif ans==14;
    Eff=abs (Eft)
  elseif ans==15
    break
  end
end
```

CHAPTER 10

Fractional Kilowatt Motors

LEARNING OBJECTIVES

After studying this chapter, students will be able to

LO1 Examine the production of rotating magnetic field and performance of various types of single-phase induction motors

LO2 Describe and analyze the performance of reluctance and hysteresis motor

LO3 Discuss the construction and working of two-phase servomotor and two-phase tachometer

LO4 Explain the performance of various types of stepper motors and ac series motors

10.1 | INTRODUCTION

So far three-phase ac motors which are used for high-power rating applications have been discussed. For reasons of economy, most homes, offices and also rural areas are supplied with single-phase ac, as the power requirements of individual load items are rather small. This has led to the availability of a wide variety of small-size motors of fractional kilowatt ratings. These motors are employed in fans, refrigerators, mixers, vacuum cleaners; washing machines, other kitchen equipment, tools, small farming appliances, etc.

Though these motors are simpler in construction as compared to their three-phase counterparts, their analysis happens to be more complex and requires certain concepts which have not been developed so far. Also the design of such motors are carried out by trial and error till the desired prototype is achieved. Because of the vast numbers in which these motors are produced, even a fractional efficiency increase or a marginal cost saving is extremely important. Nowadays, as in other fields, computers are employed for more accurate and optimum paper designs.

In this chapter these motors will first be discussed qualitatively and semi-quantitatively and then the more detailed quantitative treatment will be described, which is used for accurate design/ analysis purposes

Note: As in Chapter 9, phasor superbar will be used only for phasor equations.

Industry Relevance

- Induction machine is the most used machine in industrial areas, domestic section, educational institutions etc.
- Three phase induction motor is widely used in industrial drives while single phase induction motors are preferred for domestic applications.
- Motors of refrigerators, Water pump motors, Grinders, Mixers and Cooler motors are common applications of induction motor.

Some other applications are:

- Pumps
- Compressors
- Small Fans
- Toys
- High speed vacuum cleaners
- Electric Shavers
- Drilling Machines
- Lathes
- Blowers
- Printing Machines

10.2 | SINGLE-PHASE INDUCTION MOTORS

LO 1 Examine the production of rotating magnetic field and performance of various types of single-phase induction motors

A single-phase induction motor comprises a single-phase distributed winding on the stator and normal squirrel-cage rotor as shown schematically in Fig. 10.1 wherein for convenience the stator winding is shown in concentrated form.

Important! There are two important methods of analyzing this motor, viz. *cross-field* theory and *rotating field* theory.

As the latter is more akin to the three-phase induction machine theory advanced earlier, it will be adopted here.

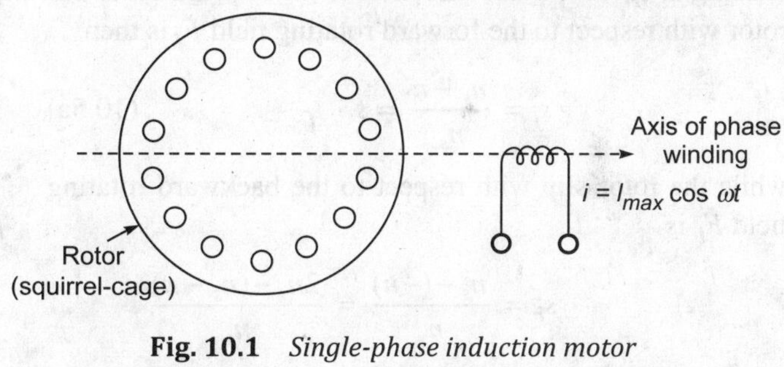

Fig. 10.1 *Single-phase induction motor*

Pulsating Field as Two Rotating Fields

Figure 10.1 gives the schematic diagram of a single-phase induction motor with one stator winding and a squirrel-cage rotor. The winding is distributed in space so that the space fundamental of mmf is the most dominant component of the actual mmf distribution. The space harmonics of mmf, as in the case of a three-phase induction motor, would then be ignored. When the winding carries a sinusoidal current, it produces a sinusoidally space-distributed mmf whose peak value pulsates with time. As seen from the axis of the winding, the mmf at any angle θ is

$$F = F_{peak} \cos \theta \tag{10.1}$$

where θ is the angle measured from the winding axis. Now

$$F_{peak} = F_{max} \cos \omega t \tag{10.2}$$

so that the mmf has both space and time distribution expressed as

$$F = F_{max} \cos \theta \cos \omega t \tag{10.3}$$

This equation can be trigonometrically manipulated into the form

$$F = \frac{1}{2} F_{\max} \cos(\theta - \omega t) + \frac{1}{2} F_{\max} \cos(\theta + \omega t) \tag{10.4}$$

Equation (10.4) tells us that a pulsating single-phase field can be considered as superposition of two rotating fields rotating at synchronous speed ($\omega = 2\pi f$ elect. rad/s) in opposite directions:

$\frac{1}{2} F_{\max} \cos(\theta - \omega t)$; the forward rotating field, F_f

$\frac{1}{2} F_{\max} \cos(\theta + \omega t)$; the backward rotating field, F_b

!Attention Both these fields have an amplitude equal to (1/2) $F_{\max}$ where $F_{\max}$ is the maximum value of the pulsating mmf along the axis of the winding. The splitting of a single pulsating field into two rotating fields rotating in opposite directions is illustrated in Fig. 10.2. This figure shows the location of the rotating fields at the time instant when the mmf along the winding axis is + $F_{\max}$.

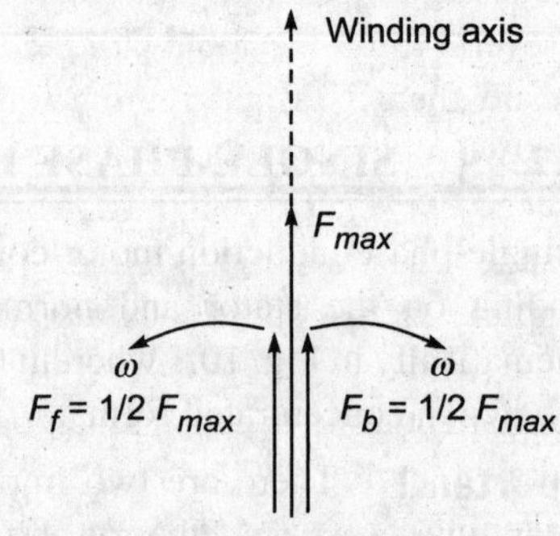

Fig. 10.2 *Two rotating field equivalent of a pulsating field*

Rotor Slip with respect to Two Rotating Fields

For the single-winding case illustrated in Fig. 10.1, Fig. 10.3 shows the forward and backward rotating fields along with the rotor which is rotating at speed n in the direction of the forward field. The slip of the rotor with respect to the forward rotating field F_f is then

$$s_f = \frac{n_s - n}{n_s} = s \tag{10.5a}$$

while the rotor slip with respect to the backward rotating field F_b is

$$s_b = \frac{n_s - (-n)}{n_s} = \frac{2n_s - (n_s - n)}{n_s}$$

$$= (2 - s) \tag{10.5b}$$

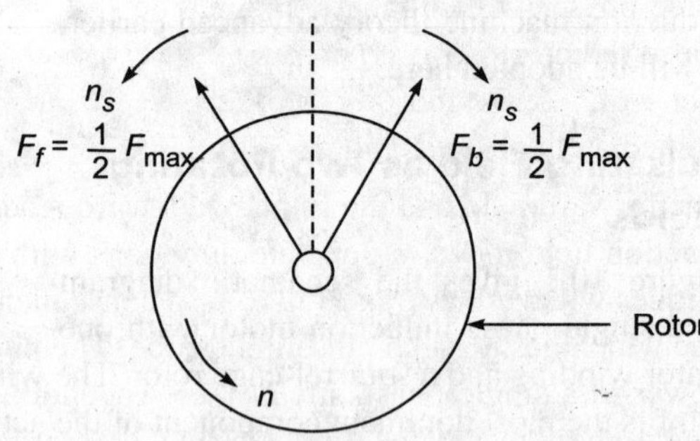

Fig. 10.3

Thus the rotor slips with respect to the two rotating fields are different and are given by Eqs (10.5a) and (10.5b).

Torque-Speed Characteristic

Qualitative Treatment Under stationary rotor condition ($n = 0$, i.e. $s = 1$), the two rotating fields slip past the rotor at the same slip, $s = 1$, (see Eqs (10.5a) and (10.5b)) inducing equal currents in the squirrel-cage rotor.

Important! The two rotating fields have the same strength and produce equal and opposite torques resulting in net starting torque of zero value. The single-winding single-phase induction motor is thus *nonself-starting*.

Further, the two rotating fields induce a resultant emf in the stator which balances the applied voltage assuming low leakage impedance of the stator winding.

If, however, the rotor is made to run at speed n in the direction of the forward field, the two slips are now s and $(2 - s)$. For normal operation $(2 - s) >> s$ and as a consequence the backward field induced rotor currents are much larger than at standstill and have a lower power factor. The corresponding opposing rotor mmf, in presence of the stator impedance, causes the backward field to be greatly reduced in strength. On the other hand, the low-slip forward rotating field induces smaller currents of a higher power factor in the rotor than at standstill. This leads to great enhancement in the forward flux wave. This reduction in the backward field and strengthening of the forward field is slip-dependent and the difference increases as slip s (with respect to the forward field) reduces or the rotor speed in the forward direction becomes close to the synchronous speed. Infact, at near about the synchronous speed, the forward field strength may be several times the backward field. As a result there is a net running torque. The two fields together must always induce the stator winding emf to balance the applied voltage. The complete torque-speed characteristics as the sum of the two (forward and backward) torque-speed characteristics is drawn in Fig. 10.4. The result of weakening of one field and simultaneous strengthening of the other leads to a toque-speed characteristic like that of a three-phase induction motor in the speed region close to synchronous. The fact of zero starting torque is immediately observed here.

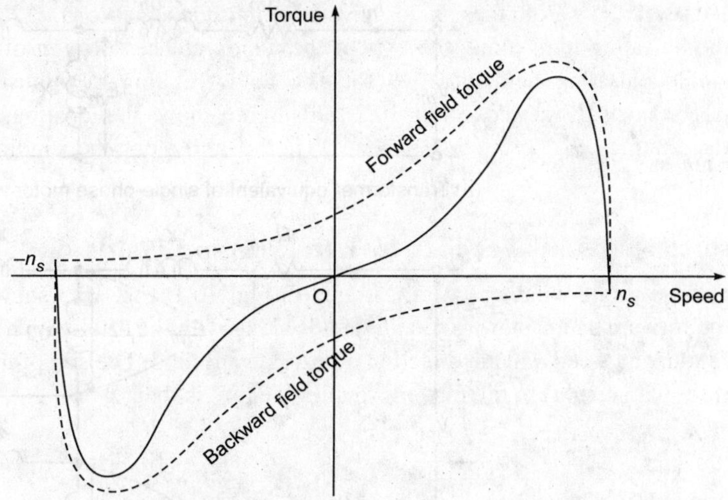

Fig. 10.4 *Torque-speed characteristic of single-winding single-phase motor*

The forward field and the rotor's backward reaction field and also the backward field and the rotor's forward reaction field move in opposite directions with relative speeds of $2n_s$ producing second harmonic pulsating torques with zero average value. As a consequence a single-phase motor is a noisier motor than a three-phase one which has no such pulsating torque. The pulsating torque in fact is a direct consequence of the pulsating power in a single-phase circuit (see Appendix I, Eq. (I-28). In fact in the torque-speed characteristic of a single-phase motor, the torque ordinate represents the average torque.

Semi-quantitative Analysis

To develop the circuit model of a single-winding (referred to as the *main winding*), single-phase motor on semi-quantitative basis, heuristic arguments will be used. The motor with a stationary rotor merely acts like a transformer with a circuit model as shown in Fig. 10.5(a). the core-loss branch having been ignored. The suffix m in the stator refers to the main winding and E_m is the stator-induced emf set up by the alternating field.

The motor is now viewed from the point of view of the rotating field theory. The resultant induced emf is composed of two equal components induced by the two oppositely rotating fields of the same strength, i.e.

$$\bar{E}_m = \bar{E}_{mf} + \bar{E}_{mb};\ \bar{E}_{mf} = E_{mb} \tag{10.6}$$

The magnetizing and rotor impedances are divided into two equal halves connected in series as shown in Fig. 10.5(b); the motor behaves like two series connected motors one corresponding to each rotating field. The circuits of the two component motors are identical under stationary condition as the rotor has the same slip with respect to each rotating field.

When the rotor is running at speed *n* with respect to the forward field, the slip is *s* with respect to it and (2 – *s*) with respect to the backward rotating field so that the circuit model now modifies as in Fig. 10.5(c). It is easily seen from this figure that $Z_f/2 \gg Z_b/2$ and so $E_{mf} \gg E_{mb}$, i.e., the forward field motor effect predominates, creating a running torque.

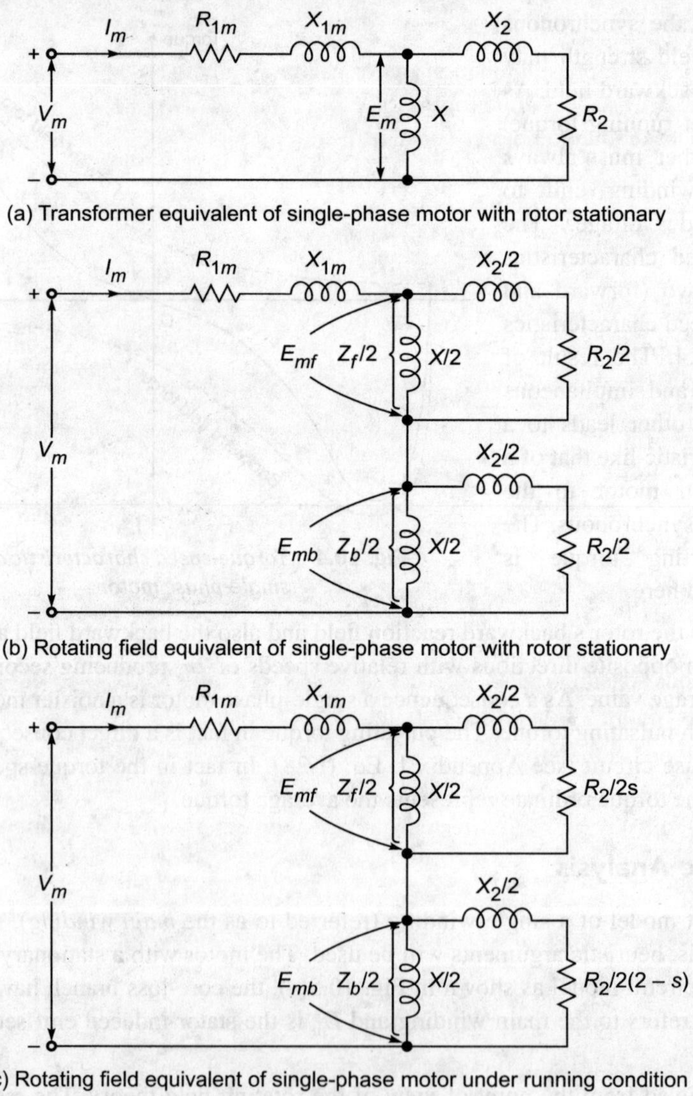

(a) Transformer equivalent of single-phase motor with rotor stationary

(b) Rotating field equivalent of single-phase motor with rotor stationary

(c) Rotating field equivalent of single-phase motor under running condition

Fig. 10.5

Important! Practical necessity dictates that the two rotating fields are made to have unequal strength under stationary conditions thereby making the motor self-starting. This requires one more winding on the motor called *auxiliary winding* which is in *space quadrature* with the main winding and comprises smaller number of turns of thinner wire.

This winding may be cut out of circuit once the motor has started except in case of the *capacitor-run* motor where it may be left connected serving the purpose of improving the overall power factor.

The detailed quantitative analysis of a single phase motor (one or two stator windings) will follow in Sec. 10.3 for the interested reader.

Performance Analysis

The performance of a single-phase induction motor can be obtained by analysis of the circuit model of the motor given in Fig. 10.5(c), as was done for the case of a three-phase induction motor. The results are similar to those for a three-phase induction motor because the circuit model is essentially the same.

The air-gap powers for the forward and backward fields are given by

$$P_{gf} = \frac{1}{2} I_m^2 R_f = \text{Air-gap power for forward field} \tag{10.7a}$$

$$P_{gb} = \frac{1}{2} I_m^2 R_b = \text{Air-gap power for backward field} \tag{10.7b}$$

where I_m is the main winding current and R_f *and* R_b are the real parts of the complex number impedances $\bar{Z}_f$ and $\bar{Z}_b$ respectively in Fig. 10.5(c).

The torques produced by the two fields can be expressed as

$$T_f = \frac{1}{\omega_s} P_{gf} \tag{10.8a}$$

$$T_b = \frac{1}{\omega_s} P_{gb} \tag{10.8b}$$

where ω_s = synchronous speed in rad/s.

Since the two fields are rotating in opposite directions the torque produced by the two oppose each other. The resultant torque developed is therefore

$$T = T_f - T_b$$

or

$$T = \frac{1}{\omega_s}(P_{gf} - P_{gb}) = \frac{I_m^2}{2\omega_s}(R_f - R_b) \tag{10.9}$$

The rotor copper losses are in general equal to slip times the air-gap power. Thus

Rotor copper-loss corresponding to forward field = sP_{gf}

Rotor copper-loss corresponding to backward field = $(2 - s)\,P_{gb}$

$$\text{Total rotor copper-loss} = sP_{gf} + (2 - s)\,P_{gb} \tag{10.10}$$

The electrical power converted to gross mechanical form is

$$P_m = (1 - s)\omega_s T$$

or

$$P_m = (1 - s)(P_{gf} - P_{gb}) \tag{10.11}$$

Equation (10.11) can also be written as

$$P_m = (1 - s) P_{gf} + [1 - (2 - s)]P_{gb} \tag{10.12}$$

This implies that the electrical power input to the motor neglecting the stator copper-loss is

$$P_{\text{elect}} = P_{gf} + P_{gb} \tag{10.13}$$

Example 10.1 *A 220-V, 6-pole, 50-Hz, single-winding single-phase induction motor has the following equivalent circuit parameters as referred to the stator.*

$$R_{1m} = 3.0\ \Omega, \qquad X_{1m} = 5.0\ \Omega$$
$$R_2 = 1.5\ \Omega, \qquad X_2 = 2.0\ \Omega$$

Neglect the magnetizing current. When the motor runs at 97% of the synchronous speed, compute the following:

(a) The ratio E_{mf}/E_{mb}
(b) The ratio V_f/V_b
(c) The ratio T_f/T_b
(d) The gross total torque.
(e) The ratios T_f/(Total torque) and T_b/(Total torque)

LO 1

Solution

$$\text{Slip} = s = 1 - 0.97 = 0.03$$

(a) From Fig. 10.5(c),

$$\frac{E_{mf}}{E_{mb}} = \frac{Z_f}{Z_b} = \frac{\left| jX \parallel \left(\dfrac{R_2}{s} + jX_2\right)\right|}{\left| jX \parallel \left(\dfrac{R_2}{2-s} + jX_2\right)\right|}$$

Since magnetizing current is neglected, $X = \infty$.

$$\therefore \quad \frac{E_{mf}}{E_{mb}} = \frac{|R_2/s + jX_2|}{|R_2(2-s) + jX_2|}$$

$$= \frac{|1.5/0.03 + j2|}{|1.5/(2 - 0.03) + j2|} = 23.38$$

(b) V_f and V_b are components of stator voltage V_m, i.e

$$\bar{V}_m = \bar{V}_f + \bar{V}_b$$

These components are defined by redrawing the circuit model of Fig. 10.5(c) in the symmetrical form of Fig. 10.6. For the purpose of this problem $X = \infty$, therefore

$$\text{Impedance offered to } V_f \text{ component} = \frac{1}{2}\left[\left(3 + \frac{1.5}{0.03}\right) + j(5+2)\right] = \frac{1}{2}(53 + j7)$$

Note: Difficulty Level ⟶ ○○● — Level 1 and Level 2 Category; ○●● — Level 3 and Level 4 Category; ●●● — Level 5 and Level 6 Category

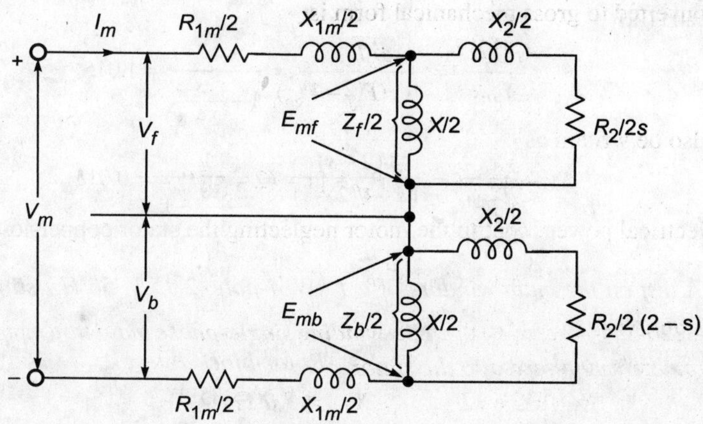

Fig. 10.6

$$\text{Impedance offered to } V_b \text{ component} = \frac{1}{2}\left[\left(3 + \frac{1.5}{1.97}\right) + j(5+2)\right] = \frac{1}{2}(3.76 + j\,7)$$

Hence
$$\frac{V_f}{V_b} = \frac{|\,53 + j7\,|}{|\,3.76 + j7\,|} = 6.73$$

(c) From Eqs (10.8a) and (10.8b)

$$\frac{T_f}{T_b} = \frac{P_{gf}}{P_{gb}} = \frac{\frac{1}{2} I_m^2 R_2/s}{\frac{1}{2} I_m^2 R_2/(2-s)} = \frac{2-s}{s} = \frac{2-0.03}{0.03} = 65.7$$

(d) Total impedance as seen from stator terminal is

$$\bar{Z}\,(\text{Total}) = \frac{1}{2}[(53 + j7) + (3.76 + j7)] = 28.38 + j\,7 = 29.2\ \angle 13.9°$$

$$I_m = \frac{220}{29.2} = 753 \text{ A}$$

$$n_s = \frac{120 \times 50}{6} = 1000 \text{ rpm}$$

$$\omega_s = \frac{2\pi \times 1000}{60} = 104.72 \text{ rad/s}$$

$$T_f = \frac{1}{\omega_s} I_m^2 \frac{R_2}{2s} \qquad \text{(i)}$$

$$T_b = \frac{1}{\omega_s} I_m^2 \frac{R_2}{2(2-s)} \qquad \text{(ii)}$$

$$T_{\text{total}} = T_f - T_b$$

$$T_{\text{total}} = \frac{I_m^2 R_2}{2\omega_s}\left(\frac{1}{s} - \frac{1}{2-s}\right) \qquad \text{(iii)}$$

$$= \frac{(7.53)^2 \times 1.5}{2 \times 104.72}\left(\frac{1}{0.03} - \frac{1}{1.97}\right) = 13.31\text{Nm}$$

(e) From Eqs (i), (ii) and (iii)

$$\frac{T_f}{T_{\text{total}}} = \frac{1/s}{1s - 1/(2-s)} = \frac{1}{1 - \dfrac{s}{2-s}} = 1.015$$

$$\frac{T_b}{T_{\text{total}}} = \frac{1/(2-s)}{1/s - 1/(2-s)} = \frac{1}{\dfrac{2-s}{s} - 1} = 0.015$$

Example 10.2 *A test on the main winding of a 1 kW, 4-pole. 2 15 V, 50 Hz, single-phase induction motor gave the following results:*

No-load test	*Rotor-blocked test*
$V_0 = 215$ V	$V_{SC} = 85$ A
$I_0 = 3.9$ A	$I_{SC} = 9.80$ A
$P_0 = 185$ W	$P_{SC} = 390$ W
$R_1 = 1.6\ \Omega$	

Given:

(a) Calculate the parameters of the circuit model assuming that the magnetizing reactance hangs at the input terminals of the model.

(b) Determine the line current power factor, shaft torque and efficiency of the motor at a speed of 1440 rpm.

LO 1

Solution

(a) Parameters of the circuit model are calculated using both no-load as well as rotor-blocked tests.

(i) No-load test: Assuming the slip to be zero, the circuit model on no-load is drawn in Fig. 10.7 with magnetizing reactance at input terminals

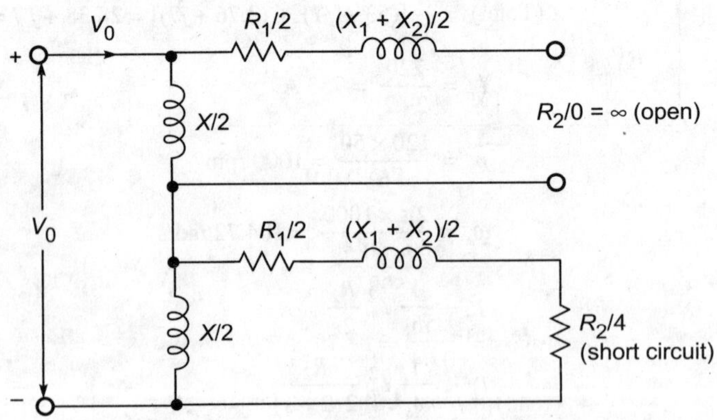

Fig. 10.7

Since the backward circuit is short-circuited for practical purposes, as X being magnetizing reactance is much larger

$$\frac{X}{2} = \frac{215}{3.9} = 55.1\ \Omega$$

Rotational loss $\quad P_0 = 185$ W

(ii) Rotor-blocked test ($s = 1$): The circuit model on rotor-blocked test is shown in Fig. 10.8.

$$390 = 85 \times 9.8 \times \cos \phi_{SC}$$

or $\quad \phi_{SC} = 62°$ lagging

With reference to Fig. 10.8

$$\bar{I}_e = \frac{V_{SC}}{jX} = \frac{85}{j2 \times 55.1} = -j\,0.77 \text{ A}$$

$$\bar{I}'_m = \bar{I}_{SC} - \bar{I}_e = 9.8\angle -62° - (-j0.77)$$
$$= 4.6 - j7.88 = 9.13 \angle -59.7°$$

$$\bar{Z}'_f = (R_1 + R_2) + j(X_1 + X_2)$$

$$\frac{V_{SC}}{I'_m} = \frac{85}{9.13 \angle -59.7°} = 9.31\angle 59.7°$$
$$= 4.7 + j\,8.04$$

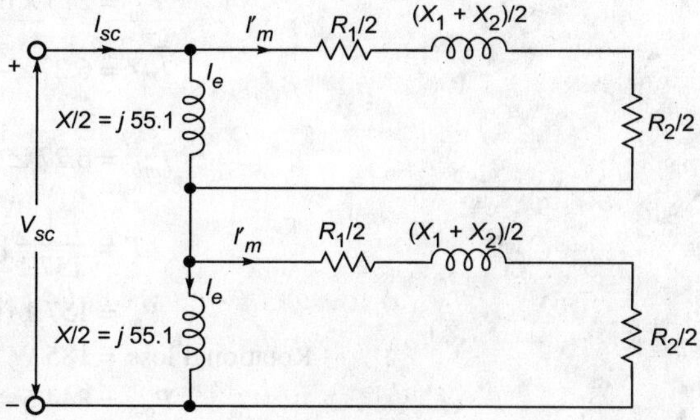

Fig. 10.8

$$R_1 + R_2 = 4.7\ \Omega$$
$$R_1 = 1.6\ \Omega \text{ (given)}$$
$$R_2 = 3.1\ \Omega$$
$$X_1 + X_2 = 8.04\ \Omega$$

The circuit model with parameter values is drawn in Fig. 10.9.

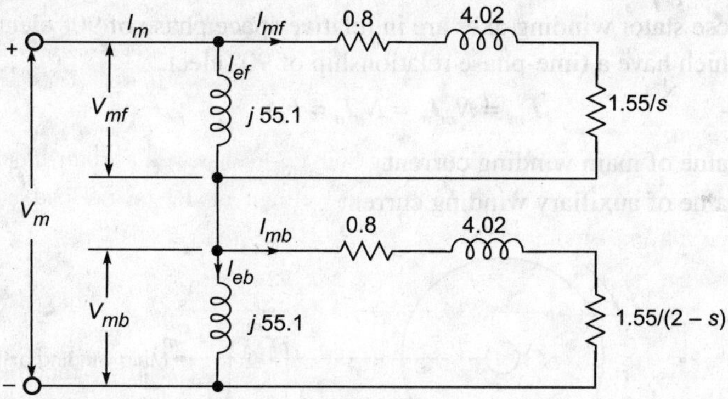

Fig. 10.9

(b)

$$s = \frac{1500 - 1440}{1500} = 0.04$$

$$1.55/s = \frac{1.55}{0.04} = 38.75;\ 1.55/(2 - s) = \frac{1.55}{1.96} = 0.79$$

$$\bar{Z}_f(\text{total}) = j55.1 \parallel (0.8 + 38.75 + j4.02)$$
$$= j55.1 \parallel (39.55 + j4.02) = 30.8 \angle 39.6° = 23.73 + j19.63$$

$$\bar{Z}_b(\text{total}) = j55.1 \parallel (0.8 + 0.79 + j4.02)$$
$$= j55.1 \parallel (1.59 + j4.02) = 4.02 \angle 70° = 1.37 + j3.78$$

$$\bar{Z}(\text{total}) = (23.73 + j19.63) + (1.37 + j3.78) = 25.1 + j23.41 = 34.3 \angle 43°$$

$$\bar{I}_m = \frac{215}{34.3 \angle 43°} = 6.27 \angle -43°$$

$$\bar{I}_L = I_m = 6.27 \text{ A}; pf = \cos 43° = 0.731$$
$$P_{in} = 215 \times 6.27 \times 0.731 = 985.4 \text{ W}$$
$$\bar{I}_{mf} = 6.27 \angle -43° \times \frac{j55.1}{39.55 + j59.12} = 4.86 \angle -9.2°$$
$$\bar{I}_{mb} = 6.27 \angle -43° \times \frac{j55.1}{1.59 + j59.12} = 5.84 \angle -41.4°$$
$$T = \frac{1}{157.1}[(4.84)^2 \times 38.75 - (5.84)^2 \times 0.79] = 5.6 \text{ Nm}$$
$$P_m = 157.1\,(1 - 0.04) \times 5.6 = 844.6 \text{ W}$$
$$\text{Rotational loss} = 185 \text{ W}$$
$$P_{out} = 844.6 - 185 = 659.6 \text{ W}$$
$$\eta = \frac{659.6}{985.5} = 66.9\%$$
$$T(\text{shaft}) = 659.6/157.1 = 4.12 \text{ Nm}$$

Single-phase 2-winding Motor

To understand the field phenomenon that contributes towards the generation of starting torque in a single-phase motor, it greatly helps to first study a balanced two-phase motor. Figure 10.10 shows a 2-winding squirrel-cage motor whose stator winding axes are in relative space phase of 90° elect. The two windings are excited with currents which have a time-phase relationship of 90° elect.

$$F_m = N_m I_m = N_a I_a = F_a \tag{10.14}$$

where I_m = rms value of main winding current

I_a = rms value of auxiliary winding current

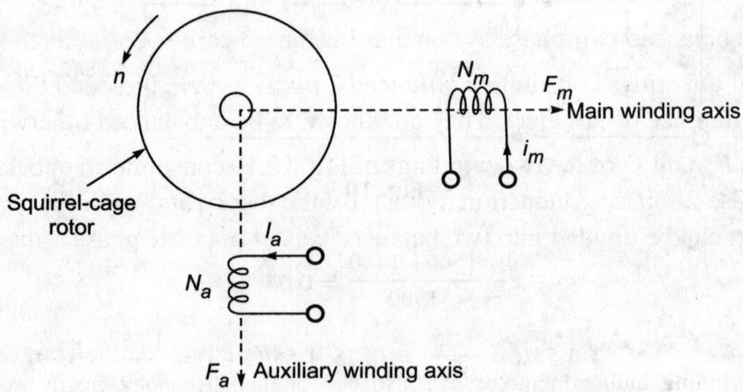

Fig. 10.10 *Two-winding single-phase motor*

In terms of the phasor relationship

$$\bar{F}_m = \bar{F}\,; \quad \bar{F}_a = -j\bar{F} \text{ (lagging, say)} \tag{10.15}$$

These windings therefore create two pulsating fields which are directed along their respective axes at 90° elect. to each other in space. The maximum value of peak field AT for each field is $F_{max} = \sqrt{2}\,F$. Figure 10.11(a) shows the time phase relationship of the pulsating main and auxiliary fields while Fig. 10.11(b) shows their space phase relationship along with their rotating components at the time instant when the main field along the winding axis has the value F_{max} while the auxiliary field has zero value in accordance with the phasor diagram of Fig. 10.11(a). Of the four rotating component fields, the two rotating in counter-clock wise direction cancel out as they are in direct opposition; while the other two rotating in the clockwise direction being coincident add up to a single rotating field of magnitude F_{max} as shown in Fig. 10.11(c).

Important! From the above it is concluded that two pulsating fields, of equal strength out of time phase by 90° elect, and oriented along axes at 90° elect, in space, produce a single rotating field which rotates in the direction of leading phase to lagging phase axis.

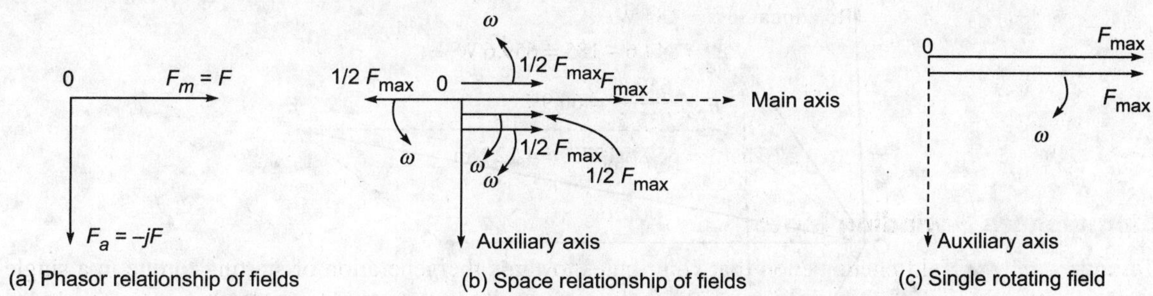

Fig. 10.11 *Production of a single rotating field by two pulsating fields*

It now easily follows that if the phase sequence of the fields is reversed, i.e. $\bar{F}_m = \bar{F}$ and $\bar{F}_a = j\bar{F}$ then the direction of rotation of the resultant field would also be reversed. Corresponding conclusions for a three-phase induction motor were proved trigonometrically* in Sec. 5.5.

Unbalanced Pulsating Fields

While the concept of a balanced two-phase system has been used earlier, let it be redefined once again.

A set of two sinusoidal quantities constitute a *balanced 2 phase system* provided they have equal amplitude and a relative phase difference of 90° elect. They are known to be unbalanced otherwise.

Let the pulsating fields F_m and F_a of the two windings of Fig. 10.10 constitute an unbalanced set (this is in fact the case in a single-phase 2-winding induction motor). By the theory of *two-phase symmetrical components,* the unbalanced field set can be divided into two balanced sets of opposite phase sequence. Thus,

$$\bar{F}_m = \bar{F}_f + \bar{F}_b \tag{10.16a}$$

*These results for a 2-winding induction motor can also be obtained trigonometrically as follows: The equations representing the fields of Fig. 10.11(b) when the time phase relation is given by Fig. 10.11(a) are

$$F_m(\theta, t) = F_{max} \cos\theta \cos\omega t$$
$$F_a(\theta, t) = F_{max} \cos(\theta - 90°) \cos(\omega t - 90°)$$

The resultant field is then

$$F_m + F_a = F_{max} \cos\theta \cos\omega t + \cos(\theta - 90°) \cos(\omega t - 90°)$$

This can be simplified as

$$F_m + F_a = F_{max} \cos(\theta - \omega t)$$

which is a rotating field of amplitude equal to F_{max}.

$$\bar{F}_a = j\bar{F}_f - j\bar{F}_b \tag{10.16b}$$

where $\bar{F}_f$ and $j\bar{F}_f$ constitute one balanced (forward) set and $\bar{F}_b$ and $-j\bar{F}_b$ constitute the other with reversed (backward) phase sequence. The operations of Eqs (10.16a) and (10.16b) are illustrated by the phasor diagram of Fig. 10.12.

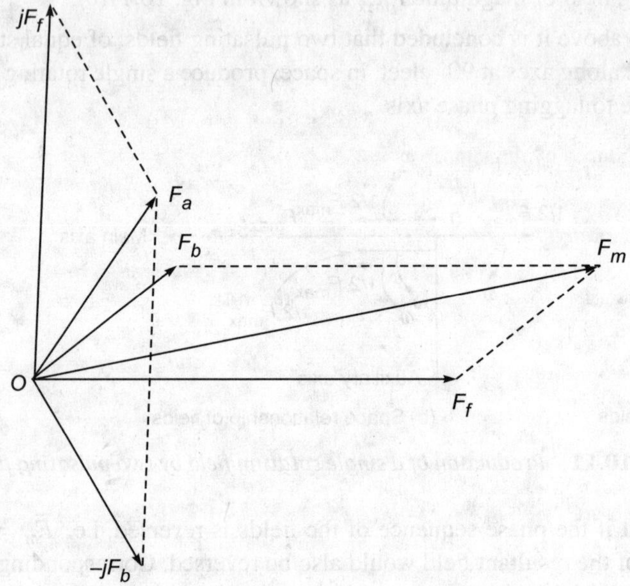

Fig. 10.12 *Two mmf phasors expressed in terms of their symmetrical components*

The inverse of the relationships (10.16a) and (10.16b) can be expressed as

$$\bar{F}_f = \frac{1}{2}(\bar{F}_m - j\bar{F}_a) \tag{10.17a}$$

$$\bar{F}_b = \frac{1}{2}(\bar{F}_m + j\bar{F}_a) \tag{10.17b}$$

from which the forward and backward symmetrical components can be computed for a set of two unbalanced fields. The phasor diagram depicting the operation of Eqs (10.17a) and (10.17b) is given in Fig. 10.13.

In fact any set of phasors (may be voltages or currents) can be similarly expressed in terms of their symmetrical components and vice versa.

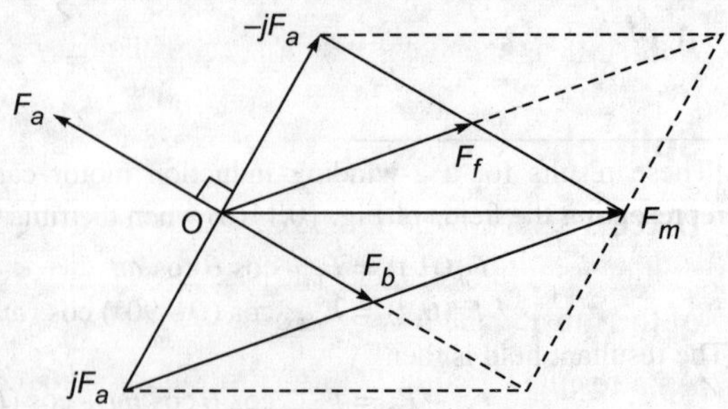

Fig. 10.13 *Symmetrical components of two-phase mmfs*

With reference to Eqs (10.16a) and (10.16b) it is observed that two fields F_m and F_a (assumed to be in space quadrature) can be split into two symmetrical component sets: $\bar{F}_f$, $j\bar{F}_f$ and $\bar{F}_b$, $-j\bar{F}_b$. Since the component fields of each set are equal in magnitude and are in both time and space quadrature, the first set produces a forward rotating field and the second set produces a backward rotating field. It is therefore concluded that two unbalanced pulsating fields in space quadrature are equivalent to two fields rotating in opposite directions.

Example 10.3 *A 2-winding single-phase motor has the main and auxiliary winding currents $I_m = 15$ A and $I_a = 7.5$ A at standstill. The auxiliary winding current leads the main winding current by $\alpha = 45°$ elect. The two windings are in space quadrature and the effective number of turns are $N_m = 80$ and $N_a = 100$. Compute the amplitudes of the forward and backward stator mmf waves. Determine the magnitude of the auxiliary current and its phase angle difference α with the main winding current if only the backward field is to be present.*

LO 1

Solution The mmf produced by the main winding

$$\bar{F}_m = N_m \bar{I}_m = 80 \times 15 \angle 0° = 1200 \angle 0°$$

The mmf produced by the auxiliary winding

$$\bar{F}_a = N_a \bar{I}_a = 100 \times 7.5 \angle 60° = 750 \angle 60°$$

From Eq. (10.17a), the forward field is given by

$$\bar{F}_f = \frac{1}{2}(\bar{F}_m - j\bar{F}_a) = \frac{1}{2}(1200 \angle 0° - j750 \angle 45°) = 334.8 - j265.2$$

$$F_f = 427.1 \text{ AT}$$

Similarly from Eq. (10.17b) the backward field is given by

$$\bar{F}_b = \frac{1}{2}(\bar{F}_m + j\bar{F}_a) = \frac{1}{2}(1200 \angle 0° + j750 \angle 45°) = 865.2 + j265.2$$

$$F_b = 904.9 \text{ AT}$$

Since the forward field is to be suppressed

$$\bar{F}_b = \frac{1}{2}(1200 \angle 0° + j100 I_a \angle \alpha) = 0$$

or

$$1200 + 100\, I_a \sin \alpha + j100\, I_a \cos \alpha = 0$$

Equating real and imaginary parts to zero

$$100\, I_a \cos \alpha = 0$$

or

$$\alpha = 90°$$

$$1200 + 100\, I_a \sin 90° = 0$$

or

$$I_a = -12 \text{ A}$$

Note: Minus sign only signifies a particular connection of the auxiliary winding with respect to the main winding.

Split-phase Motor

!Attention When a motor is provided with two windings, even though these are excited from the same voltage (supply being single-phase), the currents in the two windings can be made out-of-phase by adjustment of the impedance of the auxiliary winding in relation to the main winding.

As a result $\bar{F}_m$ and $\bar{F}_a$ constitute an unbalanced field set with 90° elect. space-phase relationship. The two symmetrical components now being unequal $F_f \neq F_b$ ($F_f > F_b$ is desired); the forward rotating field is made stronger than the backward rotating field resulting in the net production of starting torque. This is how a single-phase motor is made self-starting. In fact phase splitting can be so devised (particularly with capacitive splitting discussed soon after), wherein the backward field is reduced to zero at a specified speed resulting in a completely balanced operation. But such operation is possible only at one speed which can be optimally selected.

Two of the important methods of phase-splitting are discussed below:

Resistance Split-phase Motor The schematic diagram of the resistance split phase motor is given in Fig. 10.14(a). The motor employs an auxiliary winding with a higher *R*/*X* ratio as compared to the main winding. A high *R*/*X* ratio is achieved by using a smaller number of turns of thin wire for the auxiliary winding. The coil-sides of the auxiliary winding are sometimes placed on the top of the slots in order to reduce the reactance. This difference in the *R*/*X* ratio causes the auxiliary winding current I_a to lead the main winding current I_m by angle α as shown in the phasor diagram of Fig. 10.14(b). The fields created by the two currents also have a phase difference of α thereby constituting an unbalanced field system. The result is the production of the starting torque as explained earlier.

The torque-speed characteristic of this motor is shown in Fig. 10.14(c) which also shows the speed n_0 at which a centrifugal switch operates and thereafter the motor runs only on the main winding. The auxiliary winding need then be designed only for short-time use whereas the main winding is designed for continuous use. The value of α, the phase difference between the two currents, can at best be about 30° elect. resulting in poor starting torque as shown in Fig. 10.14(c).

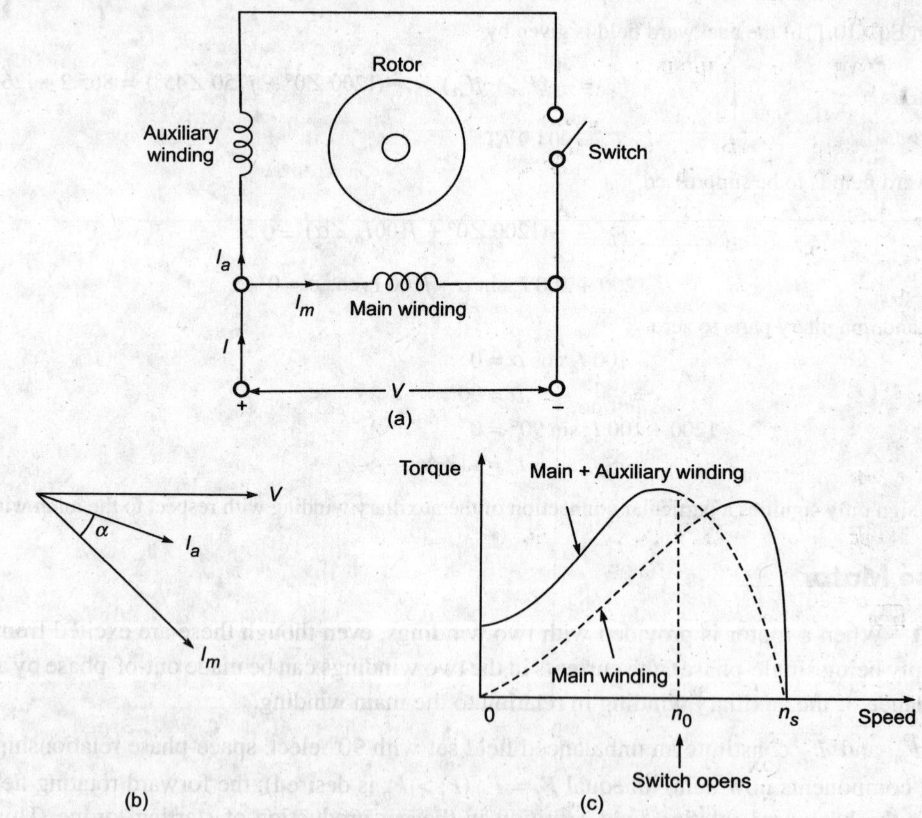

Fig. 10.14 *Resistance split-phase motor*

Capacitor Split-phase Motors

The problem of poor starting torque in a resistance split-phase motor is solved by using a capacitor in series with the auxiliary winding and thereby reaching the ideal case of $\alpha = 90°$. The auxiliary winding along with the capacitor may be disconnected after starting. However, generally the capacitor and auxiliary windings are allowed to remain connected thereby improving the overall motor performance and in particular the power factor. Thus two types of capacitor split-phase motors (also known as *capacitor motor*) exist.

(a) Capacitor-start Motor The schematic diagram of a capacitor-start motor is given in Fig. 10.15(a). The motor is so named because it uses the capacitor only for the purpose of starting. The capacitor value is usually so chosen as to give $\alpha = 90°$ elect. as shown in the phasor diagram of Fig. 10.15(b). The torque-speed characteristic with switching operation is shown in Fig. 10.15(c) which also shows that the starting torque is high. It may he noted that the capacitor need only be short-time rated. Because of the high VAR rating of the capacitor required, electrolytic capacitors must be employed. The range of capacitance is 250 μF or larger.

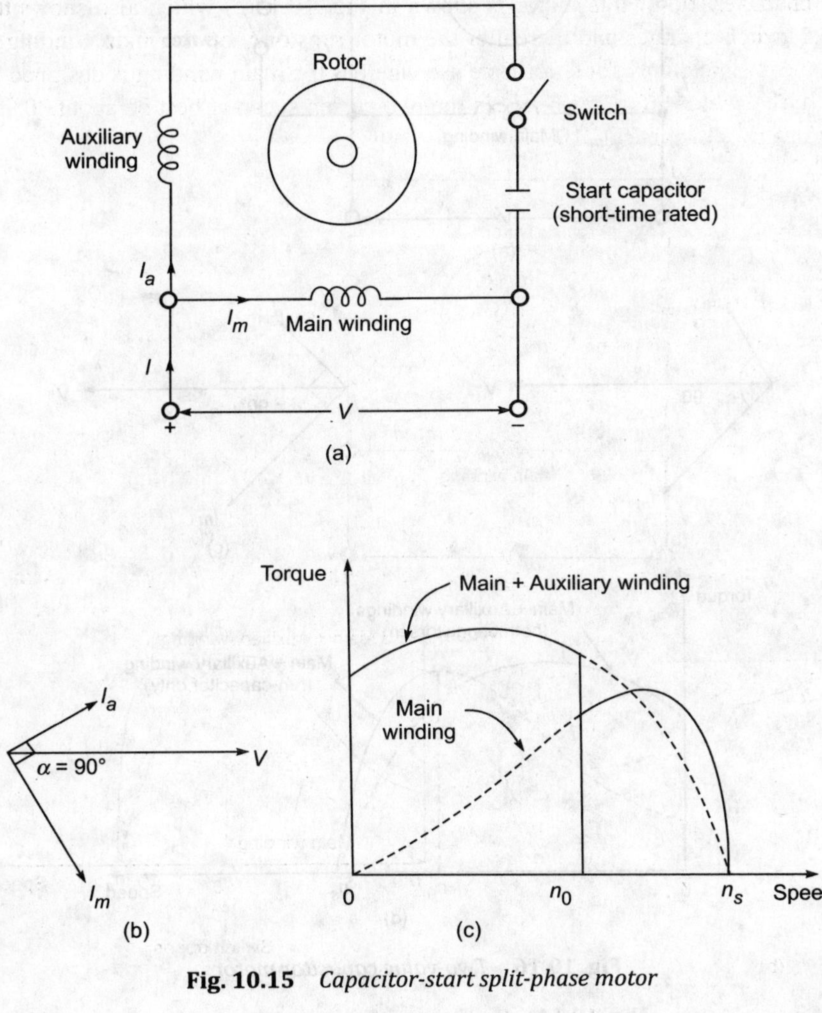

Fig. 10.15 *Capacitor-start split-phase motor*

Addition of a capacitor and accompanying switch naturally increases the cost of the motor and simultaneously reduces its reliability (because of the inclusion of extra components).

(b) Two-value Capacitor Motor The schematic diagram of a 2-value capacitor motor is given in Fig. 10.16(a). As the name suggests, the 2-value capacitor motor not only uses a capacitor for starting but also continuous (run) operation. The capacitor used permanently is called the *runcapacitor,* the use of which improves the motor running performance. Figure 10.16(b) shows the phasor diagram of currents while starting (both capacitors in circuit) where $\alpha > 90°$ elect. so that when the start-capacitor is disconnected α becomes 90° elect. as shown in Fig. 10.16(c).

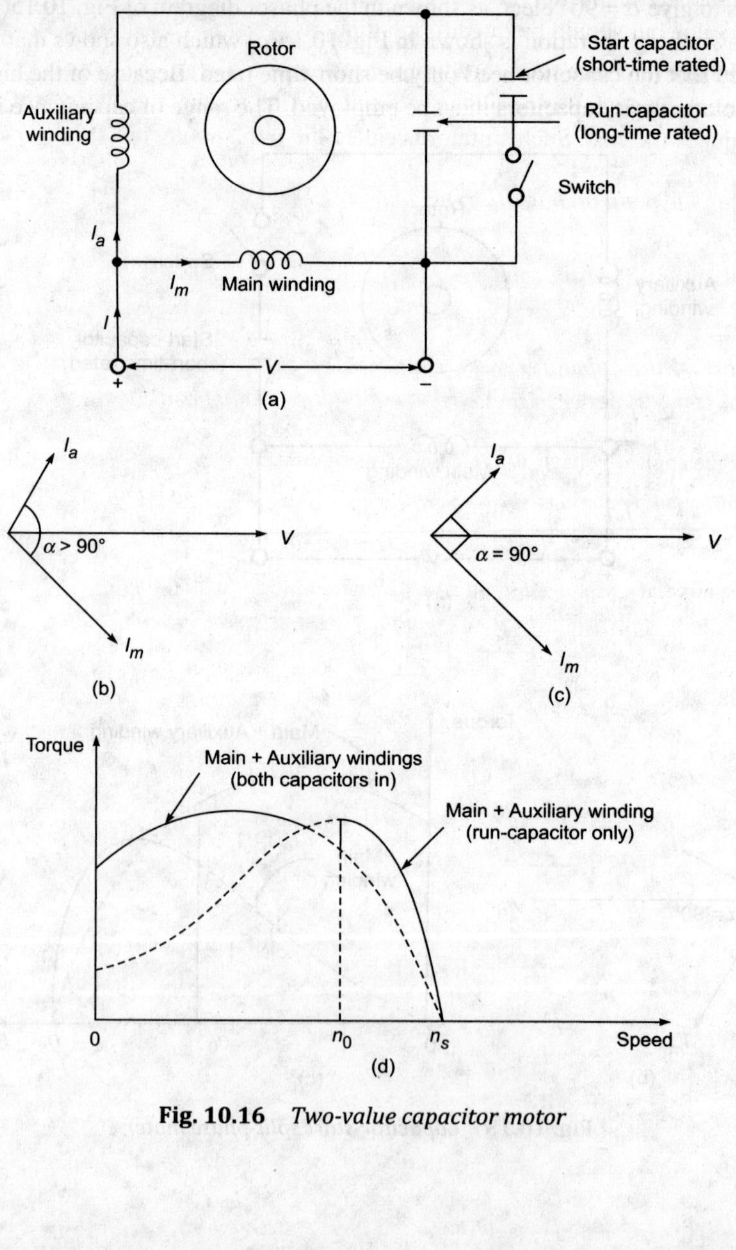

Fig. 10.16 *Two-value capacitor motor*

The quantitative circuit treatment given in Sec. 10.3 could be used for design purposes to find the capacitor values for optimum performance. The practical values of α under running condition are close to 90° elect. Figure 10.16(d) gives the torque-speed characteristics of a two-value capacitor motor. The auxiliary winding and run-capacitor can be designed to give a balanced two-phase field set at a specified speed in which case the backward rotating field does not exist, thereby improving motor efficiency. This would also eliminate the second harmonic torques making the motor smooth running. Hence, this motor would exhibit the best start and best run characteristics with optimum efficiency at an extra cost incurred for the specially designed auxiliary winding and the capacitors.

It may be noted that the start-capacitor has a large value and is rated for short-time whereas the run-capacitor required is of small value but should be rated for continuous operation entailing more expense.

It may also be noted that even if the start-capacitor is not connected at all, the motor is self-starting as the run-capacitor is still present. Though this results in reduced starting torque, the absence of the switch simplifies construction and reduces the cost. Such a motor is called the *permanent-capacitor* or *capacitor-run motor.*

Example 10.4 *The main and auxiliary winding impedances of a 50 Hz, capacitor-start single-phase induction motor are:*

$$\text{Main winding } \bar{Z}_{lm} = 3 + j2.7$$

$$\text{Auxiliary winding } \bar{Z}_{la} = 7 + j3$$

Determine the value of the capacitor to be connected in series with the auxiliary winding to achieve a phase difference of α = 90° between the currents of the two windings at start. LO 1

Solution Choose the applied voltage as a reference for phase angles.

Phase angle of the main winding current

$$\angle \bar{I}_m = -\angle \bar{Z}_{1m} = -\angle(3 + j2.7) - 42°$$

The phase angle of the auxiliary winding current with capacitor in series

$$\angle \bar{I}_a = -\angle[(7 + j3) - j/\omega C]$$

Now

$$\alpha = \angle \bar{I}_a - \angle \bar{I}_m$$

$$90° = -\tan^{-1}\left(\frac{3 - \frac{1}{\omega C}}{7}\right) - (-42°)$$

or

$$\tan^{-1}\left(\frac{3 - \frac{1}{\omega C}}{7}\right) = -48°$$

or

$$\frac{3 - \frac{1}{\omega C}}{7} = -1.11$$

for

$$\omega = 2\pi \times 50 \text{ rad/s, this yields}$$

$$C = 295.5\ \mu\text{F}$$

Shaded-pole Motor

Figure 10.17 shows a typical shaded-pole motor with a squirrel-cage rotor. A small portion of each pole is covered with a short-circuited, single-turn copper coil called the *shading coil.* The sinusoidally-varying flux created by ac (single-phase) excitation of the main winding induces emf in the shading coil. As a result, induced currents flow in the shading coil producing their own flux in the shaded portion of the pole.

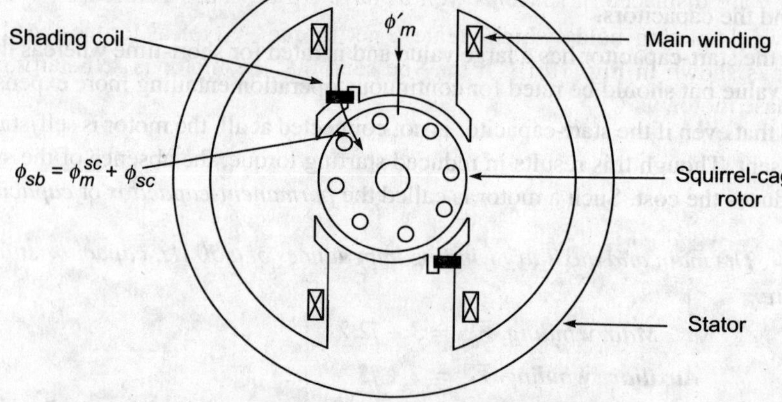

Fig. 10.17 *Shaded-pole motor*

Let the main winding flux be

$$\phi_m = \phi_{\max} \sin \omega t \tag{10.18}$$

where $\phi_m = \phi_m^{sc}$ (flux component linking shading coil)

$+ \phi'_m$ (flux component passing down the air-gap of the rest of the pole)

The emf induced in the shading coil is given by

$$e_{sc} = \frac{d\phi_m^{sc}}{dt} \text{ (since single-turn coil)}$$

$$= \phi_{\max}^{sc}\, \omega \cos \omega t \tag{10.19}$$

Let the impedance of the shading coil be

$$\bar{Z}_a = R_{sc} + jX_{sc} = Z_{sc}\angle\theta_{sc} \tag{10.20}$$

The current in the shading coil can then be expressed as

$$i_{sc} = \frac{\phi_{\max}^{sc}\, \omega \cos(\omega t - \theta_{sc})}{Z_{sc}} \tag{10.21}$$

The flux produced by i_{sc} is

$$\phi_{sc} = \frac{1 \times i_{sc}}{\mathcal{R}} = \frac{\phi_{\max}^{sc}}{Z_{sc}\, \mathcal{R}} \cos(\omega t - \theta_{sc}) \tag{10.22}$$

where $\mathcal{R}$= reluctance of the path of ϕ_{sc}

As per Eqs (10.21) and (10.22), the shading coil current ($\bar{I}_{sc}$) and flux (Φ_{sc}) phasors lag behind the induced emf ($\bar{E}_{sc}$) by angle θ_{sc}; while as per Eq. (10.19), the flux phasor $\bar{\Phi}_m^{sc}$ leads $\bar{E}_{sc}$ by 90°. Obviously the phasor $\bar{\Phi}'_m$ is in phase with Φ^{sc}_m. The resultant flux in the shaded-pole is given by the phasor sum

$$\bar{\Phi}_{sp} = \bar{\Phi}_m^{sc} + \bar{\Phi}_{sc}$$

as shown in Fig. 10.18 and lags the flux $\bar{\Phi}'_m$ of the remaining pole by angle α. The two sinusoidally varying fluxes $\bar{\Phi}'_m$ and $\bar{\Phi}'_{sp}$ are displaced in space as well as have a time phase difference (α) thereby producing forward and backward rotating fields which create a net torque. A typical torque-speed characteristic of shaded-pole motor is shown in Fig. 10.19. It may be seen that the motor is self-starting unlike a single-winding, single-phase motor.

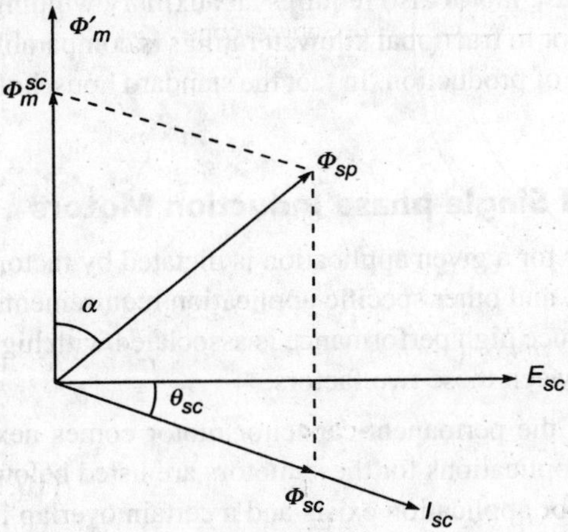

Fig. 10.18

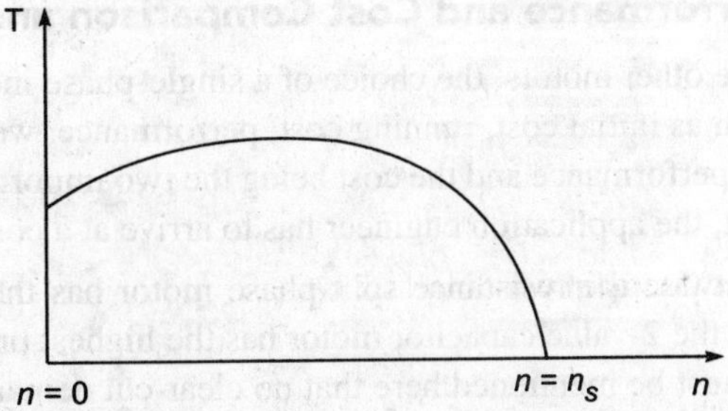

Fig. 10.19 *Torque-speed characteristic of shaded-pole motor*

It is seen from the phasor diagram of Fig. 10.18 that the net flux in the shaded portion of the pole ($\bar{\Phi}'_{sp}$) lags the flux ($\bar{\Phi}'_m$) in the unshaded portion of the pole resulting in a net torque which causes the rotor to rotate from the unshaded to the shaded portion of the pole. The motor thus has a definite direction of rotation which cannot be reversed[†].

Important! The fact that the shaded-pole motor is single-winding (no auxiliary winding), self-starting motor makes it less costly and results in rugged construction. The motor has low efficiency and low power factor and is usually available in a range of 1/300 to 1/20 kW. It is used for domestic fans, record players and tape recorders, humidifiers, slide projectors, small business machines, etc. The shaded-pole principle is used in starting electric clocks and other single-phase synchronous timing motors.

Comparison of Single-phase and Three-phase Induction Motors

Recall In a three-phase induction motor the maximum torque is independent of the rotor resistance, while the slip at which it occurs increases with the rotor resistance.

[†] Reversal of direction of rotation where desired can be achieved by providing two shading coils, one on each end of every pole and by open-circuiting one set of shading coils and short-circuiting the other set.

No such neat result is possible in a single-phase induction motor as the backward rotating field reduces the voltage available for creating the forward rotating field thereby reducing the forward torque and also the torque of the backward field reduces the net available torque. As a result the maximum torque in a single-phase induction motor reduces as the motor resistance is increased while the slip at maximum torque increases.

As a consequence of the presence of the backward field, the performance of a single-phase motor in every respect is somewhat inferior to that of three-phase motor for the same frame size. It has a lower maximum torque at higher slip and greater losses. Further, it has greater volt-ampere and watt input because of their consumption in the backward rotating field. Even the stator copper-losses are higher in a single-phase motor as a single winding is required to carry all the current. All this results in lower efficiency and higher temperature rise for single-phase motors. For a given power and speed rating, a single phase motor must therefore have a larger frame size than a three-phase motor. Further, a single-phase motor also requires an auxiliary winding. In spite of these factors, the cost of a single-phase induction motor in fractional kilowatt ratings is comparable to that of its three-phase counterpart owing to its greater volume of production. In fact the standard household power supply provides for single-phase loads only.

Performance and Cost Comparison and Choice of Single-phase Induction Motors

Like other motors, the choice of a single-phase induction motor for a given application is dictated by factors such as initial cost, running cost, performance, weight and size, and other specific application requirements, the performance and the cost being the two important factors. Since high performance is associated with high cost, the application engineer has to arrive at a compromise between these two factors.

Costwise the resistance split-phase motor has the lowest cost, the permanent-capacitor motor comes next and the 2-value capacitor motor has the highest price. Typical applications for these motors are listed below. It must be mentioned here that no clear-cut demarcation in motor application exists and a certain overlap in application is always found.

Resistance Split-phase Motor It has a low starting current and moderate starting torque. It is used for easily started loads and typical applications include fans, saws, grinders, blowers, centrifugal pumps, office equipment, washing machines, etc. These are usually available in the range of 1/20 to 1/2 kW.

Important! ***Capacitor-start Motor*** This motor has a high starting torque and therefore is used for hard starting loads, such as compressors, conveyors, pumps, certain machine tools, refrigeration and air-conditioning equipment, etc. This is the most commonly used induction motor and is available up to sizes as large as 6 kW.

Permanent-capacitor Motor It has a high starting torque but slightly lower than that of the capacitor-start motor as a result of the compromise between starting and running performances and the capacitor cost. Because of the permanent capacitor it has a better running power factor and efficiency and a quieter and smoother operation. It is used for both easy and hard to start loads. In fact in modern practice ceiling fans, air-circulators and blowers use this type of motor.

Two-value Capacitor Motor It combines the advantages of capacitor-start and permanent-capacitor motors and is used for hard to start loads. At the same time it gives a high power factor and efficiency under running conditions. Typical applications are refrigerators, compressors and stockers.

Shaded-pole Motor It is a cheap motor with a low starting torque and low power factor and efficiency during running. It is available in small sizes up to 1/20 kW. It is commonly used for fans of all kinds (particularly table fans), humidifiers, vending machines, photocopying machines, advertising displays, etc.

10.3 | CIRCUIT MODEL OF SINGLE-PHASE INDUCTION MOTOR

LO 1

The winding unbalance and the fact that both the main and auxiliary windings are fed by the same supply result in unbalanced main and auxiliary fields. The phasor along the winding axes can be split into symmetrical components $\bar{F}_f$ and $\bar{F}_b$ as given by Eqs (10.17a) and (10.17b). The forward component set $\bar{F}_f$ and $j\bar{F}_f$ produces a forward rotating field; similarly the backward component set $\bar{F}_b$ and $-j\bar{F}_b$ results in a backward rotating field.

The rotor slips with respect to the two rotating fields are s and $(2 - s)$ respectively as given by Eqs (10.5a) and (10.5b) and as a result the magnetizing and rotor circuits as seen by the two rotating fields with reference to the main winding are different and are shown in Figs 10.20(a) and 10.20(b).

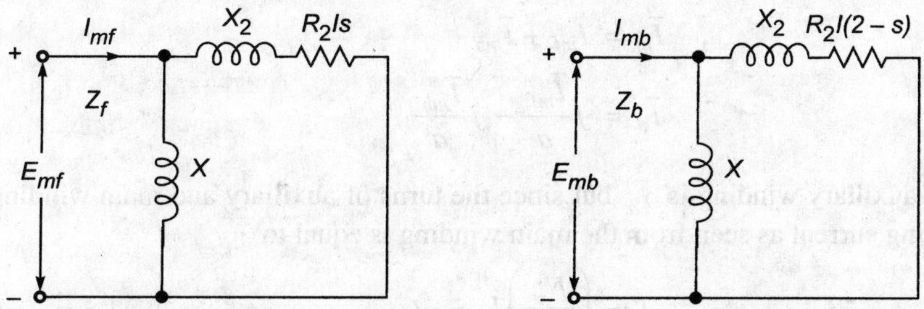

(a) Seen by forward field (b) Seen by backward field

Fig. 10.20 *Magnetizing branch and rotor circuit as seen by rotating fields with reference to the main and auxiliary winding*

It is noted that no-load losses[‡] have been neglected and therefore the core-loss conductance has not been shown in both the circuits. The impedance seen by E_{mf}, the forward field induced emf of the main winding, is

$$\bar{Z}_f = jX \,\|\, \left(\frac{R_2}{s} + jX_2\right) \tag{10.23a}$$

and the impedance seen by E_{mb}, the backward field induced emf in the main winding, is

$$\bar{Z}_b = jX \,\|\, \left(\frac{R_2}{2-s} + jX_2\right) \tag{10.23b}$$

Hence $$\bar{E}_{mf} = \bar{Z}_f \bar{I}_{mf} \tag{10.24a}$$

and $$\bar{E}_{mb} = \bar{Z}_b \bar{I}_{mb} \tag{10.24b}$$

where $\bar{I}_{mf}$ = forward component current in main winding

[‡]These losses can be considered as part of mechanical load on the motor.

$\bar{I}_{mb}$ = backward component current in main winding

Of course $\bar{I}_m$ = main winding current $= \bar{I}_{mf} + \bar{I}_{mb}$ (10.25)

Equations (10.16a) and (10.16b) will now be converted into the current form.

Let N_m = equivalent number of main winding turns

N_a = equivalent number of auxiliary winding turns

Define $$a = \frac{N_a}{N_m} \quad \text{or} \quad N_a = aN_m \tag{10.26}$$

Then from Eqs (10.16a) and (10.16b)

$$\bar{F}_m = N_m\bar{I}_m = N_m\bar{I}_{mf} + N_m\bar{I}_{mb} \tag{10.27a}$$

$$\bar{F}_a = N_a\bar{I}_a = aN_m\bar{I}_a = jN_m\bar{I}_{mf} - jN_m\bar{I}_{mb} \tag{10.27b}$$

From Eqs (10.27a) and (10.27b)

$$\bar{I}_m = \bar{I}_{mf} + \bar{I}_{mb} \tag{10.28a}$$

$$\bar{I}_a = j\frac{\bar{I}_{mf}}{a} - j\frac{\bar{I}_{mb}}{a} \tag{10.28b}$$

The current in the auxiliary winding is $\bar{I}_a$ but since the turns of auxiliary and main windings are different, the auxiliary winding current as seen from the main winding is equal to

$$I'_a = \left(\frac{N_a}{N_m}\right)I_a = aI_a \tag{10.29}$$

From Eqs (10.28a) and (10.28b), the symmetrical components of main and auxiliary winding currents with reference to the main winding can be expressed as

$$\bar{I}_{mf} = \frac{1}{2}(\bar{I}_m - ja\bar{I}_a) \tag{10.30a}$$

$$\bar{I}_{mb} = \frac{1}{2}(\bar{I}_m + ja\bar{I}_a) \tag{10.30b}$$

The forward field reaches the auxiliary winding 90° elect. ahead of the main winding and vice versa for the backward rotating field. Thus the emf's in the auxiliary winding induced by the two fields are:

$$\bar{E}_{af} = ja\bar{E}_{mf} \tag{10.31a}$$

$$\bar{E}_{ab} = -ja\bar{E}_{mb} \tag{10.31b}$$

Also let the main and auxiliary winding terminal voltages be $\bar{V}_m$ and $\bar{V}_a$ respectively. The auxiliary winding voltage is equal to (V_a/a) as seen from the main winding. This set of voltages can also be split into symmetrical components as

$$\bar{V}_m = \bar{V}_{mf} + \bar{V}_{mb} \tag{10.32a}$$

$$\frac{\bar{V}_a}{a} = j\bar{V}_{mf} - j\bar{V}_{mb} \tag{10.32b}$$

or alternatively

$$\bar{V}_{mf} = \frac{1}{2}\left(\bar{V}_m - j\frac{\bar{V}_a}{a}\right) \tag{10.33a}$$

$$\bar{V}_{mb} = \frac{1}{2}\left(\bar{V}_m + j\frac{\bar{V}_a}{a}\right) \tag{10.33b}$$

Now consider the main winding terminal voltage $\bar{V}_m$. It comprises three components: the emf induced by the forward rotating field, the emf induced by backward rotating field and the voltage drop in the winding impedance $\bar{Z}_{1m}$ owing to current I_m flowing through it. Thus

$$\bar{V}_m = \bar{I}_m\bar{Z}_{1m} + \bar{E}_{mf} + \bar{E}_{mb} \tag{10.34a}$$

Substituting for $\bar{E}_{mf}$ and $\bar{E}_{mb}$ from Eqs (10.24a) and (10.24b)

$$\bar{V}_m = \bar{I}_m\bar{Z}_{1m} + \bar{Z}_f\bar{I}_{mf} + \bar{Z}_b\bar{I}_{mb} \tag{10.34b}$$

which is represented by the circuit of Fig. 10.21(a).

Similarly the auxiliary winding terminal voltage $\bar{V}_a$ comprises three components,

i.e, $$\bar{V}_a = \bar{I}_a\bar{Z}_{1a} + \bar{E}_{af} + \bar{E}_{ab} \tag{10.35a}$$

where $\bar{Z}_{1a}$ is the winding impedance of the auxiliary winding which in general has a capacitor included in it (starting/running capacitor). Using Eqs (10.31a) and (10.31b),

$$\bar{V}_a = \bar{I}_a\bar{Z}_{1a} + ja\bar{E}_{mf} - ja\bar{E}_{mb} \tag{10.35b}$$

or $$\bar{V}_a = \bar{I}_a\bar{Z}_{1a} + ja\bar{Z}_f\bar{I}_{mf} - ja\bar{Z}_b\bar{I}_{mb} \tag{10.35c}$$

whose circuit representation is given in Fig. (10.21b).

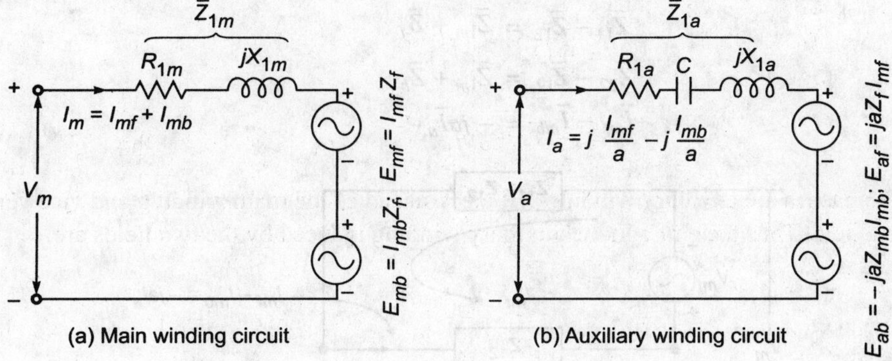

Fig. 10.21

Substituting I_m from Eq. (10.28a) in Eq. (10.34b),

$$\bar{V}_m = (\bar{Z}_{1m} + \bar{Z}_f)\bar{I}_{mf} + (\bar{Z}_{1m} + \bar{Z}_b)\bar{I}_{mb} \tag{10.36a}$$

Similarly substituting I_a from Eq. (10.28b) in Eq. (10.35c),

$$\bar{V}_a = j\left(\frac{\bar{Z}_{1a}}{a} + a\bar{Z}_f\right)\bar{I}_{mf} - j\left(\frac{\bar{Z}_{1a}}{a} + a\bar{Z}_b\right)\bar{I}_{mb} \tag{10.36b}$$

With $\bar{V}_m$ and $\bar{V}_a$ as expressed in Eqs (10.36a) and (10.36b), one obtains from Eqs (10.33a) and (10.33b)

$$\bar{V}_{mf} = \left(\frac{\bar{Z}_{1m}}{2} + \frac{\bar{Z}_{1a}}{2a^2} + \bar{Z}_f\right)\bar{I}_{mf} - \frac{1}{2}\left(\frac{\bar{Z}_{1a}}{a^2} - \bar{Z}_{1m}\right)\bar{I}_{mb} \tag{10.37a}$$

$$\bar{V}_{mb} = -\frac{1}{2}\left(\frac{\bar{Z}_{1a}}{a^2} - \bar{Z}_{1m}\right)\bar{I}_{mf} + \left(\frac{\bar{Z}_{1m}}{2} + \frac{\bar{Z}_{1a}}{2a^2} + \bar{Z}_b\right)\bar{I}_{mb} \tag{10.37b}$$

Defining

$$\bar{Z}_{11} = \left(\frac{\bar{Z}_{1m}}{2} + \frac{\bar{Z}_{1a}}{2a^2} + \bar{Z}_f\right) \tag{10.38a}$$

$$\bar{Z}_{12} = \frac{1}{2}\left(\frac{\bar{Z}_{1a}}{a^2} - \bar{Z}_{1m}\right) \tag{10.38b}$$

$$\bar{Z}_{22} = \left(\frac{\bar{Z}_{1m}}{2} + \frac{\bar{Z}_{1a}}{2a^2} + \bar{Z}_b\right) \tag{10.38c}$$

Equations (10.37a) and (10.37b) can be written as

$$\bar{V}_{mf} = \bar{Z}_{11}\bar{I}_{mf} - \bar{Z}_{12}\bar{I}_{mb} \tag{10.39a}$$

$$\bar{V}_{mb} = \bar{Z}_{12}\bar{I}_{mf} + \bar{Z}_{22}\bar{I}_{mb} \tag{10.39b}$$

Equations (10.39a), (10.39b) and (10.32a) are represented by the circuit model of Fig. 10.22. It is further noted that

$$\bar{Z}_{11} - \bar{Z}_{12} = \bar{Z}_{1m} + \bar{Z}_f \tag{10.40a}$$

$$\bar{Z}_{22} - \bar{Z}_{12} = \bar{Z}_{1m} + \bar{Z}_b \tag{10.40b}$$

$$\bar{I}_{mf} - \bar{I}_{mb} = -ja\bar{I}_a \tag{10.40c}$$

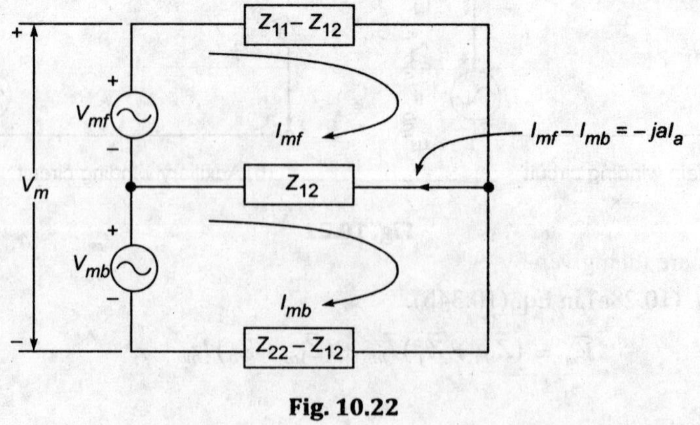

Fig. 10.22

From these the circuit model of Fig. 10.22 can be drawn in the form of Fig. 10.23. In Fig. 10.23 disconnecting the auxiliary winding under running condition is equivalently represented as the opening of switch *S*.

Once the auxiliary winding is disconnected.

$$\bar{I}_{mf} = \bar{I}_{mb} = \frac{1}{2}\bar{I}_m \qquad (10.41)$$

$$\bar{V}_m = \bar{V}_{mf} + \bar{V}_{mb} \qquad (10.42)$$

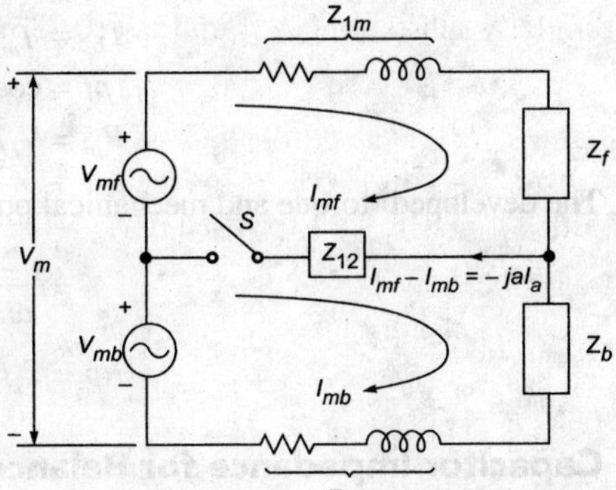

Fig. 10.23 *Circuit model of single-phase induction motor (both windings connected)*

By doubling the current and reducing the impedances to half the circuit model of Fig. 10.24 is obtained. It may be seen that this is the same circuit model as already presented in Fig. 10.5(c) on a heuristic basis.

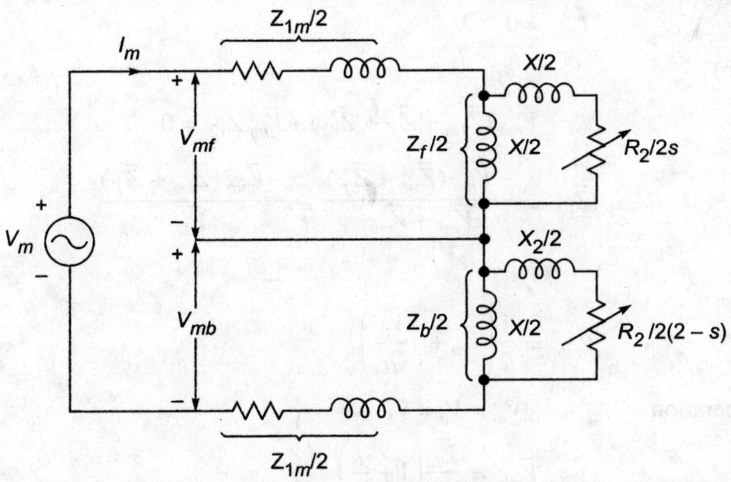

Fig. 10.24 *Circuit model of single-phase induction motor (only main winding connected)*

From Eqs (10.37a) and (10.37b),

$$\bar{I}_{mf} = \frac{\bar{V}_{mf}(\bar{Z}_{1m} + \bar{Z}_b + \bar{Z}_{12}) + \bar{V}_{mb}\bar{Z}_{12}}{(\bar{Z}_{1m} + \bar{Z}_f + \bar{Z}_{12})(\bar{Z}_{1m} + \bar{Z}_b + \bar{Z}_{12}) - \bar{Z}_{12}^2} \qquad (10.43a)$$

$$\bar{I}_{mb} = \frac{\bar{V}_{mb}(\bar{Z}_{1m} + \bar{Z}_f + \bar{Z}_{12}) + \bar{V}_{mf}\bar{Z}_{12}}{(\bar{Z}_{1m} + \bar{Z}_f + \bar{Z}_{12})(\bar{Z}_{1m} + \bar{Z}_b + \bar{Z}_{12}) - \bar{Z}_{12}^2} \qquad (10.43b)$$

The winding currents are then given by

$$\bar{I}_m = \bar{I}_{mf} + \bar{I}_{mb} \qquad (10.44a)$$

$$\bar{I}_a = \frac{j}{a}(\bar{I}_{mf} - \bar{I}_{mb}) \qquad (10.44b)$$

$$\bar{I}_L = \bar{I}_m + \bar{I}_a = I_L \angle \theta \quad (10.45a)$$

$$pf = \cos\theta \quad (10.45b)$$

$$P_{in} = V_L I_L \cos\theta \quad (10.45c)$$

The developed§ torque and mechanical power are given by

$$T = \frac{2}{\omega_s}(I_{mf}^2 R_f - I_{mb}^2 R_b) \quad (10.46)$$

$$P_m = 2(I_{mf}^2 R_f - I_{mb}^2 R_b)(1-s) \quad (10.47)$$

Capacitor Impedance for Balanced Operation at Specific Speed

For balanced single-phase operation

$$\bar{V}_L = \bar{V}_{mf} + \bar{V}_{mb}$$

$$\bar{I}_{mb} = 0$$

or from Eq. (10.43b)

$$\bar{V}_{mb}(\bar{Z}_{1m} + \bar{Z}_f + \bar{Z}_{12}) + \bar{V}_{mf}\bar{Z}_{12} = 0$$

or

$$\bar{Z}_{12} = \frac{-\bar{V}_{mb}(\bar{Z}_{1m} + \bar{Z}_f)}{\bar{V}_{mf} + \bar{V}_{mb}} = \frac{-\bar{V}_{mb}(\bar{Z}_{1m} + \bar{Z}_f)}{\bar{V}_L} \quad (10.48)$$

From Eq. (10.33b)

$$\bar{V}_{mb} = \frac{1}{2}\left(\bar{V}_m + j\frac{\bar{V}_a}{a}\right)$$

For single-phase operation $\quad V_m = V_a = V_L$

$$\therefore \quad \bar{V}_{mb} = \frac{\bar{V}_L}{2}\left(1 + \frac{j}{a}\right) \quad (10.49)$$

Substituting Eq. (10.49) in Eq. (10.48)

$$\bar{Z}_{12} = -\frac{1}{2}\left(1 + \frac{j}{a}\right)(\bar{Z}_{1m} + \bar{Z}_{1f}) \quad (10.50)$$

With $\bar{Z}_{12}$ defined in Eq. (10.38b)

$$\bar{Z}_{12} = \frac{1}{2}(\bar{Z}_{1a}/a^2 - \bar{Z}_{1m})$$

§Notice the factor 2 in Eqs (11.46) and (11.47). This is on account of the two-phase structure of a single-phase (one or two winding) motor. For the single winding case

$$I_{mf} = I_{mb} = I_m/2$$

$$T = \frac{I_m^2}{2\omega_s}(R_f - R_b)$$

which is the same as Eq. (10.9) as it should be.

or
$$\bar{Z}_{1a} = a^2(2\bar{Z}_{12} + \bar{Z}_{1m}) \tag{10.51}$$
$$= (R_C - jX_C) + (R_{1a} + X_{1a})$$

$\therefore$
$$R_C - jX_C = \bar{Z}_{1a} - (R_{1a} + jX_{1a}) \tag{10.52}$$

Example 10.5 *A (1/2) kW, 4-pole, 50 Hz, 220 V, two-value capacitor motor has the following circuit model parameters:*

$$R_{1m} = 4.2\ \Omega, \quad X_{1m} = 11.3\ \Omega$$
$$R_{1a} = 5.16\ \Omega, \quad X_{1a} = 12.1\ \Omega$$
$$X = 250\ \Omega, \quad a = 1.05\ \Omega$$
$$R_2 = 7.48\ \Omega, \quad X_2 = 7.2\ \Omega$$

Friction, windage and core losses = 45 W

(a) Calculate the starting torque and current if the two capacitors in parallel equal 70 μF.

(b) Calculate the value of the run capacitor for zero backward field when the motor is running at a slip of 0.04. What is the meaning of the associated resistance value?

(c) Calculate the motor performance for the value of the run capacitor as in part (b). Assume $R_C = 0$.

LO 1

Solution

(a) $s = 1$

$$\bar{Z}_f = \bar{Z}_b = j250 \parallel (7.48 + j7.2)$$
$$= 10.1\ \angle 45.6° = 7.07 + j7.22$$

$$\bar{Z}_{1a} = (5.16 + j12.1) - j\frac{10^6}{314 \times 70}$$
$$= 5.16 - j33.4$$
$$\bar{Z}_{1a}/a^2 = 4.68 - j30.29$$
$$\bar{Z}_{12} = \frac{1}{2}(\bar{Z}_{1a}/a^2 - \bar{Z}_{1m})$$
$$= \frac{1}{2}(4.68 - j30.29 - 4.2 - j11.3)$$
$$= 0.24 - j20.8 = 20.8\ \angle -89.3°$$

$$\bar{V}_{mf} = \frac{220}{2}\left(1 - \frac{j}{1.05}\right) = 151.9\ \angle -43.60°$$

$$\bar{V}_{mb} = \frac{220}{2}\left(1 + \frac{j}{1.05}\right) = 151.9\ \angle 43.6°$$

$$\bar{Z}_{1m} + \bar{Z}_f + \bar{Z}_{12} = \bar{Z}_{1m} + \bar{Z}_b + \bar{Z}_{12} = (4.2 + j11.3) + (7.07 + j7.22) + (0.24 - j20.8)$$
$$= 11.51 - j2.28 = 11.73\ \angle -11.2°$$

Substituting in Eqs (10.48a) and (10.48b)

$$\bar{I}_{mf} = \frac{151.9\ \angle -43.6° \times 11.73\ \angle -11.2° + 151.9\ \angle 43.6° \times 20.8\ \angle -89.3°}{(11.73)^2\ \angle -22.4° - (20.8)^2\ \angle -178.6°}$$

$$= \frac{4928\angle -49°}{561.3\angle -43°} = 8.78\angle -44.7° = 6.24 - j6.18$$

$$\bar{I}_{mb} = \frac{151.9\angle 43.6° \times 11.73\angle -11.2° + 151.9\angle -43.6° \times 20.8\angle -89.3°}{561.3\angle -4.3°}$$

$$= \frac{-1506\angle 64.6}{561.3\angle -4.3°} = -2.68\angle 68.9° = -0.96 - j2.5$$

$$n_s = 1500 \text{ rpm}, \ \omega_s = \frac{2\pi \times 1500}{60} = 157.1 \text{ rad/s}$$

$$T_s = \frac{2}{157.1} \times 7.07\{(8.78)^2 - (2.68)^2\}$$

$$= 6.31 \text{ Nm}$$

$$\bar{I}_m = \bar{I}_{mf} + \bar{I}_{mb} = 5.28 - j8.68$$

$$\bar{I}_a = \frac{j}{a}(\bar{I}_{mf} - \bar{I}_{mb}) = \frac{j}{1.05}(7.2 - j3.68) = 3.5 + j6.86$$

$$\bar{I}_L = \bar{I}_m + \bar{I}_a = 8.78 - j1.82 = 8.97\angle -11.7°$$

$$\bar{I}_L \text{ (start)} = 8.97 \text{ A}$$

(b) $s = 0.04$

$$\bar{Z}_f = j250 \,||\left(\frac{7.48}{0.04} + j7.2\right)$$

$$= j250 \,||\, (187 + j7.2) = 147\angle 38.2° = 115.5 + j90.9$$

From Eq. (10.55)

$$\bar{Z}_{12} = -\frac{1}{2}\left(1 + \frac{j}{a}\right)(\bar{Z}_{1m} + \bar{Z}_f)$$

$$= -\frac{1}{2}\left(1 + \frac{j}{1.05}\right)[(4.2 + j11.3)] + (115.5 + j90.9)]$$

$$= -108.6\angle 84.1° = -11.2 - j108$$

From Eq. (10.56)

$$\bar{Z}_{1a} = a^2(2\bar{Z}_{12} + \bar{Z}_{1m})$$

$$= (1.05)^2\,[2(-11.2 - j108) + (4.2 + j11.3)] = -20.1 - j225.7$$

From Eq. (10.57)

$$R_C - jX_C = (-20.1 - j225.7) - (5.16 + j12.1)$$

$$= -25.3 - j238$$

$$X_C = \frac{1}{314 \times C}$$

or

$$C = \frac{10^6}{314 \times 238} = 13.4\ \mu\text{F}$$

R_C being negative is unrealizable so that a completely balanced operation is not possible.

(c) $s = 0.04$

$$\bar{Z}_f = 147\angle 38.2° = 115.5 + j90.9$$

(as calculated in part (b))

$$
\begin{aligned}
\bar{Z}_b &= j250 \,||\left(\frac{7.48}{1.96} + j7.2\right) \\
&= j250 \,||\, (3.82 + j7.2) \\
&= 7.92 \angle 63° = 3.6 + j7.06 \\
\bar{Z}_{1a} &= (5.16 + j12.1) - j\frac{10^6}{314 \times 13.4} \\
&= 5.16 - j225.6 \\
\bar{Z}_{1a}/a^2 &= 4.68 - j204.6 \\
\bar{Z}_{12} &= \frac{1}{2}(4.68 - j204.6 - 4.2 - j11.3) \\
&= 0.24 - j108 = 108 \angle -89.9° \\
\bar{Z}_{1m} + \bar{Z}_f + \bar{Z}_{12} &= 4.2 + j11.3 + 115.5 + j90.9 + 0.24 - j108 \\
&= 119.9 - j5.8 = 120 \angle -2.8° \\
\bar{Z}_{1m} + \bar{Z}_b + \bar{Z}_{12} &= 4.2 + j11.3 \\
&\quad\; 3.6 + j7.06 \\
&\quad\; 0.24 - j108 \\
&= 8.04 - j89.64 = 90 \angle -84.9°
\end{aligned}
$$

Substituting in Eqs (10.48a) and (10.48b)

$$
\begin{aligned}
\bar{I}_{mf} &= \frac{151.9 \angle -43.6° \times 90 \angle -84.9° + 151.9 \angle 43.6° \times 108 \angle -89.9°}{120 \angle -2.8° \times 90 \angle -84.9° - (108)^2 \angle -179.8°} \\
&= \frac{22735 \angle -82.9°}{16177 \angle -41.6°} = 1.405 \angle -41.3° = 1.06 - j0.927 \\
\bar{I}_{mb} &= \frac{151.9 \angle 43.6° \times 120 \angle -2.8° + 151.9 \angle -43.6° \times 108 \angle -89.9°}{120 \angle -2.8° \times 90 \angle -84.9° - (108)^2 \angle -179.8°} \\
&= \frac{2506 \angle 0.24°}{16177 \angle -41.6°} = 0.155 \angle 41.8° = 0.116 + j0.103
\end{aligned}
$$

$$
T = \frac{2}{157.1} \times [(1.405)^2 \times 115.5 - (0.155)^2 \times 3.6] = 2.9 \text{ Nm}
$$

$$
\begin{aligned}
\bar{I}_m &= \bar{I}_{mf} + \bar{I}_{mb} = 1.06 - j0.927 + 0.12 - j0.103 = 1.18 - j0.824 \\
\bar{I}_a &= \frac{j}{a}(\bar{I}_{mf} - \bar{I}_{mb}) = \frac{j}{1.05}(1.06 - j0927 - 0.12 - j0.103) \\
&= \frac{j}{1.05}(0.94 - j1.13) = 1.107 + j0.895 \\
\bar{I}_L &= \bar{I}_m + \bar{I}_a = 2.25 - j0.205 \\
I_L &= 2.26; \qquad pf = 1 \text{ (almost unity)} \\
P_m &= 2[(1.405)^2 \times 115.5 - (0.155)^2 \times 3.6]\,(1 - 0.04) = 437.8 \text{ W} \\
P_{\text{out}} &= 437.8 - 45 = 392.8 \text{ W} \\
P_{\text{in}} &= 2.26 \times 220 = 497.2 \text{ W} \\
\eta &= \frac{392.8}{497.2} = 79\%
\end{aligned}
$$

Concept Check LO 1

1. Explain why a single phase single winding induction motor produces no starting torque. ○○●
2. What is the advantage of a capacitor start motor over a resistance split phase motor? ○●●
3. Why single-phase induction motor is not a self-starting one? ○●●
4. What is a fractional H.P. motor? ○○●
5. Where do we require single-phase induction motors? ○○●
6. Why centrifugal switches are provided on many 1-phase induction motors? ○●●

10.4 | SINGLE-PHASE SYNCHRONOUS MOTORS

Reluctance Motor

LO 2 Describe and analyze the performance of reluctance and hysteresis motor

Recall The three-phase reluctance motor has already been discussed as an unexcited salient-pole synchronous machine in Sec. 8.11.

The expression for reluctance torque of a single-phase device was derived in Sec. 4.3 from basic principles of energy conversion. The reluctance motor, in general, results wherever the stator produces a rotating field in space and the rotor is noncylindrical such that the reluctance of the magnetic path offered by the rotor to the rotating field is a function of the space angle.

Important! The origin of the reluctance torque lies in the tendency of the rotor to align itself in the minimum reluctance position with respect to the synchronously rotating flux of the forward field.

The motor is made self-starting by the induction principle by providing short-circuited copper bars in the projecting parts of the rotor.

Recall In a single-phase reluctance motor the rotating field can be produced by any of the phase-splitting methods discussed above.

The salient-pole structure is given to the rotor by removing some of the teeth of an induction motor rotor as shown in Fig. 10.25(a). The remaining teeth carry short-circuited copper bars to provide the starting induction torque. After starting the rotor reaches near synchronous speed by induction action and is pulled into synchronism during the positive half-cycle of the sinusoidally varying synchronous torque[¶]. This would only be possible if the rotor has low inertia and the load conditions are light. The torque-speed characteristic of a typical reluctance motor with induction start is given in Fig. 10.25(b). As seen from this figure the starting torque is highly dependent upon the rotor position because of the projecting nature of the rotor. This phenomenon is known as *cogging*. For satisfactory synchronous motor performance, the frame size to be used must be much larger than that for a normal single-phase induction motor. This accounts for the high value of starting torque indicated in Fig. 10.25(b).

[¶] At speed less than synchronous the torque alternates between positive and negative half-cycles.

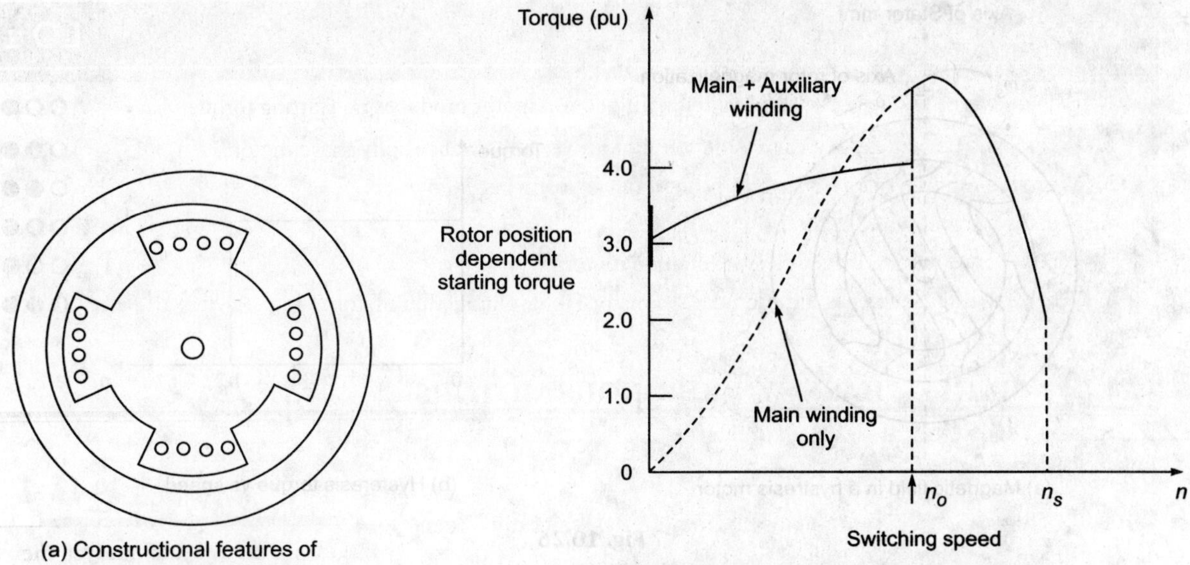

(a) Constructional features of 4-pole reluctance motor

(b) Reluctance motor torque-speed characteristics

Fig. 10.25

Hysteresis Motor

The stator of a hysteresis motor is wound with main and auxiliary windings with a permanently connected capacitor for phase splitting. The capacitor is selected to create balanced two-phase conditions.

!Attention The rotor is a smooth solid** cylinder of hard steel (this has high hysteresis loss) and does not carry any winding (no rotor bars).

Both the stator windings are distributed such as to create a rotating field with as nearly a sinusoidally space distribution as possible; this is necessary to keep down iron-loss due to space harmonics of the field.

The phenomenon of hysteresis causes the rotor magnetization to lag behind the stator-created mmf wave. As a consequence, the rotor flux lags by angle δ the stator mmf axis. Figure 10.26(a) shows the magnetic condition in the motor at any instant. As the angle δ is hysteresis-dependent, it remains constant, at all rotor speeds. The interaction torque (hysteresis torque) between stator and rotor fields therefore is constant at all speeds (Fig. 10.26(b)). Under the influence of the hysteresis torque the rotor accelerates smoothly and finally runs at synchronous speed with angle δ getting adjusted to the load torque. This is a contrast to the "pull-in" phenomenon in a reluctance motor when it synchronizes. Constancy of the hysteresis torque is demonstrated by the derivation below.

The hysteresis loss is expressed as

$$P_h = K_h f_2 B^2 \tag{10.53}$$

where B = maximum flux density, $f_2 = sf$ = rotor frequency

$$\text{Power across air-gap} = \frac{P_h}{s} \tag{10.54}$$

** To cause a saving in expensive hard steel instead a thick annular ring of hard-steel can be placed on an ordinary cylindrical steel core.

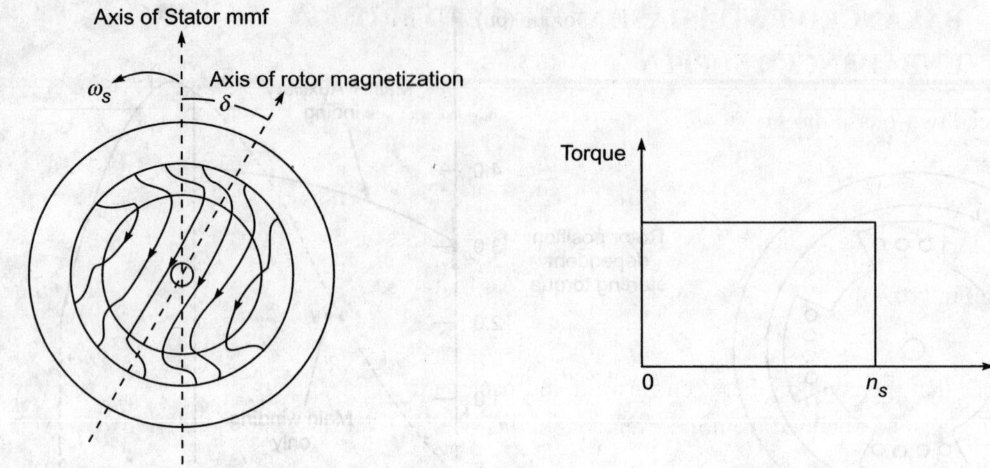

(a) Magnetic field in a hystresis motor

(b) Hysteresis torque vs speed

Fig. 10.26

$$\text{Torque developed} = \frac{P_h}{s\omega_s} = \frac{K_h sf B^2}{s\omega_s}$$

$$= \frac{K_h f B^2}{\omega_s} = \text{constant} \tag{10.55}$$

Another component of torque caused by eddy-current loss is simultaneously created in the motor. This can be derived as:

$$P_e = K_e f_2^2 B^2$$

$$= K_e s^2 f^2 B^2 \tag{10.56}$$

$$T = \frac{P_e}{s\omega_s} = \left(\frac{K_e f^2 B^2}{\omega_s}\right) s \tag{10.57}$$

As per Eq. (10.57), the eddy-current torque is highest at start and reduces linearly with slip vanishing at synchronous speed. This torque component aids the hysteresis torque at starting, endowing excellent starting characteristic to the hysteresis motor.

!Attention The hysteresis motor has a low noise figure compared to the single-phase induction motor such that the load runs at uniform speed.

This is because it operates at one speed (synchronous) and nearly balanced two-phase conditions are not disturbed (as they would in the induction motor when the slip changes with load). Further, a smooth (unslotted) rotor greatly aids in low noise performance of this motor. Multispeed operation is easily possible by arranging pole changing of stator windings; the rotor being unwound reacts to create the same number of poles as the stator. As already pointed out, the motor has excellent starting characteristics (starting torque equal to running torque). Therefore, it is well-suited to accelerate high-inertia loads.

10.5 BALANCED TWO-PHASE MOTOR FED FROM UNBALANCED SUPPLY

LO 2

In a balanced two-phase motor

$$a = \frac{N_a}{N_m} = 1$$

and

$$\bar{Z}_{1a} = \bar{Z}_{1m}$$

Hence[††] in Eq. (10.43b)

$$\bar{Z}_{12} = 0$$

This implies that the coupling between forward and backward motor circuits of Fig. 10.25 disappears and $\bar{I}_{mf}$ and $\bar{I}_{mb}$ can be obtained from independent circuits of Fig. 10.27. Notice that for convenience suffix m has been dropped.

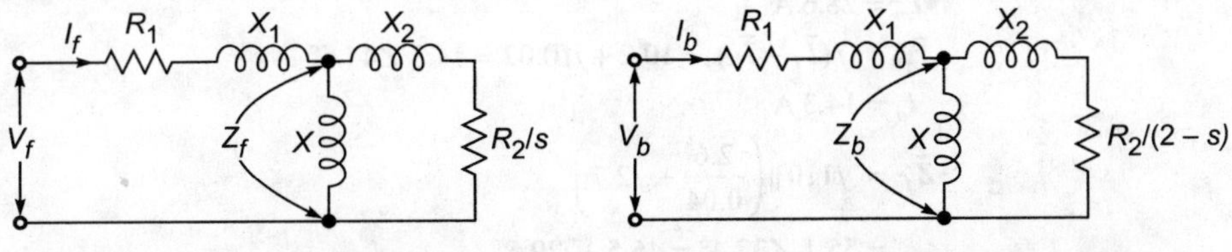

(a) Circuit model for forward component (b) Circuit model for backward component

Fig. 10.27 *Circuit model for balanced two-phase motor fed from unbalanced supply*

Example 10.6 *The circuit model parameters in Ω/phase (referred to stator) of a two-phase, 1 kW, 220 V, 4-pole, 50 Hz squirrel-cage motor are given below:*

$$R_1 = 3\ \Omega \qquad R_2 = 2.6\ \Omega$$
$$X_1 = X_2 = 2.7\ \Omega \qquad X = 110\ \Omega$$

The windage, friction and core losses equal 200 W.
The applied voltages are adjusted such that

$$\bar{V}_a = 110\ \angle 90° \text{ and } \bar{V}_m = 220\ \angle 0°$$

(a) Calculate the starting torque and starting current (in each phase).
(b) Calculate the motor performance at s = 0.04.
(c) With the motor running at s = 0.04, the phase a gets open-circuited. What voltage will be developed across this phase?

LO 2

Solution From Eqs (10.38a) and (10.38b) (a = 1)

$$\bar{V}_f = \frac{1}{2}(220 - j110\ \angle 90°) = 165 \text{ V}$$

$$\bar{V}_b = \frac{1}{2}(220 + j110\ \angle 90°) = 55 \text{ V}$$

[††] In general $\bar{Z}_{12} = 0$

requires $\bar{Z}_{1a}^2 / a^2 = \bar{Z}_{1m}$

(a) $s = 1$

$$\bar{Z}_f = \bar{Z}_b = j110 \parallel (2.6 + j2.7)$$
$$= 3.66 \angle 47.4° = 2.48 + j2.69$$
$$\bar{Z}_f\,(\text{total}) = \bar{Z}_b\,(\text{total}) = (3 + j2.7) + (2.48 + j2.69)$$
$$= 5.48 + j\,5.39 = 7.69 \angle 44.5°$$
$$\bar{I}_f = \frac{165}{7.69 \angle 44.5°} = 21.45 \angle -44.5° = 15.3 - j15.03$$
$$\bar{I}_b = \frac{55}{7.69 \angle 44.5°} = 7.15 \angle -44.5° = 5.1 - j5.01$$
$$\bar{I}_s = \frac{2}{157} \times 2.48\,[(21.45)^2 - (7.15)^2]$$
$$= 12.9 \text{ Nm}$$
$$\bar{I}_m = \bar{I}_f + \bar{I}_b = 20.4 - j20.04 = 28.6 \angle -44.4°$$
$$I_m = 28.6 \text{ A}$$
$$\bar{I}_a = j(\bar{I}_f - \bar{I}_b) = 10.2 + j10.02 = 14.3 \angle 44.5°$$
$$I_a = 14.3 \text{ A}$$

(b)

$$\bar{Z}_f = j110 \parallel \left(\frac{2.6}{0.04} + j2.7\right)$$
$$= 55.1 \angle 32.4° = 46.5 + j29.5$$
$$\bar{Z}_b = j110 \parallel \left(\frac{2.6}{2 - 0.04} + j2.7\right)$$
$$= 2.93 \angle 64.5° = 1.26 + j2.64$$
$$\bar{Z}_f\,(\text{total}) = (3 + j2.7) + (46.5 + j29.5)$$
$$= 49.5 + j32.2 = 59 \angle 33°$$
$$\bar{Z}_b\,(\text{total}) = (3 + j2.7) + (1.26 + j2.64)$$
$$= 4.26 + j5.34 = 6.83 \angle 51.4°$$
$$\bar{I}_f = \frac{165}{59 \angle 33°} = 2.79 \angle -33° = 2.34 - j1.52$$
$$\bar{I}_b = \frac{55}{6.83 \angle 51.4°} = 8.05 \angle -51.4° = 5.02 - j6.29$$
$$n_s = 1500 \text{ rpm} \quad \text{or} \quad \omega_s = 157.1 \text{ rad/s}$$
$$T_s = \frac{2}{\omega_s}(I_f^2 R_f - I_f^2 R_b)$$
$$= \frac{2}{157.1}[(2.79)^2 \times 46.5 - (8.05)^2 \times 1.26]$$
$$= 3.57 \text{ N m}$$
$$\bar{I}_m = \bar{I}_f + \bar{I}_b = (2.34 - j1.52) + (5.02 - j6.29)$$
$$= 7.36 - j7.81 = 10.73 \angle -46.7°$$
$$\bar{I}_a = j(\bar{I}_f - \bar{I}_b) = j[(2.34 - j1.52) - (5.02 - j6.29)]$$
$$= -3.5 + j8.63 = -9.31 \angle 112.1°$$

$$I_m = 10.73 \text{ A}, I_a = 9.31 \text{ A}$$

$$P_m = \omega_s(1-s)T = 157.1\,(1-0.04) \times 3.57 = 538.4 \text{ W}$$

$$P_{out} = 538.4 - 200 = 338.4 \text{ W}$$

$$P_{in}^m = 220 \times 10.73 \times \cos 46.7° = 1619 \text{ W}$$

$$P_{in}^a = 220 \times 5.47 \times \cos 29.3° = 1049 \text{ W}$$

$$P_{in} = 1619 + 1049 = 2668 \text{ W}$$

$$\eta = \frac{338.4}{2668} = 12.7\% \text{ (low because of unbalanced operation)}$$

(c)
$$\bar{I}_a = j(\bar{I}_f - \bar{I}_b) = 0$$

or
$$\bar{I}_f = \bar{I}_b \quad \text{(i)}$$

$$\bar{V}_m = \bar{V}_{mf} + \bar{V}_{mb} = 220 \text{ V} \quad \text{(ii)}$$

$$\bar{V}_a = j(\bar{V}_{mf} - \bar{V}_{mb}) = ? \quad \text{(iii)}$$

$$\bar{I}_f = \frac{\bar{V}_{mf}}{\bar{Z}_f \text{(total)}} = \frac{\bar{V}_{mb}}{\bar{Z}_b \text{(total)}} = \bar{I}_b$$

or
$$\frac{\bar{V}_{mf}}{\bar{V}_{mb}} = \frac{\bar{Z}_f \text{(total)}}{\bar{Z}_b \text{(total)}} \quad \text{(iv)}$$

$$= \frac{59\angle 33°}{6.83\angle 51.4°} = 8.63\angle -18.4° \quad \text{(v)}$$

Solving Eqs (ii) and (v)

$$\bar{V}_{mf} = 198.2\angle -1.9° = 198.1 - j6.6$$

$$\bar{V}_{mb} = 22.96\angle 16.5° = 22.01 + j6.5$$

$$\bar{V}_a = j(176.1 - j13.1)$$
$$= 13.1 + j176.1 = 176.6\angle 85.7°$$

$$V_a = 176.6 \text{ V}$$

It may be seen that the angle of V_a is slightly short of 90°.

Concept Check LO 2

1. Give reasons for the low efficiency of hysteresis and reluctance motors.
2. What are the advantages of reluctance motor?
3. What are the applications of reluctance motor?
4. What is a hysteresis motor?
5. What are the advantages of using a hysteresis motor?

10.6 | TWO-PHASE SERVOMOTOR

LO 3 Discuss the construction and working of two-phase servomotor and two-phase tachometer

For a low-power (a few hundred watts) control application, a two-phase (balanced) servomotor is ideally suited as it can be driven by means of a relatively rugged (drift-free) ac amplifier. The motor torque can be easily controlled by varying the magnitude of the ac voltage applied to the *control phase* (phase *a*) of the motor as shown in Fig. 10.28. While the second phase called the *reference phase* (phase *m*) is excited at a fixed-voltage synchronous ac voltage (both the voltages must be drawn from the same source[‡‡]). The control phase voltage is shifted in phase by 90° from the reference phase voltage by means of phase-shifting networks included in voltage amplification stages of the amplifier. The motor torque gets reversed by phase reversal of the control phase voltage.

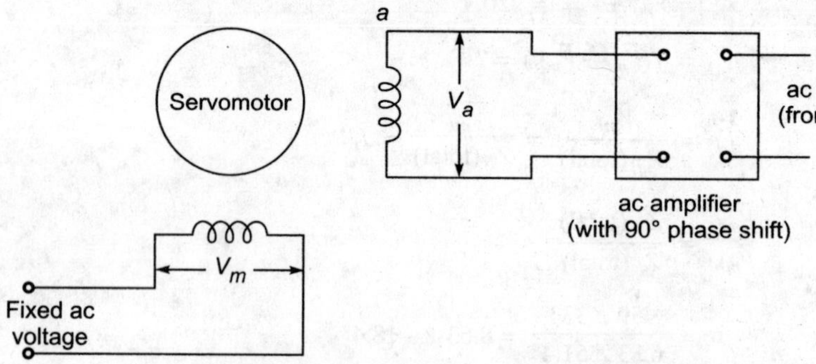

Fig. 10.28 *Control scheme for two-phase servomotor*

!Attention For linear stable operation, the torque-speed characteristic of a servo-motor must be linear with negative slope (torque reducing with increasing speed). The torque speed characteristic of a normally designed induction motor is highly nonlinear and the characteristic is unstable for normal loads in the region from zero speed to speed at breakdown torque.

This indeed is the useful region of operation of a servomotor employed in position control systems. The desired linear characteristic is obtained in a servo motor by designing a rotor with high resistance so that the maximum torque occurs at a slip of –0.5 or so. The high resistance also imparts another important desirable feature to the servomotor, i.e. it does not develop a single-phasing torque[§§] which would disturb the control characteristic of the motor.

The torque-speed characteristics of a servomotor for various per unit values of phase *a* voltage are drawn in Fig. 10.29. If the reference phase voltage is $V_m \angle 0°$, the control phase voltage is

$$\bar{V}_a = aV_m \angle 90°; \qquad a = \text{variable}$$

[‡‡] It is already known that this is a necessary condition for establishing a rotating magnetic field in the air-gap.

[§§] A single-phasing torque is given by

$$T(\text{single-phase}) = \frac{I_m^2}{2\omega_s}(R_f - R_b) \tag{10.9}$$

When the rotor resistance is large, $R_f \approx R_b$, almost eliminating the single-phasing torque.

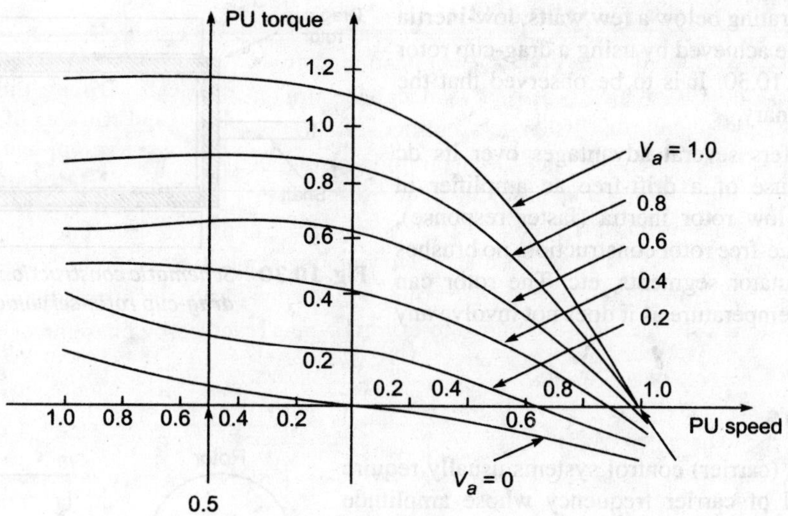

Fig. 10.29 *Torque-speed characteristics of servomotor; V_a variable, phase difference = 90° (fixed)*

Now
$$V_{mf} = \frac{V_m}{2}(1 + a) \tag{10.58a}$$

$$V_{mb} = \frac{V_m}{2}(1 - a) \tag{10.58b}$$

Thus the motor is excited with a special kind of voltage unbalance–angular phase difference of 90° is maintained while the magnitude of phase a voltage varies. The corresponding torque-speed characteristics (Fig. 10.29) are nearly linear with respect to motor speed and voltage of phase a.

Based on the linearity assumption, the following relation is obtained[¶¶]

$$T = KV_a - f\omega \tag{10.59a}$$

where ω = rotor speed

Also
$$T = J\omega + f_0\omega \tag{10.59b}$$

where J = motor inertia, and f_0 = motor viscous friction.

Laplace transforming Eq. (10.59a) and (10.59b), the motor transfer function[***] can be written as

$$G_M(s) = \frac{\omega(s)}{V_a(s)} = \frac{K}{Js + (f_0 + f)}$$

$$= \frac{K_m}{\tau_m s + 1}$$

where
$K_m = K/(f_0 + f)$ = motor gain constant
$\tau_m = J/(f_0 + f)$ = motor time-constant

[¶¶]Negative value of V_a implies phase reversal.

[***]Electrical transient is quite fast compared to mechanical dynamics and has therefore been ignored in this derivation.

In servomotors of rating below a few watts, low-inertia construction can be achieved by using a drag-cup rotor illustrated in Fig. 10.30. It is to be observed that the rotor core is stationary.

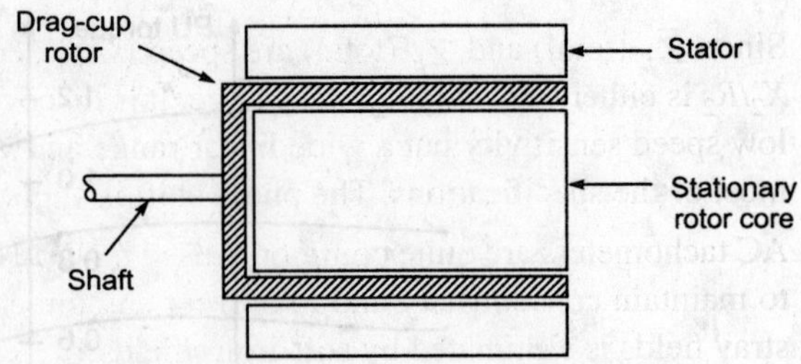

Fig. 10.30 *Schematic constructional diagram of drag-cup rotor servomotor*

ac servomotor offers several advantages over its dc counterpart—the use of a drift-free ac amplifier in control circuitry, low rotor inertia (faster response), rugged maintenance-free rotor construction, no brushes contacting commutator segments, etc. The rotor can withstand higher temperature as it does not involve any insulation.

ac Tachometers

Important! ac (carrier) control systems usually require a feedback signal of carrier frequency whose amplitude is proportional to speed. Such signals are conveniently obtained by means of ac tachometers.

An ac tachometer is nothing but a two-phase induction motor with one phase (m) excited from the carrier frequency, while the phase a winding is left open-circuited as shown in Fig. 10.31. To achieve a low-inertia, a rotor drag-cup construction is commonly employed. It is shown below that the voltage across phase a is proportional to rotor speed while it has phase shift close to 90° (see Ex. 10.5).

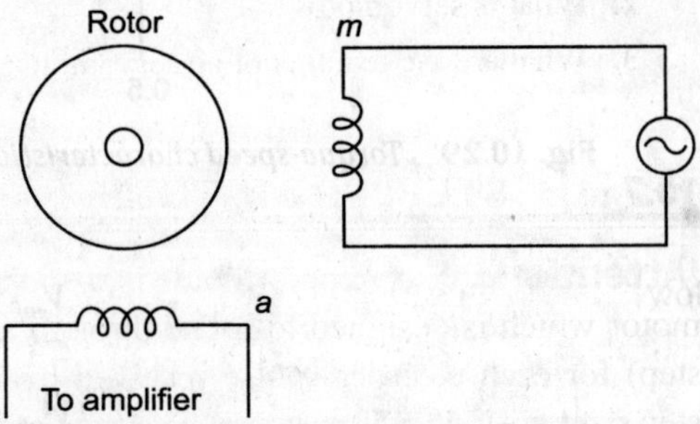

Fig. 10.31 *Schematic diagram of a two-phase tachometer*

For a balanced two-phase winding

$$\bar{Z}_{1a}/a^2 = \bar{Z}_{1m}$$

With the phase a open-circuited

$$\bar{I}_a = j(\bar{I}_f - \bar{I}_b) = 0 \tag{10.60}$$

or

$$\bar{I}_f = \bar{I}_b \tag{10.61}$$

Therefore

$$\frac{\bar{V}_{mf}}{\bar{V}_{mb}} = \frac{\bar{Z}_f\,(\text{total})}{\bar{Z}_b\,(\text{total})} = \bar{A}_z$$

Also

$$V_m \angle 0° = \bar{V}_{mf} + \bar{V}_{mb}$$

$\therefore$

$$\bar{V}_{mb} = \frac{\bar{V}_m}{1 + \bar{A}_z}$$

$$\bar{V}_{mf} = \frac{V_m \bar{A}_z}{1 + \bar{A}_z}$$

Now

$$\bar{V}_a = j(\bar{V}_{mf} + \bar{V}_{mb})$$

$$= jV_m\left(\frac{1 - \bar{A}_z}{1 + \bar{A}_z}\right)$$

Since $\bar{Z}_f$ (total) and $\bar{Z}_b$ (total) are speed (slip) dependent, so is $\bar{A}_z$. Thus V_a is speed dependent. If the rotor X_2/R_2 is either very small or very large, it is discovered that V_a - speed relationship is linear. Low X_2/R_2 gives low speed sensitivity but a wide linear range and vice versa. An intermediate value of X_2/R_2 generally meets most of the specifications. The phase shift is slightly less than 90° but is quite insensitive to speed.

AC tachometers are quite commonly used in 400-Hz control systems. High precision construction is required to maintain concentricity and to prevent any direct coupling between the two phase windings. Pick-up from stray fields is eliminated by soft-iron shields.

Concept Check LO 3

1. What are servomotors and list their characteristics? ○○●
2. What is servomotor? ○○●
3. What are the features of ac servomotor? ○●●

10.7 | STEPPER MOTORS

LO 4 Explain the performance of various types of stepper motors and ac series motors

!Attention The stepper motor is a special type of synchronous motor which is designed to rotate through a specific angle (called a step) for each electrical pulse received by its control unit. Typical step sizes are 7.5°, 15° or larger. The stepper motor is used in digitally controlled position control systems in open loop mode. The input command is in the form of a train of pulses to turn a shaft through a specified angle.

There are two advantages in using stepper motors. One is their compatibility with digital systems and secondly no sensors are needed for position and speed sensing as these are directly obtained by counting input pulses and periodic counting if speed information is needed. Stepper motors have a wide range of applications; paper feed motors in typewriters and printers, positioning of print heads, pens in XY-plotters, recording heads in computer disk drives and in positioning of work tables and tools in numerically controlled machining equipment. The range of applications of these motors is increasing as these motors are becoming available in larger power ratings and with reducing cost.

Elementary operation of a four-phase stepper motor with a two-pole rotor can be illustrated through the diagram of Fig. 10.32. Let us assume that the rotor is permanent magnet excited.

Such a rotor aligns with the axis of the stator field with torque being proportional to the sin θ, θ being the angle of displacement between the rotor axis and stator field axis. The torque-angle characteristics is drawn in Fig. 10.33(a) with phase a excited and also with phase b excited. It is easily observed that the stable position of the rotor corresponds to that angle at which the torque is zero and is positive for smaller angles and negative for larger angles. Thus with phase a excited, the stable (or locked) position is $\theta = 0°$ but not $\theta = 180°$ (unstable) and the torque has a maximum positive value at $\theta = -90°$. It is therefore easily concluded that each excitation pattern of phases corresponds to a unique position of the rotor. Therefore the excitation sequence a, b, c, d, a... causes the rotor to move in positive sequence in steps of 90°.

With rotor in position $\theta = 0$ and a and b both excited the rotor will move to 45°, which is a stable position (net torque due to a and b zero and torque-angle slope negative). So excitation sequence a, $a + b$, b, $b + c$, c...

make the rotor to move forward in steps of 45°. Patterns for phase winding excitations can be easily visualized for steps of 22.5°, 11.25°, and smaller per pulse input to the circuit. Another feature of a PM stepper motor is that, when excited, it seeks a preferred position which offers advantage in certain applications.

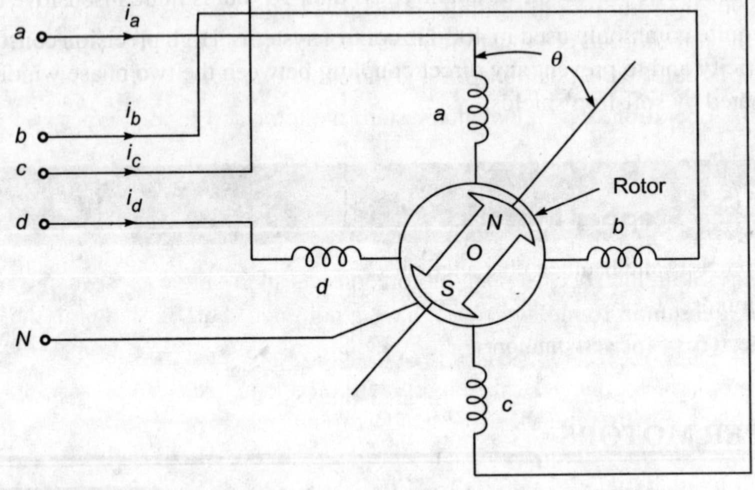

Fig. 10.32 *A 4-phase 2-pole stepper motor; elementary diagram*

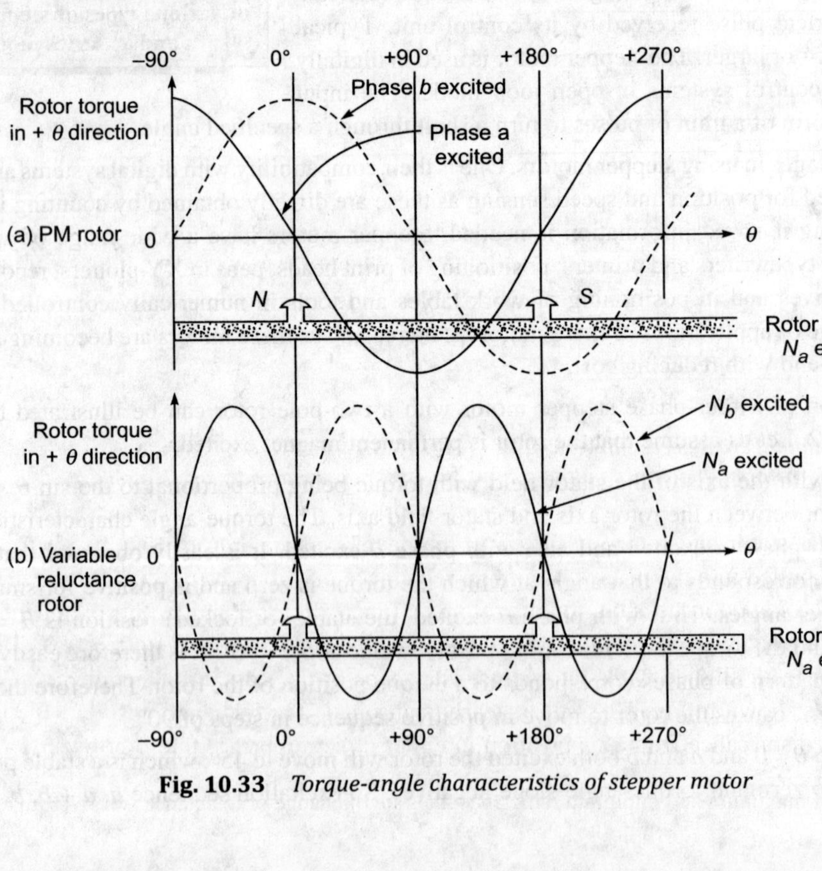

Fig. 10.33 *Torque-angle characteristics of stepper motor*

Consider now that the rotor (projecting pole) is made of just ferromagnetic material (no permanent magnet). The device now behaves as a variable-reluctance motor. The ferromagnetic rotor seeks the position which presents minimum reluctance to the stator field, i.e. the rotor axis aligns itself to the stator field axis. In Fig. 10.32 with phase a excited, this happens at $\theta = 0°$ as well as $\theta = 180°$ in which positions the torque on the rotor is zero. At $\theta = 90°$ the rotor presents infinite reluctance to the phase *'a'* axis and so the torque has also a zero[†††] there. Thus the rotor torque is a function of sin 2θ as drawn in Fig. 10.33(b). For a reluctance stepper motor there are two possible stable positions for a given excitation or set of excitations.

Having illustrated the operation of an elementary stepper motor and its two types, we shall now consider some further details.

Variable-reluctance Stepper Motors

Important! A variable-reluctance stepper motor consists of a single or several stacks of stators and rotors—stators have a common frame and rotors have a common shaft as shown in the longitudinal cross-sectional view of Fig. 10.34 for a 3-stack motor.

Both stators and rotors have toothed structure as shown in the end view of Fig. 10.35. The stator and rotor teeth are of same number and size and therefore can be aligned as shown in this figure. The stators are pulse excited, while the rotors are unexcited.

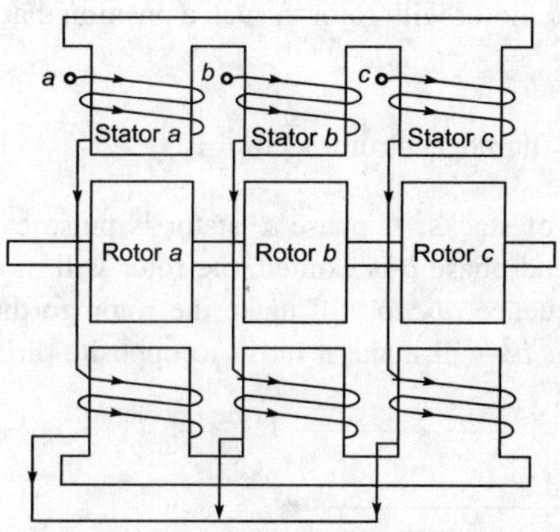

Fig. 10.34 *Longitudinal cross-sectional view of 3-stack variable reluctance motor*

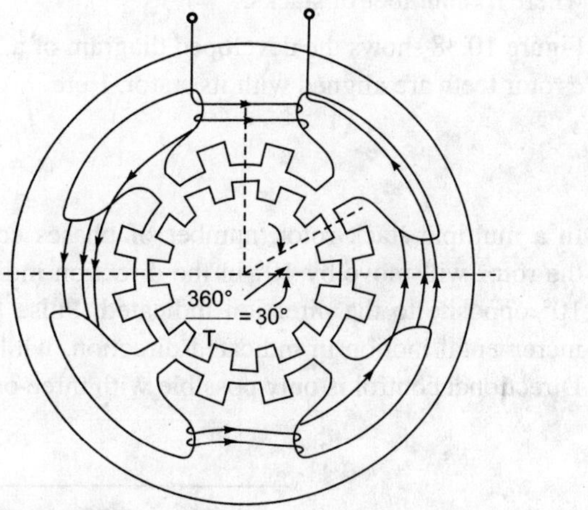

Fig. 10.35 *End view of stator and rotor (12-teeth) of a multistack variable reluctance step motor*

Consider a particular stator and rotor set shown in the developed diagram of Fig. 10.36. As the stator is excited, the rotor is pulled into the nearest minimum reluctance position; the position where stator and rotor teeth are aligned. The static torque acting on the rotor is a function of the angular misalignment θ. There are two positions of zero torque: $\theta = 0$, rotor and stator teeth aligned and $\theta = 360°/(2 \times T) = 180°/T$ (T = number of rotor teeth), rotor teeth aligned with stator slots. The shape of the static torque-angle curve of one stack of a stepper motor is shown in Fig. 10.37. It is nearly sinusoidal. Teeth aligned position ($\theta = 0$) is a stable position, i.e., slight disturbance from this position in either direction brings the rotor back to it. Tooth-slot

[†††] At these position of maximum and minimum reluctance, the rate of change of reluctance is zero (zero torque).

aligned position ($\theta = 180°/T$) is unstable i.e, slight disturbance from this position in either direction makes the rotor move away from it. The rotor thus locks into stator in position $\theta = 0$ (or multiple of $360°/T$). The dynamic torque-speed characteristic will differ from this due to speed emf induced in stator coils.

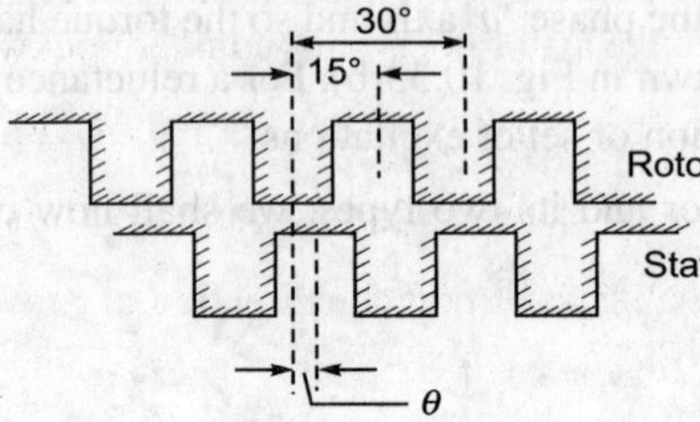

Fig. 10.36 *Developed view of teeth of a pair of stator-rotor*

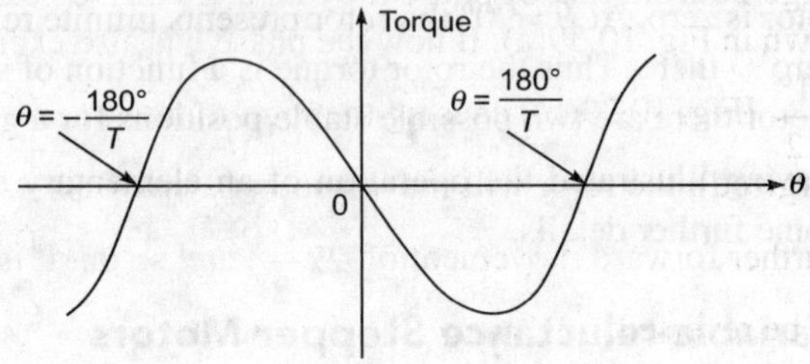

Fig. 10.37 *Static torque-angle curve of a stepper motor stack*

While the teeth on all the rotors are perfectly aligned, stator teeth of various stacks differ by an angular displacement of

$$\alpha = \frac{360°}{nT} \tag{10.62}$$

where n = number of stacks.

Figure 10.38 shows the developed diagram of a 3-stack stepper motor with rotor in such a position that stack c rotor teeth are aligned with its stator. Here

$$\alpha = \frac{360}{3 \times 12} = 10° \text{ (number of rotor teeth = 12)} \tag{10.63}$$

In a multiple stack rotor, number of phases equals number of stacks. If phase 'a' stator is pulse excited, the rotor will move by 10° in the direction indicated. If instead phase b is excited, the rotor will move by 10° opposite to the direction indicated. Pulse train with sequence *abcab* will make the rotor go through incremental motion in indicated direction, while sequence *bacba* will make it move to opposite direction. Directional control is only possible with three or more phases.

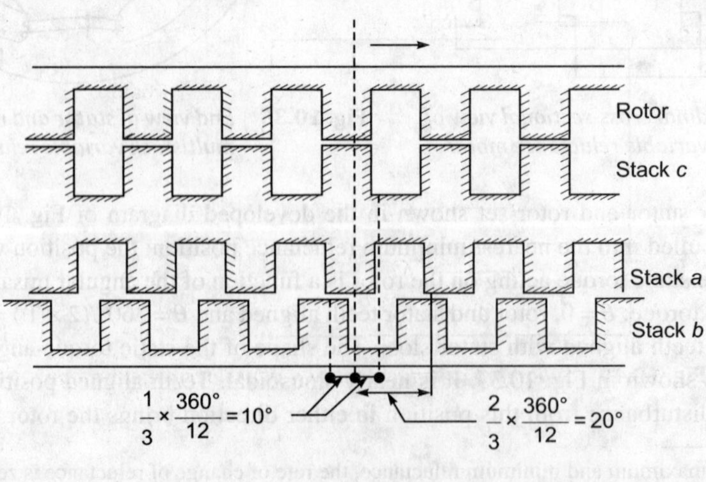

Fig. 10.38 *Developed view of rotor and stator stacks of a 3-stack stepper motor*

Permanent-Magnet Motor

Figure 10.39 shows the phases (stacks) of a two-phase, 4-pole permanent magnet stepper motor. *The rotor is made of ferrite or rare-earth material which is permanently magnetized.* The stator stack of phase *b* is staggered from that of phase *a* by an angle of 90° elect. When the phase *a* is excited, the rotor is aligned as shown in Fig. 10.39(a). If now the phase *b* is also excited, the effective stator poles shift counterclockwise by $22\frac{1}{2}^\circ$ [Fig. 10.39(b)] causing the rotor to move accordingly. If the phase '*a*' is now de-energized (with *b* still energized), the rotor will move another step of $22\frac{1}{2}^\circ$. The reversal of phase *a* winding current will produce a further forward movement of $22\frac{1}{2}^\circ$, and so on. It is easy to visualize as to how the direction of movement can be reversed.

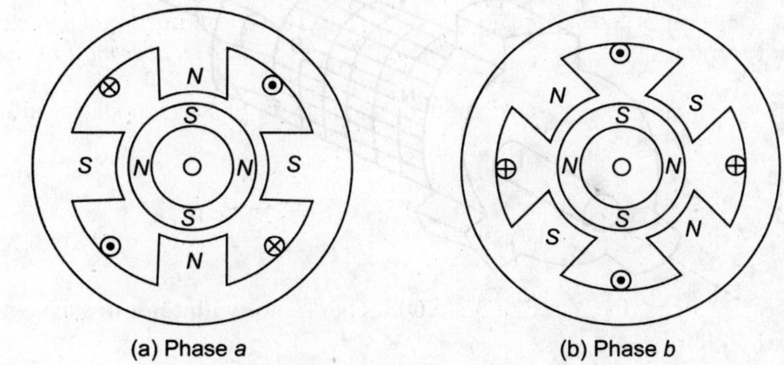

Fig. 10.39 *Permanent-magnet stepper motor*

To simplify the switching arrangement, which is accomplished electronically, double coils are provided for each phase. The schematic diagram of switching circuit is shown in Fig. 10.40.

Compared to variable-reluctance motors, typical permanent-magnet stepper motors operate at larger steps up to 90°, and at maximum response rates of 300 pps.

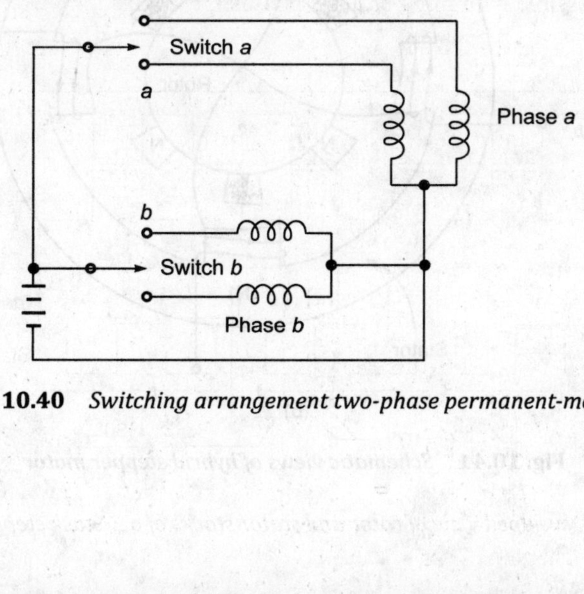

Fig. 10.40 *Switching arrangement two-phase permanent-magnet*

Hybrid Stepper Motor

This is in fact a PM stepper motor with constructional features of toothed and stacked rotor adopted from the variable-reluctance motor. The stator has only one set of winding-excited poles which interact with the two rotor stacks. The permanent magnet is placed axially along the rotor in form of an annular cylinder over the motor shaft. The stacks at each end of the rotor are toothed. So all the teeth on the stack at one of the rotor acquire the same polarity while all the teeth of the stack at the other end of the rotor acquire the opposite polarity. The two sets of teeth are displaced from each other by one half of the tooth pitch (also called pole pitch). These constructional details are brought out by Fig. 10.41(a) and (b) for the case of three teeth on each stack so that tooth pitch $\gamma_t = 360°/3 = 120°$. This motor has a two-phase, 4-pole stator.

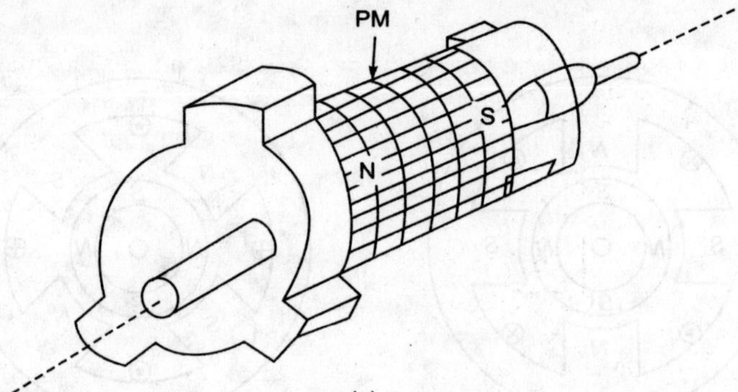

(a)

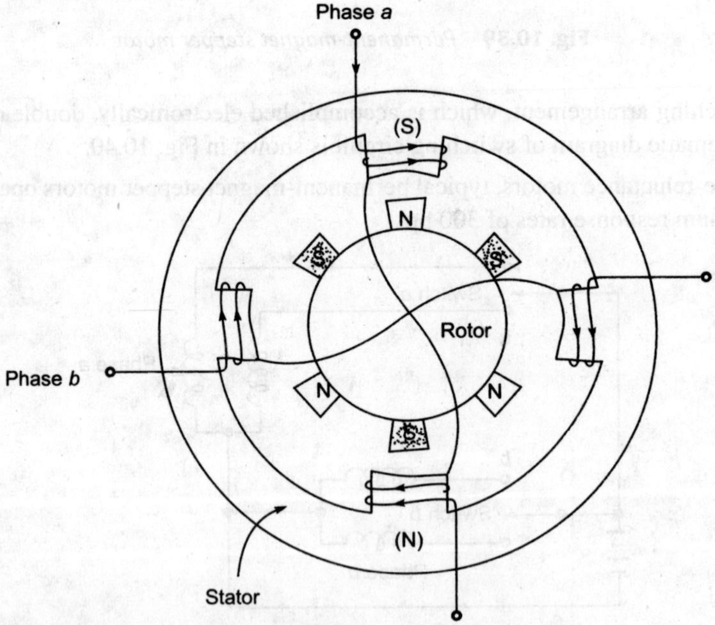

(b)

Fig. 10.41 *Schematic views of hybrid stepper motor*

Consider now that the stator phase *'a'* is excited such that the top stator pole acquires north polarity while the bottom stator pole acquires south polarity. As a result of the nearest tooth of the front stack (assumed to be of north polarity) is pulled into locking position with the stator south pole (top) and the diametrically opposite tooth of the rear stack (south polarity) is simultaneously locked into the stator north pole (bottom). The repulsive forces on the remaining two front stack teeth balance out as these are symmetrically located w.r.t. the bottom stator pole and so do the repulsive forces due to the top stator pole on the remaining two rear stack teeth. This rotor position is thus a stable position with net torque on rotor being zero.

If the excitation is shifted from phase *a* to phase *b* such that right-hand pole becomes south and left-hand pole becomes north, this would cause the rotor to turn anti-clockwise by $\gamma_t/4 = 30°$ into the new locking position. The rotor will turn clockwise by 30° if the phase *b* were oppositely excited.

If the stator excitation were to be removed, the rotor will continue to remain locked into the same position as it is prevented to move in either direction by torque because of the permanent magnet excitation. This fact is in favour of hybrid (also PM) motors. Compared to PM motor finers steps for better resolution are easily obtained in hybrid motor by increasing the number of stack teeth and also by adding additional stack pairs on the rotor for example for a seven teeth stack the steps size is $(360°/7)/4 = (90°/7)°$. Also compared to variable-reluctance motor a hybrid motor requires less stator excitation current because of the PM excited rotor.

!Attention Half stepping can be achieved in a hybrid motor by exciting phase *a* and then exciting phase *b* before switching off the excitation of phase *'a'* and so on. In fact any fractional step can be obtained by suitably proportioning the excitation of the two phases. Such stepping is known as microstepping.

Typical step angles for stepper motors are 15°, 7.5°, 2° and 0.72°. The choice of the angle depends upon the angular resolution required for application.

Torque vs Pulses Rate

As the stepping rate is increased, the rotor has less time to drive the load from one position to the next as the stator-winding current pattern is shifted. Beyond a certain pulsing rate [(pulse/s) = pps] the rotor cannot follow the command and would begin to miss pulses.

Important! The start range as shown in Fig. 10.42 is that in which the load position follows the pulses without losing steps. The slew-range is one in which the load velocity follows the pulse rate without losing a step, but cannot start, stop or reverse on command.

Obviously the maximum slew rate would increase as the load is lightened. Typical variable-reluctance motors can operate at maximum position response rates up to 1200 pps.

Step skipping may also occur if motor oscillation amplitude about the locking position is too large.

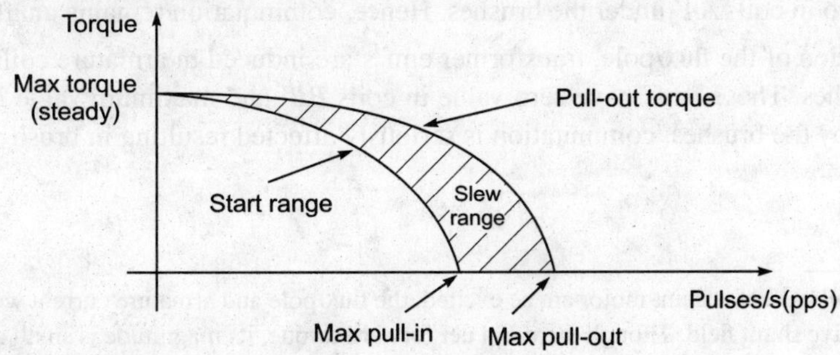

Fig. 10.42 *Torque versus pulses/s for a stepper motor*

10.8 | SERIES MOTOR—UNIVERSAL MOTOR

LO 4

A series motor can be run from either dc or ac (single-phase) supply provided that both stator and rotor cores are laminated to limit the iron-loss. Figure 10.43 is the cross-sectional view of a series motor connected to ac supply. As the armature current alternates, the field polarity alternates in phase with it as shown in Fig. 10.44; as a consequence the torque developed ($\propto \phi i_a$) is unidirectional with a pulsating component over and above the average value[‡‡‡]. The pulsating component of torque is filtered out by the rotor and load inertia so that the speed pulsations are almost negligible. The necessity of laminating the stator (poles and yoke) is indicated by the alternating flux it is required to carry.

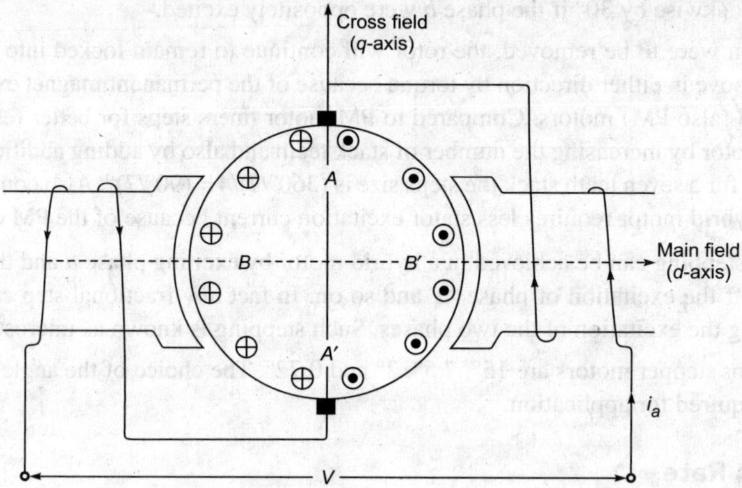

Fig. 10.43 *Series motor*

The current-carrying armature produces cross-flux along the q-axis in space quadrature with the main flux (in the d-axis). These two fluxes apart from being in space quadrature are alternating at a supply frequency. Each of these fluxes then induces rotational and static (transformer) emf's in the armature.

Emf's of Main Field (*d*-axis)

As in the dc case, rotational emf is induced in armature conductors adding to a maximum value at the q-axis brushes. The emf alternates in unison with flux/pole. The magnitude of this emf is maximum in coils BB' (Fig. 19.43) and is zero in coils AA' under the brushes. Hence, commutation remains unaffected by this emf. Caused by time-variation of the flux/pole, transformer emf's are induced in armature coils which add to net zero value at the brushes. These emfs have zero value in coils BB' and maximum value in coils AA'. Since coils AA' are shorted by the brushes, commutation is seriously affected resulting in brush sparking.

‡‡‡ When the armature and field of a shunt motor are ac excited, the flux/pole and armature current would be out-of-phase because of highly inductive shunt field. Though there is a net forward torque, its magnitude is small and the motor operation is highly inefficient.

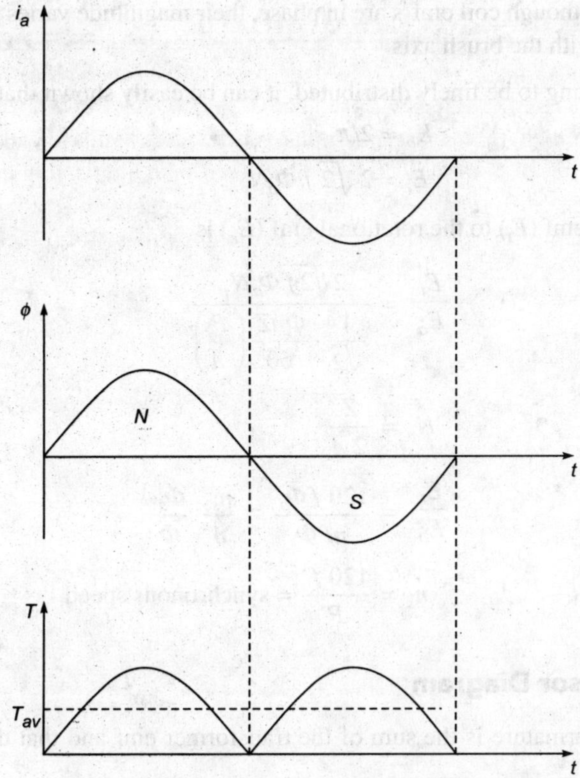

Fig. 10.44 *Torque in ac-excited series motor*

The rms value of the rotational emf caused by the main field is

$$E_a = \frac{1}{\sqrt{2}} \frac{\Phi n Z}{60} \times \left(\frac{P}{A}\right) \tag{10.64}$$

where Φ = maximum value of flux/pole

This emf is in-phase with the main flux phasor.

Emf's of Cross Field (*q*-axis Field)

The rotational emf caused by cross-flux (along *q*-axis) induces net zero emf at the brushes. This emf is zero in coils *BB*′ and maximum in coils *AA*′ thereby interfering in the commutation process.

On the other hand, the transformer emf of armature coils adds to a definite value and is alternating leading the crossflux by 90° (refer to Fig. 3.6). This emf is maximum in coils *BB*′ and zero in coils *AA*′ and so does not affect commutation.

The rms value of transformer emf E_t at the brushes is calculated below:

$$E_t = \sqrt{2}\pi K_b f \Phi_a N_p \tag{10.65}$$

where N_p = turns/parallel path; Φ_a = armature (cross) flux; proportional to I_a

K_b = space factor; though coil emf's are in phase, their magnitude varies as cosine of angle the coil axis makes with the brush axis

Assuming the armature winding to be finely distributed, it can be easily shown that

$$K_b = 2/\pi$$

Hence
$$E_t = 2\sqrt{2}\, f \Phi_a N_p \quad (10.66)$$

The ratio of the transformer emf (E_t) to the rotational emf (E_a) is

$$\frac{E_t}{E_a} = \frac{2\sqrt{2} f \Phi_a N_p}{\frac{1}{\sqrt{2}} \cdot \frac{\Phi n Z}{60}\left(\frac{P}{A}\right)}$$

But
$$N_p = \frac{Z}{2A}$$

∴
$$\frac{E_t}{E_a} = \frac{120 f \Phi_a}{n P \Phi} = \frac{n_s}{n} \cdot \frac{\Phi_a}{\Phi} \quad (10.67)$$

where
$$n_s = \frac{120 f}{P} = \text{synchronous speed}$$

Circuit Model and Phasor Diagram

The net emf induced in the armature is the sum of the transformer emf and that due to leakage inductance (x_a) and is given by

$$j\bar{I}_a x_a + j\bar{E}_t = j\bar{I}_a (x_a + \bar{E}_t/\bar{I}_a)$$

Since E_t is caused by cross-flux whose origin is in armature current, the two must be in phase. Define

$$\bar{E}_t/\bar{I}_a = x_{at}$$

Thus the net inductive emf in the armature circuit is

$$j\bar{I}_a (x_a + x_{at}) = j\bar{I}_a X_a$$

where
$$X_a = x_a + x_{at} \quad (10.68)$$

The circuit model of the series motor for ac operation can now be immediately drawn as in Fig. 10.45 where r_f, x_f = field resistance and reactance respectively.

From the circuit model[§§§] of Fig. 10.45

$$\bar{V} = \bar{I}_a (r_f + jx_f) + \bar{I}_a (R_a + jX_a) + \bar{E}_a \quad (10.69)$$

The corresponding phasor diagram is drawn in Fig. 10.46. The flux phasor $\bar{\Phi}$ (main) and $\bar{\Phi}$ (cross) are assumed in phase with motor current $\bar{I}_a$. In fact to be strictly correct $\bar{I}_a$ will lead these flux phasors by a small angle on account of hysteresis and eddy-current losses of the motor.

[§§§] This circuit model is not highly accurate because hysteresis and eddy-current losses and magnetic saturation have been ignored which in fact are significant.

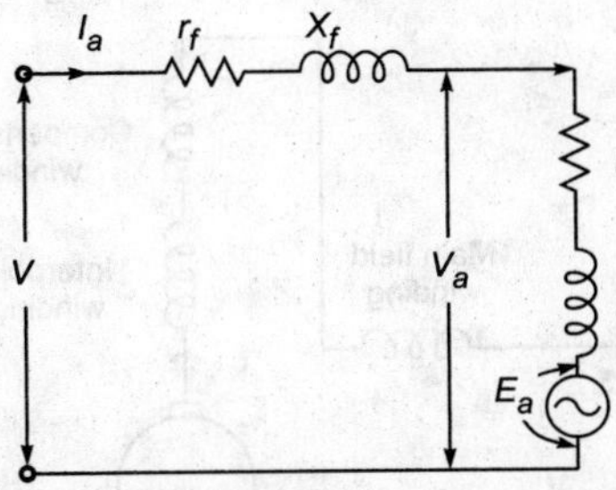

Fig. 10.45 *Circuit model of ac series motor for ac operation*

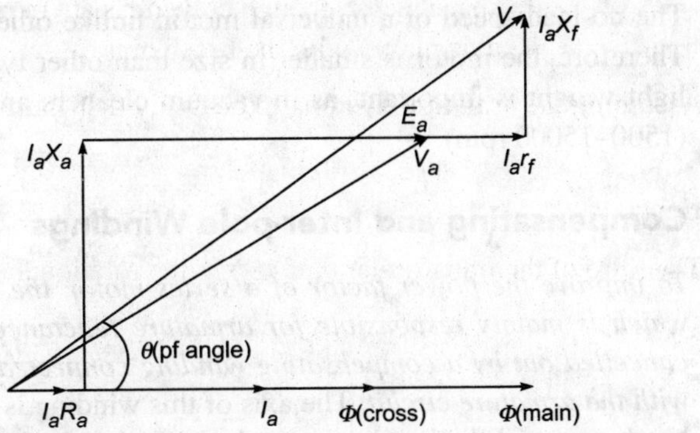

Fig. 10.46 *Phasor diagram of series motor for ac operation*

Torque Developed

As shown in the beginning, the torque developed is pulsating in nature with an average value and a predominant second harmonic component. The average torque is given by

$$T_{av}\omega_m = E_a I_a \tag{10.70}$$

Important! Being a series motor $E_a \propto I_a$ (assuming linear magnetic circuit), the torque is therefore proportional to square of armature current which becomes directly proportional at high armature current because of saturation of the magnetic circuit.

Performance Characteristics

Figure 10.47 shows the performance characteristics of an ac-operated series motor. The speed-torque characteristic has a typical series shape. The speed-torque characteristic for dc operation would lie somewhat higher as shown dotted. This is because of reactance voltage drops $[I_a(x_f + X_a)]$ in ac operation so that E_a and hence speed is lower for a given current and torque. The power factor of an ac-operated series motor is rather poor because of the large series reactance $(X_f + X_a)$.

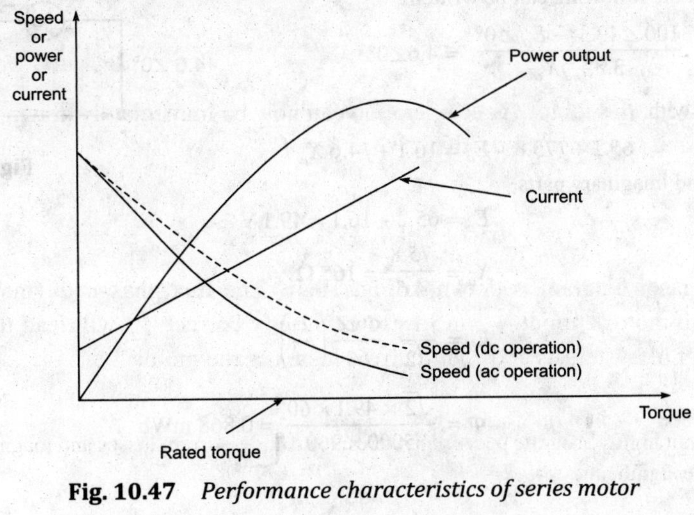

Fig. 10.47 *Performance characteristics of series motor*

The no-load speed of a universal motor, unlike other machines, is very high (of the order of 20000 rpm). Therefore, the motor is smaller in size than other types for a given load. Universal motors are used where light weight is important, as in vacuum cleaners and portable tools which usually operate at high speeds (1500–15000 rpm).

Compensating and Inter-pole Windings

To improve the power factor of a series motor, the cross-flux which is mainly responsible for armature reactance must be cancelled out by a compensating winding connected in series with the armature circuit. The axis of this winding is along the brush axis and the winding must be spread over the full pole pitch for effective neutralization of cross-flux. Like in a dc machine the interpole winding when provided must be effective over a narrow interpolar region. The schematic connection diagram of all the stator windings and the armature is shown in Fig. 10.48.

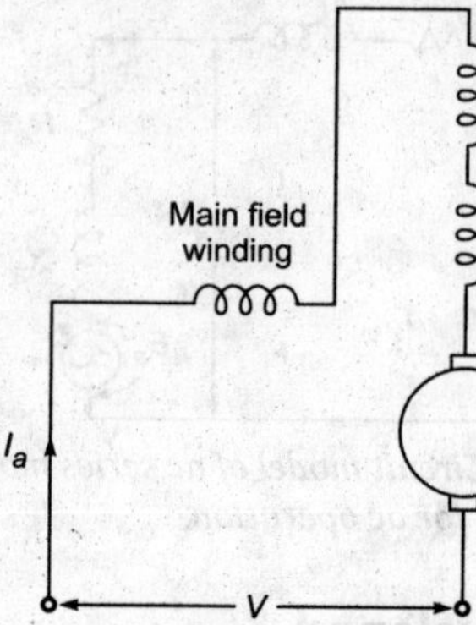

Fig. 10.48 *Arrangement of stator winding in ac-operated series motor*

Example 10.7 *A universal motor (ac-operated) has a 2-pole armature with 960 conductors. At a certain load the motor speed is 5000 rpm and the armature current is 4.6 A; the armature terminal voltage and input are respectively 100 V and 300 W.*

Compute the following, assuming an armature resistance of 3.5 Ω.

(a) Effective armature reactance

(b) Maximum value of useful flux/pole

LO 4

Solution The operating conditions in terms of voltage and current of the armature circuit are shown in Fig. 10.49.

$$100 \times 4.6 \cos \phi = 330 \text{ W}$$

or $$\phi = 49.3°$$

(lagging because of reactive nature of the circuit).

(a) From the circuit the following can be written

$$\frac{100\angle 49.3° - E_a\angle 0°}{3.5 + jX_a} = 4.6\angle 0°$$

$\bar{E}_a$ is in-phase with $\bar{I}_a$

or $$65.2 + j75.8 - E_a = 16.1 + j4.6\,X_a$$

Equating real and imaginary parts,

$$E_a = 65.2 - 16.1 = 49.1 \text{ V}$$

$$X_a = \frac{75.8}{4.6} = 165\ \Omega$$

(b) $$E_a = \frac{1}{\sqrt{2}} \cdot \frac{\Phi n Z}{60}\left(\frac{P}{A}\right)$$

$$\Phi = \frac{\sqrt{2} \times 49.1 \times 60}{5000 \times 960} = 0.868 \text{ mWb}$$

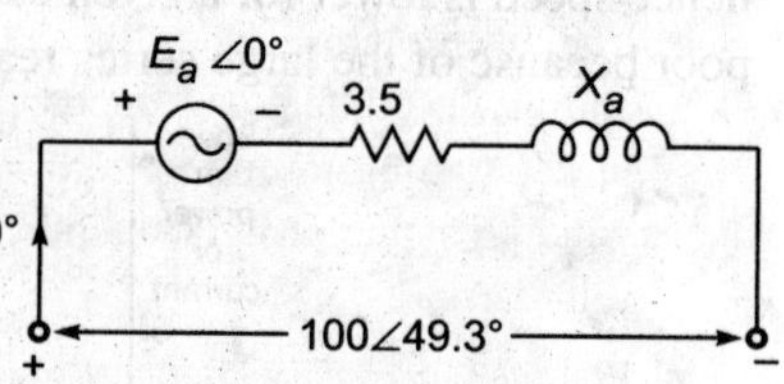

Fig. 10.49

Concept Check LO 4

1. Why are high speeds often desirable in the operation of universal motors? What limit the speed? ○●●
2. How can the speed of a linear induction motor (LIM) be controlled? ○●●
3. State the various applications of a stepper motor. ○○●
4. Mention some applications of stepper motor. ○○●
5. Define resolution. ○●●
6. What is meant by micro stepping in a stepper motor? ○●●
7. Mention few applications of ac series motors (or universal motor). ○○●
8. What modifications have to be done on dc series motor to make it work with single-phase ac supply? ●●●

SUMMARY

LO 1

- Fractional kilowatt motors are employed in fans, refrigerators, mixers, vacuum cleaners, washing machines, other kitchen equipment, tools, small farming appliances etc.
- When single phase induction motor winding carries a sinusoidal current, it produces a sinusoidally space-distributed mmf whose peak value pulsates with time.
- The two rotating fields have the same strength and produce equal and opposite torques resulting in net starting torque of zero value so the single winding single phase induction motor is thus non self-starting.
- Practical necessity dictates that the two rotating fields are made to have unequal strength under stationary conditions there by making the motor self-starting. This requires one more winding on the motor called auxiliary windings which is in space quadrature with the main winding and comprises smaller number of turns of thinner wire.
- Two value capacitor motor has high starting torque and better running power factor and efficiency and quieter and smoother operation.

LO 2

- The single phase reluctance motor working with the principle of the origin of the reluctance torque lies in the tendency of the rotor to align itself in the minimum reluctance position with respect to the synchronously rotating flux of the forward field.
- In hysteresis motor, the rotor magnetization to lag behind the stator created mmf wave, so rotor flux lags by angle of the stator mmf axis under the influence of the hysteresis.

PRACTICE PROBLEMS

10.1 A 220 V, 50 Hz, 6-pole, single-phase induction motor has the following circuit model parameters: LO 1 ○○●

$$R_{1m} = 3.6\ \Omega \qquad (X_{1m} + X_2) = 15.6\ \Omega$$
$$R_2 = 6.8\ \Omega \qquad X = 96\ \Omega$$

The rotational losses of the motor are estimated to be 75 W. At a motor speed of 940 rpm, determine the line current, power factor, shaft power and efficiency.

10.2 A(1/4) kW, 230 V, 50 Hz, 4-pole split-phase motor has the following circuit model parameters: LO 1

$R_{1m} = 10.1\ \Omega$, $X_{1m} = 11.6\ \Omega$
$R_{1a} = 40.30\ \Omega$, $X_{1a} = 9.65\ \Omega$
$X = 236\ \Omega$, $a = 0.92$
$R_2 = 9.46\ \Omega$ $X_2 = 6.86\ \Omega$

Friction, windage and core-loss = 45 W.

(a) Calculate starting torque and current of the motor.

(b) Calculate the performance of the motor at a slip of 0.035. (The auxiliary winding is open-circuited.)

10.3 A 400 W, 220 V, 50 Hz, 6-pole, permanent-capacitor motor has the following circuit model parameters: LO 1

$R_1 = 9.2\ \Omega$, $X_{1m} = 8.7\ \Omega$
$R_{1a} = 15.5\ \Omega$, $X_{1a} = 13.5\ \Omega$
$X = 138.5\ \Omega$

$\bar{Z}_C = -j257\ \Omega$ (series capacitive reactance in auxiliary winding)

$a = 1.25$

$R_2 = 14.3\ \Omega$

$X_2 = 6.84\ \Omega$

The windage friction and core-loss is 45 W

(a) Calculate starting torque and current.

(b) Calculate motor performance at $s = 0.1$.

10.4 Show that if the stator voltages of two-phase induction motor are V_m and V_a with a fixed phase difference of 90°, the starting torque is the same as for a balanced voltage of $\sqrt{V_m V_a}$ per phase. LO 1

10.5 For a two-phase servomotor with a high resistance rotor, find approximate expressions for forward and backward torques in terms of phase voltages (differing 90° in phase) and motor speed. Assume the stator impedance and rotor reactance to be negligible. LO 3

10.6 Show that in a two-phase tachometer with a high resistance rotor, the voltage induced on the open-circuited phase (*a*) is proportional to the rotor speed and leads the other phase (*m*) voltage by 90°. Neglect the stator impedance and rotor reactance. LO 3

Answers

10.1 5.22 A, 0.55 lagging, 404.3 W, 64%

10.2 (a) 1.78 N m, 12.88 A; (b) $I_L = I_m = 2.14$ A, 0.647 lagging, $P_m = 241$ W, 75.67%

10.3 (a) 3.69 N m, 6.95 A, (b) $I_L = 2.78$ A, 0.96 lagging, 3.83 N m, 68%

10.5 $T = kV_m V_a - \dfrac{V_m^2}{\omega_s}\omega_0$

MULTIPLE CHOICE QUESTIONS

1. In case of shaded pole induction motor, the direction of rotation of motor is LO 1
   (a) from main pole to shaded pole
   (b) from shaded pole to main pole
   (c) Either of the above depending on voltage
   (d) Either of the above depending on power factor

2. Direction of rotation of a split-phase induction motor can be reversed by reversing the connection of
   (a) running winding only (b) starting winding only
   (c) Either (a) or (b) (d) Both (a) and (b)
   LO 1

3. A hysteresis motor works on the principle of
   LO 2
   (a) hysteresis loss (b) magnetization of rotor
   (c) eddy current loss (d) electromagnetic induction

4. For which of the applications a reluctance motor is preferred?
   LO 2
   (a) Electric shavers (b) Refrigerators
   (c) Signaling and timing devices (d) Lifts and hoists

5. Which of the following motors is preferred for tape recorders?
   LO 2
   (a) Shaded pole motor (b) Hysteresis motor
   (c) Two-value capacitor motor (d) Universal motor

6. A hysteresis motor
   LO 2
   (a) is not a self-starting motor (b) is a constant speed motor
   (c) needs dc excitation (d) cannot be run in reverse speed

7. The most common two-phase ac servomotor differs from the standard ac induction motor because it has
   (a) higher rotor resistance (b) higher power rating
   (c) motor stator windings (d) greater inertia
   LO 3

8. One of the basic requirements of a servomotor is that it must produce high torque at all
   LO 3
   (a) loads (b) frequencies
   (c) speeds (d) voltages

9. An ac tachometer is just a/an ________with one phase excited from the carrier frequency.
   LO 3
   (a) two-phase ac servomotor (b) two-phase induction motor
   (c) ac operated universal motor (d) hybrid stepper motor

10. Which of the following motors is used in mixer grinders?
    LO 4
    (a) Repulsion motor (b) Reluctance motor
    (c) Hysteresis motor (d) Universal Motor

11. Universal motor can run only on
    LO 4
    (a) ac only (b) dc only
    (c) Either ac or dc (d) None of these

12. In ac series motor, compensating winding is employed to
    LO 4
    (a) reduce the effect of armature reaction (b) increase the torque
    (c) reduce sparking at the brushes (d) None of these

13. A stepper motor is a
    LO 4
    (a) dc motor (b) single-phase ac motor
    (c) multi-phase motor (d) two-phase motor

14. A 1.8°step, 4-phase stepper motor has a total of 40 teeth on 8 pole of a stator. The number of rotor teeth for their rotor will be
    LO 4
    (a) 40 (b) 50 (c) 100 (d) 80

Answers

1. (a)	**2.** (c)	**3.** (a)	**4.** (c)	**5.** (b)	**6.** (a)	**7.** (a)	**8.** (c)	**9.** (d)	**10.** (d)
11. (c)	**12.** (c)	**13.** (d)	**14.** (b)						

CHAPTER 11

Generalized Theory of Electrical Machines

LEARNING OBJECTIVES

After studying this chapter, students will be able to

LO1 Explain the working of basic two-pole machines

LO2 Describe the functioning of a transformer with movable secondary winding and also define Kron's primitive machine

LO3 Illustrate the various transformations in linear machines

LO4 Explain the three-phase to two-phase and rotating axis to stationary axis transformations

LO5 Interpret the physical concept of Park's transformation and derive conclusions from it

11.1 | INTRODUCTION

The objective of generalized machine theory is to establish an expression of electromagnetic torque in terms of machine variables (i and θ). This theory may be developed based on anyone of the following principles:

(i) The derivation of equivalent-circuit representation of magnetically coupled circuits

(ii) The concept of sinusoidally distributed winding

(iii) The concept of rotating air-gap Magneto Motive Force (mmf)

(iv) The derivation of winding inductances, both synchronous and asynchronous machines

Industry Relevance

Commutators are used in dc machines as well as in dc generators. High performance commutator machines find their applications in mining, aerospace, military, etc. On the other hand, induction machines are cheap, efficient, and have less maintenance cost as compared to commutator machines. Induction machines have their applications in industrial as well as domestic purposes like water pumps, electrical industries, refrigeration, etc.

11.2 | CONVENTION

In the development of the generalized machine theory, certain conventions have been adopted and they are consistently maintained throughout, for intelligent specification of theory. Some of them are listed as follows:

(i) The distribution of current and flux under one pair of poles, repeats itself under all other pairs of poles.

(ii) Each part of the actual machine forming a single circuit is represented in the idealized or basic two-pole machine by a single coil. These single coils occupy only a part of the radius on rotor and stator, the coils are shown on one side of the machine only.

(iii) The axis of the poles is called the direct axis (D_{axis}) of the machine, The axis which is 90° away from it is called the quadrature axis (Q_{axis}).

(iv) The positive direction of the current in any coil is towards the coil in the lead nearer to the centre of the diagram.

(v) The positive direction of rotation of the rotor is taken as clockwise or anti-clockwise and torque is also taken positive when acting in the sense of positive rotation.

11.3 | BASIC TWO-POLE MACHINE

LO 1 Explain the working of basic two-pole machines

In a two-pole machine representation, each part of the winding forming a single circuit is represented by a single coil. In general, the basic two-pole machine diagrams for any type of rotating electrical machine, can be drawn by knowing

(i) the stator and rotor configuration, salient pole member being taken as stationary

(ii) the winding arrangement on both stator and rotor

(iii) the position of the brushes on the rotor

Commutator Machines

All the direct current (dc) machines are invariably fitted with a commutator but in case of alternating current (ac) machines few of them have commutator winding. Fig. 11.1(a) depicts the two-pole machine representation of a commutator machine such as a dc separately excited motor. Commutator action effectively divides the armature winding along the brush axis to form two uniform current sheets which are in opposite polarity. While the armature rotates, the direction and magnitude of the armature mmf F_{QR} remain unchanged, because the direction of current under north and south poles does not vary with armature rotations.

> **Probe**
> What is the result of commutator action?

The two-pole representation of a commutator machine is shown in Fig. 11.1(b). As per magnetic considerations, the commutator winding behaves as a stationary concentrated coil (Q_r) shown in Fig. 11.1(b). The direction of the field current flow in the coil (D_s shown in Fig. 11.1(b)) should divide the poles such as North Pole (N) or South Pole (S). For total equivalence, the stationary concentrated coil (Q_r) generates the same rotational voltage as that generated by the commutator winding, when acting as a generator. In motor action, the same

> **Probe**
> What is a pseudo-stationary or quasi-stationary coil?

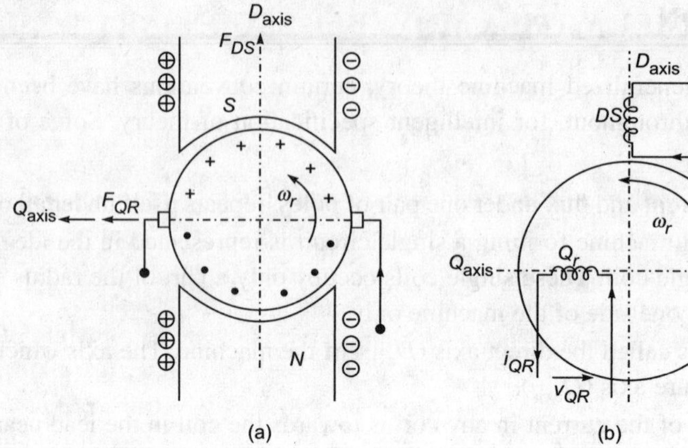

Fig. 11.1 *(a) Commutated machine (separately excited dc/motor)*
(b) Two-pole representation of commutator machine

torque is developed for the same currents in the coil (Q_r) and in the commutator winding. Such a coil is fictious and as it is to be used to represent commutator winding, it is given the identifying name of **pseudo-stationary coil** or **quasi-stationary coil.** This coil does not move along the rotor. However, voltage is induced in the coil by virtue of rotation. In general, the coil is located in the quadrature axis.

Induction Machines

The basic 3-ϕ induction machine diagram is shown in Fig. 11.2. It consists of three windings on the stator can be represented by three coils R, Y and B. These windings are 120° electrically displaced between them. Letting the winding R is taken along the d-axis for convenience. The rotor carries a short-circuited cage or a slip ring winding. The rotor windings are represented by three coils R^1, Y^1 and B^1 on the rotor.

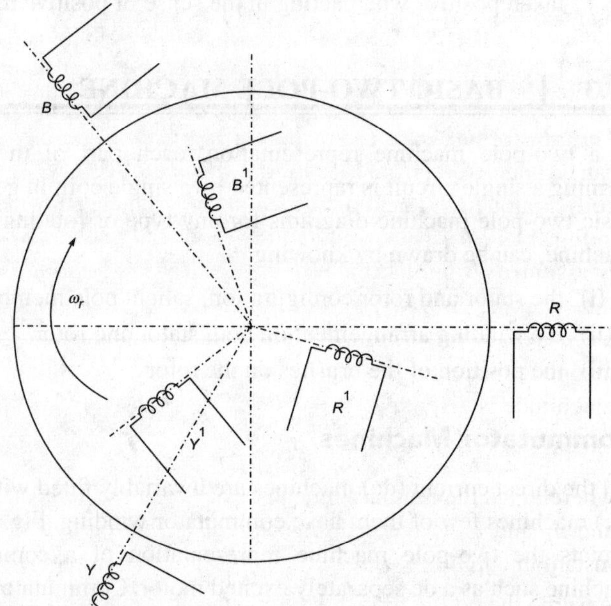

Fig. 11.2 *Basic machine diagram for an induction machine*

Concept Check — LO 1

1. What is Pseudo-stationary coil or quasi-stationary coil in commutator machine? ○○●
2. Discuss the significance of speed voltage. ○●●

Note: Difficulty Level ⟶ ○○● — Level 1 and Level 2 Category; ○●● — Level 3 and Level 4 Category; ●●● — Level 5 and Level 6 Category

11.4 | TRANSFORMER WITH A MOVABLE SECONDARY WINDING

LO 2 Describe the functioning of a transformer with movable secondary winding and also define Kron's primitive machine

An ideal transformer has been represented with two windings shown in Fig. 11.3. The secondary winding (S_y) is coincident with primary (P_y) winding shown fin Fig. 11.3(a). In Fig. 11.3(b), the secondary winding axis is θ degrees mechanically deviated with respect to the primary magnetic axis. Assume both windings have an equal number of turns. Suppose the secondary winding is coincident with the magnetic axis of the primary winding and the same amount of emf is induced in the secondary. Otherwise, the emf induced in the secondary winding given by

$$e = N\cos\theta \tag{11.1}$$

Probe

Write the emf equation induced in the secondary winding of a transformer with movable secondary winding ?

where e is the total emf induced in the secondary winding in volts; N is the number of turns and θ is the angle between magnetic axis of primary and secondary windings.

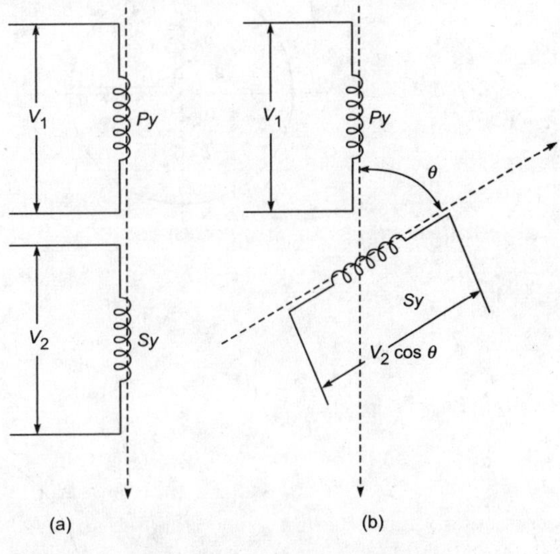

Fig. 11.3 *Transformer with primary (Py) and secondary (Sy) winding axis (a) coincident (b) at an angle of θ*

Transformer and Speed Voltage

The transformer and speed voltage of an armature winding can be understood from Fig. 11.4. It consists of coil (field coil) d_S along D_{axis}, and on the stationary element shown in Fig. 11.12(a). The magnetic flux produced by the coil is assumed to be distributed sinusoidally in space and time varying in nature. Another similar coil d_R along the d-axis is on the moving element. The alternating magnetic flux produced by i_{ds} in coil d_s links d_R and its magnitude is given by

$$\lambda_{md} = M_d i_{ds} \tag{11.2}$$

where M_d is the mutual inductance between the coils d_s and d_R and i_{ds} is the exciting current at this particular instant in coil d_s.

Assume the coil d_R is a moving element, transverses an angle θ in time, say t seconds, such that $\theta = \omega_r t$ radians shown in Fig. 11.12(b). Therefore, the flux linkage in the coil d_R after t seconds is given by

$$\lambda_{md}\cos\theta \tag{11.3}$$

Applying Faraday's law, the emf induced in the coil is written by

$$e = -\frac{d}{dt}(\lambda_{md}\cos\theta) \tag{11.4}$$

Probe

What is the speed voltage component expressed as?

i.e., $$e = +\lambda_{md}\,\omega_r\sin\theta - \cos\theta\, M_d\, p i_{ds};\ p = \frac{d}{dt},\ \text{derivative operator} \tag{11.5}$$

where $\frac{d\theta}{dt} = p\theta = \omega_r$; the first term in. Eq. (11.5) is the **speed voltage component**, because speed ω_r appears in it. The other terms is called as **transformer voltage**, since this, component involves the time derivative of current i_{ds}. Figure 11.4(c) illustrates the coil d_R coincide with q-axis, so emf induced in the coil d_R at this instant is given by

$$e_q = -\omega_r \lambda_{md}\text{; no transformer voltage, } \cos 270° = 0 \tag{11.6}$$

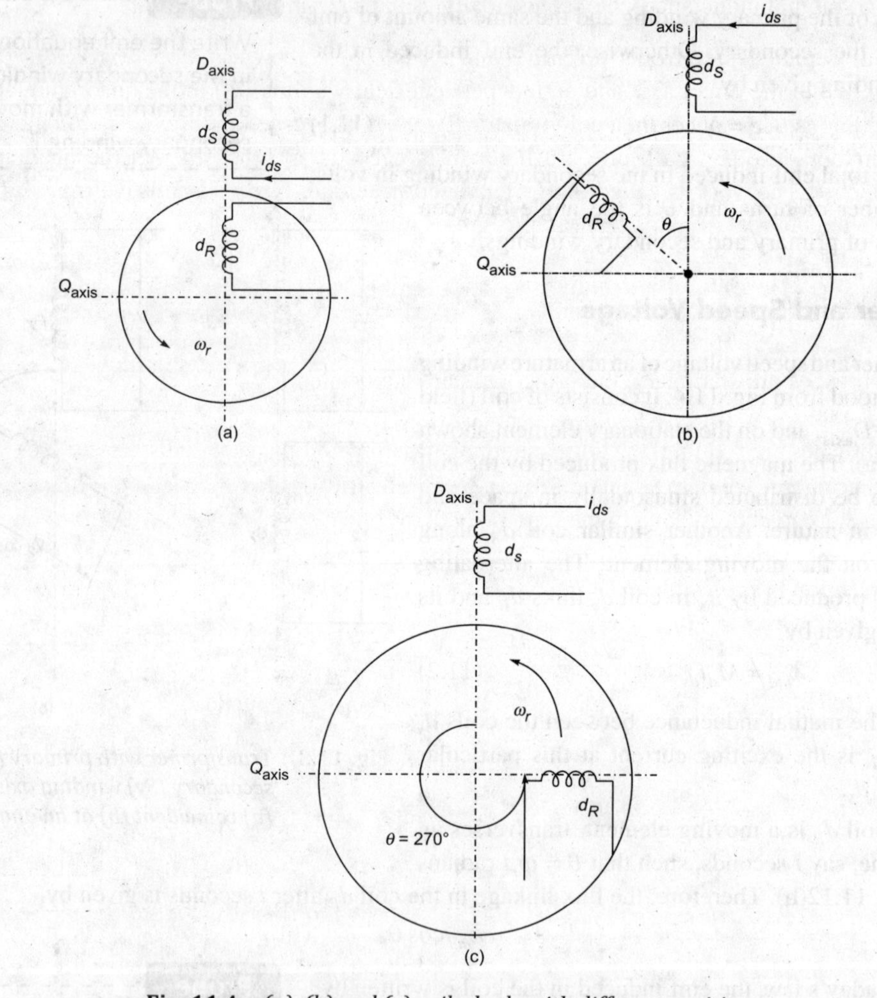

Fig. 11.4 *(a), (b) and (c) coils d_S, d_R with different positions*

11.5 | KRON'S PRIMITIVE MACHINE

LO 2

The equivalent electrical network of a two-pole machine is called Kron's primitive machine or generalized machine.

Fundamental Assumption of a Primitive Machine

(i) The MMF distributed along the air-gap periphery is sinusoidal.
(ii) Space harmonics and their effect on torque and on induced voltage are neglected.
(iii) Saturation and hysteresis effects are neglected.
(iv) No variation of inductances with the relative movement between stator and rotor (tapped by commutator brushes).
(v) Inductance variations due to slot openings are neglected.
(vi) Self and mutual inductances vary sinusoidally as the rotor moves.

Figure 11.5(a) depicted the Kron's primitive and its two-pole representation. It consists of a stationary field winding (d_s) in the direct axis and an independent field winding (q_s) in the quadrature axis. A rotating armature winding is taken out from a commutator. Two sets of brushes are provided which are magnetically perpendicular to each other, with one brush set in direct axis and the other in the quadrature axis. The equivalent electrical network is called the **generalized machine, Kron's primitive machine, generalized Model or two axis model of rotating machines.**

Note:

(i) Equivalent armature coils d_R and q_R as in Fig. 11.5(b) produce stationary flux along the brush axes.
(ii) If the brushes are stationary, the brush axes are also called **Park's axes.**
(iii) If the brushes are rotating, the corresponding armature coil fluxes will also rotate.
(iv) Single armature winding is made to act as two electrically and magnetically independent armature windings in quadrature, due to the presence of two brush sets.
(v) This representation is valid in both eclectically and magnetically linear systems and the principle of superposition is applicable.

Probe

Define Park's axes.

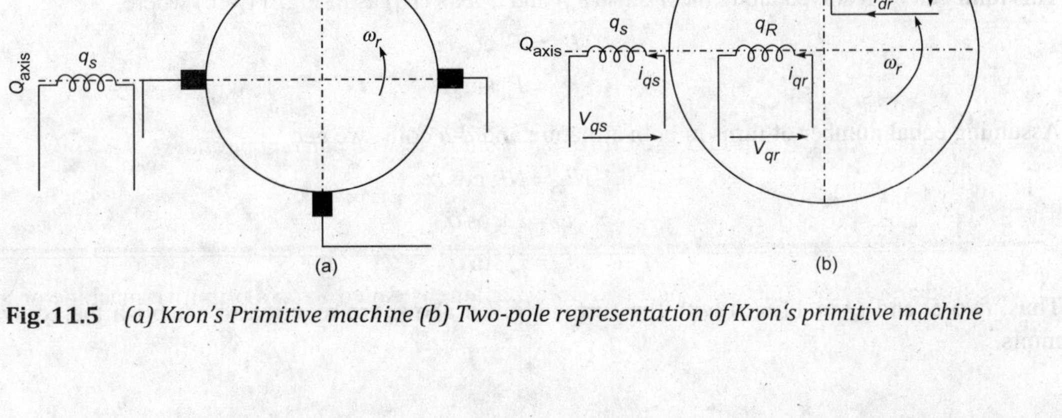

Fig. 11.5 *(a) Kron's Primitive machine (b) Two-pole representation of Kron's primitive machine*

Concept Check LO 2

1. State the fundamental assumptions of a primitive machine.

11.6 LINEAR TRANSFORMATIONS IN MACHINE

LO 3 Illustrate the various transformations in linear machines

The term 'linear transformation' means that the transformation from old to new set of variables or vice versa is governed by linear equations (An equation, each of whose terms is a first-degree term in the variable is called a linear algebraic equation). A linear equation does not involve squares, higher powers of the variables or their products.

[New variables] = [Transformation matrix] [Old variables]

or [Old variables] = [Transformation matrix] [New variables]

Invariance of Power

Probe
What do you understand by invariance of power?

This principle of maintaining the same power under transformation from an old to a new set of variables are known as invariance of power. If power is variant under transformation, power and torques can be determined only from currents expressed in the original system reference axis. In other words, if the current in the new set of variables are obtained, it will be essential to transform them back to the original set of variables before power or torque is calculated. However, if the transformations corresponds to the power invariance, power and torques 'calculated either from the new set of variables or from the old set of variables' will yield identical results.

Transformation Related to Displaced Brush-Axis

If a commutator machine has brushes displaced from *d*- or *q* axes shown in Fig. 11.6(a), a transformation is necessary from the displaced brush axis to *d*-*q* axis. Suppose one set of brushes, denoted by *AA*, makes an angle α with the *d*-axis as shown in Fig. 11.6(b). The armature establishes an mmf (F_a) along its brush axis. This mmf can be resolved along the *d*-axis (F_d) and *q*-axis (F_q) as in Fig. 11.6(c) where,

$$F_d = F_a \cos\alpha \tag{11.7}$$

$$F_q = F_a \sin\alpha \tag{11.8}$$

Assuming equal number of turns in both armature and *d*-*q* coils, we get

$$Ni_d = Ni_a \cos\alpha$$

$$i_d = i_a \cos\alpha \tag{11.9}$$

$$i_q = i_a \sin\alpha \tag{11.10}$$

Thus, two *D* and *Q* in 11.6(c) with the currents given above are required to establish the equivalence of mmfs.

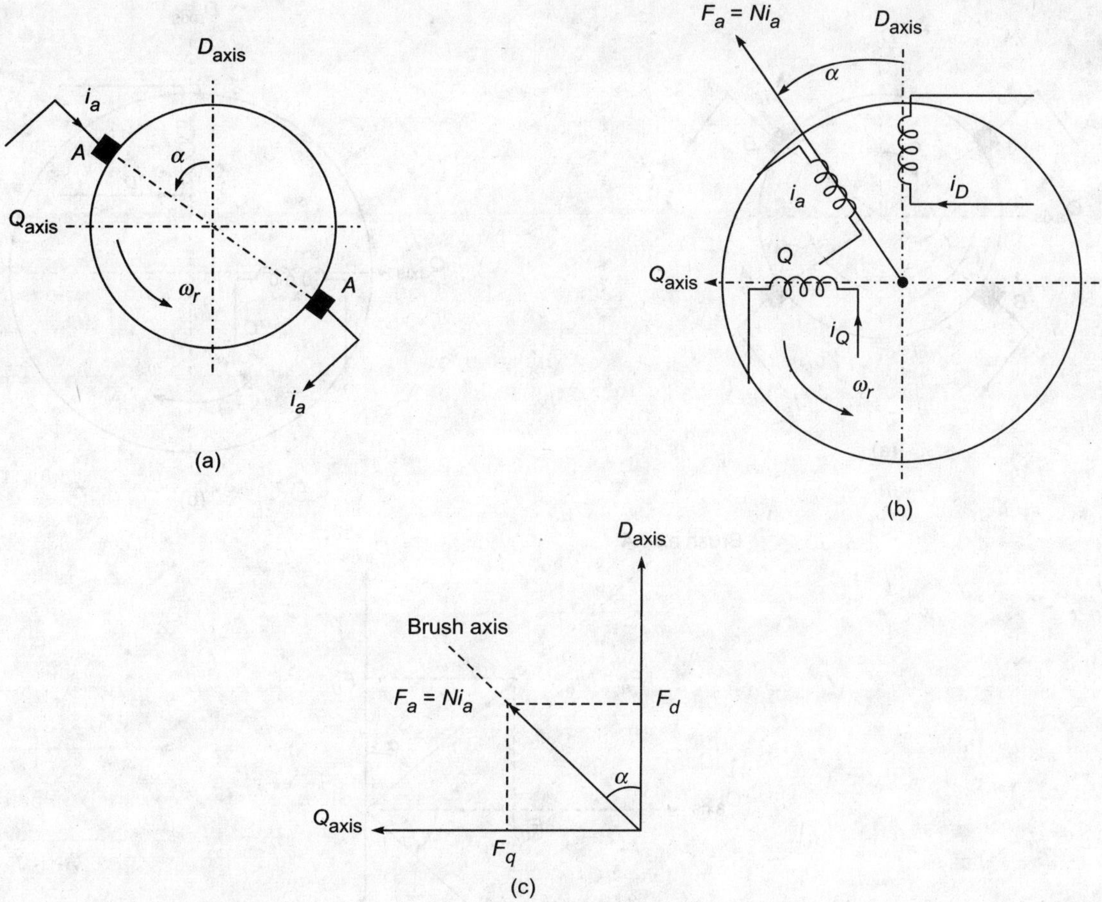

Fig. 11.6 *(a) One set of displaced brush axis, (b) Its D-Q equivalent, (c) Shifting transformation of brush*

Consider now two sets of brushes *AA* and *BB* displaced from the *D*-, *Q*-axis by angles α and β respectively as shown in Fig. 11.7(a). The equivalent *D*-, *Q*-axis coils are shown in Fig. 11.7(b).

The equivalent mmfs along *D*-axis and *Q*-axis can be written from Fig. 11.7(c) as

The mmf in *D*-axis and *Q*-axis can be written as

$$F_d = F_a \cos\alpha - F_b \sin\gamma \tag{11.11}$$

and

$$F_q = F_a \sin\alpha + F_b \cos\gamma \tag{11.12}$$

Probe

Write an expression for mmf in D- and Q-axis.

Assuming equal number of turns as above, we get

$$i_d = i_\alpha \cos\alpha - i_b \sin\gamma \tag{11.13}$$

$$i_q = i_a \sin\alpha + i_b \cos\gamma \tag{11.14}$$

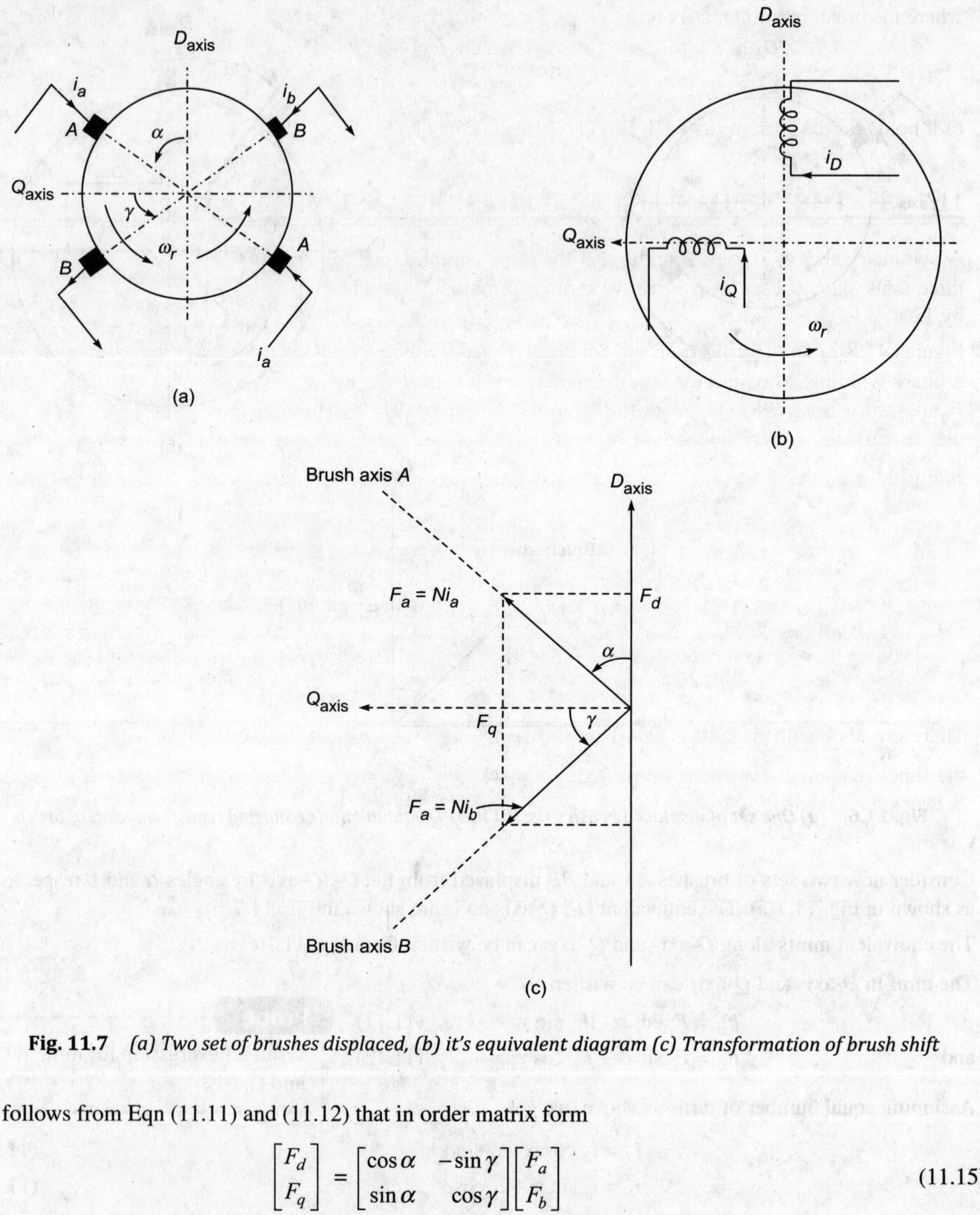

Fig. 11.7 *(a) Two set of brushes displaced, (b) it's equivalent diagram (c) Transformation of brush shift*

If follows from Eqn (11.11) and (11.12) that in order matrix form

$$\begin{bmatrix} F_d \\ F_q \end{bmatrix} = \begin{bmatrix} \cos\alpha & -\sin\gamma \\ \sin\alpha & \cos\gamma \end{bmatrix} \begin{bmatrix} F_a \\ F_b \end{bmatrix} \tag{11.15}$$

where the brush rotation matrix is

$$B(\text{Rot}) = \begin{bmatrix} \cos\alpha & -\sin\gamma \\ \sin\alpha & \cos\gamma \end{bmatrix}$$

of it holds for the currents as well; Eqs (11.14) and Eq. (11.15).

11.7 | THREE-PHASE TO TWO-PHASE TRANSFORMATION

LO 4 Explain the three-phase to two-phase and rotating axis to stationary axis transformations

A symmetrical 3-ϕ, 2-pole, windings on the rotor is represented by three coils A, B, C each of N effective turns and mutually displaced by 120°.

Figures 11.8(a) and (b) illustrate the balanced 3-phase winding and 2-phase winding. Maximum value of magnetizing forces F_a, F_b and F_c are shown along their axis-respective phases. The combined effect of these three mmfs results in a constant magnitude mmf, which rotates at a constant angular velocity depending on the poles and frequency.

In Fig. 11.8(a), the three-phase currents are written as

$$i_a = I_m \cos \omega t \tag{11.16}$$

$$i_b = I_m \cos\left(\omega t - \frac{2\pi}{3}\right) \tag{11.17}$$

$$i_c = I_m \cos\left(\omega t - \frac{4\pi}{3}\right) \tag{11.18}$$

Probe

How is three-phase current expressed as?

Therefore, the resultant mmf of constant magnitude $\frac{3\,I_m\,N}{2}$ rotates with respect to three-phase winding at the time frequency. *The space phase angle between the windings must comply with the time-phase angle between the currents.*

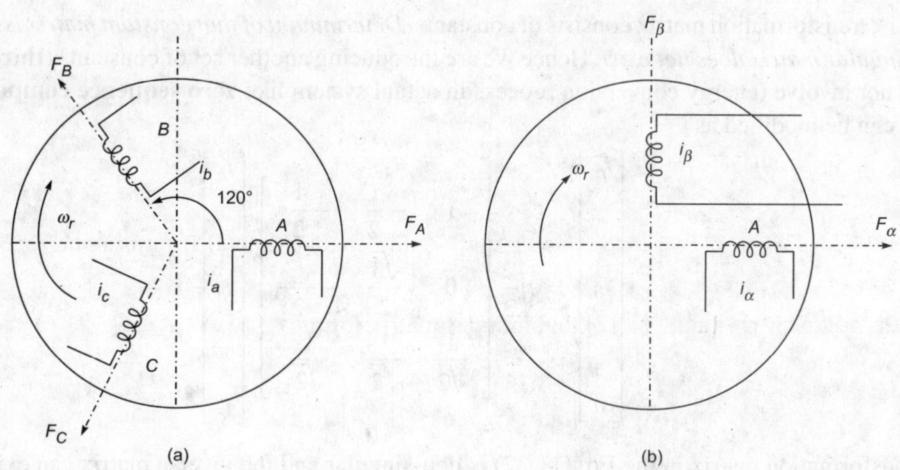

Fig. 11.8 *(a) Balanced 3-phase winding (b) 2-phase winding on the rotor*

In Fig. 11.8(b), balanced two-phase windings (α, β) are orthogonally placed on the rotor. For convenience in transformation, the axes of phase *A* and *a* are taken to be coincident.

Then the two phase currents are given by

$$i_\alpha = I_m \cos \omega t \tag{11.19}$$

$$i_\beta = I_m \cos\left(\omega t - \frac{\pi}{2}\right) = I_m \sin \omega t \tag{11.20}$$

It is flowing in the two windings, Therefore the resultant mmf of constant magnitude IN rotates with respect to two-phase winding at the time frequency of the phase currents.

If the mmfs of three-phase and two-phase systems are equal then the following modification can be made:

(i) By changing the magnitude of two-phase currents
(ii) By changing the number of turns of the two-phase windings
(iii) By changing both the magnitude of current and number of turns

Note: *Voltage magnitude of two-phase winding does not change because mmf is the product of number of turns and current* ($NI = AT$).

Significance of Zero Sequence Components in Three Phases to Two Phases Conversion

Probe

Express the transformation equation in matrix form.

The transformation equation giving the new currents i_α and i_β in terms of i_a, i_b and i_c can be written in a matrix form as follows:

$$\begin{bmatrix} i_\alpha \\ i_\beta \end{bmatrix} = \sqrt{\frac{2}{3}} \begin{bmatrix} 1 & -\frac{1}{2} & -\frac{1}{2} \\ 0 & \frac{\sqrt{3}}{2} & -\frac{\sqrt{3}}{2} \end{bmatrix} \begin{bmatrix} i_a \\ i_b \\ i_c \end{bmatrix} \tag{11.21}$$

In Eq. (11.21), transformation matrix consists of constants. *Determinant of that constant matrix is a singular, so inverse of singular matrix does not exist.* Hence we are introducing another set of constants (**third equations**) which does not involve (energy conversion process in actual system like **zero sequence component**. So the Eq. (11.21) can be modified as

$$\begin{bmatrix} i_\alpha \\ i_\beta \\ 0 \end{bmatrix} = \sqrt{\frac{2}{3}} \begin{bmatrix} 1 & -\frac{1}{2} & -\frac{1}{2} \\ 0 & \frac{\sqrt{3}}{2} & -\frac{\sqrt{3}}{2} \\ \frac{1}{\sqrt{2}} & \frac{1}{\sqrt{2}} & \frac{1}{\sqrt{2}} \end{bmatrix} \begin{bmatrix} i_a \\ i_b \\ i_c \end{bmatrix} \tag{11.22}$$

Now the transformation matrix in the Eq. (11.22) is non-singular and the inverse matrix can exists.

11.8 ROTATING AXIS (α, β, 0) TO STATIONARY AXIS (d, q, 0) TRANSFORMATION

LO 4

When the transformation is carried out from rotating axis to stationary axis, the relative position of rotating axes varies with respect to stationary or fixed axis. In view of this, the matrix giving the transformation from rotating axis to stationary axis and vice versa, must contain coefficients which are functions of the relative position of the moving (α, β) and fixed (d, q) axis. The transformation from rotating to stationary axis results in replacing the moving coils by **pseudo-stationary coils** shown in Fig. 11.9.

Probe

What happens when the transformation is carried out from rotating axis to stationary axis?

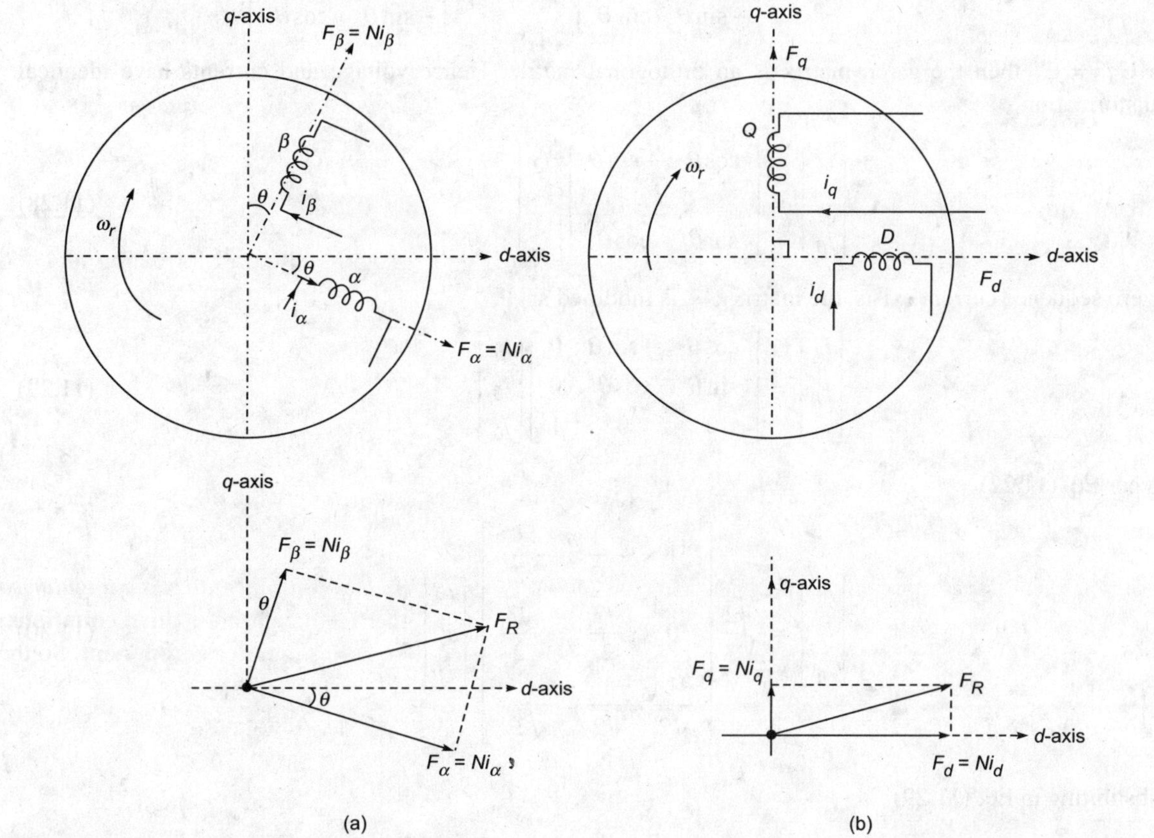

Fig. 11.9 *(a) Rotating axis α,β with rotating space phasors F_α, F_β (b) Stationary axis d, q with stationary space phasors F_d, F_q*

Zero sequence quantities are not transformed and thus the required transformation is only from (α, β) to (d, q). Assume numbers of effective turns are equal in α, β and d, q windings

$$F_d = F_\alpha \cos\theta + F_\beta \sin\theta \tag{11.23}$$

Therefore $$Ni_d = Ni_\alpha \cos\theta + Ni_\beta \sin\theta \tag{11.24}$$

or $$i_d = i_\alpha \cos\theta + i_\beta \sin\theta \tag{11.25}$$

$$i_q = -i_\alpha \sin\theta + i_\beta \cos\theta \tag{11.26}$$

In matrix form

$$\begin{bmatrix} i_d \\ i_q \end{bmatrix} = \begin{bmatrix} \cos\theta & \sin\theta \\ -\sin\theta & \cos\theta \end{bmatrix} \begin{bmatrix} i_\alpha \\ i_\beta \end{bmatrix} \tag{11.27}$$

Let $$C = \begin{bmatrix} \cos\theta & \sin\theta \\ -\sin\theta & \cos\theta \end{bmatrix} \text{ then } C^T = \begin{bmatrix} \cos\theta & -\sin\theta \\ -\sin\theta & \cos\theta \end{bmatrix} = C^{-1}$$

As $C^{-1} = C^T$ then the given matrix is an orthogonal matrix. Hence voltage and currents have identical transformation

$$\begin{bmatrix} i_\alpha \\ i_\beta \end{bmatrix} = \begin{bmatrix} \cos\theta & -\sin\theta \\ \sin\theta & \cos\theta \end{bmatrix} \begin{bmatrix} i_d \\ i_q \end{bmatrix} \tag{11.28}$$

If zero sequence current exists, the matrix C^{-1} is modified as

$$\begin{bmatrix} i_d \\ i_q \\ i_0 \end{bmatrix} = \begin{bmatrix} \cos\theta & -\sin\theta & 0 \\ \sin\theta & \cos\theta & 0 \\ 0 & 0 & 1 \end{bmatrix} \begin{bmatrix} i_\alpha \\ i_\beta \\ i_0 \end{bmatrix} \tag{11.29}$$

As per Eq. (11.22)

$$\begin{bmatrix} i_\alpha \\ i_\beta \\ i_0 \end{bmatrix} = \sqrt{\frac{2}{3}} \begin{bmatrix} 1 & -\frac{1}{2} & -\frac{1}{2} \\ 0 & \frac{\sqrt{3}}{2} & -\frac{\sqrt{3}}{2} \\ \frac{1}{\sqrt{2}} & \frac{1}{\sqrt{2}} & \frac{1}{\sqrt{2}} \end{bmatrix} \begin{bmatrix} i_a \\ i_b \\ i_c \end{bmatrix} \tag{11.30}$$

Substituting in Eq. (11.29)

$$\begin{bmatrix} i_d \\ i_q \\ i_0 \end{bmatrix} = \begin{bmatrix} \cos\theta & -\sin\theta & 0 \\ \sin\theta & \cos\theta & 0 \\ 0 & 0 & 1 \end{bmatrix} \sqrt{\frac{2}{3}} \begin{bmatrix} \cos(0°) & \cos\left(\frac{2\pi}{3}\right) & \cos\left(\frac{4\pi}{3}\right) \\ \sin(0°) & \sin\left(\frac{2\pi}{3}\right) & \sin\left(\frac{4\pi}{3}\right) \\ \frac{1}{\sqrt{2}} & \frac{1}{\sqrt{2}} & \frac{1}{\sqrt{2}} \end{bmatrix} \begin{bmatrix} i_a \\ i_b \\ i_c \end{bmatrix} \tag{11.31}$$

$$= \sqrt{\frac{2}{3}} \begin{bmatrix} \cos(0°) & \cos\left(\theta - \frac{2\pi}{3}\right) & \cos\left(\theta - \frac{4\pi}{3}\right) \\ \sin(0°) & \sin\left(\theta - \frac{2\pi}{3}\right) & \sin\left(\theta - \frac{4\pi}{3}\right) \\ \frac{1}{\sqrt{2}} & \frac{1}{\sqrt{2}} & \frac{1}{\sqrt{2}} \end{bmatrix} \begin{bmatrix} i_a \\ i_b \\ i_c \end{bmatrix} \tag{11.32}$$

Thus, the new current (or fictious current) i_d, i_q and i_0 can be expressed in terms of actual three-phase current i_a, i_b and i_c or two-phase current i_α, i_β and i_0. Similarly,

$$\begin{bmatrix} i_a \\ i_b \\ i_c \end{bmatrix} = \sqrt{\frac{2}{3}} \begin{bmatrix} \cos(0°) & -\sin(0°) & \frac{1}{\sqrt{2}} \\ \cos\left(\frac{2\pi}{3}\right) & \sin\left(\frac{2\pi}{3}\right) & \frac{1}{\sqrt{2}} \\ \frac{1}{\sqrt{2}} & \frac{1}{\sqrt{2}} & \frac{1}{\sqrt{2}} \end{bmatrix} \begin{bmatrix} \cos\theta & \sin\theta & 0 \\ \sin\theta & \sin\theta & 0 \\ 0 & 0 & 1 \end{bmatrix} \begin{bmatrix} i_d \\ i_q \\ i_0 \end{bmatrix} \tag{11.33}$$

$$= \sqrt{\frac{2}{3}} \begin{bmatrix} \cos(\theta) & -\sin(\theta) & \frac{1}{\sqrt{2}} \\ \cos\left(\theta - \frac{2\pi}{3}\right) & \sin\left(\theta - \frac{2\pi}{3}\right) & \frac{1}{\sqrt{2}} \\ \cos\left(\theta - \frac{4\pi}{3}\right) & \sin\left(\theta - \frac{4\pi}{3}\right) & \frac{1}{\sqrt{2}} \end{bmatrix} \begin{bmatrix} i_d \\ i_q \\ i_0 \end{bmatrix} \tag{11.34}$$

Concept Check LO 4

1. What is significance of considering zero sequence components in three phase to the phase conversion? ○●●
2. What are advantages of converting rotating axis to stationary axis transformation? ○○●

11.9 | PHYSICAL CONCEPTS OF PARK'S TRANSFORMATION

LO 5 Interpret the physical concept of Park's transformation and derive conclusions from it

For understanding the physical concept of Park's transformation, let the currents flowing through the winding be

$$i_a = I_m \cos(\omega t + \alpha) \tag{11.35}$$

$$i_b = I_m \cos\left(\omega t + \alpha - \frac{2\pi}{3}\right) \tag{11.36}$$

$$i_c = I_m \cos\left(\omega t + \alpha - \frac{4\pi}{3}\right) \tag{11.37}$$

where i_m is the maximum value of armature currents and α is the phase angle of i_a w.r.t. to time origin. Therefore, i_d can be written as

$$i_d = \sqrt{\frac{2}{3}}\left[I_a \cos\theta + I_b \cos\left(\theta - \frac{2\pi}{3}\right) + I_c \cos\left(\theta - \frac{4\pi}{3}\right)\right] \tag{11.38}$$

In the above equation, $\theta = \omega t$; i.e., π radian per seconds ; Substituting Equations (11.35), (11.36) and (11.37) in the Eq. 11.38,

$$i_d = \sqrt{\frac{2}{3}} I_m \left[\cos(\omega t + \alpha)\cos\omega t + \cos\left(\omega t + \alpha - \frac{2\pi}{3}\right)\cos\left(\omega t - \frac{2\pi}{3}\right) + \cos\left(\omega t + \alpha - \frac{4\pi}{3}\right)\cos\left(\omega t - \frac{4\pi}{3}\right)\right] \tag{11.39}$$

or

$$i_d = \sqrt{\frac{2}{3}} I_m \left(\frac{1}{2}\right)\left[\cos(2\omega t + \alpha) + \cos\alpha + \cos\left(2\omega t + \alpha - \frac{4\pi}{3}\right) + \cos\alpha + \cos\left(2\omega t + \alpha - \frac{2\pi}{3}\right) + \cos\alpha\right] \tag{11.40}$$

$$i_d = \sqrt{\frac{2}{3}} I_m \cos\alpha \tag{11.41}$$

Similarly, i_q can be written as

$$i_q = \sqrt{\frac{2}{3}} I_m \sin\alpha \tag{11.42}$$

Conclusion

1. Three-phase current i_a, i_b and i_c in 3-phase winding produce rotating magnetic field. The same is achieved by the existence of two *d-q* axes currents i_d, i_d in two *d-q* armature windings.
2. Three-phase currents i_a, i_b and i_c are the functions of time but i_d, i_q currents are constant (not functions of time). Machine analysis in easily carried in *d*-, *q*-currents and results converted to i_a, i_b and i_c which are function of time.

Probe

What are the conclusions drawn from Park's transformation?

MULTIPLE CHOICE QUESTIONS

1. What is the axis of the poles called? LO 1
   (a) Direct axis (b) Quadrature axis
   (c) Distribution axis (d) Imaginary axis

2. _____ action effectively divides the armature winding along the brush axis to form two uniform current sheets which are in opposite polarity. LO 1
(a) Commutator (b) Stator (c) Inductor (d) Transformer

3. A coil that is used to represent the commutator winding and is also fictious is given the name of _____ coil. LO 1
(a) pseudo-stationary (b) quad-stationary
(c) speed-voltage (d) space-hysteresis

4. The equivalent electrical network of a two-pole machine is called _____ machine. LO 2
(a) Kron's primitive (b) three-axis
(c) self-inductance (d) Park's generalised

5. The principle of maintaining the same power under transformation from an old to a new set of variables is known as _____. LO 3
(a) invariance of power (b) displaced brush transformation
(c) mutual inductance (d) mmf distribution

6. In three-phase to two-phase transformation, the _____ between the windings must comply with the time-phase angle between the currents. LO 4
(a) space phase angle (b) transverse angle
(c) displacement angle (d) space displacement angle

7. What is the product of number of turns and current? LO 4
(a) MMF (b) Voltage magnitude (c) Time-phase angle (d) Displacement

Answers

1. (a) 2. (a) 3. (a) 4. (a) 5. (a) 6. (a) 7. (a)

CHAPTER 12

Motor Control by Static Power Converters

LEARNING OBJECTIVES

After studying this chapter, students will be able to

LO1 List various solid state devices for motor control with their significance and describe their constructional features with symbols, VI Characteristics and commutation methods

LO2 Construct the block diagram of electric drive and discuss the selection of particular converter and control depending on commutation, type of source, type of load (ac or dc motor to be controlled) and quadrant of operation

LO3 Analyze the various topologies of single-phase converter, three-phase converter and chopper-fed dc motor drives with their construction and operation features

LO4 Outline the control techniques for induction motor with slip power recovery schemes and explain the principle of operation of various single-phase, three-phase inverter configurations and forced commutation methods

LO5 Illustrate the direct and indirect vector control scheme of induction motor with its algorithm and analyze the basic principle of DTC scheme

12.1 | INTRODUCTION

Power electronics is a multidisciplinary technology that encompasses power semiconductor devices, converter circuits, electric machines, signal electronics, control theory, microcomputers, very-large scale integration (VLSI) circuits, computer-aided design techniques, etc. Electrical power is processed by power electronics to make it suitable for various applications such as dc and ac regulated power supplies, electrochemical processes, heating and lighting control, electrical machine drives, induction heating, electronic welding, active power line filtering, static var compensating, etc. Today, power electronics is an indispensable tool in any country's industrial economy. An important aspect of power electronics applications is energy conservation, i.e. more efficient use of electricity. Power electronics permits generation of electrical power from environmentally clean photovoltaic, fuel cell, and wind energy sources.

Power semiconductor devices are going through a dynamic evolution in recent years. A large number of exotic power semiconductor devices have emerged in the last decade. They constitute the heart of modern power electronics [67].

A power converter acts as an interface which accepts electric power from the existing source and converts it in a controlled manner into a suitable form compatible with the particular load or the process for which it is used. The main sources for electric power are: 1 ph or 3 ph 50/60 Hz ac from utility systems and dc from batteries or solar cells. The four basic forms of power conversion required (Fig. 12.1) are: ac to dc, dc to dc, dc to ac and ac to ac. One important example where power conversion required is variable speed motor drives with variable voltage/variable frequency.

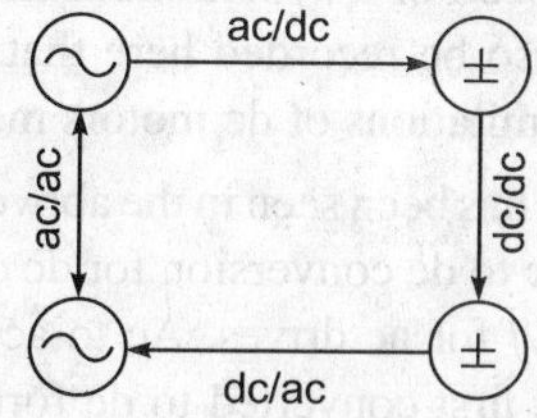

Fig. 12.1 *Four basic forms of power conversion*

The modern converters are compact, cheap, reliable, durable, flexible, completely controllable. They need reduced maintenance too. They are suitable for all the four basic forms of power conversion mentioned above through rectifiers (ac/dc), choppers (dc/dc), inverters (dc/ac) and cycloconverters or ac regulators (ac/ac). These are now conditioning electric power to meet the requirements in the industry at all levels.

Currently ac drives are in the process of replacing dc drives in many industrial applications including locomotion, machine tool drives, paper mills, electric vehicles, waste water treatment, air conditioning, and host of other applications. About 70% of the cost of an ac drive is contained in the power electronics portion of the drives versus only about 30% for the dc drive. But, technology is on the side of the ac drive and they are expected to overtake dc drives in most of the industrial applications as the technology continues to advance and mature.

Recall The control of speed-torque characteristics of motors for adjustable speed drive was discussed in Ch. 7 for dc motors and in Ch. 9 for induction motors. It was discovered that the speed of a dc motor can be controlled with the relative ease as compared to ac motors. The speed of a dc motor is controllable by the flux/pole (field control) as well as by the armature voltage. A wide range of speed is obtained by a combination of these methods. Speeds below the base speed are achieved by reducing the armature voltage while the field is kept at its maximum setting. A constant-torque load (torque corresponding to rated armature current and full field current) can be met with the load power being directly proportional to motor speed. For speeds above the base value, the armature voltage is held at constant value and the field is weakened. In this region, the torque decreases in proportion to the increase in speed and therefore the mechanical power output above the rated speed remains constant. Typical speed ranges for dc drives are 6 : 1 by armature voltage control and 3 : 1 by a field weakening control. Armature voltage control requires a power converter with adjustable voltage and capability to handle rated armature current. The speed range of a dc motor cannot be increased beyond the range mentioned as it is limited by extreme distortion of the air-gap flux of a weak field caused by the armature reaction and consequent poor commutation. The complex constructional features of dc armature also place an upper limit on the motor speed. Neither of these two limiting factors inhibit the speed of a rugged wound rotor induction motor. On the other hand, ac motors suffer from the inherent speed-constancy of the rotating field dictated by the supply frequency. Speed control in induction motors is possible from the stator side by frequency control with a rotating field flux kept constant by simultaneous variation of V such that V/f remains constant. Constant V/f supply for this control requires expensive frequency changers (power converters). The stator voltage control (fixed frequency) is ruled out except for small motors because the motor torque reduces as the square of the voltage and the motor cannot meet the load torque at low or even medium speeds. Further, the motor has to be derated to avoid overheating at low speeds. The control of speed from the rotor side is only possible for the more expensive and less rugged wound-rotor induction motor. The rotor-resistance control is too inefficient for practical use except for momentary reduction of speed. Exotic frequency changers must, therefore, be introduced in the rotor circuit for efficient control linking variable-frequency rotor to the constant-frequency supply while being capable of handling the full rotor current. The speed of a synchronous motor can be controlled by combined voltage and frequency control only. It must also be recorded here that very high speeds are only possible by using ac motors because of the inherent limitations of dc motors mentioned earlier.

It has been seen in the above discussion that speed control of both dc and ac drives requires power converters—ac to dc conversion for dc drives and ac to ac conversion, with one side having a variable frequency constant *V*/*f* for ac drives. Ac to ac power conversion normally requires an intermediate dc stage, i.e. the ac power is first converted to dc form which is then converted to a variable-frequency at form. Before the advent of static power switching devices-silicon *rectifiers* and *silicon controlled rectifiers* (*SCRs*) or *thyristors—power* conversion necessitated the use of expensive rotating machines. Motor-generator (MG) sets were used for ac to dc conversion (refer to Ward-Leonard speed control: Sec. 7.14) and ac to ac conversion required an even more elaborate and expensive rotating machine set and in fact was rarely employed. Even then a stringent upper limit existed for the high frequency end of the variable-frequency conversion imposed by high specific iron-loss.

With the availability of SCRs (thyristors) beginning in late 1950's, followed by development of other power semiconductor devices having features superior to SCRs in some respects, their commercial introduction in 1970's and with ever-increasing voltage, current and operating frequency rating of SCRs and other power devices coupled with simultaneous reduction in their cost, a virtual revolution has taken place in the field of power converters.

Important! Power ratings and speed limitation are now determined by electric motors rather than semiconductor devices.

In future large chunks of all power reaching industrial end-use equipment and domestic lighting and gadgetry would have passed through power semiconductor devices; once and even more than once. This is visualised to be the future trend as is already witnessed in most of the advanced countries.

Alongwith improvement in characteristics of power switching devices, considerable ingenuinity has been acquired in design and layout of associated circuitry. At present the bulk of size and cost of a power electronic equipment is due to passive components. Search is on for "silicon solution" for these, like the use of resonant and quasi-resonant link principle.

The power electronic systems today incorporate power semiconductor devices as well as microelectronic IC's, both of which are digital in nature (one provides the muscle and the other gives the intelligence). The ultimate goal of R&D is to put all the silicon on one chip i.e. VLSI.

Industry Relevance

In the modern robotic-industrial era almost every automated mechanism uses motors for producing either linear or rotational motion, which are controlled precisely by static power converters. The use of static power converters not only increases the efficiency of the system but also reduces the size and thereby weight. Be it a hand drilling machine, door closures of furnaces, rocket launching mechanism or pointing mechanism of satellites, precise control of mechanism is essential which is achieved by static power converter control of motors.

12.2 SOLID STATE DEVICES

LO 1 List various solid state devices for motor control with their significance and describe their constructional features with symbols, VI Characteristics and commutation methods

The main devices normally employed for motor control are diode, Silicon-Controlled Rectifier (SCR) or thyristor, triac, diac, Insulated Gate Bipolar Transistor (IGBT) and Gate Turn off (GTO) thyristors. Advancement in semiconductor devices made the motor control drive more easy and flexible. These are the materials with conductivities intermediate between insulators and metals, which can be controlled over a wide range. In this section various semiconductor devices are explained with their construction and their capability to work with high power rating motor control.

Introduction

When energy bands are empty or completely filled with electrons, there can be no movement of electrons, i.e., no current flow will take place. Otherwise some electrons in an empty band will make the current to flow. If some electrons are missing from an otherwise completely filled band, empty electron states or holes are formed, and the remaining electrons may move into these holes, thus creating other holes in the states the electrons have vacated. Thus, the holes can move in the energy band. At low temperatures the highest band containing electrons (valence band) is filled completely, while the next band (conduction band) is empty, therefore the conductivity is zero.

When an electron leaves the conduction band to fill a hole in the valence band, energy is liberated and dissipated as heat. This mechanism is called recombination and annihilates two charge carriers, i.e., one hole and one electron. The number of free electrons and holes in an ideal semiconductor is a function of temperature.

In addition to the carriers generated by thermal energy or light, further carriers can be provided by doping or adding impurities (such as phosphorus or arsenic) purposely introduced into the semiconductor. Doping can be of *n*-type, the species affecting it being called donors because they add a freely moving electron to the conduction band, or of *p*-type called acceptors, which increase the conductivity by one positively charged hole. In contrast to generation, doping with certain impurities adds only one of the two carrier types. The carriers introduced by the impurity atoms determine the conductivity in a doped region and are called majority carriers. Devices in which both types of carriers are present are called bipolar devices, as opposed to unipolar devices, in which the number of one type of carrier is much larger than the other.

Classification of Power Semiconductor Devices

Power semiconductor devices are classified based on their various operating capability as follows.

1. Uncontrolled turn on and off (Power Diode)
2. Controlled turn on and uncontrolled turn off (SCR)
3. Controlled turn on and off characteristics (Power BJT, Power MOSFET, GTO, SITH, IGBT, SIT MCT)
4. Continuous gate signal requirement (Power BJT, Power MOSFET, IGBT, SIT)
5. Pulse gate requirements (SCR, GTO, MCT)
6. Bipolar voltage-withstanding capability (SCR, GTO)
7. Unipolar voltage-withstanding capability (Power BJT, Power MOSFET, GTO, IGBT, MCT)
8. Bidirectional current capability (TRIAC, RCT)

9. Unidirectional current capability (SCR, GTO, Power BJT, Power MOSFET, MCT, IGBT, SITH, SIT, Power Diode)

High power and high voltage application SCR (thyristor) has been practically replaced by its close cousin GTO (gate turn-off thyristor) in force-commutated, voltage fed converters because of reduced overall cost and improved efficiency. These, however, have the disadvantage of low operating frequency. Other power devices that are being preferred for low and medium power applications are: Power BJT (bipolar junction transistor), Power MOSFET (metal-oxide semiconductor field effect transistor), IGBT (insulated gate bipolar transistor), SIT (Static Induction Transistor). SITH (Static Induction Thyristor) and MCT (mos-controlled thyristor). MOSFETs have higher operating frequency but lower power handling capacity compared to BJTs, IGBT which offer the compromise between power and frequency. SIT and SITH have good power and frequency characteristic but high turn-on resistance and consequent higher loss. MCT with comparatively complex structure offers superior all-round characteristics.

Power Diode

Recall The diodes (Fig. 12.2(a)) described here are based on the property of *p*- and *n*-type materials in conjunction. An ideal diode presents zero impedance to current flow in one direction and infinite impedance in the opposite direction. A practical diode has a very low forward resistance and has a voltage drop of 0.3 to 0.7 V at all current levels and a reverse current of a few milliamperes when *blocking voltage* (or *peak reverse voltage*) is around 1 kV. The *v-i* characteristic of a diode and its idealized form, commonly employed in diode circuit analysis, are shown in Figs 12.2(b) and (c) respectively. The voltage rating (maximum 5 kV), is fixed by the reverse characteristic in terms of the break-down voltage. The main parameters of a diode are: blocking voltage, average forward current and maximum operating junction temperature (nearly 125°C for most silicon rectifiers). Protection is necessary against voltage surges, overcurrents and excessive temperature rise.

Probe

List the main parameters of a diode.

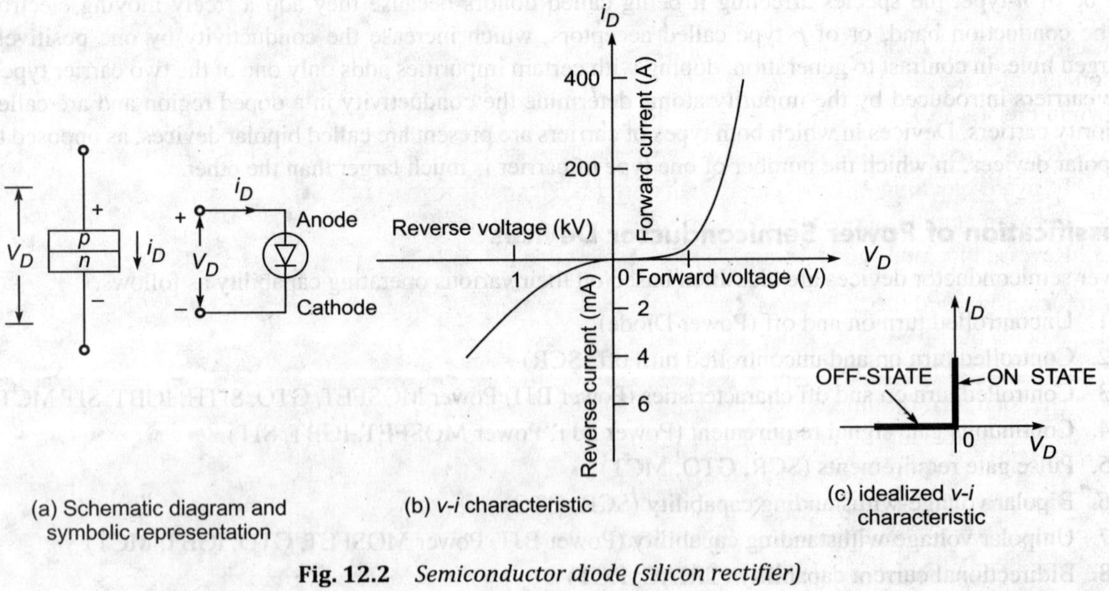

Fig. 12.2 *Semiconductor diode (silicon rectifier)*

A "free-wheeling" diode is commonly used in motor control systems to provide an alternate path for continuity of current in the inductive circuit following the switching-off of some power device between the motor and the energy source.

Recall The Zener diode, though not a true power controlling device, is quite often used as a voltage control and sensing device in many motor controllers. If conducts from cathode to anode at a definite voltage, which is then held constant.

Silicon-Controlled Rectifier (SCR)

This semiconductor switching device is also known as a *thyristor*. It is a 4-layer, 3-junction, bistable (ON and OFF) semiconductor switch. The terminal to the outer *p*-layer is the anode, the terminal to the outer *n*-layer is the cathode and the terminal to the inner *p*-layer is the gate as shown in the schematic diagram and the symbolic representation of Fig. 12.3(a). The anode is connected to a heat sink base providing a rigid platform for the SCR. The rating of the SCR has very much improved since its introduction in 1967 and now SCRs voltage rating 10 kV and an rms current rating of 3500 A with switching speed of 1 μs corresponding to a power-handling capacity of 30 MW are available. A device of such a high rating can be switched on by a low voltage supply of about 1 A and 10 W which gives us an idea of the tremendous power amplification of this device (3×10^6). Because of its compactness, high reliability and low loss, the SCR has almost replaced the earlier power switching devices—thyratron and magnetic amplifier.

Figure 12.3(b) shows the static *v-i* characteristic of the SCR which is similar to that of the diode in the reverse direction. The forward characteristic is the same up to the *turn-on* (or *breakdown*) *voltage* (V_{BD}) at which avalanche multiplication starts and the current begins to rise rapidly to a value determined by the supply voltage and the external impedance. The SCR *latches* on and this is the *conducting mode* (*state*). The SCR stays in this mode till the forward current is brought below *the holding current* level (a few milliamperes) and at this point it reverts to the forward blocking state (J_1 and J_3 forward biased and J_2 reverse biased). By increasing the gate or trigger current, both the forward break-down voltage and the holding current can be reduced. Thus the SCR can be made to fire, i.e. to conduct at a specified forward voltage, by controlling the gate current. Ideal characteristic of the SCR is shown in Fig. 12.3(c).

Probe

How to turn on the SCR?

The SCR has three operating modes:

1. Forward blocking mode (J_1 and J_3 forward biased and J_2 reverse biased). This is the off-state.
2. Forward conducting mode: the on-state.
3. Reverse blocking mode (J_1 and J_3 reverse biased and J_2 forward biased)—this is also the off-state.

When forward biased the SCR can be turned on by any of the following three ways:

1. When a positive trigger pulse is applied to the gate, the SCR switches rapidly to the conducting mode, after a brief delay. The total turn-on time which comprises delay and rise time is of the order of 0.2 and 0.1 μs and is dependent upon the character of the trigger pulse. The SCR continues to conduct after the trigger pulse has ceased provided the thyristor current has become more than the *latching current* which is slightly more than the holding current.
2. The SCR can also be turned on by increasing the forward voltage beyond the forward breakdown voltage V_{BD}.

Probe

Why a series inductor is inserted with a thyristor?

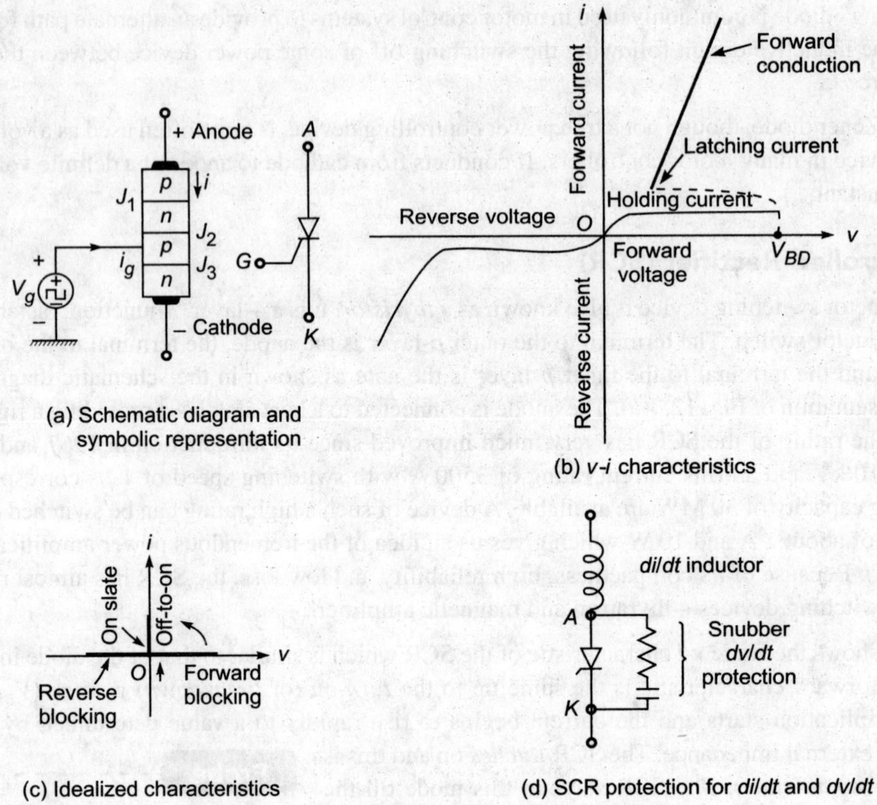

Fig. 12.3 *Silicon-controlled rectifier (thyristor)*

3. An SCR may turn on (without any gate pulse) if the forward voltage is applied suddenly. This is called as *dv/dt* turn on and it may result in improper operation of the circuit. A simple R-C snubber (see Fig. 12.3(d)) is normally used to limit the *dv/dt* of the applied forward voltage.
4. If the current in a thyristor rises at too high rate, that is, high *di/dt*, the device might be destroyed. Some inductance, therefore, must be inserted in series to keep *di/dt* below a safe limit.

The gate control method, i.e. applying a positive gate signal to the forward-biased SCR is an efficient and simple method of turning on the SCR and is commonly applied. Using gate control, the device can be triggered either by a dc gate signal or by a pulsed-gate signal. The total turn-on time depends on the anode circuit parameters, the gate signal amplitude and its rise time.

A conducting SCR can be switched off only by reducing its conduction current below the holding current (a few mA) and being reverse-biased (i.e. the cathode positive with respect to the anode) for a minimum turn-off time (10–100 μs). Also, the rate of rise of the anode voltage after turn-off must be limited, otherwise the device may start conducting again.

An important consideration in the use of SCRs in motor control is the means by which *turn-off* or *commutation* is achieved and the associated commutation circuitry—its complexity, cost and weight. In ac circuits the current goes through a natural zero value causing the device to be automatically turned off. This is known as *natural* or *line commutation.*

!Attention Here the SCR has to be triggered synchronously with or without delay with the zero crossing of the voltage across it whenever it is positively biased.

In dc circuit application where is no natural current zero, the forward current through the SCR can be reduced by shunting it with a low impedance path or by applying a reverse voltage across it forcing the forward current to zero value—this is known as *forced commutation.* Some important ways to achieve commutation will be discussed in motor control—See Section 12.6 and 12.7.

An SCR gets fired spuriously by an excessive rate of rise (*dv*/*dt*) of the anode voltage. System transient voltages of a spike waveform have to be suppressed by input filters. The necessary protective features are: (i) temperature limitation to avoid thermal run-away, (ii) protection against overcurrent (e.g. caused by a short circuit), and (iii) suppression of impulsive transient voltages and currents.

Probe

Which circuit application of SCR needs forced commutation?

There are mainly two types of SCRs for use in motor controllers—inverter grade and converter grade. The inverter grade SCRs are used in inverters, cyclo-converters, and brushless dc motor systems while the converter grade are employed in choppers, phase-controlled rectifiers, ac regulators, etc. These two grades vary in their time responses. Inverter grade SCRs are costlier than the converter grade.

Since the introduction of SCRs, their power and voltage ratings and their characteristics have considerably improved. Further several variations of thyristors such gate-assisted turnoff thyristors (GATT), asymmetrical silicon-controlled rectifier (ASCR), reverse conducting thyristors (RCT) are used in certain applications.

Light-Activated Silicon-Controlled Rectifiers (LASCR) is a recent device which is turned on by direct radiation of silicon with light. LASCRs are used in high-voltage and high-current applications, e.g. HVDC and VAR compensators. The voltage rating of an LASCR could be as high as 4 kV at 1500 A, *di*/*dt* of 250 A/μs and *dv*/*dt* of 2 kV/μs.

Triac

It is a bilateral switch with three terminals—anode, cathode and a single gate. It conducts in either direction being equivalent to two SCRs connected in antiparallel or back-to-back on a single chip of semiconductor material. It is the most widely used device for power control. Triacs with reasonably large ratings (1000 V, 2000 A) and speeds of 1 μs are now available. Figure 12.4 shows the symbolic representation and the *v-i* characteristic of a triac.

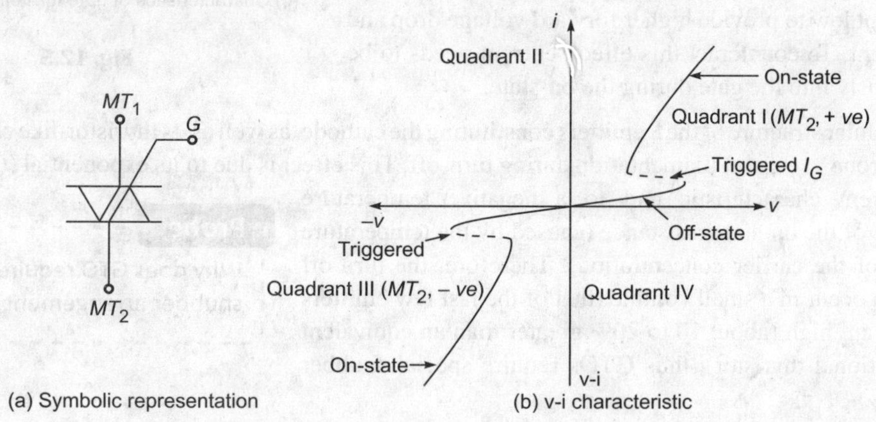

Fig. 12.4 *Triac*

A triac is economical compared to a pair of antiparallel thyristors and the control circuit is simpler. However, it is difficult to apply with inductive load and turn-off time is longer. Triacs are normally used for phase control with a resistive load (i.e. lamp dimming and heater control type applications. The state-of-the-art device voltage and current ratings are much smaller than thyristors.

Gate Turn-Off Thyristor (GTO)

The Gate Turn Off (GTO) thyristor was introduced in the late 1970s. It is mainly suitable for designing self-commutating converters of larger power ratings. The GTO is a three-terminal device. The GTO switching requires more complex gating and snubber circuits. GTO allows the extraction of holes from the *p*-base by means of a high negative current on the gate, which stops the *n* + emitter injection and turns the device off. However, since in the thyristor the lateral dimensions are bigger than its thickness, a high lateral voltage drop under the *n* + emitter would occur and a large part of it would continue injecting electrons. Therefore, the GTO *n* + emitter is configured as an array of many single and narrow stripes and to reduce the voltage drop in the gate, the *p*-base layer (which constitutes the gate) is metallised outside of the *n* emitters.

The GTO is manufactured as a whole wafer or as a die and is available in module, plastic and capsule packages. In the on-state the GTO behaves exactly as a thyristor, and, therefore, turn-on requires the application of a positive triggering current to the gate, *di*/*dt* is less critical, though, because all the emitter area is turned on quickly due to its multicell structure. As the *n* emitters are not shorted, a negative gate voltage must be applied continually during the off-state to improve the blocking behaviour at high temperatures. Shorts are introduced in the *p* emitter to improve the turn-off capability. Also, the carrier lifetime is kept low to provide higher forward voltage drop and holding current. To counteract this effect, current needs to be fed continuously into the gate during the on-state.

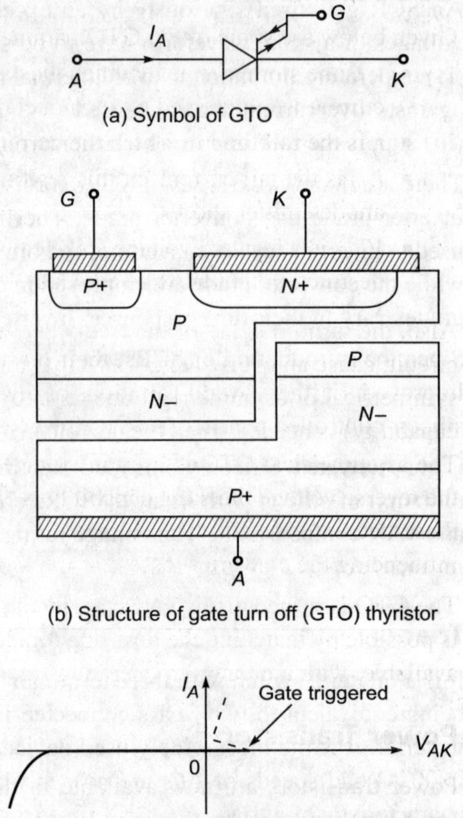

(a) Symbol of GTO

(b) Structure of gate turn off (GTO) thyristor

(c) Characteristics of gate turn off (GTO) thyristor

Fig. 12.5

Due to the cellular structure of the *n* emitters constituting the cathode, as well as its thyristor-like characteristics, the GTO is prone to current filamentation during turn-off. This effect is due to its exponential (thyristor-type) forward current characteristic and to a negative temperature characteristic of the on-state resistance (caused by the temperature dependence of the carrier concentration). Therefore, the turn-off losses, which occur in a small volume (that of the last few emitters turning off), are high (about 10 to 20% greater than an equivalent rated conventional thyristor), thus GTOs require special snubber arrangements.

Probe

Why does GTO require special snubber arrangement?

The resistor R limits the capacitor discharge current during turn-on and the R_S and C_S combination determines the minimum time required by the GTO to turn off again. For instance, for the case of a GTO with a peak voltage rating of 4.5 kV and a peak turn-off current of 4 kA, to turn the GTO safely the snubber capacitor C_S must be at least 6 mF to be able to dissipate the energy stored in the capacitor after the switching. Also the energy stored in the *di*/*dt* limiting inductor L_S at turn-off is dissipated via the discharge resistor and diode.

GTO Parameters

Given below are some of the GTO parameters:

- t_s is the storage time in which the carriers are removed from under the cathode emitters, while the load current is maintained by its inductance.
- t_f is the fall time in which the current reduces very quickly.
- i_{tail} is the tail current, in this region the remaining carriers are removed via recombination across the anode emitter shorts.
- V_{DSP} is a forward voltage spike caused by stray inductance in the snubber circuit (a possible source of destruction that must be avoided).

Also, the turn-on delay of the free-wheeling diode D_F, together with the stray inductances of the diode and circuit, cause an overvoltage VDM that must not exceed the maximum rating. The GTO can be designed for symmetrical or asymmetrical operation. A symmetrical GTO (with forward and backward blocking voltage capability) is needed when the dc voltage can reverse its polarity, as in the case of current source conversion. The asymmetrical GTO is used in voltage source conversion and permits larger power ratings. In this case, the reverse voltage blocking capability is eliminated on purpose by means of internal anode to cathode shorts, these decrease the forward voltage drop and improve the switching characteristics (two important factors influencing the converter losses).

The GTO is turned off by gate current and the turn-off gain is between 3 and 5. Series connection of devices is possible by matching the turn-off storage time of the devices. Presently a 6 kV, 6 kA GTO is commercially available, with a maximum operating frequency of 500 Hz.

Power Transistor

Power transistors are now available in ratings suitable for motor control. This device is a 3-layer *p-n-p* (or *n-p-n*) structure, as shown in Fig. 12.6. This type of transistor is known as a bipolar junction transistor (BJT).

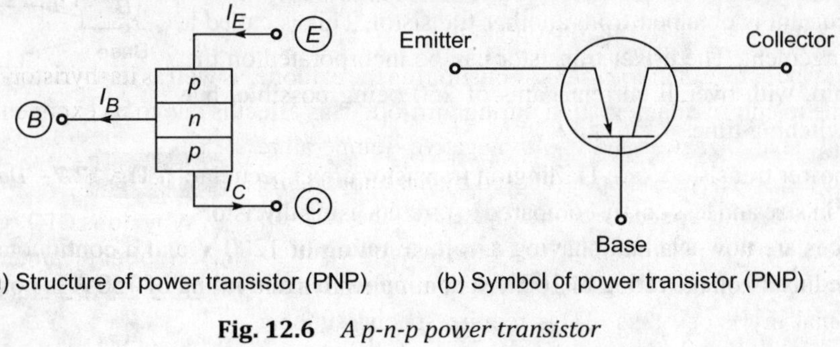

(a) Structure of power transistor (PNP) (b) Symbol of power transistor (PNP)

Fig. 12.6 *A p-n-p power transistor*

In normal operation the emitter-base junction is forward-biased and the collector-base junction reverse-biased. The collector current is almost equal but slightly less than the emitter current. The collector current can be controlled by the base current. Except for small motors, transistors are employed in the switching (on/off) mode. The common-emitter configuration is normally used owing to its higher power gain. The common-emitter saturation voltage $V_{CE(SAT)}$ for typical power transistors is from 0.3 to 0.7 V. This being sufficiently less than the on-state anode-to-cathode voltage drop of an SCR, the average power-loss in a transistor is less than that of an SCR of an equivalent power rating.

The switching of a power transistor is controlled through the base current. A continuous base current is required to keep the transistor in the saturated state, i.e on-state. In a high-current device, the base-to-collector current gain is as low as 10, and therefore a base current of several amperes is required to keep the device in the on-state; the power-loss in the base circuit may be appreciable. The rise time in turn-on may be reduced by momentarily overdriving the base current. To initiate turn-off, the excess saturation charge, which is proportional to the base current, must be removed from the base region before the transistor can begin to return to the off-state. This storage delay time can be minimized by applying reverse current to the base during turn-off.

Probe

How to control the switching of a power transistor?

Power transistors have no surge current capacity and can withstand only a low rate of change of current. There is no regenerative action in a transistor and it needs a continuous base current to continue in the on-state, whereas an SCR requires only a gate pulse to turn it on. The switching times of power transistors are, however, generally faster than those of SCRs, and the commutation problems of SCR are not present in transistors, Power transistors are finding increasing popularity in low-to-medium-power applications, where they compete successfully with thyristors and GTOs. The transistor can switch considerably faster than the thyristor, typically a switching time of less than 1 μs being possible.

Power transistor applications range from a few kW to several hundred kWs size in voltage-fed choppers and inverters with switching frequency up to 10–15 kHz. The state of the art devices are available with ratings up to 600 V, 500 A.

Power Darlington Transistor

Recall A Darlington-connected transistor pair is shown in Fig. 12.7. The current gain of the power transistor can be considerably improved if the base drive current is obtained from another transistor. This is called a Darlington arrangement. The driver transistor can be incorporated on the same silicon chip, with overall current gains of 250 being possible, but with a longer switching time.

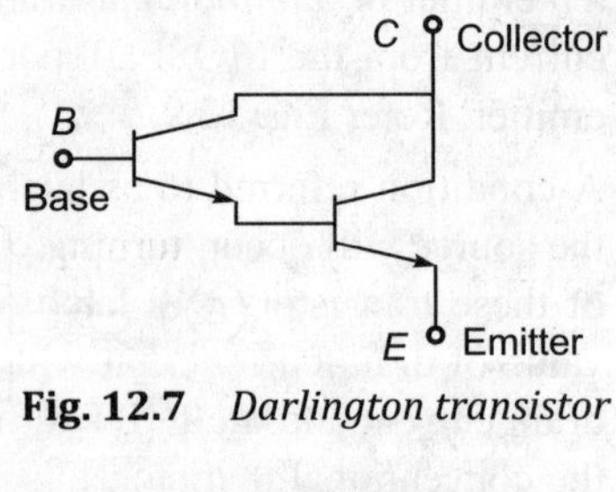

Fig. 12.7 *Darlington transistor*

Circuits using power transistors or a Darlington transistor are expected to become smaller in size and less costly compared to circuits using thyristors. Individual devices are now available having a voltage rating of 1200 V and a continuous-current rating of 300 A. With a reduced voltage rating of 100 V, a continuous-current rating of 750 A is possible.

Metal-Oxide-Semiconductor Field-Effect Transistor (Power MOSFET)

The metal-oxide-semiconductor field-effect transistor (MOSFET) is a unipolar device, (i.e., the charge flow consists only of majority carriers). The gate electrode is insulated from the *p*-substrate by a thin layer of SiO_2, which together with the surface layer of the substrate constitutes a capacitor. Two highly doped *n* + regions at the sides of the gate area form the source (cathode) and drain (anode) which are connected to a load RL and a voltage source V_{DS}. When a positive gate voltage is applied, the capacitor is charged. The current is supported only by majority carriers (electrons), which have very low relaxation time and therefore permit very fast switching. The current causes a voltage drop along the channel and, therefore, the capacitor voltage at the drain end is lower. Above a certain value of drain source voltage, the channel is pinched at the drain side and becomes shorter for increasing V_{DS} levels. High conductivity requires short channel lengths, which results in relatively low voltage blockings, a property that makes the MOSFET unsuitable for high-power and high-voltage applications.

Probe

Is MOSFET a unipolar device? If yes, state why.

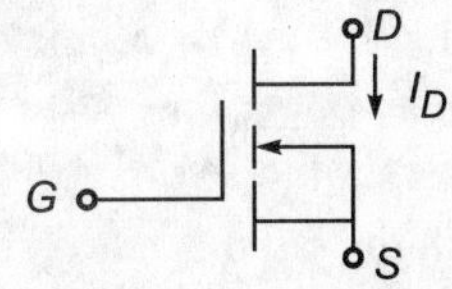

(a) Symbol of Power MOSFET

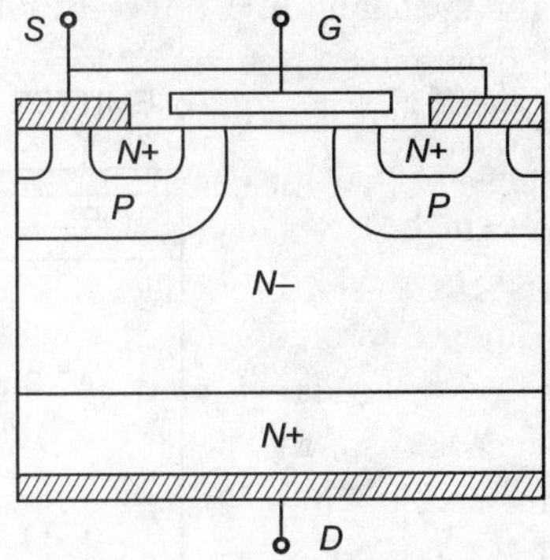

(b) Structure of Power MOSFET

Fig. 12.8

Insulated Gate Bipolar Transistor (IGBT)

A development from MOSFET technology for high power ratings, the IGBT is designed with high switching speed and low conduction loss. Its feasibility for high-voltage application was realised in the 1990s, and had already reached 6.5 kV by 2002. Although the IGBT behaves as a bipolar transistor, it has a double transistor structure, like the thyristor.

Important! The IGBT combines a high-impedance and low-power gate input with the power handling capacity of the bipolar transistors and thyristors.

The IGBT is achieved by adding a *p* + layer, this layer is referred to as the collector of the IGBT but acts as a *p* emitter of a *pnp* bipolar transistor to the MOSFET drain. The bipolar transistor is driven by the electron current from the MOSFET portion, which provides the base current, and holes are injected from the *p* + emitter. Refer Fig. 12.9.

A condition referred to as latch-up can occur at high voltage, when substantial injection of electrons from the source will occur, turning on the two transistor components (*npn* and *pnp*). Then the thyristor composed of these transistors will latch on and the IGBT will latch up. For each IGBT geometry there is a critical value of drain current that will activate this condition. Once latch-up occurs, the gate has no control of the drain current and the IGBT turn-off will have to be forced as in a conventional thyristor. However, unlike in the conventional thyristor, this will have to be done quickly to avoid device destruction by excessive power dissipation. The risk of latch-up can be reduced by controlling the doping profiles to ensure that the *npn* transistor does not turn on.

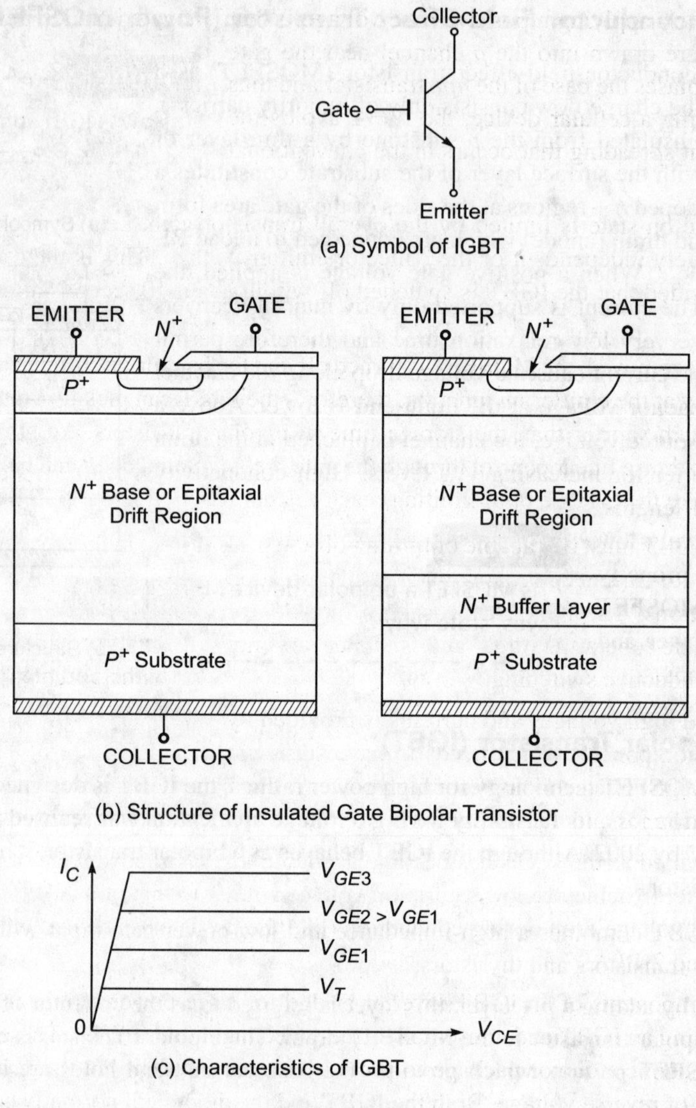

Fig. 12.9

Control is accomplished by using a pattern of MOS transistors distributed on the surface of the device. As a MOS gated device, the IGBT only needs a small amount of power to be supplied (via a voltage source) to the control gate for full turn-on and turn-off control (typically 10 to 15 V). Although the drive circuit still needs enough power to charge and discharge the gate capacitance, the circuitry required for this purpose is small compared with that of other power devices.

The IGBT is turned on by applying a positive base voltage. When the gate is made positive with respect to the emitter, *n* carriers are drawn into the *p* channel near the gate region, which forward biases the base of the *npn* transistor and thus causes the turn-on. Being a cellular device, the IGBT avoids the problem of slow current spreading that occurs in the conventional thyristor.

Probe

How is IGBT turned on?

The peak current in the on-state is limited by the overall transistor gain, i.e., for a given base current the collector current is largely independent of the collector-emitter voltage. This is the case even under short-circuit conditions, provided that the IGBT is switched off within about 10 seconds after the occurrence of a fault.

When the gate voltage is removed, the MOSFET is turned off and the base drive current cut off. A space-charge region builds up quickly at the emitter *pn* junction, therefore, the IGBT can pick up voltage immediately and its turn-off delay is much shorter than the storage time in bipolar transistors. An attractive feature of the IGBT is the ability to exercise linear control through the gate, i.e., *di*/*dt* and *dv*/*dt* can be controlled during the commutations. So, unlike the GTO, a *di*/*dt* limiting reactor is not required in the IGBT case.

During switching, the IGBT must be capable of turning off the peak current including ripple. Some additional current rating is needed for current control regulation and protection during transient conditions. The IGBT has a square turn-off (or true reverse-biased) safe operating area and snubber are not normally required for inductive switching.

Probe

What provides IGBT the ability to withstand high voltage and high current?

The ability to withstand high voltages and currents is provided by the vertical transistor component, which needs to have sufficient thickness to withstand high voltages. This vertical transistor is the major determining factor in the conductivity of the semiconductor material and the voltage drop over the device in the conducting state.

A new type of IGBT, called as the IEGT has become available that takes advantage of the effect of electron injection from the emitter to achieve a low saturation voltage similar to that of a GTO. Therefore, this device achieves a low forward voltage drop during conduction (like the thyristor). The maximum present ratings of the IGBT are 4.5 kV and 2.1 kA.

The semiconductor components of an IGBT inverter bridge for large power ratings may consist of several modules, each comprising an IGBT and anti-parallel free-wheeling diode (FWD). These two components are integrated in the same semiconductor package to provide current capability in the reverse direction and to prevent the application of reverse voltage. Both the IGBT and the diode will normally consist of a number of chips in parallel to obtain the rated current capability. Moreover, the IGBT and the diode must have the same voltage capability. In this arrangement, no ancillary components are needed, the *di*/*dt* controlling function being implemented by gate control and turn-on of the IGBT is slowed to match the allowable turn-off speed of the diode. The absence of the inductance between the inverter and the dc link results in extremely high fault currents in the event of simultaneous failure of two devices in the same phase. Thus the applicability of the topology is limited to low voltages (of about 1.5 kV). At higher voltages decoupling of link and inverter via inductance or resistance is required to limit the collateral explosion damage due to the electromagnetic force caused by the fault current.

SIT (Static Induction Transistor)

An SIT is a high-power high-frequency device essentially the solid state version of triode vacuum tube. It was commercially introduced by Tokin Corp, Japan, in 1987. The device symbol is shown Fig. 12.10(a).

The triode-like characteristics make the device suitable, both in active and switching modes. The device has been used in audio, VHF/UHF and microwave amplifiers. The reliability, noise and radiation hardness of SIT are claimed to be superior to MOSFET. The general features of SIT are given in Table 12.1. This device is normally on, that is in the absence of any base signal the SIT conducts.

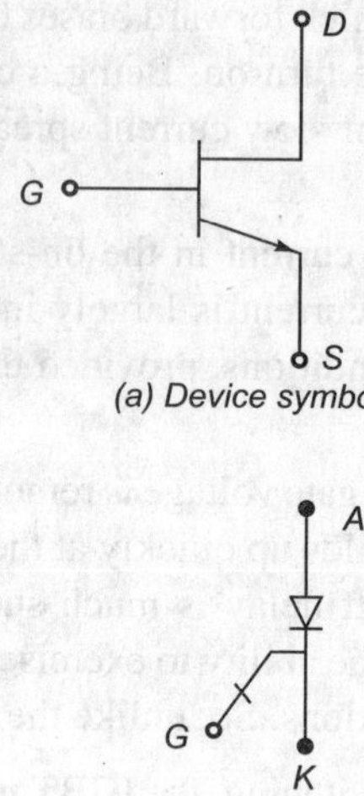

(a) Device symbol of SIT

(b) Device symbol of SITH

Fig. 12.10

Table 12.1 SIT features

•	High frequency high power vacuum triode-like device (1500 V, 180 A)
•	Switching speed faster than MOSFET (t_{ON} = 0.35 μs, t_{OFF} = 0.35 μs)
•	Lower gate-to-source capacitance and resistance
•	Expansion limited by T_j
•	Easy paralleling of devices
•	Not good for general PE applications

Probe

Why is switching loss less in SITH?

SITH (Static Induction Thyristor)

An SITH or SI-thyristor is a self-controlled GTO-like on-off device commercially introduced in Japan in 1988. Similar device, known as field-controlled thyristors (FCT) or field-controlled diode (FCD), were developed in USA. The device symbol is shown in Fig. 12.10(b). It is essentially a $P^+ NN^+$ diode. Similar to SIT, SITH is a normally on device with the *N*-region saturated with minority carrier. The turn-off behaviour of SITH is similar to that of GTO. It is a self-controlled GTO-like high power device (1200 V, 800 A). Its general comparison with GTO is given below.

1. Unlike GTO, SITH is a normally on device-asymmetric blocking.
2. The conduction drop is higher (4.0 V)
3. The turn-off current gain is lower (l to 3)
4. The switching frequency is higher (t_{ON} = 2 μs, t_{OFF} = 9 μs) and has lower switching losses.
5. The *dv*/*dt* and *di*/*dt* ratings are higher. (2 kV/μs), 900 A/μs.
6. The Expansion is improved and T_j is limited.

Integrated Gate-Commutated Thyristor (IGCT)

The need to simplify the GTO snubber circuit requirements and improve its switching performance has led to the development of the IGCT. It is a hard-driven GTO, in which during turn-off the full main current is commutated to the gate. However, the pulse is of a short duration and the energy dissipated in the gate drive are very small. Because of that, there is no current concentration during turn-off, a problem of the conventional GTO commutation. Consequently, turn-off capability is greatly improved and snubberless turnoff is possible.

The turn-off storage time is decreased to about one-tenth when compared with that of a GTO and its operating frequency can be increased. The small storage time also simplifies the series connection of IGCTs. The gain is only 1, but the gate charge is only half that of the GTO and the gate power is substantially decreased. An integrated annular gate unit on a printed circuit board achieves a very fast gate drive, enabling a current rise as high as 4 kA/_s with a gate cathode voltage of about 20 V. Thus the *di/dt* withstand can be increased by two to three times that of the GTO and, therefore, the anode reactor size can be reduced considerably.

The symbol and device structure, normally include a reverse diode. The symbol and device structure diagram of IGCT is shown in Fig. 12.11. In common with other GTO devices, the design minimises the inductance of the gate drive and cathode loop, an important parameter to achieve the required fast rising gate current. Thus, high *dv/dt* is achieved, as well as a greatly increased current turn-off capability. The energy stored in the inductor L, once the diode is commutated, is dissipated in a resistance R and a clamp capacitor C_{clamp} can be used to minimise the voltage overshoot. In the event of failure of two devices in one phase, the inductance limits the resulting current to a value $I_{fault} = V_{DC}\sqrt{C/L}$, where C is the capacitance of the dc link.

Probe

Is it possible to increase the operating frequency of IGCT? If yes, how?

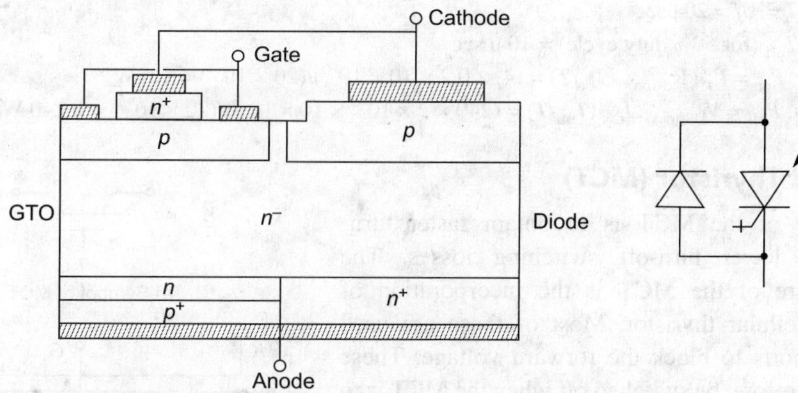

Fig. 12.11 *IGCT structure including a GCT and a reversed diode*

Electrolytic capacitors and MOSFET switches are used by the gate drivers to provide the required peak gate current at turn-off. These components reduce the reliability and increase the size of the device, a practical gate driver limits the switching frequency of the IGCT to about 500 Hz.

!Attention Like the GTO, the IGCT has high current and voltage ratings devices with 6 kV/6 kA and 6.5 kV capabilities are already commercially available. As compared with other devices, the IGCT has low on-state voltage and low total power losses (about one-half of those of the conventional GTO). Although the IGCT has overcome many of the problems of the conventional GTO, the gate driver is still complex and a large linear *di/dt* limiting inductor is needed for the anti-parallel diode.

MOS Turn-Off Thyristor (MTO)

The main aim in the development of the MTO, as compared with the previous thyristor devices, was to achieve a switch with similar power handling but faster turn-off capability, lower turn-off switching losses and a simpler gate drive. The small turn-off gate drive required by the MTO reduces the cost and improves the reliability of the device, as compared with the conventional GTO and IGCT. The storage time is comparable with that of the IGCT, which makes it suitable for series operation. Moreover, the MTO structure permits double-sided cooling. The MTO gate drive requirement for turn-on is similar to that of the GTO. A peak current of 30–60 A is applied from gate to cathode for 5–10 s, a few microseconds are needed before the anode–cathode current begins to rise and the voltage starts to fall. Following full turn-on, some gate current (about 0.5 %) must be maintained to ensure that the gate does not unlatch. Turn-off requires the application of about 15 V on second gate terminal which is held continuously during the off-state period. In the double-gated MTO, the MOSFETs are bonded in order to provide high-impedance turn-off gates on both the anode and the cathode sides. This permits the device to operate at much higher frequencies than other power devices (up to 10 kHz).

Example 12.1 *Calculate the on state and off state power loss for the MOSFET having following parameters:* $V_{ds} = 240$ V, $I_d = 4$ A, $t_r = 2$ mA, $t_f = 120$ ns, $I_{dss} = 2$ mA, $R_{ds(on)} = 0.2\ \Omega$, *duty cycle* $D = 50\%$, $F_{switching} = 50$ kHz. LO 1

Solution

(i) Total period $T = 1/f = 20$ μsec

(ii) Period $T_{on} = T_{off}$ (for 50% duty cycle) = 10 μ sec

(iii) On-state loss, $P_{on} = I_D^2 .\ R_{ds(on)} .\ (T_{on}/T) = (4^2 \times 0.2 \times 10 \times 10^{-6})/(20 \times 10^{-6}) = 1.6$ W

(iv) Off-state loss, $P_{off} = V_{DS(max)} .\ I_{dss}\ (T_{off}/T) = (240 \times 2 \times 10^{-3} \times 10 \times 10^{-6})/(20 \times 10^{-6}) = 0.240$ W

MOS Controlled Thyristor (MCT)

Again, the purpose of the MCT is to obtain faster turn-off capability and lower turn-off switching losses. The distinguishing feature of the MCT is the incorporation of MOSFETs into a cellular thyristor. Most of these are used to switch emitter shorts to block the forward voltage. These MOSFETs must, therefore, be switched off when the MCT is to be turned on. At the same time a smaller number of MOSFETs of opposite type are switched on. There are two types of MCT, i.e., the *p*-MCT, which has a *p*-type lightly doped layer to take up the off-stage voltage, and the *n*-type MCT, with an *n*-type layer as in the conventional thyristor.

In the case of the *n*-type MCT, Fig. 12.12, an *n*-type MOSFET is connected across the cathode of the *npn* transistor.

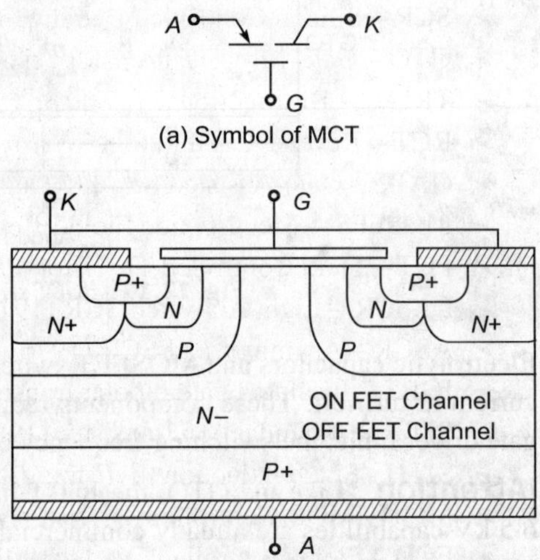

(a) Symbol of MCT

(b) Structure of MCT

Fig. 12.12

!Attention When this MOSFET is turned on by the application of a positive gate to cathode voltage, a current flow from anode to base in the lower *npn* transistor, as a result the latter turns on and produces the latched turn-on of the MCT. Similarly, a *p*-type MOSFET connected across the gate cathode of the cathode side *npn* transistor is activated to turn the device off, in this case a gate voltage must be applied to the base of the *p*-MOSFET to turn it off. The MCT structure uses many single emitter cells which can be turned on very fast. Turn-off, on the other hand, is less straightforward and requires imposing limits for the maximum controllable current. This is due to the relatively low efficiency of the MOSFET shorts, a voltage limit is also imposed due to the device tendency to filamentation. Both these limits lead to amplification of the inevitable lack of current homogeneity effect, which restricts the device application to high powers. The MCT is switched either against high voltage, with the result of a reduction in the controllable current, or at high current, with the voltage level well below that of a comparable thyristor or GTO. Thus, further development of the MCT is needed before its ideal characteristics, i.e., fast switching with low on-state and switching losses, can be exploited in high-power static conversion.

Emitter Turn-Off Thyristor (ETO)

The ETO, a promising new device for high-power applications, is reported to combine the high-voltage and high-current capability of the traditional thyristor and the easier gate controllability of the IGBT. The present voltage and current ratings are claimed to be 4500 V and 4000 A respectively. As compared with present competing technologies, the ETO offers improved performance in current conduction and turn-off speed, as well as snubberless turn-off current capability. The switching speed capability is of the order of 1 kHz and the total losses for switching frequencies of 500 and 1000 Hz are similar to those of the IGCT. A modular voltage source converter has been developed to demonstrate these benefits. Due to very high silicon utilisation and use of conventional thyristor technology, the ETO is a very cost-effective device. Further development in the device is expected to include the following capabilities: (i) built-in voltage sensing, (ii) current sensing, (iii) temperature sensing, (iv) control power self-generation, and (v) high-voltage current saturation.

Recall Various Power Semiconductor devices and their Year of Invention

- BJT – Bipolar Junction transistor (1950)
- SCR – Silicon Controlled Rectifier, thyristor (1956)
- GTO – Gate Turn-Off thyristor (1960)
- TRIAC – Bi-directional Triode Thyristor (1963)
- RCT – Reverse Conducting Thyristor (circa 1963)
- GATT – Gate-Assisted turn-off Thyristor (circa 1963)
- MOSFET – Metal-Oxide Field Effect Transistor (1975)
- FCT – Field Controlled Thyristor (1975)
- LASCR – Light Activated SCR (1976)
- ASCR – Asymmetrical SCR (1976)
- IGBT – Insulated gate bipolar transistor (1983)
- SIT – Static induction transistor (1983)
- SITH - Static induction thyristor (1983)
- MCT – MOS-Controlled Thyristors (1984)
- FCD – Field-Controlled Diode (1991)
- IGCT – Integrated Gate-Commutated Thyristor (1997)

!Attention All power semiconductor devices at present employ silicon as the basic material. However, the new type of materials such as gallium arsenide, silicon carbide and diamond show a great promise for future generation of devices. Further, super conductive power control devices based on Josephson effect also show future promise.

Table 12.2 Rating of Various Power Semiconductor Device

Device	Type	Voltage Rating (V)	Current Rating (A)	Switching time (μs)	On state Resistance (Ω)
Power	General	5000	5000	100.00	0.16 m
Diodes	Schottky	40	60	0.23	10 m
	High speed	3000	1000	2.00	1 m
Triac		1200	300	200.00	3.57 m
	Rev. blocking	5000	5000	200.00	0.25 m
Thyristor (SCR)	High speed	1200	1500	20.00	0.47 m
	GATT	1200	400	8.00	2.24 m
	LASCR	6000	1500	200.00	0.53 m
Power	Single	500	50	0.60	0.60
MOSFETs		1000	4.7	0.90	2
SITs		1200	300	0.55	1.2
IGBTs	Single	1200	1200	2.00	60 m
MCTs	Single	1200	500	2.20	18 m
GTO		6000	6000	15.00	15.00
SITH		4000	2200	6.50	6.50
IGCT		4500	3000	6.50	6.50

Concept Check LO 1

1. What are the important ratings of SCR? ○●●
2. What is commutation and explain the different ways of an SCR can be commutated? ○●●
3. How can you use triac for speed control of a three phase induction motor? ●●●
4. State the role of solid state devices in motor control. ○●●
5. Categorise solid state devices on the operating capability. ○○●
6. Differentiate between SCR and TRIAC. ○●●
7. Differentiate between MOSFET and IGBT. ○●●
8. Define latching current and holding current of SCR. ○○●
9. How will you turn on and turn off the SCR? ○●●
10. List the operating modes of SCR. ○○●
11. Is protection necessary for SCR? If yes, how will you protect it? ●●●
12. How to turn on MOSFET? ○●●
13. Give the significance of power modulator in between the source and motor in an electric drive system? ●●●
14. Compare phase control and integral control. ○●●
15. What is natural commutation and forced commutation? ○○●
16. Give the construction features of MOSFET. ○●●

Note: Difficulty Level ⟶ ○○● — Level 1 and Level 2 Category; ○●● — Level 3 and Level 4 Category; ●●● — Level 5 and Level 6 Category

17. How will you turn on IGBT?
18. State the operational capability of MCT and draw its structure.
19. How MTO differs from IGCT?

12.3 ELECTRICAL DRIVES

LO 2 Construct the block diagram of electric drive and discuss the selection of particular converter and control depending on commutation, type of source, type of load (ac or dc motor to be controlled) and quadrant of operation

Invention of thyristor led to emergence of semiconductor drives in early sixties. The improvements in converters (as discussed in Sec. 12.4) and development of new drive-control strategies, such as field oriented (or vector) control of ac drives, sliding-mode control, energy saving strategies, etc. brought about another revolution in drive technology and performance [66].

!Attention Efforts are also underway to tap alternative/renewable/new energy sources, especially solar energy. Solar drives [51] have a vital role to play in India particularly in water pumping, airconditioning, transportation and space applications.

Block diagram of an electrical drive is shown in Fig. 12.13.

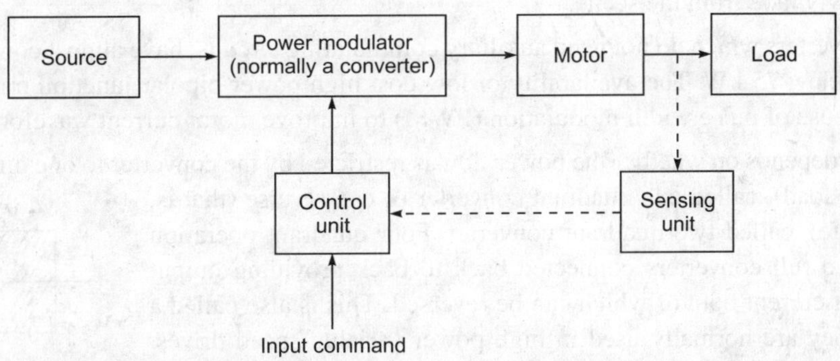

Fig. 12.13 *Block diagram of an electrical drive*

So far variable speed drive applications were dominated by dc motors. Squirrel-cage induction motor which costs approximately one-third of a dc motor of the same rating, is quite rugged, maintenance-free, can be built for higher speeds, torques and power ratings. Wound-rotor motors, though costlier than squirrel cage, also have less maintenance and available in higher power ratings. Thus, because of these advantages, ac drives have succeeded in replacing dc drives in a number of variable speed applications.

The PM synchronous motors, stepper motor and switched reluctance motor have also become popular for various applications.

Need for a converter arises when nature of the available electrical power is different than what is required for the motor. Normally, 1-phase and 3-phase 50 Hz ac supplies (except aircraft and space applications where 400 Hz ac supply is used to achieve high power to weight ratio for motors) are used. 1-ph supply is used for economy. The brushless dc motor is being used for replacing dc servomotors for fractional kW range.

Probe
What is the necessity of a sensing unit?

Next section deals with various power converters.

12.4 POWER CONVERTERS

LO 2

Since the introduction of the thyristor, the converter technology has progressed through intense technological evolution for the last four decades and has emerged into a multidisciplinary technology [69].

With basic thyristors, the converters can be naturally commutated or force commutated type. In a naturally commutated converter, the circuit has a natural built-in ability to turn-off the thyristor either by line commutation, load voltage commutation, load resonance commutation or self commutation where the turn-on of the thyristor initiates the commutation process also. In a force commutated converter, the thyristor is turned off by external intervention connecting a pulse of voltage (voltage commutation) or current (current commutation) in parallel or in series with it respectively.

As per the function, four basic categories have already been discussed (see Fig. 12.1). Out of these, ac/dc converters, cycloconverters and ac/ac voltage regulators using SCR's are of the naturally commutated type. The inverter circuits may be further categorised as per their source nature as voltage source inverters (VSI) and current source inverters (CSI). This classification depends on the type of filter and control used on the dc terminals of the converter. The VSI switches dc voltage in controlled blocks to the load whereas in CSI, the dc current is switched in blocks to the load. However, it is felt that with its somewhat inflexible topology, CSI converter will slowly fade from the scene.

The transistor converters which do not need auxiliary commutating circuits, have almost completely replaced SCR converters below 75 kW. The availability of low-cost high power bipolar junction transistors has also made possible the use of pulse width modulation (PWM) to improve motor current waveform.

Another grouping depends on whether the power flow is restricted by the converter to one direction only (that is, from supply to load), called one quadrant converter or can reverse (that is, load can regenerate), called two-quadrant converter. Four quadrant operation obtained using two full converters connected back to back providing output voltage and output current both of which can be reversed. This is also called a dual converter. They are normally used in high power variable speed drives. Figure 12.14 shown the *v-i* characteristics of a dual converter.

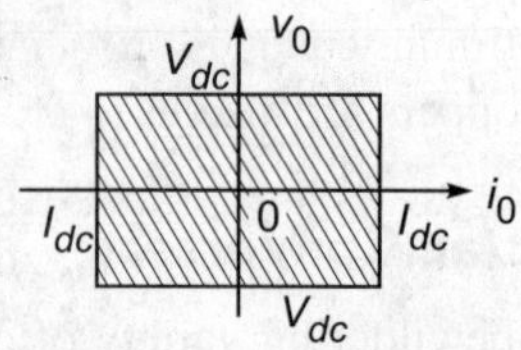

Fig. 12.14 *Four quadrant operation*

ac/dc Converters

The naturally commutated phase-controlled converter is a common type of controlled power electronic converter which has reigned supreme for the last 30 years.

Above a few tens of kW, three-phase rectifiers are used, the most common arrangements being the fully/semi-controlled bridges. Semiconverter systems, including a free-wheeling diode give better power factor than the full converter. The poor power factor is a major concern and various techniques such as asymmetrical gating, bias voltage control, etc., have been tried to improve it over the years. A number of gate-firing controllers have been used to control the firing angle in response to a variable input control voltage. Microprocessor-based control is more popular due to implementation possibility of any control law. Two phase controlled converters connected in antiparallel form a dual converter. It is also used for dc motor drives.

Probe

Which type of drive requires an ac/dc converter?

ac/ac Converters

Thyristor based ac voltage controller (same output frequency) using the phase control principle is normally used in many applications including single-phase appliance-type drives.

Important! Phase-controlled cycloconverters are used in VSCF (variable speed constant frequency) large capacity ac drives.

The inherent harmonics and lagging var problems due to phase control are also present here.

The Matrix Converter

The desirable features of MCT's and IGBT's are expected to initiate a revolution in power converter design over the next decade. The most important will probably be the so-called matrix converter [70] which with proper switching algorithm is capable of generating an almost sinusoidal waveform at input and output and can operate at lagging, unity or even leading fundamental power factor while supplying power to a lagging power factor load at a variable frequency. Four-quadrant operation can clearly be obtained, i.e. motor or regenerating operation with either leading or lagging current. The important feature of this converter is the complete absence of a "dc link" and thus no energy storage capacitor or inductor is needed to support the link voltage or current.

dc/dc Converters

Direct dc/dc chopper converters convert fixed voltage dc source into a variable voltage dc source and can be used as a dc transformer to step down (buck) or step up (boost). The force commutated thyristor chopper with voltage commutation was earlier known as oscillation chopper. Other variations are a current commutated or load commutated chopper. Choppers are normally used for traction motor control, rotor resistance control of slip ring induction motors and dc circuit breakers (with zero chopping frequency). A microprocessor based optimum control for four-quadrant operation chopper is described by Dewan and Mirbord [71]. Multiphase choppers are also used for large traction drives requiring large load current.

dc/ac Converters

A bewildering variety of direct dc/ac converters or inverters are in use today employing different auxiliary arrangements for forced commutation or new switching devices. As mentioned, they may be either voltage source or current source type. In current source inverters, the direct current is restricted to one polarity and power reversal for regeneration is carried out by reversing the polarity of the direct voltage. In voltage source inverters, the direct voltage polarity cannot change and reversal of direct current is needed for regeneration using a feedback converter. Also, according to switching, inverters can be PWM or resonant type.

Control

The expert system (ES), ANN and fuzzy control techniques are slowly being used in the control of power electronics. ES techniques have tremendous potential in automating system design, modelling, simulation, real-time control, tests and diagnostics. Fuzzy logic deals with uncertain or imprecise situations.

Probe

What is the role of an expert system in control of power electronics?

Power electronic converters are able to perform almost any type of electric power conversion or control. The present research effort is directed towards reducing the cost, size, weight and losses increasing reliability and achieving better performance of various converters

Voltage Source Inverters (VSI)

A VSI power circuit supplied from a fixed or variable dc voltage (when supplied from an ac source through a phase controlled converters or a diode rectifier followed by a chopper) with a filter capacitor and small inductor. Main applications of VSI are either in ac motor drives with variable voltage and frequency or in UPS having constant voltage and frequency.

Current Source Inverters (CSI)

The CSI is a dual of the VSI described above. A phase controlled rectifier or a diode rectifier followed by a chopper with a current loop and a large inductor provides the adjustable current source. While the major advantages of CSI over VSI are rugged and reliable power circuit, inherent four quadrant operation, etc, the limitations are lower frequency range due to commutation overlap, need of large size dc link inductance and bulky capacitors, open loop instability, torque pulsations and harmonic heating of machines when supplied from it. Various attempts[68] have been made for the last several years to obviate the limitations of CSI.

12.5 THYRISTOR MOTOR CONTROL

LO 2

A variety of thyristor control circuitry has been devised for use in motor control depending upon the type of supply (ac/dc) and the type and size of the motor.

dc Motor Control

For dc motor control, controlled dc power from a constant voltage ac supply is obtained by means of *controlled rectifiers* or *converters* employing thyristors and diodes. The control of dc voltage is achieved by firing the thyristors at an adjustable angle with respect to the applied voltage. This angle is known as the *firing angle* and the scheme of control is called the *phase control.* Another basic method of control is known as the *integral-cycle control.* Here the current is gated to flow from the ac supply for a number of complete cycles and is then quenched for a further few cycles, the process being repeated continuously. Control is applied by adjusting the ratio of on and off durations. This method is suitable for the control of fractional-kW dc motors. Line commutation is easily adopted for both these control schemes.

Phase-control and integral-cycle control methods are also applicable for ac motors wherein the converter circuitry is not required.

Control of dc motors fed from dc supply is achieved by means of a thyristor-switching circuit called the *chopper.* The chopper controller periodically opens and closes wherein the control of the average voltage is achieved by varying on and off durations. This gives an efficient and step less control of dc motors. The motor can also be made to operate in the regenerative braking mode. The chopper controller requires forced commutation of the thyristor.

ac Motor Control

For the control of ac motors fed from dc supply, thyristor, transistor or MOSFET based *inverters* are employed. These switching circuits transfer energy from dc supply to ac load of variable frequency and/or variable voltage. Because of the switching operation, the ac voltage waveforms are stepped, the harmonic contents of which are filtered out by the ac motor. As the power supply is normally ac, the complete scheme for obtaining variable-voltage and frequency power involves the use of both an inverter and a converter.

A cycloconverter is a control unit for obtaining variable-voltage and frequency-power directly from a fixed-frequency supply without the necessity of an intermediate dc stage. The mechanism of voltage and frequency control is a combination of those used in the phase-controlled converter and the pulse-modulated inverter. Cycloconverters, in spite of their attraction of direct ac to ac conversion, suffer from certain disadvantages because of which they have not gained popularity. Some of these disadvantages are: cycloconverters can produce only a sub-frequency output, they produce output with high harmonic content and have a low input power factor. Cycloconverters are used for low-speed drives and for controlling linear motors in high-speed transportation systems.

Concept Check — LO 2

1. How can the speed of a dc machine be controlled in both directions with the thyristor power supply? ○●●
2. What are the functions of power modulator? ○○●
3. List four quadrants of operation of a drive. ○○●
4. Compare VSI and CSI. ○●●
5. State the input and output of a converter, chopper and inverter. ○●●
6. Under what circumstances you need a dc/dc converter? ●●●
7. What is a matrix converter? ○○●
8. What are expert systems and what is their role in drives? ●●●
9. How will you change the direction of rotation of converter fed dc drive? ●●●

12.6 | dc MOTOR CONTROL THROUGH CONVERTERS

LO 3 Analyze the various topologies of single-phase converter, three-phase converter and chopper-fed dc motor drives with their construction and operation features

dc motor control is conveniently and efficiently achieved by phase-controlled converters wherein the ac input voltage is converted to a controlled dc output. The commutation process, the transfer of current from one thyristor to the other, in these converters is the inexpensive *natural* or *line commutation.* As, an incoming thyristor is turned-on, it reverse-biases the outgoing thyristor, turning it off. No additional commutation circuitry is therefore required. The power conversion efficiency in these converters is above 95% because of relatively low losses in thyristors. These converters are used in the speed control of fractional-kW dc motors as well as in large motors employed in variable-speed reversing drives for rolling mills with motor ratings as large as several MWs.

In certain types of converters (*semi-converters*) a diode known as *s free-wheeling diode* is connected across motor terminals to allow for dissipation of energy stored in motor inductance and to provide for continuity of motor current when the thyristors are blocked. It also provides protection against transient overvoltages.

Single-phase Converters

Half-wave Converter Figure 12.15 shows a single-phase half-wave converter for controlling a separately excited dc motor. It requires a single thyristor and a free-wheeling diode. In this circuit the motor current is always discontinuous, resulting in poor motor performance. This type of converter is employed only for motors below 400 W. It will not be described in detail here; these easily follow from the description of waveforms (voltage and current) for the semi-converter.

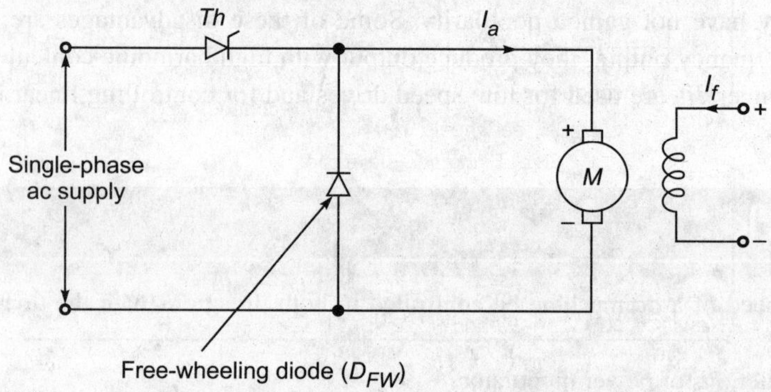

Fig. 12.15 *Single-phase, half-wave converter*

Semi-converter This is a one-quadrant converter (Fig. 12.16a) which gives voltage and current of one polarity at dc terminals. It therefore does not provide for regenerative braking, i.e. power flow from dc motor to the ac supply. Where regeneration is not required, this converter is used for reasons of economy.

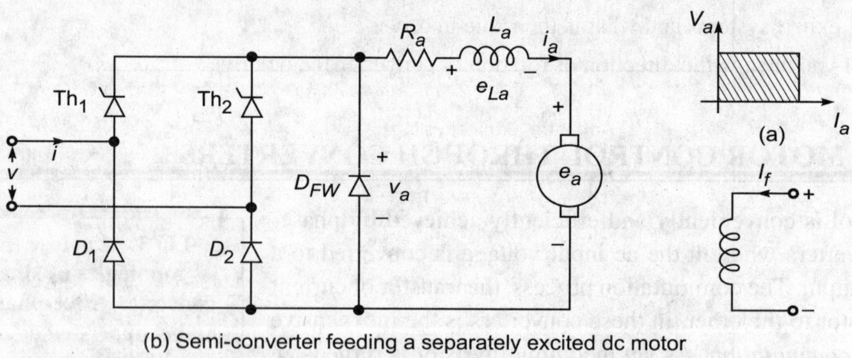

Fig. 12.16

Figure 12.16(b) gives the circuitry of a semi-converter feeding a separately excited dc motor. The armature resistance R_a and inductance L_a are shown lumped in series with an armature-induced emf e_a while the armature terminal voltage is v_a. It has two thyristors, two diodes and a free-wheeling diode (connected across motor terminals). It will be assumed that in steady-state operation the armature current is continuous over the whole range of operation. Typical steady-state voltage and current waveforms are shown in Fig. 12.17. The thyristor Th_1 is fired at angle α and Th_2 at angle $\pi + \alpha$ with respect to the supply voltage v and the process is repeated continuously.

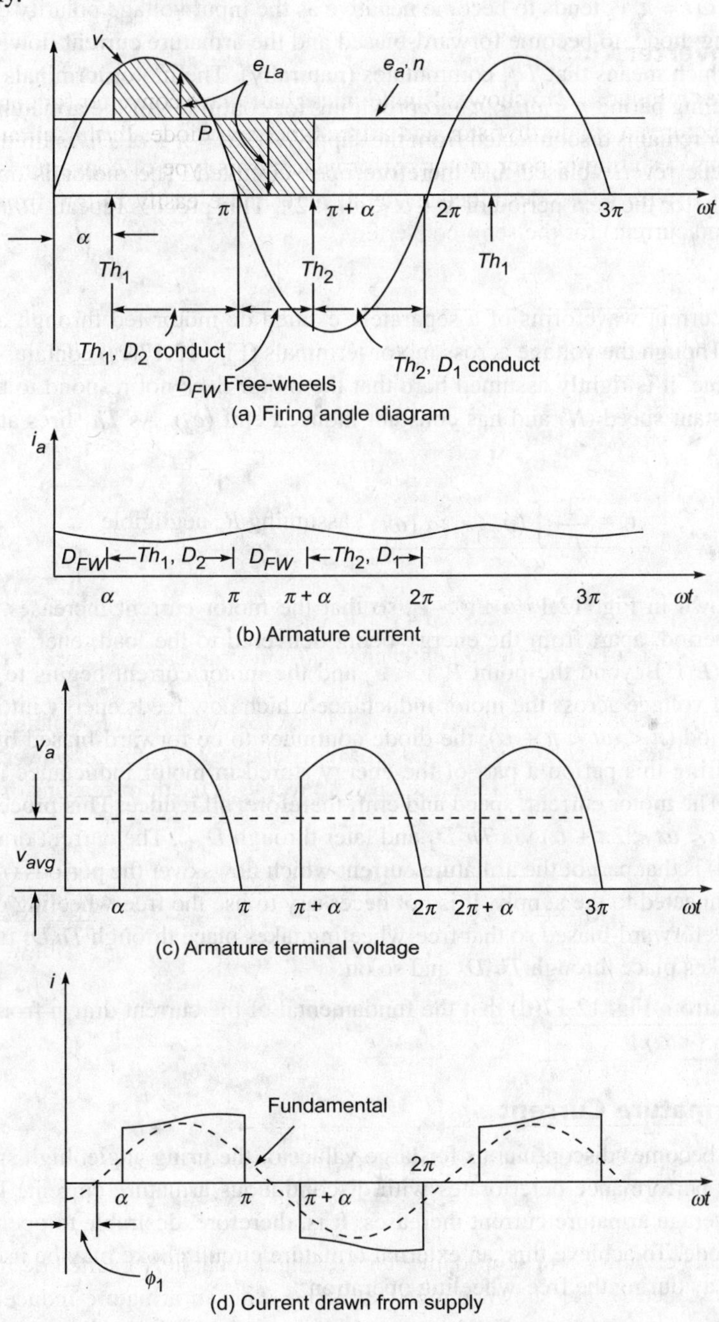

Fig. 12.17 *Voltage and current waveforms of semiconverter feeding separately excited dc motor; continuous armature current*

Operation

Under steady state conditions, as Th_1 is fired ($\omega t = \alpha$), Th_1 and D_2 conduct and the motor is connected to the supply, i.e. $v_a = v$. At $\omega t = \pi$, v_a tends to become negative as the input voltage polarity changes. This causes D_{FW}, the free-wheeling diode, to become forward-biased and the armature current flowing through Th_2D_2 is transferred to D_{FW}, which means that Th_1 commutates (naturally). The motor terminals are shorted through D_{FW} for the free-wheeling period $\pi < \omega t < \pi + \alpha$ providing for continuity of the armature current during this period when the motor remains disconnected from the supply. At $\omega t = \pi + \alpha$, Th_2 is fired and Th_2D_1 conduct, causing D_{FW} to become reverse-biased and therefore open-circuited. The motor is once again connected positively to the supply for the next period of $\pi + \alpha < \omega t < 2\pi$. This process repeats continuously.

Analysis

Various voltage and current waveforms of a separately excited dc motor fed through a semi-converter are shown in Fig. 12.17. Though the voltage across motor terminals (Fig. 12.17(c)) contains harmonics over and above a steady dc value, it is rightly assumed here that the motor does not respond to these harmonics and therefore runs at constant speed (N) and has constant induced emf (e_a). As Th_1 fires at $\omega t = \alpha$, the motor current is given by

$$i_a = \frac{1}{L_a}\int_{\alpha}^{\omega t} (v - e_a)d\,(\omega t)\text{ ; assuming } R_a \text{ negligible} \tag{12.1}$$

up to the point P shown in Fig. 12.17(a); $v > e_a$ so that the motor current increases. So does the motor emf e_a. During this period, apart from the energy being delivered to the load, energy is also being stored in motor inductance (L_a). Beyond the point P, $v < e_a$ and the motor current begins to decrease. This also implies the reversal of voltage across the motor inductance which now feeds energy into the system. During the free-wheeling period ($\pi < \omega t < \pi + \alpha$), the diode continues to be forward-biased by the reversal of the inductive voltage. During this period a part of the energy stored in motor inductance is consumed to feed the mechanical load. The motor current, speed and emf, therefore, all reduce. This process then repeats over the next period ($\pi + \alpha < \omega t < 2\pi + \alpha$) via Th_2D_1 and later through D_{FW}. The current drawn from the supply shown in Fig. 12.17(d) is that part of the armature current which flows over the periods (α, π), ($\pi + \alpha$, 2π), … when the motor is connected to the supply. It is not necessary to use the free-wheeling diode. In its absence at $\omega t = \pi$, D_1 becomes forward-biased so that free-wheeling takes place through Th_1D_1 till Th_2 is fired. At $\omega t = 2\pi$ free-wheeling takes place through Th_2D_2 and so on.

It should be observed from Fig. 12.17(d) that the fundamental of the current drawn from the mains lags the voltage by an angle ϕ_1 ($< \alpha$).

Discontinuous Armature Current

The armature current becomes discontinuous for large values of the firing angle, high speed and low values of torque. The motor performance deteriorates with discontinuous armature current. The ratio of peak to average and rms to average armature current increases. It is, therefore, desirable to operate the motor in the continuous current mode. To achieve this, an external armature circuit choke may be used, which decreases the rate of current decay during the free-wheeling operation.

Operation

The voltage and current waveforms for semi-converter with discontinuous current are shown in Fig. 12.18. The motor is connected to supply through Th_1D_2 for the period $\alpha < \omega t < \pi$. Beyond π, the motor is shorted through the free-wheeling diode D_{FW}. The armature current decays to zero at angle β (*extinction angle*) $< \pi + \alpha$, i.e. before the thyristor Th_2 is fired, thereby making the armature current discontinuous. During α to π, the conduction period through Th_1D_2, the motor terminal voltage is the same as the input voltage. During π to β, the motor terminal voltage is zero as motor terminals are shorted by the free-wheeling diode. From β to $\pi + \alpha$, the motor coasts and so its terminal voltage is the same as its induced emf.

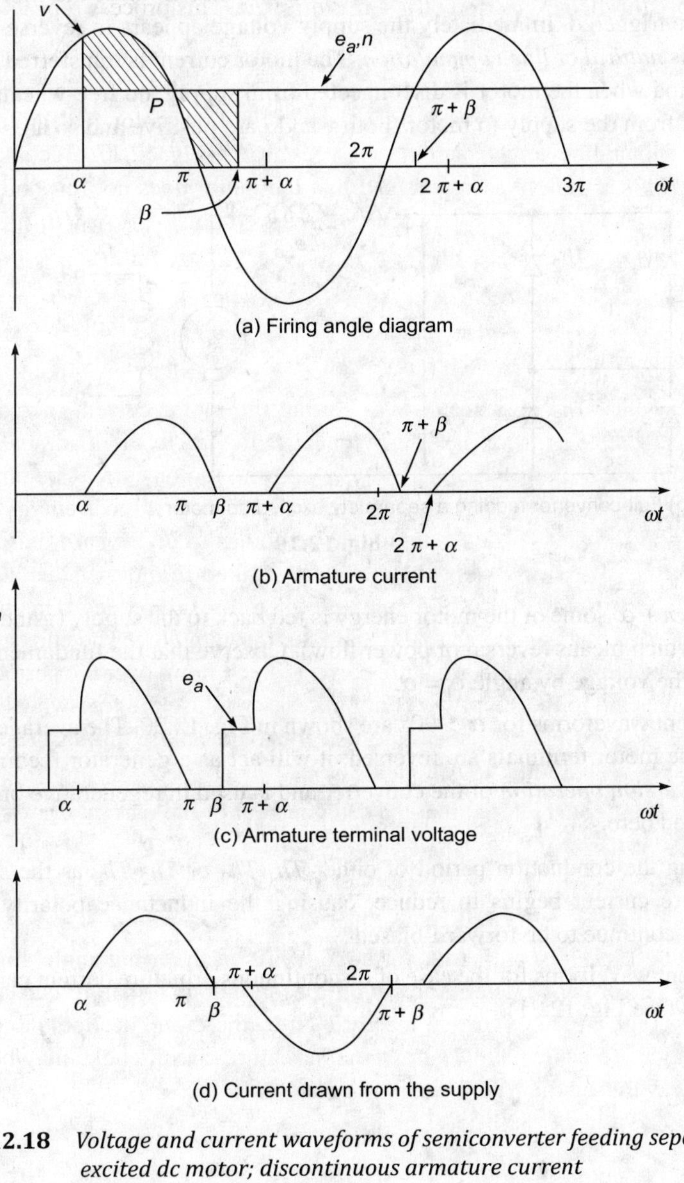

Fig. 12.18 *Voltage and current waveforms of semiconverter feeding separately excited dc motor; discontinuous armature current*

Full-Converter

A full-converter is a two-quadrant converter (see Fig. 12.19(a)) in which the voltage polarity of the output can reverse, but the current remains unidirectional because of the unidirectional thyristors. A full-converter employs four thyristors but no free-wheeling diode is required. A full-converter feeding a separately excited dc motor is shown in Fig. 12.19.

Assumption The voltage and current waveforms are shown in Fig. 12.20 with the assumption that the armature current i_a is almost constant.

Thyristors Th_1 Th_3 conduct for the interval $\alpha < \omega t < \pi + \alpha$ and connect the motor to supply. At $\pi + \alpha$, thyristors Th_2 Th_4 are triggered. Immediately the supply voltage appears in reverse-bias across Th_1 Th_3 and turns them off. This is *natural* or *line commutation.* The motor current is transferred from Th_1 Th_3 to $Th_2 Th_4$. Since there is no period when the motor is disconnected from supply, no free-wheeling is necessary. During α to π, energy flows from the supply to motor (both v and i are positive and so are v_a and i_a).

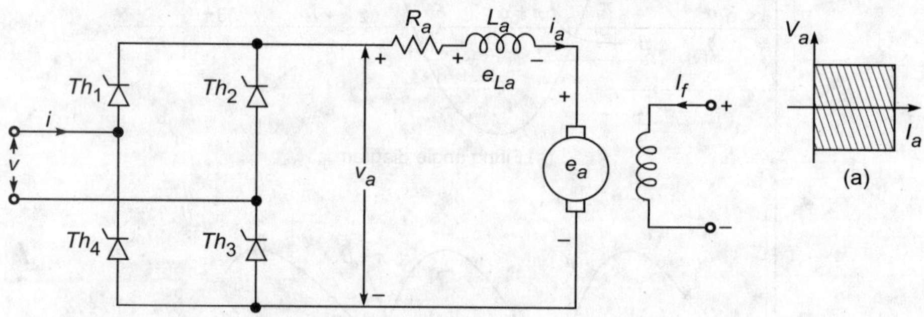

(b) Full-converter feeding a separately excited dc motor

Fig. 12.19

However, during π to $\pi + \alpha$, some of the motor energy is fed back to the supply (v and i have opposite polarity and so are v_a and i_a which means reversal of power flow). Observe that the fundamental of the current drawn from the mains lags the voltage by angle $\phi_1 = \alpha$.

The voltage and current waveforms for $\alpha > 90°$ are shown in Fig. 12.20. The average motor terminal voltage is now negative. If the motor terminals are reversed, it will act as a generator feeding power back to the ac supply. This is the *inversion operation* of the converter and is used in regenerative braking of the motor. One point needs to be noted here.

!Attention During the conduction period of either Th_1 Th_3 or Th_2 Th_4 as the supply voltage becomes negative, the armature current begins to reduce, causing the inductance polarity to reverse so that the conducting thyristors continue to be forward-biased.

The voltage and current waveforms for the case of discontinuous armature current can be found as in the case of the semi-converter (see Fig. 12.21).

Torque-Speed Characteristic

Analysis It will be assumed here that the armature current is continuous. For a semiconverter with free-wheeling diode action, the armature circuit equations are

$$v_a = v = R_a i_a + L_a \frac{di_a}{dt} + e_a \qquad \alpha < \omega t < \pi \tag{12.2}$$

$$v_a = 0 = R_a i_a + L_a \frac{di_a}{dt} + e_a \qquad \pi < \omega t < \pi + \alpha \tag{12.3}$$

For a full-converter, the armature circuit equation is

$$v_a = v = R_a i_a + L_a \frac{di_a}{dt} + e_a \qquad \alpha < \omega t < \pi + \alpha \tag{12.4}$$

Let $v = \sqrt{2}V \sin \omega t$. The average motor terminal voltages are: With a semi-converter,

$$V_a = \frac{1}{\pi}\int_{\alpha}^{\pi} \sqrt{2}\,V \sin \omega t \; d(\omega t) = \frac{\sqrt{2}\,V}{\pi}(1 + \cos \alpha) \tag{12.5}$$

With a full-converter,

$$V_a = \frac{1}{\pi}\int_{\alpha}^{\pi+\alpha} \sqrt{2}\,V \sin \omega t \; d(\omega t) = \frac{2\sqrt{2}V}{\pi}\cos \alpha \tag{12.6}$$

Figure 12.22 gives the variation of the motor terminal voltage as a function of the firing angle for both the semi-converter and full-converter. In the case of full-converter inversion operation occurs for $90° < \alpha < 180°$.

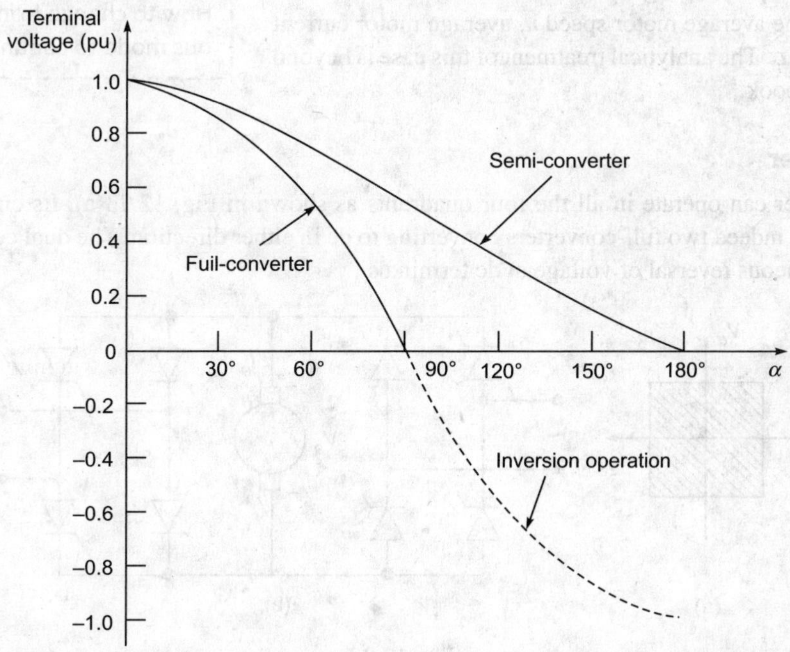

Fig. 12.22 *Variation of the converter terminal voltage with firing angle*

The motor equations for average values are

$$E_a = K_a \Phi n \tag{12.7}$$

$$T = K_a \Phi I_a \tag{12.8}$$

and

$$n = \frac{V_a - I_a R_a}{K_a \Phi} \tag{12.9}$$

Substituting for I_a from Eq. (12.8) and V_a from Eqs (12.5) or (12.6) in Eq. (12.9), the speed-torque characteristics are given as follows:

With a semi-converter,

$$n = \frac{(\sqrt{2}V/\pi)\,(1+\cos\alpha)}{K_a \Phi} - \frac{R_a}{(K_a \Phi)^2} T \tag{12.10}$$

With a full-converter

$$n = \frac{2\sqrt{2}\,V/\pi \cos\alpha}{K_a \Phi} - \frac{R_a}{(K_a \Phi)^2} T \tag{12.11}$$

The first term in Eqs (12.10) and (12.11) represents the theoretical no-load speed while the second term represents speed drop caused by armature resistance. The theoretical no-load speed can be varied by the firing angle α.

In the case of discontinuous current, the average voltage at motor terminals depends upon the angle β (*extinction angle*) which itself is dependent on the average motor speed *n*, average motor current I_a and firing angle α. The analytical treatment of this case is beyond the scope of this book.

Probe

Why is discontinuous mode of operation not preferred?

Probe

How to change from discontinuous mode to continuous mode?

Dual-Converter

The dual-converter can operate in all the four quadrants as shown in Fig. 12.23(a). Its circuit is shown in Fig. 12.23(b). It is indeed two full-converters converting to dc in either direction. The dual converter provides virtually instantaneous reversal of voltage at dc terminals.

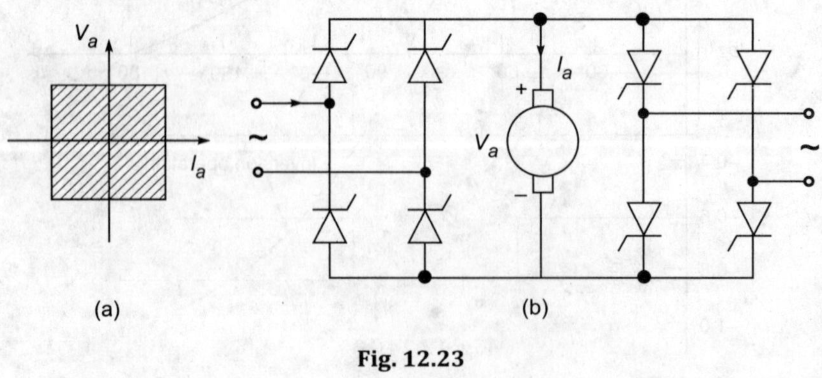

Fig. 12.23

Example 12.2 *For the single-phase converter shown in Fig. 12.24(a), assume the firing angle to be α for the thyristor.*

(a) Draw the load voltage waveform assuming the output current to be I_a, a constant.

(b) Draw the waveforms of the currents through Th_1, D_1, D_2 and D_3.

(c) Calculate the power fed to the load. Given V = 100 V, R_d = 10 Ω and firing angle, α = 45°. LO 3

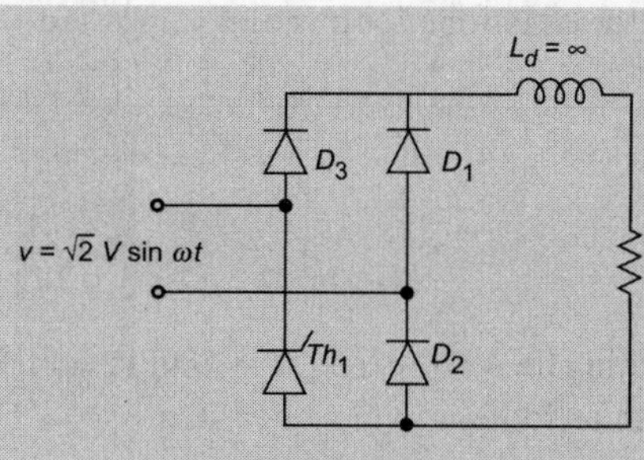

Fig. 12.24(a)

Solution

(a), (b) The various waveforms are drawn in Fig. 12.24(b).

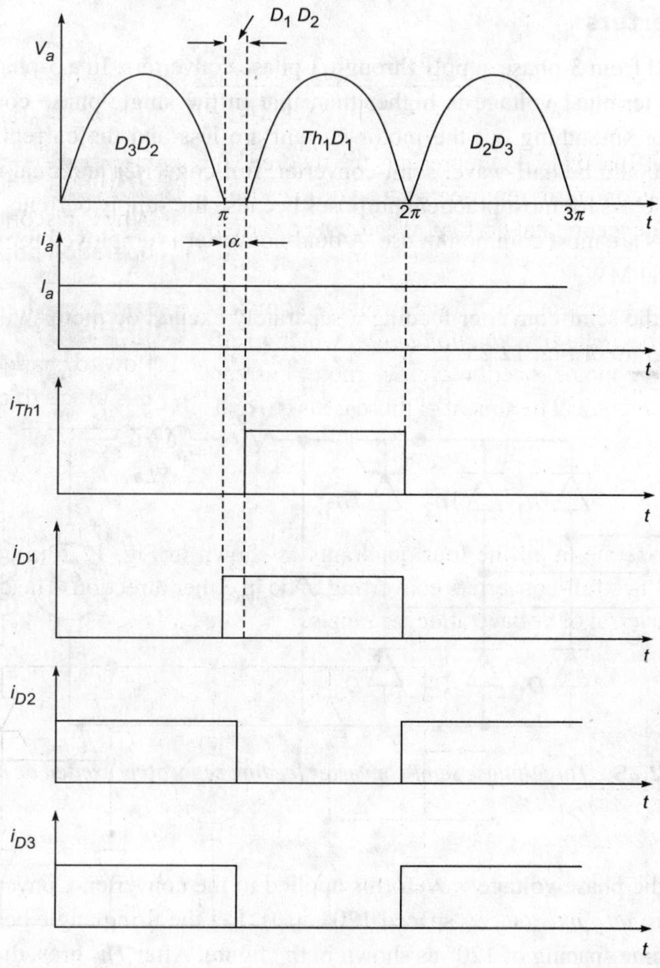

Fig. 12.24(b)

(c)
$$V_a = \frac{1}{2\pi}\left[\int_0^{\pi} \sqrt{2}V \sin \omega t \, d(\omega t) + \int_{\pi+\pi/4}^{2\pi} (-\sqrt{2}V \sin \omega t)\, d(\omega t)\right]$$

$$= \frac{V}{\sqrt{2}\pi}\left(2 + \frac{1}{\sqrt{2}}\right)$$

$$= \frac{100}{\sqrt{2}\pi}\left(2 + \frac{1}{\sqrt{2}}\right) = 60.93 \text{ V}$$

Load power, $$P_a = \frac{V_a^2}{R_d} = \frac{(60.93)^2}{10}$$

$$= 371.2 \text{ W}$$

Three-Phase Converters

Large-kW motors are fed from 3-phase supply through 3-phase converters. In a 3-phase converter, the ripple frequency of the motor terminal voltage is higher than that in the single-phase converter, as a result the filtering requirements for smoothing out the motor current are less and the current is mostly continuous. Three-phase converters could be half-wave, semi-converter, full-converter and dual-converter. The 3-phase half-wave converter is useless for most practical purposes because the supply currents contain dc component. Semi- and full-converters are most common in use. A dual-converter is employed for reversible drives with a power rating up to several MW.

For want of space only the semi-converter feeding a separately excited dc motor will be described here, as shown in the circuit diagram of Fig. 12.25.

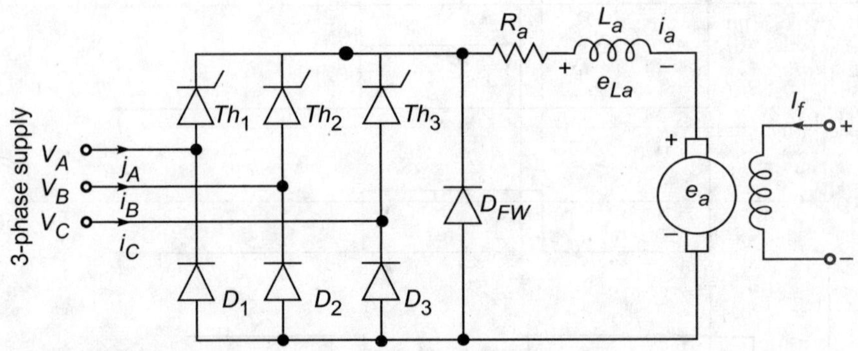

Fig. 12.25 *Three-phase semi-converter feeding separately excited dc motor*

Operation

Figure 12.26(a) shows the phase voltage waveforms applied to the converter. Convenient references for the thyristor firing angles are ωt_1, ωt_2, ωt_3 ... spaced 120° apart. Let the firing angle be α. The three thyristors fire in sequence with a time spacing of 120° as shown in the figure. After Th_1 fires, diodes D_2 or D_3 conducts depending upon which voltage is more positive of V_{AB} and V_{AC}. For the value of α chosen, it may be immediately seen from Fig. 12.26(b) that diode D_2 conducts up to 90° beyond which D_3 conducts till Th_2 fires. Then D_3

and D_1 conduct till Th_3 fires followed by D_1, D_2 conducting in sequence. This process then repeats resulting in the voltage waveform of v_a as shown in Fig. 12.26(b). For this value of α, the freewheeling diode is not called upon to conduct the armature current. The armature current i_a and line current i_A are shown in Figs. 12.26(c) and 12.26(d) respectively. The fundamental i_{A1} of the line current lags by a small angle ϕ_1 resulting in less than unity power factor.

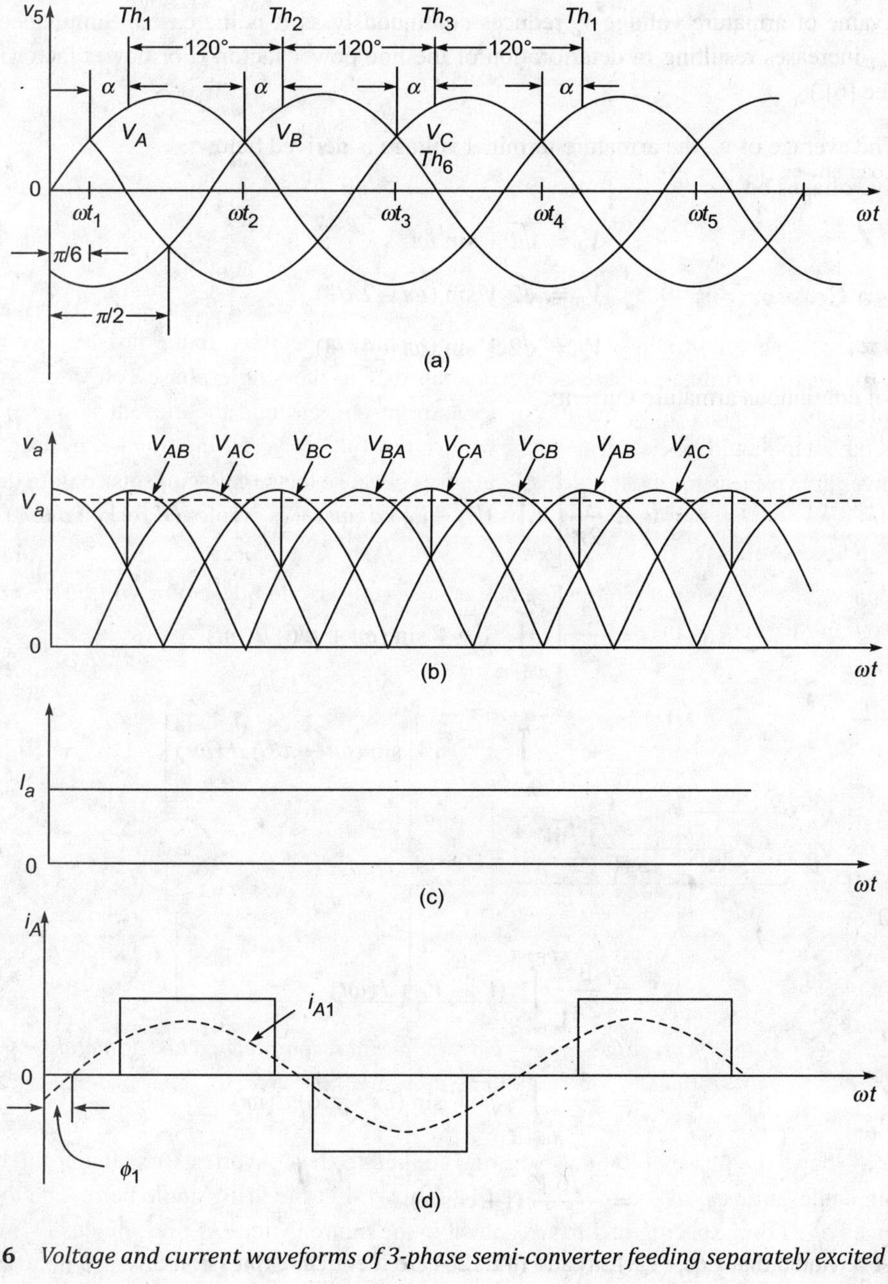

Fig. 12.26 *Voltage and current waveforms of 3-phase semi-converter feeding separately excited dc motor*

For a firing angle $\alpha = 60°$, only one diode conducts during the conduction period of a thyristor. The conduction sequence is now Th_1D_3, Th_2D_1, Th_3D_2. For larger values of α, V_{AC} becomes negative before Th_2 fires resulting in interruption of line current so that i_a flows through D_{FW}, the free-wheeling diode.

If the free-wheeling diode is not used, free-wheeling takes place through the diode and thyristor of a given branch, e.g. after D_3 stops conducting, free-wheeling occurs through Th_1D_1.

The average value of armature voltage V_a reduces continuously as α is increased. Simultaneously, the lag angle ϕ_1 of i_{A1} increases resulting in deterioration of the line power factor. (For power factor improvement techniques, see [6].)

Analysis The average of v_a, the armature terminal voltage is derived below:

Let the supply voltages be

$$V_A = \sqrt{2}\, V \sin \omega t$$

$$V_B = \sqrt{2}\, V \sin (\omega t - 2\pi/3)$$

$$V_C = \sqrt{2}\, V \sin (\omega t - 4\pi/3)$$

For the case of continuous armature current:

(i) $\alpha < 60°$

$$V_a = \frac{3}{2\pi}\left[\int_{\pi/6+\alpha}^{\pi/2} (V_A - V_B)\, d(\omega t) + \int_{\pi/2}^{\pi/6+\alpha+2\pi/3} (V_A - V_C)\, d(\omega t)\right]$$

$$= \frac{3}{2\pi}\left[\int_{\pi/6+\alpha}^{\pi/2} \sqrt{6}\, V \sin (\omega t + \pi/6)\, d(\omega t) + \int_{\pi/2}^{\pi/6+\alpha+2\pi/3} \sqrt{6}\, V \sin (\omega t - \pi/6)\, d(\omega t)\right]$$

or

$$V_a = \frac{3\sqrt{6}V}{2\pi}(1 + \cos \alpha) \tag{12.12a}$$

(ii) $\alpha \geq 60°$

$$V_a = \frac{3}{2\pi}\int_{\pi/6+\alpha}^{\pi/6+\pi} (V_A - V_C)\, d(\omega t)$$

$$= \frac{3}{2\pi}\int_{\pi/6+\alpha}^{\pi/6+\pi} \sqrt{6}\, V \sin (\omega t - \pi/6)\, d(\omega t)$$

$$= \frac{3\sqrt{6}\, V}{2\pi}(1 + \cos \alpha) \tag{12.12b}$$

It must be noted that Eqs. (12.12a) and (12.12b) are identical expressions.

The expression of Eq. (12.10) for average motor speed now modifies to

$$n = \frac{(3\sqrt{6}\,V/2\pi)(1+\cos\alpha)}{K_a\Phi} - \frac{R_a}{(K_a\Phi)^2}T \tag{12.13}$$

Example 12.3 *A resistive load is connected to 3-phase semi-converter through a large reactor. If the power output at the load terminals is 15 kW, calculate (a) the value of the firing angle assuming $R_0 = 1.5\ \Omega$ and V_{supply} (phase) = 100 V, rms, (b) the average values of currents through thyristors and diodes.* LO 3

Solution From Eq. (12.12(a)),

$$V_a(\alpha) = \frac{3\sqrt{6}\,V}{2\pi}(1+\cos\alpha) = \sqrt{P_0 R_0} \quad (\because P = V^2/R)$$

$$= \frac{3\sqrt{6}\times 100}{2\pi}(1+\cos\alpha) = \sqrt{15000\times 1.5}$$

$$\alpha = 73.6°$$

or

$$I_a = \frac{\sqrt{15000\times 1.5}}{1.5} = 100\text{ A}$$

$$I_{Th}\,(\text{av}) = I_D\,(\text{av}) = \frac{I_a}{3} = \frac{100}{3} = 33.3\text{ A}$$

Example 12.4 *An RL load with R = 10 Ω and L = 0.25 H is connected to a battery of source voltage 100 V dc. A thyristor is connected in series between battery and RL load as a switch. The latching current of thyristor is 20 mA. Will the thyristor get fired if the duration of the firing pulse is 50 μsec?* LO 3

Solution As SCR is triggered the current will rise exponentially in the inductive circuit.

$$i(t) = (V/R)\,(1 - e^{(-t/\tau)})$$

where

$$\tau = L/R = 0.25/10 = 0.025\text{ sec}$$

⇒

$$i(t) = (100/20)\,(1 - e^{(-t/0.025)})$$

At t = 50 μsec,

⇒

$$i(50\ \mu\text{sec}) = 5\,(1 - e^{(-50\times 10^{-6}/0.025)}) = 9.99\text{ mA}$$

Since, i(50 μsec) < latching current, the thyristor will not turn on.

Example 12.5 *A separately excited dc motor having 2 Ω resistance and 0.3 H inductance is supplied by a 230 V, 50 Hz supply. (i) If a full-converter interface is used between supply and motor, estimate the average motor terminal voltage for a firing angle of 120°. (ii) Estimate the average terminal voltage for the same firing angle if the full-converter is replaced with semi-converter.* LO 3

Solution

(i) Full-converter

Average motor terminal voltage

$$V_a = \frac{2\sqrt{2}V}{\pi}\cos\alpha = \frac{2\sqrt{2}\times 230}{\pi}\cos 120 = -103.52\text{ V}$$

(ii) Semi-converter
Average motor terminal voltage

$$V_a = \frac{2\sqrt{2}V}{\pi}(1+\cos\alpha) = \frac{2\sqrt{2}\times 230}{\pi}(1+\cos 120) = 103.52\text{ V}$$

So as the converter is changed, for the same firing angle, the terminal voltage is same in magnitude but polarity is negative. Hence, direction of rotation will also be reversed.

12.7 dc MOTOR CONTROL THROUGH CHOPPERS

LO 3

For controlling a dc motor operated from dc supply, the dc voltage level must be controlled. Examples can be cited of subway drives, trolley buses or battery-operated vehicles. Conventional methods of converting a fixed-voltage dc source to a variable-voltage dc source are:

Resistance Control

Recall A variable resistance is inserted between the load and the source. The method is highly wasteful of energy. Also, for a given output voltage, different values of resistance are needed for different values of load current. This method is still used for older traction installations.

Motor-generator Set

Separate generator excitation gives a voltage which can be varied from zero to rated value with either polarity. The set-up is bulky, costly, less efficient and has a sluggish response because of the generator field time-constant.

Inverter Rectifier

The dc source could be converted to ac form by an inverter (see Sec. 12.9), transformed (by a transformer) to a suitable voltage level and then rectified to dc form. Because of two stages of conversion, the set-up is bulky, costly and less efficient.

dc Chopper

As mentioned in Sec. 12.5, a chopper is essentially an electronic switch that turns on the fixed-voltage dc source for short time intervals and applies the source potential to motor terminals in series of pulses. This process controls the average voltage applied to the motor. The chopper can also be easily operated in the regenerative mode. The chopper provides efficient and stepless control of dc voltage and is less expensive and bulky compared to the three methods mentioned above. It has the added advantage of fast response and regeneration. The chopper is a relatively new technology and can be visualized as a dc-to-dc transformer with a continuously-variable ratio of conversion.

Probe
What are the limitations of two stage conversion?

Principle of Operation

A chopper is essentially a thyristor switch in series with the load as shown in Fig. 12.27(a). A shunting diode is provided across the load for free-wheeling the load current when the thyristor is off. The thyristor shown enclosed by a dotted square can be turned-on and turned-off over a time period and the cycle is repeated.

The flow of dc current through the chopper during the on-period necessitates the use of a forced commutation circuit for turning off the thyristor. The average value of the chopper output voltage waveform shown in Fig. 12.27(b) is given by

$$V_a = V\frac{t_{ON}}{t_{ON} + t_{OFF}}$$

$$= V\frac{t_{ON}}{T} = \alpha V \tag{12.14}$$

where t_{ON} = on-time over one period

t_{OFF} = off-time over one period

T = chopping period

$$\alpha = \frac{t_{ON}}{T} = \text{duty cycle}$$

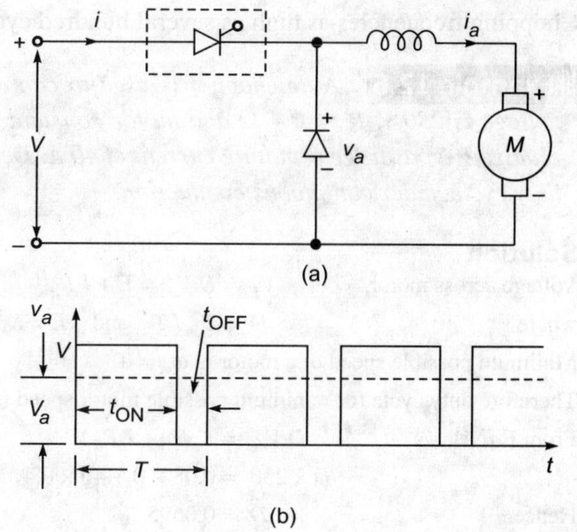

Fig. 12.27 *Basic chopper circuit and output voltage waveform*

It is therefore seen that the output voltage of a chopper is controlled by its duty cycle. It can be varied in the following ways:

1. Constant-frequency System In this system $f = 1/T$ (fixed), i.e. the chopping frequency or period are fixed and t_{ON} is varied to control α. This is *pulse-width modulation.* Figure 12.28 illustrates this system of control.

2. Variable-frequency System Here the chopping period T is varied and either t_{ON} or t_{OFF} is held fixed as illustrated in Fig. 12.29. This is *frequency modulation.* This method of control presents two difficulties: (i) Control of V_a requires variation of chopping frequency over a wide range. Filter design for variable-frequency operation is difficult. (ii) At low output voltage, a large value of t_{OFF} makes the motor current discontinuous.

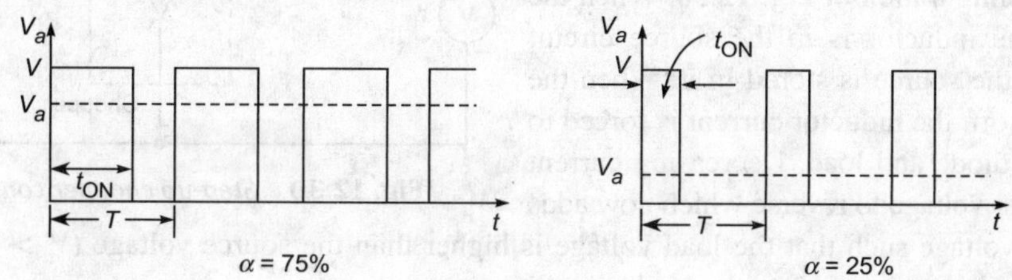

Fig. 12.28

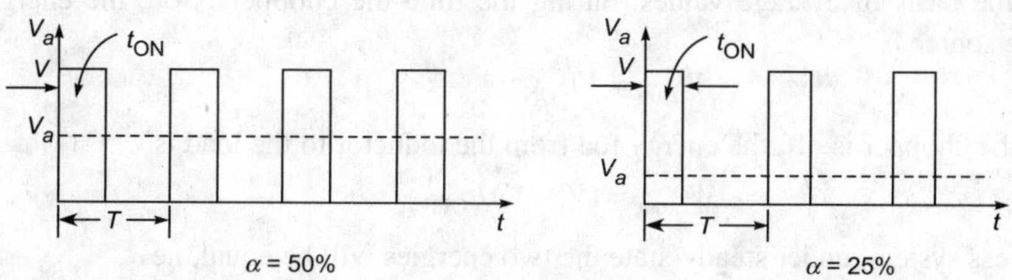

Fig. 12.29

Chopping frequencies as high as several hundred cycles/second are used in practice.

Example 12.6 *A dc chopper is used to control the speed of a separately excited dc motor. The dc voltage is 230 V, $R_a = 0.4\ \Omega$ and motor constant $K_e\phi = 0.08$ V/rpm. The motor drives a constant load requiring an average armature current of 40 A. Determine (i) the range of speed control, (ii) the range of duty cycle under continuous conduction.* LO 3

Solution

Voltage across motor, $V_L = D.V_{dc} = E + I_a R_a$ (i)

where $D = T_{on}/T$ and $E = K_e\phi\omega$

Minimum possible speed of a motor is $\omega_m = 0$

Therefore duty cycle for minimum possible motor speed ($\omega_m = 0$) is,

From Eq. (i), $D.V_{dc} = K_e\phi\omega + I_a R_a$

$$D \times 230 = 0.08 \times 0 + 40 \times 0.4$$

Hence, $D = 0.0695$

Maximum possible duty cycle of chopper is D = 1

The speed corresponding to maximum duty cycle is,

$$1 \times 230 = 0.08 \times \omega + 40 \times 0.4 \quad \text{[from Eq. (1)]}$$

On solving, $\omega = 2675$

(i) Range of speed control: $0 < \omega < 2675$

(ii) Range of duty cycle: $0.0695 < D < 1$

Step-up Chopper

The chopper configuration of Fig. 12.27(a) produces an output voltage less than the input voltage. An output voltage higher than the input voltage can be obtained by the chopper configuration of Fig. 12.30. When the chopper is on, the inductor is in the source circuit and energy from the source is stored in it. When the chopper is turned off, the inductor current is forced to flow through the diode and load. Decreasing current causes the inductor voltage to reverse which now adds up to the source voltage such that the load voltage is higher than the source voltage ($V_a > V$). The energy stored in the inductor is released into the load.

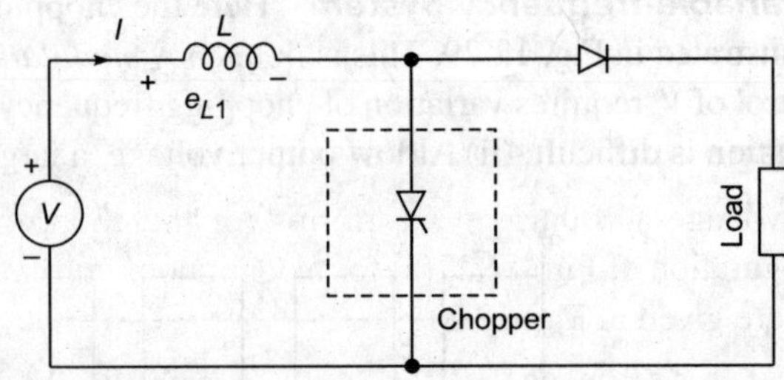

Fig. 12.30 *Step-up chopper configuration*

Analysis On the basis of average values, during the time the chopper is on, the energy stored in the inductor from the source is

$$W_{L,in} = VIt_{ON} \tag{12.15}$$

During the time the chopper is off, the energy fed from the inductor to the load is

$$W_{L,out} = (V_a - V)It_{OFF} \tag{12.16}$$

Assuming a lossless system, under steady-state the two energies will be equal, i.e

$$VIt_{ON} = (V_a - V)It_{OFF} \tag{12.17}$$

from which
$$V_a = V\frac{t_{ON} + t_{OFF}}{t_{OFF}}$$
$$= V\frac{T}{T - t_{ON}} = \frac{V}{1-\alpha} \quad (12.18)$$

For α varying between 0 to 1, the theoretical load voltage varies from V to ∞.

The step-up chopper can be employed for regenerative braking of a dc motor. In Fig. 12.30 if V_a represents the motor armature and V the dc source, power is fed from the decreasing voltage motor to a higher but fixed voltage source.

Example 12.7 *A step-up chopper has input voltage of 220 V and output voltage of 660 V. If the off time of the chopper is 100 μs, compute the pulse width of the output voltage. In case the pulse width is reduced by 50%, find the new output voltage.* LO 3

Solution

Voltage across load $V_L = V_{dc}\,[1/(1 - d)]$

$\Rightarrow$ $660 = 220[1/(1 - d)]$

Therefore, $D = 2/3 = T_{on}/(T_{on} + T_{off})$

$= T_{on}/(T_{on} + 100\ \mu s)$

Thus on solving, $T_{on} = 200\ \mu s$ and $T = 300\ \mu s$

Now, if $T_{on} = (1/2) \times 200 = 100\ \mu s$

$\Rightarrow$ $T_{off} = 200\ \mu s$, since $T = 300\ \mu s$

The new output voltage $V_L = 220[1/(1 - (1/3))]$

$V_L = 330V$

Voltage and Current Waveforms

The voltage and current waveforms for the chopper configuration of Fig. 12.27(a) for a continuous current case are given in Fig. 12.31.

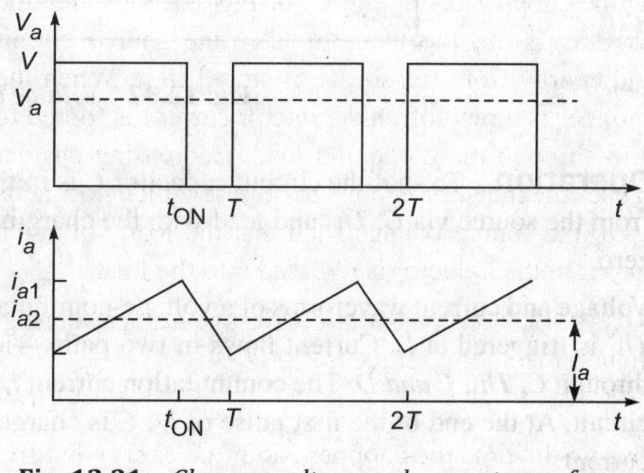

Fig. 12.31 *Chopper voltage and current waveforms*

Chopper Commutation

It has already been mentioned that the thyristor in a chopper has to be turned off by auxiliary commutation circuitry. The commutation can be broadly classified as:

1. Forced Commutation In this type of commutation, current through the thyristor is forced to become zero to turn it off. This can be accomplished in two ways.

!Attention

(a) Voltage Commutation—A charged capacitor momentarily reverse-biases the conducting thyristor to turn it off.

(b) Current Commutation—A current pulse is forced in the reverse direction through the conducting thyristor. As the net current becomes zero, the thyristor is turned off.

2. Load Commutation The load current flowing through the thyristor either becomes zero (as in natural or line commutation employed in converters) or is transferred to another device from the conducting thyristor.

Attention will be restricted to voltage commutation and load commutation of a chopper.

Voltage Commutation

This commutation circuit comprises an auxiliary thyristor Th_2, a diode *D*, inductor *L* and capacitor *C* as shown in Fig. 12.32 wherein the total chopper circuitry has been outlined with a dotted box.

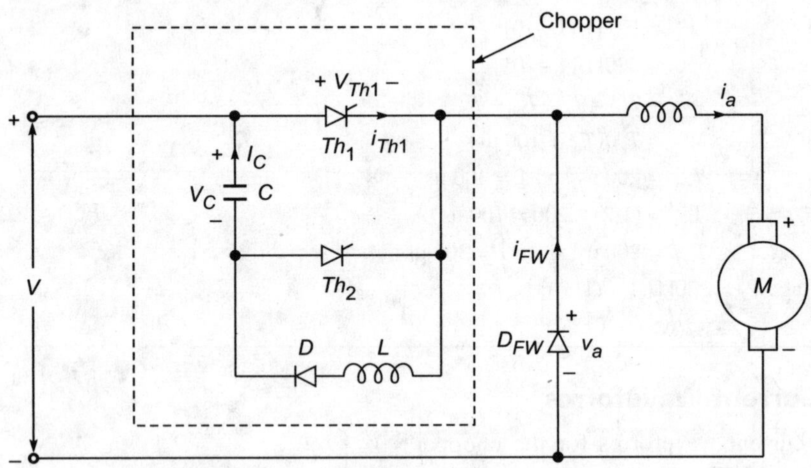

Fig. 12.32 *Voltage commutation of chopper*

Operation To start the circuit, capacitor *C* is initially charged by firing the thyristor Th_2; charging is now from the source via *C*, Th_2 and load with the charging thyristor turning off as the charging current decays to zero.

Voltage and current waveforms of a voltage-commutated chopper are shown in Fig. 12.33. The main thyristor Th_1 is triggered at t_0. Current flows in two paths—load current i_a through Th_1 and commutation current i_C through *C*, Th_1, *L and D*. The commutation current i_C is oscillatory of frequency determined by the *LC* series circuit. At the end of the first pulse of i_C, *C* is charged negatively and the diode *D* blocks-off further flow of current.

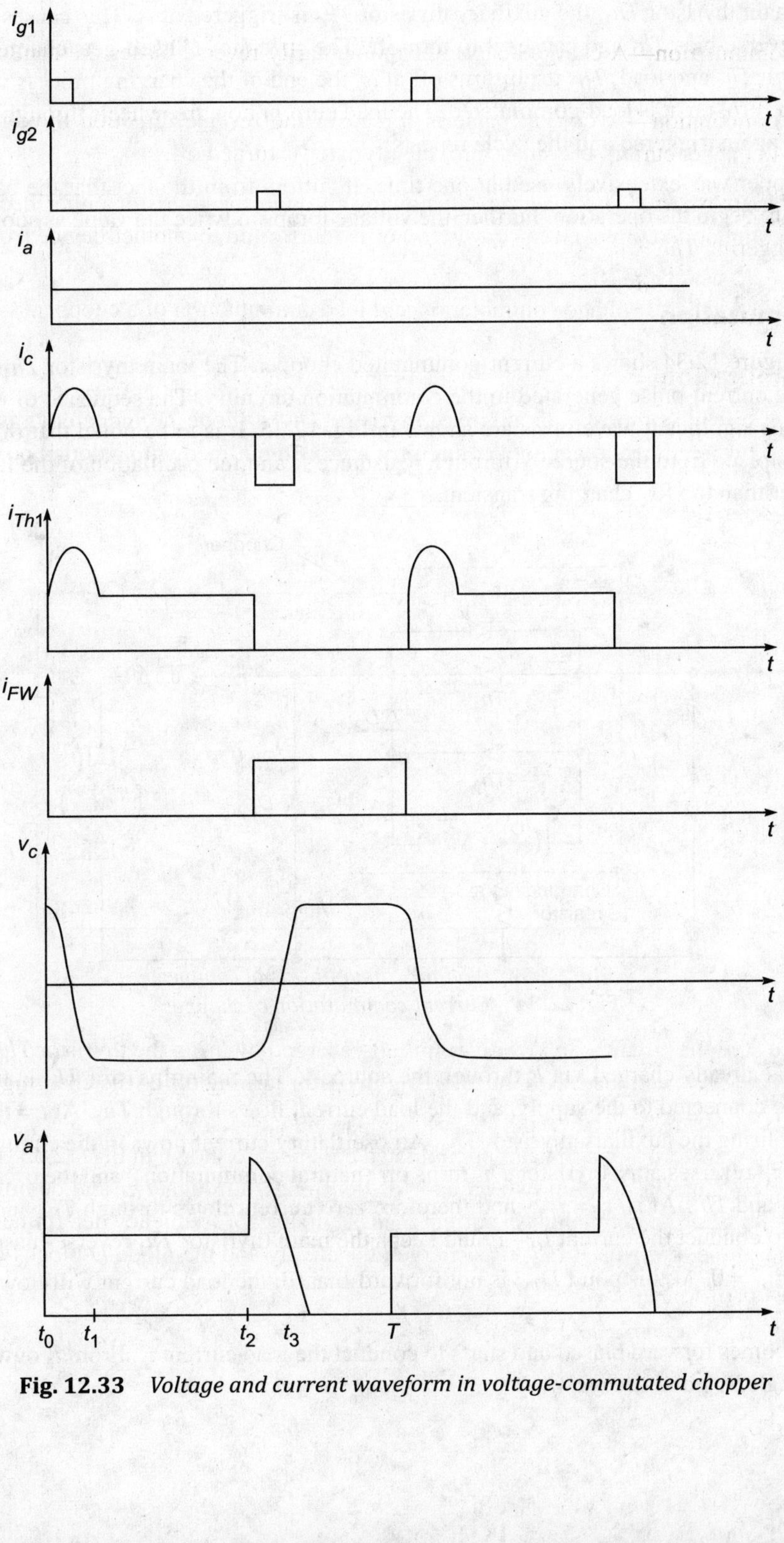

Fig. 12.33 *Voltage and current waveform in voltage-commutated chopper*

To turn-off the main thyristor Th_1, the auxiliary thyristor Th_2 is triggered at t_2. This causes the fully-charged capacitor C to reverse-bias Th_1 and causes it to turn off. The capacitor C itself gets charged positively from the source through Th_2 and load; Th_2 itself turns off at t_3, the end of the charging process when the charging current decays to zero, i.e. it is load-commutated. The load current now freewheels through D_{FW}. At $t = T$, the main thyristor is again triggered and the cycle repeats.

This simple chopper was extensively used at one time. It suffers from the fact that the capacitor has to be initially charged to begin the operation. Further, the voltage jumps to twice the value as soon as commutation is initiated by triggering Th_2.

Current Commutation

Operation Figure 12.34 shows a current-commutated chopper. The main thyristor Th_1 of the chopper is commutated by a current pulse generated in the commutation circuitry. The sequence of operation is given below, and various significant waveforms are shown in Fig. 12.35. It is to be noted that this charging of the capacitor C takes place from the source V through resistance R and the oscillation of the LC combination is set to be far faster than the RC charging transient.

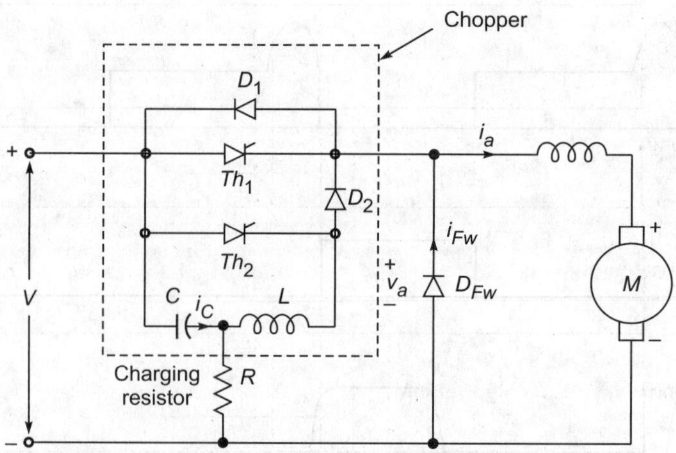

Fig. 12.34 *Current commutation of chopper*

The capacitor C is already charged via R through the source V. The main thyristor Th_1 is fired at $t = t_0$. The load terminals are connected to the supply, and the load current flows through Th_1. At $t = t_1$ commutation of Th_1 is initiated by firing the auxiliary thyristor Th_2. An oscillatory current flows in the circuit consisting of C, Th_2 and L. When i_C reverses at t_2, thyristor Th_2 turns off (natural commutation), and the oscillatory current i_C flows through D_2 and Th_1. At t_3, $i_C = i_{Th1}$, and therefore zero current flows through Th_1 and it turns off, and diode D_1 begins to conduct the current $i_C - i_a$ and keeps the main thyristor Th_1 reverse-biased.

At t_4, $i_C = i_a$ and $i_{D1} = 0$. At this point D_{FW} is not forward-biased, the load current will flow through V, C, L, D_2 and load.

At t_5, the D_{FW} becomes forward biased and starts to conduct the load current i_a. From t_6 onward, load current free-wheels through D_{FW} and decays.

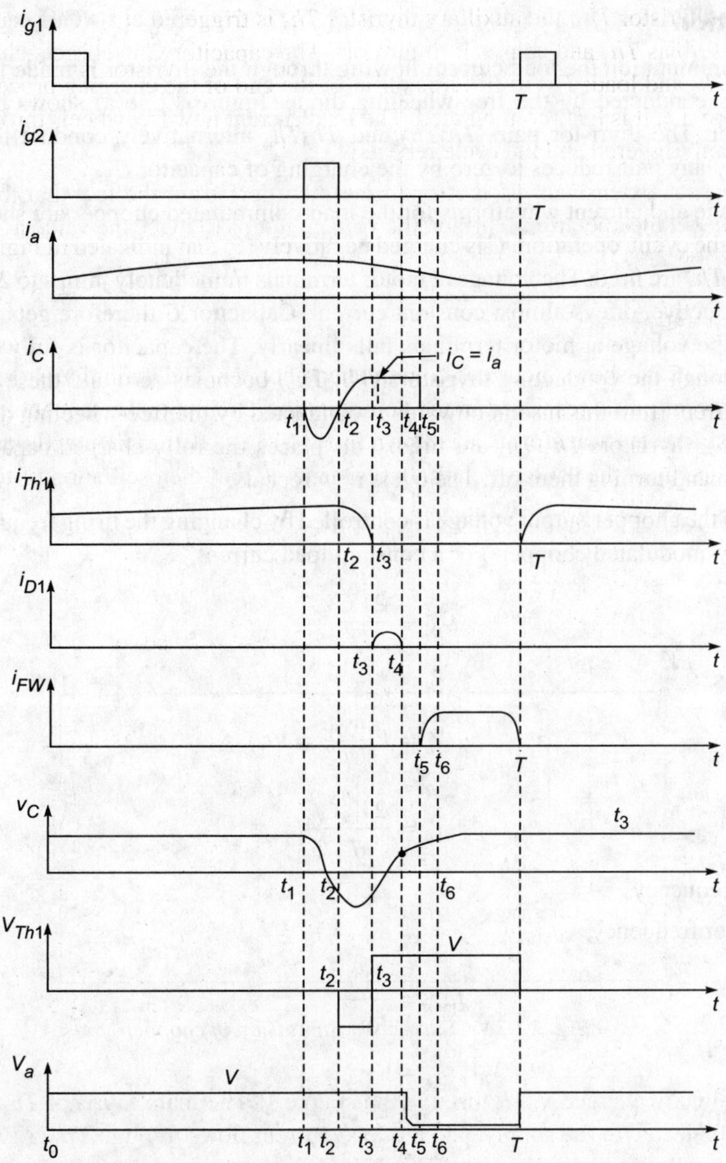

Fig. 12.35 *Voltage and current waveforms in current-commutated chopper*

At $t = T$, the main thyristor Th_1 is fired again, and the cycle repeats.

This chopper has several advantages. Commutation is reliable as long as load current is less than the peak commutation current i_C. The capacitor always remains charged with right polarity. The auxiliary thyristor Th_2 is naturally commutated as its current passes through zero value.

Load Commutation

In this method of commutation the load current flowing through the thyristor is made to become zero while the motor current is conducted by the free-wheeling diode. Figure 12.36(a) shows the circuit of a load-commutated chopper. The thyristor pairs Th_1Th_2 and Th_3Th_4 alternatively conduct the load current. The current conducted by any pair reduces to zero by the charging of capacitor C.

Operation Voltage and current waveforms for the load-commutated chopper are shown in Fig. 12.36(b). It is assumed that in the cyclic operation, C is charged negatively (to that indicated in Fig. 12.32(a)) when at $t = t_0$, the thyristors Th_1 Th_2 are fired. The voltage at motor terminals immediately jumps to 2 V. The motor circuit, because of being inductive, draws almost constant current. Capacitor C therefore gets charged positively at a uniform rate and the voltage at motor terminals falls linearly. The capacitor is fully charged positively at $t = t_1$ the current through the conducting thyristors (Th_1 Th_2) becomes zero and these go into the blocking mode. The motor current from this instant onwards is conducted by the free-wheeling diode D_{FW}. At the end of one period ($t = T$), thyristors Th_3 Th_4 are fired. This places the fully charged capacitor across Th_1 Th_2, reverse-biasing them and turning them off. The cycle now repeats.

The average value of the chopper output voltage is controlled by changing the firing frequency of the choppers. It is thus a frequency-modulated chopper. For a constant load current,

$$t_{ON} = \frac{2VC}{I_a} \tag{12.19}$$

The output voltage is

$$V_a = V\frac{t_{ON}}{T} = Vt_{ON}f$$

$$\therefore \quad V_a = \frac{2V^2Cf}{I_a} \tag{12.20}$$

where f = chopper frequency.

At maximum chopper frequency,

$$f_{max} = \frac{1}{t_{ON}}$$

Hence

$$V_a|_{f_{max}} = V \tag{12.21}$$

Then from Eq. (12.20)

$$V_a = \frac{2V^2C}{I_a}f_{max}$$

or

$$f_{max} = \frac{I_a}{2VC} \tag{12.22a}$$

The value of the capacitor is chosen for maximum load current $I_{a\,max}$. From Eq. (12.22a) this value is

$$C = \frac{I_{a\,max}}{2V f_{max}} \tag{12.22b}$$

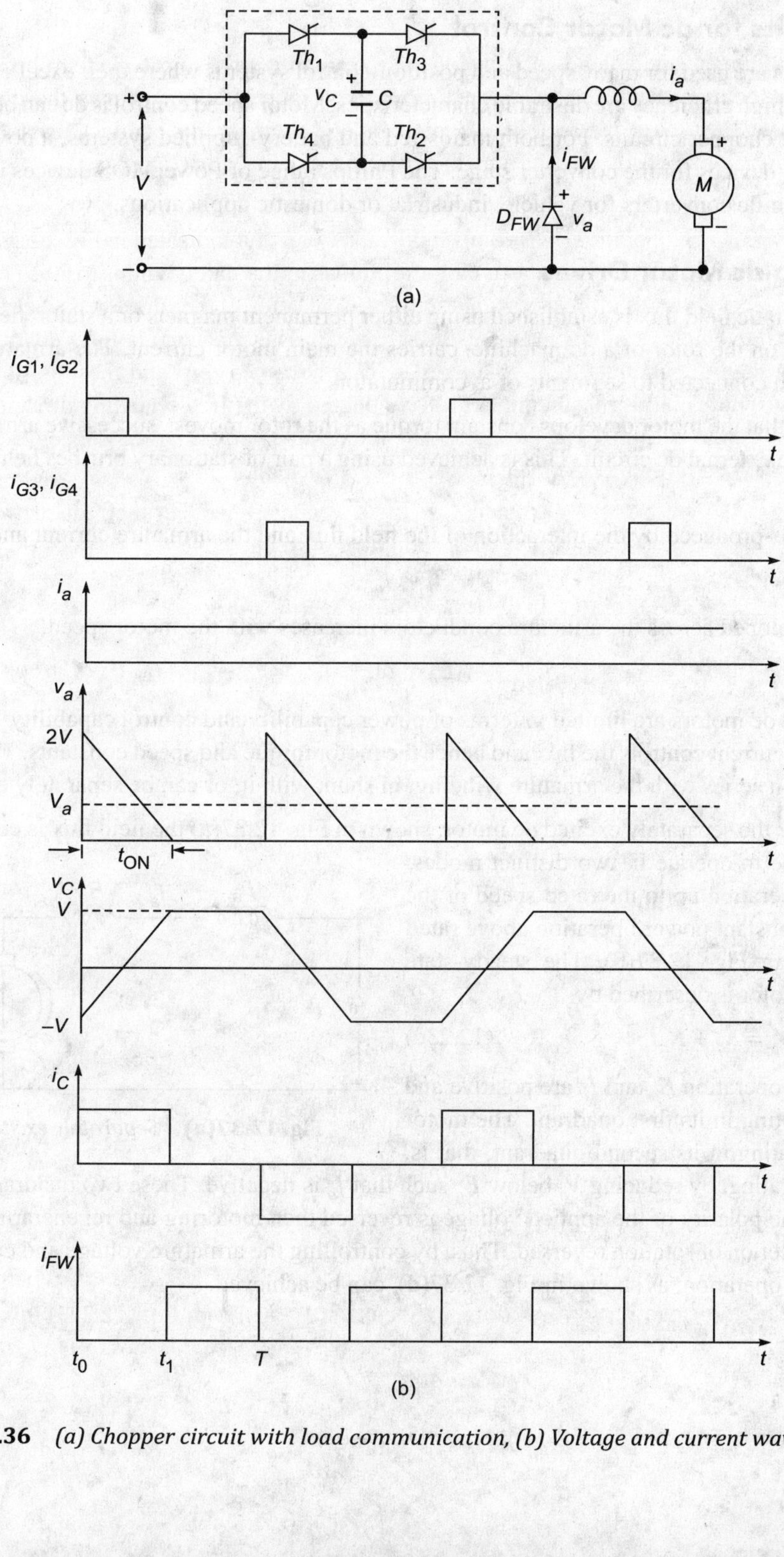

Fig. 12.36 *(a) Chopper circuit with load communication, (b) Voltage and current waveforms*

Chopper Circuits for dc Motor Control

The dc motor drives are used for many speed and position-control systems where their excellent performance, ease of control and high efficiency are desirable characteristics. Motor speed control is dc can be achieved using switch mode dc-dc chopper circuits. For both mains-fed and battery-supplied systems, a power MOSFET is the ideal switching devices for the converter stage. The Philips range of PowerMOS devices includes devices suitable for most dc-dc converters for vehicle, industrial or domestic applications.

Introduction to dc Motor Drives

In a dc motor, the static field flux is established using either permanent magnets or a stator field winding. The armature winding, on the rotor of a dc machine, carries the main motor current. The armature winding is a series of coils, each connected to segments of a commutator.

Recall In order that the motor develops constant torque as the rotor moves, successive armature coils must be connected to the external dc circuit. This is achieved using a pair of stationary brushes held in contact with the commutator.

The motor torque is produced by the interaction of the field flux and the armature current and is given by

$$T_e \propto I_a \tag{12.23a}$$

The back emf developed across the armature conductors increases with the motor speed:

$$E_a \propto \omega_m \tag{12.23b}$$

Permanent magnet dc motors are limited in terms of power capability and control capability. For field wound dc motor, the field current controls the flux and hence the motor torque and speed constants. The field winding can be connected in series with the armature winding, in shunt with it, or can be separately excited.

!Attention For the separately excited dc motor, shown in Fig. 12.37(a) the field flux is controlled and the motor can be made to operate in two distinct modes: constant torque operation up to the rated speed of the motor, and then constant power operation above rated speed, as shown in Fig. 12.37(b). The steady-state operation of the motor is described by

$$V_a = E_a + R_a I_a \tag{12.23c}$$

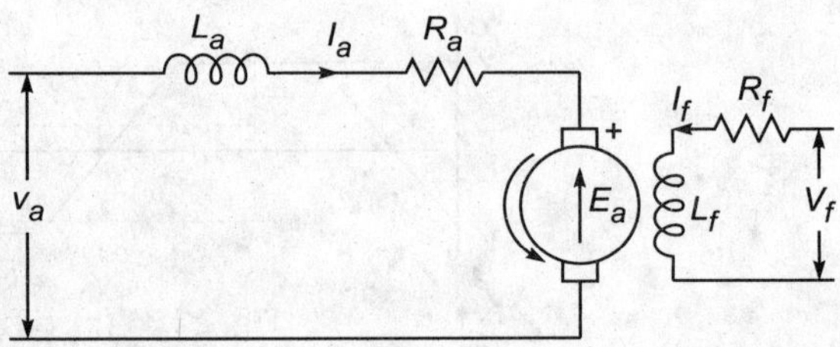

Fig. 12.37(a) *Separately excited dc motor*

For normal motor operation E_a and I_a are positive and the motor is operating in its first quadrant. The motor is said to be operating in its second quadrant, that is, braking or regenerating, by reducing V_a below E_a such that I_a is negative. These two quadrants are shown in Fig. 12.37(c). If the polarity of the applied voltage is reversed then motoring and regenerating operation can occur with the direction of rotation reversed. Thus, by controlling the armature voltage and current polarities, full four-quadrant operation, as shown in Fig. 12.37(d), can be achieved.

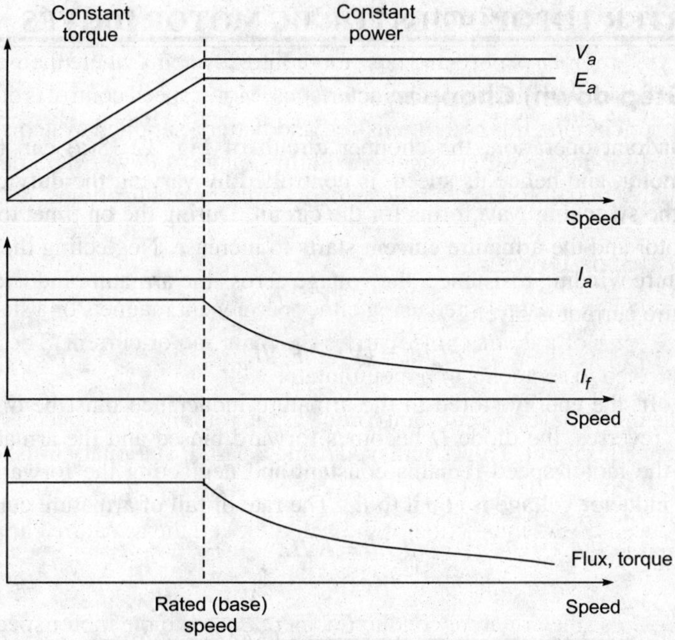

Fig. 12.37(b) *Motor operating characteristics*

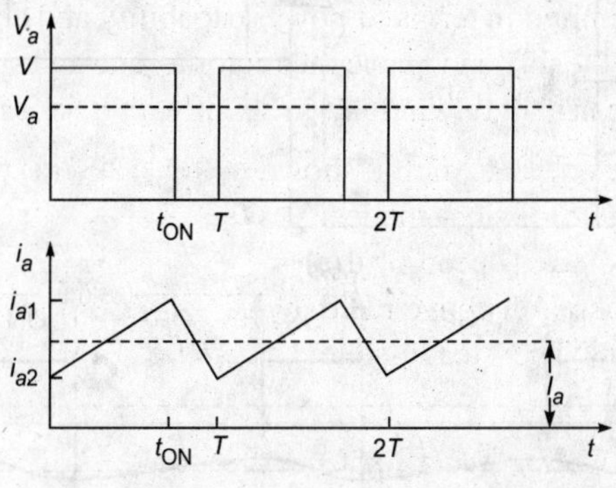

Fig. 12.37(c) *Two quadrant operation*

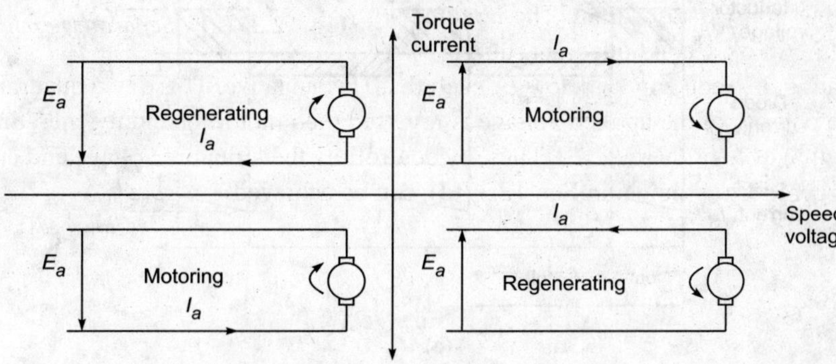

Fig. 12.37(d) *Torque–Speed Characteristics of dc Motor*

12.8 | CONVERTER TOPOLOGIES FOR DC MOTOR DRIVES

LO 3

Single Quadrant (Step-down) Chopper

Recall For single quadrant operation, the chopper circuit of Fig. 12.38(a) can be used. The average voltage applied to the motor, and hence its speed, is controlled by varying the duty cycle of the switch, *S*. Figure 12.38(b) shows the switching waveforms for the circuit. During the on time, ton, the supply voltage, V_{dc}, is applied to the motor and the armature current starts to increase. Neglecting the on-state resistance of the switch and the armature winding resistance, the voltage across the armature inductance is $V_{dc} - E_a$ and so the rate of rise of armature current is given by

$$dI_a/dt = (V_{dc} - E_a)/L_a \tag{12.24a}$$

When the switch turns off, the energy stored in the armature inductance must be dissipated. The polarity of the voltage across L_a reverses, the diode *D* becomes forward biased and the armature current continues to flow. Assuming that the motor speed remains constant and neglecting the forward voltage drop of the freewheeling diode, the inductor voltage is equal to E_a. The rate of fall of armature current is given by

$$dI_a/dt = E_a/L_a \tag{12.24b}$$

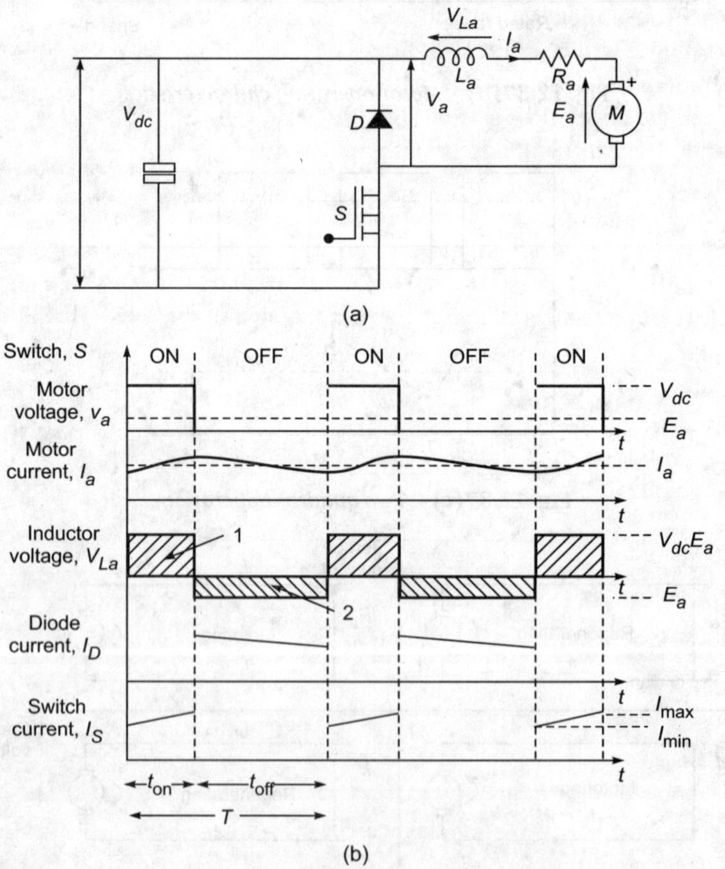

Fig. 12.38 *(a) Single quadrant chopper circuit, (b) Single quadrant chopper circuit–Switching waveforms*

If this switching sequence is repeated at same frequency then the motor voltage can be controlled by altering the relative duration of the on period and off period. Variation of the duty cycle of the switch (t_{on}/T) to control the motor voltage is referred to as Pulse Width Modulation (PWM) control. As the average voltage across the inductor over a period must be zero then

$$\int_0^T v_L\, dt = \int_0^{t_{on}} v_L\, dt + \int_{t_{on}}^T v_L\, dt = 0 \qquad (12.25a)$$

The integral of inductor voltage for the interval t_{on} corresponds to the shaded areal in Fig. 12.38(b) the integral of inductor voltage for the t_{off} interval corresponds to the shaded area in the Fig. 12.38(b). These two areas must be equal and so from Eqs. 12.24 to 12.25 or Fig. 12.38(b) the transfer function of the controller is given by

$$V_a = (t_{on}/T).\ V_{dc} \qquad (12.25b)$$

Two Quadrant, Chopper

Figure 12.39(a) shows a half-bridge circuit for two quadrant dc drive. For motoring operation S_1 and D_2 operate as described above for the single-quadrant controller.

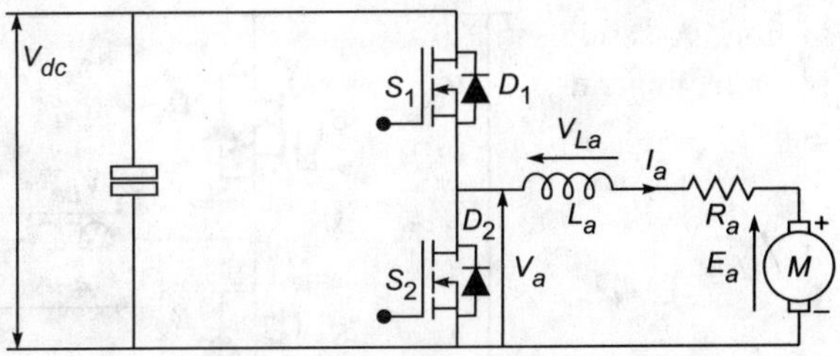

Fig. 12.39(a) *Two-quadrant half-bridge chopper circuit*

!Attention The freewheel diode D_2 may be the internal diode of a MOSFET or a discrete device.

For regenerative operation, the dc motor acts as the active power source and the power flow is from right to left in Fig. 12.39(a). The regenerating current is controlled by varying the duty cycle of S_2. When S_2 is on, the negative armature current increases through the switch and the armature inductance. When S_2 is turned off, D_1 becomes forward biased and the current regenerates into the supply. The relevant circuit waveforms are shown in Fig. 12.39(b) showing the equal areas of the inductor volt-seconds over each period of the switching cycle. During regeneration, the transfer function of the converter is given by

$$V_a = [1 - (t_{on}/T)]V_{dc} \qquad (12.25c)$$

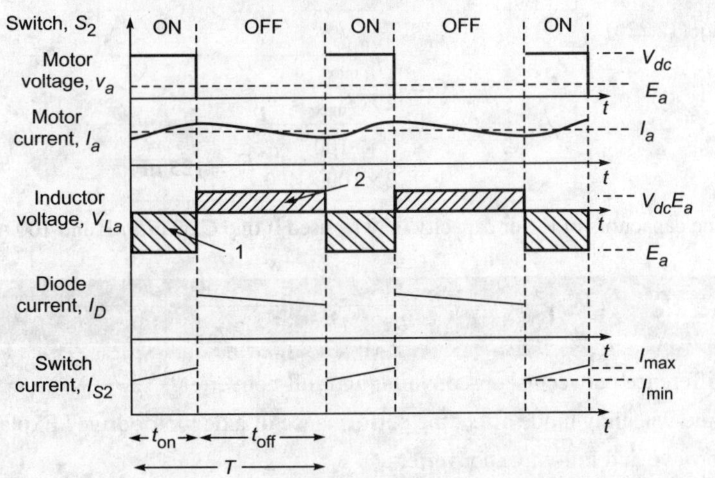

Fig. 12.39(b) *Two-quadrant half-bridge chopper circuit–switching waveforms*

Four Quadrant, Chopper

If motoring and regenerating operation are required with both directions of rotation then the full bridge converter of Fig. 12.40 is required.

Important! Using this configuration allows the polarity of the applied voltage to be reversed, thus reversing the direction of rotation of the motor.

Thus in a full-bridge converter, the motor current and voltage can be controlled independently. The motor voltage V_a is given by

$$V_a = V_{12} - V_{34} \tag{12.25d}$$

Operation where V_{12} is controlled by switching S_1 and S_2 as described above, and V_{34} by switching S_3 and S_4. The usual operating mode for a full-bridge converter is to group the switching devices so that S_1 and S_3 are always on simultaneously and that S_2 and S_4 are on simultaneously. This type of control is then referred to as bipolar control.

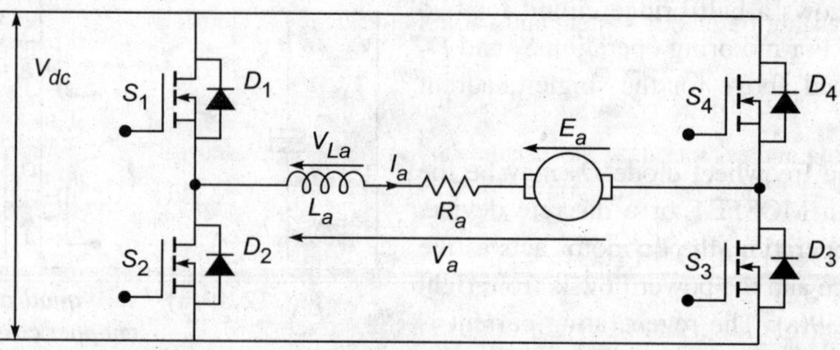

Fig. 12.40 *Four-quadrant full-bridge chopper circuit*

Example 12.8 *A battery-powered electric car is controlled by a load-commutated chopper as shown in Fig. 12.36(a). The battery voltage is 100 V; starting current is 100 A and maximum chopper frequency is 4 kHz. Calculate the value of the commutating capacitor C.*

LO 3 ○○●

Solution From Eq. (12.22b)

$$C = \frac{I_{a\,\max}}{2V f_{\max}}$$

$$= \frac{100}{2 \times 100 \times 400} F = 125\ \mu\text{F}$$

The low-loss paper type capacitors of lower capacity may be used if one C of 125 μF and 100 A rating is not available.

Concept Check

LO 3

1. What is the difference between semi-converter and full-converter? ○○●
2. How does a free-wheeling diode affect the performance of a dc motor drive? Explain. ●●●
3. Compare half-wave and full-wave converters. ○○●

4. Compare semi-converter and full-converter. ○●●
5. Is discontinuous mode of operation advantageous? ●●●
6. What is inversion mode of operation in a full-converter? ○○●
7. Why two-quadrant operation is not possible in semi-converter fed drive? ●●●
8. What do you understand by firing angle and extinction angle? ○●●
9. What is the purpose of freewheeling diode in a semi-converter? ●●●
10. Comment on the ratings of thyristor used for full-converter and semi-converter. ●●●
11. How will you assess the power loss of a converter? ●●●
12. Does the type of load (R, RL) influence the operation of converter? ●●●
13. Why a three-phase converter is required? ○○●
14. What is the role of chopper in dc motor drive? ○●●
15. How will you control the output voltage of chopper? ○○●
16. Compare constant frequency system and variable frequency system. ○●●
17. What is duty cycle? ○○●
18. What are the various forced commutation techniques used in choppers? ○●●
19. How motoring and regenerating mode is possible with chopper control? ●●●

12.9 | ac MOTOR CONTROL

LO 4 Outline the control techniques for induction motor with slip power recovery schemes and explain the principle of operation of various single-phase, three-phase inverter configurations and forced commutation methods

The only way to control the speed of a synchronous motor is to control the input frequency and voltage such as to keep V/f constant. The induction motor, on the other hand, can be controlled by the following means:

Stator Voltage Control

This is applicable for small motors and for fan-type load, where the load torque increases with speed. As already mentioned in Sec. 12.5, there are two ways of controlling the rms value of the stator voltage—phase-control and integral cycle control.

The torque-speed characteristics of an induction motor with stator voltage control are drawn in Fig. 12.41(a). The maximum torque reduces as the square of the voltage.

Figure 12.41(a) also illustrates that speed regulation is only possible for a load whose torque decreases with speed. Even then speed regulation is not achieved for low speeds—here the speed drops off sharply.

Frequency Control

Here the input frequency is varied by means of an inverter; for subnormal frequency the stator voltage is also varied to keep V/f constant which maintains the resultant flux/pole in the air-gap. A typical control system utilizing frequency control is schematically illustrated in Fig. 12.41(b). Because of large capital outlay, this system is justified only for drives wherein the rugged, maintenance-free characteristics of the induction motor are essential. Otherwise, a dc motor with converter control is the logical and cheap alternative.

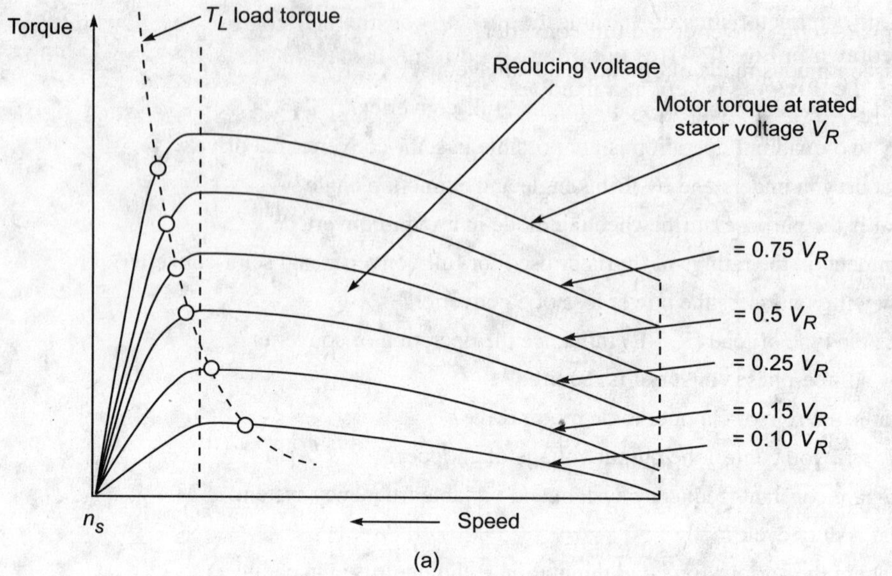

Fig. 12.41(a) *Torque-speed characteristics of induction motor with variable stator voltage*

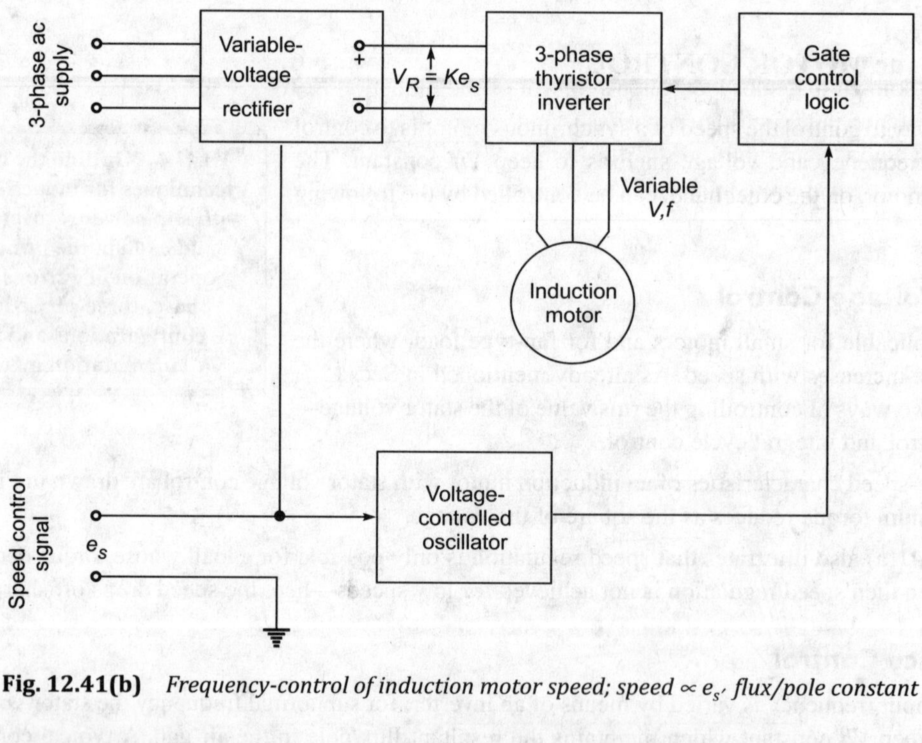

Fig. 12.41(b) *Frequency-control of induction motor speed; speed ∝ e_s, flux/pole constant*

The heart of the frequency control scheme is the 3-phase thyristor inverter. The inverter circuitry and operation will be discussed later in this section.

The torque-slip characteristics of variable-frequency, constant-*V*/*f* control were illustrated in Fig. 12.41(c). These are redrawn in Fig. 12.41(c) in torque-speed form. It may be seen that the maximum torque does not alter because the flux/pole is kept at a fixed value.

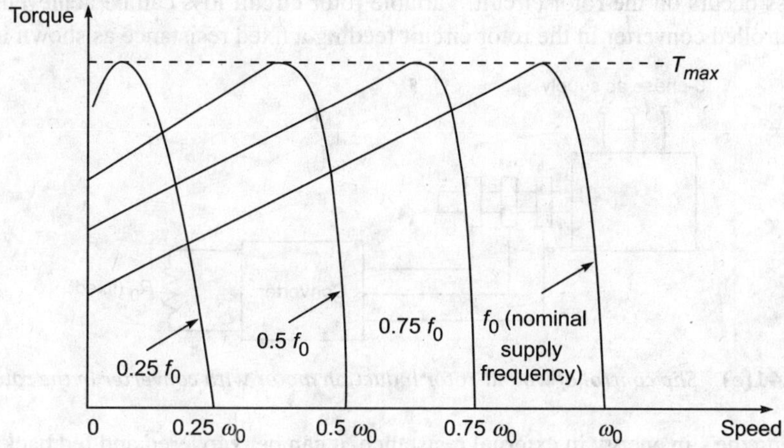

Fig. 12.41(c) *Induction motor—variable-frequency, constant-V/f operation*

Slip Control

It is already known that subsynchronous control of the induction motor speed is possible by introducing a variable resistance in the rotor circuit of a wound-rotor induction motor. This method provides a wide range of speed and a good starting torque as illustrated in Fig. 12.41(d). The maximum torque in this case remains constant. Good regulation of speed is achievable for both constant-torque and fan-type loads.

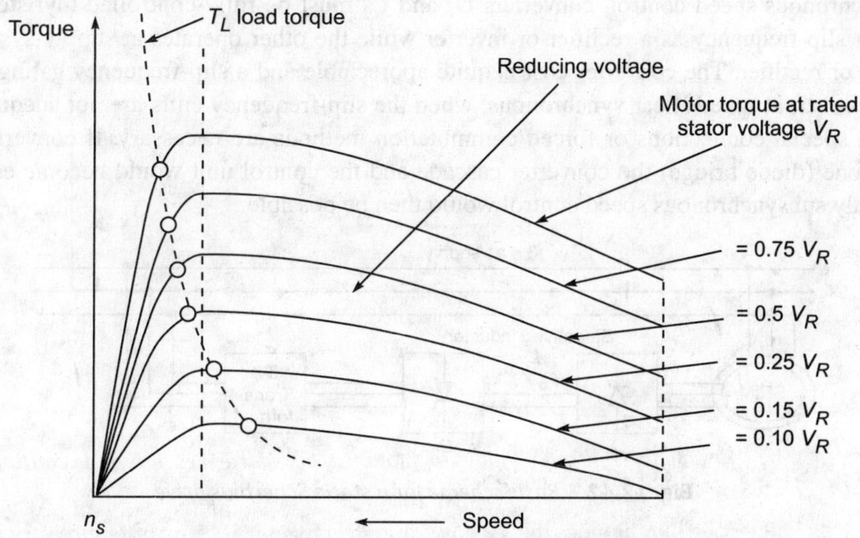

Fig. 12.41(d) *Slip control of induction motor*

The greatest draw-back of this method of speed control is its poor efficiency because of power wasted in the external rotor resistance.

The slip control method is based upon the general principle that the induction motor slip increases as more electric power loss occurs on the rotor circuit. Variable rotor circuit loss can be achieved by the scheme of introducing a controlled converter in the rotor circuit feeding a fixed resistance as shown in Fig. 12.41(e).

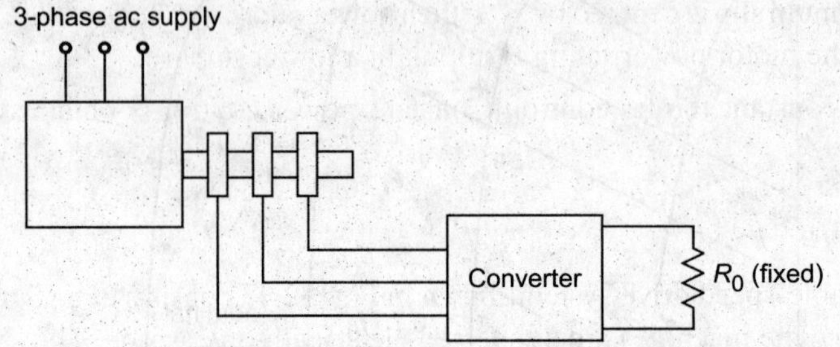

Fig. 12.41(e) *Slip control of wound-rotor induction motor with converter in the rotor circuit*

Rather than wasting the slip energy in external resistance, it can be recovered and fed back to supply, thereby removing the disadvantage of low efficiency of the slip-control method.

Slip-power Recovery Schemes

In fact, the slip power is either returned to the supply network as in Scherbius Scheme or used to drive an auxiliary motor which is mechanically coupled to the induction motor shaft, as in Kramer scheme.

Static Scherbius Scheme A basic scheme of a Scherbius static-converter cascade, that is employed for recovery of slip power in electrical form itself, is given in Fig. 12.42. For achieving both subsynchronous and supersynchronous speed control, converters C_1 and C_2 must be fully-controlled thyristor bridges, one functioning at slip frequency as a rectifier or inverter while the other operates at supply-system frequency as an inverter or rectifier. The converter cost is quite appreciable and a slip-frequency gating circuit is also required. Moreover, at speeds near synchronous, when the slip-frequency emfs are not adequate for natural commutation, special connections or forced commutation methods are necessary. If converter C_1 were an uncontrolled one (diode bridge) the converter cascade and the control unit would become economical and simple, but only subsynchronous speed control would then be possible.

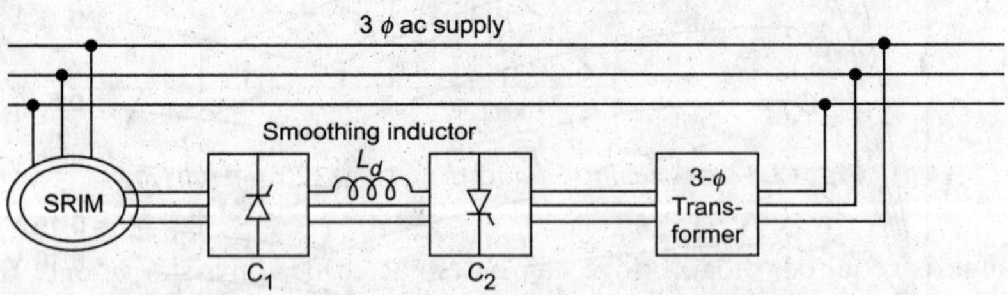

Fig. 12.42 *Basic scheme for a static Scherbius drive*

The three-phase transformer between the inverter C_2 and the supply network is required with large supply voltages, since the voltage across the sliprings is limited and may be extremely low for direct inversion into the ac supply.

The principal disadvantage of the subsynchronous cascade drive is its low power factor particularly at reduced speeds. The reactive power consumption of the inverter is largely responsible for the low power factor of the cascade drive.

The drive has applications in large power fan and pump drives which require speed control in a narrow range only.

!Attention If maximum slip is denoted by s_{max}, then power ratings of diode bridge, inverter and transformer can be just s_{max} times the motor power rating resulting in a low cost drive.

This drive provides a constant torque control. Constant power control is obtained by static Kramer drive discussed below.

Static Kramer Drive

The static Kramer variable-speed drive system shown in Fig. 12.43 consists of a slipring induction motor and a rectifier-fed dc motor. The machines are connected electrically in cascade. The slip power recovery takes place through the rectifier and the auxiliary dc motor. Torque supplied to the load is sum of torque produced by induction and dc motor. Speed control is obtained by controlling field current of dc motor.

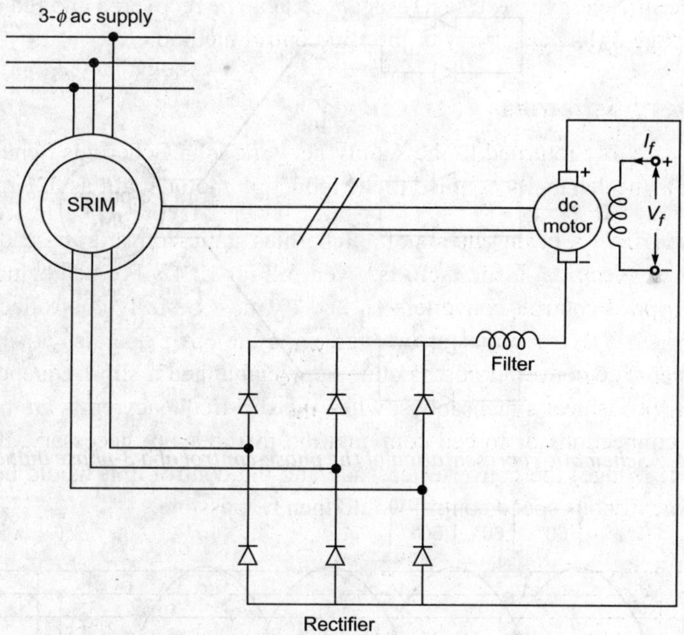

Fig. 12.43 *Schematic diagram of a static Kramer drive*

If larger speed range is required, diode bridge can be replaced by a thyristor bridge. Speed can now be controlled up to standstill.

As a static Kramer system possesses no line-commutated inverter, it causes less reactive power and smaller harmonic contents of currents than a static Scherbius system. But it has maintenance problems caused by the brushes and commutator of an auxiliary dc motor and the disadvantage of a large moment of inertia.

It is observed from the above discussion that for implementation of speed-control of the induction motor, one needs to understand the operation of the inverter which transforms the dc power into variable-frequency ac power. Before taking this up, the phase-control scheme for controlling the stator voltage will first be considered.

Phase-Control of Induction Motor

Operation Figure 12.44 shows the schematic representation of phase-control of a 3-phase induction motor. Two thyristors are connected in antiparallel in each line. The firing sequence of the thyristors for all modes of operation is 1, 3′, 2, 1′, 3 and 2′ as shown in Fig. 12.45. The interval between successive firings is 60° and the firing frequency is six times the input frequency. It is seen from Fig. 12.45 that as 3′ is fired 1 and 3′ conduct for 60°; then 2 is fired and 3′ and 2 conduct for the next 60° and so on. Each thyristor after being fired conducts for 120°. As long as a thyristor conducts the corresponding phase voltage will be known. When a phase gets open circuited, its voltage will be that induced due to currents flowing in the other two windings of the motor. As a result of phase control, the effective ac rms voltage applied to motor terminals decreases. This indeed is a method of voltage control.

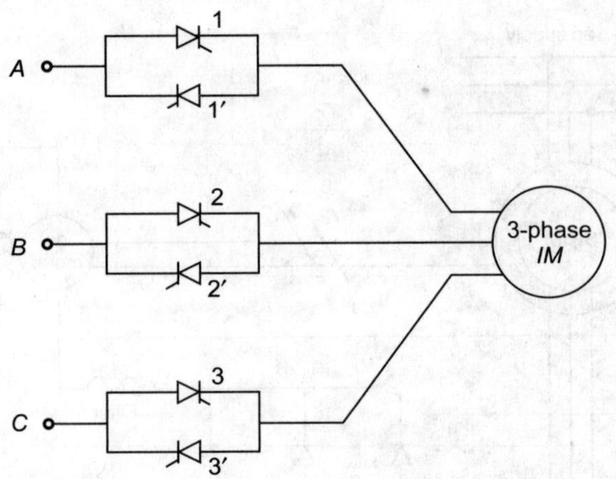

Fig. 12.44 *Schematic representation of the phase-control of a 3-phase induction motor*

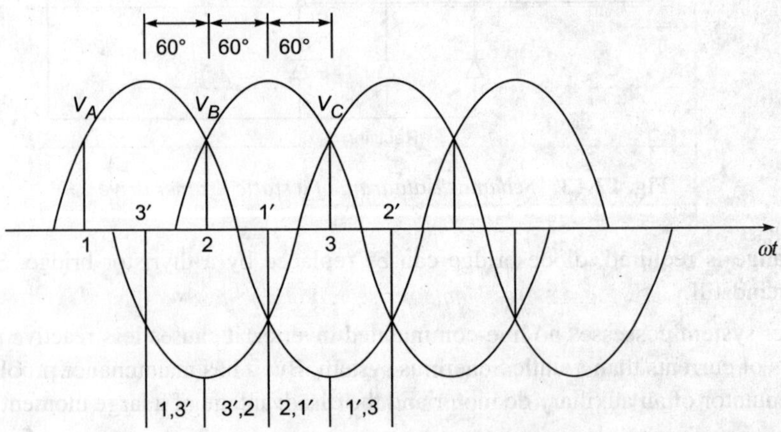

Fig. 12.45 *Firing sequence in Fig. 12.44*

12.10 | INVERTERS

LO 4

It was seen above that an inverter, which converts a dc supply into a variable frequency ac, is the basic power modulator of a variable-speed ac motor drive. Certain simple inverter configurations and their principle of operation will now be discussed.

Single-Phase Half-Bridge Inverter

Figure 12.46(a) gives the circuit configuration of a single-phase half-bridge inverter. It has two thyristors and two free-wheeling diodes. Each thyristor is gated at frequency $f = 1/T$ of the ac supply desired. The gating signals of the two thyristors have a phase angle of 180°. From Fig. 12.46(b), the output is easily seen to be rectangular ac waveform of frequency.

$$\omega = \frac{2\pi}{T} \quad \text{rad/s}$$

where T = triggering period of the thyristor.

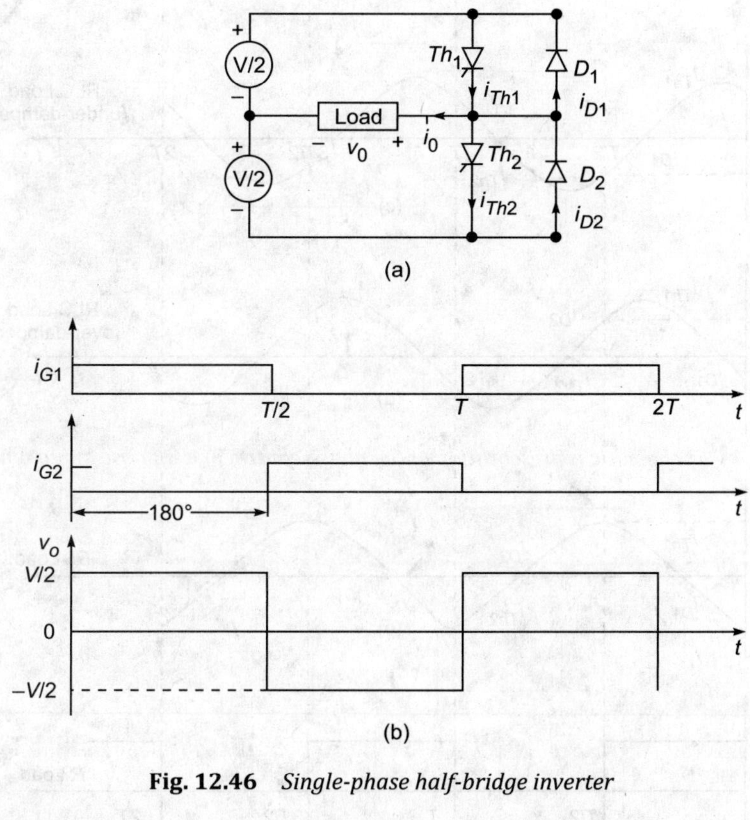

Fig. 12.46 *Single-phase half-bridge inverter*

The output waveform feeds the load which may in general comprise *RLC* components. The circuit model of the inverter is given in Fig. 12.47(a). After several cycles of source voltage v_0 have elapsed, the time variation of current settles down to periodic form such that

$$i_0 = \mp I_{01}; \qquad t = T/2$$

$$i_0 = \pm I_{01}; \qquad t = 0, T \tag{12.26}$$

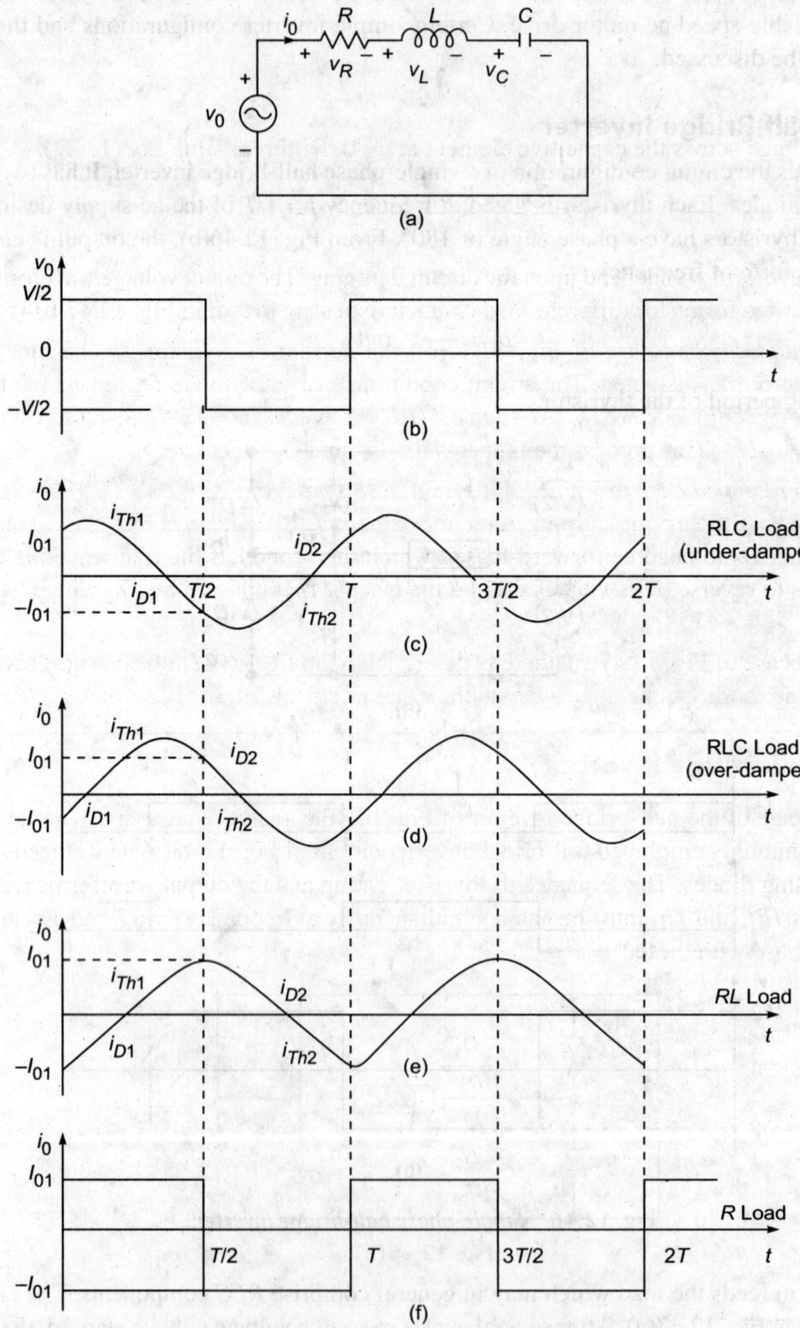

Fig. 12.47 *Circuit model and waveforms for the inverter of Fig. 12.46(a)*

During the interval $0 < t < T/2$,

$$v_0 = \frac{V}{2} = v_R + v_L + v_C$$

or

$$\frac{V}{2} = Ri_0 + L\frac{di_0}{dt} + \frac{1}{C}\int_0^t i_0\,dt + V_{C0} \tag{12.27}$$

where V_{C0} is the voltage across the capacitive element at $t = 0$. Differentiating Eq. (12.27),

$$\frac{d^2 i_0}{dt^2} + \frac{R}{L}\frac{di_0}{dt} + \frac{1}{LC}i_0 = 0 \tag{12.28}$$

The nature of the waveform will depend upon the circuit damping. The output voltage waveform (rectangular) and various current waveforms for different load characteristics are drawn in Fig. 12.47(b)-(f).

In the *RLC load underdamped case* of Fig. 12.47(c), the current of thyristor Th_1 becomes zero and the thyristor turns off before Th_2 is gated. The circuit conditions cause the diode D_1 to become forward-biased conducting the free-wheeling current i_{D1}. As soon as Th_2 is triggered, D_1 is reverse-biased and the current is transferred from D_1 to Th_2. This process then repeats for Th_2 and D_2 and so on.

In the *RLC load overdamped case* of Fig. 12.47(d) and the *RL* load of Fig. 12.47(e), Th_1 is forced to switch off (at $T/2$) while it is still carrying current. As a consequence, the voltage across the *L*-component of the load reverses, causing D_2 to become forward-biased which then conducts the free-wheeling current. As the circuit current tends to reverse, D_2 switches off and the current is conducted by Th_2 which has already been gated* at $t = T/2$.

For the ideal *R* load case of Fig. 12.47(f), there is theoretically no free-wheeling current. This does not mean that the free-wheeling diode can be done away with as any practical circuit always has some inductance.

Single-Phase Full Bridge Inverter

A serious disadvantage of the half-bridge inverter of Fig. 12.46 is that it requires a 3-wire dc supply. This is overcome by the commonly employed full-bridge inverter circuit of Fig. 12.48(a) which needs four thyristors and four free-wheeling diodes. The sequence of thyristor gating and the output waveforms are shown in Fig. 12.48(b). Thyristors Th_1, and Th_3 must be gated simultaneously at frequency $f = 1/T$ and Th_2 and Th_4 must be gated 180° out of phase with these.

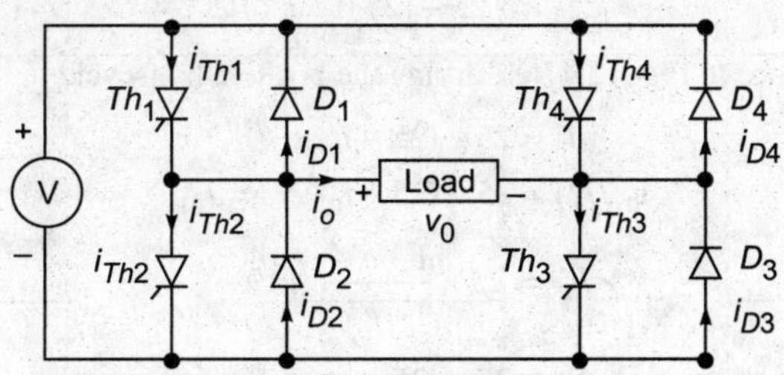

Fig. 12.48(a)

*It may be seen that the unlike converters and choppers where the thyristors are pulse triggered, in case of inverters the gate current is continuously supplied to the thyristor for one-half period after it is triggered.

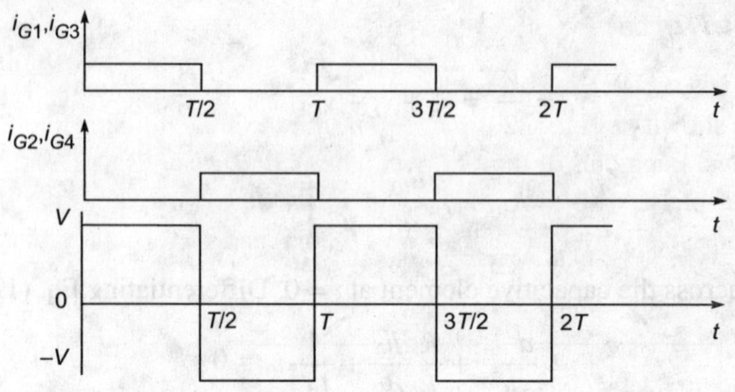

Fig. 12.48(b) *Single-phase bridge inverter*

Example 12.9 *The load connected at the output of a single-phase bridge inverter is a series combination of R = 10 Ω and L = 0.05 H. Determine the load voltage and current waveforms for the first two half-cycles with rectangular wave output at 50 Hz. The input to the inverter is V_{dc} = 250 V. Also find the expression for steady-state current in each half-cycle.* LO 4

Solution

Load time-constant, $\tau = L/R = \dfrac{0.05}{10} = 0.005$ s

$$\text{Half-cycle time} = \frac{1}{100}\text{ s}$$

For the first half-cycle, initial current is zero

$$v_0 = +250\text{ V}$$

$$i_0 = 25 - 25e^{-t/\tau}$$

$$= 25\,(1 - e^{-200t})$$

$$i_0 \text{ (at end of first half-cycle)} = 25\,(1 - e^{-2})$$

For the second half-cycle.

$$v_0 = -250\text{ V}$$

$$i_0 = -25 + [25\,(1 - e^{-2})]e^{-200t}$$

Steady-state current With reference to Fig. 12.47(e), during the positive half-cycle

$$i_0 = 25 - (25 + I_{01})e^{-200t}$$

$$i_0(T/2) = 25 - (25 + I_{01})e^{-2} = I_{01}$$

or

$$I_{01} = 25 \times \frac{1 - e^{-2}}{1 + e^{-2}} = 19$$

Hence

$$i_0 = 25 - (25 + 19)e^{-200t}$$

$$= 25 - 44e^{-200t} \text{ (+ve half-cycle)}$$

$$= -25 + 44e^{-200t} \text{ (–ve half-cycle)}$$

Three-Phase Bridge Inverter

The basic 3-phase inverter is a six-step inverter. A step is defined as a change in the firing sequence. A 3-phase thyristor bridge-inverter is shown in Fig. 12.49. Th_1 to Th_6 are the six load-carrying thyristors while D_1 to D_6 are the free-wheeling diodes. Each pair of thyristors in a branch (Th_1 and Th_4; Th_2 and Th_5; Th_3 and Th_6) are gated for $T/2$ and are out-of-phase with each other, i.e. they are never gated simultaneously. Th_1, Th_2 and Th_3 are fired out-of-phase progressively by 120° and so are Th_4, Th_2 and Th_6. This is a must to obtain three output voltages out-of-phase 120°.

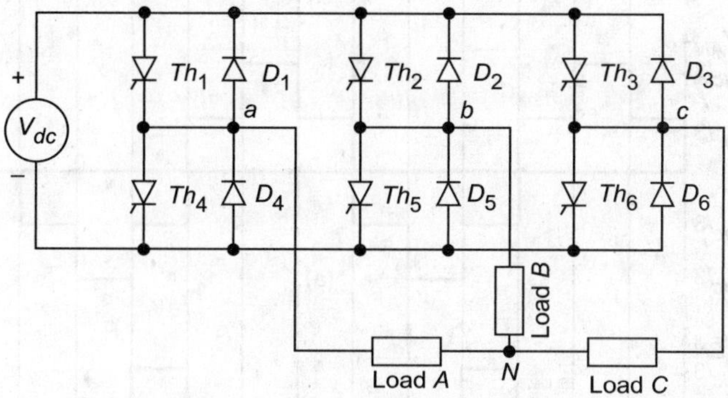

Fig. 12.49 *Three-phase bridge inverter*

Figure 12.50 shows the conducting periods of various thyristors as per the firing sequence indicated above. Over an angle of 2π, six periods (one-sixth of each cycle period) are recognized and the thyristors conducting in these periods are identified. It is noticed that in anyone period only three thyristors are conducting. The frequency of firing is six times the output frequency. The circuit models during three typical consecutive periods corresponding to the positive half-cycle of v_{AN} are drawn in Fig. 12.51 and the output voltages in terms of the input dc voltage (V_{dc}) are determined. The voltage waveforms for three phase-to-neutral voltages of the 3-phase inverter of Fig. 12.49 can be easily drawn by this procedure. It is immediately obvious that these voltages are out-of-phase by 120°. The phase sequence can be reversed by simply reversing the sequence of firing the thyristors. The line-to-line voltages are found by taking the difference of phase voltages. The waveform of $v_{AB} = v_{AN} - v_{BN}$ is illustrated in Fig. 12.50(d). It is also easily seen that the fundamental components of line-to-line (or phase-to-neutral) voltages form a balanced set. The free-wheeling diodes permit currents to flow which are out-of-phase with these voltages.

Probe

In a three phase bridge inverter, what is the conduction period of each thyristor?

Th1			Th4			Th1			Th4		
Th5		Th2			Th5			Th2			Th5
Th3	Th6			Th3			Th6			Th3	
I	II	III	IV	V	VI	I	II	III	IV	V	VI
Th1	Th1	Th1	Th4	Th4	Th4	Th1	Th1	Th1	Th4	Th4	Th4
Th5	Th5	Th2	Th2	Th2	Th5	Th5	Th5	Th2	Th2	Th2	Th5
Th3	Th6	Th6	Th6	Th3	Th3	Th3	Th6	Th6	Th6	Th3	Th3

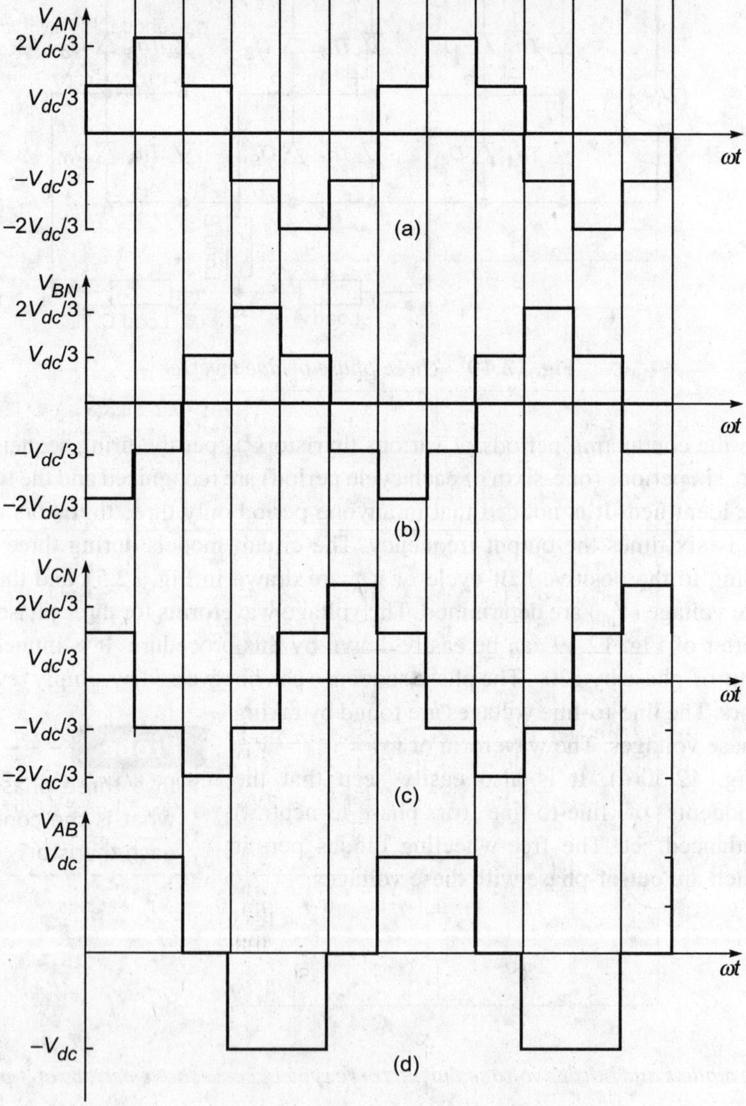

Fig. 12.50 *Sequence of conducting thyristors in 3-phase bridge inverter and output voltage waveform*

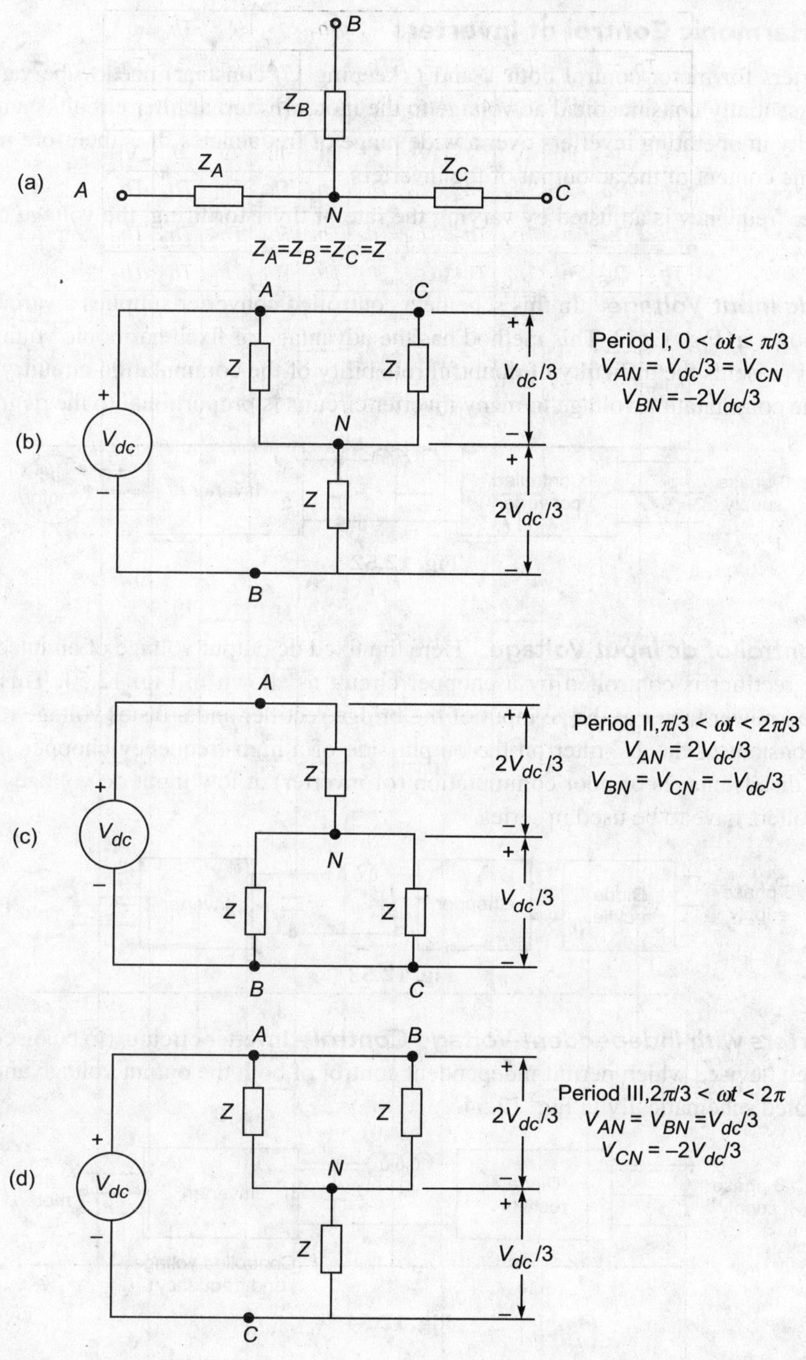

Fig. 12.51 *Circuit models and output voltage during three typical consecutive periods of 3-phase bridge inverter*

Voltage and Harmonic Control of Inverters

In applying inverters for motor control both V and f (keeping V/f constant) need to be varied. Further, the inverters apply essentially nonsinusoidal ac voltage to the motor. External filter circuits cannot be employed due to the difficulty in operating inverters over a wide range of frequencies. It is therefore necessary to keep down the harmonic content of the ac output of the inverters.

While the inverter frequency is adjusted by varying the rate of thyristor firing, the voltage can be controlled in the following ways:

1. Control of dc Input Voltage In this scheme a controlled converter supplies a variable dc voltage to the inverter as shown in Fig. 12.52. This method has the advantage of fixed harmonic voltage content in the inverter output. It presents the difficulty of doubtful reliability of the commutation circuitry at low values of the dc input as the commutation voltage in many inverter circuits is proportional to the dc input voltage.

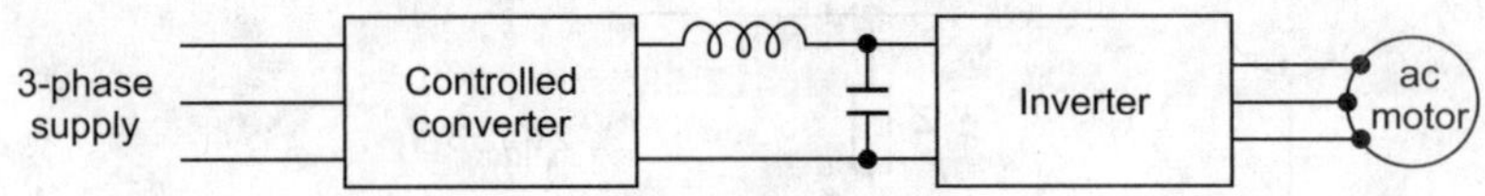

Fig. 12.52

2. Chopper Control of dc Input Voltage Here the fixed dc output voltage of an uncontrolled 3-phase full-wave bridge rectifier is controlled by a chopper circuit as shown in Fig. 12.53. This scheme has the advantage of good power factor at the ac input of the bridge rectifier and a faster voltage response owing to a smaller time-constant of the LC-filter on the output side of a high-frequency chopper. This scheme also suffers from the disadvantages of poor commutation (of inverter) at low input dc voltages and the fact that two power controllers have to be used in series.

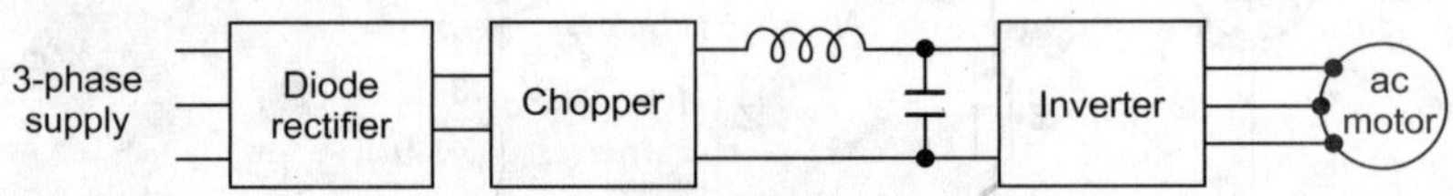

Fig. 12.53

3. Use of Inverters with Independent Voltage Control Inverter circuits (to be discussed later in this section) have been devised which permit independent control of both the output voltage and frequency. This method is illustrated schematically in Fig. 12.54.

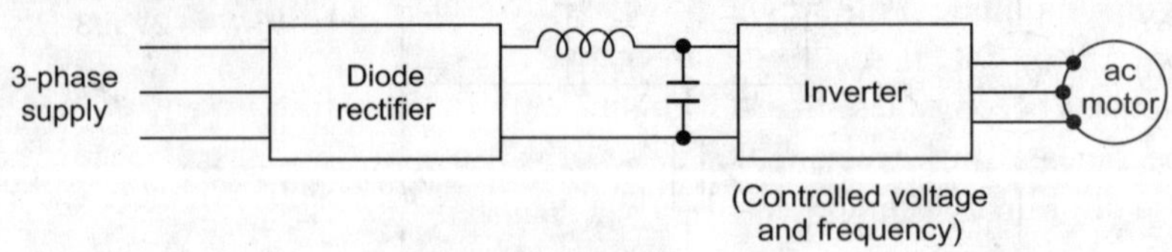

Fig. 12.54

Inverters with Pulse-Width Modulation (PWM)

In presenting the arguments here attention will be focussed on a single-phase inverter. Instead of the single rectangular pulse output of Figs 12.46 and 12.48 during each half-cycle, a commutation circuitry can be devised to produce a series of short duration pulses as output in each half-cycle as shown in Fig. 12.55. The amplitude of the fundamental of this output waveform can now be easily controlled by varying the ON/OFF periods. This method of voltage control of an inverter is known as pulsewidth modulation. It has the added advantage of low harmonic content at low output voltages. Further, the content of difficult-to-filter low harmonics in the output wave is reduced compared to the single rectangular pulse output.

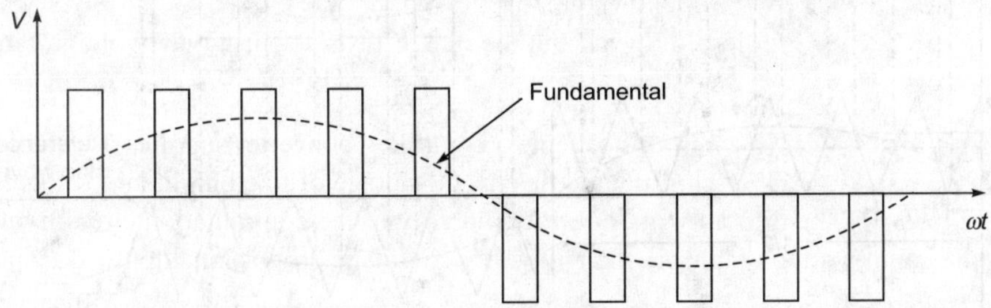

Fig. 12.55

Sinusoidal-Pulse Modulation

In sinusoidal-pulse modulation the pulse-width instead of being uniform as in the waveform of Fig. 12.55 is a sinusoidal function of its angular position with respect to a reference sine wave resulting in a reduction in the harmonic content. The control function consists of a sinusoidal wave obtained from an oscillator of variable amplitude A and of fundamental output inverter frequency $f = 1/T$ as well as a triangular wave of fixed amplitude A_p and frequency f_p with a direct component of amplitude A_p as shown in Fig. 12.56. The biased triangular waveform is reversed in polarity at the end of each half-cycle of the output voltage. Pulsed gating signals are generated by comparison of the sinusoidal and triangular waveforms—gating pulses are obtained for time intervals in which the sinusoidal signal is more positive than the triangular signal. It is easily established from the figure that the number of gate pulses (sinusoidally modulated) per half-cycle is

$$N = \frac{f_p}{2f} = \text{integer} \tag{12.29}$$

It follows from the above that the angles for turn-on and commutation of the thyristors are determined by the intersections of the two above referred signals. It can be easily seen from Fig. 12.56 that the amplitude of the fundamental voltage can be controlled by varying amplitude A of the sine wave over the range $0 < A < A_{max}$ where $A_{max} = 2A_p$. If A is made larger than A_{max}, the number of output pulses becomes less than N (Eq. (12.29)). In the limit the output becomes a single rectangular pulse of half-cycle width as in the case of ordinary (unmodulated) bridge inverter.

> **Probe**
> What do you mean by PWM control of inverters?

> **Probe**
> In a sinusoidal PWM, how many number of pulses are obtained per half cycle?

The gating signals obtained as the above are employed to fire the thyristors in the bridge inverter of Fig. 12.48. Thyristors Th_1 and Th_3 are gated during the interval of a positive pulse and Th_2 and Th_4 are gated during the negative pulse. During the zero-value intervals of the pulsed-gating signals Th_1 and Th_4 or Th_2 and Th_3 are gated resulting in zero output voltage. Thus the output voltage as shown in Fig. 12.56 is a replica of the pulsed-gating signals generated by comparison of sinusoidal and triangular waveforms fed to the commutating circuitry.

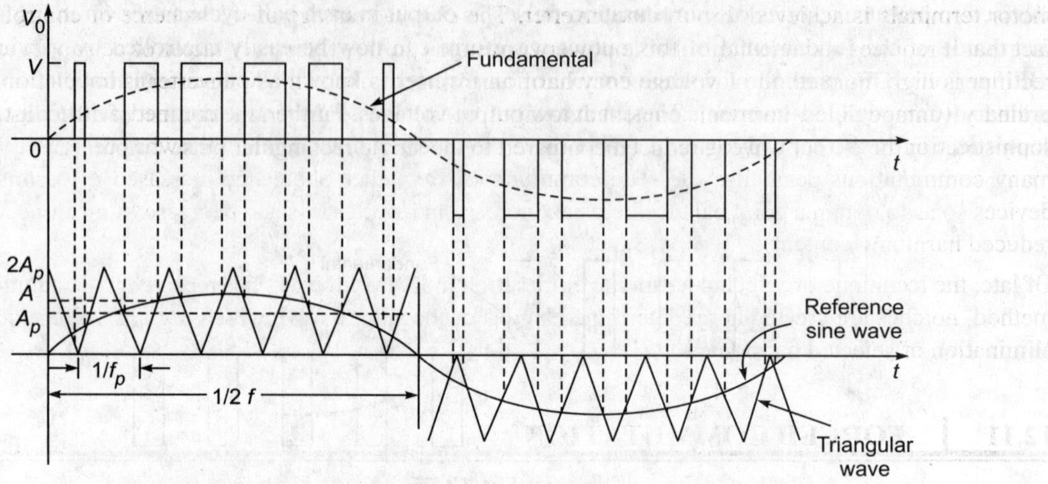

Fig. 12.56 *Output voltage with sinusoidal pulse modulation*

Examination of the circuit of the half-bridge inverter of Fig. 12.46(a) reveals that there is no way in which the load voltage can be reduced to zero. At all times, $v_0 = \pm V_{dc}/2$. The above method of sinusoidal pulse modulation, therefore, cannot be applied as such. A modified method of modulation that can be employed in the half-bridge inverter is illustrated in Fig. 12.57. Voltage control is achieved by varying the amplitude of the reference sine wave. This type of voltage control is rarely employed because of its high ratio of harmonic amplitudes to the amplitude of the fundamental component of the output voltage wave.

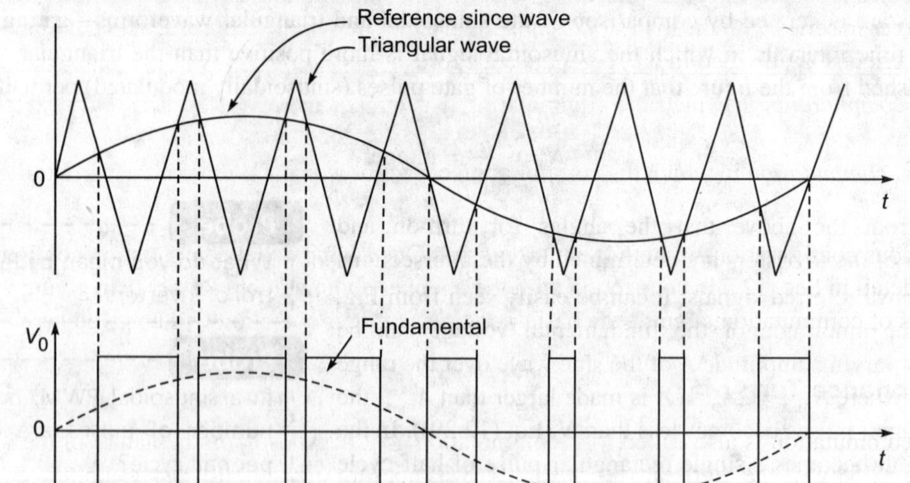

Fig. 12.57

The use of a pulse-width modulated (PWM) inverter for ac motor control is illustrated in Fig. 12.54 where the inverter is fed from a substantially constant dc voltage diode rectifier, while the control of ac voltage at motor terminals is achieved through the inverter. The chief advantage of this scheme of control lies in the fact that it requires a single controllable power conversion so that the power factor of ac supply to the diode rectifier is high. Further, the low-frequency harmonic content of the PWM inverter is lower than that of an ordinary (unmodulated inverter). These advantages of the PWM drive are obtained at the cost of greater sophistication in the control logic. In PWM drive the inverter efficiency is somewhat decreased because of many commutations per half-cycle. The commutation frequency should be increased as permitted by the devices so as to obtain a good balance between increase in inverter loss and decrease of machine loss (due to reduced harmonic content).

Of late, the technique of selected harmonic elimination in PWM inverters has received wide attention. In this method, notches are created at predetermined angles of the square wave which permits voltage control with elimination of selected harmonics.

12.11 | FORCED COMMUTATION

LO 4

Recall It was observed from the discussion on choppers and inverters that in thyristor systems fed from dc supply, the forward current does not pass through zero naturally and must therefore be forced to become zero at appropriate instants by means of auxiliary circuitry, called the commutation circuit. Apart from forcing the forward current to zero, the commutation circuit must also apply a reverse voltage to the thyristor being commutated for sufficient duration (at least equal to its turn-off time) to accelerate and complete the recovery process of the thyristor blocking state. This process of turning off the thyristor as mentioned earlier is called *forced commutation.*

In general, a forced commutation circuit consists of an oscillatory *LC* circuit, which can be considered as a voltage or current source in relation to the thyristor. The commutating voltage or current source can be introduced either in series with the thyristor or in parallel across it by means of some auxiliary device. Forced commutation circuits are divided into two general categories:

(i) *Series commutation* in which the commutating capacitor is in series with the thyristor to be turned off.

(ii) *Parallel commutation* in which the commutating capacitor is in parallel with the thyristor to be turned off.

There are various configurations in both these categories. One such parallel commutation configuration was described in detail in Sec. 12.7 in connection with the chopper commutation. Rather than giving the detailed configurations of commutation circuits, the basic principles of their operation will be given here.

Series Resonance Turn-off

This type of commutation is also called load commutation as it depends upon the nature of the load. In the circuit of Fig. 12.58(a) if the thyristor is turned on, it will not turn off naturally because of the continuity of current. To turn off the thyristor a series capacitor is included in the circuit as shown in Fig. 12.58(b) such that the circuit is underdamped. If the load inductance is negligible, a series inductor must be included in the circuit to make it oscillatory. Figure 12.58(c) shows the oscillatory current of the circuit of Fig. 12.58(b) in the absence of the thyristor. At the end of the first positive current pulse, the thyristor current falls to zero and it turns off. In the meantime capacitor *C* has been charged positively as shown which reverse-biases the thyristor to complete the turn-off process. The current through the thyristor is nearly sinusoidal so that the initial switching current has low *di*/*dt* limiting the switching loss.

This type of commutation circuits have been operated successfully at 20 to 30 kHz.

Important! The frequency limitation being on account of the thyristor turn-off time which must be less than a quarter-cycle period of the circuit oscillation.

One disadvantage of this type of commutation is that the load regulation is poor.

Figure 12.58(d) shows an alternative configuration of the series commutation circuit. Here the circuit oscillation will cause C to charge to a voltage nearly twice the supply voltage (i.e., $2V_{dc}$). The turn off is obviously accomplished in shorter time here.

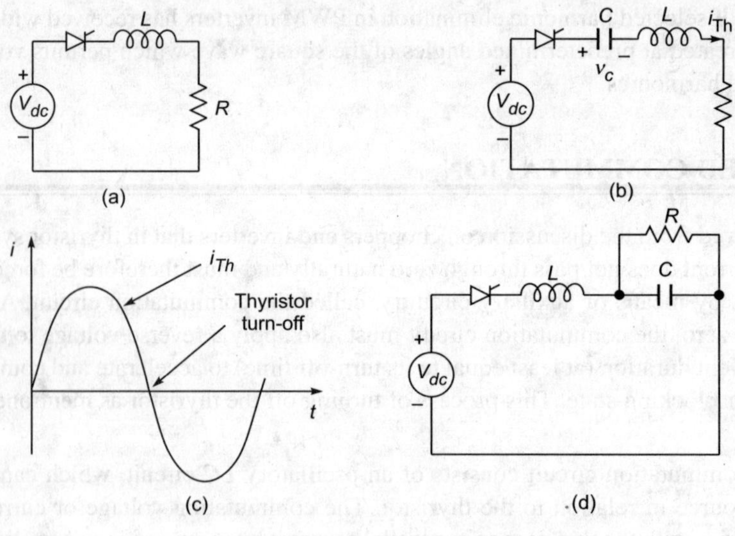

Fig. 12.58 *Series commutation*

Parallel Resonance Turn-off

Figure 12.59 illustrates the principle of parallel commutation. When the thyristor is off, C charges to V_{dc} as shown. As the thyristor is turned on, it conducts the load current I_R and the oscillatory current of the LC circuit. The current in the LC circuit reverses in the second-half oscillation (see Fig. 12.58(c)). This reduces the thyristor current to zero switching it off. The charge still remaining on C reverse-biases the thyristor after current zero.

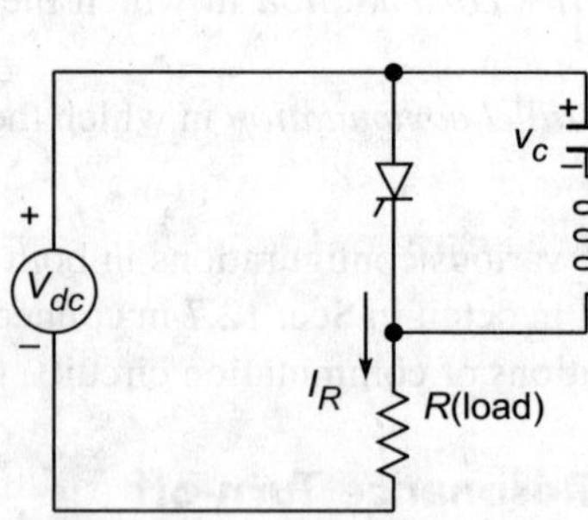

Fig. 12.59 *Parallel commutation*

The advantages of this type of commutation are good load regulation and that the voltage across the thyristor is restricted to no more than the supply voltage.

!Attention If load resistance R is increased, the charging of capacitor C will be slow, making the minimum off-time longer.

Hence this circuit is mostly restricted to loads of constant resistance.

Complementary Commutation

Figure 12.60 shows the circuit of a single-phase half-bridge inverter with the inclusion of commutating capacitors C_1 and C_2 and indictors L_1 and L_2 which are tightly coupled. This circuit provides for complementary impulse voltage commutation. In this circuit if Th_1 is on, firing Th_2 automatically transfers the load current to Th_2, turning Th_1 off. The load being assumed inductive implies that the load current I_0 is almost steady. When Th_1 is conducting, the voltage drop across L_1 is negligible as the rate of change of the inductive load current is small. Capacitance C_2 therefore charges to V_{dc}, the supply voltage. The energy stored in the magnetic field of the inductor is $\frac{1}{2}LI_0^2$, I_0 being the load current. The circuit model prior to triggering Th_2 is shown in Fig. 12.61(a).

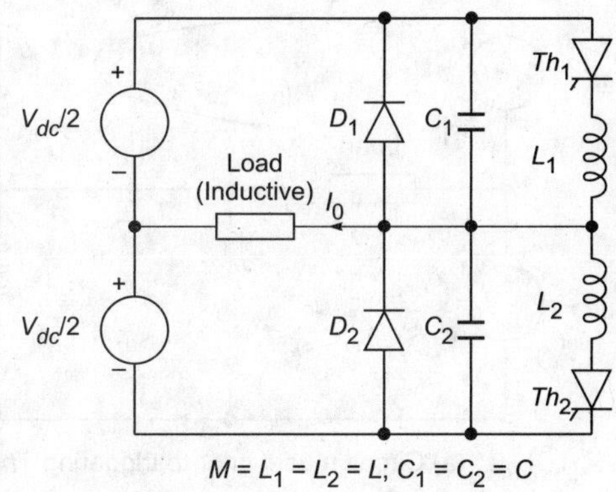

Fig. 12.60 *Complementary commutation*

Circuit Operation

As Th_2 is fired, the voltage V_{dc} of C_2 appears across L_2; as a consequence, voltage V_{dc} is induced across L_1 which reverse-biases Th_1 and turns it off. This interrupts the current I_0 in L_1 which gets immediately transferred to L_2 since the magnetic flux cannot change instantly. Further, since the inductive load current cannot get interrupted, it is fed equally ($I_0/2$) by capacitors C_1 and C_2; C_1 is getting charged and C_2 is getting discharged at the same rate. The capacitor C_2 is now placed in parallel with L_2 (as Th_2 is turned on) and an oscillatory current appears in the closed loop. At the first quarter-cycle of this oscillation, the oscillatory current rises to its maximum value I_m. At this instant Th_2 is carrying I_m which returns equally via the two capacitors. Thus current in both C_1 and C_2 is $(I_0 + I_m)/2$ as shown in the circuit model of Fig. 12.61(b).

In the process C_1 is charging to the supply voltage while the induced voltage across L_1 is reducing. This causes Th_1 to become forward-biased. At the end of the charging process as C_1 gets charged to V_{dc}, the voltage across C_2 simultaneously reduces to zero. As C_2 tends to become oppositely charged during the second quarter of the oscillatory current, the diode D_2 goes into the conducting mode and the current $(I_0 + I_m)$ immediately transfers to it. The current I_m flowing in L_2 and its associated energy dissipates through Th_2 and D_2 at the end of which Th_2 turns off. This dissipation mode is shown in the circuit model of Fig. 12.61(c). The inclusion of a small resistance in series with the diode would speed up the dissipation process.

The capacitor C_1 has by now been fully charged to V_{dc} and the load current I_0 is being fed by D_2 as shown in the circuit model of Fig. 12.61(d). After I_0 has decayed to zero via the free-wheeling diode D_2, D_2 blocks off and Th_2 is fired again or gets fired if it is still carrying the triggering current. The load current now builds in the opposite direction.

The commutation circuit of Fig. 12.60 has to be employed in all the three legs of a 3-phase bridge inverter.

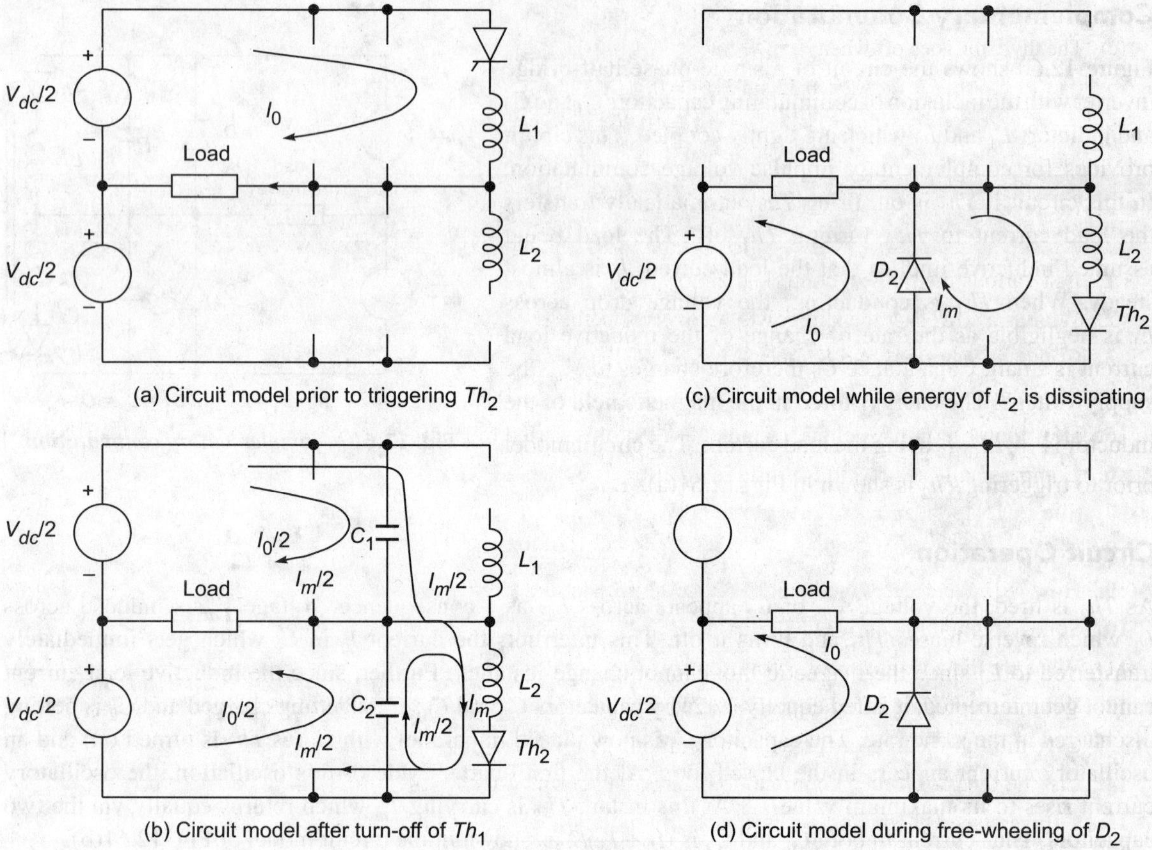

Fig. 12.61

Example 12.10 *Figure 12.62 shows a circuit used for self-commutation by resonance.*

(a) With the thyristor in the blocking state, what is the steady-state capacitor voltage and load current?

(b) Derive the expression for the load current, thyristor current and $v_C(t)$ after the thyristor is triggered at $t = 0$.

(c) Calculate the time when the thyristor would stop conducting.

LO 4

Solution

(a) $V_C = 200 \text{ V}, I_L = 0$

(b) $$v_C(t) = 200 \cos \frac{t}{\sqrt{LC}} = 200 \cos 316t$$

i_C (in the forward direction of the thyristor)

$$= -C\frac{dv_C(t)}{dt} = 6.32 \sin 316t$$

$$i_L = \frac{200}{100} = 2 \text{ A}$$

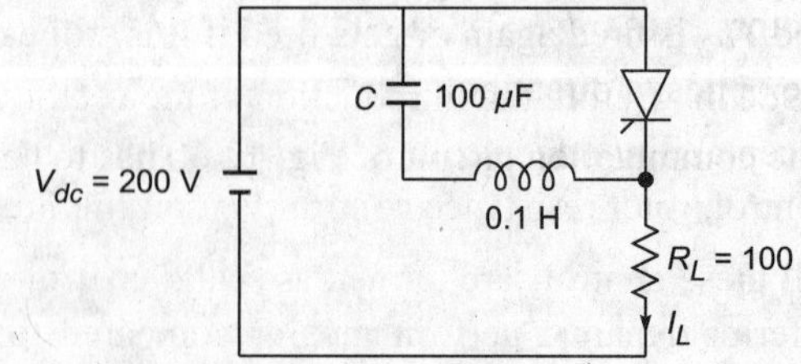

Fig. 12.62

$$i_{Th} = 2 + 6.32 \sin 316t$$

(c) The thyristor goes off when

$$2 + 6.32 \sin 316t = 0$$

$$t = 10.96 \text{ ms}$$

Concept Check — LO 4

1. List various methods to control ac motor.
2. What is a slip? List the slip power recovery schemes.
3. Compare Scherbius and Kramer drive.
4. State the applications of Scherbius drive.
5. How motor terminal voltage of induction motor is varied by phase control?
6. Compare half-bridge and full-bridge inverter.
7. Site some applications of inverter fed drives.
8. Compare 120° and 180° of operation of three-phase inverters.
9. How will you control the output voltage of inverter fed to an ac motor?
10. State a method to control the harmonic content of inverter fed ac drives.
11. What are the advantages of PWM control?
12. What is the impact of reference voltage amplitude on output voltage in PWM technique?
13. Give the significance of forced commutation.
14. List the types of forced commutation techniques for inverter.
15. Compare series and parallel resonance commutation.
16. Explain the principle of PWM of an inverter.

12.12 VECTOR CONTROL OF AN INDUCTION MOTOR

LO 5 Illustrate the direct and indirect vector control scheme of induction motor with its algorithm and analyse the basic principle of DTC scheme

The sole idea behind the vector control of an induction motor is to have an electrical drive which must offer superior performance than the widely used separately excited dc motor which has faster dynamic response as compared to the induction motor. The faster dynamic response of the dc motor lies into its being a doubly fed motor along with inherent facility of independent control of torque and flux in the motor.

Recall Before the introduction of the vector control of induction motor, the methods which enjoyed wide acceptability in controlling the speed of the induction motor drive are termed as voltage control, frequency control, rotor resistance control, *V*/*f* control, flux control, slip control, slip power recovery control, etc.

All these controls are termed as scalar control of an induction motor in which the induction motor exhibits inferior dynamic performance as compared to the separately excited dc motor. In their efforts to have a maintenance free, robust and high performance ac drive, the researcher wanted to realize a separately excited dc motor performance and characteristics with an induction motor with cage rotor. In this direction,

Blaschke introduced in 1972 the concept of vector control of an induction motor based on the principle of field orientation to realize dc motor characteristics in an induction motor drive. In vector control, the torque and flux in the motor are controlled independently.

The cage induction motor drive with vector or field-oriented control offers a high level of dynamic performance and the closed-loop control associated with this drive provides the long-term stability of the system. Despite there being no major difference between scalar and vector controls, the latter has some properties which make it favourable as a control system with high dynamic performance.

In a broad sense, control of cage induction motor such that it behaves like a fully compensated separately excited dc motor, is known as its vector. In this case, the stator currents are expressed with reference to the frame of coordinates which rotates in synchronism either with the stator or rotor mmf vector. Stator currents expressed on these coordinates are resolved into two orthogonal components which produce the flux and torque in the motor. These are similar to the dc motor in which torque and flux are controlled by controlling armature and field currents independently.

Important! To control the torque and flux (thereby speed) independently in the induction motor, there is need to control the magnitude and phase of the three stator currents (i_{as}, i_{bs}, i_{cs}) through a fast inverter. For this purpose, a normally a CC-VSI (current controlled voltage source inverter) is needed. Such a control algorithm would be highly involved.

Use is made of three to two-phase transformation (a well-known matrix operation). The two-phase currents i_{ds} and i_{qs} being in phase quadrature require the control of the two magnitudes and one phase which is what is carried out by the vector control.

Vector control techniques are classified as direct vector and indirect vector control. In direct vector control, the field angle is calculated by using terminal voltages and currents. The field angle is obtained by rotor position estimation and is known as indirect vector control scheme. Direct vector control depends on Hall sensors, search coils for sensing the flux. The magnetizing flux linkage space vector is chosen for controlling the torque. A suitable Hall sensor or search coil is used to sense/monitor the magnetizing flux linkage (ψ_m). The ψ_m can also obtained by $\psi_m = \psi_s - \psi_{s\,\text{leakage}}$ difference between stator flux linkage and stator leakage flux component.

Probe

What is the purpose of hall sensor in direct vector control scheme?

Analysis of Vector Control

It is assumed that position of rotor flux linkage phasor ψ is known at field angle θ_{st} from a stationary frame. The stator currents can be transformed into q and d axis in the synchronous reference frames by using the transformation

$$\begin{bmatrix} i_{qs} \\ i_{ds} \end{bmatrix} = \frac{2}{3} \begin{bmatrix} \sin\theta_s & \sin\left(\theta_s - \dfrac{2\pi}{3}\right) & \sin\left(\theta_s + \dfrac{2\pi}{3}\right) \\ \cos\theta_s & \cos\left(\theta_s - \dfrac{2\pi}{3}\right) & \cos\left(\theta_s + \dfrac{2\pi}{3}\right) \end{bmatrix} \begin{bmatrix} I_a \\ I_b \\ I_c \end{bmatrix} \tag{12.30}$$

Now stator current phasor is derived as $I_{st} = \sqrt{(I_{qs})^2 + (I_{ds})^2}$ and stator phase angle is $\theta_{st} = \tan^{-1}\left(\frac{I_{qs}}{I_{ds}}\right)$

where I_{qs}, and I_{ds} are q and d axis currents in synchronous reference frames that are obtained by projecting the stator current phasor on the q and d axis respectively as shown in Fig. 12.63. The current phasor I_s produces rotor flux linkage ψ and torque T. The component of current producing the rotor flux phasor has to be in phase with ψ. The component I_f is the field producing component and the perpendicular component I_p is torque producing component.

Recall Under steady state conditions both I_p and I_f have only dc components because the relative speed with respect to the rotor field is zero. The rotor flux linkage ψ is oriented towards synchronous reference frame hence flux and torque components are dc quantities. The main advantage of this method is computation and processing the dc signals is easier.

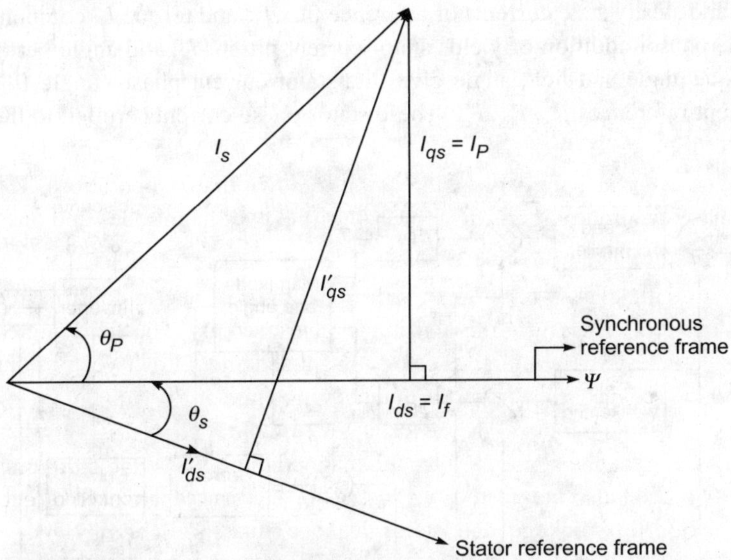

Fig. 12.63 *Phasor diagram of vector controller phasor diagram*

Vector Control Algorithm

1. Obtain the field angle.
2. Calculate the flux producing component of current I_f for a required rotor flux linkage ψ. The rotor flux linkages are controlled by field current I_f. This is similar to dc machines where field current controls the field flux
3. The required torque T from the rotor flux linkage ψ can be obtained from torque producing component of stator current I_P.

   Important! The rotor flux linkage ψ is constant, hence controlling the torque producing component current gives independent control of electromagnetic torque. This is similar to field current is maintained constant in separately excited dc machine
4. Calculate the stator current phasor magnitude I_{st} from the vector sum of I_P and I_f

5. Obtain the torque angle from flux and torque producing components of stator current.

$$\theta_P = \tan^{-1}\frac{I_P}{I_f}$$

6. Add θ_P and θ_f to obtain stator current phasor angle θ_S

Direct Vector Control Scheme

The functional block diagram of this scheme with voltage source inverter is shown in Fig. 12.64. The rotor speed ω_r is compared to the reference speed ω_r^* and the error is amplified by an error amplifier which generates a signal to limit the reference torque T^*. The rotor flux linkage reference ψ^* is derived from the rotor speed. Increasing the rotor speed beyond the base speed is done by weakening the rotor flux linkages. Hence the electromagnetic torque is reduced and by increasing the rotor speed can be controlled to produce constant power output. The torque and rotor flux linkages references are compared with T and ψ respectively. The errors are generated and steady state currents of reference flux I_P^* and torque I_f^* components are obtained by the PI controller. The phasor addition of yields stator current phasor I_S^* and angle between I_P^* and I_f^* gives θ_P^*. The sum of torque angle and field angle gives the stator current phasor angle θ_S along with I_S^* gives the stator phase current references i^*_{as}, i^*_{bs}, i^*_{cs}. These stator phase currents are fed to the inverter by current controller.

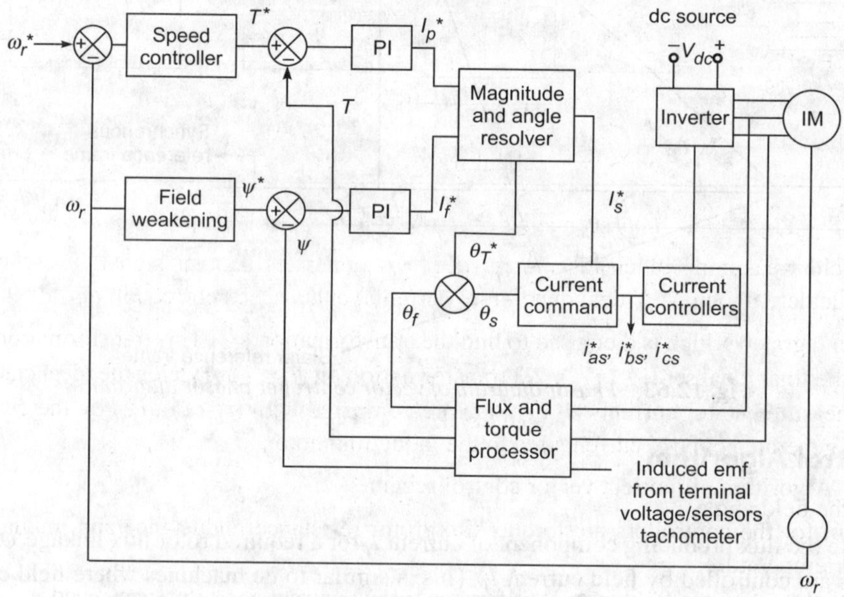

Fig. 12.64 *Direct vector control scheme*

Indirect Vector Control Scheme

The Indirect Vector Control technique depends on machine parameters. In the stationary reference frame the stator flux linkage is the integral of difference between terminal voltage and ohmic stator loss.

$$\psi_s = \int (v_s - R_s i_s)\, dt \tag{12.31}$$

where ψ_s, v_s and i_s are the space vectors of stator flux linkage, stator voltage and stator current respectively.

!Attention The main drawback in applying direct vector control to induction machine is at low frequencies the stator voltages becomes small and ohmic drop is dominant. Hence its ohmic drop affects total integration and precise control is not achieved. At low frequencies indirect vector control method is required.

The indirect vector control method in block diagram form is illustrated in Fig. 12.65. Certain crucial steps are as follows.

From the motor speed signal (ω_r) and the desired speed (ω_r^*) the error ω_e is determined. Speed controller calculates the motor torque (T_o) needed to correct the speed which is passed through a limiter to determine the torque signal T^*.

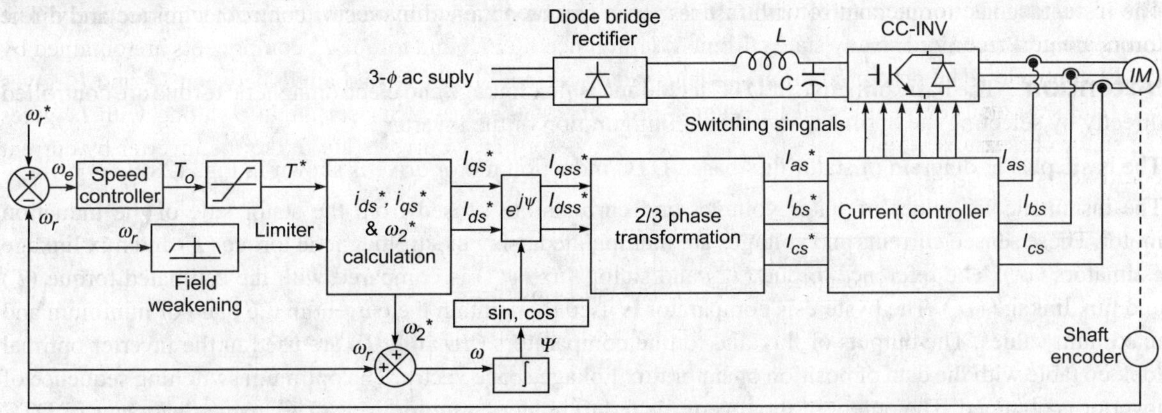

Fig. 12.65 *Indirect method of vector control of an induction motor drive*

In a parallel field weakening block the motor speed ω_2^* generates another signal. These two signals are employed to calculate i^*_{ds} and i^*_{qs} (ideal quadrature currents) and a speed correction ω_2^*.

$\omega = \omega_r + \omega_r^*$ is integrated which is then used to find the transformation $e^{j\psi}$. This transformation carried out on (i^*_{ds}, i^*_{qs}) gives the final ideal set (i^*_{dss}, i^*_{qss}). 2/3 transformation on (i^*_{dss}, i^*_{qss}) yields the ideal stator current (i^*_{as}, i^*_{bs}, i^*_{cs}). The measured stator currents (i_{as}, i_{bs}, i_{cs}) are compared with (i^*_{as}, i^*_{bs}, i^*_{cs}) by the current controller and six signals generated control currents fed to the induction motor.

Important! Algorithm of indirect vector control scheme:

⇒ The input for the controllers are torque, flux, rotor resistance, inductance and mutual inductance at nominal temperature.

⇒ Calculate the reference field current, reference torque current and reference speed.

⇒ Compute the magnitude of stator current, reference torque angle, slip angle, field angle and stator current angle.

⇒ Calculate the direct axis and quadrature axis current.

⇒ Use the *dq-abc* transformation to obtain the three phase stator current commands to the inverter.

The above is a simplified explanation of the vector control algorithms for induction motor control. It requires accurate value of inductance. But the value of inductance is not same at saturation levels and the effect on low frequency also affects the performance to control by direct vector method.

Also the cost of Hall sensor is high and at high temperatures the system perform poorly. Henceforth in industries they prefer sensorless control of induction machine.

Sensorless Estimation of Rotor Speed

In sensorless control, speed is not directly monitored but it is estimated by using intelligent techniques like fuzzy, neural network and neuro-fuzzy. The estimation of rotor speed is obtained by use of stator voltage and stator current. This estimation depends on the accuracy of the machine parameters. At low speeds, the effect of stator resistance is high since it affects stator flux linkages. The rotor flux linkage is obtained from stator flux linkage. Hence, the effect of stator resistance is important for accurate control.

Direct Torque Control

The instantaneous torque control with fast response can be obtained by vector control technique and direct torque control technique.

!Attention The basic principle of DTC technique is flux linkage and electromagnetic torque are controlled directly by selecting the optimum switching configuration of the inverter.

The basic phasor diagram of stator flux based DTC induction motor drive is shown in Fig. 12.66.

The instantaneous values of stator voltages and currents are sensed from the stator side of the induction motor. These sensed currents and voltages are manipulated using electromagnetic torque (T) and flux linkage estimators (ψ_s). The reference torque (T^*) and stator flux (ψ_s^*) is compared with the estimated torque (T) and flux linkage (ψ_s). The hysteresis comparator is used to maintain the current in the band of minimum and maximum values. The outputs of flux and torque comparators ($d\psi$ and dT) are used in the inverter optimal look up table with the data of position of stator flux linkage space vector. The optimum switching sequence of inverter is obtained. The output of the inverter is fed to induction motor drive. The main advantages of DTC are direct control of torque and flux, flux wave is sinusoidal approximately. The main demerits of DTC are during starting and low speed operation it is not effective.

In a symmetrical three-phase induction machine, the instantaneous electromagnetic torque is proportional to vector product of stator flux linkage (ψ_s) and stator current (I_s).

$$T_e = \frac{3}{2} P \bar{\psi}_s \times \bar{I}_s \tag{12.32}$$

These are represented in stationary reference frames shown in Fig. 12.66. $\psi_s = |\psi_s| e^{j\alpha_s}$, where α_s is angle of the stator flux linkage with respect to the direct axis of stator reference frame and $I_s = |I_s| e^{j\beta_s}$.

Hence electromagnetic torque is

$$T_e = \frac{3}{2} P |\psi_s||I_s| e^{j\alpha_s} e^{j\beta_s}$$

$$= \frac{3}{2} P |\psi_s||I_s| \sin(\beta_s - \alpha_s)$$

$$= \frac{3}{2} P |\psi_s||I_s| \sin\beta \tag{12.33}$$

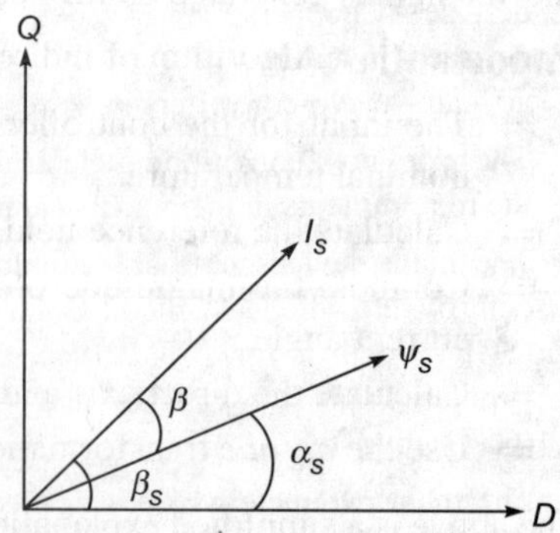

Fig. 12.66 *Phasor diagram ψ_s and I_s*

$\beta = \beta_s - \alpha_s$ is the angle between the stator flux linkage and stator current.

When three phase balanced voltages are applied to the stator of induction motor, it keeps the stator flux constant. The dynamic response of the torque is obtained quickly by giving the necessary torque command. When the motor torque is smaller than its reference value, then motor torque is increased. When motor torque is equal to reference torque then rotation is stopped.

If the stator flux linkage vector is accelerated in the forward direction then positive electromagnetic torque in reverse direction and negative electromagnetic torque is obtained. Hence, from this discussion electromagnetic torque can be changed dynamically by controlling the stator flux linkage vector which can be changed by using stator voltage generated by inverter.

Concept Check LO 5

1. List the major advantages of vector control technique. ○○●
2. What is the role of position sensors in vector control scheme? ●●●
3. Compare direct and indirect vector control scheme. ○●●
4. State the drawback of direct control scheme at low frequency operation. ○●●
5. How will you estimate rotor speed of induction motor without sensor? ●●●
6. What is direct torque control? ○○●
7. Why DTC is not effective during starting and low speed operation of induction motor? ●●●
8. Are feedback sensors must in a closed loop control of induction motor drive? ○●●
9. Can the speed of an induction motor be altered by DTC technique? ●●●

SUMMARY

LO 1

- Four basic forms of power conversions are
  ac – dc – Rectifiers
  dc – dc – Choppers
  dc – ac – inverters
  ac – ac – cyclo-converters or ac regulators

  These are now conditioning electric power to meet the requirements in the industry at all levels.
- The main devices normally employed for motor control are diode, Silicon-Controlled Rectifier (SCR) or thyristor, triac, diac, Insulated Gate Bipolar Transistor (IGBT) and Gate Turn Off (GTO) thyristor.
- The main parameters of power electronic devices are blocking voltage, average forward current, max frequency of operation and maximum operating junction temperature.
- Commutation is the process of transferring of current from one device (thyristor) to the other.
- SCR is switched ON by applying a positive gate signal to the forward biased SCR and switched OFF by reducing its conduction current below holding current and being reverse biased.
- Snubber circuit with series inductor connected across an SCR serves dv/dt and di/dt protection respectively
- GTO is a PNPN device with three terminals anode, cathode and gate; device can be triggered into conduction by a small positive gate-current pulse and can be turned off by issuing negative gate current pulse.

- Except for small motors, transistors are employed in the switching (on/off) mode. The switching times of power transistors are, however, generally faster than those of SCRs, and the commutation problems of SCR are not present in transistors.
- Power MOSFETs are suitable for applications that require low blocking voltages and high operating frequencies but IGBTs are suitable for applications with high blocking voltages and low operating frequencies.
- As compared with present competing technologies, the ETO offers improved performance in current conduction and turn-off speed, as well as snubberless turn-off current capability.

LO 2

- The advantages of ac drives such as low cost, maintenance free, and ruggedness, have succeeded in replacing dc drives in a number of variable speed applications.
- In dual converter, four quadrant operation is obtained using two full-converters connected back to back providing output voltage and output current both of which can be reversed.
- For controlling dc motor, depending on the source, either converter or chopper is used, and for ac motor control, either inverter or cycloconverter is used.

LO 3

- Single phase half-wave converter is always discontinuous, hence used only for motors below 400 W. Semi-converter is a one quadrant converter and so regenerative braking is not possible.
- Full-converter is a two quadrant converter and the average motor terminal voltage is negative if the firing angle ******⊠ is greater than 90°.
- If the polarity of the applied voltage is reversed then motoring and regenerating operation can occur with the direction of rotation reversed.
- The output voltage of the chopper is controlled by its duty cycle either by pulse width modulation system or by variable frequency system.

LO 4

- The speed of synchronous motor can be controlled by combined voltage and frequency control only.
- Rather than wasting the slip energy in external resistance of induction motor, it can be recovered and fed back to supply, thereby removing the disadvantage.
- Static Scherbius drive provides constant torque control and static Kramer drive provides constant power control.
- Full-bridge inverter is used to overcome the disadvantage of requirement of 3-wire dc supply in half-bridge inverter.

LO 5

- *Slip Control* Sub-synchronous control of the induction motor speed is possible by introducing a variable resistance in the rotor circuit of a wound-rotor induction motor. This method provides a wide range of speed and good starting torque but poor efficiency.
- Vector control is the principle of field orientation to realize dc motor characteristics in an induction motor drive. It offers a high level of dynamic performance and the closed-loop control associated with this drive provides the long term stability of the system.
- Direct torque control is a technique of controlling flux linkage and electromagnetic torque directly by selecting the optimum switching configuration of the inverter.

PRACTICE PROBLEMS

12.1 A separately excited dc motor is fed from a 230 V, 50 Hz Source via a single-phase full-converter. The motor armature resistance and inductance are respectively 2 Ω and 0.05 mH, The motor's torque constant is 1 Nm/A and the voltage constant is 1 V/rad/s. LO 3

If the current through the armature flows for 160° after the commencement of current flow at $\alpha = 60°$ (causing discontinuous current operations); calculate the average voltage across the load, the average current through the load, and the torque developed. Assume $\omega = 150$ rad/s.

12.2 A chopper circuit as shown in Fig. P12.2 is inserted between a battery, $V_{dc} = 100$ V and a load of resistance $R_L = 10\ \Omega$. The turn-off time for the main thyristor Th_1 is 100 μs and the maximum permissible current through it is 25 A. Calculate the values of the commutating components *L and* C. LO 1

Hint: With the thyristor Th_1 in the conducting state, as Th_2 is fired, the capacitor discharges via R_L and the battery (Th_1 behaves as open-circuit to C as it is conducting current in forward direction, while the voltage of C reverse-biases it). Therefore

$$v_{Th_1} = v_C = V_{dc}(1 - 2e^{-t/RC})$$

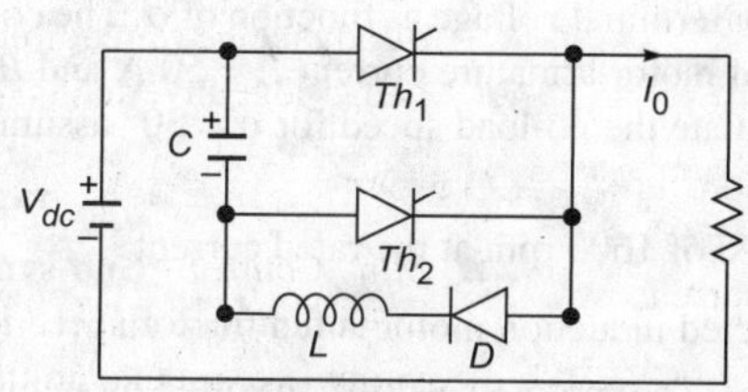

Fig. P12.2

For turn-off in 100 μs

$$v_{Th_1} = 0 = V_{dc}(1 - 2e^{-100 \times 10^{-6}/10C}) \qquad \text{(i)}$$

which gives

$$C = 14.4\ \mu\text{F}$$

which C has fully charged oppositely, Th_2 stops conductions. As Th_1 is fired again, diode D, being positively biased, becomes conducting and the LC resonant circuit oscillates via Th_1. The maximum value of the oscillatory current through Th_1 is given by

$$i_C\,(\text{max}) = V_{dc}C\omega = V_{dc}C \times \frac{1}{\sqrt{LC}}$$

$$= V_{dc}\sqrt{\frac{C}{L}}$$

Hence the maximum value of current through Th_1 is

$$i_{Th_1}(\text{max}) = I_0 + i_C\,(\text{max})$$

$$= \frac{V_{dc}}{R_L} + V_{dc}\sqrt{\frac{C}{L}}$$

Substituting values

$$L = 640\ \mu\text{H}$$

12.3 Figure P12.3 shows a 3-phase, full-converter feeding a separately excited dc motor. LO 3

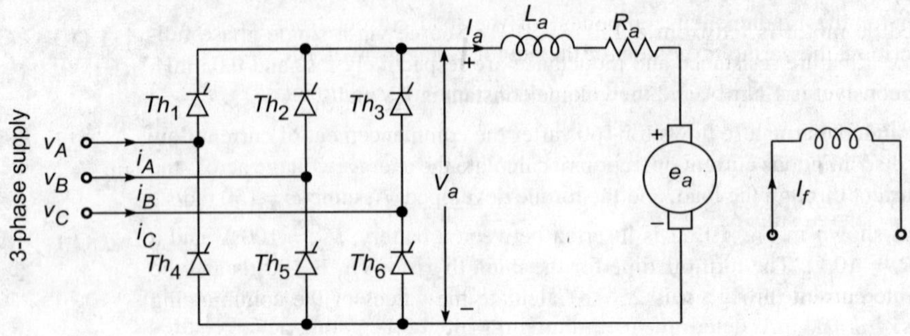

Fig. P12.3

The thyristors are fired at intervals of 60° (Th_1, Th_2, Th_3, Th_4, Th_5, Th_6). Draw the voltage and current waveforms for the firing angle $\alpha = 60°$.

Derive the expression for the average motor terminal voltage as function of α. The converter is fed from a 400 V, 50 Hz supply. The rated motor armature current $I_a = 50$ A and $R_a = 0.1$ Ω, $L_a = 6.5$ mH and $K_a\Phi = 0.3$ V/rpm. Calculate the no-load speed for $\alpha = 30°$ assuming the no-load current as 5 A.

Also calculate the value of α to obtain a speed of 1600 rpm at the rated current.

12.4 A 4-pole, 3-phase, 400 V, 50 Hz, star-connected induction motor is fed from an inverter such that the phase voltage is a six-step waveform. The motor speed is controlled by maintaining *V/f* constant at a value corresponding to the rated voltage and frequency. LO 4

(a) Determine the expression for the fundamental and harmonics of the inverter output voltage waveform.

(b) Calculate the dc input voltage required to feed the inverter for operating the motor at 60, 50 and 40 Hz.

(c) Calculate the firing angle, if the dc input voltage to the inverter is obtained from a 3-phase semi-converter fed from a 500 V (line-to-line), 50 Hz source while the inverter output corresponds to 60 Hz.

12.5 A single-phase bridge inverter with a quasi-square wave (see Fig. P12.5) output with an on-period of 5 ms feeds a load of $R = 8$ Ω in series with $L = 0.044$ H from a 200 V dc source. The output frequency of the inverter is 50 Hz. Determine: LO 4

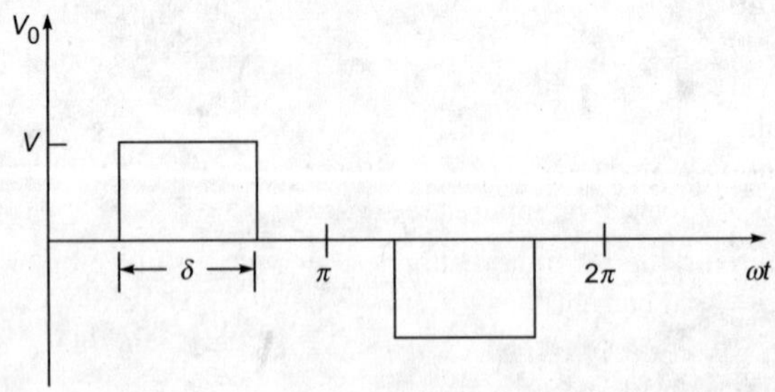

Fig. P12.5

(a) The load current waveform for the first two half-cycles. Also find expressions for the steady-state current.

(b) The expression for the fundamental component of the load current from the general expression describing the harmonic content of the output voltage waveform.

Remark This output waveform can be obtained from the bridge inverter circuit of Fig. 12.48 by the firing sequence

$Th_1\ Th_3,\ Th_1\ Th_4,\ Th_2,\ Th_4,\ Th_2,\ Th_3, \ldots$

Also $$v_0 = \frac{4V}{n\pi} \sin \frac{n\delta}{2} \sin n\,\omega t$$

12.6 A thyristor is connected in series with a battery of 48 V dc and 0.05 H inductor. If the latching current of the thyristor is 100 mA, determine the minimum width of the gating pulse required to properly turn on the thyristor. LO 1

12.7 A single-phase bridge rectifier using four SCRs feeds power to RLE load with $R = 10\ \Omega$, $L = 100$ mH and $E = 150$ V. The ac source voltage is 230 V at 50 Hz. If the firing angle delay is equal to 45°, assuming continuous conduction, calculate the average value of load current. LO 2

12.8 A single phase half-bridge inverter has a resistive load of $R = 4.8\ \Omega$ and dc input voltage is $V_s = 48$ V. Determine (a) RMS output voltage at the fundamental frequency V1, (b) output power P_0, and (c) peak current of each device. Also compute and compare the above values (a, b, and c) for a single phase full-bridge inverter. LO 4

12.9 A dc chopper operating at 230 V dc with chopping frequency of 400 Hz feeds an RL load. Determine the ON time of the chopper for the output voltage of 150V. LO 3

12.10 A 200 V, 1000 rpm and 200 A separately excited dc motor has an armature resistance of $0.02\ \Omega$ and braking resistance of $2\ \Omega$. The motor is fed from a chopper which provides both motoring and braking operations. If the source is of 250 V dc, assuming continuous conduction, calculate duty ratio of chopper for (i) motoring operation at rated torque and 450 rpm, (ii) braking operation at rated torque and 450 rpm. LO 3

MULTIPLE CHOICE QUESTIONS

12.1 A thyristor is a semiconductor switch which is LO 1
- (a) unilateral and stable
- (b) bilateral and stable
- (c) unilateral and bistable
- (d) bilateral and bistable

12.2 A dc chopper circuit can LO 3
- (a) only reduce the voltage to any value
- (b) only raise the voltage to any value
- (c) reduce or raise the voltage to any value
- (d) reduce or raise the village to half or double the input value

12.3 In a load-commutated dc chopper circuit, the maximum chopper frequency is given by (V = input voltage, C = capacitor value, I = load current). LO 3

(a) 2 *VC/I* (b) *1/2 VC* (c) *I/VC* (d) *VC/I*

12.4 In a voltage-commutated dc chopper circuit, a capacitor is LO 3
- (a) reverse charged by the load current reducing it to zero.

(b) charged by LC oscillation and is placed in series with the forward thyristor to reduce its current to zero.
(c) charged by the load current which is reduced to zero as the capacitor is fully charged.
(d) charged by LC oscillation and is placed in parallel with the forward thyristor to reduce its current to zero.

12.5 The rms value of the ac output of an inverter can be controlled LO 4
(a) by adjusting triggering times of thyristors only
(b) by varying the dc side voltage only
(c) by either (a) or (b) method above
(d) it cannot be controlled

12.6 A single-phase bridge inverter requires LO 4
(a) two thyristors, two diode and three dc lines
(b) four thyristors, two diodes, and two dc lines
(c) four thyristors, four diodes and three dc lines
(d) four thyristors, four diodes and two dc lines

12.7 In a three-phase six-step thyristor inverter, the frequency of firing is LO 4
(a) the same as output frequency (b) six times the output frequency
(c) three times the output frequency (d) independent of the output frequency

12.8 In a three-phase six-step thyristor inverter currents out of phase with the output voltage are carried by LO 4
(a) any three thyristors at a time (b) any three freewheeling diodes at a time
(c) three specific thyristors at a time (d) three specific freewheeling diodes at a time

12.9 The sinusoidal pulse modulation of a thyristor inverter gives control over LO 4
(a) output voltage only (b) both output voltage and frequency
(c) output frequency only (d) output voltage, frequency and harmonic content

12.10 In a sinusoidal pulse modulated thyristor inverter, the number of gate pulses per half cycle is (f_P = frequency of triangular biasing wave, and f = frequency of reference sine wave) LO 4
(a) f_P/f = rational number (b) $f_p/2$ = rational number
(c) f_P/f = integer (d) $f_P/2f$ = integer

12.11 Name the diode commonly used to provide continuity of current in the inductive circuit following the switching off of power device. LO 1
(a) Power diode (b) Zener diode
(c) Freewheeling diode (d) Ideal diode

12.12 The switching of power transistor is controlled through LO 1
(a) collector current (b) CE voltage (V_{CE})
(c) base current (d) BE voltage (V_{BE})

12.13 Statement 1: ac drives have succeeded in replacing dc drives. LO 2
Statement 2: ac drives are available in high power ratings and they are rugged and maintenance free.
(a) Statement 1 is true and 2 is false (b) Statement 1 and 2 are true
(c) Statement 1 and 2 are false (d) Statement 2 is true and 1 is false

12.14 Four-quadrant operation is possible with LO 2
(a) half-wave converter (b) full-converter
(c) semi-converter (d) dual converter

12.15 Name the control unit for obtaining variable voltage and variable frequency power directly from a fixed frequency ac supply voltage without any intermediate dc stage. LO 2

(a) Inverter (b) Cycloconverter
(c) Chopper (d) Integral cycle controller

12.16 In a full-converter feeding separately excited dc motor, which of the following statements is not true? LO 3

(a) Full-converter is a two quadrant converter
(b) No freewheeling diode is necessary
(c) Current is bidirectional
(d) Voltage polarity of output can be reversed

12.17 In a semi-converter, at a firing angle of 180°, the output voltage will be LO 3

(a) zero (b) maximum
(c) equal to input voltage (d) negative

12.18 Statement 1: In pulse width modulation control chopper period (T) is fixed. LO 3

Statement 2: In frequency modulation control chopper period (T) is varied.

(a) Statement 1 is true and 2 is false (b) Statement 1 and 2 are true
(c) Statement 1 and 2 are false (d) Statement 2 is true and 1 is false

12.19 In a step-up chopper, if the duty cycle is 1, the output voltage is LO 3

(a) zero (b) negative
(c) infinity (d) equal to input voltage

12.20 Which of the following methods is not an induction motor speed control technique? LO 4

(a) Stator voltage control (b) Frequency control
(c) Slip control (d) Rotor current control

12.21 In frequency control technique of speed control of induction motor LO 4

(a) frequency is varied and voltage is constant
(b) voltage is varied and frequency is maintained constant
(c) frequency and voltage are varied to keep v/f ratio constant
(d) flux/pole is varied and v/f ratio is constant

12.22 Output voltage of inverters with sinusoidal PWM control is varied by LO 4

(a) varying amplitude of reference sine wave (b) varying frequency of sinusoidal wave
(c) varying amplitude of carrier wave (d) varying frequency of carrier wave

12.23 Which of the statements are true for series and parallel commutation techniques? LO 4

In series commutation technique,

Statement 1: Commutating capacitor is in series with the thyristor to be turned off.

Statement 2: Commutating capacitor is in series with the load.

In parallel commutation technique,

Statement 3: Commutating capacitor is in parallel with the thyristor.

Statement 4: Commutating capacitor is in parallel with the load.

(a) Statements 1 and 2 (b) Statements 2 and 3
(c) Statements 1 and 3 (d) Statements 2 and 4

12.24 Assertion: Indirect vector control of induction motor at low frequencies doesn't give precise control. LO 5

Reason: When an induction motor is at low frequency, the stator voltage becomes small and ohmic drop is dominant.

(a) Both assertion and reason are true (b) Only assertion is true
(c) Only reason is true (d) Both are false

12.25 Statement 1: In the direct vector control scheme, the field angle is calculated by using terminal voltage and currents. LO 5

Statement 2: Indirect vector control scheme doesn't need rotor position estimator.

(a) Statement 1 is true and 2 is false
(b) Statements 1 and 2 are true
(c) Statements 1 and 2 are false
(d) Statement 2 is true and 1 is false

12.26 In direct torque control scheme, when motor torque is equal to reference torque the rotation is LO 5

(a) decreased
(b) increased
(c) stopped
(d) jerked

Answers

1. (c)	**2.** (c)	**3.** (b)	**4.** (d)	**5.** (c)	**6.** (a, d)	**7.** (b)	**8.** (d)	**9.** (d)	**10.** (d)
11. (c)	**12.** (c)	**13.** (b)	**14.** (d)	**15.** (b)	**16.** (c)	**17.** (a)	**18.** (b)	**19.** (c)	**20.** (d)
21. (c)	**22.** (a)	**23.** (c)	**24.** (c)	**25.** (a)	**26.** (c)				

APPENDIX I

ac Steady-State Circuit Analysis

AI.1 | PHASOR REPRESENTATION OF SINE WAVES

The sinusoidal waveform of Fig. AI.1 can be represented as

$$a = \sqrt{2}\ A \sin(\omega t + \phi)$$

where $\sqrt{2}A$ = maximum amplitude of sine wave

A = rms value of sine wave

$\omega = 2\pi f$ = frequency in rad/s

f = frequency in Hz (cycles/s)

ϕ = phase angle

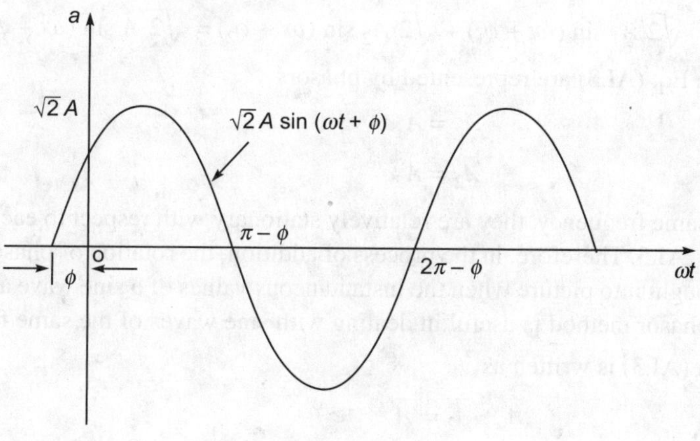

Fig. AI.1

It can also be written as

$$a = \sqrt{2}\ A \sin(\omega t + \phi) = \text{Im}\left[\sqrt{2}\ Ae^{j\phi} e^{j\omega t}\right] \quad \text{(AI.1a)}$$

$$= \text{Im}\left[\sqrt{2}\,\bar{A}e^{j\omega t}\right] \tag{AI.1b}$$

where
$$\bar{A} = Ae^{j\phi} = A\angle\phi \tag{AI.2}$$

According to Eq. (AI.1b) a sine wave can be represented by the phasor $\bar{A}$ as defined in Eq. (AI.2) where A is the magnitude of the phasor representing the rms value of the sine wave and ϕ is its phase angle. By performing the imaginary operation on $\sqrt{2}\,\bar{A}e^{j\omega t}$ as in Eq. (AI.1b), the instantaneous value of the sine wave represented by the phasor $\bar{A}$ can be found. Another powerful way is the graphical representation of a phasor as shown in Fig. AI.2 wherein the phasor $\bar{A}$ is represented by a directed line of magnitude A located at an angle ϕ from the reference axis. Anti-clockwise rotation at ω rad/s is imparted to the phasor $\bar{A}$ by $e^{j\omega t}$ in Eq. (AI.1b). It easily follows from the graphical representation that the projection of the phasor on the y-axis at time t multiplied by $\sqrt{2}$ gives the instantaneous value of the sine wave represented by the phasor*.

The phasor concept is very useful in ac steady-state circuit analysis.

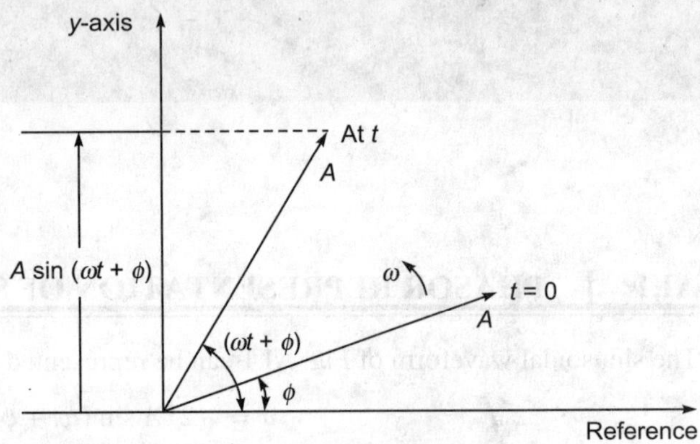

Fig. AI.2 *Phasor representation of sine wave*

AI.2 | ADDITION OF PHASORS

Addition (or subtraction) of two or more sine waves of the same frequency can be easily carried out by means of phasor representation. For example,

$$\sqrt{2}\,A_1 \sin(\omega t + \phi_1) + \sqrt{2}\,A_2 \sin(\omega t + \phi_2) = \sqrt{2}\,A \sin(\omega t + \phi) \tag{AI.3}$$

The two sine waves in Eq. (AI.3) are represented by phasors

$$\bar{A}_1 = A_1 e^{j\phi_1}$$

$$\bar{A}_2 = A_2 e^{j\phi_2}$$

Since they are of the same frequency, they are relatively stationary with respect to each other as shown in the phasor diagram of Fig. AI.3. Therefore, in the process of addition, the rotation of phasors has no significance; it needs only to be brought into picture when the instantaneous values of a sine wave are required. It is also to be observed that the phasor method is useful in dealing with sine waves of the same frequency†.

In its phasor form Eq. (AI.3) is written as

$$\bar{A}_1 + \bar{A}_2 = \bar{A} = Ae^{j\phi} \tag{AI.4}$$

The corresponding graphical phasor addition is shown in Fig. AI.3. Phasor subtraction is shown in Fig. AI.4.

---

* Phasor representing sine wave in time is also called a *time phasor*.

† The phasor concept can be extended to addition (subtraction) of two sine waves of different frequencies wherein the two phasor move relative to each other and the resultant phasor at any instant yields the instantaneous rms value and phase angle of the sum (see Sec. 3.13).

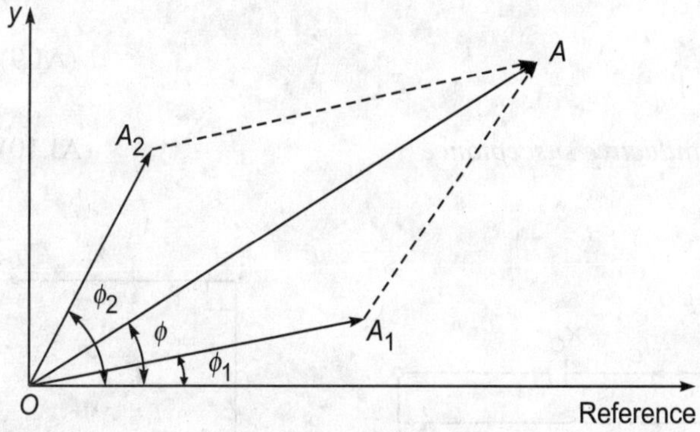

Fig. AI.3

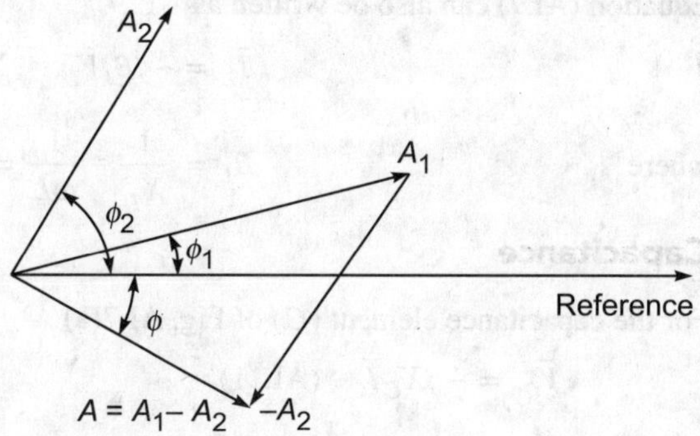

Fig. AI.4

AI.3 | AC STEADY-STATE ELEMENTAL BEHAVIOUR

A circuit excited with ac after the transients have died out, reaches steady-state wherein voltages and currents at all points in the circuit are sinusoidal provided the circuit is composed of linear elements. Therefore, ac steady-state behaviour of linear lumped elements—resistance, inductance and capacitance—can be modelled in terms of phasor voltages and currents.

Resistance

For the resistance element shown in Fig. AI.5(a),

$$\bar{V}_R = R\bar{I}_R \quad \text{(AI.5)}$$

wherein the voltage across resistance is *in phase* with current though it as shown by the phasor diagram of Fig. AI.5(b). Equation (AI.5) can also be written as

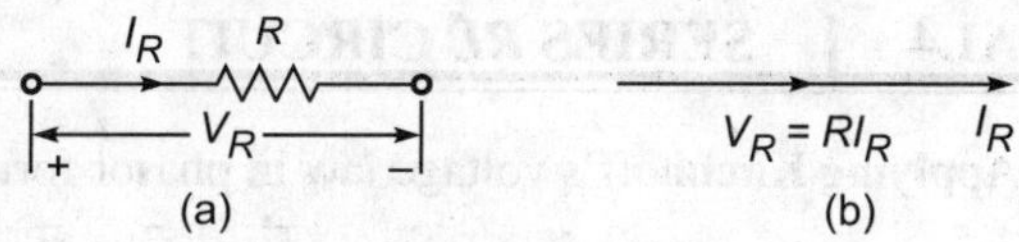

Fig. AI.5

$$\bar{I}_R = G\bar{V}_R \quad \text{(AI.6)}$$

where $G = \frac{1}{R}$ = *conductance*

Inductance

For the inductance element (L) of Fig. AI.6(a)

$$\bar{V}_L = jX_L\bar{I}_L \quad \text{(AI.7)}$$

where $X_L = \omega L = 2\pi fL$ = *inductive reactance* (AI.8)

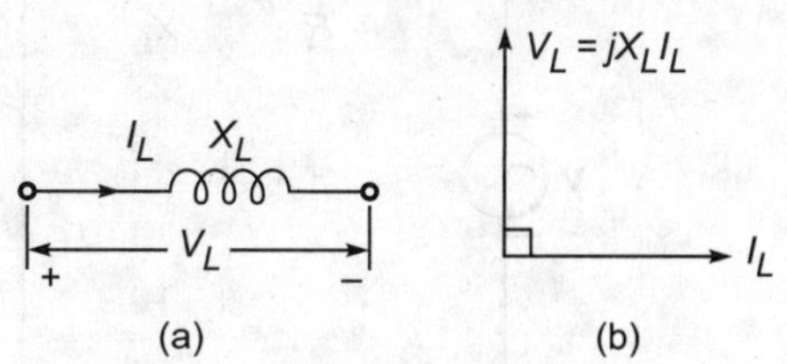

Fig. AI.6

The phasor diagram corresponding to the elemental behaviour of Eq. (AI.7) is drawn in Fig. AI.6(b) wherein the voltage across inductance and current through it are in phase quadrature with the current lagging the voltage by 90°.

Equation (AI.7) can also be written as

$$\bar{I}_L = -jB_L\bar{V}_L \tag{AI.9}$$

where
$$B_L = \frac{1}{X_L} = \frac{1}{\omega L} = \textit{inductive susceptance} \tag{AI.10}$$

Capacitance

For the capacitance element (C) of Fig. AI.7(a)

$$\bar{V}_C = -jX_C\bar{I}_C \tag{AI.11}$$

where
$$X_C = \frac{1}{\omega C} = \frac{1}{2\pi fC}$$

$$= \textit{capacitive reactance} \tag{AI.12}$$

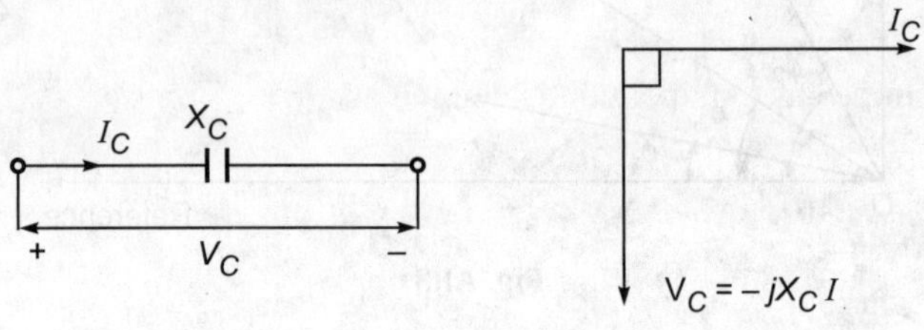

(a) (b)

Fig. AI.7

Corresponding to Eq. (AI.11), the phasor diagram is drawn in Fig. AI.7(b), wherein current leads the voltage by 90°. Equation (AI.11) can also be written as

$$\bar{I}_c = jB_C\bar{V}_C \tag{AI.13}$$

where
$$B_C = \frac{1}{X_C} = \omega C = \textit{capacitive susceptance}$$

AI.4 | SERIES *RL* CIRCUIT

Applying Kirchhoff's voltage law in phasor form to the series *RL* circuit of Fig. AI.8(a).

$$\bar{V} = \bar{V}_R + \bar{V}_L$$

$$= R\bar{I} + jX_L\bar{I} = (R + jX_L)\bar{I} \tag{AI.14}$$

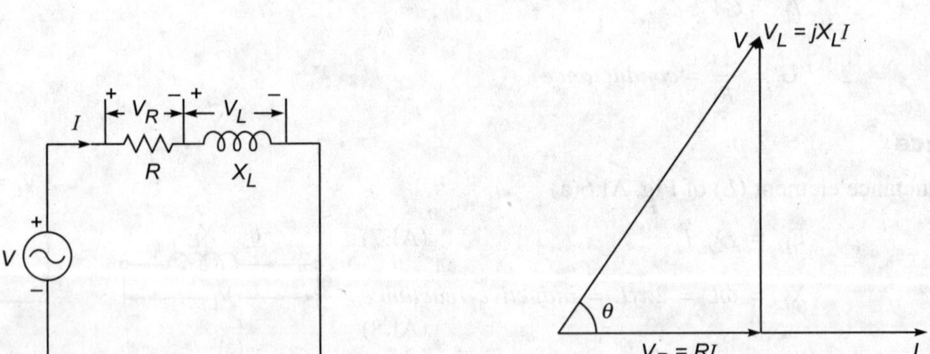

(a) (b)

Fig. AI.8

By use of elemental relations described earlier, the phasor diagram corresponding to Eq. (AI.13) is drawn in Fig. AI.8(b) from which it is observed that the current lags voltage by

$$\theta = \tan^{-1}\frac{X_L}{R} \tag{AI.15}$$

Equation (AI.14) can be written as

$$\bar{I} = \frac{\bar{V}}{R + jX_L} \tag{AI.16}$$

Impedance of a circuit is defined as

$$\bar{Z} = (R + jX_L) = \textit{impedance} \text{ (complex number)} \tag{AI.17}$$
$$= Z\angle\theta$$

where $\theta = \tan^{-1}\frac{X_L}{R}$,

which in fact is the angle by which the current lags the voltage in the circuit. The impedance of the circuit is represented as a complex number in Fig. AI.9. It is also observed that impedance is not a phasor.

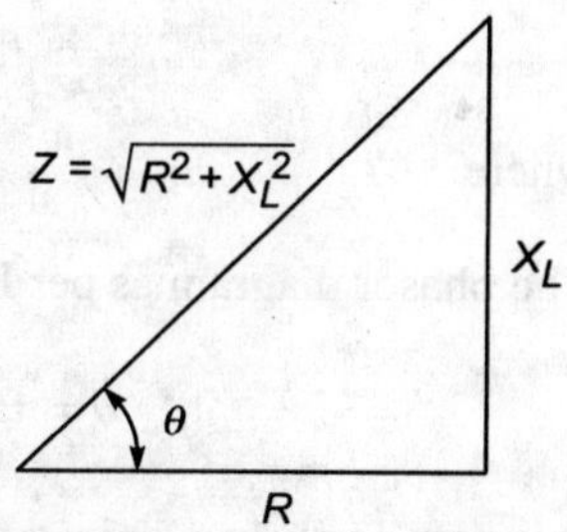

Fig. AI.9 *Impedance triangle*

Equation (AI.16) can also be written as

$$\bar{I} = \frac{\bar{V}}{\bar{Z}} = \frac{V}{Z}\angle -\theta \text{ (with } \bar{V} = V\angle 0°) \tag{AI.18}$$

which is nothing but Ohm's law in phasor form. Also

$$\bar{I} = \bar{Y}\bar{V} \tag{AI.19}$$

where

$$\bar{Y} = \frac{1}{\bar{Z}} = \textit{admittance} \text{ (complex number)} \tag{AI.20a}$$

$$= G - jB \tag{AI.20b}$$

where G = circuit conductance

B = circuit susceptance

It easily follows that in this circuit

$$G = \frac{R}{R^2 + X_L^2} \tag{AI.21}$$

$$B = \frac{X_L}{R^2 + X_L^2} \tag{AI.22}$$

AI.5 PARALLEL *RL* CIRCUIT

Consider the parallel *RL* circuit of Fig. AI.10(a) excited by voltage $\bar{V}$. The two circuit elements are described by

$$\bar{I}_R = \frac{\bar{V}}{R} = G\bar{V}; \quad \bar{I}_L = \frac{\bar{V}}{jX_L} = -jB_L\bar{V}$$

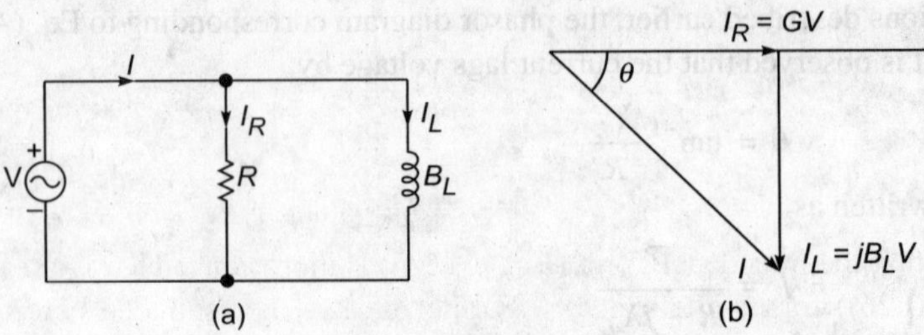

Fig. AI.10 *Parallel RL circuit*

Applying Kirchhoff's current law

$$\bar{I} = \bar{I}_R + \bar{I}_L$$

$$= (G - jB_L)\bar{V} = \bar{Y}\bar{V} \quad \text{(AI.23)}$$

where $\bar{Y} = G - jB_L$

The phasor diagram as per Eq. (AI.23) is drawn in Fig. AI.10(b) wherein

$$\theta = \tan^{-1}\frac{B_L}{G} = \text{angle by which current lags the applied voltage}$$

It is now easily recognized from Eq. (AI.20b) that the parallel equivalent of the series circuit of Fig. (AI.8) is as shown in Fig. (AI.11) wherein G and B are given by Eqs (AI.21) and (AI.22).

It is suggested that the reader find the equivalent series form for the parallel *RL* circuit of Fig. AI.10 by use of Eq. (AI.23).

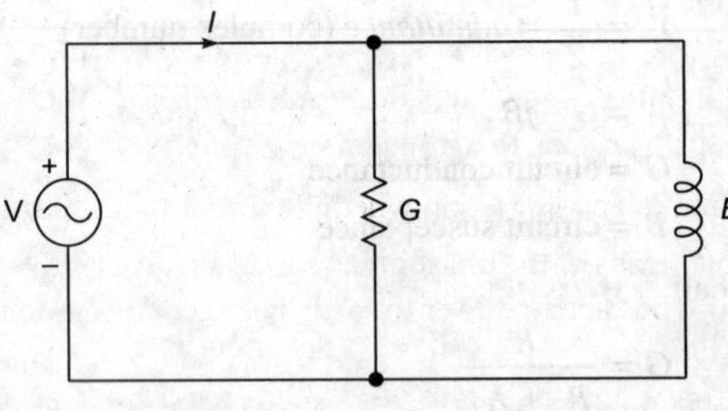

Fig. AI.11 *Parallel equivalent of the series RL circuit of Fig. AI.8*

AI.6 | IMPEDANCES/ADMITTANCES IN SERIES/PARALLEL

For impedances in series

$$\bar{Z} = \bar{Z}_1 + \bar{Z}_2 + \ldots \quad \text{(AI.24)}$$

or

$$\frac{1}{\bar{Y}} = \frac{1}{\bar{Y}_1} + \frac{1}{\bar{Y}_2} + \ldots \quad \text{(AI.25)}$$

For admittances in parallel

$$\bar{Y} = \bar{Y}_1 + \bar{Y}_2 + \dots \quad \text{(AI.26)}$$

or

$$\frac{1}{\bar{Z}} = \frac{1}{\bar{Z}_1} + \frac{1}{\bar{Z}_2} + \dots \quad \text{(AI.27)}$$

It is to be observed here that the impedance form (Eq. (AI.24)) is more convenient for series circuit and the admittance form (Eq. (AI.26)) for parallel circuits. Further, it must be reminded here that series/parallel combinations involve complex number computations.

AI.7 | POWER IN AC CIRCUIT (SINGLE-PHASE)

Consider Fig. AI.12 in which a single-phase source supplies a load of impedance $Z\angle\theta$. Then in instantaneous form

$$v = \sqrt{2}\, V \sin \omega t$$

$$i = \sqrt{2}\, I \sin (\omega t - \theta)$$

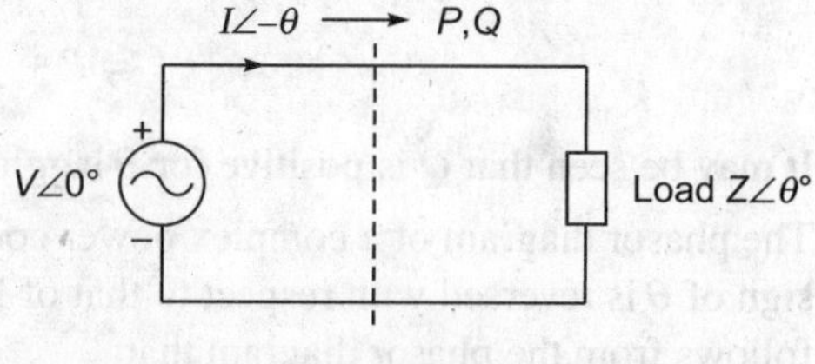

Fig. AI.12

The instantaneous power fed to the load is given by

$$p = vi = 2VI \sin \omega t \sin (\omega t - \theta)$$

$$= VI[\cos \theta - \cos (2\omega t - \theta)] \quad \text{(AI.28)}$$

It is easily seen that the instantaneous power has a constant component and an oscillating component of double frequency (2ω). The average power (*active*) being fed to the load over a complete cycle is

$$P = VI \cos \theta \text{ W} \quad \text{(AI.29)}$$

Large power can be expressed in units of kilowatts or megawatts (kW/MW). We define

$$\cos \theta = \textit{power factor} \text{ (pf)} \quad \text{(AI.30)}$$

and it could be lagging or leading depending upon whether current lags or leads the voltage.

Power can also be interpreted by means of the phasor diagram of Fig. AI.13(a) wherein the current component $I \cos \theta$ in phase with voltage supplies active power to load. The second component of current $I \sin \theta$ supplies *reactive* power to the load, i.e.

$$Q = VI \sin \theta \quad \text{VAR (VA reactive)} \quad \text{(AI.31)}$$

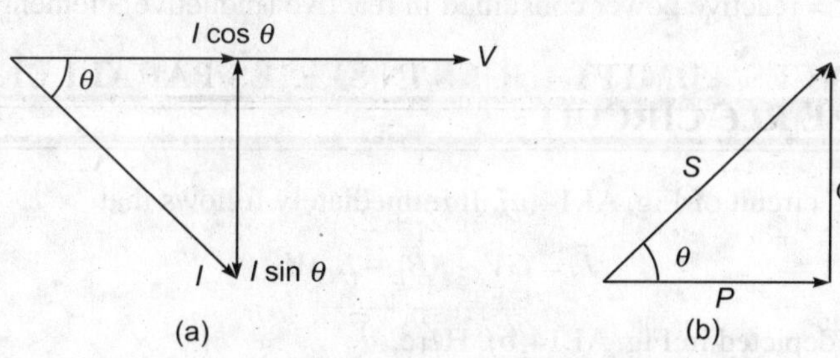

Fig. AI.13

which can also be measured in practical units of kVAR/MVAR (kilo volt-amp reactive/mega volt-amp reactive) for large power equipment. Reactive power will be taken as *positive* for *lagging* load and *negative* for *leading* load.

AI.8 | COMPLEX POWER

It is convenient to express power in complex form

$$\bar{S} = \bar{V}\bar{I}^* \tag{AI.32}$$

Let $\bar{V} = Ve^{j\phi};\ \bar{I} = Ie^{j(\phi-\theta)};\ \bar{I}$ lags $\bar{V}$ by θ

then

$$\begin{aligned}\bar{S} &= Ve^{j\phi} \cdot Ie^{-j(\phi-\theta)}\\ &= VIe^{j\theta}\\ &= VI\cos\theta + jVI\sin\theta \quad \text{(AI.33)}\\ &= P + jQ \quad \text{(AI.34)}\end{aligned}$$

It may be seen that Q is positive for θ lagging.

The phasor diagram of a complex power corresponding to Eq. (AI.34) is drawn in Fig. AI.13(b) wherein the sign of θ is reversed with respect to that of Fig. AI.13(a) because of conjugating of current in Eq. (AI.32). It follows from the phasor diagram that

$$\begin{aligned}S &= \sqrt{P^2 + Q^2} \quad \text{(AI.35)}\\ &= \text{VA delivered to load}\end{aligned}$$

$$\text{pf} = \cos\theta = \cos\tan^{-1}\frac{Q}{P};\quad \text{lagging for positive } Q, \text{ leading for negative } Q \tag{AI.36}$$

For the series *RL* circuit of Fig. AI.8

$$\begin{aligned}\bar{S} &= \bar{V}\bar{I}^*\\ &= \bar{Z}\bar{I}\bar{I}^*\\ &= (R + jX_L)I^2 = RI^2 + jX_LI^2 \quad \text{(AI.37)}\end{aligned}$$

Hence

$P = RI^2$ = real power consumed in resistive element

$Q = X_LI^2$ = reactive power consumed in reactive (inductive) element

AI.9 | PARALLEL *RLC* CIRCUIT

Consider the parallel *RLC* circuit of Fig. AI.14(a). It immediately follows that

$$\bar{I} = GV - j(B_L - B_C)V \tag{AI.38}$$

whose phasor diagram is depicted in Fig. AI.14(b). Here,

$$\theta = \tan^{-1}\left(\frac{B_L - B_C}{G}\right) \tag{AI.39}$$

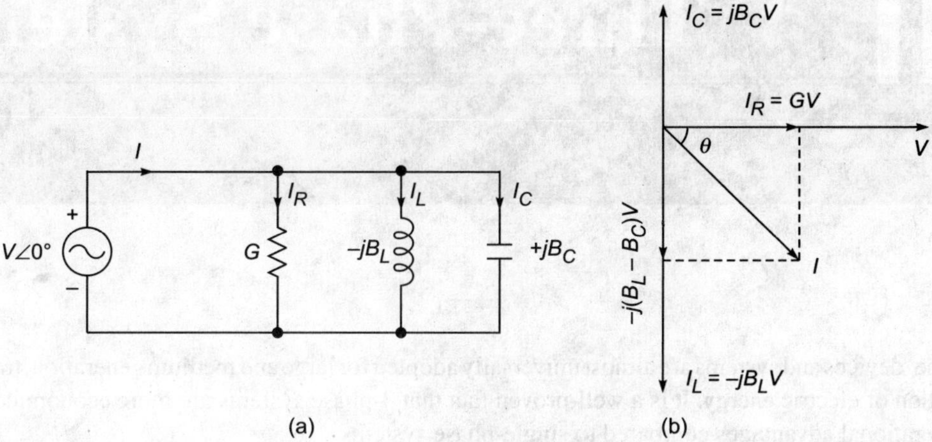

Fig. AI.14 *Parallel RLC circuit*

lagging for $B_L > B_C$ and leading for $B_L < B_C$. The power factor of the circuit is expressed as

$$\cos\theta = \cos\tan^{-1}\left(\frac{B_L - B_C}{G}\right) \tag{AI.40}$$

It is easily seen that for a given load (G and B_L), the power factor can be adjusted by B_C of the capacitance. The complex power consumed by the circuit is

$$\bar{S} = GV^2 + j(B_L - B_C)V^2 \tag{AI.41}$$

Thus with the real power fixed (GV^2), the reactive power can be controlled by adjusting the capacitance (B_C) which, in other words, means that the circuit power factor can be controlled.

APPENDIX II Three-Phase Systems

Three-phase* devices and systems are almost universally adopted for large and medium generation, transmission and utilization of electric energy. It is a well-proven fact that 3-phase systems are more economical and have definite operational advantages compared to single-phase systems.

AII.1 THREE-PHASE VOLTAGES AND CURRENTS

A set of three voltages/emfs which are equal in magnitude and have a relative progressive phase difference† of $2\pi/3(=120°)$ constitute *balanced* 3-phase voltages/emfs. The 3-phase voltages are otherwise said to be *unbalanced.* The same definition applies to a set of three currents. Figure AII.1 shows three windings (*phases*) whose voltages/emfs form a balanced set. In instantaneous form these can then be expressed as

$$e_{aa'} = \sqrt{2}\, E \sin \omega t \tag{AII.1a}$$

$$e_{bb'} = \sqrt{2}\, E \sin (\omega t - 120°) \tag{AII.1b}$$

$$e_{cc'} = \sqrt{2}\, E \sin (\omega t - 240°) \tag{AII.1c}$$

The waveforms of these voltages are drawn in Fig. AII.2. In the case being considered the voltage of phase *a* leads that of *b* and the voltage of phase *b* leads that of *c*; a fact which is remembered by saying that the 3-phase voltages have a *phase sequence abc.* The phase sequence could alternatively be *acb*, i.e. *a* lags *b* lags *c*.

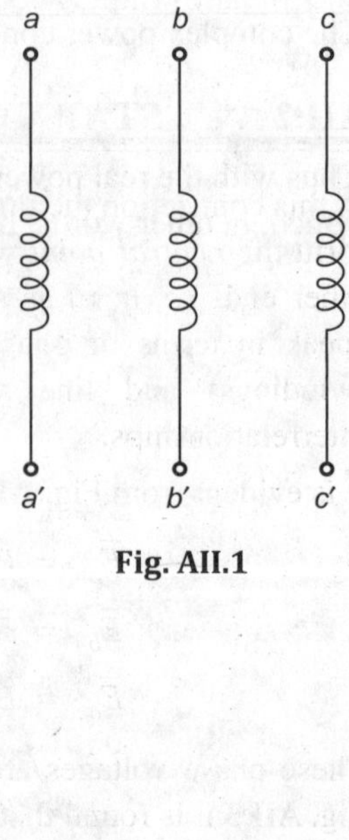

Fig. AII.1

In phasor form Eqs (AII.1a, b, c) can be written as

$$\bar{E}_{aa'} = E\angle 0° \tag{AII.2a}$$

$$\bar{E}_{bb'} = E\angle -120° \tag{AII.2b}$$

$$\bar{E}_{cc'} = E\angle -240° \tag{AII.2c}$$

*For the last decade interest is also being expressed in 6-phase systems.

†A balanced set of *q*-phase voltages have a relative phase difference of

$$2\pi/q(q \neq 2) \text{ and } \pi/2 \text{ for } q = 2.$$

Consider now the instantaneous voltage sum

$$e_{aa'} + e_{bb'} + e_{cc'} = \sqrt{2}E[\sin \omega t + \sin(\omega t - 120°) + \sin(\omega t - 240°)]$$

or $$e_{aa'} + e_{bb'} + e_{cc'} = 0 \tag{AII.3}$$

It is therefore seen that the sum of balanced 3-phase voltages (or currents) is zero at all time. This is also evident from the waveforms of Fig. AII.2. In phasor form this implies

$$\bar{E}_{aa'} + \bar{E}_{bb'} + \bar{E}_{cc'} = 0 \tag{AII.4}$$

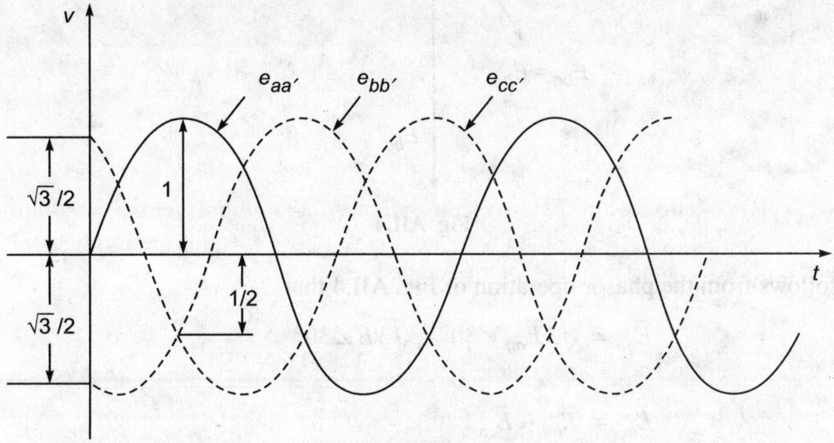

Fig. AII.2

The 3-phase windings of Fig. AII.1 can be connected in two ways illustrated below.

AII.2 | STAR CONNECTION

In this connection the three similar polarity ends (say a', b', c') of the phasor windings are joined together to form the *neutral point* while three lines are taken from the other ends (a, b, c) as shown in Fig. AII.3. One can now speak in terms of phase voltages and currents in phases (windings) and line voltages and currents and their interrelationships.

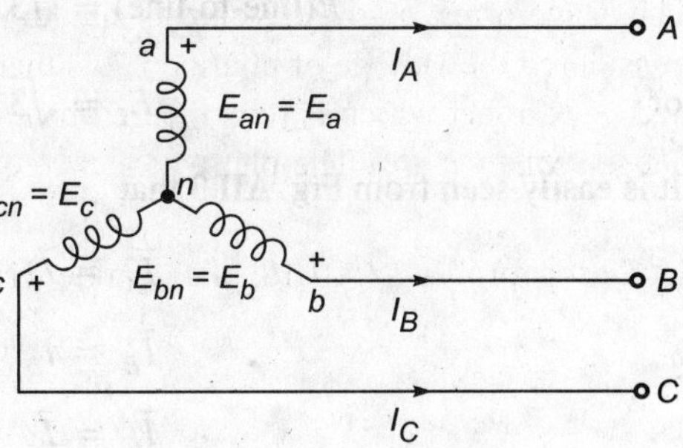

Fig. AII.3 *Star connection*

It is evident from Fig. AII.3 that

$$\bar{E}_{an} = \bar{E}_a = \bar{E}_{aa'} = E\angle 0° \tag{AII.5a}$$

$$\bar{E}_{bn} = \bar{E}_b = \bar{E}_{bb'} = E\angle -120° \tag{AII.5b}$$

$$\bar{E}_{cn} = \bar{E}_c = \bar{E}_{cc'} = E\angle -240° \tag{AII.5c}$$

These phase voltages are indicated in the phasor diagram of Fig. AII.4. From the connection diagram of Fig. AII.3 it is found that the voltage between a and b (or A and B) is

$$\bar{E}_{ab} = \bar{E}_{an} - \bar{E}_{bn}$$

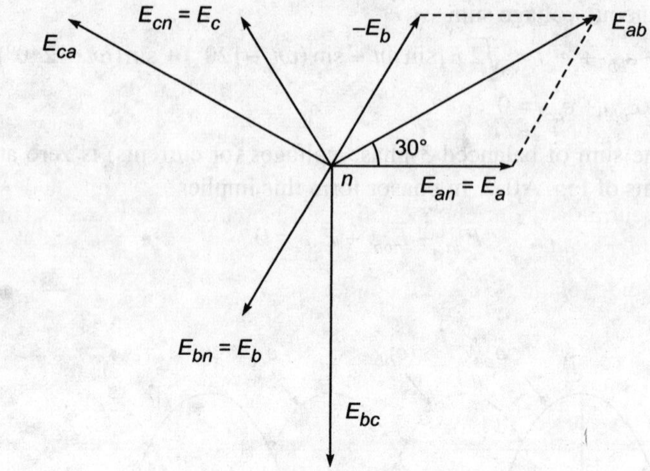

Fig. AII.4

It immediately follows from the phasor operation of Fig. AII.4 that

$$\bar{E}_{ab} = \sqrt{3}\,\bar{E}_{an} \angle 30° = \sqrt{3}\,E \angle 30° \tag{AII.6a}$$

Similarly,

$$\bar{E}_{bc} = \bar{E}_{bn} - \bar{E}_{cn}$$
$$= \sqrt{3}\,\bar{E}_{bn} \angle 30° = \sqrt{3}\,E \angle -90° \tag{AII.6b}$$

$$\bar{E}_{ca} = \bar{E}_{cn} - \bar{E}_{an}$$
$$= \sqrt{3}\,\bar{E}_{cn} \angle 30° = \sqrt{3}\,E \angle 150° \tag{AII.6c}$$

It is seen from the complete phasor diagram of Fig. AII.4 that for balanced phase voltages, line voltages also form a balanced set advanced in phase by 30° from the phase voltages. In terms of magnitudes

$$E(\text{line-to-line}) = \sqrt{3}\,E\,(\text{phase})$$

or

$$E_L = \sqrt{3}\,E_P \tag{AII.7}$$

It is easily seen from Fig. AII.3 that

$$\bar{I}_A = \bar{I}_a \tag{AII.8a}$$

$$\bar{I}_B = \bar{I}_b \tag{AII.8b}$$

$$\bar{I}_C = \bar{I}_c \tag{AII.8c}$$

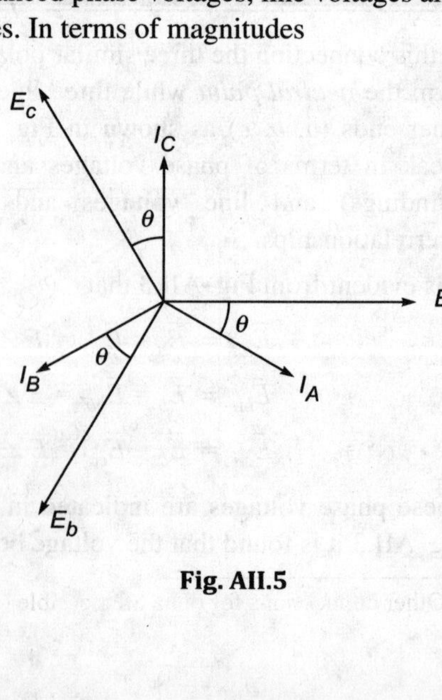

Fig. AII.5

Thus if the phase currents are a balanced set, the line currents also constitute a balanced set as shown in Fig. AII.5. In a balanced system, θ is the phase angle difference between the voltages and currents of corresponding phases. In terms of magnitudes

$$I(\text{line}) = I(\text{phase})$$

or
$$I_L = I_P \tag{AII.9}$$

Being a balanced set

$$\bar{I}_A + \bar{I}_B + \bar{I}_C = 0 \tag{AII.10}$$

However, with isolated neutral in a star connection, even when currents are unbalanced, Eq. (AII.10) must hold.

To summarize, the magnitude relations in a star connection:

$$E_L = \sqrt{3}\, E_P; \qquad I_L = I_P \tag{AII.11}$$

AII.3 | DELTA CONNECTION

The phase windings of Fig. AII.1 can be connected[‡] in delta by joining *ac′*, *ba′* and *cb′* and taking out three lines from the nodes so formed. It is immediately seen that the voltage round the closed loop formed by the delta connection is

$\bar{E}_{aa'} + \bar{E}_{bb'} + \bar{E}_{cc'} = 0$ (being a balanced set of phase voltages and hence delta does not constitute a short circuit).

In delta connection, phase and line currents are related as shift (Fig. AII.6)

$$\bar{I}_A = \bar{I}_{ba} - \bar{I}_{ac} = \sqrt{3}\,\bar{I}_{ba} \angle -30° \tag{AII.12a}$$

$$\bar{I}_B = \bar{I}_{cb} - \bar{I}_{ba} = \sqrt{3}\,\bar{I}_{cb} \angle -30° \tag{AII.12b}$$

$$\bar{I}_C = \bar{I}_{ac} - \bar{I}_{cb} = \sqrt{3}\,\bar{I}_{ac} \angle -30° \tag{AII.12c}$$

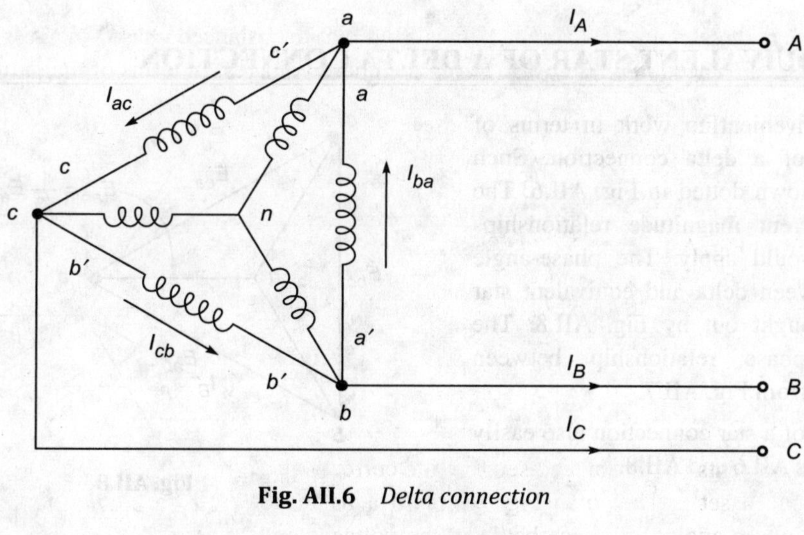

Fig. AII.6 *Delta connection*

[‡]Other connections for delta are possible (see Sec. 3.13).

These relationships are also shown in the phasor diagram of Fig. AII.7. Obviously when phase currents are balanced, so are line currents and vice versa. In terms of magnitudes

$$I_L = \sqrt{3}\, I_P \tag{AII.13}$$

It is easily concluded that

$$E_L = E_P \tag{AII.14}$$

The reader should compare these magnitude relationships (Eqs (AII.13) and (AII.14)) with those of star connection (Eq. (AII.11)).

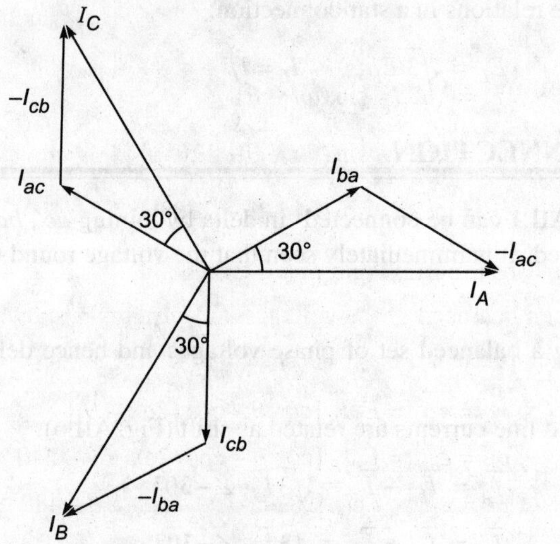

Fig. AII.7

AII.4 | EQUIVALENT STAR OF A DELTA CONNECTION

It is usually convenient to work in terms of equivalent star of a delta connection. Such equivalence is shown dotted in Fig. AII.6. The voltage and current magnitude relationships already given would apply. The phase-angle relationship between delta and equivalent star voltages are brought out by Fig. AII.8. The corresponding phase relationship between currents follows from Fig. AII.7.

Equivalent delta of a star connection also easily follows from Figs AII.6 and AII.8.

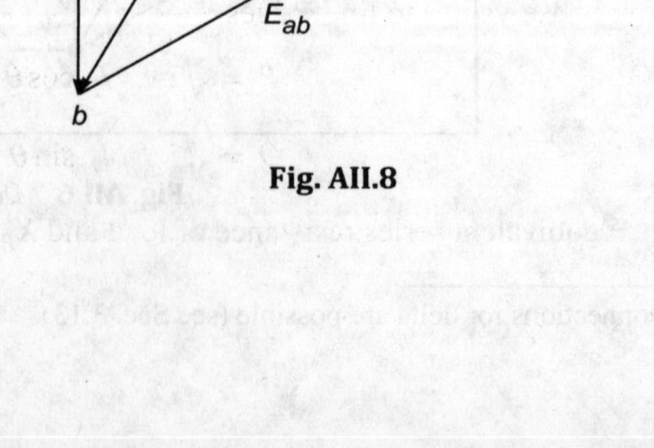

Fig. AII.8

AII.5 POWER

Phase voltages and currents in a balanced 3-phase circuit (star or delta) can be written in the instantaneous form as

$$e_a = \sqrt{2}\, E_P \sin \omega t \quad \text{(AII.15a)}$$

$$e_b = \sqrt{2}\, E_P \sin (\omega t - 120°) \quad \text{(AII.15b)}$$

$$e_c = \sqrt{2}\, E_P \sin (\omega t - 240°) \quad \text{(AII.15c)}$$

and

$$i_a = \sqrt{2}\, I_P \sin (\omega t - \theta) \quad \text{(AII.16a)}$$

$$i_b = \sqrt{2}\, I_P \sin (\omega t - \theta - 120°) \quad \text{(AII.16b)}$$

$$i_c = \sqrt{2}\, I_P \sin (\omega t - \theta - 240°) \quad \text{(AII.16c)}$$

where θ = phase angle between phase current and phase voltage.

The phasor relationship between balanced phase voltages and currents is shown in Fig. AII.5.

The instantaneous power in each phase is

$$p_a = e_a i_a = E_P I_P [\cos \theta - \cos (2\omega t - \theta)] \quad \text{(AII.17a)}$$

$$p_b = e_b i_b = E_P I_P [\cos \theta - \cos (2\omega t - \theta - 240°)] \quad \text{(AII.17b)}$$

$$p_c = e_c i_c = E_P I_P [\cos \theta - \cos (2\omega t - \theta - 480°)] \quad \text{(AII.17c)}$$

The total instantaneous power is

$$p = p_a + p_b + p_c = 3 E_P I_P \cos \theta \quad \text{(AII.18)}$$

Notice that the sum of the three cosine terms, which have a progressive phase difference of 120°, is zero. From Eq. (AII.18) it is found that the total instantaneous power in a balanced 3-phase system is constant and is equal to three times the average power per phase. This observation is in contrast to power in a single-phase system (Eq. (AI.28)) which has a double-frequency oscillating component. Constancy of instantaneous power in a balanced 3-phase system affords the important advantage of uniform torque in rotating machines with the absence of pulsating torque which would be present in single-phase machines. This is one of the many but important reasons in universal adoption of 3-phase system except in low power applications.

Thus in star/delta balanced system

$$P = 3\, E_P I_P \cos \theta \quad \text{(AII.19)}$$

Converting phase values to line values

$$P = \sqrt{3}\, E_L I_L \cos \theta = 3 I_p^2 R_p \quad \text{(AII.20)}$$

Also

$$Q = \sqrt{3}\, E_L I_L \sin \theta = 3 I_p^2 X_p \quad \text{(AII.21)}$$

where R_P = equivalent series resistance of load and X_P = equivalent series reactance of the load.

The phase angle of the load is given by

$$\theta = \tan^{-1}\frac{X_P}{R_P} = \cos^{-1}\frac{R_P}{Z_P} = \tan^{-1}\frac{Q}{P} \tag{AII.22}$$

In a balanced 3-phase system each phase has the same power factor.

AII.6 THREE-PHASE CIRCUITS

Figure AII.9 shows a typical balanced 3-phase system—source, transmission line and load. The source and load are assumed to be star-connected though either or both of them could be delta-connected. The two neutrals are isolated; even if these are connected (shown by dotted line) the current in the neutral connection will be zero by virtue of the balanced currents $(\bar{I}_n = \bar{I}_a + \bar{I}_b + \bar{I}_c = 0)$. For the reference phase a (loop shown)

$$\bar{E}_a = (\bar{Z}_{TL} + \bar{Z}_L)\bar{I}_a \tag{AII.23}$$

Because of zero current in the neutral connection, if present, its impedance would not appear in Eq. (AII.23). The currents and voltages in other phases would have the same magnitude as given byEq. (AII.23) but would have a progressive shift in phase of 120°. Equation (AII.23) corresponds to the single-phase circuit shown in Fig. AII.10 whose solution completely determines the solution of the 3-phase circuit of Fig. AII.9. If the load or any part of the system were unbalanced, one must proceed on a 3-phase basis by writing three loop equations.

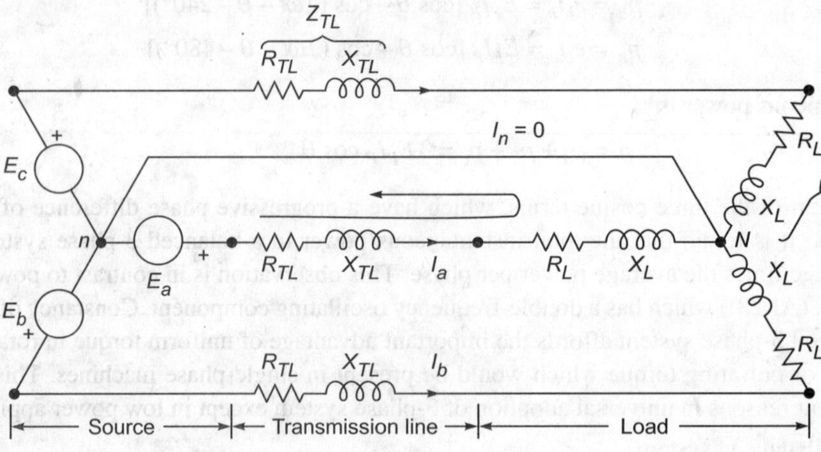

Fig. AII.9 *Balanced 3-phase system*

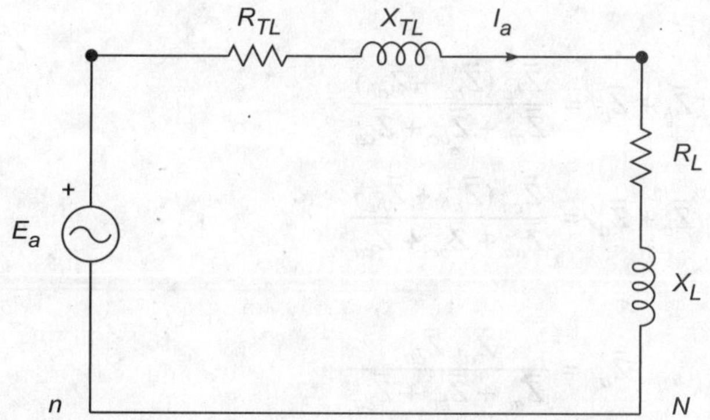

Fig. AII.10

Analysis of a balanced 3-phase system is thus extremely simple and proceeds on a single-phase basis. Where deltas are involved, one can either proceed directly or convert these to equivalent star. This procedure will be illustrated in Ex. AII.1.

AII.7 STAR-DELTA CONVERSION

Figure AII.11 shows three delta-connected impedances whose equivalent star, shown dotted, is required to be found, by equating the impedances between all three 2-terminal pairs by turn. Between terminals *a* and *b*.

$$\bar{Z}_a + \bar{Z}_b = \bar{Z}_{ab} \,||\, (\bar{Z}_{bc} + \bar{Z}_{ca})$$

$$= \frac{\bar{Z}_{ab}\,(\bar{Z}_{bc} + \bar{Z}_{ca})}{\bar{Z}_{ab} + \bar{Z}_{bc} + \bar{Z}_{ca}} \qquad \text{(AII.24)}$$

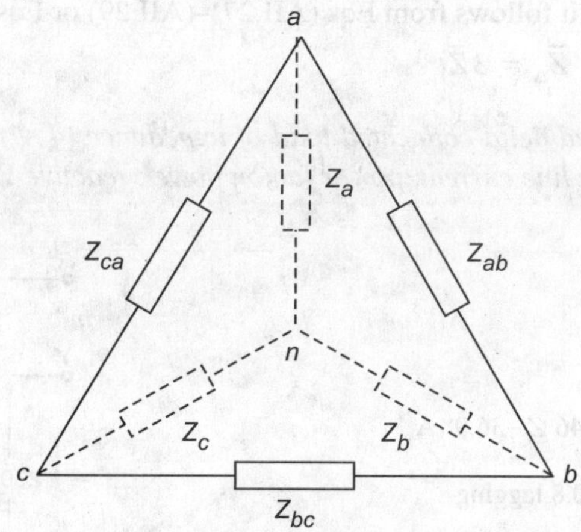

Fig. AII.11

Similarly

$$\bar{Z}_b + \bar{Z}_c = \frac{\bar{Z}_{bc}(\bar{Z}_{ca} + \bar{Z}_{ab})}{\bar{Z}_{ab} + \bar{Z}_{bc} + \bar{Z}_{ca}} \tag{AII.25}$$

$$\bar{Z}_c + \bar{Z}_a = \frac{\bar{Z}_{ca}(\bar{Z}_{ab} + \bar{Z}_{bc})}{\bar{Z}_{ab} + \bar{Z}_{bc} + \bar{Z}_{ca}} \tag{AII.26}$$

Solving

$$\bar{Z}_a = \frac{\bar{Z}_{ab}\,\bar{Z}_{ca}}{\bar{Z}_{ab} + \bar{Z}_{bc} + \bar{Z}_{ca}} \tag{AII.27}$$

$$\bar{Z}_b = \frac{\bar{Z}_{ab}\,\bar{Z}_{bc}}{\bar{Z}_{ab} + \bar{Z}_{bc} + \bar{Z}_{ac}} \tag{AII.28}$$

$$\bar{Z}_c = \frac{\bar{Z}_{ca} + \bar{Z}_{bc}}{\bar{Z}_{ab} + \bar{Z}_{bc} + \bar{Z}_{ac}} \tag{AII.29}$$

Solving Eqs (AII.24)–(AII.26), equivalent delta impedance of a star can be obtained. The results are reproduced below:

$$\bar{Z}_{ab} = \frac{\bar{Z}_a\bar{Z}_b + \bar{Z}_b\bar{Z}_c + \bar{Z}_c\bar{Z}_a}{\bar{Z}_c} \tag{AII.30}$$

$$\bar{Z}_{bc} = \frac{\bar{Z}_a\bar{Z}_b + \bar{Z}_b\bar{Z}_c + \bar{Z}_c\bar{Z}_a}{\bar{Z}_a} \tag{AII.31}$$

$$\bar{Z}_{ca} = \frac{\bar{Z}_a\bar{Z}_b + \bar{Z}_b\bar{Z}_c + \bar{Z}_c\bar{Z}_a}{\bar{Z}_b} \tag{AII.32}$$

If the impedances are *balanced*, it follows from Eqs (AII.27)–(AII.29) or Eqs (AII.30)–(AII.32) that

$$\bar{Z}_\Delta = 3\bar{Z}_Y \tag{AII.33}$$

Example AII.1 *A balanced delta-connected load of impedance (4 + j3) Ω/phase is connected to a 3-phase 230 V supply. Find the line current, power factor, power, reactive VA and total VA.*

Solution

With reference to Fig. AII.2

$$V_P = V_L = 230 \text{ V}$$

$$I_P = \frac{230\angle 0^\circ}{4 + j3} = 46\angle -36.9^\circ \text{ A}$$

$$pf = \cos 36.9^\circ = 0.8 \text{ lagging}$$

$$I_L = \sqrt{3}\,I_P = \sqrt{3} \times 46 = 79.6 \text{ A}$$

$$P = \sqrt{3}\,V_L I_L \cos\theta$$

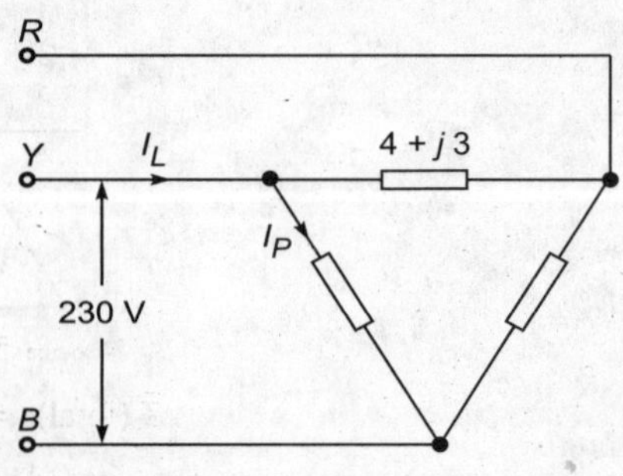

Fig. AII.12

$$= \sqrt{3} \times 230 \times 79.6 \times 0.8 = 25.36 \text{ kW}$$

$$Q = \sqrt{3}\, V_L I_L \sin\theta = \sqrt{3} \times 230 \times 79.6 \times 0.6 = +19.02 \text{ kVAR}$$

$$S = \sqrt{P^2 + Q^2} = 31.7 \text{ kVA}$$

Solution by equivalent star

With reference to Fig. AII.13

$$\bar{Z}_Y = \frac{1}{3}(4 + j3)$$

$$V_P = 230/\sqrt{3} = 132.8 \text{ V}$$

$$\bar{I}_L = \bar{I}_P = \frac{132.8\angle 0^\circ}{\frac{1}{3}(4 + j3)} = 79.6\angle -36.9^\circ$$

All other results now follow.

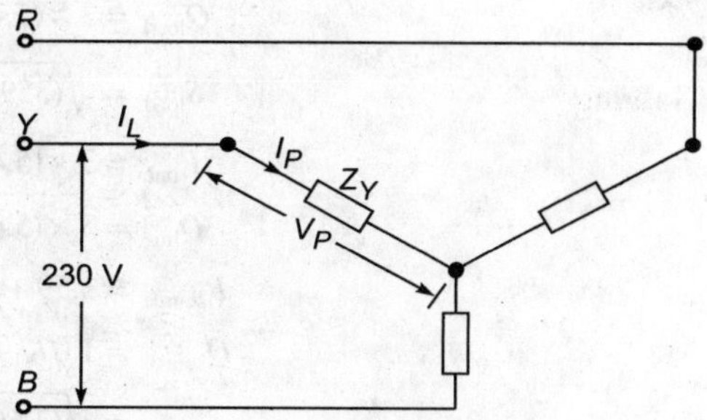

Fig. AII.13

Example AII.2 *A 3-phase, 400 V source (terminal voltage assumed constant independent of load) supplies a load with an equivalent-star impedance of (40 + j 12) Ω/phase through a transmission line impedance (0.2 + j 0.8) Ω/line. Compute (a) the line current; (b) the load voltage; (c) the power, reactive power and VA consumed by the load; (d) the power and reactive power loss in the line; and (e) the power, reactive power and VA supplied by the source.*

Solution

Figure AII.14 is the per phase circuit of the system (star basis)

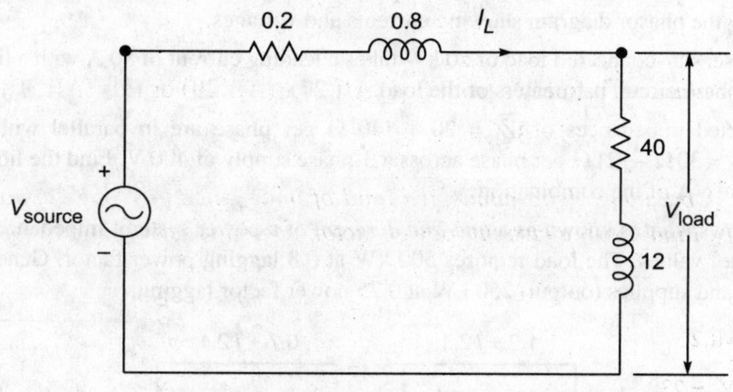

Fig. AII.14

$$V_{\text{source}} = 400/\sqrt{3} = 231 \text{ V}$$

$$\bar{Z}\text{ (total)} = (0.2 + j0.8) + (40 + j12) = 40.2 + j12.8 = 42.2\angle 17.7^\circ$$

$$I_L = \frac{231}{42.2} = 5.47 \text{ A}$$

$$\bar{Z}_{load} = 40 + j12 = 41.8 \angle 16.7°$$

$$V_{load} = 5.47 \times 41.8 = 228.6 \text{ V} \quad \text{(phase-to-neutral)}$$

$$= 396 \text{ V} \quad \text{(line-to-line)}$$

$$P_{load} = 3 \times (5.47)^2 \times 40 = 3.59 \text{ kW}$$

$$Q_{load} = 3 \times (5.47)^2 \times 12 = 1.077 \text{ kVAR}$$

$$S_{load} = \sqrt{(3590)^2 + (1077)^2} = 3.748 \text{ kVA}$$

$$P_{line} = 3 \times (5.47)^2 \times 0.2 = 18 \text{ W}$$

$$Q_{line} = 3 \times (5.47) \times 0.8 = 72 \text{ VAR}$$

$$P_{source} = 3.59 + 0.018 = 3.608 \text{ kW}$$

$$Q_{source} = 1.077 + 0.072 = 1.149 \text{ kVAR}$$

$$S_{source} = \sqrt{(3.608)^2 + (1.149)^2} = 3.786 \text{ kVA}$$

PRACTICE PROBLEMS

AII.1 A balanced star-connected load is supplied from a symmetrical 3-phase, 400 V system. The current in each phase is 30 A and lags 30° behind the phase voltage. Find (a) the phase voltage, and (b) power and reactive VA drawn by the load. Draw the phasor diagram.

AII.2 A balanced delta-connected load of $15 + j12$ Ω/phase is connected to a 3-phase. 230 V supply. Find the line current, power factor, power, reactive VA and total VA.

AII.3 A symmetrical 3-phase, 400 V system supplies a balanced delta-connected load. The current in each delta branch is 25 A and its phase angle is 30° lagging (with respect to voltage across the branch). Find the line current and the total power. Draw the phasor diagram showing currents and voltages.

AII.4 A balanced 3-phase, star-connected load of 20 kW takes a leading current of 50 A with a line voltage of 400 V, 50 Hz. Find the per phase circuit parameters of the load.

AII.5 Three-star-connected impedances of $\bar{Z}_1 = 20 + j40$ Ω per phase are in parallel with three delta-connected impedances of $\bar{Z}_2 = 30\,\Omega - j21\,\Omega$ per phase across a 3-phase supply of 400 V. Find the line current, power factor, power and reactive VA of the combination.

AII.6 Figure APII.6 shows what is known as a *one-line diagram* of a power system. Impedances indicated on the lines are per line (phase) values. The load requires 500 kW at 0.8 lagging power factor. Generator G_1 has a terminal voltage of 10 kV and supplies (output) 250 kW at 0.75 power factor lagging.

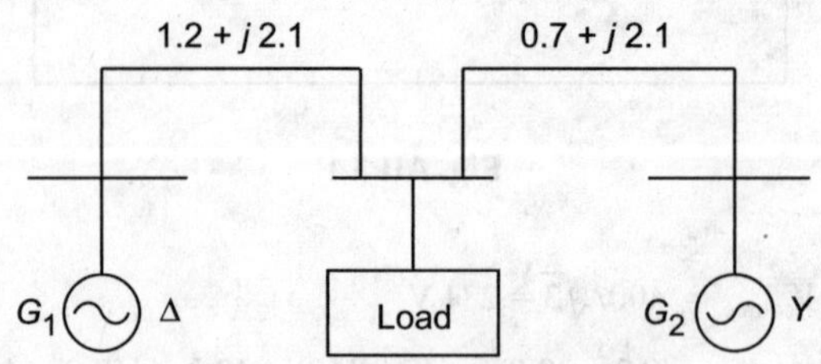

Fig. APII.6

Find the load voltage and terminal voltage and power and reactive VA supplied by G_2.

APPENDIX III

Special Topics in Transformers

AIII.1 | FREQUENCY RESPONSE

In electronic amplifiers, transformers are used at the output stage with frequency being variable over a band. In such variable-frequency applications, the frequency response of a transformer is of great significance. Frequency response is the V_2/V_1 behaviour for fixed V_1 and load but variable frequency. A flat frequency response over the frequency band of interest is most desirable. The corresponding variation of the phase angle of the load voltage (V_2) with respect to the source voltage (V_1) is called the phase response. A small phase angle is acceptable.

The complete frequency range can be divided into three parts—low frequency (LF) range, intermediate frequency (IF) range and high frequency (HF) range. The transformer is so designed that the equivalent circuit of Fig. 3.14(c) is applicable over the IF range. Here the magnetizing reactance of the transformer is large and does not affect its voltage-ratio behaviour and the series reactances of the windings are small enough so that V_2/V_1 remains substantially constant as shown in Fig. AIII.1. In the LF range the transformer draws large magnetizing current causing heavy voltage drop in the primary winding such that V_2/V_1 reduces continuously with decreasing frequency. The output voltage V_2 is of course zero for dc ($f = 0$). In the HF range the effect of inter-turn, turn-to-ground and stray capacitances can no longer be neglected.

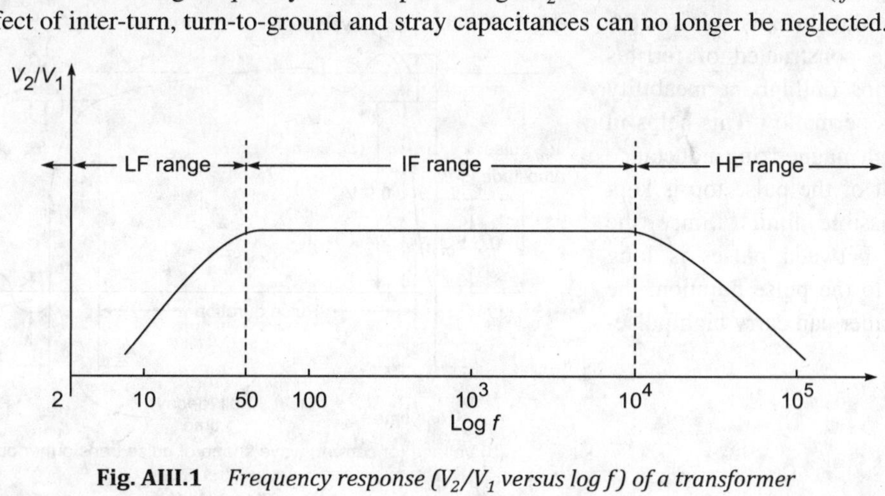

Fig. AIII.1 *Frequency response (V_2/V_1 versus log f) of a transformer*

The combined effects of increasing series inductive reactance and decreasing shunt capacitive reactance with increasing frequency cause the output voltage to drop sharply. The complete frequency response of a transformer on a logarithmic frequency scale is illustrated in Fig. AIII.1.

AIII.2 | PULSE TRANSFORMERS

Transformers used in pulse or digital circuits are known as pulse transformers. These normally operate from a discontinuous source of excitation because of which the distributed capacitance of the transformer winding must be taken into account.

For a pulse transformer, the average power is normally very low, but the peak power is quite large. However, recently, medium- and low-peak power pulse transformers have been developed for many electronic and control applications, such as in the gate firing circuits of SCRs. It is important that the transformer reproduces the input pulse (normally rectangular) as accurately as possible at its secondary terminals. Figure AIII.2(a) depicts a square wave input pulse. The pulse width varies from a fraction of a microsecond to about 25 μs and quite sometime will elapse before the next pulse appears. To find the output waveform, a transient analysis needs to be resorted to. Figure AIII.2(b) shows the corresponding output pulse wherein various pulse transformer parameters are defined.

The analysis is carried out normally by solving the differential equations in a piecewise manner using the equivalent circuit of a transformer including stray capacitance. Owing to the stray capacitance, there will usually be damped oscillations. Leakage inductance is kept to a minimum value if the shortest rise time is desired.

Pulse transformers are quite small in size and have comparatively few turns so that the leakage inductance is minimized. The cores are constructed of ferrites or wound strips of high permeability alloys such as permalloy. This helps in achieving a high magnetizing inductance so that the tilt of the pulse top is kept within permissible limits. Since the time interval between pulses is long as compared to the pulse duration, the pulse transformer can carry high pulse-power levels.

(a) Rectangular pulse input voltage (or current)

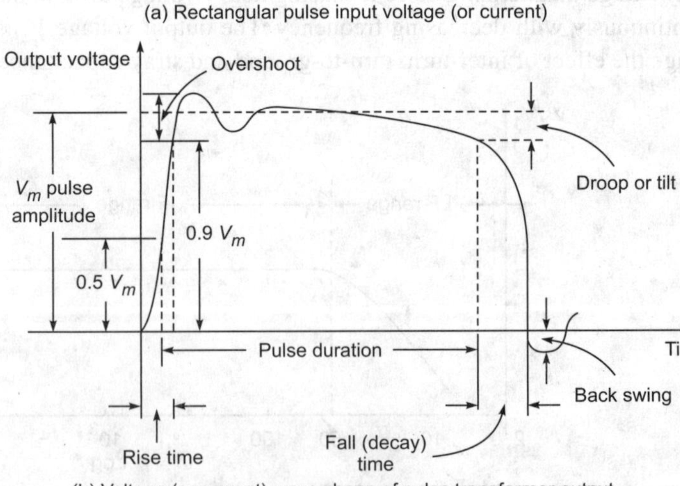

(b) Voltage (or current) wave shape of pulse transformer output

Fig. AIII.2

APPENDIX IV Cross-Field Machines

At the output stage of a control system, a low power electrical control signal is required to control large electric power necessitating large power amplification. Presently, SCR power amplifiers serve this purpose excellently. Prior to the advent of SCRs, large power amplification with low associated time constants were devised by means of a family of dc machines called cross-field machines. Though an ordinary dc generator is a power amplifier—a small field power controls large power at armature terminals—their power amplification is not sufficiently high and their response is sluggish. Cross-field machines which overcome both these drawbacks of dc generators have windings on *d*- and *q-axes* of the stator and brush pairs on the rotor forming both *q*- and *d-axes* mmf directions fully utilizing the potentiality of the dc machine. Cross-field machines have two brush sets per pair of poles. Thereby they use the armature mmf to provide most of the excitation and achieve high power gains. In fact, cross-field machines behave like two dc generators in cascade accommodated in a single machine frame. There are several types of cross-field generators—metadyne, amplidyne and regulex. Here metadyne will be studied first and then the most commonly used form, the amplidyne, will be dealt with.

An ordinary dc generator is modified into a metadyne by providing an additional pair of brushes in the *d-axis* as shown in the schematic diagram of Fig. AIV.1. The *q-axis* brushes are shorted, while the output is obtained from the *d*-axis brushes. Control field winding mmf, F_f induces the *q-axis* emf which causes the short-circuit *q*-axis current I_q to flow. The *q-axis* mmf F_q thus established induces the *d*-axis emf causing the *d*-axis current I_d, to flow through the load. The *d*-axis reaction mmf F_d, by Lenz's law, opposes the cause F_f. For a given F_f and load resistance, the system settles down to certain steady values of I_q and I_d. Any increase in load (decrease of load resistance) increases F_d which neutralizes more of F_f effectively reducing *d*-axis induced emf preventing the load current from increasing. Thus for a given F_f (control excitation), the output current of the metadyne remains substantially constant over a wide range of load variation, i.e. metadyne behaves as a constant current generator.

To convert a constant-current metadyne to *constant-voltage amplidyne,* a more useful machine, the *d*-axis mmf must be cancelled by means of a *compensating winding* located in the *d*-axis on the stator as shown in the schematic diagram of Fig. AIV.2 The *d*-axis output current flowing through the compensating winding develops an mmf, F_c, in opposition to that of the *d*-axis armature mmf. The negative feedback effect of the load current is therefore minimized. The *d*-axis flux is now mainly dependent on the field winding current, giving it wider control over the net *d*-axis flux. Because of the difference in the manner of space distribution of F_d and F_c, exact neutralization of F_d is never possible.

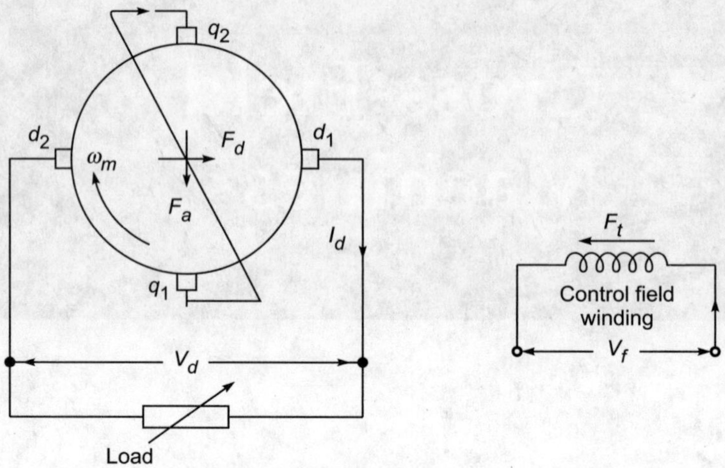

Fig. AIV.1 *Schematic diagram of a metadyne generator*

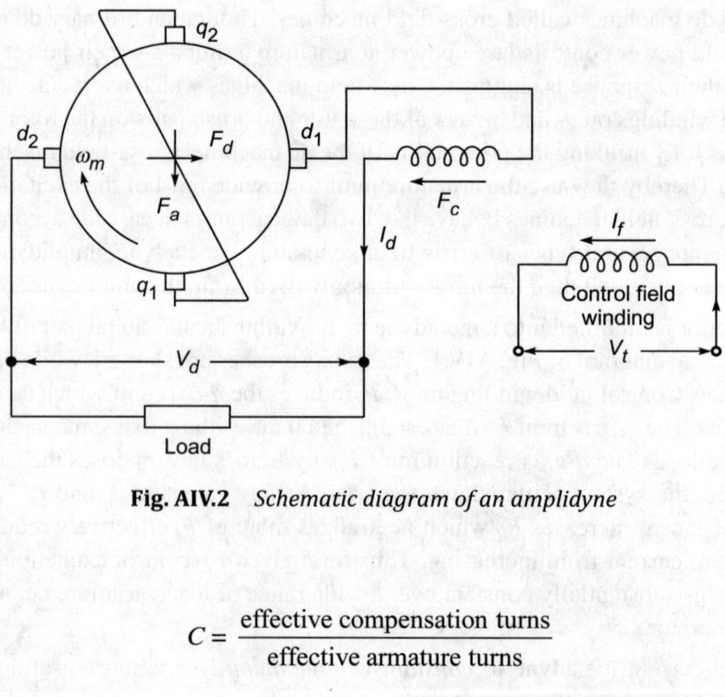

Fig. AIV.2 *Schematic diagram of an amplidyne*

By definition

$$C = \frac{\text{effective compensation turns}}{\text{effective armature turns}}$$

The effect of compensating factor C on the load characteristic of machine is depicted in Fig. AIV.3. For metadyne, $C = 0$ and for an amplidyne, $C = 1$, giving an amplidyne a characteristic like that of a separately excited generator. It is clear from Fig. AIV.3 that a metadyne should never be operated on open-circuit and an amplidyne on short-circuit.

To create a given F_q with smaller I_q, a series connected q-axis winding is usually provided on the stator of an amplidyne. The d- and q-axes currents flowing on the armature of an amplidyne tend to cancel each other considerably increasing the amplidyne rating over that of a dc generator for the same frame size. This fact also results in large power amplification ($V_d I_d / V_f I_f$) in the range of 10^5.

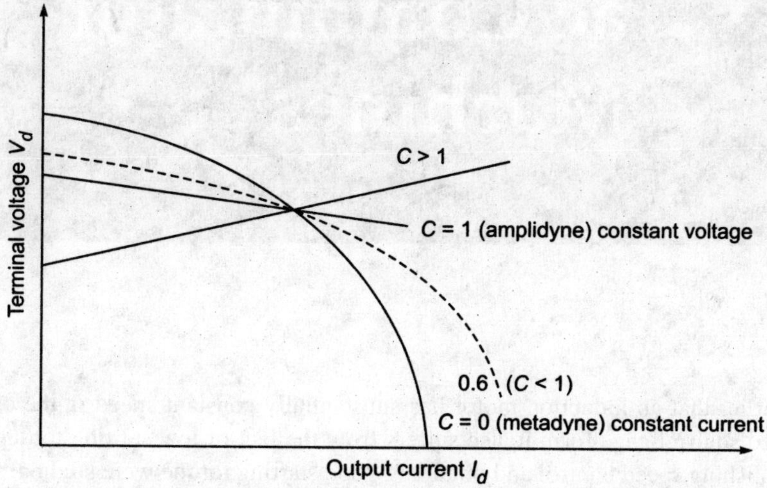

Fig. AIV.3 *External characteristics of a cross-field generator*

Amplidynes are used to provide controlled power amplification in a variety of control systems, e.g. as the voltage regulator in the excitation systems of large ac generators to introduce a buck or boost voltage in series with the field winding of the main exciter. An amplidyne may be used as a generator in the Ward-Leonard speed or position control system. In modern control systems these functions of an amplidyne have been taken over by SCR circuitry.

APPENDIX V

ac Commutator Machines

It has been seen earlier that an induction motor has substantially constant speed in the operating range. In its cheap and rugged squirrel-cage form it also suffers from the fact of low starting torque. In Ch. 9 several methods of accomplishing speed control and obtaining good starting torque were studied. Another answer to these problems of the induction machine is the ac commutator machine. Because of the inherent problems of ac commutation, ac commutator motors have been manufactured in medium sizes only. With the advent of thyristors, the frequency-converting action of the commutator has now been taken over by SCR circuitry. Though ac commutator machines can still be found in older installations, they have for all practical purposes become a part of the history of electrical machines.

The underlying principles and basic constructional features of two once commonly used types of ac commutator machines—the Repulsion motor (single-phase) and Schrage motor (three-phase)—are described below.

AV.1 REPULSION MOTORS

The repulsion motor is like a dc series motor with the rotor energized inductively. The stator slots house a single-phase exciting winding similar to the main winding of a single-phase induction motor. The rotor carries an ordinary distributed dc winding connected to the commutator at one end. The brushes are short-circuited to provide a closed current path as shown in Fig. AV.1.

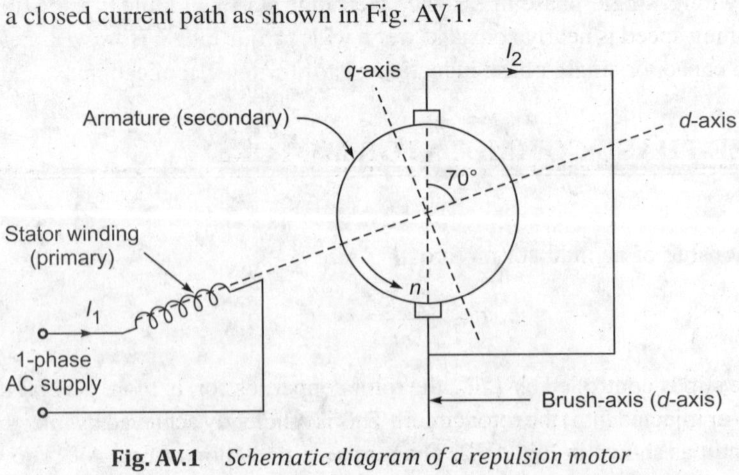

Fig. AV.1 *Schematic diagram of a repulsion motor*

If the brushes are located in the *d*-axis, the transformer emf in the effective armature coil will be maximum and so would be the short-circuit current I_2; but no torque will be produced as stator and rotor fields are aligned. If the brushes are placed in the *q*-axis, the transformer emf induced in the armature would add up to zero and so current $I_2 = 0$ and hence no torque is produced. As it is essential that both an armature current should flow and an angular displacement must exist between the two fields, the brushes are placed in an intermediate position making a large angle of about 70° with the *d*-axis. In a futile attempt to reduce copper losses, the rotor runs in a direction to orient itself with the *q*-axis, i.e. the direction of rotation is counter-clockwise for brushes displaced from the *d*-axis as shown in Fig. AV.1. This appears like repulsion between the two fields and hence the name. The direction of rotation will reverse if the brushes are displaced on the other side of the *d*-axis.

The unmodified repulsion motor has speed-torque characteristic similar to that of an ac series motor. It has a large starting torque and a tendency to overspeed at light load.

Repulsion-Induction Motor

The repulsion principle is often used merely to provide a high starting torque at a relatively low starting current. When the machine has accelerated close to the operating speed, it may be converted to a single-phase induction motor to provide essentially constant speed running characteristics. This is achieved if, in addition to an outer commutator winding, the rotor carries an inner squirrel-cage winding. The torques developed then are additive. Such a motor is called a *repulsion-induction motor*. At starting, the cage current is limited by its relatively high reactance and the commutator winding develops a major part of the torque, but during running conditions the cage torque increases, compensating for the natural drop in the speed-torque characteristic of the commutator winding.

Repulsion-Start Induction-Run Motor

Constancy of speed and high starting torque can be achieved in a repulsion motor by providing a centrifugal force-operated device to short-circuit all commutator segments when the motor has accelerated to approximately half the speed. This short-circuited condition produces the equivalent of a squirrel-cage rotor for normal-speed running.

Such repulsion-induction or repulsion-start induction motors are very much suitable for use in the rural areas that may be served by long, single-phase lines, since their high starting torque is available without excessive current, and their running speed is nearly constant over a wide torque range. However, these motors have now been replaced by two capacitor single-phase induction motor for most applications.

AV.2 | THREE-PHASE COMMUTATOR MOTORS

Principle

The torque-slip relationship of an inductor motor is

$$T = \frac{3I_2^2 R_2}{s\omega_s} \tag{AV.1}$$

For a given torque the slip is controlled by $I_2^2 R_2$, the rotor copper loss or, in more general terms, the electrical power removed from (or injected into) the rotor circuit. This is efficiently achieved by injecting a slip-frequency emf into the rotor circuit as shown in Fig. AV.2. The torque-slip relation of Eq. (AV.1) now generalizes to

$$T = \frac{3(I_2^2 R_2 \pm E_i I_2 \cos\phi_2)}{s\omega_s} \quad \text{(AV.2)}$$

$$s\bar{E}_2 \pm \bar{E}_i = \bar{I}_2 \bar{Z}_2$$

or

$$\bar{I}_2 = \frac{s\bar{E}_2 \pm \bar{E}_i}{\bar{Z}_2} \quad \text{(AV.3)}$$

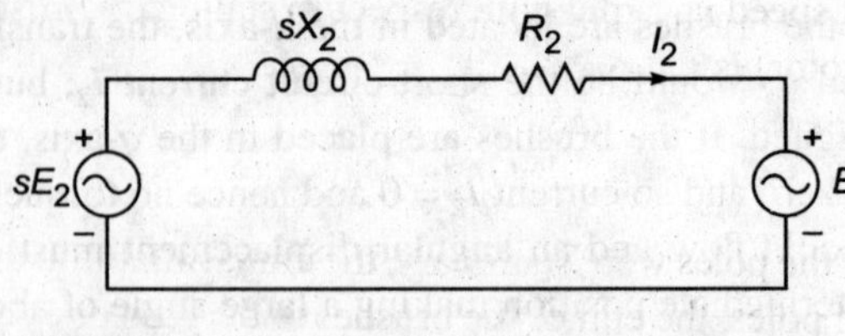

Fig. AV.2 *Slip control by injected slip-frequency emf into rotor circuit*

When the injected emf E_i is in-phase with the rotor induced emf sE_2, the electrical power $E_2 I_2 \cos\phi_2$ is removed (recovered) from the rotor circuit and the plus sign applies in Eq. (AV.2). On the other hand, when E_i is in phase opposition to sE_2, the electrical power $E_2 I_2 \cos\phi_2$ is injected into the rotor circuit and the minus sign applies in Eq. (AV.2). It easily follows from Eq. (AV.2) that for a given torque, assuming rotor current I_2 to remain nearly constant*, as the magnitude of the in-phase emf E_i is increased, the slip increases (speed decreases). When the phase of E_i is reversed and as E_i is increased in magnitude, the numerator in Eq. (AV.2) becomes negative and so does the slip, i.e. a super-synchronous speed is achieved.

If instead the phase of the injected emf is changed with the magnitude of E_i remaining fixed, control over the rotor current power factor and therefore over that of the primary input current is achieved. A leading phase angle of E_i would result in pf improvement. This is illustrated in Fig. AV.3 for E_i leading sE_2 by 90°. To a rough approximation the component of E_i in-phase (or in phase opposition) controls motor speed and the quadrature component controls the motor power factor.

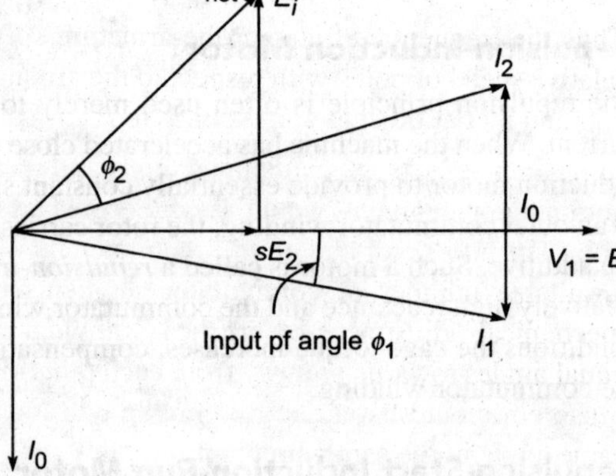

Fig. AV.3 *Power factor control by rotor injection*

Generation of Slip-Frequency EMF

Since the injected emf must have slip frequency, a power converter of variable frequency (and voltage) is needed which interchanges electrical power with the fixed-frequency mains. This would maintain high overall efficiency of the set.

Presently, this is accomplished by means of thyristor frequency converters. An older arrangement used for this purpose was the commutator. The principle of generalized frequency conversion by means of a commutator is explained below with the help of Fig. AV.4.

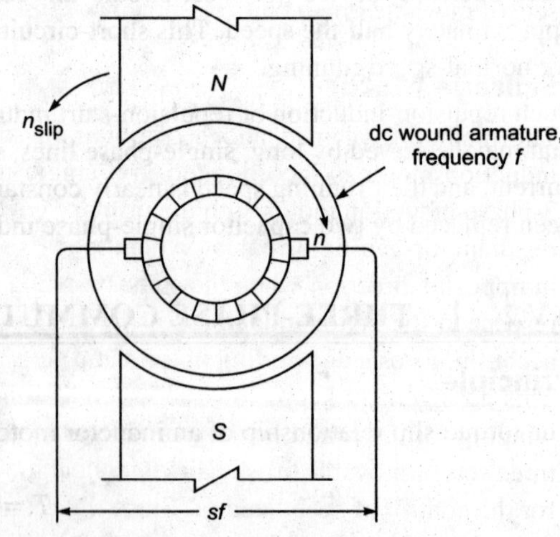

Fig. AV.4 *Generalized frequency conversion*

A dc-wound armature is running at speed n rpm, while the machine poles are made to rotate in the opposite direction

*For large values of slip, the rotor power factor ($\cos\phi_2$) decreases and the current for a given torque increases.

at speed n_{slip} rpm with respect to stationary brushes as shown. The frequency of emf induced in the armature (rotor) is

$$f = \frac{(n + n_{slip})P}{120} = \frac{n_s P}{120} \quad \text{(AV.4)}$$

If the poles were stationary, the emf available at the brushes would be dc (zero frequency). For every revolution of poles, the emf of the brushes undergoes one complete alteration (for the 2-pole case illustrated). In general, the frequency of emf at brushes would then be

$$f_{brushes} = \frac{n_{slip} P}{120} = \left(\frac{n_{slip}}{n_s}\right) f = sf \quad \text{(AV.5)}$$

Thus the frequency of emf in the armature corresponds to the relative speed of poles with respect to the armature, i.e. $n_s = n + n_{slip}$ and that at brushes corresponds to n_{slip}.

If $n_s = n + n_{slip}$ is maintained constant, the rms emf per commutator segment would remain constant for fixed flux/pole. The brush emf magnitude could then be varied by tapping a suitable number of commutator segments by symmetrical mechanical adjustment of brushes—moving both brushes by equal angles towards or away from each other. The phase of the brush emf could be varied by moving both the brushes by the same angle in the same direction.

A set of balanced 3-phase emf's of adjustable magnitude and phase could be tapped from the commutator by placing three sets of movable brush pairs as shown in Fig. AV.5.

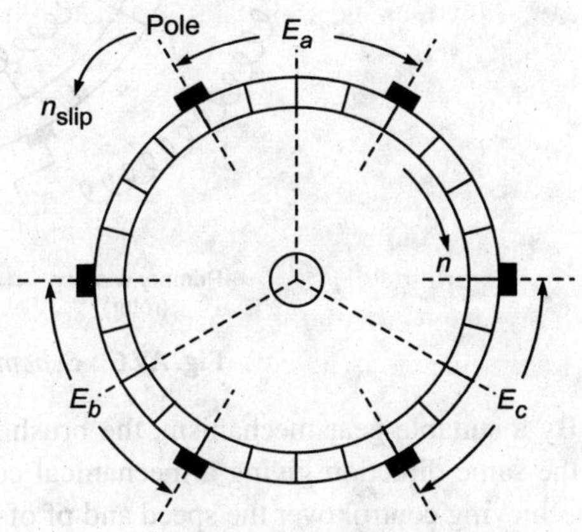

Fig. AV.5 *Tapping 3-phase emf's from generalized frequency converter*

Schrage Motor

The method explained above of obtaining slip-frequency emf and injecting it into the secondary of an induction motor is incorporated in a single machine called the Schrage motor. It is basically an inverted 3-phase induction motor with the primary winding on the rotor and the secondary winding on the stator. The resultant air-gap field runs at synchronous speed n_s with respect to the rotor and the rotor runs at speed n in opposite direction when the stator winding is shorted. As a consequence, the field runs at slip speed with respect to the stator inducing slip frequency current in it which causes torque production. The rotor also accommodates a dc winding in the same slots as the primary conductors, but placed nearer the slot openings. This arrangement keeps low the leakage inductance of the dc winding coils facilitating commutation. This winding (dc) is called *tertiary* or *regulating winding*. It is connected to a commutator on which are placed three sets of movable brush pairs to collect the 3-phase emf's for injection into the secondary (stator) winding for the control of speed and pf. Since the rotating field moves at slip speed (n_{slip}) with respect to brushes, the frequency of brush emf's is always the slip-frequency. The schematic diagram of a 2-pole Schrage motor is shown in Fig. AV.6.

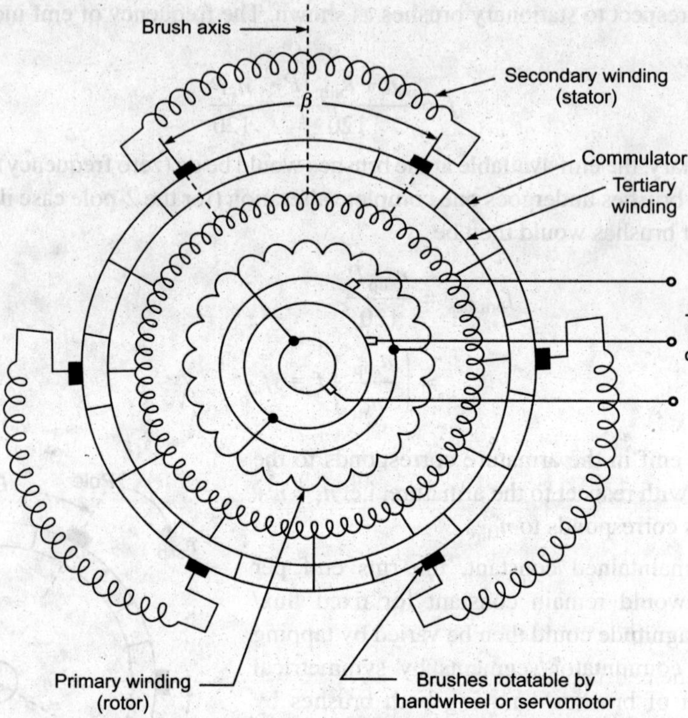

Fig. AV.6 *Schematic diagram of a two-pole Schrage motor*

By a suitable gear mechanism, the brush pairs can be moved towards or away from each other or both in the same direction giving a mechanical control over the magnitude and phase of the injected emf thereby achieving control over the speed and pf of the motor.

When the brushes are moved so that the injected emf opposes the secondary current, the speed is decreased (positive slip) and for movement in the opposite direction the speed is increased, i.e. the slip is negative. When the brushes are on the same commutator segment, i.e. the brush separation angle is zero, the secondary winding is short-circuited in itself and the motor runs at its natural slip as an inverted induction motor. For an electrical brush separation β, the no-load motor speed is

$$n \approx n_s (1 - C \sin 0.5\, \beta) \qquad \text{(AV.6)}$$

where C is a constant whose value depends upon the turn ratio of the stator and rotor commutator windings and β = elect. brush separation angle. For a given brush separation, the motor power pf can be controlled by moving each brush of brush pairs in the same direction. Starting can be effected by direct switching with brush separation of 180° elect., i.e. one pole-pitch, in the lowest speed position (emf opposing rotor current).

The Schrage motor speed can be varied over the range $0.4n_s$ to $1.4n_s$. The efficiency of the machine on load is slightly lower than that of a corresponding induction motor as the Schrage machine has additional components in the form of commutator and brushes and there are additional losses in the third winding. Commutation problems and design considerations limit the rated output to around 25 kW/pole and placing of the primary winding on the rotor keeps the supply voltage restricted to around 600 V.

The Schrage motor is used as a constant torque drive and the power developed varies directly with speed. It is usually applied in printing presses, textile mills, paper mills, cranes, etc.

APPENDIX VI

Resistance

AVI.1 | CHANGE OF RESISTANCE WITH TEMPERATURE

The dc resistance loss can be found out if the dc resistance is measured. For copper conductors

$$\frac{R_{t1}}{R_{t2}} = \frac{234.5 + t_1}{234.5 + t_2} \tag{AVI.1}$$

where R_{t1} and R_{t2} are the resistances at t_1 and t_2 °C, respectively.

The R_{t1}/R_{t2} versus temperature plot is shown in Fig. AVI.1 for the case $t_2 = 0$°C.

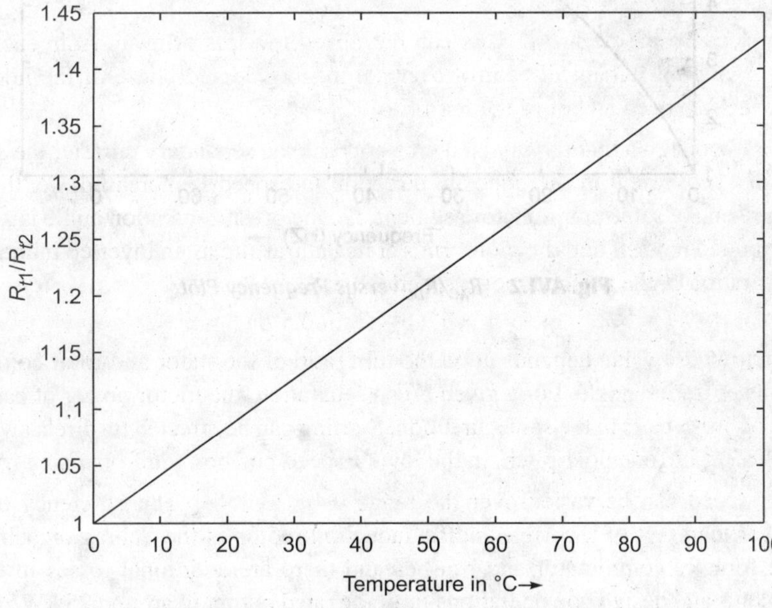

Fig. AVI.1 *R_{t1}/R_{t2} versus Temperature Plot*

AVI.2 | CHANGE OF RESISTANCE WITH FREQUENCY

ac and dc resistances of copper conductor are related as

$$\frac{R_{AC}}{R_{DC}} = \rho(f) = \text{function of frequency} \tag{AVI.2}$$

When ac is flowing through a conductor, the current is non-uniformly distributed over the cross-section. This effect becomes more pronounced as frequency is increased. This phenomenon is called *skin effect*. It causes larger power loss for a given rms ac than the loss when the same value of dc is flowing through the conductor. Consequently, the effective conductor resistance is more for ac than for dc.

Apart from skin effect, non-uniformity of current distribution is also caused by *proximity effect*. Thus the ratio of R_{AC}/R_{DC} is a function of frequency. R_{AC}/R_{DC} versus frequency plot for copper is shown in Fig. AVI.2.

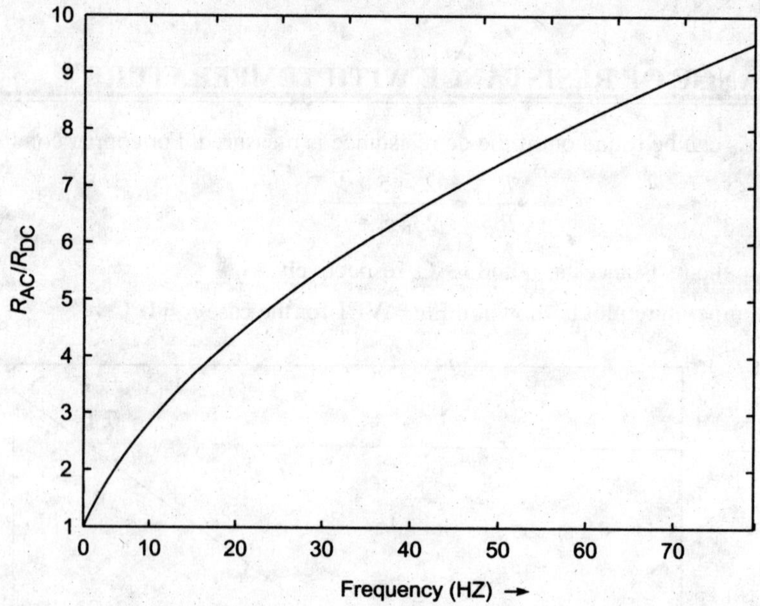

Fig. AVI.2 *R_{AC}/R_{DC} versus Frequency Plot*

APPENDIX VII

Table of Constants and Unit Conversion

CONSTANTS

Permeability of free space, $\mu_0 = 4\pi \times 10^{-7}$ H/m

Permittivity of free space, $\varepsilon_0 = 8.854 \times 10^{-12}$ F/m

Gravitational acceleration, $g = 9.807$ m/s^2

UNIT CONVERSION

Quantity	*SI Unit*	*English Unit*	*CGS Unit*
Length	1 m	3.281 ft = 39.37 in	100 cm
Mass	1 kg	2.205 lb (mass); 35.27 ounces (oz)	1000 gm
Force	1 N	0.2248 lb = 7.23 poundals	10^5 dynes
Torque	1 N m	0.738 pound-feet; 141.6 oz.in	10^7 dyne cm
Energy	1 J = 1 Ws = 9.84×10^{-4} BTU = 0.239 Calories	0.738 ft-lb; 2.778×10^{-7} kWh	10^7 ergs
Power	1 W	1.341×10^{-3} hp (1 hp = 746 W) 3.412 BTU/hr	
Moment of inertia	1 kg m^2 = 0.738 slug-ft^2	23.73 lb-ft^2	10^7 g cm^2; 141.6 oz.in sec^2
Magnetic flux	1 Wb	10^8 lines	10^8 maxwells
Magnetic flux density	1 Wb/m^2 = 1T	64.5 kilo lines/in^2	10^4 gauss
Magnetic field intensity	1 A/m	0.254 A/in	0.004 π oersted
Pressure	1 Pa (N/m^2)	4.45×10^{-4} lbf/in^2	9.86×10^{-6} atm

APPENDIX VIII

Electric Drives

AVIII.1 | INTRODUCTION TO ELECTRIC DRIVES

To improve manufacturing performance and flexibility, automation control, motion control and machine automation systems are used. Energy consumption can be reduced by engineering assistance with machine safety, energy efficiency, and motion control by braking system automation concepts, global application expertise, and support. Efficient use of new and integrated approaches to complex engineering challenges helps in improved worker's safety. A production machine requires high precision movement, or just simple positioning. This architecture can be implemented using integrated motor-drive technology. It distributes functions such as motion control, safety and predictive maintenance to individual drives.

A modern controller can be used to control the transients and dynamics of the electrical or mechanical machine. Its response is then applied to the loads. It enables self-starting of the motor and can also be used for providing electronic commutation.

AVIII.2 | DRIVES

Drives are defined as a system which is used to employ for motion control. They are basically classified into five categories as follows:

A. On the Basis of Mode of Operation

(i) ***Continuous duty drives*:** In these drives, to reach temperature equilibrium, the motor works at a constant load for enough time.

(ii) ***Short time duty drives*:** In these drives, the time of operation is very low and the heating time is much lower than the cooling time. The motor operates at constant load and these types of motors are used in crane drives.

(iii) ***Intermittent duty drives*:** In these drives, the starting current has little effect on temperature. Temperature equilibrium is never reached by using these drives.

---

* For detailed discussion please refer to the book on *Electric Drives* by D P Kothari and Rakesh Singh Lodhi published by I K Int New Delhi 2016.

B. On the Basis of Means of Control

(i) ***Manual control drives***: These drives can be either simple like a room fan or complicated like a push button starter.

(ii) ***Semi-automatic control drives:*** These drives consist of a manual device for giving a command and automatic device in response to that command.

(iii) ***Automatic control drives:*** These drives do not consist of a manual device but have a control gear.

C. On the Basis of Number of Machines

(i) ***Individual drive***: In these drives, a separate motor drives each individual machine.

(ii) ***Group drive***: These drives are also called shaft drives. They consist of one motor, which rotate/control one or more line shafts supported on bearings. The line shaft may be fitted with pulleys, belts, and gears.

(iii) ***Multi-motor drive***: In these drives, a separate motor takes care of each operation of the mechanism. There are several drives in this drive system; it serves to make one of the working parts of drive mechanism error free.

D. On the Basis of Dynamics and Transients

(i) Independent transient period
(ii) Dependent transient period

E. On the Basis of Methods of Speed Control

(i) Reversible and non-reversible independent constant speed
(ii) Reversible and non-reversible step speed control
(iii) Variable position control
(iv) Reversible and non-reversible smooth speed control

AVIII.2 | ELECTRICAL DRIVES

An electrical drive is an electromechanical energy conversion (EMEC) device used for transforming electrical energy into mechanical energy. It is used in various kinds of process control to impart motion to different machines and mechanisms.

An electrical drive system has the following components (Fig. AVIII.1)

1. Electrical machines and loads
2. Motor
3. Power Modulator
4. Sources
5. Control unit
6. Sensing unit

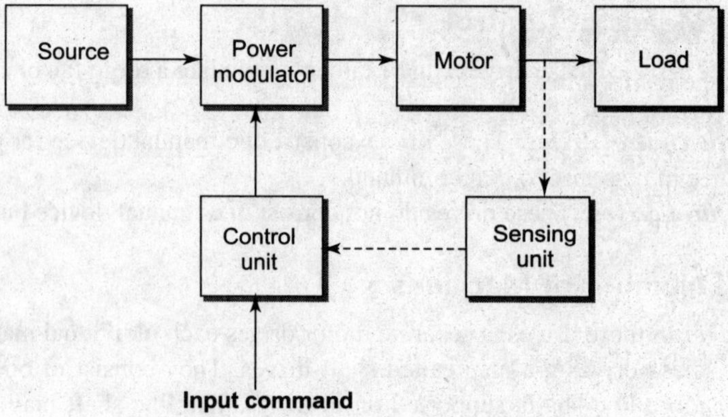

Fig. AVIII.1 *Block diagram of electrical drives*

Types of Power Modulators

In an electric drive system, the power modulators are of the following types:

1. Controlled rectifiers (ac to dc converters)
2. Inverters (dc to ac converters)
3. ac voltage controllers (ac to ac converters)
4. dc choppers (dc to dc converters)
5. Cycloconverters (Frequency conversion)

Classification of Electrical Drives

The schematic representation of classification of electrical drives is shown in Fig. AVIII.2.

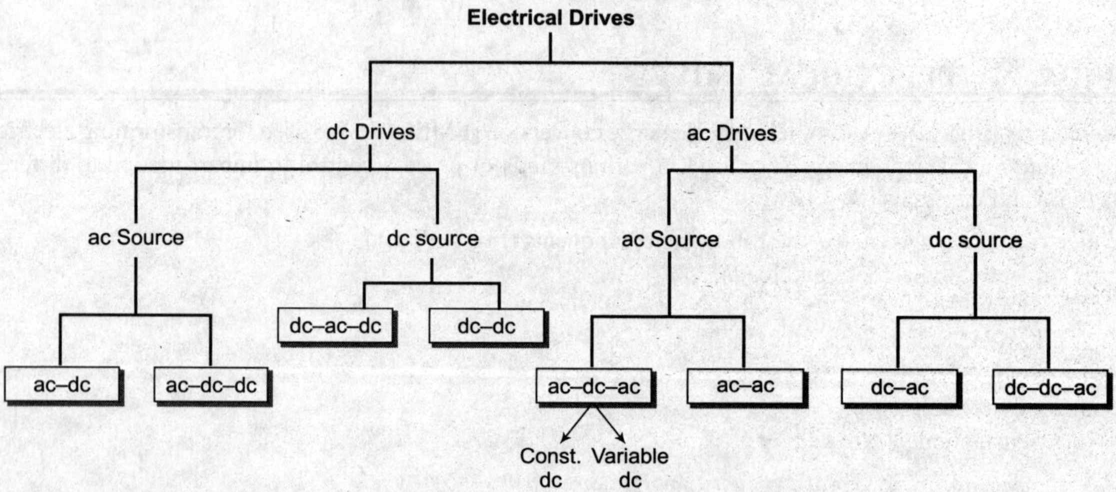

Fig. AVIII.2 *Classification of Electrical Drives*

Main Parts of a dc Machine

Figure AVIII.3 depicts the main parts of a 4-pole dc machine. They are described as follows:

1. ***Stator*:** It is the stationary part of the machine. The field winding is placed on stator side. The main parts of a stator are: (i) yoke or frame, (ii) pole core, (iii) pole shoe, (iv) lifting eye, and (v) feet.
2. ***Rotor*:** It is the rotating or movable part of the machine. The armature winding is placed on rotor side. The main parts of a rotor are: (i) armature core (ii) commutator (iii) shaft, and (iv) armature winding.

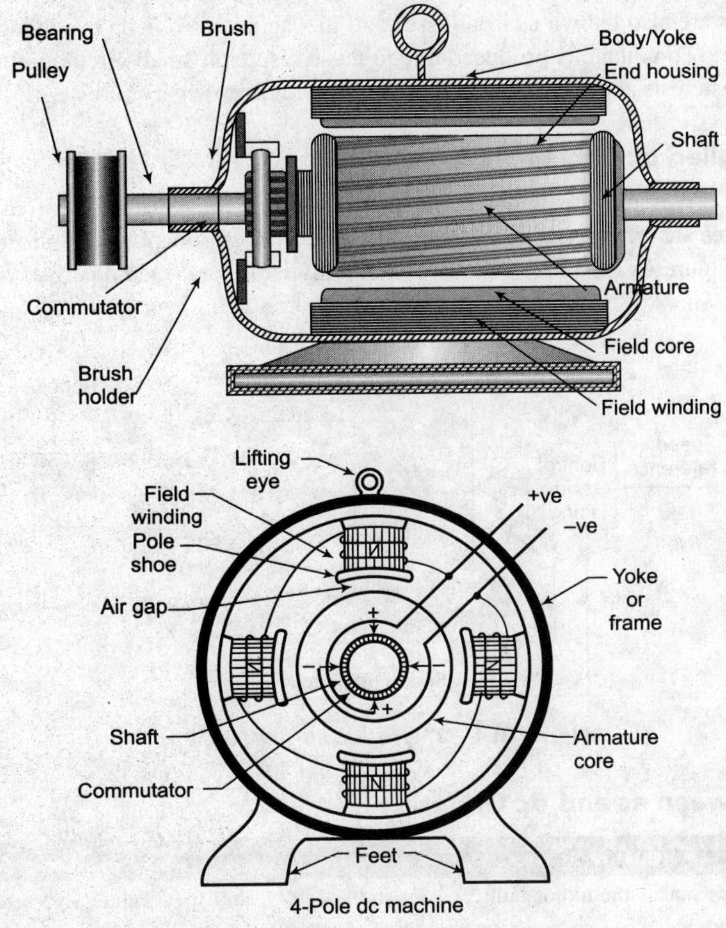

Fig. AVIII.3 *Construction of a dc Machine**

Losses in dc Motor

1. ***Friction and Windage Losses:*** These losses are also known as mechanical losses. These include bearing friction, brush friction, and windage. They vary with changes in speed but remain constant at a given speed. Power losses due to windage increase as the cube of the speed; and those due to friction, increase as the square of the speed.

*For construction of dc machines, refer Chapter 7.

2. ***Armature Copper Losses:*** These are the I^2R losses of the armature circuit which includes armature winding, commutator and brushes. They vary as the square of currents and directly with the resistance.
3. ***Field Copper Losses:*** These are I^2R losses of series and shunt field winding, interpole windings and any shunts used in connection with these windings. They vary as the square of currents and directly with the resistance.
4. ***Core Losses:*** Core losses are basically the eddy current and hysteresis losses in the armature. An expenditure of energy is required with continuous change of direction of flux in the armature iron. These losses are also known as iron losses and magnetic losses. Iron is conducting and moves in a magnetic field so voltage is produced due to these results in small circulating currents called eddy currents. Hysteresis and eddy current losses vary with flux density and speed.

Thyristor Controlled dc Drives

Thyristor controlled drives operate under switching frequency equal to supply frequency according to converter. These drives are employed in low performance, low cost, low power applications as well as high power applications. Figure AVIII.4 shows the circuit of a thyristor controlled dc drive.

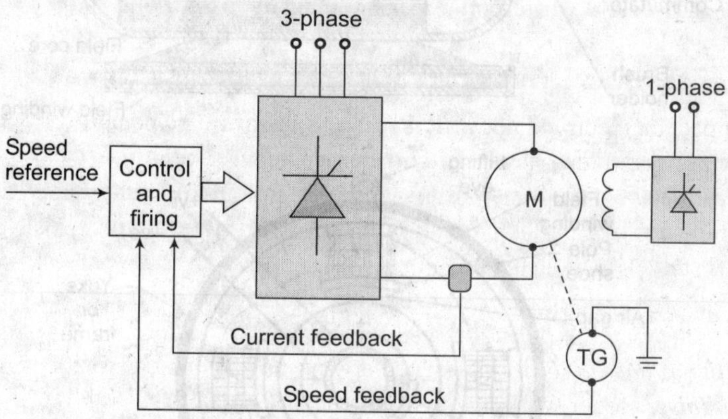

Fig. AVIII.4 *Thyristor Controlled dc Drive*

Comparison between ac and dc Drives

S. No.	dc Drives	ac Drives
1.	The commutator makes the motor bulky, costly and heavy.	Motors are inexpensive, particularly squirrel cage motor.
2.	The converter technology is well established. The power converter is simple and inexpensive.	The inverter technology is still being developed. The power circuit of the converter and its control are simple.
3.	Line commutation of the converter is used	Forced commutation is used with induction motors. Sometimes machine commutation may be used with synchronous motors.
4.	Fast response and wide speed range smooth control.	With solid state converter, the speed range is wide. With conventional methods, it is stepped and limited.
5.	Small power/weight ratio.	Large power/weight ratio.
6.	Cost does not depend on the solid-state converter	Solid-state converter employed also decides the cost.

Closed Loop Operation of dc Motor

To reach the required velocity, a closed loop controller regulates the power delivered to the motor. If the motor needs to run faster than the required velocity then the controller will deliver less power to the motor.

Advantages of Control Scheme

- It is cost effective.
- It is easy on hardware design.
- It can be retrofitted on systems in which motor does not come along with the encoder.
- Small motors with gear used for hobby robotics usually run at high speed (gear runs at low speed, gear reduction) and can easily use this scheme.

Disadvantages of Control Scheme

- It cannot be used to control motors running at extremely low speeds as motors do not generate sufficient back emf in such cases.
- It cannot be used in system where torque ripple cannot be tolerated. Torque ripple is caused by stopping the PWM for back emf measurement.
- The back emf measurement circuit becomes complicated if the motor is configured to run in both directions using H-bridge (differential measurement).

Feature and Benefits of ac Drives or Variable Frequency Drives

- ***Soft start:*** Impact on electrical network is reduced; there are no penalties from utility; reduced stress on motor, coupling and load, hence life time extension is given.
- ***Precise speed and torque control:*** Enhanced product quality; better cost of ownership; improved protection of motor (e.g., stall protection and load); stable product quality despite input power variations and sudden load changes.
- ***Wide speed control range:*** Better efficiency compared to traditional flow control methods, e.g., damper control and throttling; lower maintenance.
- ***High reliability and availability:*** Lessened downtime and enhanced process availability.
- ***Low audible noise:*** Better working environment for operators.
- ***Capability for speed reversal/regenerative braking:*** Desired torque during braking, therefore better product quality; enhanced braking characteristics; advanced efficiency.
- ***Flux optimization (motor flux automatically adapted to load):*** Better motor efficiency and reduced motor noise.
- ***Power loss ride through:*** Diminished number of drive trips and improved process availability.
- ***Automatic start (drive can catch a spinning load):*** Reduced waiting time and downtime.
- ***Energy saving:*** ac drives can be retrofired to standard induction motors to provide substantial energy savings.

Breaking Methods of Induction Motor Drive

There are three methods of braking in an induction motor. These are as follows:

Generating mode When the energy stored in the rotating mass is dissipated in an electric resistance then the dynamic braking of electric motors occurs. This requires the motor to operate as a generator in order to convert this stored energy into electrical energy.

Regenerative breaking mode When the motor speed exceeds the synchronous speed, then occurs the regenerative braking. In regenerative braking, motor works as a generator and electrical power is delivered back to the electrical system. This method of breaking is called regenerative braking.

Plugging The operation shifts from motoring to plugging region, when the phase sequence of supply of motor running at a speed is reversed by interchanging the connections of any two phases of stator, with respect to supply terminals.

Comparison of Induction Motor Starters

Description of Starter	% of line voltage applied	Starting current (I_s) compared with		Starting torque (T_s) compared with	
		D.O.L current (I_{dol})	Full load current (I)	D.O.L Torque (T_{dol})	Full load torque (T)
D.O.L Starter	100	$I_s = I_{dol}$	$I_s = 6\,I$	$T_s = T_{dol}$	$T_s = 6\,T$
Star delta starter	57.7	$I_s = (1/\sqrt{3})^2\,I_{dol}$	$I_s = 2\,I$	$T_s = (1/\sqrt{3})^2\,T_{dol}$	$T_s = 2/3\,T$
Auto transformer starter	80	$I_s = (0.8)^2\,I_{dol}$	$I_s = 3.84\,I$	$T_s = (0.8)^2\,T_{dol}$	$T_s = 1.28\,T$
	60	$I_s = (0.6)^2\,I_{dol}$	$I_s = 2.16\,I$	$T_s = (0.6)^2\,T_{dol}$	$T_s = 0.72\,T$
	40	$I_s = (0.4)^2\,I_{dol}$	$I_s = 0.96\,I$	$T_s = (0.4)^2\,T_{dol}$	$T_s = 0.32\,T$
Reactance-resistance starter	64	$I_s = (0.64)^2\,I_{dol}$	$I_s = 2.5\,I$	$T_s = (0.425)^2\,T_{dol}$	$T_s = 0.35\,T$

Speed Control of Induction Motors

Induction motors are of two types, namely—squirrel-cage motor and wound-rotor motor (slip ring rotor motor). Given below are the various types of speed control methods of induction motor.

- Pole changing
- Stator voltage control
- Supply frequency control
- Eddy-current coupling
- Rotor resistance control
- Slip power recovery

Stator Voltage Control

The applications of the stator voltage control method are applicable for torque demand reduced with speed. Examples are fan and pump drives.

Merits of Stator Voltage Control Method

- Simple control circuitry
- Compact size
- Fast response time

Demerits of Stator Voltage Control Method

- Voltage and current waveform gets highly distorted due to harmonics, and therefore affects the efficiency of motor.
- Performance is poor under running condition at low speeds.
- Operating efficiency is low and resistance losses are high.
- Maximum torque of the motor decreases with reduction in stator voltage.

Few Important Electric Drives

1. Static Kramer Drive

An approach has been made to feed the slip energy wastage back to the ac line by converting it into an appropriate signal so as to recover the wastage. Two methods in support of this statement are as follows:

(i) ***Static Kramer Drive:*** This method is used to obtain an injected voltage that is in phase with the rotor current.

(ii) ***Static Scherbius Drive:*** These drives are used for bi-directional power flow so both positive and negative voltages injected are possible either in phase with or opposing the rotor current.

2. Diesel-electric Drives

These drives use electrical transmission and motor drives to convert movable mechanical power from a diesel engine to traction mechanical power at the train wheels.

Vector Control of Induction Motor

There are two methods of vector control. They are as follows:

1. Direct or feedback method
2. Indirect or feed forward method

Salient Features of Vector Control

- The rotor frequency, ω_e, is not directly controlled as in scalar control. The motor is essentially self-controlled where rotor frequency as well as phase is controlled indirectly with the help of unit vector.
- There is no fear of instability by crossing the operating point beyond the electromagnetic torque, T_{em}, just like in scalar control. Limiting the total I_s within the safe limit automatically limits the operation within stable region.
- Transient response is fast in vector control as dc machines because torque is control by I_{qs}. It does not affect the flux. Signal processing and parameter variation are effected due to delays in converter, so ideal vector control is not possible.
- Speed control is possible in four quadrants as dc machines without using additional control elements. If the torque T_e is negative in forward motoring condition, then the drive initially goes to regenerative braking mode and the speed slows down. At zero speed, phase sequence of the unit vector is reversed, giving reverse motoring operation

Figure AVIII.5 gives a schematic representation of vector control in an induction motor.

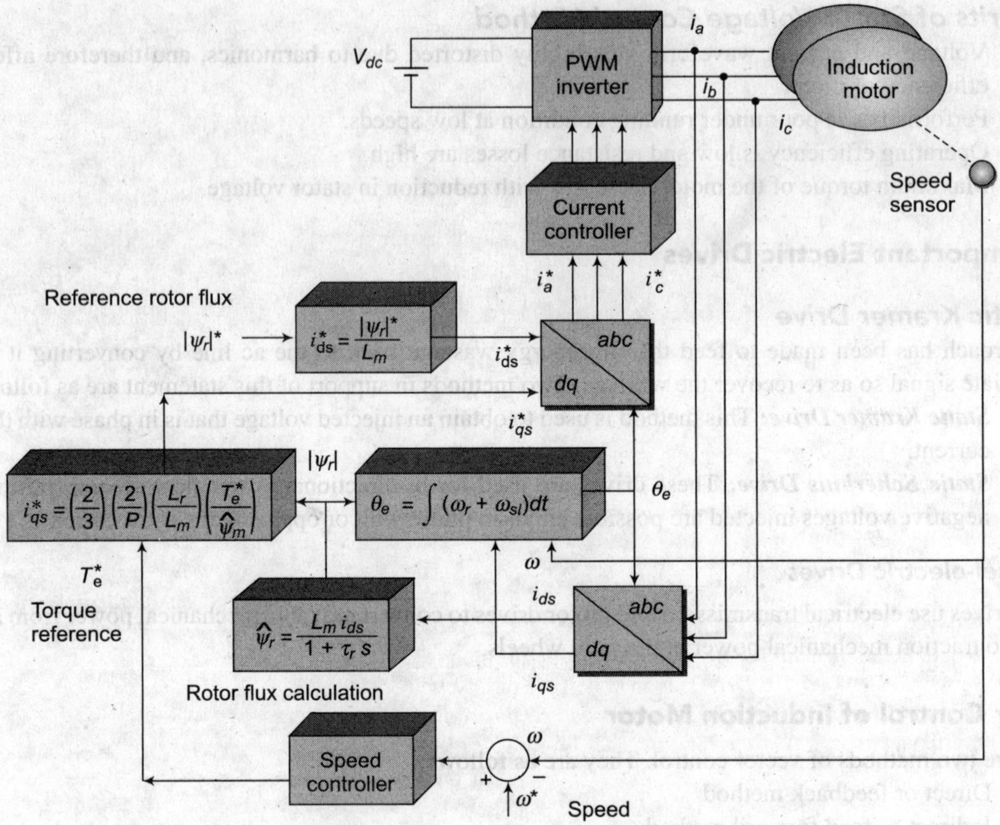

Fig. AVIII.5 *Vector control in induction motor*

References

1. Fitzgerald, A E, C Kingsley and S D Umans, *Electric Machinery,* 7th edition, McGraw-Hill, New York, 2013.
2. Say, M G, *Alternating Current Machines,* 5th edition, Sir Isaac Pitman and Sons Ltd, 1984.
3. Murthy, S S, B P Singh, D P Kothari and C S Jha (Eds), *Laboratory Manual for Electromechanics,* Wiley Eastern, New Delhi, 1982.
4. George McPherson, Robert D. Laramore, *An Introduction to Electrical Machines and Transformers*, John Wiley & Sons; 2nd Revised edition, 1989.
5. Wood, Peter, *Switching Power Conservators,* Van Nostrand Reinhold Co., New York, 1981.
6. Sen, P C, *Thyristor DC Drives,* John Wiley & Sons, New York, 1981.
7. Nagrath, I J and D P Kothari, *Power System Engineering,* 2nd edition, Tata McGraw-Hill, New Delhi, 2007.
8. Pearman, Richard A, *Power Electronics Solid State Motor Control*, Reston Pub. Co. Inc., Reston, Virginia, 1980.
9. Say, M G and E O Taylor, *Direct Current Machines,* John Wiley, New York, 1980.
10. Slemon, G R and A Straughen, *Electric Machines,* Addison Wesley, Reading, Mass, 1980.
11. Deltoro, V, *Principles of Electrical Engineering,* 2nd edition, Prentice-Hall of India, New Delhi, 1979.
12. Nasar, S A and L E Unnewehr, *Electromechanics and Electric Machines,* John Wiley, New York, 1979.
13. Shanmugasundaram, A, G Gangadharan and R Palani, *Electrical Machine Design Data Book,* Wiley Eastern, New Delhi, 1979.
14. Sarma M S, *Synchronous Machines,* Gordon & Breach Science Publishers, New York, 1979.
15. Fink, D G and H W Beaty, *Standard Handbook of Electrical Engineers,* 11th edition, McGraw-Hill, New York, 1999.
16. Richardson, D V, *Rotating Electric Machinery and Transformer Technology,* Prentice Hall, 1997.
17. Elgerd, O I, *Basic Electric Power Engineering,* Addison Wesley, Reading, Massachusetts, 1977.
18. Ramamoorthy, M, *An Introduction to Thyristors and their Applications,* Palgrave, 2014.
19. Dewan, S B and A Straughenm, *Power Semiconductor Circuits,* John Wiley, New York, 1975.
20. Mazda, F F, *Thyristor Control,* Newnes-Butterworths, 1973.
21. Murphy, J M D, *Thyristor Control of AC Motor,* Pergamon Press, London, 1973.
22. Ramshaw, R S, *Power Electronics Semiconductor Switches,* 2nd edn. Chapman and Hall, London, 1993.
23. General Electric Co, *General Electric SCR Manual,* 5th edition, Syracuse, USA, 1972.
24. Harnden, J D and F B Golden, *Power Semiconductor Applications,* Vol. I, IEEE Press, 1972.
25. Kosow, I L, *Electric Machinery and Transformers,* 2nd edition, Prentice-Hall Inc., Englewood, Cliffs, NJ, 1991.
26. Moltgen, G, *Line Commutated Thyristor Converters,* Pitman Publishing, London, 1972.
27. Kusko, A, *Solid-State AC Motor Drives,* The MIT Press, Cambridge, Mass, 1971.
28. Pelly, B R, *Thyristor Phase Controlled Converter and Cyclo-converters,* John Wiley, New York, 1971.
29. Lord, H W, "Pulse transformers," *IEEE Trans. on Magnetics,* Vol. Mag-7, No. 1, New York, pp. 17–28, March 1971.

30. Openshaw Taylor, E, *The Performance and Design of AC Commutator Motors,* A H Wheeler & Co. (P) Ltd., Allahabad, 1971.
31. Alger, P L, *Induction Machines*, 2nd edition, Gordon and Breach, New York, 1970.
32. Chilkan, M, *Electric Drive,* MIR Publishers, Moscow, 1970.
33. Westinghouse Electric Corp, *Westinghouse SCR Designers Handbook,* 2nd edition, Youngwood, Pa, 1970.
34. Kusko, A, *Solid-State DC Motor Drives,* MIT Press, Cambridge, Mass, 1969.
35. Daniels, A R, *The Performance of Electrical Machines,* McGraw-Hill, London, 1968.
36. Humphrey, A J, "Inverter commutation circuits," *IEEE Trans. I.G.A.*, Vol. IGA-4, No. 1, pp. 104–110, January-February, 1968.
37. Nagrath, I J, J R Handa and S C Rastogi, *Electric Circuits and Machines* (*Direct Current*), Jain Brothers, New Delhi, 1968.
38. Draper, A, *Electric Machines,* 2nd edition, Longman, London, 1967.
39. Grossner, N R, *Transformers for Electronical Circuits,* McGraw-Hill, New York, 1967.
40. Thaler, G J and M L Wilcox, *Electric Machines: Dynamics and Steady State,* John Wiley, New York, 1966.
41. Schaefer, J, *Rectifier Circuits. Theory and Design,* John Wiley, New York, 1965.
42. Bedford, B D and R G Hoft, *Principles of Inverter Circuits,* John Wiley, New York, 1964.
43. Heumann, K, *Pulse Control of dc and ac Motors by SCRs, IEEE Trans. on Communication and Electronics*, pp. 390–399, 1964.
44. Puchestein, A F, T C Lloyd and A G Conrad, *Alternating Current Machines,* Asia Publishing House, Bombay. 1964.
45. Philip Kemp, *Alternating Current Electrical Engineering,* Macmillan and Co. Ltd., London, 1963.
46. Clayton, A E and N H Hancock, *Performance and Design of DC Machines,* ELBS Pitman edition, 1962.
47. Veinott, C, *Theory and Design of Small Induction Motors,* McGraw-Hill, New York, 1959.
48. Langsdorf, E H, *Theory of Alternating Current Machinery,* McGraw-Hill, New York, 1999.
49. *Electrical Transmission and Distribution Reference Book,* 4th edition, Westinghouse Electric Corporation, 1964.
50. Staff, E E, MIT, *Magnetic Circuits and Transformers,* John Wiley, New York, 1965.
51. Dube, G K, *Fundamentals of Electrical Drives,* CRC Press, 2002.
52. Rashid, M H, *Power Electronics, Circuits Devices & Applications,* 3rd edition, Pearson Education India, 2009.
53. Pillai, S K, *Analysis of Thyristor Power Conditioned Motors,* Universities Press (India) Pvt. Ltd, Hyderabad, 1992.
54. Berde, M S, *Electric Motor Drives,* Khanna Publishers, Delhi, 2008.
55. Murphy, J M D and F G Turnbull, *Power Electronic Control of D.C. Motors,* 1st edition, Pergamon Press, Oxford, 1988.
56. Sen, P C, *Principles of Electric Machines & Power Electronics,* 2nd edition, John Wiley & Sons, New York, 2005.
57. Dubey, G K, *Power Semiconductor Controlled Drives,* Prentice-Hall, New Jersey, 1989.
58. Bose, B K, *Power Electronics and D.C. Drives,* Prentice-Hall, New Jersey, 1986.
59. Lander, C W, *Power Electronics,* McGraw-Hill, International edition, Singapore, Third edition, 1993.
60. Sen, P C, *Power Electronics,* Tata McGraw-Hill Publishing Company Ltd., New Delhi, 1987.
61. Mohan, Underland and Robbins, *Power Electronics Converters, Applications and Design,* John Wiley & Sons, 2006.
62. Leonhard, W, *Control of Electrical Drives,* Springer-Verlag, Berlin, Heidelberg, 1985.
63. Dubey, G K and C R Kasarabada (edition), *Power Electronics & Drives,* Tata McGraw-Hill Publishing Company Ltd., 1993.
64. Peter Vas, *Vector Control of D.C. Machines,* Clarendon Press, Oxford, 1990.
65. Syed A Nasar, *Electric Machines and Power System,* Vol. 1, *Electric Machines,* McGraw-Hill, New York, 1995.
66. Revankar, G N and S K Pillai, *Recent Advances in A.C. Drives, Institution of Engineers (India),* Vol. 70, pp. 79–87, June 1989.
67. Bimal K Bose, *Recent Advances in Power Electronics, IEEE Transactions on Power Electronics,* Vol. 7, No. 1, pp. 1–7, January 1992.

68. Thomas A Lipo, *Recent Progress in the Development of Solid State A.C. Motor Drives, IEEE Transactions on Power Electronics,* Vol. 3, No. 2, pp. 105–116, April 1988.
69. Chattopadhyay, A K, *Power Electronic Converter Technology—A State of the Art Review, Institution of Engineers (India),* Vol. 70, pp. 33–45, June 1989.
70. Quicoe, J E and S B Dewan, *A damped a.c.-a.c. Frequency Converter for Induction Heating, IEEE Transactions on Industry Applications,* Vol. 1A-22, November-December 1986.
71. Dewan, S B and A Mirbord, *Microprocessor Based Optimal Control for Four Quadrant Chopper, IEEE Transactions on Industry Applications,* Vol. IA-17, January-February 1981.
72. Ryff, P F, *Electric Machinery,* Prentice-Hall, New Jersey, 1988.
73. Ryff, P T, D Platnick and J A Karnas, *Electric Machines and Transformers: Principles and Applications,* Prentice-Hall, New Jersey, USA, 1987.
74. Miller, T J E, *Brushless Permanent Magnet and Reluctance Motor Drives,* Oxford Science Publications, Oxford, 1989.
75. Singh, M and D P Kothari, *Starting of a Three-phase Induction Motor Connected to a Single-phase Supply System, I.E. (India),* Vol. 76, pp. 63–67, May 1995.
76. Kothari, D P and I J Nagrath, *Electric Machines*, Sigma Series, Tata McGraw-Hill, New Delhi, 2006.
77. Kothari, D P and I J Nagrath, *Modern Power System Analysis,* 4th edition, Tata McGraw-Hill, New Delhi, 2011.
78. Agarwal, S C, M Singh, and D P Kothari, *Analysis of Induction Generator for Wind Energy Conversion System at Constant Output Voltage, I.E. (India),* Vol. 77, pp. 182–185, February 1997.
79. Bhatia, C M and D P Kothari (edition.), *Power Electronics-Laboratory Manual,* Galgotia Book Source, New Delhi, 1988.
80. Indulkar, C S and D P Kothari, *Power System Transients: A Statistical Approach,* 2nd Edition, Prentice-Hall of India, New Delhi, 2010.
81. Kothari, D P and I J Nagrath, *Basic Electrical Engineering,* 3rd edition, Tata McGraw-Hill, New Delhi, 2010.
82. Kothari, D P and I J Nagrath, *Theory and Problems of Basic Electrical Engineering,* 2016, Prentice-Hall of India, New Delhi, 2016.
83. Chapman, S J, *Electric Machinery Fundamentals,* 3rd edition, McGraw-Hill, Singapore, 1999.
84. Guru, B S and H R Hiziroglu, *Electric Machinery and Transformer,* 3rd edition, Oxford University Press, N.Y. 2004.
85. Anuradha, Bhim Singh and D P Kothari, *Generalised Concepts for Balancing Single Phase Load Fed from Three Phase Supply, Int. J. of Electric Machines and Power Systems,* Vol. 27, No. 1, pp. 63–78, 1999.
86. Das, D, S K Aditya and D P Kothari, *Dynamics of Diesel and Wind Turbine Generators on Isolated Power Systems, Int. J. of Elect Power and Energy System,* Vol. 21, No. 3, pp. 183–189, March 1999.
87. Suryawanshi, H M, K L Thakre, S G Tarnekar, D P Kothari and A G Kothari, *Power Factor Improvement and Closed Loop Control of an AC-DC Resonant Converter, IEE Proc. Elect Power Appl,* Vol. 149, pp. 101–110, March 2002.
88. Bansal, R C, T S Bhatti and D P Kothari, *A Bibliographical Survey on Induction Generators for Application of Non-Conventional Energy Systems, IEEE Trans. on EC,* Vol. 18, No. 3, pp. 433–439, September 2003.
89. Bhim Singh, A Pande, D P Kothari et al, *A Review of 1 ϕ Improved Power Quality AC-DC Convertors, IEEE Trans. Industrial Electronics,* Vol. 50, 962–981, October 2003.
90. Bimal K Bose, (Ed.) *Power Electronics and Variable Frequency Drives IEEE Press,* 1997.
91. Peter Vas, *Sensorless Vector and Direct Torque Control,* Oxford University Press, N.Y. 1998.
92. Dr. Bimbhra, P S, *Electrical Machinery*, Seventh Edition, Khanna Publishers, New Delhi, 2009.
93. Ashfaq Hussain, *Electrical Machines*, Dhanpat Rai Publications, 2nd edition, 2014.
94. Ashish Pandey, D P Kothari and Bhim Singh, *A Novel Regulation Band Approach for Fast Voltage Regulation in Single-phase PFC Converters, IETE Journal of Research*, Vol. 49, No. 6, pp. 445-451, November-December 2003.
95. Bhim Singh, Brij N Singh, Ambrish Chandra, Kamal Al-Haddad, Ashish Pandey and Dwarka P Kothari, *A review of Three-Phase Improved Power Quality AC-DC Converters, IEEE Trans. Industrial Electronics,* Vol. 51, No. 3, June 2004.
96. Ashish Pandey, D P Kothari, Ashok K Mukerjee and Bhim Singh, *Modelling and Simulation of Power Factor Corrected AC-DC Converters, International Journal of Electrical Engineering Education*, Vol. 41, No. 3, pp. 244-264, July 2004.

97. Pandey, A, Bhim Singh, D P Kothari, *Comparative Evaluation of Single-phase Unity Power Factor AC-DC Boost Converter Topologies, I.E. (I) Journal–EL*, Vol. 85, pp. 102-109, September 2004.
98. Bansal, R C, T S Bhatti and D P Kothari, *A Novel Mathematical Modelling of Induction Generator for Reactive Power Control of Isolated Hybrid Power Systems, International Journal of Modelling and Simulation*, Vol. 24, No. 1, pp. 1-7, December 2004.
99. Bansal, R C, T S Bhatti, D P Kothari, *Automatic Reactive Power Control of Isolated Wind-Diesel Hybrid Power Systems for Variable Wind Speed/Slip, Electric Power Components and Systems*, Vol. 32, pp. 901-912, 2004.
100. Ashish Pandey, Bhim Singh, Brij N Singh, Ambrish Chandra, Kamal Al-Haddad, Dwarka P Kothari, *A Review of Multilevel Power Converters, Journal of Institution of Engineers (I)*, Vol. 86, pp. 220-231, March 2006.
101. Paul, R N, H K Verma, L D Arya and D P Kothari, *Design Aspects of Three-Phase Squirrel – Cage A.C. Traction Motor with Aluminum Alloy Rotor Bars*, Proc. 16th AUPEC – 2006, Melbourne, 10th-13th December 2006.
102. Bhim Singh., Jayaprakash P & D P Kothari, *A Three-phase Four-wire DSTATCOM for Power Quality Improvement, International Journal of Power Electronics*, Korea, July 2008.
103. Bhim Singh, Jayaprakash P, Somayajulu R and D P Kothari, *Reduced Rating VSC with a Zig-zag Transformer for Power Quality Improvement in Three-phase Four-wire Distribution System, IEEE Trans. Power Delivery*, Vol. 24, No. 1 pp. 249-259, January 2009.
104. Bhim Singh, Jayaprakash P and D P Kothari, *Star/Hexagon Transformer Based Three-Phase Four-Wire DSTATCOM for Power Quality Improvement, International Journal of Emerging Electric Power Systems*, December 2008.
105. Bhim Singh, Jayaprakash P and D P Kothari, *A T-connected Transformer and Three-Leg VSC based DSTATCOM for Power Quality Improvement, IEEE Transaction Power Electronics*, Vol. 23, pp. 2710-2718, November 2008.
106. Bandyopadhyay, MN, *Electrical Machines*, Prentice-Hall of India Pvt. Ltd., 2007.
107. D P Kothari and J S Dhillon, *Power System Optimization*, Prentice-Hall of India, 2004, 2nd Edition, 2011.
108. D P Kothari & B S Umre, *Laboratory Manual for Electric Machines*, I.K. International Publishing House Ltd, 2014.
109. D P Kothari and I J Nagrath, *Control System Engineering Theory and Problems*, New Age International, New Delhi, 2015.
110. D P Kothari and Rakesh Singh Lodhi, *Electric Drives*, I.K. International Publishing House Ltd, 2015.
111. D P Kothari & I J Nagrath, *Basic Electrical & Electronic Engineering*, McGraw Hill Education (India) Private Limited, 2014.
112. D P Kothari, Mahima Jain and Shefali Jagwani, *Electrical and Electronics Material*, Narosa Publishing House Pvt Ltd, New Delhi, 2015.
113. D P Kothari, I J Nagrath and HS Kalsi, *Electrical Engineering and Instrumentation*, McGraw Hill Education (India) Private Limited, New Delhi, 2015.
114. Majid Jamil, M. Rizwan and D P Kothari, *Grid Integration of Solar Photovolatic Systems*, Taylor and Francis, 2017.
115. N R Sree Harsha Anupama Prakash D P Kothari, *The Foundations of Electric Circuit Theory*, IOP publishers, UK, 2017.
116. Akhilesh A. Nimje and D P Kothari, *Electrical Circuit Analysis and Synthesis*, New Age International Publishers, New Delhi, 2017.
117. Shobhana D Langade, D. P. Kothari, *A Closed Loop Analysis of Z-Source Inverter Fed Induction Motor Drive with Variable Load Torque*, IRJET, vol 2, 4th July 2015.
118. D P Kothari, *Energy & Environment Problems Facing the Third World and Their Probable Solutions for Sustainable Development, Tech-Chronicle*, Vol 1: Issue 1, April 2015, PP- 45 - 53.

Index